Adjustable Speed AC Drive Systems

OTHER IEEE PRESS BOOKS

Optical Fiber Technology, II, *Edited by C. K. Kao*
Protective Relaying for Power Systems, *Edited by S. H. Horowitz*
Analog MOS Integrated Circuits, *Edited by P. R. Gray, D. A. Hodges, and R. W. Brodersen*
Interference Analysis of Communication Systems, *Edited by P. Stavroulakis*
Integrated Injection Logic, *Edited by J. E. Smith*
Sensory Aids for the Hearing Impaired, *Edited by H. Levitt, J. M. Pickett, and R. A. Houde*
Data Conversion Integrated Circuits, *Edited by Daniel J. Dooley*
Semiconductor Injection Lasers, *Edited by J. K. Butler*
Satellite Communications, *Edited by H. L. Van Trees*
Frequency-Response Methods in Control Systems, *Edited by A.G.J. MacFarlane*
Programs for Digital Signal Processing, *Edited by the Digital Signal Processing Committee*
Automatic Speech & Speaker Recognition, *Edited by N. R. Dixon and T. B. Martin*
Speech Analysis, *Edited by R. W. Schafer and J. D. Markel*
The Engineer in Transition to Management, *I. Gray*
Multidimensional Systems: Theory & Applications, *Edited by N. K. Bose*
Analog Integrated Circuits, *Edited by A. B. Grebene*
Integrated-Circuit Operational Amplifiers, *Edited by R. G. Meyer*
Modern Spectrum Analysis, *Edited by D. G. Childers*
Digital Image Processing for Remote Sensing, *Edited by R. Bernstein*
Reflector Antennas, *Edited by A. W. Love*
Phase-Locked Loops & Their Application, *Edited by W. C. Lindsey and M. K. Simon*
Digital Signal Computers and Processors, *Edited by A. C. Salazar*
Systems Engineering: Methodology and Applications, *Edited by A. P. Sage*
Modern Crystal and Mechanical Filters, *Edited by D. F. Sheahan and R. A. Johnson*
Electrical Noise: Fundamentals and Sources, *Edited by M. S. Gupta*
Computer Methods in Image Analysis, *Edited by J. K. Aggarwal, R. O. Duda, and A. Rosenfeld*
Microprocessors: Fundamentals and Applications, *Edited by W. C. Lin*
Machine Recognition of Patterns, *Edited by A. K. Agrawala*
Turning Points in American Electrical History, *Edited by J. E. Brittain*
Charge-Coupled Devices: Technology and Applications, *Edited by R. Melen and D. Buss*
Spread Spectrum Techniques, *Edited by R. C. Dixon*
Electronic Switching: Central Office Systems of the World, *Edited by A. E. Joel, Jr.*
Electromagnetic Horn Antennas, *Edited by A. W. Love*
Waveform Quantization and Coding, *Edited by N. S. Jayant*
Communication Satellite Systems: An Overview of the Technology, *Edited by R. G. Gould and Y. F. Lum*
Literature Survey of Communication Satellite Systems and Technology, *Edited by J. H. W. Unger*
Solar Cells, *Edited by C. E. Backus*
Computer Networking, *Edited by R. P. Blanc and I. W. Cotton*
Communications Channels: Characterization and Behavior, *Edited by B. Goldberg*
Large-Scale Networks: Theory and Design, *Edited by F. T. Boesch*
Optical Fiber Technology, *Edited by D. Gloge*
Selected Papers in Digital Signal Processing, II, *Edited by the Digital Signal Processing Committee*
A Guide for Better Technical Presentations, *Edited by R. M. Woelfle*
Career Management: A Guide to Combating Obsolescence, *Edited by H. G. Kaufman*
Energy and Man: Technical and Social Aspects of Energy, *Edited by M. G. Morgan*
Magnetic Bubble Technology: Integrated-Circuit Magnetics for Digital Storage and Processing, *Edited by H. Chang*
Frequency Synthesis: Techniques and Applications, *Edited by J. Gorski-Popiel*
Literature in Digital Processing: Author and Permuted Title Index (Revised and Expanded Edition), *Edited by H. D. Helms, J. F. Kaiser, and L. R. Rabiner*
Data Communications via Fading Channels, *Edited by K. Brayer*
Nonlinear Networks: Theory and Analysis, *Edited by A. N. Willson, Jr.*
Computer Communications, *Edited by P. E. Green, Jr. and R. W. Lucky*
Stability of Large Electric Power Systems, *Edited by R. T. Byerly and E. W. Kimbark*
Automatic Test Equipment: Hardware, Software, and Management, *Edited by F. Liguori*
Key Papers in the Development of Coding Theory, *Edited by E. R. Berkekamp*
Technology and Social Institutions, *Edited by K. Chen*
Key Papers in the Development of Information Theory, *Edited by D. Slepian*
Computer-Aided Filter Design, *Edited by G. Szentirmai*

Adjustable Speed AC Drive Systems

Edited by
Bimal K. Bose
Corporate Research and Development
General Electric Company

A volume in the IEEE PRESS Selected Reprint Series,
prepared under the sponsorship of the
IEEE Industry Applications Society.

IEEE PRESS

The Institute of Electrical and Electronics Engineers, Inc., New York

PRINTED IN THE UNITED STATES OF AMERICA

Sole Worldwide Distributor (Exclusive of IEEE):

JOHN WILEY & SONS, INC.
605 Third Ave.
New York, NY 10016

Wiley Order Numbers: Clothbound: 0-471-09395-5
Paperbound: 0-471-09396-3

IEEE Order Numbers: Clothbound: PC01404
Paperbound: PP01412

Library of Congress Catalog Card Number 80-27789

Library of Congress Cataloging in Publication Data
Main entry under title:

Adjustable speed Ac drive systems.
(IEEE Press selected reprint series)
Bibliography: p.
Includes indes.
1. Electric driving. I. Bose, Bimal K.
TK4058.A3 621.46'2 80-27789
ISBN 0-87942-145-2
ISBN 0-87942-146-0 (pbk.)

Contents

Preface

THE TECHNOLOGY OF solid-state speed control of ac machines made great strides during the last one and a half decades. Traditionally, ac machines were considered suitable for constant speed applications though complex, inefficient, and expensive methods of speed control were known before the advent of the solid-state era. For a long time, dc motors were the work horses in the industry for adjustable speed applications. Therefore, when the thyristors were introduced in the late nineteen-fifties, these found immediate application in the converters for dc drives. In the early nineteen-sixties, solid-state inverters employing elegant commutation techniques were introduced by several notable contributors. This resulted in the realization of practical and efficient static variable-voltage variable-frequency power supplies required for the speed control of ac motors.

The ac machines, especially the cage-type induction motors, seem to possess several distinct virtues in comparison with dc machines. These relate to lower cost and weight, lower inertia, higher efficiency, improved ruggedness and reliability, low maintenance requirement, and the capability to operate in a dirty and explosive environment due to the absence of commutators and brushes. Some of these virtues are of paramount importance, which makes the ac drive mandatory in several areas of application. In addition, when precision speed control is needed or close speed tracking in multimotor drives is required, the synchronous machines seem to be an obvious choice.

In spite of the many virtues of ac machines, the cost of the converters and complexity of control requirement are the main factors which are at present impeding the widespread application of ac drives in competition with dc drives. The research and development effort in recent years has been focused mainly in those directions. The price of thyristors has practically levelled off, though the performance, including the voltage and current ratings, is continuously improving. The power transistor ratings have recently increased significantly and they are finding application in low to medium power drives. The power MOS devices are also on the horizon. The light fired thyristors are beginning to get attention for converter applications. The replacement of discrete semiconductors by integrated power modules will help to substantially reduce the semiconductor packaging cost and the ultimate reality of mounting the integrated converter system on the machine frame is possibly not far away. The cost of control electronics has been substantially reduced because of the advent of LSI technology. The microcomputer implementation of the control system not only provides simplification of hardware and improvement of reliability but permits performance optimization of the drive system which could not be possible by hardwired control. The ac machines are being designed progressively for improved efficiency without much penalty of additional cost. The permanent magnet synchronous motors are being widely used recently because of the absence of field power supply and the attendant copper loss. The modeling and computer simulation of inherently complex ac drive systems have received wide attention in the literature, and the control performance of ac machines is now almost comparable to that of dc machines.

Considering the present trend of the technology, widespread application of ac drives seems inevitable in the near future. There is at present a great awakening in the industrial community with an intense urge to understand the basics of ac drives technology. Unfortunately, the literature on ac drives has grown immensely and has proliferated in different directions, so that a motivated reader trying to study the literature gathers only frustration. No suitable textbook has yet been published in this area.

This book is intended for the engineers in industry who are willing to study ac drives and eventually to use the knowledge for development and design of the drive systems. The content of the book may also be considered as a one semester graduate course in a university. The readers, of course, are assumed to have basic knowledge in converters, machines, and control theories. The book essentially consists of fifty carefully selected reprinted papers from a vast literature on ac drives, but also includes a section specially contributed by the editor. The criteria for selection of the papers were such that they should reflect the mainstream of technology and the write-up should be clear and concise with unity of thoughts and expressions. Often a well written review paper is considered as an essential ingredient. However, even with the above criteria, the selection process could not be easy because of so many excellent papers in the literature. Primarily, the papers from IEEE publications have been considered, though international publications in English have been reviewed, and only the papers which are of special reader interest have been considered.

The content of the book is classified principally into two parts: the first part deals with induction motor drives whereas the second part includes synchronous motor drives. Attempt has been made to organize the papers so that the theory, control, design, and application appear in sequence. Selected bibliography has been added at the end of the book. For the readers who are relatively new in this area, the papers may appear somewhat difficult to comprehend. For that reason, Section 1, which is the editor's own contribution, comprehensively reviews ac drives technology. It is a simpleminded and nonmathematical description of different types of ac drives which hopefully will provide the background to enable one to successfully launch into the other sections. Section 2 describes induction motor drives with phase-controlled ac power supply to the stator. This type of drive has simple converter configuration and is used in low to medium power applications. Section 3 describes variable-voltage variable-frequency inverter drives of induction motors which is the most common type of ac drive today. The literature in this area is vast and therefore maximum numbers of papers are included in this section. Section 4 deals with variable-frequency current-controlled induction motor drives which are receiving wide attention today because of several attractive features. Section 5 describes induction motor drives with slip power regulation, where wound rotor machines are used with converters in the rotor circuit. This type of drive is favoured in large power and limited speed range applications. Section 6 describes synchronous motor drives using current-controlled inverters. This type of drive normally requires shaft position sensitive signals to generate the firing pulses of the inverter and is commonly known as commutatorless dc motor (CLM) or electronically commutated dc motor (ECM). Section 7 deals with cycloconverter-controlled synchronous machines drives. The book does not include any paper on cycloconverter-fed induction motor or voltage-fed synchronous motor, though these are by no means unimportant. Single-phase ac drives which are so common in low power applications are also excluded. Only papers which exclusively relate to ac drives have been considered. Of course, exception has been made in several cases.

Before concluding, I must acknowledge the help of several of my professional colleagues. Prof. R. G. Hoft of the University of Missouri, Columbia, has given valuable input to this book. The discussion with Prof. T. A. Lipo, University of Wisconsin, from time to time was very helpful. Thanks are also due to Mr. F. G. Turnbull and Dr. R. L. Steigerwald, who are my colleagues in General Electric Company.

Bimal K. Bose
General Electric Company
Corporate Research and Development
Schenectady, New York 12301

October 8, 1980

Section 1
Introduction to AC Drives

BIMAL K. BOSE
General Electric Company
Schenectady, NY 12301

There are many industrial applications where variable speed drives of electrical machines are needed. This requirement can be met either by dc or by ac machines. The speed of a dc machine can be varied by controlling the armature voltage through a phase-controlled rectifier, or by a dc-dc converter if the input power supply is dc. In a dc machine, the torque is related to the product of field flux and dc armature mmf which remain stationary in space. In an ac machine, a three-phase ac power supply produces a rotating magnetic field in the airgap which reacts with the rotor mmf wave to develop the torque. The rotor mmf in a synchronous machine is created by a separate field winding which carries dc current, whereas in an induction motor it is created by the stator induction effect. The speed of an ac machine is related to stator supply frequency which produces the synchronously rotating magnetic field. If frequency is increased to increase speed of the machine, the magnitude of air-gap flux is reduced due to increased magnetizing reactance, and correspondingly the developed torque is reduced. For this reason, an ac machine normally requires variable-voltage variable-frequency power supply for speed control. This type of power supply can be obtained by a dc link converter system which consists of a rectifier followed by an inverter, or by a cycloconverter. The machine can be excited by a current source instead of a voltage source. The current source operation has several advantages but somewhat complicates the machine operation. Since the voltage or current waves are fabricated by solid state switches of a converter, the harmonics associated with the waves cause problems of harmonic heating and torque pulsation. The converter which supplies power to the machine is expensive, and is designed with a limited peak power rating which may be substantially less than that of the machine. This influences the machine performance when compared to its performance with the utility supply. An ideal voltage-fed inverter should offer zero Thevinin impedance at the machine terminal, and similarly an ideal current-fed inverter should have infinite Thevinin impedance. Such conditions are difficult to meet in practice because of cost considerations. The finite source impedance can be shown to have an effect on the harmonic performance and stability condition of the drive system. In this section, the various performance and control features of converter-fed ac drive systems have been discussed in detail. The background knowledge gained in this section will help to successfully explore the other sections which describe all the features in considerable detail.

1. Induction Motor Drive

The speed of an induction motor is determined by the synchronous speed and slip of the rotor. The synchronous speed is related to the supply frequency and the slip can be controlled by the regulation of voltage or current supplied to the motor. There are several mechanisms for controlling the speed of induction motor. These are: (1) variable-voltage constant-frequency or stator voltage control, (2) variable-voltage variable-frequency control, (3) variable-current variable-frequency control, and (4) regulation of slip power. The operation principle and characteristics of these methods are described as follows.

1.1 Stator Voltage Controlled Drive

This is a simple and economical method of speed control of a cage type induction motor. In this scheme, the stator supply voltage is controlled at line frequency

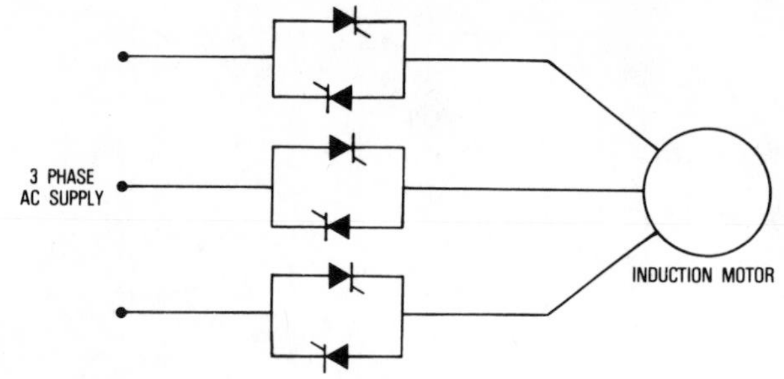

Fig. 1. Induction motor speed control with phase-controlled ac power supply.

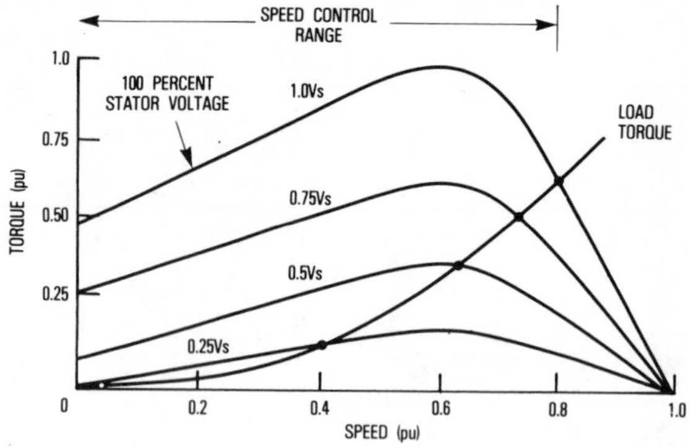

Fig. 2. Torque-speed curves of induction motor with variable stator voltage.

by symmetrically controlling the trigger angles of three-phase, line commutated antiparallel thyristors as shown in Fig. 1. In low power application, triacs can be used giving further simplification to the power and control circuits. The stator voltage can be varied steplessly between zero and the full value within the trigger angle range $0 < \alpha < 120°$, giving the circuit a "solid-state Variac" characteristic. Of course, the load and the supply lines carry rich harmonics at integral frequencies which are generated by the converter. If the machine is star-connected with isolated neutral, the triplen harmonics are eliminated. The line side power factor is poorer than that with Variac control because of the added harmonics and the additional reactive power taken by the converter due to phase control.

Though the converter is very simple, the drive system is characterized by poor performance. It is used in low to medium power applications, especially with pump or blower type loads, where the torque is low at starting but increases typically as square of the speed. Single-phase fractional horsepower drives with phase control have been widely used in appliance type control, where efficiency and power factor considerations have been sacrificed for the advantage of low cost. Another area of recent application is the solid state starter for medium to high horsepower induction motors. Here, the motor is generally started at no load with stator voltage control to limit the inrush current. At normal running condition with full voltage and load, the thyristors are often bypassed by mechanical contactors to eliminate the device standby loss.

The motors with high slip (typically 10 to 12 percent) are used in this method of speed control which correspondingly causes higher copper loss in the machine. Fig. 2 explains the speed control characteristics of a blower type drive. As the stator voltage is reduced, the developed torque is reduced correspondingly, because the torque varies with square of the voltage at a fixed slip. The stable speed at different stator voltage is given by the point of intersection as shown in the figure. Evidently, the range of speed control will be diminished if the machine is designed with low slip. On the other hand, by designing the machine with high slip (i.e., high rotor resistance), the constant torque type load can be controlled in the full range of speed. In this method of speed control, the torque per ampere of stator current is reduced as the speed is decreased by stator voltage control. The airgap flux which is related to the ratio of stator voltage and frequency decreases as the voltage is decreased. Therefore, the magnitude of current increases for the same value of torque at reduced voltage resulting in a poor torque/amp characteristic. It can be shown that for the blower type of load ($T_1 \alpha N^2$), the stator current reaches maximum value at 66.7 percent of synchronous speed; and for a machine designed with 12 percent full load slip, the current is 25 percent higher than the full load rated current. The harmonic currents, of course, add to the above theoretical values and the resulting high copper loss at reduced speed may cause severe machine heating problem. This type of drive is amenable to plugging braking by reversal of phase sequence, but the additional dissipation of rotor power compounds the heating effect. Because of the above complexities, the machine-converter system is to be carefully designed for each application with the given performance specifications.

Though many configurations of phase control circuit are possible, two additional practical circuits are shown in Figs. 3 and 4. In the delta-connected controller, a thyristor pair is connected in series with each delta winding and each phase constitutes as an independent single-phase load to the line. Once the phase current wave is determined, the line current wave can be constructed by superposition. One apparent advantage of this connection is that the thyristors have to handle less fundamental current, but the harmonic currents become higher offsetting this advantage. The neutral-connected controller in Fig. 4 is simplest of all, but the harmonic penalty is very severe. At a reduced speed, the power consumed by the load can be as high as 100 percent greater than that with sine wave voltage control.

1.2 Voltage-Fed Square-Wave Inverter Drive

The voltage-fed inverters are generally classified into two types: square-wave inverter and pulse-width modulated inverter. This type of inverter was intro-

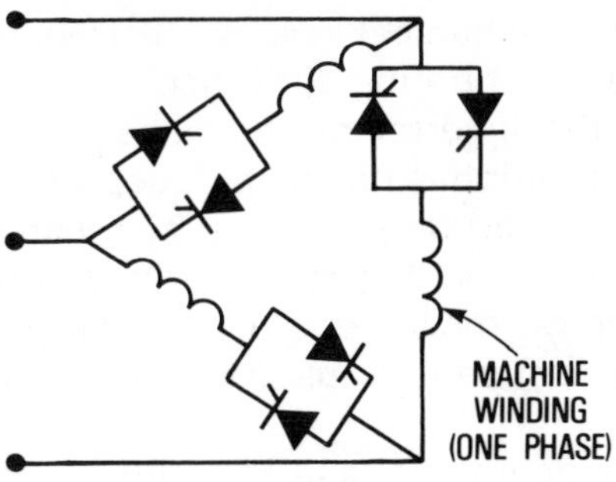

Fig. 3. Delta-connected controller.

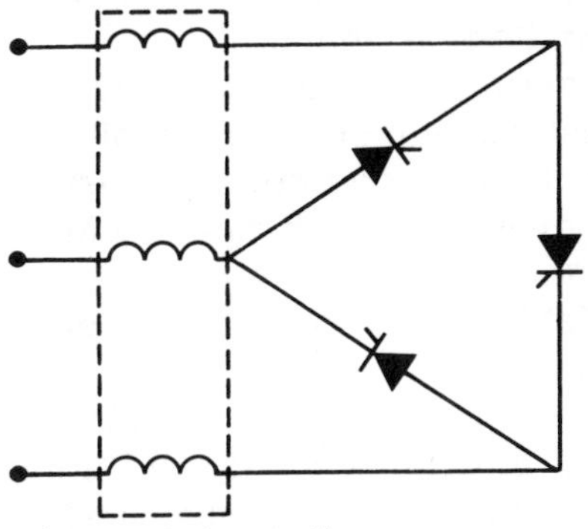

Fig. 4. Neutral connected controller.

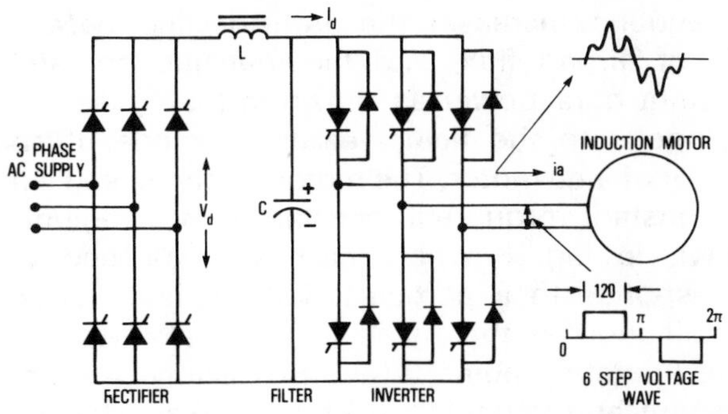

Fig. 5. Variable-voltage variable-frequency square wave induction motor drive (inverter forced commutation is not shown).

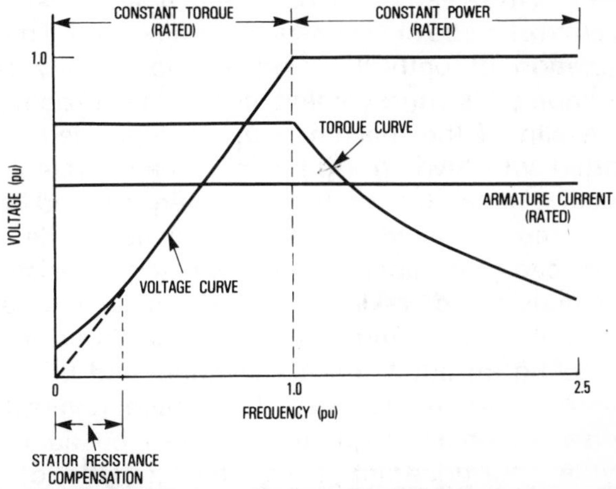

Fig. 6. Voltage-frequency relation of induction motor.

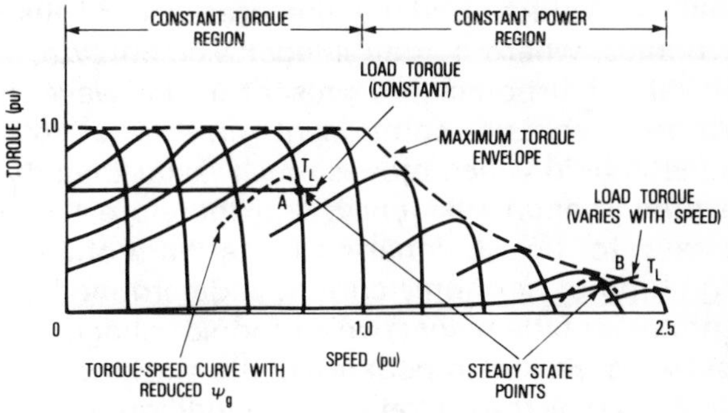

Fig. 7. Torque-speed curves of induction motor with variable-voltage variable-frequency power supply.

duced from the beginning of nineteen-sixties when elegant force-commutation techniques were introduced. Figure 5 shows the conventional power circuit of a square-wave inverter drive. A three-phase bridge rectifier converts ac to variable-voltage dc, which is impressed at the input of a force-commutated bridge inverter. The inverter generates a variable-voltage variable-frequency power supply to control the speed of the motor. The inverter is called voltage-fed because a large filter capacitor provides a stiff voltage supply to the inverter and the inverter output voltage waves are not affected by the nature of the load. Normally, each thyristor of an inverter leg conducts for 180° to generate a square-wave voltage at the machine phase with respect to the fictitious center point of the dc supply. The line to line voltage can be shown to be a six-stepped wave as shown in the figure. Since the induction motor constitutes a lagging power factor load, the inverter thyristors require forced commutation. The feedback diodes help circulation of load reactive power with the filter capacitor and maintain the output voltage waves clamped to the level of dc link voltage. The diodes also participate in the commutation and braking process which will be explained later.

The theory of variable-voltage variable-frequency speed control method can be explained with the help of Figs. 6 and 7. The motor used in this type of drive has low slip characteristic which results in improvement of efficiency. The speed of the motor can be varied by simply varying its synchronous speed, i.e., by varying the inverter frequency. However, as the frequency is increased, the machine airgap flux falls causing low developed torque capability. The airgap flux can be maintained constant as in a dc shunt motor if the voltage is varied with frequency so that the ratio remains constant. Figure 6 shows the desired voltage–frequency relationship of the motor. Below the base (1.0 PU) frequency, the airgap flux is maintained constant by the constant volts/Hz ratio which results in constant torque capability. At very low frequency, the stator resistance dominates over the leakage inductance, and therefore additional voltage is impressed to compensate this effect. At base frequency, the full motor voltage is established as permitted by fully advancing the rectifier firing angle. Beyond this point, as frequency increases, the torque declines because of loss of airgap flux, and the machine operates in constant horsepower as shown in the figure. This is analogous to the field weakening mode of speed control of a dc motor. The motor torque-speed curves for constant torque and constant power regions are shown in Fig. 7 where each torque-speed curve corresponds to a particular voltage and frequency combination at the machine terminal. Two steady state operating points A and B which correspond to constant and variable load torque, respectively, are shown in the figure. The machine can accelerate from zero speed at maximum available torque and then approach the steady state points either at constant flux slip control mode or at constant slip flux control mode. Regulation of both flux and slip for steady state operation adds improvement of machine efficiency.

The slip of the machine can be controlled to be negative which will make the machine operate as an induction generator and then the energy stored in the system inertia can be pumped back to the dc side. This energy can be dissipated into a switched-in resistor to provide dynamic braking of the drive system, or else, regenerative breaking can be made possible by connecting an additional phase-controlled bridge in antiparallel to the rectifier. The speed reversal by reversal of phase sequence can also be added to provide four-quadrant drive characteristic of the system.

The six-stepped voltage wave at the machine terminal is characterized by the presence of $6n \pm 1$ harmonics, where $n =$ an integer, i.e., 5th, 7th, 11th, 13th, etc., harmonics are present in the wave. Each harmonic voltage component creates a rotating magnetic field which moves much faster than that of the fundamental frequency, and therefore the rotor appears to be stationary to the harmonics. The magnitude of harmonic current is determined by the magnitude of the respective harmonic voltage and the passive equivalent impedance of the machine, i.e., it is not affected by the operating slip condition. Therefore, at light load operation, the ripple current may exceed the fundamental current. With variable-voltage variable-frequency supply, the harmonic currents tend to remain constant, but as the machine enters into constant power region, these attenuate because of increasing frequency resulting in improvement of total copper loss. The harmonic heating effect should be taken into consideration to determine the frame size of the machine for a specified horsepower output. An additional harmful effect of harmonics is the torque pulsation on the machine. It can be shown that the harmonic torque at the lowest frequency ($6f$) may be caused by interaction of fundamental airgap flux and 5th and 7th harmonic currents. The effect of harmonic torque tends to be filtered by the large inertia of the drive system. However, if the frequency of the torque is near the mechanical resonance frequency of the system, severe hunting may occur. The lower harmonics of the inverter output voltage can be eliminated by multiphasing technique, i.e., 12 or 24 stepped-wave can be generated by mixing the phase shifted inverter voltages through a multiwinding transformer or machine. Such a complex and expensive system can be justified only for high power drive.

Since the dc link voltage is generated by a phase-controlled rectifier, the input line power factor tends to deteriorate as the dc voltage is reduced. The line current wave is ideally six-stepped and therefore its harmonic pattern is identical with inverter voltage wave. The harmonics in the line current wave can be improved by the multiphasing technique as explained before. A simple diode bridge rectifier followed by a dc-to-dc converter can substantially improve the line power factor.

The thyristors in a voltage-fed inverter need to be force-commutated as mentioned before, and in Fig. 5 the inverter requires six commutations per cycle of fundamental frequency. In forced commutation, a precharged capacitor creates a voltage or current transient around the conducting thyristor. This causes diversion of the main current and an inverse voltage impressed across the device helps to turn it off. Many different commutation techniques are possible and only a few common types are shown in Fig. 8. The McMurray, Verhoef, and Brown Boveri commutation schemes operate on an auxiliary commutation principle, i.e., auxiliary thyristors are used to turn off the conducting thyristors. The inverters using this type of commutation adapt well to pulse-width modulation. The McMurray-Bedford inverter uses a complementary commutation method, i.e., switching in of a main thyristor causes the conducting device to turn off. The McMurray and Brown Boveri inverters are also known as current commutated inverters, because the bypass diode carries the commutation resonance current and helps to turn off the main thyristor by the limited diode drop only. The other two types are known as voltage commutated where large capacitor voltage is impressed across the conducting thyristor.

Since the commutation capacitor is usually charged by the dc link voltage, the commutating capability deteriorates when the dc voltage decreases. Therefore, in Fig. 5, the inverter soon reaches its commutation limit as dc link voltage is reduced. Due to this reason, the inverter is normally provided with an auxiliary constant voltage dc supply for commutation purpose.

The voltage-fed square-wave drives are normally used in low to medium horsepower industrial applications where the speed ratio is usually limited to 10:1. Recently, this type of drive has largely been superseded by PWM drives which will be described in

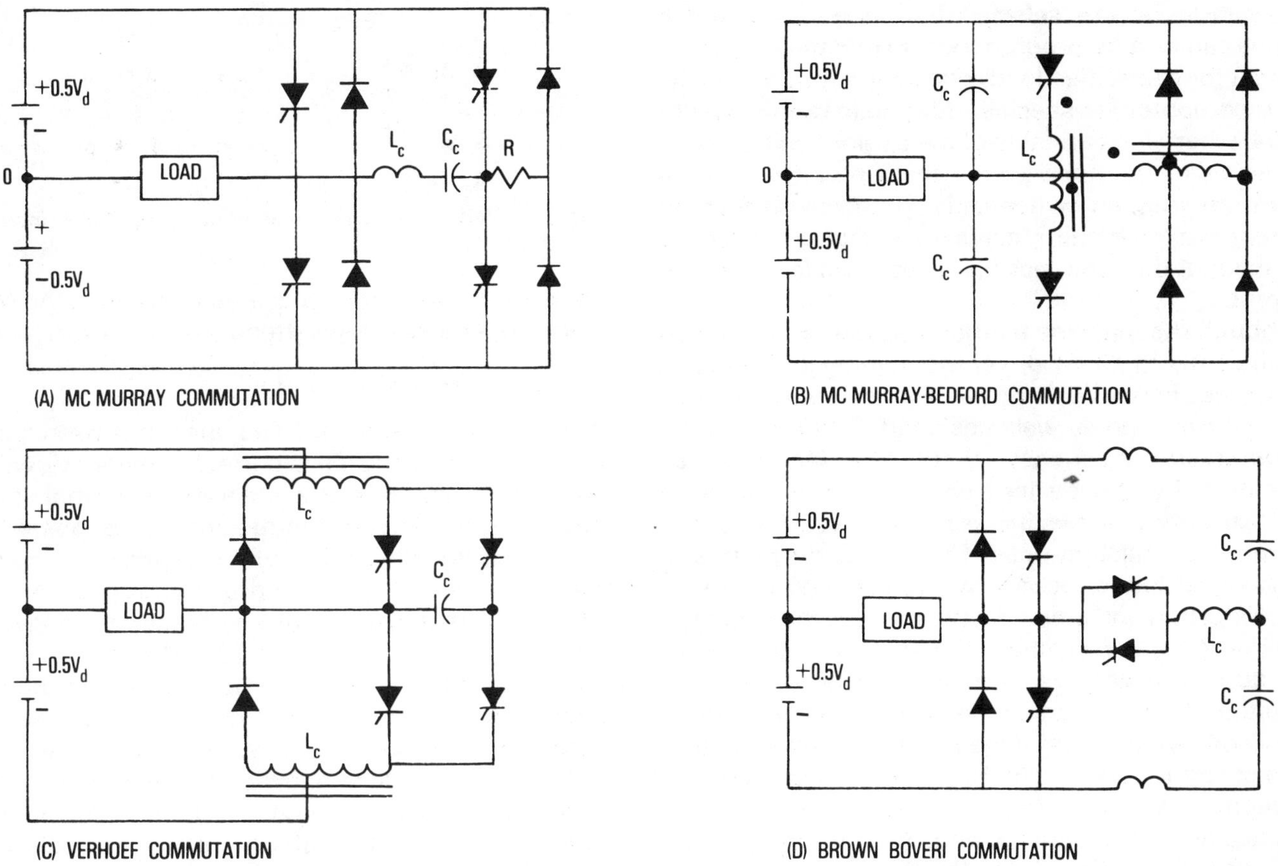

Fig. 8. Several commutation circuits with half-bridge inverter.

next section. The voltage-fed inverters are easily adaptable to multimotor drives where speed of a number of induction motors can be closely tracked.

1.3 PWM Inverter Drive

In the variable-voltage variable-frequency inverter drive described in the previous subsection, the dc link voltage can be kept uncontrolled by a diode rectifier and the fundamental frequency output voltage can be controlled electronically within the inverter by using a pulse-width modulation technique. In this method, the thyristors are switched on and off many times within a half-cycle to generate a variable-voltage output which is normally low in harmonic content. Among several PWM techniques, sinusoidal PWM is common and it is explained in Fig. 9. An isosceles triangle carrier wave is compared with the sine wave signal and the crossover points determine the points of commutation. Except at low frequency range, the carrier is synchronized with the signal, and an even integral (multiple of three) ratio is maintained to improve the harmonic content. The fundamental output voltage can be varied by variation of the modulation index. It can be shown that if the modulation index is less than unity, only carrier frequency harmonics with fundamental frequency related side bands appear at the output. Such a waveform causes considerably less harmonic heating

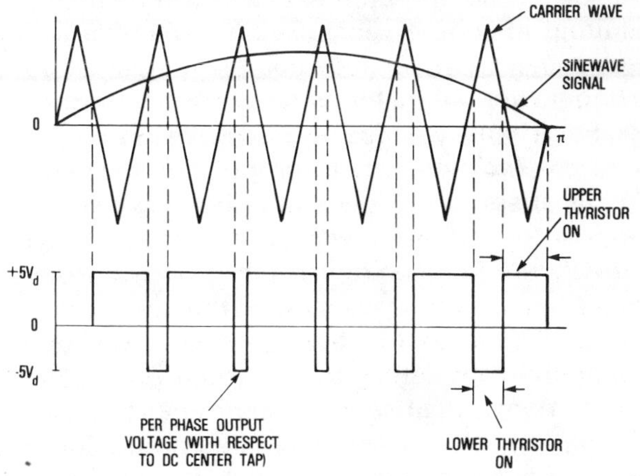

Fig. 9. Principle of pulse width modulation.

and torque pulsation compared to that of a square wave. The voltage can be increased beyond the modulation index of unity until maximum voltage is obtained in square wave mode. Therefore, PWM voltage control is applicable in the constant torque region (see Fig. 6), whereas in the constant power region the operation is identical to that of square-wave drive.

The technique of selected harmonic elimination PWM has received wide attention recently. In this method, notches are created at predetermined angles of the square-wave which permits voltage control with elimination of the selected harmonics. The notch angles can also be programmed so that the rms ripple current for a specified load condition is minimum. The microcomputer is especially adaptable to this type of PWM, where look up table of the angles can be stored in the ROM memory. In the bang bang method of PWM, the inverter switching is controlled so that the current wave remains confined within a hysteresis band about the reference wave resulting in low ripple current.

Though the machine harmonic losses are improved significantly in PWM drive, the inverter efficiency is somewhat lessened because of many commutations per halfcycle. In a well designed PWM drive, the commutation frequency should be increased as permitted by the devices so as to obtain a good balance between the increase of inverter loss and decrease of machine loss. The use of a simple and economical diode rectifier in the front end improves line distortion and power factor, reduces the filter size, and improves the reliability of system operation. Since the dc link voltage is relatively constant, the commutation of thyristors is satisfactory in the whole range of fundamental voltage. In addition, low harmonics and minimal torque pulsation in the low frequency region permit wide range speed control practically from standstill with the full torque capability of the machine. Since the dc link voltage is not controlled, a number of inverters with independent control can be operated with a single rectifier supply resulting in considerable saving of rectifier cost. The drive system can be made uninterruptible for possible ac line power failure by switching in a battery in the dc link. For a battery or dc powered drive system, such as, electric vehicle or subway propulsion, the supply can directly absorb the regenerative braking power.

In low to medium power applications, the thyristors in the inverter can be replaced by power transistors. A power circuit with transistors in the inverter is shown in Fig. 10. The cost of transistors may be somewhat higher than thyristors for a certain power handling requirement, but absence of commutation circuits and elimination of corresponding commutation losses may make the transistor circuit more economical and efficient. Since the transistor switches at faster speed, pulse-width modulation is possible at higher frequency cutting down the harmonic losses of the machine further. The state of the art single chip power transistors are available typically with the ratings, 500V and 200A. Recently, General Electric Company introduced integrated power Darlington Modules (450V, 200A) which were used in electric vehicle drive system. For higher power applications, the devices can be connected in series-parallel configuration adding cost and complexity of the inverter. For low power and higher frequency applications, power MOS inverter looks attractive because of reduced losses and low

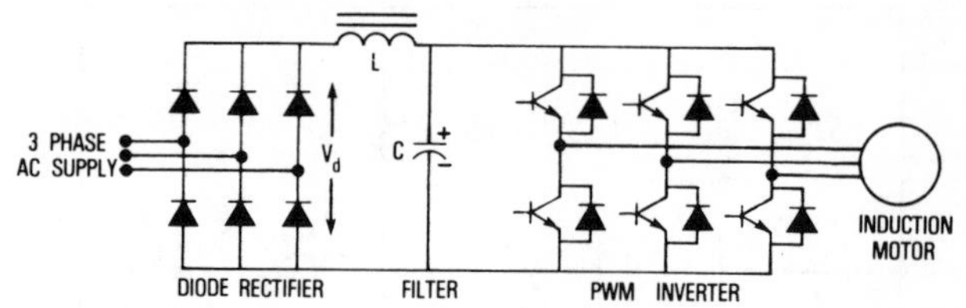

Fig. 10. Variable-voltage variable-frequency PWM induction motor drive.

gate drive requirement. It is anticipated that the MOS devices will receive wide attention in the future.

1.4 Current-Fed Inverter Drive

During the last several years, there has been a wide interest in the controlled current inverter drives. A current-fed inverter likes to see a stiff dc current source (ideally infinite Thevinin inductance) as opposed to a voltage-fed (ideally zero Thevinin impedance) inverter which was described in the previous subsection. Fig. 11 shows a schematic diagram of current-fed inverter drive and Fig. 12 gives the typical phase voltage and current waves. A phase-controlled rectifier generates variable dc voltage which is converted to a current source by connecting a large inductor in series. A diode rectifier followed by a dc chopper can also constitute the variable voltage dc source. The thyristors of the inverter steer the current source I_d symmetrically to the three phases of the machine to generate a variable-frequency, six-stepped current wave. Since the source is considered stiff, the wave is not affected by the nature of the load, i.e., it is dual to the voltage wave of a voltage-fed inverter. It can be shown that the machine terminal voltage is nearly sinusoidal with superimposed voltage spikes due to commutation.

The motor torque-speed characteristics at different current but at a fixed frequency is shown in Fig. 13. If, for example, the machine is operated with rated current ($I_d = 1.0$ PU), the starting developed torque will be very low compared to the voltage-fed machine, because airgap flux will be low due to low machine impedance. As the motor speed increases, the terminal voltage rises due to higher motor impedance and, as a result, torque increases with higher airgap flux. If saturation of the machine is neglected, the torque rises to a high value as shown by the dotted line, and then decreases to zero with steep slope at synchronous speed. In a practical machine, however, the saturation will limit the developed torque as shown by the solid line. A torque curve with rated voltage is also shown in Fig. 13, where the portion with negative slope can be considered to have stable operation with the rated airgap flux. This curve intersects $I_d = 1.0$ PU torque curve at the point A. The machine can be operated either at A or B for the same torque demand. At the point B, the rotor current is lower but the airgap flux is somewhat higher with partial saturation

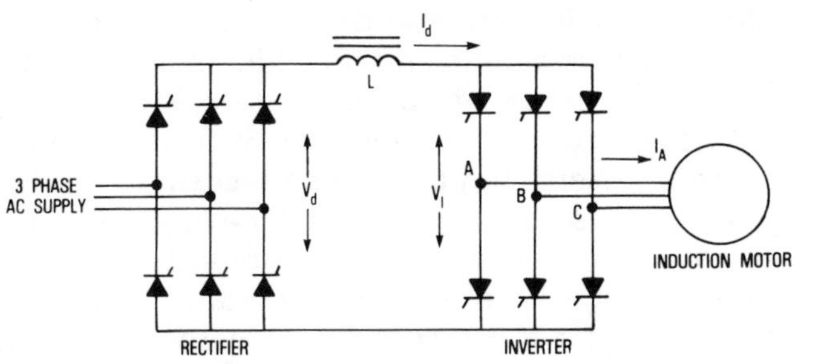

Fig. 11. Variable-current variable-frequency induction motor drive (inverter forced commutation is not shown).

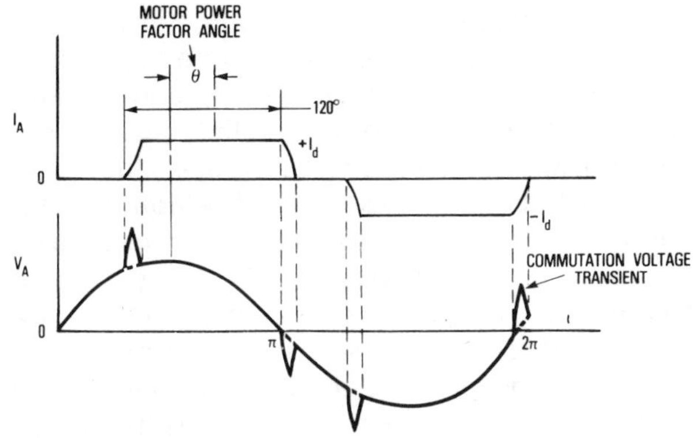

Fig. 12. Machine phase voltage and current waves.

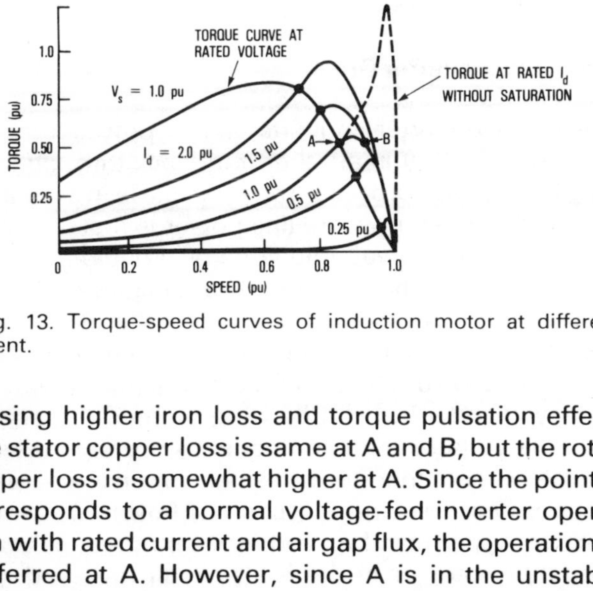

Fig. 13. Torque-speed curves of induction motor at different current.

causing higher iron loss and torque pulsation effect. The stator copper loss is same at A and B, but the rotor copper loss is somewhat higher at A. Since the point A corresponds to a normal voltage-fed inverter operation with rated current and airgap flux, the operation is preferred at A. However, since A is in the unstable region of the torque curve, close loop operation of the drive system is mandatory. The torque at the rated flux can be varied by varying the current and slip as shown in the figure so that the operation lies on the negative slope of the equivalent voltage torque curve. The various operation points in the torque speed plane which may lie in the constant torque or constant power regions, can be established by a variable-current variable-frequency power supply similar to that in Fig. 7.

The six-step current wave in a current-fed inverter may cause problem of torque oscillation at low speed operation. The fifth and seventh harmonic currents reacting with fundamental flux causes the sixth harmonic torque oscillation as explained before. This can be minimized by modulating the dc link current, or by introducing some amount of pulse-width modulation in the inverter. The six-stepped current wave also causes higher harmonic heating of the machine. If the machine horsepower is high, two inverters can be operated with phase shift to generate a 12-step current wave. The harmonic heating and torque pulsation effect of a 12-step inverter are substantially less.

Since the induction motor constitutes a lagging power factor load, the thyristors of the inverter need forced commutation. Several common force commutated inverter types are shown in Figs. 14, 15, and 16. The auto-sequentially commutated inverter (ASCI), as shown in Fig. 14, is very commonly used. The six capacitors and six diodes constitute the forced commutation circuit. The diodes tend to isolate the capacitors from the load and help to store energy for commutation. If, for example, the thyristor 3 is conducting in the beginning, it can be commutated by firing the incoming thyristor 5. The thyristor 3 is turned off by the reverse capacitor voltage, and subsequently the current is transferred from phase B to phase C with a resonant overlap angle which is determined by the magnitude of dc current, commutation capacitance and subtransient inductance of the machine. The commutation process causes transient overvoltage at the machine terminal as shown in Fig. 12. The designed capacitance value depends on the compromise of transient overvoltage and highest operating frequency of the inverter. Fig. 15 shows an inverter with individual auxiliary commutation. Here, an auxiliary thyristor bridge with the help of three capacitors permits independent commutation of each thyristor of the inverter. Since the commutation takes place through the auxiliary device, the duration of commutation tends to widen causing significant torque reduction at higher frequency. Fig. 16 shows the simplest commutation circuit where a capacitor is connected between the netural of the machine and the fourth leg of the inverter. The auxiliary thyristor A is responsible for commutation of the upper half of the bridge, whereas B commutates the lower half of the bridge. In a full cycle of operation, the capacitor voltage alternates six times and that is why this circuit is called third harmonic commutated inverter. The circuit, though simple, is not normally used for induction motor drive, because torque reduction due to commutation is substantial at higher frequency.

The current-fed inverter drive has several good features which can be summarized as follows. The power circuit is rugged and reliable, and there is no possibility of shoot-through fault as in a voltage-fed inverter. The inverter may also recover from occasional commutation failure. Any fault on the inverter side causes slow rise of fault current which can be cleared by rectifier gate suppression. In addition to

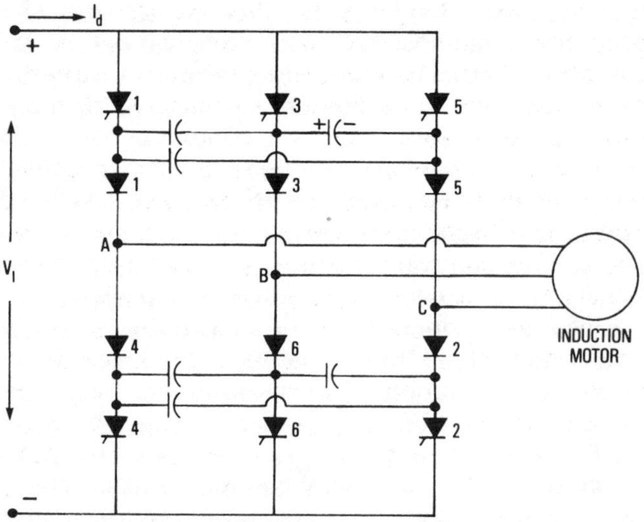

Fig. 14. Current inverter with autosequential commutation (ASCI).

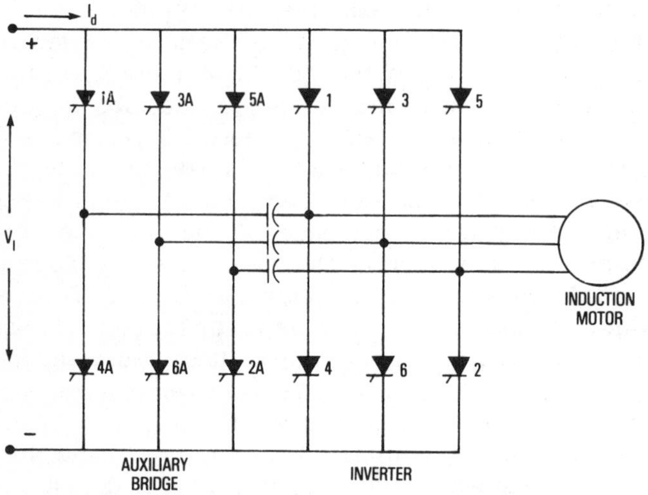

Fig. 15. Current inverter with individual commutation.

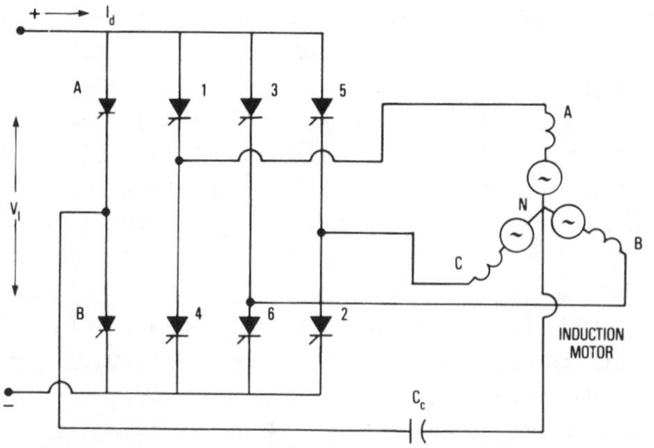

Fig. 16. Current inverter with fourth leg commutation.

less number of components in the inverter, the commutation loss is somewhat lower. Since only six thyristors are to be controlled, the control circuit is simpler and more reliable. The inverter can be designed with low speed rectifier grade thyristors and diodes permitting cost saving. One important feature of the current-fed inverter drive is that the regeneration process is simple and no additional component is needed in the power circuit. During regeneration, the machine runs as a generator with negative slip, and the inverter firing angle with respect to machine voltage wave is adjusted such that the input voltage V_I reverses in polarity. For the same direction of I_d, the rectifier voltage V_d is reversed by retarding the firing angle such that the power flows in the reverse direction.

In spite of several merits as mentioned above, the current-fed inverter drive has several limitations. The frequency range of the inverter is somewhat lower and it cannot operate at no load, i.e., some minimum load current is required to satisfactorily commutate the inverter. The large size of the dc link inductance and the commutation capacitors make the inverter somewhat bulky and expensive. The commutation is dependent on machine subtransient inductance which adds large transient overvoltage at the machine terminal. The response of the drive is somewhat sluggish and tends to give a stability problem at the light load high speed condition. The current-fed inverters are used in single motor drives in medium to high horsepower range. The multimotor operation is somewhat difficult but it is recently receiving considerable attention.

1.5 Cycloconverter Drive

A cycloconverter converts ac line power from one frequency to that in another frequency through a one step conversion process. The principle is in contrast to dc link conversion where line power is first converted to dc and then to variable-voltage variable-frequency source through an inverter. The technique of cycloconversion was known in Germany in the 1930's, when mercury arc rectifiers were used to convert power from 50 Hz to 16 2/3 Hz for railway transportation.

A cycloconverter can be programmed to generate variable-voltage variable-frequency power to drive an ac motor and therefore the performance characteristics as described in subsection 1.2 hold true here. The cycloconverter can be used in the stator circuit of an induction rotor or in the rotor circuit if wound rotor machine is used. The latter is known as static Scherbius drive and will be described in subsection 1.7. Figure 17 shows three-phase, three-pulse cycloconverter drive using 18 thyristors, and Fig. 18 gives the phase voltage and current waves. The power circuit in each phase consists of positive (P) and negative (N) groups of thyristors which permit bidirectional flow of phase current. The cycloconverter is said to operate in a noncirculating mode because no circulating current is permitted between the positive and the negative groups. The circuit operates in phase-control line-commutation mode and the firing angles of the thyristors are modulated to synthesize a mean

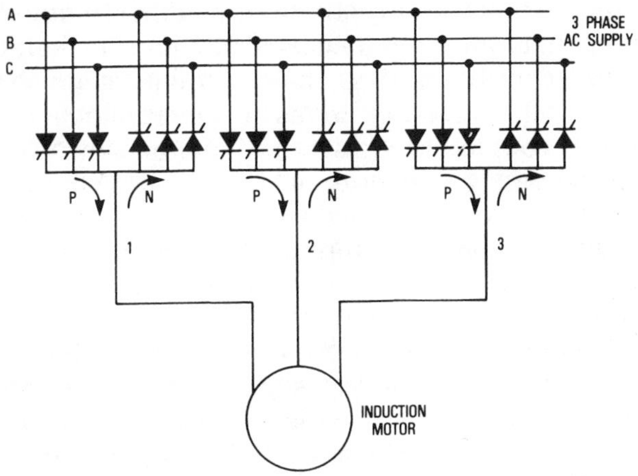

Fig. 17. Three-phase, three-pulse cycloconverter drive (showing noncirculating mode of operation).

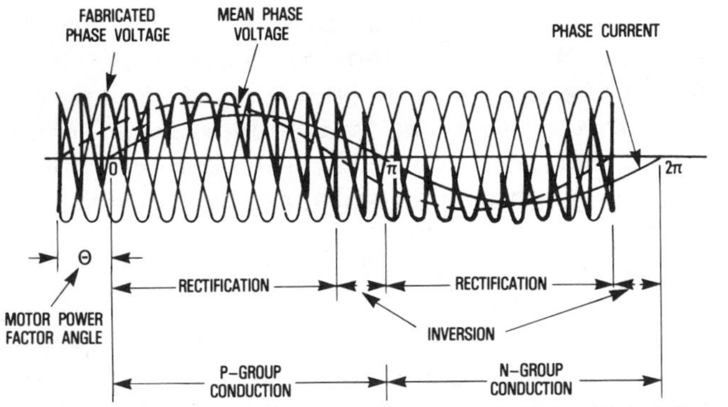

Fig. 18. Machine voltage and current waves with cycloconverter drive.

sine wave voltage as shown in Fig. 18. The raw fabricated voltage wave, when impressed across the machine terminal, results in nearly sinusoidal current at a lagging power factor. The positive phase current which is shared by the P-group of thyristors has two regions of operation, i.e., rectification and inversion. During rectification, the firing angle α varies between 0° and 90° when active power flows to the load. During inversion, on the other hand, α varies between 90° and 180° when reactive power is fed back to the source. The operation of the N-group is similar except that it shares the negative half-cycle of line current. The cycloconverter is normally operated as a step-down frequency changer and the output frequency is restricted typically to 1/3 or 1/2 of the line frequency.

The output voltage of a cycloconverter contains complex harmonic patterns which can be given by $k_1 n f_i \pm k_2 f_o$, where f_i = input frequency, f_o = output frequency, n = pulse number and k_1, k_2 are integers. The harmonic attenuation by the machine inductance is normally adequate. The harmonic currents are also reflected to the input line. Because of the phase control principle, the cycloconverter always presents lagging reactive power at the input irrespective of power factor of the load. The input power factor worsens if the fabricated output voltage is decreased and load power factor is reduced. The power factor can of course be improved by forced commutation which adds complexity and reduces efficiency of the converter. One advantage of cycloconverter drive is that regeneration is simple and the system can be designed easily for four-quadrant operation.

The cycloconverter drives are normally used in very large horsepower applications. The cost and complexity of power and control circuits make them uncompetitive with other classes of drives in general applications. They have been used in diesel electric locomotives, where a high frequency alternator coupled to the engine shaft provides power at the input. Several cycloconverter motor drive units, each operating from the same source, can independently power the axles of the locomotive. Another area of cycloconverter application is the gearless cement mill or ball mill drives. Here, the mill drive is directly connected to the motor which is supplied by a low frequency cycloconverter.

1.6 Static Kramer Drive

This class of drives requires a wound rotor induction motor compared to the other types discussed so far which used simple and rugged cage motors. The principle of speed control of a wound rotor machine by mechanically varying the rotor circuit resistance was known long ago. The equivalent rotor resistance can also be varied electronically either by an ac phase controlled circuit or by a diode rectifier and a chopper. The inefficiency of the drive system because of large slip power dissipation can be overcome by the static Kramer drive which is shown in Fig. 19. In this scheme, the slip power of the rotor is rectified by a diode rectifier and is then pumped back to the ac line through a line-commutated inverter. It is the recent version of the older scheme which used auxiliary machines in the rotor circuit.* Since the drive system operates only in the subsynchronous speed range, it is also known as a drive with subsynchronous static converter cascade. At any operating condition of the motor, the power input to the stator is distributed between the shaft power output and the power fed back to the line. The torque-speed characteristics of the machine at different cos α, where α = firing angle of the inverter, are shown in Fig. 20. The machine always operates with nearly constant airgap flux which is established by the fixed stator voltage and frequency condition. It can be shown that the torque is proportional to dc link current I_d and the machine torque characteristic is similar to that of a separately excited dc motor. If the machine runs at synchronous speed, the developed torque will be zero making V_d and I_d of the dc link also zero, and correspondingly α of the inverter will be set to 90°. If now the load torque is applied, the machine will tend to slow down increasing V_d and I_d until developed

*Many authors in the literature confuse Static Kramer drive with Static Scherbius drive.

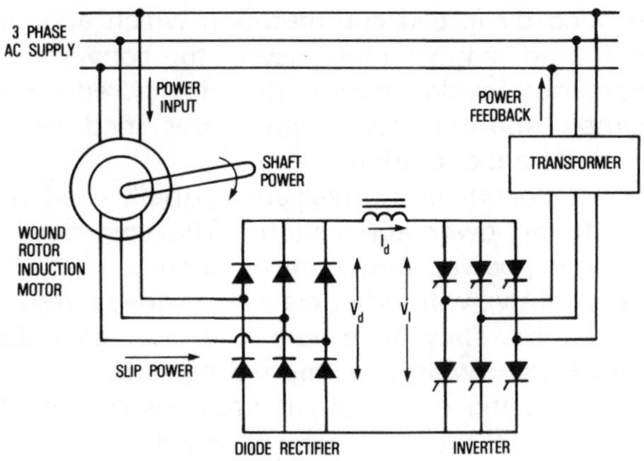

Fig. 19. Static Kramer drive system.

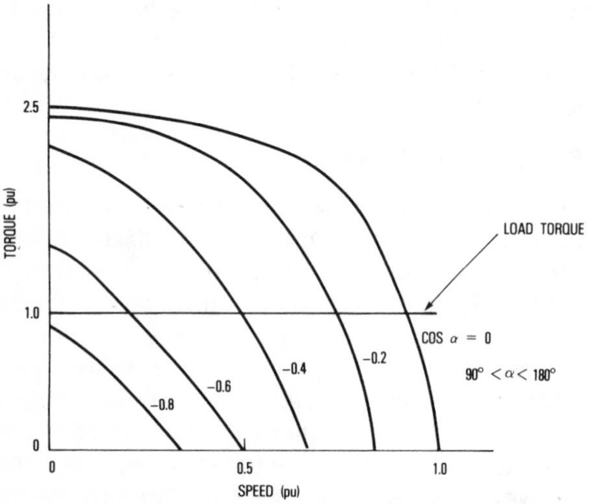

Fig. 20. Torque-speed curves at different firing angles of inverter.

torque balances the load torque. If load torque, i.e., the corresponding I_d is maintained constant, the speed can be decreased by increaseing cos α, such that V_d increases to match the V_I for the same I_d. Since the converter group handles only slip power, its power rating can be low, if subsynchronous speed control range is restricted. Obviously, if speed control up to standstill is desired, the converters are to be rated for the full machine power rating. The current inrush at the start up condition can be avoided in the converters by connecting separate starting resistances across the slip rings. The power factor of the drive system is poor because lagging reactive power is drawn by the machine to establish airgap flux and by the inverter for line commutation. At full load and rated speed, the power factor optimistically may be around 0.7, but decreases as torque and speed conditions are decreased. The power factor can be improved by designing the inverter so that α lies nearer to 180° for the designed speed control range. This requires a step-up transformer at the inverter output as shown in Fig. 19. At the lowest speed, V_d is maximum and correspondingly $\alpha = 180°$, so that the rotor induced voltage matches with the transformer induced voltage. The power factor of the inverter can of course be improved by using forced commutation. Another method of power factor improvement used in the recent literature is called commutatorless static Kramer system. In this scheme, the inverter drives a synchronous motor which is coupled to the same shaft. The synchronous machine with the inverter acts like a commutatorless dc machine (described later) and the reactive power required for inverter commutation is controlled by the field current.

The machine rotor current is typically a six-stepped wave as determined by the diode rectifier. Since the rotor voltage is low at low slip condition, the rectifier commutation overlap angle is large. This involves additional reactive current requirement at the stator input. Obviously, the harmonic currents are reflected to the ac line both from the machine and the inverter sides. The torque pulsation and additional heating effect due to harmonics should be considered in designing the drive system. Another disadvantage is that regeneration and speed reversal are not possible in the drive system. The static Kramer drives are used in large horsepower pump and blower type applications where limited range of speed control is required.

1.7 Static Scherbius Drive

In the static Kramer drive described in the previous subsection, the diode rectifier can be replaced by a thyristor bridge permitting slip power to flow in either direction. The speed of such a doubly-fed machine can then be controlled in both subsynchronous and supersynchronous regions. If the slip power is fed back to the line, the machine will operate in the subsynchronous region and the mode of operation will correspond to static Kramer drive. If, on the other hand, the slip power is pumped in the rotor by reversal of rectifier and inverter operation, the motor will operate in the super-synchronous region. The scheme is the static version of the older Scherbius system where rotory converter, dc motor and induction generator were used in the rotor circuit. The advantage of such a drive system is that full range (±50 percent at about synchronous speed) speed control is possible with converter rating only half that of the machine. The commutation of the machine side converter may become a problem at low slip and may therefore need forced commutation.

The dual converter system as explained above can be replaced by a phase-controlled line-commutated cycloconverter as shown in Fig. 21. The additional cost and complexity of the cycloconverter can be justified because of several advantages. The problem of commutation near synchronous speed disappears and the near sinusoidal current wave in the machine substantially improves torque pulsation and harmonic heating effects. The line current waveform is also improved correspondingly. A step-up transformer can be connected to the line side of the cycloconverter to

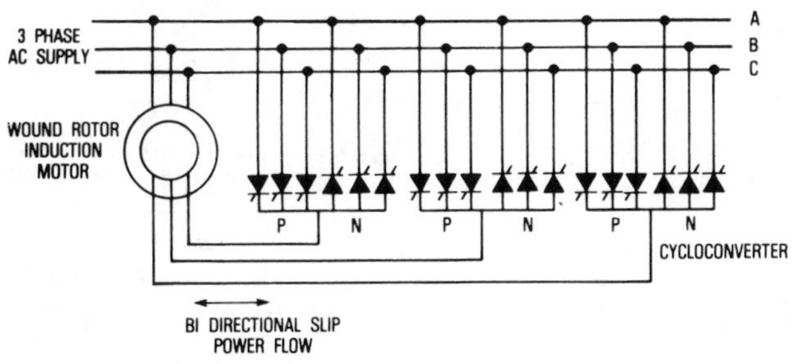

Fig. 21. Static Scherbius drive system.

reduce the voltage rating of the thyristors and to improve the line power factor. The drive system can have regeneration but speed reversal is not possible. The static Scherbius drive is used in very high horsepower pump and blower type applications.

2. Control of Induction Motor Drive

There are many possible methods for the formulation of induction motor drive control systems. The complexity of a particular method and the corresponding pay-off in performance are to be determined by the nature of application. While a typical industrial drive can be identified as a speed control system, there are many applications in which torque, current or position is controlled in the outer loop. A control system is characterized by the hierarchy of control loops, i.e., the external loop generates the command for the next inner loop which correspondingly generates command for the next loop and so on. The inner loops are designed to be progressively faster and the commands of all the loops are normally clamped to have limited excursion of the respective variables. Though the primary objective of the control system is to design transient response and steady state error of the control parameters within the desired specifications, it should take into consideration the efficiency and harmonic improvement, and fault conditions in the converter system.

2.1 Modeling and Simulation

The dynamics of the ac machine drive control system is extremely complex and the subject is receiving wide attention in the recent years. The complexity arises because of the nonlinearity and discrete time nature of the converter-machine system. For this reason, when a control strategy is developed, it is the usual practice to simulate the drive system on the computer and study the performance in detail before proceeding to build a breadboard. The electrical dynamics of an induction motor can be represented by a fourth-order nonlinear differential equation which may be either in a stationary reference frame (Stanley equation), or in a synchronously rotating reference frame. The advantage of the later model is that the steady state sinusoidally varying parameters appear as dc quantities. The rectifier and inverter can either be represented by the exact switching topology, or may be transformed and merged with the machine model neglecting their discrete time behavior. The converter-machine model with the proposed configuration of feedback control system can be simulated on a computer and control parameters may be fine tuned until desired performance is obtained. An alternative approach of modeling is that the converter-machine system can be linearized about a steady state operating point using the small signal perturbation principle. The corresponding transfer function model can be used to study stability of the control system by using the classical Nyquist, Bode, or root locus technique. Once the simulation performance is found to be satisfactory, the breadboard can be built and tested, which will show correlation with the simulation result.

A few words about the type of computer for simulation study may be appropriate here. A digital computer has been commonly used for drive system simulation, but an analog computer can also be used for this purpose. Since the machine and the converters are basically analog in nature, instantaneous access to the variables is possible in analog simulation. In a hybrid computer, which consists of analog and digital computers, the system can be appropriately partitioned for analog and digital simulation. Normally, the converters and the machine are simulated on analog computer, but the control can be simulated either in analog or digital computer. A microcomputer based control system, which is discrete time in nature and where complex computation and decision making processes are involved, is simulated in the digital computer. Recently, the parity simulation technique has appeared on the horizon and shows a great promise for the future. In this method, analog and digital simulation principles are combined so that individual components in the drive system maintain their equivalence, i.e., a machine simulation looks like a machine at the terminal except that the voltages and currents are appropriately scaled. Such a simulation with modular building blocks can be done quickly by an inexperienced engineer. Real time parity simulation of the power components can be driven by breadboard control hardware which subsequently can be used as a prototype.

2.2 Stator Voltage Speed Control

The block diagram of a stator voltage speed control system using thyristor ac controller is shown in Fig. 22. The command speed ω^* is compared with the measured speed ω from a tachogenerator, and the resulting error passes through the speed regulator to generate the firing angle signals of the controller. The speed controller usually consists of a gain and a compensator function. The compensator can be designed as lag-lead type $(1 + T_1S)/(1 + T_2S)$ or

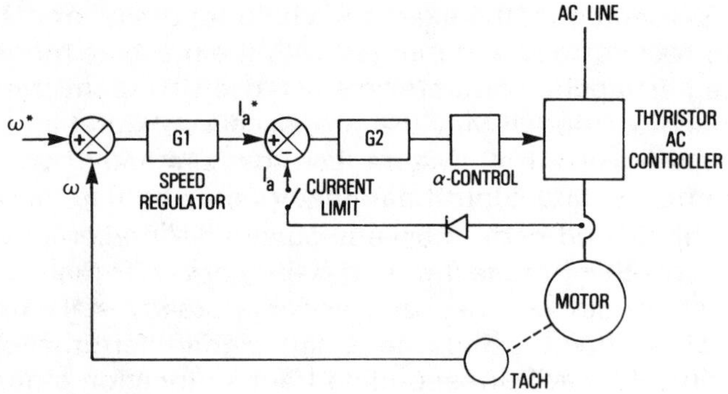

Fig. 22. Stator voltage speed control with current limit.

proportional-integral type $K_1 + K_2/S$, but the later has the advantage that the steady state error tends to be zero. An inner current limit control loop may be added if start up current transient tends to exceed the capability of the controller. Restricting starting current, however, delays the machine acceleration.

2.3 Voltage-Fed Inverter Speed Control

A simple form of speed control of a voltage-fed inverter drive is the constant volts/Hz control as shown in Fig. 23. The speed loop error signal controls the inverter frequency which also generates the voltage signal through a proportional gain factor G3. A constant offset V_o is added to the voltage command to compensate the stator resistance effect at low frequency. The constant volts/Hz control maintains approximately constant airgap flux, as discussed before. Initially, if a step speed command ω^* is applied, the machine accelerates at constant torque with a current limit control until steady state condition is reached, when I_a falls to a value as determined by the load. If ω exceeds base speed of the machine, the inverter frequency increases but the rectifier voltage remains fixed at the maximum value, and the machine is said to enter into field weakening constant power region. One advantage of the voltage-fed control is that the drive system can operate with all the loops open, but in such a system, the airgap flux may vary widely due to supply voltage fluctuation, and the finite filter impedance effect may cause stability problem at low frequency region.

A block diagram of constant volts/Hz control with slip regulation is shown in Fig. 24. The slip frequency ω_{s1} which is proportional to torque is regulated by the speed loop errors. The ω_{s1} signal is added with the speed signal ω to generate the inverter frequency $\omega_e{}^*$. The voltage control signal $V_s{}^*$ is generated from $\omega_e{}^*$ through a function generator. The drive system accelerates with the clamped value of slip which may correspond to the maximum torque, and then settles down to a value in steady state condition as dictated by the load torque (Fig. 7). If the command speed ω^* is reduced at steady state condition, the slip becomes negative and the drive system goes into dynamic or regenerative braking mode as designed in the system.

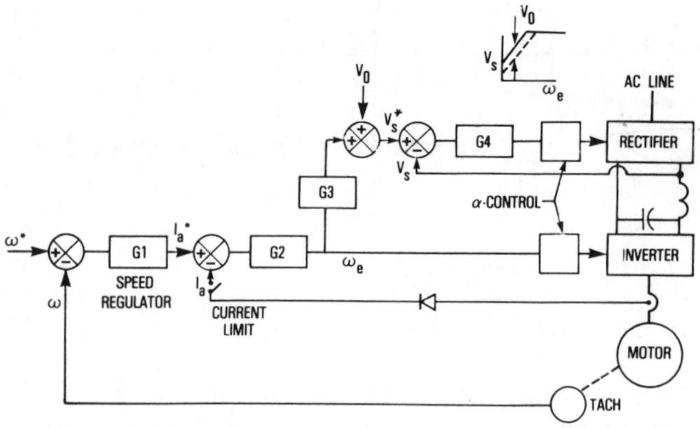

Fig. 23. Volts/Hz speed control with current limit.

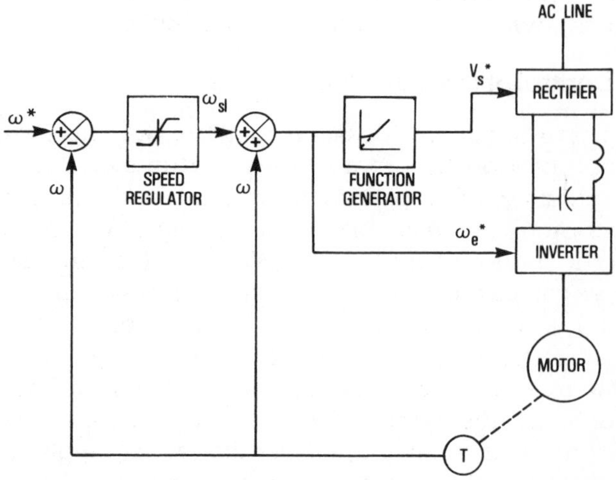

Fig. 24. Constant volts/Hz speed control with slip regulation.

Instead of regulating slip, it can be maintained constant and the speed loop error may control the dc link voltage. The variation of volts/Hz ratio causes variation of airgap flux and correspondingly the developed torque is regulated. A slightly modified control system incorporates close loop armature current or power factor control which closely resembles the diagram of Fig. 23. It can be shown that the current or power factor is uniquely related with the slip of machine. In Fig. 24, if the PWM inverter is used, obviously both the voltage and frequency are controlled within the inverter.

A very improved control system can be designed where the flux and torque are close loop controlled as shown in Fig. 25. The torque loop error generates the slip command which is added with the speed to generate the frequency command. The airgap flux ψ_g may either be maintained constant as in a dc shunt motor or programmed as a function of torque for steady state efficiency optimization. The flux loop error generates the armature current command $I_s{}^*$, and the machine is operated continuously in current control mode. A set of three-phase sinusoidal reference current waves is generated and the inverter switching devices are controlled such that the actual current profile remains confined within a hysteresis

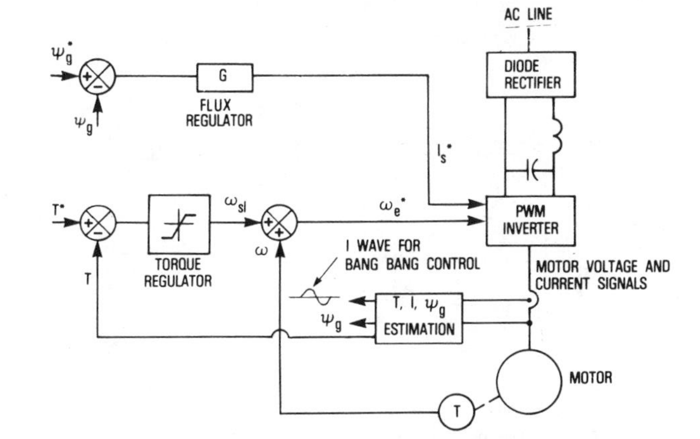

Fig. 25. Independent torque and airgap flux regulation with machine operating in current control mode.

band (bang bang control). Such a PWM control scheme tightly controls the ripple in the machine current wave. The feedback airgap flux and torque signals can be estimated from the machine terminal voltages and currents. The accuracy of estimation determines the precision within which the parameters can be controlled. The airgap flux can also be measured by inserting flux coils in the machine airgap.

The concept of field oriented control which was introduced by Siemens Company several years ago is recently receiving wide attention. This is also known as transvector control, because the control implementation is based on vector transformation from rotating to stationary reference frame and vice versa. The field oriented control method permits optimum transient response of the drive and can be explained as follows. It can be shown that generically ac and dc machines are analogous, and this analogy becomes somewhat obvious if the stator voltages and currents of an ac machine are converted from d-q stationary reference frame to the corresponding synchronously rotating reference frame. In a dc machine, if armature reaction and saturation effects are neglected, the developed torque is given by $T = K_t I_f I_a$, where I_f = field current and I_a = armature or torque producing component of current. The parameters I_f and I_a are mutually decoupled and best transient torque response can be obtained by controlling I_a with the rated value of I_f. In an induction motor the active and reactive power at the machine terminal correspond to the output and excitation power, respectively, and these can be translated into the corresponding torque component and field component of current at a given frequency. This is explained in Fig. 26 with the help of phasor diagrams in d-q rotating reference frame. The developed torque across the airgap is given by $T = K_t \psi_g i_{qs}^e$, where $i_{qs}^e = I_s \sin\theta$, as shown in the figure. If the rotor leakage inductance is neglected, the airgap flux ψ_g is contributed by the component $i_{ds}^e = I_s \cos\theta$. Therefore, the torque can be expressed as $T = k_t' i_{ds}^e i_{qs}^e$, where i_{ds}^e = field component of current and i_{qs}^e = torque component of current. The torque equation is analogous to that of a dc machine. The parameters i_{ds}^e and i_{qs}^e are mutually decoupled and can be independently varied as shown in Fig. 26.

The transvector control method implementation for a PWM current controlled inverter is shown in Fig. 27. The primary control parameters i_{ds}^e and i_{qs}^e are converted from d-q rotating reference frame to the corresponding stationary reference frame with the help of cos $\omega_e t$ and sin $\omega_e t$ signals generated by the PLL as shown in the figure. The resulting stationary frame signals i_{ds}^s and i_{qs}^s are then converted to a-b-c axes to generate the reference current waves of the inverter. The principle of vector transformation is shown in Fig. 28. The outer loops of flux and torque can be added to have precision control of the above parameters. Various refinements of field oriented control are possible but these are outside the scope of this section. It may be mentioned here that the field oriented control may be difficult to implement with dedicated hardware but adapts well to microcomputer control.

Some additional control methods have been described in the next section the concepts of which are applicable to voltage-fed inverter drives. The control methods described so far can be easily extended to cycloconverter drive, and therefore no additional description will be included.

2.4 Current-Fed Inverter Speed Control

A simple form of speed control in which the slip ω_{s1} is maintained constant and the speed is controlled by variation of dc link current I_d is shown in Fig. 29. The performance of the drive can be understood by referring to the torque-speed curves shown in Fig. 13. The command slip is fixed in the positive slope region of torque-speed curves, where the airgap flux remains within saturation. When a step command speed ω^* is applied, the machine accelerates with high I_d which is proportional to the torque, and then settles down to a low value as steady state condition is approached. One demerit of the system is that the airgap flux will vary widely at different operating points giving poor performance of the drive. The scheme can operate in the constant torque region, and it has been shown that as the rectifier voltage saturates, the system oscillates at low frequency. It may be noted that a current-fed drive system cannot be operated in open loop as in a voltage-fed drive, because the operation needs dynamic stabilization for a statically unstable point on torque-speed curve.

An improved method of control in which the slip ω_{s1} is varied as a function of I_d to maintain approximately constant airgap flux is shown in Fig. 30. The functional relationship between ω_{s1} and I_d can be precomputed for the given parameters of the machine and that the operating point remains on curve joining the dots as shown in Fig. 13. The feature of regeneration control has also been added in the diagram. When the error in

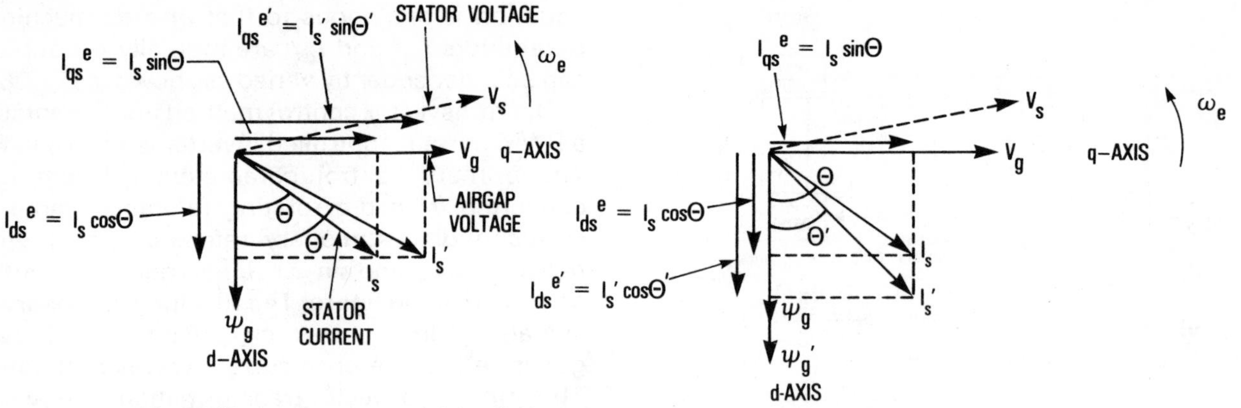

Fig. 26. Phasor diagrams explaining field oriented control method (rotor leakage effect neglected).

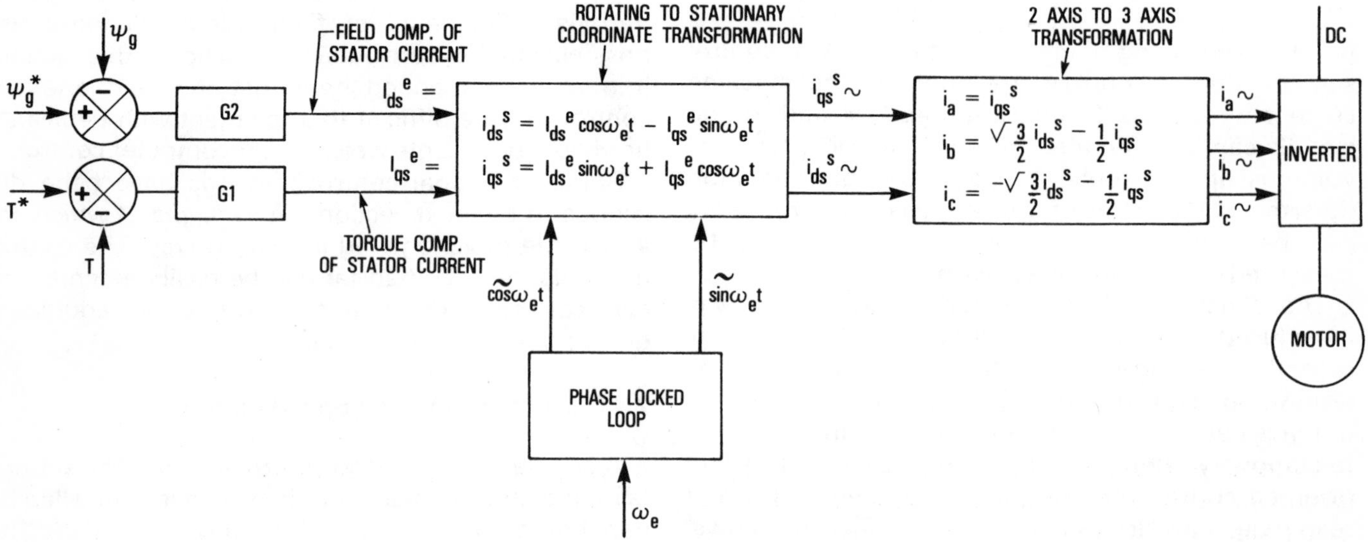

Fig. 27. Basic implementation of field oriented control.

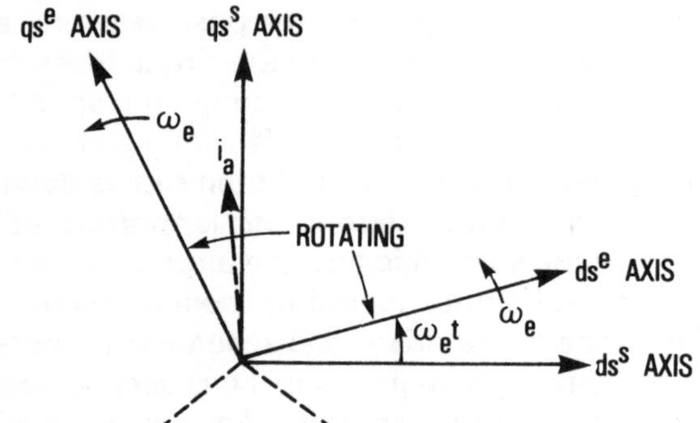

Fig. 28. Principle of vector transformation.

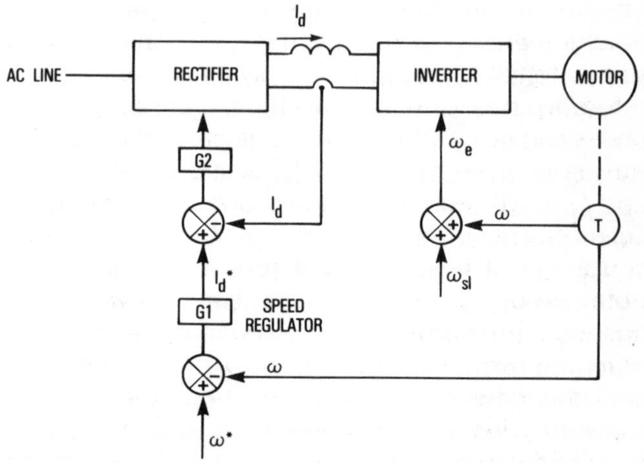

Fig. 29. Speed control at constant slip.

speed control loop becomes negative demanding regenerative torque, the polarity of ω_{s1} is reversed causing the machine to regenerate. The full four-quadrant capability of the drive can be obtained by adding phase sequence reversal to the inverter. The drive accelerates at constant I_d and ω_{s1} as determined by the clamped value of I_d, and then settles down in the active region at steady state condition. A minimum value of I_d is maintained so that the inverter can commutate satisfactorily. The scheme shown is suitable for operation in the constant torque region. To operate in constant power region, the slip must be increased with increasing speed to force constant current operation. This can be accomplished by adding an additional steady state slip command

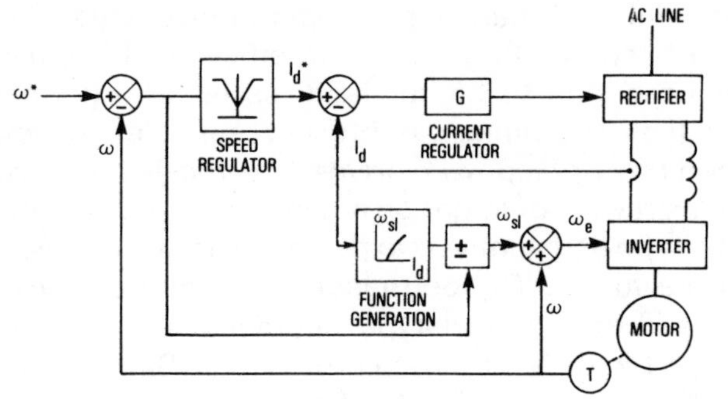

Fig. 30. Speed control at programmable slip.

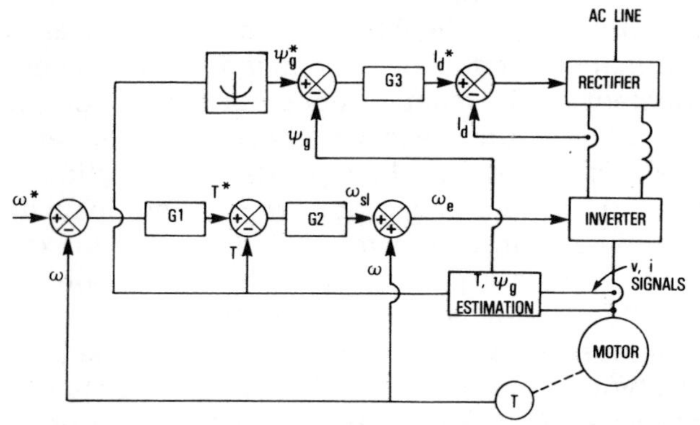

Fig. 31. Speed control by regulation of torque and airgap flux.

proportional to speed during the saturated voltage region.

A further improved method of control in which the torque and airgap flux are directly controlled by close loop regulation is shown in Fig. 31. In this scheme, the torque is controlled by the current, though the exchange of outer control loops is perfectly possible. A fixed functional relationship is maintained between torque and airgap flux to improve steady state efficiency. Operation in the constant power region is obtained by modifying the control system with a flux command which is inversely proportional to motor speed, when the rectifier voltage is saturated.

A sophisticated control method which can be visualized as polar form of field oriented control is shown in Fig. 32. The polar form has the same transient performance as the cartesian form but it does not require complex coordinate transformation. If commutation effect is neglected, the dc link current I_d can be shown to have unique relation with the stator current I_s, and therefore the torque component $I_s \sin\theta$ and field component $I_s \cos\theta$ as shown in Fig. 26 can be directly controlled in polar form through the dc link. The $\sin\theta$ control loop which controls the inverter frequency via the inner slip control loop can be considered as a phase-locked loop system. For a fixed ψ_g command, the torque is controlled by I_d and $\sin\theta$ control as shown in the figure. For a fixed T command, the flux is controlled by controlling I_d^* through an adder, and $\sin\theta^*$ through a function generator and multiplier. If torque is zero, for example, $\sin\theta^*$ becomes zero and the flux is corrected by adding to I_d^* signal only. If regeneration is desired, T^* is reversed in polarity which maintains I_d positive but $\sin\theta$ and ω_{s1} become negative.

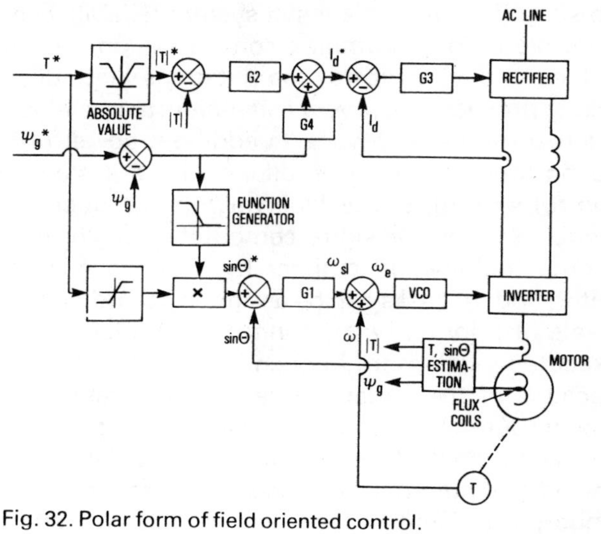

Fig. 32. Polar form of field oriented control.

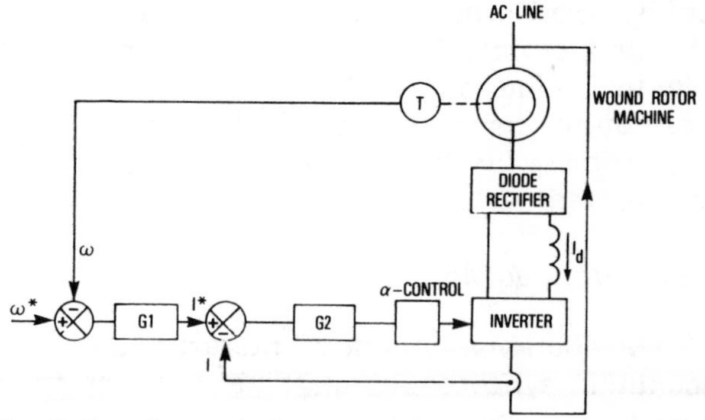

Fig. 33. Speed control of static Kramer drive.

2.5 Slip Power Recovery Speed Control

The speed control of static Kramer drive system is shown in Fig. 33. As explained before, this type of drive has characteristics similar to a dc machine and therefore the control configuration is analogous to phase-controlled dc machine drive system. The error in the speed control loop constitutes the inverter current command, which is proportional to developed torque. The current I is then controlled by firing angle control of the inverter. The airgap flux remains approximately constant as determined by the stator supply voltage and frequency. Evidently, the control method is considerably simpler than the conventional method of induction motor control, and the stability problem is practically nonexistent.

2.6 Microcomputer Control

The control block diagrams described so far can be implemented either by dedicated hardware or by microcomputer software. The microcomputer implementation has the advantages that the hardware can be simplified and as a result system reliability can be improved. For a complex control system which involves many intricate computations and decision taking processes, it is very convenient to implement by microcomputer software. In addition, powerful diagnostic tools can also be added to the system. In a control system, the tasks involving execution of the feedback loops and signal computation can be done in a time multiplexed manner by a single microcomputer, and the tasks can be assigned according to their levels of priority, i.e., the inner control loops can be executed much more frequently than the outer loops. Because of the serial nature of computation, a microcomputer is generally slower than the dedicated hardware control. However, this problem can be solved by multi-microprocessor control, where judicious partitioning of tasks can significantly enhance the execution speed. The advent of the fast and sophisticated processors, such as Intel 8086, Zilog 8000 or Motorola 68000, has given immense possibility to the future of ac drive control systems. In the near future, possibly a single VLSI chip will take responsibility for the entire control system, however complicated.

3. Synchronous Motor Drive

A considerable amount of interest has developed recently in synchronous machine drives as competitors to induction motor drives. The speed of a synchronous machine is exclusively determined by the stator supply frequency, i.e., either it must run at synchronous speed or will not run at all. When a load torque is impressed on the machine shaft, the corresponding developed torque is established almost instantaneously to maintain stability of the machine. It was explained before, that the excitation of an induction motor is supplied from the stator side which therefore requires that the machine must draw lagging reactive power at the input. In a synchronous machine, on the other hand, the excitation is provided by dc excited magnetic field which permits the machine operation at any power factor, leading, lagging or unity. The operation near unity power factor not only saves armature copper loss, but the inverter size is also reduced. The field mmf requirement is somewhat larger than that of induction motor because of armature reaction effect, and it has to be supplied by a separate rectifier.

The synchronous machine can be classified into several types. The conventional wound field machines are normally large in size. The dc excitation current for the rotor can be supplied either by a static exciter through slip rings and brushes, or by an ac exciter, or rotating transformer followed by a rotating diode rectifier, or by an induction generator-rectifier system. In an inductor homopoler or claw pole type of machine, the field winding can be transferred from the rotor to the stator permitting higher speed of operation, but the machine size becomes larger and the efficiency is somewhat reduced due to higher field losses. The permanent magnet machines, at present suitable for low to medium horsepower drives, are receiving considerable amount of attention nowadays. The application of this type of machine is expected to grow tremendously in the future. The permanent magnet machines may have drum- or disk-type geometry and can be designed with ferrite, Alnico or Cobalt–Samarium (rare earth) magnets. The Co–Sm machines are used in special applications because of higher cost though there is advantage in the size and weight reduction. The efficiency of the PM machine is higher because of elimination of the field copper loss. However, the constant field excitation does not permit compensation of armature reaction effect in the airgap flux. There is also the problem of demagnetization effect due to armature reaction mmf. The inability to manipulate airgap flux makes it difficult to run in high speed constant power mode. The synchronous reluctance motor is simplest of all and its torque is developed due to saliency of the rotor. The poles of this type of machine are induced from the stator side, and as a result, the machine suffers from poor power factor. Reluctance machines are favoured in multimotor drives though recently they are being replaced by PM machines.

The synchronous motor drives have essentially two different modes of operation. One is the true synchronous mode in which the machine speed is controlled by an independent oscillator. The other mode is called dc commutatorless motor (CLM) or electronically commutated motor (ECM), where the inverter firing signals are derived from rotor shaft position sensors. In this mode, the analogy with the dc machine needs some explanation. Internally, a dc machine can be viewed as an ac synchronous machine in which the field is stationary but the armature with multiphase ac winding is rotating. The ac power to the armature is derived from the input dc supply through commutators and brushes, which can be considered as a mechanical shaft position sensitive inverter. An equivalent dc motor operation is possible with a synchronous machine, where the field is rotating but the armature is stationary and it is supplied by a shaft position controlled electronic inverter. The advantages for replacement of mechanical commutators and brushes in a dc machine by the electronic commutation are obvious.

3.1 Voltage-Fed Inverter Drive

A typical power circuit of voltage-fed inverter drive as shown in Fig. 34 has already been described in subsection 1.2. The inverter may operate in square-wave or PWM mode, and the inverter switches may be transistors, thyristors or gate-turnoff devices (GTO).

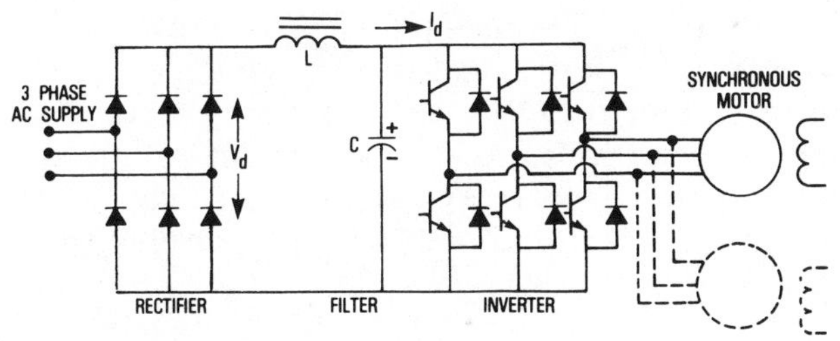

Fig. 34. Voltage fed synchronous motor drive.

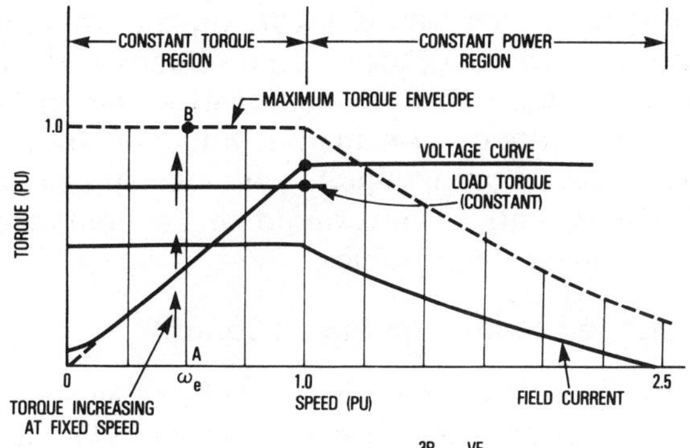

$$T = \frac{3P}{4\omega_e} \frac{VE}{X_s} \sin\delta$$

$$= \frac{4P}{3\omega_e} \psi_g I_s \sin\Theta$$

Fig. 35. Torque-speed curves of synchronous machine.

The inverter is normally controlled to have constant volts/Hz power supply. The torque-speed curves of the machine with volts/Hz control is shown in Fig. 35. For a fixed voltage and frequency combination, the torque increases in a vertical line from the point A as the armature current increases until it reaches to the pull out torque defined by the point B. Different vertical lines are described by different voltage and frequency combinations. Beyond the base speed, the machine enters into constant power region and the maximum torque decreases as volts/Hz ratio decreases. At any operating condition, the field current can be adjusted independently to maintain the desired power factor at the machine terminal.

The discussion on regeneration, harmonic heating, and torque pulsation as given previously for induction motor drives, in general, holds true here. Voltage-fed inverter drive with multiple reluctance or PM machines are normally used in fiber spinning mill type applications where precision tracking of speeds is required. Here, a crystal controlled oscillator determines the inverter frequency, which makes the speed of all machines exactly identical. Another group of machines operating on a separate inverter can be connected to the common rectifier but the speed may vary by a fixed ratio from the previous group. A single machine in such an application can be started and brought into synchronism by a cage winding which also helps to maintain its stability. A single motor voltage-fed drive can be used typically in transportation type application, where the primary supply is dc. Here, the machine is controlled in CLM mode which adds improved stability to the drive system.

3.2 Current-Fed Inverter Drive

The power circuit of current-fed inverter drive is shown in Fig. 36. The performance characteristics of the drive in regard to regeneration, harmonic heating, and torque pulsation are similar to induction motor drive, and have already been discussed in subsection 1.4. One important feature of synchronous machine is that it can be operated in leading power factor and therefore the inverter can be operated with load commutation.

The operation principle of load-commutated inverter is explained with the phasor diagram and waveforms of Fig. 37. The commutation is initiated by firing the incoming thyristor 2 at an advance angle β shown in the figure. This causes transfer of line current from thyristor 1 to thyristor 2 during the overlap angle θ. Subsequently, the reverse voltage is impressed across the outgoing thyristor 1 for the duration γ which helps it to turn off. Since the angle θ depends on the magnitude of the current and the machine subtransient inductance, the advance angle β is to be carefully predicted so that the minimum turn-off time is maintained under all frequency and load conditions. The absence of a commutation circuit makes the inverter simple, reliable, and more efficient. The thyristors of the inverter can be slow speed rectifier type saving further cost. However, the inverter is not capable of load commutation below a certain speed (typically 10 percent) because of inadequate counter emf. The low speed start up operation using forced commutation by a fourth leg is shown in Fig. 36. The operation of the circuit has been explained previously in subsection 1.4. Another method of starting uses the principle of dc link current interruption to commute the thyristors, and it is popularly used in Europe. In this method, shaft position sensitive signals are used to fire the thyristors so that torque is developed in one direction only. Instead of load commutation, forced commutation can be used to maintain machine power factor at unity condition. Elimination of reactive power results in minimal size of inverter-machine system at the cost of a small size commutation circuit.

The load commutated inverter drive operating in CLM mode is widely used in high horsepower applications, such as pumps and compressor drives. It has been recently popular in pumped storage hydro and gas turbine start up applications. As a general purpose solid state starter of a large synchronous machine, the drive has three modes of operation. At a low speed, the drive is started in forced commutation mode, then as adequate counter emf is developed the drive enters into load commutation mode. As the

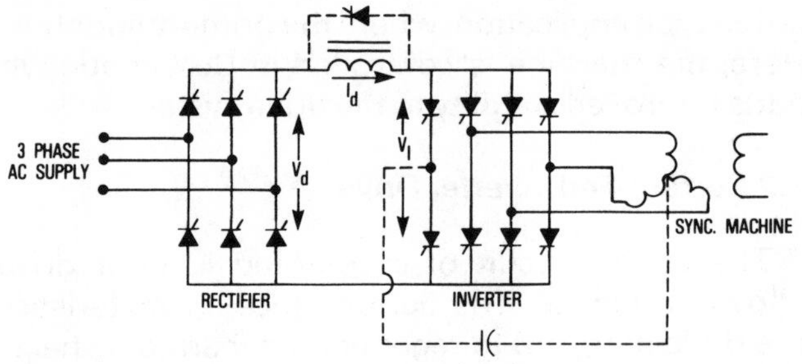

Fig. 36. Current-fed inverter drive of synchronous machine.

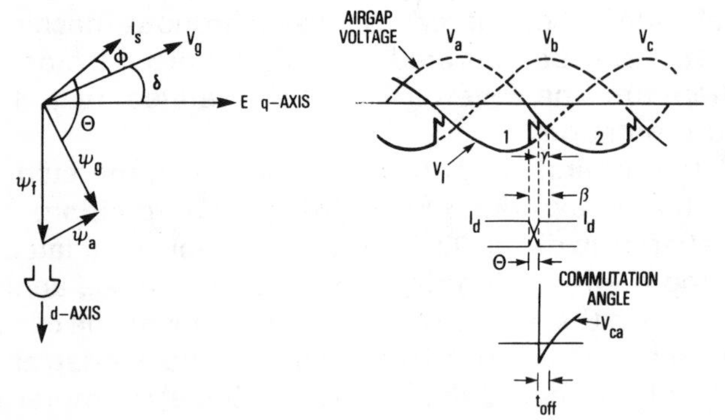

Fig. 37. Phasor diagram and waveforms for load commutation.

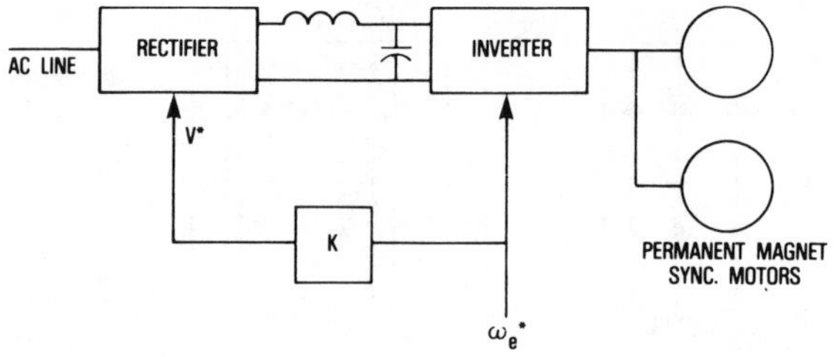

Fig. 38. Constant volts/Hz speed control for multimotor drive.

voltage, frequency, and phase conditions are matched, the starter is taken out and the machine is switched into the utility line. The solid state starter, though expensive, can be used in an installation where one starter is common to a number of machines.

3.3 Cycloconverter Drive

The cycloconverter fed synchronous machine drives are normally used in high horsepower (above 10,000 HP) and low speed, typically in gearless slow running mill drives. The cycloconverter is normally operated with line commutation, but can have load commutation if the output frequency approaches or exceeds the line frequency. The firing pulses of cycloconverter are normally derived from shaft position sensors and the machine terminal power factor is maintained at unity by field excitation control. In this self-controlled mode, the drive is sometimes defined as an ac commutatorless motor (AC CLM). The operation and performance characteristics of this type of drive have already been explained in subsection 1.5.

4. Control of Synchronous Motor Drive

The synchronous machine drive systems may have many different control configurations and only a few sample and representative types will be reviewed here. Of course, the complexity of control method will depend on the nature of application. The general discussion on modeling and simulation given in subsection 2.1 is also applicable for synchronous machine drives. Due to the presence of dc field winding and saliency of rotor poles, the only practical model of the machine is described in d-q axis synchronously rotating reference frame. Including the effect of the amortisseur winding, the machine electrical dynamics are represented by a fifth-order state equation, known as Park's equation. As in the induction motor, the model is based on negligible core loss and saturation nonlinearity and is considered to be unaffected by temperature and harmonics.

4.1 Voltage-Fed Inverter Speed Control

A simple form of open loop constant volts/Hz speed control for multimotor drive is shown in Fig. 38. The speed of the machines is uniquely related to the command frequency, and frequency to voltage proportionality makes the maximum torque available for the machines. When the dc link voltage reaches maximum ($\alpha = 0$), the machine enters into constant power region, i.e., further increase of frequency decreases the available torque due to reduction of airgap flux. The drive system may show some sign of instability in the low frequency region of operation.

A more sophisticated form of close loop control system is shown in Fig. 39. The motor is shown to operate in self-controlled mode (CLM), i.e., the inverter firing pulses are derived from the shaft position sensors. Since the field flux (ψ_f) vector remains fixed with respect to the rotor position, this method permits unique positioning of stator voltage vector with respect to ψ_f giving the drive a dc machine characteristic (in a dc machine ψ_f and armature current remain at fixed 90° phase relation) at the input terminal of the inverter. The phase relationship can be varied for different stator current (see Fig. 37) so that the machine power factor remains constant at a predetermined value. The power factor control can also be done in a closed loop manner as shown by computing the machine terminal power factor. The airgap flux may be maintained constant independently by field current control. The self-controlled synchronous machine, either voltage-fed or current-fed, cannot fall out of step and rarely shows a stability problem. Its dynamic behaviour can be shown to approximately correspond to that of a dc machine.

4.2 Current-Fed Inverter Speed Control

A self-controlled synchronous machine with current-fed inverter drive is shown in Fig. 40. The shaft position sensors have been eliminated, but the

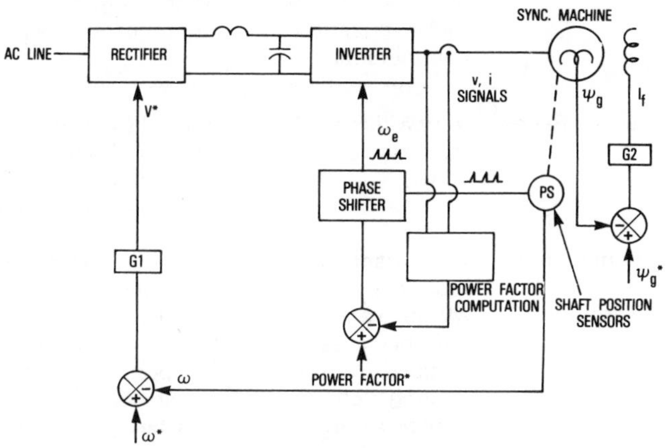

Fig. 39. Self controlled synchronous machine with close loop airgap flux and power factor control.

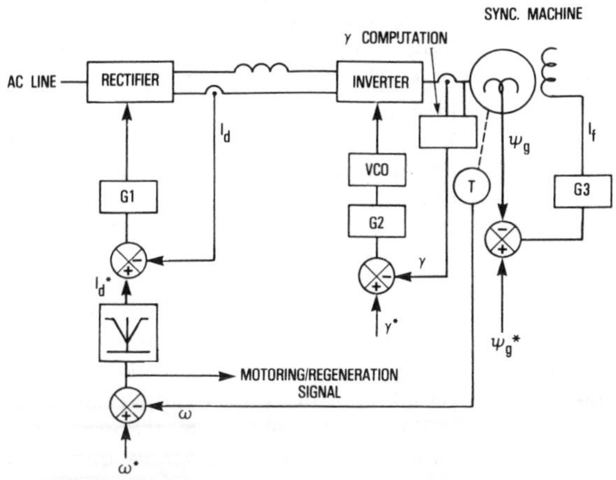

Fig. 40. Load commutated inverter drive with constant turn-off angle (γ) control.

equivalent signals for firing the inverter thyristors have been derived from the phase position of machine terminal voltages. This is because the synchronous machine with the inverter can be considered as symmetrical with the line side converter with the supply source, where the firing pulses are synchronized with the line voltage waves. In Fig. 40, the phase angle γ between the stator current and terminal voltage has been maintained constant by phase-locked loop control. In a load-commutated inverter drive, γ corresponds to the fixed turn-off angle as shown in Fig. 37, which is required for satisfactory commutation. If, on the other hand, the inverter is force-commutated, the γ angle may be such that the terminal power factor is unity. The airgap flux ψ_g is controlled by field current regulation, and it is reduced in constant power region when dc link voltage reaches to the maximum value. The machine behaves like a current-controlled separately excited dc motor at the inverter input terminal. The torque of the machine is controlled by I_d whose minimum and maximum values are clamped. In motoring condition, the polarity of speed error remains positive, but as the error becomes negative, the drive goes into regeneration mode and γ angle is advanced to be near 180°.

At this condition, the rectifier and inverter functions are exchanged, and the ac line side converter firing angle is shifted automatically to limit the commanded I_d value.

5. Summary Comparison of AC Drives

	5.1 Induction Motor with Stator Voltage Control	5.2 Induction Motor with Square-Wave Inverter
1. Speed Range	Subsynchronous speed range with limited lower speed.	Medium to high speed range, typically 10:1 ratio.
2. Regeneration	Dynamic braking possible by plugging.	Dynamic braking possible. Regeneration needs extra converter on line side.
3. Harmonics	Rich harmonics in line and machine. Problem of motor heating.	Six-step inverter voltage and ac line current waves. Motor harmonic heating high at lower frequency.
4. Torque Pulsation	High pulsating torque frequency may not be a problem.	May be a problem at low speed.
5. Power Factor	Poor line power factor due to phase control.	Low line power factor due to phase control.
6. Cost	Low cost, simple control circuit.	High converter cost but may be less with transistors.
7. Efficiency	Poor.	Medium but improves with transistors.
8. Application	Low to medium power pump and blower type drives. Large motor starter. Low power appliance type drives.	General purpose industrial drive. Low to medium power. Multi-motor drives possible.
9. General Comments	Requires relatively high slip cage motor. Application is limited.	Open loop control possible but may have stability problem at low speed.

	5.3 Synchronous Motor with Square-Wave Inverter	5.4 Induction Motor with PWM Inverter
1. Speed Range	Medium to high speed range with limited lower speed.	Very wide speed range up to zero speed possible.
2. Regeneration	Dynamic braking possible. Regeneration needs extra converter on line side.	Dynamic braking possible. Regeneration possible if primary supply is dc; otherwise need extra converter.
3. Harmonics	Six-step inverter voltage and ac line current waves. Motor harmonic heating high at lower frequency.	Machine voltage and current waves nearly sinusoidal. Six-step ac line current wave.
4. Torque Pulsation	May be a problem at low speed.	Minimal.
5. Power Factor	Low line power factor due to phase control. Machine power factor can be improved by field control.	Line power factor close to unity.

6. Cost	High converter cost but may be less with transistors.	Moderately high but improves with transistors.
7. Efficiency	Moderately good but improves with transistors and PM motor.	Good but improves with transistors.
8. Application	General purpose industrial drive. Low to medium power. Multimotor drives possible.	General purpose industrial drive. Low to high power range. High HP transportation drive. Can use transistors in low to medium power. Multimotor drives possible.
9. General Comments	Open loop control possible but may have stability problem at low speed. CLM mode is very stable. Starting by cage winding or by open loop frequency control.	Open loop operation possible. Fast transient response. Control complex except with microcomputer. Very popular drive.

	5.5 Synchronous Motor with PWM Inverter	5.6 Induction Motor with Current-Fed Inverter
1. Speed Range	Very wide speed range up to zero speed possible.	Very wide speed range up to zero speed possible.
2. Regeneration	Dynamic braking possible. Regeneration possible if primary supply is dc; otherwise needs extra converter.	4-quadrant operation is simple.
3. Harmonics	Machine voltage and current waves nearly sinusoidal. Six-step ac line current wave.	Machine and ac line current waves six-stepped. Harmonic heating high.
4. Torque Pulsation	Minimal.	May be a problem at low speed.
5. Power Factor	Line power factor close to unity. Machine power factor can be improved by field control.	Low line power factor due to phase control.
6. Cost	High but improves with transistors.	High cost of inverter.
7. Efficiency	Good but improves with transistors and PM motor.	Moderately good.
8. Application	General purpose industrial drive. Low to high power range. High HP transportation drive. Can use transistors in low to medium power. Multimotor drives possible. Single machine drive normal in CLM mode.	Medium to high power industrial drive. Multimotor drive difficult.
9. General Comments	Open loop operation possible. Fast transient response. Control complex except with microcomputer. Very popular drive. Can be started by cage winding or by open loop frequency control.	Simple control. Commutation relatively simple. Cannot be operated at open loop. Problem of voltage spike.

	5.7 Synchronous Motor with Current-Fed Inverter (forced commutation)	5.8 Synchronous Motor with Load-Commutated Inverter
1. Speed Range	Very wide speed range up to zero speed possible.	Typically above 10-percent of base speed, but forced commutation may extend up to zero speed.
2. Regeneration	4-quadrant operation is simple.	4-quadrant operation is simple.
3. Harmonics	Machine and ac line current waves six-stepped. Harmonic heating high.	Machine and ac line current waves six-stepped. Harmonic heating high.
4. Torque Pulsation	May be a problem at low speed.	May be a problem at low speed.
5. Power Factor	Low line power factor due to phase control. Machine power factor can be improved by field control.	Line power factor low due to phase control. Machine power factor slightly leading.
6. Cost	High cost of inverter.	Converter cost medium.
7. Efficiency	Moderately good but improves with PM motor.	Good but improves with PM motor.
8. Application	Low to high power drive in CLM mode.	Medium to very high power drive in CLM mode, e.g., gas turbine starting, pumped-hydro turbine starting, pump and blower drives.
9. General Comments	Simple control. Commutation relatively simple. Cannot be operated at open loop. Problem of voltage spike. Improved stability in CLM mode.	Simple control. Commutation relatively simple. Cannot be operated at open loop. Problem of voltage spike. Needs additional starting method. Popular drive at high HP.

	5.9 Induction Motor with Stator Side Cycloconverter	5.10 Synchronous Motor with Line-Commutated Cycloconverter
1. Speed Range	Zero to maximum speed which typically corresponds to 1/3 or 1/2 of line frequency.	Zero to maximum speed which typically corresponds to 1/3 or 1/2 of line frequency.
2. Regeneration	4-quadrant operation is simple.	4-quadrant operation is simple
3. Harmonics	Voltage and current waves are nearly sinusoidal.	Voltage and current waves are nearly sinusoidal.
4. Torque Pulsation	Minimal.	Minimal
5. Power Factor	Line power factor poor.	Line power factor somewhat better because machine power factor can be made unity.
6. Cost	High cost of converter with complex control.	High cost of converter with complex control.
7. Efficiency	Good	Good

8. Application	High power low speed drives, such as ball and cement mills. Diesel electric locomotive.	High power low speed drives, such as ball and cement mills. Diesel electric locomotive. Operates in ac CLM mode.
9. General Comments	Fast Response.	Fast Response.

	5.11 Synchronous Motor with Load-Commutated Cycloconverter	5.12, Static Kramer Drive
1. Speed Range	Zero to maximum speed which may correspond to above the line frequency.	Subsynchronous speed range. Limit lower speed to save converter size.
2. Regeneration	4-quadrant operation is simple.	Regeneration and speed reversal not possible.
3. Harmonics	Voltage and current waves are nearly sinusoidal	Six-stepped current waves.
4. Torque Pulsation	Minimal.	Torque pulsation effect is present.
5. Power Factor	Line power factor low.	Line power factor poor.
6. Cost	High cost of converter with complex control.	Converter cost moderate but machine cost is higher.
7. Efficiency	Good.	Good.
8. Application	High power pump and blower type drives.	Large pump and fan type drives within limited speed range.
9. General Comments	Started at low frequency in line commutation mode. Fast response.	Requires separate starting resistors.

	5.13 Static Scherbius Drive
1. Speed Range	Speed range between subsynchronous and supersynchronous. Limit range to save converter size.
2. Regeneration	Regeneration is possible but not speed reversal.
3. Harmonics	Nearly sinusoidal waves with cycloconverter.
4. Torque Pulsation	Minimal.
5. Power Factor	Line power factor low.
6. Cost	High converter and machine cost, complex control.
7. Efficiency	Good.
8. Application	Large pump and fan type drives within limited speed range.
9. General Comments	Requires separate starting resistors.

Conclusion

A simple but comprehensive review of different features of adjustable speed ac drive systems has been presented in this section. The review is expected to provide a background of knowledge which will help the reader in successful launching into the other sections of the book. All the concepts which have been briefly discussed here will appear in considerable detail in the papers included in the book. For a relative newcomer in the ac drives area, understanding a paper in the recent literature may be difficult, and often the difficulty is compounded due to rigour of mathematical implementation. For this reason, mathematics has been intentionally avoided in this section and total emphasis is given to the description of physical principles of operation. Since many features are in common to both induction and synchronous machine drives, the part of the descriptions which is common has not been repeated in the latter case. The literature on ac drives is so vast and proliferated that only salient features could be included in the limited number of pages of this section. Some general comments about the present research and development trend have also been included. The subsection on summary comparison of ac drives will help the reader to review collectively the capabilities and limitations of different types of ac drives. This will also help him or her to make a preliminary selection of a drive system if he or she has a particular application in his mind. The final selection, of course, will depend on an in-depth evaluation of many tradeoff considerations.

Part I
Induction Motor Drives

Section 2
Phase Controlled Converter Drives

Induction Motor Speed Control by Stator Voltage Control

DEREK A. PAICE, SENIOR MEMBER, IEEE

Abstract—**This paper establishes the fundamental laws relating to the speed control of induction motors by simple voltage control and emphasizes the problems that may be caused by excessive input currents which cause stator overheating. Eight different thyristor voltage control circuits have been tested to determine a *best* control circuit for three-phase motors and test results are given. A practical speed control for a 4-hp motor is described and the way in which a special rotor design can minimize the problem of excessive stator losses is convincingly demonstrated.**

Introduction

The SQUIRREL cage induction motor is basically a simple, cheap, and reliable machine which can provide excellent characteristics where a constant shaft speed is required. For example, the speed of a typical motor fed from a constant frequency supply may vary less than 5 percent over the range of normal load and supply voltage variations. This basic characteristic of the motor, to run at a virtually constant speed close to synchronous, has for long been a challenge to designers who have sought to devise variable speed schemes.

The advent of silicon power thyristors around 1960 added new impetus to the development of variable speed schemes by making variable frequency power supplies a practical possibility and by providing a versatile and economic solid state power control element. Numerous low power motor speed controls using simple voltage control are already in use and currently more efficient controls for motors in the thousands of horsepower are under consideration.

The object of this paper is first to define those fundamental principles which are inviolate to induction motor performance; second, some possible methods of voltage control by means of thyristors are evaluated to determine a best approach as far as the motor is concerned; finally, a practical 4-hp variable speed fan drive is described in detail.

Motor Theoretical Considerations

Power Characteristics

In the first case, sinusoidal variations of supply voltage, currents, etc., will be assumed to delineate the basic operating principles of the squirrel cage induction motor will be assumed.

A rotating magnetic field is set up by the voltage applied to the stator windings and the speed of rotation is uniquely determined by the number of poles and supply frequency such that

$$N_s = \frac{120 \times f}{p} \tag{1}$$

Paper 31 PP 67–76, recommended and approved by the Rotating Machinery Committee of the IEEE Power Group for presentation at the IEEE Winter Power Meeting, New York, N. Y., January 29–February 3, 1967. Manuscript submitted August 10, 1966; made available for printing October 4, 1967.

The author is with the Research and Development Center, Westinghouse Electric Corporation, Pittsburgh, Pa. 15235

where N_s is the synchronous speed in r/min; f is the frequency in Hz; and p is the number of poles. The rotating field induces currents in the squirrel cage rotor which interact with that field so as to produce a torque. This torque accelerates the rotor until finally the rotor revolves at some speed N_r, less than the synchronous speed N_s. Now the rotor requires a torque to satisfy the load torque and this same torque is experienced by the rotating (air gap) field. Thus a situation occurs, similar to a slipping clutch, in which two rotating members experience the same torque but have a different speed. Therefore, a power difference exists between the motor stator power input $\omega_s T$ and rotor output $\omega_r T$ where T is the common torque and ω_s and ω_r are the angular speeds of the air gap field and rotor, respectively. This power difference, called slippage power, is fundamental to all induction motors and if the speed of a motor is controlled by varying the rotor speed ω_r without varying the synchronous speed ω_s, then the slippage power must be recognized. In the case of squirrel cage motors the slippage power $(\omega_s - \omega_r)T$ is dissipated as heat in the rotor and it is seen immediately therefore that, even with sinusoidal voltages and currents, speed control of the squirrel cage induction motor fed from a fixed frequency supply is inevitably an inefficient process. Fig. 1 illustrates the power flow diagram for such a motor.

The power flow chart shows that as well as slippage power there are other losses to be considered and in particular the stator copper losses are likely to create difficulties with simple voltage control schemes in conjunction with standard class B motors. This fact will be evident if the variations to be expected in the motor input current are examined as the speed varies.

Input Current and I^2R Losses

For simplicity, the ensuing analysis will neglect the magnetizing current required by the stator winding. Although the magnetizing current may be comparatively high at rated voltage (e.g., 50 percent of full load current), as the speed is reduced by reducing the voltage the magnetizing current becomes of little importance in determining the total input current.

Assuming sinusoidally varying quantities, the average torque produced by the stator field reacting with the rotor current is given by the following proportionality:

$$T_m \propto \frac{I^2R}{s} \tag{2}$$

where

- T_m motor torque
- R rotor resistance, assumed constant
- I rotor current
- s fractional slip and equals (synchronous speed − rotor speed)/synchronous speed.

Neglecting the magnetizing current associated with the stator winding, the input current is proportional to the rotor current.

Reprinted from *IEEE Trans. Power App. Syst.*, vol. PAS-87, pp. 585–590, Feb. 1968.

Consideration of how the rotor current varies with slip will therefore indicate how the rotor and stator I^2R losses may vary.

If the load torque is related to the square of the motor speed as is approximately true for a fan load, this proportionality occurs:

$$T_L \propto (\text{rotor speed})^2 \propto (1 - s)^2 \tag{3}$$

where T_L is the load torque.

Now at the steady operating condition the motor and load torques are equal and combining (2) and (3),

$$I \propto \frac{(1 - s)(s)^{1/2}}{R^{1/2}}. \tag{4}$$

This proportionality indicates the input current will be maximum when $s = 1/3$ and also that the peak input current is inversely proportional to (rotor resistance)$^{1/2}$.

A plot of $(1 - s)\, s^{1/2}$ is given in Fig. 2 and the input current of a 5-hp motor and fan in which the slip was increased by lowering the supply voltage by means of a variac is included in this figure. There is seen to be a good agreement between the simple theory and practice.

It is evident from Fig. 2 that if the normal full load slip of the motor is low, e.g., 2–8 percent, then if the speed is reduced, a very large increase of input current can be expected. A conventional motor would not normally tolerate this magnitude of current and might be expected to *burn out* the stator windings.

Further tests were carried out with single-phase fractional horsepower fan motors and similar current versus slip characteristics were obtained. On the smaller motors tested the full load slip was about 12 percent and speed control of these motors by voltage control, without stator overheating, is possible and has in fact been achieved.

Fig. 3 has been drawn to illustrate the increase to be expected in the line input currents of fan motors controlled by varying the supply voltage. Using the curve in Fig. 3 it is seen for example that a motor with a normal full load slip of 12 percent will have a maximum input current about 25 percent greater than the normal full load current. Also note that the rotor and stator losses will increase by about 56 percent.

Fig. 4 has been drawn to illustrate the motor and fan performance as slip varies, and the efficiency curve represents the best that can be achieved since it only considers the rotor losses. Correction can be made for the stator I^2R losses in a practical case since it has been shown that these vary in the same way as the rotor losses if the rotor resistance is assumed constant.

The way in which the input current will vary with slip when the load is basically constant torque can be determined by substituting $T_m = K$ in (2). In this case

$$I \propto \left(\frac{s}{R}\right)^{1/2}. \tag{5}$$

As in the previous example, the input current is inversely proportional to (rotor resistance)$^{1/2}$ but there is no maximum value as for the case of a fan load. The plot of $(s)^{1/2}$ in Fig. 5 indicates how the input current may be expected to vary with slip for the constant torque load case.

A measurement obtained on a small single-phase motor is inuded in Fig. 5. To obtain the measured results, a closed-loop achometer feedback scheme was used in conjunction with a hyristor voltage control. In this test the motor voltage was erefore a chopped sine wave but there is still a fair agreement between the form of the theoretical and measured characteristics.

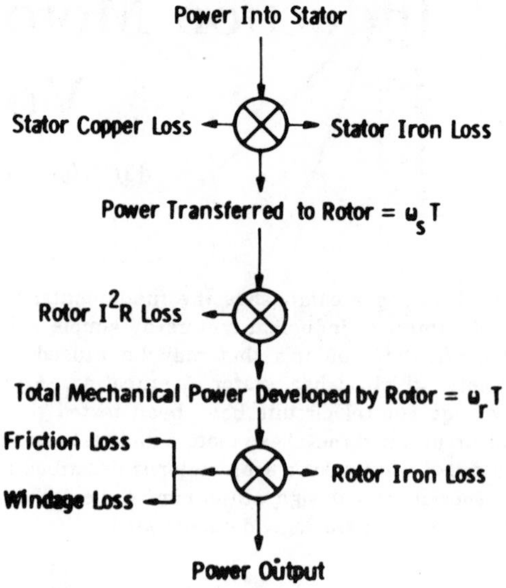

Fig. 1. Induction motor power flow.

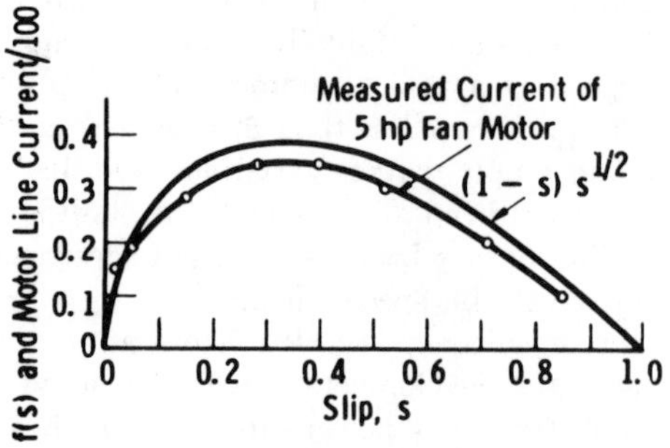

Fig. 2. Theoretical and practical form of motor line current–fan load.

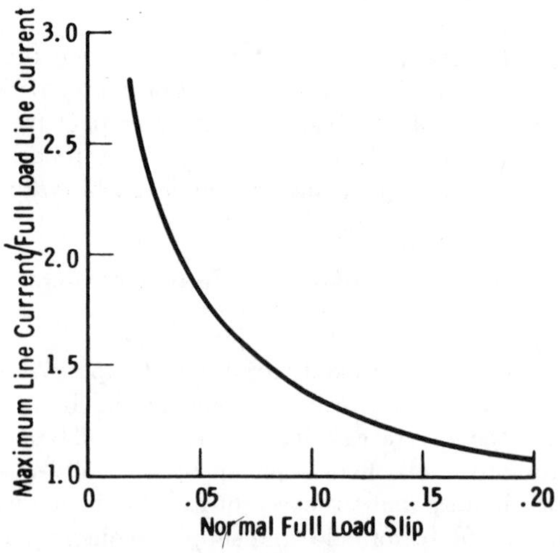

Fig. 3. Relating motor full load slip to peak line current–fan load.

The conclusions to be drawn from the foregoing theory can be summarized as follows. If the rotor resistance is substantially constant and the motor full load slip is low, then heating problems of the stator winding may be expected as the speed is reduced because of the large increase of input current. Practical solutions to this problem can be obtained by using machines with high resistance rotors, and hence high full

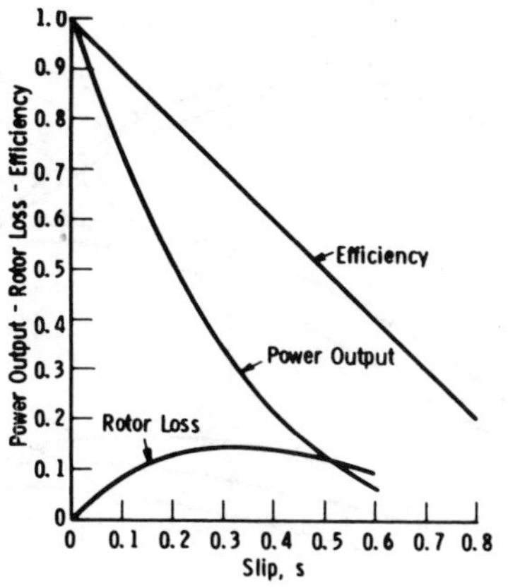

Fig. 4. Fan motor performance as slip varies.

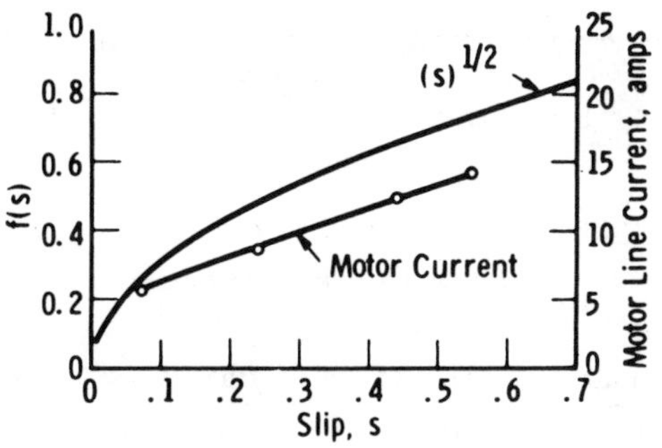

Fig. 5l Theoretical and practical form of motor line current–constant torque load.

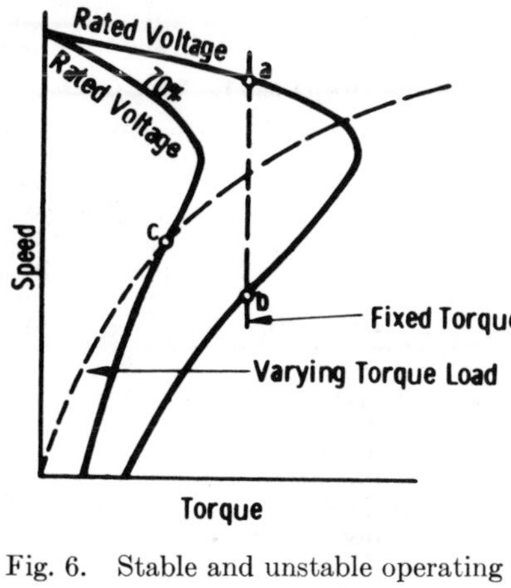

Fig. 6. Stable and unstable operating points.

load slip, or better still special machines with variable resistance rotor characteristics can be employed. This latter type of machine was used in the fan speed control scheme described later.

The above conclusions have been drawn from a very simple theory based on sinusoidal voltages. The conclusions do indicate however the first-order effects to be considered in relation to speed control by voltage control. The harmonics introduced by, for example, chopped sine wave control by thyristors are likely to degrade the motor performance further.

Stability

To ensure that an induction motor will operate with a stable speed, it is necessary to ensure that at the required operating point the motor torque increases less than the load torque if the speed increases, and decreases less than the load torque if the speed decreases. Fig. 6 shows a typical motor characteristic and points a and c represent stable operating points. Point b would not be a stable condition without additional control means. However, in general, operation at points such as b can be obtained by the use of a feedback loop which continuously adjusts the motor voltage so as to maintain the required output speed. Such a system in which a tachometer is used for speed sensing is readily applied to thyristor voltage control circuits and schemes of this type are in use.

The small signal transfer function of the motor is very dependent upon the set speed. For example, at point a the motor speed/voltage relationship can be approximated by an integrating term, whereas at point b the characteristic is that of a nonminimum phase term with low-frequency phase shift of 180°. A detailed analysis of the motor used in this way is beyond the scope of this paper, but it would seem that this is an area in which little work has been published and yet is worthy of attention.

Thyristor Power Control Circuits

Single-phase and three-phase thyristor control circuits are already well described in the literature,[1]–[3] but the emphasis does not seem to have been placed on restraining the motor input current with a view to minimizing stator losses. Further development was carried out therefore to check this point especially.

There are many thyristor power circuit variations that can be used and eight of those tested are shown in Fig. 7. This drawing will be referred to in the text and the circuit configuration will be referred to by number rather than by name. The load controlled by the thyristors was in all cases a three-phase induction motor with a special solid iron rotor and a nominally 5-hp fan load. Depending on whether the coils per phase of the stator were connected in parallel or series, the phases have been shown as n or $2n$, respectively. Furthermore, the supply voltage was adjusted in each circuit to provide constant volts per turn at full speed.

The results of tests on these circuits are illustrated in Fig. 8. The peak of input current which was so pronounced with a class B rotor machine is not apparent with a solid iron rotor machine which has an effective rotor resistance which varies inversely with air gap voltage.

In each case voltage, current, and power were measured by three instruments and where only one value is given in the results it represents the arithmetic mean of three readings. The ammeters were *square law* instruments and gave good agreement with a thermocouple type despite the chopped waveforms occurring in some circuits. Speed was measured by a tachometer coupled to the motor shaft. Further discussion of the test circuits follows.

Test Circuit 1

In this circuit it was found that excessive current was required through two lines. The peak line current was 28 amperes and occurred at approximately 40 percent synchronous speed.

Test Circuit 2

This circuit has the merit of extreme simplicity and since the motor stalls when operated single phase, the speed can be controlled down to zero. It was found, however, that excessive current was required through the two unregulated lines. The results of tests on this circuit are given in Table I.

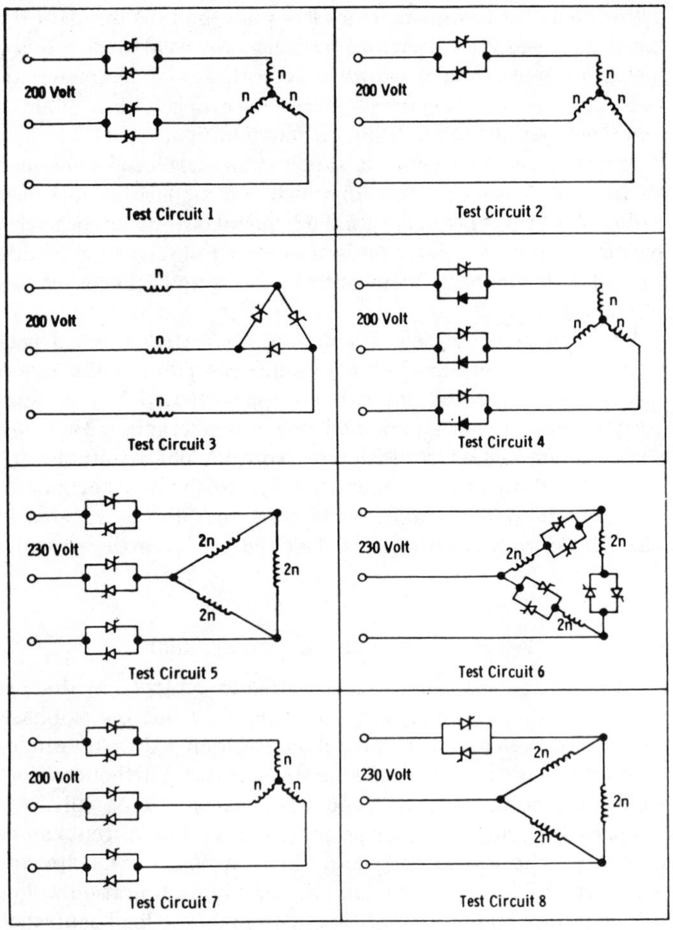

Fig. 7. Thyristor power test circuits.

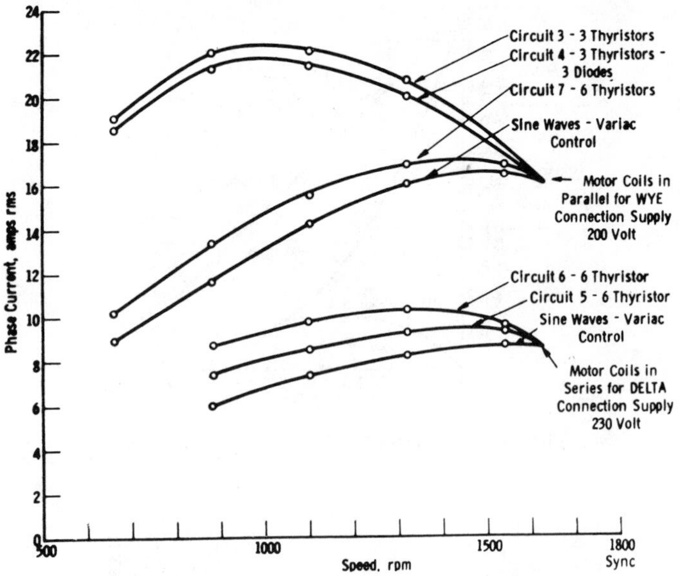

Fig. 8. Performance of thyristor power control circuits.

TABLE I

PHASE CURRENT, AMPERES

Speed r/min	Phase 1	Phase 2	Phase 3
1600	16.5	16.8	17.1
1560	14	17.5	21.6
1490	11.9	21	25.6
1430	11.0	23	27.5
1380	10.0	24	28.5

TABLE II

Test Circuit	Description	Percentage Increase of Current at 33 Percent Slip Compared to Sine Wave Control
3	3 thyristor	43.4
4	3 thyristor–3 diodes	38.2
5	6 thyristor–delta load	14
6	6 thyristor–open delta load	30.8
7	6 thyristor–wye load	8

Test Circuit 3

This circuit requires only three thyristors and presents the best possibility of the *simple* circuits.

It was considered that the peak input current of 22.4 amperes was excessive. At a speed of 1120 r/min the power consumed by the motor was 100 percent greater than for the sine wave control.

Test Circuit 4

This circuit is almost identical to Test Circuit 3 in performance.

Test Circuit 5

This circuit, using six thyristors, represents a conventional balanced three-phase control approach and gives a good performance. At 33 percent slip the input current is 14 percent greater than for sine wave control. It is apparently not as good as Test Circuit 7 in which the load is wye connected. It is possibly not so good because any third harmonic voltages developed by the motor back EMF can cause harmonic circulating currents when the load is delta connected.

Test Circuit 6

This circuit is a conventional approach using six thyristors, but the results are not as good as for Test Circuit 5.

The circuit operation was modified by shorting two of the bilateral switches and controlling with one third only. With this switch open, the motor ran in open delta and the speed control range was from 1600 to 1320 r/min only. At the low speed condition the winding current was 60 percent greater than the full load value.

Test Circuit 7

This test is of a balanced circuit using six thyristors and gave a very good performance. At 33 percent slip, for example, the input current is only 8 percent greater than for sine wave control.

Test Circuit 8

This test is similar to Test Circuit 2 and causes excessively large currents in the motor windings.

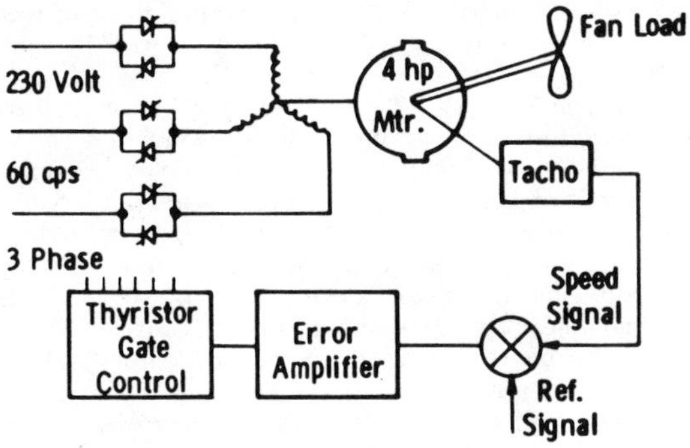

Fig. 9. Induction motor speed control with closed-loop feedback.

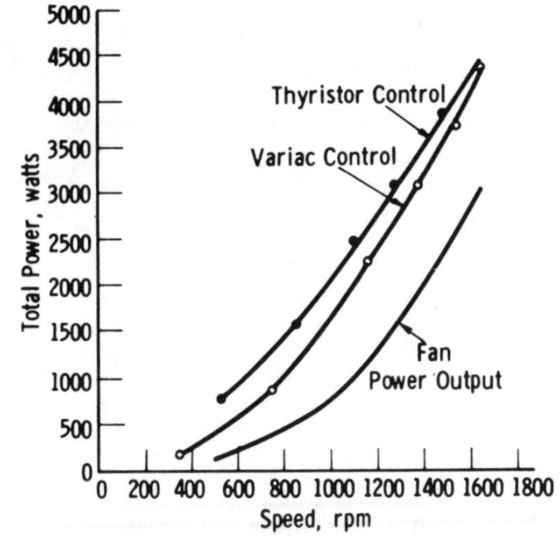

Fig. 10. Comparison of motor power for variac and thyristor control.

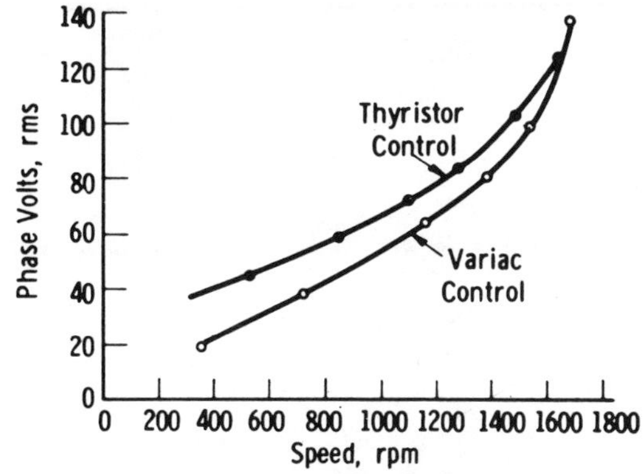

Fig. 11. Comparison of motor speed for variac and thyristor control.

The performance of the *balanced* control circuits has been compared to that of a sine wave at a slip of 33 percent at which the rotor losses for the sine wave case are maximum, and the results are given in Table II.

The foregoing tests indicated that Test Circuit 7 represented the best control approach as far as the motor performance was concerned and the rotor power requirements were not greatly different from the sine wave tests. Since the input current is not significantly increased over the sine wave control, this thyristor circuit does not unduly increase the stator loss.

The evidence obtained, however, indicates that in power applications a six thyristor control circuit and star connected load gives the *best* control approach; however, it may be that in some applications the motor temperature rise is not the limiting factor and other thyristor control circuits, such as Test Circuit 2, could be used because of their simplicity.

Application of Thyristors to a 4-hp Fan Speed Control

The solid rotor motor used to determine the characteristics of the various thyristor voltage control circuits was not as efficient as had been hoped and some additional work was carried out to improve its efficiency before completing a control scheme. A significant improvement was obtained by using the known technique of enlarging the stator-rotor air gap[5] and in conjunction with a three-phase control circuit, with six thyristors, as shown in Test Circuit 7, good results were obtained. Tachometer feedback into a simple transistor error amplifier as shown in Fig. 9 enabled good closed-loop performance to be achieved and a speed range of 5 to 1 was possible. The motor speed remained within ± 1 percent of the set value for ± 15 percent variations of supply voltage.

Tests were made of the system performance in terms of total power input versus speed and Fig. 10 compares the performance of motor control by variac and by the six thyristor circuit. Fig. 11 shows variation of speed with voltage.

Conclusions

1) The importance of input current in simple induction motor speed control systems is shown theoretically and demonstrated practically. The way in which special rotor designs, which exhibit varying resistance, can be beneficial is demonstrated.

2) A six thyristor power control circuit in conjunction with a wye connected motor was found to be the least demanding on the motor windings.

3) The experimental results obtained from a practical scheme show that speed control by simple voltage control is possible, especially when the motor has a *fan type* load.

Acknowledgment

The author is indebted to the Westinghouse Electric Corporation for permission to publish this paper and especially wishes to thank Dr. L. A. Finzi, Consultant to Westinghouse, for his advice and encouragement.

References

[1] W. Shepherd and J. Stanway, "The polyphase induction motor controlled by firing angle adjustment of silicon controlled rectifiers," *1964 IEEE Internat'l Conv. Rec.*, vol. 12, pt. 4, pp. 135–154.

[2] L. R. Foote, "Adjustable speed control of ac motors," *Elec. Engrg.*, vol. 78, pp. 840–843, August 1959.

[3] J. G. Petersen, "Thyristor drives for ac cranes and hoists," presented at the ASME-IEEE Conf., Pittsburgh, Pa., October 1965.

[4] L. A. Finzi and D. A. Paice, "Analysis of the solid iron rotor induction motor for solid-state speed controls," this issue, pp. 590–596.

[5] W. J. Gibbs, "Induction and synchronous motors with unlaminated rotors," *J. IEE* (London), vol. 95, pt. 2, pp. 411–420, 1948.

Discussion

John R. M. Szogyen (Fair Haven, N. J.): I should like to congratulate the author of this interesting and timely paper. Speed control on variable torque drives can be attractively simple by stator voltage control. Magnetic amplifiers and saturable reactors have been used with some success for stator voltage control of such speed controlled drives. The smaller bulk and space requirements of

Manuscript received February 24, 1967.

thyristor controls offer obvious advantages over the magnetic devices. Reliability and overload capability and freedom from detrimental harmonics may yet have to be demonstrated to be better than, or as good as, the magnetic devices. I think it would be particularly valuable if the author could comment on these factors.

The experimental evaluation of different circuits is particularly interesting. It is gratifying to see the old established rule of "avoid parallel circuits and delta connections to avoid extra losses due to unknown harmonics" confirmed.

It is known that the presence or absence of rotor *end rings* and that the resistance of these rotor end rings have a marked influence on the performance of induction motors with solid iron rotors. Torque per unit of current improves very much as copper end rings are attached to solid iron rotors. It is thought that depending on the type of thyristor circuit arrangement, the combined effect of harmonics and end rings might be used to find optimum end rings. I should like to ask the author if he has considered different end rings that might yield as much improvement as different thyristor circuits did.

References

[6] Fidelity Instrument Corp., York, Pa., Bulletin 459-1.

[7] W. Leonhard, "Elements of reactor-controlled reversible induction motor drives," *Trans. AIEE (Applications and Industry)*, vol. 78, pp. 106–115, May 1959.

[8] H. M. McConnell and E. F. Sverdrup, "The induction machine with solid iron rotor," *Trans. AIEE (Power Apparatus and Systems)*, vol. 74, pp. 343–349, June 1955.

[9] G. Angst, "The polyphase induction motor with solid rotor effects of saturation and finite length," presented at the Fall General Meeting, Detroit, Mich., October 15–20, 1961.

D. A. Paice: Mr. Szogyen has pointed out two of the major advantages of thyristor controls, the small size and weight, but he asked to what extent these and other potential advantages of solid state controls can be exploited. In particular, how do the thyristor controls compare with magnetic types as regards reliability overload capability and harmonics?

Manuscript received June 21, 1967.

I believe that the use of thyristors for stator voltage control represents a *natural* application area and one in which the thyristor is able to compete very successfully with existing techniques. For example, thyristor crane control drives are available which exhibit significant advantages over the older reactor types: reduced weight and volume and improved performance.[1] In most of these ac applications slip ring motors are used and slip power is dissipated outside the machine. This gives additional advantages in that larger motors can be controlled with simpler thyristor circuits such as shown in Test Circuit 1.

The simplicity of the thyristor voltage control and inherent high reliability of the devices themselves provides opportunity for an extremely reliable equipment design. Inrush current capability has to be designed into the thyristor equipment to start the motor properly and this is achieved readily in practice with present day thyristors which also are well able to tolerate locked rotor currents. The range of motor sizes that can be controlled depends greatly upon application; with varying torque loads and squirrel cage machines controls up to 50 hp can be evisaged; with slip ring motors controls (with convection cooled single cell thyristors) to 125 hp are already available suitable even for constant torque loads.

Thyristor and magnetic amplifier controls are similar as regards production of harmonic voltages which may modify the motor torque capabilities. The thyristor by virtue of its very fast switching action may also generate radio frequency noise. However, where necessary this can readily be corrected.

Mr. Szoygen's remarks pertaining to the effect of *harmonics and end rings* with a possible optimum configuration are intriguing. The end rings could be used to filter out harmonic currents, but I have no practical information to prove the feasibility of this. From Fig. 10 the total power loss of the motor caused by harmonics in the supply can be estimated and the possible benefit of an optimum *end ring configuration*, gauged in the case of a six thyristor control circuit.

[1] See Bulletin DB 22–501, Crane Drive Group, Systems Control Division, Westinghouse Electric Corporation, Buffalo, N. Y.

A Comparative Study of Symmetrical Three-Phase Circuits for Phase-Controlled AC Motor Drives

WILLIAM McMURRAY, SENIOR MEMBER, IEEE

Abstract—The speed of an induction motor in certain types of drives can be adjusted by phase control of the applied voltage. Three pairs of inverse-parallel thyristors are required for three-phase bipolar symmetry, but several different arrangements of the thyristors and motor windings are possible. A direct series connection of thyristors and windings in each phase can be made; the three phases can then be connected in delta, wye, or wye with neutral return. Alternatively, a delta connection of thyristor modules can be inserted in the opened neutral junction of wye-connected windings. Representation of the motor as a counter EMF in series with leakage reactance allows a simplified analysis of the current waveforms in terms of the thyristor conduction angle. With a given fundamental current, the current harmonics in the motor windings and in the supply lines for each of the circuits can then be compared. The results are equally applicable to thyristor-controlled inductive loads such as may be used for reactive power adjustment.

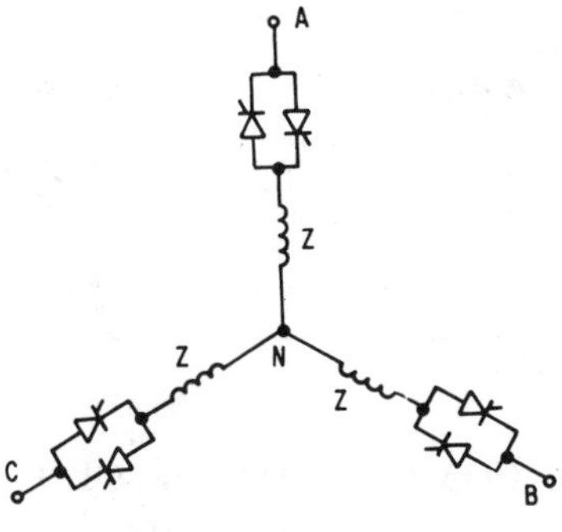

Fig. 1. Motor in wye, thyristors in wye (Circuit 1).

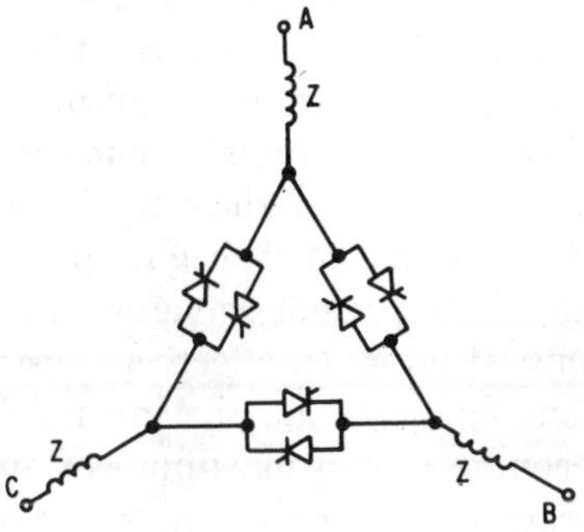

Fig. 2. Motor in wye, thyristors in delta (Circuit 2).

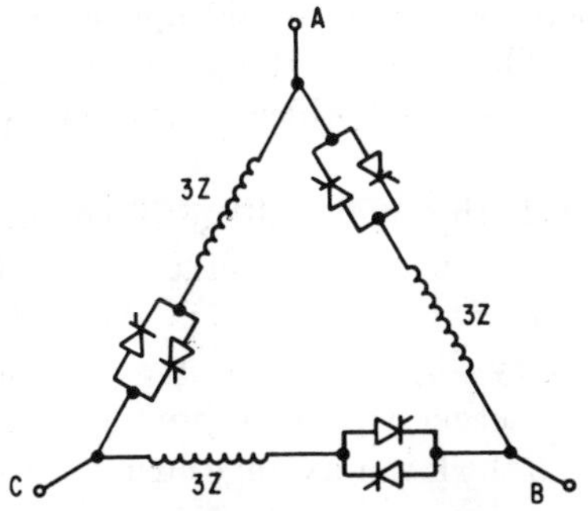

Fig. 3. Motor in delta, thyristors in delta (Circuit 3).

INTRODUCTION

THIS STUDY is restricted to the circuit arrangements shown in Figs. 1–3 that have complete three-phase bipolar symmetry. Since they have been given no widely accepted descriptive names, they will be referred to as Circuits 1, 2, and 3, respectively. The results of the study are easily extended to cover certain similar circuit configurations. The variation of Circuit 1 in which the motor is connected in delta involves a simple wye–delta transformation of the motor parameters with the possibility of third-harmonic current circulation around the delta windings, but it does not affect the thyristors. Similarly, the performance of Circuit 1 with the motor neutral N connected to the neutral of a four-wire supply system is easily obtained by delta–wye transformation from Circuit 3, and third-harmonic currents flow in the neutral wire.

A number of studies [1]–[6] have been made of these circuits by different methods. The motor and line currents in Circuit 2 are the same as in Circuit 1; only the thyristor currents are different. Thus, a complete solution of the behavior of Circuit 2 includes the solution for Circuit 1, so that it is only necessary to detail the analyses of Circuits 2 and 3 to cover all the symmetrical cases.

Unsymmetrical arrangements, such as Circuit 1 with only one or two phases controlled by thyristors, Circuit 1 with one thyristor in each phase replaced by a diode, or Circuit 2 with three of the thyristors omitted, have been proposed [5]–[7] for low-cost applications, but their performance is too poor for large drive systems. With the phase-back control being asymmetric, the motor and line currents have a considerably higher harmonic content, including even-order components. Also, the total thyristor rating may not be reduced. The three-thyristor version of Circuit 2, for example, requires each of the three devices to carry three times the average current that each of the six devices in the symmetrical circuit carry, when both circuits are fully on, so that the total rating is increased by 50 percent.

Paper TOD-73-90, approved by the Static Power Converter Committee of the Industry Applications Society for presentation at the 1973 Eighth Annual Meeting of the Industry Applications Society, Milwauk~~ Wis., October 8–11, 1973. Manuscript released for publication January 15, 1974.

The author is with the Research and Development Center of the General Electric Company, Schenectady, N. Y. 12301.

Reprinted from *IEEE Trans. Ind. Appl.*, vol. IA-10, pp. 403–411, May/June 1974.

TABLE I
COMPARISON OF THYRISTOR RATING PARAMETERS WITH FULL-ON CONDUCTION, BASED ON PEAK VALUES IN CIRCUIT 3.

Circuit Number	Motor Connection	Thyristor Connection	Conduction Angle	Thyristor Current			Initial di/dt	Peak Thyristor Voltage	
				Peak	rms	Average		Normal	Fault
1	Y	Y	180°	$\sqrt{3} = 1.732$	$\frac{\sqrt{3}}{2} = 0.866$	$\frac{\sqrt{3}}{\pi} = 0.552$	$\sqrt{3}$	$\frac{\sqrt{3}}{2}$	1
2	Y	Δ	120°	$\frac{3}{2} = 1.5$	$\sqrt{\frac{1}{2} - \frac{3\sqrt{3}}{8\pi}} = 0.541$	$\frac{\sqrt{3}}{2\pi} = 0.276$	$\sqrt{3}$	1	1
3	Δ	Δ	180°	$1 = 1$	$\frac{1}{2} = 0.5$	$\frac{1}{\pi} = 0.318$	1	1	1

GENERAL COMPARISON OF CIRCUITS 1–3

When the thyristors are fully on, the motor currents are sinusoidal in all cases, assuming no saturation and a design that avoids third harmonics in the case of the delta of Circuit 3. For Circuits 1 and 3, the current in each thyristor is a half sine wave, differing in magnitude by the usual factor of $\sqrt{3}$. However, the thyristor currents in Circuit 2 are nonsinusoidal because of their nonlinear conducting impedance and the timing of their gate pulses.

The actual thyristor current waveform in Circuit 2 will be further discussed later. In effect, a substantial triplen harmonic current flows around the inner delta such that the thyristors conduct for a maximum of only 120° each cycle. The average thyristor current is reduced to a minimum, but the rms value is slightly increased over that in Circuit 3. The ideal current distribution may be slightly altered if there is significant linear impedance in series with the thyristors. The relative thyristor duty for the three circuits when fully on under ideal conditions is summarized in Table I.

Since the peak thyristor voltage is equal to the crest value E_{Lm} of the line-to-line voltage in both Circuits 2 and 3, the rating of the thyristors is about the same. Circuit 2 may be very slightly better or worse than Circuit 3, depending on whether the peak junction temperature is more sensitive to the lower average current or to the higher peak and rms values.

The volt–ampere rating of the thyristors in Circuit 1 will be the same as in Circuit 3 if their peak voltage is reduced by the same wye–delta transformation factor of $\sqrt{3}$ by which their current is increased. However, this is not true because of the nonlinear switching operation. When the thyristors in Circuit 1 are phased back, there will be an interval in which phase A, for example, is nonconducting while phases B and C are on. With no motor back EMF, as at zero speed, the potential of neutral point N and the inner terminal of the phase A thyristors will be midway between the potentials of lines B and C during this interval. The peak voltage from A to N is, then, $\sqrt{3}/2$ times the peak-to-line voltage E_{Lm}, occurring when the potentials of B and C are equal.

Under abnormal conditions, the voltage across the thyristors in Circuit 1 may be still higher, equal to the whole line voltage E_{Lm}, the same as for the other two circuits. Suppose that, when all phases are intended to be off, the thyristors in one phase are rendered conductive by misfiring or by device failure. Then, full line-to-line voltage is applied to the blocking thyristors in the other two phases. Since the thyristors should be designed to withstand such conditions, the volt–ampere rating of the thyristors in Circuit 1 will be $\sqrt{3}$ times the rating for Circuit 3.

Another disadvantage of Circuit 1 is the need for two thyristors to be fired simultaneously in order to establish a conductive path between two supply lines. Hence, double-pulse gate signals spaced 60° apart, or else a single long gate pulse exceeding 60° in duration are required. For the thyristors in Circuits 2 and 3, single short gate pulses are sufficient.

SIMPLIFIED EQUIVALENT CIRCUIT FOR MOTOR

The choice between Circuits 2 and 3 will depend largely on their relative performance under phased-back conditions. The comparison is best made in terms of the harmonic content of the motor currents and of the line currents. In both cases, third-harmonic components are not present in the line, but Circuit 3 produces third-harmonic currents in the motor windings. Thus, Circuit 2 would seem to be best, unless the other harmonic components are significantly higher. In order to effect a comparison of harmonics without a complete detailed analysis of both cases, some simplification of the motor behavior is desirable.

For steady-state sinusoidal operation, an induction motor load on a supply system is often represented as an equivalent passive inductance and resistance. However, such a representation is grossly inaccurate when harmonic components are involved. It is better to represent induction motors, as well as synchronous motors or generators, as an equivalent counter EMF and a small series impedance. For simplicity, we will assume that the counter EMF is sinusoidal (i.e., pure fundamental) despite the harmonic currents and neglect the series resistance. The remaining series reactance is produced by the leakage inductance of the machine.

With the motor represented in this simplified manner, Fig. 4 shows the equivalent circuit for phase BC in Fig. 3. The counter EMF e_{bc}' is in series with an inductance $3L$, where L is the leakage inductance referred to the wye configuration. As shown in the phasor diagram Fig. 5, the counter emf e_{bc}' is assumed to lag the corresponding supply line voltage e_{bc} by some arbitrary phase angle and to be somewhat lower in magnitude. When the thyristors conduct, the phasor difference voltage e_{bc}'' appears across the series inductance $3L$ and produces a current i_{bc} having a fundamental component that lags the phasor voltage e_{bc}'' by 90°. The conduction interval of the thyristors depends upon the timing of their gate pulses with respect to the phase of the voltage e_{bc}'', and the current waveform can

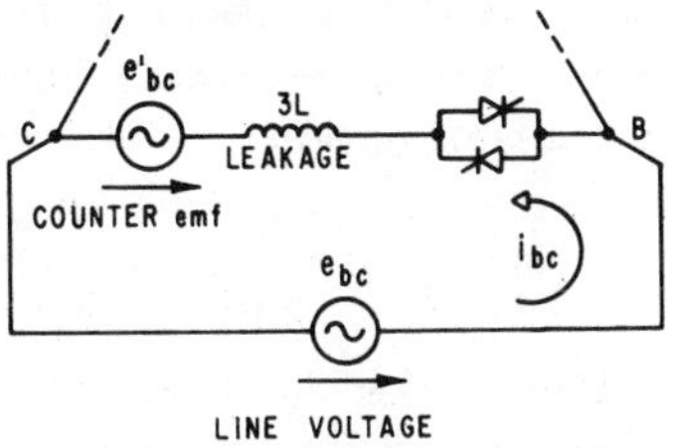

Fig. 4. Phase *BC* of Circuit 3 with motor represented by a counter EMF and leakage inductance.

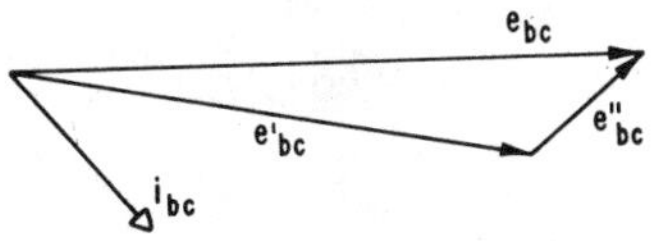

Fig. 5. Phasor diagram for Fig. 4.

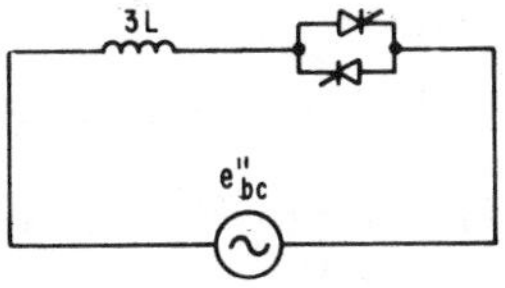

Fig. 6. Equivalent circuit for analysis of current waveform.

be determined by considering the behavior of the reduced equivalent circuit in Fig. 6.

With each phase treated in this fashion, it is seen that Figs. 2 and 3 become polyphase versions of Fig. 6 if the source voltages are replaced by the phasor differences between the actual supply voltages and the counter EMF's of the motor, while the complex motor impedances Z are replaced by pure leakage inductances. The peak motor current in the delta configuration with the thyristors full on is given by $I_m = E_{Lm}''/(3L\omega)$, and all currents in the analyses that follow will be defined in terms of this base quantity.

Note that, while it is necessary to know the magnitude and phase of the counter EMF in order to find the values of the peak voltage difference E_{Lm}'' and hence I_m, the shape and relative harmonic content of the current waveforms depend only on the thyristor conduction angle θ or the hold-off angle γ, which is directly related to θ. Of course, the thyristor firing angle α, measured from the line voltage zero crossing, that is required to achieve the assumed value of θ will also depend upon the counter EMF.

Since the speed–torque performance of the motor is largely determined by the fundamental components of voltage and current, the counter EMF generated under a given load condition will be the same, whichever circuit configuration is employed. Thus, E_{Lm}'' and I_m will be the same for each circuit, as well as the required fundamental component of current. Therefore, the alternative circuits can be compared if the conduction angle in each case is chosen to produce the same per unit value of the fundamental current, based on I_m. In particular, the harmonic current spectra can be compared.

The purely inductive impedances of the simplified equivalent circuits result in current waveforms that are readily deduced and can be expressed mathematically by displaced cosine functions. With reasonable effort, these functions yield to direct Fourier analysis, and the resulting Fourier series coefficients can be evaluated by digital computer. This procedure is followed for each of the circuits in turn in the following sections. The accuracy of the analysis and computation has been checked by verifying that evaluation of the sum of the Fourier series yields an approximation of the actual time function.

CURRENT WAVEFORMS IN CIRCUIT 3

In the simplified equivalent circuit for Fig. 3, the currents in the thyristors and the stator windings have the waveforms shown in Fig. 7(a), drawn for a particular value of the conduction angle θ. In this circuit, θ can vary over the range $0 < \theta < \pi$ and the hold-off angle γ is $\gamma = \pi - \theta$. It is convenient to express the waveforms of Fig. 7 in terms of half the conduction angle, $\beta = \theta/2$. The thyristor firing angle with respect to the zeroes of the voltage wave e_L'' is $\alpha'' = \pi - \beta$.

If one selects the time origin at the instant when the current in a particular phase (*ab* in Fig. 7a) is maximum, the waveform can be expressed as a Fourier cosine series of odd-order terms

$$i = \sum_{m=1}^{\infty} i_n = \sum_{m=1}^{\infty} a_n \cos n\omega t \tag{1}$$

where $n = 2m - 1$ and

$$a_n = \frac{4}{\pi} \int_0^{\pi/2} i(\omega t) \cos n\omega t \, d(\omega t) \tag{2}$$

where $i(\omega t)$ is the actual instantaneous time function defined over the interval $0 < \omega t < \pi/2$.

As seen in Fig. 7(a),

$$\begin{aligned} i(\omega t) &= I_m(\cos \omega t - \cos \beta), && \text{for } 0 < \omega t < \beta \\ i(\omega t) &= 0, && \text{for } \beta < \omega t < \pi/2. \end{aligned} \tag{3}$$

Performing the necessary integrations and simplifying the results, the following Fourier series is obtained, in two alternative forms:

$$\frac{i}{I_m} = \frac{2\beta - \sin 2\beta}{\pi} \cos \omega t + \sum_{m=2}^{\infty} \frac{2}{\pi n} \cdot \left[\frac{\sin (n-1)\beta}{n-1} - \frac{\sin (n+1)\beta}{n+1} \right] \cos n\omega t \tag{4}$$

$$= \frac{\theta - \sin \theta}{\pi} \cos \omega t + \sum_{m=2}^{\infty} \frac{1}{\pi n} \cdot \left[\frac{\sin (m-1)\theta}{m-1} - \frac{\sin m\theta}{m} \right] \cos n\omega t. \tag{5}$$

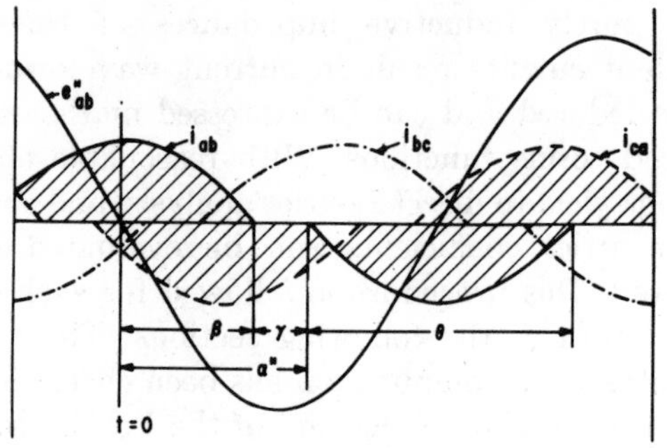

(a)

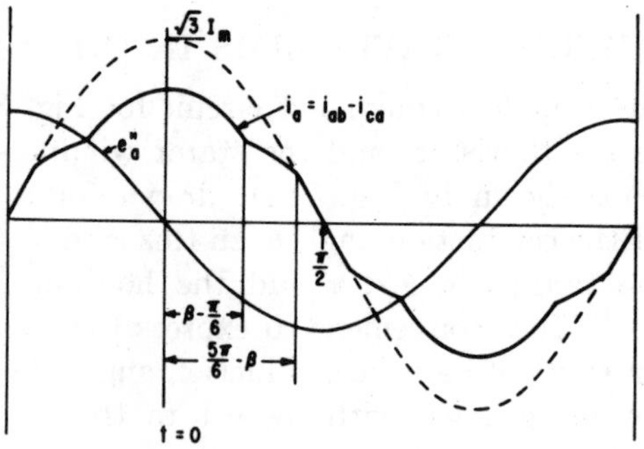

(b)

Fig. 7. Currents in Circuit 3 for $2\pi/3 < \theta < \pi$. (a) Thyristor currents. (b) Line A current.

The average current I_a in a pair of thyristors and the windings is given by

$$I_a = \frac{2}{\pi}\int_0^{\pi/2} i(\omega t)\, d(\omega t)$$

$$= \frac{2}{\pi} I_m \int_0^{\beta} (\cos x - \cos\beta)\, dx$$

$$= I_m \frac{2}{\pi}(\sin\beta - \beta\cos\beta). \tag{6}$$

The rms current I_e is obtained by a similar integration

$$I_e = \left(\frac{2}{\pi}\int_0^{\pi/2} [i(\omega t)]^2\, d(\omega t)\right)^{1/2}$$

$$= I_m\left(\frac{1}{\pi}\left[\beta(2 + \cos 2\beta) - \frac{3}{2}\sin 2\beta\right]\right)^{1/2} \tag{7}$$

The average and rms currents in an individual thyristor are 1/2 and $1/\sqrt{2}$ times these values, respectively.

The current in line A is given by $i_a = i_{ab} - i_{ca}$, which results in the waveform shown in Fig. 7(b). The Fourier series for this current can easily be obtained from the previous result (4). For the nth-harmonic component i_n' of the line current, with the time origin shifted to coincide with the center of the positive line current wave,

$$i_n' = a_n \cos n(\omega t + \pi/6) + a_n \cos n(\omega t - \pi/6)$$

$$= 2a_n \cos(n\omega t)\cos(n\pi/6)$$

$$= 2\cos(n\pi/6) i_n. \tag{8}$$

Thus, each term in the series (4) is multiplied by $2\cos(n\pi/6)$, which has the following values, as expected for a delta–wye transformation:

n	1	3	5	7	9	11	13
$2\cos(n\pi/6)$	$\sqrt{3}$	0	$-\sqrt{3}$	$-\sqrt{3}$	0	$\sqrt{3}$	$\sqrt{3}$.

The same modified series should result from integration of the segment equations for the current waveform in Fig. 7(b) over the interval $0 < \omega t < \pi/2$. However, the applicable segments depend upon β as follows. When $\pi/3 < \beta < \pi/2$ or $2\pi/3 < \theta < \pi$ (as drawn in Fig. 7(b)),

$$i'(\omega t) = I_m(\sqrt{3}\cos\omega t - 2\cos\beta), \quad \text{for } 0 < \omega t < \beta - \pi/6$$

$$i'(\omega t) = I_m[\cos(\omega t - \pi/6) - \cos\beta], \quad \text{for } \beta - \pi/6 < \omega t < 5\pi/6 - \beta \tag{9}$$

$$i'(\omega t) = I_m\sqrt{3}\cos\omega t, \quad \text{for } 5\pi/6 - \beta < \omega t < \pi/2$$

when $\pi/6 < \beta < \pi/3$ or $\pi/3 < \theta < 2\pi/3$,

$$i'(\omega t) = I_m(\sqrt{3}\cos\omega t - 2\cos\beta), \quad \text{for } 0 < \omega t < \beta - \pi/6$$

$$i'(\omega t) = I_m[\cos(\omega t - \pi/6) - \cos\beta], \quad \text{for } \beta - \pi/6 < \omega t < \beta + \pi/6$$

$$i'(\omega t) = 0, \quad \text{for } \beta + \pi/6 < \omega t < \pi/2 \tag{10}$$

when $0 < \beta < \pi/6$ or $0 < \theta < \pi/3$,

$$i'(\omega t) = 0, \quad \text{for } 0 < \omega t < \pi/6 - \beta$$

$$i'(\omega t) = I_m[\cos(\omega t - \pi/6) - \cos\beta], \quad \text{for } \pi/6 - \beta < \omega t < \pi/6 + \beta$$

$$i'(\omega t) = 0, \quad \text{for } \pi/6 + \beta < \omega t < \pi/2. \tag{11}$$

The average and rms values of the line current can be obtained from a numerical evaluation of the integrals of $i'(\omega t)$ and its square. An approximate value of the rms currents in all cases can be calculated by summing the squares of the Fourier coefficients for a large number of terms.

CURRENT WAVEFORMS IN CIRCUIT 2

The thyristor currents in the simplified equivalent circuit for Fig. 2 have the waveforms sketched in Fig. 8(a) for a value of conduction angle within the range $\pi/3 < \theta < 2\pi/3$ or $\pi/6 < \beta < \pi/3$. To preserve the nomenclature used by Lipo [4], the hold-off angle γ is defined here as the interval of zero line current, so that $\gamma = 2\pi/3 - \theta$. The nonconducting angle of the thyristors is, of course, $\pi - \theta$ and their firing angle with respect to the voltage e_L'' is still $\alpha'' = \pi - \beta$.

For the current in the thyristors between phases A and B in Fig. 8(a) defined over the interval $0 < \omega t < \pi/2$

with the indicated origin, the segment equations can be shown to be

$$i(\omega t) = \tfrac{3}{2}I_m[\cos \omega t - \sqrt{3}\cos(\beta + \pi/6)],$$
$$\text{for } 0 < \omega t < \pi/3 - \beta$$
$$i(\omega t) = \sqrt{3}I_m[\cos(\omega t + \pi/6) - \cos(\beta + \pi/6)],$$
$$\text{for } \pi/3 - \beta < \omega t < \beta$$
$$i(\omega t) = 0, \qquad \text{for } \beta < \omega t < \pi/2. \tag{12}$$

Fourier analysis of this function yields the following series:

$$\frac{i}{I_m} = \frac{3}{2\pi}[2\beta - \sin(2\beta + \pi/3)]\cos \omega t + \sum_{m=2}^{\infty} \frac{\sqrt{3}}{\pi n}$$
$$\cdot\left[\frac{\cos(n-1)(\pi/3 - \beta) - 2\cos[\pi/3 + (n-1)\beta]}{n-1}\right.$$
$$\left. + \frac{\cos(n+1)(\pi/3 - \beta) - 2\cos[\pi/3 - (n+1)\beta]}{n+1}\right]$$
$$\cdot \cos n\omega t \tag{13}$$

where $n = 2m - 1$, or the alternative form in terms of $\theta = 2\beta$ and m

$$\frac{i}{I_m} = \frac{3}{2\pi}[\theta - \sin(\theta + \pi/3)]\cos \omega t + \sum_{m=2}^{\infty} \frac{\sqrt{3}}{2\pi n}$$
$$\cdot\left[\frac{\cos(m-1)(2\pi/3 - \theta) - 2\cos[\pi/3 + (m-1)\theta]}{m-1}\right.$$
$$\left. + \frac{\cos m(2\pi/3 - \theta) - 2\cos(\pi/3 - m\theta)}{m}\right]\cos n\omega t. \tag{14}$$

The average value of the current in a pair of thyristors is

$$I_a' = I_m\frac{\sqrt{3}}{\pi}[\sin(\beta + \pi/6) - (\beta + \pi/3)\cos(\beta + \pi/6)]. \tag{15}$$

The rms value of the current has been determined by numerical methods.

When the conducting angle θ is small, within the range $0 < \theta < \pi/3$ or $0 < \beta < \pi/6$ the instantaneous current has the form

$$i(\omega t) = \tfrac{3}{2}I_m(\cos \omega t - \cos \beta), \qquad \text{for } 0 < \omega t < \beta$$
$$i(\omega t) = 0, \qquad \text{for } \beta < \omega t < \pi/2. \tag{16}$$

This function is seen to be 1.5 times the function (3) for the case of Circuit 3, so that the Fourier series is 1.5 times (4) or (5) and the average and rms currents are 1.5 times (6) and (7), respectively.

The harmonic components of the motor current, which is the same as the line current, are easily obtained by the transformation (8). The actual waveform is sketched in

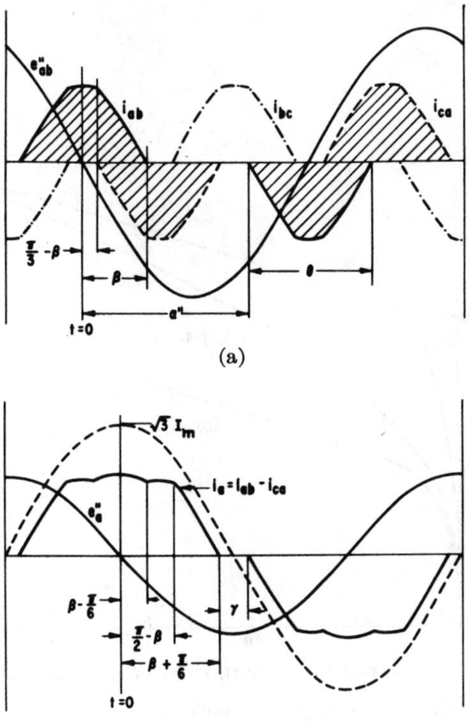

Fig. 8. Currents in Circuit 2 for $\pi/3 < \theta < 2\pi/3$. (a) Thyristor currents. (b) Line A current.

Fig. 8(b), for which the segment equations are

$$i'(\omega t) = I_m\sqrt{3}[\cos \omega t - 2\cos(\beta + \pi/6)],$$
$$\text{for } 0 < \omega t < \beta - \pi/6$$
$$i'(\omega t) = I_m\tfrac{3}{2}[\cos(\omega t - \pi/6) - \sqrt{3}\cos(\beta + \pi/6)],$$
$$\text{for } \beta - \pi/6 < \omega t < \pi/2 - \beta$$
$$i'(\omega t) = I_m\sqrt{3}[\cos \omega t - \cos(\beta + \pi/6)],$$
$$\text{for } \pi/2 - \beta < \omega t < \beta + \pi/6$$
$$i'(\omega t) = 0, \qquad \text{for } \beta + \pi/6 < \omega t < \pi/2. \tag{17}$$

The average and rms values of the line current have been determined by numerical evaluation of the integral of the function (17) and its square.

When the conducting angle θ is in the range $0 < \theta < \pi/3$ the applicable function for the line current is 1.5 times the function (11). The average value of line current is 3 times (6) and the rms value is $1.5\sqrt{2}$ times (7).

EVALUATION OF COMPUTED CURRENT HARMONICS

The computer programs for analyzing the current waveforms in the two circuits were written with the fundamental component of current as the independent variable, in normalized form as a percentage of the current with full conduction. The performance of each circuit at a given point can then be compared directly. For each circuit, the thyristor conduction angle required to obtain the specified fundamental current is determined by numerical solution of a transcendental equation for θ, given

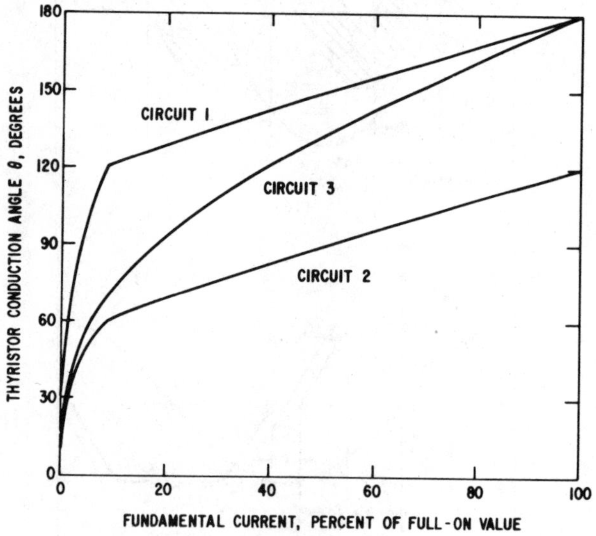

Fig. 9. Thyristor conduction angle as a function of the desired fundamental current.

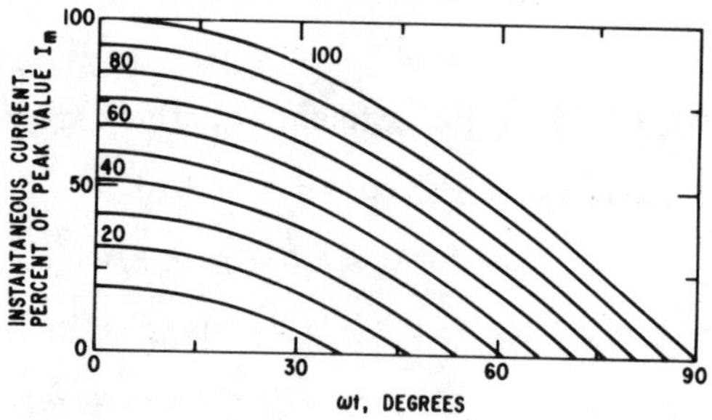

Fig. 10. Thyristor current waveforms in Circuit 3 with percent fundamental current as a parameter.

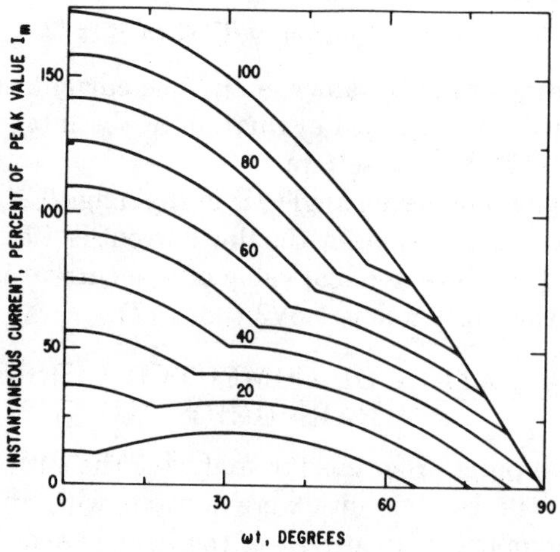

Fig. 11. Line current waveforms in Circuit 3 with percent fundamental current as a parameter.

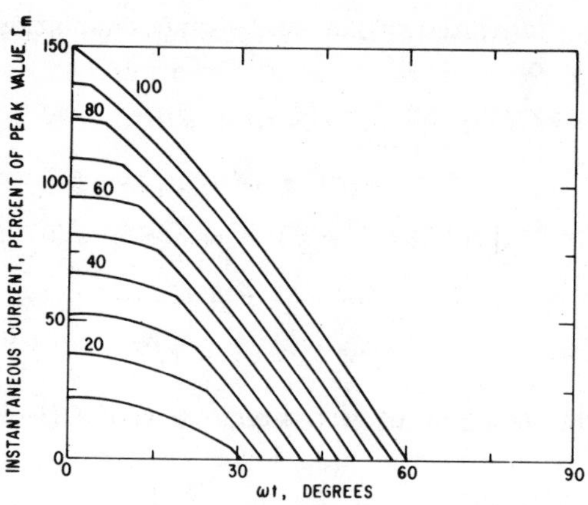

Fig. 12. Thyristor current waveforms in Circuit 2 with percent fundamental current as a parameter.

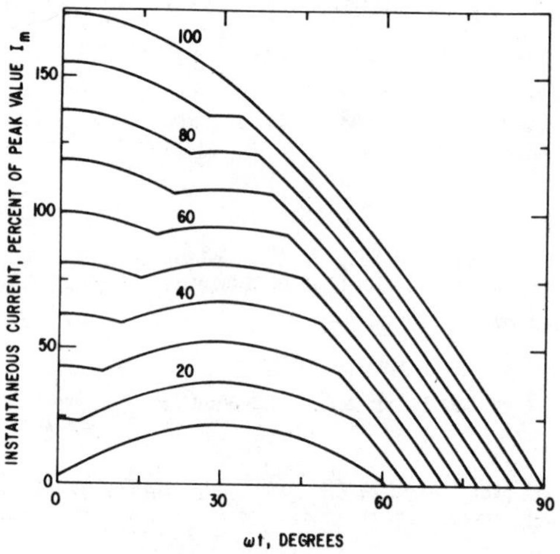

Fig. 13. Line current waveforms in Circuit 2 or Circuit 1 with percent fundamental current as a parameter.

by the coefficient of the series (5) or (14) for $n = 1$. The results are plotted in Fig. 9. For currents down to 20 percent of the full-on value, the current is an almost linear function of the conduction or firing angle.

Quarter-cycle waveforms for the thyristor current and line current in each of the two circuits are presented in Figs. 10–13, for values of the fundamental component from 10 percent to 100 percent of the full-on current in 10 percent steps. The total rms values of the harmonic content in these waveforms are shown in Fig. 14 as a percentage of the full-on current, and are presented again in Fig. 15 as a percentage of the actual fundamental current at each point. The individual harmonic components, up to the ninth, are plotted on a logarithmic scale in Figs. 16–19. Note that the triplen harmonics appear only in the thyristors and are eliminated from the line currents, whereas the other components have the same relative magnitude in both locations. It is seen that the harmonics in Circuit 2 are generally greater than in Circuit 3.

However, in Circuit 3 where the motor windings are delta connected in series with the thyristors, the windings also carry the triplen harmonic currents. These increase the total harmonic content of the motor current despite the lower values of the other components relative to Circuit 2. Hence, Circuit 2 is preferred where motor heating is the major concern. Where distortion of the line current is more important, Circuit 3 has the advantage.

In this analysis, the effect of line impedance has been neglected. For Circuit 2, the effective source reactance is directly additive to the motor reactance and will not change the circuit behavior, although the conduction angle required to achieve a given current will be increased. However, inclusion of source reactance in Circuit 3 tends to change the configuration towards that of Circuit 2, and the behavior will be modified if the source reactance becomes significant with respect to the motor leakage reactance. The analysis also assumes that the motor generates a sinusoidal counter EMF. This may not be valid when the conduction angle is small, but the performance in this region is not critical. If the motor is designed to produce a nonsinusoidal counter EMF, the circuit behavior will be altered.

TRANSIENT CONSIDERATIONS

There is a significant difference between the three circuits in the stresses imposed upon the thyristors by transients on the power lines, such as produced by switching or lightning strokes. In all circuits, the motor windings must be designed to withstand the transient voltage stress. The thyristors are partially protected by the motor in Circuit 2. For Circuit 1 as drawn, the thyristors are subjected to incoming line transients. However, if the locations of the thyristors and the motor windings are interchanged, then Circuit 1 becomes similar to Circuit 2 with respect to transient stresses. In the arrangement of Circuit 3, both the motor windings and the thyristors must be prepared to withstand transient voltages appearing on the supply lines. Therefore, Circuit 3 is at a disadvantage compared with the other two circuits.

There should be no significant difference between the three arrangements with respect to electromagnetic interference (EMI) generation. In all cases, the firing and blocking of the thyristors should produce about the same noise due to dv/dt and di/dt. However, propagation of the noise into the supply lines will be reduced if the motor windings are located between the thyristors and the lines, as in Circuit 2 or Circuit 1 with the previously mentioned interchange of windings and thyristors.

CONCLUSIONS

It has been shown that if an ac motor can be represented as a counter EMF in series with an inductance, then the stator and line current waveforms obtained in a phase-controlled drive system are similar to the waveforms produced by a polyphase thyristor-controlled pure inductive load, such as may be used for reactive power adjustment in a power factor correction scheme, for example. The performance with different arrangements of the thyristors can then be readily analyzed and compared. The optimum circuit configuration depends largely upon whether motor

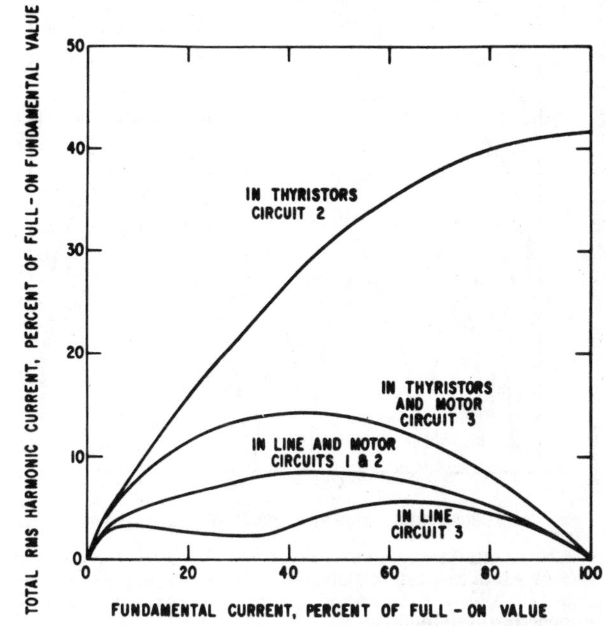

Fig. 14. Total rms values of harmonic currents as a percentage of full-on fundamental value.

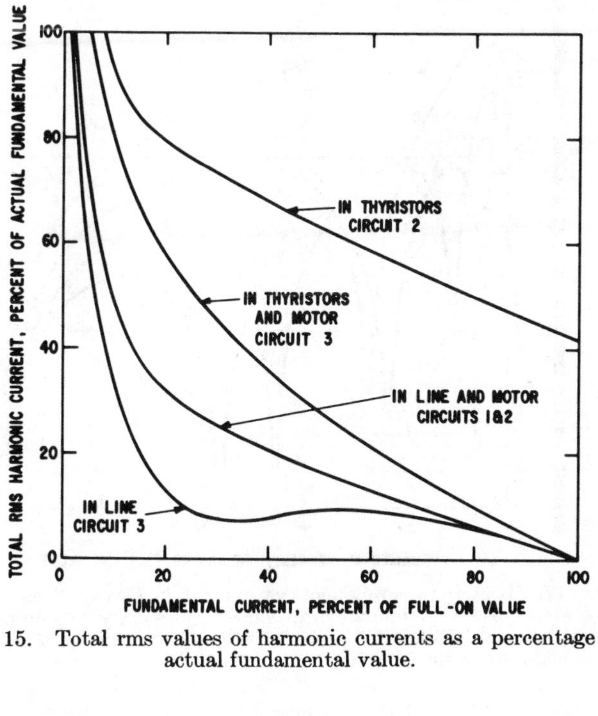

Fig. 15. Total rms values of harmonic currents as a percentage of actual fundamental value.

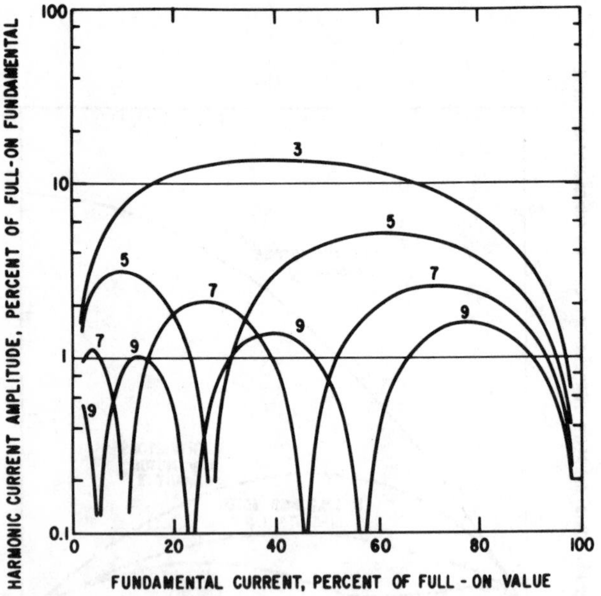

Fig. 16. Harmonic components of current in Circuit 3 as a percentage of fundamental full-on value. The curves are numbered with the order of the harmonic. *Note:* The triplen harmonics do not appear in the line current.

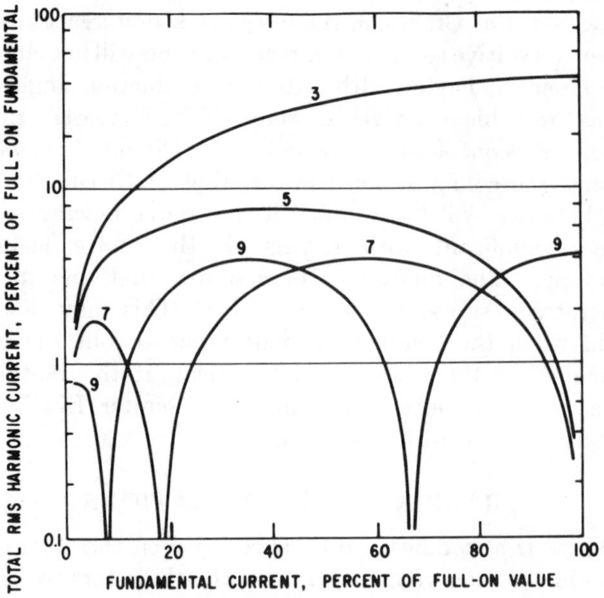

Fig. 18. Harmonic components of current in Circuit 2 or 1 as a percentage of fundamental full-on value. The curves are numbered with the order of the harmonic. *Note:* the triplen harmonics do not appear in the line current.

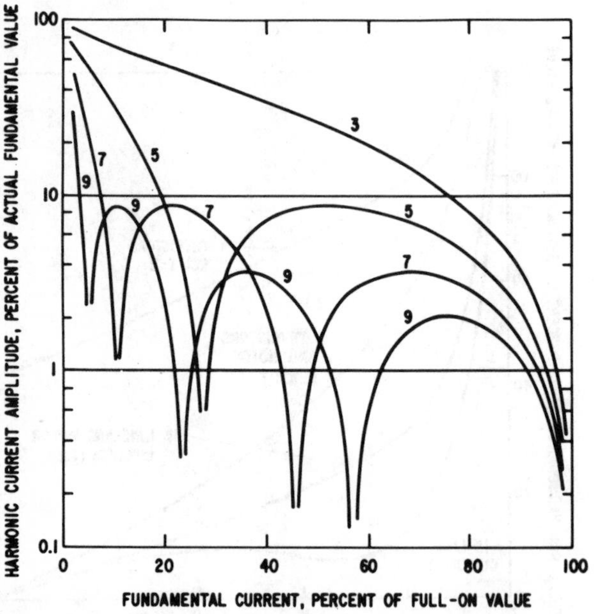

Fig. 17. Harmonic components of current in Circuit 3 as a percentage of actual fundamental value. The curves are numbered with the order of the harmonic. *Note:* the triplen harmonics do not appear in the line current.

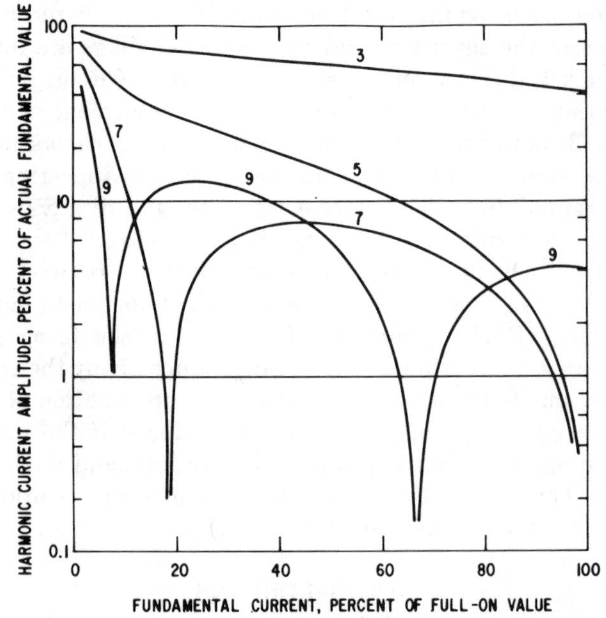

Fig. 19. Harmonic components of current in Circuit 2 or 1 as a percentage of actual fundamental value. The curves are numbered with the order of the harmonic. *Note:* the triplen harmonics do not appear in the line current.

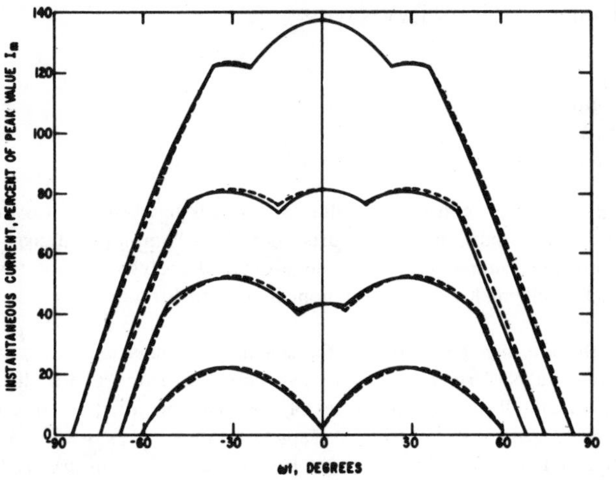

Fig. 20. Comparison of line current waveforms in Circuit 1 or 2. —— Accurate equivalent circuit for motor. - - - Approximate equivalent circuit for motor.

current harmonics or line current harmonics are of greater concern in the particular application.

To substantiate the validity of the simplified equivalent circuit of the motor, Lipo made several test runs of his previously developed [4] computer program for the analysis of Circuit 1 with an accurate equivalent circuit for an induction motor. Using the same motor parameters [4], a 0.833 per unit rotor speed was selected for all cases. A set of hold-off angles was specified to correspond with the 10 percent, 30 percent, 50 percent, and 80 percent curves in Fig. 13, and the computed current waveforms were scaled to match the peak values of those curves. The results are presented in Fig. 20. It is seen that the accurate and approximate current waveforms are in close agreement. This implies that a typical motor is able to produce a near-sinusoidal counter EMF despite considerable distortion of the stator current waveform. It also suggests that other types of converters, such as inverters or cycloconverters, should be analyzed with a counter EMF-inductance load instead of the commonly assumed series resistance–inductance load or a sinusoidal current sink. The behavior of the converter deduced under counter-EMF load conditions will be closer to its actual behavior in a motor drive system, so the effort of analysis will be more profitably spent.

REFERENCES

[1] W. Shepherd and J. Stanway, "The polyphase induction motor controlled by firing-angle adjustment of silicon controlled rectifiers," in *1964 IEEE Int. Conv. Rec.*, vol. 12, pt. 4, pp. 135–154.

[2] W. Shepherd, "On the analysis of the three-phase induction motor with voltage control by thyristor switching," *IEEE Trans. Ind. Gen. Appl.*, vol. IGA-4, pp. 304–311, May/June 1968.

[3] T. J. Tacheuchi, *Theory of SCR Circuit and Application to Motor Control.* Tokyo: Tokyo Electrical Engineering College Press, 1968.

[4] T. A. Lipo, "The analysis of induction motors with voltage control by symmetrically triggered thyristors," *IEEE Trans. Power App. Syst.*, vol. PAS-90, pp. 515–525, Mar./Apr. 1971.

[5] "Trinistor three-phase ac regulators," Westinghouse Brake and Signal Co. Ltd., London, England, Application Rep. 9, 1962.

[6] D. A. Paice, "Induction motor speed control by stator voltage control," *IEEE Trans. Power App. Syst.*, vol. PAS-87, pp. 585–590, Feb. 1968.

[7] W. Shepherd and J. Stanway, "Unbalanced voltage control of three-phase loads by the triggering of silicon controlled rectifiers," *IEEE Trans. Ind. Gen. Appl.*, vol. IGA-1, pp. 206–216, May/June 1965.

THE ANALYSIS OF INDUCTION MOTORS WITH VOLTAGE CONTROL BY SYMMETRICALLY TRIGGERED THYRISTORS

T. A. Lipo
General Electric Company
Schenectady, N. Y.

ABSTRACT

The application of thyristor switching to induction motor speed control has resulted in a number of unconventional supply systems. One such technique, which has been successfully employed in a number of applications, is variable voltage control by means of symmetrically triggered thyristors in the stator phases of a wye-connected machine. In this paper, an analytic method for predicting the steady-state performance of such a system is presented. The solution is obtained in closed form without the necessity for iteration to obtain the proper boundary conditions. In addition, it is shown that the symmetry of the solution permits an additional reduction in the computation time. Since matrix techniques are utilized throughout the analysis, the equations can be easily implemented into a digital computer program. Hence, the method is well suited to the evaluation of proposed motor designs when used in conjunction with thyristor voltage control. Measured torque-speed characteristics of a typical drive system are included and the results compared to an analytical solution. It is demonstrated that a computed solution will favorably predict the performance of an actual system.

INTRODUCTION

With the development of the thyristor, a variety of schemes have been evolved incorporating these devices for the purpose of controlling the speed of induction motors. Perhaps the most straightforward application, from a conceptual point of view, is the use of thyristors in the supply lines of an induction machine to adjust the effective voltage applied to the stator terminals. Since the output torque varies as the square of the air-gap emf, variation of the terminal voltage is basically an inexpensive and reliable means of adjusting the motor speed. When a relatively high rotor resistance is incorporated in the design of the induction machine, speed ranges of 5 to 1 can be readily obtained.[1]

In many cases, stator voltage speed control is not practical owing to the inherently poor efficiency of the scheme. At low speeds, as much as 18-20% of the full load power is dissipated as heat within the machine, and considerable care must be exercised to properly match the machine design to the application. Efficiency, however, must be weighed with a number of other important factors including initial cost, system power factor, reliability, and ease of maintenance. Hence, the method is being increasingly employed in low and medium power applications wherein the load torque varies as the square of the motor speed. Such applications are typically fan or pump loads in the 5 – 150 hp range.

Thyristors can be connected to induction machines in a variety of ways for the purpose of adjusting the terminal voltage. However, it has been demonstrated that a circuit containing a pair of thyristors connected back-to-back in each line of a wye-connected motor is superior to all other conventional connections if average torque per rms stator ampere is used as the performance criterion.[1] Although thyristor voltage control of induction motors has been in use for many years[2] and component parts, as well as completely designed systems, are commercially available, a detailed analysis and solution of such a scheme has not, as yet, appeared in the literature.

Despite the simplicity of the scheme, an analysis of even its steady-state performance is extremely complex owing to the difficulty of establishing a suitable set of boundary or initial conditions needed to generate a solution. An approximate analysis of a system with a delta-connected primary has been reported.[3] In this paper, an expression for average torque using the first harmonic component of stator current was derived. In this case, the harmonic components were measured directly from an actual system. Only relatively poor correlation of the computed and measured torque-speed curves was obtained for various firing angles. An exact analysis of the same delta-connected configuration has been attempted.[4] However, a solution of the resulting equations was not carried out, apparently owing to their complexity. In a recent book,[5] Takeuchi has analyzed a version of the wye-connected machine. The average torque, voltages, and currents are approximated as an infinite series of symmetrical components and computation of the resulting equations is tedious. Since the voltage across an open-circuited phase is neglected, results are variable and become inaccurate for large delay angles when the rotor is near maximum speed.

Implicit solutions by analog or digital simulation techniques have also been obtained.[6,7,8] However, such an approach does not easily lend itself to system design aspects which generally require a repeated number of steady state solutions.

These analyses seem to demonstrate that a precise solution of the stator and rotor currents is required for accurate prediction of torque-speed curves, power factor, efficiency and other important performance characteristics. In this paper, the steady-state solution of a wye-connected induction machine with back-to-back thyristors connected in the lines is developed in detail. State variable techniques[9] are utilized to generate, without iteration, a set of initial conditions which will yield the steady-state solution. The analysis differs considerably from related work[10] in that the equations defining the appropriate initial conditions are obtained by simple matrix algebra. Also, the open circuit conditions which occur are defined without reducing the rank of the resulting matrix differential equation. Hence, it is not necessary to define different sets of equations for each open-circuit condition. Since the solution is carried out entirely with matrices, the method is well suited for digital computer programming. It is shown that when attention is given to the symmetry of the system variables, it is possible to significantly reduce the computational effort required to obtain the solution.

BASIC ASSUMPTIONS

A simplified diagram of the system considered in this paper is given in Fig. 1. The system consists of a three-phase power source, three pairs of identical thyristors (silicon controlled-rectifiers), connected back-to-back in series with the phases of a three-wire wye-connected induction machine. The thyristor gate control signals used to trigger the thyristors are derived from zero crossings of the three-phase power source. Stator voltage control is achieved by alternately open-circuiting the three stator phases at instants of zero current.

Paper 70 TP 661-PWR, recommended and approved by the Rotating Machinery Committee of the IEEE Power Group for presentation at the IEEE Summer Power Meeting and EHV Conference, Los Angeles, Calif., July 12-17, 1970. Manuscript submitted February 16, 1970; made available for printing May 8, 1970.

Reprinted from *IEEE Trans. Power App. Syst.*, vol. PAS-90, pp. 515–525, Mar./Apr. 1971.

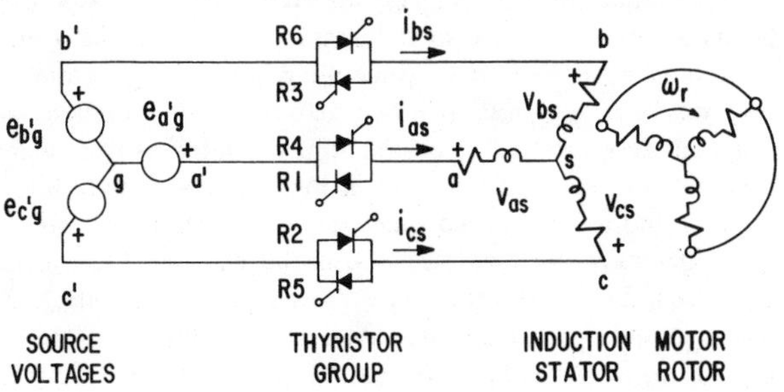

Fig. 1. System studied.

A frequently-used variation of this system incorporates three pairs of delta-connected thyristors in the neutral of the machine.[5] Performance of this system is identical to that in Fig. 1 and the analysis contained herein is applicable to this configuration with proper interpretation of the results.

In this paper it is assumed that:

(a) The power source may be considered as a set of balanced, sinusoidal three-phase voltages having zero source impedance.

(b) The six thyristors have identical characteristics, are symmetrically triggered and can be considered as a device which presents an infinite impedance in the blocking mode when the forward or anode-to-cathode voltage changes from positive to negative. The impedance changes to zero whenever a trigger pulse is applied, provided the forward voltage is positive.

(c) The induction machine is an idealized machine in which the stator and rotor windings are distributed so as to produce a single sinusoidal MMF wave in space when balanced sets of currents flow in the stator and rotor circuits.

(d) All parameters of the machine are assumed to be constant and saturation of the magnetic circuit is neglected.

(e) The system is in the steady-state. In particular, the rotor speed of the induction machine is assumed to be constant.

CONSIDERATIONS OF SYMMETRY

Because of the symmetrical nature of the voltages applied to the induction machine and the symmetry of the resulting line currents, a considerable reduction in the effort required to obtain a steady-state solution is possible. A typical steady-state solution of the voltage-controlled induction motor scheme of Fig. 1 is given in Fig. 2, which serves to set forth the notation used and to illustrate the symmetry of the solution.

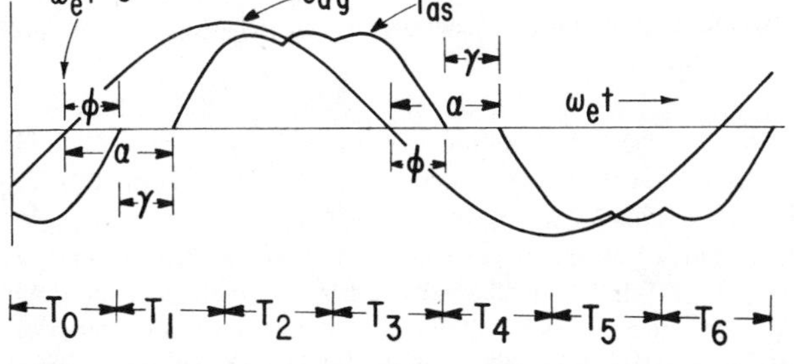

Fig. 2. Typical steady-state solution, α=60°, γ = 30°, ϕ = 30°.

In the steady-state, an induction motor operates as an inductive load. Hence, a phase current typically lags its respective line-to-neutral voltage by a phase angle ϕ. The delay from a point of zero phase voltage to the conduction of the succeeding thyristor in that phase is generally termed the delay angle α. The delay from the instant the phase current reaches zero to the firing of the succeeding thyristor in that phase is defined in this paper as the current delay angle or hold-off angle γ. These angles are illustrated in Fig. 2. Since the thyristors are fired symmetrically, γ is the same for all phases.

At any instant of time, the differential equations defining the state of the system are linear since the rotor speed of the induction motor has been assumed constant. In addition, the voltages fed to the six controlled-rectifiers form a balanced polyphase set. All controlled-rectifiers are triggered in a symmetrical fashion with respect to a current zero and the induction motor is itself symmetrical. Hence, the stator line-to-neutral voltages and phase currents form balanced sets. That is to say, the voltage v_{bs} across phase bs of the induction machine and the current i_{bs} in this phase are identical in form to the corresponding voltage and current in phase as and phase delayed by $2\pi/3$ radians. Similarly, the voltage v_{cs} and i_{cs} are identical in form to v_{bs} and i_{bs} and phase delayed by a further $2\pi/3$ radians. Furthermore, since the hold-off angle for each of the two thyristors in a given phase is identical, each voltage and current is half-wave symmetric.

Because of this symmetry, not only the hold-off angle γ, but also the phase angle ϕ and the delay angle α associated with the turn-off of each thyristor are identical. In order to facilitate the analysis, it is convenient to divide a full period into six equal intervals, T_1, T_2, ..., T_6, each having a duration of $\pi/3$ radians. These intervals are also noted in Fig. 2. Intervals 1 to 6 are initiated when the thyristors R1 to R6, as indicated in Fig. 1, enter the blocking mode. An extra interval T_0, occurring one interval prior to T_1, has been included for purposes of analysis.

On the basis of three-phase symmetry and half-wave symmetry, the following conclusions can be obtained.

(a) Only five system states can exist for a wye-connected three-wire machine as follows:

(1) All three phases connected to the source voltage; i_{as}, i_{bs}, and i_{cs} are non-zero.

(2) Phase as disconnected from the source voltages; $i_{as} = 0$.

(3) Phase bs disconnected; $i_{bs} = 0$.

(4) Phase cs disconnected; $i_{cs} = 0$.

(5) All three phases disconnected from the source voltages; $i_{as} = i_{bs} = i_{cs} = 0$.

(b) A solution for any $\pi/3$ interval is sufficient to uniquely define all variables over an entire period.

(c) By virtue of (b), only three system states need be investigated in detail. If interval T_1 is chosen as the interval over which a solution is desired, these states are (1), (2), and (5) as defined in (a) above.

(d) Only two distinct modes of operation exist for a wye-connected, three-wire machine and these modes are defined by the hold-off angle γ, as follows:

(1) First mode: $0 < \gamma < \pi/3$. When γ ranges between 0 and $\pi/3$, either three or two windings of the induction motor are connected to the power source. The sequence of system states during intervals T_1 to T_6 are 2-1-4-1-3-1 as noted in (a).

(2) Second mode: $\pi/3 < \gamma < 2\pi/3$. During this mode, either two or one winding of the induction motor is connected to the power source. Since the machine is wye-connected, all three stator currents are zero for the latter connection. The sequence of system states during a full period beginning with interval T_1 is then 5-2-5-4-5-3.

A trivial case exists for $2\pi/3 < \gamma < \pi$, when either one winding or zero windings are connected to the source. For either connection, the three stator currents are zero; and, hence, this mode is not of interest. However, if the induction motor has a neutral return path to the power source, it is clear that three modes of operation will exist.

At this juncture, it is possible to proceed towards a solution utilizing either the delay angle or hold-off angle as the basic parameter. However, the two operating modes are related directly to the hold-off angle. It will be shown that a closed form of solution for a set of initial conditions can be found when γ and ϕ are selected as the basic parameters, whereas the use of α and ϕ results in a need for an iterative solution. If necessary, results can be readily converted from one form to another once a solution has been obtained.

STATE VARIABLE ANALYSIS

In the analysis of problems involving induction machinery, it has proven useful to transform the equations which describe the behavior of the machine to d-q axes fixed either on the stator or the rotor, or rotating in synchronism with the applied voltages. When actual phase voltage and currents are used as variables, time-varying coefficients appear in the resulting differential equations owing to the sinusoidal variation of mutual inductance with displacement angle. However, a transformation of voltage and current variables to d-q axes results in a set of constant coefficients in the resulting differential equations when the rotor speed is constant.

In the case of the induction motor with stator voltage control, switching of phase currents takes place in the stator of the machine. Hence, it is desirable to fix the d-q axes in the stator. These axes were originally termed the α-β axes by Stanley.[11] However, since the subscript α is also used to denote the delay angle, the notation of Krause and Thomas[12] will be employed. These equations, expressed in per unit, wherein the d-q axes of the arbitrary reference frame are fixed in the stator (d^s-q^s) axes are given in matrix form by

$$\begin{bmatrix} v_{qs}^s \\ v_{ds}^s \\ 0 \\ 0 \end{bmatrix} = \begin{bmatrix} r_s + \frac{p}{\omega_b} x_s & 0 & \frac{p}{\omega_b} x_m & 0 \\ 0 & r_s + \frac{p}{\omega_b} x_s & 0 & \frac{p}{\omega_b} x_m \\ \frac{p}{\omega_b} x_m & \frac{\omega_e - \omega_r}{\omega_b} x_m & r_r' + \frac{p}{\omega_b} x_r' & \frac{\omega_e - \omega_r}{\omega_b} x_r' \\ -\frac{\omega_e - \omega_r}{\omega_b} x_m & \frac{p}{\omega_b} x_m & -\frac{\omega_e - \omega_r}{\omega_b} x_r' & r_r' + \frac{p}{\omega_b} x_r' \end{bmatrix} \times \begin{bmatrix} i_{qs}^s \\ i_{ds}^s \\ i_{qr}'^s \\ i_{dr}'^s \end{bmatrix} \quad (1)$$

or in the corresponding vector-matrix form as

$$\bar{v} = \bar{X} \frac{p}{\omega_b} \bar{i} + \bar{R}\bar{i} \quad (2)$$

In these equations, the superscript s is employed to denote that the d-q axes have been fixed in the stator. A p denotes the operator d/dt. Although six equations are generally required to completely define the machine response, the two zero-sequence equations have been immediately omitted since the sum of stator as well as rotor currents are zero. Also, in Eq. 1, ω_r is the electrical angular velocity of the rotor and is assumed constant, ω_e is the electrical angular velocity of the source voltages and ω_b is the base electrical angular velocity used to obtain the per unit machine parameters. The voltages v_{ds}^s and v_{qs}^s are an equivalent set of voltages, often written v_α and v_β,[11] related to the phase voltages by the equations

$$v_{qs}^s = v_{as} \quad (3)$$

$$v_{ds}^s = \frac{1}{\sqrt{3}}(v_{cs} - v_{bs}) \quad (4)$$

where in order to obtain Eq. 3 it has been noted that the sum of the stator line-to-neutral voltages of a three wire-wye-connected machine is zero.[12]

The $d^s - q^s$ stator currents are related to the stator phase currents by

$$i_{qs}^s = i_{as} \quad (5)$$

$$i_{ds}^s = \frac{1}{\sqrt{3}}(i_{cs} - i_{bs}) \quad (6)$$

The stator referred d-q rotor currents are related to the rotor phase currents by

$$i_{qr}'^s = \frac{N_r}{N_s}\left[i_{ar}\cos\theta_r + \frac{1}{\sqrt{3}}(i_{cr} - i_{br})\sin\theta_r\right] \quad (7)$$

$$i_{dr}'^s = \frac{N_r}{N_s}\left[-i_{ar}\sin\theta_r + \frac{1}{\sqrt{3}}(i_{cr} - i_{br})\cos\theta_r\right] \quad (8)$$

Again, the sum of the stator currents and the sum of the rotor currents have been assumed to be zero. In eqs. 7 and 8, N_r/N_s is the effective rotor to stator turns ratio, θ_r denotes the relative displacement in electrical radians of the ar rotor axis with respect to the as axis.

When peak rated line-to-neutral voltage and peak rated line current are chosen as base quantities, the electromagnetic torque expressed in terms of the d^s-q^s variables is[12]

$$T_e = x_m (i_{qs}^s i_{dr}'^s - i_{ds}^s i_{qr}'^s) \quad (9)$$

In addition to the equations defining operation of the induction machine, equations expressing the ac source voltages may be written in differential equation form. Since the machine is supplied by a balanced, polyphase voltage source, the per unit source voltages may be expressed

$$e'_{ag} = V_m \sin \omega_e t \quad (10)$$

$$e'_{bg} = V_m \sin(\omega_e t - 2\pi/3) \quad (11)$$

$$e'_{cg} = V_m \sin(\omega_e t + 2\pi/3) \quad (12)$$

where V_m denotes the peak value of the source voltages expressed in per unit. The electrical angular velocity of the source voltages,

ω_e, has been assumed equal to the base angular velocity throughout this study.

The source voltages e'_{ag}, e'_{bg} and e'_{cg} may, in turn, be obtained from the pair of first order differential equations

$$\frac{p}{\omega_e} e_1 = e_2 \tag{13}$$

$$\frac{p}{\omega_e} e_2 = -e_1 \tag{14}$$

where $e_1(0) = 0$, $e_2(0) = V_m$. In matrix form, Eqs. 13 and 14 become

$$\frac{p}{\omega_e} \bar{e} = \begin{bmatrix} 0 & 1 \\ -1 & 0 \end{bmatrix} \bar{e} \tag{15}$$

where

$$\bar{e} = [e_1, e_2]^T \tag{16}$$

and T denotes the transpose. The source voltages are related to the e_1, e_2 set of voltages by

$$e'_{ag} = e_1 \tag{17}$$

$$e'_{bg} = -\frac{1}{2} e_1 - \frac{\sqrt{3}}{2} e_2 \tag{18}$$

$$e'_{cg} = -\frac{1}{2} e_1 + \frac{\sqrt{3}}{2} e_2 \tag{19}$$

Eqs. 1 and 9 are sufficient to properly express the behavior of the induction machine provided that the d^s_s-q^s_s axis voltages are known. The identity of these voltages depend upon the system state. Although five system states exist, it has been noted that only system states 1, 2 and 5 need be defined to completely specify the solution. Hence, it is necessary to relate the d^s_s-q^s_s axes voltages only to these system conditions.

System state 1; i_{as}, i_{bs} and i_{cs} nonzero

In state 1, the machine is simply supplied by a conventional polyphase source. The line-to-neutral voltages across the terminals of the machine are

$$v_{as} = e'_{ag} - v_{sg} \tag{20}$$

$$v_{bs} = e'_{bg} - v_{sg} \tag{21}$$

$$v_{cs} = e'_{cg} - v_{sg} \tag{22}$$

However, since the sum of the stator line-to-neutral voltages and the sum of the source voltages are zero, it is readily noted that

$$v_{sg} = 0 \tag{23}$$

Utilizing Eqs. 3 and 4, the direct quadrature axis voltages are given by

$$v^s_{qs} = e_1 \tag{24}$$

$$v^s_{ds} = e_2 \tag{25}$$

The voltage forcing function vector $\bar{v}$ in Eq. 2 can now be related to the source voltage vector $\bar{e}$ by the matrix equation

$$\bar{v} = \begin{bmatrix} 0 & 1 & 0 & 0 \\ 1 & 0 & 0 & 0 \end{bmatrix}^T . \bar{e} \tag{26}$$

which, for convenience, can be expressed in the form

$$\bar{v} = \bar{C}_1 \bar{e} \tag{27}$$

Hence, the differential equations of the system for state 1 may be expressed in matrix notation as

$$\frac{p}{\omega_b} \bar{i} = -\bar{X}_1^{-1} \bar{R} \bar{i} + \bar{X}_1^{-1} \bar{C}_1 \bar{e} \tag{28}$$

$$\frac{p}{\omega_e} \bar{e} = \begin{bmatrix} 0 & 1 \\ -1 & 0 \end{bmatrix} \bar{e} \tag{29}$$

It will be shown that the reactance matrix $\bar{X}$ as defined by Eq. 2 is valid only for system state 1. Thus, the subscript 1 has been appended to the reactance matrix in Eq. 28 although the corresponding matrices in these two equations are identical.

By partitioning, these equations may be written as a single matrix equation

$$\frac{p}{\omega_e} \begin{bmatrix} \bar{i} \\ \bar{e} \end{bmatrix} = \begin{bmatrix} -\bar{X}_1^{-1} \bar{R} & \bar{X}_1^{-1} \bar{C}_1 \\ \bar{0} & \begin{bmatrix} 0 & 1 \\ -1 & 0 \end{bmatrix} \end{bmatrix} \times \begin{bmatrix} \bar{i} \\ \bar{e} \end{bmatrix} \tag{30}$$

where 0 is a 2 x 4 matrix of zeros and it has been noted that in this application $\omega_e = \omega_b$. Eq. 30 is expressed in the standard state-variable format familiar to control engineers.[10] It is normally described by the single compact equation

$$\frac{p}{\omega_e} \bar{x} = \bar{A} \bar{x} \tag{31}$$

In this application, the column vector $\bar{x} = [\bar{i}^T, \bar{e}^T]^T$ and contains the system state variables. $\bar{A}$ is the 2 x 2 partitioned system matrix of Eq. 30.

System state 2; $i_{as} = 0$

When the stator current i_{as} reaches zero, thyristor R_1 enters the blocking state and the voltage which appears across phase as is an induced emf owing to mutual coupling with the other stator and rotor phases. By virtue of Eq. 5 when $i_{as} = 0$, then $i_{qs} = 0$, and the open circuit voltage may be expressed in the d^s_s-q^s_s axes as[12]

$$v^s_{qs} = \frac{p}{\omega_e} x_m i'^s_{qr} \tag{32}$$

Since the bs and cs phases remain connected, the d^s_s axis voltage is again given by

$$v^s_{ds} = \frac{1}{\sqrt{3}} (e'_{cg} - e'_{bg}) = e_2 \tag{33}$$

When Eq. 32 is substituted into the general induction machine expression, Eq. 1, and is cancelled with the similar term on the right-hand side of the equation, the system differential equations expressed in state-variable form for system state 2 may be written

$$\frac{p}{\omega_e}\begin{bmatrix}\bar{i}\\ \\ \bar{e}\end{bmatrix} = \begin{bmatrix}-\bar{X}_2^{-1}\bar{R} & \bar{X}_2^{-1}\bar{C}_2 \\ 0 & \begin{bmatrix}0 & 1\\ -1 & 0\end{bmatrix}\end{bmatrix} = \begin{bmatrix}\bar{i}\\ \\ \bar{e}\end{bmatrix} \tag{34}$$

where $\bar{X}_2$ is the reactance matrix

$$\bar{X}_2 = \begin{bmatrix} x_s & 0 & 0 & 0 \\ 0 & x_s & 0 & x_m \\ x_m & 0 & x_r' & 0 \\ 0 & x_m & 0 & x_r' \end{bmatrix} \tag{35}$$

and $\bar{C}_2$ is the voltage connection matrix

$$\bar{C}_2 = \begin{bmatrix} 0 & 1 & 0 & 0 \\ 0 & 0 & 0 & 0 \end{bmatrix}^T \tag{36}$$

Eq. 34 may now be represented compactly by the state variable equation

$$\frac{p}{\omega_e}\bar{x} = \bar{B}\bar{x} \tag{37}$$

where $\bar{B}$ is the 2 x 2 partitioned matrix of Eq. 34.

System state 5; $i_{as} = i_{bs} = i_{cs} = 0$

In cases where the hold-off angle γ is large, two of the stator phase currents, will be zero whenever two of the controlled-rectifiers are simultaneously in the forward-biased blocking mode. Since the machine is wye-connected, no currents flow in the stator winding during this interval. However, since currents continue to flow through the short-circuited rotor windings, voltages may appear across the three stator phases owing to mutual coupling. By virtue of Eqs. 5 and 6, when the three stator phase currents are zero, then both i_{qs}^s and i_{ds}^s are zero. The voltages which appear across the stator phase may be represented in the d^s-q^s axes as

$$v_{qs}^s = \frac{p}{\omega_e} x_m i_{qr}'^s \tag{38}$$

$$v_{ds}^s = \frac{p}{\omega_e} x_m i_{dr}'^s \tag{39}$$

It is readily established that the system differential equation for system state 5 is

$$\frac{p}{\omega_e}\begin{bmatrix}\bar{i}\\ \\ \bar{e}\end{bmatrix} = \begin{bmatrix}-\bar{X}_5^{-1}\bar{R} & \bar{0}^T \\ \bar{0} & \begin{bmatrix}0 & 1\\ -1 & 0\end{bmatrix}\end{bmatrix} x \begin{bmatrix}\bar{i}\\ \\ \bar{e}\end{bmatrix} \tag{40}$$

where

$$\bar{X}_5 = \begin{bmatrix} x_s & 0 & 0 & 0 \\ 0 & x_s & 0 & 0 \\ x_m & 0 & x_r' & 0 \\ 0 & x_m & 0 & x_r' \end{bmatrix} \tag{41}$$

Hence, the state variable equation for the condition $i_{as} = i_{bs} = i_{cs} = 0$ is

$$\frac{p}{\omega_e}\bar{x} = \bar{C}\bar{x}$$

where $\bar{C}$ is the 2 x 2 partitioned system matrix of Eq. 40.

Solution for mode 1; $0 < \gamma < \pi/3$

When the hold-off angle ranges between 0 and $\pi/3$, either three windings of the induction machine are connected to the source voltages or two windings are connected while the third is disconnected. If t_o denotes a time reference in interval T_o defined such that $\phi + \gamma - \pi/3 \leqslant \omega_e t_o \leqslant \phi$, by referral to Fig. 2, the condition of the system can be noted as state 1. The solution to Eq. 31 for any time t where $\omega_e t_o \leqslant \omega_e t \leqslant \phi$, expressed in terms of the system initial conditions at $t = t_o$, is

$$\bar{x}(\omega_e t) = \epsilon^{\omega_e(t-t_o)\bar{A}}\,\bar{x}(\omega_e t_o) \tag{43}$$

The matrix exponential function $\exp[\omega_e(t-t_o)\bar{A}]$ appearing in the solution is generally termed the state transition matrix by control engineers. Computation of this quantity will be discussed in a subsequent section.

When $\omega_e t = \phi$, thyristor R_1 enters the blocking mode and the system is in state 2. The solution to Eq. 37, expressed in terms of the system condition at $\omega_e t = \phi$, for $\phi \leqslant \omega_e t \leqslant \phi + \gamma$ is

$$\bar{x}(\omega_e t) = \epsilon^{(\omega_e t-\phi)\bar{B}}\,\bar{x}(\phi) \tag{44}$$

During the period $\phi + \gamma \leqslant \omega_e t \leqslant \omega_e t_o + \pi/3$, the system is again in state 1 and the solution in terms of the system conditions at $\omega_e t = \phi + \gamma$ is

$$\bar{x}(\omega_e t) = \epsilon^{(\omega_e t-\phi-\gamma)\bar{A}}\,\bar{x}(\phi+\gamma) \tag{45}$$

Eqs. 43, 44, and 45 serve to completely define the operation of the system over the interval $\omega_e t_o \leqslant \omega_e t \leqslant \omega_e t_o + \pi/3$. However, the solution is not complete until the initial condition vector $x(\omega_e t_o)$ is known. This task may be carried out by utilizing the symmetry of the solution.

Since Eq. 44 is valid for the condition $\omega_e t = \phi$, and Eq. 45 is valid for $\omega_e t = \phi + \gamma$

$$\bar{x}(\phi) = \epsilon^{(\phi-\omega_e t_o)\bar{A}}\,\bar{x}(\omega_e t_o) \tag{46}$$

and

$$\bar{x}(\phi+\gamma) = \epsilon^{\gamma\bar{B}}\,\bar{x}(\phi) \tag{47}$$

Hence, Eq. 45 can be written

$$\bar{x}(\omega_e t) = \epsilon^{(\omega_e t-\phi-\gamma)\bar{A}}\,\epsilon^{\gamma\bar{B}}\,\epsilon^{(\phi-\omega_e t_o)\bar{A}}\,\bar{x}(\omega_e t_o) \tag{48}$$

where, $\phi + \gamma \leqslant \omega_e t \leqslant \omega_e t_o + \pi/3$. In particular, when $\omega_e t = \omega_e t_o + \pi/3$, the solution of Eq. 48 is

$$\bar{x}(\omega_e t_o+\pi/3) = \epsilon^{(\omega_e t_o-\phi-\gamma+\pi/3)\bar{A}}\,\epsilon^{\gamma\bar{B}}\,\epsilon^{(\phi-\omega_e t_o)\bar{A}}\,\bar{x}(\omega_e t_o) \tag{49}$$

It is shown in the Appendix that regardless of the form of the solution,

$$\bar{x}(\omega_e t_o + \pi/3) = \bar{T}\,\bar{x}(\omega_e t_o) \tag{50}$$

where $\bar{T}$ can be regarded as a time translation matrix which relates the solution at any two instants which are phase displaced by $\pi/3$ radians. The expression for the translation matrix $\bar{T}$ is given by Eq. 78. Utilizing Eq. 50, Eq. 49 can be expressed in the form

$$\left[\bar{T} - \epsilon^{(\omega_e t_o - \phi - \gamma + \pi/3)\bar{A}}\,\epsilon^{\gamma\bar{B}}\,\epsilon^{(\phi - \omega_e t_o)\bar{A}}\right]\bar{x}(\omega_e t_o) = \bar{0} \tag{51}$$

where $\bar{0}$ in this case is a 6 x 1 column vector of zeros. Because of the required symmetry of the solution, Eq. 51 is a necessary condition for steady-state operation. In addition, this equation may be utilized to solve for the unknown initial condition $\bar{x}(\omega_e t_o)$.

Thus far, Eq. 51 is valid for any time t_o where $\omega_e t_o$ is in the interval T_o and $\phi + \gamma - \pi/3 \leqslant \omega_e t_o \leqslant \phi$. If t_o is selected to be the particular initial time $\omega_e t_o = \phi$ or the initiation of interval T_1, the first element of the state vector $\bar{x}$ must be zero since at this instant i_{as}, and thus i_{qs}^s, is identically equal to zero. With this choice of initial time, Eq. 51 reduces to

$$\left[\bar{T} - \epsilon^{(\pi/3-\gamma)\bar{A}}\,\epsilon^{\gamma\bar{B}}\right]\bar{x}(\phi) = \bar{0} \tag{52}$$

or for simplicity

$$\bar{W}\,\bar{x}(\phi) = \bar{0} \tag{53}$$

Eq. 53 can be written in partitioned form as

$$\begin{bmatrix} \bar{W}_1 & \bar{W}_2 \\ \bar{W}_3 & \bar{W}_4 \end{bmatrix} \times \begin{bmatrix} \bar{i}(\phi) \\ \bar{e}(\phi) \end{bmatrix} = \bar{0} \tag{54}$$

It has been noted by Fath[7] that matrices $\bar{W}_3$ and $\bar{W}_4$ are identically zero and hence cannot be utilized further. When the vector $\bar{i}(\phi)$ is solved in terms of the source voltage vector $\bar{e}(\phi)$ using the upper two matrices

$$\bar{i}(\phi) = \bar{W}_1^{-1}\,\bar{W}_2\,\bar{e}(\phi) \tag{55}$$

or simply

$$\bar{i}(\phi) = \bar{Y}\,\bar{e}(\phi) \tag{56}$$

Since $i_{qs}^s = 0$ when $\omega_e t_o = \phi$, the first row of Eq. 56 must equal zero. If y_{11} denotes the (1,1) element and y_{12} the (1,2) element in the 4 x 2 matrix $\bar{Y}$ defined by Eq. 55, then

$$y_{11}e_1 + y_{12}e_2 = 0 \tag{57}$$

However, by virtue of Eqs. 13 and 14, $e_1(\phi) = V_m \sin\phi$ and $e_2(\phi) = V_m \cos\phi$. Substituting these two expressions in Eq. 57 yields the final result

$$\phi = \tan^{-1}(-y_{12}/y_{11}) \tag{58}$$

Having computed ϕ, the remaining unknown currents in the vector $i(\phi)$ are readily obtained by means of Eq. 56.

Equation 56, together with 58, generate the required set of initial conditions needed to obtain the steady-state solution. It is of interest to note the added significance of the equations which have been derived. Although numerous switchings occur over a complete cycle, the system equations for any period are always linear. A necessary requirement for a linear system is that a linear relationship be maintained between the forcing function (source voltages) and the response (resulting currents). That is to say, if the terminal voltage is doubled, the machine currents should double as well. It is immediately apparent from Eq. 56 that this condition is satisfied for the instant $\omega_e t = \phi$. It follows that the result is also true for all successive time instants. Also, it is recalled that for a linear system, the phase angle between voltage and current is independent of the magnitude of the forcing function and depends only upon system parameters. Eq. 58 verifies this requirement.

Having obtained one point on the steady-state solution curve, the solution for any successive instant is readily evaluated by means of Eqs. 44 and 45. If n_2 is the number of solution points desired during the hold-off period γ of T_1, then for $\phi \leqslant \omega_e t \leqslant \phi + \gamma$

$$\bar{x}(\phi + n\gamma/n_2) = \epsilon^{\bar{B}\gamma/n_2}\,\bar{x}[\phi + (n-1)\gamma/n_2] \tag{59}$$

where $n = 1, 2, \ldots, n_2$;

During the remainder of time interval T_1, $\phi + \gamma \leqslant \omega\, t \leqslant \phi + \pi/3$ and all three machine terminals are connected to the source. If n_1 is the number of solution points desired during system state 1

$$\bar{x}[\phi+\gamma+(\pi/3-\gamma)n/n_1] = \epsilon^{\bar{A}(\pi/3-\gamma)n_1}\,\bar{x}[\phi+\gamma+(\pi/3-\gamma)(n-1)/n_1] \tag{60}$$

where $n = 1, 2, \ldots, n_1$.

Having computed the solution for the $\pi/3$ interval T_1, the solution for the remaining five intervals is immediately obtained by means of the recursion relationship, Eq. 50. These solution points require no actual computation and only involve systematic rotation of the values obtained for the first interval.

Although a d^s-q^s axes solution is often sufficient, it is generally desirable to obtain the actual phase voltages and currents. The required inverse relationships needed to transform system voltage from the d^s-q^s variables to the phase variables are

$$v_{as} = v_{qs}^s \tag{61}$$

$$v_{bs} = -\frac{1}{2}v_{qs}^s - \frac{\sqrt{3}}{2}v_{ds}^s \tag{62}$$

$$v_{cs} = -\frac{1}{2}v_{qs}^s + \frac{\sqrt{3}}{2}v_{ds}^s \tag{63}$$

In system state 1, the d^s-q^s axes voltages are related to the source voltage vector $\bar{e}$ by Eqs. 24 and 25. In system state 2, the as phase is open circuited and

$$v_{qs}^s = x_m \frac{p}{\omega_e} i'^s_{qr} \tag{64}$$

The quantity $(p/\omega_e)\, i'^s_{qr}$ is readily evaluated by computing the third row of the state variable equation defining the open circuited condition, Eq. 37. A set of transformation equations, analogous to Eqs. 61-63, is used to obtain the stator phase currents. The rotor currents can be obtained, if desired, by using the inverse expression for Eqs. 7 and 8.

Solution for mode 2; $\pi/3 < \gamma < 2\pi/3$

When the hold-off angle γ becomes greater than $\pi/3$, the three terminals of the machine are no longer simultaneously connected to the source voltages. Either one or two terminals are connected and Eqs. 37 and 42 describe the operation of the machine during the first time interval T_1. If an initial time t_o is again chosen such that $\omega_e t_o = \phi$, it can be readily shown that a necessary condition for steady-state operation is

$$\left[\bar{T} - \epsilon^{(2\pi/3-\gamma)\bar{B}}\,\epsilon^{(\gamma-\pi/3)\bar{C}}\right]\bar{x}(\phi) = \bar{0} \qquad (65)$$

where $\bar{T}$ is again the time translation matrix defined by Eq. 78. As was the case for mode 1, $i_{qs}^{s} = 0$ when $\omega_e t_o = \phi$ so that Eqs. 53-58 can again be employed to solve for a suitable initial condition vector. In this case, $\bar{W}$ is the bracketed 6 x 6 matrix of Eq. 65.

COMPUTER SOLUTION OF THE SYSTEM EQUATIONS

In implementing a computer solution of the equations which have been derived, four matrix exponential functions must be evaluated in order to complete the solution. When γ is less than $\pi/3$, these expressions are: exp $[(\pi/3-\gamma)\bar{A}]$ exp $(\gamma\bar{B})$, exp $(\gamma\bar{B}/n_2)$ and exp $[(\pi/3-\gamma)\bar{A}/n_1]$ from Eqs. 52, 59 and 60, respectively. Although the computation of these functions is straightforward, a familiarity with numerical analysis is required. Briefly, the matrix exponential is defined by a power series expansion of the matrix argument similar in form to the familiar power series expansion of the scalar exponential ϵ^a. Accuracy can be obtained to an arbitrary number of significant digits by calculating enough terms in the power series. However, the number of terms required for a specified accuracy depends greatly on the matrix argument and if convergence is slow, round-off errors may introduce erroneous results. Fortunately, when the induction machine parameters are expressed in per unit, as in this paper, the system matrix is conditioned so that convergence is very rapid. When six digit accuracy is desired, the number of terms required in the power series expansion for the matrix exponentials used in Eq. 52 is typically ten. For the same accuracy, five terms are typically required for the matrix exponentials used in Eqs. 59 and 60 when n_1 or n_2 is approximately ten.

An excellent introduction to the computation of the matrix exponential is given by Liou.[13] A more efficient method used to obtain the computer results described herein is given by Fath.[14]

COMPARISON OF COMPUTED AND TESTED RESULTS

In order to verify the equations which have been developed, the solution obtained from a computer analysis was compared to the tested results of an actual system. The induction machine used for purposes of comparison was a 1/3 hp, 220 volt, 50 Hz, 4 pole, wound-rotor machine. Since both rotor and stator of the machine was random-wound, this choice represents a severe test of the accuracy of the analytical method. The measured parameters of the machine expressed in per unit using rated voltage as base volts and 375 watts as base power are: $r_s = 0.0566$, $r_r' = 0.069$, $x_s = 1.0318$, $x_r' = 1.0318$, $x_m = 0.969$. In order to more closely approximate an induction machine which might be utilized in a practical application, an external resistance $r_{ext}' = 0.0562$ was connected to the machine via slip rings. Hence, the total effective rotor resistance was increased to 0.1252 pu. Rated voltage at rated frequency was applied to the machine so that $V_m = 1.0$ and $\omega_e = 314$ rad/s.

In Fig. 3, a comparison of the actual machine phase voltage and current with the results obtained from a digital computer solution is shown. In order to facilitate a comparison, the calculated results have been converted from the per unit form to actual units. The hold-off angle γ was set at 45° ($\pi/4$) so that the induction machine was operating in mode 1. Also, for this case, the load torque $T_L = 1.0$ N-m, $\omega_r = 1325$ r/min. Although not available for measurement in the physical system, the instantaneous electro-magnetic torque obtained from the computer solution is also shown. A close comparison of computed and tested results is clearly evident.

In Fig. 4, the hold-off angle was increased to 70°. For this case, the machine was unable to overcome even its internal losses so that $T_L \cong 0$, and $\omega_r/\omega_e = 0$. Again, a close comparison is apparent.

Calculated torque-speed curves are given in Fig. 5 for constant values of γ. Again, actual rather than per unit values have been plotted. Measured values of average torque are denoted by an x or Δ. Due to the characteristic of the loading device, it was not possible to reliably measure torque beyond the effective breakdown values. Also, the current rating of the thyristors used prevented checking high load conditions for small γ. However, satisfactory correlation is evident over the range of operation measured. It can be noted that negligible torque is generated when γ is greater than 60° and mode 2 operation of the induction machine is clearly of little practical importance.

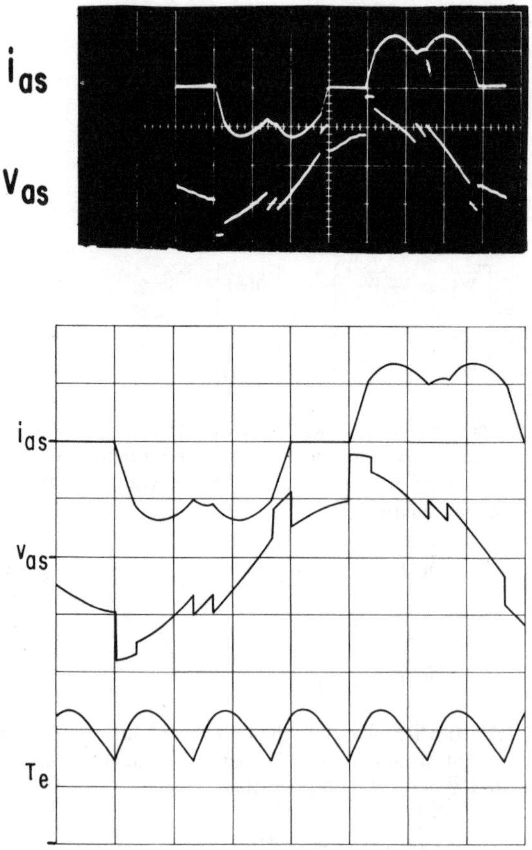

(b) Computed results

Fig. 3. Comparison of computed and measured results, γ = 45°, ω_r = 1325 r/min. Scale: i_{as} - 1.0 A/division, v_{as} - 100 V/div., T_e - 0.5 N·m/div.

CONCLUSIONS

This paper has developed a direct, straightforward method of calculating the steady-state behavior of an induction motor operating with thyristors in the stator phases. The method is free from

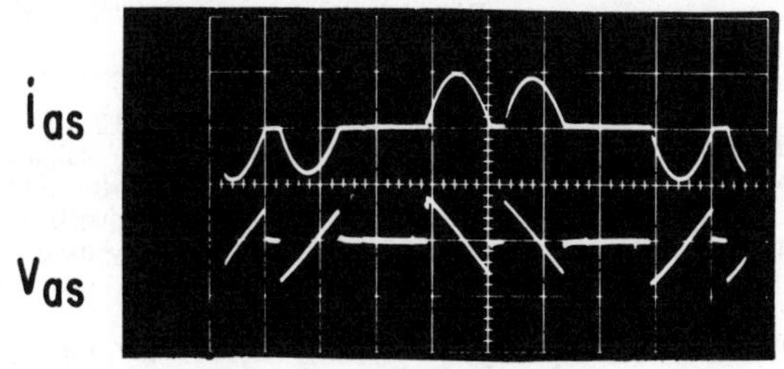

(a) Actual system

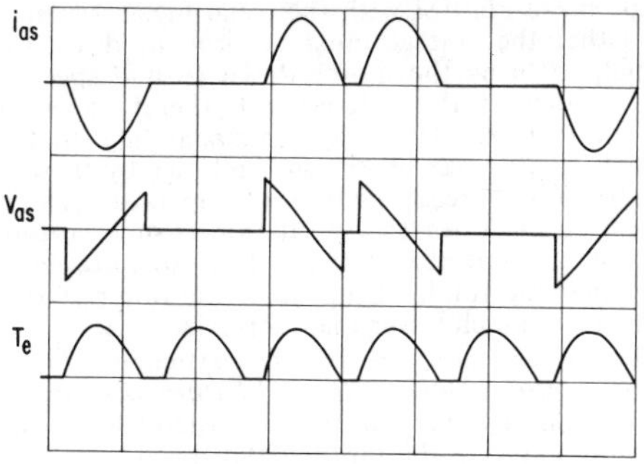

(b) Computed results

Fig. 4. Comparison of computed and measured results, γ *= 70°,* ω_r*=0. Scale:* i_{as} *- 1.0 A/division,* v_{as} *- 100 V/div.,* T_e *- 0.1 N·m/div.*

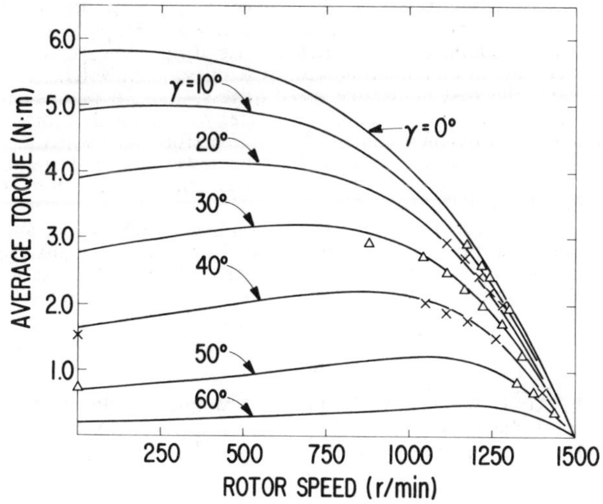

Fig. 5. Torque-speed curves for changes in γ*.*

the maze of complex components which has characterized the solution of this and similar problems in the past.

Two distinct modes of operation are shown to exist for a wye-connected, three-wire machine. These modes are independent of the machine parameters and depend only on the hold-off angle γ. Complete solutions have been obtained for both modes of operation. However, on the basis of this study, the second mode of operation does not appear to be of practical importance. The results of a digital computer solution are plotted and compared with the results obtained from an actual system. Good correlation is shown to exist over a wide range of delay angles.

Modern state variable techniques have been utilized throughout this analysis, and it is shown that this technique permits a closed form solution for a suitable set of initial conditions. Since matrix methods are exclusively employed, the analysis is well suited to a digital computer solution. This problem and solution vividly demonstrates that state variable methods provide the analyst with a new and powerful tool for the study of modern static ac and dc drives. It is expected that this technique will find increasing use in the analysis of such systems wherein the complex switching of the power converter defines numerous modes of operation.

ACKNOWLEDGEMENTS

The author wishes to thank the University of Manchester Institute of Science and Technology for facilities provided. Acknowledgement is made to the U. K. Science Research Council for the award of a grant in support of research in the area of variable speed drives. The author is also greatly indebted to Mr. J. Hindmarsh for his assistance in obtaining the experimental results.

REFERENCES

(1) D. A. Paice, "Induction motor speed control by stator voltage control," IEEE TRANS. POWER APPARATUS AND SYSTEMS, vol. PAS 87, pp. 585-590, February 1968.

(2) M. C. Spencer, "Squirrel-cage motor speed control system," ELECTRONICS, vol. 28, pp. 126-129, August 1955.

(3) W. Shepherd and J. Stanway, "The polyphase induction motor controlled by firing angle adjustment of silicon controlled rectifiers," IEEE INTERNATIONAL CONV. REC., vol. 12, pt. 4, pp. 135-154, 1964.

(4) W. Shepherd, "On the analysis of the three-phase induction motor with voltage control by thyristor switching," IEEE TRANS. INDUSTRY AND GENERAL APPLICATIONS, pp. 304-311, May/June 1968.

(5) T. J. Takeuchi, Theory of SCR Circuit and Application to Motor Control. Tokyo: Tokyo Electrical Engineering College Press, 1968.

(6) E. Gerecke and H. Badr, "Asynchronmaschine mit primärseitig eingebauten steuerbaren Ventilen" (in German), Neue Technik (Switzerland), no. 3, pp. 125-134, 1962.

(7) P. C. Krause, "A Constant Frequency Induction Motor Speed Control," Proc. National Electronics Conf., vol. 20, pp. 361-365, 1964.

(8) K. P. Kovacs, "Uber die genaue und vollständige Simulation des am Ständer mit steuerbaren Siliziumtrioden geregelten Drehstrom-Asynchronmotors" (in German), Acta Tech. (Hungary), vol. 48, no. 3-4, pp. 445-459, 1964.

(9) J. T. Tou, Modern Control Theory. New York: McGraw-Hill, 1964.

(10) D. Novotny and A. F. Fath, "The analysis of induction machaines controlled by series connected semiconductor switches," IEEE TRANS. POWER APPARATUS AND SYSTEMS, vol. PAS 87, pp. 597-605, February 1968.

(11) H. C. Stanley, "An analysis of the induction machine," AIEE TRANS., vol. 57, pp. 751-575, 1938.

(12) P. C. Krause and C. H. Thomas, "Simulation of symmetrical indnction machinery," IEEE TRANS. POWER APPARATUS AND SYSTEMS, vol. PAS 84, No. 11, pp. 1038-1053, 1965.

(13) M. L. Liou, "A novel method of evaluating transient response," PROCEEDINGS OF IEEE, vol. 54, No. 1, pp. 20-23, January 1966.

(14) A. F. Fath, "Evaluation of a matrix polynomial," IEEE TRANS. AUTOMATIC CONTROL, vol. AC-13, No. 2, pp. 220-221, April 1968.

APPENDIX

When the stator currents i_{as}, i_{bs} and i_{cs} are half-wave symmetric

$$i_{as}(\omega_e t + \pi) = - i_{as}(\omega_e t) \tag{66}$$

$$i_{bs}(\omega_e t + \pi) = - i_{bs}(\omega_e t) \tag{67}$$

$$i_{cs}(\omega_e t + \pi) = - i_{cs}(\omega_e t) \tag{68}$$

The assumption of three-phase symmetry implies

$$i_{bs}(\omega_e t + 2\pi/3) = i_{as}(\omega_e t) \tag{69}$$

$$i_{cs}(\omega_e t + 2\pi/3) = i_{bs}(\omega_e t) \tag{70}$$

$$i_{as}(\omega_e t + 2\pi/3) = i_{cs}(\omega_e t) \tag{71}$$

Increasing the argument of Eqs. 69-71, and substituting the result in 66-68, yields

$$i_{as}(\omega_e t + \pi/3) = - i_{bs}(\omega_e t) \tag{72}$$

$$i_{bs}(\omega_e t + \pi/3) = - i_{cs}(\omega_e t) \tag{73}$$

$$i_{cs}(\omega_e t + \pi/3) = - i_{as}(\omega_e t) \tag{74}$$

The equivalent d^s-q^s stator currents at time instants $\omega_e t$ and $\omega_e t+\pi/3$ are related to the phase currents by Eqs. 6 and 7.

Eliminating the phase variables from the resulting six equations by means of the symmetry equations 72-74, the following equations for the variables are readily obtained.

$$i_{qs}^s(\omega_e t + \pi/3) = \frac{1}{2}\, i_{qs}^s(\omega_e t) + \frac{\sqrt{3}}{2}\, i_{ds}^s(\omega_e t) \tag{75}$$

$$i_{ds}^s(\omega_e t + \pi/3) = -\frac{\sqrt{3}}{2}\, i_{qs}^s(\omega_e t) + \frac{1}{2}\, i_{ds}^s(\omega_e t) \tag{76}$$

It is clear that a similar development applies for the stator phase voltages and referred rotor currents since transformation equations similar to 72-74 also apply for these variables. In addition to the machine variables, the source voltages e'_{ag}, e'_{bg}, and e'_{cg} are also half-wave and three-phase symmetric. Hence, a general symmetry relationship can be constructed which relates any of the machine currents or source voltages to these same variables phase displaced by $\pi/3$ radians. This relationship is expressed in matrix form as

$$\bar{x}(\omega_e t + \pi/3) = \bar{T}\,\bar{x}(\omega_e t) \tag{77}$$

where $\bar{x}$ is defined as $[i_{qs}^s, i_{ds}^s, i'^s_{qr}, i'^s_{dr}, e_1, e_2]^T$ and

$$\bar{T} = \begin{bmatrix} \frac{1}{2} & \frac{\sqrt{3}}{2} & 0 & 0 & 0 & 0 \\ -\frac{\sqrt{3}}{2} & \frac{1}{2} & 0 & 0 & 0 & 0 \\ 0 & 0 & \frac{1}{2} & \frac{\sqrt{3}}{2} & 0 & 0 \\ 0 & 0 & -\frac{\sqrt{3}}{2} & \frac{1}{2} & 0 & 0 \\ 0 & 0 & 0 & 0 & \frac{1}{2} & \frac{\sqrt{3}}{2} \\ 0 & 0 & 0 & 0 & -\frac{\sqrt{3}}{2} & \frac{1}{2} \end{bmatrix} \tag{78}$$

Discussion

Philip L. Alger (1758 Wendell Ave., Schenectady, N.Y. 12308): This paper presents for the first time a sound procedure for calculating the currents and voltages of a three-phase induction motor supplied through SCRs with delayed firing. This use of SCRs for voltage control is growing in importance and promises to be widely used in the future, particularly for induction motors driving fans and pumps, where feedback control of motor speed is required.

While I have not followed the analysis in the paper, I accept its conclusions as correct, based on the demonstrated agreement with test results and after talking with the author. A chief objection to this scheme of speed control with the usual high-resistance squirrel-cage motor is that the voltage must be brought down to less than 50% of normal, to bring the motor down to half speed; and in so doing the harmonics in the voltage and, therefore, in the motor current are quite large. It is my understanding that the high-frequency currents produced in the squirrel cage by these harmonics create high losses and increase the motor temperature, so that special provisions must be made for the ventilation of such motors. Also, in order to hold the current at half speed down to a low enough value, it is necessary that the full-load slip be of the order of 10%, making the motor efficiency much lower than normal.

As shown by several recent papers, particularly *Speed Control of Wound Rotor Motors with SCRs and Saturistors**, these difficulties can be overcome by the addition of Alnico bars in the slots of a squirrel-cage motor, giving the motor a flat speed-torque curve and a low starting current. Such a motor, when operated with SCR phase angle control, can be designed for a normal full-load slip of about 5%, and will come down to half speed when the effective voltage is reduced to perhaps 60% of rated voltage. Under these conditions, the harmonics in the motor voltage are much smaller for a given speed than in the case of the standard motor, and the induced high-frequency rotor currents are still smaller, because of the higher reactance introduced by the Alnico bars. Thus, as shown in the paper cited above, the effective currents and heating of the rotor are practically the same with SCR phase angle control as with sinusoidal voltage.

It is my opinion, therefore, that the use of squirrel-cage motors with Alnico bars in combination with SCR phase angle voltage control provides the best adjustable speed drive system for pumps and fans. It is true that the frequency converter with a standard motor provides better efficiency and much better performance when high torque at low speeds is required, but the lower cost and greater simplicity of the delayed firing SCRs make this system preferable for motors of moderate size.

Since the Saturistor impedance is non-linear with current and, when referred to primary, is independent of the speed, the analysis of its performance is quite different from that given in the author's paper. I hope that the author will extend his studies to this new case.

Manuscript received June 5, 1970.

*Philip L. Alger, William A. Coelho, and Mukund R. Patel, IEEE Trans. on Industry and General Applications, Vol. IGA-4, No. 5, Sept./Oct. 1968, pp. 477-485.

Ed. Gerecke (Freiestrasse, 212, 8032 Zurich, Switzerland): Dr. Lipo has reported an interesting analysis of the voltage controlled induction motor. However, in view of the considerable amount of work conducted here at the E.T.H. (Swiss Federal Institute of Technology) at Zurich/Switzerland, in the field of simulating induction motors on electronic computers, it seems that an information gap may exist between Schenectady and Zurich.

In 1956 we began to control induction motors by 6 thyratrons using the same configuration as in Fig. 1 of the paper of Dr. Lipo. In 1958, I asked Mr. Badr to simulate this system on an electronic computer. This was just at the end of the development of FORTRAN, and we had the impression that the time for programming and computation with a slow digital computer of that period would be too long. We chose, therefore, a small DONNER analog computer. Because simulation on this computer is executed by integration, we had to arrange the differential equations in such a way that on the left side of each equation appears the first derivatives of the unknown stator and rotor currents. This corresponds exactly to the today so-called state transition equations. Therefore, the differential equations on pages 41-51-72-79-80 of Mr. Badr's Doctor Thesis [15]

Manuscript received July 22, 1970; revised September 3, 1970.

for the three-phase motor and the two axis system (d-q) correspond with equation (1) of Dr. Lipo's Paper. We developed a simulation of a 5 kW motor with good accuracy and obtained the steady state solution for the stator currents, the stator and valve voltages, the torque $T_e(t)$ and the steady-state torque-speed curves for the induction motor and generator as a function of the firing angle α (Fig. 49). They are similar to the torque-speed curves on Fig. 5 of Dr. Lipo's paper. However, because he has chosen as parameter the hold-off angle γ, an exact comparison is not possible. Dr. Lipo should indicate that for every angle γ a corresponding angle α exists. Using a method similar to the OSSANA circle diagram for induction motors. Dr. Badr has also drawn the current loci in the complex plane for the first harmonic of the stator current. The cos ϕ factor goes down with firing control.

A very interesting feature is the transient behavior of the controlled induction motor, accelerating through or braking below synchronism, since the machine works during a short time as generator and has then a negative torque. A large oscillating torque is superposed over the steady state torque. The analog computer showed a great capability to display both these phenomena, as well as the influence of changes in the parameters and the electromagnetic torque $T_e(t)$, which cannot be measured directly.

Because Dr. Badr had connected the neutral points of the motor to the neutral point g of the generator (see Fig. 1), a small "ZERO-TORQUE" appeared. This fact was pointed out by Dr. Kovacs, who was a guest professor at our institute.

Fig. 5 of my paper [16] on the "Fifth International Analog Computation Meetings" [AICA] at Lausanne 1967, shows the block diagram of an induction machine in a closed control loop. Figs. 14 to 18 represent the steady state solution of the currents and the torque $T_e(t)$ for full load and no load; therefore, I can confirm the Figs. 3 and 4 of Dr. Lipo's paper. These were obtained by digital computation on the CDC-1604A computer of Control Data Corporation. The computing time for a period of 20 ms was 12 to 20 seconds. At slow motor speed, the computer did not arrive in a useful time at the steady state. Oscillations occur as a form of subharmonic.

The stimulation of iron saturation presents no special difficulty. Mr. Stürzinger [17] has simulated on the analog computer PACE-231R of EAI the behavior of the induction motor fed from a D.C. source by a self-commutating inverter.

Mr. Fatton [18] has simulated the transient state of a three-phase synchronous generator together with its water-turbine on a digital computer in the interval from no load to rated load as a nonlinear system.

As third IFAC president I quite agree with Dr. Lipo that the state variable methods of control engineers provide the analyst with a new and powerful tool. They give him a clear directive how to solve dynamic problems. We have applied it to the digital simulation of a high voltage D.C. power line together with the emitter and the receptor station. Every station has a three-phase thyristor bridge with 6 thyristors, which results in 64 + 64 = 128 different system states for the conducting and blocking valves. The power line of 1000 km length was simulated as a distributed system. Different kinds of transient phenomena could be studied such as switching on full load, no load and short-circuit, the behavior of automatic control and many kinds of defects of the valves.

When simulating such large power systems, we have always found 3 kinds of equations:

a) algebraic linear and nonlinear equations, i.e. from the 2 Kirchhoff laws,

b) linear and nonlinear first order differential equations in connection with the energy stores,

c) logical equations for the firing and the extinction of the valves.

In many cases, a hybrid computer consisting of an analog and a digital part, together with a memory CRO, has distinct advantages. In general, simulation of large power systems demand much skill and experience as well as a large high speed computer.

References

[15] H. A. Badr, "Primary-side thyratron controlled three-phase induction machine," ETH Doctor Thesis Nr. 3060, Juris Verlag, Zürich, 1960

[16] Ed. Gerecke/Thomas Müller," Untersuchung von thyristorgesteuerten Drehstrohm-Käfig-läufer-Motoren mit dem Analogrechner," 5th International Analog Computation Meetings, Lausanne, AICA 1967, pages 709-718. Presses acadèmiques Bruxelles

[17] Peter Stürzinger, "Drehstromasynchronmaschine und Pulswechselrichter gesteuert durch Zweipunktregelung des Statorstromes," ETH Doctor Thesis Nr. 4199, Juris Verlag Zürich 1969

[18] R. G. Fatton, "Dynamisches Grossbereichverhalten des nichtlinearen Zweifachregelkreises Wasserturbine-Synchrongenerator mit frequenzunabhängiger Last im Isnelbetrieb," ETH Doctor Thesis Nr. 4201, Juris Verlag Zürich 1969

P. C. Krause (Purdue University, Lafayette, Ind.): Dr. Lipo has solved a problem which has eluded other researchers. Moreover, he accomplishes this task directly with the simplicity of solution which cannot be equaled by any other method.

There are two features of this paper worth discussion. First, Dr. Lipo employs the d-q axis variables fixed in the stator. He does not complicate his analysis by using phase variables directly as is often suggested.[19,20] I attempted to point out the advantages of the d-q axis approach for this problem in my discussions of Refs. 7, 19 and 20. This paper by Dr. Lipo will place all other approaches to this problem in their proper perspective.

The second feature of this paper, which is actually linked to the first, is the concept of applying the open-circuit voltage to the d-q model for the purpose of maintaining a current zero. This concept enables one to develop the open-circuit voltage and to maintain a current zero without changing the parameters in the d-q model of the machine. Dr. Lipo and I have employed this idea in numerous analog computer simulations. The theory was first set forth in Ref. 9 and it was first employed in Ref. 21 to study the dynamic performance of an induction machine with series controlled rectifiers. Dr. Lipo has applied this idea to advantage in his analysis of steady-state performance.

Finally, if such information is readily available, it would be helpful to know how the speed-torque curves appear when α is used as a parameter rather than γ.

References

[19] S. D. T. Robertson and K. M. Hebbar, "A Digital Model for Three-Phase Induction Machines," IEEE TRANS. POWER APPARATUS AND SYSTEMS, vol. 88, pp. 1624-1634, November, 1969.

[20] A. K. De Sarkar and G. J. Berg, "Digital Simulation of Three-Phase Induction Motors," IEEE TRANS. POWER APPARATUS AND SYSTEMS, vol. 89, pp. 1031-1037, July/August 1970.

[21] P. C. Krause, "A Constant Frequency Induction Motor Speed Control," *Proc.' National Electroncis Conf.,* vol. 20, pp. 361-365, 1964.

Manuscript received July 28, 1970.

W. Shepherd and H. Gaskell (University of Bradford, Bradford BD7 1DP, England): Dr. Lipo is to be congratulated on obtaining an analytic form of solution to a hitherto intractable problem. It is perhaps poetic justice that the thyristor extinction angle, which proved the stumbling block in an earlier attempt at solution (reference 4 of the paper), is used as a key parameter in the state-space approach.

The author's assumption of constant rotor speed with symmetrical phase-angle control is justified for machines of normal per-unit inertia and large conduction angle. For low inertia motors the double supply frequency torque harmonic which is usually obtained with phase-angle triggering causes speed ripple. Even for machines of normal per-unit inertia the use of large retardation of firing-angle has been found in some cases to cause second harmonic torque pulsations of such severity as to produce slight speed ripple. The performance of Fig. 3 which combines torque pulsations with constant, non-zero speed may therefore imply a motor of rather high inertia. It would be of interest in control applications to investigate at what value of the "figure of merit", torque/inertia ratio, the assumption of constant speed ceases to give accurate values of torque.

We would like to commend the attention of the author to the further class of problems whereby induction motor voltage control is obtained by integral-cycle triggering of thyristors. When the phase-current is interrupted for several supply cycles there is definite speed droop unless the motor has very large inertia. Such problems appear to be quite intractable of analytic solution but can be tackled digitally. The state-space approach, with appropriate transformations of variables, may provide a solution here also.

Manuscript received July 31, 1970.

T. A. Lipo: I wish to thank the discussors for their stimulating comments which have contributed both to the history of this problem and to trends in its future application.

Professor Alger has emphasized the use of Saturistors in the rotor circuit of an induction motor when operating from a variable voltage thyristor power supply. The use of nonlinear devices can, indeed, produce a significant reduction in high frequency losses for high slip conditions and an improved torque per rms ampere. However, work thus far has centered about the wound rotor motor in which the Saturistor is externally connected to the rotor current.* In order that the technique be practical, it would be necessary, as Professor Alger has noted, to incorporate the Saturistor in the rotor slots. Unfortunately, this introduces added machining and assembly operations to the construction of a squirrel cage rotor. It has not yet been established whether the improvement in performance warrants the added cost of material and manufacture.

Professor Gerecke has provided an impressive summary of his research on the same problem. I am aware of the simulation work carried out at E.T.H., Zürich. Since little information has passed between Zürich and Schenectady, I agree that an "information gap" exists. However, having pioneered in the simulation of ac machinery,[22,23,24], General Electric has developed the capability of simulating many of the problems described by Dr. Gerecke. Hence, an "information gap" should not be considered as a "technology gap."

Digital or analog simulation methods, indeed, provide the analyst with an alternative means to obtain a solution. In fact, the author completed a similar study of the same problem on an analog computer several years ago. The difference between simulation techniques and the state variable method used in this paper centers around how the solution is obtained.

Simulation is basically an implicit formulation of the system differential equations in which the equations are simply rearranged to accommodate the computer to be used. Initial conditions are guessed, and the steady-state solution is generated as a by-product of the transient solution obtained either by feedback or iteration. In this paper, an explicit analytic solution has been obtained whereby the initial conditions needed to generate the steady-state have been expressed directly in terms of system parameters (terminal voltage amplitude, slip, hold-off angle, and machine constants). The system currents are then solved as explicit functions of time.

Professor Gerecke has stated that computing time for simulating one cycle of the steady-state solution was 12 to 20 seconds. It should be noted that since the initial conditions needed to achieve steady state are not known a priori in a simulation, 10 to 20 cycles must typically be computed to pass through the initial transient. Hence, 120 seconds or more of computing time would be a more accurate figure for an entire computing run.

Using the explicit solution, machine currents are expressed as functions of time. Hence, the number of solution points computed per cycle for display purposes is arbitrary. Since Professor Gerecke does not mention the step size of the integration algorithm used for his simulation, a direct comparison is difficult. However, when as many as 200 solution points are computed per cycle, the time needed to obtain a full cycle of the steady-state solution is less than one second. The computer used during the study was a GE 600 series.

I concur with the statements of Professor Krause. It appears that phase coordinates are currently in favor, having been suggested for use in synchronous machine modeling[25] as well as for induction machines.[19,20] When space harmonics are neglected, I feel that d-q axis techniques continue to be the most straightforward and efficient method of analyzing ac machinery.

Professor Krause has suggested that speed-torque curves using the delay angle α as a parameter rather than γ would be useful. These curves are given in Fig. 6. The curves were obtained by extrapolating between solutions for constant γ. This technique is straightforward, and it should be emphasized that the procedure is not the same as iterating boundary conditions in order to arrive at a solution. A less rapid change in the characteristic with α than with γ can be noted. Rather than all curves terminating at zero torque and synchronous speed as was the case in Fig. 6, the curves intersect the curve for zero delay denoted by $\alpha = \phi$.

Dr. Shepherd and Dr. Gaskel have pointed out the necessity of constant rotor speed in the formulation of the solution. Such an assumption is, of course, required in order to arrive at a permissible set of linear differential equations. In this paper, it has been assumed that the hold-off angle γ associated with each of the six thyristors is identical. In this case, the lowest torque harmonic is six times supply

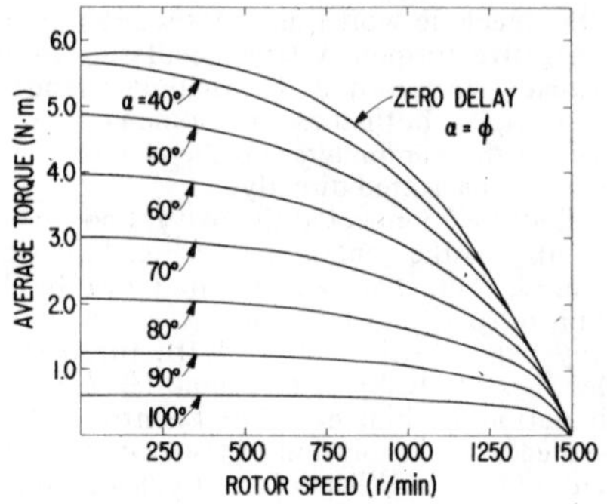

Fig. 6. Torque-speed curves for changes in α.

frequency, and the assumption of constant rotor speed is valid even for machines of relatively low inertia.

Imperfections in the thyristor gate circuitry or in the thyristors themselves can sometimes cause unsymmetrical firing. A second harmonic or even a fundamental component of electromagnetic torque can appear. The case of unsymmetrical firing is clearly not within the scope of this paper. However, the analysis techniques employed can be readily adapted to such a study. Since constant speed would again be assumed, it would then be quite important to establish a minimum torque/inertia ratio for each unbalanced condition. With an integral-cycle or burst firing type of speed control, speed deviation is of utmost importance, and a constant speed assumption is valid as over only a small portion of the speed range. At the present time, it appears that this problem is best studied using simulation techniques.

References

[22] F. J. Maginniss and N. R. Schultz, "Transient performance of induction motors," Trans. AIEE, vol. 63, pp. 641-646, Sept. 1944.

[23] K. G. Black and R. J. Noorda, "Analog computer study of wind-tunnel drive," AIEE Trans. (Communications & Electronics), vol. 76, pp. 745-750, January 1958.

[24] F. P. de Mello and G. W. Walsh, "Reclosing transients in induction motors with terminal capacitors," Trans. AIEE (Power Apparatus & Systems), vol. 80, pp. 1206-1213, Feb. 1961.

[25] M. Ramamoorty and G. R. Slemon, "Analysis of Synchronous Machine Transients Using Phase Parameters," IEEE Paper 31CP67-469.

Manuscript received September 3, 1970.

Three-Phase AC Power Control Using Power Transistors

ALEXANDER MOZDZER, JR., AND BIMAL K. BOSE, MEMBER, IEEE

***Abstract*—A three-phase ac power control circuit using power transistors which operate in a high-frequency chopping mode is described. The circuit is capable of handling several kilowatts of power at any lagging load power factor angle and the output voltage can be smoothly controlled from zero to full supply voltage. The circuit has inherently fast response and the high-frequency ripple at the output is easily filtered. The development, study, and experimental evaluation of the circuit with resistive and induction motor loads is described.**

INTRODUCTION

IN SOLID-STATE ac power control applications, such as light dimming, heater control, and induction motor speed control, thyristor ac switches are usually employed. A thyristor ac switch is essentially an inverse-parallel connected pair of thyristors, operating in a phase control line commutated mode. Although large amounts of ac power can be controlled economically by this technique, it has several limitations due to the inherent characteristics of phase control. A phase control circuit presents a lagging power factor at the input even if the load power factor is unity and, in general, the input power factor angle is always greater than the load power factor angle. For a fixed firing angle, the load voltage waveform is determined by the load power factor and usually contains harmonics at multiples of the supply voltage frequency which become excessive if discontinuous conduction occurs. Harmonic currents flowing through the machine windings cause undesirable heating and contribute to pulsating developed torque. The response of the phase-controlled circuit is slow due to the inherent dead time lag. In special applications where the supply frequency is above several kilohertz, the minimum turnoff time requirement of thyristors prevents the successful operation of a phase control scheme.

To overcome some of the problems discussed above, a single-phase and a three-phase ac power controller employing power transistors which operate in a high-frequency time-ratio mode were designed. Transistor ac switches connected in series and in shunt are controlled so that a lagging reactive load of any power factor angle can be satisfactorily handled. Power transistors are at present costly and have limited power handling capacity. For higher power ratings, transistor switches can be replaced by thyristors operating in the chopping mode at the expense of additional commutation circuitry. However, considering the current trend of development, higher power, less expensive transistors should be available in the near future.

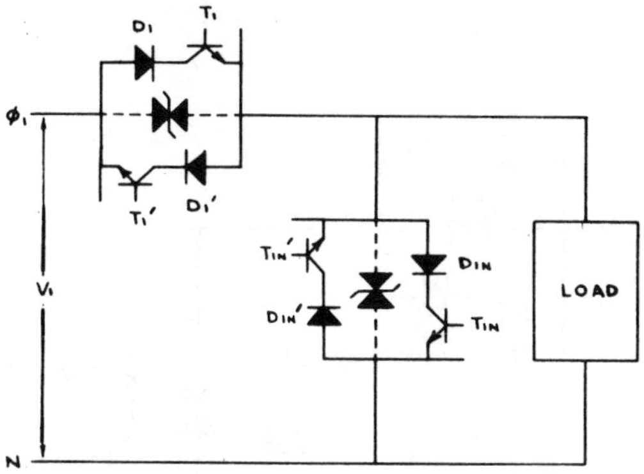

Fig. 1. Single-phase power control circuit.

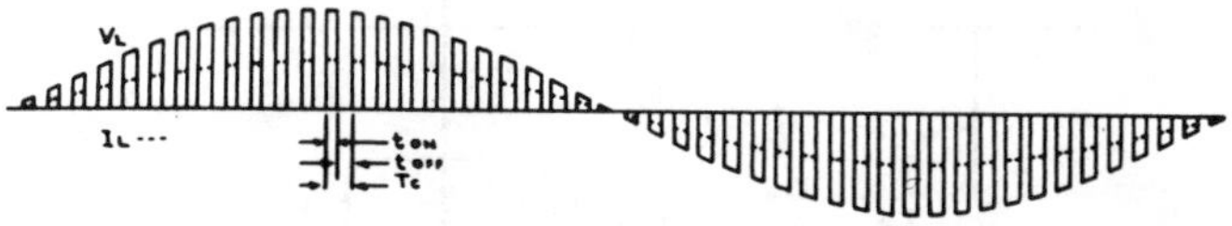

Fig. 2. Waveforms with resistive load.

DESCRIPTION OF SINGLE-PHASE CIRCUIT

The operation of the power controllers can be understood best by considering the single-phase version of the circuit, shown in Fig. 1, first. The circuit consists of two transistor ac switches, one connected in series and the other in parallel with the load as shown. Each ac switch is a pair of inverse-parallel connected n-p-n transistors, a diode in series with each to block reverse voltage. The circuit operates as a bidirectional chopper. Series transistors T_1 and T_1' chop in the positive and negative half cycles of the supply voltage, respectively, with the corresponding shunt transistor providing the free-wheeling path. The fundamental or rms voltage can be controlled by varying the time ratio, $\tau = t_{on}/(t_{on} + t_{off})$, of the chopping transistors. With a resistive load, the operation of the shunt transistors is redundant and they can be disconnected from the circuit. Fig. 2 shows the load voltage and current waveforms with a resistive load. The circuit can be used for light dimming and heater control applications, but an RFI filter is required at the input.

Now consider the circuit operation with an inductive load. Fig. 3 shows the load voltage and current waveforms. A complete cycle of supply voltage is subdivided into two modes. During mode I, which corresponds to the positive half-cycle, the transistor T_1 is chopping, T_{1n} is open, and T_1' and T_{1n}' are closed. In mode II, corresponding to the negative half-

Paper TOD-74-19, approved by the Power Semiconductor Committee of the IEEE Industry Applications Society for presentation at the 1975 Tenth Annual Meeting of the IEEE Industry Applications Society Atlanta, GA, September 28–October 2, 1975. Manuscript released for publication March 20, 1976. This work was supported in part by a grant from the General Electric Company.

A. Mozdzer, Jr., is with the Stanley Elevator Company, Inc., Long City, NY 11101.

B. K. Bose was with Rensselaer Polytechnic Institute, Troy, NY. He is now with the General Electric Co., Corporate Research and Development, Schenectady, NY 12345.

Reprinted from *IEEE Trans. Ind. Appl.*, vol. IA-12, pp. 499–505, Sept./Oct. 1976.

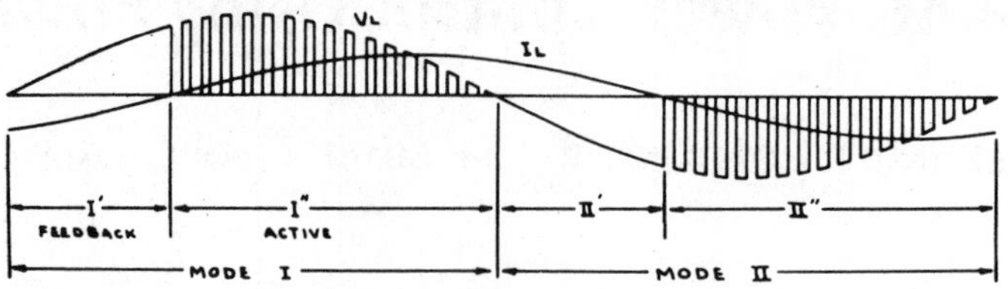

Fig. 3. Waveforms with inductive load.

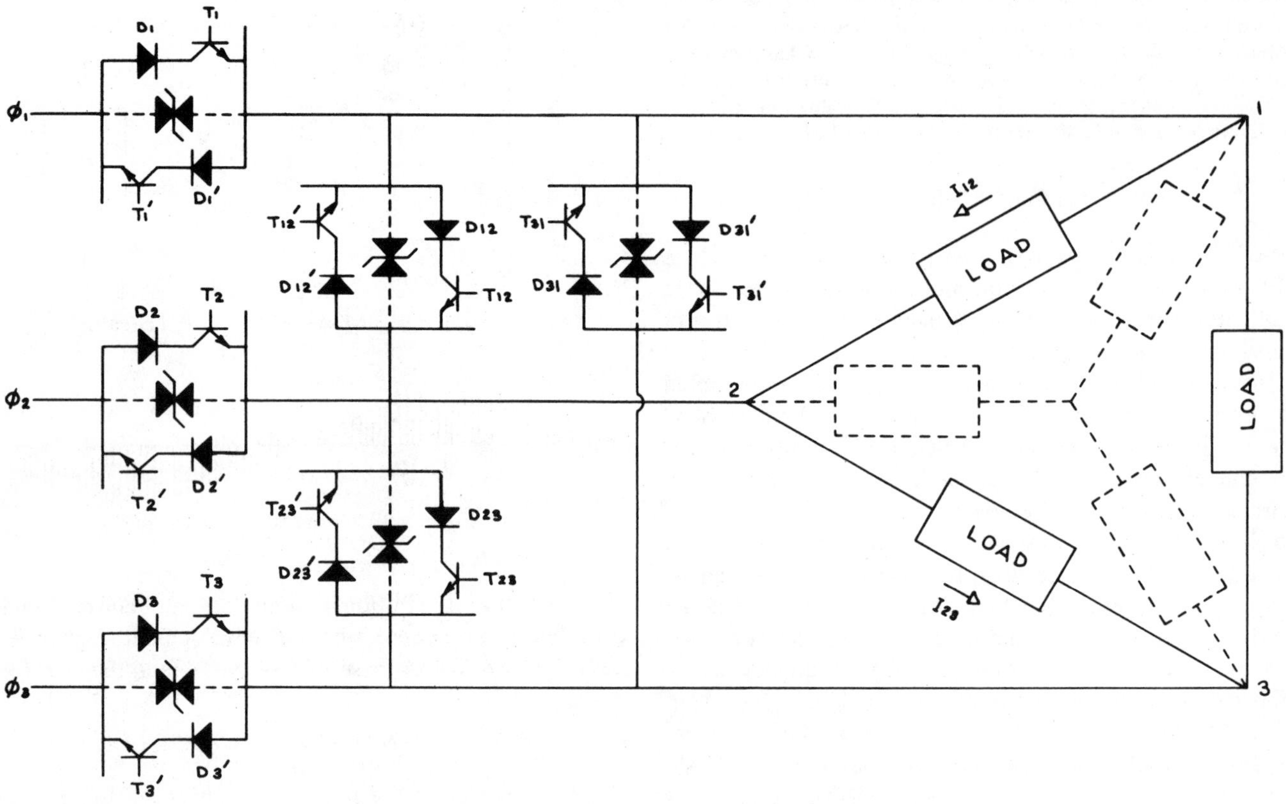

Fig. 4. Three-phase power control circuit.

cycle, T_1' is chopping, T_{1n}' is open and T_1 and T_{1n} are closed. In submode I$'$ supply voltage is positive, but load current is negative. The load feeds power back to the source through the closed transistor T_1'. A block of positive supply voltage is impressed across the load and the chopping becomes redundant. As the polarity of the current reverses, the circuit becomes active, entering into submode I$''$. The circuit operates in this submode as a dc chopper where transistor T_1 is chopping and T_{1n}' is free-wheeling. Mode II operation is similar. In the transition from mode I to mode II, a shoot through fault can occur due to the finite turnoff time of the transistors. This can be avoided by suitable control circuit design, discussed later in detail. The load voltage can be controlled smoothly irrespective of load power factor by varying the time ratio of the chopping transistors.

DESCRIPTION OF THREE-PHASE CIRCUIT

The principles of the single-phase power controller described can be extended to the three-phase version. For a star connected load with a neutral, the circuit consists of three identical single-phase controllers operating independently. If the load is delta- or star-connected without a neutral, the mode of operation of the circuit becomes quite different. Fig. 4 shows the circuit of a three-phase power controller without a load neutral. Since delta and star loads are mutually convertible, and the mode of circuit operation is the same in either case, only a delta load will be described. The circuit consists of six transistor ac switches, three connected in series and three in shunt as shown. The series transistors either act as chopping switches in the active mode or as closed switches in the feedback mode as explained before. The shunt transistors provide a free-wheeling path in the active mode or remain open if the corresponding line voltage is positive. With unity power factor loads, the shunt transistors are redundant and can be removed.

A summary of states of all the transistors in different modes is shown in Fig. 5. A complete cycle of supply voltage is divided into six equal, 60° modes. As shown in Fig. 4, each phase of the load is connected to the supply through two transistor ac switches, one in series with each line. One transistor in each switch is in the chopping mode to control the load voltage; the other kept closed to allow feedback current flow. Under ideal conditions, the composite control of the three-

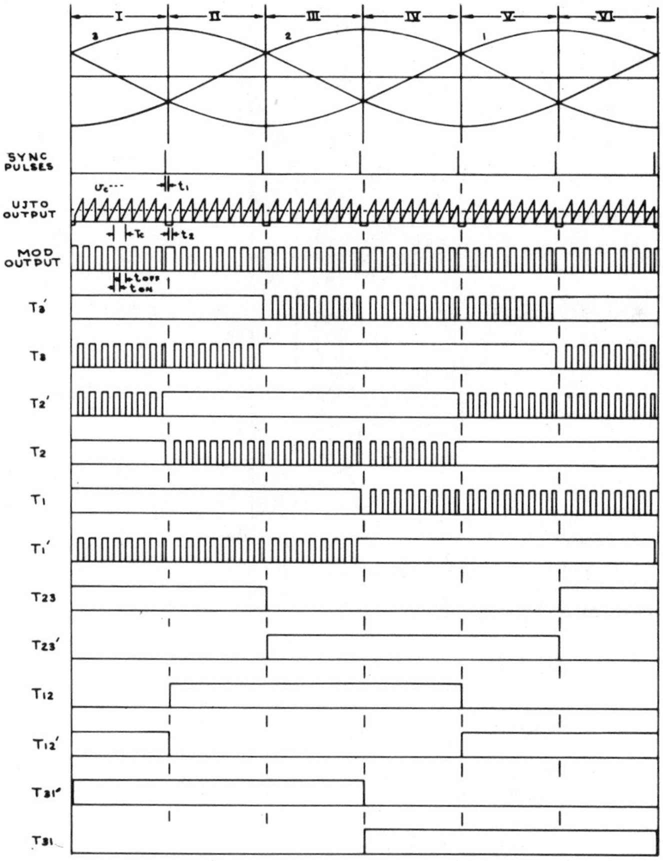

Fig. 5. Waveforms illustrating transistor states in different modes.

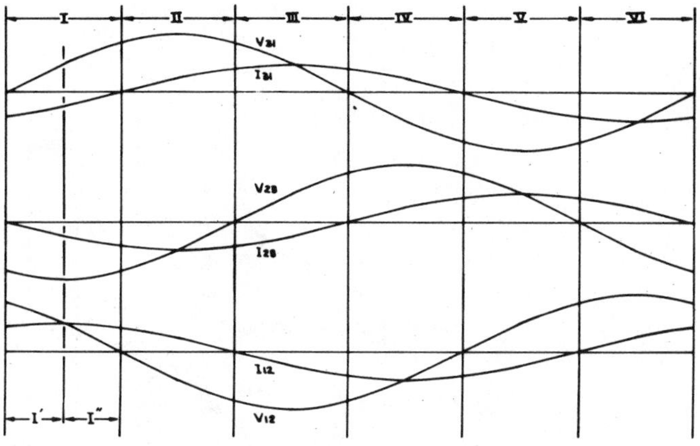

Fig. 6. Supply voltage and load current waveforms (for lag angle of 60°).

phase load can be obtained by chopping transistors between the positive and negative envelopes of the phase voltages. However, as shown in Fig. 5, the chopping interval of each transistor is extended by 60° at the leading edge resulting in a 50-percent duty cycle. This allows some simplification of the control circuit, but excludes operation with a leading power factor load and makes one of the three chopping transistors in any mode redundant. The chopping period of each transistor is given an additional small leading phase shift to prevent misoperation of the circuit near mode transitions.

Consider, for example, the circuit operation in detail during mode I. The series transistors T_3, T_2', T_1' are chopping and T_3', T_2, T_1 are closed. The shunt transistors T_{23}, T_{12}, T_{31}' are closed and T_{23}', T_{12}, T_{31} are open. Load phase 31 is chopped by T_3, T_1' and free-wheeled by T_{31}'; phase 23 is chopped by T_3, T_2' and free-wheeled by T_{23}; and phase 12 is chopped by T_2' and free-wheeled by T_{12}'. Transistor T_1' chopping is redundant. Up to a maximum load power factor angle of 90°, reactive power feedback is permitted in phase 31 through T_1, T_3' and in phase 23 through T_2, T_3'. Since phase 12 is in the trailing 60° interval of its half-cycle, no feedback path is provided. As the voltage V_{12} changes polarity in mode II, the reactive current in phase 12 is allowed to flow through T_1 and T_2'. T_2' is closed in lead time t_1 to prevent transient voltage due to inadvertent interruption of reactive current. After the mode transition, there will be a brief conduction overlap between T_{12} and T_{12}' due to turn off delay, and a shoot through fault can occur through T_2 and T_1'. To prevent this, a time gap t_2 is provided after the transition point. In addition, time margin t_1 is provided for T_2 before the transition point to allow for drift of the zero crossing point due to supply voltage imbalance. The circuit operation is similar for all the remaining modes.

Operation with a reactive load will be considered next. Fig. 6 shows line voltage and phase current waveforms (assuming perfect filtering) for a load power factor angle of 60°. In mode I, phase 31 is feeding back and phases 23 and 12 are active. For $|I_{31}| > |I_{23}|$, submode I', the current distribution during the off period of the chopping transistors T_3, T_2' and T_1' is shown in Fig. 7(a). All the load phases act as current sources. Current $I_{31} - I_{23}$ is fed back to the line V_{31}. The rest of I_{31}, in series with I_{23}, adds to I_{12} and free-wheels through T_{12}'. Under these conditions, the load voltages are $V_{L31} = V_{L23} = V_{31} = V_{31}$ and $V_{L12} = 0$. When the chopping transistors are on, the load phase voltages equal their respective line voltages. For $|I_{23}| > |I_{31}|$, submode I'', the current distribution pattern changes to that shown in Fig. 7(b). Phase 31 no longer feeds back to the source. I_{31}, in series with a portion of I_{23}, adds to I_{12} and free-wheels through T_{12}. The remaining current $|I_{23} - I_{31}|$ free-wheels through T_{23}. This results in zero load voltage for all three phases. The voltage waveforms for a lagging power factor angle of 60° is shown in Fig. 8(b). If the power factor angle becomes 30° or less, submode I' disappears, that is, there is no feedback region and the load voltage waveform is similar to that with a resistive load (Fig. 8(a)). For an extreme power factor angle of 90°, submode I' extends over the whole of mode I as shown in Fig. 8(c). The load voltage and current waveforms of phase 31 for different power factor angles are shown in Fig. 9.

CONTROL CIRCUIT

The states of the transistors in different modes as summarized in Fig. 5 can be obtained through appropriate logic circuits. The shunt transistors operate as open-closed switches on a 50-percent duty cycle of supply frequency, remaining open when their collector voltage is positive. Fig. 10 shows the schematic diagram of the shunt transistor control circuit. Line voltages are stepped down and isolated through three single-phase transformers, the primaries of which are connected in

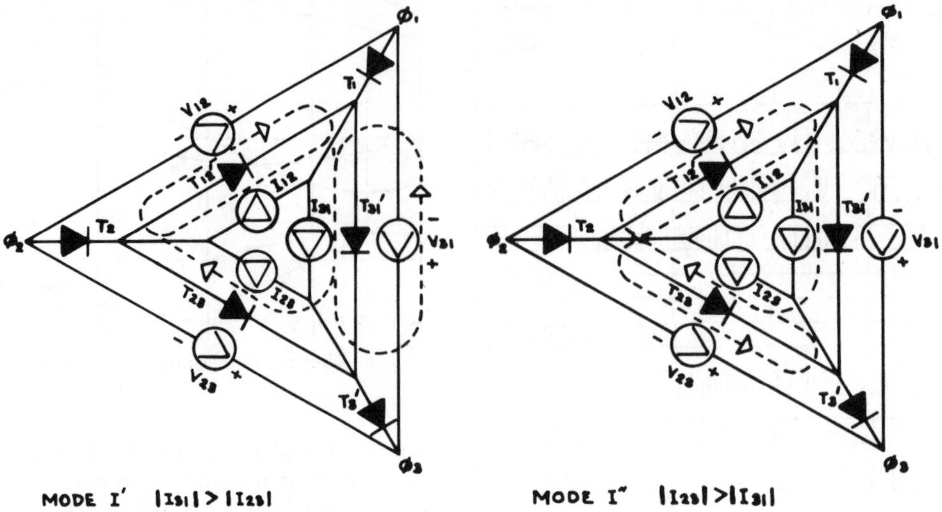

Fig. 7. Current paths in mode I (for lag angle of 60°).

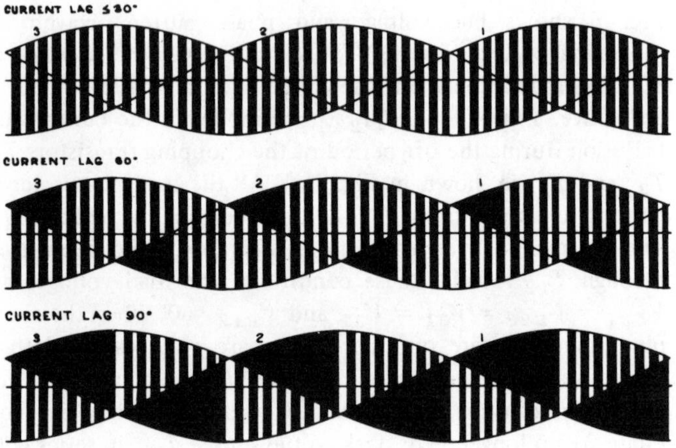

Fig. 8. Voltage waveforms at load terminals for different lag angles.

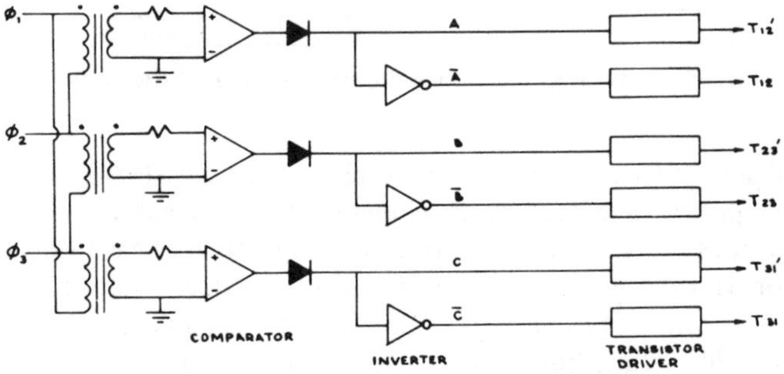

Fig. 10. Shunt transistor control circuit.

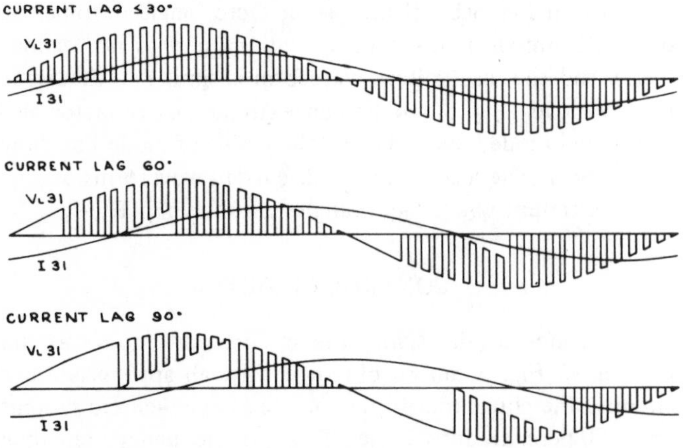

Fig. 9. Load voltage and current waveforms of phase 31 for different lag angles.

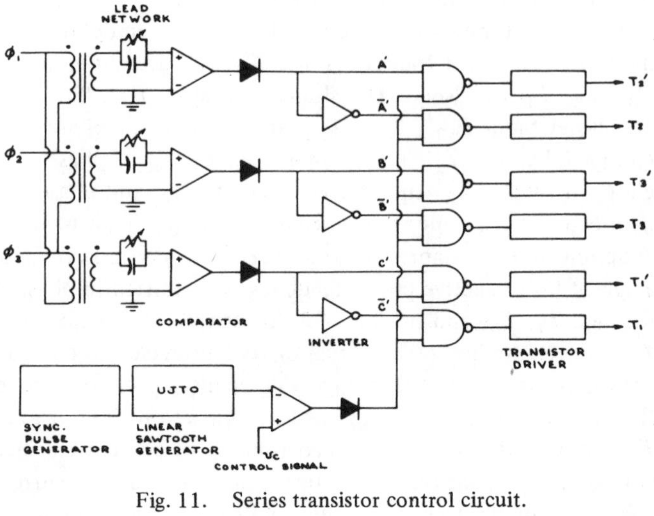

Fig. 11. Series transistor control circuit.

delta. The secondary of each transformer is connected to a comparator which converts the sine wave to a square wave with the same phase. A series diode suppresses the negative amplitude and drives the transistor T_{12}' through a driver circuit. An inverter generates a complementary voltage waveform to drive the inverse transistor T_{12}. Fig. 11 shows the schematic diagram of the series transistor control circuit. The waveforms are explained in Fig. 5. The circuit is similar to Fig. 10 except that it is coupled to a pulse width modulator circuit and the transformer secondaries are connected through adjustable RC lead networks to provide the margin time t_1 shown in Fig. 5. Outputs at A', A', etc., are each coupled to the modular through a two-input NAND gate, providing the chopping signals for the series transistors. The modulator consists of a linear sawtooth UJT oscillator synchronized near the transition of each mode and a comparator with a dc control signal v_c as shown. Synchronization pulses are generated by a set of NAND gates (Fig. 12), differentiated and used to trigger a one-shot which maintains the UJTO capacitor discharged (slightly negative) for the interval $t_1 + t_2$. The sawtooth waveform is compared with the dc signal v_c to provide the modulator output. The NAND gates complement the modulator output which is then fed to the transistor drivers. As the amplitude of v_c decreases, the time ratio $\tau = t_{on}/(t_{off})$ increases. For $\tau = 1$, the load voltage equals the supply voltage. However, the snychronization gap $t_1 + t_2$ is maintained at each mode transition.

The drive circuit for each power transistor is shown in Fig. 13. Single-phase ac is transformer isolated, rectified and filtered to provide a bipolar power supply for the drive circuit. The logic signal is coupled to the base of transistor $T-1$ which drives an optical coupler. For a logic 1 condition, the phototransistor saturates turning transistors $T-2$ on and $T-3$ off. This in turn saturates the power transistor. For a logic 0 condition, $T-3$ is on and $T-2$ off; the reverse base voltage V_z cutting off the power transistor. Each pair of power transistors is protected by a Thyrector surge suppressor in case of an inadvertent voltage transient in the collector circuits.

DESIGN AND TESTS

A complete model of the power control circuit was designed and tested in the laboratory. The experimental results agreed well with the theory. Voltage and current ratings of the power transistors determine the power handling capability of the circuit. Delco type DTS-430 transistors ($V_{CEO} = 400$ V, $I_c = 5$ A) were used in the model, allowing operation from a 208-V, three-phase supply with a peak load current of 5 A. The chopping frequency was arbitrarily selected to be 6 kHz, but could be chosen within wide limits.

The circuit was tested first with a resistive load, varying the control voltage v_c to smoothly control the load voltage from zero to the full supply voltage. Fig. 14(a) shows the load voltage and current waveforms per phase for a star connected resistive load with $\tau = 1$. The small discontinuities in the waveforms at the mode transition points (60° intervals) are due to the sync. signals as explained before. The gap intervals

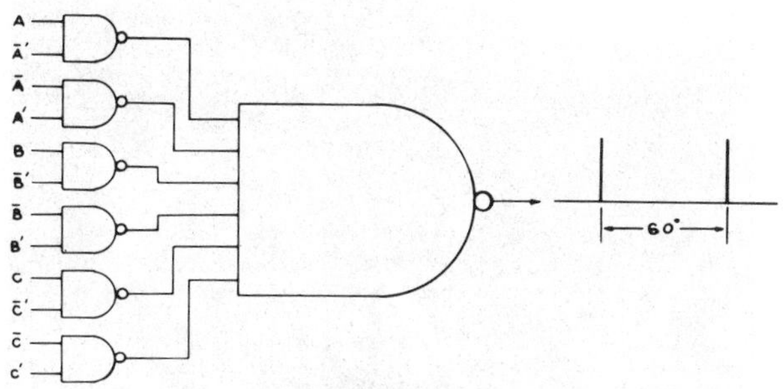

Fig. 12. Sync pulse generator circuit.

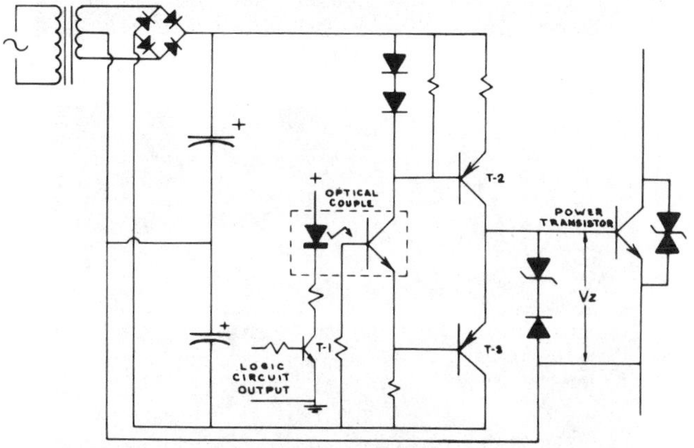

Fig. 13. Power transistor driver circuit.

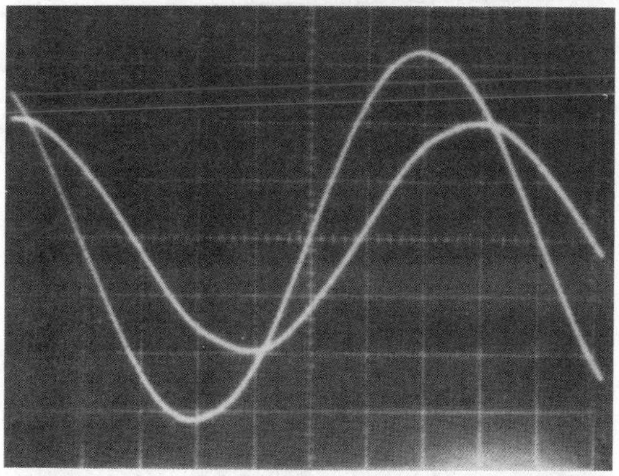
(a)

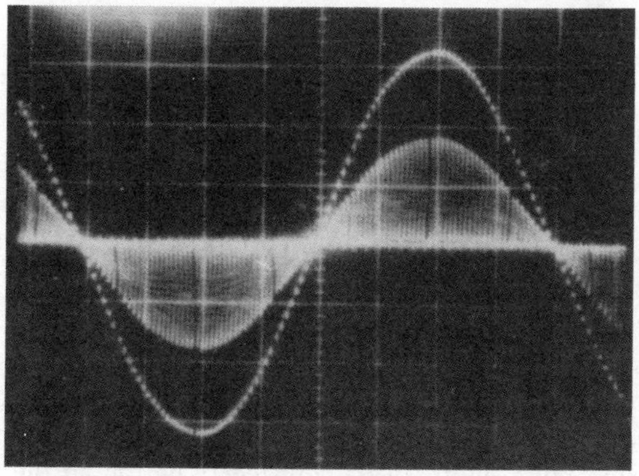
(b)

Fig. 14. Waveform with resistive load. (a) $T = 1$. (b) $T = 0.5$.

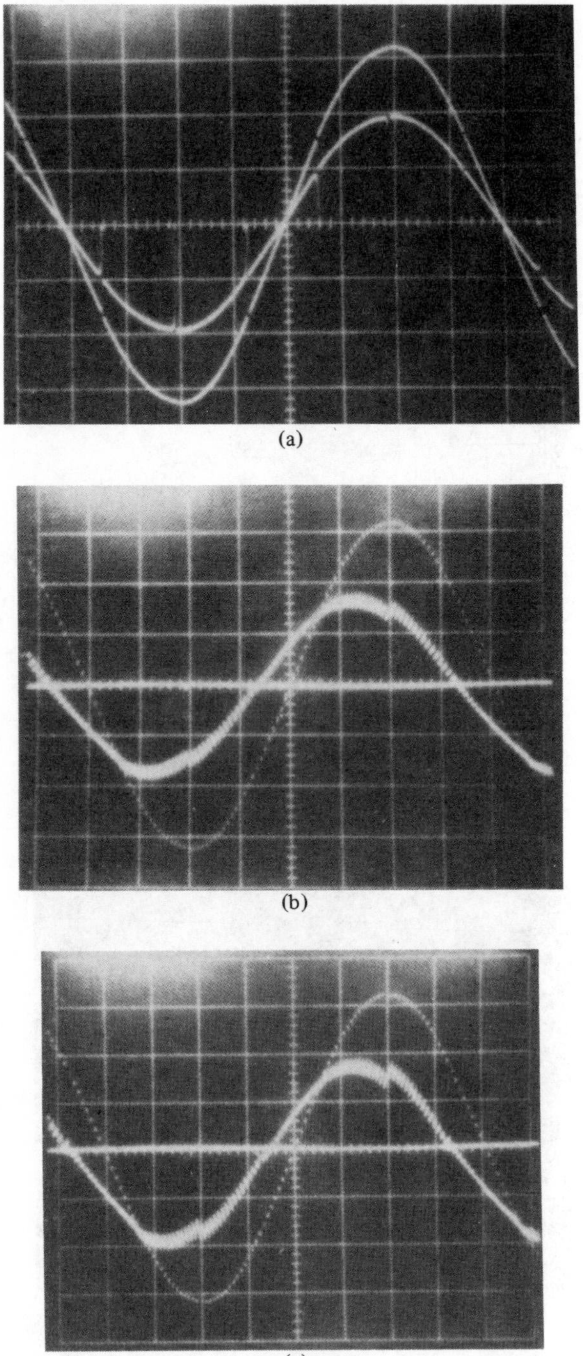

Fig. 15. Waveform with induction motor load. (a) $T = 1$. (b) $T = 0.25$. (c) $T = 0.15$.

t_1 and t_2 (see Fig. 5) are adjustable, and were both set a nominal value of 20 s. Fig. 14(b) shows the waveforms $\tau = 0.5$.

The circuit was tested next with 220-V, three-phase, 1.5-hp squirrel cage induction motor with star connected stator windings. Fig. 15 shows the load voltage and current waveforms per phase for different values of τ. Since the circuit cannot withstand high starting inrush current, the motor was started with no load and gradually increasing stator voltage. The sync pulse widths were adjusted to be low in the motor tests and do not appear as prominent due to the filtering action of the load. In Fig. 15(b) and (c), the load power factor angle is near 30° and the load voltage waveform is identical to that for a resistive load as predicted. Due to the filtering action of the load, the current waveforms contain little ripple. The torque-speed characteristics of the motor were studied and the performance compared to that obtained with an adjustable autotransformer voltage settings (Fig. 16). As expected with voltage control of an induction motor, the scheme can only be employed where load torque requirements are low at low speed, for example, blower speed control.

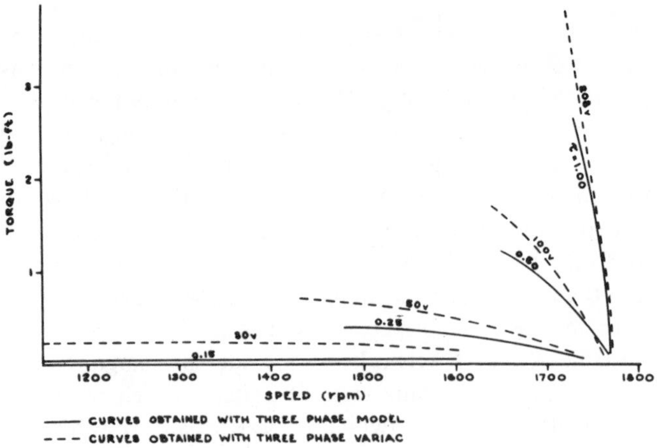

Fig. 16. Experimental torque-speed characteristic curves.

CONCLUSION

A three-phase ac power controller employing power transistors which operate in a high-frequency chopping mode is discussed in detail. The load voltage is smoothly controlled from zero to the full supply voltage by varying the time ratio τ in response to a dc control signal v_c. Several kilowatts of power can be handled at any lagging power factor angle. The circuit features a fast response, low load current ripple with a nominal value of load inductance and the capability of instantaneous circuit interruption if a fault occurs.

The circuit has been systematically studied and experimentally evaluated in the laboratory; the results agreeing well with the theory. A laboratory model was constructed to operate on a 208-V, three-phase ac supply, and was tested with a resistive load and a 1.5-hp induction motor load. Designing the circuit to other specifications is limited only by the power handling capabilities of the transistors.

At present, power transistors are expensive and their power handling capability small in comparison to thyristors, but with the present trend of development the situation should improve in the not too distant future. Regardless, the cost can be justified in part by considering the advantages of the transistor circuit over an equivalent circuit employing thyristors.

TABLE OF SYMBOLS

ϕ_1, ϕ_2, ϕ_3	Three phase supply lines.
V_{12}, V_{23}, V_{31}	Line voltages.
$V_{L12}, V_{L23}, V_{L31}$	Load voltages.
I_{12}, I_{23}, I_{31}	Load currents.

T_1, T_1', T_2, T_2', etc.	Series circuit transistors.
$T_{12}, T_{12}', T_{23}, T_{23}'$, etc.	Shunt circuit transistors.
D_1, D_1', D_2, D_2', etc.	Series circuit diodes.
$D_{12}, D_{12}', D_{23}, D_{23}'$, etc.	Shunt circuit diodes.
t_1	Time margin before mode transition to prevent transient over voltage.
t_2	Time margin after mode transition to prevent shoot through.
t_{on}	Chopper on time.
t_{off}	Chopper off time.
T_c	Chopper time period.
τ	t_{on}/T_c.
v_c	Control voltage.

REFERENCES

[1] K. Nohara, "A switched-mode modulator for power circuit," *Elec. Eng.* (Japan), vol. 90, pp. 17-25, 1970.

[2] A. Mozdzer, "Three-phase ac power control using power transistors," M.S. dissertation, Rensselaer Polytechnic Institute, Troy, NY, Aug. 1973.

Triac Speed Control of Three-Phase Induction Motor with Phase-Locked Loop Regulation

WILLIAM L. KENLY, AND BIMAL K. BOSE, MEMBER, IEEE

Abstract—**The speed control of a three-phase squirrel cage induction motor by employing triacs in the lines which operate in the normal phase control mode is described. The inherent limitation of low circuit commutated (dv/dt), which makes the triac circuit somewhat unreliable, has been overcome by suitably designing the snubber circuit. The closed-loop speed regulation of the system is then investigated by employing digital phase-locked loop scheme. Compared to the conventional method with antiparallel SCR's and analog servo, the present scheme is somewhat simpler, more economical, and extreme precision in speed control is possible. A complete model of the speed control system has been designed and experimentally evaluated in the laboratory.**

INTRODUCTION

SOLID STATE speed control of induction motors by employing thyristors has been increasingly popular during the recent years for various industrial drive applications. Between the two principal schemes of motor speed control, i.e., variable voltage-variable frequency, the former is simpler, more economical, and is commonly used for pump and blower type speed control where starting torque requirement is low and load torque usually increases with the speed. In such a scheme, normally three pairs of antiparallel SCR's are installed, one in each line, and the firing angles are symmetrically controlled to smoothly regulate the stator voltage of the motor. A triac can be conveniently substituted for a pair of antiparrallel SCR's, but because of the integral construction in the same pellet, commercial triacs are available with limited voltage, current, turnoff time, and (dv/dt) ratings and are normally used in a phase control mode with power line frequency. However, the state-of-the-art triac ratings are such that three-phase drive systems in a considerable range of horsepower can be built wiithout difficulty.

DESCRIPTION OF POWER CIRCUIT

The triac power control circuit for a three-phase star-connected motor load is shown in Fig. 1. The basic operation mode of the circuit is similar to that employing SCR's, and it has been widely discussed in the literature [1]-[5]. The circuit operation will, however, be briefly reviewed here.

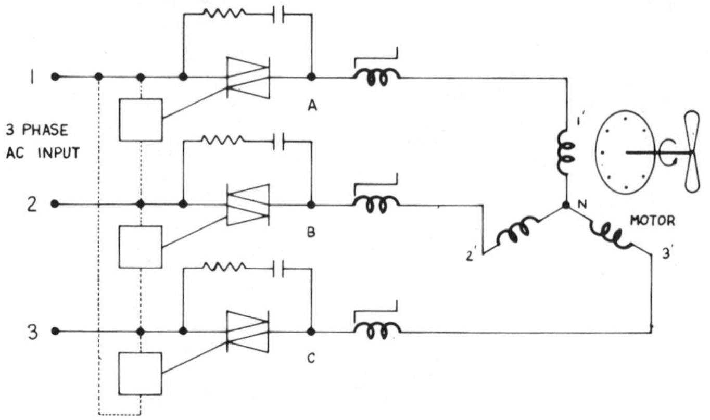

Fig. 1. Triac power circuit for motor speed control.

In the present scheme, the triacs are phase controlled symmetrically in the positive and negative half-cycles of the respective phase voltages with long pulse-bursts triggering between 0° to 180°, so as to smoothly regulate the stator voltage of the motor. Fig. 2 shows the voltage waves for different phase control angle α. Since the load has no neutral, at least two triacs must conduct simultaneously to establish the stator current of the machine. Consider, for example, Fig. 2(a) which shows waveforms for $\alpha = 25°$. During the interval, the traics B and C are on, the stator voltage $V_{2'3'}$ equals the line voltage V_{23}. At the instant t_1, all the triacs are conducting, when phase 1 current is positive and phases 2 and 3 currents are negative. Then, at the instant t_2, the phase voltages V_1 and V_3 are equal and opposite, and phase 2 current will tend to go to zero trying to commutate the triac B off. However, if the load is inductive, the phase 2 current will be continued until the instant t_3, describing the overlap angle μ. The triac B is again fired at the instant t_4 (corresponds to angle α) in the positive half-cycle and the line voltage is impressed across the load. During the interval $t_4 - t_3$, the stator voltage $V_{2'3'}$ is given by the sum of the mean voltage between phases 1 and 3 and the induced voltage across the stator phase 2'. A complete cycle of voltage $V_{2'3'}$ can be described in this way which will have notches as shown in the figure. The principle can be extended to explain the waveforms at other firing angles. Table I summarizes the triac conduction states per 60° interval at resistive load and at inductive load for different ranges of α. Evidently α can be controlled smoothly between 0° to 120° with resistive load and between μ° to 120° with inductive load.

Paper TOD-75-89, approved by the Static Power Converter Committee of the IEEE Industry Applications Society for presentation at the 1975 Tenth Annual Meeting of the IEEE Industry Applications Society, Atlanta, GA, September 28–October 2, 1976. Manuscript released for publication October 20, 1975. This work was supported in part by a grant from Lutron Electronics Company, Coopersburg, PA.

W. L. Kenly is with PCI Ozone Corporation, West Caldwell, NJ 07006.

B. K. Bose was with Rensselaer Polytechnic Institute, Troy, NY. He is now with the General Electric Co., Corporate Research and Development, Schenectady, NY 12345.

Reprinted from *IEEE Trans. Ind. Appl.*, vol. IA-12, pp. 492–498, Sept./Oct. 1976.

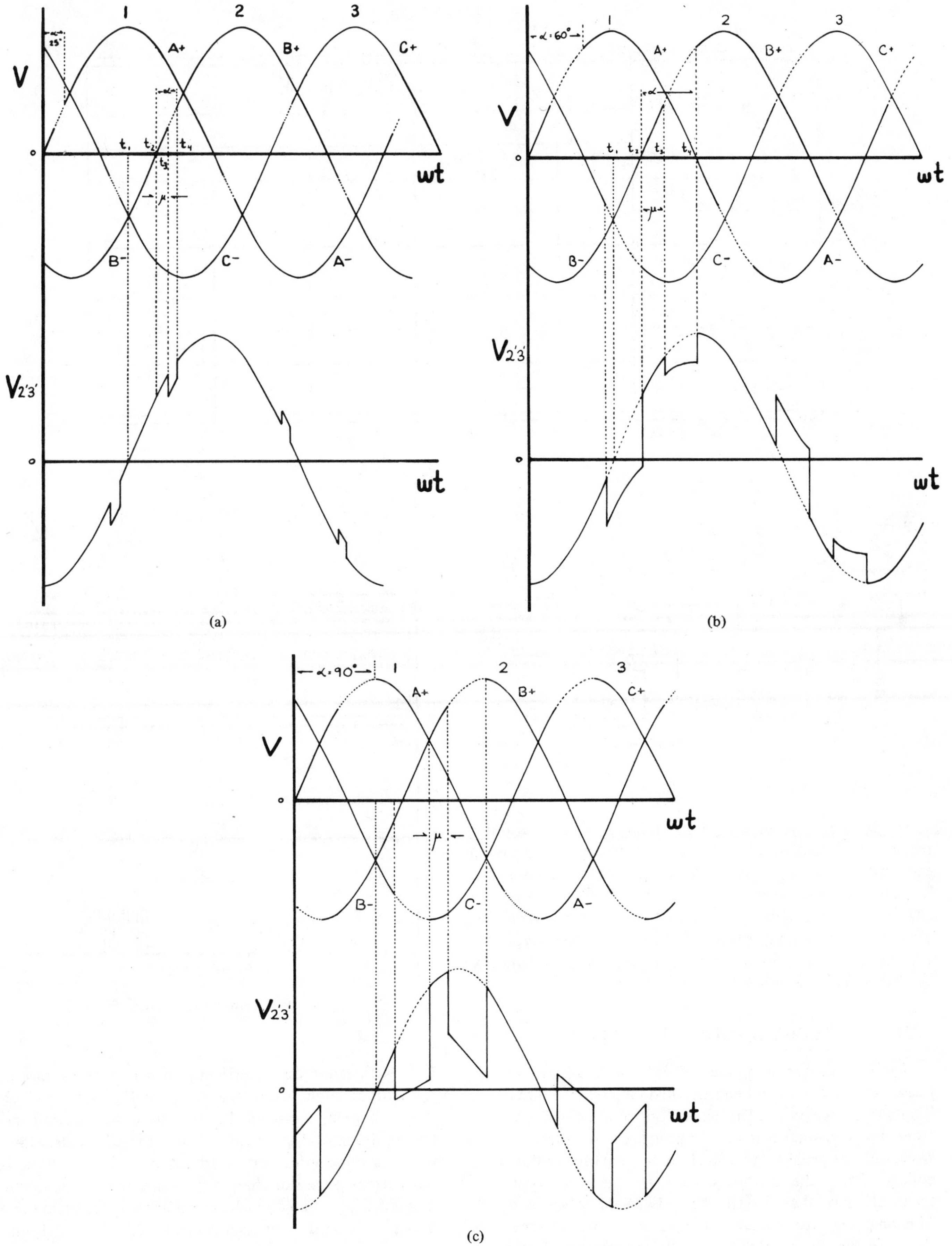

Fig. 2. Theoretical load voltage waves. (a) $\alpha = 25°$. (b) $\alpha = 60°$. (c) $\alpha = 90°$.

TABLE I
SUMMARY OF TRIAC CONDUCTION STATES PER 60° INTERVAL

FIRING ANGLE α	Number of Triacs On: R Load, 3	R Load, 2	R Load, None	RL Load, 3	RL Load, 2	RL Load, None
$0 \leq \alpha \leq \mu^{\circ}$	$60-\alpha$	α	--	60	--	--
$\mu^{\circ} \leq \alpha \leq 60^{\circ}$				$60-(\alpha-\mu)$	$\alpha-\mu$	
$60^{\circ} \leq \alpha \leq 90^{\circ}$	--	60	--	μ	$60-\mu$	--
$90^{\circ} \leq \alpha \leq (90^{\circ}+\mu)$	--	$150-\alpha$	$\alpha-90$	$(90+\mu)-\alpha$	$\alpha-30-\mu$	--
$(90+\mu)^{\circ} < \alpha < 120^{\circ}$				--	$150-\alpha+\mu$	$\alpha-(90+\mu)$
$120^{\circ} \leq \alpha \leq 180^{\circ}$	--	--	60	--	--	60

μ = overlap angle with inductive load.

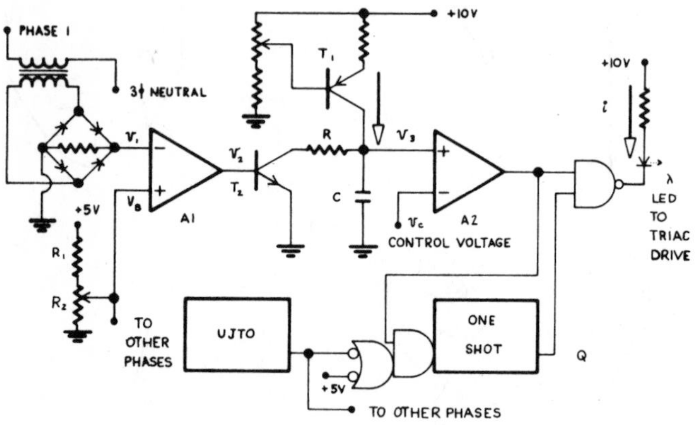

Fig. 3. Control circuit for phase 1.

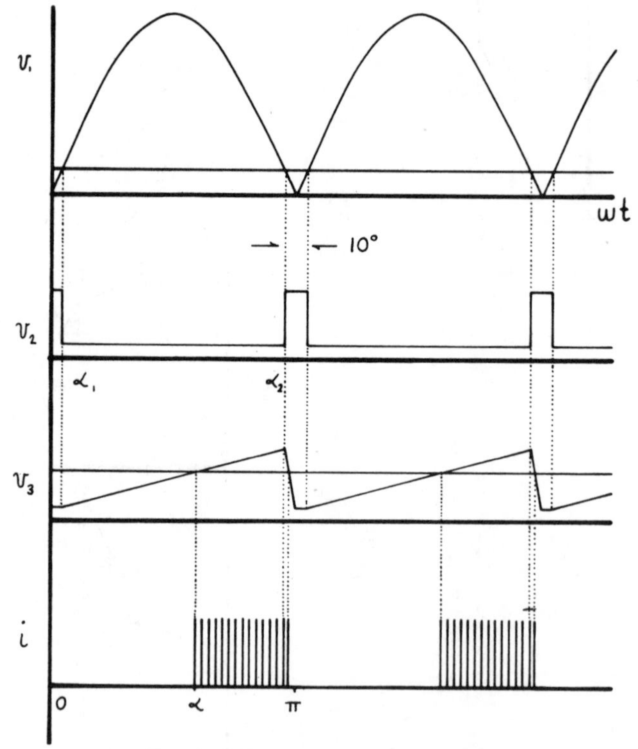

Fig. 4. Waveform for the control circuit.

An extra feature in the power circuit is a small saturable reactor employing square-loop core, connected in series with each triac. It aids the *RC* snubber circuit to overcome the limitation of low circuit commutated *dv/dt* (4 to 5 V/s) and enables reliable circuit operation in case of adverse transients. The reactor essentially sustains the inverse voltage across the triac during commutation and helps to recombine the minority carriers in the base region.

DESCRIPTION OF CONTROL CIRCUIT

The control circuit of the system can be subdivided into phase control circuit and phase-locked loop scheme for closed-loop speed regulation. The phase control circuit of a triac is to be designed such that the firing angle can be smoothly controlled in the range 0° to 180° in either polarity of the phase voltage. Since the control circuits for the three triacs are identical, only that for triac *A* will be described in detail. The simple control circuit employed in the present scheme is shown in Fig. 3, and Fig. 4 shows the explanatory waveforms. The phase 1 voltage is transformered to isolate and step down, which is then full-wave rectified to generate the voltage wave v_1. It is then compared with a small bias V_B, and the synchronizing voltage v_2 is generated at the output of the comparator. The amplitude of V_B is adjusted such that the pulsewidth of the v_2 wave is approximately 10°. A common base type constant current generator charges the capacitor *C* linearly during the interval α_1 and α_2 which is then discharged quickly by the transistor T_2 with short time constant *RC*. The synchronized saw tooth wave v_3 is then compared with the dc control signal v_c to generate the logic interval for the firing pulses as

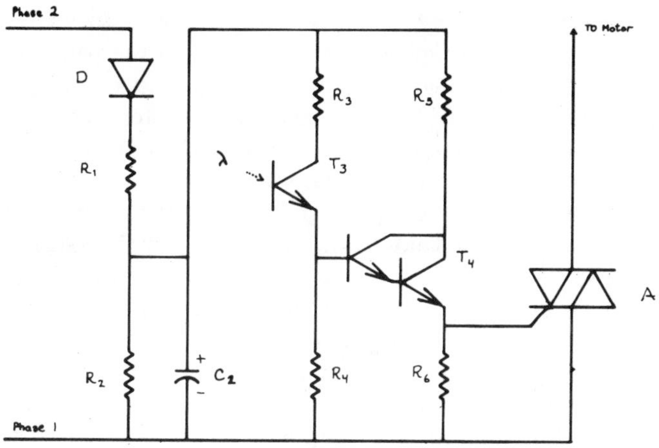

Fig. 5. Triac drive circuit.

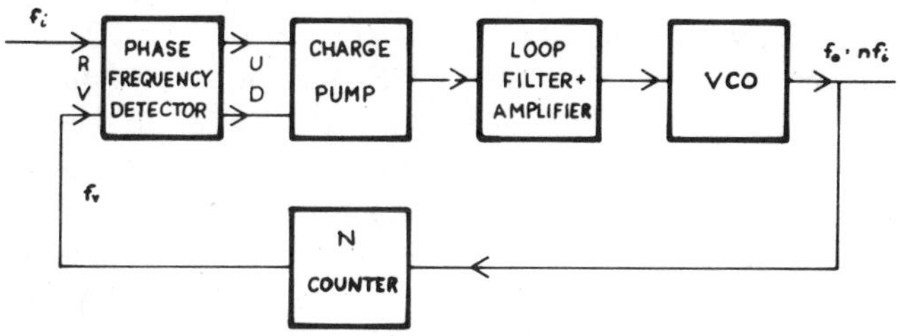

Fig. 6. Block diagram of phase-locked loop frequency synthesizer.

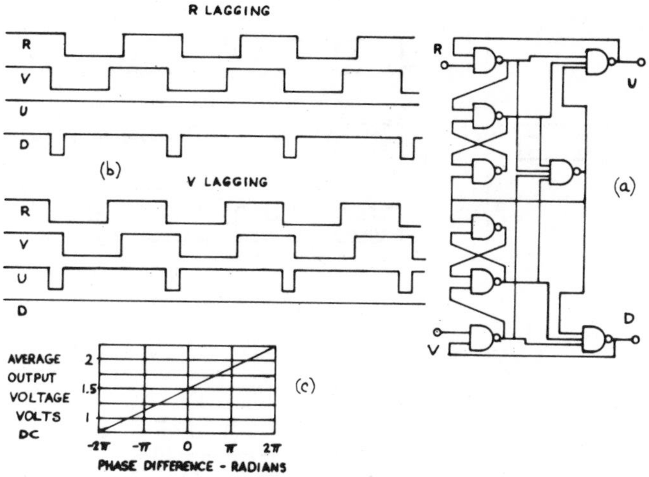

Fig. 7. Phase-frequency detector and its characteristics.

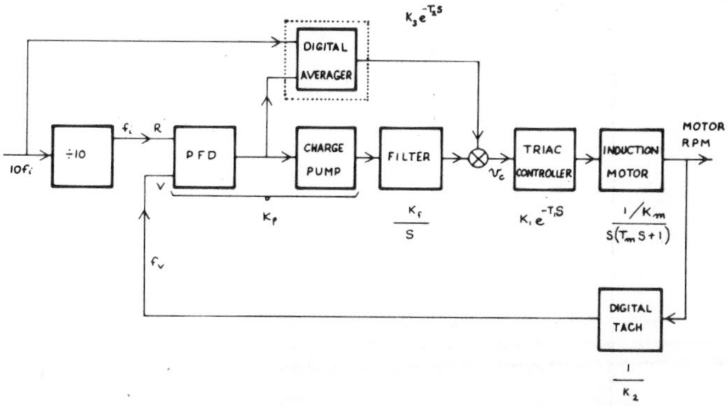

Fig. 8. Closed-loop motor control system.

shown. The firing pulses of the triacs are generated by a common UJT oscillator. The pulses are stretched by a one shot to desired interval and are then converted to optical pulses to feed the triac drive circuit. The connection as shown assures the generation of a firing pulse near the leading edge of angle.

The drive circuit for the triac A is shown in Fig. 5. The dc supply of the circuit is obtained by half-wave rectification and filtering of the voltage between phases 1 and 2 of the supply. The diode also provides the necessary isolation. The circuit has been designed such that the total drive energy required per cycle can be supplied by the capacitor C_2 without appreciable voltage drop. The train of optical pulses from Fig. 3 couple to the phototransistor T_3, which is then amplified by Darlington pair T_4 and fed to the gate of the triac. Obviously, the triac is triggered in the modes I+ and III+ and the particular triac employed in the circuit has good sensitivity in both the quadrants.

PHASE-LOCKED LOOP REGULATION

The speed of the three-phase induction motor has been close loop controlled by employing digital phase-locked loop scheme and it can be shown that this technique can provide extreme precision in speed control. The concept of phase-locked loop was known long ago, but only recently these have been used extensively because of their availability in the form of inexpensive digital IC's.

One of the most common applications of phase-locked loop is the design of programmable frequency synthesizer. A block diagram of frequency synthesizer is shown in Fig. 6. It generates an output frequency $f_0 = Nf_i$ which can be programmed either by varying the input frequency f_i or by setting the modulus N of the counter. Essentially, it is a digital feedback system in which the reference frequency and feedback frequency are compared in the phase-frequency detector and an analog error signal, proportional to the phase difference of the input waves, is generated at the output of the loop filter. The amplified error signal actuates a voltage-controlled oscillator (VCO) to generate the desired frequency. If the output wave tends to fall back in phase (or frequency), the error voltage builds up to correct the VCO output such that the reference and feedback waves always lock together in phase under stable condition.

The key element in the phase-locked loop frequency synthesizer is the phase-frequency detector which is explained in Fig. 7. Essentially, it consists of NAND logic gate combination that behaves in a manner similar to that of flip-flop. As shown in the figure, if the feedback wave V lags the reference wave R, the phase error signal is present on the U (up) output line. Conversely, if R lags V, the error signal appears on the D (down) output line. The circuit responds only to negative transitions of the input signals and locks the loop (zero phase difference) when the negative transitions on both input lines coincide. The charge pump converts the signals at U and D into a single width modulated wave such that the transfer characteristics at the output of the filter is given by Fig. 7(c). This type of phase-frequency detector (for example, MC4344/4044) is insensitive to harmonics, i.e., locking does not occur at multiples of the reference frequency. In addition, the operation is unaffected by changing duty cycle of the input signals.

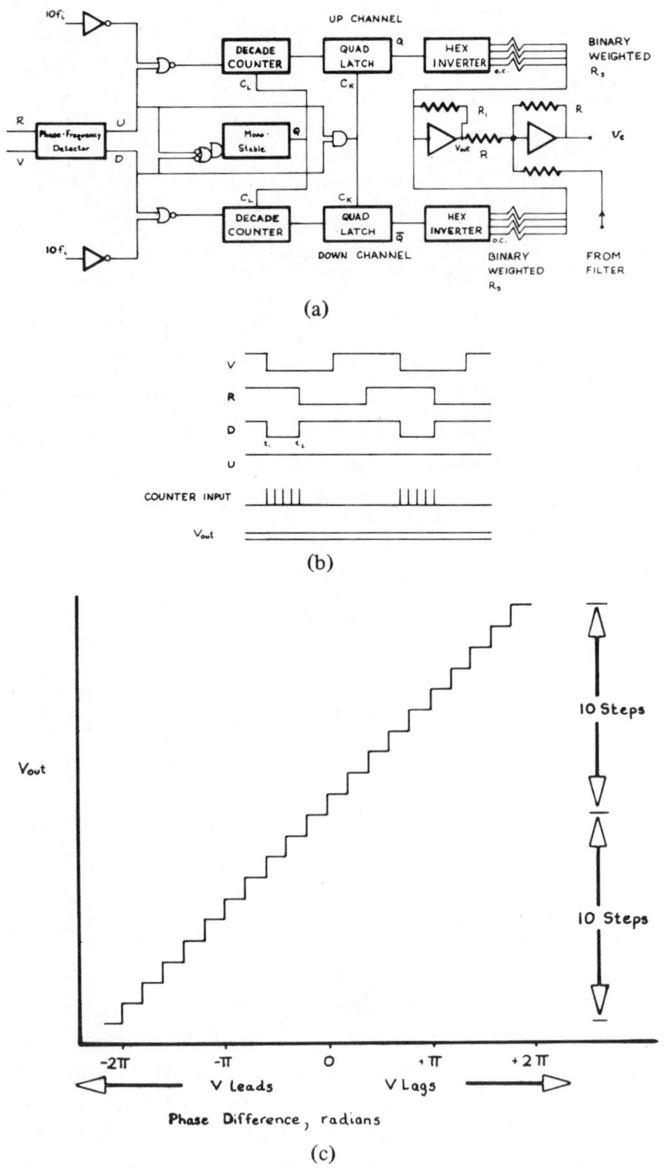

Fig. 9. Digital averager. (a) Block diagram. (b) Waveform. (c) Transfer characteristic.

The block diagram of the phase-locked loop speed control system is shown in Fig. 8. Here, the feedback frequency f_y is derived from a digital tachometer coupled to the motor shaft. Essentially, it consists of a slotted disk and LED-phototransistor sensor which generates pulses at frequency proportional to r/min. The pulse train is shaped through a monostable and fed to the phase-frequency detector input. The operation principle of the system is similar to that of frequency synthesizer, except that the motor and triac phase controlled substitute the VCO. The performance characteristics of the system, however, can be different because the composite transfer characteristic between control voltage v_c and motor rpm is nonlinear, and the system bandwidth is smaller because of the sluggish response characteristic of the motor. A digital averager (shown by dotted lines) is incorporated to improve stability of the system.

The block diagram shows the approximate transfer function of the elements. The induction motor transfer function is given by $(1/k_m)/(1 + T_m s)$ with respect to spend or frequency. However, phase, not frequency, is the parameter of primary interest, and since phase is equal to the integral of frequency, a term $1/S$ is included in the transfer function.

The digital averager provides the faster feed-forward path of the error signal over the sluggish response of the filter. It supplies the necessary poles and zeros to the composite loop gain function and helps to stabilize the system. Fig. 9(a) shows the schematic diagram of the averager and (b) and (c) explain its operation. It consists of UP-channel and DOWN-channel which receive signal from U and D, respectively, of the phase frequency detector. A common D/A converter at the end translates the digital error information into equivalent analog signal. Consider, for example, that V leads with respect to R, resulting in the D and U signals as shown in Fig. 9(b). This activates the DOWN-channel, whereas the UP-channel remains inoperative. At the edge t_1, the counter is cleared by a short pulse from the monostable which then fills in by frequency $10\,f_i$ during the time slot $t_2 - t_1$. At the edge t_2, the content of the counter is read in the latch, which is then twice inverted and converted to analog signal by the D/A converter. When V lags with respect to R, the UP-channel becomes activated, inhibiting the DOWN-channel, the operation principle remaining otherwise similar. The transfer characteristic of the averager in the full range is shown in Fig. 9(c).

DESIGN AND TESTS

A complete model of the speed control unit with phase-locked loop regulation was designed and tested in the laboratory, and the experimental results were found to agree well with the theory. The triac type MAC 10-7 (I_T = 10 A, V_{DRM}/V_{RRM} = 500 V) was employed in the power circuit which controlled the speed of Rotron 318 QS (1.5 hp) blower type motor from 220-V, 3-phase supply.

The circuit was first tested in the open loop condition by varying the control voltage v_c. Fig. 10 shows the experimental load voltage (line to line) and line current waves for firing angles $\alpha = 25°$, $60°$, $90°$, and $105°$, respectively. These waves are found to agree well with the theoretical waves shown in Fig. 2. The triac snubber circuit was designed with nominal values of R = 100 Ω and c = 0.1 μF. In addition, a toroidal square-loop core of Delta max type T4168 with 60 turns was connected in series with triac to improve the commutating capability. Fig. 11 shows the voltage across the saturable reactor and the triac, respectively, under a typical operating condition.

Next, the experimental performance characteristics of the drive system were determined systematically by varying the firing angle of the triacs. These were then compared with those under variac control. Fig. 12 shows the comparison of the performance characteristics. Dynamometer type instruments were used to measure voltage, current and power, which give reasonable accuracy under the harmonic conditions. The two sets of performance curves have similar shapes. However, at lower voltage range, the triac controlled curves are shifted to the right because of the harmonic effect. For example, at a certain motor terminal voltage, the developed torque is higher

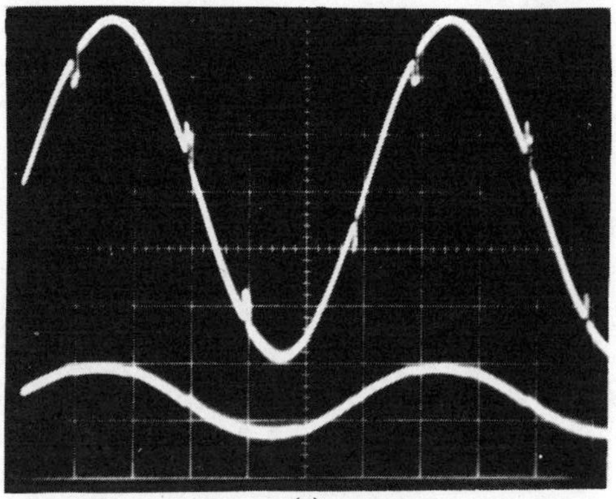

(a)

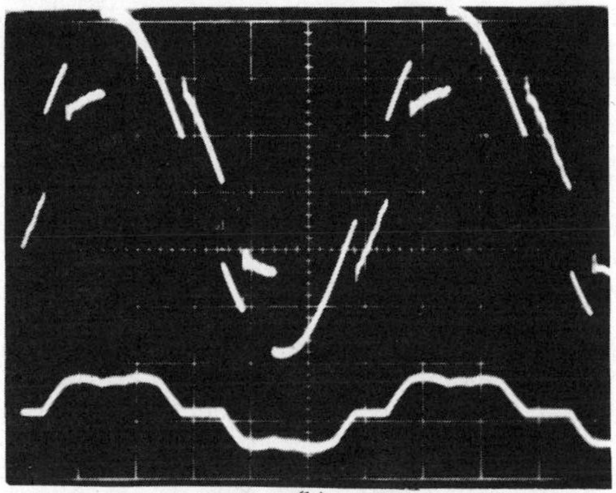

(b)

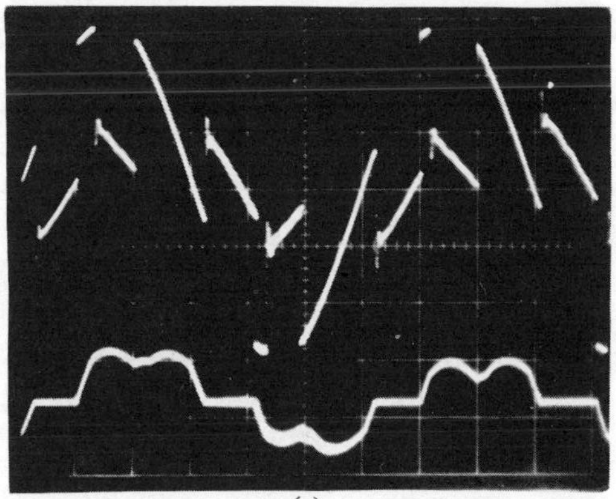

(c)

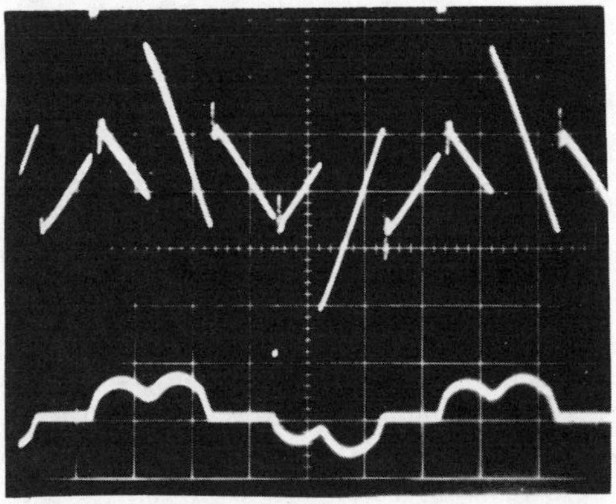

(d)

Fig. 10. Experimental load voltage and current waves. (a) $\alpha = 25°$. (b) $\alpha = 60°$. (c) $\alpha = 90°$. (d) $\alpha = 105°$.

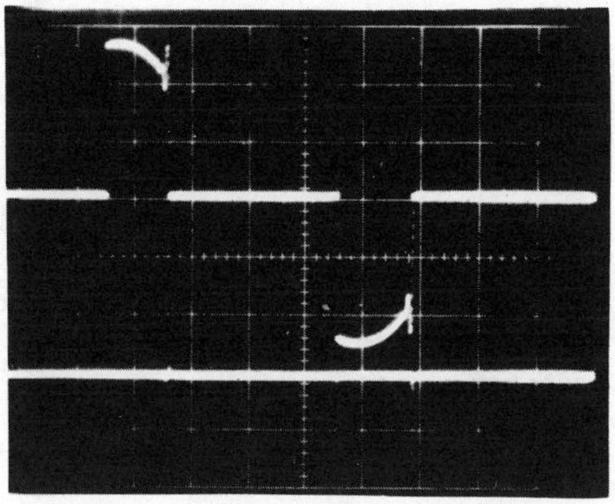

Fig. 11. Voltage waves cross saturable reactor and triac.

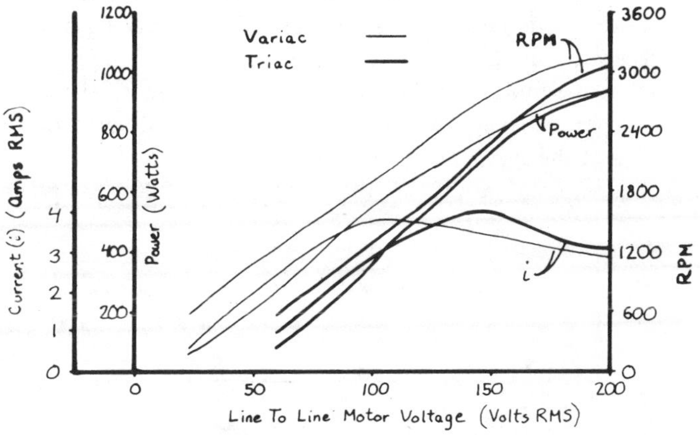

Fig. 12. Experimental performance characteristics in comparison with variac control.

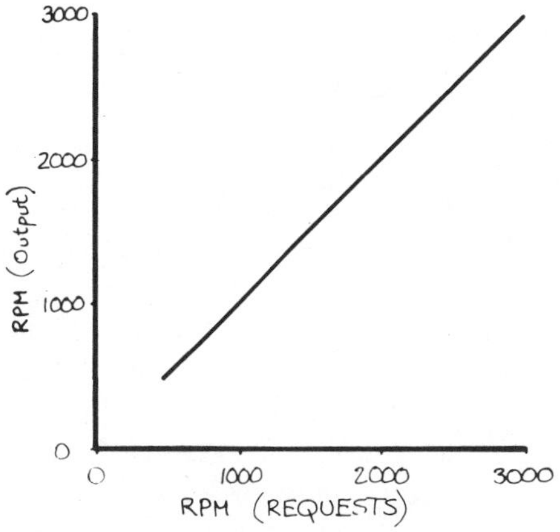

Fig. 13. Closed-loop performance characteristics.

with variac control than that with triac and, therefore, the speed is higher. For the same reason, higher voltage is needed for starting under triac control.

The speed control unit is then tested in the whole range under closed-loop condition with an external oscillator which sets the reference speed of the motor. Fig. 13 shows the performance characteristic under closed-loop condition. The speed was found to track accurately with the oscillator frequency, and at a particular setting, the supply voltage variation by +20 percent did not show any perceptible change in the motor speed. The change of external torque on the motor shaft did also have no effect. However, in the present status of experimentation, the machine has a slight instability in a low speed, which is under present investigation.

CONCLUSION

The speed control of squirrel cage motor driving a blower type load has been discussed in the paper by employing triacs in the lines instead of antiparallel SCR's. The low circuit commutated (dv/dt) of a triac and the consequent problem of unreliable commutation has been solved by designing snubber circuit with an additional square-loop core saturable reactor in series. Closed-loop speed control is then discussed by employing digital phase-locked loop scheme. The design and performance criteria of the system is then briefly reviewed.

A complete model of the system with 220-V, 3-phase, 1.5 hp motor has been designed and systematically tested in the laboratory, and the experimental results agree well with the theory. The scheme can be easily extended to higher power range, and it being limited only by the present power handling capability of triac. Compared to the conventional system with antiparallel SCR's and analog servo, the present scheme is somewhat simpler, cheaper, and provides speed control with extreme precision. It can be employed to a number of motors requiring closely tracked speed control. The scheme is expected to be popular in many industrial drive applications.

REFERENCES

[1] D. A. Paice, "Induction motor speed control by stator voltage control," *IEEE Trans. Power App. Syst.*, vol. PAS-87, pp. 585–590, Feb. 1968.

[2] W. Shepherd and J. Stanway, "The polyphase induction motor controlled by firing angle adjustment of silicon controlled rectifiers," *IEEE Int. Conf. Rec.*, vol. 7, pt. 4, pp. 135–154, 1964.

[3] B. C. Krause, "A constant frequency induction motor speed control," *Proc. NEC*, vol. 20, pp. 361–365, 1964.

[4] W. McMurray, "A comparative study of symmetrical three-phase circuits for phase controlled ac motor drives," *in Proc. IEEE-IAS Ann. Meeting*, pp. 765–774, 1973.

[5] T. A. Lippo, "The analysis of induction motors with voltage control by symmetrically triggered thyristors," *IEEE Trans. Power App. Syst.*, vol. PAS-90, pp. 515–525, Mar./Apr. 1971.

[6] A. W. Moore, "Phase locked loops for motor speed control," *IEEE Spectrum*, pp. 61–67, Apr. 1973.

[7] G. Nash, "Phase locked loop design fundamentals," Appl. Note An-535, Motorola Semiconductor Products, Inc., Phoenix, AZ, 1970.

DESIGN AND APPLICATION OF A SOLID STATE AC MOTOR STARTER

JOHN MUNGENAST
Power Semiconductors, Incorporated
Devon, Connecticut

Abstract

This paper will explore the design, application and field operating experience of the first complete line of 25 HP to 1000 HP solid state A.C. motor starters. Contents are as follows:

A. The basic requirements of A.C. motor starting.
B. Power circuitry -- a brief review of solid state.
C. A complete solid state A.C. motor starter with provision for reduced voltage starting.
D. Application problems, advantages and disadvantages of solid state.
E. The "Hybrid" motor starter.
F. New dimensions for squirrel cage motors using solid state control.
G. Limited range speed control using variable voltage, fixed frequency.
H. The universal replacement. Is it practical?

Forward

Since the birth of the thyristor it has been predicted that the days of the electromechanical motor starter were numbered!

Needless to say this has not happened -- the conventional AC motor starter continues to dominate the market. However, as in so many areas, solid state is complementing the electro mechanical starter in thousands of difficult applications. Indeed, certain hybrid approaches promise a fruitful and peaceful coexistence.

Let us first define the advantages and disadvantages of solid state for AC motor starting and we can judge its place for ourselves.

Definition

For purposes of this paper, we will define the solid state AC motor starter as a three phase controller suitable for across-the-line or reduced voltage connection of a conventional squirrel cage motor.

The least expensive circuit is shown in FIG. 1, however it imposes costly additional heating conditions on the motor and makes complete motor protection difficult. The circuit shown in FIG. 3 is the most widely used, while control with only three thyristors is possible with thyristors in delta-windings in the wye, provided that all six motor leads are brought out. The McMurray paper listed in the bibliography gives an excellent insight into circuit considerations.

Note that coordination with motor characteristics becomes necessary even at the earliest stages of design. There is no such thing as a universal replacement and the author doubts if there ever will be. Interface with motor, circuit and thyristor is imperative as we examine the next area:

Horsepower Rating

Assuming the circuit in FIG. 1, what is the maximum capacity of a solid state motor starter today? The answer lies in FIG. 2, showing five years progress in power thyristors. The 28mm and 33mm were the largest available from 1964 to 1968. In just the last few years we have available sufficient power handling capability for across-the-line starting to 400 HP and reduced voltage starting over 1200 HP at 480-3-60. The comparison below illustrates this progress and, we hope, aids in clearing away some confusion over semiconductor ratings. (For reference purposes, one LINE AMP equals approximately 1.2 Horsepower at 480-3-60.)

Year	1966	1969	1974
Thyristor Wafer Diameter	28mm	52mm	76mm
Thyristor Nominal Rating			
Amps RMS	500A	1600A	3300A
Amps Ave	320A	1000A	2100A
"USABLE"* LINE CURRENT			
Free Convection Cooling	235A RMS	540A RMS	720A RMS
Forced Air Cooling	465A	1000A	1700A

Reprinted from *IEEE/IAS 1974 Annual Meeting* pp. 861-866, 1974.

Water Cooling	770A	2200A	4900A
Intermittent Ratings, Forced Air Cooling			
1 minute	800A RMS	1700A RMS	2800A RMS
30 seconds	910A	2000A	3400A
1 second	1500A	3000A	5000A

* Continuous line current in a three phase AC circuit, with conventional cooling systems and ambient temperature of 50°C air and 25°C water. Coolant velocities of 1000 LFPM and IGPM.

Note that the Thyristor Nominal Ratings given above refer to semiconductor industry definitions and are absolute values, under impractical conditions, used for comparative purposes. More significant are the "USABLE" ratings, which relate directly to continuous load capability.

Equally significant are the Intermittent Ratings, which illustrate the improvement in thyristor stamina -- the overcurrent capability for those instances where the trigger logic misfires for some reason and the thyristors must survive an unintended across-the-line surge. The one minute rating also gives an insight into the ability to withstand locked rotor conditions.

Thus we can see that the Solid State AC motor starter is no longer a toy. Declining thyristor prices and increased capabilities give us an entirely new tool for sophisticated motor control.

Modes of Control

Across-the-line starting is possible and practical, of course, but the cost structure of solid state AC motor starters makes this somewhat unlikely as over 60% of the cost is generally in the thyristor power control, with only 40% in the thyristor regulator, or firing circuit. A more detailed schematic is shown in FIG. 3.

Reduced voltage starting is the great strength of solid state as phase control is inherent in the thyristors. Three systems are in widespread use.

A. Ramp Control

The motor voltage is increased at an adjustable, pre-set rate, without feedback from the motor.

B. Current Limit

The regulator limits motor current to an adjustable, pre-set level -- feedback is obtained from current sensors in the line. (FIG. 4)

C. Linear Acceleration and Deceleration

Motor acceleration and/or deceleration is maintained at an adjustable, pre-set level with feedback coming from a standard tachometer (FIG. 5)

Each mode of operation has distinctive features. Some general applications are as follows:

A. Ramp Control

Certain specialized applications where the user wishes to assure that full voltage is applied to the motor after a given time interval under any circumstance -- with no regard to load current or shaft speed. This mode corresponds to conventional schemes which have no option but timed control.

B. Current Limit

The most popular mode to date, used in all types of single purpose machines such as compressors, extruders, pumps, presses where the requirement is to limit the KVA demand on the electrical system.

C. Linear Acceleration

Linear Acceleration is widely used on ski-lifts and conveyors and in some cases has allowed replacement of a DC motor drive with a squirrel cage motor. It is also used on process machines which must be accelerated in a precise time, such as wire drawing machines. Linear deceleration provides the additional function of controlled rate of deceleration -- provided that the coast-to-rest time of the machine is less than the pre-set deceleration time.

FIG. 6 is a comparison of starting characteristics for solid state vs. conventional systems.

Starter-Motor Coordination

The solid state starter is generally used with a NEMA design B motor. In some applications, such as high-inertia or high breakaway torque loads, NEMA design C on D motors are used, as they would be with other types of reduced-voltage motor starters.

Overtemperature protection of the motor is important with controlled acceleration. Since losses are very high during starting, and since the time of acceleration can be controlled at will, it is not difficult to establish a condition that will allow the motor to burn out. Normal overload protection will not be adequate; it is strongly recommended that an overtemperature detector be imbedded in the motor winding.

Another word of caution -- while the starter itself is capable of providing adjustable voltage for continuous speed control, this system should not be considered for this function other than for the short duration of the starting period. The normal squirrel cage motor is not intended for continuous reduced speed operation and the most careful integration is necessary between motor, control and load characteristics.

Such integration has been accomplished in a number of applications involving both variable torque applications such as fans, blowers, and pumps and constant torque applications such as hoists and elevators.

Changing patterns in the cost and availability of squirrel cage vs. DC motors will result in increased interest in this subject; it is certainly forseeable that the solid state starter will, in some instances, evolve into a special purpose motor speed control.

The Hybrid Motor Starter

At this point the reader is familiar with some of the advantages and disadvantages of solid state vs. electro mechanical motor starters. A most valuable review is the paper by R. Locke listed in the bibliography.

As we consider these points, it becomes apparent that a combined approach has much merit;

A. Solid State has the inherent ability to modulate by means of phase control, allowing reduced voltage starting.

B. Solid State has no wear-out mechanism when properly designed, thus repetitive operations pose no problems. This holds true during current interruption, also.

C. Mechanical switching, on the other hand, has far lower heat dissipation due to the much lower voltage drop across contacts vs. thyristors.

D. Mechanical switching can also offer isolation and positive disconnection impossible with solid state.

Combining the two results in an interesting hybrid offering most of the best features of either.

A simple system is shown in FIG. 9, sequence is as follows:

A. Solid State Switch 1 phases into full conduction, providing reduced voltage starting.

B. When full voltage is reached, Contact C1 closes, load current is now conducted through contacts.

C. If "soft stop" is desired, Contact C1 opens with SS1 fully phased on. SS1 then phases back at the desired rate until circuit is deenergized.

Miscellaneous

A. A more detailed description of operation of the typical starter shown in FIG. 3 is as follows.

Schematic shows circuit for standard current-regulated starter. Motor current is measured by three current transformers and the combined rectified signal is compared to an adjustable reference signal from the "current limit set" potentiometer.

The resulting error signal causes the trigger circuit to control the firing angle of the SCRs between 0 and 180 degrees. When the "start" button is pressed relay 1CR sequentially unclamps the SCR gating circuits and then the reference signal. This insures that proper regulator voltage levels are achieved before the SCRs are permitted to turn on.

The opposite sequence on stopping assures smooth current delay before turn-off. Under control of the trigger circuit, as guided by the error signal, the firing angle of the SCRs is advanced until the motor reaches full speed. At this point the regulator amplifier turns full on, applying full sine-wave voltage to the motor.

B. The question often arises as to isolation and safety aspects of solid state, for sufficient leakage flows through the thyristors even in an OFF state to constitute a potential hazard. While standards are still being formulated, it is prudent to view the solid state starter in the same light as an electromechanical starter with a questionable control circuit. That is, positive and visible isolation must be used in the form of a disconnect switch or molded case circuit breaker.

Acknowledgments

The author acknowledges the assistance of the Electric Regulator Corporation of Norwalk, Connecticut, manufacturer of the first complete line of solid state AC motor starters.

Bibliography

1. McMurray, W., "A Comparative Study of Symmetrical Three-Phase Circuits for Phase Controlled AC Motor Drives," IEEE IAS Conference Record, 1973.
2. Locke, R., "Design and Application of Industrial Solid State Contactors," IEEE IAS Conference Record, 1973.
3. Reimers, E., "Hybrid Electric Propulsion Utilizing Reconnectible Motor Windings in Wheels," IEEE IAS Conference Record, 1973.
4. Fox, J. "Use Solid State Unit for Smooth Starting," POWER Magazine, McGraw-Hill, 1973.

Addenda

Operating Experience

Over a thousand solid state AC motor starters have been in field operation for period up to five years. Results have been most encouraging; many users found that improved flexibility and smoothness of operation was very valuable. Other reversing applications have returned to feasibility because, for the first time, the machine operator need not replace large starter contacts every month. Reliability has been excellent for most starter designs have been based on solid state DC drives with years of field experience. The few field problems have generally been due to small electronic component failure in the regulator section. Tighter incoming inspection and circuit burn-in solved these problems, and modular replacement proved invaluable in minimizing down-time.

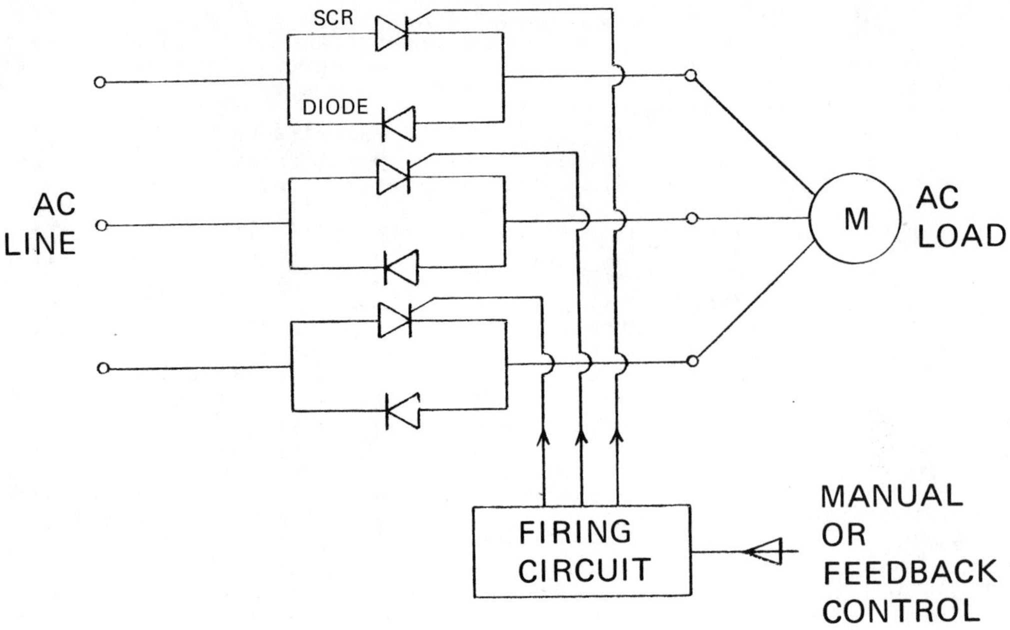

FIG. 1 BASIC AC MOTOR STARTER CIRCUIT

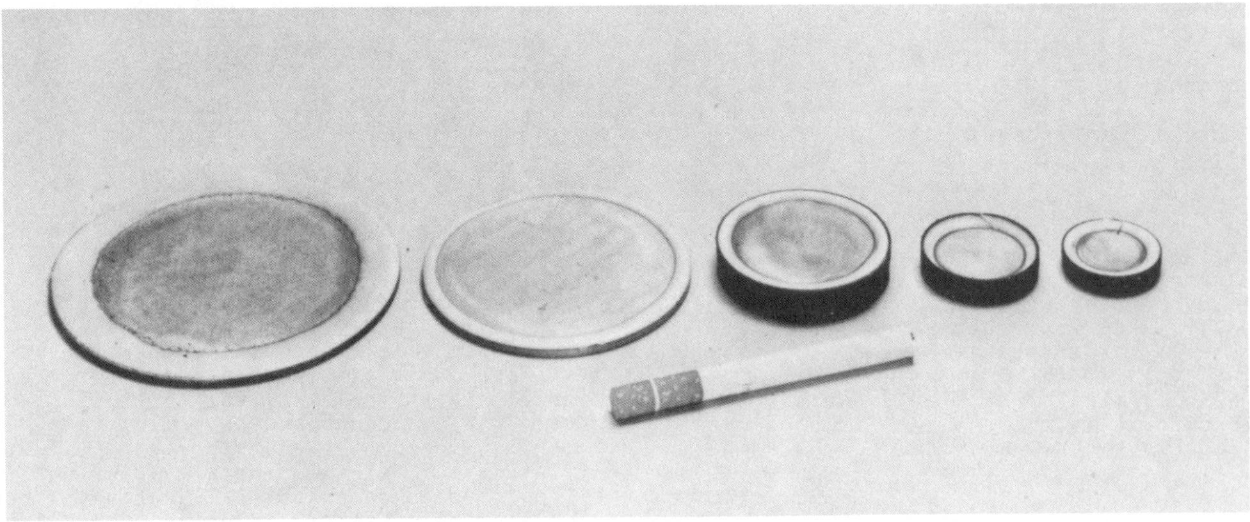

FIG. 2 Progress in Thyristor Capability, 102mm, 76mm, 52mm, 33mm, 28mm Thyristor wafers

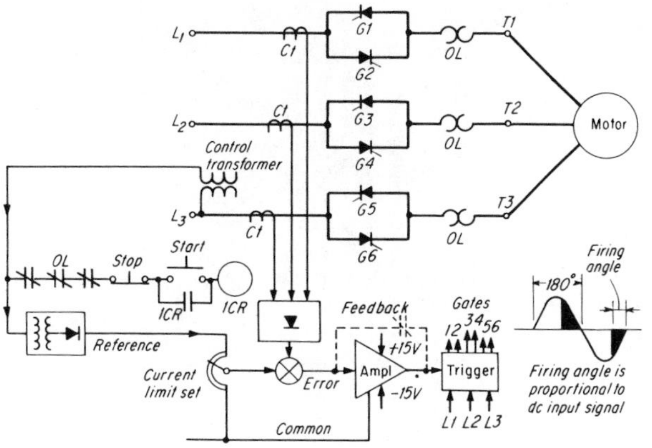

FIG. 3 Schematic of AC motor starter

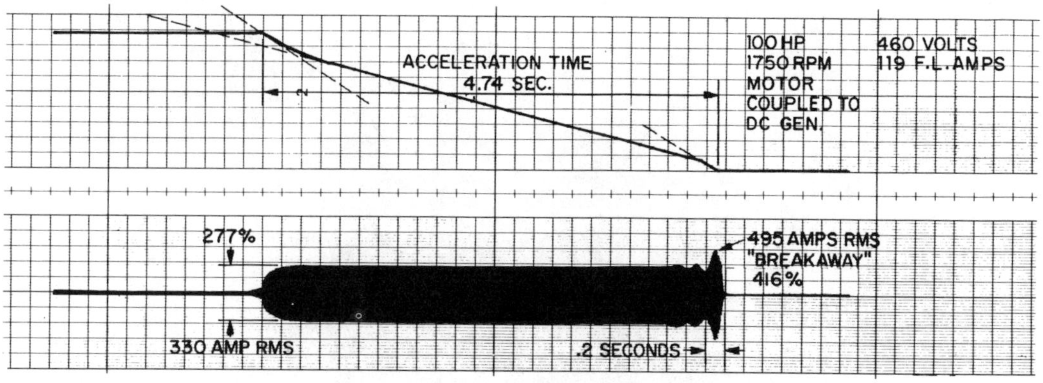

FIG. 4 Motor Current during Current Limit Mode

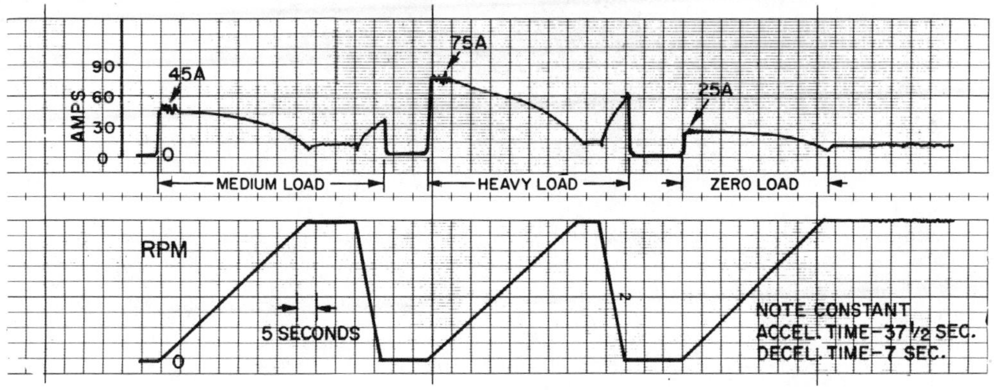

FIG. 5 Motor Current during Constant Acceleration Mode

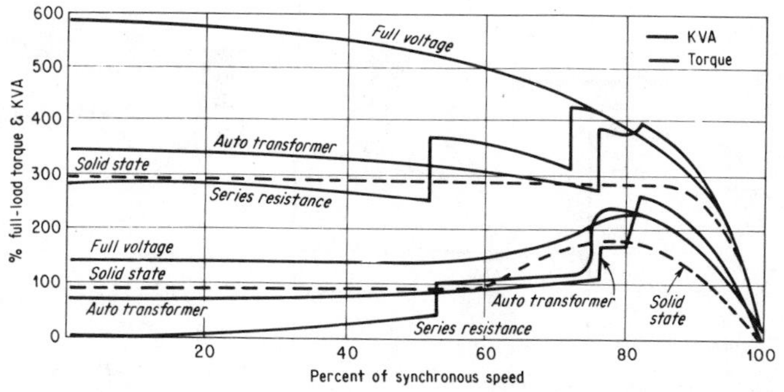

FIG. 6 Comparison of Solid State Reduced Voltage Starting with Conventional Methods

FIG. 7 A 400 HP Solid State AC Motor Starter

FIG. 8 A Typical Thyristor Firing Circuit

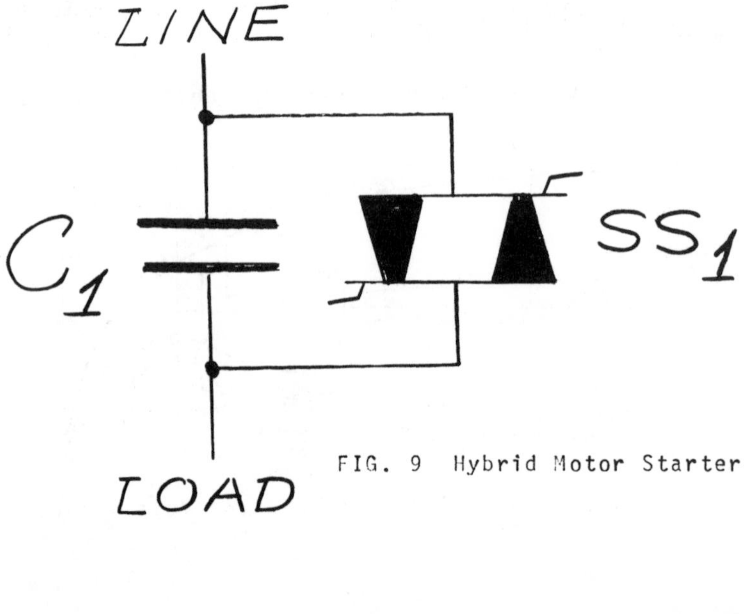

FIG. 9 Hybrid Motor Starter Concept

Section 3
Voltage Fed Inverter Drives

A Wide-Range Static Inverter Suitable for AC Induction Motor Drives

PAUL M. ESPELAGE, MEMBER, IEEE, JACOB A. CHIERA, AND FRED G. TURNBULL, MEMBER, IEEE

***Abstract*—A dc to ac, variable voltage, variable frequency, three-phase, external impulse commutated inverter, rated at 200 kVA, is described that controls the speed of an induction motor that is suitable for traction applications. The inverter is supplied from a variable voltage dc source for the load circuit and a fixed voltage dc source for the commutation circuit. Isolation between the two sources is provided by a high-frequency pulse transformer. The output ac voltage is controlled from 0 to 250 volts; the output frequency is controlled from 4 to 400 Hz. The load is a low-slip, 12 000 r/min induction motor with a continuous rating of 15 hp over a 10:1 speed range, and 60 hp over a 2:1 speed range. The theory of operation and design equations for the inverter power circuit, the design and construction of the motor, and the test results of the inverter–motor combination are presented.**

INTRODUCTION

OF THE MANY techniques for controlling the ac output voltage from a static dc to ac inverter, control of the dc input voltage is the most straightforward [1]–[15]. The line-to-line output voltage from a three-phase full-wave inverter circuit with 180° conduction of the inverter SCRs is the familiar 120° waveform shown on Fig. 1. The output voltage is shown for three voltages and frequencies, such that the volt–second area of each half-cycle is constant. Control of the ac voltage is provided by controlling the magnitude of the 120° voltage waveform. Control of the output frequency is provided by controlling the inverter frequency. The harmonic voltages present in the line-to-line voltage are composed of all of the odd harmonics of the fundamental except those divisible by three. The two lowest harmonics are the fifth and the seventh. The total harmonic distortion in the output voltage is a constant, and the percentage of each harmonic with respect to the fundamental is also constant over the entire range of fundamental frequency voltage control.

A diagram of a complete system is shown in Fig. 2. Control of the output frequency is provided in the dc to ac inverter; control of the output voltage is provided in the ac to variable dc voltage part of the system. It is relatively easy to provide the fixed dc voltage and variable dc voltage with a common negative with any of the three techniques shown in Fig. 2. With the proper control technique, the speed or torque of either an ac induction or an ac synchronous motor can be controlled.

THEORY OF OPERATION

One phase of the three-phase inverter is shown in Fig. 3. The fixed dc voltage for commutation and the variable dc voltage for the load circuit are shown. There are two inverters shown in Fig. 3, a square-wave inverter for the load, and a sine-wave inverter used to provide a source of commutating current impulses for the square-wave inverter. The square-wave inverter operates in a conventional manner; SCR_1 and SCR_2 are supplied with gating signals for approximately 180° of the fundamental output frequency. The oncoming SCR at the beginning of a half-cycle is not supplied with a gate signal until after the off-going SCR has been successfully commutated or turned off. Diodes D_1 and D_2 provide a path for the return of reactive current, (which is associated with the lagging power factor of the motor load) to the dc source. Therefore, point A (Fig. 3) is connected to either the positive dc bus through SCR_1 or D_1 or the negative dc bus through SCR_2 or D_2. Reactors L_3 and L_4 control the rate of change of SCR current during the first few microseconds after the SCR begins conduction. Capacitors C_3 and C_4, and transformer winding T_{1S} do not influence the operation of the inverter except during the commutating interval. When SCR_1 or D_1 is conducting, point B, Fig. 3, is approximately at the positive dc potential; when SCR_2 or D_2 is conducting, point B is approximately at the negative dc potential.

The sine-wave inverter shown as a source for commutating current pulses is a conventional series capacitor commutated inverter [1]–[3]. This portion of the circuit operates as follows: SCR_{1A} is turned on and a half sine wave of current flows through SCR_{1A}, L_1, T_{1P}, and divides equally into capacitors C_1 and C_2. This series L–C circuit is underdamped and the current pulse naturally drops to zero. The peak value of the current pulse is determined by the circuit constants (L and C) and the value of voltage on the series capacitors (C_1 and C_2). The width of the current pulse is determined by the circuit constants (L_1 in series with C_1 and C_2 in parallel). Since the circuit is underdamped, the common point of capacitors C_1 and C_2 is charged to a voltage above the positive input bus on one half-cycle and below the negative bus on the next half-cycle. At the end of the next half cycle of the output fre-

Paper 69 TP 16-IGA, approved by the Petroleum and Chemical Industry Committee of the IEEE IGA Group for presentation at the 1968 IEEE Industry and General Applications Group Annual Meeting, Chicago, Ill., September 29–October 3. Manuscript received February 11, 1969. This work was supported in part by the U.S. Army Mobility Equipment Research and Development Center under Contract DA-44-099-AMC-1820(T).

P. M. Espelage and F. G. Turnbull are with the General Electric Company, Schenectady, N.Y. 12305.

J. A. Chiera is with the General Electric Company, Erie, Pa.

Reprinted from *IEEE Trans. Ind. Gen. Appl.*, vol. IGA-5, pp. 438–445, July/Aug. 1969.

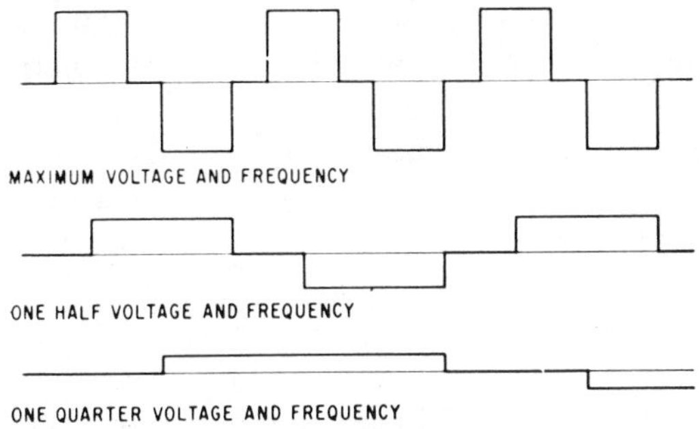

Fig. 1. Inverter line-to-line output voltage.

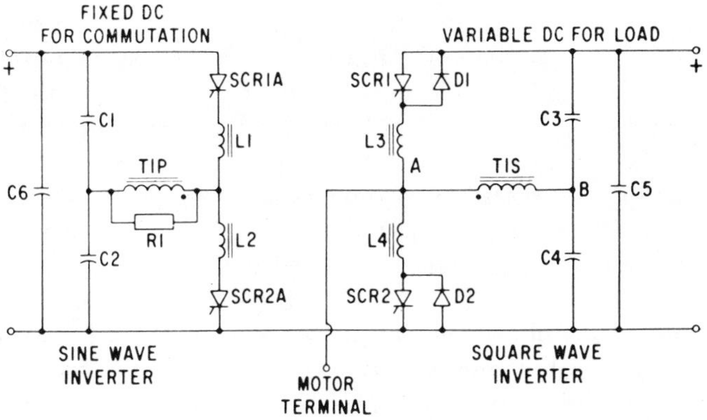

Fig. 3. Schematic diagram of one phase.

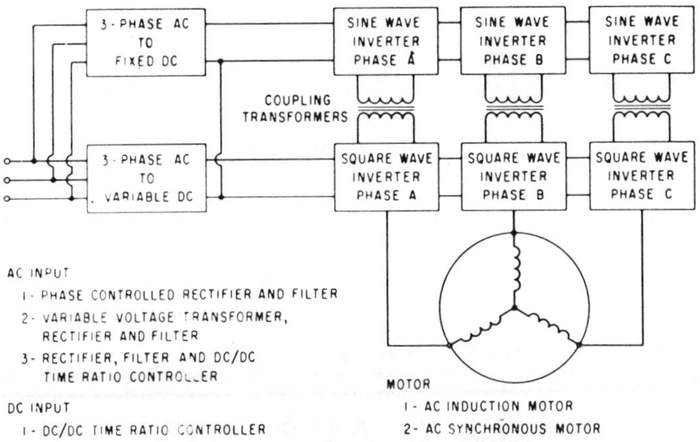

Fig. 2. Block diagram of system.

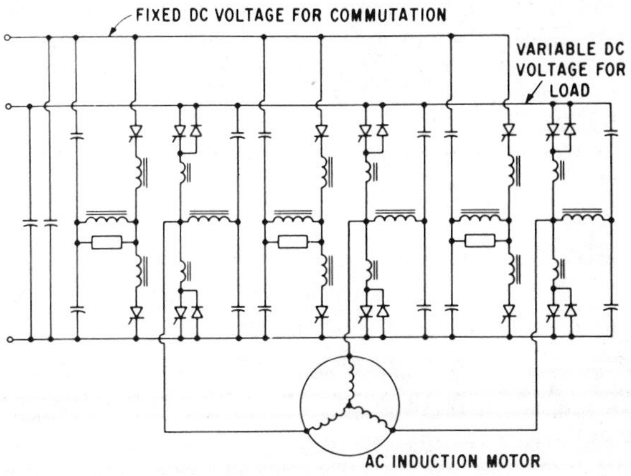

Fig. 4. Schematic diagram of three-phase system.

quency, SCR_{2A} is turned on and a half sine-wave pulse of current flows from capacitors C_1 and C_2, through T_{1P} in the opposite direction, L_2 and SCR_{2A}. These pulses of current flowing through T_{1P} are coupled into the square-wave inverter by means of a secondary winding, T_{1S}. Transformer T_1 is connected such that current flowing into the dot end of T_{1P} causes current to flow out of the dot end of T_{1S}. Assume that SCR_1 is conducting load current, SCR_{1A} is turned on and the commutating current builds up in T_{1P} and T_{1S}. When the instantaneous current out of the dot end of T_{1S} equals the instantaneous load current, the current in SCR_1 has been forced to zero. The current out of T_{1S} is designed to be greater than the largest value of instantaneous load current, and this excess current flows through diode D_1. The voltage drop of diode D_1 reverse biases SCR_1 for the appropriate length of time in order that SCR_1 can recover its forward voltage blocking capability. Capacitors C_3 and C_4 provide a path for the commutating current. Half of the current from T_{1S} flows into C_4 and half into C_3. When the instantaneous current supplied by T_{1S} again equals the instantaneous load current, which was assumed to be constant during the commutating interval, diode D_2 starts to conduct and supply the current demanded by the load. At this point in the sequence, SCR_2 is supplied with gate signals. However, with an inductive load it does not begin to conduct until later in the half-cycle at a time determined by the load power factor. During one commutation, transformer T_1 has experienced only a unidirectional half-cycle of current. At the end of the commutation interval, the transformer attempts to reset rather than wait for a commutating current pulse in the opposite direction at the end of the next half cycle of the fundamental output frequency. A circuit for reset exists in the combination of the open circuit inductance of transformer T_1 and capacitors C_3 and C_4. This reset manifests itself in a voltage oscillation on the nonconducting main controlled rectifier, for example, SCR_1. This relatively low frequency oscillation changes the voltage on the commutating capacitors (C_1 and C_2) so that the peak current in transformer T_1 is a function of the output frequency, undergoing successive maxima and minima. Resistor R_1 is used to damp out these oscillations in order that the peak current in the commutating inverter not be a function of the output frequency. The resistor also serves as a method for controlling the Q of the series L-C circuit and provides a path for current in the event of a fuse opening in the main inverter circuit. Transformer T_1 does not carry the load current, and its wire size can be calculated based on the rms value of the commutating current rather than on the commutating and the load currents. Its volt–second rating is based on the fixed supply voltage, the circuit Q, and the SCR turn-off time. The leakage reactance of transformer T_1 and the parallel combination of C_3 and C_4 are designed to provide an impedance equal to the equivalent load resistance calculated in (1). Since the sine-wave inverter operates as a current source, the peak current in the transformer secondary T_{1S} is not a function of the magnitude of the variable dc source voltage. Therefore, the peak commutating current is constant over the range of output frequencies (4–400 Hz) and the range of the variable input dc voltage (0–336 volts). The variable dc supply could even be greater than the fixed dc supply without a reduction in peak commutating current. The pulsewidth of the half-sinusoid of commutating current is also independent of supply voltage and output frequency. Three identical circuits similar to that shown in Fig. 3 were provided to supply a three-phase motor load. A complete power circuit schematic diagram is shown in Fig. 4.

Commutating Inverter Design

The basic design equations for a series capacitor commutated inverter are contained in [2] and [3]. In normal operation, the series capacitor commutated inverter is operated at a frequency close to the circuit resonant frequency. However, in this application the frequency of operation of the series capacitor commutated inverter is much lower than the circuit resonant frequency. The design equations for the peak currents and voltages can be used directly from the references, the equations for rms and average values of current and voltage must be modified in order to take into account the low repetition rate. Several requirements of the system must be known before the design can begin. The most important of these requirements are: 1) the maximum current that the main inverter SCRs must commutate, 2) the main SCR turn-off time at this value of peak current, 3) the maximum value of forward voltage rating of the auxiliary SCRs and 4) the maximum and minimum values of the fixed dc input voltage. In the particular case at hand the maximum main SCR current at commutation was specified as 570 amperes, the main SCR turn-off time was specified as 30 μs, the maximum forward voltage on the auxiliary SCRs was specified as 1100 volts, and the maximum dc input voltage was 450 volts, with a minimum value of 336 volts.

The peak current in the sine-wave inverter must be greater than the largest value of load current at the time of commutation for a length of time at least as long as the main SCR turn-off time. Suitable safety factors should include the effects of transformer magnetizing current and the current flowing through the damping resistor R_1 (Fig. 3).

The maximum current to be commutated by the main inverter SCRs may occur at the minimum value of the fixed dc input voltage (336 volts). Providing a half-sinusoid of current with a base width of twice the required main SCR turn-off time results in a value of peak current equal to $(\sqrt{2})$ (570) or approximately 800 amperes. Assuming that the peak current is a linear function of dc supply voltage results in a peak current of 800 (450/336) or approximately 1100 amperes at 450 volts.

The equivalent load resistance can be calculated from [3] as follows:

$$R = \frac{(2)(E_{dc})}{(1.11)(\pi)(i_{peak})}. \tag{1}$$

Substituting the values of E_{dc} and i_{peak} into (1) results in an effective load resistance of 0.234 ohm.

The circuit Q can be calculated as follows:

$$Q = \sqrt{\left(\frac{(e_{SCR})(\pi)^2}{(4)(E_{dc})}\right) - 1}. \tag{2}$$

Substituting the values of e_{SCR} and E_{dc} into (2) results in a Q of 1.65.

The resonant frequency is related to the SCR turn-off time requirement of 30 μs as follows:

$$f_r = \frac{1}{(2)(2)(\text{turn-off time})(10^{-6})}. \tag{3}$$

Substituting the value of turn-off time into (3) results in a resonant frequency of 8.34 kHz.

Once the effective load resistance, the Q, and the resonant frequency are known, the values of circuit constants can be calculated. The value of capacitance can be calculated from the following equation:

$$C = \frac{10^6}{(2\pi)(Q)(R)(f_r)}. \tag{4}$$

Substituting the values of Q, f_r, and R into (4) results in a value of C of 49.5 μF. The actual value of C is divided into two units C_1 and C_2 shown on Fig. 3. C_1 and C_2 are each equal to 25 μF, since they are in parallel and the total capacitance must equal approximately 50 μF from (4).

The value of inductance required can be calculated from the following equation:

$$L = \frac{(Q)(R)}{(2\pi)(f_r)} 10^6. \tag{5}$$

Substituting the values of Q, R, and f_r into (5) results in a value of L of 7.4 μH. The actual value of L_1 and L_2 shown in Fig. 3 was approximately 7.5 μH. No coupling was provided between these two air core inductors.

The peak capacitor voltage can be calculated from the following equation

$$V_{c\ peak} = (Q)(i_{peak})(R). \tag{6}$$

Substituting the values of Q, i_{peak}, and R into (6) results in a value of peak capacitor voltage of 426 volts. This is the ac voltage that is superimposed on the dc volt-

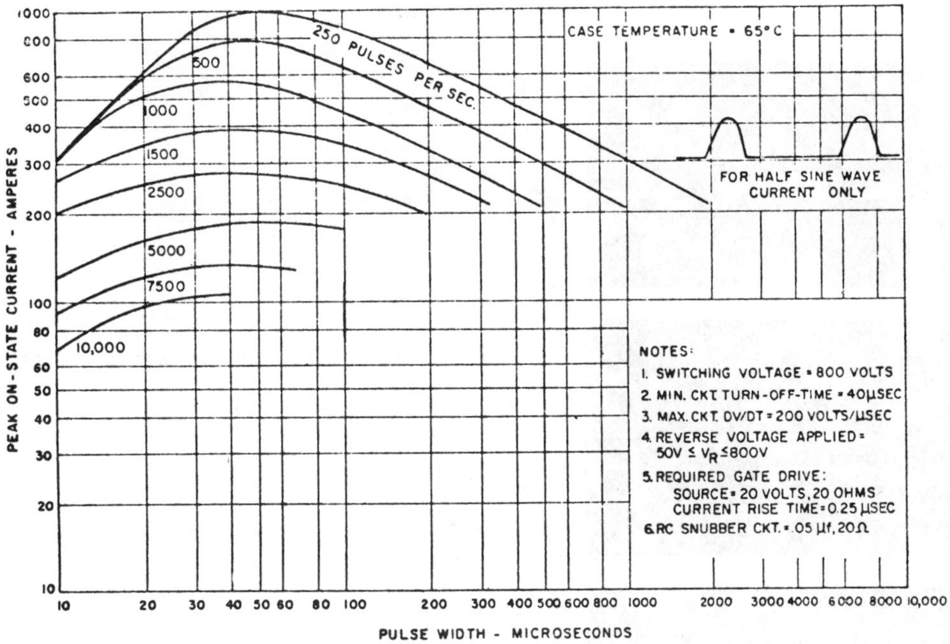

Fig. 5. SCR rating curve.

age equal to one-half of the supply voltage, and therefore, the center tap of C_1 and C_2 oscillates between +651 volts and −201 volts when the dc supply is equal to +450 volts.

The rms values of currents can be used to calculate wire sizes for the inductors and transformers, and to aid in the capacitor selection. The selection of a controlled rectifier can be accomplished by using a chart similar to that shown in Fig. 5 for the C158/C159 SCR. The chart shows the rating of this particular SCR when it is operated at various peak half-sinusoidal currents, base widths, and repetition rates for a particular SCR case temperature. This particular chart together with an energy loss per pulse chart and the design value of voltage, current, turn-off time, and other dynamic requirements enables the circuit designer to select a SCR to meet the circuit requirements.

The value of the damping resistor was chosen experimentally to be approximately twenty times the effective load resistance. This was a negligible loss over the entire range of output frequency at rated output. The energy storage capacitor C_6 (Fig. 3) was selected to have an rms current rating equal to one-half of the rms commutating capacitor current.

The pulse transformer T_1 (Fig. 3) must be rated to operate at the maximum value of voltage for one-half-cycle of the resonant frequency (8.34 kHz). Since this frequency is relatively high, the transformer can be made quite small. A one-to-one turns ratio between primary and secondary was selected since it was sufficient and made the transformer easier to construct with the desired close coupling between primary and secondary. A split C core was chosen because of its high value of magnetizing re-

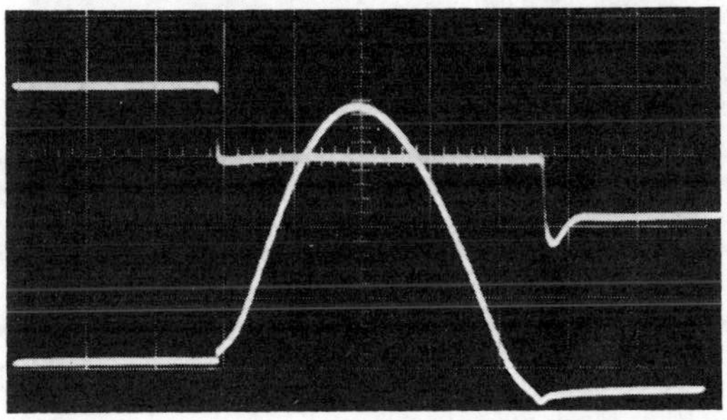

Fig. 6. Sine-wave inverter SCR voltage and current.

actance. This high value of reactance reduces the oscillation frequency of the transformer and capacitors C_3 and C_4 after the conclusion of the commutation interval.

The capacitors shown as C_3 and C_4 (Fig. 3) are required in order to provide a path for the commutating current which is transformer coupled from the sine-wave inverter. These capacitors should be sized so that they, in conjunction with the impedance of L_3 or L_4 and the leakage reactance of T_{1S}, provide an impedance equal to the effective load resistance at the resonant frequency. The units selected were 80 μF each. The auxiliary SCR current and voltage during the commutating interval are shown in Fig. 6.

Square-Wave Inverter Design

The square-wave inverter is composed of main SCRs (SCR_1 and SCR_2), feedback diodes D_1 and D_2, and rate of change of current reactors L_3 and L_4. The voltage rating of the SCRs and diodes should be equal to the maximum dc source voltage plus safety factors for overshoot caused by the inductors L_3 and L_4. The current ratings of the SCRs

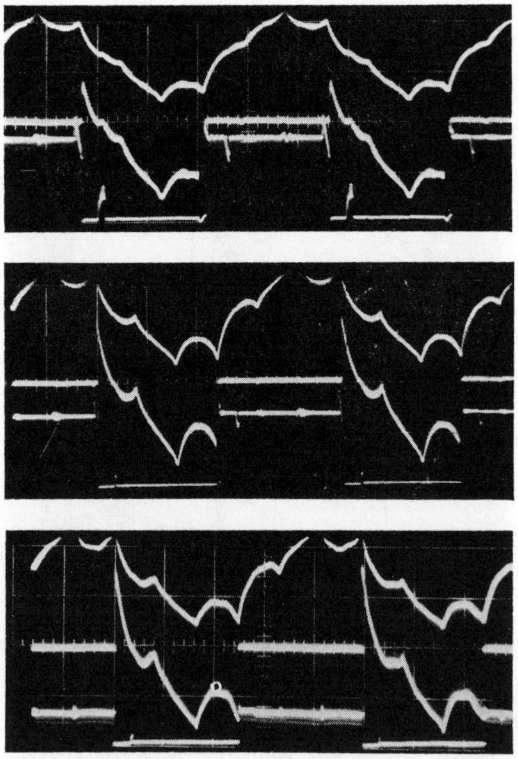

Fig. 7. Square-wave inverter SCR voltage and current and motor line current.

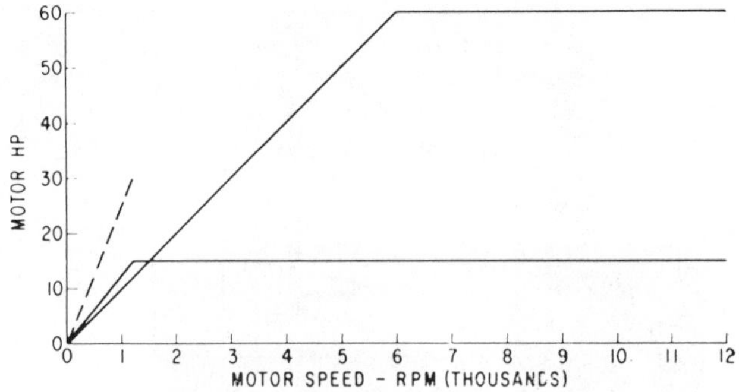

Fig. 8. Motor horsepower versus motor speed profile.

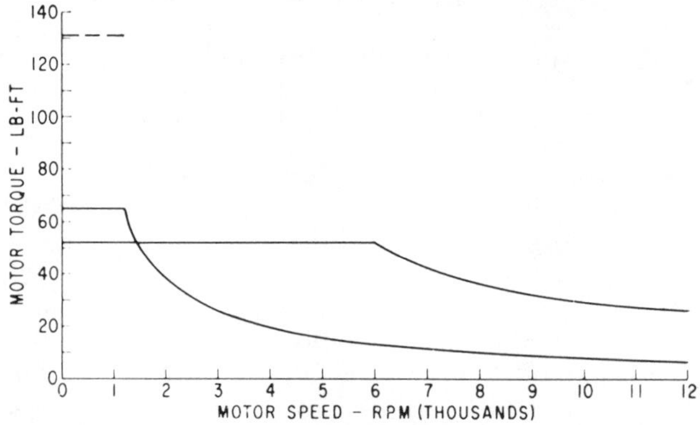

Fig. 9. Motor torque versus motor speed profile.

are determined by the load current and load power factor specifications. For a balanced resistive load, each SCR conducts load current for a 180° interval. For balanced loads with a power factor equal to zero lagging or zero leading, each SCR conducts for a 90° interval and its associated feedback diode conducts for the other 90° interval in a given half-cycle. With an induction motor load, the current waveform and SCR conduction intervals are more complex due to the difference in motor impedance to the various harmonics in the voltage waveform in addition to the nonlinearities caused by motor saturation. Several examples of SCR and feedback diode voltages and currents together with motor line current are shown in Fig. 7.

Motor Design Requirements

The motor was designed in accordance with U.S. Army Mobility Equipment Research and Development Center (MER&DC) requirements. A brief summary of important specification requirements are as follows:

1) horsepower envelope (Figs. 8 and 9):
   15hp at $40 \leq f \leq 400$ Hz
   60hp at $200 \leq f \leq 400$ Hz,
2) motor shall be capable of stalled operation for 2 minutes while delivering 131 lb-ft torque and not exceeding 380 amperes,
3) motor frequency range 4–400 Hz, maximum speed = 12 000 r/min,
4) motor shall have 3 phases, 4 poles,
5) maximum voltage 250 line–line rms,
6) life (minimum) 1500 hours,
7) temperature rise $\leq 150°$C (class H insulation) in an ambient of 125°F,
8) motor must be submersible,
9) motor must be water cooled,
10) within the constant horsepower range, efficiency (minimum) 82 percent and power factor (minimum) 78 percent,
11) motor must incorporate a digital speed indicating device,
12) motor target weight 180 pounds.

This motor was accepted by MER&DC early in 1968 after having performed successfully in conformance with their requirements.

Design and Construction

The stator assembly (Fig. 10) consisted of a core assembly shrunk into a steel barrel in which double-lead spiral grooves were cut for stator water cooling channels. The core assembly, made up of 14-mil laminations due to high-frequency operation, was held together with welds across the back of the punchings terminating in end rings at each end. The stator coils completing the stator assembly were form wound in the interests of weight and size reduction as well as for

Fig. 10. Motor stator.

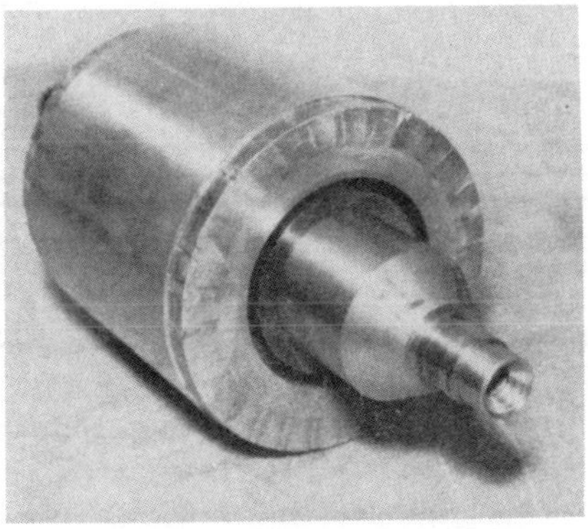

Fig. 11. Motor rotor.

Fig. 12. Complete motor assembly.

TABLE I

INVERTER AND MOTOR EFFICIENCY VERSUS MOTOR HORSEPOWER AND SPEED

Frequency (Hz)	Horsepower (hp)	Motor Efficiency (percent)	Inverter Efficiency (percent)
400	60	82.2	86.7
300	60	84.7	88.5
200	60	86.1	91.2
250	45	84.4	95.9
40	30	63.2	78.3
40	12	78.4	77.2
4	0 (locked rotor)	0.0	41.1

heat transfer and end winding strength considerations. The rectangular conductors were insulated with Kapton tape, while the polycoil had Nomex for ground insulation. Kapton string was used to lash all of the coils together to minimize end winding vibration since a wide spectrum of magnetic exciting forces is encountered in the frequency range of this application. Following assembly, the stator was vacuum pressure impregnated with varnish.

The rotor assembly (Fig. 11) consisted of a core assembled on a fabricated hollow shaft with a shrink fit. Rotor core punchings were held together by zirconium copper end rings which were brazed to copper bars. The entire rotor core surface was machined to a smooth finish to minimize windage and surface core losses.

Water circulation was also employed in the rotor to cool the core and bearings. Water entered the end of the rotor shaft opposite the drive and through the inlet pipe, passed into the hollow shaft by means of a nozzle, and then reverse flowed out of the shaft through an annular passage around the nozzle. A carbon face type rubbing seal was used to prevent water from entering the machine. It has proved capable of effective sealing at a pressure of 70 psig which is almost double the pressure required for maximum rotor flow ($2^1/_2$ gal/min). The machine was equipped with a built in tachometer consisting of a digital speed pickup gear on the shaft and a stationary magnetic speed pickup head mounted in the connection end framehead. Both the connection end and spline drive end frameheads were constructed of aluminum in order to minimize weight.

There are two 25-mm ball bearings in the machine. Precision bearings were selected because of the high speed. The floating end bearing at the motor connection end was preloaded (about 20 pounds) in order to prevent skidding at high speeds and acceleration. Bearings were lubricated with Andok C grease. The complete motor assembly can be seen in Fig. 12.

INVERTER–MOTOR TEST DATA

Table I summarizes the inverter and motor efficiency at the extremes of the horsepower and torque versus speed profile.

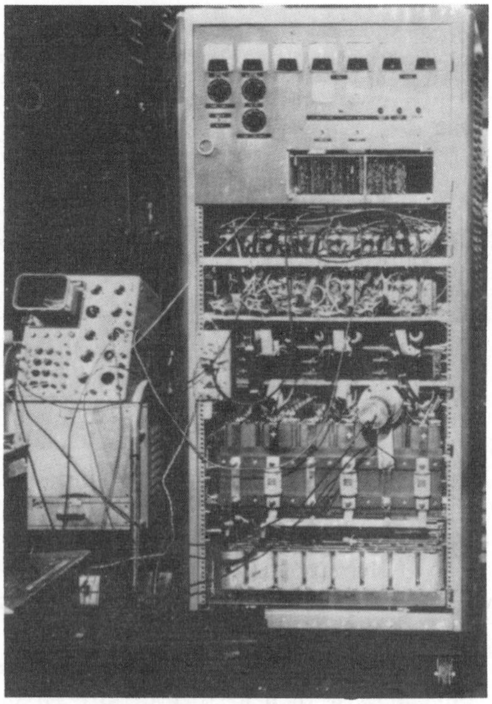

Fig. 13. Inverter showing input capacitor, commutating capacitors, pulse transformers, and square-wave semiconductors.

Fig. 14. Inverter showing sine-wave semiconductors.

As shown in Table I, the motor efficiency was comparable to the inverter's efficiency. At the 400-Hz 60-hp load point, the overall efficiency from dc power input to low-speed shaft output was 72 percent. Numerous load points were obtained at various frequencies, voltages, and load torques in order to characterize the system and determine the optimum conditions for the complete system.

At the 400-Hz 60-hp load point, the inverter output voltage was 250 volts rms. At the 40-Hz 30-hp load point, the inverter output current was 465-amperes rms. These two outputs determine the maximum capability of the inverter, which is approximately 200 kVA.

Figs. 13 and 14 show the inverter during test and evaluation. Fig. 13 shows the input capacitor, commutating capacitors, pulse transformers, and square-wave inverter semiconductors from bottom to top. Fig. 14 shows the sine-wave inverter semiconductors. The total weight of the electrical components is approximately 300 pounds.

Conclusions

The paper has described an external impulse commutated inverter that can commutate a fixed value of load current when supplied from a variable source of dc voltage and operated over a 100:1 range in frequency. A fixed voltage supply is provided for commutation. The two dc power supplies are coupled through a high-frequency pulse transformer; there is no direct current path between the two power sources. The design equations and SCR selection criteria are described and the component selection for a 200-kVA system is presented. The design and construction details of the low slip induction motor are described. Test data of inverter and motor efficiency is presented during conditions of operation over a wide range of speed, torque, and horsepower. The system is useful for ac motor drive applications requiring a wide range in frequency, voltage, speed, and torque control.

References

[1] B. D. Bedford and R. G. Hoft, *Principles of Inverter Circuits*. New York: Wiley, 1964.

[2] F. W. Gutzwiller, Ed., *Silicon Controlled Rectifier Manual*, 4th ed. Syracuse, N.Y.: General Electric Co., 1967.

[3] N. Mapham, "An SCR circuit to produce high frequency power," General Electric Co., Application Note 200.24, June 1962.

[4] I. M. MacDonald, "A static inverter, wide range, adjustable speed drive," *1964 IEEE Internatl. Conv. Rec.*, vol. 12, pt. 4, pp. 34–41.

[5] K. Heumann, "Pulse control of dc and ac motors by silicon controlled rectifiers," *IEEE Trans. Communications and Electronics*, vol. 83, pp. 390–399, July 1964.

[6] D. A. Bradley, C. D. Clarke, R. M. Davis, and D. A. Jones, "Adjustable frequency inverters and then application to variable speed drives," *Proc. IEE* (London), vol. 111, pp. 78–83, November 1964.

[7] A. J. Humphrey, "Precise speed control with inverters," *1965 IEEE Industrial Static Power Conversion Conf. Rec.*, pp. 78–83.

[8] E. Ohno and M. Akamatsu, "Variable frequency SCR inverter with an auxiliary commutation circuit," *IEEE Trans. Magnetics*, vol. MAG-2, pp. 25–30, March 1966.

[9] F. G. Turnbull, "Wide range impulse commutated, static inverter with a fixed commutation circuit," *Conf. Rec. 1966 IEEE IGA Group Annual Meeting*, pp. 475–482.

[10] A. J. Humphrey, "Inverter commutation circuits," *IEEE Trans. Industry and General Applications*, vol. IGA-4, pp. 104–110, January/February 1968.

[11] H. Stemmler, "The use of static converters to vary the speed of three phase drives without losses," *Brown Boveri Rev.*, vol. 54, pp. 217–232, M[illegible]ne 1967.

[12] J. T. Salihi, P. D. [illegible]wal, and G. S. Spix, "Induction motor control scheme for a battery-powered electric car (GM-Electrovair I)," *IEEE Trans. Industry and General Applications*, vol. IGA-3, pp. 463–469, September/October 1967.

[13] C. D. Beck and E. F. Chandler, "Motor drive inverter ratings," *IEEE Trans. Industry and General Applications*, vol. IGA-4, pp. 589–595, November/December 1968.

[14] P. D. Agarwal, "The GM high-performance induction motor drive system," *IEEE Trans. Power Apparatus and Systems*, vol. PAS-88, pp. 86–93, February 1969.

[15] R. W. Johnston, "Modulating inverter system for variable speed induction motor drive (GM Electrovair II)," *IEEE Trans. Power Apparatus and Systems*, vol. PAS-88, pp. 81–85, February 1969.

Analysis and Simplified Representations of a Rectifier-Inverter Induction Motor Drive

PAUL C. KRAUSE, SENIOR MEMBER, IEEE, AND THOMAS A. LIPO, MEMBER, IEEE

Abstract—**Simplified representations of a rectifier-inverter induction motor drive system are established and verified by comparing the results obtained from a computer study using these representations to those obtained using a detailed simulation of the system. It is shown that when all harmonic components are neglected the static drive system may be conveniently represented in the synchronously rotating reference frame. The computer simulation resulting from this type of representation can be readily implemented, and in many cases it will predict the system performance with sufficient accuracy. Also, in the analysis leading to these simplified representations, the operation of the inverter is analytically expressed in the synchronously rotating reference frame with the harmonic components due to the switching in the inverter included. These equations of transformation may be used to advantage in describing the interaction between the filter and the induction motor.**

Introduction

Cycloconverters, inverters, and rectifier-inverter systems are being used with ac machines in an increasing number of variable speed applications. There are many aspects of these static drive systems which offer interesting and challenging problems. However, most variable frequency drive systems are quite complex, and it is often difficult to predict their dynamic performance without the aid of a computer. Consequently, the design of control systems associated with static drive systems may be facilitated by appropriate use of either an analog or a digital computer.

In earlier publications the analog computer simulation of a rectifier-inverter induction motor drive system was developed and verified [1], [2]. This simulation was recently used to verify the stability analysis of the static drive system at low frequencies [3]. However, this analog computer simulation is involved and requires a substantial amount of computing equipment. Therefore, it is desirable to establish simplified representations which yield a more direct means of simulating the drive system while maintaining the salient features which determine its performance. The analytical development and the verification of simplified representations of the rectifier-inverter induction motor drive system is the subject of this paper.

Paper 68 TP 640-PWR, recommended and approved by the Rotating Machinery Committee of the IEEE Power Group for presentation at the IEEE Summer Power Meeting, Chicago, Ill., June 23–28, 1968. Manuscript submitted February 12, 1968; made available for printing May 6, 1968. This work was supported by the U. S. Army Mobility Equipment Research and Development Center, Fort Belvoir, Va., under Contract DAAK02-67-C-0073.

P. C. Krause is with the Department of Electrical Engineering, University of Wisconsin, Madison, Wis. 53706.

T. A. Lipo was with the Department of Electrical Engineering, University of Wisconsin, Madison, Wis. He is now with the Department of Electrical and Electronic Engineering, University of Manchester Institute of Science Technology, Manchester, England.

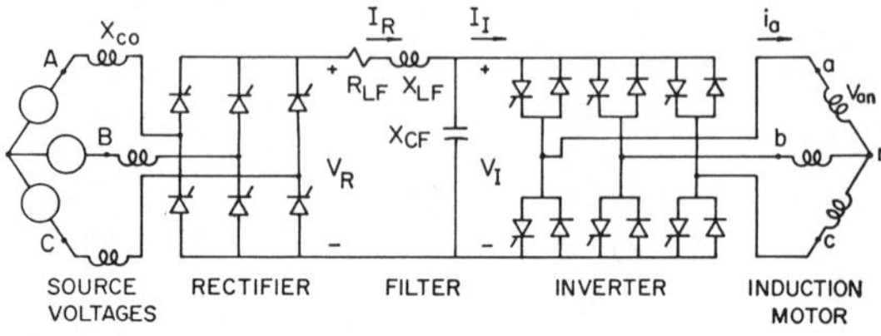

Fig. 1. System studied.

The first simplified system representation is developed by neglecting the harmonic components due to the switching in the rectifier. Next, the operation of the inverter and induction motor is expressed in a reference frame which rotates in synchronism with the fundamental component of the stator applied voltages. This analysis yields equations of transformation for the inverter and establishes an equivalent circuit which describes the static drive system in a synchronously rotating reference frame with only the harmonic components due to the switching in the rectifier neglected. This equivalent circuit conveniently describes the interaction between the filter and the motor. Moreover, a markedly simplified representation results from this equivalent circuit if the harmonic components due to the switching in the inverter are also neglected.

The analysis set forth in this paper yields two simplified system representations which may be simulated on the analog computer more directly than the detailed computer simulation of the complete system. As mentioned, the first of these result from the system representation wherein the harmonic components due to rectifier switching are neglected. The second is obtained by neglecting all harmonic components and employing the representation of the system in the synchronously rotating reference frame. The validity and limitations of each of these simplified representations are established by comparing the results obtained from the analog computer simulation of these representations with those obtained from the detailed analog computer simulations of the complete system.

System Description and Basic Equations

A simplified diagram of the rectifier-inverter drive system is given in Fig. 1. Similar systems were studied in two previous papers and in a companion paper [1], [3], [4]. This system consists of a three-phase power source, a six-phase rectifier, a filter, an inverter, and a three-phase symmetrical induction machine. Although there are various converter configurations and control systems being employed, the system shown in Fig. 1 forms the basis of many of the present-day rectifier-inverter drive systems.

The equations which describe the symmetrical induction machine in the arbitrary reference frame may be expressed [5]–[7] as follows:

Reprinted from *IEEE Trans. Power App. Syst.*, vol. PAS-88, pp. 588–596, May 1969.

$$
\begin{bmatrix} v_{qs} \\ v_{ds} \\ v_{qr}' \\ v_{dr}' \end{bmatrix} = \begin{bmatrix} r_s + (p/\omega_b)X_s & (\omega/\omega_b)X_s & (p/\omega_b)X_m & (\omega/\omega_b)X_m \\ -(\omega/\omega_b)X_s & r_s + (p/\omega_b)X_s & -(\omega/\omega_b)X_m & (p/\omega_b)X_m \\ (p/\omega_b)X_m & (\omega - \omega_r/\omega_b)X_m & r_r' + (p/\omega_b)X_r' & (\omega - \omega_r/\omega_b)X_r' \\ -(\omega - \omega_r/\omega_b)X_m & (p/\omega_b)X_m & -(\omega - \omega_r/\omega_b)X_r' & r_r' + (p/\omega_b)X_r' \end{bmatrix} \times \begin{bmatrix} i_{qs} \\ i_{ds} \\ i_{qr}' \\ i_{dr}' \end{bmatrix} \tag{1}
$$

where

$$X_s = X_{ls} + X_m \tag{2}$$

$$X_r' = X_{lr}' + X_m. \tag{3}$$

The electromagnetic torque, expressed positive for motor action, is

$$T = \left(\frac{n}{2}\right)\left(\frac{P}{2}\right)\left(\frac{X_m}{\omega_b}\right)(i_{qs}i_{dr}' - i_{ds}i_{qr}') \tag{4}$$

where

- p operator d/dt
- r_s stator resistance
- r_r' rotor resistance (referred to the stator winding)
- X_{ls} stator leakage reactance
- X_{lr}' rotor leakage reactance
- n number of phases
- P number of poles.

The base electrical angular velocity ω_b is introduced in the machine equations for the purpose of converting inductances to inductive reactances whereupon a per unit system may be conveniently employed. It is clear that $v_{qr}' = v_{dr}' = 0$ if the machine is singly fed.

In (1) the electrical angular velocity of the arbitrary rotating reference frame is denoted as ω. If it is desirable to express the equations of the induction machine in a stationary reference frame, ω is set equal to zero. For a reference frame fixed in the rotor ω is set equal to the electrical angular velocity of the rotor ω_r. The equations in the synchronously rotating reference frame are obtained by setting ω equal to the electrical angular velocity of the fundamental frequency components of the applied stator voltages herein denoted as ω_e.

The notation employed in this paper differs slightly from that used in earlier publications involving induction machines [5], [8]. In these papers, as in this paper, ω_e is used to denote the speed of the synchronously rotating reference frame which corresponds to the frequency of the stator applied voltages. During constant rated frequency operation ω_e would generally be equal to ω_b. However, during variable frequency operation ω_e varies with the frequency of the applied voltages. The notation employed in [5] was selected without consideration for variable frequency operation. That is, ω_e was used in the induction machine equations to denote the constant base electrical angular velocity (herein denoted as ω_b), and it was also used to denote the speed of the synchronously rotating reference frame.

In the discussion of Jordan's paper, it is demonstrated that if the applied stator voltages form a balanced-set variable-frequency operation can be simulated in the synchronously rotating reference frame by continuously changing the speed of the reference frame to correspond to the frequency of the applied voltages [8]. The previous use of ω_e as a constant to convert an inductance to an inductive reactance is misleading during variable frequency operation. However, this situation may be easily resolved if all ω_e appearing in (54)–(68), (79), (80), and (103)–(126), as well as in the computer representation shown in Fig. 6 of [5] are changed to ω_b. The computer representation of the symmetrical machine in the arbitrary reference frame convenient for variable-frequency applications is given in Fig. 2. Therein

$$\psi_{qs} = X_{ls}i_{qs} + \psi_{mq} \tag{5}$$

$$\psi_{ds} = X_{ls}i_{ds} + \psi_{md} \tag{6}$$

$$\psi_{qr}' = X_{lr}'i_{qr}' + \psi_{mq} \tag{7}$$

$$\psi_{dr}' = X_{lr}'i_{dr}' + \psi_{md} \tag{8}$$

$$\psi_{mq} = X_{mq}\left(\frac{\psi_{qs}}{X_{ls}} + \frac{\psi_{qr}'}{X_{lr}'}\right) \tag{9}$$

$$\psi_{md} = X_{md}\left(\frac{\psi_{ds}}{X_{ls}} + \frac{\psi_{dr}'}{X_{lr}'}\right) \tag{10}$$

$$X_{ls} = \omega_b L_{ls}, \text{ etc.} \tag{11}$$

$$X_{mq} = X_{md} = \frac{1}{1/X_m + 1/X_{ls} + 1/X_{lr}'}. \tag{12}$$

The torque is expressed

$$T = \left(\frac{n}{2}\right)\left(\frac{P}{2}\right)\left(\frac{1}{\omega_b}\right)(\psi_{qr}'i_{dr}' - \psi_{dr}'i_{qr}'). \tag{13}$$

Hereafter, ω_e will be reserved to denote only the electrical angular velocity of the synchronously rotating reference frame.

The voltage equation for the filter during continuous operation ($I_R > 0$) may be written

$$V_R = V_I + [(p/\omega_b)X_{LF} + R_{LF}]I_R \tag{14}$$

$$V_I = (\omega_b/p)X_{CF}(I_R - I_I). \tag{15}$$

It is clear that the rectifier current I_R cannot be negative. When V_I exceeds V_R and I_R is forced to zero, V_R becomes equal to V_I, that is, when

$$I_R = 0 \tag{16}$$

$$V_R = V_I. \tag{17}$$

A detailed analog computer simulation of the complete system shown in Fig. 1 is described in [1] and [2], and results obtained from a computer study are compared with test results. This detailed simulation will not be repeated. However, the computer results obtained using this simulation will be compared with those obtained using the simplified representations presented in the following sections.

In regard to the development of the detailed computer simulation, it is important to mention that C. H. Thomas, G. E. Gareis, and R. A. Hedin were instrumental in the development of the simulation of the rectifier. Also L. T. Woloszyk contributed to the simulation of the rectifier-filter-inverter combination [1], [2].

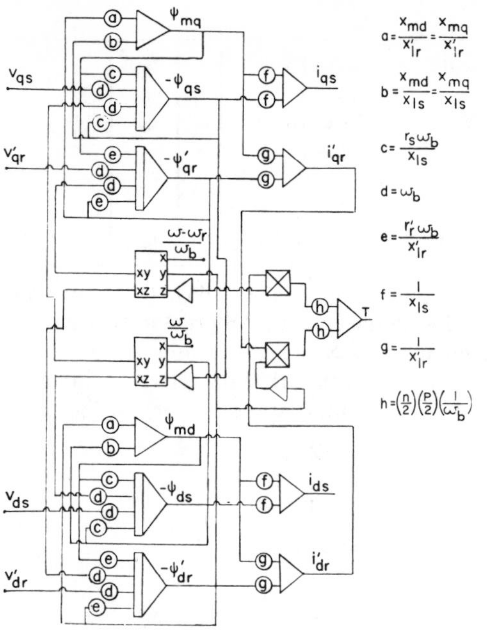

Fig. 2. Computer representation of symmetrical induction machine; arbitrary reference frame.

Simplified Representation of Rectifier with Machine Simulated in Stationary Reference Frame

The complete computer simulation of the static drive system is rather involved and requires a substantial amount of computing equipment [1], [2]. However, by employing a functional representation for the rectifier the complexity of the simulation may be reduced while retaining many of the salient features of the actual system. A suitable functional representation of the rectifier was reported in a recent paper [9]. In particular, an equation which expresses the average output voltage of a six-phase converter in a d–q axis is set forth and verified. This relationship was established by substituting equations of transformation into the equations which describe the operation of the converter. If the commutating reactance X_{co} is small compared to the filter reactance X_{LF}, the average output voltage V_R may be expressed as follows:

$$V_R = (3\sqrt{3}/\pi)V_S \cos\alpha - (3/\pi)X_{co}I_R \tag{18}$$

where V_S is the magnitude of the line-to-neutral source voltage and α the delay angle. It is clear that (18) is the familiar expression for the average output voltage of a six-phase converter. However, a slightly modified interpretation is possible. That is, if the ac voltages at the rectifier are appropriately transformed to a synchronously rotating reference frame which corresponds to the frequency of these voltages, the q-axis voltage is equal to V_S while the d-axis voltage is maintained at zero. This transformation can be readily simulated if the source voltages (e_{GA}, e_{GB}, and e_{GC}) are independent of load current. However, if it is necessary to include source impedance other than the commutating reactance, the simulation is more involved [9]. Source impedances other than the commutating reactance will not be considered in this paper.

The control system which establishes the delay angle α may also be incorporated using techniques described in [9]. Consequently (18) describes the average rectifier voltage which may be related to the reference frame which rotates in synchronism with the electrical angular velocity of the source voltages e_{GA}, e_{GB}, and e_{GC}. It is clear that all variables (voltages and currents) of the dc system (filter) may be interpreted directly, without transformation.

In this simulation the inverter is represented as described in [1] and [2]. In particular, the operating frequency of the inverter is established from a variable-frequency sine–cosine oscillator. This oscillator is used to establish a three-phase set of voltage signals displaced 120 electrical degrees. These voltage signals are then used to operate three comparator relays which simulate the switching of the machine terminals to the appropriate capacitor terminal. Commutating circuitry is not included in the simulation of the inverter, that is, it is assumed that commutation occurs instantaneously. In this type of static drive system, the applied phase voltages of the induction machine will be of stepped waveform. When the applied voltages of the induction machine are not a balanced sinusoidal three-phase set, it is desirable to represent the machine in the stationary reference frame ($\omega = 0$) [5].

In summary, this method of simulation differs from the complete simulation of the system in that the rectifier is replaced by a functional representation. The filter, inverter, and the induction machine are simulated using the same techniques employed in the complete simulation [1], [2].

Representation in the Synchronously Rotating Reference Frame

The stepped voltages applied to the stator phases of the induction machine may be approximated by Fourier series expansions. For example, during normal balanced operation, phase A of the three-phase set of stepped voltages may be expressed [6] as follows:

$$v_{as} = (2V_I/\pi)(\cos\omega_e t + \tfrac{1}{5}\cos 5\omega_e t - \tfrac{1}{7}\cos 7\omega_e t - \cdots). \tag{19}$$

It is clear that during normal operation the stator-phase voltages may be considered as a series of three-phase sets formed by the fundamental and the 5th, 7th, 11th, 13th, ... harmonic components. Moreover, the amplitude of each of these balanced three-phase sets is determined by the instantaneous value of the capacitor voltage V_I. That is, the fundamental and the harmonic components of the phase voltages each may be considered as sinusoidal functions modulated by the instantaneous value of V_I.

If the stepped-phase voltages are transformed to a reference frame which rotates in synchronism with the fundamental frequency of the applied stator voltages v_{qs}^e and v_{ds}^e become [3], [6]

$$v_{qs}^e = (2V_I/\pi)(1 + \tfrac{2}{35}\cos 6\omega_e t - \tfrac{2}{143}\cos 12\omega_e t + \cdots) \tag{20}$$

$$v_{ds}^e = (2V_I/\pi)(\tfrac{12}{35}\sin 6\omega_e t - \tfrac{24}{143}\sin 12\omega_e t + \cdots). \tag{21}$$

In the above equations the angular relationship between the q axis and the magnetic axes of the stator and rotor phases has been selected so that these axes coincide at time zero [5]–[7]. It is clear that the speed of the synchronously rotating reference frame ω_e is determined by the frequency of the fundamental component of the applied voltages which in turn is determined by the frequency at which switching (commutation) is caused to occur in the inverter.

If it assumed that there is no power loss in the inverter,

$$V_I I_I = \tfrac{3}{2}(v_{qs}^e i_{qs}^e + v_{ds}^e i_{ds}^e). \tag{22}$$

If (20) and (21) are substituted into (22), the inverter current may be expressed as follows:

$$I_I = (3/\pi)i_{qs}^e(1 + \tfrac{2}{35}\cos 6\omega_e t - \tfrac{2}{143}\cos 12\omega_e t + \cdots) + (3/\pi)i_{ds}^e(\tfrac{12}{35}\sin 6\omega_e t - \tfrac{24}{143}\sin 12\omega_e t + \cdots). \tag{23}$$

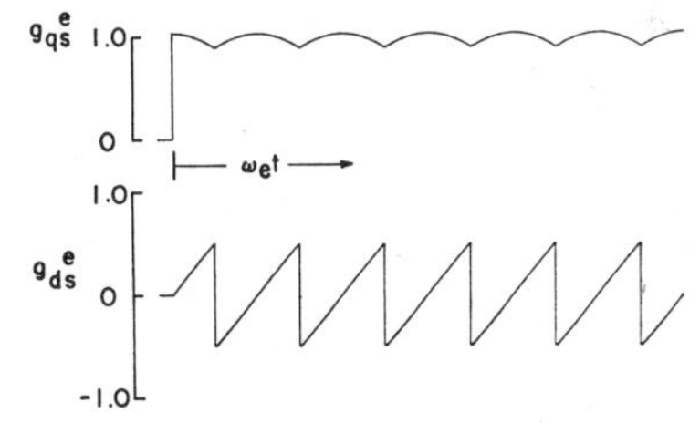
Fig. 3. g_{qs}^e and g_{ds}^e.

It is convenient to define

$$g_{qs}^e = 1 + \tfrac{2}{35}\cos 6\omega_e t - \tfrac{2}{143}\cos 12\omega_e t + \cdots \tag{24}$$

$$g_{ds}^e = \tfrac{12}{35}\sin 6\omega_e t - \tfrac{24}{143}\sin 12\omega_e t + \cdots. \tag{25}$$

Equations (20), (21), and (23) may now be written

$$v_{qs}^e = (2/\pi)V_I g_{qs}^e \tag{26}$$

$$v_{ds}^e = (2/\pi)V_I g_{ds}^e \tag{27}$$

$$I_I = (3/\pi)(i_{qs}^e g_{qs}^e + i_{ds}^e g_{ds}^e). \tag{28}$$

The above expressions of g_{qs}^e and g_{ds}^e are Fourier series expansions of the functions shown in Fig. 3. The maximum value of g_{qs}^e and g_{ds}^e is $\pi/3$ and $\pi/6$, respectively. It is interesting to note that the waveform of g_{qs}^e is analogous to the output of an ideal six-phase rectifier which is operating without phase delay and with zero commutating reactance. Similarly, the waveform of g_{ds}^e is analogous to the negative of the continuous output of an ideal six-phase rectifier with a 90-degree phase delay and with zero commutating reactance.

It is desirable to incorporate the following substitute quantities which, in effect, refers the variables to the stator winding of the induction machine.

$$V_I' = (2/\pi)V_I \tag{29}$$

$$I_I' = (\pi/3)I_I \tag{30}$$

$$R_{LF}' = (6/\pi^2)R_{LF} \tag{31}$$

$$X_{LF}' = (6/\pi^2)X_{LF} \tag{32}$$

$$X_{CF}' = (6/\pi^2)X_{CF}. \tag{33}$$

Substituting (29) and (30) into (22) yields

$$\tfrac{3}{2}V_I'I_I' = \tfrac{3}{2}(v_{qs}^e i_{qs}^e + v_{ds}^e i_{ds}^e). \tag{34}$$

The following substitute variables are introduced for the purpose of referring the rectifier variables to the stator winding of the induction machine:

$$V_R' = (2/\pi)V_R \tag{35}$$

$$I_R' = (\pi/3)I_R. \tag{36}$$

Thus (14) and (15) may be written

$$V_R' = V_I' + [(p/\omega_b)X_{LF}' + R_{LF}']I_R' \tag{37}$$

$$V_I' = (\omega_b/p)X_{CF}'(I_R' - I_I'). \tag{38}$$

The simplified representation for the rectifier developed in the previous section is also used to describe the operation of the rectifier in the representation developed in this section. With the appropriate substitute variables introduced, (18) may be expressed

$$V_R' = V_{RO}' - X_{co}'I_R' \tag{39}$$

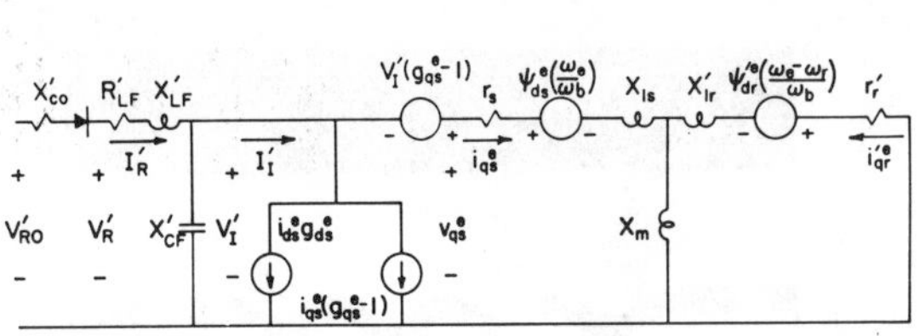
(a)

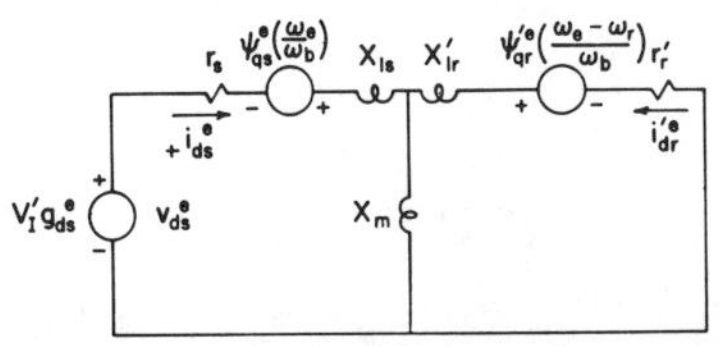
(b)

Fig. 4. Equivalent circuit in synchronously rotating reference frame with harmonic components due to rectifier switching neglected. (a) q axis. (b) d axis.

where

$$V_{RO}' = (6\sqrt{3}/\pi^2)V_S\cos\alpha \tag{40}$$

$$X_{co}' = (18/\pi^3)X_{co}. \tag{41}$$

Incorporating (24)–(41) with the equation which describes the induction machine in the synchronously rotating reference frame yields the equivalent circuit shown in Fig. 4. This equivalent circuit describes the operation of the rectifier-inverter induction motor drive system in the synchronously rotating reference frame with the harmonic components due to the rectifier switching neglected. The electrical angular velocity of the synchronously rotating reference frame ω_e corresponds to the frequency of the fundamental component of the applied voltages which is determined by the frequency at which switching occurs in the inverter. Thus changing ω_e in Fig. 4 corresponds to changing the frequency at which switching occurs in the inverter. In Fig. 4 it is clear that

$$\psi_{ds}^e = X_{ls}i_{ds}^e + X_m(i_{ds}^e + i'^e_{dr}) \cdots. \tag{42}$$

Although the equivalent circuit shown in Fig. 4 is of importance in that it permits one to describe conveniently the interaction between the filter and the motor, it does not yield a computer simulation which is more convenient or more readily implemented than the one described in the preceding section. If, however, the harmonic components due to the switching of the inverter are neglected, the equivalent circuit and thus the computer representation are markedly simplified. That is, if

$$g_{qs}^e = 1 \tag{43}$$

$$g_{ds}^e = 0 \tag{44}$$

then

$$v_{qs}^e = V_I' \tag{45}$$

$$v_{ds}^e = 0 \tag{46}$$

$$I_I' = i_{qs}^e. \tag{47}$$

Equations (45)–(47) describe the operation of an idealized inverter which supplies the induction machine with only a fundamental set of balanced three-phase variable-frequency voltages. Therefore, if the speed of the synchronously rotating reference frame appearing in the equations for the induction machine is changed to correspond to the frequency of this balanced three-phase set ($\omega = \omega_e$), the applied voltages in this synchronously rotating reference frame are related directly to the capacitor voltage.

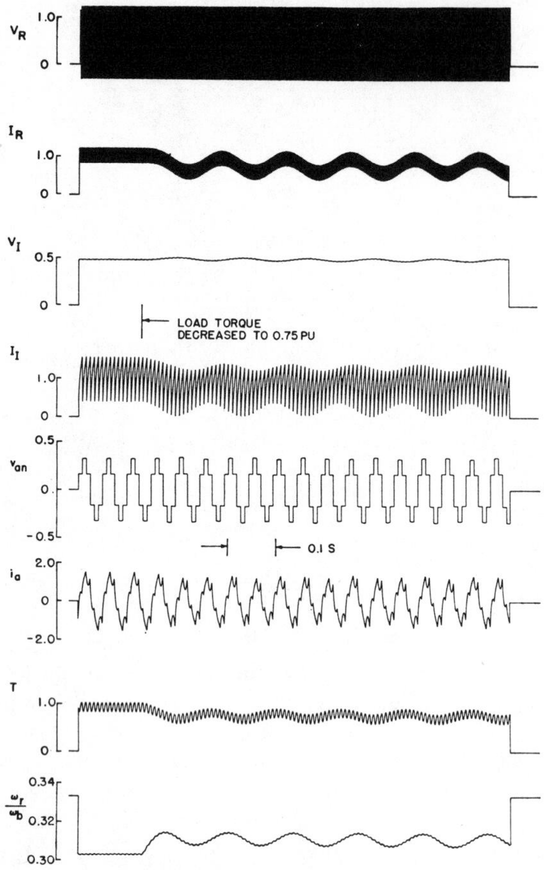

Fig. 5. Load torque switching from 0.925 to 0.75 pu; operation at 20 Hz. Detailed system representation.

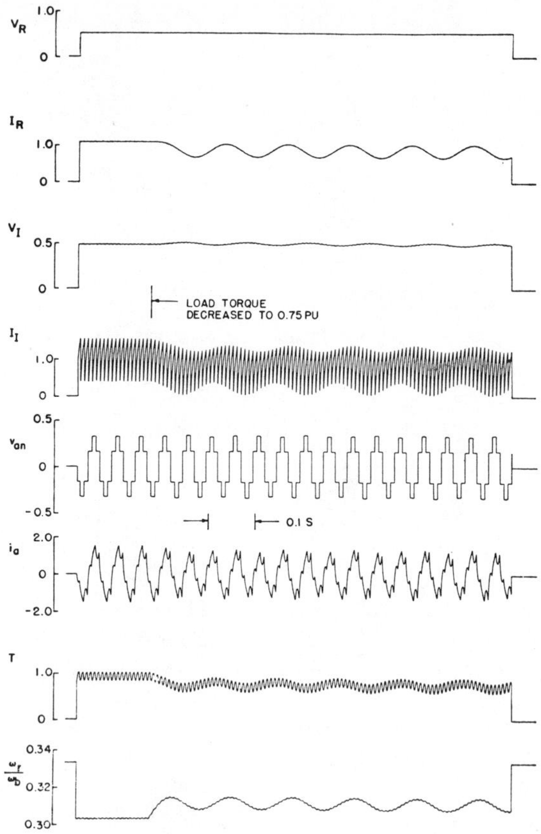

Fig. 6. Load torque switching from 0.925 to 0.75 pu; operation at 20 Hz. Functional representation of rectifier.

It is clear, however, that in general both i_{qs}^e and i_{ds}^e will be made up of a constant and a series of harmonic components. Due to the selection of the time-zero position of the axes, the constant component of i_{ds}^e corresponds to the magnitude of the magnetizing or reactive component of the fundamental phase current. It is important to note that the current i_{ds}^e appears only as the coefficient of the 6th, 12th, ... harmonics in the expression for the inverter current (23). Thus the fundamental magnetizing current is supplied to the machine by the harmonic components (predominantly the 6th harmonic) of the inverter current. Therefore, if the harmonic components are neglected, as in (43)–(47), the effect of the magnetizing current flowing in the machine is not included in the inverter current. Consequently, neglecting all harmonic components may, at first, appear as an invalid means of approximating the performance of this static drive system. However, since the electromechanical performance of this system is determined primarily by the real power transferred through the inverter rather than the reactive power exchange, many of the dominant performance characteristics are preserved even though all harmonics are neglected.

Comparisons of System Representations—Computer Study

In order to investigate the validity of two of the simplified representations set forth in the previous sections, the results obtained from the analog computer simulations of these representations of the static drive system are compared to those obtained from the detailed analog computer simulation of the complete system. The simplified representations considered herein are 1) the representation where only the switching of the

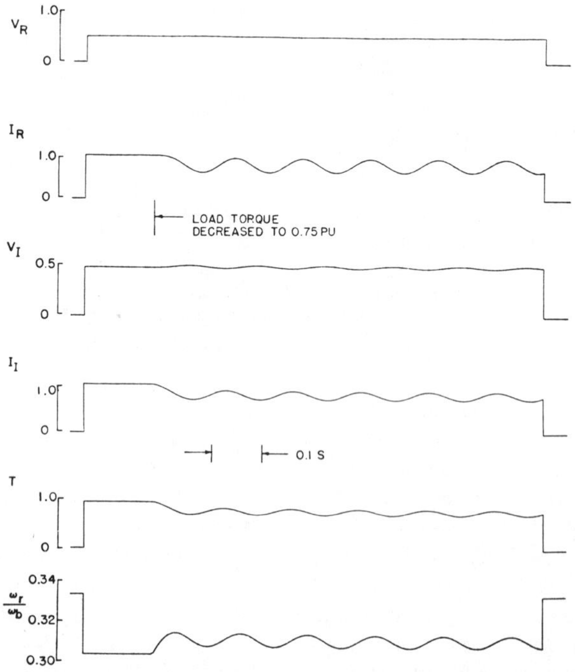

Fig. 7. Load torque switching from 0.925 to 0.75 pu; operation at 20 Hz. System represented in synchronously rotating reference frame with harmonics neglected.

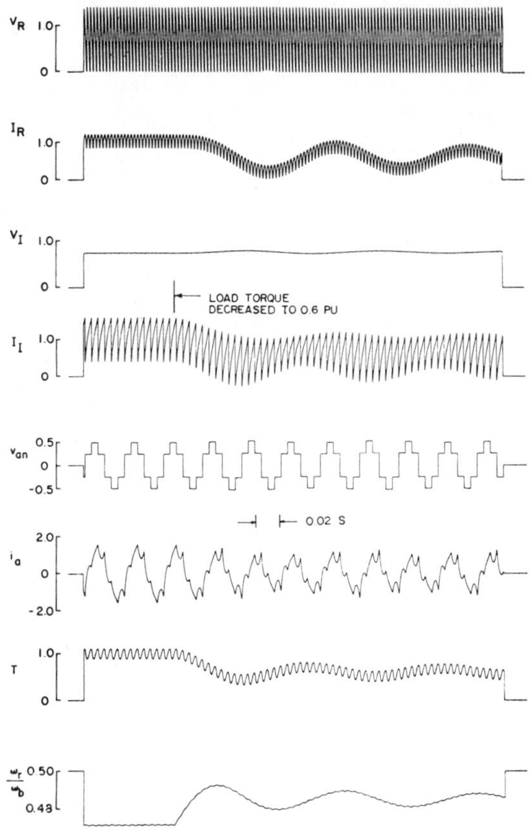

Fig. 8. Load torque switching from 1.0 to 0.6 pu; operation at 30 Hz. Detailed system representation.

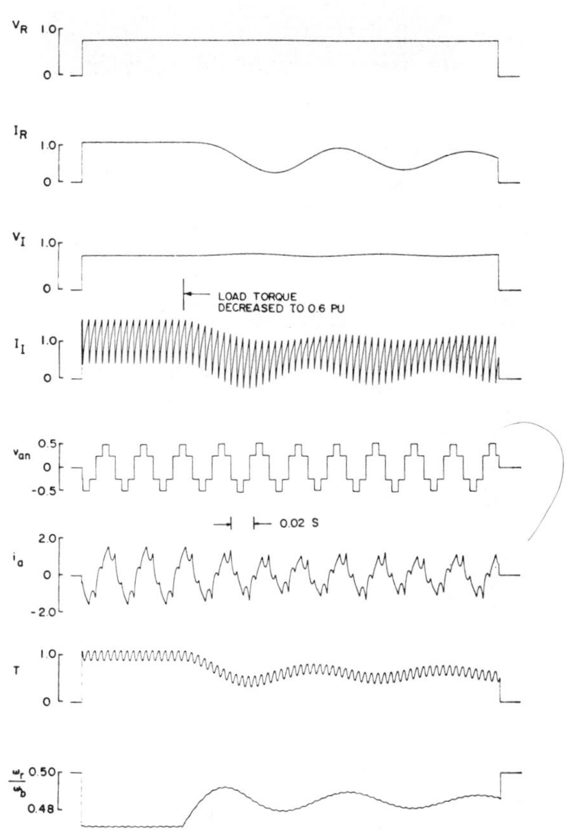

Fig. 9. Load torque switching from 1.0 to 0.6 pu; operation at 30 Hz. Functional representation of rectifier.

rectifier is neglected (18) while the inverter and the induction motor are represented in the stationary reference frame as in the detailed simulation, and 2) the representation of the static drive system in the synchronously rotating reference frame where the harmonic components due to the rectifier and the inverter are neglected (18) and (45)–(47).

The per unit parameters of the induction motor, filter, and commutating reactance are

$$X_{co} = 0.016 \qquad R_{LF} = 0.025 \qquad r_s = 0.025$$

$$X_{CF} = 0.0141 \qquad X_{LF} = 0.5 \qquad X_s = 2.075$$

$$r_r' = 0.020 \qquad X_m = 2.0$$

$$X_r' = 2.075 \qquad H = 0.2 \text{ s.}$$

The above parameters are based on a 7.5-hp induction motor having a base impedance of 9.45 ohms. The base frequency ω_b is assumed to be the rated frequency of the induction motor which is 60 Hz. In variable-frequency systems the amplitude of the applied stator voltages is decreased, in some manner, as frequency decreases. In the studies reported in this paper the amplitude of the fundamental component of the open-circuit inverter voltage is decreased linearly with frequency with 1.0-pu voltage occurring at 60 Hz [3]. In a previous paper it was shown that with the parameters and voltage–frequency relation given above, the static drive system will become unstable during operation at low frequencies [3]. For example, at 20 Hz the system is unstable if the load torque is equal to or less than 0.75 pu.

The computer traces given in Figs. 5, 6, and 7 show the system performance when the load torque is changed from a

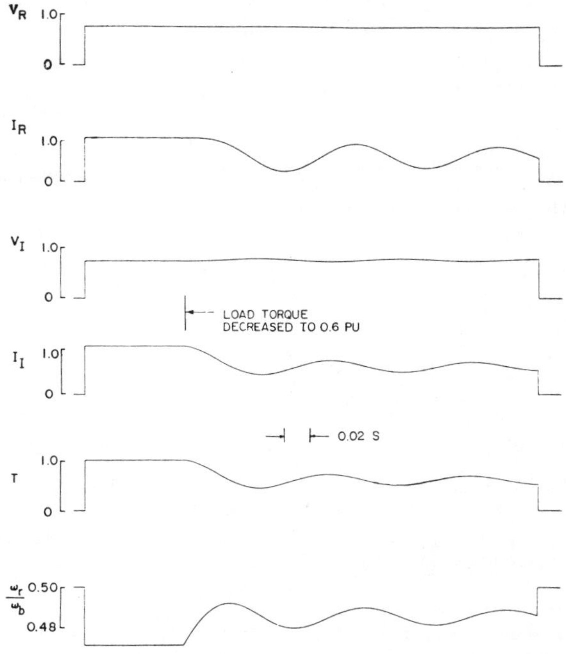

Fig. 10. Load torque switching from 1.0 to 0.6 pu; operation at 30 Hz. System represented in synchronously rotating reference frame with harmonics neglected.

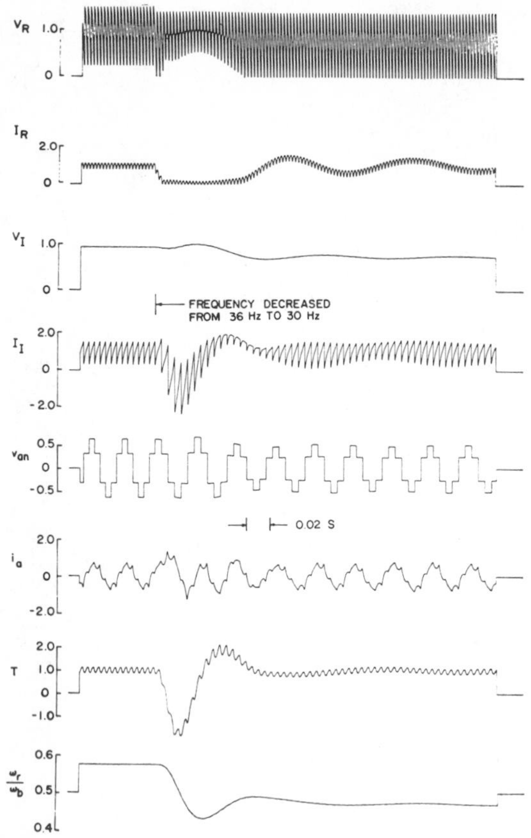

Fig. 11. Load torque held at 1.0 pu; frequency stepped from 36 to 30 Hz. Detailed system representation.

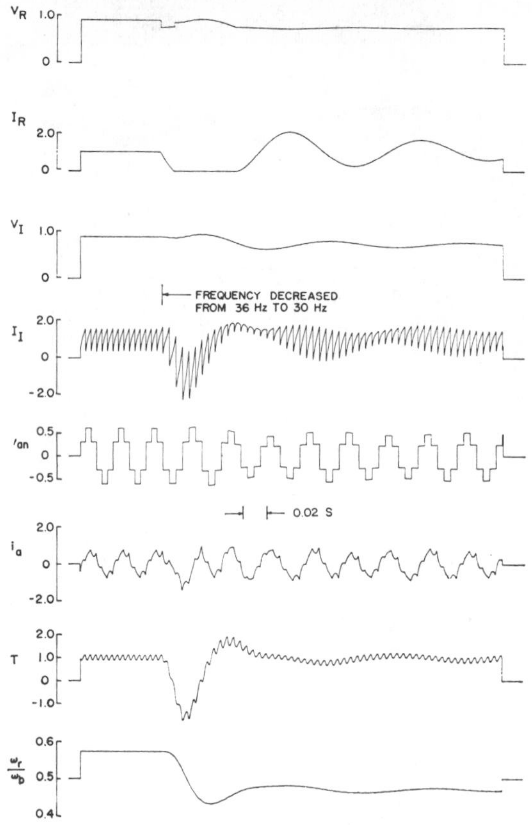

Fig. 12. Load torque held at 1.0 pu; frequency stepped from 36 to 30 Hz. Functional representation of rectifier.

stable to an unstable load point at an operating frequency of 20 Hz. In particular, with the machine operating at 0.925-pu torque, a stable operating point, the load torque is switched to 0.75 pu, an unstable operating point. The computer recordings shown in Fig. 5 were obtained using the detailed simulation of the complete system. Computer tracings shown in Fig. 6 were obtained using the representation of the system wherein a functional (average value) representation of the rectifier is employed (18). The traces given in Fig. 7 were obtained using the representation in the synchronously rotating reference frame with all harmonic components neglected (18) and (45)–(47). In Figs. 5 and 6 the following system variables are recorded:

V_R	rectifier output voltage
I_R	rectifier current
V_I	capacitor voltage
v_{an}	line-to-neutral stator voltage
i_a	phase current
T	electromagnetic torque
ω_r/ω_b	per unit electrical angular velocity of the rotor.

In Fig. 7, only V_R, I_R, V_I, T, and ω_r/ω_b are recorded. However, it is clear that in this case

$$V_I = (\pi/2)v_{qs}^e, \quad I_I = (3/\pi)i_{qs}^e.$$

A comparison of the computer traces given in Figs. 5–7 reveals that the average response of the system variables is identical. In particular, the computer recordings of the system variables obtained using the representation in the synchronously rotating reference frame (Fig. 7) are the average of the corresponding system variables recorded from the detailed simulation (Fig. 5)

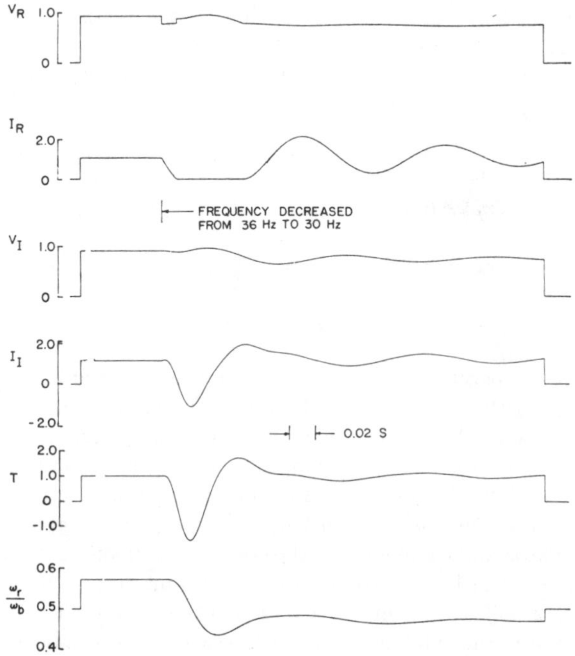

Fig. 13. Load torque held at 1.0 pu; frequency stepped from 36 to 30 Hz. System represented in synchronously rotating reference frame with harmonics neglected.

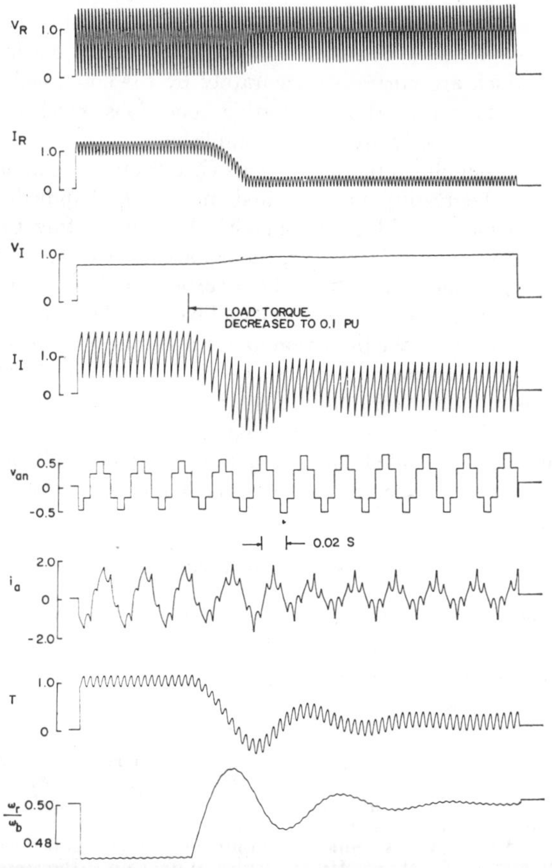

Fig. 14. Load torque switching from 1.0 to 0.1 pu; operation at 30 Hz. Detailed system representation.

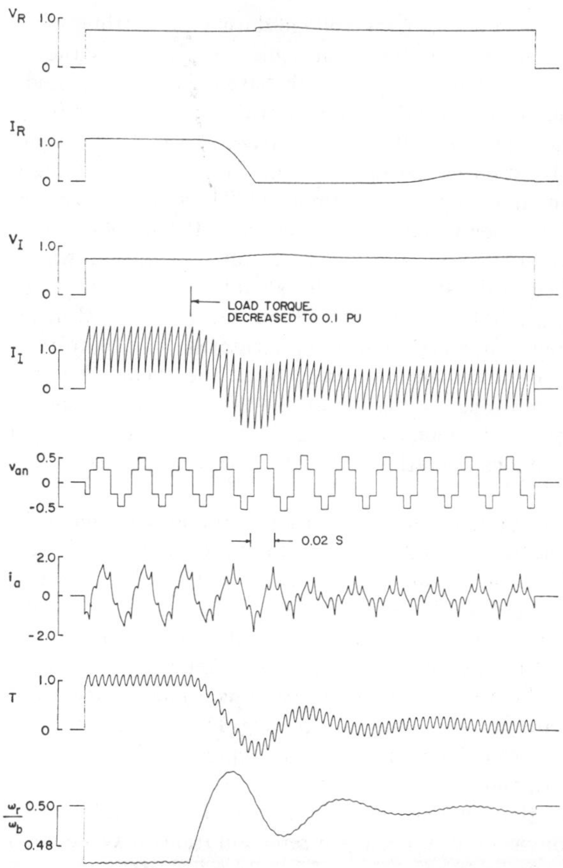

Fig. 15. Load torque switching from 1.0 to 0.1 pu; operation at 30 Hz. Functional representation of rectifier.

and those obtained using the functional representation of the rectifier (Fig. 6).

The computer traces given in Figs. 8–10 permit a comparison of the representations during load-torque switching at an operating frequency of 30 Hz. With the system operating initially at 1.0-pu load torque, the load torque is decreased (stepped) to 0.6 pu. Although the rotor speed is lightly damped at small load torques with the system parameters used, the system is stable for all load torques less than the breakdown value [3]. The average response of the system variables given in Figs. 8–10 is identical for the three types of representations.

Figs. 11–13 show the system response using the three representations during a step change in the frequency of the fundamental component of the voltages applied to the induction machine. In particular, the system is initially operating at 36 Hz with 1.0-pu load torque. The frequency is then stepped to 30 Hz, and simultaneously the delay angle of the rectifier is increased so that the open-circuit inverter voltage is decreased linearly with frequency. A comparison of Figs. 11–13 reveals an excellent correlation between the system performance obtained using the three methods of representation.

Although the simplified representations enable one to predict accurately the performance of the static drive system for many operating conditions, these representations do not properly account for all modes of discontinuous rectifier operation. This feature is demonstrated in Figs. 14–16, where the system is operating at 30 Hz and the load torque is switched from 1.0 to 0.1 pu. The recording of V_I, in Fig. 14, shows an increase in the capacitor voltage during the sustained, periodic, discontinuous operation of the rectifier at the light load. This relatively slow

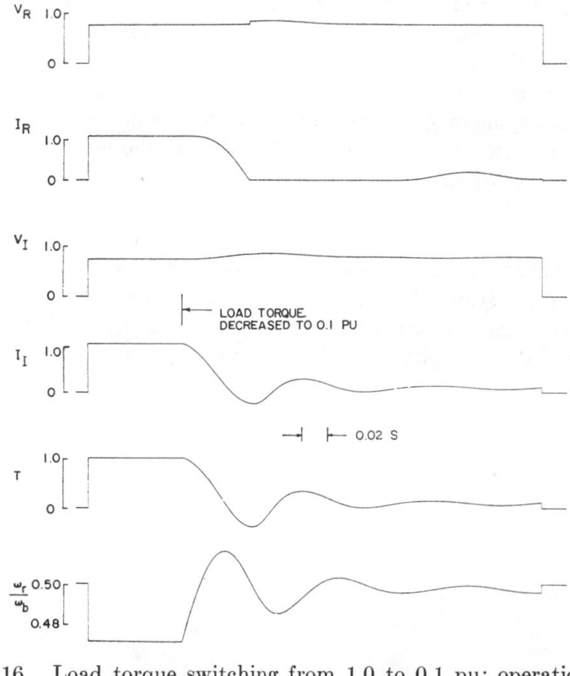

Fig. 16. Load torque switching from 1.0 to 0.1 pu; operation at 30 Hz. System represented in synchronously rotating reference frame with harmonics neglected.

increase in the capacitor voltage during the settling out of the system occurs due to the harmonic components of the rectifier voltage (predominately the 6th harmonic). This periodic discontinuous operation will persist at the 6th harmonic frequency, and the capacitor voltage will increase to a value determined by a combination of the torque load, the amplitude of the source voltage, and the system parameters. The harmonic components of the rectifier voltage are neglected in the simplified representations. Therefore, continuous rectifier operation will be established in the steady state for all values of load torque greater than zero but less than the breakdown value. Although the functional (average value) representation of the rectifier will exhibit discontinuous operation when the average value of the capacitor voltage is larger than the average value of the rectifier voltage, the harmonic charging of the capacitor due to periodic discontinuous operation of the rectifier will not occur (Figs. 15 and 16).

The computer traces show that at the load torque of 0.1 pu, the capacitor voltage increases to approximately 0.9 pu in the actual system (detailed simulation, Fig. 14), whereas the simplified representations give a capacitor voltage of about 0.78 pu (Figs. 15 and 16). Although discontinuous rectifier operation occurred briefly during the change in frequency shown in Figs. 11–13, the time interval during discontinuous operation was small, and the capacitor voltage (Fig. 11) increases only slightly above that predicted by functional representation of the rectifier (Figs. 12 and 13).

Since the simplified representations may be readily simulated and conveniently used, these representations may be employed to advantage in conducting feasibility and preliminary control studies. The simplicity of the simulation in the synchronously rotating reference frame is indeed a desirable feature, and in many cases the results obtained using this representation will predict the system performance with sufficient accuracy. However, in modes of operation where the exact system performance during sustained discontinuous operation is of importance the detailed simulation must be used in order to portray accurately the operation of the system. Perhaps in some cases it may be desirable to incorporate a detailed representation of the rectifier with the synchronously rotating reference frame representation of the inverter and induction motor wherein the harmonics due to inverter switching are neglected.

Conclusions

Simplified representations of the rectifier-inverter induction motor drive system have been developed. A computer study has been conducted wherein the system performances predicted by two of these simplified representations are compared to that obtained from a detailed computer simulation of the complete static drive system. The modes of operation of the static drive system which are portrayed accurately by the simplified representations as well as those operating conditions which can not be duplicated exactly are clearly established.

Since the simplified representations yield computer simulations which can be readily implemented and easily handled, these representations should be invaluable when conducting preliminary control studies. Moreover, the equations of transformation which express the operation of the inverter in the synchronously rotating reference frame can be employed to describe conveniently the effects of the induction motor upon the filter. Although the harmonic components in these equations of transformation are neglected when establishing the most simple representation of the static drive system, the complete equations, wherein the harmonics are included, offer a new method of analyzing the steady-state behavior of filter-inverter-induction motor combination.

Acknowledgment

The computer studies were performed at the University of Wisconsin Hybrid Computer Laboratory, using equipment provided in part by an NSF grant.

References

[1] P. C. Krause and L. T. Woloszyk, "Comparison of computer and test results of a static ac drive system," *IEEE Trans. Industry and General Applications*, vol. IGA-4, pp. 583–588, November/December 1968.

[2] L. T. Woloszyk, "An analog computer study of a static ac drive system," M. S. thesis, University of Wisconsin, Madison, 1967.

[3] T. A. Lipo and P. C. Krause, "Stability analysis of a rectifier-inverter induction motor drive," *IEEE Trans. Power Apparatus and Systems*, vol. PAS-88, pp. 55–66, January 1969.

[4] T. A. Lipo, P. C. Krause, and H. E. Jordan, "Harmonic torque and speed pulsations in a rectifier-inverter induction motor drive," this issue, pp. 579–587.

[5] P. C. Krause and C. H. Thomas, "Simulation of symmetrical induction machinery," *IEEE Trans Power Apparatus and Systems*, vol. PAS-84, pp. 1038–1053, November 1965.

[6] P. C. Krause, "Method of multiple reference frames applied to the analysis of symmetrical induction machinery," *IEEE Trans. Power Apparatus and Systems*, vol. PAS-87, pp. 218–227, January 1968.

[7] P. C. Krause and C. H. Thomas, "The arbitrary reference frame applied to the analysis of symmetrical induction machinery," presented at the IEEE Winter Power Meeting, New York, N. Y., January 28–February 2, 1968.

[8] P. C. Krause, Jr., discussion of H. E. Jordan, "Analysis of induction machines in dynamic systems," *IEEE Trans. Power Apparatus and Systems*, vol. PAS-84, pp. 1085–1088, November 1965.

[9] H. A. Peterson and P. C. Krause, Jr., "A direct- and quadrature-axis representation of a parallel ac and dc power system," *IEEE Trans. Power Apparatus and Systems*, vol. PAS-85, pp. 210–225, March 1966.

Stability Analysis of a Rectifier–Inverter Induction Motor Drive

THOMAS A. LIPO, MEMBER, IEEE, AND PAUL C. KRAUSE, SENIOR MEMBER, IEEE

Abstract—**A stability study of a rectifier–inverter induction motor drive system is performed by neglecting the harmonic content of the stator voltages and applying Nyquist stability criterion to the small-displacement equations obtained by linearization about an operating point. This investigation reveals that system instability can occur over a wide speed range if the system parameters are improperly selected. It appears that the method of analysis presented in this paper is sufficient to predict stability of practical rectifier–inverter induction motor drive systems. Also, with slight modifications, this method of analysis can be applied to rectifier–inverter systems which supply reluctance-synchronous machines or synchronous machines.**

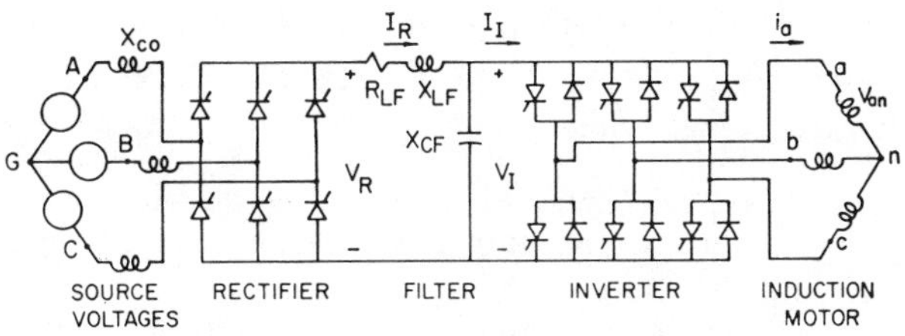

Fig. 1. System studied.

INTRODUCTION

RECTIFIER–inverter drive systems are used in a variety of applications for the purpose of controlling the speed of ac machines by adjusting the frequency of the applied voltages. There are many problems associated with this type of variable frequency drive which are of current interest. Perhaps the problem of system instability at low-frequency operation is one of the major concerns in the design of rectifier–inverter drive systems which are to operate over a wide speed range.

In earlier papers [1], [2], the authors have shown that during low-frequency operation, the reluctance-synchronous machine as well as a synchronous machine can demonstrate continuous oscillations in speed and may even become unstable. These regions of instability occur with balanced sinusoidal applied stator voltages which are independent of load current (zero-impedance source). Consequently, the presence of machine instability is not necessarily due to an interaction between the reluctance-synchronous machine or the synchronous machine and a static converter.

In a paper by Rogers [3], it is shown that speed disturbances of an induction motor may be lightly damped during operation at low frequencies. However, the symmetrical induction machine in most cases will not demonstrate sustained oscillations when supplied from a balanced set of voltages which are independent of load current.

The rectifier–inverter drive system is generally equipped with a filter between the rectifier and the inverter. Energy can be stored in the inductance and capacitance of the filter. The interchange of energy between the filter components and magnetic field and rotor of the machine can cause the rectifier–inverter induction motor drive system to become unstable at low operating frequencies. The stability analysis of this type of drive system is the subject of this paper.

In the analysis presented in this paper the harmonics in the terminal voltages are neglected and the machine is analyzed in the synchronously rotating reference frame. The small-displacement equations are established from these machine equations and from the equations which predict the average value of the converter variables. A stability analysis is then performed by employing the Nyquist stability criterion to the small-displacement equations. The influence on system stability due to changes in several machine parameters, filter parameters, and rectifier commutating reactance are presented.

SYSTEM STUDIED

A simplified diagram of the system studied is shown in Fig. 1. This system is comprised of a three-phase power source, a six-phase rectifier with filter, an inverter, and a three-phase induction machine. Practical static drive systems may be equipped with various feedback control systems for the purpose of maintaining prescribed modes of operation. For example, control loops may be incorporated to regulate the capacitor voltage or the rate of acceleration. Regardless of the control systems which may be incorporated, the system shown in Fig. 1 forms the basis of many of the present-day rectifier–inverter drive systems.

The purpose of this paper is to establish a method of predicting system stability by employing Nyquist stability criterion to the equations which describe the behavior of the system during small excursions about a steady-state operating point. This method of analysis is verified by comparing the predicted system performance obtained by a digital computer solution to the performance obtained from an analog computer simulation. The analog computer simulation differs from the analysis which follows in that the nonlinear equations which describe the behavior of the system are simulated in detail.

Although the analog computer simulation is secondary to the purpose of this paper, it seems appropriate to discuss briefly the method of simulation while describing the system components and to cite references wherein the simulation techniques are set forth. Simulation of the three-phase source voltages e_{GA}, e_{GB}, and e_{GC} is readily accomplished by implementing a constant frequency

Paper 68 TP 109–PWR, recommended and approved by the Rotating Machinery Committee of the IEEE Power Group for presentation at the IEEE Winter Power Meeting, New York, N.Y., January 28–February 2, 1968. Manuscript submitted September 25, 1967; made available for printing December 13, 1967. This work was supported by the U.S. Army Mobility Equipment Research and Development Center, Fort Belvoir, Va., under Contract DAAK02–67–C–0073.

T. A. Lipo was with the Department of Electrical Engineering, University of Wisconsin, Madison, Wis. He is now with the Department of Electrical and Electronic Engineering, University of Manchester Institute of Science Technology, Manchester, England.

P. C. Krause is with the Department of Electrical Engineering, University of Kansas, Lawrence, Kans., on one year's leave of absence from the Department of Electrical Engineering, University of Wisconsin, Madison, Wis.

Reprinted from *IEEE Trans. Power App. Syst.*, vol. PAS-88, pp. 55–66, Jan. 1969.

oscillator. The simulation of the rectifier is given in [4] and [5]. Briefly, each of the controlled rectifiers is represented by clamping circuitry which allows tube current when the forward voltage as well as the firing signal are positive. Although the commutating reactance X_{co} is included in the simulation, the resistance of the commutating inductance and the voltage drop across the tubes during conduction are neglected. These features may be readily included with only minor modifications of the computer simulation. Control of the amplitude of the rectifier voltage V_R is achieved by varying the firing signal. Although continuous voltage control may be incorporated in the computer simulation, in this paper the constant volts-per-hertz type of operation is considered.

The symmetrical induction machine is simulated using techniques discussed in [7] and [8]. The equations which describe the induction machine in the stationary reference frame are used in this simulation.

Simulation of the filter and the inverter is given in [6]. The switching of the inverter is simulated without regard to the commutating circuitry. That is to say, it is assumed that commutation occurs instantaneously in the inverter.

A comparison of analog computer results and test results obtained from an actual static drive system is presented in [9]. The drive system considered in [9] is nearly identical to the system shown in Fig. 1. The two systems differ only in the rectifier configuration.

System Equation

The equations which describe the symmetrical induction machine in the arbitrary reference frame may be expressed [7], [10]

$$\begin{bmatrix} v_{qs} \\ v_{ds} \\ 0 \\ 0 \end{bmatrix} = \begin{bmatrix} r_s + \frac{p}{\omega_b} X_s & \frac{\omega}{\omega_b} X_s & \frac{p}{\omega_b} X_m & \frac{\omega}{\omega_b} X_m \\ -\frac{\omega}{\omega_b} X_s r_s & +\frac{p}{\omega_b} X_s & -\frac{\omega}{\omega_b} X_m & \frac{p}{\omega_b} X_m \\ \frac{p}{\omega_b} X_m & \frac{\omega - \omega_r}{\omega_b} X_m & r_r' + \frac{p}{\omega_b} X_r' & \frac{\omega - \omega_r}{\omega_b} X_r' \\ -\frac{\omega - \omega_r}{\omega_b} X_m & \frac{p}{\omega_b} X_m & -\frac{\omega - \omega_r}{\omega_b} X_r' & r_r' + \frac{p}{\omega_b} X_r' \end{bmatrix} \times \begin{bmatrix} i_{qs} \\ i_{ds} \\ i_{qr}' \\ i_{dr}' \end{bmatrix} \tag{1}$$

where

$$X_s = X_{ls} + X_m \tag{2}$$

$$X_r' = X_{lr}' + X_m. \tag{3}$$

In the above equations p is the operator d/dt; r_s the stator resistance; r_r' the rotor resistance (referred to the stator winding); X_{ls} and X_{lr}' the stator and rotor leakage reactances, respectively. The base electrical angular velocity ω_b is a constant selected convenient to the rating of the machine. The electrical angular velocity of the arbitrary rotating reference frame is denoted as ω. If it is desirable to express the equations of the induction machine in a stationary reference frame, ω is set equal to zero in (1). For a reference frame fixed in the rotor, ω is set equal to the electrical angular velocity of the rotor ω_r. The equations in the synchronously rotating reference frame are obtained by setting ω equal to the electrical angular velocity of the fundamental frequency components of the applied voltages herein denoted as ω_e. The machine equations expressed in the synchronously rotating reference frame are used in the following analysis, thus with $\omega = \omega_e$ (1) becomes

$$\begin{bmatrix} v_{qs} \\ v_{ds} \\ 0 \\ 0 \end{bmatrix} = \begin{bmatrix} r_s + \frac{p}{\omega_b} X_s & f_R X_s & \frac{p}{\omega_b} X_m & f_R X_m \\ -f_R X_s & r_s + \frac{p}{\omega_b} X_s & -f_R X_m & \frac{p}{\omega_b} X_m \\ \frac{p}{\omega_b} X_m & f_R S X_m & r_r' + \frac{p}{\omega_b} X_r' & f_R S X_r' \\ -f_R S X_m & \frac{p}{\omega_b} X_m & -f_R S X_r' & r_r' + \frac{p}{\omega_b} X_r' \end{bmatrix} \times \begin{bmatrix} i_{qs} \\ i_{ds} \\ i_{qr}' \\ i_{dr}' \end{bmatrix} \tag{4}$$

where

$$S = \frac{\omega_e - \omega_r}{\omega_e} \tag{5}$$

$$f_R = \frac{\omega_e}{\omega_b}. \tag{6}$$

The electromagnetic torque expressed in per unit is

$$T_e = X_m(i_{qs} i_{dr}' - i_{ds} i_{qr}'). \tag{7}$$

In [7], [8], [10], and [11] a superscript e is used to denote variables in the synchronously rotating reference frame. This superscript is omitted in (4), since only the synchronously rotating reference frame is used in the analysis presented in this paper.

The stepped wave voltages applied to the terminals of the machine may be expressed in the synchronously rotating reference frame

$$v_{qs} = \frac{2V_I}{\pi}\left(1 + \frac{2}{35}\cos 6\omega_e t - \frac{2}{143}\cos 12\omega_e t + \cdots\right) \tag{8}$$

$$v_{ds} = \frac{2V_I}{\pi}\left(\frac{12}{35}\sin 6\omega_e t - \frac{24}{143}\sin 12\omega_e t + \cdots\right) \tag{9}$$

where V_I is the per unit capacitor voltage, Fig. 1. In the above equations the angular relationship between the q-axis and the magnetic axes of the stator and rotor phases has been selected so that these axes coincide at time zero [7], [10].

Since system stability is to be established by small excursions about a steady-state operating point, it is sufficient to consider only the fundamental component of the applied voltages [8], [12].

Although the fundamental component of the applied stator voltages together with the harmonics combine to produce a steady-state operating point, the contribution of the voltage harmonics on the electromechanical behavior of the machine are small for most operating conditions. The change in the average steady-state operating point and the perturbation about this operating point due to the harmonics are neglected. If the inertia of the mechanical system is relatively small, the harmonics in voltage waveform may have secondary effects upon system stability during operation at very low frequencies. These effects are not considered in this paper.

Neglecting the harmonics in the voltage waveform, the *qs* and *ds* voltages in the synchronously rotating reference frame become [(8) and (9)]

$$v_{qs} = \frac{2V_I}{\pi} \tag{10}$$

$$v_{ds} = 0. \tag{11}$$

The voltage equations for the filter during continuous operation ($I_R \neq 0$) may be written

$$V_R = V_I + \left(\frac{p}{\omega_b} X_{LF} + R_{LF}\right) I_R \tag{12}$$

$$V_I = \frac{\omega_b}{p} X_{CF}(I_R - I_I). \tag{13}$$

In a recent paper, Krause [13] developed the equation which expresses the average output voltage of a six-phase converter in a *d–q* axis. This relation was established by substituting the equations of transformation into the equations which describe the operation of the converter. If the commutating reactance X_{co} is small compared to the filter reactance X_{LF}, the average output voltage V_R may be expressed

$$V_R = \frac{3\sqrt{3}}{\pi} V_S \cos\alpha - \frac{3}{\pi} X_{co} I_R \tag{14}$$

where V_S is the magnitude of the line-to-neutral source voltage and α the delay angle. It is shown that an additional term is necessary if the commutating reactance is comparable to X_{LF} [13]. This refinement was not necessary in this study; however, the modifications necessary to include the additional term are straightforward.

Since at time zero the position of the synchronously rotating reference frame is selected so that v_{ds} is always zero (neglecting harmonics), the following instantaneous power balance equation is readily established

$$V_I I_I = \frac{3}{2} v_{qs} i_{qs}. \tag{15}$$

Equations (4)–(7) and (10)–(15) are the expressions from which the small-displacement equations will be derived and system stability investigated.

Application of Small-Displacement Theory to a Static Drive System

Small-displacement theory has had a long and productive history in analysis of electric machinery [14]–[17]. This theory enables one to establish linear equations which describe operation of a system for small changes about an operating point. Although the small-displacement equations are not valid for large changes of system variables, these relationships can be utilized to establish system stability when used in conjunction with Nyquist's or Routh's criterion

$$\begin{bmatrix} v_{qso} \\ 0 \\ 0 \\ 0 \end{bmatrix} = \begin{bmatrix} r_s & f_R X_s & 0 & f_R X_m \\ -f_R X_s & r_s & -f_R X_m & 0 \\ 0 & f_R S_o X_m & r_r' & f_R S_o X_r' \\ -f_R S_o X_m & 0 & -f_R S_o X_r' & r_r' \end{bmatrix} \times \begin{bmatrix} i_{qso} \\ i_{dso} \\ i_{qro}' \\ i_{dro}' \end{bmatrix}. \tag{16}$$

In the previous section, the equations defining the behavior of the rectifier–inverter system have been established. During balanced, steady-state operation wherein the voltage harmonics are not considered, the machine variables (voltages and currents) referred to the synchronously rotating reference frame are constants. Bearing in mind that in this analysis harmonics are neglected, the variables in the dc portion of the system are also constant in the steady state. The equations which describe the steady-state mode of operation may be obtained by setting the time rate-of-change of all currents equal to zero. The resulting steady-state equations are expressed where

$$v_{qso} = \frac{2}{\pi} V_{Io}. \tag{17}$$

Also

$$T_{eo} = X_m(i_{qso} i_{dro}' - i_{dso} i_{qro}') \tag{18}$$

$$V_{Ro} = V_{Io} + R_{LF} I_{Ro} \tag{19}$$

$$I_{Ro} = I_{Io} \tag{20}$$

$$V_{Ro} = \frac{3\sqrt{3}}{\pi} V_S \cos\alpha - \frac{3}{\pi} X_{co} I_{Ro}. \tag{21}$$

In the above equations the letter *o* has been added to the subscript so as to denote steady-state variables. If all variables are allowed to change by a small amount about an initial operating point and if the terms which describe the steady-state mode are eliminated, (4) becomes [1]

$$\begin{bmatrix} \Delta v_{qs} \\ 0 \\ (X_m i_{dso} + X_r' i_{dro}') \dfrac{\Delta\omega_r}{\omega_b} \\ -(X_m i_{qso} + X_r' i_{qro}') \dfrac{\Delta\omega_r}{\omega_b} \end{bmatrix} = \begin{bmatrix} r_s + \frac{p}{\omega_b} X_s & f_R X_s & \frac{p}{\omega_b} X_m & f_R X_m \\ -f_R X_s & r_s + \frac{p}{\omega_b} X_s & -f_R X_m & \frac{p}{\omega_b} X_m \\ \frac{p}{\omega_b} X_m & f_R S_o X_m & r_r' + \frac{p}{\omega_b} X_r' & f_R S_o X_r' \\ -f_R S_o X_m & \frac{p}{\omega_b} X_m & -f_R S_o X_r' & r_r' + \frac{p}{\omega_b} X_r' \end{bmatrix} \times \begin{bmatrix} \Delta i_{qs} \\ \Delta i_{ds} \\ \Delta i_{qr}' \\ \Delta i_{dr}' \end{bmatrix} \tag{22}$$

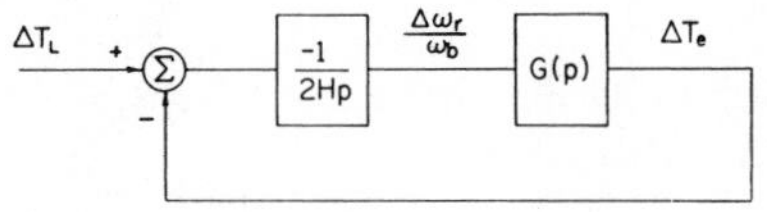

Fig. 2. Small-displacement closed-loop system.

where

$$\Delta v_{qs} = \frac{2}{\pi} \Delta V_I. \tag{23}$$

Also

$$\Delta T_e = X_m(i_{qso}\Delta i_{dr}' + i_{dro}'\Delta i_{qs} - i_{dso}\Delta i_{qr}' - i_{qro}'\Delta i_{ds}) \tag{24}$$

$$\Delta V_R = \Delta V_I + \left(\frac{p}{\omega_b} X_{LF} + R_{LF}\right)\Delta I_R \tag{25}$$

$$\Delta V_I = \frac{\omega_b X_{CF}}{p} (\Delta I_R - \Delta I_I) \tag{26}$$

$$\Delta V_R = -\frac{3}{\pi} X_{co}\Delta I_R. \tag{27}$$

If ΔV_R and ΔI_R are eliminated from (25)–(27) one obtains

$$\Delta V_I = -X_{CF}\left[\frac{\frac{p}{\omega_b} X_{LF} + \left(R_{LF} + \frac{3}{\pi} X_{co}\right)}{\left(\frac{p}{\omega_b}\right)^2 X_{LF} + \left(\frac{p}{\omega_b}\right)\left(R_{LF} + \frac{3}{\pi} X_{co}\right) + X_{CF}}\right]\Delta I_I. \tag{28}$$

However, since changes in the average value of inverter voltage and current are related to the quadrature axis by

$$\Delta V_I = \frac{\pi}{2} \Delta v_{qs} \tag{29}$$

$$\Delta I_I = \frac{3}{\pi} \Delta i_{qs} \tag{30}$$

(28) can be written as

$$\Delta v_{qs} = -\frac{6}{\pi^2} X_{CF} \times \left[\frac{\left(\frac{p}{\omega_b}\right) X_{LF} + \left(R_{LF} + \frac{3}{\pi} X_{co}\right)}{\left(\frac{p}{\omega_b}\right)^2 X_{LF} + \left(\frac{p}{\omega_b}\right)\left(R_{LF} + \frac{3}{\pi} X_{co}\right) + X_{CF}}\right] \Delta i_{qs} \tag{31}$$

or simply

$$\Delta v_{qs} = -Z_F(p)\Delta i_{qs}. \tag{32}$$

If (32) is substituted into (22) one obtains

$$\begin{bmatrix} 0 \\ 0 \\ (X_m i_{dso} + X_r' i_{dro}') \frac{\Delta\omega_r}{\omega_b} \\ -(X_m i_{qso} + X_r' i_{qro}') \frac{\Delta\omega_r}{\omega_b} \end{bmatrix} = \begin{bmatrix} r_s + \frac{p}{\omega_b} X_s + Z_F(p) & f_R X_s & \frac{p}{\omega_b} X_m & f_R X_m \\ -f_R X_s & r_s + \frac{p}{\omega_b} X_s & -f_R X_m & \frac{p}{\omega_b} X_m \\ \frac{p}{\omega_b} X_m & f_R S_o X_m & r_r' + \frac{p}{\omega_b} X_r' & f_R S_o X_r' \\ -f_R S_o X_m & \frac{p}{\omega_b} X_m & -f_R S_o X_r' & r_r' + \frac{p}{\omega_b} X_r' \end{bmatrix} \times \begin{bmatrix} \Delta i_{qs} \\ \Delta i_{ds} \\ \Delta i_{qr}' \\ \Delta i_{dr}' \end{bmatrix}. \tag{33}$$

It is clear from the form of (33) that the four machine currents can be expressed in terms of $\Delta\omega_r/\omega_b$ by inverting the 4×4 matrix. If the results are substituted into (24), it is evident that one can obtain an expression which has the form

$$\Delta T_e = G(p) \frac{\Delta\omega_r}{\omega_b}. \tag{34}$$

Although the procedure is straightforward, it is a formidable task to solve for $G(p)$. It is more direct and useful to solve for $G(j\nu)$ where p is set equal to $j\nu$. The derivation of the expression $G(j\nu)$ is given in the Appendix.

Equation (34) formulates a linear, small-displacement relationship between electromagnetic torque and rotor speed. An additional small-displacement equation can be obtained by considering the dynamics of the mechanical system. In the steady-state, the load torque T_{Lo} is equal to the electromagnetic torque T_{eo}. If ΔT_L is a small positive change from the steady-state torque then

$$T_{Lo} + \Delta T_L = T_{eo} + \Delta T_e - \frac{2H}{\omega_b} p\Delta\omega_r \tag{35}$$

where H is inertia constant of the induction machine expressed in seconds. Eliminating the steady-state terms

$$\Delta T_L = \Delta T_e - 2Hp\left(\frac{\Delta\omega}{\omega_b}\right). \tag{36}$$

Equations (34) and (36) suggest the block diagram representation shown in Fig. 2. It is noted that the technique of small displacements has permitted the problem of determining system stability to be recast in the form of a simple feedback control system. Hence, the problem of system stability is amenable to any of the approaches commonly employed in linear feedback control theory. Stability could be established employing the Routh test or the root locus method; however, the Nyquist criterion will be used in this development. It is clear from Fig. 2 that the open-loop transfer function is

$$F(p) = \frac{G(p)}{2Hp}. \tag{37}$$

The stability of the closed-loop system can be determined by setting $p = j\nu$ and observing the locus of $F(j\nu)$ as ν varies from $-\infty$ to ∞. When $F(p)$ does not have poles with positive real parts, the closed-loop system is stable if and only if the locus of $F(j\nu)$ does not pass through or encircle the $(-1,0)$ point [18].

Stability Studies

The linearized equations which have been developed offer a convenient means of predicting the behavior of a rectifier–inverter induction motor drive system for any operating frequency.

TABLE I

Figure	r_s	r_r'	X_s	X_r'	X_m	H	V_m	X_{CF}	X_{LF}	R_{LF}	X_{co}
3–6	0.025	0.020	2.075	2.075	2.0	0.2	1.0	0.0141	0.5	0.025	0.016
7	0.025	0.020	2.075	2.075	2.0	varied	1.0	0.0141	0.5	0.025	0.016
8	0.025	0.020	varied	varied	varied	0.2	1.0	0.0141	0.5	0.025	0.016
9	0.025	0.020	2.075	2.075	2.0	0.2	varied	0.0141	0.5	0.025	0.016
10	0.025	0.020	2.075	2.075	2.0	0.2	1.0	varied	0.5	0.025	0.016
11	0.025	0.020	2.075	2.075	2.0	0.2	1.0	0.0141	varied	0.025	0.016
12	0.025	0.020	2.075	2.075	2.0	0.2	1.0	0.0141	0.5	varied	0.016
13	0.025	0.020	2.075	2.075	2.0	0.2	1.0	0.0141	varied	varied	0.016
14	0.025	0.020	2.075	2.075	2.0	0.2	1.0	0.0141	0.5	0.025	varied

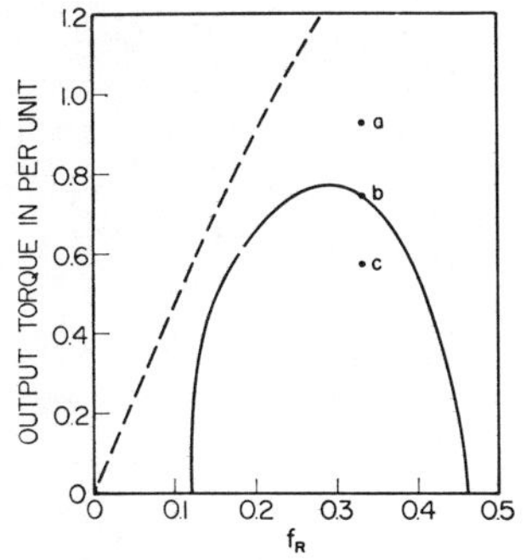

Fig. 3. Region of instability of a rectifier–inverter induction motor drive system.

A change in frequency can be taken into account by appropriate changes in the value of f_R. In this study, 60 Hz is assumed to be base frequency and the per unit system is based on operation at this frequency (ω_b = 377 rad/s). Thus the frequency ratio $f_R = \omega_e/\omega_b$ can be interpreted as the steady-state operating frequency expressed in per unit.

In variable speed systems, the amplitude of the applied stator voltages is decreased as frequency decreases. In some applications, it may be desirable to adjust the terminal voltage so as to maintain a constant breakdown torque. However, in this study it is assumed that the amplitude of the fundamental component of the open-circuit inverter voltage is decreased linearly with frequency. That is

$$v_{qso} = f_R V_m \tag{38}$$

where V_m corresponds to rated voltage (1.0 pu voltage). In order to develop this open-circuit voltage across the machine terminals, it is clear that the average value of the rectifier voltage is to be set at

$$V_{Ro} = \frac{\pi}{2} f_R V_m. \tag{39}$$

By utilizing the equations developed in the previous section, a digital computer can be programmed to compute Nyquist contours for the entire range of possible operating conditions. If the system is unstable, the region of instability can be readily determined. A region of instability of a typical rectifier–inverter drive system is given in Fig. 3. The per unit parameters of the induction motor, filter, and source reactance of the system are given in Table I. The parameters are based on a 7.5-hp induction motor having a base impedance of 9.45 ohms.

The dashed line shown in Fig. 3 indicates breakdown or maximum steady-state torque at various operating frequencies. The continuous contour shown in Fig. 3 forms the boundary between stable and unstable regions of operation. Specifically, the contour connects all operating points for which the locus of $F(j\nu)$ passes through the $(-1,0)$ point. Three operating points indicated as a, b, and c are shown in Fig. 3. These operating points occur at a frequency ratio of 0.333. In this case, the applied stator voltages vary at 20 Hz. With T_{eo} = 0.75 pu, point b, the locus of $F(j\nu)$ passes through the $(-1,0)$ point. At the initial operating point wherein T_{eo} = 0.575 pu, point c, the plot of $F(j\nu)$ encircles the $(-1,0)$ point. The system is unstable at these operating conditions. With T_{eo} = 0.925, point a, the locus of $F(j\nu)$ fails to encircle the $(-1,0)$ point as ν varies from $-\infty$ to $+\infty$; the system is stable.

In order to determine the actual system performance in the region of instability, the equations which describe the performance of the rectifier, filter, inverter, and induction motor were simulated on the analog computer. The computer traces of the system variables during load torque switching is shown in Figs. 4–6. With the machine initially operating at 0.925 pu torque, the load is switched to 0.75 pu (Fig. 4). This switching corresponds to a change from point a, a stable point of operation to point b, an unstable operating point. The traces shown in Fig. 4 illustrate the sustained oscillations which occur with the load torque of 0.75 pu. The traces in Fig. 5 demonstrate the increasing system oscillations which occur when the load is switched from 0.75 pu, point b, to 0.575 pu, point c, an operating condition inside the region of instability. In Fig. 6 the load is switched from point c back to point a, a stable operating point. Positive system damping is evident at the stable operating point.

The results of the stability study based on the theory of small displacements indicate that sustained oscillations will occur only at operating points on the contour; oscillations will be damped for operation outside the region of instability, and negative damping will exist in the unstable region. One might expect that operation at a point within the region of instability will cause unbounded increases in oscillation causing the motor to reach its breakdown torque, subsequently resulting in a braking of the machine. However, positive damping occurs during that part of the oscillation which includes operating points outside the region of instability. Also, as the system oscillations increase, the rectifier current assumes an intermittent zero mode of operation due to the unidirectional current constraint of the rectifier. It was found, for the system studied, that this discontinuous action of the rectifier current provides system damping. Sustained oscillations will occur when the equivalent damping over a complete cycle is zero. It can be concluded from this study that sustained oscillations of the machine variables usually occur rather than a breakdown of the machine. However, breakdown may occur if the region of instability is sufficiently large.

The contours shown in Figs. 7–14 illustrate the effect upon machine stability due to changes in system and machine con-

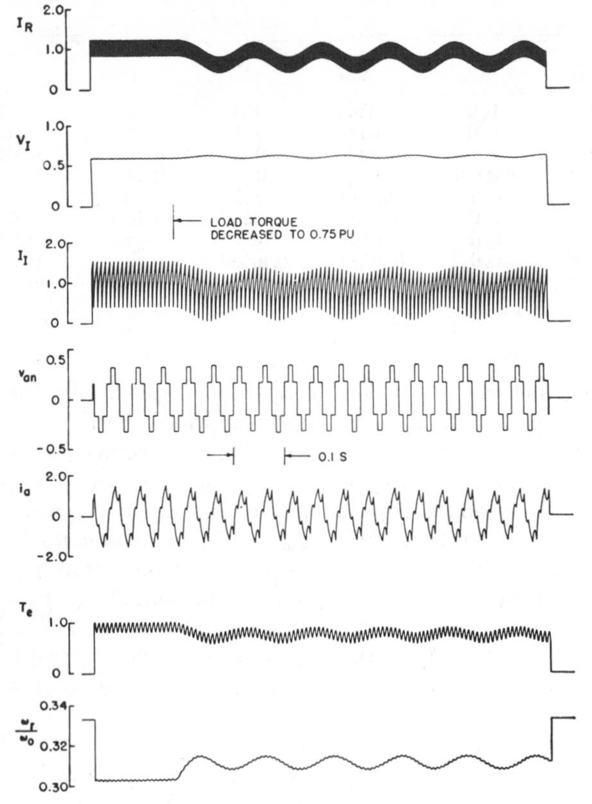

Fig. 4. Load torque switching from 0.925 pu to 0.75 pu (analog computer study).

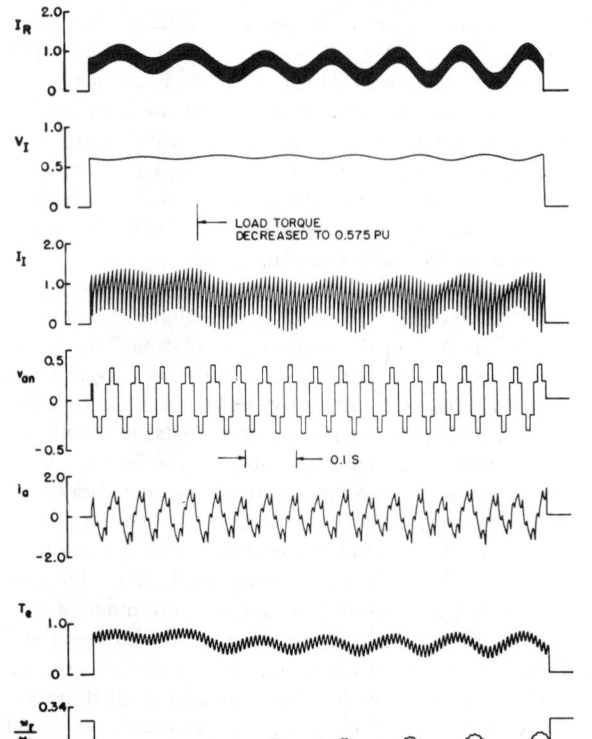

Fig. 5. Load torque switching from 0.75 pu to 0.575 pu (analog computer study).

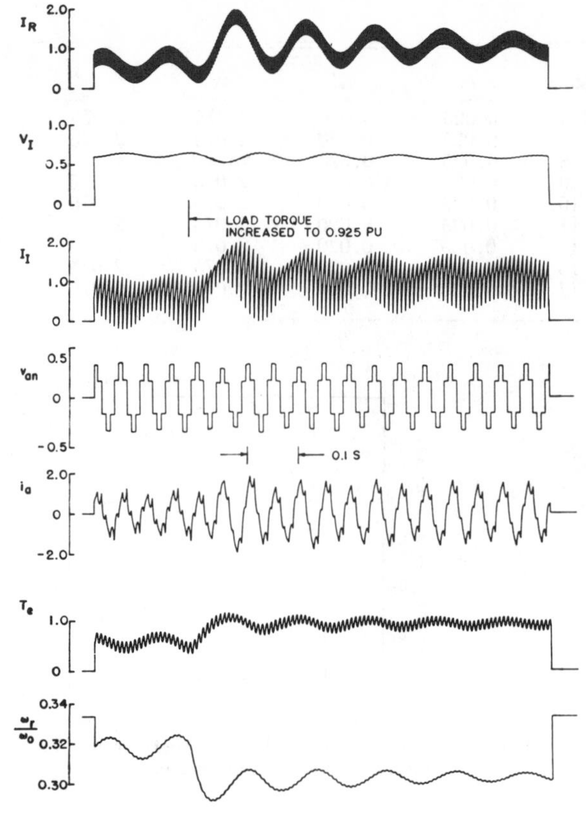

Fig. 6. Load torque switching from 0.575 pu to 0.925 pu (analog computer study).

stants. To facilitate a direct comparison, the contour given in Fig. 3 is also included in Figs. 7–14. The values of the system parameters and the quantities which were varied in each case are given in Table I.

Variation in the region of instability due to a change in system inertia is given in Fig. 7. Regions of instability are shown for $H = 0.075$, 0.2, 0.5, and 1.0 s. It is interesting to note that for this system an intermediate value of $H = 0.2$ produces the largest area of instability. This system is stable over all operating regions for $H = 0.05$ and 1.4 s.

In order to enhance inverter commutation, it may be desirable to make the magnetizing reactance large thereby reducing the magnetizing current. The curves shown in Fig. 8 indicate that an increase in magnetizing reactance increases the region of instability.

The contours given in Fig. 9 show an increase in the region of instability with an increase in the amplitude of the fundamental of the applied stator voltages. Hence an increase in breakdown torque is accompanied by a larger region of instability. The applied stator voltage fundamental, it is recalled, is decreased linearly with f_R. That is, $v_{qso} = f_R V_m$, where the voltage V_m is normally 1.0 pu. Fig. 9 shows the contours for $V_m = 0.8$, 1.0, and 1.2 pu.

The increase in the region of instability due to an increase in capacitive reactance (decrease in capacitance) is shown in Fig. 10. Regions of instability are given for values of $X_{CF} = 0.0094$, 0.0141, and 0.0282 pu. With a base impedance of 9.45 ohms and $\omega_b = 377$ rad/s a capacitive reactance of 0.0141 pu corresponds to 20 000 μF. When $X_{CF} = 0.00705$, the system is stable over all

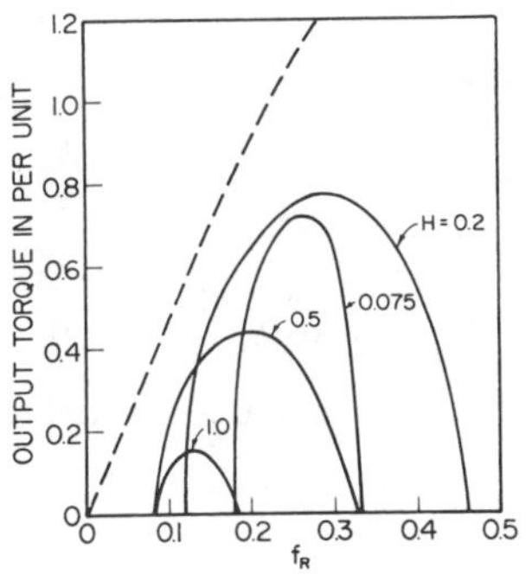

Fig. 7.   Regions of instability for changes in system inertia.

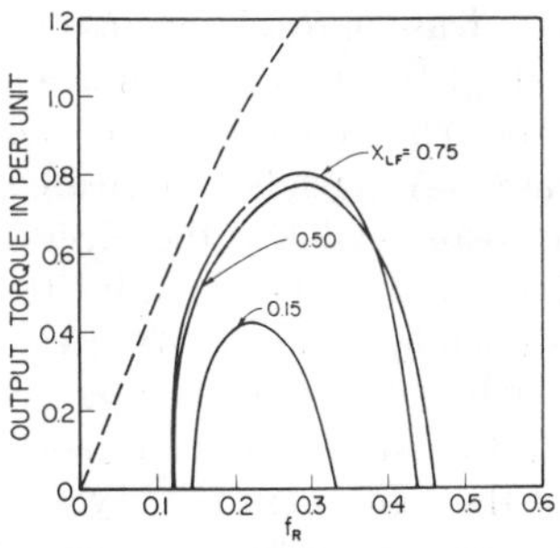

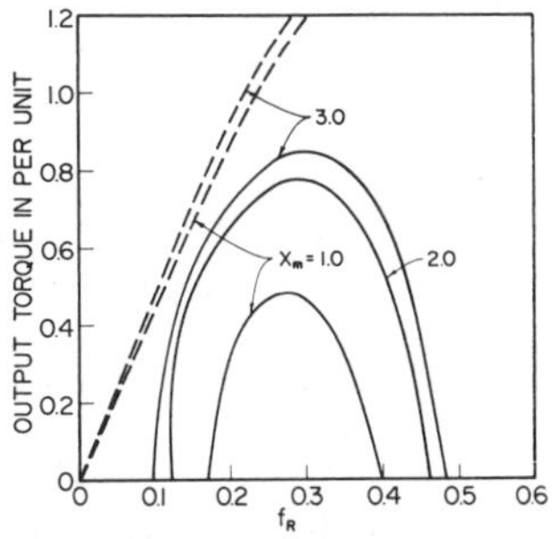

Fig. 8.   Regions of instability for different values of magnetizing reactance.

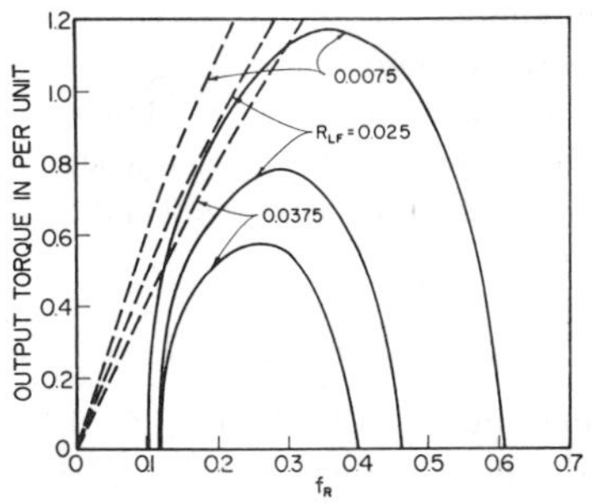

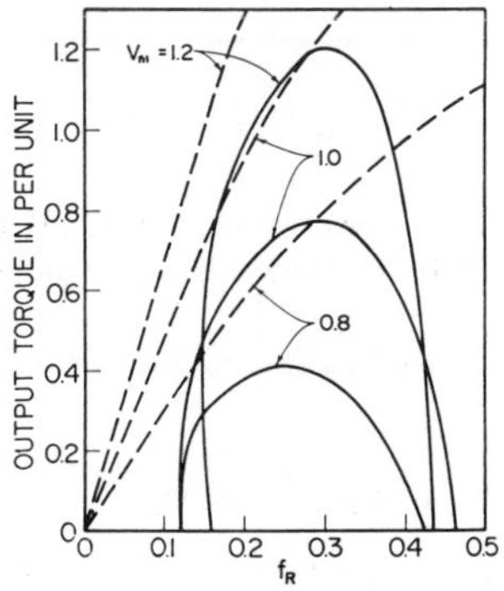

Fig. 9.   Regions of instability for changes in V_m.

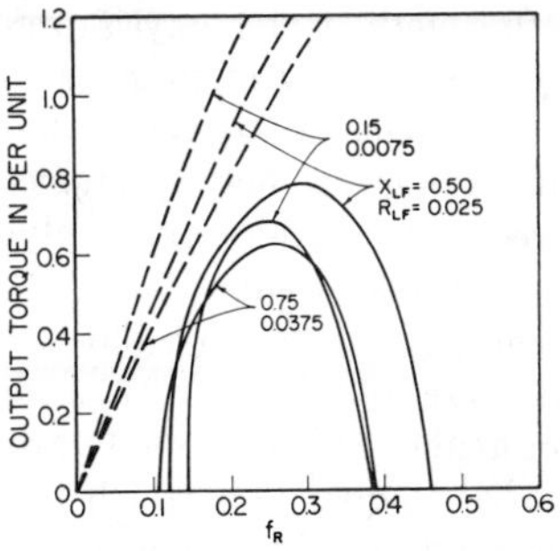

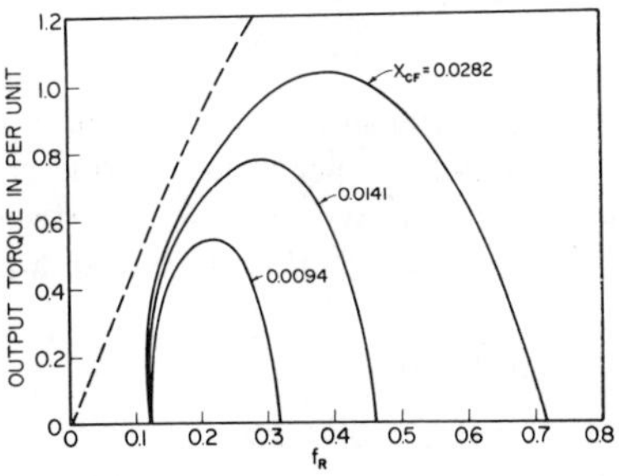

Fig. 10.   Regions of instability for various values of filter capacitance.

Fig. 11.   Regions of instability for several values of filter inductance.

Fig. 12.   Regions of instability for changes in R_{LF}.

Fig. 13.   Regions of instability for changes in both filter inductance and resistance.

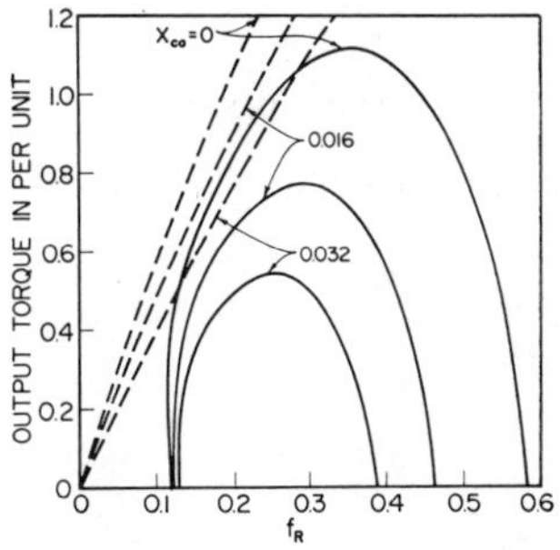

Fig. 14.   Regions of instability for different values of commutating reactance.

regions of operation. Regions of instability for several values of filter inductive reactance and resistance is given in Figs. 11 and 12. A decrease in inductive reactance tends to promote system stability whereas a decrease in resistance results in an increase in the region of instability. In Fig. 13 the resistance of the filter is maintained at five percent of the inductive reactance. In this case, since an increase in X_{LF} tends to decrease stability whereas an increase in R_{LF} increases stability, the effects tend to cancel.

The contours given in Fig. 14 illustrate the effect of commutating reactance upon system stability. Regions of instability are given for $X_{co} = 0.0$, 0.016, and 0.032 pu. Since a change in X_{co} has a similar effect on system stability as R_{LF}, an increase in X_{co} tends to decrease the region of instability.

Throughout this study, it has been assumed that the rectifier current is always positive (nonzero). If the filter inductance is decreased to a sufficiently small value, it is possible that at large phase delay angles (low frequencies) discontinuous conduction of the rectifier current I_R may occur. This effect is not incorporated in the analysis presented herein. Also, the harmonic content of the stator voltages may affect the operation of the drive system at very low frequencies. However, for the system studied, harmonic components did not significantly change the instability region predicted when considering only the fundamental components.

Conclusions

A method of determining the stability of a rectifier–inverter induction motor drive system has been set forth. In this method of analysis the harmonics of the stator voltages are neglected and the equations which describe the induction machine in the synchronously rotating reference frame are related to the equations which express the average value of the converter variables.

It appears that the method of analysis presented herein is adequate to predict the stability of practical inverter–rectifier induction motor drive systems. Moreover, with only slight modifications, this method of analysis may be applied to a drive system consisting of a rectifier–inverter unit supplying a reluctance-synchronous or a synchronous machine.

Regions of instability predicted by the method employing Nyquist's criterion were obtained from a digital computer study. Results of an analog computer study provide a means of interpreting the modes of operation which occur within these regions of system instability. The regions of instability depend upon various system parameters. For example, the amplitude of the applied stator voltages, machine parameters, filter parameter, and commutating reactances of the rectifier all affect system stability.

Appendix

Equation (33) can be written as

$$\Delta\bar{v} = z\Delta\bar{\imath} \tag{40}$$

where

$$\Delta\bar{v} = [0,0,X_m i_{dso} + X_r' i_{dro}', -(X_m i_{qso} + X_r' i_{qro}')]^T \frac{\Delta\omega_r}{\omega_b} \tag{41}$$

$$\Delta\bar{\imath} = [\Delta i_{qs}, \Delta i_{ds}, \Delta i_{qr}', \Delta i_{dr}']^T \tag{42}$$

and z is the 4×4 impedance matrix in (33) and T indicates the transpose.

If p is set equal to $j\nu$ and the elements of z are separated into real and imaginary parts, (40) becomes

$$\Delta\bar{v} = (r + jx)\Delta\bar{\imath}. \tag{43}$$

If the inverse of z is defined as

$$g + jb = (r + jx)^{-1} \tag{44}$$

then

$$(r + jx)(g + jb) = I + j0 \tag{45}$$

where I is the identity matrix and 0 is the null matrix. Thus

$$rg - xb = I \tag{46}$$

and

$$rb + xg = 0. \tag{47}$$

If x possesses an inverse, then (47) can be solved for g as

$$g = -x^{-1}rb. \tag{48}$$

The matrix x is given by

$$x = \begin{bmatrix} \frac{\nu}{\omega_b}(X_s + X_F) & 0 & \frac{\nu}{\omega_b}X_m & 0 \\ 0 & \frac{\nu}{\omega_b}X_s & 0 & \frac{\nu}{\omega_b}X_m \\ \frac{\nu}{\omega_b}X_m & 0 & \frac{\nu}{\omega_b}X_r' & 0 \\ 0 & \frac{\nu}{\omega_b}X_m & 0 & \frac{\nu}{\omega_b}X_r' \end{bmatrix} \tag{49}$$

where

$$\frac{\nu}{\omega_b}X_F = \mathrm{Im}[Z_F(j\nu)]. \tag{50}$$

The matrix x will not possess an inverse if the determinant of x is zero. It can be readily shown that this condition is satisfied if

$$X_s X_r' - X_m^2 = 0 \tag{51}$$

or

$$(X_s + X_F)X_r' - X_m^2 = 0. \tag{52}$$

The first condition is unimportant since this restriction implies a perfectly coupled machine. The second condition may be satisfied if

$$X_F = -\frac{X_s X_r' - X_m^2}{X_r'}. \tag{53}$$

Although this isolated situation is possible, it is certainly improbable and has not been experienced in this study. If filter parameters have been chosen such that (53) is satisfied, a solution for z^{-1} is still possible although the procedure is possibly more involved [19].

Substituting (48) into (46)

$$-rx^{-1}rb - xb = I \tag{54}$$

or

$$b = -(x + rx^{-1}r)^{-1} \tag{55}$$

and

$$\boldsymbol{g} = \boldsymbol{x}^{-1}\boldsymbol{r}(\boldsymbol{x} + \boldsymbol{r}\boldsymbol{x}^{-1}\boldsymbol{r})^{-1}. \tag{56}$$

Hence

$$\Delta\bar{\imath} = (\boldsymbol{x}^{-1}\boldsymbol{r} - j\boldsymbol{I})(\boldsymbol{x} + \boldsymbol{r}\boldsymbol{x}^{-1}\boldsymbol{r})^{-1}\Delta\bar{v}. \tag{57}$$

The transfer function $G(j\nu)$ can now be readily written as

$$G(j\nu) = X_m[i_{dro}', -i_{qro}', -i_{dso}, i_{qso}](\boldsymbol{x}^{-1}\boldsymbol{r} - j\boldsymbol{I})(\boldsymbol{x} + \boldsymbol{r}\boldsymbol{x}^{-1}\boldsymbol{r})^{-1}$$

$$\times \begin{bmatrix} 0 \\ 0 \\ X_m i_{dso} + X_r' i_{dro}' \\ -(X_m i_{qso} + X_r' i_{qro}') \end{bmatrix}. \tag{58}$$

ACKNOWLEDGMENT

The computer studies were performed at the University of Wisconsin, Hybrid Computer Laboratory, using equipment provided in part by a National Science Foundation grant. The authors wish to thank L. T. Woloszyk for his counsel.

REFERENCES

[1] T. A. Lipo and P. C. Krause, "Stability analysis of a reluctance-synchronous machine," *IEEE Trans. Power Apparatus and Systems*, vol. PAS-86, pp. 825–834, July 1967.
[2] —— "Stability analysis for variable frequency operation of synchronous machines," *IEEE Trans. Power Apparatus and Systems*, vol. PAS-87, pp. 227–234, January 1968.
[3] G. J. Rogers, "Linearised analysis of induction-motor transients," *Proc. IEE* (London), vol. 112, pp. 1917–1926, October 1965.
[4] R. A. Hedin, "The dynamic behavior of a synchronous generator with rectifier load," M.S. thesis, University of Wisconsin, Madison, 1964.
[5] H. A. Peterson, P. C. Krause, J. F. Luini, and C. H. Thomas, "An analog computer study of a parallel ac and dc power system," *IEEE Trans. Power Apparatus and Systems*, vol. PAS-85, pp. 191–209, March 1966.
[6] L. T. Woloszyk, "An analog computer study of a static ac drive system," M.S. thesis, University of Wisconsin, Madison, 1967.
[7] P. C. Krause and C. H. Thomas, "Simulation of symmetrical induction machinery," *IEEE Trans. Power Apparatus and Systems*, vol. PAS-84, pp. 1038–1053, November 1965.
[8] P. C. Krause, Discussion of [12], *IEEE Trans. Power Apparatus and Systems*, vol. PAS-84, pp. 1085–1088, November 1965.
[9] P. C. Krause and L. T. Woloszyk, "Comparison of computer and test results of a static ac drive system," *IEEE Trans. Industry and General Applications*, vol. IGA-4, pp. 583–588, November/December 1968.
[10] P. C. Krause and C. H. Thomas, "The arbitrary reference frame applied to the analysis of symmetrical induction machinery," presented at the IEEE Winter Power Meeting, New York, N.Y., January 28–February 2, 1968.
[11] P. C. Krause, "Method of multiple reference frames applied to the analysis of symmetrical induction machinery," *IEEE Trans. Power Apparatus and Systems*, vol. PAS-87, pp. 218–227, January 1968.
[12] H. E. Jordan, "Analysis of induction machines in dynamic systems," *IEEE Trans. Power Apparatus and Systems*, vol. PAS-84, pp. 1080–1088, November 1965.
[13] H. A. Peterson and P. C. Krause, "A direct- and quadrature-axis representation of a parallel ac and dc power system," *IEEE Trans. Power Apparatus and Systems*, vol. PAS-85, pp. 210–225, March 1966.
[14] R. H. Park, "Two-reaction theory of synchronous machines—II," *AIEE Trans.*, vol. 52, pp. 352–355, June 1933.
[15] C. Concordia, "Synchronous machine damping torque at low speeds," *AIEE Trans.*, vol. 69, pp. 1550–1553, 1950.
[16] J. W. Lynn, "Tensor analysis of electrical machine hunting," *Proc. IEE* (London), vol. 105, pt. C, pp. 420–431, 1958.
[17] A. S. Aldred and G. Shackshaft, "A frequency-response method for the predetermination of synchronous-machine stability," *Proc. IEE* (London), vol. 107, pt. C, pp. 2–10, 1960.
[18] H. Nyquist, "Regeneration theory," *Bell Sys. Tech. J.*, vol. 11, pp. 126–147, January 1932.
[19] C. Froberg, *Introduction to Numerical Analysis*. Reading, Mass.: Addison-Wesley, 1965, p. 99.

Discussion

Linos J. Jacovides and **Paul D. Agarwal** (General Motors Research Laboratories, Warren, Mich. 48090): The paper provides a stability analysis of a system which is becoming increasingly important in variable speed electric drives. With the advent of low-cost silicon controlled rectifiers, an induction motor supplied with variable frequency may be expected to replace the dc motor in many traction as well as servo applications in the near future.

The method used is particularly suitable for analyzing such a complex system. Although the result does not guarantee stability for large changes in operating conditions in practice, all should be well if the system operates at a point far removed from the stability boundary.

The authors' assumption of decreasing the inverter open-circuit voltage linearly with frequency results in a system with very low torque capability at low speeds, as shown in Fig. 3. The analysis would be of greater practical value if the motor voltage was so varied as to maintain constant airgap induction. This would make the breakdown torque constant over the entire speed range, Fig. 6 [20].

A second point that may be raised is the experimental confirmation of instability in a practical system. The results shown in Figs. 7–14 indicate that the system is always unstable over a part of the speed range in spite of wide changes in parameters. Systems like the one described in the paper have been built in a variety of sizes and speed ranges, both in this country and abroad. To our knowledge, no open-loop instability has been reported in the literature except in the present paper.

REFERENCES

[20] P. D. Agarwal, "The GM high-performance induction motor drive system," this issue, pp. 82–89.

Manuscript received February 19, 1968.

George H. Studtmann and **Raymond J. Yarema** (Borg-Warner Corp., Roy C. Ingersoll Research Center, Des Plaines, Ill. 60018): This paper handles a difficult problem which is of interest to those involved with ac motor speed control. Until the present time, little analytical work has been published which treats a system of this nature. The described approach presents a new analytical tool for analyzing variable frequency motor drive systems.

Several interesting and important trends relating to stability of the system are predicted. In Fig. 3 a region of instability is shown and three points (a, b, c) are marked. As their analog computer results have verified, operation at point a outside the region in question is stable. When the load torque was reduced to point b, sustained oscillations were observed with the analog computer which indicated the edge of the region of instability. The sustained oscillations are explained from a quasi-steady state point of view by saying that during the part of the oscillation which includes operating points outside the region of instability, positive damping occurs, and that negative damping occurs during the part of the oscillation which is inside the region of instability. It follows that as the load torque is reduced further, the magnitude of the oscillation should increase. The authors also point out that as the system's oscillations increase, it has been observed on the computer that the rectifier current assumes an intermittent zero mode of operation which provides system damping. Sustained oscillations occur when the net damping over a complete cycle is zero. Thus when the load torque is reduced from point b to point c, growing oscillations were observed on the computer. If the oscillations were allowed to continue, intermittent zero mode operation of the rectifier current would be observed which would limit the growing oscillation. It is implied by the authors that for all values of load torque below that value at point c the system

Manuscript received February 15, 1968.

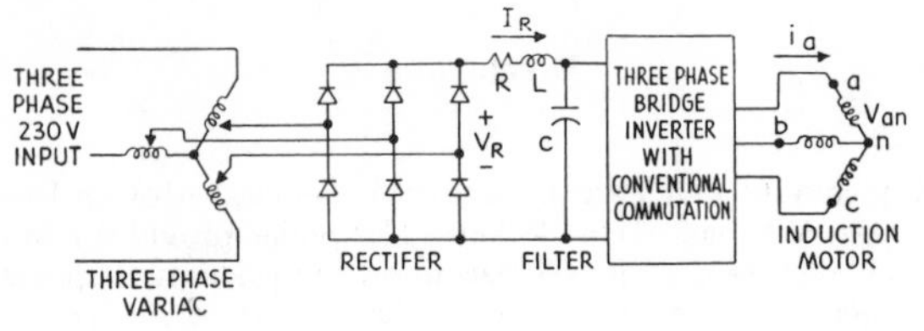

Fig. 15. System studied.

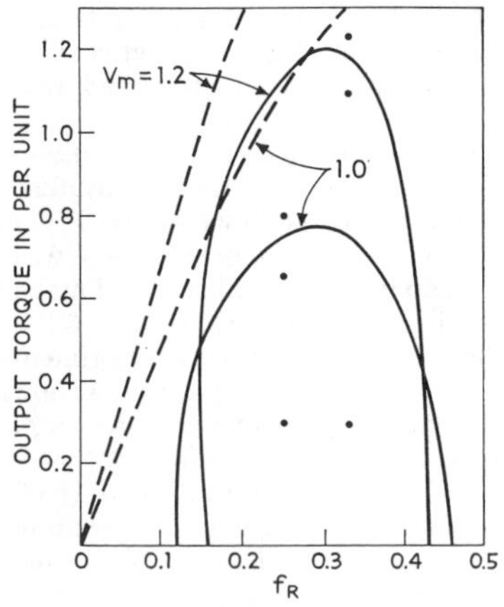

Fig. 16. Regions of instability with test points.

TABLE II

Per Unit Quantities

	Lipo and Krause	Studtmann and Yarema
r_s	0.025	0.0363
r_r'	0.020	0.0328
X_s	2.075	1.304
X_r'	2.075	1.304
X_m	2.0	1.25
H	0.2	0.223
V_m	1.0	1.0
X_{Cf}	0.0141	0.0137
X_{Lf}	0.5	0.5
R_{Lf}	0.025	0.0163
X_{Co}	0.016	

would be in an unstable operating region in which the oscillations would become sustained due to intermittent conduction of the rectifier.

In an attempt to verify the described computer results, a system similar to the authors' was tested in the laboratory. The system used is shown in Fig. 15. Input voltage for the system is obtained from a three-phase variac whose output is rectified by a three-phase diode bridge to obtain the voltage V_R. From the authors' paper it does not appear that the different type of bridge rectifier input we have used would change the results they obtained. The inverter is a three-phase bridge inverter with conventional commutation. No attempt has been made in the laboratory results to account for inverter impedance or losses, since it was not known how to include these factors. The load is a 1.5-hp induction motor which has a synchronous speed of 1800 r/min at 60 Hz. An attempt was made to use approximately the same per unit quantities as the authors and run under the same operating conditions (e.g., constant V/f, fixed rectifier voltage V_R). A comparison of the per unit quantities used by the authors to those used in the laboratory is shown in Table II. The experimental quantities were calculated using a base impedance of 43 ohms.

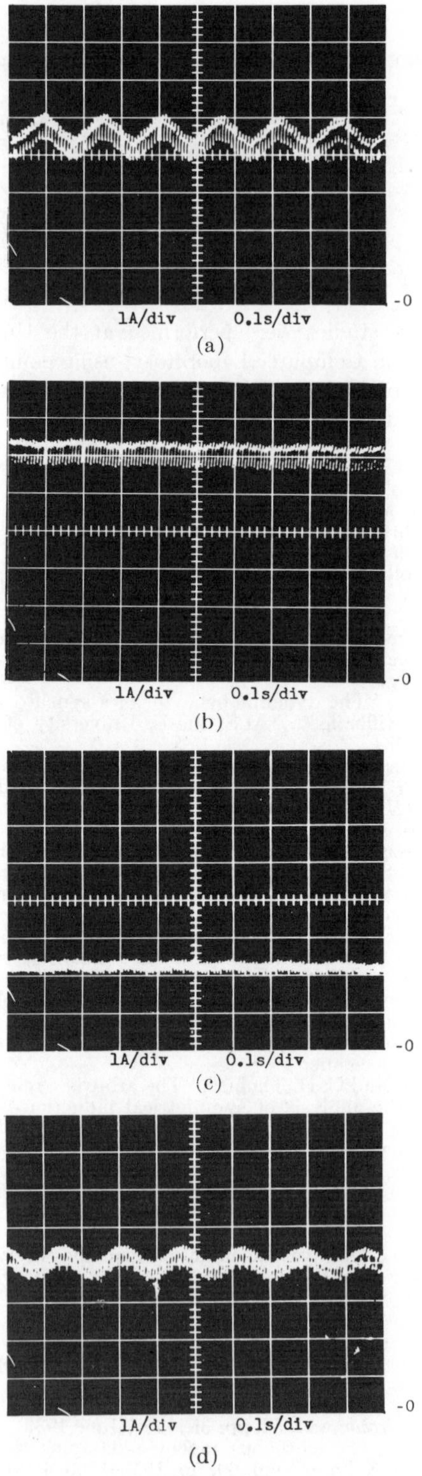

Fig. 17. Current I_R through filter choke at $f_R = 0.25$. (a) $T_L = 0.295$ pu; $V_m = 1.0$ pu. (b) $T_L = 0.655$ pu; $V_m = 1.0$ pu. (c) $T_L = 0.80$ pu; $V_m = 1.0$ pu. (d) $T_L = 0.80$ pu; $V_m = 1.2$ pu.

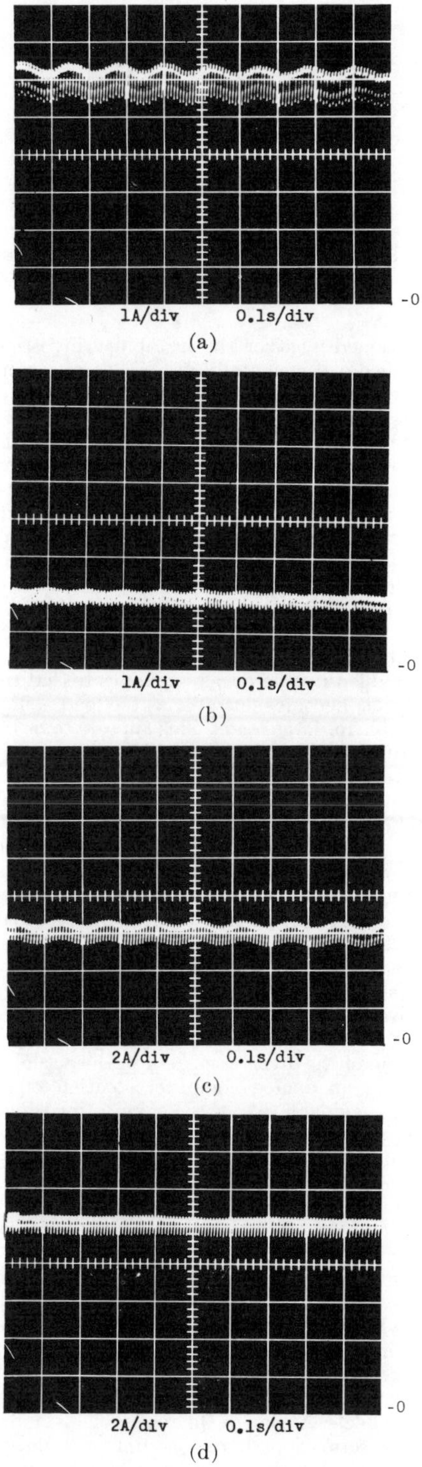

Fig. 18. Current I_R through filter choke at $f_R = 0.33$. (a) $T_L = 0.28$ pu; $V_m = 1.0$ pu. (b) $T_L = 1.09$ pu; $V_m = 1.0$ pu. (c) $T_L = 1.23$ pu; $V_m = 1.0$ pu. (d) $T_L = 1.34$ pu; $V_m = 1.2$ pu.

From the test system two general observations have been made. First, limited oscillations in regions close to those predicted by the authors have been observed. It is found that these oscillations are limited by some other factor than intermittent nonconduction of the rectifiers in the input bridge. Second, in most cases, load torque must be applied to the motor to obtain noticeable oscillations. The results of tests at 15 Hz and 20 Hz which show these points are presented in this discussion.

The test points which have been chosen for the purpose of discussion are shown in Fig. 16 superimposed on curves showing regions of instability calculated by the authors for the parameters used by the authors. All tests were performed by starting with the motor initially unloaded and then continually increasing the load. At $f_r = 0.25$ (15 Hz) and with $V_m = 1.0$ pu no oscillations were observed when the motor was unloaded. When the load was increased to 0.295 pu, as shown in Fig. 17(a), still no oscillations were observed. When the load was increased above 0.295 pu, oscillations were observed which grew in amplitude as the load was increased. An oscillation of maximum amplitude shown in Fig. 17(b) was found at a load torque of 0.655 pu. When the load torque was increased further, the amplitude of the oscillation decreased until at a load torque of 0.8 pu, as shown in Fig. 17(c), no oscillations were apparent. At this point it was felt that the upper bound of the region of instability predicted by the authors had been crossed. To verify this, the voltage V_m was raised to 1.2 pu which, according to the authors, should increase the region of instability and hence make the system unstable again. As can be seen in Fig. 17(d), when the voltage was raised to 1.2 pu, limited oscillations again occurred. Results for a similar test at 20 Hz are shown in Fig. 18. When the load was increased to 0.28 pu small oscillations were observed. Increasing the load further, oscillations of maximum amplitude were found at a load torque of 1.09 pu. Larger motor loads caused the oscillations to diminish in amplitude until at a load torque of 1.23 pu, which was very near the stall torque of the motor, the oscillations were not present as can be seen in Fig. 18(c). Again the voltage was raised to 1.2 pu and the oscillations returned as seen in figure 18(d). It should be noted that at no time during either of the tests did the rectifier current assume an intermittent zero mode of operation.

From the analysis by the authors and the experimental results which have been presented, it is not understood what factor or factors limit the amplitude of the oscillations which have been observed. It is clearly some factors other than nonconduction of the input diodes. Also it is not understood why the motor must be loaded to obtain oscillations. It may be that there is only one factor or it may be that there is more than one factor which raises these questions. The cause or causes of the discrepancies which have been observed may be particular to our system. On the other hand, assuming that the analysis is correct it may be that additional factors must be considered in an analysis of this nature in order to obtain closer agreement between theoretical and experimental results.

As a final point of the discussion, two questions relating to the application of the results are posed to the authors. First, in the analysis presented the rectified voltage V_R was set. In most applications the motor terminal voltage V_m would be adjusted so as to maintain breakdown torque. How could regulating V_m instead of V_R be handled in an analysis of this nature? Second, the results which have been presented apply to a particular set of parameters. Can results from this type of analysis be extended with additional work to apply to a general system and hence divorce the computer from the analysis of the problem?

T. A. Lipo and **P. C. Krause:** We wish to thank the discussers for their practical and thus, valuable contributions to our paper.

Dr. Jacovides and Dr. Agarwal question the choice of decreasing the inverter open circuit voltage as a linear function of frequency. Although this choice is satisfactory for frequencies above 30 Hz, due to the losses within the system it does tend to deplete the breakdown torque of the machine at low frequencies.

Manuscript received April 3, 1968.

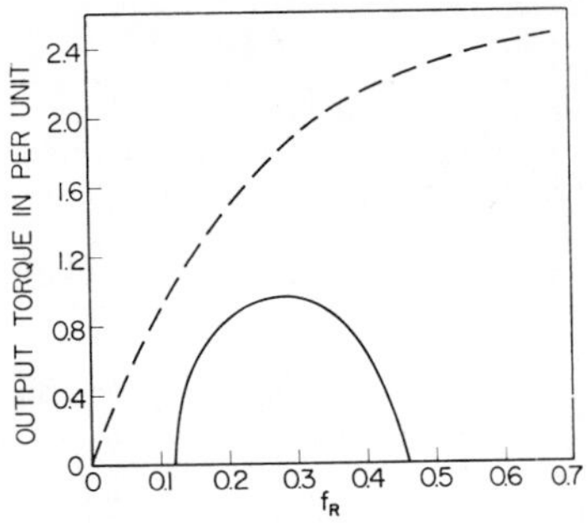

Fig. 19. Region of instability for $V_{I0} = (\pi/2)f_R$.

In practical systems, a regulator may be required to compensate for these system losses. This can be accomplished by various means. For example, the inverter dc voltage could be varied linearly with frequency and maintained constant for an increase in load, or the fundamental component of line to neutral stator voltage could be varied linearly with frequency and maintained constant with an increase in load. Finally, as mentioned in the discussion, the flux level in the machine could be maintained constant. Each technique requires a different form of regulator.

For example, if it is desired to maintain the ac stator volts per hertz constant, a feedback signal which is related to ac voltage is required. If it is required to keep the flux level in the machine constant, a feedback signal which is a function of ac voltage and stator phase current is required. In this paper, the open-loop behavior of a rectifier–inverter drive was investigated. A linear change of inverter open-circuit voltage appeared to be the most logical choice if the system is operated without feedback.

For purposes of comparison, the effect of compensation for system losses can be investigated if the regulator time constants are much greater than the important electrical time constants of the system. It can then be assumed that the regulator serves to adjust the steady-state quantities (16)–(21), but will not affect the small displacement quantities (33).

A region of instability for a system in which the fundamental component of stator voltage is varied linearly with frequency but maintained constant with an increase in load is given in Fig. 19. This has been accomplished by adjusting the average value of rectifier voltage so that

$$V_{I_0} = \frac{\pi}{2} f_R V_m \tag{59}$$

where $V_m = 1.0$ pu. A linear variation in inverter voltage can be implemented by setting X_{co} and R_{LF} to zero in the computer program which was written to obtain steady-state quantities for a linear variation in rectifier voltage. The voltage drops due to commutating reactance and filter resistance which are then readily established by (19) and (21); hence the rectifier voltage V_{ro} required to yield (59) can be obtained. The parameters of the system are identical to the parameters chosen for Figs. 3–6. Because of the stator losses in the induction machine, the breakdown torque again ultimately approaches zero as the frequency approaches zero.

In Fig. 20 the rectifier voltage has been adjusted so that

$$X_m \sqrt{i_{qso}^2 + i_{dso}^2} = 1.0 \tag{60}$$

The constraint given by (60) may be incorporated into the computer program in a manner similar to that used to obtain (59). In this case the flux level in the machine is constant and the break-down torque of the induction motor is independent of frequency. In either case, the shape of the region of instability does not differ significantly from the open-loop system studied in the paper.

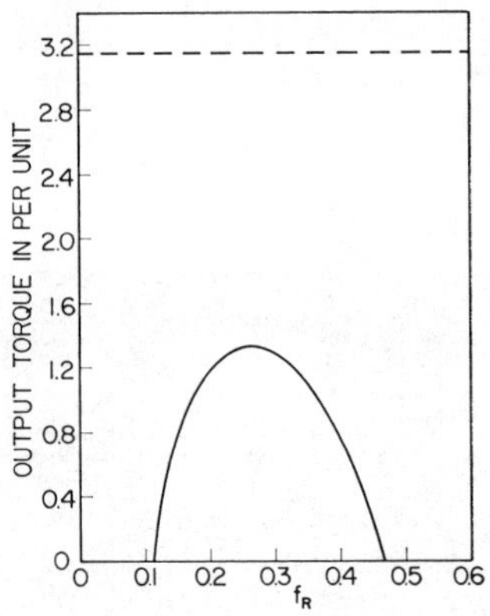

Fig. 20. Region of instability for constant breakdown torque operation.

It is also questioned whether the instability investigated in the paper can be verified experimentally in a practical system. The fact that such a system indeed can become unstable is clearly demonstrated by Mr. Studtmann and Mr. Yarema.

Mr. Studtmann and Mr. Yarema have provided a most interesting and important verification of our work. Although the parameters of the filter network were selected so as to approximate the system studied by the authors, the parameters of the induction machine differ appreciably from our parameters. Thus a one-to-one correlation cannot, of course, be expected. In particular, the relative decrease in stator and rotor leakage reactance (33 percent) tends to increase the region of instability. The relative increase in stator resistance (14 percent) tends to move the region of instability to the right (increasing f_R). A qualitative comparison can, however, be expected.

The figures provided by the discussers indicate that the system is stable for light loads, becomes unstable as the load is increased, and again becomes stable for large loads. This appears to be in contrast to the theoretical results which predict an instability at no load. In order to correlate the results, the discussers provided one of the authors the opportunity to visit the Borg-Warner facility and observe a second set of experimental results. At this time it was apparent that the behavior of the experimental system, identical with that previously employed, differed materially from that reported in the discussion. At frequencies of 15 and 20 Hz ($f_R = 0.25$ and 0.333), a sustained system oscillation was now evident from no load operation to the large load condition originally reported in the discussion. These qualitative results were thus in good agreement with that predicted in the paper. However, it was evident that the magnitude of the oscillations were not as severe as that predicted by the analog computer simulation of the idealized system (Fig. 5).

Upon comparison of the inverter current in the actual system with that obtained from an analog computer solution of an idealized inverter, it became evident that the commutating circuitry of the practical inverter provides a certain amount of damping not realized by an idealized inverter. This damping tends to reduce the system oscillations and, if the total negative damping provided by other system components is small, may stabilize the system. This appears to have been the case during the original experimental work carried out by Mr. Studtmann and Mr. Yarema.

In view of the number and complexity of the commutating circuits employed in present day inverters, it may be extremely difficult to include a term which would take the damping of a commutating circuit properly into account. Other sources of damping such as saturation, windage, and friction damping also have an effect upon stability. However, it is again very difficult to properly predict and account for these effects. Although the region of instability obtained by neglecting these terms is perhaps sometimes pessimistic, it yields a worst cast situation which is usually preferable in engineering design.

Pulsewidth Modulated Inverter Motor Drives with Improved Modulation

JACOB ZUBEK, MEMBER, IEEE, ALBERTO ABBONDANTI, MEMBER, IEEE, AND CRAIG J. NORDBY, MEMBER, IEEE

Abstract—**The performance of inverter drives, which use the pulsewidth modulated (PWM) technique to control motor applied voltage and frequency, are critically influenced by the choice of the modulation policy used in the control circuits. This paper deals with practical inverter drives for squirrel cage induction motors and presents some basic considerations on modulation requirements. The advantages and limitations of popular modulation methods are discussed, and an improved modulation scheme, which allows us to extend the practical speed range of PWM ac drives, is presented.**

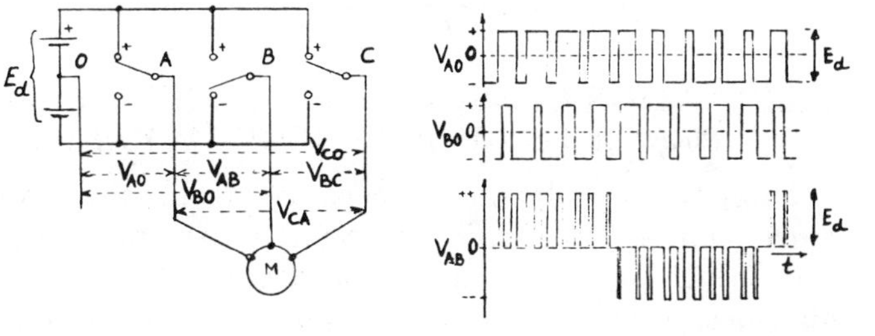

Fig. 1. Conceptual diagram and typical waveforms of a PWM inverter motor drive.

INTRODUCTION

THE PULSEWIDTH modulated (PWM) inverter drive for ac induction motors has the distinctive feature that it obtains its input power from a dc source (the "dc link") of roughly constant magnitude. The output stage can be conceptually represented by the set of single-pole double-throw switches shown in Fig. 1. Through proper activation of the switches, the fixed dc link voltage is applied for short time intervals with either polarity across the motor terminals, in such a way as to result in a fundamental output voltage of adjustable magnitude and frequency. The set of rules determining the sequence and the timing of the switch activations is termed the modulation policy.

The use of the PWM technique in motor drive applications is considered advantageous in many ways. For traction ac drives fed by a dc input power source, the PWM inverter is a practical solution which only involves a single power conversion. For industrial applications the PWM drive obtains its dc input through simple uncontrolled rectification of the commercial ac line and is favored for its good power factor, good efficiency, its relative freedom from regulation problems, and mainly for its ability to operate the motor with nearly sinusoidal current waveforms. At very low frequency this results in smooth speed and torque performance and freedom from cogging down to zero speed during reversal of the direction of rotation. Therefore, on the basis of overall performance the PWM inverter is able to compete favorably with the dc drive. The PWM drive is often preferred for its compatibility with the use of a single centralized dc rectifier facility providing a common dc bus serving a number of inverters, which in large installations is considered a favorable arrangement.

The potential advantages of the PWM drive over rival inverter techniques (namely converter-inverter or chopper-inverter systems using adjustable dc link voltage or current concepts) are obtained at the cost of greater sophistication in the control logic. This sophistication is required to properly implement the mechanism of modulation in a manner that minimizes the effects of unwanted harmonics or subharmonics of the motor terminal voltage. The presence of these undesirable voltage components is detrimental in a number of ways: it can cause torque and speed disturbances (cogging or frequency beats); it can cause motor overheating due to excessive copper or iron losses; and it can cause motor current transients which overtax the inverter commutation ability. Unless the modulation approach is carefully selected to contain within reasonable limits these harmful side effects of the modulation process, full practical advantage of the favorable features of the PWM drive cannot be taken.

This paper discusses some aspects of the modulation problems encountered in the development of a practical thyristor ac drive without regard to a particular power circuit configuration, and which, at nominal dc link voltage, meets the requirements of constant flux operation (constant V/f) to 80 Hz (base speed), and constant voltage operation above base speed (field weakening) to 160 Hz. This frequency range of operation corresponds to a 0–4800 r/min speed range for a four-pole traction motor. Workable conceptual solutions are presented which have been successfully reduced to practice.

Paper TOD-75-47, approved by the Static Power Converter Committee of the IEEE Industry Applications Society for presentation at the 1974 IEEE Industry Applications Society Annual Meeting, Pittsburgh, Pa. Manuscript released for publication May 5, 1975.

J. Zubek and A. Abbondanti are with the Westinghouse Electric Corporation, Pittsburgh, Pa. 15235.

C. J. Nordby is with the Westinghouse Electric Corporation, Buffalo, N. Y.

Reprinted from *IEEE Trans. Ind. Appl.*, vol. IA-11, pp. 695–703, Nov./Dec. 1975.

TABLE I

TRUTH TABLE SHOWING THE NECESSARY CORRELATION BETWEEN POLE VOLTAGE LEVELS AND LINE VOLTAGE LEVELS IN A 3-PHASE BRIDGE INVERTER

Pole voltage level			Line voltage level		
V_{AO}	V_{BO}	V_{CO}	V_{AB}	V_{BC}	V_{CA}
(+)	(+)	(+)	0	0	0
(-)	(+)	(+)	(--)	0	(++)
(+)	(-)	(+)	(++)	(--)	0
(+)	(+)	(-)	0	(++)	(--)
(-)	(-)	(+)	0	(--)	(++)
(+)	(-)	(-)	(++)	0	(--)
(-)	(+)	(-)	(--)	(++)	0
(-)	(-)	(-)	0	0	0

MODULATION CONSTRAINTS IMPOSED BY THE POWER CIRCUIT CONFIGURATION

From the modulation approach viewpoint, the power circuit imposes two basic constraints on the modulation rules which are found in varying degrees of severity in all known commutation circuits. First, there is an upper limit on the usable rate of commutation which restricts the number of pulses/s in the PWM waveform (or the number of activations/s of each switch in Fig. 1). This limit is dictated by thermal considerations. At each cycle, the commutation capacitors exchange their stored charge, and in the process, joule, dielectric, and eddy current effects generate losses in the commutation circuit components. At high rates of commutation the problem of heat evacuation becomes difficult to solve with certain types of enclosures. Besides, the increased current ratings requirements of the components and devices have an adverse effect on the price. In practice it becomes counterproductive to consider commutation rates in excess of 500 Hz.

The other restriction in the modulation process imposed by the power circuit follows from the existence of a minimum delay that must elapse between the instant a given main thyristor is fired and the instant it can be commutated. This minimum conduction time must be respected to ensure safe communtation. The gate drive logic to the thyristor includes the necessary circuitry to ensure that whenever a given main thyristor is fired, the mechanism for its commutation remains disabled for the duration of the minimum conduction time t_m. This restricts the freedom to maneuver the switches of Fig. 1. The commutation circuit must be selected and designed in order to minimize t_m. For small drives (15 hp), t_m will typically be as low as 200 μs. For a 500 hp peak ac traction drive, t_m can be as large as 460 μs. It is desirable that virtually no circuit modification of the modulation controls be required to accommodate a wide range of values of t_m.

MODULATION CONSTRAINTS IN A 3-PHASE BRIDGE

In the literature one can find a number of analytical waveform studies aimed at defining a waveform suited for use in PWM systems. Often these studies ignore one fundamental handicap encountered in inverter systems having the 3-phase bridge configuration. This handicap stems from the interaction between output phases which restricts the freedom one has of shaping at will the waveform of one pole, lest the other two be seriously distorted. The constraints imposed by the 3-phase bridge configuration on the freedom of the output waveform to assume any given level are summarized in Table I. First, pole voltage will be defined as the voltage V_{AO}, Fig. 1, between the pole output and a fictive center tap on the dc link supply E_d, and line voltage as the voltage V_{AB} between two output lines. We see that a pole voltage can only assume two levels, called $(+)$ and $(-)$ on Fig. 1, whereas a line voltage can assume three levels, called 0, $(++)$ and $(--)$. In Table I, all possible pole voltage combination levels are listed and the corresponding combination levels of the line voltages. We see that three line voltages must, at any given instant, assume levels which are different from each other, except for the trivial all-zero combination, which can only occur when all poles are at the same level.

Another constraint is imposed by the need to respect the 3-phase symmetry of the output fundamental line voltages. Fig. 2 shows the relationship between the 3 fundamental line sine waves. One can see that an output cycle can be divided into six identical segments such as the one represented between the XX and YY axes. In each such segment, one line voltage has a peaking course, i.e., rides through its crest with a given polarity, whereas the other two have a zero crossing course, i.e., rise from zero or decline to zero with the opposite polarity. Thus, in the modulation process, whenever a line voltage is being manipulated so that its fundamental has a peaking course of a given polarity, the other two line voltages must be manipulated according to rules that will confer a zero crossing course of opposite polarity to their fundamentals.

One can combine these constraints with another rule not necessarily imposed by the 3-phase configuration. This is the "pulse polarity consistency rule," which says that the string of pulses in a modulated line voltage should have the same polarity through an entire half cycle of the output cycle. All the pulses should be positive for the positive fundamental half cycle. Fig. 3 shows two waveforms, corresponding to the same fundamental amplitude and frequency and the same number of pulses per output cycle. The top waveform conforms to the pulse polarity consistency rule; the bottom does not conform. A harmonic analysis would confirm the intuitively evident fact that the nonconforming waveform contains many more unwanted high frequency components.

Combining the above constraints and rules, one can convince oneself that in each one of the six identical output cycle sements, there is only one permissible pole commutation sequence, that is, only one sequence according to which the three switches of Fig. 1 may be activated. (See Table II.) Typically, in a segment in which line voltage V_{AB} has a (positive) peaking course, the sequence would start with the three switches (A, B,

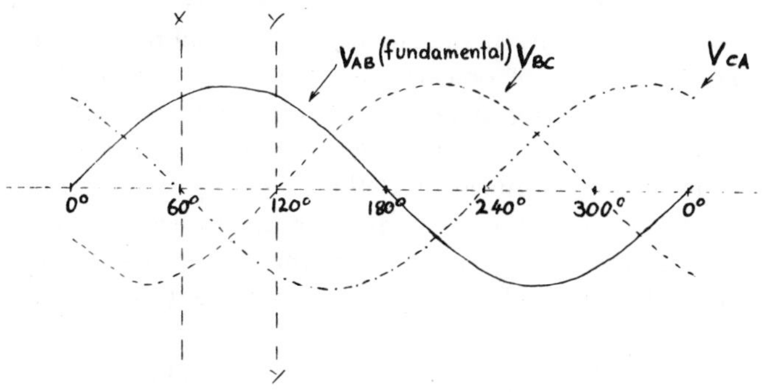

Fig. 2. Relationship between the three fundamental line voltages in one of six identical segments of an output cycle.

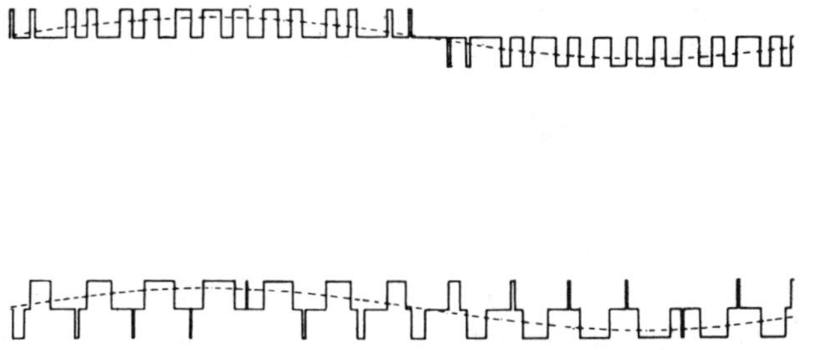

Fig. 3. Two PWM line voltage waveforms of same fundamental amplitude. Only the upper one has a consistent pulse polarity through a half cycle.

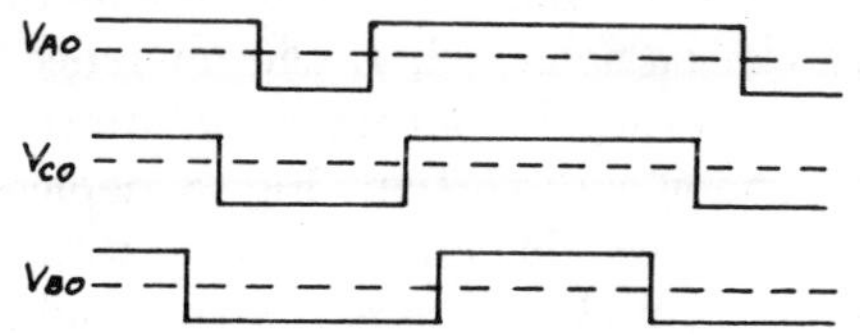

Fig. 4. "Encased" pole voltage waveform aspect, typical of the optimum commutation sequence.

TABLE II

TABLE SHOWING THE OPTIMUM COMMUTATION SEQUENCE VALID THROUGH A GIVEN SEGMENT OF THE OUTPUT CYCLE

Line Voltage having a peaking course in the considered segment	Polarity of the line voltage crest in the considered segment	Optimum pole commutation sequence
V_{AB}	+	B↘ ,C↘ ,A↘ ,A↗ ,C↗ ,B↗
	-	A↘ ,C↘ ,B↘ ,B↗ ,C↗ ,A↗
V_{BC}	+	C↘ ,A↘ ,B↘ ,B↗ ,A↗ ,C↗
	-	B↘ ,A↘ ,C↘ ,C↗ ,A↗ ,B↗
V_{CA}	+	A↘ ,B↘ ,C↘ ,C↗ ,B↗ ,A↗
	-	C↘ ,B↘ ,A↘ ,A↗ ,B↗ ,C↗

B↘ means: pole voltage V_{BO} commutates from the (+) level to the (-) level

A↗ means: pole voltage V_{AO} commutates from the (-) level to the (+) level

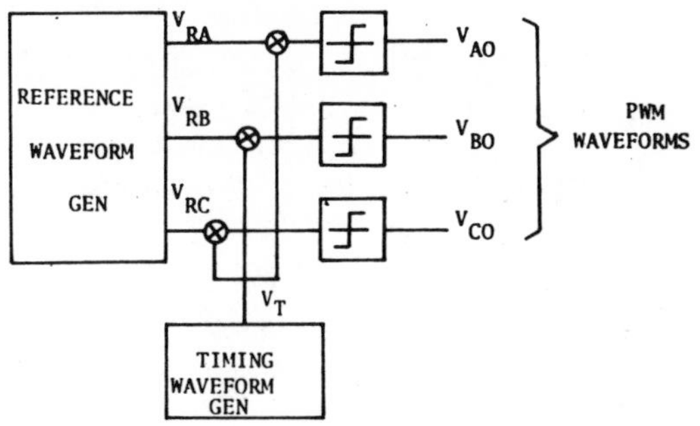

Fig. 5. Block diagram of a modulator using the triangulation method.

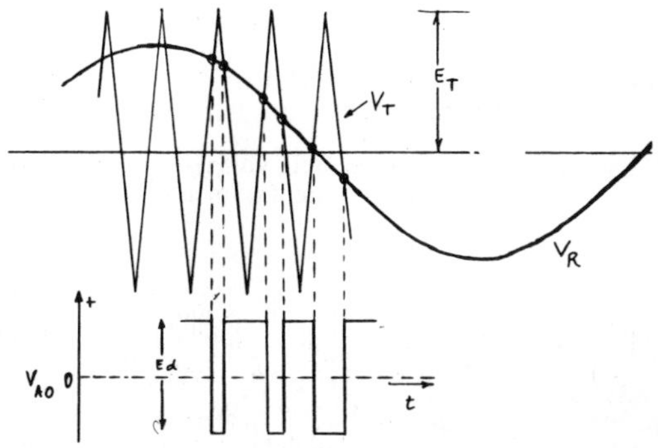

Fig. 6. Modulation by the triangulation approach.

C) in the same (+) state. The switches then toggle in the following sequence: B, C, A, ending with all switches again in the same state, opposite to the initial one. The switches then retoggle in the reverse sequence: A, C, B, ending in the initial combination. The timing of the togglings depends on the output fundamental amplitude requirements and is a variable. However, the sequence is imposed and must repeat itself through the segment of output cycle. This results in the "encased" aspect of the pole voltages represented in Fig. 4. Whatever modulation approach is used, this pole commutation sequence rule must be satisfied.

THE TRIANGULATION METHOD

The described inverter drive uses a known modulation method which inherently satisfies the optimum pole switching sequence rule and is referred to as the triangle interception method [1], or the subharmonic method [2]. Referring to Fig. 5, a set of three reference signals V_R in a symmetric 3-phase relationship is supplied by a reference generator subunit. These reference signals should preferably be sinusoidal or exhibit little harmonic distortion. Their frequency should be adjustable within the range of the desired inverter output frequency, and their common level should be controllable over a wide range independent of frequency. Each reference signal is compared to a common triangular wave V_T (said "timing wave") of fixed amplitude and frequency. The output V_O of each comparator has a "high" level whenever the instantaneous input reference level exceeds the timing wave level and a "low" level when the reference is exceeded by the timing wave, resulting in a PWM waveform, as shown in Fig. 6.

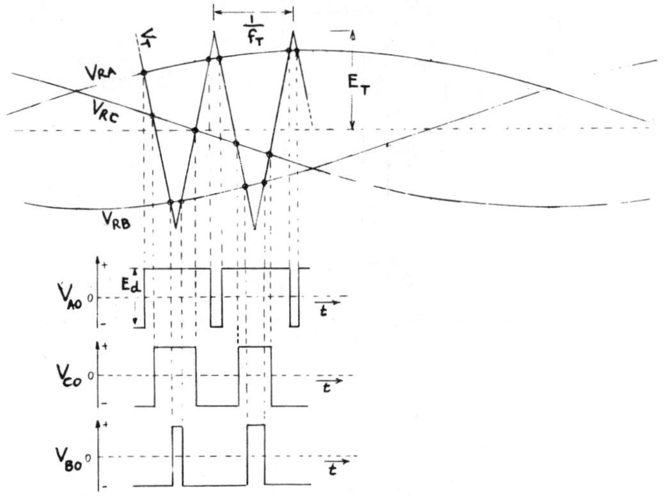

Fig. 7. Relationship between pole voltages in a 3-phase modulator using the triangulation method.

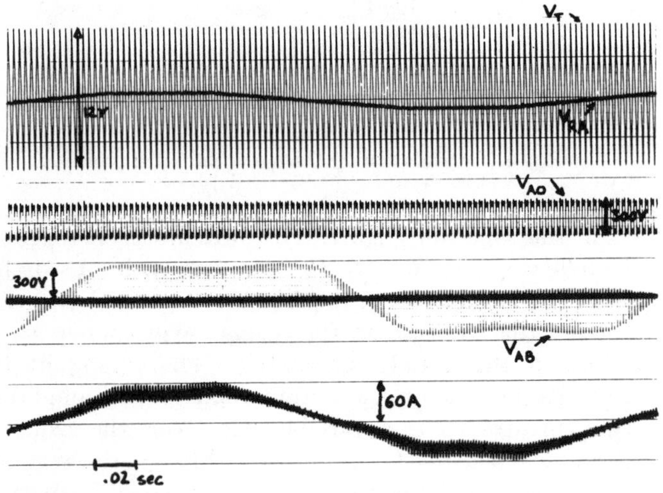

(a)

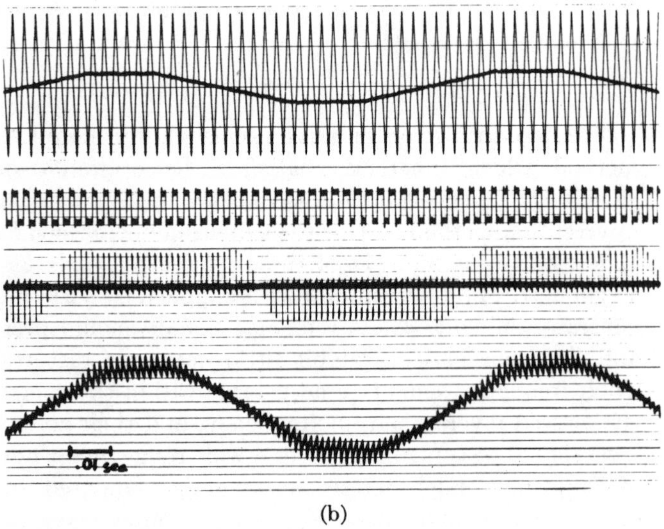

(b)

Fig. 8. (a) Oscillograph recordings showing the triangulation method applied at an inverter output frequency of 3 Hz. (b) Oscillograph recordings showing the triangulation method applied at an inverter frequency of 10 Hz. Upper traces: superposition of the timing waveform (360 Hz) and one reference signal. Second trace: pole voltage. Third trace: line voltage. Lowest trace: motor current.

The comparator outputs are used to generate gate drive signals to the main and commutation thyristors of each pole in such a manner that the two discrete levels of a comparator output are made to correspond to the two possible states of one of the switches of Fig. 1. As a result, the inverter pole voltage waveform reproduces the comparator output at the high level within the approximation of the commutation delays. The 3-phase relationship between the reference signals and the use of a single timing wave for the system are the conditions that ensure the fulfillment of the optimum pole switching sequence rule (Fig. 7).

The triangulation method produces a pole output voltage waveform whose fundamental has the same frequency and same phase as the corresponding reference signal, and whose amplitude E_1 is a linear function of the modulation ratio, i.e., of the reference amplitude E_R relative to the timing waveform amplitude E_T according to

$$E_1 = \frac{E_d E_R}{2E_T}, \qquad E_d = \text{dc link voltage}. \tag{1}$$

For a sinusoidal reference, the output waveform harmonic distortion is low if the ratio between the timing waveform frequency and the reference frequency is high. In this case not only the distortion is low, but the residual harmonics all have such a high order that they are virtually without influence on the motor torque continuity. This makes the method ideally suited for low frequency applications where a smoothness of rotation superior to that of a dc drive can be achieved. If the reference signal is not sinusoidal, however, the output waveform reproduces the reference harmonic distortion. Fig. 8 depicts this situation when a trapezoidal reference waveform is used. The figures are multitrack recordings of the modulator signals, the output voltages, and one output line current in a typical drive operating at nominal load at 3 and 10 Hz. One notes the high number of pulses per output cycle in the line voltage, resulting in a smooth motor current reproducing the distortion of the trapezoidal reference signal. If the latter would be made perfectly sinusoidal, the motor current would also become a sinusoid with a finely jagged outline. If the reference signal would be made a square wave, the output waveform would reproduce the characteristics of known "multipulse" modulation approaches [3] with high carrier frequency to inverter frequency ratios. Such systems produce an output waveform which is, at best, equivalent to a low frequency, low voltage square wave. The resulting motor behavior at low frequency exhibits cogging effects practically absent when a triangulation method with near sinusoidal reference is used.

As the ratio between the timing waveform frequency f_T and the reference or inverter output frequency f_R decreases for higher motor speeds, the pulses in each output cycle become fatter and fewer [Fig. 8(b)], and the harmonic distortion becomes more pronounced. One must compromise between waveform quality and inverter losses, since at any given frequency the distortion is reduced if the timing waveform frequency is increased. This produces an increase in the rate of commutation with associated thermal problems. Usually, at relatively high frequency, the presence of harmonics in the output voltage waveform has less significant effects on the torque smoothness of the machine; therefore, from the viewpoint of the harmonic distortion, one could tolerate rather low f_T/f_R ratios and low commutation rates. However, another side effect of the triangulation method becomes important in that case, namely, the generation of subharmonic voltage components or frequency "beats". This occurs because the phase relationship between reference signals and timing waveform is not fixed. Consequently, the pulse pattern does not repeat itself identically from cycle to cycle, but small changes continually occur at a rate related to the difference between the timing waveform frequency and some multiples of the reference frequency. It is said that the pulse pattern is "free-running," as opposed to the synchronized pattern obtained through other approaches. As long as the number of pulses per output cycle is high, and the magnitude of the output fundamental is small, the free running pulse pattern has no detrimental effect on the motor operation. At higher frequencies, however, the simultaneous occurrence of a reduction of the number of pulses and an increase in their width (to produce more output voltage) make the system much more sensitive to small changes of the pattern. From cycle to cycle the output voltage amplitude undergoes a small but cumulative change, which causes low rate fluctuations in the machine excitation and characteristic current, torque, and speed beats. To contain the beats within tolerable limits, the f_T/f_R ratio cannot be made too small. In practice f_T/f_R should not be made much lower than nine.

The main drawback of the triangulation method is the limitation of the frequency range of its applicability. Theoretically the method is applicable for modulation ratios E_R/E_T comprised between zero and one with the condition $f_T/f_R > 9$. From (1) this leads to a maximum pole voltage fundamental amplitude

$$E_{1m} = \frac{E_d}{2}. \tag{2}$$

The above maximum output is lower than the peak fundamental pole voltage E_{1p} possible in a three phase bridge inverter of given dc link voltage E_d. This absolute maximum output amplitude is obtained if the inverter is operated with 120° displaced, unmodulated square waves as pole voltages (V_{AO}, V_{BO}, and V_{CO}, Fig. 9). The line voltages V_{AB}, V_{BC}, and V_{CA} then have the familiar six step pseudo square wave shape. In this case the pole voltage fundamental amplitude is

$$E_{1P} = \frac{2}{\pi} E_d. \tag{3}$$

By comparison with (2), one can see that the use of the triangulation method prevents full utilization of the inverter voltage capability.

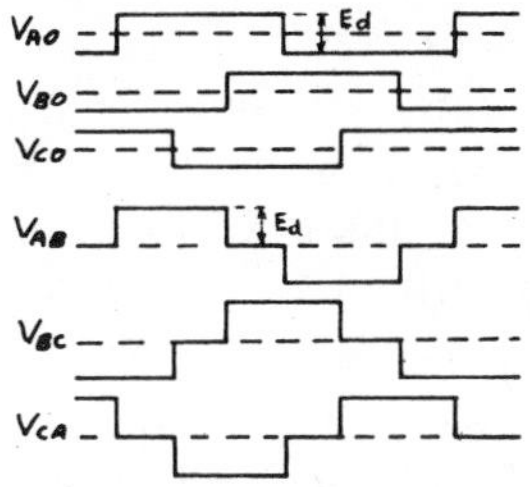

Fig. 9. Unmodulated output square waves yielding the maximum output voltage possible with a given dc link voltage magnitude E_d.

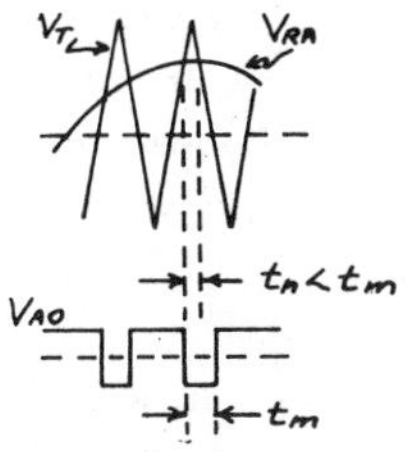

Fig. 10. Case of a modulation ratio exceeding the critical level of activation of the minimum pulsewidth clamp.

In practice, moreover, the underutilization is more severe than shown by (2) and (3). The linear relationship between reference amplitude and output level, expressed by (1), does not hold for modulation ratio values approaching one. This would imply that infinitely narrow notches would be present in the pole voltage waveform, and that a minimum pulsewidth clamp circuitry exists to prevent that occurrence in order to safeguard the commutation ability. When the amplitude of the reference waveform exceeds a given critical percentage of the timing waveform peak level (Fig. 10), the minimum pulsewidth clamp is enabled, and the normal modulation process is overridden so that a notch width at least equal to the minimum conduction time t_m is ensured. The critical level depends on the power circuit characteristics which impose t_m and is related to the chosen timing waveform frequency and the desired output volt/hertz characteristic.

Under conditions in which the minimum conduction time clamp is activated, two things happen. First, the output level ceases to be a linear function of the reference level, which results in a lower possible maximum output than is indicated by (2). However, the most harmful consequence is that beat effects develop as the free-running pulse pattern becomes disturbed in correspondence with the crests of the reference. The beat effects rise rapidly in severity as the modulation ratio is increased above the level of activation of the minimum pulsewidth clamp, to the extent that this modulation ratio level can be considered as the upper limit of the range of applicability of the triangulation method. In a practical system, such as the described inverter drive with a timing wave frequency of 360 Hz, such level is reached at about 40 Hz output frequency. This corresponds to the case in which a standard 60 Hz motor is used having a nameplate voltage equal to the commercial line voltage feeding the inverter front end rectifier. The output volt/hertz characteristic is such that the motor is driven at constant nominal flux over the modulation range. The inverter would have the voltage capability to drive the motor at nominal flux up to 60 Hz, if a proper modulation approach would be applied. The frequency could be increased well above 60 Hz if field weakening operation with a constant amplitude output square wave would be considered. This gives an idea of the degree of underutilization of the inverter capabilities which results when the modulation policy is restricted to the use of the triangulation method.

THE MULTIMODE APPROACH

Modulation techniques resulting in pulse patterns which are phase locked or synchronized [1], [4] to the output fundamental are inherently free of beat effects and are well suited for application at frequencies beyond the reach of the triangulation method; that is, the frequency span in which the modulation process is gradually phased out until the output becomes a full amplitude square wave. No inverter underutilization problem exists with such methods. Unfortunately, they generally do not perform as well as the triangulation method at the low end of the frequency range. This follows from the fact that they work at a fixed number of pulses per cycle, resulting inevitably in excessive motor current ripple if the inverter frequency is made low enough, or alternatively, they use a "multipulse" approach in which the waveform quality is at best equivalent to a six step square wave system.

To retain the low frequency performance typical of the triangulation method and benefit of the good high frequency characteristics of a synchronized method, one can consider switching the modulation strategy at some suitable point of the inverter operational frequency range. This is the approach presented here, by which several modes of operation of the modulator are applied, one of which results in the triangulation method, the other resulting in fixed pulse pattern methods, particularly suited to the frequency span in which they are used.

In order to minimize the complexity of the controls, it is desirable that the same circuit arrangement necessary to implement the modulation process in one of the modes can be put to use when the modulator operates in the other modes. With this in mind, particular fixed pattern waveforms are adopted, so that they can be generated from the circuits and signals involved in the modulator block diagram of Fig. 5 primarily intended for implementation of the triangulation approach. The only requirement necessary to convert Fig. 5 into a fixed pattern waveform generator is to properly alter some characteristics of the common timing waveform signal applied to the three comparators. This alteration must be done in such a manner that after the change the modulator will still retain its original property of inherently satisfying the optimum pole commutation sequence rule.

These conditions are fulfilled if the following modulation strategy is used. At low frequency the modulation operates in mode 1, characterized by the application of a triangulation technique with a triangular timing waveform of fixed amplitude, fixed frequency, and independent phase

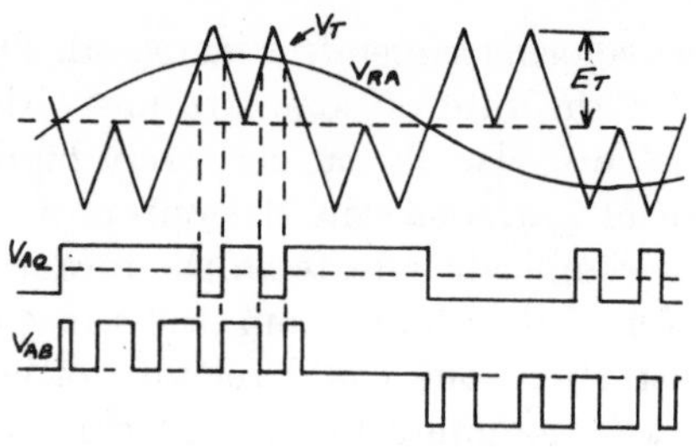

Fig. 11. Mode 2 waveforms.

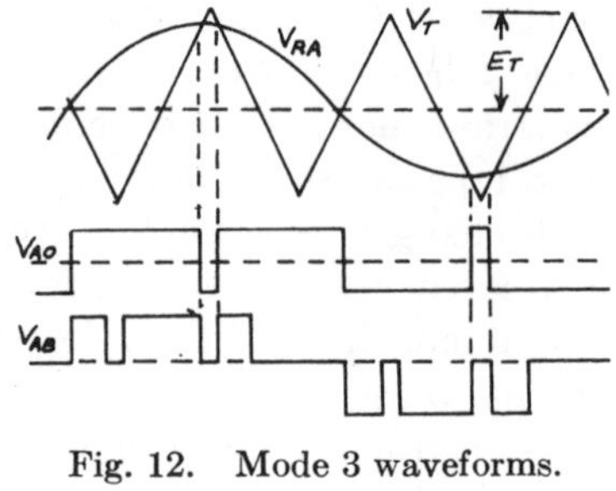

Fig. 12. Mode 3 waveforms.

with respect to the reference signals. When a preset inverter operating frequency is detected, for which it is known that the modulation ratio approaches the critical upper level limiting the modulator linear range, the system is switched to mode 2, described by Fig. 11. The timing waveform V_T has the same amplitude as in mode 1, but its frequency (or slope) changes. It becomes a multiple of the reference signal frequency, meaning that in case of reference signal frequency variations, the timing waveform frequency must track. Moreover, the shape of the timing waveform ceases to be triangular. At the zero crossings corresponding to the reference signal angular positions $(1 + 2k)\pi/6$ (i.e., 30°, 90°, 150°, etc.), it undergoes abrupt slope inversions. One can convince oneself that the figure depicts a relationship valid for each of the three reference signals.

The PWM waveform resulting from the new situation is shown on Fig. 11. The line voltage has five pulses per half cycle, the pole voltage half cycle has two equal notches, equidistant from the 90° position. The amplitude E_1 of the pole voltage fundamental is a nearly linear function of the reference signal amplitude E_R relative to the peak level E_T of the timing waveform, according to

$$E_1 = \frac{2E_d E_R}{\pi E_T}. \tag{4}$$

Compared to (1), one can see that at the moment of the switching from mode 1 to mode 2, the reference amplitude E_R must be reduced by a factor $4/\pi$ to ensure that the transition occurs without discontinuity in the amplitude of the output waveform (see Fig. 16).

Because of this amplitude reduction at the switching moment, and because of the reduced slope of V_T, the notch width, which was approaching the critical value just before switching, widens considerably in mode 2. However, if the output frequency is further increased, together with the output amplitude to keep the volt/hertz characteristic constant, the notches narrow rapidly, and a frequency is reached for which the notch width becomes again nearly equal to t_m. This frequency represents the upper boundary of the frequency region in which mode 2 can be used. If the frequency is increased without switching to a new mode the minimum pulsewidth clamp is enabled, and the output fundamental cannot further increase. In fact it would start decreasing with frequency, because the clamped notches would grow wider in terms of electrical degrees of the reference signal.

To conserve the ability of keeping the machine flux constant for frequencies beyond the range of mode 2, at some preset inverter frequency the system is switched in mode 3 by manipulating again the timing waveform as shown on Fig. 12. The amplitude E_T is still unchaged, and V_T remains synchronized to the reference signal. However, the shape of the timing waveform is modified to become a triangle again, with a phase relationship to the fundamental such that the zero crossings occur at angular positions $k(\pi/3)$ (for example, 0°, 60°, 120°, etc.). This results in a pole voltage waveform having a single notch per half cycle, centered at the 90° angular position, and a line voltage made up of 3 pulses per half cycle. Because of the lower frequency of the timing waveform (the slope steepness is reduced by a factor of two compared to mode 2), the single notch of the new pole waveform is wider than each of the two notches in the old waveform; therefore, further voltage control is possible. This can be done by varying the amplitude of the reference signal. A nearly linear relationship between reference amplitude E_R and output fundamental amplitude E_1 applies, approximated by (4). No reference amplitude adjustment is required upon switching from mode 2 to mode 3: automatic continuity of the output amplitude through the transition results by nature.

At some higher frequency the central notch of the mode 3 waveform will have narrowed to the critical width t_m. This is the upper boundary frequency of the region of the application of mode 3. For further frequency increases, operation in mode 3 must be abandoned, because the output fundamental amplitude would start decreasing with frequency as soon as the minimum pulsewidth clamp is enabled. Thus, at a preset frequency level the system is switched to mode 4 which is characterized by a pole-to-pole output where one of the notches in each half cycle in mode 3 is removed and the remaining notch in each half cycle is controlled to provide the proper fundamental amplitude (see Fig. 13). After the transition to mode 4, the single notch in each half cycle of pole output voltage must be twice as large as either of the two notches in mode 3 to provide the proper amplitude of fundamental components. The increase in notch width is accomplished by increasing the timing wave amplitude in mode 4 by approximately a factor of two. The removal of one notch in each half cycle of line-to-line output is accomplished by either exaggerating the amplitude of the reference waveform on the negative half cycle so as to avoid a transition with the timing waveform as shown in Fig. 13, or by logically overriding the intersection of reference and timing wave on each negative half cycle. Modulation in this

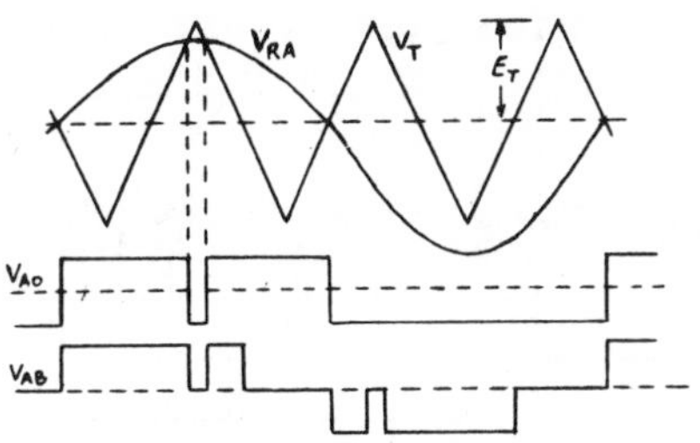

Fig. 13. Mode 4 waveforms.

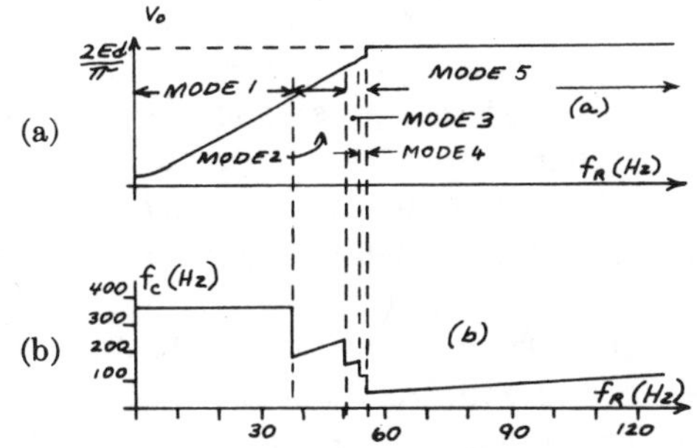

Fig. 14. (a) Typical output voltage/frequency characteristic obtainable with the multimode modulation approach. (b) Typical variations of the rate of commutation (number of commutations per second of a given main thyristor).

mode can proceed until the notch in each half cycle of line-to-line output again reaches the critical width t_m. Thus again, at a preset frequency level the system is switched to mode 5, which is characterized by the absence of modulation. The pole output voltage is a full conduction square wave and the line voltage is the six step quasi-square wave of Fig. 9. This transition is accomplished by a last manipulation of the timing wave, whose amplitude is suddenly lowered so as to avoid any intersection with the reference outside the zero axis.

In mode 5 the pole fundamental voltage amplitude E_1 becomes uncontrollable. It has its absolute maximum value given by (3). For further frequency increases, the machine enters the field weakening region of operation. It is therefore not possible to avoid a step variation of the output amplitude (and the machine flux) at the mode 4 to mode 5 transition. This step, which in a practical system has a magnitude of about 7 percent of the motor nameplate voltage, is usually a tolerable inconvenience.

Applied to the described PWM inverter drive, the multimode modulation approach allows a constant volt/hertz operation up to a frequency close to the motor nameplate frequency of 60 Hz. Fig. 14 shows the obtainable voltage/frequency characteristic, with indication of the boundary frequencies for the various modes. A plot of the rate of commutation versus operating frequency is also given, showing how the rate never exceeds the low frequency value of 360 Hz and drops by stages at each mode switching.

At constant torque in the constant volt/hertz range of ac motor operation, the peak current that the inverter thysistors must commutate increases as the fundamental frequency increases in modulation mode 1, and decreases as the fundamental frequency increases in modes 2, 3, and

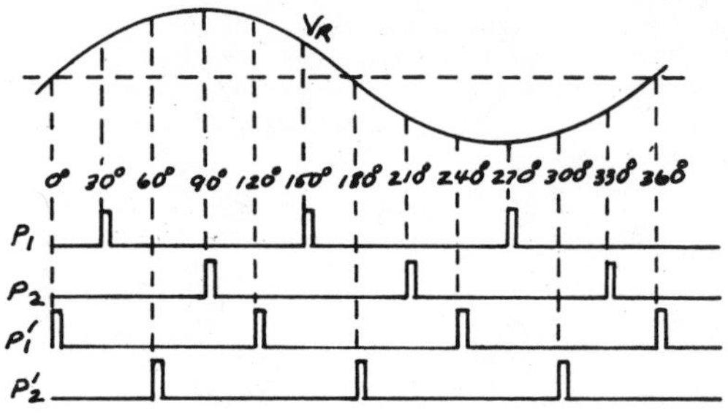

Fig. 15. Synchronizing pulses at fixed angular positions of the reference signal.

4, since the harmonic voltages applied to the motor and the resultant harmonic motor currents increase as the fundamental frequency increases in mode 1 and decrease as the fundamental frequency increases in modes 2, 3, and 4.

The transition from mode 1 to mode 2 is dictated by the necessity to minimize subharmonic beat currents in the motor. These increase rapidly when the fundamental frequency and voltage applied to the motor reach a point at which the inverter output is constrained not to follow the commands of the triangular and reference wave intersections, due to the inverter commutating circuit minimum conduction time requirements.

Modulation is switched to mode 2 before the minimum condition time limit is reached in mode 1. Typically, for the same torque, the current that the inverter thyristors must commutate at the highest fundamental frequency in mode 1 is less than at the lowest fundamental frequency in mode 2, 3, and 4. The peak currents that must be commutated at the lowest fundamental frequency in modes 2, 3, and 4 (the worst case points) are typically comparable.

MODE SWITCHING COORDINATION

The boundary frequencies for which mode switching is desired can be detected through level comparators processing an analog signal, proportional to the inverter frequency and controlling the mode of operation of the timing waveform generator circuit. However, if the modifications in timing waveform shape are initiated at random without method and organized time relationship with the reference signals, uncontrolled motor current transients at the beginning of a new mode are likely to occur. A first requirement is the introduction of some hysteresis in the comparison level (e.g., ± 1 Hz) to avoid repeated back and forth mode switching when the frequency signal happens to hover around the detection level. Another important condition is that any new timing waveform shape should be initiated at a reference angular position where it normally crosses zero. For this, one should obtain the pulse trains P_1, P_2, P_1', P_2' from the reference signal generator circuit (Fig. 15) in the shown phase relationship with the reference.

A mode 1 to mode 2 transition (in both directions) should be enabled only in coincidence with pulses P_1 or P_2 (P_1 initiates the timing waveform with a down-going slope, P_2 with an up-going slope). Mode 2 to mode 3 or mode 3 to mode 4 transitions should be enabled only in coincidence with pulses P_1' or P_2'. Besides their mode switching coordination function, these pulses play a role in ensuring proper synchronization of the timing waveform in modes 2, 3, 4, and 5.

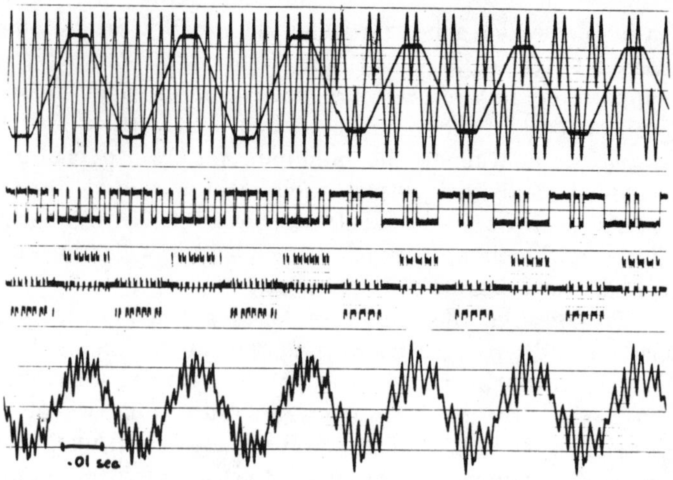

Fig. 16. Recording of modulator waveforms and motor voltage and current waveforms at a mode 1-mode 2 transition. Upper traces: superimposed reference and timing wave. Second trace: pole voltage. Third trace: line voltage. Lowest trace: motor current.

The effectiveness of the proposed coordination approach is illustrated by the oscillograph recording of Fig. 16 showing the absence of motor line current transients through a mode 1 to mode 2 transition (15 hp motor under nominal load, 37 Hz boundary frequency). One can observe on the figure how the free-running pulse pattern becomes fixed after mode switching and how the number of pulses fall with corresponding increases in current waveform roughness.

CONCLUSIONS

The multimode modulation approach discussed in this paper is a technique which combines the advantages of two opposed modulation concepts (free-running pattern and fixed pattern) and eliminates their respective drawbacks. It has been found of easy practical implementation using straightforward circuitry. Predicted results have been experimentally verified with prototype ac motor drives intended either for traction applications or for industrial use. Satisfactory performance was obtained over an overall frequency range of 1–160 Hz, including the capability of direction reversal. While recognizing that there is still room for improvement in this aspect of the PWM drive technology, the authors feel that the described concept is a practical modulation strategy for PWM drives applicable over a wide frequency range.

REFERENCES

[1] K. Heintze *et al.*, "Pulse width modulating static inverters for the speed control of induction motors," *Siemens-Z.*, vol. 45 (3), pp. 154-161, 1971.

[2] A. Schonung and H. Stemmler, "Static frequency changers with subharmonic control in conjunction with reversible variable speed ac drives," *Brown-Boveri Review*, pp. 555-577, Aug./Sept. 1964.

[3] R. D. Adams and R. S. Fox, "Several modulation techniques PWM inverter," in *Conf. Rec. IGA 1970 IEEE Fifth Annu. Meet. of IEEE Ind. Gen. Appl. Group*, pp. 687–693.

[4] S. B. Deman and J. B. Forsythe, "Harmonic analysis of a synchronized pulsewidth modulated three phase inverter," in *Conf. Rec. IGA 1971 Sixth Annu. Meet. of IEEE Ind. Gen. Appl. Group*, pp. 327–332.

Generalized Techniques of Harmonic Elimination and Voltage Control in Thyristor Inverters: Part I—Harmonic Elimination

HASMUKH S. PATEL AND RICHARD G. HOFT

Abstract—This paper considers the theoretical problem of eliminating harmonics in inverter-output waveforms. Generalized methods are developed for eliminating a fixed number of harmonics in the half-bridge and full-bridge inverter-output waveforms, and solutions are presented for eliminating up to five harmonics. Numerical techniques are applied to solve the nonlinear equations of the problem on the computer. The uneliminated higher order harmonics can be easily attenuated by using filter circuits in the output stage of the inverter. The results show the feasibility of obtaining practically sinusoidal output waveforms, which are highly desirable in most inverter applications.

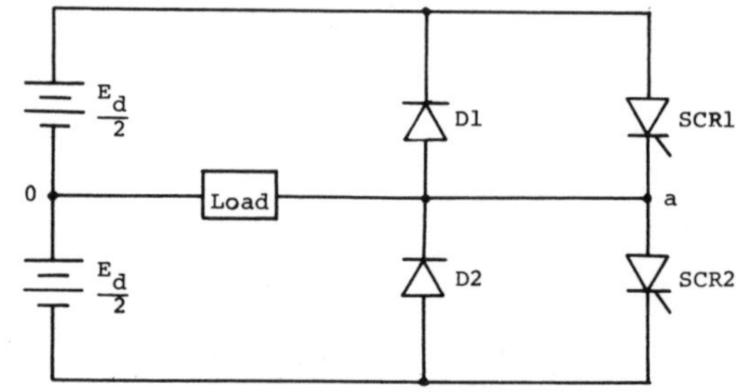

Fig. 1. Single-phase half-bridge inverter.

INTRODUCTION

SINCE THE advent of the thyristor family of semiconductors, including the SCR, tremendous interest has been renewed in inverter technology. In recent years the SCR-device technology has also made significant progress, enabling the creation of sophisticated inverter circuits for a wide variety of applications. The availability of SCRs in high power ratings, having turn-off times in the range of a few microseconds, has increased the feasibility of achieving a practically sinusoidal output by employing sophisticated switching patterns in inverter circuits. The derivation of optimal switching patterns to obtain a harmonic-free sinusoidal output is the subject of this paper.

The trends of modern integrated circuit technology are favorable in considering the implementation of the theoretical results. It is foreseen that the techniques developed should be practical as well as economical considering the scope of applications. Desirability of very low output waveform distortion in standby static power conversion equipment, favors the implementation of the theoretical techniques developed. The results for eliminating two harmonics in the output agree with those derived by Turnbull [2], [3].

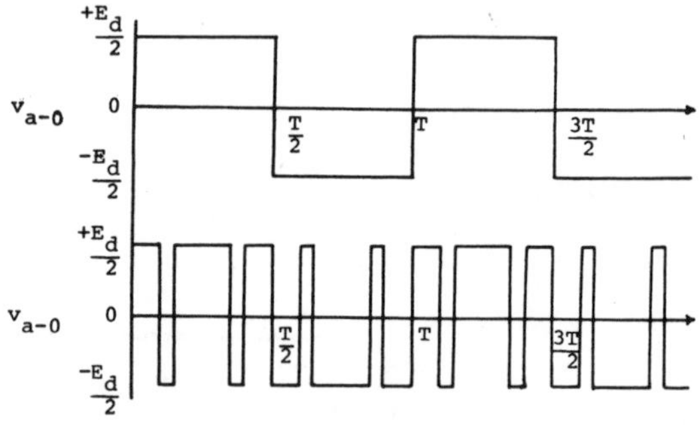

Fig. 2. Typical waveforms for circuit of Fig. 1.

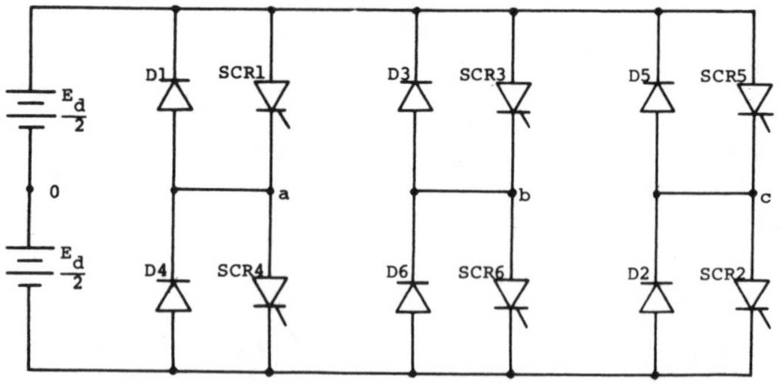

Fig. 3. Basic configuration of three-phase half-bridge inverter. 0 is theoretical source center tap.

Half-Bridge or Center-Tapped DC-Source Inverter

Fig. 1 shows the basic configuration of this type of single-phase inverter circuit (commutation and firing circuits not shown). SCR1 and SCR2 are alternately turned on to connect point a to the positive and negative lines, respectively. Theoretically, there are four states of operation for the circuit. The state when both SCRs are turned on obviously short circuits the dc supply. The voltage at point a for both SCRs off depends on the nature of the load and the current in the circuit prior to the initialization of this state. Thus the inverter has only two fully controllable states that can be utilized to generate an alternating voltage across the load; with SCR1 on and

Paper TOD-72-136, approved by the Static Power Converter Committee of the IEEE Industry Applications Society for presentation at the 1972 IEEE Industry Applications Society Annual Meeting, Philadelphia, Pa., October 9–12. Manuscript released for publication December 15, 1972.

H. S. Patel is with the Electrical Equipment Division, Electrotechnology Department, Mobility Equipment Research and Development Center, Fort Belvoir, Va. 22060.

R. G. Hoft is with the Department of Electrical Engineering, University of Missouri, Columbia, Mo.

Reprinted from *IEEE Trans. Ind. Appl.*, vol. IA-9, pp. 310–317, May/June 1973.

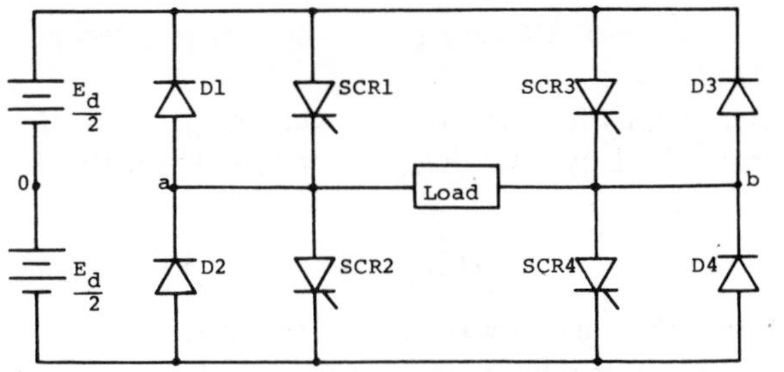

Fig. 4. Single-phase full-bridge inverter. 0 is theoretical source center tap.

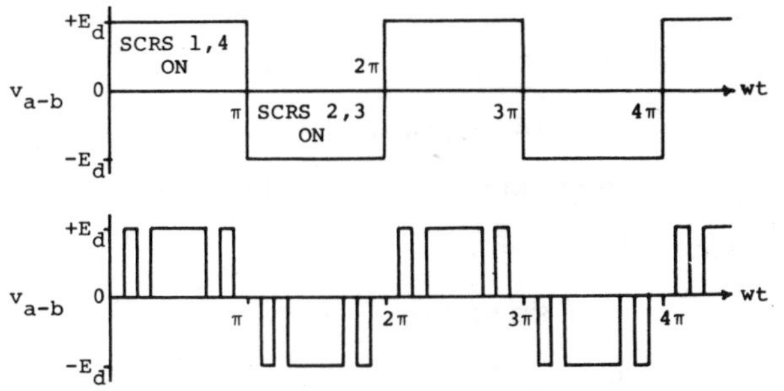

Fig. 5. Typical waveforms for circuit of Fig. 4.

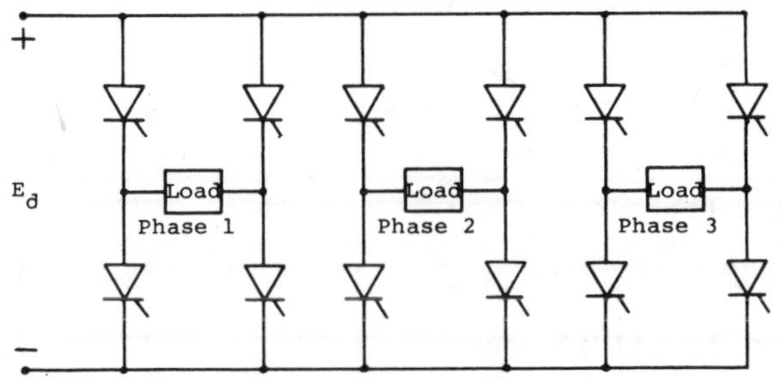

Fig. 6. Three-phase full-bridge inverter (feedback diodes and commutation circuits not shown).

SCR2 off, v_{a-0} is positive, and with SCR1 off and SCR2 on, v_{a-0} is negative. The circuit can be used to generate any waveform having only the two states mentioned previously. Fig. 2 shows two examples of such periodic waveforms that can be generated by the circuit. Three such building blocks can be used to create a three-phase half-bridge inverter as shown in Fig. 3.

Full-Bridge Inverter

Fig. 4 shows the basic circuit configuration of a single-phase full-bridge inverter, also known simply as the bridge inverter. The bridge is, in fact, derived from the half-bridge inverter. There are four SCRs as compared to two in the half-bridge circuit, hence it has $2^4 = 16$ different possible combinations of switching. Only four of these combinations are useful for obtaining an alternating waveform across the load. The four states referring to Fig. 4 are as follows.

Conducting SCRs	Load Voltage v_{a-b}
SCR1, SCR4	$+E_d$
SCR2, SCR3	$-E_d$
SCR1, SCR3	0
SCR2, SCR4	0

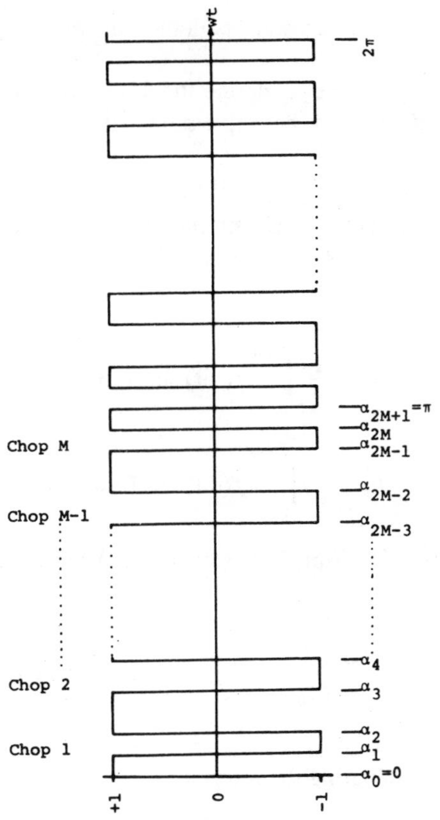

Fig. 7. Generalized output waveform of half-bridge inverter (magnitude normalized).

Thus there are only three possible states for the load voltage v_{a-b}. A number of periodic waveforms can be generated using three states. Fig. 5 shows two examples of such waveforms that can be generated by the circuit.

The voltage v_{a-b} can also be looked upon as

$$v_{a-b} = v_{a-0} - v_{b-0} \tag{1}$$

where 0 is the theoretical center tap of the dc supply E_d, and v_{a-0} and v_{b-0} are two-state waveforms as in the half-bridge inverter of Fig. 1. Three such circuits can be used as building blocks to create a three-phase full-bridge inverter as shown in Fig. 6.

A Generalized Method of Harmonic Elimination in the Half-Bridge Inverter

The two-state output waveform of the single-phase half-bridge inverter is approached from an analytical viewpoint, and a generalized method for theoretically eliminating any number of harmonics is developed in this section. The basic square wave output is "chopped" a number of times, and a fixed relationship between the number of chops and possible number of harmonics that can be eliminated is derived.

Fig. 7 shows a generalized output waveform with M chops per half-cycle. It is assumed that the periodic waveform has half-wave symmetry and unit amplitude. Therefore

$$f(\omega t) = -f(\omega t + \pi) \tag{2}$$

where $f(\omega t)$ is a two-state periodic function with M chops per half-cycle.

Let $\alpha_1, \alpha_2, \cdots, \alpha_{2M}$ define the M chops as shown in Fig. 7. The waveform can be represented by a Fourier Series as follows:

$$f(\omega t) = \sum_{n=1}^{\infty} [a_n \sin(n\omega t) + b_n \cos(n\omega t)] \tag{3}$$

where

$$a_n = \frac{1}{\pi}\int_0^{2\pi} f(\omega t)\sin(n\omega t)\,d(\omega t). \tag{4}$$

$$b_n = \frac{1}{\pi}\int_0^{2\pi} f(\omega t)\cos(n\omega t)\,d(\omega t). \tag{5}$$

Substituting for $f(\omega t)$ in (4) and using the half-wave symmetry property

$$a_n = \frac{2}{\pi}\sum_{k=0}^{2M}(-1)^k\int_{\alpha_k}^{\alpha_{k-1}} \sin(n\omega t)\,d(\omega t) \tag{6}$$

where $\alpha_0 = 0$, $\alpha_{2M+1} = \pi$, and $\alpha_0 < \alpha_1 < \alpha_2 \cdots < \alpha_{2M+1}$. From (6), evaluating the integral

$$a_n = \frac{2}{n\pi}\sum_{k=0}^{2M}(-1)^k[\cos(n\alpha_k) - \cos(n\alpha_{k+1})]$$

$$= \frac{2}{n\pi}[\cos n\alpha_0 - \cos n\alpha_{2M+1} + 2\sum_{k=1}^{2M}(-1)^k \cos n\alpha_k] \tag{7}$$

but $\alpha_0 = 0$ and $\alpha_{2M+1} = \pi$. Hence

$$\cos n\alpha_0 = 1 \tag{8}$$

$$\cos n\alpha_{2M+1} = (-1)^n. \tag{9}$$

Therefore, (7) reduces to

$$a_n = \frac{2}{n\pi}[1 - (-1)^n + 2\sum_{k=1}^{2M}(-1)^k \cos n\alpha_k]. \tag{10}$$

Similarly

$$b_n = -\frac{4}{n\pi}\sum_{k=1}^{2M}(-1)^k \sin n\alpha_k. \tag{11}$$

Utilizing the half-wave symmetry property of the waveform, $a_n = 0$ and $b_n = 0$ for even n. Therefore, for odd n, from (10) and (11)

$$a_n = \frac{4}{n\pi}\left[1 + \sum_{k=1}^{2M}(-1)^k \cos n\alpha_k\right] \tag{12}$$

$$b_n = \frac{4}{n\pi}\left[-\sum_{k=1}^{2M}(-1)^k \sin n\alpha_k\right]. \tag{13}$$

Equations (12) and (13) are functions of $2M$ variables, $\alpha_1 \cdots \alpha_{2M}$. In order to obtain a unique solution for the $2M$ variables, $2M$ equations are required. By equating any M harmonics to zero, $2M$ equation are derived from equations (12) and (13).

The M equations derived by equating $b_n = 0$ for M values of n, are solved by assuming quarter-wave symmetry for $f(\omega t)$, i.e.,

$$f(\omega t) = f(\pi - \omega t). \tag{14}$$

From the quarter-wave symmetry property the following relations are obvious, with regard to Fig. 7:

$$\alpha_k = \pi - \alpha_{2M-k+1}, \quad \text{for } k = 1, 2, \cdots, M. \tag{15}$$

Therefore, using (15)

$$\sin n\alpha_k = \sin n(\pi - \alpha_{2M-k+1})$$

$$= [\sin n\pi \cos n\alpha_{2M-k+1}$$

$$- \cos n\pi \sin n\alpha_{2M-k+1}], \quad \text{for } k = 1, 2, \cdots, M. \tag{16}$$

For odd n

$$\sin n\pi = 0, \quad \cos n\pi = -1.$$

Substituting in (16)

$$\sin n\alpha_k = \sin n\alpha_{2M-k+1}, \quad \text{for } k = 1, 2, \cdots, M. \tag{17}$$

Substituting (17) in (13)

$$b_n = \frac{4}{n\pi}\sum_{k=1}^{M}(\sin n\alpha_k - \sin n\alpha_{(2M-k+1)}) = 0. \tag{18}$$

From (15)

$$\cos n\alpha_k = \cos n(\pi - \alpha_{(2M-k+1)}), \quad \text{for } k = 1, 2, \cdots, M. \tag{19}$$

For odd n, (19) becomes

$$\cos n\alpha_k = -\cos n\alpha_{(2M-k+1)}, \quad \text{for } k = 1, 2, \cdots, M. \tag{20}$$

Substituting (20 in (12)

$$a_n = \frac{4}{n\pi}\left[1 + 2\sum_{k=1}^{M}(-1)^k \cos n\alpha_k\right]. \tag{21}$$

THEOREM

For a two-state waveform of the type shown in Fig. 7, any M harmonics can be eliminated by solving the M equations obtained from setting (21) equal to zero. The waveform is chopped M times per half-cycle and is constrained to possess odd quarter-wave symmetry.

An analytical proof of this theorem has not been devised. However, the theorem has been applied to a wide variety of two-state waveforms and shown to be correct, using numerical techniques to solve the equations involved. The problem as defined previously involves solving M equations of the type given in (21) for M different values of n; i.e., setting M harmonics equal to zero. These equations are nonlinear as well as transcendental in nature. There is no general method that can be applied to solve such equations. Moreover, an analytical method is highly improbable unless the equations involved are relatively simple with well-behaved nonlinearities. The transcendental nature of the equations involved suggests a possibility of multiple solutions. The practical method of solving these equations is a trial and error process. Taking all the factors into account, a numerical technique is the best approach in solving the equations.

A Numerical Method for Solving a System of Nonlinear Equations [4], [5]

The system of nonlinear equations in M variables can be represented as

$$f_i(\alpha_1, \alpha_2, \cdots, \alpha_M) = 0, \qquad i = 1, 2, \cdots, M. \tag{22}$$

These M equations are obtained for the problem by equating (21) to zero for any M harmonics desired to be eliminated.

Equation (22) is written in vector notation as

$$f(\alpha) = \mathbf{0} \tag{23}$$

where

$$f = [f_1\, f_2 \cdots f_M]^T, \qquad \text{an } M \times 1 \text{ matrix}$$

$$\alpha = [\alpha_1\, \alpha_2 \cdots \alpha_M]^T, \qquad \text{an } M \times 1 \text{ matrix.}$$

Equation (23) can be solved by using a linearization technique, where the nonlinear equations are linearized about an approximate solution. The steps involved in computing a solution are as follows.

1) Guess a set of values for α; call them

$$\alpha^0 = [\alpha_1^0, \alpha_2^0 \cdots \alpha_M^0]^T.$$

2) Determine the values of

$$f(\alpha^0) = f^0. \tag{24}$$

3) Linearize (23) about α^0

$$f^0 + \left[\frac{\partial f}{\partial \alpha}\right]^0 d\alpha = \mathbf{0} \tag{25}$$

where

$$\left[\frac{\partial f}{\partial \alpha}\right]^0 = \begin{bmatrix} \frac{\partial f_1}{\partial \alpha_1} & \frac{\partial f_1}{\partial \alpha_2} & \cdots & \frac{\partial f_1}{\partial \alpha_M} \\ \frac{\partial f_2}{\partial \alpha_1} & \frac{\partial f_2}{\partial \alpha_2} & \cdots & \frac{\partial f_2}{\partial \alpha_M} \\ & & \vdots & \\ \frac{\partial f_M}{\partial \alpha_1} & \frac{\partial f_M}{\partial \alpha_2} & \cdots & \frac{\partial f_M}{\partial \alpha_M} \end{bmatrix}$$

evaluated at α^0 and $d\alpha = [d\alpha_1\, d\alpha_2 \cdots d\alpha_M]^T$.

4) Solve (25) for $d\alpha$.

5) Repeat 1)-4) using, as improved guesses,

$$\alpha^1 = \alpha^0 + d\alpha. \tag{26}$$

The process is repeated until (23) is satisfied to the desired degree of accuracy. If the previous method converges, it will give a solution to (23). In case of divergence from the initial guess, it is necessary to make a new initial guess. The process is a trial and error method. The correct solution must satisfy the condition

$$0 < \alpha_1 < \alpha_2 < \cdots < \alpha_M < \pi/2. \tag{27}$$

In solving a set of nonlinear equations numerically, the primary concern is the convergence of the method used. Unlike solving a single nonlinear equation, where there are many methods of obtaining *a priori* information on the location of the root, the convergence itself is a serious problem in solving a set of nonlinear equations. It is usually a trial and error process, and no general method exists that can guarantee convergence to a solution.

Problem Formulation and a Generalized Algorithm for Obtaining a Solution

A computer algorithm, implementing the numerical technique discussed previously, is developed here to solve the M-nonlinear transcendental equations obtained from (21).

Let $n_1, n_2, \cdots, n_M$ be the M harmonics to be eliminated; then from (21) the following equations are obtained:

$$f_1(\alpha) = 1 + 2 \sum_{k=1}^{M} (-1)^k \cos n_1 \alpha_k = 0$$

$$f_2(\alpha) = 1 + 2 \sum_{k=1}^{M} (-1)^k \cos n_2 \alpha_k = 0$$

$$\vdots$$

$$f_M(\alpha) = 1 + 2 \sum_{k=1}^{M} (-1)^k \cos n_M \alpha_k = 0. \tag{28}$$

The derivative matrix, $\partial f/\partial \alpha$ of (25) is obtained from (28)

$$\frac{\partial f}{\partial \alpha} = \begin{bmatrix} 2n_1 \sin n_1 \alpha_1 & -2n_1 \sin n_1 \alpha_2 & \cdots \pm 2n_1 \sin n_1 \alpha_M \\ 2n_2 \sin n_2 \alpha_1 & -2n_2 \sin n_2 \alpha_2 & \cdots \pm 2m_2 \sin n_2 \alpha_M \\ & & \vdots \\ 2n_M \sin n_M \alpha_1 & -2n_M \sin n_M \alpha_2 & \cdots \pm 2n_M \sin n_M \alpha_M \end{bmatrix}. \tag{29}$$

The elements of the last column of the matrix in (29) are positive if M is odd, and negative if it is even. Using the numerical method discussed previously the algorithm of Fig. 8 is obtained. The computer program implementing the algorithm is given in [1].

In order to solve the M-linear equations (25), the $M \times M$ matrix of (29) must be nonsingular. This condition is violated if any one of $\alpha_1, \alpha_2, \cdots \alpha_M$ is equal to zero, assuming the domain of the solution is the closed interval $[0, \pi/2]$. Also, if any two α are equal, two columns of the matrix are identical, except for the sign, in case they are opposite. The rank of the matrix in that case is reduced to $M - 1$, and the matrix is singular. The condition of (27) insures the nonsingularity of the matrix as well as a meaningful solution to (21).

Results

The algorithm developed previously was implemented on the computer to obtain solutions for eliminating up to five harmonics. From the practical viewpoint, the lowest existing har-

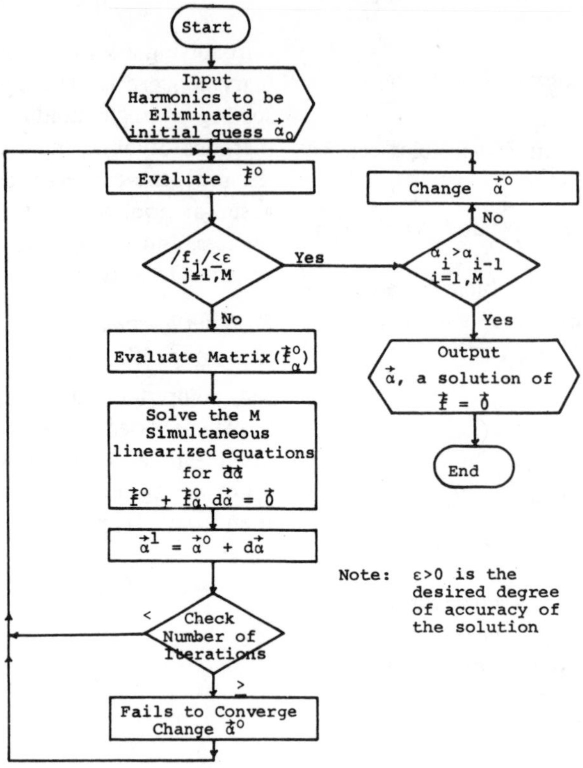

Fig. 8. Computer algorithm for numerical solution of problem.

(a) Solution for No Fifth and Seventh Harmonic Waveform

$\alpha_1 = 16.2448°$; $\alpha_2 = 22.0630°$

Order of harmonic	Absolute value of the harmonic coefficients	Absolute value of harmonic as % of the fundamental
1 (fundamental)	1.1879	100.00
3	0.2070	17.43
5	0.0000	0.00
7	0.0001	0.01
9	0.1086	9.14
11	0.2412	20.31
13	0.3223	27.13
15	0.3084	25.96
17	0.2030	17.09
19	0.0514	4.33
21	0.0825	6.94

(b) No Fifth and Seventh Harmonic Waveform

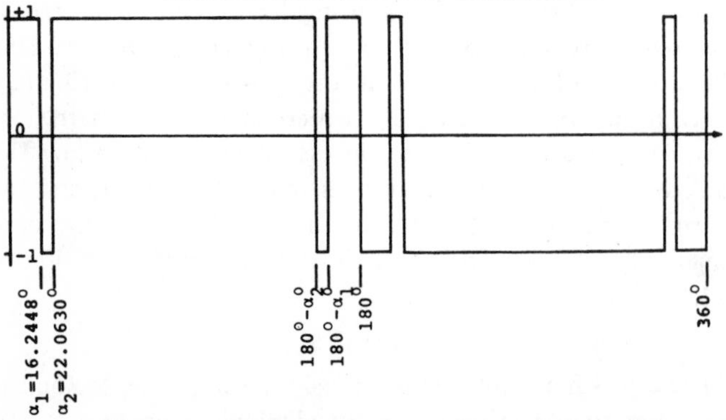

Fig. 9. Solution and waveform for eliminating fifth and seventh harmonics.

(a) Solution for No 5th, 7th, 11th, 13th and 17th Harmonic Waveform

$\alpha_1 = 6.7952°$; $\alpha_2 = 17.2962°$; $\alpha_3 = 21.0252°$;

$\alpha_4 = 34.6566°$; $\alpha_5 = 35.9840°$

Order of harmonic	Absolute value of the harmonic coefficients	Absolute value of harmonic as % of the fundamental
1 (fundamental)	1.1663	100.00
3	0.1748	14.99
5	0.0000	0.00
7	0.0000	0.00
9	0.0130	1.11
11	0.0000	0.00
13	0.0000	0.00
15	0.0216	1.85
17	0.0000	0.00
19	0.1190	10.20
21	0.2825	24.22

(b) No 5th, 7th, 11th, 13th and 17th Harmonic Waveform

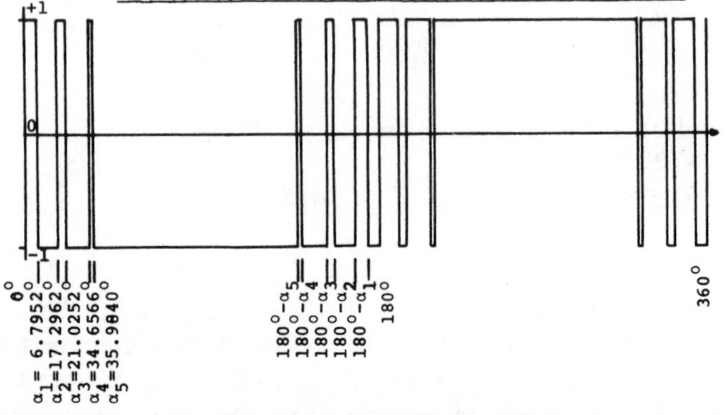

Fig. 10. Solution and waveform for eliminating fifth, seventh, eleventh, thirteenth, and seventeenth harmonics.

monics are the most undesirable. Thus the solutions are presented for eliminating these harmonics. The higher harmonics can be easily attenuated using filter circuits. Figs. 9 and 10 show the harmonic content up to the twenty-first harmonic and the waveforms for eliminating two and five harmonics, respectively. The triplen harmonics are absent in a three-phase system; thus they are not eliminated in the single-phase waveforms shown. Detailed computer results and solutions for more problems are given in [1].

A Generalized Method of Harmonic Elimination in the Full-Bridge Inverter

The full-bridge inverter as discussed previously, has three states of operation. Fig. 11 shows a generalized waveform that can be generated by the full-bridge inverter. Instead of chopping the square wave as in the half-bridge waveform, identical but opposite polarity pulse trains are generated in each half-cycle. It is shown in this section that it is possible to eliminate as many harmonics as the number of pulses per half-cycle of the waveform by constraining the size and position of the pulses.

One advantage of generating the waveform of Fig. 11 over that of Fig. 7 is a reduction in the number of commutations per cycle required to eliminate the same number of harmonics. Let M be the number of chops per half-cycle or the number of pulses per half-cycle in Figs. 7 and 11, respectively.

Then the number of commutations N_1 per cycle of the waveform of Fig. 7 is given as

$$N_1 = 2(2M+1) = 4M+2. \tag{30}$$

For the waveform of Fig. 11, the number of commutations N_2 per cycle is

$$N_2 = 2(2M) = 4M. \tag{31}$$

Thus the half-bridge inverter waveform of Fig. 7 requires two extra commutations per cycle as compared to the full-bridge inverter waveform of Fig. 11. As the number of chops or pulses increases, the relative advantage of less commutations in Fig. 11, decreases.

Assuming odd quarter-wave symmetry for the unit height waveform of Fig. 11, the Fourier series coefficients are given by for odd n

$$a_n = \frac{4}{\pi}\int_0^{\frac{\pi}{2}} f(\omega t)\sin(n\omega t)\, d(\omega t) \tag{32}$$

and for even n

$$a_n = 0 \tag{33}$$

and

$$b_n = 0 \tag{34}$$

for all n. From (32)-(34), the Fourier series is given as

$$f(\omega t) = \sum_{n=1}^{\infty} a_n \sin(n\omega t). \tag{35}$$

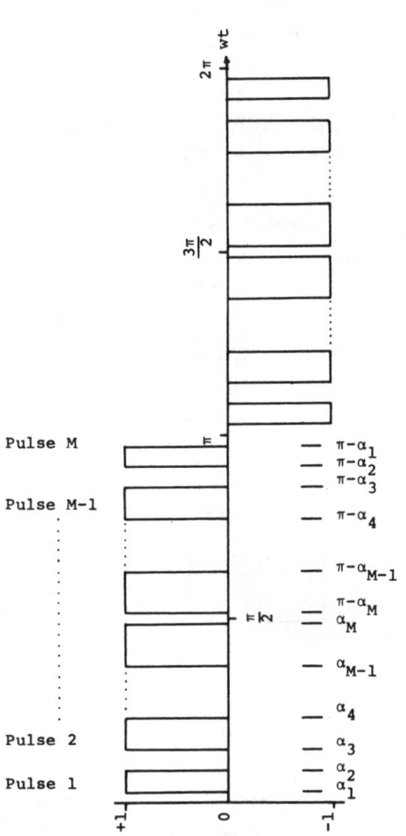

Fig. 11. Generalized output waveform of full-bridge inverter (magnitude normalized).

From Fig. 11 and (32), for odd n, and odd M

$$a_n = \frac{4}{\pi}\left[\int_{\alpha_1}^{\alpha_2} \sin n\omega t\; d\omega t + \int_{\alpha_3}^{\alpha_4} \sin n\omega t\, d\omega t + \cdots + \int_{\alpha_M}^{\frac{\pi}{2}} \sin n\omega t\, d\omega t\right]$$

$$= \frac{4}{n\pi}\sum_{k=1}^{M} (-1)^{k+1}\cos n\alpha_k \tag{36}$$

since $\cos n(\pi/2) = 0$ for odd n.

For odd n and even M

$$a_n = \frac{4}{\pi}\left[\int_{\alpha_1}^{\alpha_2} \sin n\omega t\; d\omega t + \int_{\alpha_3}^{\alpha_4} \sin n\omega t\; d\omega t + \cdots + \int_{\alpha_{M-1}}^{\alpha_M} \sin n\omega t\; d\omega t\right]$$

$$= \frac{4}{n\pi}\sum_{k=1}^{M} (-1)^{k+1}\cos n\alpha_k. \tag{37}$$

Therefore, since (36) and (37) are the same, for any M and odd n, a_n is given by

$$a_n = \frac{4}{n\pi}\sum_{k=1}^{M} (-1)^{k+1}\cos n\alpha_k \tag{38}$$

where $0 < \alpha_1 < \alpha_2 \cdots < \alpha_M < \pi/2$.

(a) Solution for Eliminating Fifth and Seventh Harmonics

$\alpha_1 = 15.4226^\circ; \alpha_2 = 87.3949^\circ$

Order of Harmonic	Absolute Value of Harmonic Coefficient	Absolute Value of Harmonic as % of Fundamental
1 (Fundamental)	1.1698	100.000
3	0.3501	29.931
5	0.0000	0.000
7	0.0000	0.000
9	0.1621	13.855
11	0.0590	5.045
13	0.1456	12.477
15	0.0000	0.000
17	0.0618	5.282
19	0.0768	6.563
21	0.0000	0.000

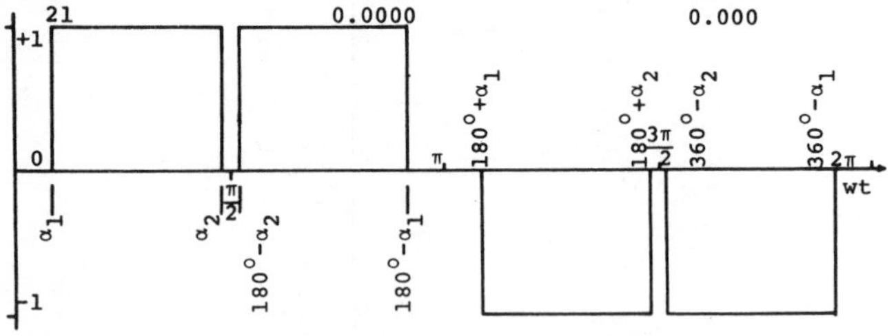

(b) No Fifth and Seventh Harmonic Waveform

Fig. 12. Solution for eliminating fifth and seventh harmonics from full-bridge inverter output.

(a) Solution for Eliminating 5th, 7th, 11th, 13th and 17th Harmonics $\alpha_1 = 11.3490^\circ; \alpha_2 = 17.2616^\circ;$

$\alpha_3 = 23.8017^\circ; \alpha_4 = 34.8708^\circ; \alpha_5 = 37.2567^\circ$

Order of Harmonic	Absolute Value of Harmonic Coefficient	Absolute Value of Harmonic as % of Fundamental
1 (Fundamental)	1.1657	100.00
3	0.1739	14.92
5	0.0000	0.00
7	0.0000	0.00
9	0.0124	1.06
11	0.0000	0.00
13	0.0000	0.00
15	0.0183	1.57
17	0.0000	0.00
19	0.0848	7.28
21	0.1701	14.59

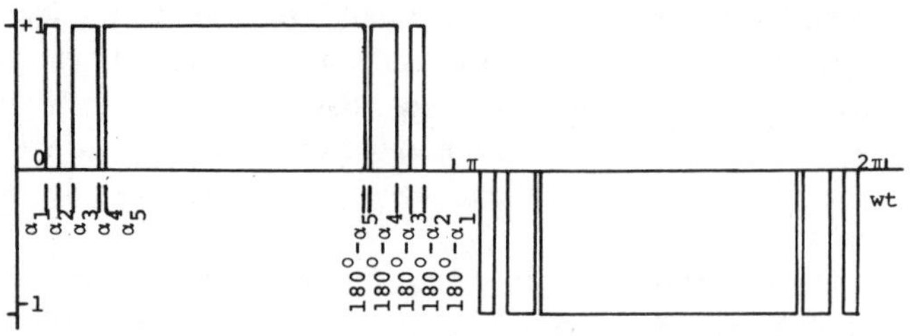

(b) No 5th, 7th, 11th, 13th and 17th Harmonic Waveform

Fig. 13. Solution for eliminating fifth, seventh, eleventh, thirteenth, and seventeenth harmonics from full-bridge inverter output.

The equations resulting from (38) equated to zero for any M harmonics, give the equations, whose solution $(\alpha_1, \alpha_2 \cdots \alpha_M)$ eliminates the M harmonics. Thus

$$f_i(\alpha) = \frac{4}{n_i \pi} \sum_{k=1}^{M} (-1)^{k+1} \cos n_i \alpha_k = 0$$

for

$$i = 1, 2, \cdots M \tag{39}$$

where n_i, $i = 1, 2, \cdots M$, are the harmonics to be eliminated.

Equation (39) is similar to (28). The same numerical method using the linearization technique is applied to (39) to obtain solutions for eliminating one–five harmonics.

Figs. 12 and 13 show the resultant waveforms, together with the solutions and harmonic amplitudes up to the twenty-first harmonic for eliminating two and five harmonics, respectively. The lowest existing harmonics in a three-phase system are eliminated in each case. Thus the triplen harmonics which are absent in a three-phase system are not eliminated in the single-phase waveforms shown. Of course any desired set of harmonics can be eliminated and the solutions obtained. From the preceding results it is concluded that it is possible to eliminate as many harmonics as the number of pulses per half-cycle of the waveform of Fig. 11.

Conclusions

It has been shown in the preceding sections that it is possible to eliminate as many harmonics as chops or pulses per half-cycle of the half-bridge or full-bridge inverter-output waveform, respectively. Solutions for more than five variables can be easily obtained using the same approach. The implementation of the solutions is an involved problem. Complex logic circuits are needed to generate the desired waveforms accurately. Digital integrated circuits including counters, shift registers, and logic gates can be economically combined for the thyristor-firing circuits required to generate the waveforms.

If the solutions, which are calculated with relatively great precision, are approximated, the logic circuitry can be simplified. This will of course increase the tolerance limits on the harmonics to be eliminated. To get a rough idea about the effect of approximating the solutions, Fig. 14 gives the harmonic content of the half-bridge waveform with five harmonics eliminated, where the solution is approximated to the nearest degree. It is observed that the harmonics intended to be eliminated are quite small so that this approximation would be acceptable in many applications. The logic circuits required to

Approximation to No 5th, 7th, 11th, 13th and 17th Harmonic Waveform

$\alpha_1 = 7^\circ; \alpha_2 = 17^\circ; \alpha_3 = 21^\circ; \alpha_4 = 35^\circ; \alpha_5 = 36^\circ$

Order of harmonic	Absolute value of the harmonic coefficients	Absolute value of harmonic as % of the fundamental
1 (fundamental)	1.1701	100.00
3	0.1764	15.08
5	0.0157	1.34
7	0.0306	2.62
9	0.0115	0.98
11	0.0021	0.18
13	0.0133	1.14
15	0.0294	2.51
17	0.0073	0.63
19	0.1303	11.14
21	0.2840	24.27

Fig. 14. Approximate solution for limiting five harmonics.

generate the SCR-triggering signals will be simpler and less expensive. Thus depending on the tolerance limits on the harmonics, a suitable approximation to the solution can be made.

The technique of harmonic elimination presented in this paper can be applied to inverters used to supply constant frequency, constant voltage, and sinusoidal output. The output filter required to attenuate the remaining harmonics is much smaller as the lowest existing harmonic frequency is relatively high.

References

[1] H. S. Patel, "Thyristor inverter harmonic elimination using optimization techniques," Ph.D. dissertation, Dep. Elec. Eng., Univ. Missouri, Columbia, 1971.

[2] F. G. Turnbull, "Selected harmonic reduction in static dc-ac inverters," *IEEE Trans. Commun. Electron.*, vol. 83, pp. 374–378, July 1964.

[3] B. D. Bedford and R. G. Hoft, *Principles of Inverter Circuits*. New York: Wiley, 1964.

[4] R. W. Hamming, *Numerical Methods for Scientists and Engineers*. New York: McGraw-Hill, 1962.

[5] A. Ralston, *A First Course in Numerical Analysis*. New York: McGraw-Hill, 1965.

Generalized Techniques of Harmonic Elimination and Voltage Control in Thyristor Inverters: Part II—Voltage Control Techniques

HASMUKH S. PATEL, MEMBER, IEEE, AND RICHARD G. HOFT, SENIOR MEMBER, IEEE

Abstract—**Theoretical techniques of voltage control for the half-bridge and full-bridge inverters are derived based on the results in [1]. Detailed analytical results for the symmetrical pulsewidth modulation method of voltage control are also presented. Voltage control techniques are derived whereby harmonic elimination is possible in variable-frequency variable-voltage three-phase inverter circuits. The technique for the half-bridge inverter is optimized subject to the constraint of switching frequency of the SCR's, using the concepts of modern control theory. Variable-frequency variable-voltage sinusoidal output in three-phase inverters is possible by employing the techniques developed. The methods show great promise in application to variable-speed ac motor drive systems.**

INTRODUCTION

IN MANY inverter applications, it is desirable to obtain an ac output voltage with variable frequency and amplitude. The variable-speed ac motor drive system is one major area of application for the variable-frequency inverter. Sinusoidal voltages are obviously desirable in these applications. With the present inverter technology, inverter circuits with variable-frequency ac output are available, although it is often difficult to achieve a distortion-free sinusoidal output. Many techniques have been developed to reduce the harmonics in the inverter output. The pulsewidth modulation technique is presently the most popular and economical method of voltage and frequency control [3]. Methods employing multiple inverters in parallel have also been developed [4]. Some versions incorporate harmonic elimination schemes to achieve a nearly sinusoidal output [5], [6].

SYMMETRICAL PULSEWIDTH VOLTAGE CONTROL IN BRIDGE INVERTERS

The technique of symmetrical multiple pulsewidth voltage control can easily be applied to the three-phase circuits of [1, Figs. 3 and 6], the half-bridge and full-bridge three-phase inverters, respectively. In the half-bridge circuit the line-to-line voltage is limited to a quasi-rectangular waveform as the maximum conduction interval for each half-cycle is 120°. To vary the voltage from this maximum value, the output voltage waveform is divided into a number of symmetrical pulses whose width is varied to achieve the desired voltage control. For the full-bridge inverter the same method is used, but the maximum voltage is available when the output is a square wave.

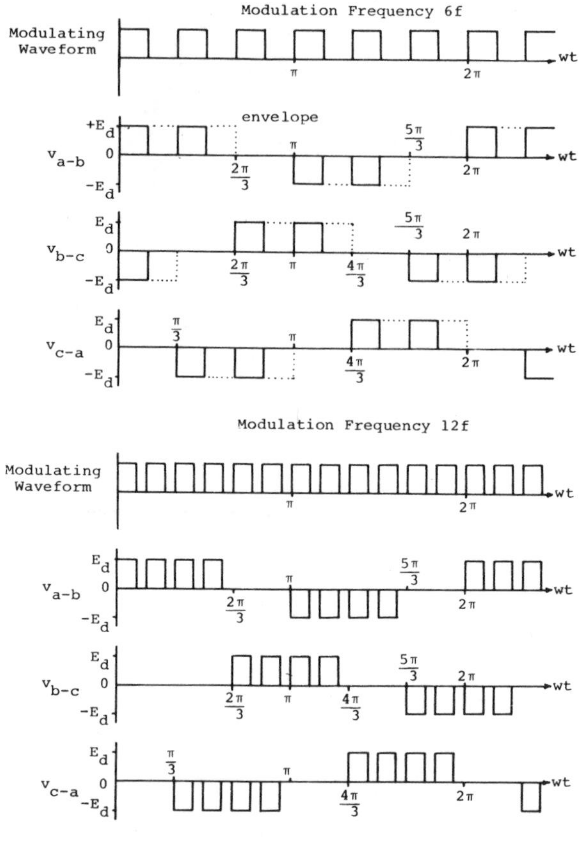

Fig. 1. Symmetrical pulsewidth modulated voltage waveforms for the three-phase half-bridge inverter.

To obtain symmetrical pulses in the three-phase output of the half-bridge inverter, the modulating frequency is constrained to be $6mf$, where $m = 1,2,\cdots$ and f is the desired output frequency. Fig. 1 shows the modulated three-phase output voltages for modulation frequencies of $6f$ and $12f$, for the circuit of [1, Fig. 3]. The number of pulses per half-cycle of the line-to-line voltages is given by $2m$, where m is as just defined. The output voltage is controlled by varying the widths of the pulses symmetrically. In some applications the number of pulses per half-cycle

Paper TOD-72-137, approved by the Static Power Converter Committee of the IEEE Industry Applications Society for publication in this TRANSACTIONS. Manuscript released for publication May 1, 1974.

H. S. Patel was with the Department of Electrical Engineering, University of Missouri, Columbia. He is now with the General Electric Company, Philadelphia, Pa.
R. C. Hoft is with the Department of Electrical Engineering, University of Missouri, Columbia, Mo.

Reprinted from *IEEE Trans. Ind. Appl.*, vol. IA-10, pp. 666–673, Sept./Oct. 1974.

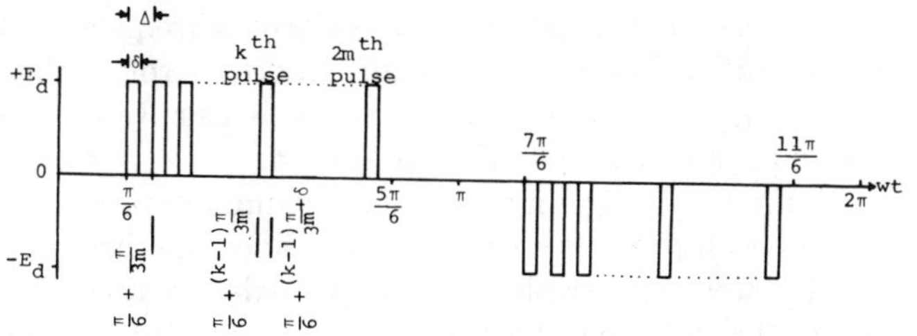

Fig. 2. Generalized line–line voltage waveform for symmetrical multiple pulsewidth voltage control in the three-phase half-bridge inverter.

is kept fixed at all voltages. More complex schemes allow for increasing the number of pulses per half-cycle at lower output voltages to reduce the harmonic content of the waveform. The amplitudes of the harmonics in the waveforms with different numbers of pulses per half-cycle are derived as

$$M = \frac{\text{modulation frequency}}{6 \text{ fundamental frequency}} \tag{1}$$

$$\text{number of pulses per half-cycle} = 2m. \tag{2}$$

Fig. 2 shows the generalized line-to-line voltage waveform with $2m$ symmetrical pulses per half-cycle. The theoretical maximum pulsewidth Δ is given by

$$\Delta = \frac{\pi}{3m}. \tag{3}$$

Let δ be the variable pulsewidth in radians.

The waveform of Fig. 2 is represented by the Fourier series

$$f(\omega t) = \sum_{n=1}^{\infty} [a_n \sin (n\omega t) + b_n \cos (n\omega t)]. \tag{4}$$

From the odd half-wave symmetry of the waveform

$$a_n = 0, \quad \text{for even } n$$

$$b_n = 0, \quad \text{for even } n.$$

The coefficients a_n for odd n are

$$a_n = \frac{2E_d}{\pi} \int_0^{\pi} f(\omega t) \sin (n\omega t)\, d(\omega t), \quad \text{for } n = 13, \cdots \tag{5}$$

$$= \frac{2E_d}{\pi} \sum_{k=0}^{2m-1} \left[\int_{\pi/6+k\pi/3m}^{\pi/6+k\pi/3m+\delta} \sin (n\omega t)\, d(\omega t) \right] \tag{6}$$

$$= \frac{2E_d}{n\pi} \sum_{k=0}^{2m-1} \left[\cos n \left(\frac{\pi}{6} + \frac{k\pi}{3m} \right) - \cos n \left(\frac{\pi}{6} + \frac{k\pi}{3m} + \delta \right) \right]. \tag{7}$$

Similarly,

$$b_n = \frac{2E_d}{n\pi} \sum_{k=0}^{2m-1} \left[\sin n \left(\frac{\pi}{6} + \frac{k\pi}{3m} + \delta \right) - \sin n \left(\frac{\pi}{6} + \frac{k\pi}{3m} \right) \right]. \tag{8}$$

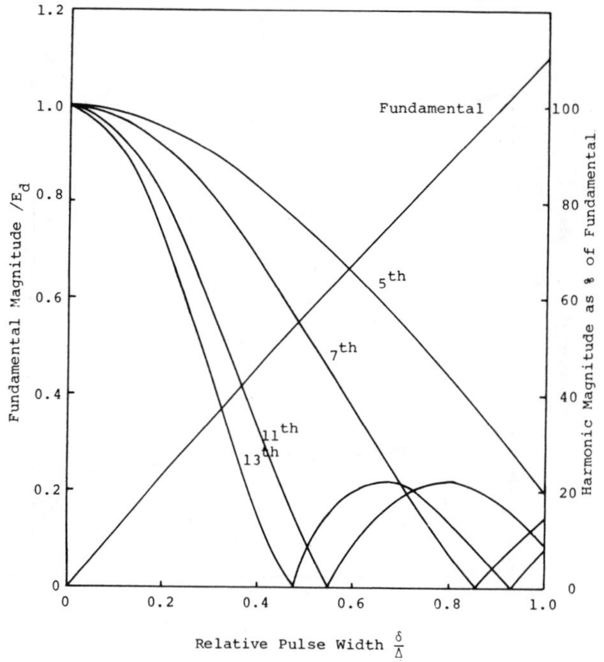

Fig. 3. Harmonic content of the line–line waveform of Fig. 2—two pulses per half-cycle.

Define c_n as

$$c_n = (a_n^2 + b_n^2)^{1/2}. \tag{9}$$

The Fourier series of the waveform $f(\omega t)$ is

$$f(\omega t) = \sum_{n=1}^{\infty} c_n \sin (n\omega t + \phi_n) \tag{10}$$

where

$$\phi_n = \tan^{-1} (a_n/b_n). \tag{11}$$

Define relative pulsewidth as

$$\text{rpw} = \frac{\delta}{\Delta}. \tag{12}$$

Fig. 3 shows the curve of the fundamental component plotted against the relative pulsewidth for two pulses per half-cycle of the waveform. The four lowest existing harmonics—the fifth, seventh, eleventh, and thirteenth—are plotted as a percentage of the fundamental against the relative pulsewidth. Figs. 4 and 5 show the same curves for four and eight pulses per half-cycle, respectively. The triplen harmonics are absent in the line-to-line voltages as they are eliminated by the three-phase star without neutral or delta load connections.

From the curves of Figs. 3–5, it is concluded that as the number of pulses per half-cycle increases, the ratio of the harmonics to the fundamental approaches the value for an equivalent square wave; i.e., for the fifth harmonic it approaches one-fifth; for the seventh harmonic, one-seventh, etc. Theoretically the harmonic content of the waveform with an infinite number of pulses approaches that of a square wave at all values of the relative pulsewidth. Thus, the harmonics in the output waveform of

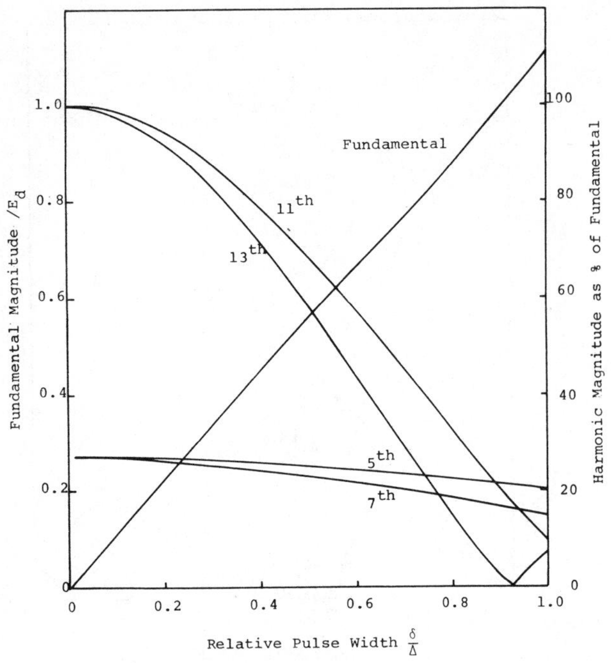

Fig. 4. Harmonic content of the line–line waveform of Fig. 2—four pulses per half cycle.

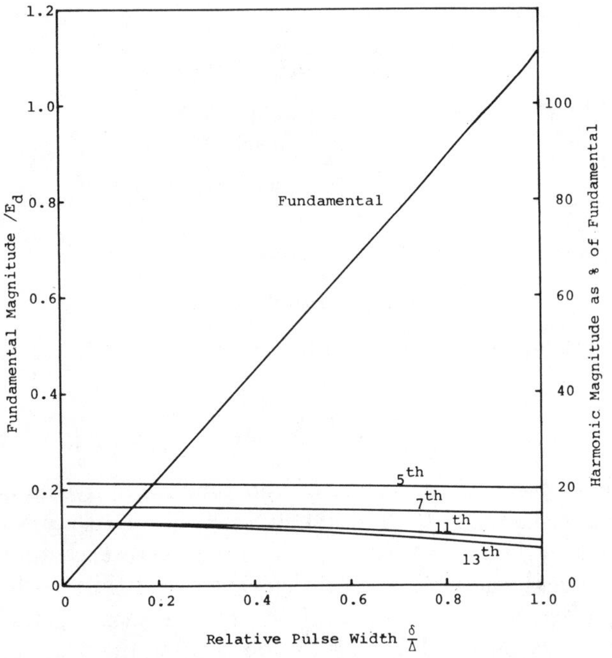

Fig. 5. Harmonic content of the line–line waveform of Fig. 2—eight pulses per half cycle.

the inverter can be limited by increasing the number of pulses per half-cycle. The lower harmonics tend to approach the limit before the higher harmonics.

The number of pulses per half-cycle for a particular application can be decided upon by the restrictions on the harmonic content. Usually harmonics higher than the eleventh or thirteenth are easily filtered out by using an output filter circuit. The number of pulses per half-cycle, in practice, is limited by the turnoff time requirements of the devices that constrain their switching frequency. The commutation losses also play an important role at higher switching frequencies. A compromise is usually necessary for an economical and efficient system.

In the full-bridge inverter, the maximum voltage available in each phase is a square wave. Since the three phases are isolated, the triplen harmonics exist in these waveforms. There is no restriction on the modulation frequency, and it is an integral multiple of the output frequency of the inverter. The voltage control technique, using symmetrical pulses, is similar to the technique used in the half-bridge circuit. It is practically the same except for the increased magnitude of the fundamental and the harmonics, due to the full square wave as the maximum output.

OPTIMAL VOLTAGE AND FREQUENCY CONTROL IN THREE-PHASE HALF-BRIDGE INVERTER

Theoretical mathematical models for variable-voltage variable-frequency inverters are derived in this section. The voltage control scheme is optimally derived, based on the concepts of modern control theory. Continuous voltage control of the output at all frequencies is obtained, together with continuous elimination of a fixed number of harmonics. The higher harmonics that are present in the output can be easily attenuated by an output filter circuit, giving essentially a variable sinusoidal output.

Problem Formulation

The control scheme is derived for the circuit of [1, Fig. 3]. In [1] the single-phase output waveforms were determined for this circuit, eliminating up to five harmonics.

In the output waveform of [1, Fig. 7], there is no way of controlling the fundamental component if the method of [1] is used to eliminate the harmonics. Referring to [1, eq. (28)], it is observed that M equations are required to solve for the M variables $\alpha_1, \alpha_2, \cdots, \alpha_M$. The problem as formulated does not possess a degree of freedom whereby the fundamental component can be controlled.

Voltage control of the inverter output is possible if a single degree of freedom is introduced in the problem; thus, to eliminate M harmonics and at the same time control the fundamental component, $(M + 1)$ variables are needed. With the degree of freedom introduced, it is possible to vary the output voltage from a maximum value to zero continuously, eliminating the M harmonics.

The problem of maximizing the magnitude of the fundamental component is solved, applying modern control theory concepts, based on the calculus of variations.

Fundamental Maximization

The waveform of [1, Fig. 7] is chopped M times per half-cycle, and it is assumed to possess odd quarter-wave symmetry. Then the Fourier series coefficients for the waveform are given by [1, eq. (21)]. It is desired to solve for $\alpha_1, \alpha_2, \cdots, \alpha_M$ to maximize the magnitude of the fundamental component and eliminate $(M - 1)$ harmonics.

The performance index for the problem is obtained from [1, eq. (21)], substituting $n = 1$, and is given by

$$L'(\boldsymbol{\alpha}) = \frac{4}{\pi}[1 + 2 \sum_{k=1}^{M} (-1)^k \cos \alpha_k] \tag{13}$$

where $\alpha_1 \cdots \alpha_M$ must satisfy

$$0 < \alpha_1 < \alpha_2 \cdots < \alpha_M < \frac{\pi}{2}. \tag{14}$$

Equation (13) is rewritten as

$$L(\boldsymbol{\alpha}) = 1 + 2 \sum_{k=1}^{M} (-1)^k \cos \alpha_k. \tag{15}$$

Any solution that maximizes the magnitude of L', also maximizes L.

To clarify the notations used, all vector quantities are indicated

$$\boldsymbol{\alpha} = [\alpha_1, \alpha_2, \cdots, \alpha_M]^T.$$

Matrices are represented by brackets.

The $(M - 1)$ constraint relations as obtained by equating $a_n = 0$, in [1, eq. (21)] are given by

$$f_i(\boldsymbol{\alpha}) = 1 + 2 \sum_{k=0}^{M} (-1)^k \cos n_i\alpha_k = 0 \tag{16}$$

for $i = 1,2,\cdots,(m - 1)$, where n_i, $i = 1,2,\cdots,(M - 1)$ are the $(M - 1)$ harmonics to be eliminated.

Equation (16) is rewritten in vector notation as

$$\boldsymbol{f}(\boldsymbol{\alpha}) = \boldsymbol{0}. \tag{17}$$

A maximum or minimum value of $L(\boldsymbol{\alpha})$, subject to the constraints of (17), gives a maximum value for the magnitude of the fundamental component, since $L(\boldsymbol{\alpha})$ is a periodic function.

Define an H function as

$$H(\boldsymbol{\alpha},\boldsymbol{\lambda}) = L(\boldsymbol{\alpha}) + \boldsymbol{\lambda}^T \boldsymbol{f}(\boldsymbol{\alpha}) \tag{18}$$

where $\boldsymbol{\lambda}^T = [\lambda_1 \lambda_2 \cdots \lambda_{M-1}]$ is a set of constant multipliers, often referred to as Lagrange multipliers. The H function is the Hamiltonian in control theory terminology.

Definition: "A stationary value of a function is such that the function has either a maximum, minimum, or a saddle point at that value."

The necessary and sufficient conditions for a stationary value of $L(\boldsymbol{\alpha})$ are

$$\boldsymbol{f}(\boldsymbol{\alpha}) = \boldsymbol{0} \tag{19}$$

and

$$H_{\boldsymbol{\alpha}} = \boldsymbol{0}$$

where

$$H_{\boldsymbol{\alpha}} = \left[\frac{\partial H}{\partial \alpha_1} \frac{\partial H}{\partial \alpha_2} \cdots \frac{\partial H}{\partial \alpha_M}\right]^T. \tag{20}$$

Equations (19) and (20) give $(2M - 1)$ equations in $(2M - 1)$ variables $\boldsymbol{\alpha}$ and $\boldsymbol{\lambda}$. Solution of the $(2M - 1)$ equations gives a stationary value of the function $L(\boldsymbol{\alpha})$. To determine the nature of the stationary value, it is necessary to investigate the behavior of second or higher order differential changes in the L, f, and H functions [7].

To second order, the differential changes of $L(\boldsymbol{\alpha})$, $f(\boldsymbol{\alpha})$, and $H(\boldsymbol{\alpha},\boldsymbol{\lambda})$ away from a nominal point $(\boldsymbol{\alpha})$ are

$$dL = L_{\boldsymbol{\alpha}}^T d\boldsymbol{\alpha} + \tfrac{1}{2} d\boldsymbol{\alpha}^T (L_{\boldsymbol{\alpha}\boldsymbol{\alpha}}) d\boldsymbol{\alpha} \tag{21}$$

$$df = f_{\boldsymbol{\alpha}}^T d\boldsymbol{\alpha} + \tfrac{1}{2} d\boldsymbol{\alpha}^T (f_{\boldsymbol{\alpha}\boldsymbol{\alpha}}) d\boldsymbol{\alpha} \tag{22}$$

$$dH = dL + \boldsymbol{\lambda}^T df = H_{\boldsymbol{\alpha}}^T + \tfrac{1}{2} d\boldsymbol{\alpha}^T (H_{\boldsymbol{\alpha}\boldsymbol{\alpha}}) d\boldsymbol{\alpha} \tag{23}$$

where

$$L_{\boldsymbol{\alpha}} = \left[\frac{\partial L}{\partial \alpha_1} \frac{\partial L}{\partial \alpha_2} \cdots \frac{\partial L}{\partial \alpha_M}\right]^T$$

$$L_{\boldsymbol{\alpha}\boldsymbol{\alpha}} = \begin{bmatrix} \frac{\partial^2 L}{\partial \alpha_1^2} & \frac{\partial^2 L}{\partial \alpha_1 \partial \alpha_2} & \cdots & \frac{\partial^2 L}{\partial \alpha_1 \partial \alpha_M} \\ \frac{\partial^2 L}{\partial \alpha_2 \partial \alpha_1} & \frac{\partial^2 L}{\partial \alpha_2^2} & \cdots & \frac{\partial^2 L}{\partial \alpha_2 \partial \alpha_M} \\ \cdots & \cdots & \cdots & \cdots \\ \frac{\partial^2 L}{\partial \alpha_M \partial \alpha_1} & \frac{\partial^2 L}{\partial \alpha_M \partial \alpha_2} & & \frac{\partial^2 L}{\partial \alpha_M^2} \end{bmatrix} \quad (M \times M \text{ matrix}).$$

Similarly,

$$f_{\boldsymbol{\alpha}} = \frac{\partial f}{\partial \boldsymbol{\alpha}} \qquad H_{\boldsymbol{\alpha}} = \frac{\partial H}{\partial \boldsymbol{\alpha}}$$

$$f_{\boldsymbol{\alpha}\boldsymbol{\alpha}} = \frac{\partial}{\partial \boldsymbol{\alpha}}\left(\frac{\partial f}{\partial \boldsymbol{\alpha}}\right)^T \qquad H_{\boldsymbol{\alpha}\boldsymbol{\alpha}} = \frac{\partial}{\partial \boldsymbol{\alpha}}\left(\frac{\partial H}{\partial \boldsymbol{\alpha}}\right)^T.$$

Equation (22) must be interpreted as applying to each component of $\boldsymbol{f}(\boldsymbol{\alpha})$. From (23), the following relation for dL is obtained

$$dL = H_{\boldsymbol{\alpha}}^T d\boldsymbol{\alpha} + \tfrac{1}{2} d\boldsymbol{\alpha}^T (H_{\boldsymbol{\alpha}\boldsymbol{\alpha}}) d\boldsymbol{\alpha} - \boldsymbol{\lambda}^T df. \tag{24}$$

Now if the nominal point $(\boldsymbol{\alpha})$ is a stationary value, $H_{\boldsymbol{\alpha}} = \boldsymbol{0}$ and with $df = \boldsymbol{0}$, (24) reduces to

$$dL = \tfrac{1}{2} d\boldsymbol{\alpha}^T (H_{\boldsymbol{\alpha}\boldsymbol{\alpha}}) d\boldsymbol{\alpha}. \tag{25}$$

Since $L_{\boldsymbol{\alpha}} = \boldsymbol{0}$ is a necessary condition even with constraints, (21) and (24) imply that when the constraints are satisfied

$$\left(\frac{\partial^2 L}{\partial \boldsymbol{\alpha}^2}\right) = \left(\frac{\partial^2 H}{\partial \boldsymbol{\alpha}^2}\right). \tag{26}$$

If the second derivative matrix $(H_{\boldsymbol{\alpha}\boldsymbol{\alpha}})$ is negative definite, then the solution gives a maximum. A positive definite $(H_{\boldsymbol{\alpha}\boldsymbol{\alpha}})$ gives a minimum value of $L(\boldsymbol{\alpha})$. It gives a saddle point if it is neither, but is a nonsingular matrix. In case it is singular, no information about the stationary value is available. Higher derivatives of $H(\boldsymbol{\alpha},\boldsymbol{\lambda})$ are needed to ascertain the nature of the solution. A direct solution of (19) and (20) is practically impossible except for a second-order system involving only two variables. Numerical methods are necessary to obtain solutions for the problem with more than two variables. The numerical method that is used to solve the problem and the computer program and results are given in [2].

Solving for Optimal Voltage Control

The solution of the preceding problem gives the maximum magnitude of the fundamental component, satisfying the constraints of (16). To obtain the optimal voltage control curve, the numerical method of [1] is used. For various values of the fundamental component, starting from zero to the maximum magnitude of steps, the solutions ($\boldsymbol{\alpha}$) are determined. The initial guess for each step is the solution from the previous step. The starting point is the solution for the zero magnitude. The step size is dependent on the curvature of the functions solved and is adjusted until a convergence is obtained. The computer program for evaluating the voltage control solutions is given in [2].

Solution to Two-Variable Problem—Eliminating Fifth Harmonic

For $M = 2$, the fundamental component function from (15) is

$$L(\alpha_1,\alpha_2) = 1 - 2\cos\alpha_1 + 2\cos\alpha_2. \tag{27}$$

The constraint function for eliminating the fifth harmonic, as derived from (16) is

$$f_1(\alpha_1,\alpha_2) = 1 - 2\cos 5\alpha_1 + 2\cos 5\alpha_2 = 0. \tag{28}$$

The H function of (18) becomes

$$H(\alpha_1,\alpha_2,\lambda_1) = L(\alpha_1,\alpha_2) + \lambda_1 f_1(\alpha_1,\alpha_2). \tag{29}$$

For a stationary value of L subject to $f_1 = 0$, the necessary and sufficient conditions as obtained from (19) and (20) are

$$\frac{\partial H}{\partial \alpha_1} = 0 \qquad \frac{\partial H}{\partial \alpha_2} = 0 \qquad f_1(\alpha_1,\alpha_2) = 0. \tag{30}$$

Therefore, from (27)–(29)

$$\frac{\partial H}{\partial \alpha_1} = 2\sin\alpha_1 + 10\lambda_1 \sin 5\alpha_1 = 0 \tag{31}$$

$$\frac{\partial H}{\partial \alpha_2} = -2\sin\alpha_2 - 10\lambda_1 \sin 5\alpha_2 = 0. \tag{32}$$

Solutions of (28), (31), and (32) for α_1,α_2 and λ_1 give the stationary value of $L(\alpha_1,\alpha_2)$.

A stationary solution of (27) subject to constraint (28) as obtained from the computer is

$$\alpha_1 = 0^\circ \qquad \alpha_2 = 12^\circ.$$

The value for the fundamental component of the unit height waveform obtained with the preceding solution is given by (13) and is

$$L'(\alpha_1,\alpha_2) = \frac{4}{\pi} L(\alpha_1,\alpha_2) = 1.21712. \tag{33}$$

To ascertain the nature of the stationary value, the curvature matrix $H_{\boldsymbol{\alpha}\boldsymbol{\alpha}}$ of (26) is evaluated at the solution $\boldsymbol{\alpha}$. The curvature matrix in this case is negative definite; hence, the stationary value is a maximum.

Fig. 6 shows the curves for the voltage control solution. Chopping angles α_1 and α_2 are plotted against the fundamental component expressed as a percentage of the maximum magnitude as given in (33). The seventh and eleventh harmonics expressed as a percentage of the fundamental component for the given α_1 and α_2 are also plotted against the percentage fundamental. As the triplen harmonics are absent for a three-phase connection they need not be eliminated. The components of $\boldsymbol{\alpha}$ for a smooth control of the voltage vary fairly linearly with the fundamental component. The fifth harmonic is always absent as desired.

Solution to Three-Variable Problem—Eliminating Fifth and Seventh Harmonics

The equations for the three-variable problem as obtained from (15) and (16) are

$$L(\boldsymbol{\alpha}) = 1 - 2\cos\alpha_1 + 2\cos\alpha_2 - 2\cos\alpha_3 \tag{34}$$

$$f_1(\boldsymbol{\alpha}) = 1 - 2\cos 5\alpha_1 + 2\cos 5\alpha_2 - 2\cos 5\alpha_3 = 0 \tag{35}$$

$$f_2(\boldsymbol{\alpha}) = 1 - 2\cos 7\alpha_1 + 2\cos 7\alpha_2 - 2\cos 7\alpha_3 = 0. \tag{36}$$

Applying the same procedures the stationary value is

$$\alpha_1 = 0^\circ$$
$$\alpha_2 = 16.241^\circ$$
$$\alpha_3 = 22.0601^\circ.$$

The maximum magnitude of the fundamental component for the unit height waveform is

$$L'(\boldsymbol{\alpha}) = -1.18791.$$

The curvature matrix is positive definite, hence the stationary value is a minimum as desired.

Fig. 7 shows the plots of the optimal voltage control solution. The lowest two existing harmonics for a three-phase system, the eleventh and thirteenth, are also plotted as a percentage of the fundamental component for the given $\boldsymbol{\alpha}$, against the percentage fundamental.

Discussion of Results

It is obvious from the preceding problems that voltage control in the three-phase half-bridge inverter can be optimally achieved, eliminating a desired number of harmonics. The frequency of the output waveform can be varied as required, as long as the chopping angles $\boldsymbol{\alpha}$ remain the same. If constant volt-seconds are desired at all frequencies, as in a motor-drive application, the fundamental

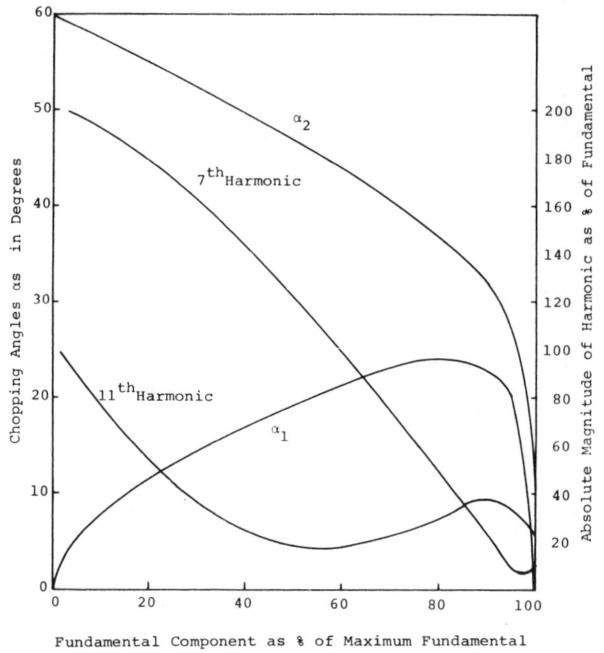

Fig. 6. Voltage control solution for no fifth harmonic.

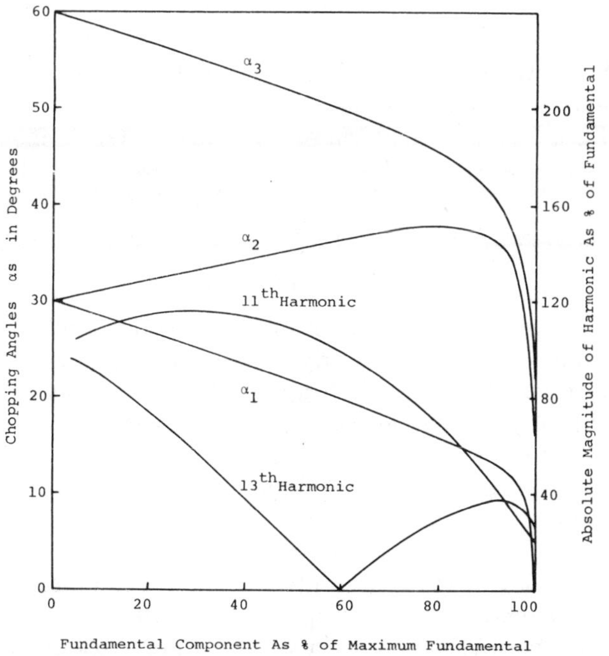

Fig. 7. Voltage control solution for no fifth and seventh harmonic.

component is varied in direct proportion to the frequency; i.e., if the frequency is halved, then the fundamental component is also halved. The chopping angles for eliminating the desired number of harmonics, as a function of the fundamental component, are readily available from curves of the type shown in Fig. 6.

The higher harmonics that exist in the waveforms are easier to eliminate using filter circuits. Moreover, the higher the lowest existing harmonic, the smaller and less expensive is the filter required. Thus, a practically sinusoidal output with variable frequency and voltage is obtainable if the preceding technique is employed. The logic circuits needed to generate the SCR firing signals will be more complex. However, the technique is feasible with the modern technology of integrated circuit logic components. Depending on the type of application, the curves of Figs. 6 and 7, which are fairly linear over a wide range, can be linearly approximated. This would lead to less complex circuitry for the inverter control system. A compromise between acceptable harmonic distortion in the output and the complexity of the circuitry is needed for a practical application of the technique.

PHASE SHIFT METHOD OF VOLTAGE AND FREQUENCY CONTROL IN THREE-PHASE FULL-BRIDGE INVERTER

In this section the phase shift method of voltage control for the three-phase full-bridge inverter of [1, Fig. 6], is derived using the results of [1]. The method is illustrated, for the single-phase version of [1, Fig. 4] and is easily extended to the three-phase circuit.

In [1, Fig. 4], the load voltage v_{a-b} is given as

$$v_{a-b} = v_{a-0} - v_{b-0} \tag{37}$$

where 0 is the theoretical dc source center-tap.

As discussed in [1], the voltages v_{a-0} and v_{b-0} are two state waveforms. In [1] the two state waveforms with a number of harmonics eliminated were derived. The waveforms v_{a-0} and v_{b-0} are generated in the same manner, eliminating a desired number of harmonics.

The voltage v_{a-b} is controlled continuously by varying the phase shift between v_{a-0} and v_{b-0}. In [1, Fig. 4] for a dc source voltage of E_d V, the voltage v_{a-0}, having M chops per half-cycle and possessing odd quarter-wave symmetry is derived from [1, eqs. (3) and (21)] as

$$v_{a-0} = \sum_{n=1}^{\infty} a_n \sin n\omega t \tag{38}$$

where

$$a_n = \frac{4E_d}{2\pi n}\left[1 + 2\sum_{k=1}^{M} (-1)^k \cos n\alpha_k\right] \tag{39}$$

for odd n only. For even n, $a_n = 0$. The waveform is chopped M times per half-cycle, and the solution $\boldsymbol{\alpha}$ to eliminate M harmonics is obtained as in [1].

Let v_{b-0} be the same as v_{a-0}, except phase shifted by ϕ rad with respect to v_{a-0}.

Then v_{b-0}, from (38) is

$$v_{b-0} = \sum_{n=1}^{\infty} a_n \sin n(\omega t - \phi). \tag{40}$$

From (38) and (40)

$$\begin{aligned} v_{a-b} &= v_{a-0} - v_{b-0} \\ &= \sum_{n=1}^{\infty} a_n[\sin n\omega t - \sin n(\omega t - \phi)]. \end{aligned} \tag{41}$$

Using $\sin C - \sin D = 2 \sin (C - D/2) \cos (C + D/2)$, (40) reduces to

$$v_{a-b} = \sum_{n=1}^{\infty} \left[2a_n \sin n \frac{\phi}{2} \cdot \cos n \left(\omega t - \frac{\phi}{2} \right) \right]. \quad (42)$$

Define the Fourier series coefficients of v_{a-b} in (42) as

$$c_n = 2a_n \sin n \frac{\phi}{2}. \quad (43)$$

Then

$$v_{a-b} = \sum_{n=1}^{\infty} c_n \cos n \left(\omega t - \frac{\phi}{2} \right). \quad (44)$$

The fundamental component of v_{a-0} is given by (43), substituting $n = 1$. It is derived as

$$c_1 = 2a_1 \sin \frac{\phi}{2} \quad (45)$$

where a_1 as obtained from (39) is

$$a_1 = \frac{2E_d}{\pi} [1 + 2 \sum_{k=1}^{M} (-1)^k \cos \alpha_k]. \quad (46)$$

Hence, by varying the phase shift angle ϕ, the fundamental component can be controlled from a maximum of $2a_1$ at $\phi = \pi$, to zero at $\phi = 0$.

If by the technique of [1], M harmonics are eliminated in v_{a-0} and v_{b-0}, then the same harmonics are also eliminated in v_{a-0} as seen from (43), which reduces to zero if a_n is zero. The remaining harmonics vary with the phase shift angle, and the relationship is given by (43).

The output frequency can be varied as desired, as long as the chopping angles α remain unaltered. Then the eliminated harmonics are always absent. The voltage in each phase of a three-phase full-bridge inverter is controlled by the technique discussed. The phase shift angle ϕ is the same in each phase, and the voltage control is symmetrical.

For constant volt-seconds operation, as desired in induction motor speed control inverters, the fundamental component of the output is varied in direct proportion to the frequency; that is

$$c_1 = k \times \text{frequency}, \qquad k = \text{constant}. \quad (47)$$

Therefore, the fundamental component is maximum at the maximum frequency of operation. For variable-frequency output with constant volt-seconds operation, the desired relationship is

$$\frac{\text{frequency}}{\text{maximum frequency}} = \frac{f}{f_{\max}} = \frac{c_1}{c_{1_{\max}}}. \quad (48)$$

From (45), c_1 is a maximum for $\phi = \pi$, or

$$c_{1_{\max}} = 2a_1. \quad (49)$$

Equation (48) reduces to

$$\frac{f}{f_{\max}} = \frac{2a_1 \sin \phi/2}{2a_1} = \sin \frac{\phi}{2}. \quad (50)$$

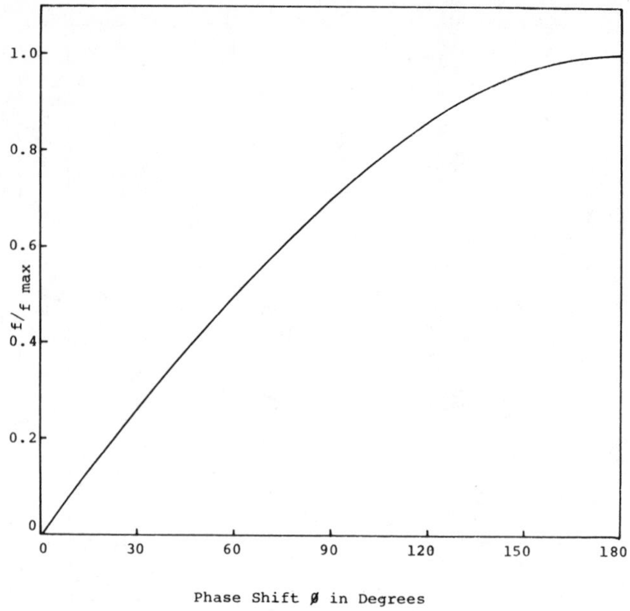

Fig. 8. Phase shift versus frequency ratio for constant volt-seconds operation. (Note: Desired volt-seconds at maximum frequency and $\phi = 180°$.)

The frequency is maximum for $\phi = \pi$, and decreases with ϕ as expressed by (50). Fig. 8 shows the curve of ϕ versus frequency ratio for constant volt-seconds operation. The solutions for α, for eliminating certain harmonics, are given in detail in [1].

Figs. 9 and 10 show curves of fundamental component versus phase shift for waveforms with four and five harmonics eliminated, respectively. The lowest two existing harmonics are also plotted against phase shift.

CONCLUSIONS

The preceding results show that theoretically it is feasible to achieve a nearly sinusoidal output in bridge inverters. The generalized approach used in solving the problems makes it an easy task to obtain solutions for eliminating as many harmonics as required.

The voltage control techniques derived make it feasible to generate a variable-frequency variable-voltage sinusoidal output in an optimal manner. The voltage control problem is solved to constantly eliminate a certain number of harmonics. The technique has considerable practical use for variable-speed ac motor drive applications, where harmonics in the output of the inverter pose serious problems in the motor performance.

The voltage control technique for the full-bridge inverter has a clear advantage of ease of voltage control over the half-bridge inverter. The chopping angles are constants and not a function of the fundamental voltage. Thus, the control circuits are easier to realize in practice. Moreover, the remaining harmonics as a percentage of the fundamental are quite high in the half-bridge inverter at reduced output voltages.

The techniques as presented in this paper provide solutions with a relatively high degree of precision. Approximations to the solutions can be made to achieve a practical

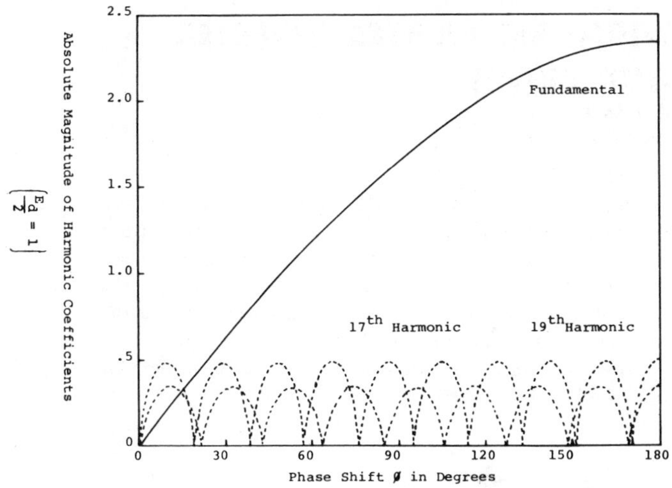

Fig. 9. Fundamental voltage control by phase shift method—No 5th, 7th, 11th and 13th harmonic waveform for v_{a-0} and v_{b-0}.

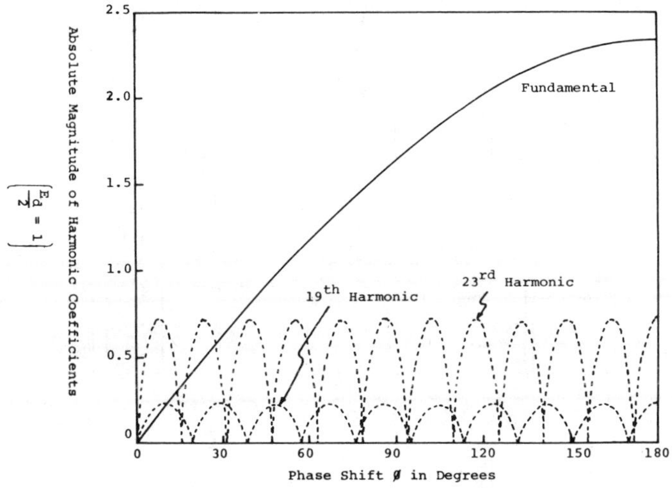

Fig. 10. Fundamental voltage control by phase shift method—No 5th, 7th, 11th, 13th, and 17th harmonic waveform for v_{a-0} and v_{b-0}.

system. The specifications for particular applications would define the allowable approximations.

The results presented should provide techniques that are practical and economical, considering the availability of low cost and sophisticated integrated circuit components for the inverter control circuits.

REFERENCES

[1] H. S. Patel and R. G. Hoft, "Generalized techniques of harmonic elimination and voltage control in thyristor inverters: Part I—Harmonic elimination," *IEEE Trans. Ind. Appl.*, vol. IA-9, pp. 310–317, May/June 1973.

[2] H. S. Patel, "Thyristor inverter harmonic elimination using optimization techniques," Ph.D. dissertation, Dep. Elec. Eng., Univ. of Missouri, Columbia, Mo., 1971.

[3] B. Mokrytzki, "Pulse width modulated inverters for ac motor drives," *IEEE Trans. Ind. Gen. Appl.*, vol. IGA-3, pp. 493–503, Nov./Dec. 1967.

[4] C. W. Flairty, "A 50 kW adjustable frequency 24-phase controlled rectifier inverter," presented at AIEE Industrial Electronics Symposium, Boston, Mass., Sept. 20–21, 1961.

[5] B. D. Bedford and R. G. Hoft, *Principles of Inverter Circuits.* New York: Wiley, 1964.

[6] F. G. Turnbull, "Selected harmonic reduction in static dc–ac inverters," *IEEE Trans. Commun. Electron.*, vol. 83, May 1964.

[7] A. E. Bryson and Y. C. Ho, *Applied Optimal Control.* Waltham: Mass.: Ginn Blaisdell, 1969.

HARMONIC EFFECTS IN PULSE WIDTH MODULATED INVERTER INDUCTION MOTOR DRIVES

GERALD B. KLIMAN
Transportation Technology Center
General Electric Co., Erie, Pa.

Abstract

When induction motors are driven by electronic inverters the applied voltage waveforms are quite non-sinusoidal. The fundamental component of the current is controlled by the load. The harmonic currents are limited principally by the motor leakage inductance and are independent of load. These harmonic currents can lead to increased motor heating and to increased peak currents.

Several inverter waveforms are analyzed and applied to a typical induction motor. Harmonic losses and peak currents are calculated for various operating conditions.

Introduction

The spectra of typical voltage and current waveforms provided by a saturating, chopped inverter to an induction motor will be found. The presence of large amplitude harmonics will increase motor heating and peak currents and may affect torque production.

The general arrangement of the circuit and the waveforms are shown in Fig. 1. The inverter, in effect, switches each line (A, B, and C) between ground and V_{DC}. The neutral is not connected. Without chopping, the resultant phase-neutral voltages are the familiar six-step square wave.

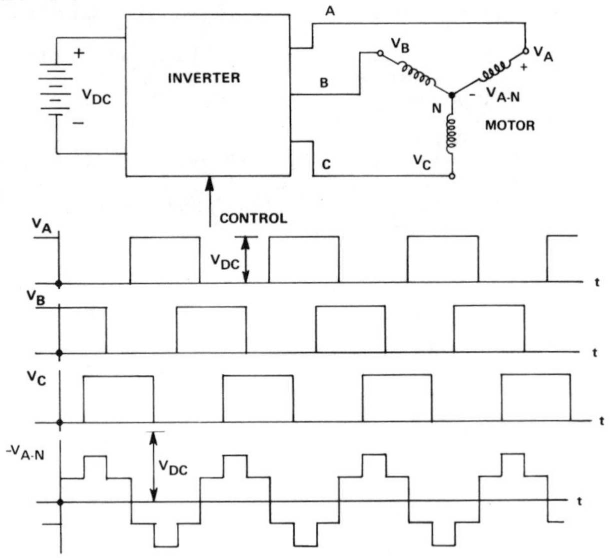

FIG. 1 GENERAL CIRCUIT ARRANGEMENT AND SIX-STEP SQUARE WAVE

The particular situation analyzed here is sub-harmonic pulse width modulation (PWM) proposed by Schonung and Stemmler[1] except that a minimum pulse width is prescribed and pulses are dropped as the modulation index increases beyond the linear modulation range.

In this method the reference signal (motor) frequency is compared to a triangle wave at a higher (chopping) frequency. The switching times are set by the intersection of the reference and triangle waves.

In this scheme, illustrated in Fig. 2, whenever the signal sinewave and the triangle wave have the same value, the inverter is switched from ground to V_{DC} or from V_{DC} to ground. The case shown is for a chopping ratio N = fc/f = 6 and a chopping index A = sinewave amplitude/triangle wave amplitude = 1.00. A chopping ratio of 6 (or some multiple of 3) is used for synchronized chopping since this results in identical waveforms on all three lines except that each waveform is displaced by 120° of the fundamental (that is the motor frequency f). The phase to neutral voltage may be found in the same way as for the six-step square wave except that now there are many more switches and the resultant phase-neutral waveforms are far more complex.

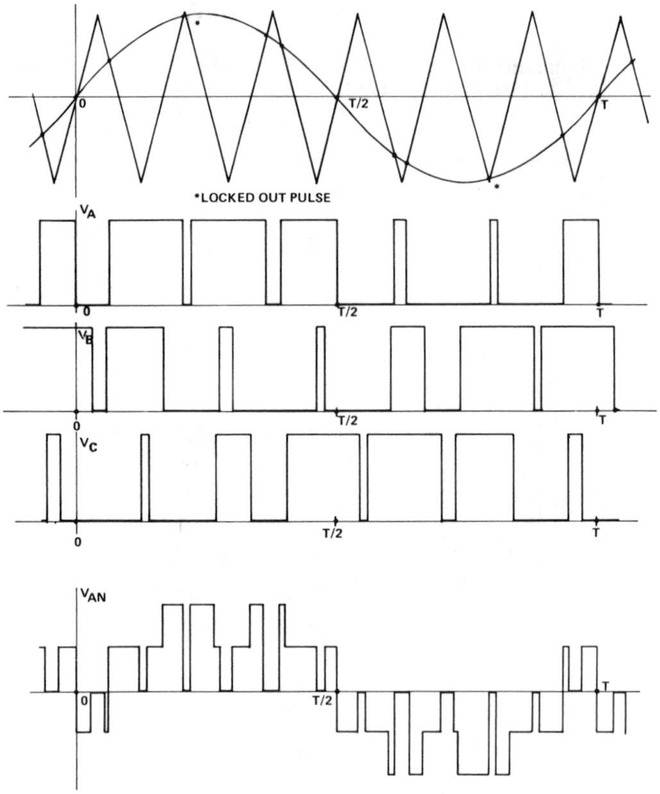

FIG. 2 INVERTER VOLTAGE A = 1, N = 6, L = 0.015T

Before attempting to analyze these waveforms, another complication that arises from limitations on the speed with which the SCR's in the PWM can switch must be added to the picture.

In Fig. 2, for N = 6 and A = 1.00, a short interval is being called for near the middle of each half period. As A gets large, this pulse becomes very short. To avoid such short intervals, a minimum pulse width is set. This "lockout" holds a constant pulse width until a zero length interval is called for. At that point the pulse (or chop) is dropped out entirely. Phase A voltages are shown in Fig. 3 for various indices A as pulses are dropped. When all the chops are dropped at high index A, the resultant waveform is again a six-step square wave.

Waveforms and Spectra

It will be assumed that the signal and chopping reference voltages are synchronized such that the chopping frequency is a triple of the signal frequency (f_c = N f). Fig. 3 shows the resultant waveforms when the modulation index takes on various values. Fig. 2 shows how the

Reprinted from *IEEE/IAS 1972 Annual Meeting*, pp. 783–790, 1972.

switch times are generated. Notice in Fig. 3 that the modulated phase voltage is just a square wave at the chopping frequency for "zero" modulation and progresses through varying the pulse widths and dropping of pulses to again a square wave at motor or signal frequency.

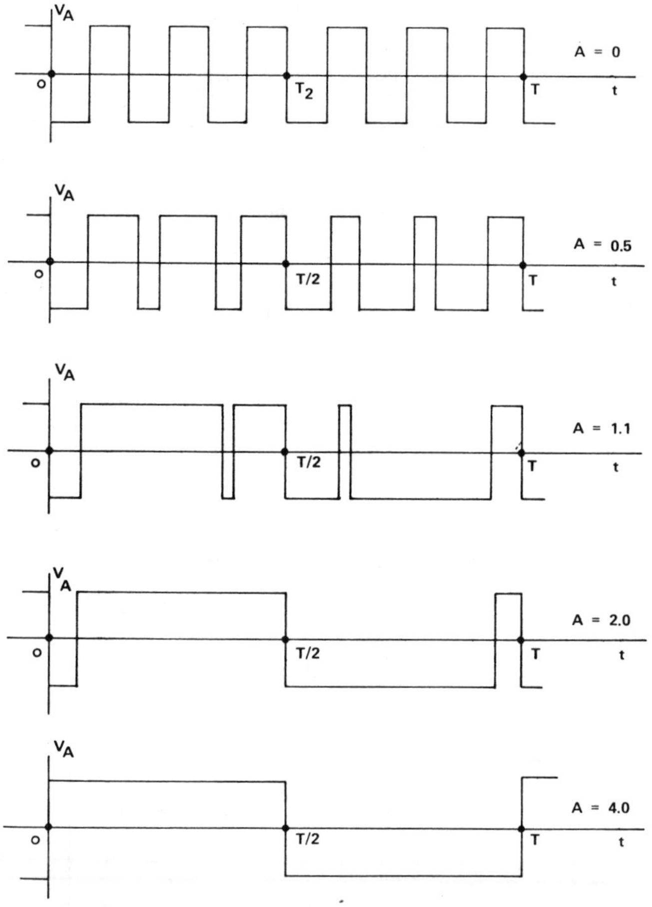

FIG. 3 PHASE A VOLTAGE WAVEFORMS AT VARIOUS MODULATION INDEX

Three such waveforms are produced, each one delayed in time by one-third of a signal period yielding "three phase" voltage, and applied to an induction motor. In the typical inverter situation these waveforms are produced by switching between zero (ground) and some DC supply voltage V_{DC}. Thus the waveforms illustrated in Fig. 3 are actually $V_{phase} - V_{DC}/2$. The waveform of interest is, however, the phase-neutral voltage ($V_{p\text{-}n}$). This is a quite complex time function which is tedious to find and to analyze. (Fig. 2) (See also Schonung and Stemmler[1], Fig. 7)

If the motor can be considered linear or represented adequately by a linear model, the phase-ground voltages may be combined to find the phase-neutral voltages. In the same way (by superposition) the phase-ground spectra may be added to find the phase-neutral spectra.

Thus to find the voltages in Fig. 4

$$V_N = 1/3\,(V_A + V_B + V_C) \quad (1)$$

$$V_{A-N} = 1/3\,(2\,V_A - V_B - V_C) \quad (2)$$

And by superposition the spectra may also be found.

$$V_{A-N}(k) = 1/3\,[2V_A(k) = V_B(k) - V_C(k)] \quad (3)$$

where k is the harmonic number.

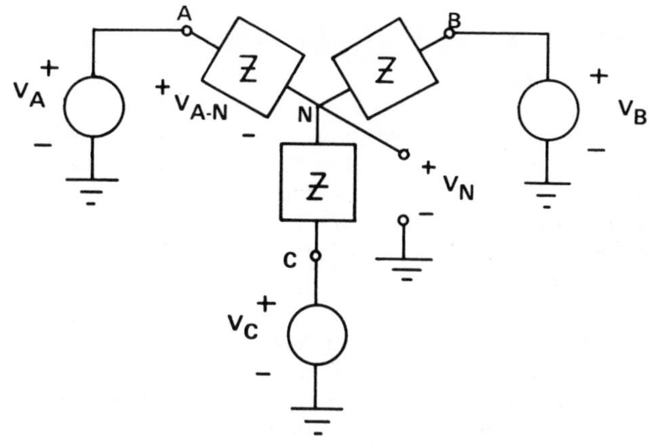

FIG. 4 LINEAR REPRESENTATION OF A MOTOR

Since the waveforms are identical in each phase, the spectra will be the same except for a phase shift of ± K 2π/3 in the Kth harmonic.

$$V_{A-N}(k) = 1/3\,V_A(k)\,[2\;\angle 0 - 1\;\angle{-k\,2\pi/3} - 1\;\angle{+k\,2\pi/3}] \quad (4)$$

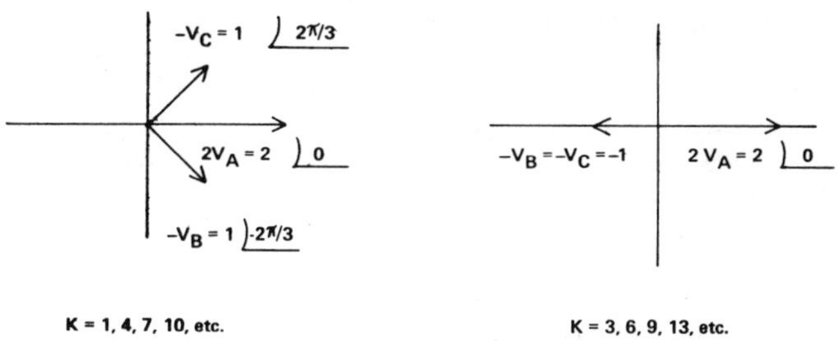

FIG. 5 PHASOR DIAGRAM FOR EACH HARMONIC

Considering then the phasor diagrams (Fig. 5) for each phase-neutral voltage at each harmonic, it may be seen that the term in the brackets has a magnitude of zero or three and causes positive or negative phase rotation as follows:

Harmonic	$V_{A\text{-}N}(k)$	Rotation
1,4,7,10,13, etc.	V_A (k)	+
2,5,8,11,14, etc.	V_A (k)	−
3,6,9,12,15, etc.	0	

This results in the spectrum of the phase-neutral voltage being identical to the corresponding phase-ground voltage except that triple multiples of the fundamental frequency are missing (the "triplen" harmonics).

The problem then reduces to the simpler one of finding the spectrum $V_A(k)$.

In order to proceed further the times at which the switches occur must be known. Referring to Fig. 2 it may be seen that the problem is just that of finding the intersection of a straight line with a sinusoid. That is the switch (or chop) times, T_M, are given by the solution of

$$A \sin 2\pi\, T_M/T = (-1)^M\, 8N\left(T_M/T - \frac{M}{4N}\right) \qquad (5)$$

where M takes on integer values from 1 to N−1 in the first half period. The switch times in the second half period are identical to the first half period but reversed in time except for the pulse width lockout. Clearly t = 0 and t = T/2 are also intersections for any N. A is the modulation index and may vary from 0 to about 4 when the limiting square wave is reached.

In the vicinity of A = 1, for the case illustrated, the time between the second and third switches is becoming small. Due to thyristor problems a minimum pulse width L is prescribed until a pulse width of zero is called for by the switching rule whereupon the pulse is dropped entirely. This process is called "lockout and drop". All these properties may be embodied in a computer algorithm for calculating the switch times.

To calculate the spectrum of the line-neutral voltage, it is only necessary to calculate the spectrum of the line-ground voltage deleting DC and all harmonics that are multiples of 3f (the "triplen" harmonics). This is easy to implement since the waveform varies only between ± 1/2 V_{DC} at the chop times so that an elaborate Fourier transform technique is not needed.

In analyzing the phase voltage note that the waveform will be antisymmetric about the 0° and 180° points so that only a sine series* is required and that both odd and even harmonics will be present. Starting at t = 0 the first segment is negative with switching in strict alternation of sign at each chop time. Putting this into the transform integral

$$V_A(k) = \frac{2}{T}\int_0^T V_A(t) \sin k\, 2\pi t/T\, dt \qquad (6)$$

there results:

$$V_A(k) = \frac{2}{k\pi}\left\{\frac{1}{2}\left[(-1)^{k+n+1} - 1\right] + \sum_{M=1}^{N-1} (-1)^{M+1} \cos 2\pi k\, T_M/T\right\} \qquad (7)$$

and the spectral density function is then

$$V_A(\omega) = V_A(k)\, U_o(\omega - k\,\omega_1) \quad k \neq 0 \text{ or triples} \qquad (8)$$

For A = 0 this reduces the spectrum of a square wave of frequency fc and for A ≳ 3.8 to a square wave of motor frequency f.

A computer program has been written to find the chop times and evaluate the Fourier transform.

The magnitude of various harmonics are shown as a function of modulation index in Fig. 6 for N = 6. Several features should be noted:

1. unlike the six-step square wave in which the most important harmonics are 5 and 7, the most important harmonics are 4, 8 (side bands of fc) and 11, and 13 (side bands of 2 fc);*

2. 4 and 8, and 11 and 13, etc. are identical until the first lockout;*

3. the fundamental is exactly V_{DC} A/2 until the first lockout;*

4. for A > 2 the harmonics do not change very much until A ≃ 3.8 where a sudden transition is made to six-step square wave;

5. The remaining harmonics do not become significant until A > 0.7.

Side bands of 3 fc (16 and 20, 1 and 22) are also of significant magnitude but are not shown.

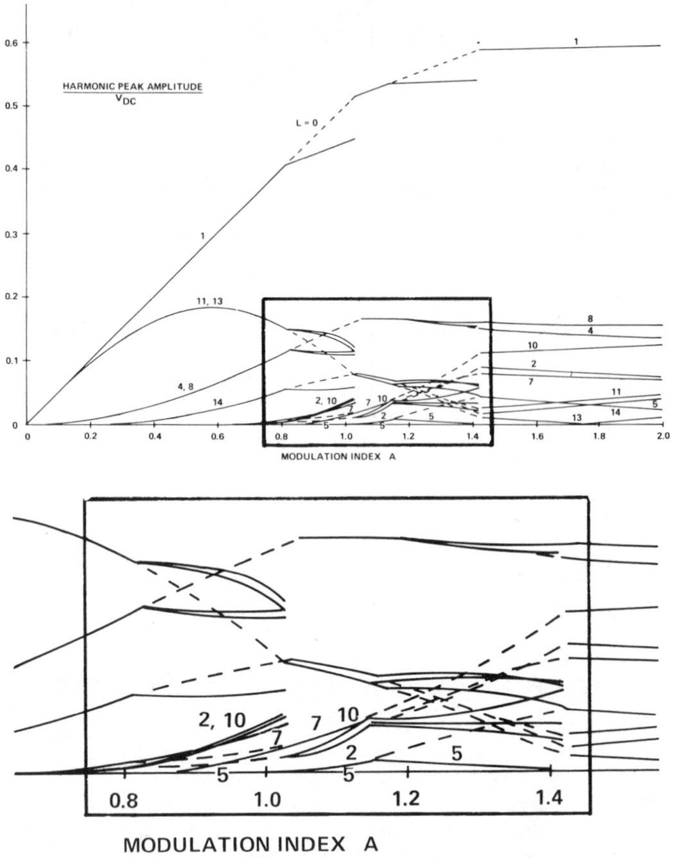

FIG. 6 HARMONIC PEAK AMPLITUDE VS. MODULATION INDEX N = 6, L = 0.015T (---L = 0)

Induction Motor Response

The major contribution to the harmonic effects in an induction motor should then arise in the side band harmonics 11 and 13, 4 and 8, or 2 and 10 depending on the modulation index in the PWM system or in the 5th and 7th harmonics in square wave operation. To estimate the harmonic effects more closely, the harmonic voltages were applied to linear equivalent circuits of an induction motor[2], Fig. 7.

It is assumed that the parameters of the linear circuit model are independent of frequency except for deep bar effect in the rotor. The stator and referred rotor inductances and resistances are divided into frequency dependent and independent parts to account for deep bar effects. Stray load and core losses will be neglected.

Thus, if L_f and R_f are the "DC" values of the frequency dependent parts of the rotor leakage inductance and resistance

*The lockout causes some dissymmetry which would, strictly speaking, require evaluating the cosine series as well but the contribution to each harmonic is negligible.

*Note: This agrees with Schonung and Stemmler[1] for A< 1, L = 0.

$$L_{2k} = L_o + L_f\, J_1 \qquad (9)$$

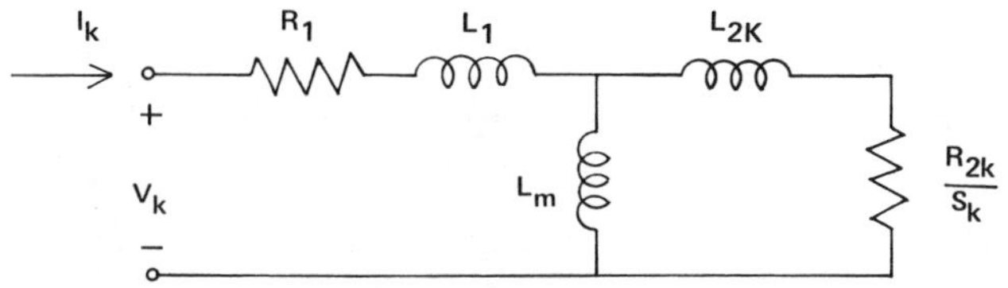

FIG. 7 INDUCTION MOTOR EQUIVALENT CIRCUIT

$$R_{2k} = R_o + R_f\, K_1 \qquad (10)$$

where for rectangular rotor bars, for example, the K1 and J1 factors are given by Alger[3].

$$K_1 = \alpha_k d \,\frac{\sinh 2\alpha_k d - \sin 2\alpha_k d}{\cosh 2\alpha_k d - \cos 2\alpha_k d} \qquad (11)$$

$$J_1 = \frac{\sinh 2\alpha_k d + \sin 2\alpha_k d}{\cosh 2\alpha_k d - \cos 2\alpha_k d} \qquad (12)$$

Hence, α_k is the reciprocal of the rotor bar skin depth and is given by

$$\alpha_k = 2\sqrt{\frac{k\, f\,|S_k|}{10^9}} \qquad (13)$$

and d is the height of a (rectangular) rotor bar.

The resistivity ρ is given in ohm-cm and the bar height d is given in cm.

The harmonic slip S_k requires some elaboration. The shaft is assumed to be running at a constant speed close to synchronous for the fundamental. The synchronous speeds corresponding to each harmonic are proportional to the harmonic number k and are rotating in either positive or negative sense so that the slip for the Kth harmonic (assuming S_1 is small) is given by:

$$S_k = \frac{k \mp 1}{K} \qquad k \neq 1 \qquad (14)$$

or

$$f_r = f\,(k \mp 1) \qquad k \neq 1 \qquad (15)$$

where the ± signs refer to the sense of rotation.

The range of slips is shown in Fig. 8 (not to scale). All of the slips will fall in the range of 0 to + 2. Note that, due to these high slips, harmonic currents are controlled by the leakage reactance.

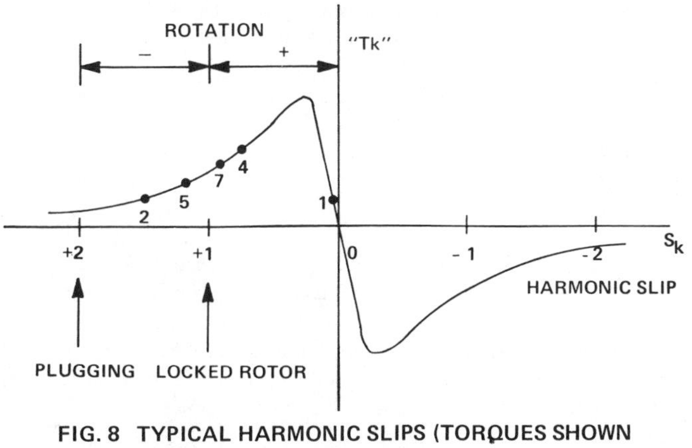

FIG. 8 TYPICAL HARMONIC SLIPS (TORQUES SHOWN ARE NOT TO THE SAME SCALES)

The harmonic outputs of the Fourier Analysis program are applied to a program evaluating the response of the equivalent circuit, Fig. 7. Torques and losses for a typical motor design were evaluated out to the 50th harmonic at the rated point. Average torque was due almost entirely to the fundamental with less than 0.3% variation due to harmonics in the PWM and six-step square. Losses turned out higher than expected. A summary of results is shown in Fig. 9. The data shown is for rated conditions in all cases.

4 POLE, 50 HERTZ INDUCTION MOTOR

$X_l = 0.113$ $\quad X_m = 1.24$ $\quad$ (PER UNIT)

$R_1 = 0.010$ $\quad R_2 = 0.010$

SIX-STEP SQUARE WAVE	STATOR	ROTOR	TOTAL
FUNDAMENTAL LOSS	.0168	.0108	.0276
HARMONIC LOSS	.0025	.0096	.0121
RATIO*	1.15	1.89	1.44

PULSE WIDTH MODULATED A = 1, N = 6	STATOR	ROTOR	TOTAL
FUNDAMENTAL LOSS	.0166	.0107	.0272
HARMONIC LOSS	.0094	.0286	.0437
RATIO*	1.57	3.67	2.61

* $\dfrac{\text{FUNDAMENTAL + HARMONICS}}{\text{FUNDAMENTAL}}$

FIG. 9 LOSS CALCULATION SUMMARY FOR VARIOUS WAVEFORMS APPLIED TO A TYPICAL MOTOR

Electrical efficiency dropped 1.4% due to harmonics in the six-step square wave and 5.1% in PWM under these conditions. Since the harmonic content depends on the modulation index, the degradation of efficiency due to harmonics will also vary with modulation index.

It should be noted that the factors by which the fundamental losses are multiplied to account for harmonic losses are greater than those cited in previously published work[4, 5, 6]. The magnitude of these factors will be quite variable in any case since harmonic losses are reasonably independent of fundamental slip (and therefore load) except through the influence of saturation.

These results are generally consistent with those previously published but predict generally larger harmonic losses (Fig. 10). The cause of these differences is partly in the rather simple models used for calculation but the wide variability in calculated and measured harmonic loss factors points out that harmonic currents and losses are extremely sensitive to motor design. In particular leakage reactance, rotor resistance and rotor bar height are critical parameters.

Harmonic loss factors for six-step square wave operation range from a low of 1.1 to a high of 1.38. Some of this variation can be rationalized. In this report a machine of conventional design is used as an example. The rotor and stator resistances are approximately equal and the rotor bars are slender resulting in severe deep bar effects at all harmonic frequencies. Information on motors and inverters is exceedingly sparse in the literature but some comments can be made. Largiader[5] uses a motor designed for use with inverters and having a very low rotor resistance. He also introduces some external inductance.

SIX-STEP SQUARE WAVE

DATA SOURCE	TOTAL LOSS/FUNDAMENTAL LOSS AT RATED POWER	
1. THIS REPORT	1.35	CALC.
2. KLINGSHIRN AND JORDAN (5)	1.19 (1.23 AT REDUCED FREQUENCY)	MEAS.
3. LARGIADER (6)	1.17 (SEPARATE CIRCUITS WITH 3 PHASE CHOKES)	CALC.
4. CHALMERS AND SARKAR (7)	1.10 (WITH EXTRA LEAKAGE REACTANCE AND POOR WAVEFORM)	MEAS.

PULSE WIDTH MODULATION

1. THIS REPORT	2.13 A = 1, N = 6		CALC.
2. KLINGSHIRN AND JORDAN (5)	1.60 } UNKNOWN (EQUAL WIDTH ?)	N = 24	MEAS.
	1.87	N = 12	
	3.21	N = 6	

FIG. 10 COMPARISON OF LOSS CALCULATIONS WITH PUBLISHED DATA

Chalmers and Sarkar[6] describe their inverter as having a "poor" waveform, low in harmonics. In addition, their motor was driven through an available transformer thus introducing extra reactances.

The losses associated with harmonics will vary with frequency, modulation index and chopping ratio but not (except through saturation) with load. Accordingly, losses and torques for a typical voltage-frequency schedule were evaluated for the same motor as described in Fig. 9. A typical operating schedule is shown in Fig. 11. The first part of the schedule is constant volts/hertz with the modulation index A increasing linearly with frequency up to f = 50 Hz. A is rapidly increased in the range from 50 Hz to 60 Hz where square wave operation is reached. Fundamental, harmonic and total losses are shown in Fig. 12. Fig. 13 displays a breakdown of the harmonic loss into stator and rotor I^2R losses. The voltage schedule shown is ideal in the sense that a feedback control system is required to alter the voltage (or modulation index) and slip to meet the torque or current requirement for the drive system.

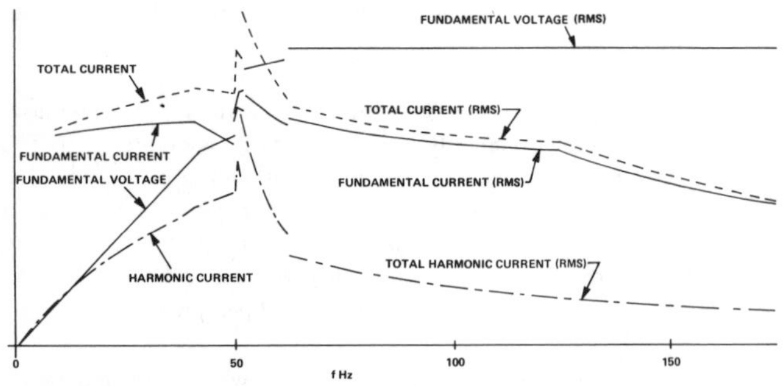

FIG. 11 TYPICAL PERFORMANCE SCHEDULE

Waveforms

Once the voltage or current (magnitude and phase) is available for each harmonic, it is straightforward to synthesize the waveforms. The motor analysis program reads out a file of 50 harmonics. These are

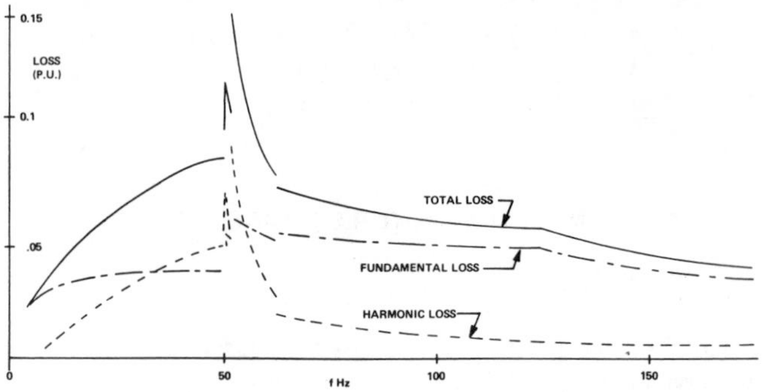

FIG. 12 COMPARISON OF FUNDAMENTAL AND HARMONIC LOSSES

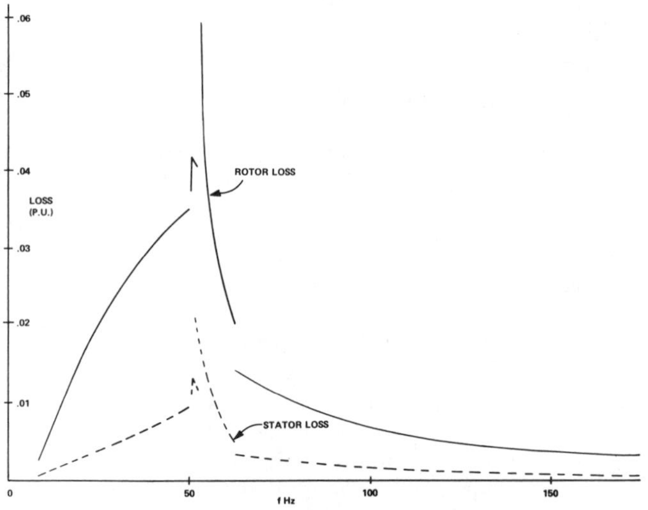

FIG. 13 ROTOR AND STATOR HARMONIC LOSSES

added together at 100 points in time using appropriate magnitude and phase. The magnitudes of the harmonics were adjusted according to Lanczos' formula to reduce the Gibbs phenomenon[7].

Fig. 14 displays the line voltage and current of a typical motor driven by the PWM inverter at a point close to its rated current and voltage. The voltage waveform was also synthesized as a check on program operation. Fig. 15a shows the line voltage at a somewhat higher modulation index at a point near the peak of the harmonic currents in Fig. 11. The line current for a somewhat lower impedance motor exhibiting very large peak currents relative to the fundamental is shown in Fig. 15b. Fig. 16a is for the same voltage but with line currents for a motor similar to that of Fig. 14. Fig. 16b shows the effect of increasing the chopping frequency 50% so that N = 9 instead of N = 6 with the same fundamental frequency. Fig. 17 illustrates the effect of utilizing a phase shift inverter on the waveforms of Fig. 15.

Conclusion

A flexible and efficient computation technique has been developed for analyzing the performance of a pulse width modulated inverter – induction motor variable speed drive in the steady state. Harmonic currents, losses and current waveforms have been calculated for a variety of input voltages and motors demonstrating the range of information that can be obtained with this technique.

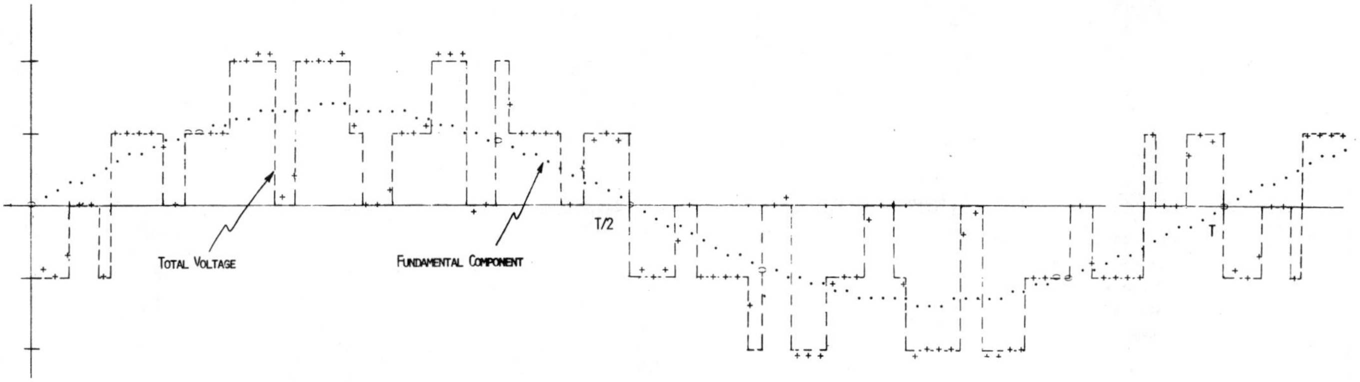

FIG. 14A LINE–TO–NEUTRAL VOLTAGE (PHASE A) A = 1, N = 6, L = 0.015T

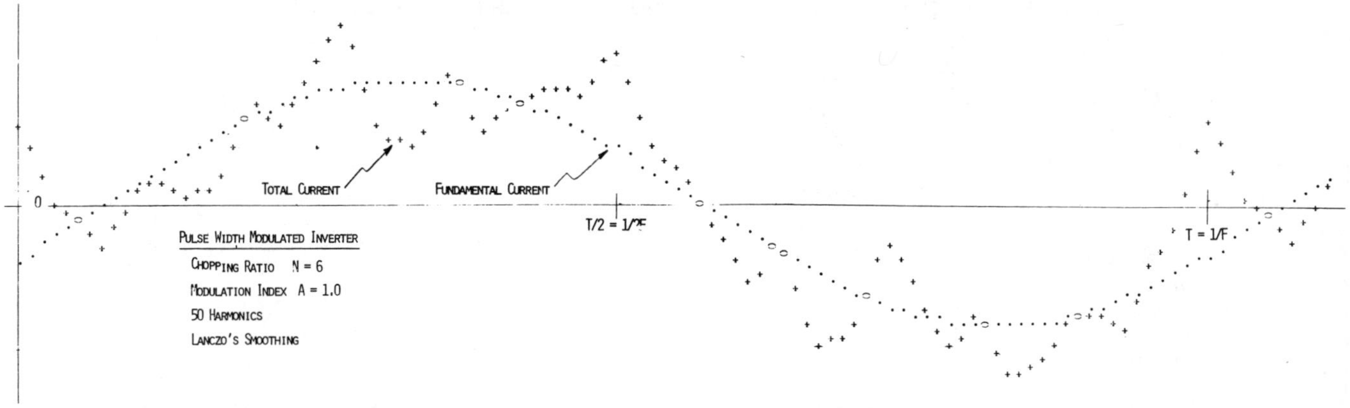

FIG. 14B LINE CURRENT FOR A TYPICAL MOTOR

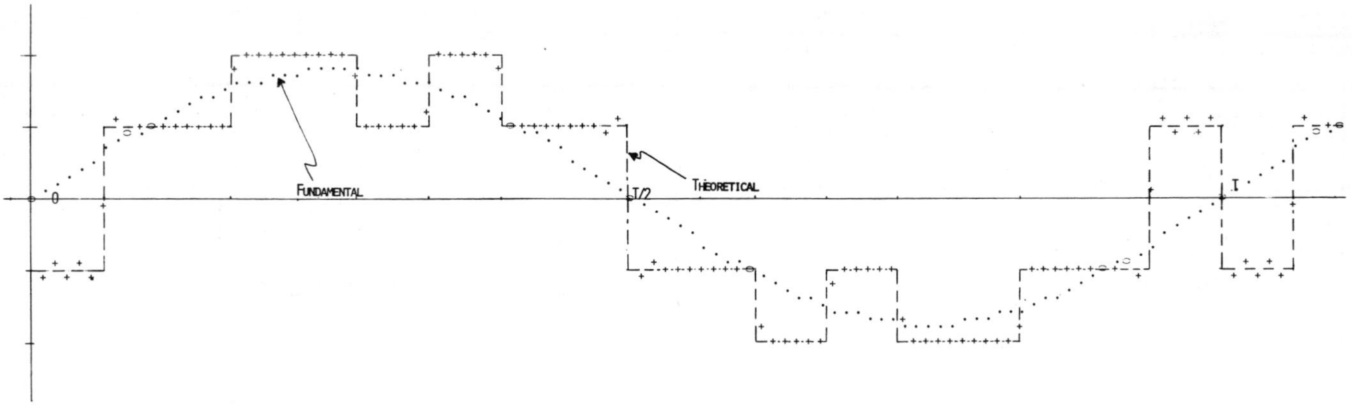

FIG. 15A LINE–NEUTRAL VOLTAGE A = 1.46, N = 6, L = 0.016T

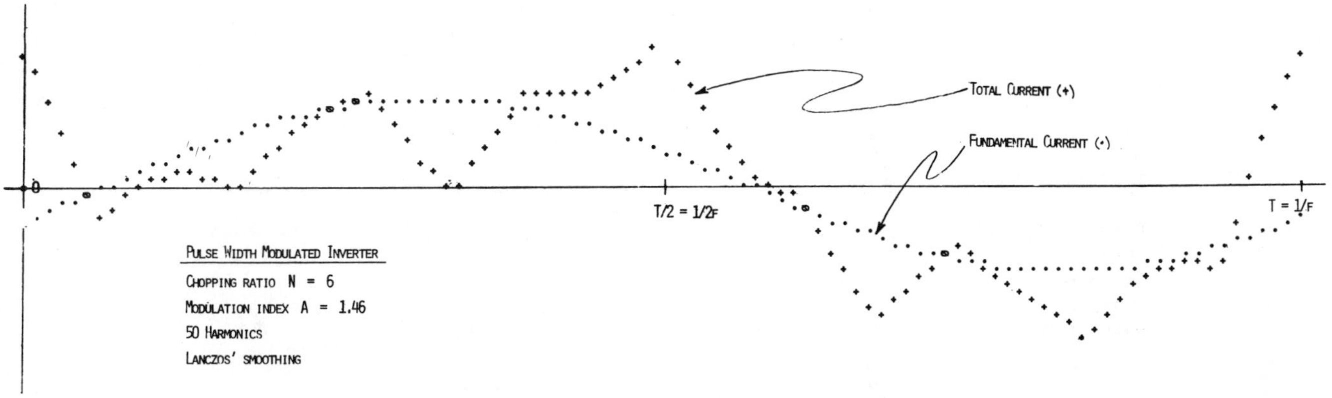

FIG. 15B LINE CURRENT FOR A LOW IMPEDANCE MOTOR

7 1/8 x 9 15/16 PRINT SU

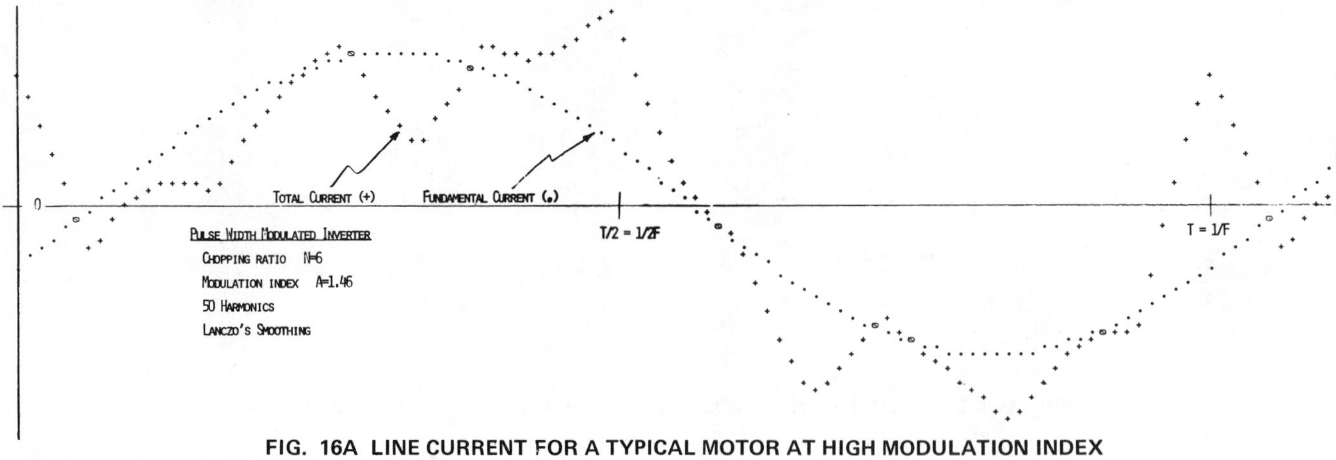

FIG. 16A LINE CURRENT FOR A TYPICAL MOTOR AT HIGH MODULATION INDEX

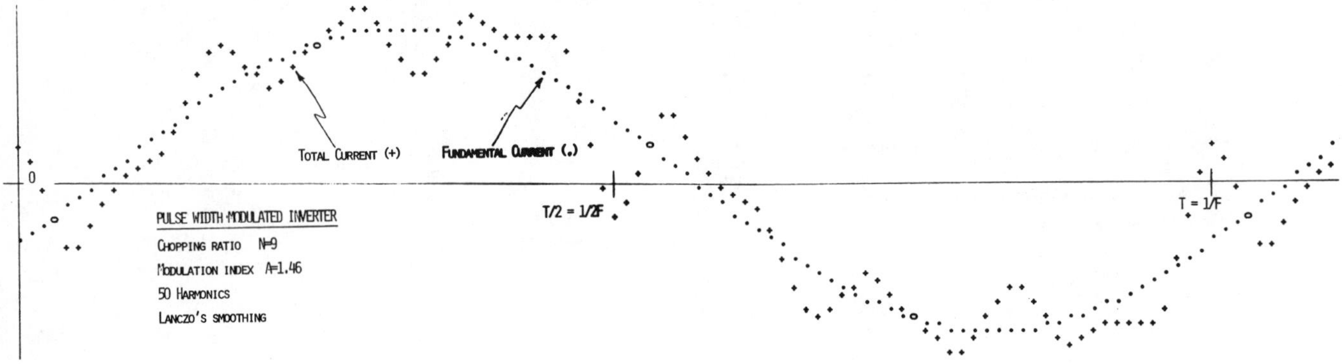

FIG. 16B LINE CURRENT FOR A TYPICAL MOTOR AT INCREASED CHOPPING RATIO

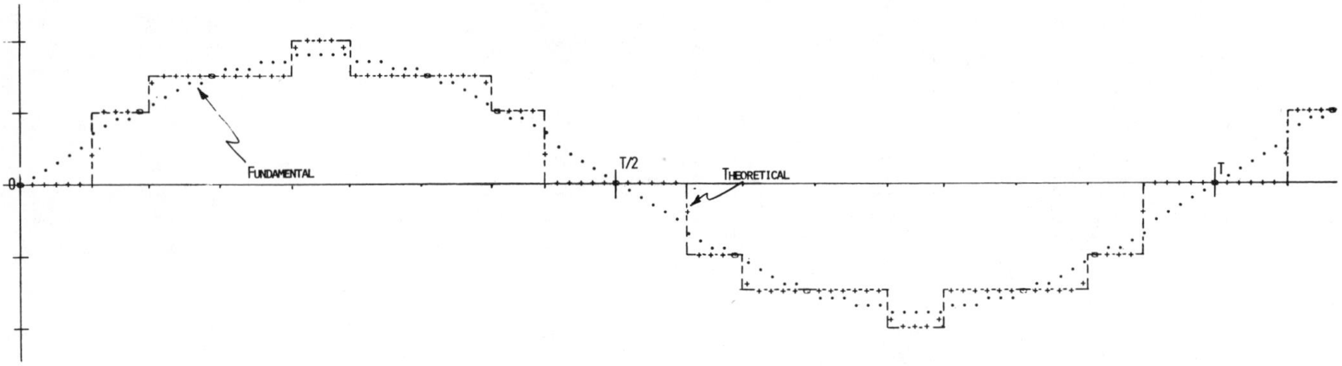

FIG. 17A LINE–NEUTRAL VOLTAGE – PHASE SHIFT INVERTER

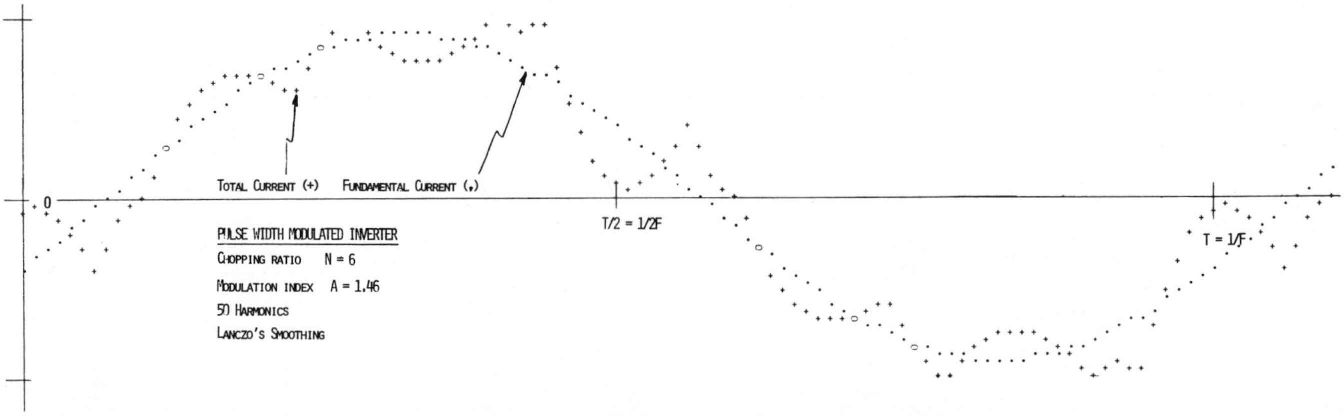

FIG. 17B LINE CURRENT PHASE SHAFT INVERTER – LOW IMPEDANCE MOTOR

Bibliography

1. Schonung, A. and Stemmler, D., "Static Frequency Changers with "Subharmonic" Control in Conjunction with Reversible Variable Speed AC Drives", The Brown-Boveri Review, Aug/Sept 1964.

2. Jain, G.C., "The Effect of Wave-Shape on the Performance of A 3-Phase Induction Motor", Trans. PAS–IEEE, Vol. 83, No. 6, pp. 561-566, June 1964.

3. Alger, P.L., "Induction Machines", Gordon and Breach Science Publishers, 2nd Ed., New York 1970, pp. 265-272.

4. Klingshirn, E.A. and Jordan, H.E., "Polyphase Induction Motor Performance and Losses on Non-Sinusoidal Voltage Sources", Trans. PAS–IEEE, Vol. 87, No. 3, pp. 624-631, March 1968.

5. Largiader, H., "Design Aspects of Induction Motors for Traction Applications with Supply Through Static Frequency Changers", The Brown-Boveri Review, April 1970.

6. Chalmers, B.J. and Sarkar, B.R., "Induction-Motor Losses Due to Non-Sinusoidal Supply Waveforms", Proc. IEE, Vol. 115, No. 12, Dec. 1968, pp. 1777-1782.

7. Hamming, R.W., "Numerical Methods for Scientists and Engineers", McGraw-Hill Book Co., Inc., New York 1962.

Torque Pulsations in Induction Motors with Inverter Drives

STUART D. T. ROBERTSON, MEMBER, IEEE, AND K. M. HEBBAR

Abstract—**Torque pulsations in induction motors would have an average value of zero and therefore are generally neglected by inverter induction machine system designs. A simple equivalent circuit method is presented for estimating the magnitude of torque fluctuations under steady-state operating conditions based on single-phase equivalent circuits. The method indicates that torque fluctuations are due mainly to the interaction of fundamental flux in the air gap at harmonic rotor currents. Inverters producing a 6-stepped voltage waveform, the predominant pulsating torque is at the sixth harmonic and the magnitude of the fluctuation is independent of operating frequency, provided that constant volts per Hz is maintained. The method is extendable to PWM inverter-machine systems with similar conclusions with respect to variation of torque over the input frequency range. However, for the PWM inverter with a fixed number of pulses per cycle, the torque pulsations can approach or exceed full-load average torque. The method outlined is simple and amenable to hand calculations.**

Introduction

INVERTERS apply essentially nonsinusoidal voltages to induction machines. The fundamental of the voltages produces the average output torque, and the harmonics produce increased losses and torque fluctuations. The contribution of the harmonics to the average torque is very small [1], [2], approximately a 4-percent reduction. For an applied six-stepped voltage waveform, where the voltages do not depart appreciably from sinusoidal, the torque fluctuations remain small. However, certain inverters, such as the pulsewidth-modulated type, produce dominant harmonics at certain frequencies, and the torque fluctuations due to such harmonics may be large. An estimation of the amplitudes of these torque fluctuations under steady state will be useful in a preliminary study of a given inverter machine derive for a particular load. Such a preliminary study will involve the calculation of machine currents, losses, torque–slip characteristics, and other similar operating features in addition to the calculation of torque fluctuations. A single method capable of obtaining all of these quantities of interest will be very useful in the preliminary study. The machine $d - q$ models have been used to calculate the torque fluctuations by Krause [3] and by Ward, Kazi, and Farkas [4]. However, the model normally used for studying the steady-state behaviour of the machine is the single-phase equivalent circuit [1], [2], [5]. Klingshirn and Jordan [1] have detailed the method of calculating the currents, losses, and other performance characteristics under steady state, using the single-phase equivalent circuit and a balanced nonsinusoidal source. Torque fluctuations have not been obtained by them using the circuit. Sandis [2] has indicated the quantities from the single-phase equivalent circuit upon which the torque fluctuations depend. A method of using the single-phase equivalent circuit model of the machine for evaluating the torque fluctuations with nonsinusoidal applied voltages is detailed in this paper.

Paper 69 TP 18-IGA, approved by the Petroleum and Chemical Industry Committee of the IEEE IGA Group for publication in this TRANSACTIONS. Manuscript received October 30, 1970.

S. D. T. Robertson is with the University of Toronto, Toronto, Ont., Canada.

K. M. Hebbar was with the University of Toronto, Toronto, Ont., Canada. He is now with the Department of Electrical Engineering, Kornadnatack Regional Engineering College, Mysore, India.

Single-Phase Equivalent Circuit

The single-phase equivalent circuit shown in Fig. 1 is used as a basis for the calculations. The effect of the presence of harmonics in the applied voltages on the parameters of the equivalent circuit has been investigated in detail by Kingshirn and Jordan [1]. It will be seen that the effect of machine resistances in the harmonic equivalent circuit is negligible, except at low frequencies, where the resistances and harmonic leakage reactances become comparable.

Phasor Diagrams: Fundamental

Fig. 2 is the phasor diagram for the induction machine for the fundamental of the applied voltage. The phasors ϕ_1^m and I_1^r, representing the mutual flux and the rotor current, can also be considered as vectors which rotate in space at an angular velocity of ω_s, the synchronous speed of the machine. These vectors will then represent the combined effect of all three phases. ϕ_1^m will represent the combined mutual flux and I_1^r the rotor magnetomotive force (MMF).

Phasor Diagrams: Harmonic

In an ideal symmetrical three-phase machine, the third harmonic cannot produce an air-gap flux, and hence it is necessary to consider harmonics only of the order $(6n \pm 1)$. Furthermore, the inverter applies normally balanced voltages to the machine. Then the $(6n - 1)$th harmonic produces a flux rotating in the opposite direction to that produced by the fundamental, while the $(6n + 1)$th harmonic produces a flux rotating in the same direction as the fundamental. Fig. 3(a) shows the phasor diagrams for the $(6n - 1)$th harmonic and Fig. 3(b) for the $(6n + 1)$th harmonic. In Fig. 3(a) the phasor rotation is made clockwise so that the flux and the rotor current phasors rotate in the same direction as the flux and rotor MMF

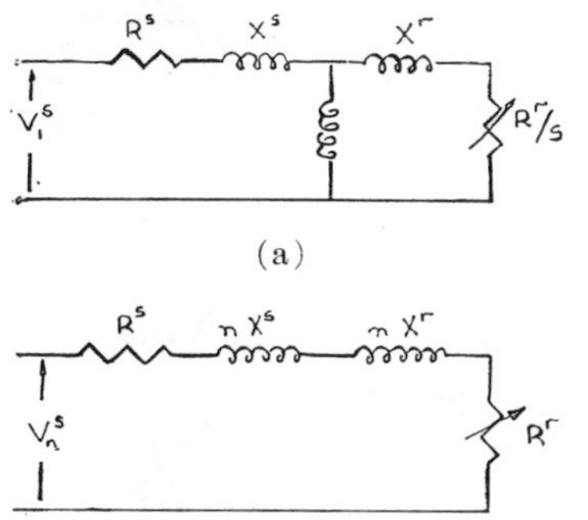

(a)

(b)

Fig. 1. Single-phase equivalent circuits for induction motor. (a) Fundamental equivalent circuit. (b) nth harmonic equivalent circuit.

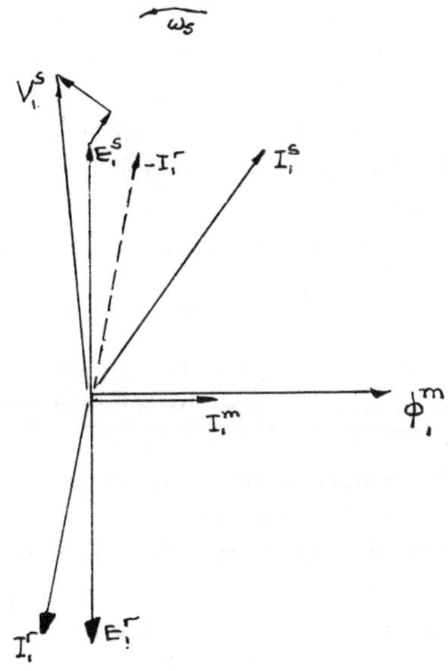

Fig. 2 Phasor diagram, fundamental.

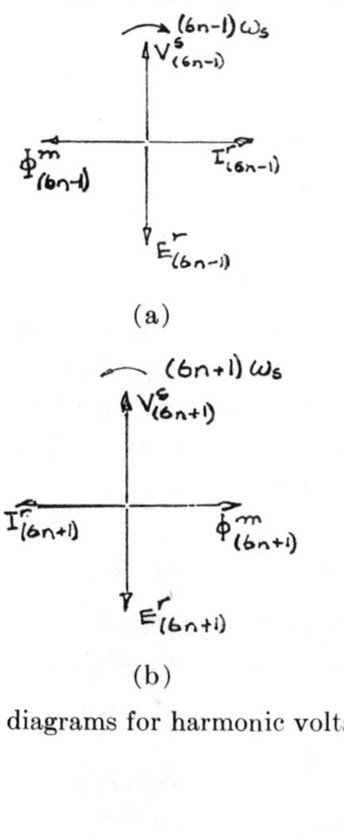

(a)

(b)

Fig. 3. Phasor diagrams for harmonic voltages applied.

vectors in space. $\phi_{(6n-1)}{}^m$ and $I_{(6m-1)}{}^r$ can now represent the flux and rotor MMF vectors at the $(6n - 1)$th harmonic.

Torque Calculations

Torque can be considered to be produced by the interaction between the mutual flux and the rotor MMF vectors. Torque will then have a magnitude $\phi^m I^r \sin \psi^{mr}$ where ψ^{mr} is the angle between ϕ^m and I^r at any instant. Since the angle between ϕ^m and I^r of the same harmonic remains constant, only steady torque is produced as a result of their interaction. Torque pulsations are produced when the angle ψ^{mr} varies. When flux of one frequency interacts with the rotor MMF of another frequency, the angle ψ^{mr} varies at a rate which is the difference between the speeds of the two rotating vectors, and pulsating torques are produced. These torques can therefore be evaluated by superimposing the flux and rotor MMF phasor vectors of various frequencies in a single diagram. The method is illustrated in the following two examples.

Example 1: Torque Fluctuations in Induction Machines Fed by Six-Stepped Voltage Waveforms

The six-stepped voltage waveform shown in Fig. 4(a) is one of the most common waveforms applied by an inverter to a machine. The Fourier components at the fundamental, fifth, and seventh harmonics of this waveform are shown in Fig. 4(b). Fig. 5 shows all three of these voltage phasors in a single diagram, rotating at their own synchronous speeds. The instant chosen in the phasor diagram is the instant t_0 of Fig. 4, when all the voltages are zero and are increasing in the positive direction. The phasor-vector diagrams can now be completed for all these three frequencies, and the flux and the rotor MMF vectors can be extracted to obtain the diagram of Fig. 6(a). In obtaining Fig. 6(a), it is assumed that stator induced voltage E_1^r at the fundamental is in phase with V_1^s, and that the effect of machine resistances at the fifth and seventh harmonics is negligible. These assumptions are normally valid. Where they are not, the modifications in the following results are straightforward. Fig. 6(b) is obtained by giving the whole diagram a clockwise rotation at ω_s to make the fundamental vectors stationary for simplicity. Fig. 6(b) shows a sixth-harmonic torque fluctuation due to 1) ϕ_1^m with I_5^r and I_7^r and 2) ϕ_5^m and ϕ_7^m with I_1^r. From this diagram, the net torque fluctuations at the sixth-harmonic frequency in the anticlockwise direction can be obtained as

$$T_6 = -\phi_1^m(I_5^r - I_7^r)\sin 6\omega_5 t + I_1^r[\phi_5^m \sin(\psi_1^{mr} - 6\omega_5 t) + \phi_n^m \sin(\psi_1^{mr} + 6\omega_5 t)].$$

Under normal operating conditions, ψ_1^{mr} will be nearly 90 degrees, in which case

$$T_6 = -\phi_1^m(I_5^r - I_7^r)\sin 6\omega_5 t + I_1^r(\phi_5^m + \phi_7^m)\cos 6\omega_5 t. \tag{1}$$

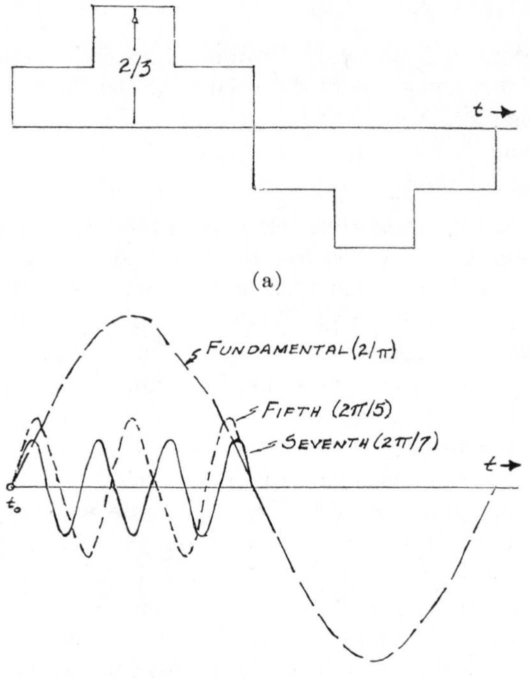

(a)

(b)

Fig. 4. Six-stepped waveform and its Fourier components. (a) Six-stepped voltage waveform. (b) Fundamental, fifth, and seventh harmonics.

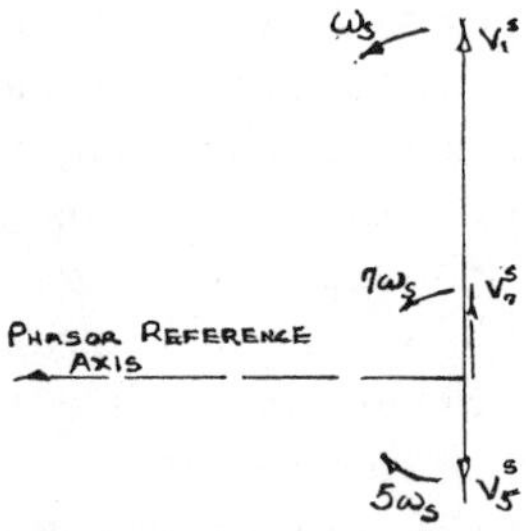

Fig. 5. Fundamental, fifth, and seventh harmonic voltage phasors.

It is seen from (1) that the sixth-harmonic torque fluctuation consists of two terms in phase quadrature, the relative magnitudes of which are dependent upon the relative magnitudes of fluxes and rotor MMFs at the various frequencies.

Relative Magnitude of Fluxes

For normal operation the fundamental flux will be about 1 pu. The fifth-harmonic voltage applied is 0.2 pu. For the harmonics, the stator and rotor drops are nearly equal, and hence gap voltage will be about 0.1 pu. Since the frequency is five times, the fifth-harmonic gap flux is 0.02 pu based upon similar considerations, the seventh-harmonic gap flux is 0.01 pu.

Relative Magnitudes of Rotor MMFs

I_1^r will vary depending upon the load on the machine, from approximately 0 to about 1 pu. I_5^r and I_7^r are dependent upon the locked-rotor impedance of the machine and the harmonic voltages applied. Blocked-rotor impedance is normally 0.1–0.2 pu at the fundamental frequency, and the harmonic voltages are 0.2 pu for the fifth and 0.16 pu for the seventh harmonics. Based upon these figures, I_5^r will be about 0.3 pu, and I_n^r will be 0.15 pu. These calculations indicate that the second term of (1) has an amplitude of less than one fifth of the amplitude of the first term and since it is in phase quadrature, the amplitude of the torque fluctuations can be obtained approximately by the amplitude of the first term:

$$T_6 = \phi_1^m(I_5 - I_7). \tag{2}$$

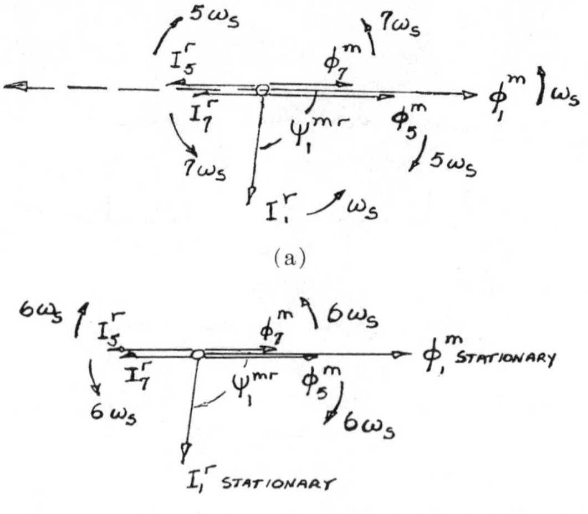

(a)

(b)

Fig. 6. Superimposed phasor-vector diagrams.

Sixth-harmonic torque fluctuations are produced by the interaction of fluxes and MMFs at other harmonics differing by six, such as the fifth and eleventh harmonics. However, by considerations similar to the ones already described, it can be shown that their relative magnitudes are small. The torque fluctuations in a machine with the parameters shown later in the section are presented in Table I. The calculations are based upon (2), while the computed results are obtained by digital solution of the complete machine equations. The following are machine parameters obtained from a 1-hp, 2-pole 110-volt 6.5-ampere three-phase 60-Hz machine:

$R^s = 0.041$ pu
$X^s = 0.061$ pu
$X^r = 0.061$ pu
$X^m = 1.69$ pu
$R^r = 0.045$ pu.

The results show that the sixth-harmonic torque fluctuations remain essentially unchanged with changing load or changing frequency. This is because ϕ_1^m, I_5^r, and I_7^r remain substantially constant under these conditions. Figs. 7–9 show the nature of torque fluctuations obtained from these computations.

Torque Fluctuations at Other Harmonic Frequencies

Following a similar method, it can be shown that the torque fluctuation at the twelfth harmonic is

$$T_{12} \simeq \phi_1^m(I_{11}^r - I_{13}^r)$$
$$\simeq \phi_1^m(I_5^r) \times 0.05$$

TABLE I

Frequency (Hz)	Fundamental Voltages (pu)	Machine Speed (rad/s)	Machine Variables $\phi_1{}^m$ (pu)	$I_1{}^r$ (pu)	$I_5{}^r$ (pu)	$I_7{}^r$ (pu)	Steady Torque: Simplified Calculation	Steady Torque: Digital Computation	Sixth-Harionic Torque: Simplified Calculation	Sixth-Harionic Torque: Digital Computation
60	1.0	370	0.950	0.392	0.328	0.164	0.577	0.583	0.258	0.278
60	1.0	360	0.928	0.937	0.328	0.164	1.400	1.380	0.252	0.272
60	1.0	350	0.903	0.430	0.328	0.164	2.110	2.080	0.247	0.265
30	0.5	180	0.929	0.457	0.328	0.164	0.665	0.728	0.252	0.264
30	0.5	175	0.908	0.720	0.328	0.164	1.045	1.050	0.247	0.256
15	0.25	90	0.931	0.232	0.293*	0.150*	0.320	0.340	0.238	0.230
15	0.25	80	0.861	0.712	0.293*	0.150*	0.980	0.990	0.204	0.193

* Stator and rotor resistances are considered in these current calculations as they become comparable to the leakage reactances at this frequency.

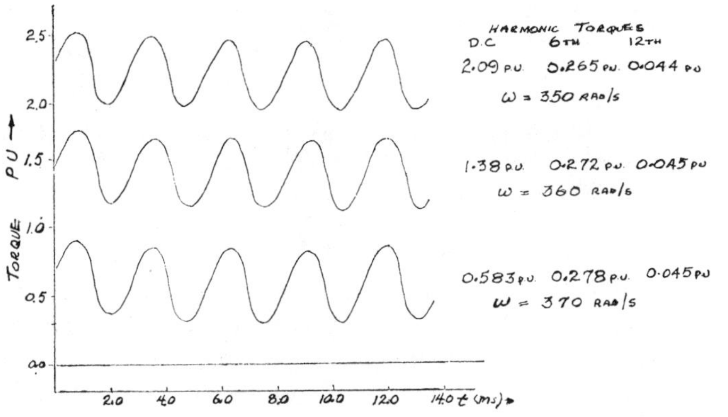

Fig. 7. Torque pulsations by digital computation. Six-stepped voltage: fundamental voltage—1.0 pu; frequency—60 Hz, 1.0 pu.

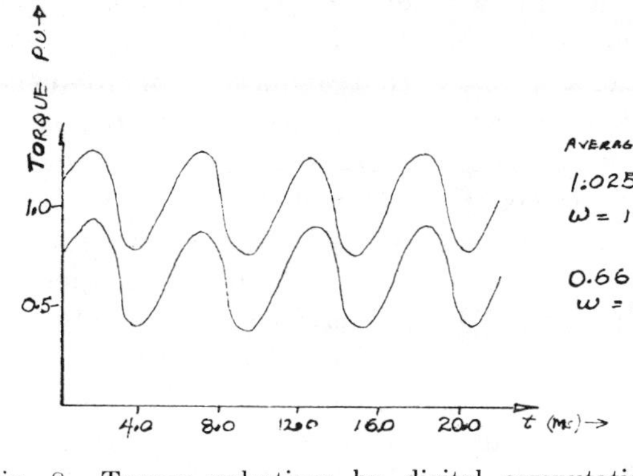

Fig. 8. Torque pulsations by digital computation. Six-stepped voltage: fundamental voltage—0.5 pu; frequency—30 Hz, 0.5 pu.

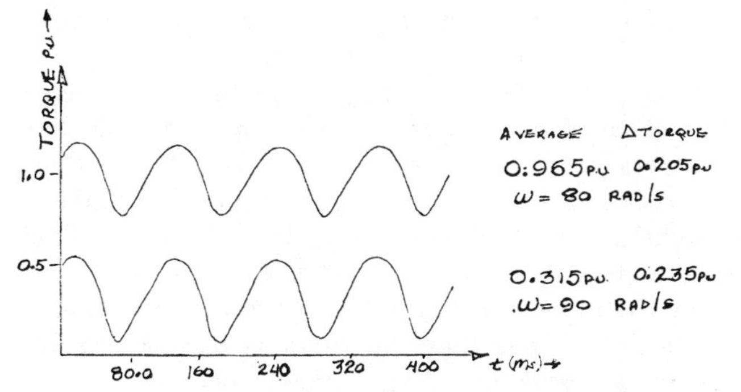

Fig. 9. Torque pulsations by digital computation. Six-stepped voltage: fundamental voltage—0.25 pu; frequency—15 Hz, 0.25 pu.

since

$$I_{11}{}^r = \frac{5^2}{11^2} I_5{}^r$$

$$I_{13}{}^r = \frac{5^2}{13^2} I_5{}^r.$$

This can be seen to be negligible when compared to the sixth-harmonic fluctuations which are approximately equal to

$$\phi_1{}^m I_5{}^r \times 0.5$$

since

$$I_7{}^r \simeq \frac{5^2}{7^2} I_5{}^r.$$

It can similarly be shown that torque fluctuations at other higher harmonics, i.e., the eighteenth, twenty-fourth, etc., will be still smaller. Hence with a six-stepped voltage wave, the main torque fluctuations are at the sixth-harmonic frequency.

Example 2: Torque Fluctuations in Induction Machines Driven by Pulsewidth-Modulated Inverters

There are different types of pulsewidth-modulated inverters. For this example, a pulsewidth-modulated inverter producing waveforms of the type shown in Fig. 10(a) is considered. It can be shown that the amplitude of the mth-harmonic voltage in such a waveform is given by

$$V_m{}^s = \frac{a}{m} \frac{\sin(\pi k m/p)}{\sin(\pi m/p)} \sin(m\omega_5 t) \qquad (3)$$

where a_1 is the amplitude of the fundamental for the six-stepped wave, enveloping given waveform, K is the ON/(ON + OFF) time, $\omega_5 = 2\pi f$, where f is fundamental frequency of voltage wave, and p is the number of pulses per cycle.

Inspection of (3) reveals that the amplitudes of voltages at the $(p - 1)$th and $(p + 1)$th frequency are very nearly the same as the amplitude of the fundamental. This gives rise to large currents at these frequencies and hence large torque pulsations at the pulsing frequency. The number of pulses per cycle p is usually a multiple of six; then the $(p - 1)$th harmonic would produce clockwise rotating

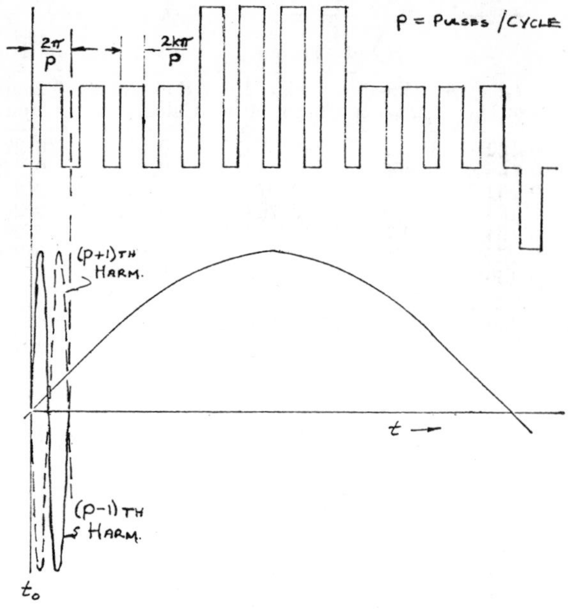

Fig. 10. Voltage waveforms of pulsewidth inverter.

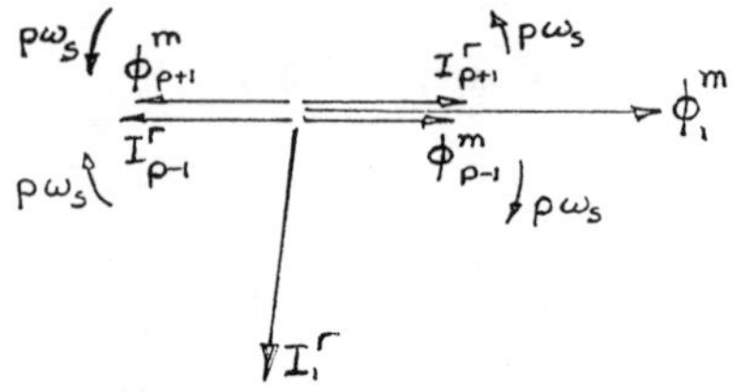

Fig. 11. Phasor-vector diagram for calculating torque fluctuations in pulsed inverters.

flux and MMFs. The fundamental and these $(p - 1)$th- and $(p + 1)$th-harmonic flux and rotor MMFs could be superimposed, as in Example 1, to obtain Fig. 11. The torque fluctuations are at the pth-harmonic frequency and consist of two component fluctuations due to (1) $\phi_1{}^m$, $I_{(p-1)}{}^r$, and $I_{(p+1)}{}^r$ and (2) $\phi_{(p-1)}{}^m$, $\phi_{(p+1)}{}^m$, and $I_1{}^r$. As in Example 1, the fluctuations due to (2) will be negligible, and hence the amplitude of torque fluctuations at the pth harmonic can be written from Fig. 11 as

$$T_p = \phi_1{}^m(I_{p-1}{}^r + I_{p+1}{}^r). \tag{4}$$

Fig. 12 shows the torque fluctuations computed for the machine of Example 1 under the following conditions: p is the number of pulses per cycle (24), K is the ON/(ON + OFF) time ratio (1/6), f is the frequency (10 Hz = 1/6 pu), and the angular velocity of the rotor is 50 rad/s. Fig. 12 indicates torque fluctuations at the sixth harmonic also which can be calculated in the manner of Example 1. The torque fluctuation at the pulsing frequency has a triangular or sawtooth waveform. Comparing the fundamental (which is at the pth harmonic of the input frequency) of this waveform with the result from (4), we have torque fluctuations at the twenty-fourth harmonic: by computation 0.854 pu, by calculation using (4) 0.888 pu; torque fluctuations at the sixth harmonic: by computation 0.173 pu, by calculation 0.181 pu.

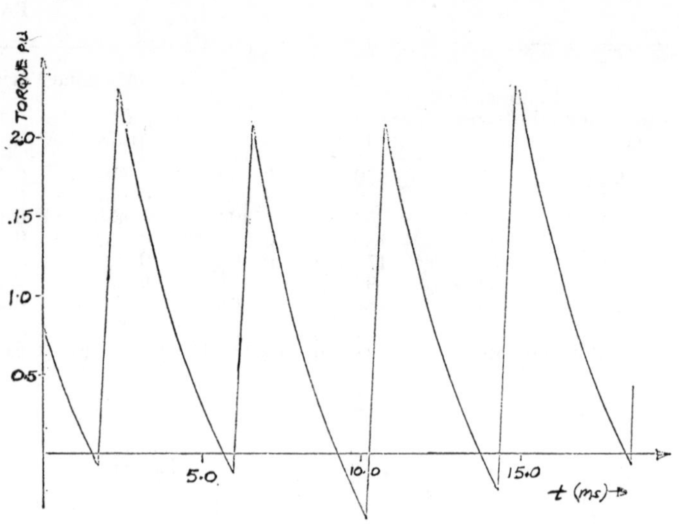

Fig. 12. Torque pulsations by digital computation. Pulsewidth-fed induction motor: frequency 10 Hz, ON/(ON + OFF) = 1/6, ω = 50 rad/s.

Discussion of Results

The results of the two examples given show that the calculation of the torque fluctuations can be done by considering the fundamental flux and the harmonic currents, as seen from (2) and (4). In normal operating conditions the fundamental flux is maintained at 1 pu. Furthermore, the harmonic currents do not vary with load since the machine presents nearly the blocked-rotor impedance to the harmonics at normal operating conditions. Hence it can be seen that the harmonic torque fluctuations are constant for normal load changes. The results of Table I show that reduction of frequency reduces the harmonic torque pulsations slightly for six-stepped waves but this reduction is seen to be due to reduction in the flux. For Example 1 the ratio V/f is kept constant. This increases the stator drop at low frequencies and hence reduces $\phi_1{}^m$. The torque fluctuations are therefore reduced. This reduction will not be obtained if $\phi_1{}^m$ is kept constant instead of V/f. Furthermore, at low frequencies, the machine resistances also appear in the calculations of harmonic currents. This may reduce the harmonic currents and consequently the torque fluctuations.

Use of pulsewidth-modulated inverters in the mode of constant number of pulses per cycle produces large torque pulsations of the order of 1 pu at the pulsing frequency. Equation (3) shows that large voltages are available at harmonic frequencies adjacent to the pulsing frequency at low values of K; that is, at conditions corresponding to low operating frequencies. Equation (4) indicates that the harmonic currents $I_{p-1}{}^r$ and $I_{p+1}{}^r$ are added to produce the fluctuations at the pth harmonic. Hence the large fluctuations in torque at this harmonic. As the operating frequency reduces, the frequency of these fluctuations also reduces and hence these fluctuations become capable of producing larger and larger effects on the mechanical rotation of the machine. However, if the number of pulses per cycle is large, these fluctuations will be less due to the reduction in the harmonic currents. The number of pulses per cycle can be increased effectively at lower frequencies by keeping the number of pulses per second constant and large enough to keep the torque pulsations within the desired limit. From considerations of torque and speed fluctuations, this mode of operation at constant pulsing rate, rather than constant pulses per cycle, is preferable.

In the calculations presented, the rotor speed is assumed to be constant. With pulsating torques produced by the machine, the rotor speed may vary which may further amplify the torque fluctuations. The machine input voltage may not be the ideal as has been assumed in these calculations. In a recent paper [7] these aspects have been discussed for a six-stepped inverter, and it is shown that the pulsations at the sixth-harmonic frequency with a six-stepped voltage could reach 0.75 pu. However, by chosing the machine and the inverter system parameters properly, these fluctuations can be reduced. The fluctuations of 0.25 pu shown in this paper for the machine exist because of the harmonic voltages in the input, without any amplifying effects due to speed or input-voltage variations from the ideal.

Conclusion

A method of estimating torque fluctuations under steady state, using the results from the single-phase equivalent circuit has been presented. This circuit is normally used for the steady-state analysis of machine performance, and the proposed method extends its usefulness. The method indicates that the torque fluctuations are due mainly to the fundamental flux and harmonic currents. With inverters producing six-stepped waveforms, the predominant torque fluctuations are at the sixth harmonic, while with pulsed inverters, the fluctuations at the pulsing frequency predominate. Since the fundamental flux is normally 1 pu, the fluctuations in torque can be reduced by reducing the harmonic currents. With pulsed inverters, this can be done by increasing the pulsing frequency or by keeping the pulsing frequency large and independent of the output frequency.

References

[1] E. A. Klingshirn and H. E. Jordan, "Polyphase induction motor performance and losses on nonsinusoidal voltage sources," *IEEE Trans. Power App. Syst.*, vol. PAS-87, Mar. 1968, pp. 624–631.

[2] J. P. Landis, "Static inverter drives," presented at the 1963 IEEE Textile Industry Tech. Conf.

[3] P. C. Krause, "Method of multiple reference frames applied to the analysis of symmetrical induction machinery," *IEEE Trans. Power App. Syst.*, vol. PAS-87, Jan. 1968, pp. 218–227.

[4] E. E. Ward, A. Kazi, and R. Farkas, "Time domain analysis of the inverter-fed induction machine," *Proc. Inst. Elec. Eng.*, vol. 114, Mar. 1967.

[5] G. C. Jain, "The effect of voltage waveshape on the performance of a 3-phase induction motor," *IEEE Trans. Power App. Syst.*, vol. 83, June 1964, pp. 561–566.

[6] K. M. Hebbar, Ph.D. dissertion, Dep. Elec. Eng., University of Toronto, Toronto, Ont., Canada, 1969.

[7] T. A. Lipo, P. C. Krause, and H. E. Jordan, "Harmonic torque and speed fluctuations in a rectifier–inverter induction motor drive," *IEEE Trans. Power App. Syst.*, vol. PAS-88, May 1969, pp. 579–587.

The Controlled Slip Static Inverter Drive

BORIS MOKRYTZKI, MEMBER, IEEE

Abstract—The ability to use the versatile characteristics of the semiconductor inverter to achieve optimized control of a squirrel cage induction motor is demonstrated. The technique has been made possible during the past few years due to the development of fast, powerful, and efficient inverters. These, in turn, are the direct result of advances in semiconductor technology, which have produced thyristors with improved qualities and the advent of integrated electronics. If the slip of an induction motor is constrained and controlled to values below breakdown, high efficiency, high power factors, and moderate currents result in performance comparable to that of a dc machine. General expressions defining torque and involving the quantities of slip and excitation are easily derived. Excitation can be expressed in terms of volts per cycle, current, or flux. Torque can be controlled by adjusting slip or excitation or both in combination. The ability to control slip and escitation precisely and accurately depends on the inverter which is used. The pulse width modulated (PWM) inverter is an extremely effective motor controller accomplishing voltage and frequency adjustment in a single circuit. It is a fast, linear device; its response is virtually instantaneous. As a power amplifier it is comparable to the duel converter of dc systems, and its speed of response makes it applicable to virtually any feedback loop. In considering the mating of the motor and inverter, several principal factors are involved in the optimization of the system. In most cases, the inverter is the most expensive part of the system. Its cost is determined primarily by its kVA output capability. In some applications, over-sizing the motor reduces the necessary inverter requirements.

Introduction

THIS paper demonstrates the ability to use the versatile characteristics of the semiconductor inverter to achieve optimized control of a squirrel cage induction motor. The induction motor has traditionally been thought of as a constant speed device since it usually operates from a fixed frequency power source. The disadvantages of operating in such a mode are exemplified when starting where high inrush currents must be experienced to achieve reasonable starting torques. The advent of the thyristor inverter as a source of variable voltage and frequency, coupled with the technique of controlled slip, transforms the ac motor to a variable speed device comparable to a dc motor. Much has already been said about the relative desirability of the ac drive which includes its ability to operate at high speeds and in adverse environments, its lightness, its compactness, etc. This paper defines the state-of-the-art as it explains the hardware and control techniques.

Manuscript received March 4, 1968.

The author is with the Reliance Electric Company, Cleveland, Ohio.

Theory

Before any drive system can be applied effectively, the output quantities of the motor must be expressed in terms of its controller's output quantities. These, in turn, must be realized from and related to controller inputs. The operation of the familiar shunt wound dc motor is defined by the speed and torque transfer functions.

$$T = K_t I_A$$

$$E = K_e S$$

where

- T torque
- S speed
- K_t torque constant
- K_e induced voltage constant
- E induced armature voltage
- I_A armature current.

A similar set of relationships can be developed for the induction motor.

When a set of polyphase voltages are applied to the stator windings of an ac machine, a rotating magnetic field is created. A magnetized rotor placed in this field tends to spin in synchronism, thereby forming a synchronous motor. The speed of the motor is dependent on the speed of the field which in turn is a function of the frequency of the applied voltage. As the frequency is changed to adjust the speed it is necessary to maintain the flux near some optimum level.

Since

$$V = N \frac{d\phi}{dt},$$

$$\phi = \frac{Vt}{N} = K \frac{V}{\omega}$$

Reprinted from *IEEE Trans. Ind. Gen. Appl.*, vol. IGA-4, pp. 312–317, May/June 1968.

where

ϕ flux
V applied voltage
ω frequency.

Thus, to control the action of ac machines, the adjustment of voltage and frequency must be coordinated.

Construction

The squirrel cage induction motor is composed of a stator incorporating a magnetic structure and a set of polyphase windings. The rotor is composed of a set of magnetic punchings or laminations into which has been cast (or fabricated) a simple set of aluminum or copper bars and end rings to form a squirrel cage winding. The rotor is essentially a solid chunk of material containing no electrical insulation other than that which is inherent on the surface of the laminations and conductors. As a result, it is mechanically rugged and able to operate at higher speeds and temperatures. At the same time, the operation of the machine can be efficient as demonstrated by the following analysis.

Variable Speed Induction Motor Drive Analysis

If excitation losses and stator voltage drops are neglected, a polyphase voltage $V \sin \omega t$, applied to the stator windings of a motor, produces a rotating magnetic field in the stator whose angular speed is $\omega_s = 2\omega/p$ (p being the number of poles) and whose magnitude is[1]

$$\phi_s \sim V/\omega.$$

If the speed of the rotor is ω_r, then

$$\omega_d = \omega_s - \omega_r \tag{1}$$

where ω_d is the slip speed.

The voltage induced in the rotor is

$$V_r \sim \phi_s \omega_d. \tag{2}$$

From this, a rotor current I_r is produced where

$$I_r = \frac{V_r}{Z_r} = \frac{V_r}{\sqrt{R_r^2 + (\omega_d L_r)^2}} \tag{3}$$

where

Z_r rotor impedance
R_r rotor resistance
L_r rotor reactance.

This, in turn, produces a flux

$$\phi_r \sim I_r \sim \frac{\phi_s \omega_d}{Z_r}. \tag{4}$$

This flux lags ϕ_s by an angle

$$\theta = \frac{\pi}{2} + \tan^{-1} \frac{(\omega_d L)}{R_r} = \frac{\pi}{2} + \gamma. \tag{5}$$

The motor torque is given by

$$T = \phi_r \phi_s \sin \theta = \phi_r \phi_s \cos \gamma. \tag{6}$$

Therefore, from (1), (4), and (5)

$$T \sim \phi_s \cdot \phi_s \omega_d \cdot \frac{1}{Z_r} \cdot \frac{R_r}{Z_r} \tag{7}$$

$$\phi_s^2 \frac{\omega_d R_r}{Z_r^2} = T \sim \left(\frac{V}{\omega}\right)^2 \frac{\omega_d R_r}{R_r^2 + (\omega_d L_r)^2}. \tag{8}$$

Thus, independent of the excitation frequency and speed, torque is dependent only on slip ω_d and volts per cycle V/ω.

Differentiating the expression for torque with respect to slip and setting it equal to zero yields

$$\omega_{dm} = R_r/L_r. \tag{9}$$

This is known as the breakdown slip associated with the breakdown point.

Controlled Slip

Operating beyond breakdown yields high-current low power factors and high losses. If the slip is constrained and controlled to values below breakdown, high efficiency, high power factors, and moderate currents result in an operation comparable to that of a dc machine. Furthermore, this holds true even when the speed of the system is varied by adjusting the stator frequency provided that, as in any well designed machine, the flux is kept at a level high enough to obtain good performance, but low enough to control saturation losses. The direction of slip relative to the stator determines the direction of power flow while the voltage determines the level of torque and, hence, along with the frequency, the level of power flow.

Thus, (8) is a general expression defining torque involving the quantities of slip and excitation. Excitation can be expressed in terms of volts per cycle, current, and flux. Torque can be controlled by adjusting excitation, or slip, or both in combination. The choice of the particular mode of control depends on the load, the motor, and the controller.

Forcing torques, which are a multiple of rated values, can be applied briefly by increasing excitation and/or slip. The resulting increase in losses can be tolerated because of the relatively long thermal time constant of the motor.

The Inverter

While other means of producing ac, such as the cycloconverter, are available, the inverter which operates from dc is more flexible and, hence, more popular. The number of different types of high power inverters available today use thyristors which act as power switches. These thyristors must be provided with a means of turn-off, almost universally employing charged capacitors, in a technique known as forced commutation.[2] The ability of an inverter to provide a given amount of turn-off time is

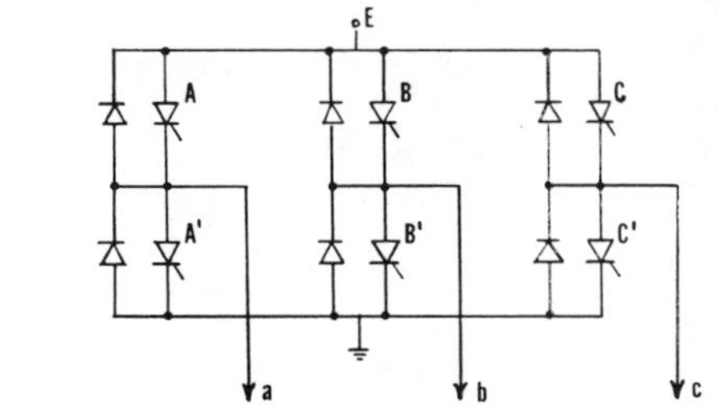

Fig. 1. The three-phase thyristor bridge.

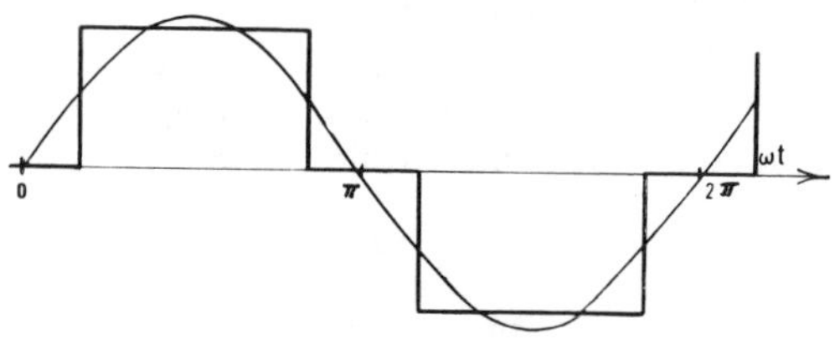

(a)

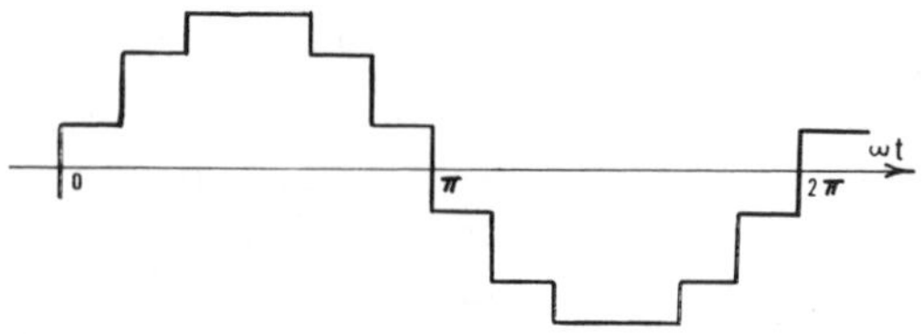

(b)

Fig. 2. Inverter waveforms. (a) Six-step. (b) Twelve-step.

proportional to the size of the commutation capacitors and the voltage to which they are charged, and inversely proportional to the current to be commutated.[3] The circuit of Fig. 1 is a three-phase thyristor bridge with the commutation circuitry removed. The parallel opposed rectifiers carry reactive current. The thyristors of each phase conduct for alternate 180° intervals with the sequencing of each phase displaced by 120°. A set of three-phase output voltages are formed with the "six-step" wave-shape of Fig. 2. These have been used successfully with motors. This emphasizes the fact that the motor does not need a sine wave and can tolerate or reject a surprisingly large amount of harmonic content. The harmonics, principally the 5th and 7th, produce about 20 percent–30 percent additional losses over the sine wave in a standard motor.

Variable Voltage Inverter

One popular way to achieve voltage control is to operate the inverter bridge from a variable voltage dc source such as a phase-controlled rectifier or a chopper. Some filtering is required to smooth the dc. The disadvantage of this system lies in the need for two power controllers in cascade, the time delay introduced by the filter, and the harmonics of the waveform. If the outputs of two or more inverters are combined, the stepped waveform of Fig. 2 can be synthesized. The harmonics of such waveforms can be neglected, except at very low frequencies where they impart a cogging or a stepping motion to the motor.

Pulse Width Modulated Inverter

The pulse width modulated (PWM) inverter is an extremely effective motor controller accomplishing voltage and frequency adjustment in a single circuit.[4] Proper sequencing of the bridge in Fig. 1 produces an output characterized as composed of discrete pulses of fixed height and frequency, and adjustable width, as shown in Fig. 3. The inverter voltage is controlled by the average width of the pulses and the harmonics are controlled by modulating to a sinusoidal envelope. If a sine wave envelope is used, the harmonics of the output waveform are principally those associated with the modulating frequency. These can be set high enough to be rejected by the motor.

The PWM inverter is a fast, linear device; its response is virtually instantaneous. As a power amplifier it is comparable to the dual converter of dc systems and its speed of response makes it applicable to virtually any feedback loop. It can be considered as a "black box" with power inputs and outputs, as well as two control inputs, one for frequency, and one for voltage.

The Inverter–Motor System

To best combine the motor and inverter, several principal factors are involved. In most cases, the inverter is the most expensive part of the system. Its cost is determined primarily by its kVA output capability. In some applications, oversizing the motor reduces the necessary inverter requirements. For example, reducing the modulating frequency also lowers inverter cost and increases motor losses. This frequency should be lowered to the point where motor heating is noticed and perhaps to the point where an extra frame size, or better grade of insulation, is required.

The four basic types of machine losses are important factors in optimization. They are

1) resistance losses (fundamental current)
2) slip losses (slip × torque)
3) magnetization losses
4) harmonic losses.

The latter two are affected principally by the level of excitation. Conventional machines are designed to sustain full load loses continuously but, as in dc machines, are usually provided with external cooling or derated at low speeds because of diminished fan cooling.

The motor has a rather long thermal time constant, relative to that of the much smaller thyristor-heatsink combination. When the motor experiences higher losses, as it produces high torques in load forcing, its tolerance to this high heating can be measured in minutes. The ability of a thyristor to deliver over currents is measured in seconds, while the maximum instantaneous output current is limited at the maximum current which can be turned off, or commutated, by the inverter. Thus, for short time overloads within the thermal limits of the thyristor, commutation limit is the determining factor, and a current limit is generally required. The steady-state rating of the system is ultimately set by motor heating and can be monitored by thermal protectors.

Control Technique

The application of dc drives is normally divided into two classical modes: the constant torque and constant horsepower ranges of operation. The term constant torque implies operating below some base speed with constant field flux or excitation. Torque is controlled by adjusting armature current which determines armature flux. The torque is variable below some maximum value which, as a limit, is constant throughout this range of operation. However, in reality, this limit may be a multiple of the rated machine torque and, consequently, may have to be modified according to the commutation ability of the machine at various speeds.

In dc, the constant horsepower range is obtained by a weakening of the shunt field flux. This produces a higher speed which varies as the inverse of the field strength. The maximum torque is proportional to the field strength, hence, the horsepower or product of speed and torque remains constant. The maximum speed under field weakening can be greater than several times the base speed. The armature flux is normally controlled by current forcing. The armature voltage must also be limited by coordinating the control of the field. Torque can be adjusted by the control of field current with the armature voltage maintained at some limit.

In both of the classical operating modes, the essence of the task is to adjust the levels of two interacting fluxes to control speed and torque. In the ac induction motor, the behavior of stator flux can be analogous to the field of a dc machine. In the constant torque range, it can be maintained by controlling V/ω and current-slip factors. However, speed is a function of applied frequency, and the flux or field "weakening" is a consequence of operating above base frequency at a fixed voltage.

Rotor flux is formed as the product of stator flux and slip. It can be varied to set the torque by adjusting slip and/or excitation, just as the armature flux is controlled by adjusting current in a dc machine. To maintain the maximum value of rotor flux at some level, the slip must be increased as the stator flux is weakened in the constant horsepower mode. A representative control system is shown in Fig. 4 which includes a slip control loop and a stator flux regulator brought about by a shaped current reference that produces indirect stator flux control. The method is simple, since the sense of the shaping deviation is the same for regeneration and motoring while the magnitude bears a close resemblance to the slip function. Direct flux control using magnetic sensors is also possible.

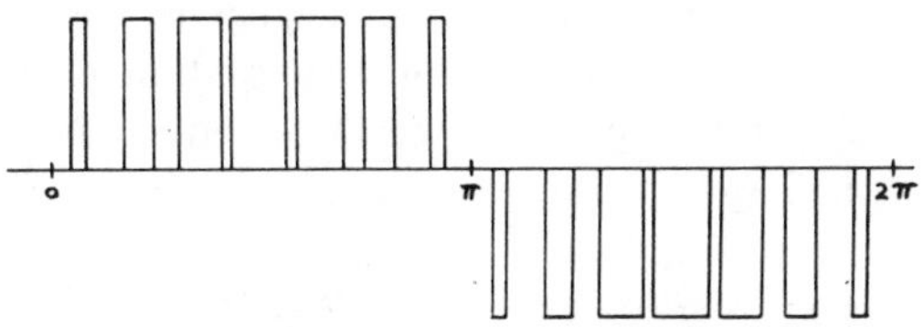

Fig. 3. Pulse width modulated inverter output.

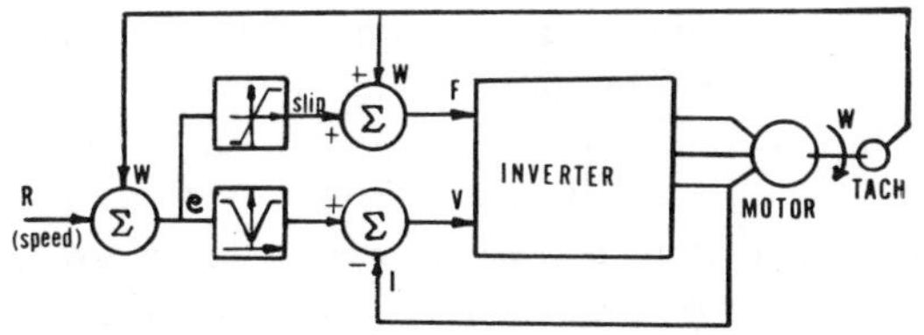

Fig. 4. Controlled slip drive system.

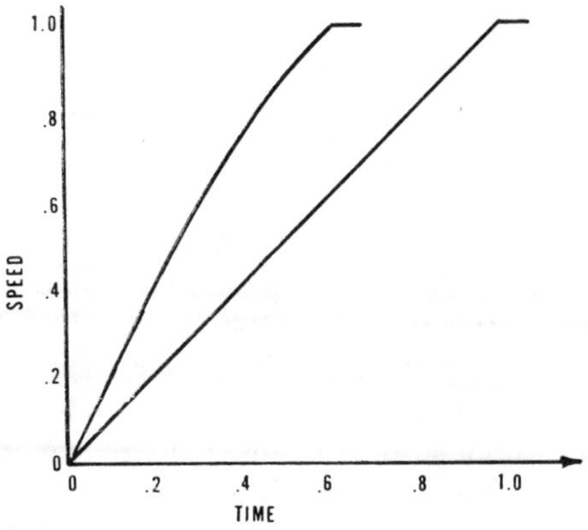

Fig. 5. Acceleration transients.

The constant stator flux method is not necessarily optimum since, in the process of load forcing in the constant torque range, for a given current and speed, the torque is proportional to the applied voltage squared. This implies that the excess torque is a function of an excess stator flux at some maximum rotor flux as measured by stator current. The high flux has the tendency to produce high magnetization losses in steady state in anticipation, so to speak, of the moment when it is necessary to produce brief forcing torque. However, slip losses appear only when the load is being driven.

On the other hand, the presence of harmonics in the waveform, while not affecting increased magnetization losses, does induce rotor losses because their relative slip is high and is virtually uneffected by the speed and loading of the machine. Hence, their contribution to motor loss is roughly proportional to the square of the stator flux. Such losses would soar under conditions of over-excitation. Thus, in the constant torque range, an optimized system can operate with a fixed value of slip. Torque is controlled by adjusting the level of stator flux. As mentioned before, this can yield a relaxation with respect to the minimum allowable carrier frequency. Such a system calling for a variable stator flux can also be obtained by forcing that flux with a shaped voltage or flux reference since

$$T \sim (V/\omega)^2 s,\ I^2/s,\ \phi_s{}^2 s.$$

Application

Operating above base speed in the constant horsepower range is limited by the maximum practical slip obtainable from the motor. This is normally about twice-rated slip, hence the constant horsepower range for a motor extends to about twice base speed. Operation above this speed can be obtained with derating or by experiencing greater than rated current. Obtaining a given horsepower at a speed of three or four times base speed requires an oversized motor, or an oversized inverter, or both.

The task of accelerating a large mass is an example of the use of an oversized motor. Fig. 5 displays the typical straight line acceleration of an inverter operating at maximum current to obtain maximum torque from a motor. This implies operation at some fixed slip s_1 during the time of acceleration T_1. To accelerate the motor in a shorter time, it is more economical to retain the inverter size at the expense of motor cost. Doubling the motor size, and connecting or winding it to be rated at twice the maximum inverter voltage output, means that the system operates in the constant horsepower mode above half speed. It also means that the motor produces twice the torque per ampere of its low voltage connection and initially it can produce twice the acceleration in the load. This again is accomplished by operating at rated slip. At half speed, the voltage becomes constant along with the horsepower as the current is maintained by increasing slip. The load continues to accelerate, and torque drops off as the reciprocal of speed until, at full speed, the torque is half its original value at twice rated slip. The acceleration time is about two-thirds of the original time as shown in Fig. 5.

Fig. 6. Thyristor packages.

The Future

The ac system can compete technically with conventional general purpose drives. But before wide scale applications or acceptance of these systems is realized, they must be competitive in cost. Today, applying the differences in cost between the dc and ac motor to the increased cost of the ac inverter still leaves the economic balance in favor of dc drives. The cost of the inverter is principally a function of semi-conductor cost and, to some extent, control circuit cost. Significant advances in thyristors is evident with the new "Pow-R-Disc" or "Press-pak" packages as shown in Fig. 6. Unlike the conventional stud mounted units, heat can propagate from two surfaces of the pellet rather than one. This yields increased ratings in addition to physical mounting

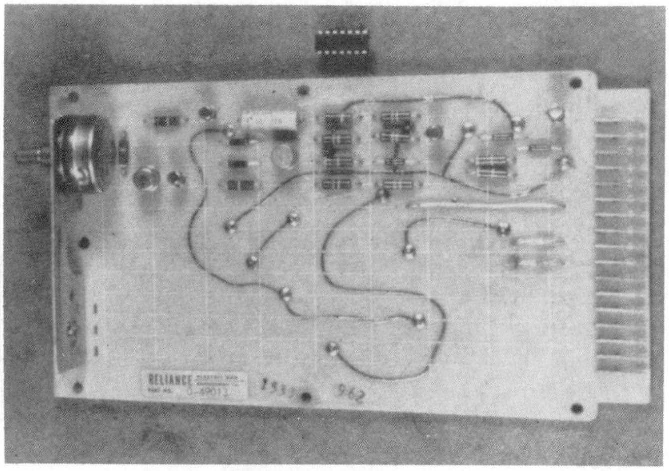

Fig. 7. Discrete and integrated circuits.

flexibility. The possibility of improved performance with respect to turn-off time and other parameters should tend to further reduce costs.

The control functions required with PWM inverters were quite costly a year or two ago. Today, the widespread use of integrated circuitry has reduced control function cost and increased reliability of these units. This is exemplified by the operational amplifier shown in Fig. 7. It is shown in its printed circuit form with discrete components as well as its integrated circuit form. The integrated circuit provides an order of magnitude improvement in gain, stability, and speed, at a fraction of the cost. Such factors make the complex coordination in ac systems of voltage, frequency, current, speed, and slip, routine and inexpensive.

References

[1] C. J. Amato, "Variable speed with controlled slip induction motor," presented at IEEE Industrial Static Power Conf., November 1965.

[2] A. J. Humphrey, "Inverter commutation circuits," *Conf. Rec. 1966 IEEE Industry and General Applications Group Annual Meeting*, pp. 97–108.

[3] B. Mokrytzki, "Solid state ac motor drives," presented at the 13th Annual IEEE Pulp and Paper Conf., May 3–5, 1967.

[4] ——, "Pulse width modulated inverters for ac motor drives." *1966 IEEE Internat'l Conv. Rec.*, vol. 14, pt. 8, pp. 8–23.

A CURRENT-CONTROLLED PWM TRANSISTOR INVERTER DRIVE

A.B. Plunkett
General Electric Company
Schenectady, New York 12305

Abstract - A method of controlling a pulse width modulated transistor inverter to minimize the peak output current will be discussed. The method adaptively modulates the inverter switching times so as to produce the maximum possible torque in the attached motor for the minimum ac peak current. As the method involves a transition for operation of the inverter as a current source to operation as a voltage source, a method of system control must be applied that is suitable for use in either mode of operation. A motor angle controller is used to supply the primary system stabilization. The described method of control generates the appropriate angle command and current amplitude signals by using feedback signals of motor torque and flux.

A hybrid computer simulation of the described method of control has been performed and the results are presented. A transistor pulse width modulated dc link inverter driving an induction motor is modeled. The method of inverter control allows for a relatively simple adaption to the use of either induction or synchronous machines and renders the inverter relatively insensitive to specific machine parameters.

Introduction

In the process of designing a transistor inverter ac induction motor drive system, it has become apparent that the development of an improved method of pulse width modulation (PWM) control is desirable. The control system must achieve several important objectives:

- Minimization of motor loss so that conventional motors can be used without derating
- Minimization of peak transistor current
- Simplification of the control to reduce cost

These objectives can be met by employing an instantaneous current control system with independent current control on each inverter output phase. The resulting system has operating characteristics similar to both current inverters and voltage inverters. However, the use of such a hybrid inverter control system leads to a number of problems. A current inverter system operates very well at low speeds but is limited in performance at high speeds (high output frequencies). A voltage inverter tends to perform poorly at low speeds. The present system eliminates these problems by acting as a hybrid of the two types; it retains the good low speed performance of the current inverter and the good high speed performance of the voltage inverter.

The method of accomplishing this desirable end and the results of a complete simulation on a hybrid analog-digital system are presented below.

Description of the Drive System

The drive system consists of a transistor inverter using power Darlington transistors and an induction motor optimized for maximum efficiency on inverter waveforms. The inverter is a simple three half bridge arrangement with feedback diodes, thus giving a defined voltage on the ac output lines at all times. Figure 1 shows the inverter's circuit diagram. Note that the inverter consists of six half-phase sections. Figure 2 is a photograph of a half-phase module, six of which are required to make a complete inverter. Figure 3 shows the ac induction motor referred to above, along with a General Electric dc motor of equivalent power, voltage and speed range rating. The induction motor designed by Dr. G.B. Kliman of General Electric contains an integral 3:1 gear box to reduce the maximum motor speed of 15,000 rpm to a maximum shaft speed of 5,000 rpm. The motor is a four-pole design and thus has a maximum frequency of 500 Hz.

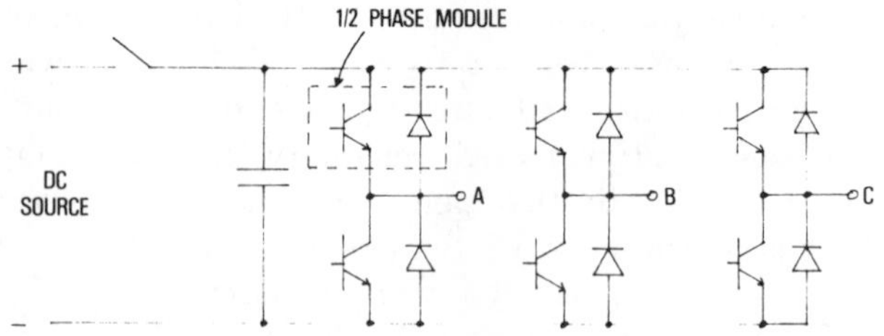

Fig. 1. Transistor Inverter Circuit

Fig. 2. Inverter Half Phase Module

Fig. 3. AC Induction Motor (left) and Equivalent DC Commutator Motor

Reprinted from *Conf. Rec. IEEE/IAS 1979 Annual Meeting,* pp. 785-792, 1979.

The motor is designed to supply constant torque with constant flux up to a speed of 5000 rpm and constant power to maximum speed with constant voltage applied. The leakage inductance is maximized consistent with the constant power requirements.[1] Maximizing the leakage inductance minimizes harmonic current but limits the pull-out torque and thus the maximum speed for constant power output at constant applied voltage. Figure 4 shows the torque, ac voltage, and ac current as a function of speed.

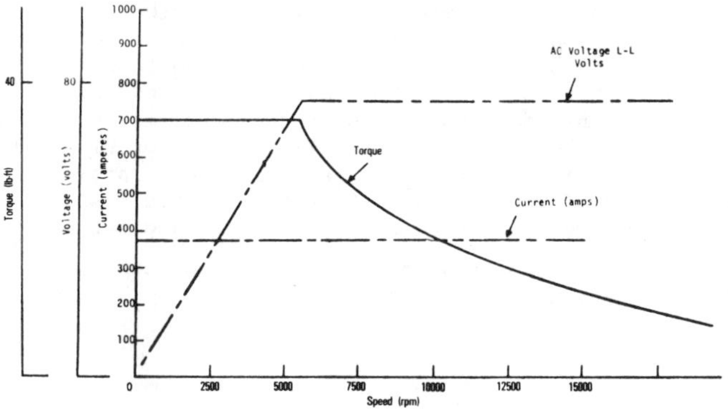

Fig. 4. Induction Motor Torque, Voltage, and Current

The Appendix shows the equivalent circuit and parameters used in the analysis.

Voltage Control

The inverter will operate from a nominally fixed dc source voltage and thus will require some sort of voltage control capability to appropriately vary the ac motor voltage. There are a number of pulse width modulation methods which could be used.[2,3,4,5]

One method presently in wide use dates back to the original paper of Schonung and Stemmler.[2] Figure 5 is an illustration of the method of generating the inverter switching waveforms and the resultant output waveform across the motor line to neutral connection. The resulting output waveform causes motor current ripple which adds to the motor losses and requires extra inverter current handling capability. In addition, because the waveform is a voltage waveform, a small imbalance in the motor voltage, due to the small motor stator resistance, can cause a relatively large current imbalance in the motor current. An imbalance in phase voltages will translate to a motor frequency ripple in the dc link and since the motor frequency varies from 0 to 500 Hz, any resonances in the dc link filter tend to be excited. In addition, it is possible to generate in the motor dc currents which will cause extra heating and torque ripple in the motor output.

All the pulse width modulation methods in References 2,3,4, and 5 have in common two disadvantages. The first, that these methods are voltage control systems; the second, that a special transition mode of operation involving the synchronization of the reference waveform with the desired output waveform is required. The requirements that the waveforms have very little imbalance and a very small dc offset impose a requirement for a high degree of precision on the waveform generation. This means that the control electronics must precisely generate a waveform with a fairly smooth sinusoidal output.

A second problem arises when an inverter output voltage greater than about 80% of maximum is required, as at high speed. For maximum output voltage (V line to neutral = .45 Vdc), the inverter must operate in the square wave mode without notches in the output

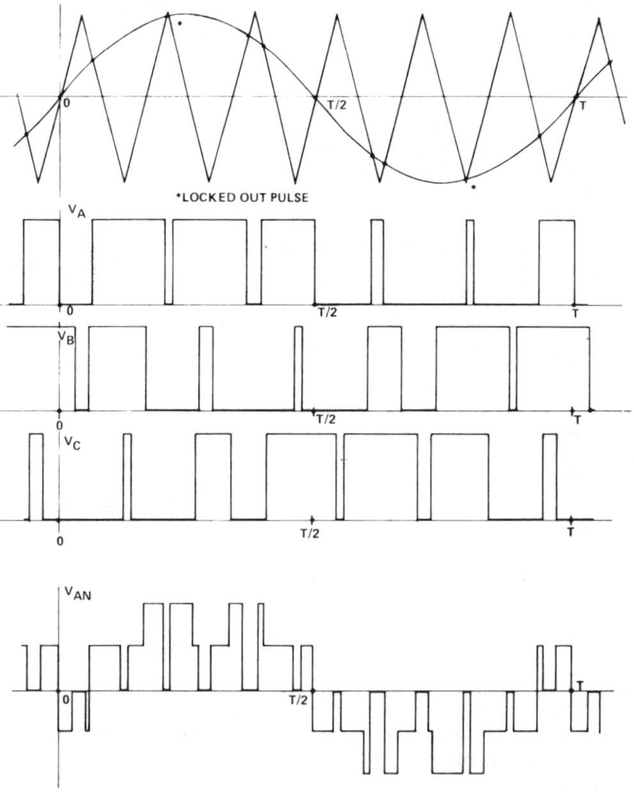

Fig. 5. Sine Wave - Triangle Wave Pulse Width Modulation

waveform. One normally uses a transition mode of pulse width modulation which synchronizes the chopping reference waveform to the output frequency waveform. References 3,4, and 5 examine some of these methods.

Reference 4 describes a method used as a transition mode of PWM. The method is to place the voltage controlling notches in the inverter output waveform near the zero crossings of the ac voltage cycle rather than in the middle of the wave. The effect of eliminating a minimum width pulse at the end of an ac cycle is almost negligible since the contribution to the fundamental frequency output waveform is small. Of course, to prevent phase jumps when a pulse is dropped, the pulses must be dropped symmetrically so that even harmonics are never developed. The transistor inverter is able to switch much faster than an equivalent thyristor inverter so the transition problem should be less significant. Nevertheless, this problem must be considered.

Current-Controlled PWM

A method has been devised to eliminate these two problems. The method involves generating a 3-phase reference sine wave at the desired output frequency and current amplitude and comparing this reference with the actual motor current. If the motor current is more positive than the reference, the inverter will switch in the negative direction and vice versa. The frequency of the inverter chopping can be controlled by introducing a small amount of hysteresis into the comparison so that in effect the amount of current ripple is regulated. Figure 6 is a block diagram of the basic concept while Figure 7 shows the current control circuit, including the hysteresis and a 25 μsec time lockout on the inverter switching frequency.

A problem one must face is the maximum operating frequency of the inverter, which the chopping frequency determines at low speeds. The inverter switching losses are proportionate to frequency and must be controlled; thus the chopping frequency should, on the average, be

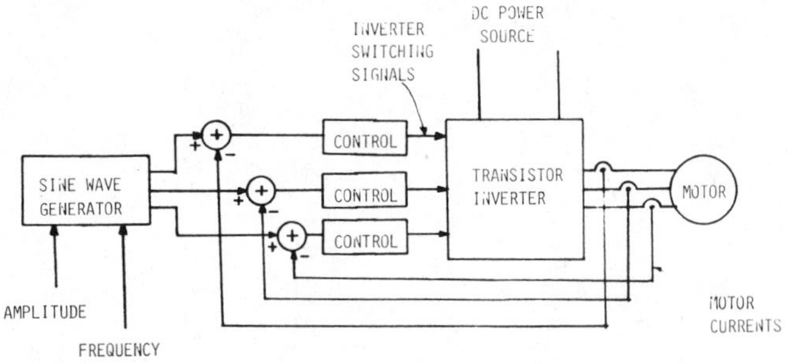

Fig. 6. Current-Controlled PWM Block Diagram

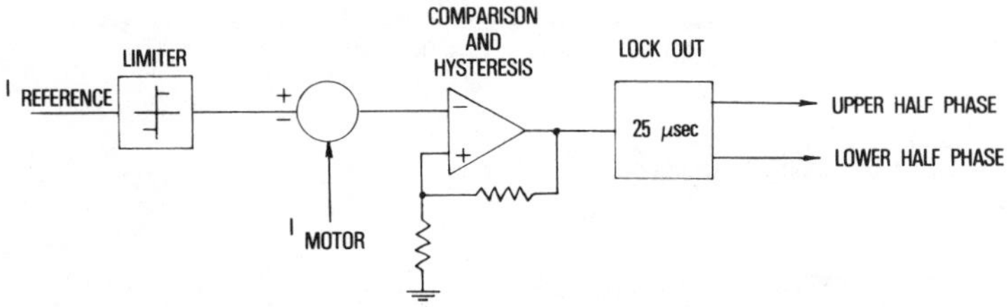

Fig. 7. Current Control Circuit for One Phase

limited. The chopping frequency is automatically controlled by the comparator hysteresis, the motor stator plus rotor leakage inductance, and the dc link voltage by the equation

$$V = -L \frac{di}{dt}$$

which, however, applies only approximately since the motor leakage inductance is a function of frequency and the voltage is actually the dc link voltage minus the motor back emf. However, the net result is that, in the PWM range, the average chopping frequency is relatively constant and not greatly affected by speed.

An additional feature of the current-controlled PWM system is its instantaneous current limit, easily achievable by limiting the maximum value of the current waveform reference generator. Limiting the reference generator's output creates a more square current reference which limits the transistor current but (up to a point) not the motor torque producing capability sinc since the fundamental frequency torque producing component of the current in a square wave is about 11% higher than in a sine wave. Of course, torque ripple at the fifth and seventh harmonics of the fundamental frequency will appear when the current waveform becomes flat-topped.

The method of controlling inverter current in effect causes the transistor inverter to become the same as a current inverter, with all the inherent advantages, stability, and problems of the current inverter at low speeds. However, at high speed the system can operate without pulse width modulation and thus has the control characteristics of a voltage inverter and does not suffer the inherent frequency limits of the current inverter. As the motor speed increases, the output voltage of the inverter will increase as necessary to maintain the motor current. Eventually the maximum output voltage capability of the inverter will be achieved. The result is a relatively smooth transition from PWM to square wave operation of the inverter, as far as amplitude of the current and voltage is concerned. A problem arises because at low speeds before the transition, the reference wave represents motor current which in an induction motor may lag the voltage by about 30° at full load to 90° at no load. After the transition, the reference represents motor voltage. Thus, as pulses are dropped (chops in the ac output waveform eliminated), jumps in the phase of the reference are required to eliminate torque transients. In addition, some sort of stabilizing control is required for low speed current inverter operation which should ideally be compatible with the voltage inverter operation at high speed.

Some known methods of controlling current inverters for motor stability include some form of slip frequency control combined with current regulation when induction motors are used,[6,7] or the well-known shaft position sensing when synchronous motors are used. Both these methods are difficult to use to handle the transition from current-controlled PWM to voltage control. An alternative stabilizing control suitable for both current and voltage control, is described in Reference 8 which shows a method of motor electrical angle control applicable to both synchronous and induction motors. This control method measures and regulates the motor internal angle between the motor air gap flux vector and the stator (inverter) current vector. Figure 8 is an illustration of the variation of angle as a function of load. Figure 9 from Reference 8 shows an example relating the motor electrical angle $\sin \theta_{eq}$ to the motor slip frequency and torque for constant flux operation.

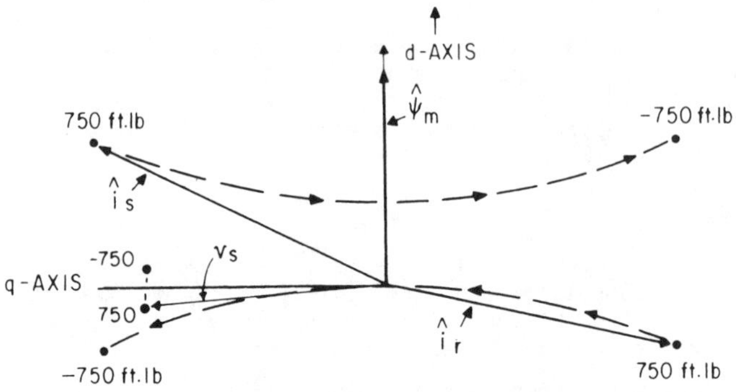

Fig. 8. Vector Diagram Showing Definition of Motor Angle

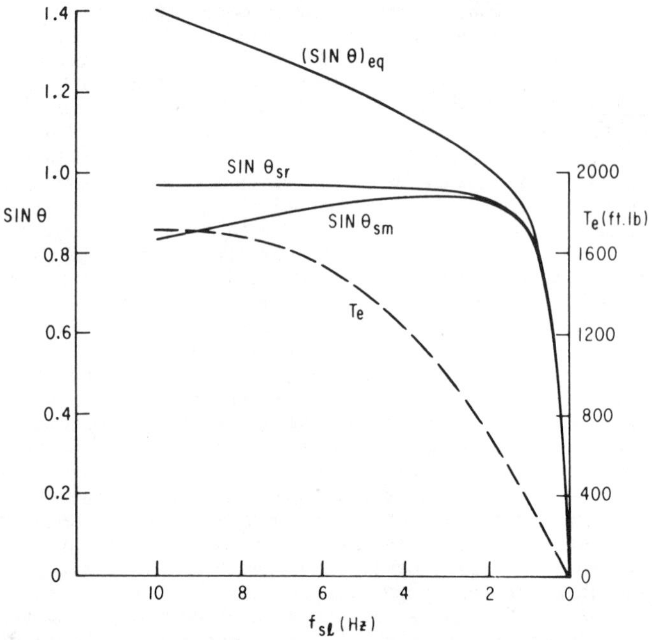

Fig. 9. Induction Motor Angle and Torque as a Function of Slip Frequency

There are two angle signals shown in Figure 9. The actual motor angle $\sin\theta_{sm}$ reaches a maximum of about 60° at about one half the induction motor pull-out torque. A modified angle signal, $\sin\theta_{eq}$, which does not have this distressing (for feedback control) property, is actually used in the control system. For a synchronous motor, either angle signal is satisfactory for control system use. However, use of the $\sin\theta_e$ signal will reduce the control gain variation as a function of the steady state angle.

Use of this control method will stabilize the drive system in current inverter operation and allow for a feedback-controlled transition between current control and voltage control. Note that the reference waveform generator phase angle will be varied as pulses are dropped to maintain the motor operation in a satisfactory manner.

The method of measuring the motor's operating angle involves measuring the motor stator current and the motor air gap flux as vectors. The air gap flux can be inferred from the motor terminal voltage but will suffer some inaccuracy for very low speed operation due to the stator resistance voltge drop; or can be measured directly by air gap flux sensing coils wound around a motor stator tooth in a manner similar to that described in Reference 10. The method of making and inserting the sensing coils without disturbing the main motor winding was developed by J. Franz of General Electric (U.S. patent No. 4,011,489). A block diagram of the motor current, flux, torque, and angle circuits is shown in Figure 10 for the situation in which flux sensing coils (the simplest circuits) are used. The inputs to the circuit are V_{ma}, V_{mb}, V_{mc}, the airgap voltages; and i_{as}, i_{bs}, and i_{cs}, the ac line currents. A block diagram of the basic drive system is shown in Figure 11, with the angle control loop indicated. The commanded inputs to the drive system shown in Figure 11 are the magnitude of motor current and the angle between the air gap flux and stator current vectors. These inputs are satisfactory for complete control of the drive system but do not allow for an easily generated operator command signal. For example, to increase torque, both the angle command and the current command must be increased. However, the relative amount of increase of these signals is a non-linear function of the desired motor flux level. The flux level in the motor will affect the speed of response to an increasing torque command as the motor may need to increase flux to be able to produce more torque.

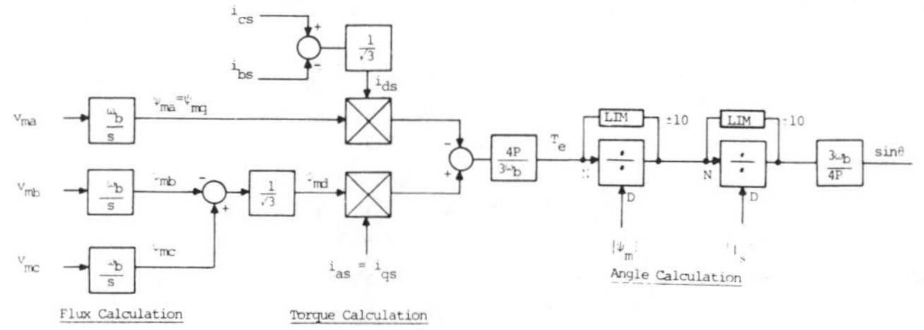

Fig. 10. Motor Flux, Torque, and Angle Sensing Block Diagram

Extra current is required to increase flux level in the induction motor so that a conflict may occur between the increased torque requirement and the necessity of increased flux. This conflict will thus tend to slow the drive system response to an increased torque command if the motor flux level is too low before the commanded torque increase. Also the gain of the control system is affected by motor flux level. Therefore, at least one, and preferably two, outer control loops are desirable to enable the use of simple operator

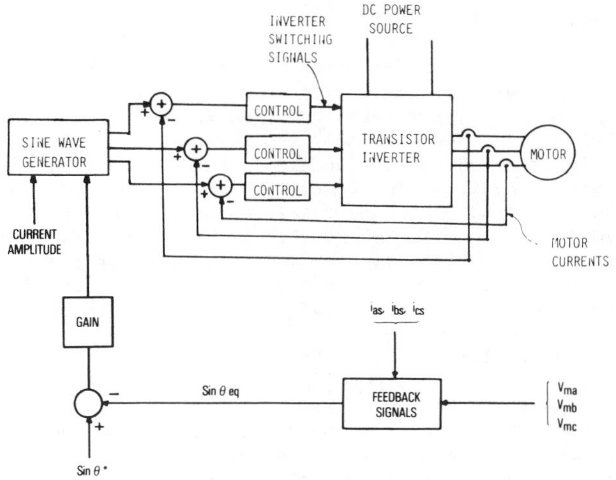

Fig. 11. Block Diagram of Drive System with Angle Loop Control

commands without the use of arbitrary non-linear functions. The outer loops chosen are a torque regulator operating through the inverter frequency input to regulate torque in order to obtain the fastest transient response and to retain torque regulation at high speed when the inverter is in the square wave mode of operation and current control is lost due to control loop saturation. At low speeds in the PWM mode of operation, some form of voltage regulation is desired. Regulation of the air gap flux level in the motor by varying the magnitude of motor current is an ideal way of controlling the effective inverter voltage output. Both of the feedback signals of torque and flux are available as by-products of the motor angle sensing circuits. Use of these signals eliminates the need of precise mechanical tachometers or the shaft position sensors required for many previously used high performance ac motor drives. A block diagram of the complete control system emphasizing the torque and flux control system is shown in Figure 12.

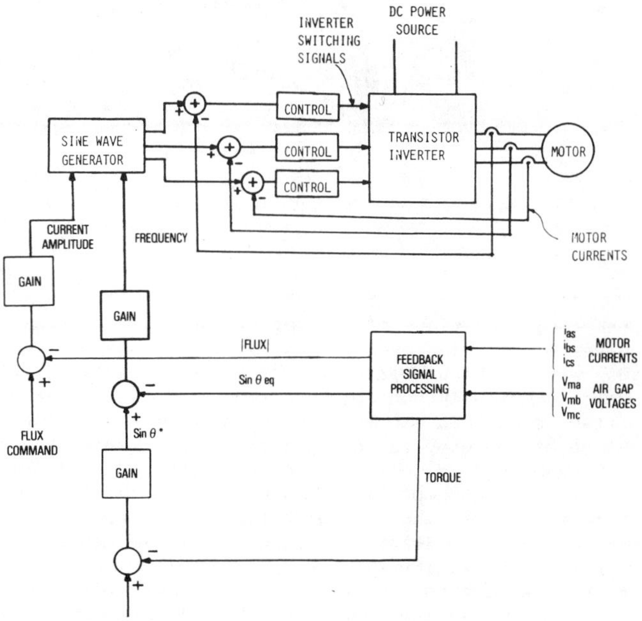

Fig. 12. Overall Drive System Block Diagram

Simulation of Current-Controlled PWM

The drive system described has been simulated on a hybrid digital and analog computer. The control system, including the reference sine wave generator is implemented in essentially the same way that it would be implemented using standard digital and analog integrated circuits. The power transistor inverter and the induction motor are simulated using standard techniques such as those described in Reference 11. The Appendix gives the simulated induction motor equivalent circuit. The inverter is simulated using a set of six ideal switches that simulate the effect of the parallel transistor and diode in the one-half-phase module of Figure 2. The detailed effect of the snubbers in this inverter circuit is not important to the control system operation and is not included in the simulation. However, the effect of the dc link inductance and capacitance is important and is included in the simulation. The results of the operation of the control system are presented below in a series of oscillograph traces as a function of time.

Figure 13 shows a typical example of operation in the current controlled PWM mode of operation. Note that the motor line current is quite sinusoidal and that the current is constrained to be within 50 A of the reference waveform, the hysteresis value chosen in the control system. The pulse width modulator has a built-in 25 μsec lock out to prohibit the generation of conflicting transistor switching signals, but for the case shown, the lock out has no detrimental effect.

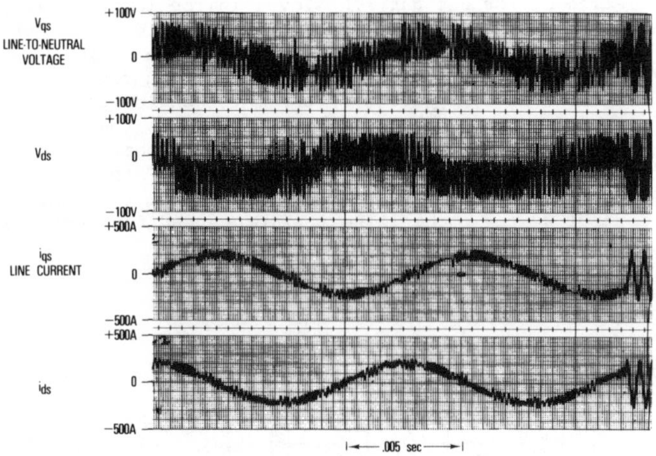

Fig. 13. Steady State Operation in PWM at Rated Motor Torque, Flux, and Current at One-Half Rated Speed

Figure 14 shows steady state operation at the rated operating conditions of the motor. Note that a few pulses still exist even though the design is for full square wave output voltage. The current regulated PWM system adaptively controls the type of waveform generated to cause a smooth transition between full PWM and square wave operation by causing the remaining chops to be near the ac voltage zero crossings. The effect of these extra chops is to minimize the peak motor current while retaining the capability for essentially full ac output voltage. As motor speed and hence frequency increase, the extra chops disappear due to the improved smoothing of the inverter chopped voltage wave by the motor leakage inductance at high ac output frequency. The system operation in steady state shows that the current-controlled PWM method controls the motor current ripple exactly as desired and, in addition, generates an optimum waveform in the operating region between PWM and square wave.

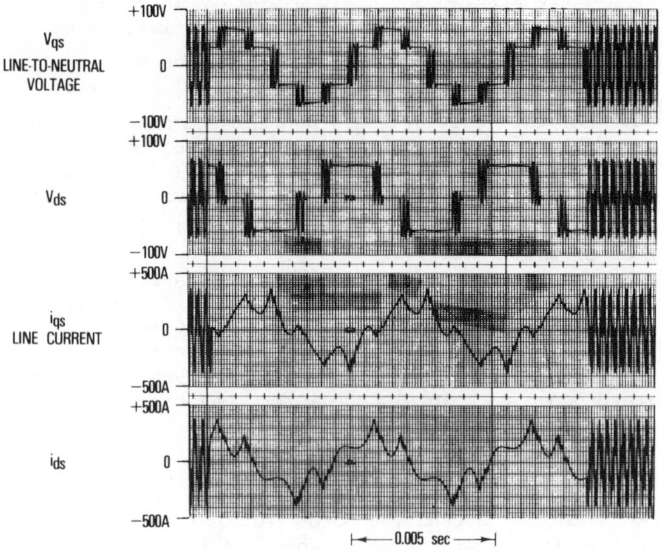

Fig. 14. Steady State Operation in PWM at Rated Motor Torque, Flux, and Current at Rated Motor Speed

Next, the transient performance of the control loops is examined to determine if satisfactory regulation of motor operating conditions such as flux level and torque can be obtained. Figure 15 shows the drive system performance, including the effect of the flux regulation loop which controls motor flux level by varying the magnitude of ac current. Note that the angle control loop controlling motor frequency is also present as is always required in PWM operation to ensure stable operation of the drive system at low speeds in the current-controlled mode of operation. Step changes of the flux command signal cause increases in ac current to change flux level at essentially constant motor angle; then the ac current drops back to a lower level as soon as the desired motor flux is obtained.

An example of sudden step changes in the current command signal upon the inverter waveforms and upon the motor torque is shown in Figure 16. Note that as the current command increases, the angle control detects the change in motor angle and compensates the reference waveform generator phase to minimize disturbances. Note that for a constant angle command, increasing the current command causes an increase in motor torque output. As the inverter waveform changes to square wave operation, no net change in average torque occurs but a six times frequency motor torque ripple appears.

The final drive system configuration includes both torque and flux regulation. The performance of the complete system in the braking mode of operation is shown in Figure 17. Both the torque and flux commands are maintained constant as the speed decreases. Note that the inverter waveform varies between quasi-square wave operation and full PWM operation with virtually no change in motor torque. The motor current waveform maintains the sinusoidal shape all the way to a standstill at which speed the current becomes dc since for safety reasons the waveform generator does not allow speed reversal. The motor torque declines to zero as the flux disappears at standstill.

The simulation results show that the complete current-controlled PWM drive system with the torque and flux regulators has very satisfactory dynamic performance at any speed, handles the transition from current-controlled PWM to square wave operation in a graceful manner, and controls the inverter current to minimize the inverter peak current requirement.

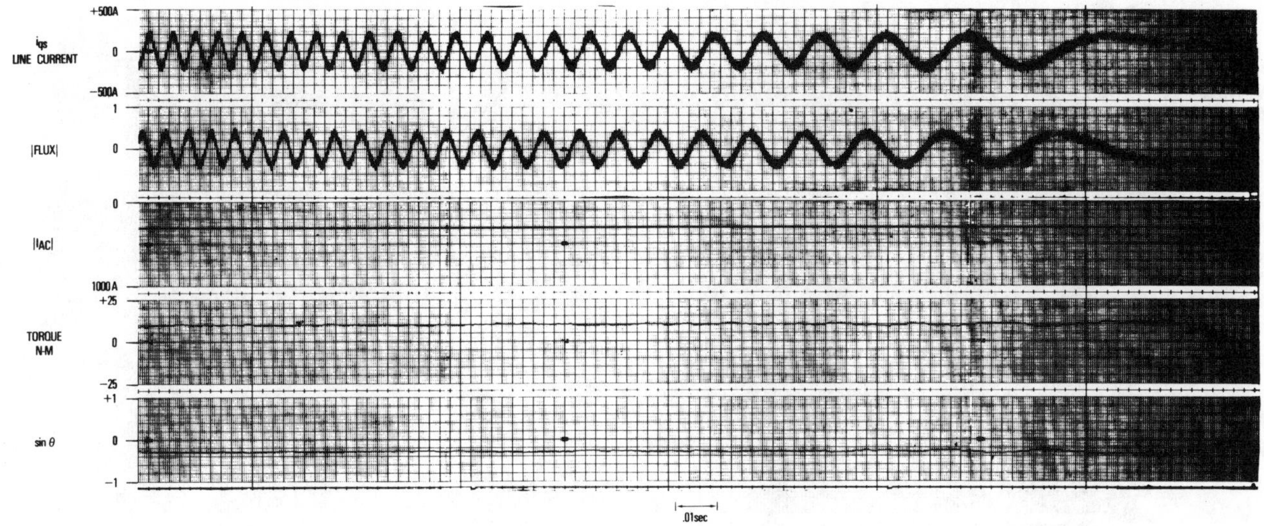

Fig. 15. Performance of the Flux and Angle Regulators for Step Changes in Flux Command and Constant Angle Command

CONCLUSIONS

The described current-controlled PWM system combines the advantages of both current and voltage inverters. The major disadvantages of the two systems are eliminated in the implementation shown here. Very fast dynamic performance with good motor current waveform control is available at any speed of operation. Motor losses are minimized so that it is possible to use conventional induction motors in this drive system with only a small penalty in increased motor heating. The use of current control eliminates the need for as critical a balancing between phases as required in a voltage inverter system and allows relatively easy and non-critical implementation of the control electronics. The current-controlled PWM system also eliminates the need to include a transition mode of PWM normally needed in order to maintain maximum inverter output power capability. If extremely good torque regulation near zero speed is not required, then the motor terminal voltage can be sensed instead of the air gap flux voltage and used in the same manner to develop the stabilizing angle control feedback signal as well as the magnitude of flux and torque signals.

REFERENCES

1. A.B. Plunkett and D.L. Plette, "Inverter-Induction Motor Drive for Transit Cars," IEEE Trans. Ind. Appl., vol. IA-13, No. 1, pp. 26-37 Jan./Feb. 1977.
2. A. Schonung and D. Stemmler, "Static Frequency Changers with Subharmonic Control in Conjunction with Reversible Variable Speed AC Drives," The Brown-Boveri Review, pp. 555-577 Aug./Sept. 1974.
3. A. Abbondanti, J. Zubek and C.J. Nordby, "Pulse Width Modulated Inverter Motor Drives with Improved Modulation," Conf. Proc. 1974 9th IAS Meeting.
4. G.B. Kliman and A.B. Plunkett, "Development of a Modulation Strategy for a PWM Inverter Drive," IEEE Trans. Ind. Appl., vol. IA-15, no. 1, pp. 72-79, Jan./Feb. 1979.
5. Von Konrad Heintze, Hermann Tappeiner and Manfred Weibelzahl, "Pulswechselrichter zur Drehzahlsteuerung von Asynchronmaschinen," Siemens Review, vol. 45, no. 3, p. 154, 1971.
6. K.P. Phillips, "Current Source Converter for AC Motor Drives," IEEE Publication 71-C1-IGA, Conf. Rec. 1971 IGA 6th Annu. Meet., pp. 385-392, Oct. 18-21, 1971.
7. E.P. Cornell and T.A. Lipo, "Modeling and Design of Controlled Current Induction Motor Drive Systems," Conf. Rec. IEEE-IAS 1975 10th Annu. Meet., Sept. 28-Oct. 2, 1975.
8. A.B. Plunkett, J.D. D'Atre and T.A. Lipo, "Synchronous Control of a Static AC Induction Motor Drive," Conf. Rec. 1977 IAS Annu. Meet., p. 609, Oct. 1977.
9. A.B. Plunkett, "Direct Torque and Flux Regulation in a PWM Inverter-Induction Motor Drive," IEEE Trans. Ind. Appl., vol. IA-13, no. 2, pp. 139-146, Mar./Apr. 1977.
10. T.A. Lipo, D.W. Novotny, A.B. Plunkett, and V.R. Stefanovic, "Dynamics and Control of AC Drives," course notes, Univ. of Wisconsin Extension, Nov. 3-5, 1976.
11. R.H. Nelson, T.A. Lipo, and P.C. Krause, "Stability Analysis of a Symmetrical Induction Machine," IEEE Transactions on Power Apparatus & Systems, vol. PAS-88, no. 11, pp. 1710-1717, Nov. 1969.

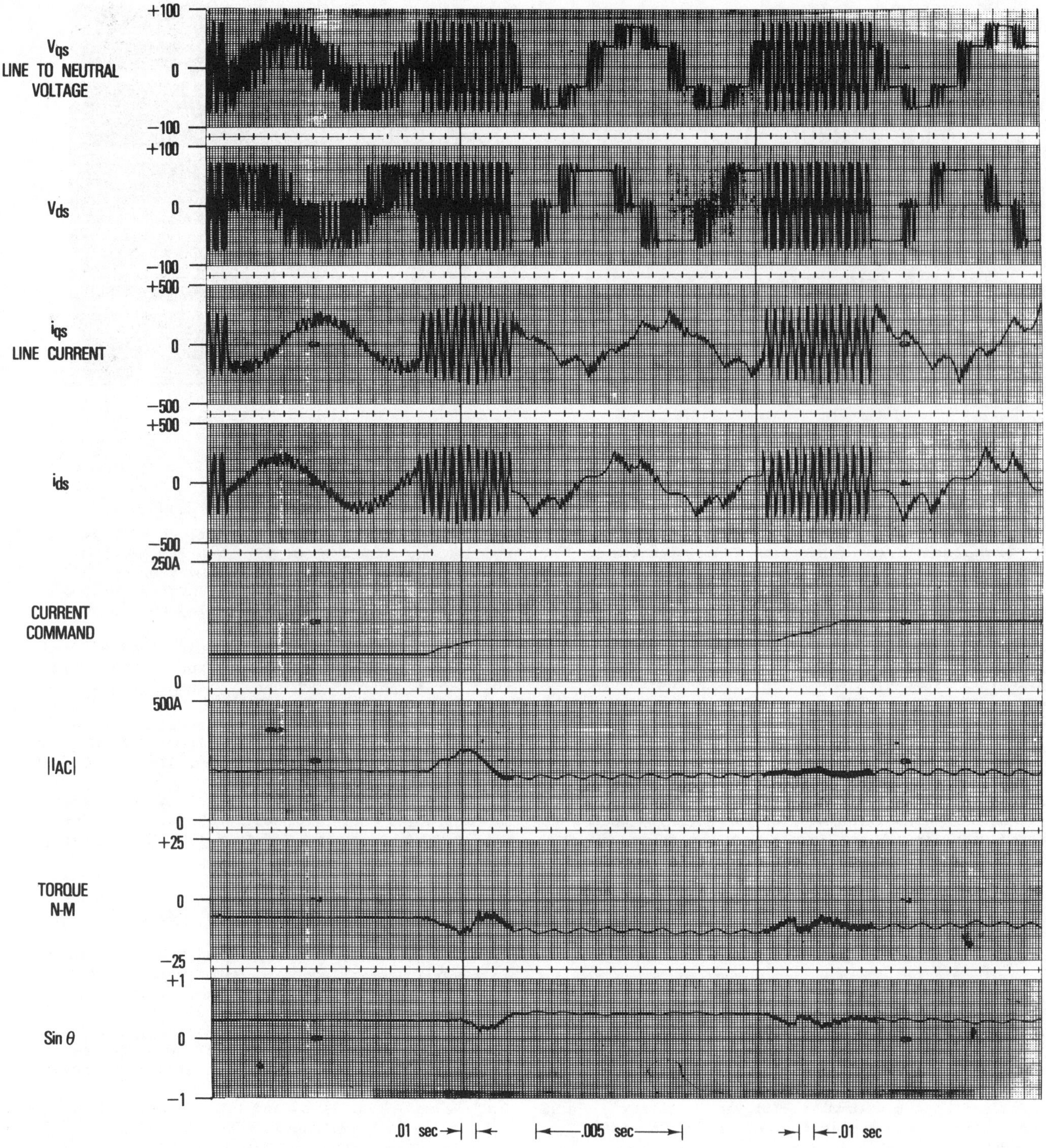

Fig. 16. The Effect of Increasing the Current Command Signal Showing the Automatic Optimum Waveform Control

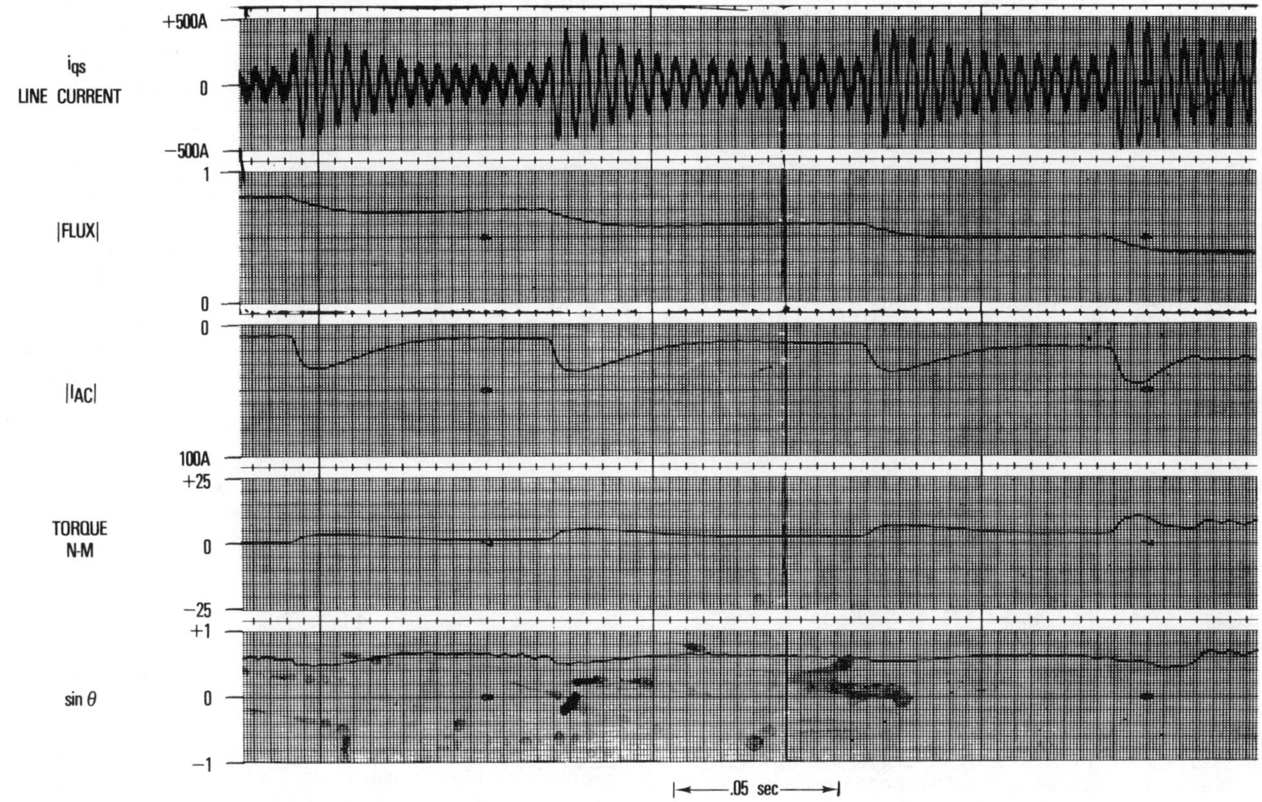

Fig. 17. Performance of the Complete Drive System When Braking to a Standstill

APPENDIX

The induction motor equations and equivalent circuit used in the analysis describe the instantaneous operation of the inverter and induction motor. These are given in the simple to use form used here in Reference 11. The motor equivalent circuit is reproduced below for reference. The factors λ^e_{ds}, λ^e_{dr}, λ^e_{qs}, and λ^e_{qr} are found simply by computing the flux linkages across the inductors as shown in Figure 18. Note that

$$\lambda = \frac{d}{dt}\ (Li).$$

The parameters of the motor analyzed are given below:

r_s = .0039Ω/phase

$L_{\ell s}$ = .0106 mhy

L_m = .320 mhy

$L'_{\ell r}$ = .0101 mhy

r'_r = .0033Ω

w_b = 2π 180 rad/sec

V_{qs}(rated) = 45 v rms/phase

P(number poles) = 4

Rated Power = 17 hp output

I_{rated} = 173.9 amps rms

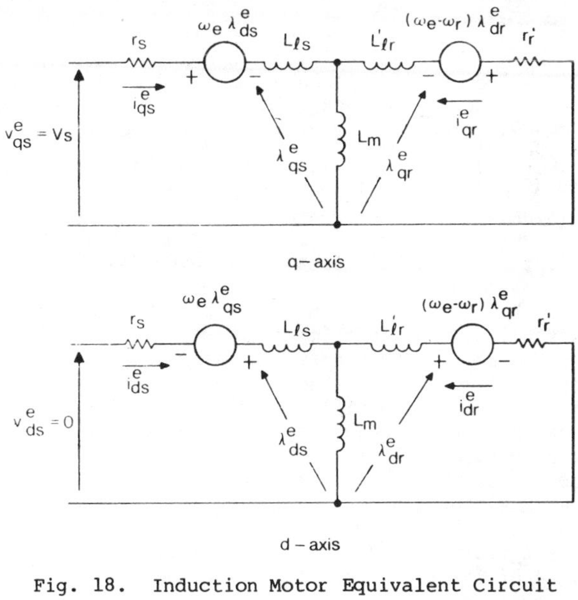

Fig. 18. Induction Motor Equivalent Circuit

METHOD OF FLUX CONTROL IN INDUCTION MOTORS DRIVEN BY VARIABLE FREQUENCY, VARIABLE VOLTAGE SUPPLIES

A. Abbondanti

Westinghouse Electric Corporation
Pittsburgh, Pennsylvania

Summary

This paper describes a technique for controlling the flux level in induction motors fed by variable frequency, variable voltage static converters, thereby optimizing motor performance over a wide speed and load range. The flux level is sensed and regulated without use of flux transducers and independently from stator resistance thermal variations.

Introduction

This paper describes a technique by which the motor flux level, in variable frequency induction motor drives, can be measured and regulated, and in particular, kept constant at its optimum level over any practical frequency range, independently from load and frequency variations. Besides being more accurate and more widely applicable than other known methods, the described approach is not dependent upon the knowledge of the value of the stator resistance, nor is it affected by stator resistance thermal variations. Furthermore, the method is based on analog operations on motor voltage and current and does not make use of transducers or probes to measure the flux.

1. The problem of Flux Control in Induction Motor Drives

When a multiphase induction motor is fed by a variable frequency multiphase supply, such as an inverter or a cycloconverter, in which both the output frequency ω_1 and the output voltage V_1 can be independently adjusted, it is usually desired to ensure that the machine flux attains and keeps, for given operating conditions, a given preferred magnitude. In particular, a frequent requirement is that the flux level remains constant and close to its nominal value* as the motor operating conditions (i.e., the speed and the torque) vary. This is referred to as the constant flux mode of operation. The fulfilment of this requirement ensures the ability of generating the highest possible torque per ampere of stator current and therefore results in the best possible utilization of the available current capability of the drive. Constant flux operation also provides consistency in motor response parameters, a must in high performance servo drives. Although other strategies of flux control are conceivable, let us primarily consider here the constant flux mode of operation as a typical illustration of the flux regulation requirements in induction motor drives.

In order to operate at constant flux level, the two controllable output quantities of the supply, namely ω_1 and V_1, will have to be adjusted for each particular operating condition. But since ω_1 is closely related to the shaft speed and is, therefore, imposed by the motor speed requirements, the only parameter independently available for flux regulation is V_1.

By considering the motor equivalent circuit of Figure 1, it is evident that for a given operating condition defined by a frequency ω_1 and a load that

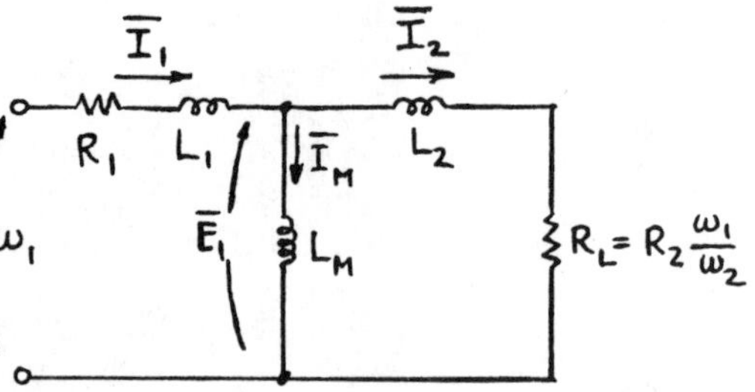

Figure 1. Adopted equivalent circuit of one phase of the induction motor.

corresponds to a stator current I_1, the machine flux level (i.e., the magnitude of the magnetizing current I_M) is determined by the value of V_1. If the nominal flux is desired, for instance, a value of V_1 should be used, such that when the $I_1 (R_1 + j\omega_1 L_1)$ drop is subtracted from it, the resulting "air gap voltage" E_1 satisfies the relation

$$E_1 = j\omega_1 L_M I_M$$

with I_M = nominal magnetizing current.

If V_1 has been adjusted to the desired magnitude for the given operating condition and if we assume that the load undergoes a variation (i.e., that I_1 varies) without variation of ω_1, we see that E_1 will vary because of the variation of the drop across R_1 and $\omega_1 L_1$. To restore E_1 to its desired value, i.e., to ensure constant flux operation, V_1 should be varied as a function of load. Similarly, if we assume a variation of the operating conditions such that ω_1 and only ω_1 varies, we see that the conservation of the flux magnitude requries that E_1 and, therefore, V_1 be varied as a function of the frequency.

This shows that in order to obtain the desired flux regulation, a precise dependence of V_1 on the frequency and the stator current must be provided.

The exact relation between V_1, ω_1, and I_1 which

*The nominal flux level is the flux level attained when the motor operates at nameplate frequency, with nameplate voltage and zero load.

Reprinted from *1977 IEEE/IAS Intl. Semi. Power Conv. Conf.*, pp. 177–184, 1977.

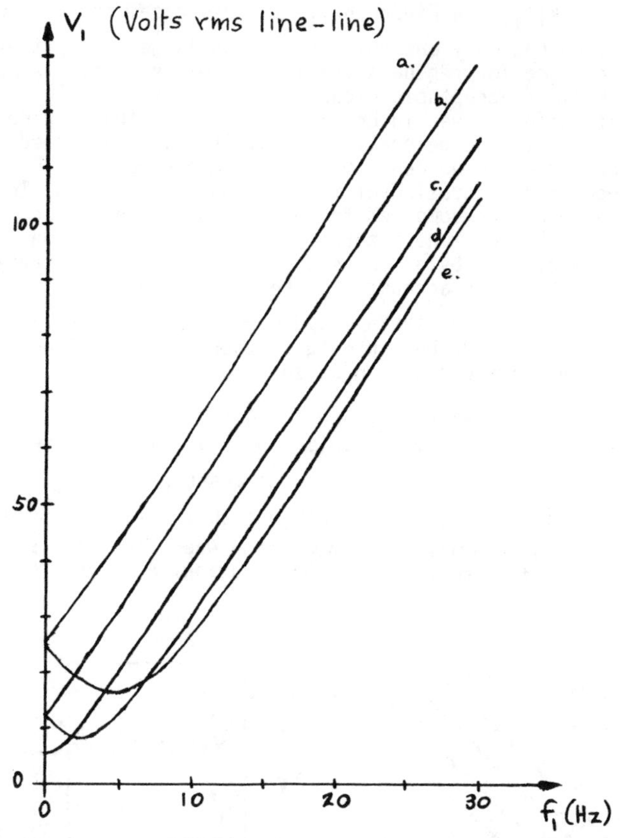

Figure 2. Relationship between stator voltage V_1 and frequency f_1 required to maintain the motor's flux at its nominal level under varying operating conditions characterized by different values of the stator current I_1.

a. I_1 = 200% of rated value, motoring.
b. I_1 = 100%, motoring.
c. Zero load; I_1 = nominal magnetizing current.
d. I_1 = 100%, regenerating.
e. I_1 = 200%, regenerating.

(Motor: Westinghouse 680B101G45, 15 hp, 1200 rpm)

fulfills the condition of constant flux can be derived from the equivalent diagram of Figure 1 and represented by a family of curves[1] as shown for a typical motor of Figure 2. The examination of these curves suggests the following qualitative considerations. The voltage should generally increase with frequency in a close to linear fashion, except for regenerative operation at low frequency. For a given frequency, the voltage should increase when motoring load is applied and decrease (except at low frequency) when regenerative load is applied. The amount of voltage increase when motoring at fixed frequency should roughly be proportional to load and fairly independent from the operating frequency. When regenerating, the required amount of voltage variation is related to the load by a non-proportional law, which is fairly constant when the frequency is high, but which changes drastically in the range of low frequencies. At relatively high frequency, when voltage decrease with regenerative load is required, the amount of decrease should always be smaller than the corresponding amount of increase required to handle the same load in motoring conditions. The voltage variation required for a given regenerative load changes sign below a given frequency which is different for each load level.

The key to the solution of the regulation problem in a drive operating in the constant flux mode is to implement, in the mechanism that regulates V_1, the proper dependence of V_1 from ω_1 and I_1, approaching as much as possible the ideal one. This paper deals with one way of automatically achieving this proper dependence.

2. Prior Art

Several approaches have been used to confer to V_1 a law of variation in function of frequency and load which would more or less approximate the ideal one required for constant flux operation. Let us briefly review some of these approaches and point out their limitations.

2.1 Programmed Volt/Hertz Characteristics

A popular control policy for induction motors consists of ignoring the stator drops in the equivalent circuit, i.e., neglecting the presence of L_1 and R_1 on Figure 1. Under this approximation, the flux resulting from the application of a given V_1 can be regarded as independent from load and dependent only from ω_1. On this base, constant flux operation

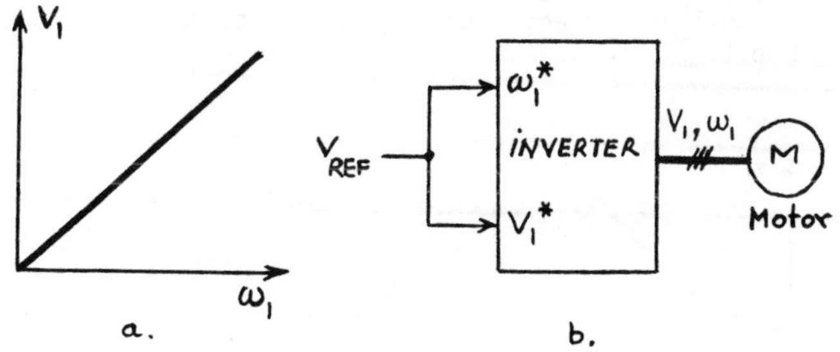

Figure 3. Constant Volt/Hertz control policy.
a. Voltage/frequency characteristic.
b. Control scheme.

is ensured simply by keeping the amplitude of V_1 strictly proportional to ω_1 (Figure 3a). This could be accomplished, for instance, by using the same signal as frequency reference ω_1^* and as voltage reference V_1^* in the converter control (Figure 3b). For relatively high frequencies of operation, this approach is valid and leads only to moderate machine underexcitation and overexcitation with motoring and regenerative loads, respectively. At the low end of the frequency range, however, the neglected stator resistive drop becomes so important compared to V_1 that ignoring it leads to severe underexcitation and intolerable loss of torque capability. To correct this deficiency, the $V_1 = f(\omega_1)$ characteristic can be programmed to follow a law other than proportional. For instance, one can deliberately accept zero load overexcitation of the machine by using a shifted linear relationship as in Figures 4a and 4b or by "boosting" the voltage at low frequency only, as in Figures 5a and 5b, or by combining these

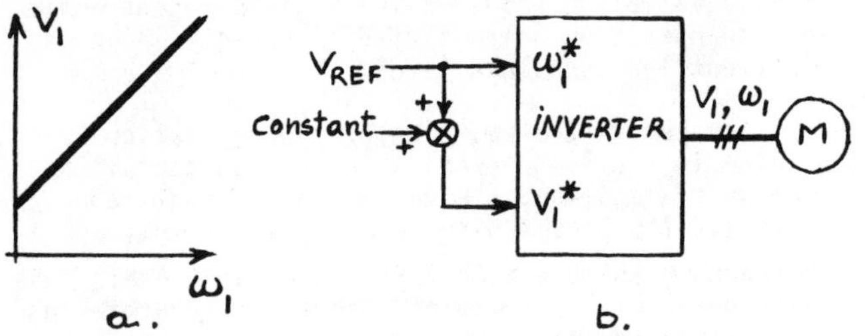

Figure 4. Constant Volt/Hertz control policy with linear shift to correct underexcitation effects at low speed.

a. Voltage/frequency characteristic.
b. Control scheme.

approaches. These techniques allow to achieve high torques at low speeds. However, when the motor is not loaded, the boost is still present and causes the machine to be overexcited and heavily saturated, particularly at low frequency. With saturation, motor heating problems develop, together with high pitch acoustical noise (in PWM inverter drives particularly). At low frequency, the steady state no-load current is comparable in magnitude to the peak overload current capability required from the drive, with resulting motor overheating. In addition, if the drive uses a current limit technique in which overload is controlled by use of a current-dependent, speed controlling feedback signal, this loop is severely disturbed at low frequency by the high level of no-load current caused by the programmed voltage boost.

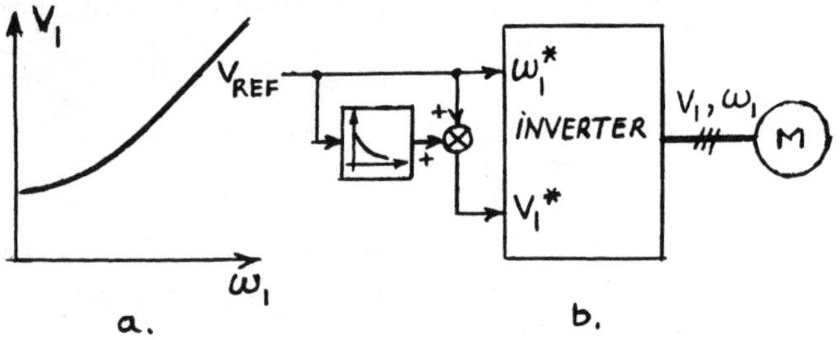

Figure 5. Constant Volt/Hertz control policy modified by low speed voltage "boost".

a. Voltage/frequency characteristic.
b. Control scheme.

2.2 RI Drop Compensation Techniques

In these approaches, it is recognized that for a fixed V_1 and at a fixed frequency, the flux decreases with increasing motor load because of the increasing voltage drop across the stator resistance R_1 caused by the stator current I_1. To neutralize the effect of the drop, the voltage V_1 is caused to increase proportionally to a load dependent parameter v_L (Figure 6a). Several approaches exist, differing by the way v_L is derived.

In one approach (current-dependent boost), v_L is obtained by full-wave rectifying the stator current ($v_L = k|I_1|$, in Figure 6b). In this case, the compensation has only one polarity, which is generally the wrong one for regenerative load. However, this is sometimes acceptable because it always results in saturation, never in underexcitation, of the machine, and the torque generating capability is safeguarded. The current-dependent boost technique somewhat reduces the saturation of the machine at zero load, low frequency, as compared to the previously mentioned approach. But the saturation effect and its inconveniences are by no means suppressed, since the technique fails to account for the V_1, I_1 phase relationship and treates the no-load magnetizing current as normal load current, providing therefore a substantial voltage boost even at zero load.

In another approach (in-phase current-dependent boost), v_L is obtained by phase sensitive rectification of I_1 using V_1 as phase reference ($v_L = kI_1 \cos \phi$, Figure 6c). In this case, the compensation automatically changes sign when the load is changed from motoring to regeneration. At high frequen-

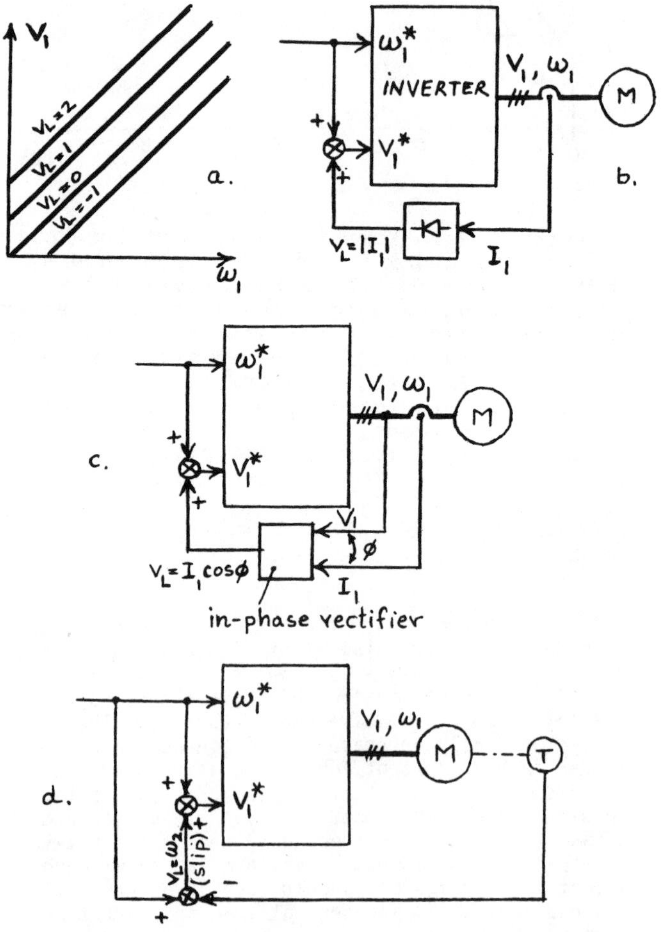

Figure 6. Control policy applying the load-dependent voltage boost concept.

a. Voltage/frequency characteristic.
b. Scheme in which boost is proportional to current magnitude.
c. Boost is proportional to magnitude of real component of stator current.
d. Boost is proportional to slip.

cies, this provides a better approximation of the ideal voltage behavior, although the compensation remains inaccurate in magnitude, if not in sign. At low frequency, the technique fails to properly handle the regenerative loads. The machine saturation at zero load, low frequencies is further reduced but still exists, because the technique fails to vectorially compensate the stator I_1R_1 drop.

It has been found that both of the above described current-dependent boost approaches exhibit poor stability characteristics whenever an attempt is made to compensate close to 100% of the stator voltage drop.

In a third approach (slip-dependent boost, Figures 6a and 6d), v_L is obtained from an analog slip frequency signal ($v_L = k\omega_2$). In this case, the compensation changes sign when required (except for low frequencies), no saturation problem exists at zero load and the stability is good even at 100% compensation level. However, this technique requires the use of a slip transducer T (Figure 6d). In its simplest implementation, this can be a dc tachometer, delivering a speed signal which, subtracted from the analog frequency reference signal, provides the slip. When the tachometer is not needed for speed regulation purposes, its use only for flux control makes the solution expensive and often impractical for mechanical reasons, considering also that a high degree of accuracy and low level of ripple are required for this use of the tachometer. Other rotary slip transducers can be devised, but their use is generally impractical in industrial drives. Besides, this technique still fails to provide the proper compensation when regenerating near zero frequency and therefore leaves the drive unable to assume any sizeable regenerative load at low speed.

2.3 Closed-Loop Flux Regulation Techniques

In these techniques, the flux Φ or a quantity related to it is sensed and compared to a reference, and the voltage is adjusted to nullify the error (Figure 7a). Since these approaches regulate the flux to its desired level in any situation, the resulting $V_1 = f(\omega_1, I_1)$ characteristics are not mere approximations of the ideal ones, but exact duplicates of the curves of Figure 2. As a result, these approaches have none of the drawbacks of the previously described methods and provide the best solutions to the flux regulation problem. The various possible approaches in this category differ by the method used for obtaining a flux related signal.

Direct Flux Probing[2]

It has been suggested that flux sensitive devices (e.g., Hall elements) can be imbedded in the motor poles to provide a flux signal. This approach is considered more costly and perhaps impractical, because of the special mechanical work required on the motor. It would make the drive unable to handle standard commercial motors.

Use of Flux Rate-of-Charge Probing Coils[3]

Small secondary stator windings could be used to sense the flux rate of change, obtaining a signal proportional to the air gap voltage E_1. This signal would then be regulated to have an amplitude strictly proportional to frequency. Although simpler to implement than the above approach, this solution also is incompatible with the use of standard motors.

a.

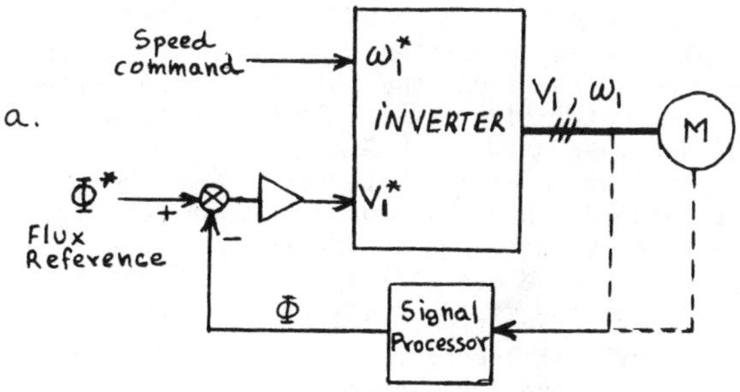

b.

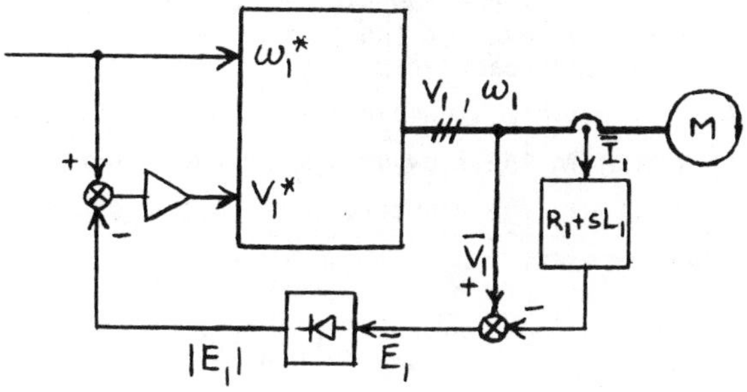

Figure 7. Closed loop flux regulation techniques.

a. Conceptual scheme.
b. Method based on the synthesis of an air gap voltage signal.

Synthesis of An Air Gap Voltage Signal[4]

Instead of measuring the air gap voltage E_1 through probing coils, one can synthesize it from V_1 if the magnitude of the stator drop terms R_1I_1 and $\omega_1L_1I_1$ are known. Since R_1 and L_1 are well defined parameters of the motor and I_1 is measurable, the synthesis of E_1 is possible by performing operations on the instantaneous values of the motor voltage and current as indicated on Figure 7b. The synthesized air gap voltage E_1 is rectified and the dc signal representing its magnitude is regulated through a feedback loop, so as to be proportional to the operating frequency ω_1.

This approach duplicates almost perfectly the characteristics of Figure 2. It has the inconvenience, however, of relying on the constancy of R_1. Unfortunately, this parameter varies substantially with the temperature of the motor, and the compensation valid at room temperature becomes inadequate, mainly at low frequency, after motor warm-up. Although this effect could be somewhat neutralized by probing the motor temperature or by other methods providing an indirect monitoring of R_1 during operation, it makes the approach somewhat impractical for industrial drives.

3. Proposed Method[5]

This paper proposes a method for regulating the

motor flux which can be classified among the methods of Paragraph 2.3 It makes use of a closed-loop regulation scheme to adjust the flux magnitude to a desired level in any situation. It differs from the other methods by the technique used to derive information on the motor flux level. Similarly to the approach described on Figure 7b, a signal related to the flux level is synthesized from the available information on the motor terminal voltage and motor line current, but the particular operation accomplished on these quantities does not involve the knowledge of the stator resistance R_1, which can, therefore, vary with temperature without affecting the derivation of the flux level.

The flux-related signal used to obtain information on the level of the flux is defined here as the motor phase "reactive power" W_X. If the motor is supplied with a sinusoidal voltage $\overline{V}_1 = V_1 \sin \omega_1 t$, resulting in the flow of a sinusoidal current $\overline{I}_1 = I_1 \sin(\omega_1 t - \phi)$, the reactive power in one phase is defined here as:

$$W_X = \frac{1}{2} V_1 I_1 \sin\phi$$

It is the power delivered to a resistive system which, when energized by a source $\overline{V}_1$, would draw from this source a current whose magnitude is equal to the magnitude of the motor quadrature current.

The quantity W_X can be derived from V_1 and I_1, by classical wattmetric techniques. A preferred one will be described in Section 5.1.

The basic concept of the proposed flux regulation method rests in the recognition that W_X is a function of, and only of:

a. The magnetizing current I_M (and therefore of Φ).

b. The line current I_1.

c. The motor inductive parameters L_1, L_M, L_2.

d. The frequency ω_1.

As will be established in **Section** 4, this function is

$$W_X = \frac{\omega_1}{2(L_M+2L_2)}\left[(L_M^2+L_M L_2)I_M^2 + (L_M L_2 + L_M L_1 + 2L_1 L_2)I_1^2\right] \qquad (1)$$

with the parameter definitions used in Figure 1.

Since ω_1 and the inductive parameters are known and since I_1 and W_X are measurable, I_M and Φ can be derived from (1) with proper processing of known and measured quantities. One can, for instance, consider the following rearrangement of equation (1).

$$\frac{L_M(L_M+L_2)}{2(L_M+2L_2)}\,\omega_1 I_M^2 = \left[\frac{W_X}{\omega_1} - \frac{1}{2} I_1^2\left(L_1 + \frac{L_M L_2}{L_M+2L_2}\right)\right]\omega_1 \qquad (2)$$

A signal proportional to the right hand side of (2) can be obtained as output of a signal processor SP (described in more detail in Section 5.2), to which information on $\overline{V}_1$ and $\overline{I}_1$ is fed and which is properly

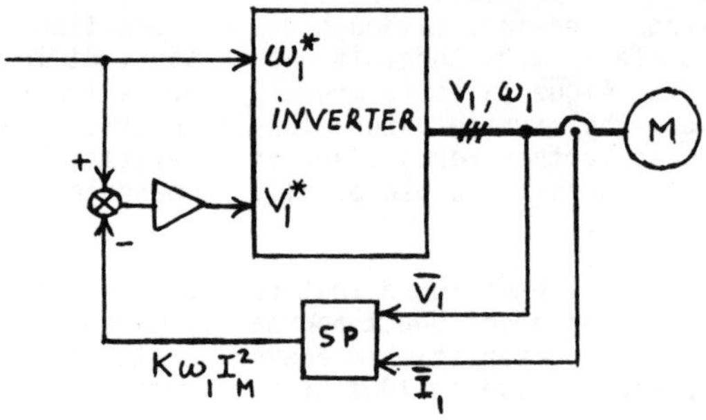

Figure 8. Conceptual scheme of the proposed new approach: a flux regulation loop based on the synthesis of a signal related to the magnetizing current.

programmed with the motor inductive constants L_M, L_1, L_2. The output of SP can then be regulated to be strictly proportional to the frequency, using the converter voltage control to maintain this proportionality, according to the scheme of Figure 8. By keeping the right-hand side of (2) proportional to ω_1, it is evident that the magnetizing current, and therefore the flux, will be maintained at a constant level, dependent only on the chosen coefficient of proportionality. By proper design, this coefficient can be selected to always ensure nominal flux level. Since the knowledge of R_1 or R is not needed to compute the right-hand side of (2), the regulation process is not dependent on the magnitude or the variation of these parameters.

4. Relationship Between Motor Reactive Power and Motor Flux Level

By applying the definition $W_X = \frac{1}{2} V_1 I_1 \sin\phi$ to the network of Figure 1, and with the aid of the phasor diagram of Figure 9, the validity of the following relation can be established:

$$W_X = \frac{1}{2}\omega_1 L_1 I_1^2 + \frac{1}{2}\omega_1 L_2 I_2^2 + \frac{1}{2}\omega_1 L_M I_M^2 \qquad (3)$$

Since $\frac{1}{2}\omega_1 L I^2$ is nothing else than the reactive power developed in inductor L by an ac current I at frequency ω_1, we can say that relation (3) expresses that the total reactive power developed in one phase of the motor is the exact sum of the reactive power developed in each one of the motor reactive components. The presence of resistive components does not contribute to any of the terms of W_X.

Equation (3) shows that the magnetizing current I_M is related to W_X by a stator current dependent term and a rotor current dependent term. The first term is calculable, since we can measure I_1. The second term is not readily available. However, the relationship between $\overline{I}_1$, $\overline{I}_M$ and $\overline{I}_2$ is such that the term I_2^2 can be eliminated from (3).

From Figure 9, we derive:

$$I_1^2 = (I_M + I_{2X})^2 + I_{2Y}^2$$
$$= I_M^2 + 2I_M I_{2X} + I_2^2$$

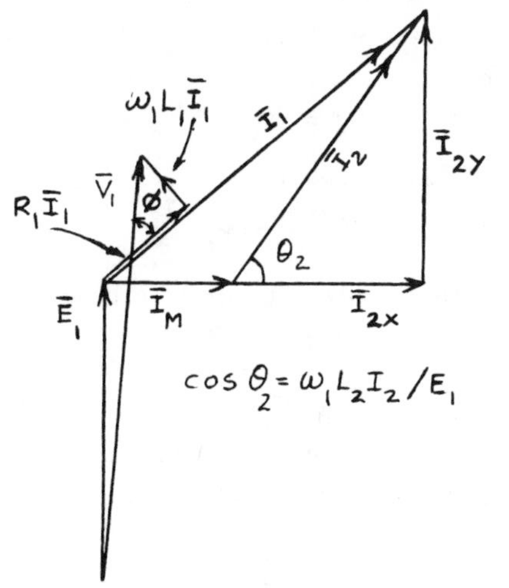

Figure 9. Phasor diagram showing the relationships between voltages and current in the motor's equivalent circuit shown on Fig. 1.

Observing that $I_M = \dfrac{E_1}{\omega_1 L_M}$ and $I_{2x} = I_2^2 \dfrac{\omega_1 L_2}{E_1}$,

we have $I_1^2 = I_M^2 + 2I_2^2\dfrac{L_2}{L_M} + I_2^2$ or

$$I_1^2 - I_M^2 = \frac{L_M + 2L_2}{L_M} I_2^2 \qquad (4)$$

This allows us to substitute I_2^2 in (3) by its expression from (4). We therefore obtain,

$$W_X = \frac{1}{2}\omega_1 \frac{L_M(L_M+L_2)}{L_M+2L_2} I_M^2 + \frac{1}{2}\omega_1 \frac{L_M L_2}{L_M+2L_2} I_1^2 + \frac{1}{2}\omega_1 L_1 I_1^2 \qquad (5)$$

This relation, identical to (1), shows that W_X can be fully defined without the knowledge of $\bar{I}_2$.

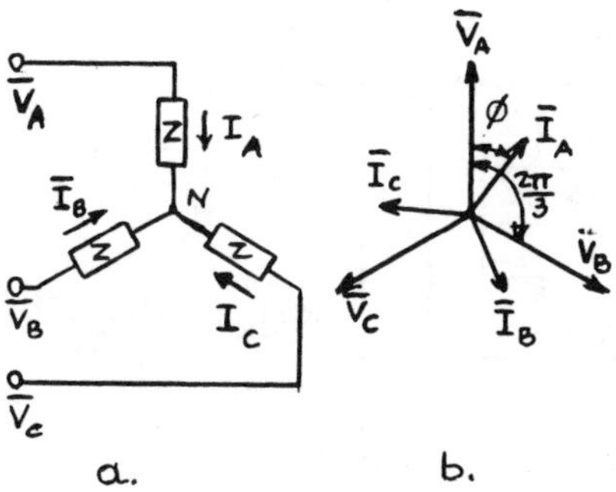

$\bar{V}_A = V_1 \sin\omega_1 t$

$\bar{V}_B = V_1 \sin(\omega_1 t - \frac{2\pi}{3})$

$\bar{V}_C = V_1 \sin(\omega_1 t + \frac{2\pi}{3})$

$\bar{I}_A = I_1 \sin(\omega_1 t - \phi)$

$\bar{I}_B = I_1 \sin(\omega_1 t - \frac{2\pi}{3} - \phi)$

$\bar{I}_C = I_1 \sin(\omega_1 t + \frac{2\pi}{3} - \phi)$

$Z = \frac{V_1}{I_1}(\cos\phi + j\sin\phi)$

Figure 10. Three-phase phasor diagram of motor.
a. Balanced three phase system representing motor.
b. Phasor diagram of stator voltages and currents.

5. Signal Processing Circuitry

5.1 Derivation of a Reactive Power Signal

Given a balanced 3-phase system described by the circuit of Figure 10a, the phasor diagram of Figure 10b and the adjacent equations, one can define a system of voltages $\bar{V}_d$, $\bar{V}_q$, and currents $\bar{I}_d$, $\bar{I}_q$, the two components of each set having equal amplitude and being in quadrature with each other, as represented by the phasor diagrams of Figures 11a and 11b respectively and the adjacent equations.

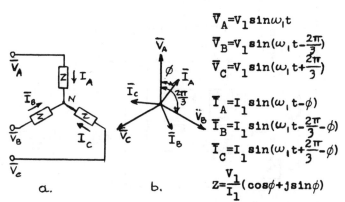

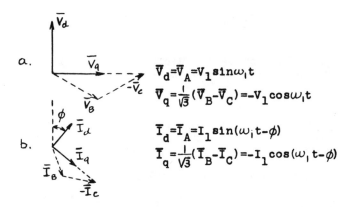

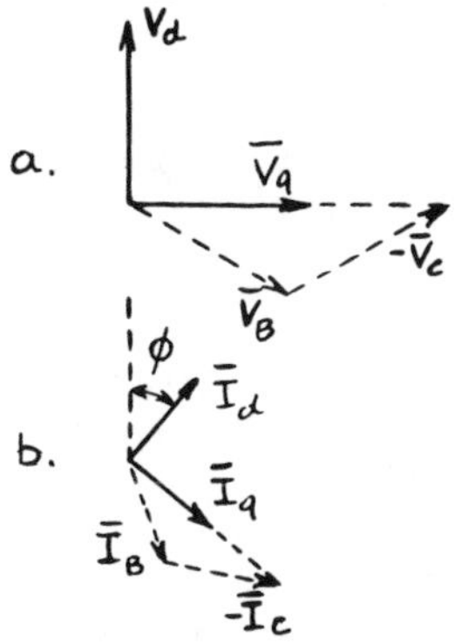

$\bar{V}_d = \bar{V}_A = V_1 \sin\omega_1 t$

$\bar{V}_q = \frac{1}{\sqrt{3}}(\bar{V}_B - \bar{V}_C) = -V_1 \cos\omega_1 t$

$\bar{I}_d = \bar{I}_A = I_1 \sin(\omega_1 t - \phi)$

$\bar{I}_q = \frac{1}{\sqrt{3}}(\bar{I}_B - \bar{I}_C) = -I_1 \cos(\omega_1 t - \phi)$

Figure 11. Two-phase phasor diagrams of motor (d-q diagrams).
a. Stator voltages.
b. Stator currents.

The product $\bar{V}_d\bar{I}_q$ can be expressed as:

$$\bar{V}_d\bar{I}_q = -VI\sin\omega_1 t(\cos\omega_1 t - \phi)$$
$$= -\frac{1}{2}VI\sin\phi - \frac{1}{2}VI\sin(2\omega_1 t - \phi) \qquad (6)$$

We see that this product contains a dc term $\frac{VI}{2}\sin\phi$ which according to our definition, is precisely equal to the reactive power W_X supplied to one phase of the system. However, the considered product contains also an ac term of amplitude $\frac{VI}{2}$ and of twice the frequency of the applied voltages.

We can also consider the product $\bar{V}_q\bar{I}_d$ and develop it as

$$\bar{V}_q\bar{I}_d = -VI\cos\omega_1 t(\sin\omega_1 t - \phi)$$
$$= \frac{1}{2}VI\sin\phi - \frac{1}{2}VI\sin(2\omega_1 t - \phi) \qquad (7)$$

This product also contains the dc term $\frac{VI}{2}\sin\phi$, and the ac term at twice the frequency of the applied voltages. Whereas the ac term in (6) is of the same amplitude and phase as the ac term in (7), the dc terms in these equations are of opposite sign. Therefore, if we take the difference between the products, we double the dc component and cancel the ac components. We obtain,

$$\bar{V}_q\bar{I}_d - \bar{V}_d\bar{I}_q = VI\sin\phi = 2W_X$$

On this basis, a scheme for obtaining a signal proportional to the motor reactive power in a 3-phase

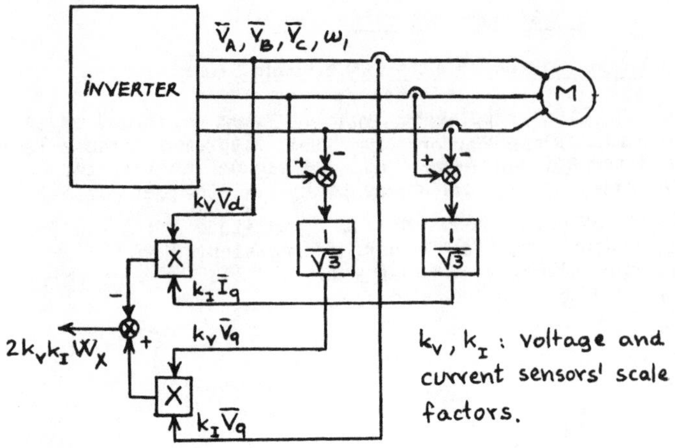

Figure 12. Scheme for derivation of a signal representative of the flux level.

converter-motor system can be represented by the block diagram of Figure 12, making use of motor voltage and current sensors and two analog multipliers.

5.2 Derivation of a Flux Dependent Signal

If we consider equations (2) or (5), we see that the quantity which is strictly related to the magnetizing current amplitude and which, therefore, can provide useful information on the flux level is not the reactive power W_X itself, but W_X reduced by a stator current dependent term W_X' given by:

$$W_X' = \frac{1}{2}\,\omega_1 I_1^2 \left(L_1 + \frac{L_M L_2}{L_M + 2L_2}\right)$$

This term has the dimensions of a reactive power, and is in fact the reactive power that would be developed in an inductor $L' = L_1 + L_M L_2/(L_M + 2L_2)$, if the stator current I_1 would flow through it.

If the motor equivalent circuit of Figure 1 would be fictitiously replaced by the circuit of Figure 13, with same input voltage, Z'' being an undefined impedance, such that the input current I_1 has the same amplitude and phase as in Figure 1, the input reactive power W_X would be unchanged, the term

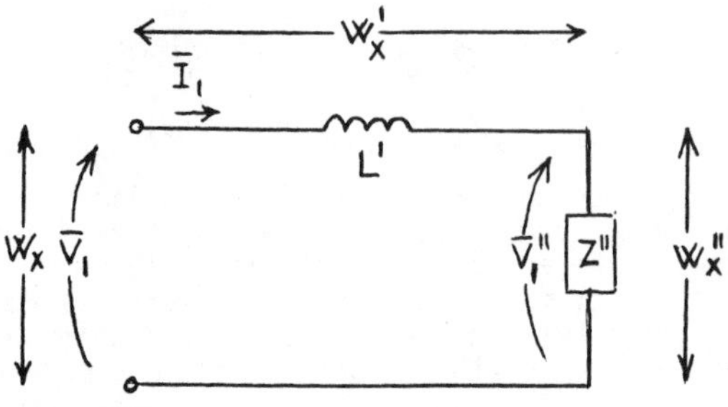

Figure 13. Equivalent circuit illustrating the relationship between the motor reactive power W_X and a term W_X'' directly representative of the flux level.

W_X' would represent the reactive power developed in the series inductance L' and the reactive power across Z'' would be

$$W_X'' = W_X - W_X'$$

By virtue of (2), we would have:

$$W_X'' = K\omega_1 I_M^2$$

If we desire to compute the reactive power W_X'', we can use the same technique as in 5.1, provided that the product terms are obtained using the voltage $\bar{V}''$ instead of the voltage $\bar{V}_1$. This voltage V_1'' can be derived from $\bar{V}_1$ by correcting the latter for the voltage drop across the series inductor L'.

$$\bar{V}_1'' = \bar{V}_1 - j\omega_1 L' \bar{I}_1$$

Therefore, the term $K\omega_1 I_M^2$ of equation (2) can be obtained from the motor voltages and currents by using a scheme similar to the one of Figure 12, with the difference that the multipliers voltage inputs should be corrected for a stator voltage drop across a fictitious stator reactance $\omega_1[L_1 + L_M L_2/(L_M + 2L_2)]$ so as to apply to the multipliers:

$$\bar{V}_d'' = \bar{V}_d - j\omega_1\left(L_1 + \frac{L_M L_2}{L_M + 2L_2}\right)\bar{I}_d$$

$$\bar{V}_q'' = \bar{V}_q - j\omega_1\left(L_1 + \frac{L_M L_2}{L_M + 2L_2}\right)\bar{I}_q$$

instead of $\bar{V}_d$, $\bar{V}_q$.

On this basis, the proposed flux regulation scheme takes the form of the block diagram of Figure 14 or the circuit diagram of Figure 15.

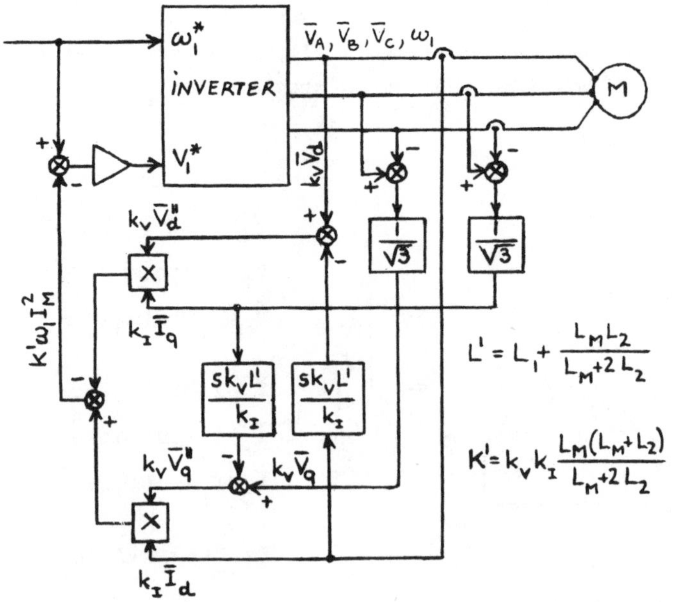

Figure 14. Functional diagram of a flux regulating loop according to the proposed method.

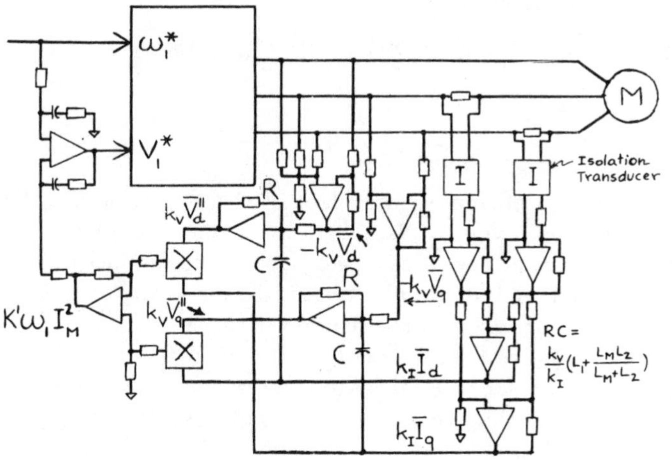

Figure 15. Circuit diagram of a control system regulating the motor's flux according to the proposed method. The derivation does not rely on the knowledge of the stator resistance R_1, a thermally sensitive parameter.

Conclusions

The practical feasibility of the proposed flux regulation scheme has been demonstrated by implementing it on a 15 hp PWM inverter-motor system. It has performed satisfactorily down to frequencies of 1 Hz, with 200% rated load, motoring or regenerative. It has shown performances superior to any other scheme known to the author. It is believed that the approach of using a reactive power signal to sense and regulate the flux provides a solution to a number of hitherto unsolved regulation problems in operating variable frequency ac drives with demanding performance requirements.

References

1. K. Heumann and A. C. Stumpe, "Thyristoren", B. G. Teubner, Stuttgart (1969), p. 250.
2. F. Blaschke, "The Principle of Field Orientation as Applied to the New Transvektor Closed-Loop Control System for Rotating-Field Machines", Siemens Review, Vol. 39, No. 5, pp. 217-220 (1972).
3. A. B. Plunkett, "Direct Flux and Torque Regulation in a PWM Inverter-Induction Motor Drive", Conference Record of IEEE-IAS 1975 Annual Meeting, pp. 591-597.
4. F. Blaschke, "Apparatus for Providing the Pilot Values of Characteristics of an Asynchronous Three Phase Machine", U.S. Patent No. 3,593,083 (1971).
5. A. Abbondanti, "Flux Control System for Controlled Induction Motors", U. S. Patent No. 3,909,687 (1975).

The Principle of Field Orientation as Applied to the New TRANSVEKTOR Closed-Loop Control System for Rotating-Field Machines

By Felix Blaschke

When rotating-field machines are employed as drive motors, the question of torque generation and control requires special consideration. It is, for instance, possible to use the vector of the stator voltage or the vector of the stator current as the manipulated variable for the torque, depending on whether the static converter supplying the motor provides a variable voltage or a variable current. This paper describes the principle of field orientation – a new closed-loop control method for rotating-field machines [1 to 4] – by way of reference to an induction motor. It is shown how these manipulated variables must be influenced to provide instantaneous and well-damped adjustment of the torque independently of the inherent characteristics of an induction motor.

Field orientation with current control

The principle of field orientation can best be explained by reference to the characteristics of a d.c. motor. Fig. 1 shows a d.c. motor of the non-salient-pole type. Arranged in the stator perpendicular to each other are two windings 1 and 2. Owing to the action of the commutator, the rotating armature winding 3 produces the effect of a stationary winding. If a current i_1 is passed through field winding 1, a magnetic field Ψ builds up in the motor (Fig. 2, left). For the generation of a torque, a current i_3 must also be passed through the armature winding. The armature current and field now set up forces in the directions shown. Since the axis of the armature winding is perpendicular to the field, the forces are applied with maximum leverage to the shaft. Hence, this position of the armature winding is the most favourable one for torque generation. The armature winding also builds up a field that is superimposed on the original field and is perpendicular to it. This effect is undesirable, since it turns the field out of the optimal position. For this reason, the armature field is compensated by a compensating winding 2 arranged in the stator in the same plane as the armature winding and carrying the same current, but in the opposite direction ($i_2 = -i_3$). This stator winding and the field produce in the stator a reaction torque which acts against the armature. The currents and the field may be represented by the vector diagram* shown on the right in Fig. 2. In a d.c. machine, therefore, current i_1 forms the field and currents i_2 and i_3, together with the field, form the torque.

Dipl.-Ing. Felix Blaschke, Siemens Aktiengesellschaft, Measurement and Process Engineering Division, Erlangen

* For a definition as space vectors due to KOVACS and RACZ see [5]. Space vectors and matrices are denoted by boldface letters.

In an induction motor, the place of the commutator-fed armature winding is taken by a short-circuited winding which may, for instance, consist of conductor bars distributed uniformly round the periphery and connected by two short-circuiting rings at the ends (Fig. 3). The current required in this winding for the setting up of a torque can only be generated by induction, i.e. by field change. Again a field is set up by a current i_1 in winding 1. If now a current i_2 is suddenly

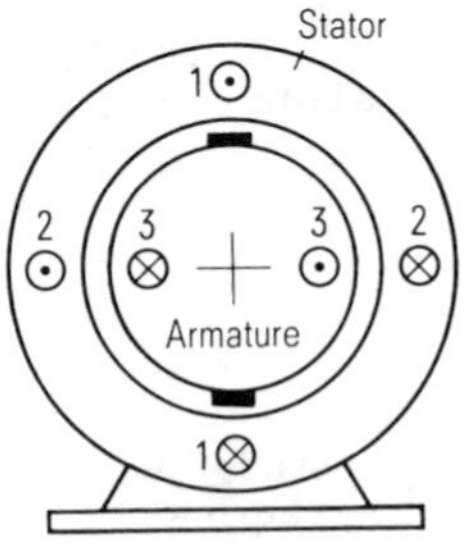

1 Field winding
2 Compensating winding
3 Armature winding

Fig. 1 Representation of a d.c. motor

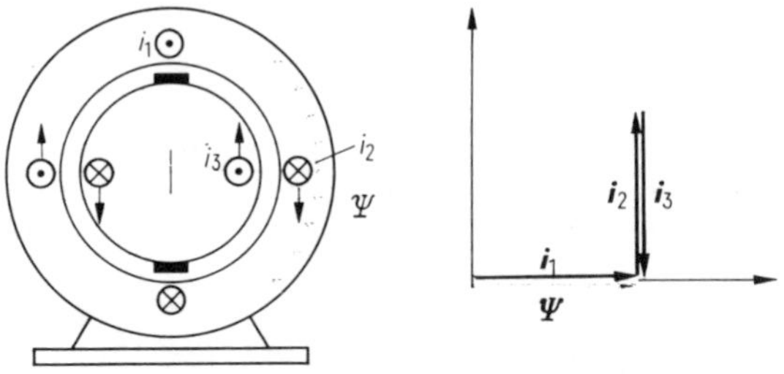

Field and currents — Vector diagram

Fig. 2 State of field and currents in a d.c. motor

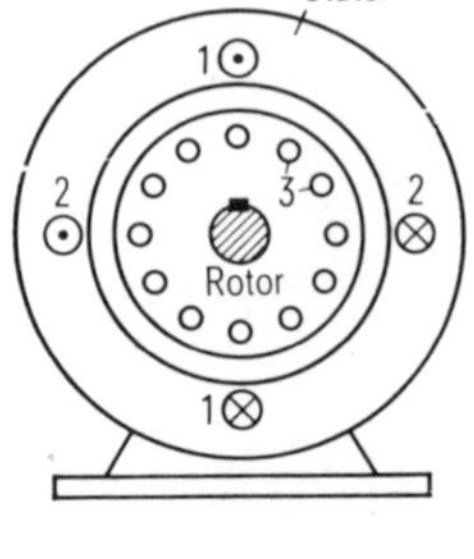

1, 2 Stator windings
3 Rotor winding

Fig. 3 Representation of an induction motor

Reprinted with permission from *Siemens Review,* vol. 34, pp. 217-220, May 1972.

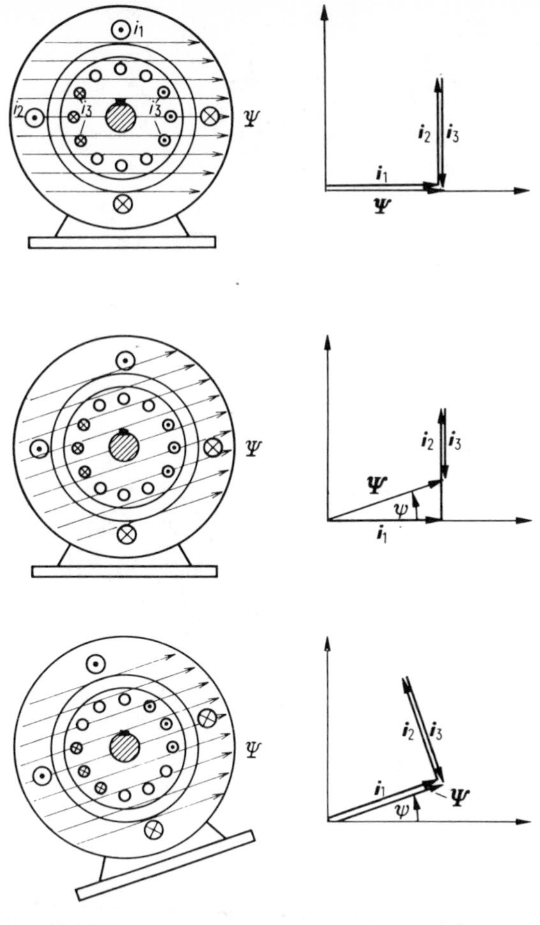

Left: Field and currents
Right: Vector diagrams

Fig. 4 States of field and currents in an induction motor

injected into winding 2, previously acting as a compensating winding, it causes an opposing current i_3 to be induced in the rotor. In the first instant this is exactly equal to stator current i_2 but opposite. The conditions obtaining in this first instant are therefore identical to those in a d.c. machine (Fig. 4a). However, since the induced rotor current requires a field change, the vector diagram is changed after a certain time to that shown in Fig. 4b. For the sake of simplicity, the rotor is assumed to be locked, as is indicated by the unchanged position of the shaft. The field has rotated slightly and the original alignment of the current vectors $\boldsymbol{i}_1$ and $\boldsymbol{i}_2$ with respect to the field has been lost. If – fictitiously – the stator is now turned until $\boldsymbol{i}_1$ and the field are again parallel (Fig. 4c), the orientation is restored. If the field is not allowed to move from the direction of $\boldsymbol{i}_1$, but the stator is turned continually with the field rotation, rigid orientation of the currents across the field is obtained. Consequently, the conditions at any instant are the same as in a d.c. motor. Rotation of the stator is, of course, merely an aid in representation.

In reality the stator and the windings α and β remain stationary (Fig. 5). In the measure shown, it is only the current vector $\boldsymbol{i}$ formed from $\boldsymbol{i}_1$ and $\boldsymbol{i}_2$ that is important (Fig. 6). Instead of producing this rotating vector from rotating windings 1 and 2 with constant currents i_1 and i_2, it is now necessary to produce it from stationary windings α and β with variable currents i_α and i_β. Fig. 6 shows which values the currents i_α and i_β must assume at any instant. i_α and i_β depend not only on the freely selectable values i_1 and i_2, which may, for instance, be constant, but also on the angle of the field ψ with respect to the stator axis α:

$$i_\alpha = i_1 \cos\psi - i_2 \sin\psi,$$

$$i_\beta = i_1 \sin\psi + i_2 \cos\psi. \qquad (1)$$

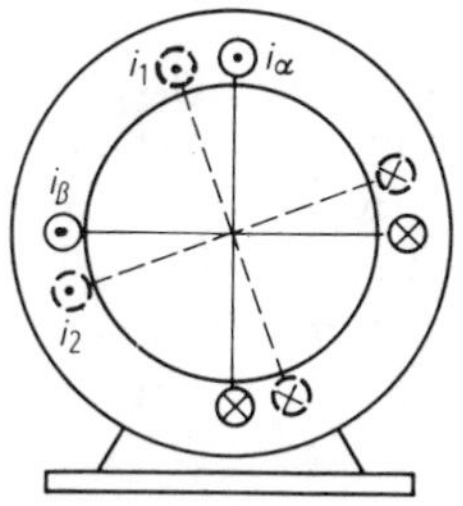

Fig. 5 Conversion of the required currents into attainable currents

This relationship can be realized by the computation circuit shown in Fig. 7 which is called a vector rotator VR, since it causes the current vector to be rotated by the angle of the field. Fig. 8 shows the application of this vector rotator for field-orientated control in an induction motor. In the vector rotator the required positioning values $i_\alpha^{(*)}$ and $i_\beta^{(*)}$ are formed from the setpoint values i_1^* and i_2^* and the angle functions of the field angle ψ. These values are fed to a variable-current static converter U as manipulated variables for the currents i_α and i_β in the corresponding stator windings. The information on the field angle ψ required by the vector rotator is obtained by measuring the field vector in the motor [6]. The two components of the field vector are measured by Hall generators arranged at different angles in the air gap and are converted to the required angle functions $\sin\psi$ and $\cos\psi$ by a vector analyzer [1].

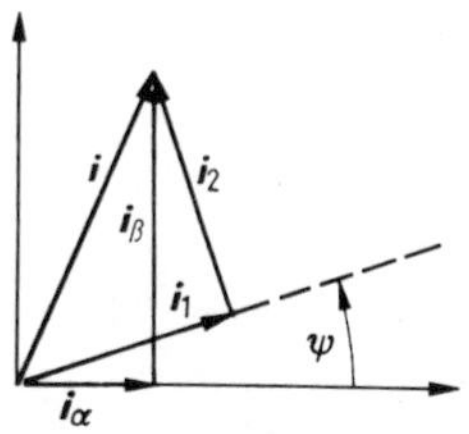

Fig. 6 Vector diagram for Fig. 5

This arrangement, which effects field orientation of the stator current, provides separate access to the field through i_1 and to the torque-producing current through i_2. Thus it is possible to operate an induction motor in the same manner as a d.c. motor with current control.

To simplify the above explanation, it has been assumed that the rotor is locked. The same result would, however, also be obtained with a rotating rotor. The only difference between rotation of the field to which orientation is being carried out and that obtaining with a locked rotor lies in the rotation of the rotor.

Vector representation of field orientation

For the following it will be expedient to employ vectors to express the relationships established at the beginning. The relationship between i_1, i_2 and i_α, i_β, shown in equation (1), can be looked upon as a co-ordinate transformation of current vector $\boldsymbol{i}$ from the field-oriented to the stator-oriented co-ordinate system. If the current vector in the field co-ordinate system is defined as $\boldsymbol{i}_\psi = \begin{pmatrix} i_1 \\ i_2 \end{pmatrix}$ and the current vector in the stator co-ordinate system as $\boldsymbol{i}_\mathrm{s} = \begin{pmatrix} i_\alpha \\ i_\beta \end{pmatrix}$, this co-ordinate transformation can be expressed as follows with the aid of the rotational matrix $\boldsymbol{D}\,(+\psi) = \begin{pmatrix} \cos\psi & -\sin\psi \\ \sin\psi & \cos\psi \end{pmatrix}$:

$$\boldsymbol{i}_\mathrm{s} = \boldsymbol{D}\,(+\psi)\,\boldsymbol{i}_\psi. \qquad (2)$$

Here, the vector rotator shown in Fig. 7 contains the transformation function of the rotational matrix.

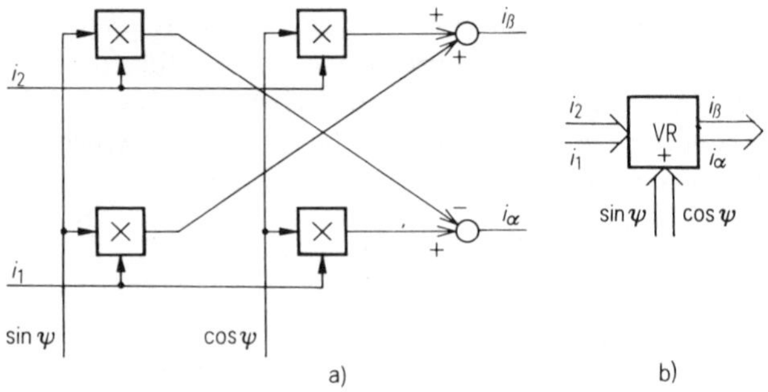

a) Computation circuit b) Symbol

Fig. 7 Vector rotator for transformation from the field co-ordinate system to the stator co-ordinate system

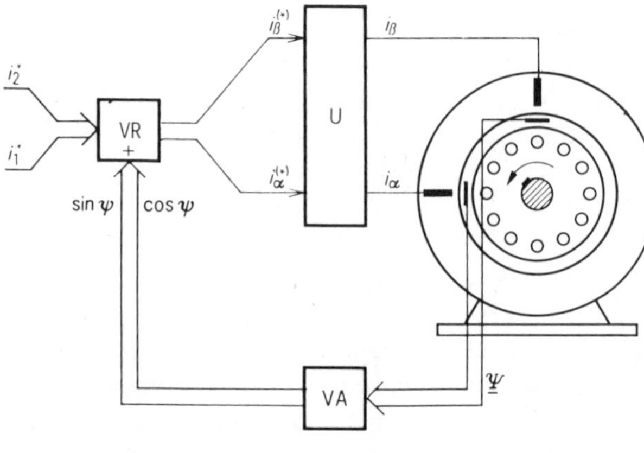

Fig. 8 Application of a vector rotator for field orientation in an induction motor

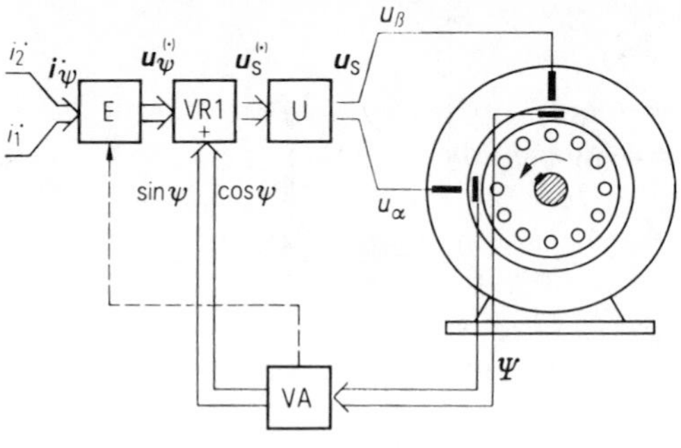

a) Open-loop control

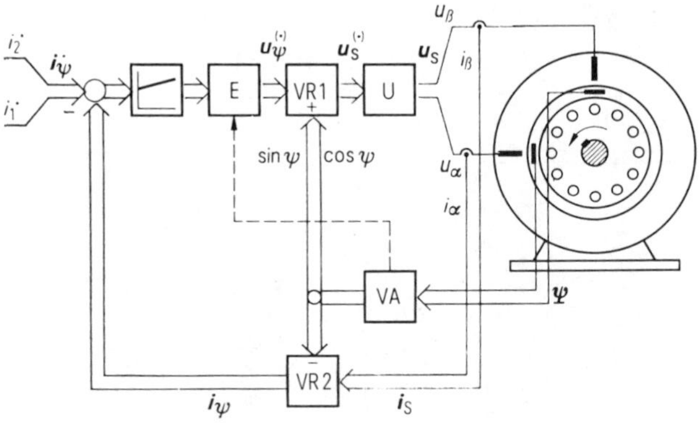

b) Closed-loop control

Fig. 9 Field orientation in an induction motor with voltage control

Field orientation with voltage control

In the case of a variable-voltage static converter, the voltages u_α and u_β are available across the windings α and β through manipulated variables $u_\alpha^{(*)}$ and $u_\beta^{(*)}$. To achieve field orientation, it is therefore necessary here, in contrast to current control, to determine the voltage positioning values $u_\alpha^{(*)}$ and $u_\beta^{(*)}$ necessary for the current setpoint values i_1^* and i_2^*. The relationship is established in two steps. First of all, the voltage vector $\boldsymbol{u}_\psi^{(*)}$ in the field co-ordinate system is formed from the current vector $\boldsymbol{i}_\psi^*$ which is in turn formed from i_1^* and i_2^*. This vector consists of the vectors for the resistive and inductive voltage drops of the current and the vector for the back e.m.f. in the motor. This relationship is established in a computation circuit E (Fig. 9a), which requires information from the motor and contains a simulation of the structure of the motor [1]. In a second step the result is transformed to the voltage vector in the stator co-ordinate system $\boldsymbol{u}_\mathrm{s}^{(*)}$ by a co-ordinate transformation

$$\boldsymbol{u}_\mathrm{s}^{(*)} = \boldsymbol{D}\,(+\psi)\,\boldsymbol{u}_\psi^{(*)},$$

which corresponds to that for current control. This transformation is carried out by the vector rotator VR 1 (Fig. 9a). The components $u_\alpha^{(*)}$ and $u_\beta^{(*)}$ of this vector are then fed to the static converter as manipulated variables.

The motor current set up by the applied voltage depends, among other things, on the resistance of the motor. This in turn depends on the operating temperature which cannot generally be taken into account in the computation circuit. Consequently, the current vector $\boldsymbol{i}_\psi$ deviates from its setpoint value $\boldsymbol{i}_\psi^*$ and it is necessary to superimpose closed-loop control of the current vector $\boldsymbol{i}_\psi$ on this control (Fig. 9b). Since the components of $\boldsymbol{i}_\psi$ remain steady under steady-state operating conditions, the deviation of $\boldsymbol{i}_\psi$ from the required value $\boldsymbol{i}_\psi^*$ caused by resistance change can be corrected by integral-action controllers. The actual value $\boldsymbol{i}_\psi$ required for the control is obtained by measurement of the stator-oriented current vector $\boldsymbol{i}_\mathrm{S}$ and subsequent transformation of the field co-ordinate system. The transformation specification is obtained by inversion of equation (2):

$$\boldsymbol{i}_\psi = \boldsymbol{D}^{-1}(+\psi)\,\boldsymbol{i}_\mathrm{S} = D(-\psi)\,\boldsymbol{i}_\mathrm{S}. \quad (3)$$

The vector rotator VR 2 required for this is shown in Fig. 9b. It has the same structure as that shown in Fig. 7, it merely being necessary to substitute $-\psi$ in place of $+\psi$; this leads to a sign change of the crossed loops.

The arrangement described thus also makes field orientation of the stator current of an induction motor possible with a static converter for voltage control.

Hence, the principle of field orientation ensures dynamically high-grade control for induction motors, independently of the type of static converter employed.

References

[1] Blaschke, F.: Das Prinzip der Feldorientierung, ein neues Verfahren zur Regelung der Asynchronmaschine. Siemens-Forschungs- u. Entwicklungsber. 1 (1971) No. 2

[2] Böhm, K.; Wesselak, F.: Variable-Speed A.C. Drives with Static Frequency Converter Feed. Siemens Rev. XXXIX (1972) pp. 126 to 129

[3] Flöter, W.; Ripperger, H.: Die TRANSVEKTOR-Regelung für den feldorientierten Betrieb einer Asynchronmaschine. Siemens-Z. 45 (1971) pp. 761 to 764

[4] Bayer, K.-H.; Waldmann, H.; Weibelzahl, M.: Field-Oriented Closed-Loop Control of a Synchronous Machine with the New TRANSVEKTOR Control System. Siemens Rev. XXXIX (1972) pp. 220 to 223

[5] Kovács, K.P.; Rácz, I.: Transiente Vorgänge in Wechselstrommaschinen. Budapest: Publishing House of the Hungarian Academy of Sciences 1959

[6] Langweiler, F.; Richter, M.: Flußerfassung in Asynchronmaschinen. Siemens-Z. 45 (1971) pp. 768 to 771

[7] Weh, H.: Elektrische Netzwerke und Maschinen in Matrizendarstellung. Mannheim, Zürich: Bibliogr. Inst. 1968

[8] Naunin, D.: Ein Beitrag zum dynamischen Verhalten der frequenzgesteuerten Asynchronmaschine. Berlin, Techn. Univ., Diss. 1968

[9] Hasse, K.: Zur Dynamik drehzahlgeregelter Antriebe mit stromrichtergespeisten Asynchron-Kurzschlußläufermaschinen. Darmstadt, Techn. Hochsch., Diss. 1969

[10] Blaschke, F.; Ripperger, H.; Steinkönig, H.: Regelung umrichtergespeister Asynchronmaschinen mit eingeprägtem Ständerstrom. Siemens-Z. 42 (1968) pp. 773 to 777

Field-Oriented Control of a Standard AC Motor Using Microprocessors

RUPPRECHT GABRIEL, WERNER LEONHARD, MEMBER, IEEE, AND CRAIG J. NORDBY, MEMBER, IEEE

Abstract—**Field orientation has emerged as a powerful tool for controlling ac machines such as inverter-supplied induction motors. The dynamic performance of such a drive is comparable to that of a converter-fed four quadrant dc drive. The complex functions required by field-oriented control may be executed by microprocessors on line, thus greatly reducing the necessary control hardware. It is shown that the flux signals may be derived from sensing coils or, with some compromise in performance, from the stator voltages and currents. The speed signal is obtained from a digital tachometer. Results from a 2-kW experimental drive are given.**

INTRODUCTION

VARIABLE-SPEED ac drives employing induction motors and static converters have been developed in recent years to the point of meeting the high standards of performance set by dc drives with line-commutated converters. Continuous-speed control down to standstill and operation at high efficiency in all quadrants of the torque–speed plane has been demonstrated. However, despite the strong incentives to use the less expensive ac motors with contact-free low-inertia rotors of rugged construction, ac drives have not supplanted variable-speed dc drives except in special circumstances. This is due to the still considerably higher cost and complexity of the complete ac drive, Fig. 1.

Since an inverter supplying an induction motor cannot operate with natural commutation, forced commutated circuits must be employed requiring many additional components and fast-switching thyristors. Also the control schemes are quite complex since the ac motor is a highly coupled nonlinear multivariable control plant, as opposed to the dc motor with its much simpler decoupled control structure. Therefore, considerable simplifications are still required before the induction motor variable-speed drive can become an economical alternative to dc drives [1].

At the inverter side there seems to be a general tendency towards the impressed current scheme with interphase commutation; also, improved thyristors, thyristor modules, and thyristor heat-sink combinations indicate gradual but steady progress. For smaller drives power transistors now look very promising.

Paper ID 79-16, approved by the Industrial Drives Committee of the IEEE Industry Applications Society for presentation at the 1979 Industry Applications Society Annual Meeting, Cleveland, OH, September 30–October 4. Manuscript released for publication November 5, 1979. This work was supported by a grant from Deutsche Forschungsgemeinschaft.

R. Gabriel and W. Leonhard are with the Institut fur Regelungstechnik, Technische Universitat Braunschweig, 3300 Braunschweig, West Germany.

C. J. Nordby is with the Westinghouse Electric Corporation, Buffalo, NY.

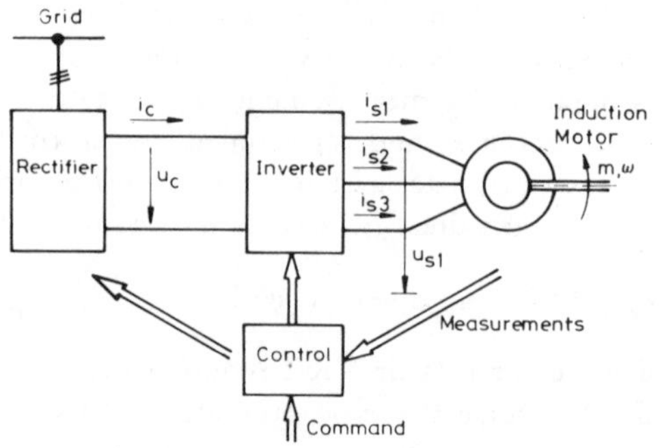

Fig. 1. AC motor drive with intermediate dc loop.

At the control end the situation is characterized by a variety of solutions designed for special applications such as traction, where speed of response is not the primary objective. The concept of field orientation as proposed by Blaschke [2], [3] stands out as a fundamental method of controlling ac machines, essentially transforming their dynamic structure into that of dc machines. Unfortunately, the scheme requires many electronic components such as sensors, amplifiers, or multipliers which increase the cost and are difficult to adjust and maintain.

This situation is now changing rapidly due to the availability of microelectronics, with the cost of control hardware being no longer a major consideration. Of course, a complex control function requires a highly flexible structure. In this paper it is shown that even presently available microprocessors can perform this task. With the rapid development in the microprocessor field there is little doubt that many improvements will be possible in the near future.

At first the theoretical background of ac motor control by field orientation is described as was done in another recent paper [10], then additional results from an experimental drive are presented.

MATHEMATICAL MODEL OF THE INDUCTION MOTOR

The dynamic properties of an induction motor as a control plant can be described by a set of nonlinear differential equations linking the stator and rotor currents and voltages with the mechanical quantities torque, speed, and angular position. The rotational symmetry of an induction motor may be taken advantage of by a vectorial representation in three-phase or two-phase coordinates. Considerable simplifications result without noticeably affecting the validity of the control model if the stator and rotor windings are assumed to produce sinusoidal magnetomotive force (MMF) waves in the air gap of the machine, disregarding spatial harmonics of

Reprinted from *IEEE Trans. Ind. Appl.*, vol. IA-16, pp. 186–192, Mar./Apr. 1980.

the windings as well as slot and end effects. Additional minor inaccuracies are the assumption of infinite permeability of the iron core and the neglect of iron losses and eddy currents in the conductors.

The symmetry of the motor construction gives rise to complex current and voltage vectors [4] defined in a plane perpendicular to the motor axis:

$$i_S(t)=i_{S1}(t)+i_{S2}(t)e^{j2\pi/3}+i_{S3}(t)e^{j4\pi/3}, \tag{1a}$$

$$i_R(t)=i_{R1}(t)+i_{R2}(t)e^{j2\pi/3}+i_{R3}(t)e^{j4\pi/3}. \tag{1b}$$

The values $i_{S\nu}(t)$ are the stator-phase and $i_{R\nu}(t)$ the rotor-phase currents; they are arbitrary functions of time, as long as

$$\sum_1^3 i_{S\nu}=0, \qquad \sum_1^3 i_{R\nu}=0$$

holds as dictated by the isolated neutrals of the windings. The instantaneous magnitudes and angles of the vectors indicate the amplitude and direction of the pertinent MMF waves.

The vector of the stator terminal voltages is defined accordingly:

$$u_S(t)=u_{S1}(t)+u_{S2}(t)e^{j2\pi/3}+u_{S3}(t)e^{j4\pi/3}. \tag{2}$$

The stator and rotor inductances per phase are, converted to equal number of turns,

$$L_S=(1+\sigma_S)L_h, \qquad L_R=(1+\sigma_R)L_h, \tag{3}$$

with

$$\sigma=1-1/(1+\sigma_S)(1+\sigma_R) \tag{4}$$

being the total leakage factor. R_S and R_R are the winding resistances per phase, and L_h is the main inductance.

With θ as the total inertia of the drive, m_{el} and m_L the driving and load torques, respectively, ω the angular velocity, and ϵ the angle of rotation, the equations of the wound-rotor two-pole induction machine with shorted rotor windings may be written as follows [4], [5]

$$R_S i_S+L_S\frac{di_S}{dt}+L_h\frac{d}{dt}(i_R e^{j\epsilon})=u_S, \tag{5}$$

$$R_R i_R+L_R\frac{di_R}{dt}+L_h\frac{d}{dt}(i_S e^{-j\epsilon})=0, \tag{6}$$

$$\theta\frac{d\omega}{dt}=m_{el}-m_L=\tfrac{2}{3}L_h \operatorname{Im}[i_S(i_R e^{j\epsilon})^*]-m_L, \tag{7}$$

$$\frac{d\epsilon}{dt}=\omega, \tag{8}$$

i^* is the conjugate complex of the vector i.

Equations (5)-(8) constitute a dynamic extension of the well-known stationary model of the induction machine, operating with constant speed off the symmetrical three-phase line. With the sinusoidal stator currents described by constant complex phasors,

$$i_{S1}(t)=\frac{\sqrt{2}}{2}[\tilde{I}_S e^{j\omega_1 t}+\tilde{I}_S{}^*e^{-j\omega_1 t}], \qquad \tilde{I}_S=I_S e^{j\alpha},$$

$$i_{S2}(t)=\frac{\sqrt{2}}{2}[\tilde{I}_S e^{j(\omega_1 t-2\pi/3)}+\tilde{I}_S{}^*e^{-j(\omega_1 t-2\pi/3)}],$$

$$i_{S3}(t)=\frac{\sqrt{2}}{2}[\tilde{I}_S e^{j(\omega_1 t-4\pi/3)}+\tilde{I}_S{}^*e^{-j(\omega_1 t-4\pi/3)}], \tag{9}$$

the current vector, defined in (1a),

$$i_S(t)=\frac{3\sqrt{2}}{2}\tilde{I}_S e^{j\omega_1 t} \tag{10}$$

exhibits constant magnitude and angular velocity, indicating a constant circumferential MMF wave moving at constant speed. By introducing phasors also for the sinusoidal stator voltages and rotor currents, (5)-(8) assume the form

$$(R_S+j\omega_1 L_S)\tilde{I}_S+j\omega_1 L_h\tilde{I}_R=\tilde{U}_S, \tag{11}$$

$$(R_R/s+j\omega_1 L_R)\tilde{I}_R+j\omega_1 L_h\tilde{I}_S=0, \tag{12}$$

$$m_{el}=3L_h \operatorname{Im}[\tilde{I}_S\tilde{I}_R{}^*]=\text{const}, \tag{13}$$

$$\omega=\text{const}, \tag{14}$$

where $s=(\omega_1-\omega)/\omega_1$ represents the rotor slip. This corresponds to the well-known single-phase equivalent circuit in stationary condition, Fig. 2.

Of course, when designing an inverter-supplied ac drive where neither the currents are sinusoidal nor, aiming at high dynamic performance, the speed may be assumed to be constant, the dynamic model of (5)-(8) will have to be used. When the motor is fed from an inverter, irregular movements are superimposed on the circular motion of the current vectors; however, $i_S(t)$, $i_R(t)$ are still continuous due to the leakage inductances.

FIELD-ORIENTED CONTROL

An ac induction motor as described by (5)-(8) is a much more difficult control plant than, for example, a dc motor. This is not only due to the fact that three ac voltages and currents of varying amplitude, frequency, and phase have to be generated instead of two direct currents, but also because there is an intricate coupling between all the control inputs and the inner quantities, flux, and electric torque. In addition, the rotor currents are inaccessible with squirrel-cage motors.

It has been shown by Blaschke [2], [3] that these problems may be overcome by field orientation, essentially reducing the control dynamics of an ac motor to those of a separately excited compensated dc motor. This is achieved by defining a time-varying vector which corresponds to a sinusoidal flux wave moving in the airgap of the machine. When referring the MMF wave of the stator currents described

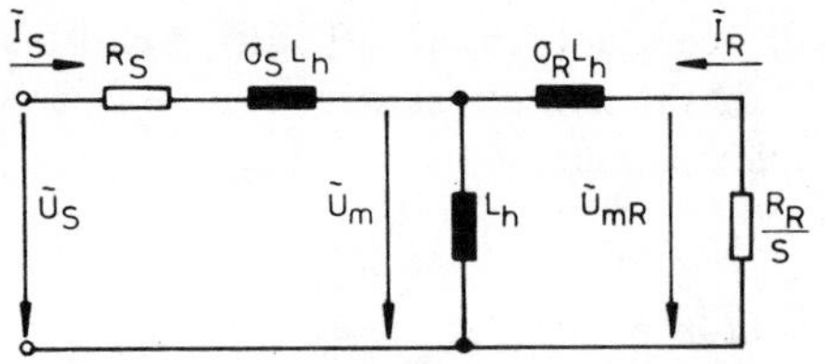

Fig. 2. Single-phase equivalent circuit of induction motor in stationary condition.

by $i_S(t)$ to this flux wave, it is realized that only the quadrature component of $i_S(t)$ is contributing to the torque, whereas the longitudinal or direct component affects the magnitude of the flux. Hence, the stator current vector is defined in a frame of reference defined by the time-varying field or in field coordinates. This indicates a close correspondance to dc machines with the direct component of the stator-current vector being analogous to the field current and the quadrature component to the armature current. The flux-oriented frame of reference may also be compared to the *d-q* coordinates commonly used with synchronous machines; however, now the reference is no longer determined by rotor position. The concept of field orientation thus provides a clue to a very effective method of controlling ac motors.

A modified magnetizing current vector representing the flux reference is defined as

$$i_{mR}(t) = i_S(t) + (1+\sigma_R) i_R(t) e^{j\epsilon} = i_{mR} e^{j\rho}. \tag{15}$$

Under stationary conditions this is the magnetizing current inducing the voltage U_{mR} in Fig. 2.

The instantaneous angular velocity of this vector is the stator frequency

$$\frac{d\rho}{dt} = \omega_1(t). \tag{16}$$

With the vector $i_{mR}(t)$, the direct and quadrature components may be defined according to Fig. 3:

$$i_S(t) e^{-i\rho} = i_{Sd} + j i_{Sq}, \tag{17}$$

$$u_S(t) e^{-j\rho} = u_{Sd} + j u_{Sq}. \tag{18}$$

At constant speed and torque the transformed variables are constant except for ripple caused by inverter operation.

Inserting the definitions (15)-(18) into the model of the motor ((5)-(8)) and separating real and imaginary parts results in six real differential equations for the transformed variables. With $T_S = L_S/R_S$ and $T_R = L_R/R_R$ we find

$$\sigma_S T_S \frac{di_{Sd}}{dt} + i_{Sd} = \frac{u_{Sd}}{R_S} + \omega_1 \sigma T_S i_{Sq} - (1-\sigma) T_S \frac{di_{mR}}{dt}, \tag{19}$$

$$\sigma_S T_S \frac{di_{Sq}}{dt} + i_{Sq} = \frac{u_{Sq}}{R_S} - \omega_1 \sigma T_S i_{Sd} - (1-\sigma)\omega_1 T_S i_{mR}, \tag{20}$$

$$T_R \frac{di_{mR}}{dt} + i_{mR} = i_{Sd}, \tag{21}$$

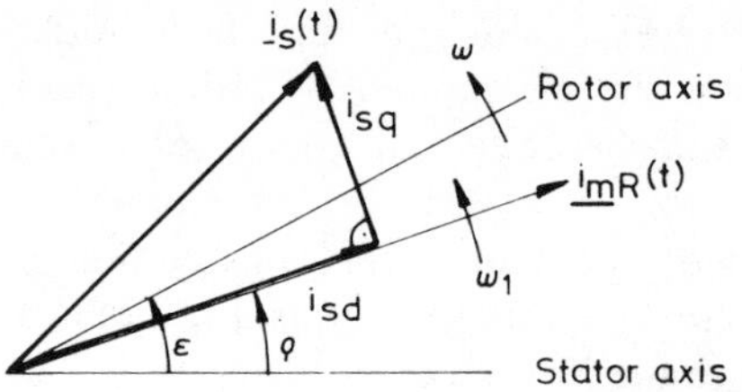

Fig. 3. Angular relations of current vectors.

$$\frac{d\rho}{dt} = \omega_1 = i_{Sq}/T_R i_{mR} + \omega, \tag{22}$$

$$\theta \frac{d\omega}{dt} = m_{el} - m_L = k i_{mR} i_{Sq} - m_L; \qquad k = \tfrac{2}{3} \frac{L_h}{1+\sigma_R} \tag{23}$$

$$\frac{d\epsilon}{dt} = \omega. \tag{24}$$

Equations (19) and (20) describe the transition from the field-oriented voltage components u_{Sd} and u_{Sq} to the current components i_{Sd} and i_{Sq} which involves small leakage time constants and some interactions, which can be balanced by suitable control.

The slow dynamics in the direct axis are governed by (21), with i_{Sd} acting as control input. The angular position of the flux vector is determined by (22) with i_{Sq} serving as a convenient input in order to control the torque (23) as the product of i_{mR} and i_{Sq}.

Analogous to the control of dc motors, it is the best policy to maintain i_{mR} at its maximum level, limited either by iron saturation or, above base speed, the ceiling voltage of the inverters. This also removes some of the nonlinear interactions.

It is noted that the control plant could be simplified by employing quasi-impressed stator currents obtained through fast current control of the inverter [5], [6]. This would all but eliminate (19) and (20). However, at higher stator frequency and with thyristor inverters having limited clock frequency, sufficient bandwidth of the control loops is difficult to achieve.

The transformed machine equations are seen in graphical form in the upper part of Fig. 4, which also show the transformation in two steps from the three terminal voltages u_{S1}, u_{S2}, and u_{S3} to the field-oriented voltages u_{Sd} and u_{Sq}. The transformation involves four multiplications with cos ρ, sin ρ.

The three-phase inverter shown at the left of Fig. 4 supplies three independently controlled voltages, usually pulsewidth modulated (PWM) square waves. The control dynamics are approximated by a small delay T_D, which is determined by the clock frequency.

In the lower part of Fig. 4 a general control scheme in field coordinates is drawn. It contains an inverse coordinate transformation by modulating the dc output quantities of the controllers with sin ρ, cos ρ in order to generate the ac voltage references for the inverters.

The main feature of field-oriented control is this: if the delay T_D of the inverter can be considered to be small in

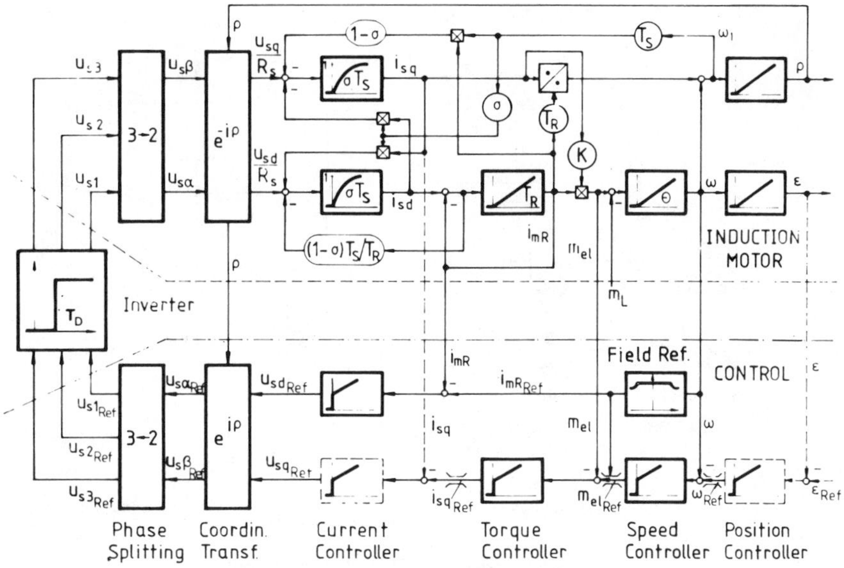

Fig. 4. Induction motor control using field coordinates.

relation to the control dynamics of the drive, the complete left portion of Fig. 4 drops out and $u_{Sd\ \text{ref}}$, $u_{Sq\ \text{ref}}$ become the input signals to a dc control plant.

If i_{mR} is maintained constant at maximum level, as explained before, the nonlinear interactions in the motor are reduced. The quadrature current controller is redundant if the torque is governed by a torque controller; there is a superimposed speed controller and, if position control is called for, a position controller. All the intermediate reference quantities can be limited as is common with cascade control [5].

It is seen that no controllable oscillator is needed, as with most ac motor control schemes. Instead the motor controls its stator frequency through feedback from the flux signal; therefore, no pull-out effect exists. If the motor is overloaded or if the speed reference is changed too quickly for the motor to follow, the speed controller will saturate and the motor generates constant maximum torque. Thus the drive can operate safely in all four quadrants of the speed-torque plane as is the case with a dc motor supplied for a reversible converter.

Despite this straightforward design, the control scheme in Fig. 4 is of considerable complexity, and it is difficult to see at first how it can be reduced to simple hardware. With conventional components it would indeed be difficult, but with microprocessors these problems can be very effectively overcome. This is seen in the next Section.

DESCRIPTION OF EXPERIMENTAL DRIVE

The field-oriented control system was tested on a system consisting of the microprocessor controller, a 2.5-kW peak-rated PWM transistor inverter, and a 2.0-kW nominal-rated squirrel-cage induction machine.

The control portion was implemented with two Intel 8085 microprocessors, each having a dedicated hardware multiplier. The microprocessor output signals, two sampled quadrature ac voltage references, serve as the input signals for the three-phase inverter, the control circuit of which employs a triangle-intersection technique implemented with discrete logic. Speed feedback is derived from a two-phase pulse tachometer which enables the control to discern between forward and reverse rotation.

Fig. 5 delineates the functions handled by each processor. Processor 1 also takes care of the keyboard, display, and service routines. All control functions are computed once every millisecond except for the speed controller and the field controller which, because of their lower dynamic requirements, are processed on 5- and 10-ms intervals, respectively.

A program flow diagram is shown in Fig. 6 along with approximate execution times for the main functions. Both processors basically work in an interrupt mode on the 1-ms intervals set by an external timer chip. Total memory requirement for the system is approximately 5.3 kbyte, of which 2 kbyte are tables for coordinate transformation and display functions.

The functions shown in the block diagram of Fig. 5 are implemented with modular software. For example, all proportional-integral (PI) controllers use the identical subroutine with the parameters substituted as required by the plant dynamics. The following algorithm has been employed

$$y_\nu = K\left[x_\nu + \frac{\Delta t}{T}\sum_{-\infty}^{\nu} x_\mu\right], \tag{25}$$

where y_ν and x_ν are the output and input variables, respectively, at the time $\nu\,\Delta t$, and K is the gain, T is the controller time constant, and Δt the sampling period.

For the field-oriented method it is necessary to know both the magnitude and angle of the i_{mR} vector. This vector can either be reconstructed from the stator variables by a variety of techniques [7], [13] or measured [11]. Both approaches have been followed here.

For the measurement version three sensing coils were placed in the stator slots above the wedges, spanning a full pole pitch each and thus eliminating spatial harmonics due to the distributed stator windings and slot effects. The induced

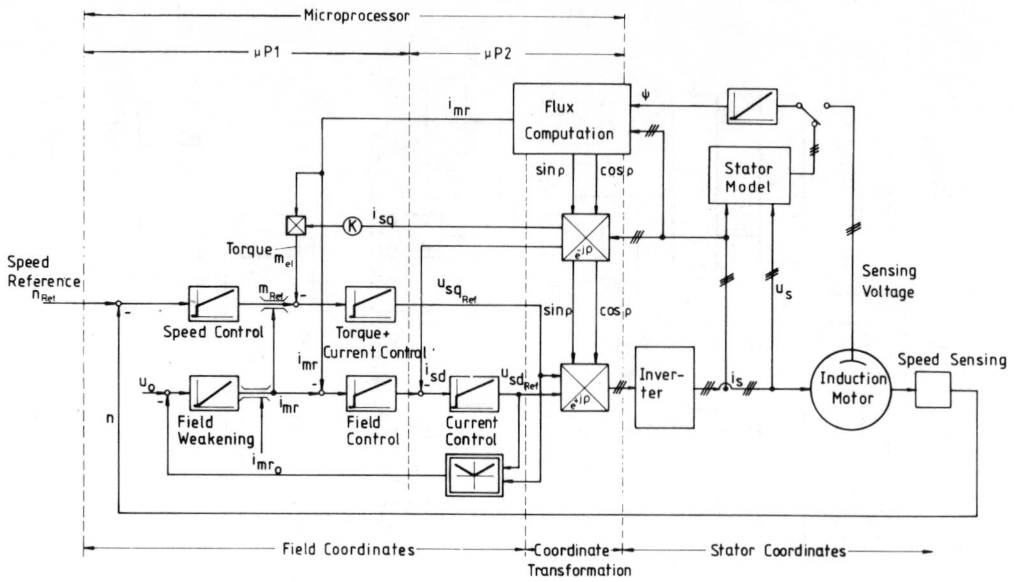

Fig. 5. Microprocessor controlled induction motor.

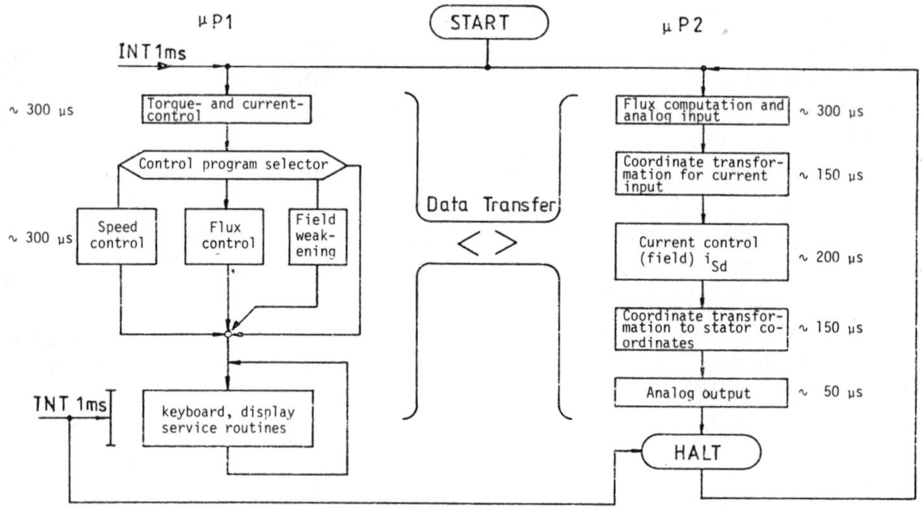

Fig. 6. Program flow diagram.

voltages were integrated, providing a quite ideal flux measurement down to the lower cutoff frequency of the integrators of about 0.5 Hz.

The coils placed at the front end of the stator slots provide a measurement of the main flux vector

$$\boldsymbol{\Psi}_m(t) = L_h(\boldsymbol{i}_S + \boldsymbol{i}_R e^{j\epsilon}), \tag{26}$$

which is then corrected according to

$$\boldsymbol{\Psi}_{mR}(t) = (1+\sigma_R)\boldsymbol{\Psi}_m(t) - \sigma_R L_h \boldsymbol{i}_S(t) \equiv L_h \boldsymbol{i}_{mR}(t), \tag{27}$$

in order to obtain the modified rotor flux as required by (15).

In order avoid the sensing coils a substitute flux signal was also constructed from the stator voltages and currents, thus making the control scheme applicable to standard ac motors as well. For this purpose, an analog model of the stator differential equation (5) was used. The drawback of this method is of course that the stator resistance, varying with temperature, must be compensated, which is of particular importance at low speed.

The analog flux signals are converted to digital form and processed by the second 8085 unit as seen in Fig. 5, resulting in the magnitude i_{mR} and the angular functions $\sin\rho$, $\cos\rho$.

The flux computation is essentially a rectangular to polar conversion. Although one can conceive of many algorithms to implement this function, the following one was chosen as a compromise between speed, memory usage, and resolution. Given inputs of

$$x = M\cos\rho, \qquad y = M\sin\rho,$$

the value $S = x^2 + y^2$ is computed; with this result M and $1/M$ are determined from look-up tables. The trigonometric functions are computed by normalizing the inputs

$$\cos\rho = x/M, \qquad \sin\rho = y/M.$$

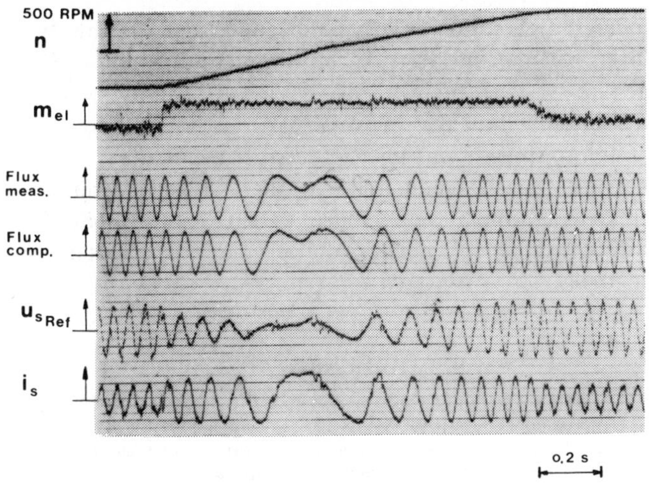

Fig. 7. Speed reversal computed flux.

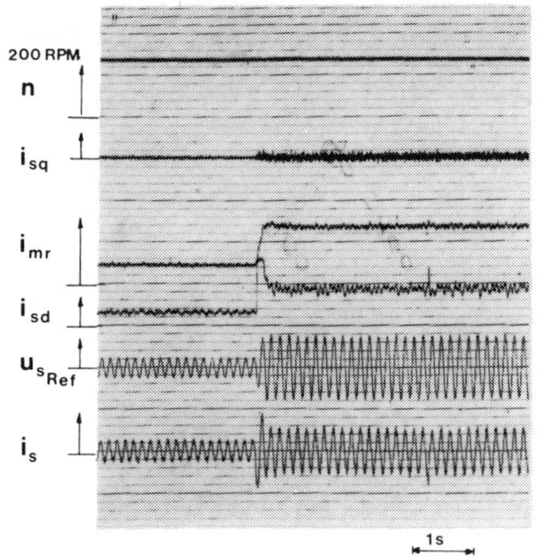

Fig. 9. Step change in field reference.

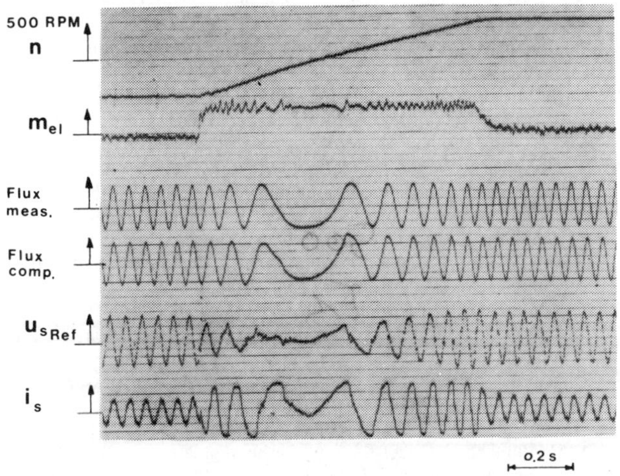

Fig. 8. Speed reversal measured flux.

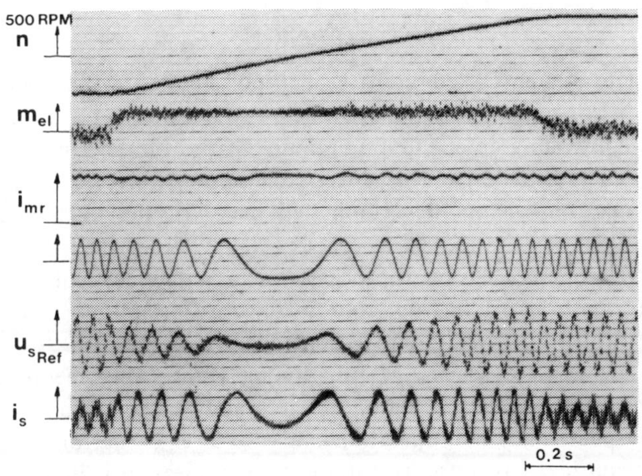

Fig. 10. Speed reversal at low commutation frequency.

The angle information is used to demodulate the stator currents into the direct and quadrature components i_{Sd} and i_{Sq}. These signals directly relate to the magnitude of the field and the torque, which enables the control to be decoupled into two loops, one for the field control and one for the torque control. The outputs from the controllers are modulated by sin ρ, cos ρ in order to produce stator voltage references for the inverter.

These coordinate transformations are defined as follows:

$$\begin{bmatrix} u_{S\alpha} \\ u_{S\beta} \end{bmatrix}_{\text{ref}} = \begin{bmatrix} \cos\rho & -\sin\rho \\ \sin\rho & \cos\rho \end{bmatrix} \begin{bmatrix} u_{Sd} \\ u_{Sq} \end{bmatrix}_{\text{ref}}.$$

Similar relations exist for the currents.

RESULTS

Oscillographs of system performance for various conditions are seen in Figs. 7–10. Fig. 7 shows a speed reversal with the flux feedback signal reconstructed from the stator voltages and currents, while Fig. 8 shows the same transient when the flux signal is derived from the sensing coils in the machine. In general, for reasonably high stator frequencies (above 10 Hz) there are only insignificant differences. At lower frequencies, however, the reconstructed flux signal deteriorates primarily because of unavoidable errors of the $i_S R_S$ compensation. As the frequency is reduced this term becomes increasingly significant, since the induced voltages are small. Thus, any difference between the actual machine resistance and the nominal resistance value assumed for the model can create substantial measurement errors. Fig. 9 shows the response of the system to a step change in the field magnitude reference.

All of the above tests were carried out with modulation frequency of the inverter of 2.5 kHz. In order to show the reaction of the field-oriented control method to currents with high harmonics content, as would be the case with a motor fed by a thyristor inverter, the modulation frequency was lowered to 250 Hz. A speed reversal for this condition using sensed flux is shown in Fig. 10. It is seen that there is no serious degradation of performance. In Fig. 11 a speed transient is shown where automatic field weakening based on inverter voltage feedback takes place.

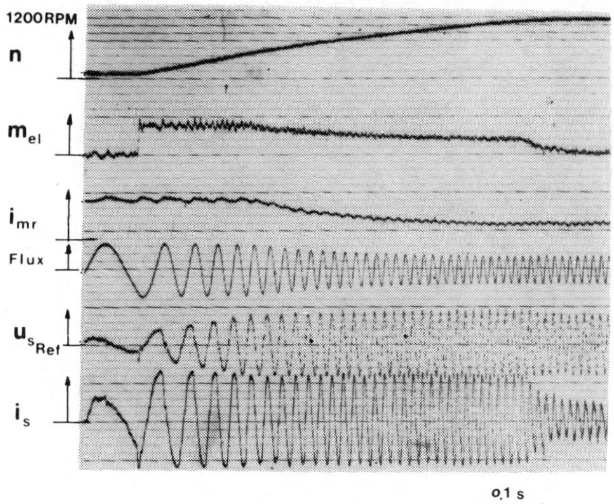

Fig. 11. Speed reversal including field weakening.

CONCLUSION

The field-oriented control method coupled with microprocessor technology has been shown to be an effective means for controlling an ac induction motor. Although field-oriented control can be implemented using analog hardware, the advent of microprocessors has eliminated many of the difficulties and complexities as well as cost associated with this approach. Even present day microprocessors can rather easily handle the functions, primarily coordinate transformations, which are required for field-oriented control.

The control scheme is functioning with a flux feedback signal based only on stator voltages and currents; however, the speed range of operation and the level of performance are considerably improved if the flux is measured directly—for example using sensing coils. Other techniques for computing the flux vector will have to be evaluated in an attempt to find the best and most practical solution. Here again microprocessors can play an important part.

Although the described system was initially designed for a PWM inverter with a high modulation frequency, there is no inherent restriction limiting the application to this type of inverter. Other inverters, such as impressed current or variable dc-link voltage types could be used, possibly, with some reduction in dynamic performance.

Improvements in microprocessor technology will make possible further reductions of the hardware required and create additional options such as self-adaptive control. This will continue to enhance the promise of adjustable-speed ac drives.

REFERENCES

[1] *Control in Power Electronics and Electrical Drives 2.* Oxford: Pergamon, 1978; IFAC Symp. Düsseldorf, Germany, 1977.

[2] F. Blaschke, "Das verfahren der feldorientierung zur regelung der asynchronmaschine," *Siemens Forschungs-und Entwicklungsberichte I*, 1972.

[3] "Das verfahren der feldorientierung zur regelung der drehfeldmaschine," Dissertation, TU Braunschweig, 1974.

[4] K. P. Kovacz, and J. Racz, *Transiente Vorgänge in Wechselstrommaschinen.* Budapest, 1959.

[5] W. Leonhard, *Regelung in der Elektrischen Antriebstechnik.* Stuttgart: Teubner, 1973.

[6] ——"Introduction to ac-motor control using field coordinates," in *Simp. Sulla Evoluzione Nella Dinamica Delle Macchine Elettriche Rotanti*, Tirrenia, p. 370, 1975.

[7] F. Blaschke, "Verfahren der felderfassung bei der regelung stromrichtergespeister asynchronmaschinen," presented at the Control in Power Electronics and Electrical Drives IFAC Symp., Düsseldorf, 1974.

[8] R. Gabriel, "Antriebsregelung einer thyristorgespeisten asynchronmaschine durch einen mikrorechner," Diplomarbeit Institut für Regelungstechnik, TU Braunschweig, 1978.

[9] H. Waldmann, "Koordinatentransformationen bei der mehrgrössenregelung von wechsel- und drehstromsystemen," Dissertation TU Braunschweig, 1978.

[10] R. Gabriel, W. Leonhard, and C. Nordby, "Microprocessor control of induction motors employing field coordinates," IEE, 2. Int. Conf. on "Electrical Variable-Speed Drives," London, September 1979.

[11] A. B. Plunkett, "Direct flux and torque regulation in a PWM inverter-induction motor drive," *IEEE Trans. Ind. Appl.*, vol. IA-13, no. 2, 1977.

[12] W. Leonhard, "Field oriented control of a variable speed alternator connected to the constant frequency line," presented at the IEEE Conf. on Control of Power Systems, College Station, TX., 1979.

[13] A. Abbondanti, "Method of flux control in induction motors driven by variable frequency, variable voltage supplies," in *Proc. IEEE/IAS Int. Semiconductor Power Conf.*, p. 177, Mar. 1977.

Inverter–Induction Motor Drive for Transit Cars

A. B. PLUNKETT, MEMBER, IEEE, AND D. L. PLETTE, SENIOR MEMBER, IEEE

Abstract—The advent of large power semiconductors has made it possible to apply inverters and ac motors to traction applications. Either synchronous or induction motors and several types of power converters can be considered. The induction motor and the pulsewidth modulated (PWM) inverter are selected as favorable for application to a transit car drive. A general method of sizing the PWM inverter and induction motor in terms of the car performance requirements is outlined. This method results in a minimum size inverter and allows optimization of system weight and cost. A discussion of wheel size effects and the optimization of regenerated energy is included.

INTRODUCTION

THE LARGE majority of transit cars in service today are driven from a fixed voltage dc third rail by dc motor drives. The motors are controlled by resistance type controllers to provide the required car acceleration. These systems generally also include dynamic braking to decelerate the car and have friction brake systems to supplement or back up the dynamic braking systems. For the most part, these systems provide entirely adequate car performance and reasonable operating costs. However, the advent of large power semiconductors has made it practical to consider more advanced propulsion systems using choppers and inverters, which offer promise of reduced operating costs. This paper describes an ac propulsion system employing inverters and induction motors and discusses its design and application considerations.

The inverter ac motor system consists of a solid state inverter, which converts the dc third rail voltage to variable voltage, variable frequency ac, and induction motors to drive four axles (see Fig. 1). The inverter supplies the motors with the proper ac voltage, current, frequency, and waveshape to provide the needed car acceleration and deceleration.

This ac propulsion system provides a high degree of regenerative braking with only minor penalty in equipment size. The amount of regenerative power savings is dependent upon many factors such as station spacing and traffic density; however, computer studies have shown the savings may be as high as 40 to 50 percent compared to equivalent cars with resistance type controllers and dynamic braking. The rising costs of electric power and the limitations of our national energy supplies make it highly desirable that future cars utilize a propulsion system which can make these savings possible.

DC motor maintenance costs are also increasing. By eliminating the maintenance and inspection costs associated with brushes and mechanical commutators, the ac induction motor can offer substantially reduced costs to the operator. Let us now consider the application requirements and how they influence the design of an ac propulsion system.

Paper IOD-75-58, approved by the Land Transportation Committee of the IEEE Industry Applications Society for presentation at the 1974 IEEE Industry Applications Society Annual Meeting, Pittsburgh, PA, October 7–10. Manuscript released for publication October 6, 1976.

A. B. Plunkett is with the General Electric Co., Schenectady, NY 12345. He was with the General Electric Co., Erie, PA.

D. L. Plette is with the General Electric Co., Erie, PA. 16501.

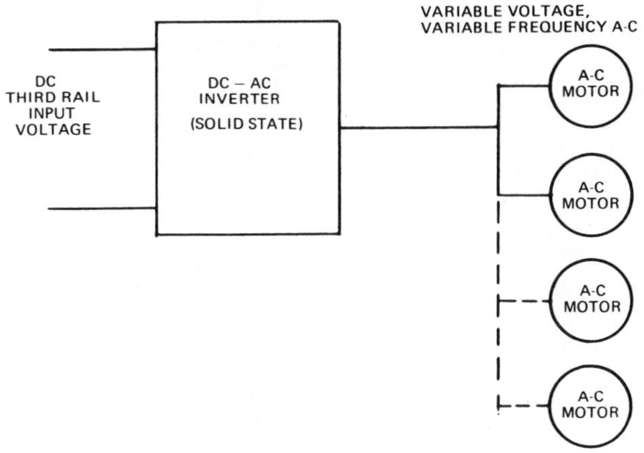

Fig. 1. System block diagram.

PERFORMANCE REQUIREMENTS

A typical transit car duty cycle is shown in Fig. 2. The car accelerates, at a decreasing rate, to the desired speed, and then runs at nearly constant speed. It will then brake at a constant rate to a stop. After a dwell time, the cycle is repeated. The amount of time spent in each operating region, the variation in station spacing, speed limits, and the presence of grades and curves will all affect the duty cycle and determine the equivalent continuous motor torque which sets the physical size of the motor.

The general form of a car performance curve is shown in Fig. 3. This figure shows acceleration and train resistance as a function of car speed. The motoring curve includes a constant acceleration section to speed $V1$, a constant horsepower section from $V1$ to $V2$, and a section from $V2$ to maximum speed where torque falls off in accordance with the natural motor characteristic. The constant horsepower section is included for two reasons; it constitutes the maximum line current draw required from the distribution system, and it determines, along with braking requirements, the size of the components in the propulsion equipment.

The braking curve of Fig. 3 has two sections. In the speed range from zero to speed $V3$, a constant deceleration rate is held. From $V3$ to maximum speed, the rate is tapered to avoid exceeding wheel to rail adhesion capability and to limit the peak power requirements of the propulsion equipment.

The effect of train resistance is to subtract from the motoring tractive effort to give an acceleration somewhat lower than would be indicated by the curve. During braking,

Reprinted from *IEEE Trans. Ind. Appl.*, vol. IA-13, pp. 26–37, Jan./Feb. 1977.

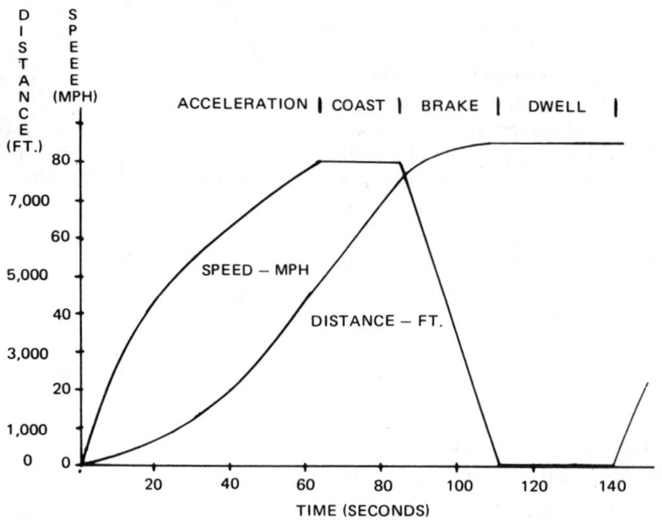

Fig. 2. Sample duty cycle.

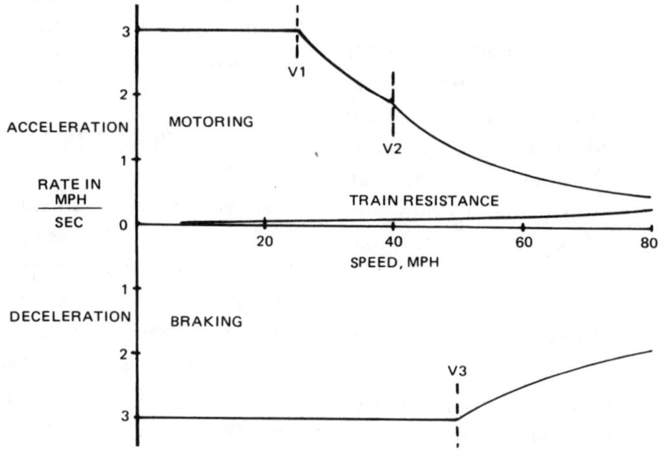

Fig. 3. Car performance curves.

FACTOR	INDUCTION MOTOR	SYNCHRONOUS MOTOR
COMPLEXITY	LOW	MEDIUM
RELIABILITY	HIGH	MEDIUM
MAINTAINABILITY	HIGH	MEDIUM
WEIGHT	MEDIUM TO LOW	MEDIUM TO LOW
COST TO MANUFACTURE	LOW	MEDIUM
INVERTERS REQUIRED PER CAR	ONE	ONE/MOTOR
EFFICIENCY	HIGH	HIGH
SENSITIVITY TO WHEEL DIAMETER DIFFERENCE	MEDIUM (UNLESS USING ONE INVERTER/MOTOR)	NONE (REQUIRES ONE INVERTER PER MOTOR)

Fig. 4. AC motor selection factors.

Fig. 5. Induction motor rotor.

train resistance adds to the braking rate and the deceleration will be greater than that provided by electric and friction brakes alone.

AC MOTOR SELECTION

There are two basic types of ac motors which could be considered for the transit application–synchronous and induction. The squirrel cage induction motor is a proven industrial workhorse which can be given the required torque/speed characteristics when operated from the variable voltage and frequency of the inverter.

The synchronous motor, in a transit application, requires a means to excite the field at all speeds, including zero speed. One method of excitation, which is often used, is to excite a coaxial transformer from the stator, to rectify the voltage induced in the secondary with rotating diodes, and to supply the rectified output to the rotating field winding. An alternate method using a solid rotor inductor or Lundell type generator is sometimes considered because of its stationary field winding, but the higher weight and rotor construction problems usually cause this alternate to be less attractive.

Fig. 4 compares the induction motor with the synchronous motor of the type employing coaxial transformer, rotating diodes, and wound rotating field.

The induction motor was selected for application since it was judged to provide lower cost, higher reliability, and better maintainability for the overall system. A major factor in the comparisons is the need for only one inverter per car with the induction motor. For further consideration of the effects of wheel diameter differences on parallel motors driven from the same inverter, see Appendix A.

Induction motors with 2, 4, and 6 poles were compared to determine an optimum design. With maximum motor speeds of 6000 to 7000 r/min, the four-pole motor is optimum since it provides minimum motor weight and reasonable inverter switching frequencies.

The rotor of an induction motor designed for transit cars is shown in Fig. 5. The lack of commutators, insulated windings, and rotating semiconductors contribute to its basic ruggedness and reliability.

INVERTER SELECTION

A full comparison of all possible inverter configurations requires extensive detailed analysis which is beyond the scope of this paper. The most important parameters for comparison are those affecting inverter cost, weight, and size. These are the number of power semiconductors, relative peak currents which must be commutated, and the need for heavy magnetic components. Effect on motor heating due to nonsinusoidal waveforms and the resulting low speed torque pulsations were also evaluated. Regenerative capability was considered as a firm requirement. Fig. 6 shows several of the systems which were considered and gives them a relative rating.

The phase shift inverter 1 eliminates a number of the possible output harmonics by the use of a phase shift reactor/transformer between inverters. A minimum of two inverters is required. Wheel diameter tolerance must be controlled on a total car basis assuming no more than two inverters per car. The principal disadvantage is the heavy weight of the reactor/transformer.

The variable voltage square wave inverter controls the dc link voltage in order to control the ac voltage. A bilateral chopper is required to permit regeneration. A second filter is likely to be required between the chopper and inverter. All power is converted twice. Cost, weight, and size of this approach are all high.

The full pulsewidth modulation (PWM) inverter provides voltage control and low frequency wave shaping by the multiple switching action of the inverter. Square wave operation of the inverter is employed at higher speeds to minimize commutation losses. This inverter can be used as shown with two inverters per car with independent frequency control or with one inverter per car driving four motors.

The controlled current inverter [2] is quite heavy due to the inverter input reactor and the need for commutating capacitors in both chopper and inverter. It could also be used with one inverter per car. Regeneration does not require a bilateral chopper. With this approach, all power is converted twice.

The full PWM inverter (*c*) was selected because of its relatively light weight, lower cost, and its need to handle the power conversion only once. More details of this approach are described in the next section.

PWM INVERTER POWER CIRCUIT

The simplified circuit for the inverter is shown in Fig. 7. For a 600-V system, each semiconductor shown is actually two thyristors or diodes in series in order to obtain a sufficient voltage range to withstand line voltage transients. This circuit is the standard McMurray inverter [1] modified by the addition of reset resistor $R1$ and diode $D1$. These serve to discharge the excess voltage which appears on commutating capacitor $C1$ when commutating the load current. Thyristor $M1$ is the main current path for motoring (power flowing from the dc to ac side) and diode $D2$ is the main current path for braking (power flowing from the ac to the dc side). Thyristor $A1$, capacitor $C1$, inductor $L1$, and diode $D2$, comprise the commutating current path. This path diverts the load current from the main SCR for about 60 μs to allow it to shut off and recover voltage blocking capability.

Fig. 8 shows the line filter, contactors, and braking resistors in their relationship to the inverter power circuit. The line

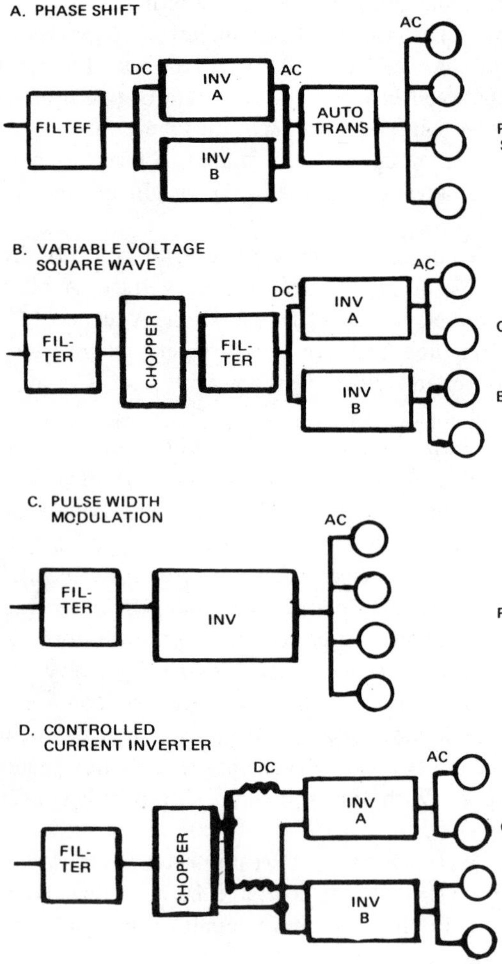

VOLTAGE CONTROL	PEAK CURRENTS*	MOTOR HEATING	STARTING (TORQUE PULSATIONS)	REGENERATION	MAGNETICS	COMPARATIVE NUMBER OF SEMICON-DUCTORS
PWM AND PHASE SHIFT	1.04	GOOD	GOOD	YES	YES+	1
CHOPPER	1.24	FAIR	POOR	NO	NO	1.3
BILATERAL CHOPPER	1.24	FAIR	POOR	YES	NO	1.5
PWM	1.37	FAIR	GOOD	YES	NO	1
CHOPPER	NA	FAIR	POOR**	YES	YES	1

*THE NUMBER GIVEN IS THE RATIO OF TOTAL PEAK TO SINE WAVE PEAK

Fig. 6. Inverter alternates.

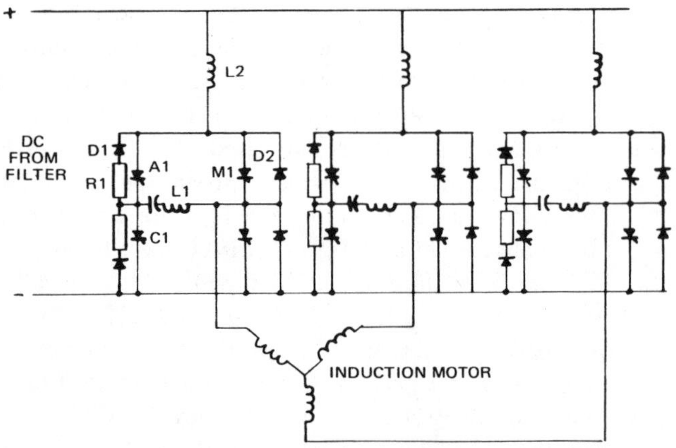

Fig. 7. Inverter power circuit.

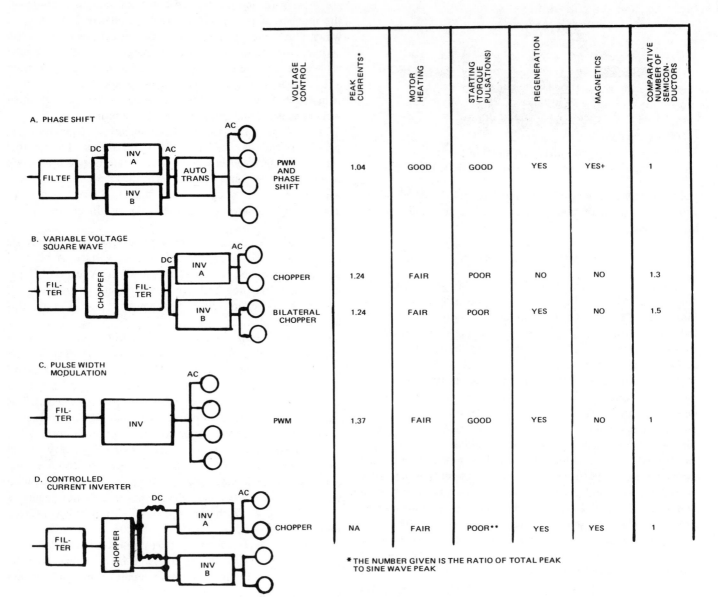

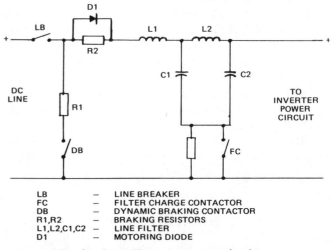

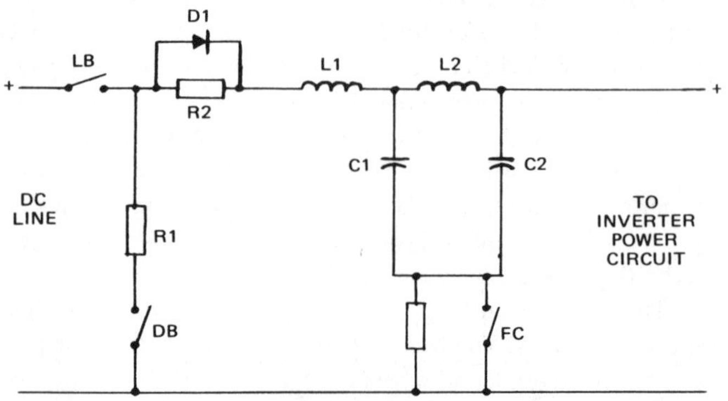

LB – LINE BREAKER
FC – FILTER CHARGE CONTACTOR
DB – DYNAMIC BRAKING CONTACTOR
R1,R2 – BRAKING RESISTORS
L1,L2,C1,C2 – LINE FILTER
D1 – MOTORING DIODE

Fig. 8. Input filter, contactors, and resistors.

filter supplies the low source impedance desired by the inverter power circuit and serves to eliminate interference with the signalling system from pulsating currents drawn by the inverter. Contactor *FC* and its associated resistor are used when initially charging the line filter in order to limit the inrush current and voltage overshoot.

The braking resistors R_1 and R_2 are in two sections. The resistor R_2 is in the circuit every time braking occurs. Its function is to allow the dc inverter voltage to be higher than the dc line voltage so as to obtain a higher braking power than motoring power from the same sized inverter. This is explained in more detail in the section of this paper on *Motor and Inverter Design and Rating.* Resistor R_1 is only used in the dynamic braking mode and is switched into the circuit by contractor *DB.*

It is necessary to calculate the currents, voltages, and frequencies that each inverter current path must handle in order to determine the inverter requirements in terms of car performance parameters.

INVERTER WAVEFORMS

In order to control the fundamental ac voltage applied to the motors and to minimize low frequency torque pulsations, a method of pulse width modulation that results in an equivalent sine wave of voltage is used as shown in Fig. 9. A sine wave reference voltage is compared with a triangular wave with the intersections determining whether the inverter phase output is plus V_{dc} or ground, thus creating the line-to-ground and line-to-neutral voltages as shown. This method of control is employed during motoring from zero speed to approximately 80 percent of speed V_1 (see Fig. 3). A modified switching mode is used during the transition from this speed (0.8 V_1) to full square wave operation which occurs at speed V_1. The transition occurs in a similar but reverse fashion from square wave to PWM operation at a somewhat higher speed (V_4) during electric braking.

The PWM chopping action results in nonsinusoidal current and flux in the motor contributing to motor losses. This subject has been treated in more detail in previous papers [3], [4]. For a given motor inductance, a higher chopping frequency will produce lower harmonic currents. This must be weighed against the disadvantages of more commutations per second which increase inverter losses and affect component ratings. The larger the motor inductance the lower the harmonic currents for a given applied voltage waveform. System considerations prevent the use of high inductance motors as described in the section on *Motor and Inverter Design and Rating.*

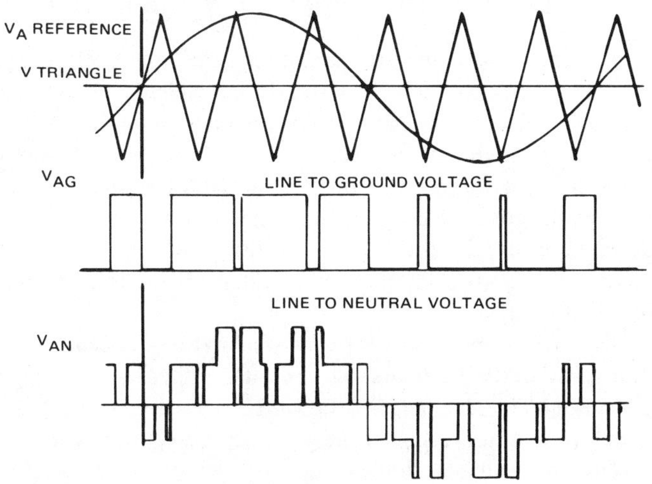

Fig. 9. PWM inverter waveforms.

MOTOR CONTROL FOR DESIRED PERFORMANCE

A general method of sizing an ac propulsion system, in terms of car performance parameters, is required in order to determine the necessary system size for a given car performance. The motor and inverter must be designed together in order to maximize the system performance while minimizing weight and cost. The range of sizes of dc/ac inverters is limited to discrete steps due to the few types of suitable power semiconductors available. The motor can usually be designed for any performance necessary within maximum size, weight, and gearbox restrictions.

In order to see how best to meet the desired torque speed characteristics, the simplified motor torque equations should be examined. Torque in any motor is proportional to the rotor current and the flux linking that current.

$$T \propto \Phi_{AG} \cdot I_R, \tag{1}$$

where Φ_{AG} is the air gap flux, I_R is the rotor current.

To the first approximation, the air gap flux in an induction motor is proportional to the applied stator voltage (V_a) divided by stator frequency (f),

$$\Phi_{AG} \propto \frac{V_a}{f}. \tag{2}$$

The rotor current at low values of slip is proportional to the air gap flux and the amount of slip (fs) betwen the stator flux wave and the speed of the rotor,

$$I_R \propto \Phi_{AG} fs. \tag{3}$$

Substituting equation (2) in equation (3),

$$I_R \propto \left(\frac{V_a}{f}\right) fs. \tag{3a}$$

Substituting equations (2) and (3a) in equation (1) yields

$$T \propto \left(\frac{V_a}{f}\right)^2 fs. \tag{4}$$

From (4), it can be seen how the slip of the motor must vary in order to provide a particular torque versus speed curve. Fig. 10 shows a general torque-speed relationship which could

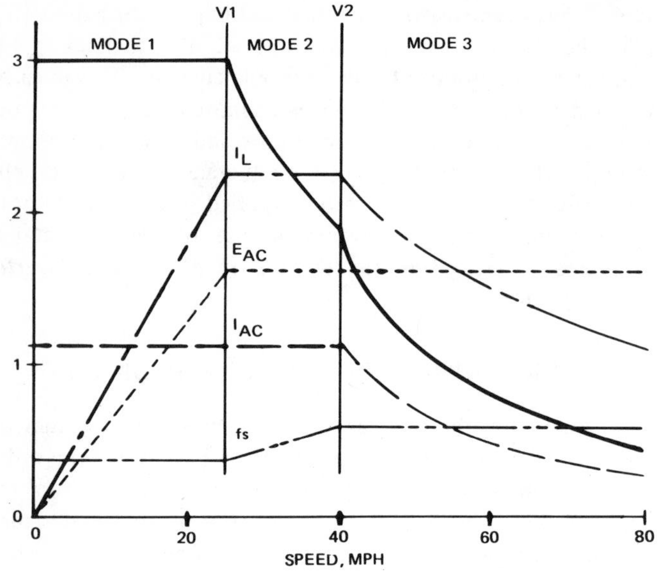

Fig. 10. Control modes, motoring.

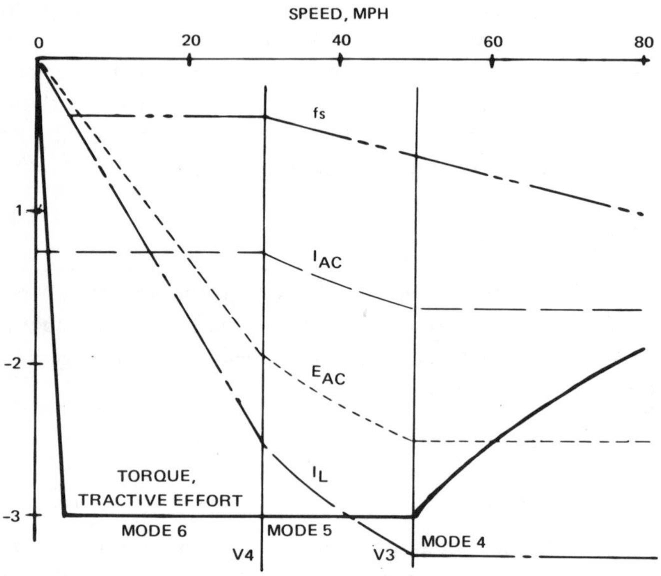

Fig. 11. Control modes, braking.

be used for accelerating the vehicle. In mode 1, constant torque can be achieved by holding constant air gap flux (V_a/f) and constant slip (fs). In modes 2 and 3, the applied voltage (E_{ac}) is constant since the inverter is in square wave operation. The motor flux will decrease in these modes inversely proportional to speed. In mode 2, constant horsepower, constant motor current (I_{ac}) operation can be achieved by increasing the motor slip proportional to speed. Above the desired maximum speed (V_2) for constant horsepower, the slip is held constant, motor current reduces inversely with speed, and torque decreases inversely as the speed squared.

In Fig. 11, similar relationships are shown for a typical braking characteristic. At speeds above V_3 (mode 4), the motor voltage, current, and hence power, are held constant to match the inverter maximum power capability. From (3a), it can be seen that the slip (fs) must therefore be proportional to speed (f) during this portion of the braking cycle.

From $V3$ to $V4$ (mode 5), torque is held constant, power varies directly with speed, and both motor voltage and current vary approximately as the square root of speed. Equation (4) will show that again the slip must be proportional to speed during this mode of operation.

Mode 6 operation is similar to mode 1 (motoring) with flux, torque and slip held constant until finally at very low speeds the system electric braking is allowed to fade.

Fig. 10 also shows the normal motoring dc line current (I_L), ac motor current (I_{ac}), and motor voltage (E_{ac}) as a function of speed.

The dc line current and the fundamental component of current will depend on the line voltage and the motoring cornerpoint speed, V_1. To obtain the dc current, the power developed at the rail is equated to the power drawn from the dc line. At speed V_1, the motoring power is maximum; thus,

$$I_L = \frac{200\, WA_1 V_1}{E_L \times \eta c} \text{ amperes,} \quad (5)$$

where

W	car weight (tons),
A_1	motoring acceleration at V_1 (mi/h/s)
V_1	motoring cornerpoint (mi/h),
E_L	dc line voltage (V),
ηc	combined efficiency of motor, inverter, gears, etc.

In calculating the ac currents in motoring, only the efficiency and power factor (PF) angle between voltage and current of the motor need to be taken into account.

Thus at V_1

$$I_{ac} = \frac{148\, WA_1 V_1}{E_L (PF) \eta m}, \quad (6)$$

ηm	motor efficiency,
PF	power factor of motor operation.

The equations for I_L and I_{ac} in braking are similar except that losses add to braking effort, whereas they subtract from motoring effort.

MOTOR AND INVERTER DESIGN AND RATING

The car performance requirements determine both inverter and motor designs. For example, the car weight, the motoring cornerpoint speed V_1, and the acceleration rate at speed V_1 define the maximum motoring power. Two principal factors determine the motor size: the motor leakage inductance, and the continuous rated torque. The motor inductance is determined by the high speed motoring and braking requirements of the car. The continuous rated torque is determined by the duty cycle.

The inverter peak power point occurs at high speed braking. The commutation circuits may be rated either by motoring or braking conditions. The motor voltage increases linearly with speed to the motoring cornerpoint speed V_1 at which the maximum available voltage will be attained. This voltage profile is chosen to minimize the inverter current and thus its size. However, this minimum power inverter size will be increased because the high speed motor pull-out torque requires a low motor inductance. The low inductance causes significant harmonic currents which the inverter must supply at speeds near V_1.

In determining the motor inductance, the motoring condition which occurs at the end of the constant power region of operation (speed V_2) must be considered. Here the motor pull-out torque must meet or exceed the torque requirement. With a three-phase four-pole induction motor, the pull-out torque is

$$T_{PO} = \frac{0.0475}{LT} \left(\frac{E_{ac}}{f}\right)^2 \text{ lb ft,} \tag{7}$$

where

L_T total leakage inductance including cables (hy),
E_{ac} applied ac voltage,
$0.45\, E_L$ for square wave,
f applied frequency (Hz).

This value takes approximate motor and gear losses into account. The ratio of required motor pull-out torque at speed V_1 to the torque required for acceleration is the speed ratio V_2/V_1.

The total motor leakage inductance for all motors in parallel, needed to meet the motoring requirement is

$$L_T \leqslant 2.04 \times 10^{-4} \left(\frac{E_L{}^2\, V_{max}}{WA_1 V_1 V_2 f_{max}}\right). \tag{8}$$

The braking requirement for a transit application will usually determine the minimum motor torque capability and, thus the motor size. The profile that is used in braking is to allow both the motor voltage and current to increase with speed, thus allowing maximum utilization of the inverter commutation capability which inherently increases with voltage. This allows the amount of regeneration to be greatly increased over a similar sized system which attempts to supply braking without a voltage increase.

Fig. 12 shows two possible dynamic braking requirements. The design point is where the maximum braking power is required. Fig. 12(a) shows this point occuring at maximum speed. The design point is that at which the constant power torque intersects the pull-out torque. At this point, the motor voltage will be maximum. Fig. 12(b) shows a steeper torque (brake) taper. The limiting point is now the V_3 point (Fig. 11) which is again at maximum motor voltage. Another way of describing the limiting point is to say that it will always be at the maximum motor voltage which will always be at the maximum power point.

The motor slip must increase with speed in dynamic braking; thus, the motor will reach pull-out torque at maximum speed without regard for the actual braking rate required.

To obtain the maximum braking capability from a given inverter, the voltage in braking must be below a safe upper limit, and the ratio of motor current in braking to the motor current in motoring should not exceed the ratio of the braking dc voltage to the motoring dc voltage. The inverter current capability increases directly with E_{dc}; thus the power handling capacity increases as the voltage squared. The maximum value of speed V_3 for a given inverter voltage capability is given by

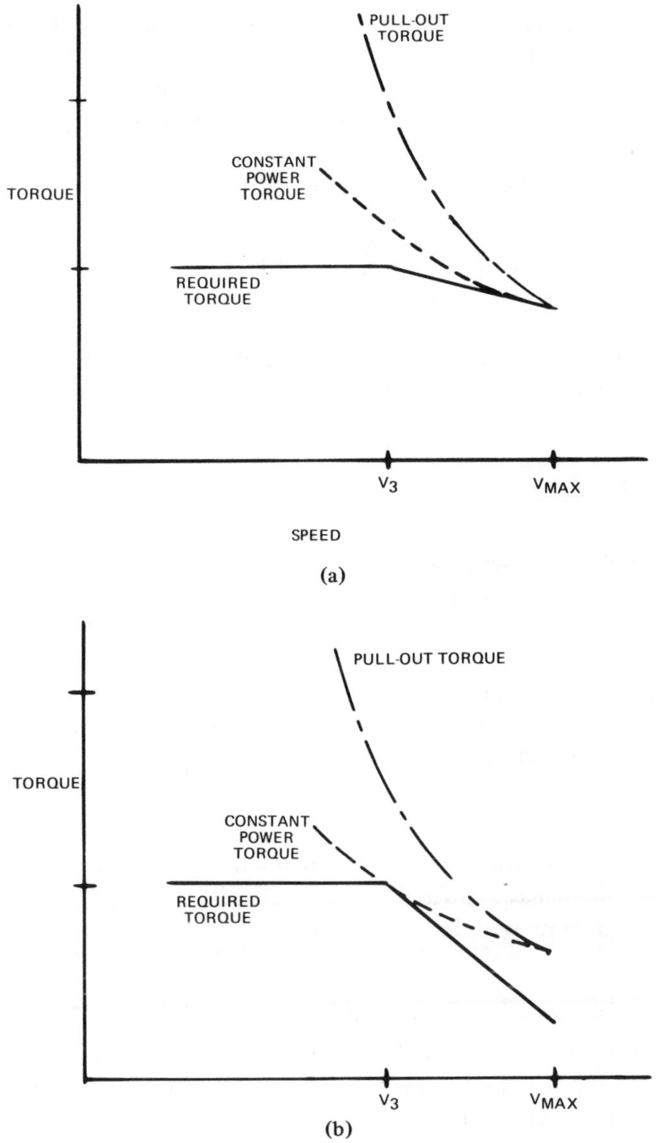

Fig. 12. (a) Brake taper less than constant power. (b) Brake taper greater than constant power.

$$V_3 = \left(\frac{E_{dc\,max}}{E_L}\right)^2 \frac{V_1 A_1}{A_3}, \tag{9}$$

Where A_3 is equal to the deceleration at V_3 (mi/h/s), neglecting the efficiency sign change between motoring and braking. The maximum inverter voltage ($E_{dc\,max}$) is limiting in braking. Thus the motor inductance requirement as determined by braking is

$$L_T \leqslant 2.94 \times 10^{-4} \left(\frac{(E_{dc\,max})^2}{WA_3 V_3 f_{max}}\right) \tag{10}$$

and A_3V_3 represents the maximum braking power regardless of brake taper.

For a given motor, speeds V_2 and V_3 are uniquely related; thus only one need be specified. In the case of a transit car. V_3 is almost always limiting. Equation (11) gives the relationship between V_2 and V_3 for any motor inductance,

$$V_3 = \left(\frac{1}{\eta c}\right)^2 \left(\frac{E_{dc\,max}}{E_L}\right)^2 \frac{A_1}{A_3} \frac{V_1 V_2}{V_{max}}. \tag{11}$$

To use the maximum inverter capability in braking, the motor must be designed so that

$$V_2 = \frac{V_{max}}{(\eta c)^2}. \tag{12}$$

With V_2 determined for maximum inverter utilization in braking, the motor pull-out torque ratio varies inversely with V_1.

The result is that the motor peak torque capability (approximately proportional to air gap area) must be increased over that for the minimum motoring requirement in order to supply the braking power.

The inverter has three basic rating factors other than maximum voltage. These are

1) commutation SCR peak current limit (to prevent mechanical fatigue),
2) commutation SCR thermal limit,
3) main SCR thermal limit.

Peak Commutation Current Rating

The inverter requires a peak commutation current at nominal line voltage of two times the motor peak current. This will allow full current operation down to a dc voltage of 83 percent of nominal line voltage.

The critical inverter section involved in peak currents is the commutation network. Any time the load current exceeds the critical level, a commutation failure will occur. The commutation SCR is the most critical component involved; thus the objective will be to calculate the maximum peak current for this device.

The peak commutation current is the sum of the fundamental and harmonic currents.

$$I_C = 2I_{ac}\left[\sqrt{2} + 1.76 \times 10^3 \frac{K_N V_2}{V_1}\right], \tag{13}$$

where K_N is a harmonic current factor depending on the type of PWM chosen. K_N is 2.05×10^{-4} for a chopping (commutation) frequency six times the fundamental and is 1.37×10^{-4} for nine times chopping frequency.

Commutation SCR Thermal Rating

The commutation SCR power dissipation is the product of the SCR energy dissipated per pulse (from the data sheet) and the chopping frequency. For the purposes of design, a nominal energy dissipated per pulse will be assumed at the approximate expected commutation current (I_c) and then will be scaled by the ratio of actual I_c to assumed I_c. This is valid over a limited range from 50 percent to 150 percent of the assumed value. If an assumed value of a commutation current on the high side of the value is used, a conservative design will result. The equation for commutation SCR dissipation is

$$P_{comm} = Kf_{chop}I_{comm}$$
$$= \frac{NK_1V_1{}^2WA_1f_{max}}{E_LV_{max}}\left[\sqrt{2} + 1.76 \times 10^3 \frac{K_NV_2}{V_1}\right], \tag{14}$$

where

$$K_1 = \frac{\text{Ws/pulse}}{\text{assumed } I_{comm}}.$$

K represents the commutation SCR current carrying capacity. The chopping frequency to fundamental frequency ratio N is assumed to apply at a speed of $0.8\,V_1$.

The value of I_{comm} is that required to meet the motoring current requirements. The peak SCR current is the expected I_{comm} at $E_{dc\,max}$ braking.

Main SCR Thermal Rating

The permitted main SCR dissipation is determined from the SCR data sheet using the average current (I_{ac}). The main SCR power dissipation will be

$$P_{main} = 87.3K_2 \frac{WA_1V_1}{E_L} \tag{15}$$

for 180° conduction. K_2 is the power dissipated per ampere at the average expected SCR current.

By using (13), (14), and (15), the inverter and power semiconductors can be sized to meet a given car performance. On the other hand, once an inverter design exists, the same equations are useful in applying it to different application requirements.

The following is an example of how a new application of an existing inverter can be explored. Assuming that maximum braking performance is limited by inverter voltage, a set of curves can be drawn which relate car weight and performance as limited by SCR capability. See Figs. 13 through 16.

Other assumptions are

V_{max}	80 mi/h,
f_{max}	200 Hz,
K_1	3×10^{-4},
K_2	1.2,
A_1	2.5 mi/h/s,
E_L	600 V,
V_2	$\frac{V_{max}}{1.44} = 55.6$ mi/h,
V_3	$2.78\,V_1$,
$I_{comm\,max}$	6667 A.

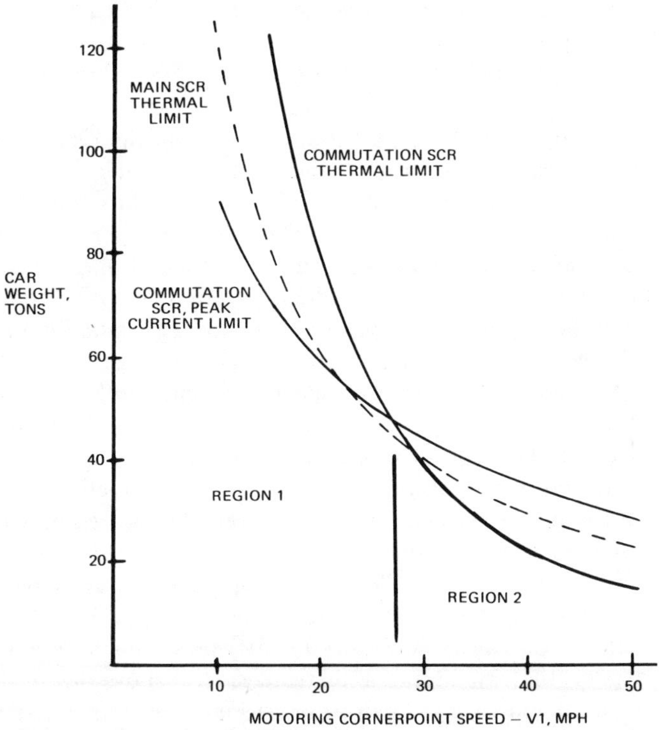

Fig. 13. SCR limits versus car performance.

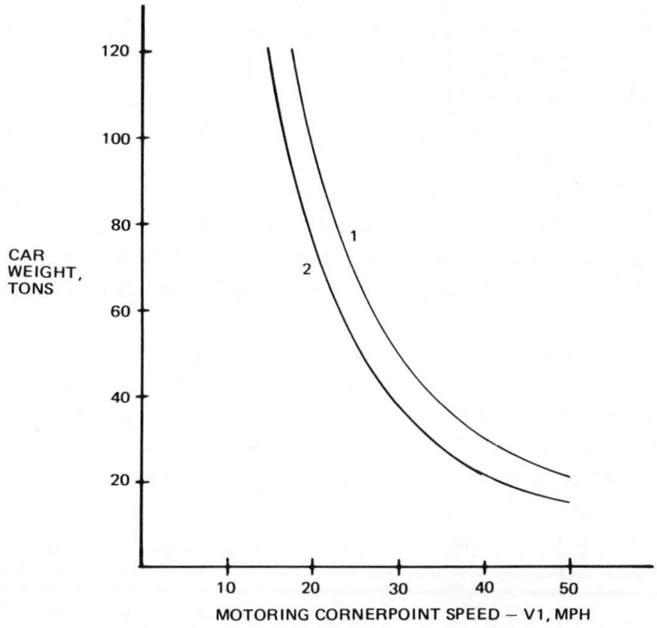

Fig. 15. Commutation SCR thermal limits.

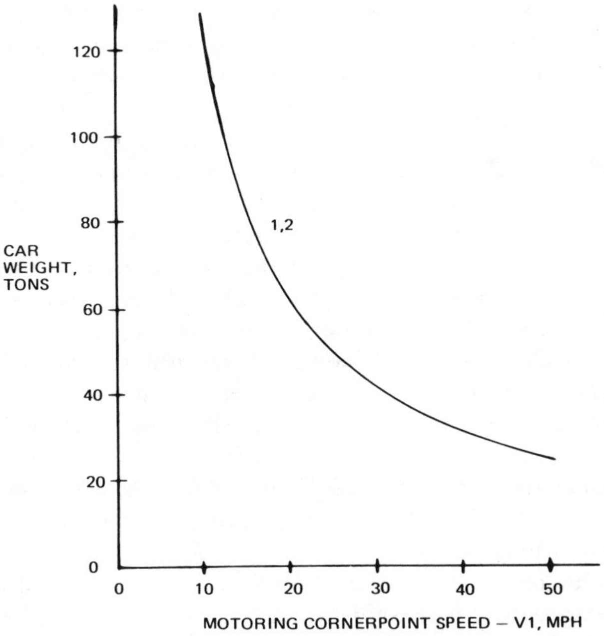

Fig. 14. Main SCR thermal limit.

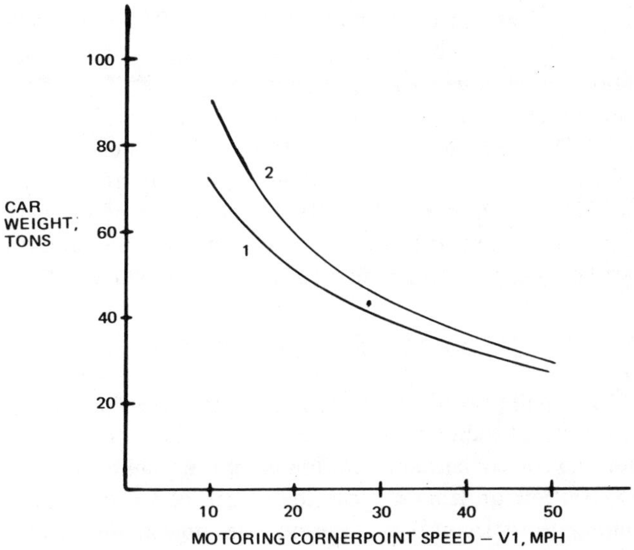

Fig. 16. Commutation SCR peak current limit.

The curves are plotted for a constant value of SCR dissipation (550 W) implying a constant inverter size. Fig. 13 shows the combined rating curves for

1) commutation SCR peak current,
2) commutation SCR thermal duty,
3) main SCR thermal duty.

The safe operating range is below all three curves. In Region 1, the performance limit is set by the comm SCR because the peak commutation current required above this value will cause too great a short-time temperature excursion inside the silicon wafer. In Region 2, the performance is limited by the comm SCR thermal dissipation rating. There are ways to relax these restrictions for a particular case by varying the $E_{\mathrm{dc\,max}}/E_L$ ratio, or the maximum motor frequency or the chopping ratio N.

Fig. 14 shows the main SCR thermal limit set by the average current. The only way to increase this limit is to improve device cooling, use larger SCR's, or parallel SCR's.

Fig. 15 shows the commutation SCR thermal limit curve for two different cases. Case 1 is for six times chopping; Case 2 is for nine times chopping. Comparing curves 1 and 2 shows that the reduced peak current with nine times chopping (2) does not compensate for the increased chopping required. Other methods of pulsewidth modulation may improve the comm SCR dissipation if it is limiting. Another way to reduce the comm SCR dissipation is to reduce the maximum motor frequency by either reducing the number of poles or reducing the gear ratio.

Fig. 16 shows the comm SCR maximum peak current which tends to vary directly with the peak load current. Thus any means to reduce peak load current, such as using a higher chopping frequency, will remove this limitation.

In conclusion, the power semiconductors are an important rating parameter in meeting car performance. The designer has the option of changing operating modes and selection of larger devices to extend the inverter rating. The motor size is determined by the continuous rated torque and the torque ratio. The torque ratio is proportional to V_2/V_1, and the continuous rated torque is inversely proportional to the gear ratio. Thus reducing the maximum inverter frequency strongly increases motor size while varying V_1 weakly affects motor size. The design method allows determination of car performance and shows the direction that changes to the propulsion system must go to meet the desired car performance.

REGENERATION FACTORS

The pulsewidth modulated inverter has the desirable capability of bidirectional power flow. Without any reconnection, power can be caused to flow from third rail to the car for acceleration or from the car to the source for deceleration. During electric braking, the reverse power flow can be returned to the third rail so that a large portion of the kinetic energy of decelerating cars can be returned to the distribution system. The amount of regeneration power savings which are achievable from a fleet of ac propulsion cars requires extensive system analysis and is dependent upon a number of factors, such as the following:

1) motor-gear-inverter combined efficiency including losses in line matching resistors,
2) third rail and running rail resistance,
3) maximum permissable voltage at the third rail during regeneration,
4) proximity of motoring trains and other loads to the regenerating train,
5) maximum inverter and motor voltage during braking.

Due to the practical limitations of third rail and running rail resistances, a voltage rise is required at the regenerating train to drive current through the rail impedance to the point of use by an accelerating train or other load. The more current being regenerated and the longer the distance to the load, the greater the required voltage rise. In the application of a regenerative propulsion system to a transit system, the maximum third rail voltage must be determined by analysis and testing.

Factors 1 and 5 also provide the designer with trade-offs which affect the regeneration savings obtainable plus the size, weight, and cost of the propulsion system.

An inverter which delivers ac power to an induction motor during motoring and receives power during regenerative braking has a current commutating capability which varies with the input voltage. During the regenerative braking mode of operation, the input voltage is increased by inserting a resistance in series between the dc power source and the inverter, thereby increasing the voltage level above that of the dc power source. The resulting increased current commutation capacity and increased voltage provides for greater power generation and braking efforts without increasing the size of the inverter or motor. During motoring, the resistance is removed to provide for the free flow of current to the inverter.

In practice, the greatest power flow requirements occur in braking and hence, this mode sizes the system. By allowing the inverter voltage to rise perhaps 55 percent above the line voltage during braking, the inverter becomes properly sized for both the motoring and braking requirements with minimal penalty in weight, size, and cost above motoring requirements. The extra voltage capability is required in any case to safely withstand the high line voltage transients that occur. With such a system, however, some of the power is dissipated in the line matching series resistor during regeneration.

An alternate system could be designed which did not employ a matching resistor between the inverter and line. Braking power would then be provided at line voltage with 55 percent more current. This system would require larger inverter commutation circuits and larger power semiconductors, thereby substantially adding to cost, size, and weight of car-borne equipment.

The bar graph of Fig. 17 shows a comparison of two inverter designs with respect to regenerative capability. Inverter (1) is designed to meet the motoring requirements, but does not have any additional capacity to handle braking.. This design has a maximum braking current of 1.0 per unit which limits the braking power such that over 58 percent of the braking energy must be absorbed by friction brakes.

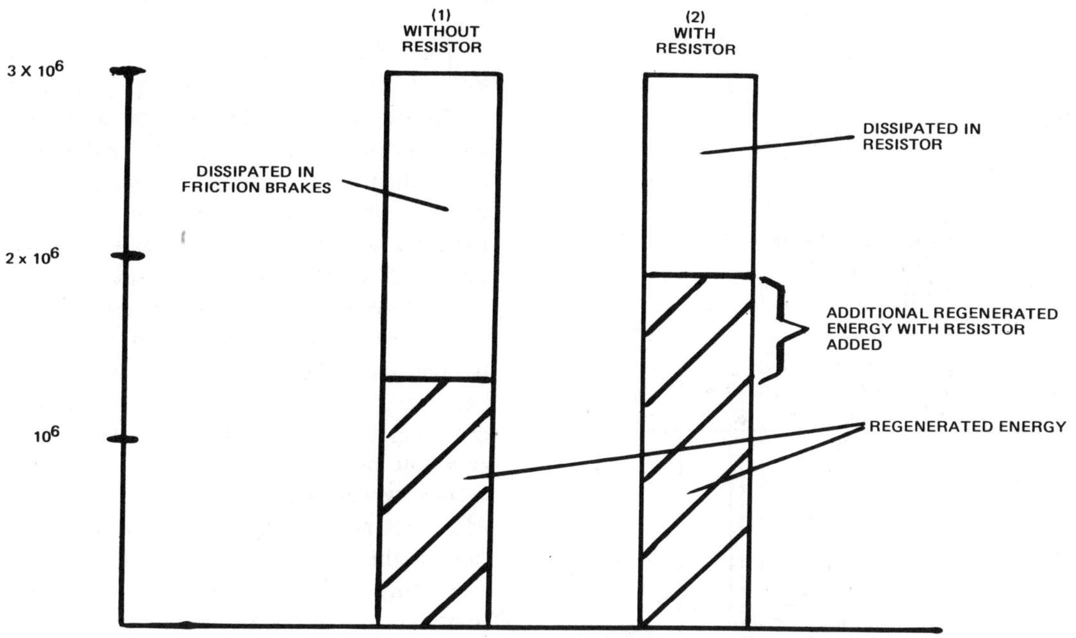

Fig. 17. Regeneration comparison.

Inverter (2) uses the series resistors which allows the inverter voltage and power rating to be increased to 155 percent and 240 percent respectively without appreciable change in component sizes over inverter (1). The result is that regenerated current is 30 percent greater, and the energy which is not returned to the line is dissipated in the series resistor. Friction brake wear is thus reduced to a very low value.

An additional consideration in the use of inverter regenerative braking is that of brake blending under conditions of less than 100 percent line receptivity. If there are insufficient motoring cars near enough to absorb the regeneration currents without exceeding the voltage limits, it is necessary to provide additional braking effort either in friction or dynamic braking. Blending friction brakes with regenerative brakes has the advantage of eliminating the weight and space requirements of mounting dynamic braking resistors on the car. However, the additional friction brake wear which results may not be acceptable to the operator. Blending regenerative brakes with dynamic brakes requires that resistors be provided and that some blending means be provided to proportion the current between the line and the braking resistors. One blending means is by stepping resistors in stages as line receptivity conditions change.

It should be noted that dynamic braking resistors can be avoided on the cars by use of regenerative braking with receptive substations. Substations can be made receptive either by use of controlled thyristor inverting rectifiers, or by the use of wayside resistors. In the former case, the inverting action of the substation transmits the power back into the ac power system. In the latter case, resistors are automatically inserted across the line in response to increasing line voltage to absorb the excess regenerated current which cannot be used by accelerating cars. This latter method is suitable with diode rectifier substations, and essentially amounts to a blending of regenerative and dynamic braking where the resistors are located on the wayside instead of on the car.

CONCLUSION

The power semiconductors and auxiliary components required to build a propulsion inverter are now available. The simple, reliable induction motor can be controlled to provide the needed accelerating and decelerating torques to meet car performance requirements. It is therefore practical to provide inverter-induction motor drives for transit cars. These drives are expected to exhibit a high degree of power savings over dc motor resistance controlled drives, primarily as a result of regeneration. They also promise to yield much lower motor maintenance as an added benefit to the user.

APPENDIX A

Effects of Unequal Wheel Diameters

The most economical and reliable ac inverter drive for transit cars is one which uses one inverter per car, driving all induction motors in parallel. Existing transit properties are generally in the practive of maintaining relatively close tolerance in wheel diameters between axles on a truck (i.e., less than 0.5 in difference). Many properties do not presently control wheel diameters between trucks on a car. On several properties, measurements at random have shown that perhaps 20 percent of the cars have maximum differences in the 1.5 to 3.0 in range. These are probably the result of truck replacement which normally is performed without regard for wheel matching.

The performance effect of unequal wheel diameters can be best understood by considering the torque versus slip curve of Fig. 18. The curve shows that when the stator frequency applied to a motor exceeds the rotor rotational speed a positive slip occurs which produces positive or motoring torque. When the applied stator frequency is reduced below the rotational speed, negative slip occurs and braking or negative torque is produced.

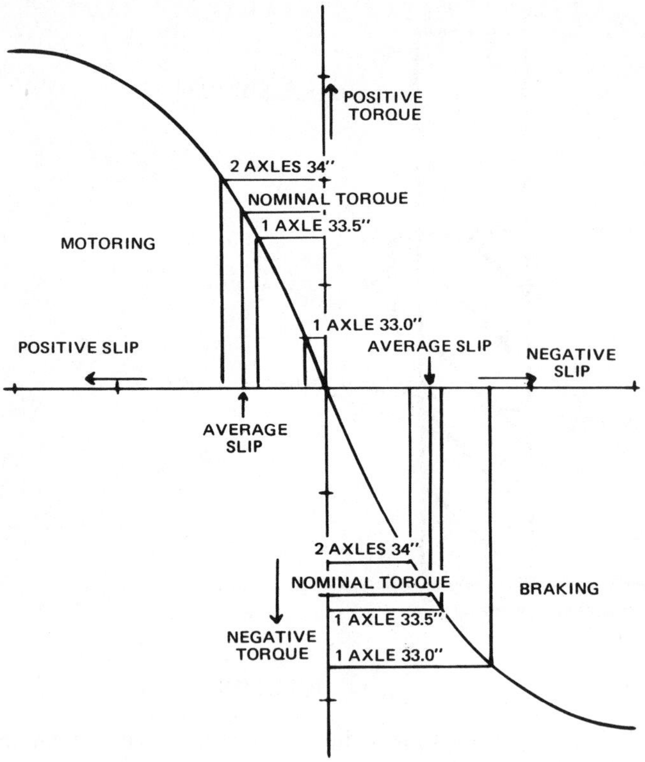

Fig. 18. Effects of unequal wheel diameters.

The example of unequal wheel diameters shown is the worst case for one in difference on a total car with one-half in maximum difference on a given truck. The two axles with 34 in wheels draw more than nominal current and supply more than nominal torque in motoring. The motors driving the smaller wheels, particularly the 33 in wheels, are supplying less motoring torque than nominal. In braking, the situation reverses and the smaller wheel motors supply more torque and current. The motor heating effect of unequal diameters is minimized where motoring and braking periods are somewhat balanced. Some additional thermal capacity is generally required in the motor, however.

The effect of varying induction motor rotor resistance is to shift the slip of the torque-slip curve of the motor. A low resistance rotor (low slip) provides a high efficiency motor but accentuates the wheel diameter problem. A high resistance rotor (high slip) reduces the sensitivity of the motors to unequal wheel diameters but increases motor losses. There is therefore a trade-off between motor weight, power consumption, and wheel diameter tolerance which must be made for overall lowest system operating cost.

A further factor in limiting wheel diameter differences is that the adhesion limit of a mismatched wheel may be exceeded. If this occurs in motoring to an extra large wheel, the resultant slippage will tend to be self-correcting. If the adhesion limits are exceeded in braking on a smaller than average wheel (a more likely occurrence in transit cars), the wheel to rail slippage will tend to make the small wheel smaller, thus aggravating the situation.

Practical efficiency considerations limit the rotor resistance to an upper limit. Calculations indicate that wheel diameters will need to be controlled to within 0.75 to 1.25 in on a total car in order to provide expected performance. A testing program is required to determine the final point at which adhesion considerations put a practical upper limit on this tolerance. The true added cost to maintain wheels within closer tolerances will vary with maintenance practices. This added cost, if required, will be small compared to the benefits of reduced motor maintenance and power savings offered by the ac propulsion system.

REFERENCES

[1] B. D. Bedford and R. G. Hoft, *Principles of Inverter Circuits.* New York: Wiley, 1964.

[2] K. P. Phillips, "Current source converter for AC motor drives," Conf. Proc. IAS Meeting, 1971, P. 385. (Also Transactions)

[3] G. B. Kliman, "Harmonic effects in pulse width modulated inverter induction motor drives," Conf. Proc. IAS Meeting, 1972, P. 783.

[4] G. B. Kliman and A. B. Plunkett, "Measurement of harmonic effects in PWM inverter drives," Conf. Proc. IAS Meeting, 1973, P. 881.

Some Guidelines for the Application of Adjustable-Speed AC Drives

JERRY J. POLLACK, MEMBER, IEEE

Abstract—**Constant-torque adjustable-speed phase-controlled SCR dc drives are compared with inverter drives from the standpoint of technical similarities and differences, application limitations, and relative costs. Various inverter types are discussed and compared. Finally, guidelines as to the factors influencing the ac versus dc decision are summarized and tabulated.**

INTRODUCTION

USING the rugged polyphase ac motor traditionally considered a source of constant-speed power, today's static adjustable-speed ac drive systems have emerges as an excellent solution to many industrial drive applications. When compared to its more familiar dc counterpart, the adjustable-speed ac drive often offers some unique advantages, both economically and from the standpoint of performance.

Adjustable-speed alternating current, however, does not offer the optimum in drive performance or economy on a universal basis. Depending on the application and its requirements, adjustable-speed alternating current may or may not be a sound decision when compared to direct current.

With a basic understanding of its operation it will be possible to compare the adjustable-speed ac drive with the dc drive and establish a few guidelines for its sensible application.

DRIVE SYSTEM REQUIREMENTS

The thyristor, introduced in the 1950's, provides an economical solid-state power device allowing efficient high-speed power-circuit switching with a minimum of control current. When used in a simple phase-controlled rectifier, as shown in Fig. 1, the thyristor allows direct conversion of plant ac to adjustable-voltage dc power. It involves pulsing the thyristor "on" by means of a low energy pulse at its control terminal or gate, during the positive half-cycle of the sine wave. The characteristics of the thyristor are such that it will remain in conduction until the polarity reverses or the sine wave goes to the negative half-cycle. Therefore, selecting the time during the positive half-cycle to activate the thyristor will determine a specific power pulse duration. It should be apparent that in varying the duration of the power pulse, the average dc voltage output from the rectifier may be controlled. A typical three-phase controlled rectifier is shown in Fig. 2.

Paper TOD-73-16, approved by the Rubber and Plastics Industry Committee of the IEEE Industry Applications Society for presentation at the 24th Annual Conference of Electrical Engineering Problems in the Rubber and Plastics Industries, Akron, Ohio, April 10–11, 1972. Manuscript released for publication February 10, 1973.

The author is with the Industrial Systems Division, Reliance Electric Company, Cleveland, Ohio.

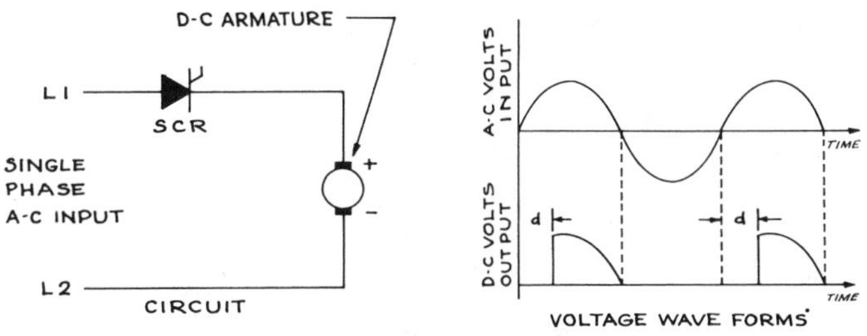

Fig. 1. Simple phase-controlled rectifier.

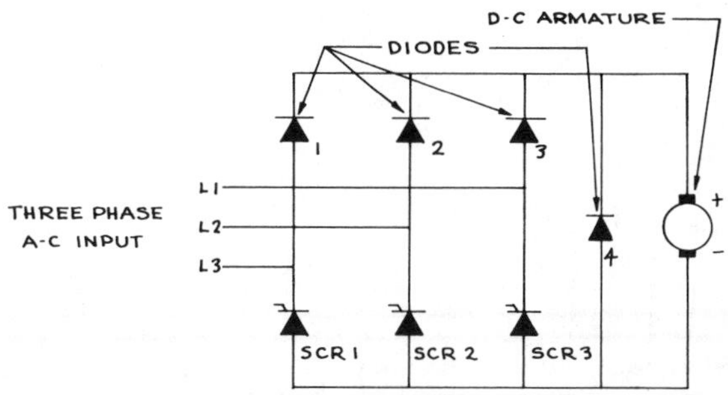

Fig. 2. Three-phase controlled rectifier.

If a shunt-wound dc motor is supplied with fixed-voltage excitation to its shunt field and adjustable dc voltage for armature power, its shaft speed may be controlled by adjusting the value of applied armature voltage.

The relationship between shaft speed and applied armature terminal voltage may be expressed as

$$\text{r/min} = \frac{V_t - IR}{K}$$

where V_t is the value of applied armature voltage, K is a constant of proportionality, and IR represents a voltage drop due to resistive losses in the armature circuit. Since armature current in a dc machine is proportional to applied torque load, the IR component represents motor slip or regulation due to loading. A typical shunt-wound dc machine will exhibit 8–15 percent of base speed slip from no load to full load.

In comparison, the speed of a squirrel-cage induction ac motor may be expressed as

$$\text{r/min} = \frac{120f}{p} - \text{slip}$$

where f represents the power supply frequency in hertz, p indicates the number of motor poles, and

$$\text{slip} = (\text{r/min at no load}) - (\text{r/min at running load}).$$

Reprinted from *IEEE Trans. Ind. Appl.*, vol. IA-9, pp. 704–710, Nov./Dec. 1973.

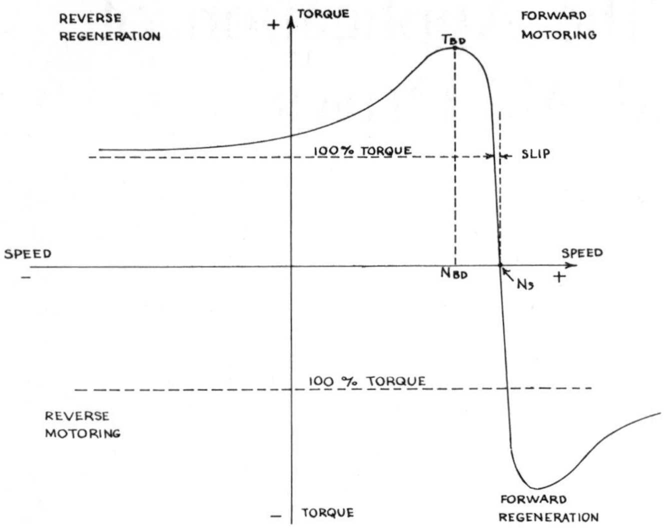

Fig 3. AC induction motor speed versus torque.

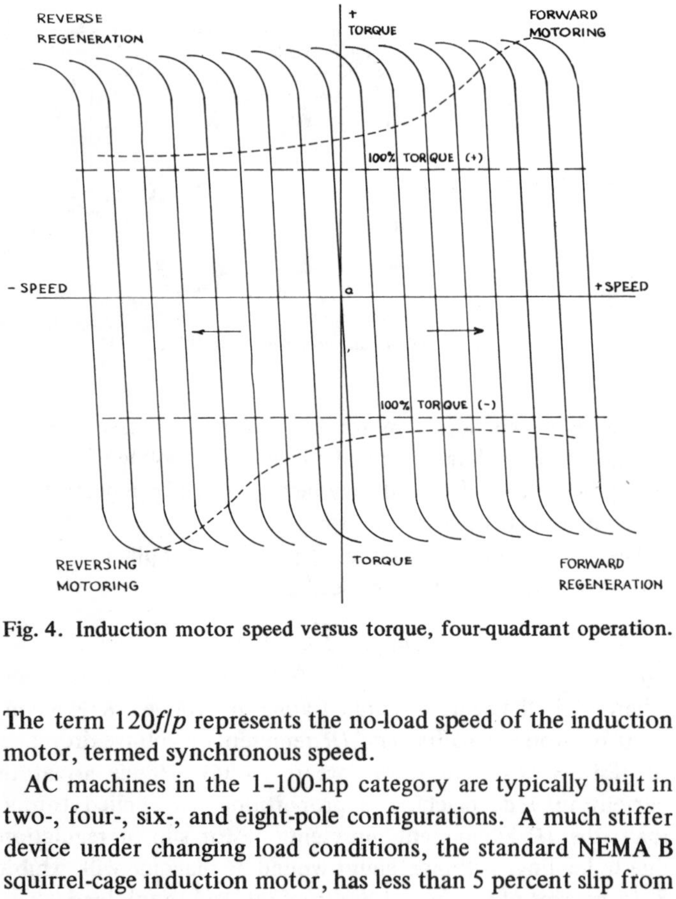

Fig. 4. Induction motor speed versus torque, four-quadrant operation.

The term $120f/p$ represents the no-load speed of the induction motor, termed synchronous speed.

AC machines in the 1-100-hp category are typically built in two-, four-, six-, and eight-pole configurations. A much stiffer device under changing load conditions, the standard NEMA B squirrel-cage induction motor, has less than 5 percent slip from no load to full load.

The speed-torque curve of Fig. 3 shows a typical ac induction motor characteristic. With the curve extending into both the motoring and regenerative quadrants of operation, it is important to recognize that the direction of slip determines the direction of power flow. With the speed of the rotor less than synchronous speed N_s, the flow of power is from the motor to the load. The rotor speed could also be greater than synchronous speed if the load were overhauling. This would result in regeneration and the flow of power from the load to

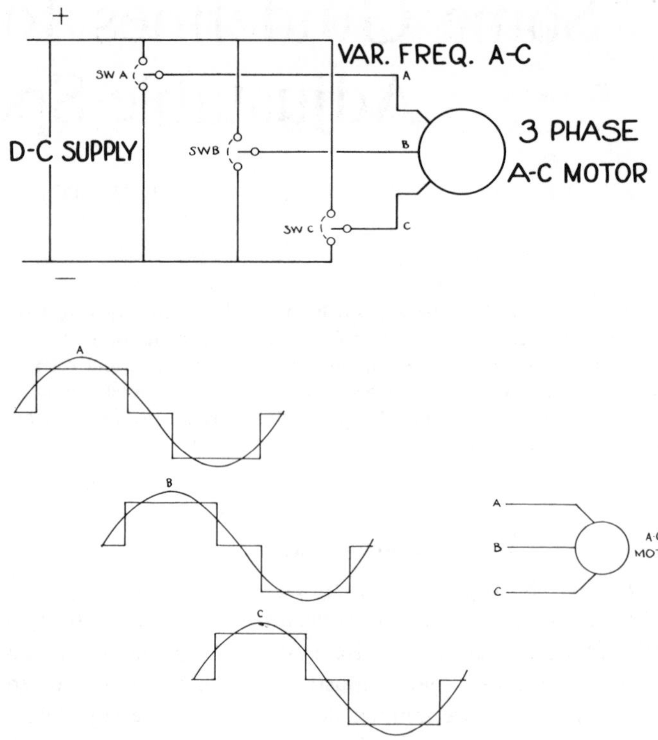

Fig. 5. Basic inverter operation.

the motor. Note also that slip is indicative of motor regulation under load. For a given load, motor slip will remain constant, although its synchronous speed ($120f/p$) may be changed.

It is evident that the synchronous speed of the ac motor can be altered either by varying the frequency of the power source, or by changing the number of poles in the motor. In the adjustable-speed ac system the frequency of the power supply is varied to provide a smooth continuous control of motor speed.

Only the speed of the ac motor should change as the frequency of the power supply is varied. The other motor capabilities should remain the same. An adjustable-speed ac drive should have a family of characteristic curves as shown in Fig. 4. For constant-torque drives it is particularly important that the torque characteristic of the motor remain the same, regardless of how the synchronous speed of the motor may change.

AC motor torque is produced by interaction of rotor and stator flux. This flux is created by current flow in both the stator and rotor. In order to maintain a constant torque, current flow should be constant. Since motor impedance decreases with applied frequency due to a reduction of inductive reactance, the applied voltage should vary in a direct relationship with the applied frequency, or

$$\text{volts/hertz} = \text{constant}.$$

The preceding discussion of the needs of an ac motor for adjustable-speed constant-torque operation has pointed out two basic requirements for the adjustable-speed ac power unit: 1) adjustable frequency output to provide control of motor speed, and 2) adjustable controlled output voltage to maintain constant motor torque over the speed range.

The past decade has seen the development of many techniques suitable for adjustable speed control of ac motors. However, they are built on a common base identified as the rectifier–inverter system. This system first converts fixed-frequency plant ac power to either fixed- or variable-voltage dc power which is subsequently chopped up or "inverted" to adjustable-frequency ac power.

A typical inverter chops the dc input voltage into a three-phase square-wave output. Fig. 5 shows a functional means of accomplishing the ac conversion from the dc power source. The fundamental action is depicted by means of switches which cause the load terminals to be alternately switched from the positive to the negative dc bus in a programmed sequence. The resulting square waves are shown superimposed on the normal ac motor sine wave input.

The switching action is actually accomplished with thyristors in the practical inverter circuit. It may be seen that by properly timing the switching of the inverter thyristors, the required three-phase excitation can be synthesized to produce a rotating magnetic field for polyphase motor operation. By controlling the rate at which this switching takes place, the inverter output frequency may be adjusted.

Inverter Types

There are numerous techniques employed to effect adjustable inverter output voltage. Since the form of voltage control used may well limit the speed of response of the inverter system or determine its effective usable speed range, an understanding of voltage control fundamentals is essential.

In general, output voltage is controlled by adjusting the amplitude of the inverter square-wave output or by controlling its duty cycle or pulse width. The inverter systems can be designated as

1) variable-input transformer (VIT);
2) variable-output transformer (VOT);
3) variable-voltage input (VVI) (phase-controlled rectifier or chopper);
4) pulse-width (phase shift);
5) pulse-width modulated (PWM).

Comparative block diagrams of these inverter types are shown in Fig. 6.

The variable-input transformer (VIT) inverter operation is based on accepting the ac power line voltage and transforming it by a controllable ratio as a function of output frequency. The transformer output voltage is subsequently rectified and then changed by the inverter to provide an adjustable-voltage adjustable-frequency output. Note that two controlled power conversion stages are required (transformer and inverter) and that the value of dc voltage present on the input side of the inverter section is variable as a function of frequency.

The variable-output transformer (VOT) inverter operates from the incoming ac supply through a fixed diode rectifier and the inverter. The output of the inverter is constant-amplitude adjustable-frequency power. The variable transformer adjusts the amplitude with output frequency to maintain proper voltage-to-frequency ratio. The output transformer, having an adjustable frequency and a constant voltage input,

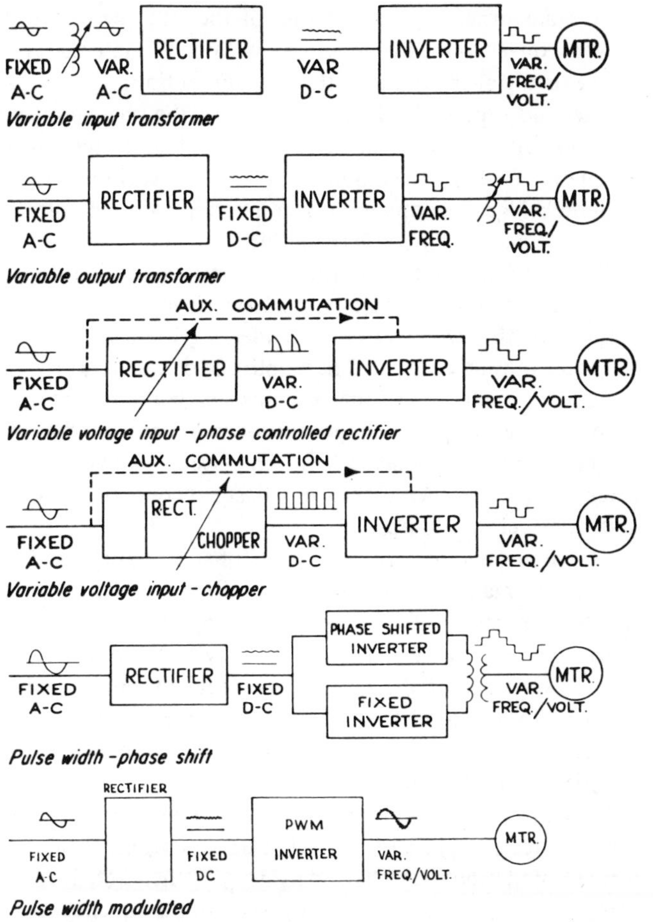

Fig. 6. Inverter system classification by type of voltage control employed.

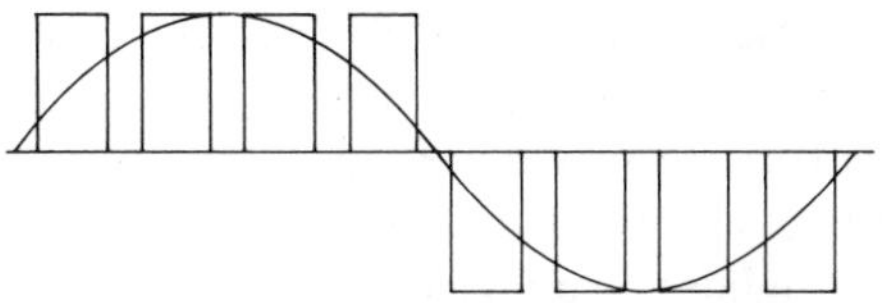

Fig. 7. Typical single-phase PWM wave shape.

must be sized to operate at rated load and minimum frequency. Consequently, the transformer will be large compared to the VIT. Again note that two controlled power conversion stages are used, but that the value of direct current at the inverter input in this design is constant, regardless of output frequency.

The VIT and VOT inverters are not completely static drives in the strictest sense as they include a moving part (variable transformer). Since the speed of response of these systems is limited by the transformer adjustment speed, the VIT and VOT inverter systems lend themselves only to relatively slowly changing drive applications.

The variable-voltage input (VVI) is an all solid-state system and is capable of good dynamic response. The most common type of VVI consists of a phase-controlled thyristor bridge which rectifies the incoming ac voltage to an adjustable level of direct current to maintain the desired voltage-to-frequency ratio. VVI inverter operation parallels that of the VIT inverter with the phase-controlled rectifier replacing the VITs input transformer. As in the case of the VIT, two controlled power conversions are required, with a variable-voltage dc level existing at the input of the inverter section.

A second type of VVI inverter replaces the phase-controlled rectifier with a fixed diode rectifier and a dc regulator or chopper. This system requires three power sections in series, a diode bridge, a chopper, and an inverter. A fixed-voltage dc bus at the chopper input is the result.

The pulse width (phase shift) inverter is an all solid-state system and involves a minimum of two inverters. The output voltages of the two inverters are added algebraically to provide the total output voltage. The amplitude of the output voltage is varied by phase shifting one inverter with respect to the other. The output voltage approaches zero as the phase displacement approaches 180°. The need for two complete inverter power stages makes this design economically unsuitable at or below 100 hp.

The circuit complexity of all types of static variable-speed ac drives is primarily in the firing logic of the inverter stage. It is feasible, therefore, to expand this firing circuitry to permit both frequency and voltage control by the inverter section. The power stage of the inverter is effectively unchanged, and the need for a second controlled power stage to control output voltage (variable transformer, chopper, or phase-controlled rectifier) is eliminated.

Thus in the pulse-width modulated (PWM) inverter, plant power is rectified to a fixed level of direct current by means of a fixed diode rectifier. This dc supply is then acted upon by the inverter section which provides both frequency control and voltage control with the same set of power thyristors. The output of the PWM inverter, as shown in Fig. 7, consists of a series of adjustable-width pulses in place of the conventional square-wave output. By means of rapidly switching the thyristors in the inverter section under regulator control, the pulse-to-notch ratio may be varied, thus controlling the effective output voltage of the inverter.

The most modern approach to static adjustable-frequency power conversion, the PWM inverter offers a list of features unobtainable with any one of the previously offered inverter types.

1) It is a completely solid-state device, offering fast accurate response and efficient power conversion.

2) Only one controlled power stage is required, eliminating the need for phase-controlled rectifiers, choppers, or transformers and their associated regulators.

3) As opposed to the phase-controlled VVI inverter, the PWM with its fixed diode rectifier taking power smoothly from the plant supply, operates at a 96 percent power factor, regardless of frequency.

4) The PWM inverter operates from a fixed-voltage dc supply, allowing multi-inverter operation from a common dc bus. Since the inverter bridge is itself inherently capable of regenerating power back to the dc bus, the bus acts as a common energy-source sink for multi-inverter systems. Motoring inverters will draw power from the bus while inverters whose motors are holding back or braking an overhauling load will pump power to the bus. The net power difference, i.e., work load plus machine losses, is drawn from plant power. (Without the ability to be operated from a common bus, any inverter with an overhauling load would require a special regenerative rectifier section to allow pumping power back to the plant line.)

The fixed-voltage dc bus also lends itself to support by means of energy storing capacitors or batteries to provide unaffected inverter operation during plant power outages.

5) The modulated wave shape greatly improves low-frequency motor operation. Using existing modulation techniques to generate low-frequency voltage waveshapes more closely resembling a sine wave, induction and synchronous reluctance motors may be operated at very low frequencies without the cogging and excessive heating inherent with conventional square-wave inverters.

System Comparison

The ac motor has long been accepted as a dependable workhorse for constant-speed applications. Actually, the ac motor, with the proper power supply and minor design considerations, is also highly suited for controlled adjustable-speed drive service.

The squirrel-cage induction machine, as described and modeled in a preceding section, offers some distinct electrical and mechanical advantages. Electrically, the motor has no slip rings or brushes requiring electrical maintenance, and needs no additional source of excitation. Both the rotating and stationary magnetic fields are supported from one set of three-power leads. Mechanically, the motor offers a rugged cast rotor capable of mechanical integrity to very high speed. Coupled with the absence of the mechanical commutator present in dc machines, this allows the ac motor to achieve far greater maximum speeds than it is possible with dc machinery.

The induction motor, when compared with direct current, is far less expensive, smaller, weighs less, and has a lower rotor inertia for better dynamic response.

The myriad of external mechanical modifications on fixed-frequency ac motors, such as special shafts, enclosures, etc., are available at moderate cost on adjustable-frequency machines as well.

When compared with the dc motor, the ac motor becomes even more advantageous from an economic standpoint when asked to perform in adverse or explosive environments. Economics favor the ac motor over the dc equivalent as base speed is reduced, as well.

As previously noted, the slip or softness of the standard NEMA B design induction motor is under 5 percent from no load to full load. When compared to the 8–15 percent speed droop exhibited by a typical dc machine under an increasing load, it can be seen that the ac induction motor is much more adaptable to open-loop control schemes. In applications where increased load sensitivity is desired, as when driving helper rolls or driven idlers, NEMA D designs with 5–8 percent or 8–13 percent slip may be specified.

Significantly, the use of the synchronous reluctance motor is possible with an adjustable-frequency ac system, offering drive possibilities not paralleled by dc equipment. The synchronous reluctance machine, by virtue of a magnetically asymmetrical rotor configuration, runs at exactly synchronous speed ($120f/p$) exhibiting zero slip from no load to well over full load. Thus shaft speed is uniquely and exactly determined by applied power source frequency.

This motor design must not be confused with old-style dc excited synchronous motors (slip-ring type) used for power factor correction. The synchronous reluctance machine, from all outward appearances, is identical to its induction counterpart, and is available with all induction motor modifications from fractional horsepowers to over 100 hp at only a slight premium in price.

The adjustable-speed ac power conversion unit and regulator, on the other hand, is a more costly and more complicated device than that required to power and control a dc motor. A number of reasons exist for this difference.

The power unit required to power and control an adjustable-speed dc drive system is a single-stage device. By using a phase-controlled rectifier described earlier, adjustable-voltage dc power may be directly converted from single- or three-phase plant alternating current. This conversion may be accomplished with as few as one thyristor and one diode for fractional-horsepower single-phase powered drives, with a conventional three-phase powered drive (5–150 hp) using three thyristors and four diodes.

The thyristors are turned on by gate pulses from the regulator whose only function is to adjust the conduction angle of the thyristors. The thyristor, once thus turned on, will conduct until automatically turned off (commutated) by the reversal of the ac plant line.

The ac power conversion unit, however, is inherently a two-stage device. Plant alternating current must be converted to direct current by some rectifying means, and then inverted by a second power stage to controlled-frequency alternating current.

The thyristors in the inverter stage of the power unit, moreover, are supplied by either fixed or variable dc voltage. Once turned on, there is no automatic commutation as with the dc drive. The conducting thyristor must be forced off, or commutated, by an auxiliary circuit containing still more thyristor devices. A typical VVI type inverter using a phase-controlled rectifier input requires three thyristors and three power diodes in the rectifier section with an additional twelve thyristors and six power diodes in the inverter, for a total of fifteen thyristors and nine power diodes. An equivalent PWM inverter would not require the thyristors in the rectifier, yielding a total of twelve thyristors and twelve power diodes.

Along with having to control a greater number of thyristor devices, the regulator for an adjustable-speed ac drive has more functions to perform than its dc counterpart. The regulator must control the firing rate to obtain adjustable frequency, the firing sequence to maintain phase coordination, the output voltage to produce the required voltage-to-frequency ratio, and the commutation.

The adjustable-speed ac power unit does, however, possess advantages not present in the typical dc drive power unit against which it was compared above. It is inherently a regenerative device capable of power flow both into and out of the drive motor. In cases where the process load is overhauling, energy removed from the process will be transferred to the dc bus where it may be burned up in a resistive load placed across the bus itself (snubbing), returned to the plant line by means of a regenerative type rectifier, or used by other inverters sharing the bus.

Contactorless reversing ability is also inherent in the static ac power unit. Since direction of rotation of a polyphase ac motor is controlled by the phase sequence of applied power, reversing can take place by modifying the order in which inverter thyristors are fired. This function is controlled at milliwatt power levels in the firing logic, without the need for power contactors. Typically, inverter frequency is decreased in a linear fashion to near zero, the phase sequence is flipped, and the inverter frequency ramped back upward, allowing smooth controlled deceleration and reacceleration in the reverse direction.

Both regenerative and contactorless reversing capabilities may be had in dc power conversion units, but at extra cost with additional complexity.

In the final analysis, the development of adjustable-speed ac and dc drives has advanced along the same lines. There are instances, to be sure, where the ac system affords the only electrical drive solution; i.e., where machine limitations do not allow the larger heavier dc motor; where inaccessibility of the drive motors or adverse environments rule out the use of brushes and commutators; and where high speeds are beyond the limits of commutation for the dc system.

But in the majority of situations, both drive types will seem applicable. The ac drive system can parallel the dc drive in nearly all functions. It is available with jog, thread, reversing, braking, and virtually every magnetic function available on dc drives. It can be operated over a wide speed range and is offered with a variety of regulator types.

In these instances, drive system cost is normally the deciding factor. It is in this area that the static ac system is perhaps the most misunderstood and misapplied.

Comparative Economics

As a first example, let us consider a single-motor 50-hp drive. The motor is a 1750-r/min base speed machine, foot mounted, and dripproof. It must operate over a wide speed range in response to the setting of an operator's potentiometer with speed regulation of no more than 5 percent of base speed. This system is depicted in Fig. 8 in both the dc and ac configuration. The bar graph to the right of the system sketch illustrates relative cost. Note that the ac system is operated open loop, without the need for tachometer feedback, through the use of a NEMA B design motor. The dc system requires a tachometer for speed feedback.

As illustrated by the cost comparison, however, although the ac system lacks the tachometer feedback requirement and uses a motor much less costly than the dc system, the ac drive becomes more expensive as a system due to the power unit dollars involved.

If the application required the use of an explosionproof machine, however, and if base speed were specified as 850 r/min, Fig. 9 would properly depict the comparative costs. Although the system concept remained exactly as in the

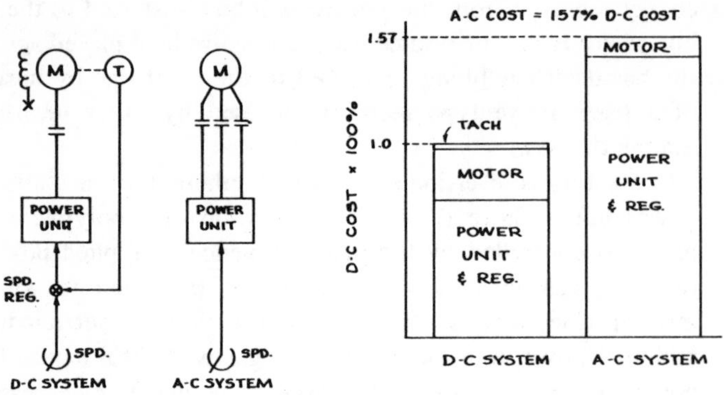

Fig. 8. Economic comparison of single-motor 50-hp 1750-r/min dripproof drive.

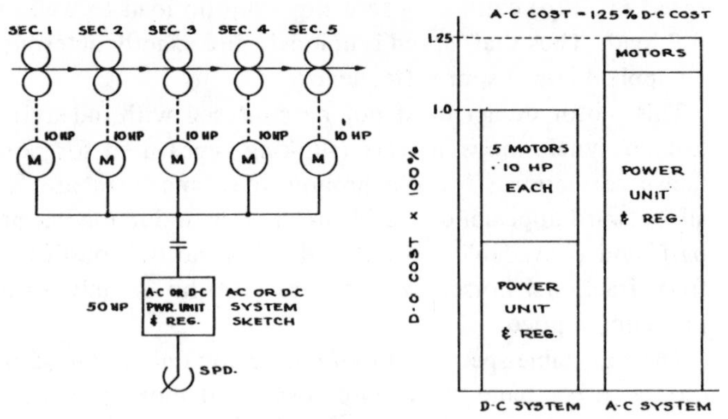

Fig. 11. Five-section machine with 4:1 speed range and 10-percent tracking.

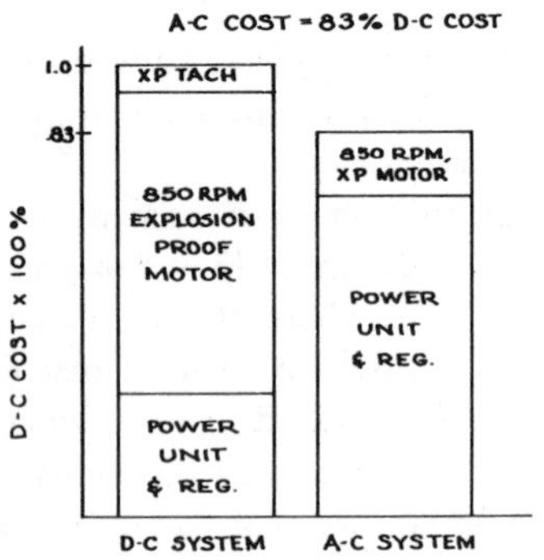

Fig. 9. Requirement of low-base-speed explosionproof machine.

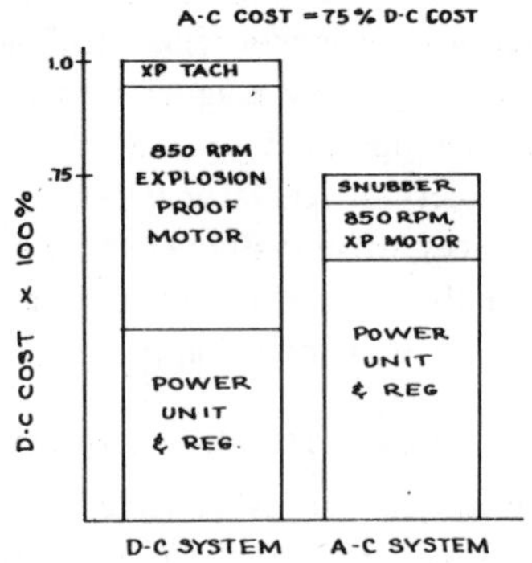

Fig. 10. Additional requirement of frequent slowdowns and reversals.

previous example, the additional cost of the low-base-speed explosionproof dc motor made the ac adjustable speed system more economical. It should be noted that either of the two additional motor requirements, explosionproof or low base speed, alone would have shown the same result.

Considering this same low-base-speed single-motor drive system operating in an explosive environment, and adding the need for frequent controlled slowdown and reversal, adjustable-speed ac drive economics become even more appealing (see Fig. 10). The addition of a static snubber to dissipate regenerative energy delivered to the dc bus by the inverter section during slowdown is the only required addition to the ac drive. The inverter itself is regenerative, and contactorless reversal is done in the regulator at an insignificant price addition. The dc drive system, however, required the use of a regenerative power conversion unit in place of the single-quadrant motor-forward-only type used in the preceding examples.

In order to compare adjustable-speed ac and dc drives as applied to multisectional industrial machinery, let us consider a five-section processing line powered by five separate motors. Each motor will be 10 hp at 1750-r/min base speed with a dripproof enclosure. The line will operate over a 4:1 speed range with required sectional tracking of 10 percent. No relative speed adjustment between adjacent sections is required.

As shown in Fig. 11, the drive system using ac or dc equipment would appear identical. A common power conversion unit feeding all five motors would be satisfactory for either scheme. The accompanying bar graph proves the dc approach to be more economical.

Increasing the speed range requirement and tightening the tracking requirements, however, yields a quite different result (see Fig. 12). Each of the dc motors requires speed regulation to meet the tracking specification, using tachometer feedback and a separate power conversion unit and regulator for each section. The ac system will remain as sketched in the preceding example (see Fig. 11), with the 5 percent tracking ability inherent in the NEMA B induction motors.

Furthermore, if the motors in sections one through four are required to do any braking or "holding back" on the web (if the web is being stretched in the process), those sections' power units, in the dc system, will have to be made regenerative. In the case of the ac drive, however, all five of the ac motors can transfer energy along their ac feeder bus, with the inverter supplying only net motoring horsepower to the load.

Whether or not the regenerative requirement exists for this example, however, the ac scheme requiring only one power unit and regulator is a more economical solution.

Going one step further, let us require of this process line absolute speed regulation and absolute tracking between machine sections. The dc system would appear as shown for the previous example with minor exceptions. In the dc system, each of the five power unit regulators would now receive a digital input reference signal and a pulse-type feedback signal from motor-mounted pulse tachometers. Speed would be set by means of an appropriate thumbwheel switch.

The ac drive would still require only one power unit (as shown in Fig. 11), with the inverter frequency regulator operating open loop. The pulse train established by the thumbwheel switch would be fed directly into the inverter firing logic, thereby establishing inverter frequency directly.

The use of synchronous reluctance motors in place of the induction machines used in previous examples would assure absolute tracking between sections, with each motor's speed ($120f/p$) exactly determined by inverter frequency.

As indicated by the bar graph in Fig. 13, the ac system is by far the most appropriate solution to the problem.

Granted, the preceding examples were "hand selected" to prove various points, but the economic guidelines which these examples were selected to illustrate are very useful over a broad range of industrial drive situations.

Economic Guidelines

The adjustable-speed ac drive system uses a power conversion device which, when compared to its dc counterpart, is more complex and considerably more costly. The savings in using an ac drive system occur through the use of the inexpensive ac motor, with the ac system *tending to* show an economic edge over the dc system in the following instances.

1) When the use of low base speed motors is required.

2) When the environment requires explosionproof or totally enclosed motors.

3) When multiple motors may be used without the need for individual speed trimming from the same ac power unit.

4) When the need for absolute shaft-to-shaft synchronization calls for synchronous motor use.

5) When frequent reversing and/or controlled slowdown would require the use of regenerative dc power units.

Since these are only rules of thumb, an economic comparison of ac and dc drive equipment should be a requirement in instances where both drive types are technically applicable.

Summary

The adjustable-speed ac drive is a useful drive tool, broadening the applicability of electrical adjustable-speed drives in industry.

Depending upon the application and its requirements, however, adjustable-speed alternating current may or may not be an economically sound decision when compared to direct current and careful economic evaluation should be performed when both drive types are technically applicable.

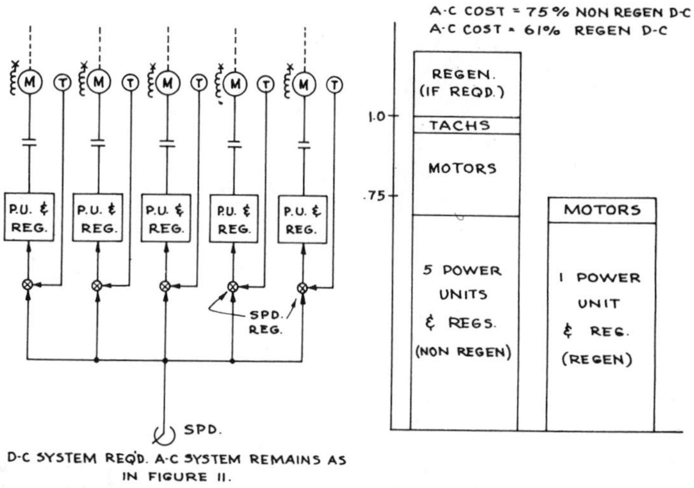

Fig. 12. Five-section machine with 10:1 speed range and 5-percent tracking.

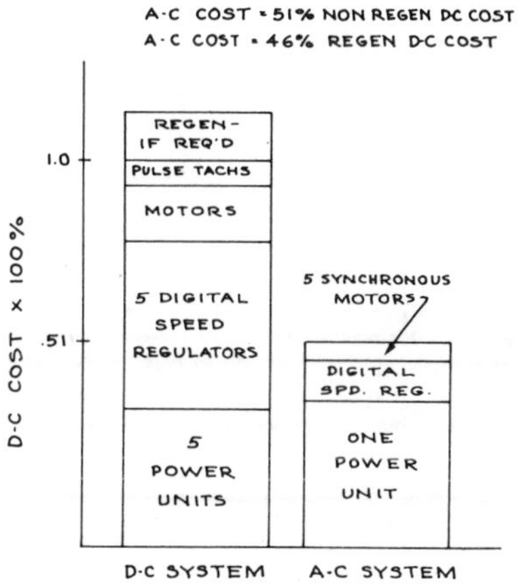

Fig. 13. Five-section machine with 10:1 speed range and absolute tracking.

References

[1] D. F. Grubb, "Theory and application of a-c adjustable speed drives," Reliance Electric Co., Reprint L-5039.

[2] R. P. Veres, "New inverter supplies for high horsepower drives," *IEEE Trans. Ind. Gen. Appl.*, vol. IGA-6, pp. 121–127, Mar./Apr. 1970.

Section 4
Current Fed Inverter Drives

Current-Source Converter for AC Motor Drives

KENNETH P. PHILLIPS

***Abstract*—The employment of current-source concepts in thyristor converters to obtain adjustable frequency and adjustable current waveforms is presented. The use of a dc filter choke and a current feedback loop to produce a regenerative current source is explained. The simplified inverter commutation circuit made possible by the current-source technique is also discussed. Finally, a brief review of the inherently rugged current-source converter's ability to provide wide range control of an ac induction motor is given.**

Introduction

THE STATIC ac motor drive described in this paper utilizes a combination of power and control techniques which achieve ac drive performance features heretofore not available. The excellent inherent characteristics of the squirrel-cage induction motor (i.e., ruggedness, simplicity, low-inertia rotor, low cost, smaller size, etc.) make it a desirable source of mechanical power. However, this source of mechanical power has not been fully utilized on adjustable speed applications.

Nearly all static ac drives built today can be described as voltage source converters (VSC) which must be designed with full application knowledge so that all possible load requirements are considered. These VSC's merely change line voltage and frequency to a setable adjustable voltage and frequency which is applied to the motor. The motor is then free to respond within its speed/torque parameters as the application dictates. Unusual load variations can, and often do, push the motor to its breakdown point or cause regeneration to occur in the converter by overhauling the motor. In either case, an untimely shutdown or damage to the motor or converter results.

The current-source converter described in this paper is slip regulated for both speed and torque control. The current-source/slip-regulated converter (CS/SR) concept controls operation of the ac motor at its optimum torque/ampere rating point while maintaining and enhancing its rugged simplicity to achieve torque limit, fast response, regeneration, and wide speed range.

Current-Source Converter Power Circuit

General

An ac drive employing the current-source converter is shown in Fig. 1. Basically, it consists of a controlled rectifier, a dc link filter choke (without a capacitor bank),

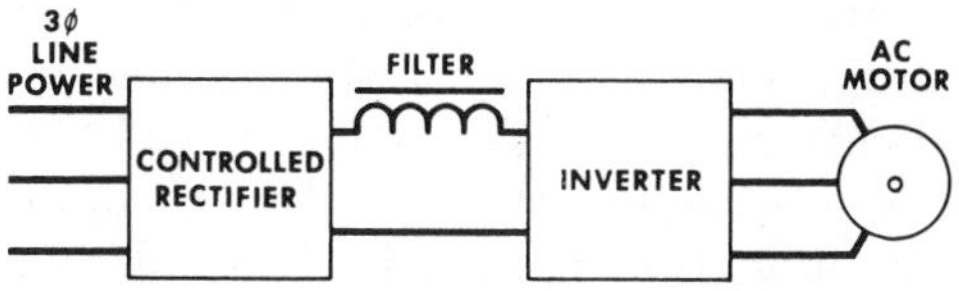

Fig. 1 Simplified CS/SR converter.

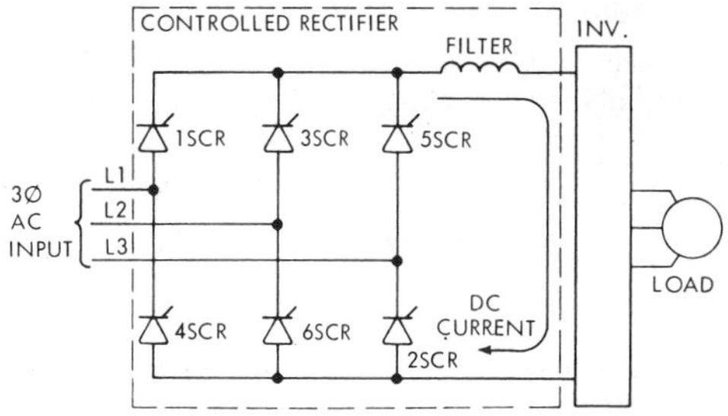

Fig. 2. Controlled rectifier and filter schematic.

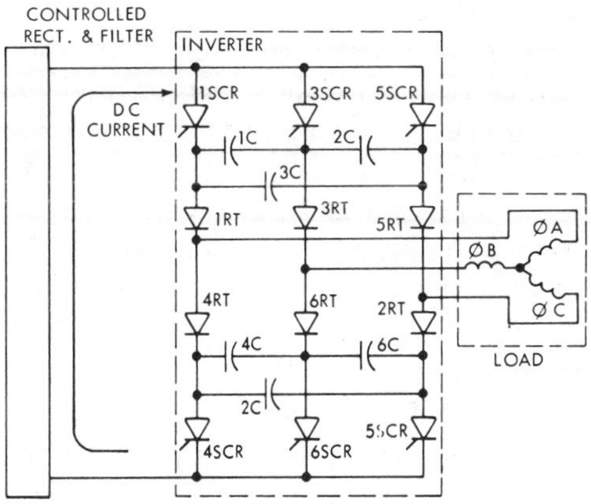

Fig. 3. Inverter power circuit.

and a current-mode inverter. The controlled rectifier and filter choke (Fig. 2) combine to form a dc current regulator which supplies a regulated dc current to the current mode inverter. As could be expected, an inverter designed to operate from a high-impedance dc current source is quite different from an inverter which operates from a low-impedance dc voltage source. Fig. 3 is a simplified schematic of a six-step current-mode inverter power circuit. Thyristors (SCR's) 1–6 switch the load current at a rate determined by the inverter control to establish output frequency. Capacitors $1C$–$6C$ provide the energy storage necessary for commutation, while the series diodes 1RT–6RT isolate the capacitors from the load. Only two thyristors are on at any one time, with each one conducting the dc link current Id for 120°. A thyristor is commutated (turned off) by the firing of the adjacent thyristor in the next phase.

Paper TOD-72-32, approved by the Static Power Converter Committee of the IEEE Industry Applications Society for presentation at the 1971 IEEE Industry and General Applications Group Annual Meeting, Cleveland, Ohio, Oct. 18–21. Manuscript released for publication July 24, 1972.

The author is with the Louis Allis Company, Milwaukee, Wis. 53201

Reprinted from *IEEE Trans. Ind. Appl.*, vol. IA-8, pp. 679–683, Nov./Dec. 1972.

Current-Mode Inverter Firing Sequence

Fig. 4(a) illustrates schematically the conditions in the inverter just prior to firing thyristor 3. Current from the controlled rectifier (dc supply) flows through thyristor 1, 1RT, motor phase A, motor phase C, 2RT, thyristor 2, and back to the supply. The commutating capacitor 1C between thyristors 1 and 3 has been charged with the polarity shown. When thyristor 3 is fired, the full capacitor voltage is applied to thyristor 1 which is then commutated. The load-current path is shown in Fig. 4(b). This current, which now goes through the capacitor 1C, linearly reverses the capacitor voltage until it is sufficient to cause 3RT to conduct. At this point, [see Fig. 4(c)] the current begins to transfer from 1RT and phase A to 3RT and phase B until the current in 1RT has been driven to zero, and phase B assumes the full value of dc current Id as shown in Fig. 4(d). Table I illustrates what then takes place in sequence.

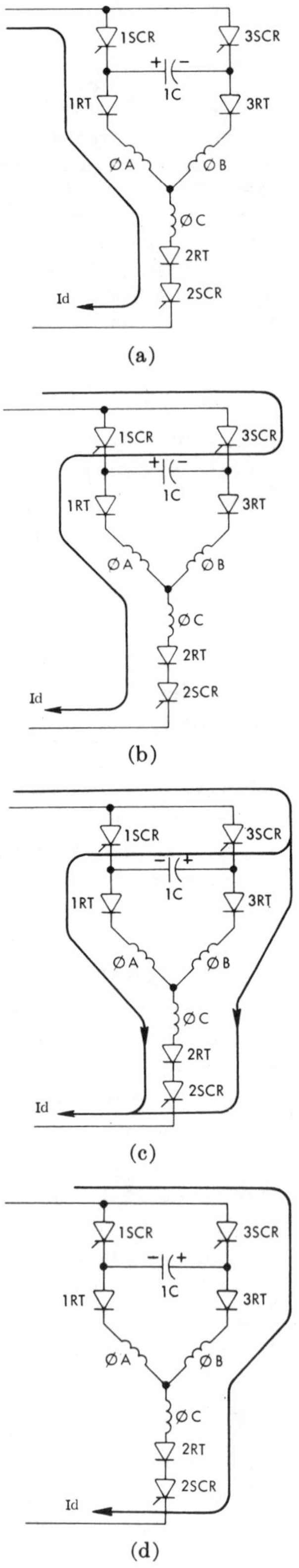

Fig. 4. Current flow through inverter and motor.

Current-Source Converter Output Waveforms and Power Flow

The typical current supplied to each phase of the motor is shown in Fig. 5. The motor voltage is determined by the response of the motor and load to the applied current.

The magnitude of current at any point in the current mode inverter is always controlled by the regulated current source (the controlled rectifier and filter). Even when there are two parallel paths for current flow, the sum of these two currents can never be greater than Id. The inverter only controls the time in which current is applied to a particular phase of the motor. At the dc terminals of the inverter, the average dc voltage will vary with the power demand of the motor. If the motor is unloaded, the dc voltage will be near zero. If the motor is supplying rated horsepower, the dc voltage will be at some maximum positive value. If the motor load is overhauling (power demand), the dc voltage will reverse polarity and power will be returned to the ac line (regeneration) through the controlled rectifier. (Refer to Fig. 6.) Note that the regulated dc link current Id between the controlled rectifier and the inverter is always in the same direction. During regeneration, the thyristors of the controlled rectifier are commutated by the alternation of the ac line.

Note that the current waveshape in the current-source converter (Fig. 5) is identical to the classical six-step voltage waveshape. With the current-source converter, the load voltage is determined by the response of the load to applied current. The harmonic content of the current waveform is well defined and has a known and favorable harmonic distribution with regard to the motor heating.

Commutation

An immediate consequence of the current-source converter concept is the simplification of the power circuitry that is required to commutate the thyristors. In Fig. 3, note that the commutation circuit consists only of

TABLE I

Turn on of SCR Number	Commutates SCR Number	Current-Path Source
4	2	3SCR–3RT–ϕB–ϕA 4RT–4SCR
5	3	5SCR–5RT–ϕC–ϕA 4RT–4SCR
6	4	5SCR–5RT–ϕC–ϕB 6RT–6SCR
1	5	1SCR–1RT–ϕA–ϕB 6RT–6SCR
2	6	1SCR–1RT–ϕA–ϕC 2RT–2SCR
3	1	3SCR–3RT–ϕB–ϕC 2RT–2SCR

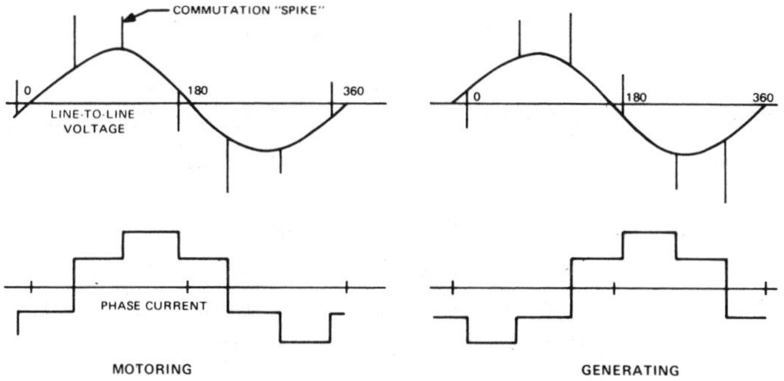

Fig. 5. CS/SR current and voltage waveforms.

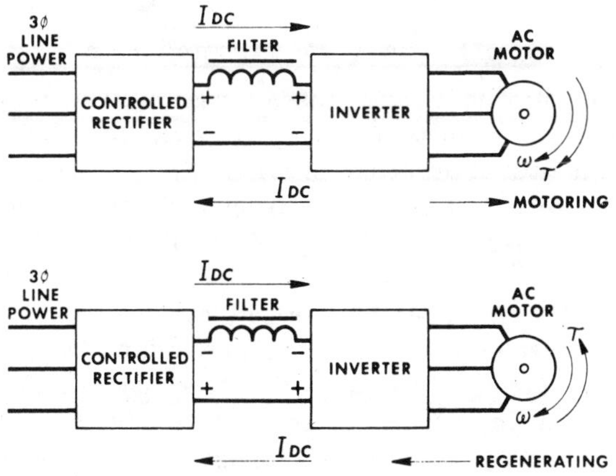

Fig. 6. Power flow: motoring and regenerating.

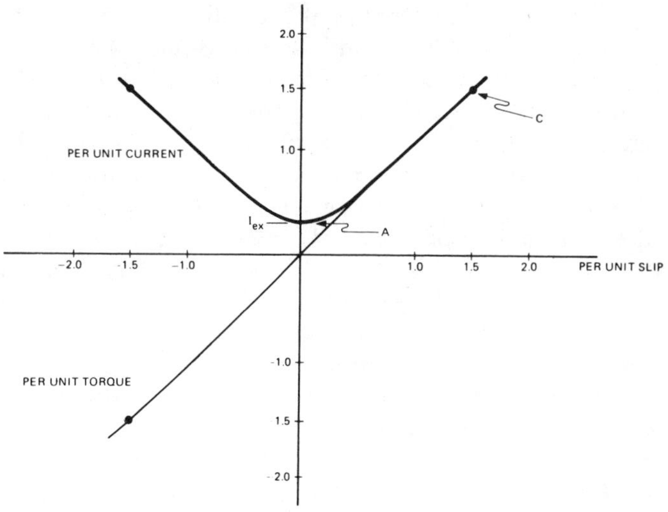

Fig. 7. Induction motor characteristics.

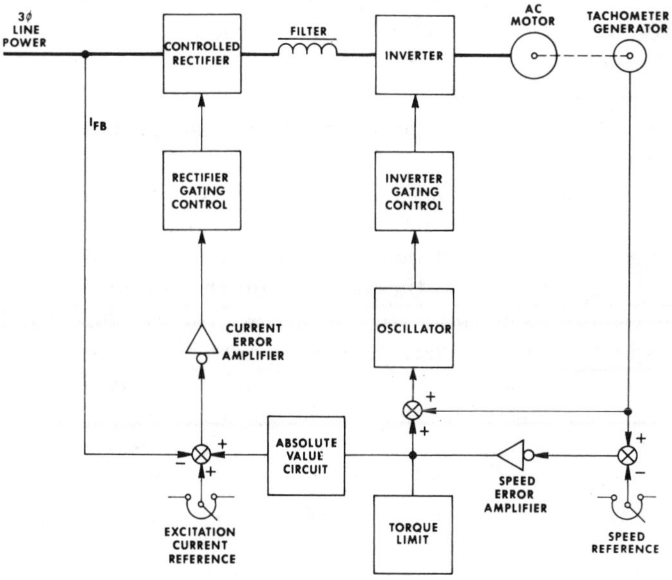

Fig. 8. Basic system.

capacitors and diodes. Commutation transformers or reactors are eliminated except for a small value of inductance in series with each thyristor to limit di/dt to a safe value. Elimination of commutation reactors not only allows increased operating frequency but also drastically reduces audible mechanical noise levels in the operating equipment. The commutation capacitor is designed to limit peak voltage on devices, rather than to provide adequate turnoff time for the thyristor. As a result, the turnoff time available is sufficient to permit the use of standard thyristors rather than special "fast turnoff" thyristors throughout the current-source converter. Because there is no requirement for special commutating transformers and premium-grade thyristors, the current-mode inverter design is economically attractive at all horsepower levels.

Current-Source Converter Control Circuits

General

Fig. 7 depicts the torque, current, and slip r/min relationship of an induction motor supplied from a fixed voltage. These relationships apply for any synchronous speed as long as the air gap volts/cycle remains the same. The motor requires some minimum value of current $I_{\min}$ to maintain synchronous speed with zero load torque. As the load (torque requirement) increases, motor current must also increase. This is accomplished by the basic speed regulator shown in Fig. 8 which consists of a "current loop" and a "speed loop." The current loop includes a current feedback, a "current error amplifier," a controlled rectifier, and filter. The speed loop includes a "speed feedback," a speed-error amplifier, an oscillator, an inverter, and a motor. Correct interaction between the two loops, regardless of the direction of the load, is provided by the absolute value circuit.

With the motor running at a condition represented by point A in Fig. 7, the load is zero, the current is I_{min} (set by the *excitation current reference*), the speed is as set by the *speed reference*, the speed error is zero, and the speed-error amplifier output is approximately zero. Assume a load equal to 100 percent of rated torque is applied to the motor. The motor slowdown causes a speed-error signal which is detected and amplified by the speed-error amplifier. The resulting output signal of the speed-error amplifier increases the oscillator frequeny and also increases the current-error output to the controlled rectifier and filter through the absolute value circuit.

The speed-error amplifier output continues to increase until the motor is operating at full rated current, slip r/min, and torque. This condition is represented by point C on Fig. 7. The full-load torque demand is now satisfied and motor speed is maintained within the regulation limit of the speed regulator.

Note that whenever there is no speed-error signal, the oscillator receives its only signal directly from the motor tachometer. Thus, whenever the shaft speed matches the speed reference signal, the oscillator drives the inverter at an electrical frequency corresponding to motor shaft speed.

If the load torque had been negative (an overhauling load), the speed-error amplifier signal would have been of reversed polarity which would decrease the oscillator frequency. However, the polarity of the signal from the absolute value circuit to the current-error amplifier would still have been such as to increase current. The dc voltage at the dc terminals of the inverter would have reversed polarity and the drive would be regenerating.

Current and Torque Limit (Fig. 8)

Limiting the speed-error amplifier output, as indicated by the *torque limit reference* adjustment, accomplishes two very useful functions. First, it limits the magnitude of the current reference, thereby limiting the current supplied to the current-mode inverter to a safe value—regardless of load variations or transient conditions within the inverter. Second, it limits the slip r/min of the motor, thereby limiting the motor torque to a desired value. Suppose this limit corresponded to 150 percent of rated slip and 150 percent of rated current. As the motor load is increased, the current and slip are increased until the speed-error amplifier is clamped at the value corresponding to point C in Fig. 7. Increasing the load beyond 150 percent of full-load torque would increase neither current nor slip. As the motor continued to slow down, the tachometer signal would cause the oscillator to decrease the frequency of the inverter proportional to the decrease in motor speed. Slip r/min and current are maintained at the torque limit value, preventing the motor from "breaking down" or operating at high slip where the torque/ampere is very low. If the overload is maintained, motor speed may decrease to zero, but the motor will still be supplying 150 percent of rated torque to the load. If the motor load is reduced to something less than 150 percent the motor will accelerate to the desired speed at a rate proportional to the difference between the load torque demand and the set torque limit.

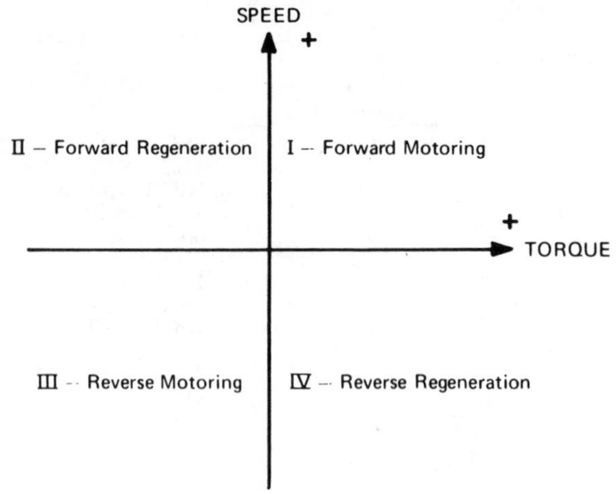

Fig. 9. Four quadrants of operation.

As a result of this torque limit control, the induction motor can be safely accelerated at the maximum rate up to any desired speed, without the fear of pulling out the motor or drawing excessive currents. In addition, any overload-torque demand beyond the capability of the drive will not cause the drive to shut down, but will merely result in a speed reduction until the overload is relieved.

CS/SR Operating Characteristics

Regeneration and Dynamic Response and Wide Speed Range

As mentioned earlier, the average voltage at the dc terminals is a function of the power supplied to the load. Power delivered to the motor is a product of the dc voltage and dc current. This is true for either direction of power flow. If the load should overhaul the motor, the induction motor becomes an induction generator and pumps power back to the current-source converter. Because the inverter and rectifier design permit the power flow to reverse in the first and fourth quadrants while maintaining a unidirectional dc current *Id*, the regenerative capability is inherent and no additional components—such as back-to-back rectifiers, are required. The inverter dc voltage reverses its polarity as the motor power factor becomes negative but the polarity of the dc link current *Id* remains unchanged. However, the direction of power flow in the dc link reverses as shown in Fig. 6. Motoring and generating waveforms are found in Fig. 5. Since the controlled rectifier and filter choke seeks to maintain constant current, its dc voltage will automatically reverse, thereby, returning power to the ac line. The ability of the dc section to operate in the first and fourth quadrant gives the total drive the ability to operate in the first and second quadrant, that is, positive and negative torque. Electronic reversal of the phase sequence of the inverter permits motor operation in the third and fourth quadrants. This is shown graphically in Fig. 9. Consequently, full four-quadrant motor operation can be achieved with only 12 power switching devices.

Since reversing can be easily achieved by electronically reversing the phase sequence in the inverter control, this ac drive is capable of extremely rapid speed reversals. The ability of the current-source converter to produce positive or negative torque is limited only by the time required to increase the dc current. Therefore, this drive can respond to a full positive- or negative-torque command within a few milliseconds. Rapid response to speed reference changes as well as transient load changes result. Since the torque to inertia ratio of an ac motor is generally three or more times greater than that of dc machines of the same horsepower, this static ac drive exhibits even faster response than its dc counterpart. Also, the cast-rotor higher speed design of the ac motor can be utilized in high-speed applications.

The ability of the inverter to supply a specific value of slip r/min and current to the motor insures that the drive will develop the desired torque at any speed from standstill to rated motor r/min. During all conditions of operation—whether starting, stopping, or reversing—the motor only sees the magnitude of current determined by the torque limit reference. The torque limit reference can be set for any value up to a point near the breakdown torque point of the motor. Since the current/torque relationship within the operating range of the ac motor is not linear, a given percentage of torque will be developed with motor current at some level below that percentage. For example, with torque limit set at 200 percent torque, the motor will experience less than 200 percent current under all operating conditions and the drive will not deliver greater than 200 percent torque to the load.

Ruggedness

The current limit feature gives this drive a truly rugged nature. By limiting the dc current to a safe value, it is impossible for the inverter components to experience excessive current. Events, such as noise firing, misfires, and momentary power interruptions—that usually cause ac drives to blow fuses, destroy devices, or at a minimum, shut down—result in only momentary loss of torque for the duration of the occurrence. Even a three-phase short across the inverter output terminals, although somewhat hard on the motor, does not harm the current-source converter. The ability of this drive to continue to operate in spite of the aforementioned conditions, not only permits troubleshooting while the unit is operating, but in many cases allows the operator to repair a component failure in the unit at some later, more convenient time.

Conclusion

From an historical perspective, the ac motor drive has not lived up to its potential. All the advantages the ac motor has possessed for years have been negated by limitations within the power conversion unit. Most ac power conversion units have been merely adjustable frequency power supplies and not truly general-purpose drives. When applying motors to these conversion units, success has been measured by a fine balance between inverter design, application knowledge, and operational safety factors. Under industrial load conditions, this balance has been all too fine to result in adequate operation in a wide range of applications.

The concept of a CS/SR converter supplying an induction motor results in the first ac motor drive which enhances all the superior characteristics of the ac motor, and maintains a rugged and simple motor drive. The resulting drive can be applied as a industrial general-purpose ac motor drive system.

Acknowledgment

The author wishes to thank C. E. Rettig, Manager of Advanced Products and L. D. Beer, Senior Engineer, Advanced Products, the Louis Allis Company, for their assistance.

References

[1] L. D. Beer, "Static induction motor drive: an ideal drive for the paper industry," presented at the 1971 IEEE Pulp and Paper Conf.

[2] R. B. Maag, "Characteristics and application of current source/slip regulated ac induction motor drives," in *1971 IEEE IGA Group Annu. Meeting*, pp. 411–413.

Power Converters for Feeding Asynchronous Traction Motors of Single-Phase AC Vehicles

WOLFGANG LIENAU, ADOLF MÜLLER-HELLMANN, AND HANS-CHRISTOPH SKUDELNY

***Abstract*—Asynchronous induction motors are very well suited to powerful traction drives. For electric locomotives and motor coaches with single-phase supply, a power conversion on the vehicle is necessary. Three different types of static power converters for this application are described. Indirect ac converters with direct voltage link have been developed successfully in the past. For indirect ac converters with direct current link some problems are discussed. At last a self-commutated direct ac converter with suppressed dc link is presented.**

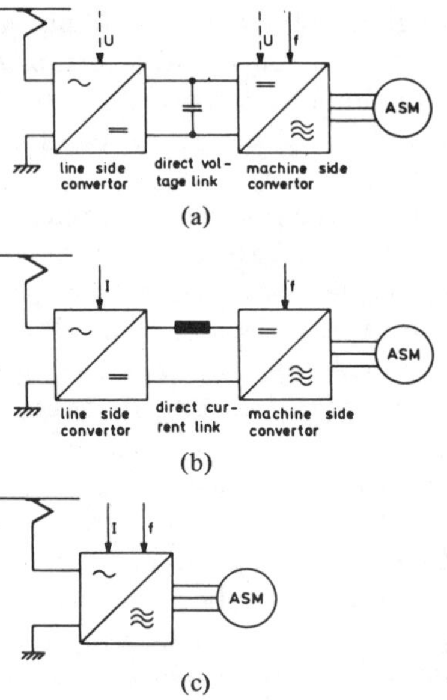

Fig. 1. Basic converter configurations. (a) Indirect converter with direct voltage link. (b) Indirect converter with direct current link. (c) Direct converter with suppressed direct current link.

INTRODUCTION

At PRESENT single-phase ac locomotives and motor coaches normally have either single-phase series wound motors with tap changer control or rectifier fed dc motors as a traction drive. Both motor types have mechanical commutators, and therefore there are several restrictions for design and operation. Asynchronous squirrel-cage induction motors are more suited to traction applications [1]. They have high reliability and no need of maintenance. The weight of asynchronous traction motors is about 2/3 of the weight of equivalent dc motors. For traction applications, the most important characteristics of asynchronous motors are a high range of operation and a high overload capability. We expect that future high-speed electric locomotives normally will be designed with asynchronous motors. As an additional advantage, it will be possible to use the same locomotives for either heavy trains or fast trains.

In our paper an investigation of converter configurations for controlling asynchronous motors shall be presented. There are two subjects to be investigated: line-side interaction and machine-side interaction. It is well-known that static converters may generate inductive power and harmonic power on both the line and the machine side.

On the machine side fundamental inductive power is necessary for proper operation of the asynchronous machine. Harmonic power may cause torque oscillations and additional losses and therefore is a disadvantage.

The power-factor λ on the line side should be as close as possible to unity, since both reactive and harmonic power contribute to additional losses and voltage distortion.

Paper SPCC 77-25, approved by the Static Power Converter Committee of the IEEE Industry Applications Society for presentation at the 1977 International Semiconductor Power Converter Conference, Orlando, FL, March 28–31. Manuscript released for publication July 18, 1979.

The authors are with the Lehrstuhl und Institute für Stromrichtertechnik und Electrische Antriebe, RWTH Aachen, Jägerstrasse 17–19, West Germany.

BASIC CONVERTER CONFIGURATION

For feeding asynchronous motors, converters of the self-commutated type are necessary. Two basic principles of self-commutated converters are known.

1) Direct voltage converters which connect a direct voltage source to an ac load generating rectangular ac-side voltage [2]. The voltage may be pulsewidth modulated in order to vary the mean value. Pulsewidth modulation (PWM) is also used for the reduction of low-order voltage harmonics [3].

2) Direct current converters which connect a direct current source to an ac load generating rectangular ac side current [2]. Pulsewidth modulation may be used in order to reduce low-order current harmonics [4]. Current modulation for varying the mean value has not been used so far.

These two converter types lead to two differenc ac–ac converter systems (see Fig. 1(a) and (b)). A third system is to be developed out of the second (see Fig. 1(c)).

INDIRECT AC CONVERTER WITH DIRECT VOLTAGE LINK

The principal configuration is shown in Fig. 1(a). The line-side converter feeds into a direct voltage link of constant voltage. The dc link contains a capacitor which takes the harmonic components of the converter line-side current.

Reprinted from *IEEE Trans. Ind. Appl.*, vol. IA-16, p. 103–110, Jan./Feb. 1980.

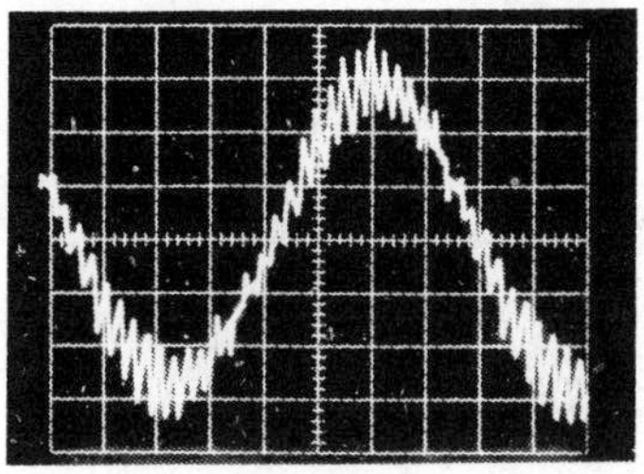

Fig. 2. Current waveform of a PWM converter.

Additional storage devices may be useful. The machine-side converter must be capable of varying both machine voltage and machine frequency. This is normally done by PWM operation of the converter.

Using the PWM technique the machine currents can be controlled to be almost sinusoidal (see Fig. 2).

Investigations have been performed on this drive system for traction [5]. They have resulted in a self-commutated line-side converter which operates very similarly to the machine-side converter; it has been called a "four quadrant controller" [6]. Using this type of converter, the line-side current can be controlled almost sinusoidally and the power factor is greater than 0.99 [7]. This drive system has been the first to be built into electric locomotives [8]. We will not discuss this system as there is a special report on it [9].

INDIRECT AC CONVERTER WITH DIRECT CURRENT LINK

This converter system consists of a line-side converter, which feeds adjustable direct current into the dc link, and a motor-side converter, which converts the direct current to adjustable frequency three-phase currents (see Fig. 1(b)). The difference of the dc side voltages of the two converters is magnetizing the series reactor of the dc link.

There are various configurations of self-commutated motor-side converters with different methods of commutation, e.g., phase commutation, consecutive phase commutation, and common commutation. A well-known configuration of a self-commutated six-pulse bridge circuit is shown in Fig. 3. This circuit operates with consecutive phase commutation and therefore does not need separate commutation thyristors. For example, the commutation of the current from thyristor T4 to thyristor T6 is illustrated in Fig. 4. Normally, the next commutation of this commutating group will be T6 → T2. However, it is also possible to perform a commutation from T6 back to T4 and thereafter again to T6. By switching between the two phases, the motor current I_1 is pulse modulated as shown in Fig. 5. The rule of modulation is that in any instant one phase of the upper commutation group and one phase of the lower commutation group is carrying current. During switching in one commutation group no commutation occurs in the other group. This leads to a symmetric control of the converter. Of course, it is not necessary to make nine current pulses per half cycle. The number of pulses, theoretically, can be any odd figure between one and infinity.

It is necessary to mention that the motor leakage reactance largely influences the commutation process. In practice the

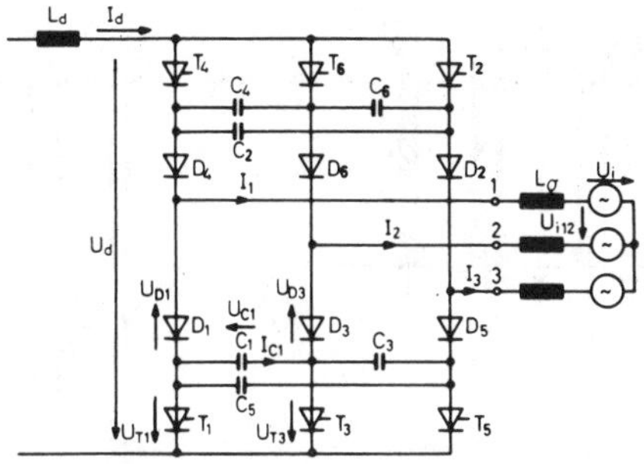

Fig. 3. Self-commutated six-pulse bridge converter.

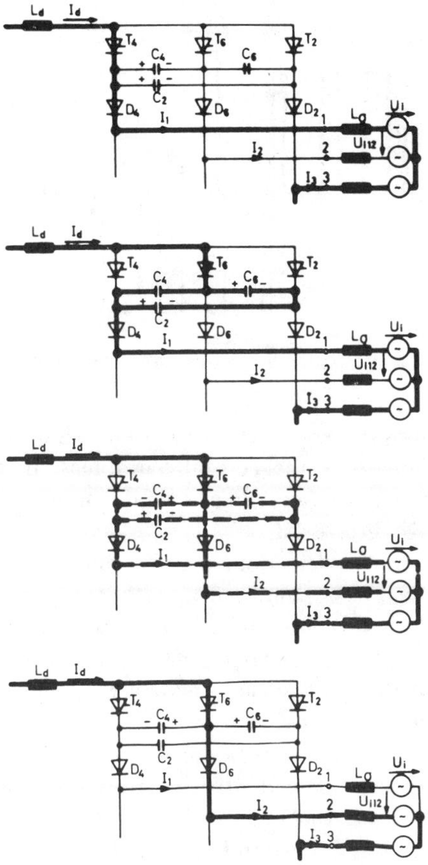

Fig. 4. Steps of commutation.

commutation takes some time. Therefore, at increasing inverter frequency the commutation time in proportion to the period is increasing as well. When the commutation time reaches 1/3 of the period, the maximum frequency of normal operation is reached. It is possible to increase the operation frequency further, but in this region of operation multiple commutations and harmonic oscillations occur.

Normally the operation region of multiple commutation is avoided by proper design of the commutation capacitors. The maximum commutation time defines the limit of the operation frequency. It also restricts the minimum duration of the current pulses.

Pulse modulation of the current is favorable with respect to harmonic torque reduction. It is well known [10] that the current harmonics together with the field fundamental wave

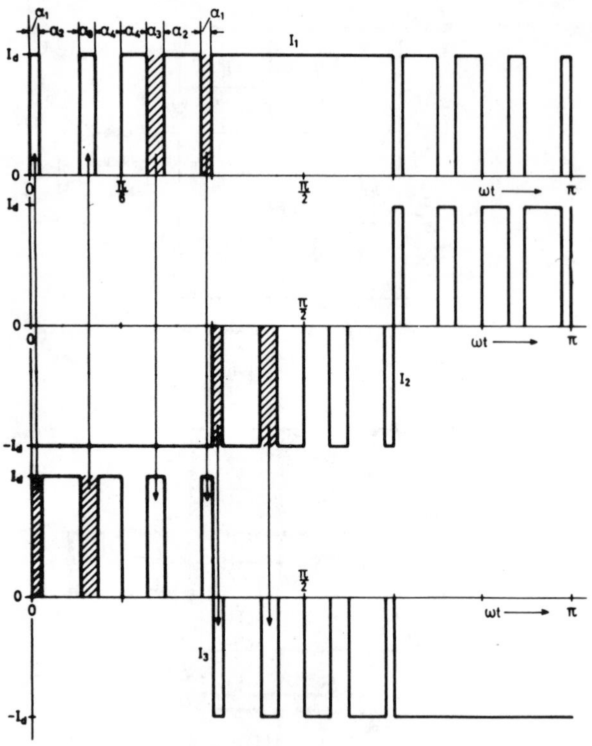

Fig. 5. Pulse modulation of the motor current..

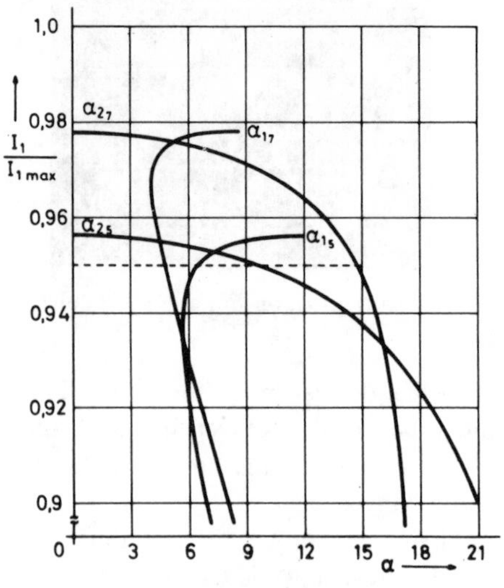

Fig. 6. Control angles to be chosen to eliminate certain current harmonics for pulse operation.

generate torque harmonics of an order which differs by one from the order of the current harmonics. In the case of a rectangular current waveshape torque harmonics of the order 6, 12, 18, 24, etc., exist.

These torque oscillations may be dangerous to the vehicle drive system. When the fundamental frequency of the motor currents is varied the frequency of the torque oscillations also varies. This leads to resonances between the harmonic torque and the mechanical transmission system for low supply frequencies.

Using pulse current modulation the low-order current harmonics, and consequently the torque harmonics, can be suppressed. As a result of calculations, Fig. 6. depicts how the control angles as defined in Fig. 5. are to be chosen to suppress certain current harmonics for five-pulse operation.

For an arbitrary fundamental current ratio of $I_1/I_{1\max} = 0.95$, the curves show that α_1 must be 4.8° and α_2 must be 14.8° in order to eliminate the seventh harmonic. The fifth harmonic can be eliminated by the combination $\alpha_1 = 6.5°$ and $\alpha_2 = 9.5°$.

There is one set of control angles that eliminates both the fifth and seventh current harmonics. This is $\alpha_1 = 5.82°$, $\alpha_2 = 16.25°$, and $I_1/I_{1\max} = 0.933$.

Similar figures can be presented for seven-pulse operation and for higher pulse operation of the converter. As a result, Table I shows for one- to nine-pulse operation how the harmonics up to the order 25 are affected. It can be seen that the reduction of the fundamental current is less than eight percent. The transition from one pulse number to the next one causes stepwise increase of the current fundamental. The steps can be reduced or avoided if harmonics are accepted for a short time during transition.

TABLE I
REDUCTION OF THE CURRENT HARMONICS UP TO THE ORDER 25

m	k		5	7	11	13	17	19	23	25
9	$\frac{I_K}{I_1}$	0,923	0	0	0	0,05	0,007	0,136	0,259	0,188
7		0,925	0	0	0	0,097	0,27	0,231	0,029	0,001
5		0,934	0	0	0,186	0,248	0,152	0,033	0,121	0,099
1		1	0,2	0,143	0,091	0,077	0,059	0,053	0,044	0,04

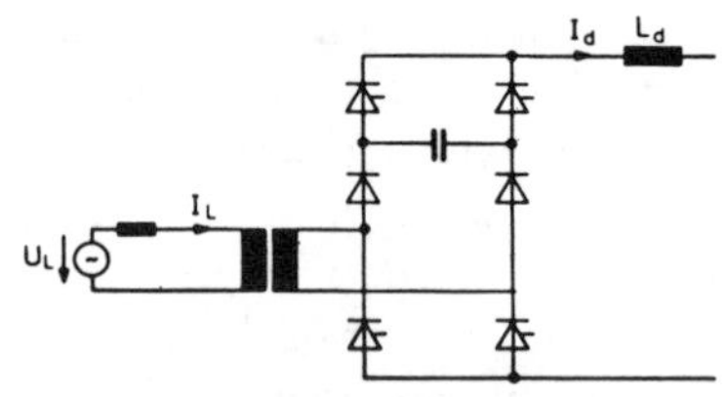

Fig. 7. Self-commutated single-phase line-side converter.

As mentioned above, the frequency of the motor current is controlled by the inverter. The motor current amplitude is adjusted by a controllable line-side converter. In order to provide a high power factor on the line side, this converter should be self-commutated as well. It can be a single-phase version of the same type of converter as used for the motor-side three-phase converter. Fig. 7 shows the circuit diagram. It is not necessary in this case to provide means of self-commutation for both parts of the bridge. In rectifier operation the thyristors of the lower part switch on the line-side current, and the self-commutated thyristors of the upper part switch off the line-side current starting free wheeling operation of I_d. In inverter operation the function of the two commutating groups is reversed.

The line-side current is composed of blocks in every half cycle, the beginning and end of which is determined by the control angle of the respective commutation groups. The control angles can be chosen such that no fundamental reactive power is supplied on the line side. However, no influence on the harmonic power is possible. The power factor λ

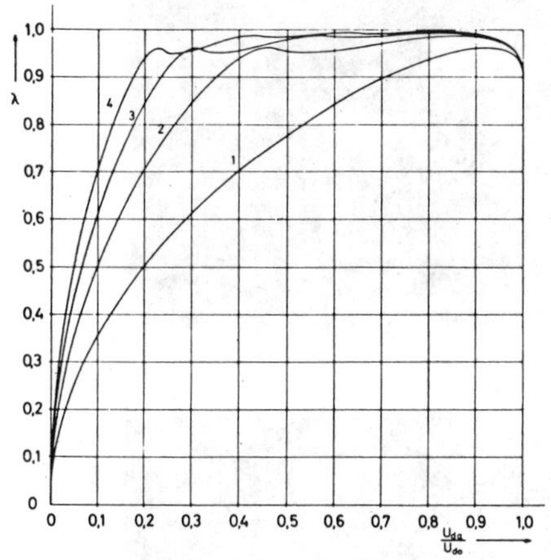

Fig. 8. Dependence of the power factor on the control ratios for multiple step control.

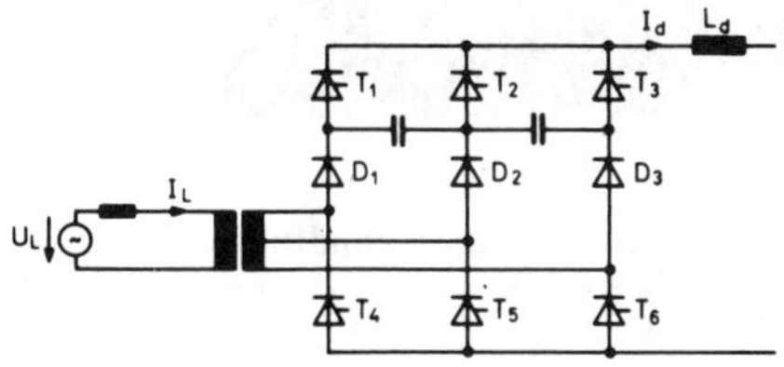

Fig. 9. Self-commutated single-phase line-side converter for two-step control.

is less than unity. Its dependence on the control ratio is shown in Fig. 8. The curves have been calculated on the basis of smooth dc, the commutation phenomena being neglected. A detailed analysis taking care of the commutation phenomena and of the direct current ripple will be published later. If the control is chosen so that no fundamental reactive power occurs it can be shown that there is not much error in the curves given in Fig. 8.

In order to improve the power factor, additional branches which are connected to transformer taps can be added to the line-side converter. To give an example, Fig. 9 shows a converter circuit for two-step control of the line-side current. The improved power factor can be read out from curve 2 of Fig. 8. This figure also indicates which further improvement of the power factor results from three-step control and from four-step control, curves 3 and 4, respectively. The entire converter system is shown in Fig. 10 [11]. It consists of a self-commutated line-side converter with two-step control and a machine-side converter with a common direct current link. Figs. 11 and 12 give oscilloscope records of the currents and voltages of the line side as well as the machine side.

DIRECT AC CONVERTER WITH SUPPRESSED DIRECT CURRENT LINK

Direct ac converters as shown in Fig. 1(c) seem to be very simple for the application discussed. However, considering the objectives pointed out in the first section, the converter shall have approximately sinusoidal voltages and currents on both input and output sides. This means different instantaneous power in the input and in the output. The converter must have inherent energy storage devices.

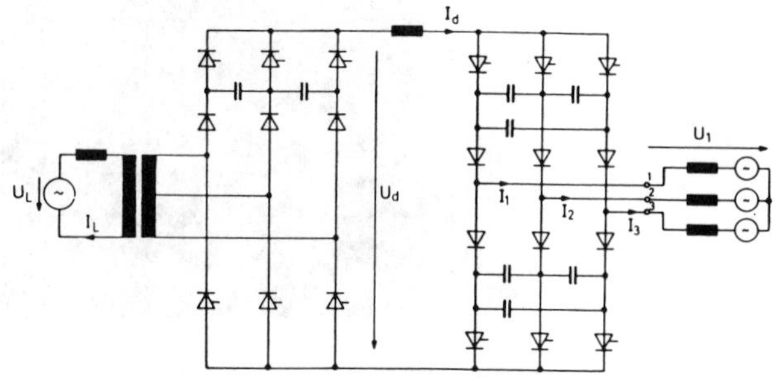

Fig. 10. AC converter system with direct current link.

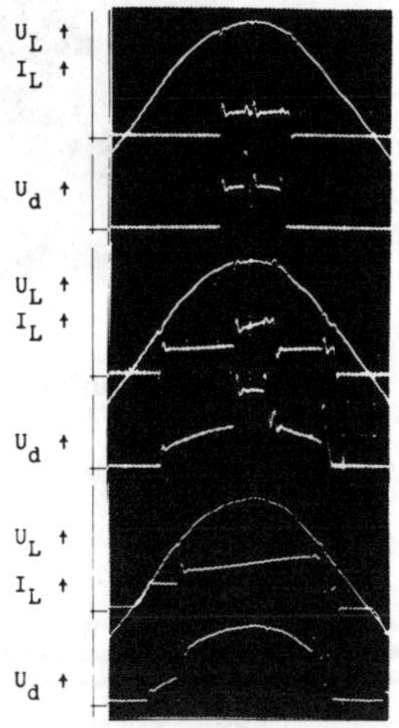

Fig. 11. Voltage and current waveform on the line side for two-step control at different control ratios.

In order to provide a sufficient range of operation, the machine frequency must be variable from zero to values above line frequency. Therefore, the converter must be capable of this frequency variation.

The basic idea of this converter as shown in Fig. 13 is well-known [13]. Two six-pulse bridge converters are connecting a single-phase supply and a three-phase supply. The reactors shown in the figure are magnetically coupled as indicated and, therefore, act as interlinked smoothing chokes. The reactors could be replaced by machine-side reactors.

As this converter is commutated from either supply side, the supply voltages must be able to deliver reactive power. The three-phase supply preferably could be a synchronous machine.

We will discuss a self-commutated version of this converter. This is capable of feeding induction machines and of improving the line-side power factor. In Fig. 14 a converter is presented with commutating circuits of the same kind as discussed in the previous section. In order to have the advantages of self-commutation on both the machine side and the line side, commutation equipment has been duplicated.

The capacitors C1 ··· C6 together with diodes D1 ··· D6 make possible the self-commutation on the machine side. Their function is exactly the same as discussed in the last section.

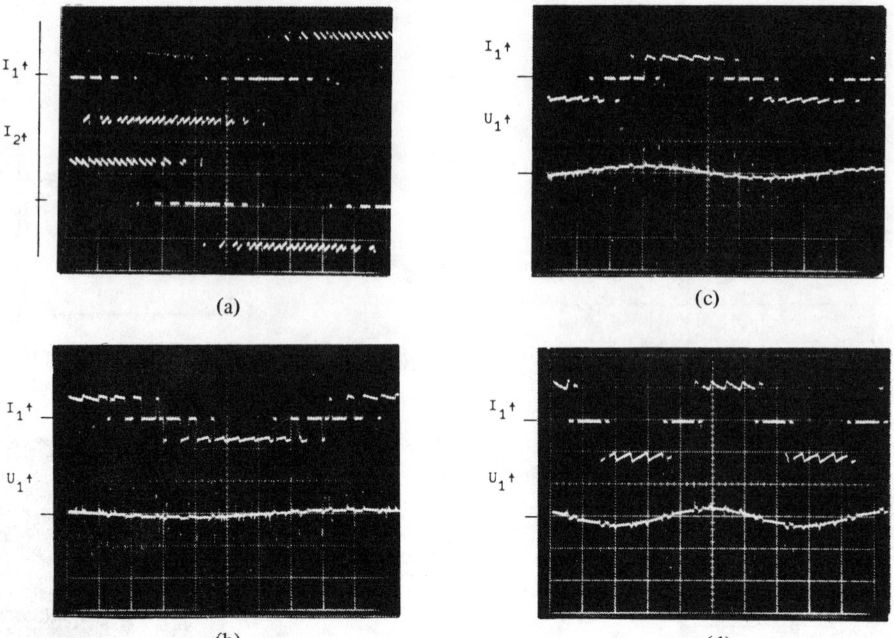

Fig. 12. Voltage and pulse current waveform on the machine side for different load and frequency conditions. (a) $m = 9$, $t = 50$ ms/cm, $\hat{I}_1, \hat{I}_2 = 19$ A, $\cos \tau = 0$. (b) $m = 7$, $t = 20$ ms/cm, $\hat{U}_1 = 10$ V, $\hat{I}_1 = 19$ A, $\cos \tau = 0$. (c) $m = 5$, $t = 20$ ms/cm, $\hat{U}_1 = 15$ V, $\hat{I}_1 = 19$ A, $\cos \tau = 0$. (d) $m = 3$, $t = 20$ ms/cm, $\hat{U}_1 = 20$ V, $\hat{I}_1 = 28$ A, $\cos \tau = 0.5$.

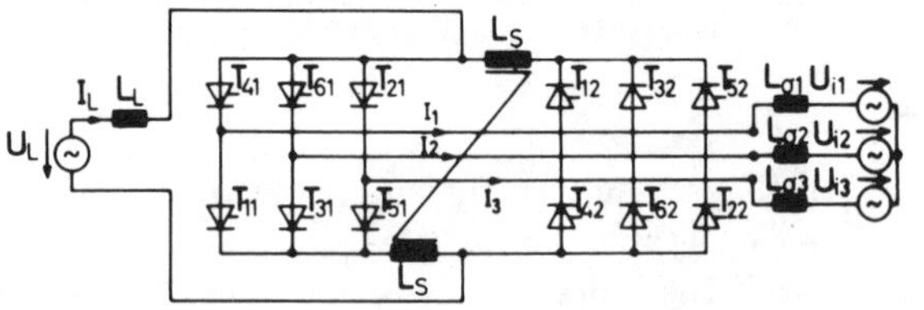

Fig. 13. Machine commutated direct converter with suppressed direct current link.

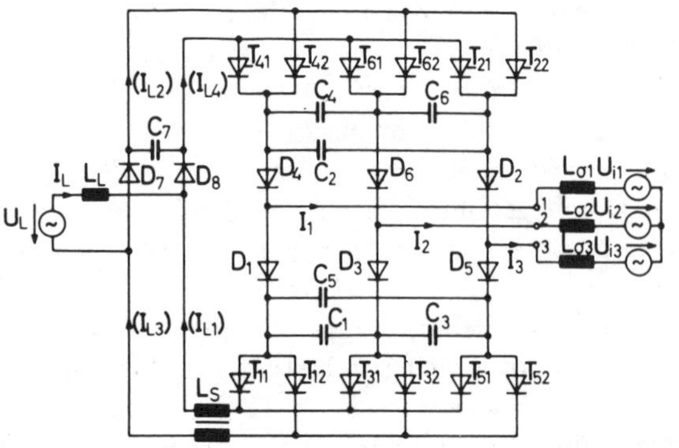

Fig. 14. Self-commutated direct converter with suppressed direct current link.

The specific property of pulse modulation of the machine current is also maintained. Thus, in order to reduce torque oscillations, the machine-side current is pulse modulated for machine-frequencies below 15 Hz.

Capacitor C7 together with diodes D7 and D8 enable self-commutation on the line side which can be used for improving the line-side power factor, as discussed the the previous section. The corresponding values of the line-side power factor are those given in Fig. 8, curve 1.

It is evident that the commutations occur exactly as discussed in the previous section, unless machine-side commutation and line-side commutation are at the same time. In the latter case, there is an interaction of the commutations which may cause a commutation failure of the converter. This can be avoided by a proper sequence of the control pulses.

Let us assume, according to Fig. 15(a), that in the beginning T4.2 and D4 in the upper group and T5.1 in the lower group are carrying the current. This means free wheeling operation with respect to the ac source.

Now assume that UL is positive and T4.1 is fired in order to start a line commutation between I_{L3} and I_{L1} (see Fig. 15(b)). Because of the line-side reactance X_L and the leakage reactance of L_1 a certain time of overlap is needed.

During that overlap, T6.1 is fired in order to start a machine-side self-commutation between phases 1 and 2. A commutation failure is now avoided by firing T6.2 together with T6.1. Thereby, the line commutating currents are changed very quickly from T4.1 and T4.2 to T6.1 and T6.2 (see Fig. 15(c)).

While the line commutation from T6.2 to T6.1 is going on, the self-commutation on the machine side continues,

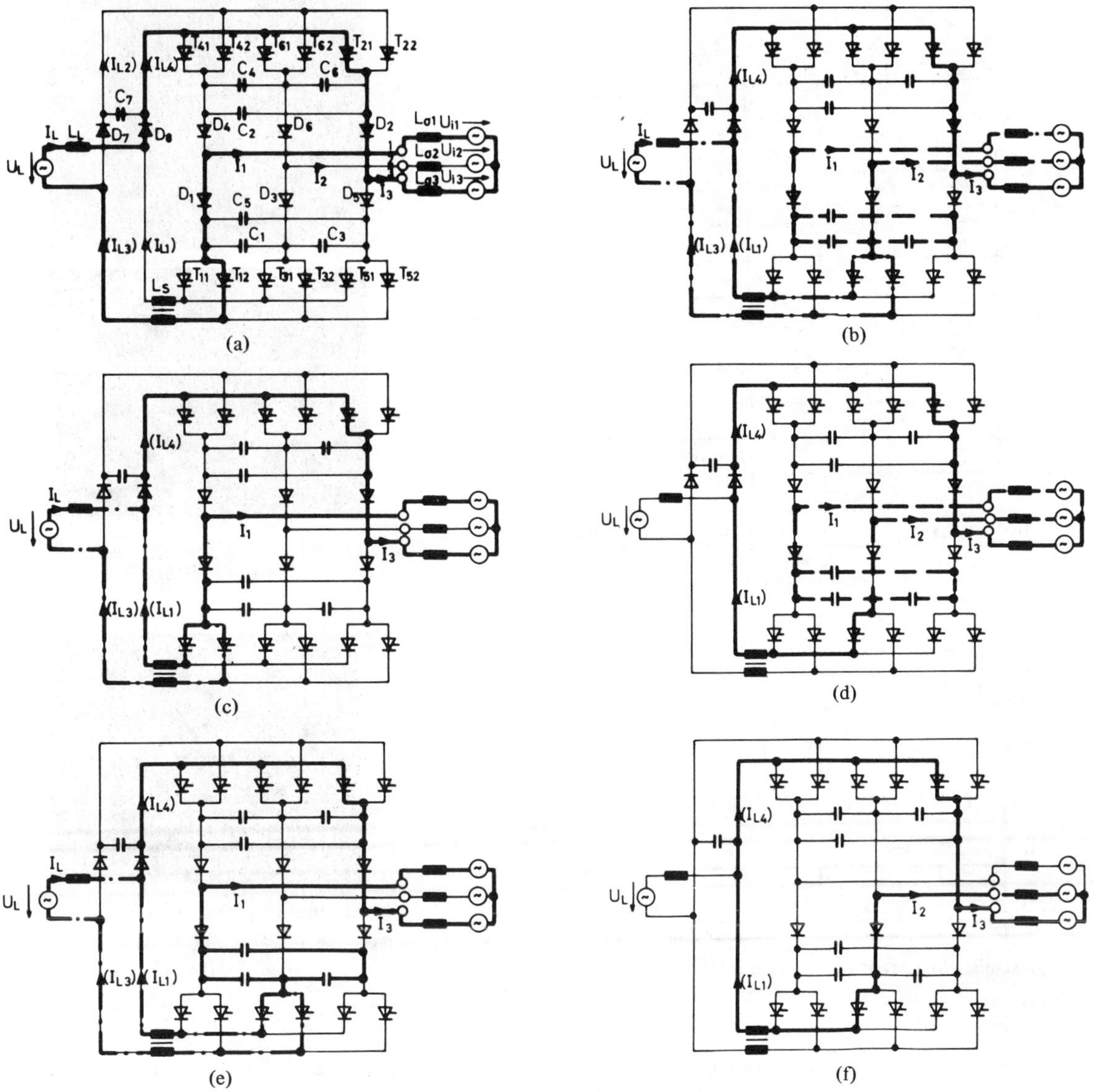

Fig. 15. Steps of commutation.

as shown in Fig. 15(d). It is finished when the current through D4 has become zero (see Fig. 15(e)).

After the machine-side commutation has been completed, the line-side commutation continues unaffected. At the end of the commutation the currents are as indicated in Fig. 15(f).

The operation of the entire converter can be followed in Fig. 16, where the currents on the line side as well as on the machine side are drawn for one point of operation. It can be seen from the figure that at this point of operation the machine frequency is above the line frequency. Furthermore, it can be seen that the control ratio of the line-side converter is below unity; that means the voltage of the suppressed dc link is less than maximum. In case a line-side commutation is started during a machine-side commutation, no disturbance will occur.

As mentioned before, the smoothing reactor L_1 consists of a laminated core with two windings. The reactance must be sufficient with respect to the ripple factor of the current in the suppressed dc link. However, the two windings should be coupled as close as possible in order to avoid additional reactance in the line-side commutation circuit.

The current of the reactor windings is pulse shaped and therefore has an rms value of $\sqrt{2}$ as compared with smooth dc of the same mean value. For this reason the design power of the two reactors is $\sqrt{2}$ as compared with the reactor of the indirect ac converter described before.

It is possible to replace the reactors by three machine-side reactors with two windings each. In this case the design power is increased by the factor $\sqrt{3/2}$ above the design power of the indirect ac converter.

The design of the semiconductor elements of the direct converter as compared with the indirect dc converter is shown in Table II.

The data of the ac supply are

$$U_L = 145\ V, \qquad f_L = 50\ \text{Hz},$$

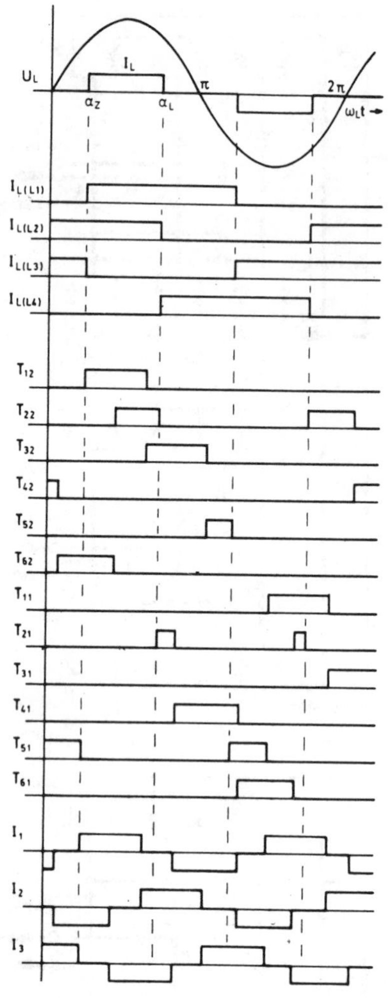

Fig. 16. Designed currents of the line and machine sides.

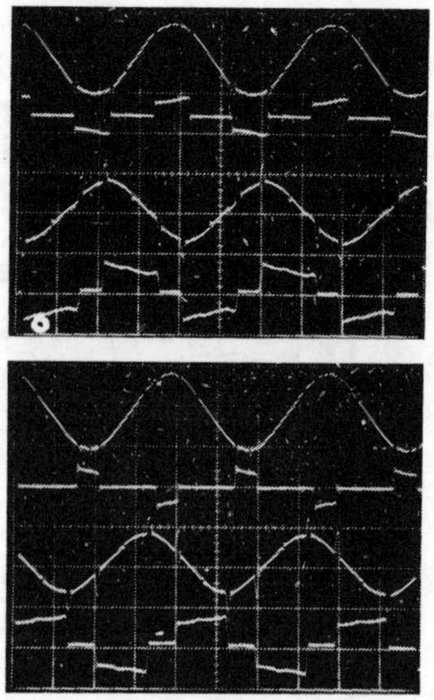

Fig. 17. Current waveforms on both line and machine sides for motoring and regenerating operation at 50-Hz line and machine frequency.

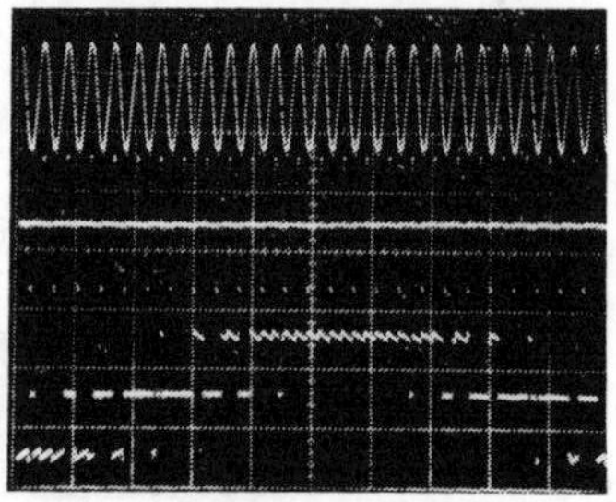

Fig. 18. Pulse current waveform for nine-pulse operation (machine frequency 0.9 Hz).

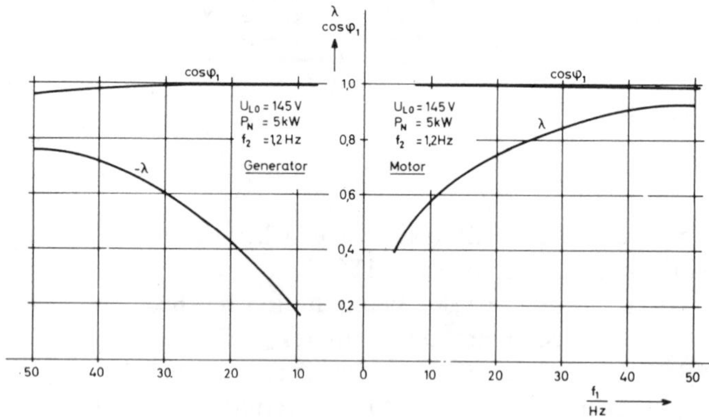

Fig. 19. Measurement of the power factor λ and the displacement factor cos τ_1 on the machine frequency f_1.

TABLE II
COMPARISON OF THE SEMICONDUCTOR ELEMENTS BETWEEN THE DIRECT CONVERTER AND THE INDIRECT CONVERTER

	indirect convertor	direct convertor
thyristors	10	12
diodes	8	8
capacitors	7	7
series connected semiconductor elements in the circuit	7	5
RMS value of the thyristor current	$4 \times \bar{I}_d/2$ $6 \times \bar{I}_d$	$12 \times \bar{I}_d/2$
RMS value of the diode current	$2 \times \bar{I}_d/2$ $6 \times \bar{I}_d$	$2 \times \bar{I}_d/2$ $6 \times \bar{I}_d$

and the data of the asynchronous machine, the load, are

$$P_N = 5\ \text{KW}, \quad U_{12} = 90\ \text{V}, \quad I_N = 44\ \text{A}, \quad f_{1N} = 100\ \text{Hz}.$$

Figs. 17 and 18 are oscilloscope records of the line voltage U_L, line current I_L, machine voltage U_{12}, and machine current I_1 for two different frequencies of the machine. In Fig. 19 are shown the results of the measurement of the power factor λ and displacement factor cos φ_1 on the machine frequency f_1.

CONCLUSION

As we have shown at the beginning of this report it is desirable to build advanced traction drives with asynchronous traction motors. According to the state of the art of power electronics three different drive systems are anticipated to be suited to this application.

Indirect ac-ac converters with a constant voltage dc link and pulsewidth modulated inverter already have been tested and introduced into production. It has been proven by experience that this drive system has excellent traction characteristics.

We have investigated two versions of an ac-ac converter with (suppressed) direct current link. The analysis of this system and the results of the first laboratory tests are shown.

We have started an extended investigation on a medium sized experimental setup of approximately 100 kW. The final result of this investigation is expected to be published in 1978.

REFERENCES

[1] J. Körber, "Grundlegende gesichtspunkte fur die auslegung elektrischer triebfahrzeuge mit asynchronen fahrmotoren," *Elektrische Bahnen*, vol. 45, pp. 52–59, 1974.

[2] K. W. Kanngiesser, "Umrichterverfahren und ihre anwendung," *BBC-Nachrichten*, vol. 46, pp. 609–624, 1964.

[3] A. Schönung and H. Stemmler, "Geregelter drehstrom—Umkehrsantrieb mit gesteuertem umrichter nach dem unterschwingungsverfahren," *BBC-Nachrichten*, vol. 46, pp. 699–721, 1964.

[4] W. Lienau and A. Müller-Hellmann, "Moglichkeiten zum betrieb van stromeinpragenden wechselrichtern ohne niedrefrequente oberschwingungen," *Elektrot. Zeitschrift Ausg. A*, vol. 97, pp. 663–667, 1976.

[5] J. Brenneisen, E. Futterlieb, E. Muller, and M. Schulz, "A new converter drive system for a diesel electric locomotive with asynchronous traction motors," *IEEE Trans. Ind. Appl.* vol. IA-9, no. 4, pp. 482–491, 1973.

[6] H. Kehrmann, W. Lienau, and R. Nill, "Vierquadrantensteller—eine netzfreundliche einspeisung fur triebfahrzeuge mit drehstromantrieb," *Elektrische Bahnen*, vol. 45, pp. 135–141, 1974.

[7] E. Becker and R. Gammert, "Drehstromversuchsfahrzeug—DE 2500 mit steuerwagen-systemerprobung eines drehstromantriebes an 15 kV 16²/3 Hz," *Elektrische Bahnen*, vol. 47, pp. 18–23, 1976.

[8] J. Körber and W. Teich, "Three phase motors for diesel and electric traction," *Railway Gazette Int.*, pp. 64–71, 1975.

[9] H. Kielgas and R. Nill, "Converter propulsion system with three phase induction motors for electric traction vehicles," presented at the IEEE 1977 Int. Semiconductor Power Convertor Conf.

[10] P. G. Sperling, "Die umrichtergespeiste asynchronmaschine im betrieb mit eingepragten rechteckströmen," *Siemens-Zeitschrift*, vol. 45, pp. 5085514, 1971.

[11] W. Lienau and A. Müller-Hellmann, "Drehstromantrieb mit stromeinprägendem zwischenkreisumrichter," *Elektrot. Zeitschrift Ausg. A*, vol. 97, pp. 84–86, 1976.

[12] E. Kern, "Der dreiphasen—stromrichtermotor und seine steuerung bei betrieb als umkehrmotor," *Elektrot. Zeitschrift Ausg. A*, vol. 59, pp. 467–470, 494–497, 1938.

[13] R. Stokes, "Three phase traction: problems and prospects," *Railway Gazette Int.*, pp. 419–422, 1976.

[14] H. J. Köpcke and W. Lienau, "Netzrückwirkungsarmer umrichter mit unterdrücktem zwischenkreis zur speisung von asynchronmotoren in der traktion," *Elektrot. Zeitschrift Ausg. A*, vol. 99, pp. 142–146, 1978.

A COMPARATIVE ANALYSIS OF TWO COMMUTATION CIRCUITS FOR ADJUSTABLE CURRENT INPUT INVERTERS FEEDING INDUCTION MOTORS

M. B. Brennen

Westinghouse Research and Development Center

Pittsburgh, Pennsylvania

ABSTRACT

A hybrid mathematical/graphical technique for quick determination of instantaneous voltages on components in two types of adjustable current input (ACI) inverters feeding a standard induction motor is presented with practical examples. A comparison between the synthesized voltage waveforms of the two inverter types, auxiliary thyristor and auto-commutated, reveals better overall performance for the auto-commutated inverter.

1. INTRODUCTION

Among several techniques applicable for variable speed induction motor drives, adjustable current (ACI) inverters seem to offer certain advantages. These inverters are characterized by a dc link input which behaves as a variable dc current source, through the action of a front end converter and a current regulation loop. The inverter thyristors, fired and commutated according to a programmed sequence, simply route the dc link current through the machine windings in a given sequence. An important advantage of such inverters is the elimination of thyristor protecting fuses. Since the current is supplied by the converter through a relatively large inductor, the dc link can be short circuited without excessive current build-up in the case of a commutation failure in the inverter. Further on, it has been found that an ACI inverter has the capability to ride through misfirings and tends to resume normal operation without the need of restarting after a temporary commutation failure. Another advantage of ACI inverters is their inherent ability to handle regenerative power. If the dc link current is set according to the slip of the motor and its value is kept constant by the action of the full wave front end converter, true four quadrant operation of the induction motor can be obtained.

The price of this ruggedness and versatility is the need of having a choke in the dc link with about a ten times higher inductance value than the phase leakage inductance of the motor. Also, the voltage stresses on the semiconductors in the inverter can be higher than those in an inverter with a constant dc voltage link, or adjustable input voltage (AVI) inverter.

The purpose of this paper is to present the mechanism of commutation process in the two, presently most encouraging ACI-inverter schemes, and to develop a technique for the calculation of worst case voltage stresses on the inverter components and motor terminals for comparison.

The first circuit to be analyzed has auxiliary commutating thyristors and is a Siemens patent application from 1968; the second one with commutating capacitors and series diodes has been known from the literature for at least 8 years.

The analysis was performed in conjunction with a 6-pole 230 V, 15 hp standard Westinghouse induction motor. The results obtained can be applied for the calculation of inverter device ratings with motors of different voltage and horsepower.

Prior to the analysis, it was necessary to determine the phase angle relationships between motor currents and motor terminal voltages in the 60 Hz speed range and with torques varying from +200% to -200% of rated torque. This was done with a computer program. It was found that the variation of voltage-current phase angle tends to decrease the relative voltage stresses in the circuit with increasing torque, compared to operation with a fictitious motor having no phase angle change for varying torque; and therefore, the phase angle should be considered in the analysis.

Since the worst case voltage stresses occur at the highest motor rpm it was sufficient to perform the stress analysis at 60 Hz, our highest operating frequency considered.

2. DESCRIPTION OF OPERATION OF THE THREE-PHASE THYRISTOR ACI INVERTER WITH AUXILIARY COMMUTATING CIRCUIT (TYPE 1)

The operation of the inverter with symmetrical L-R load is described in [1]. Unfortunately, the results given there cannot be readily translated for operation where the load is an induction motor and the magnitude of induced voltages is not negligible. The inverter circuit is shown in Figure 1.

Reprinted from *IEEE 1973 Power Elec. Spec. Conf.*, pp. 201–212, 1973.

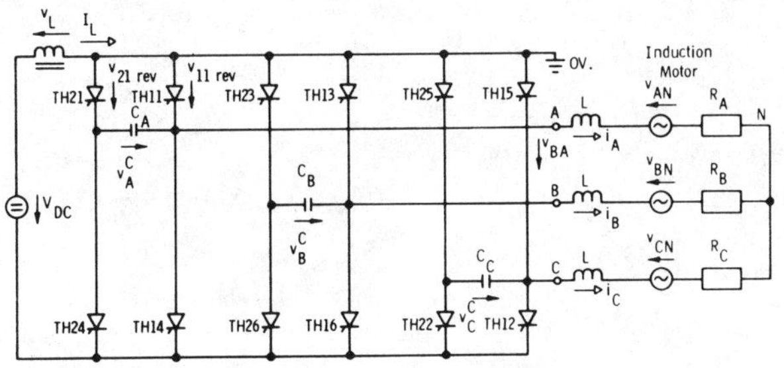

Fig. 1. Adjustable current input inverter with auxiliary commutating circuit

The "main" thyristors of the inverter are marked with a two digit number subscript starting with 1. The main thyristors are fired in a sequence shown by the second digit of their subscript. Each main thyristor has its commutating (auxiliary) thyristor which is marked with a first digit 2 and a second digit bearing the number of the main thyristor which it commutates.

The following table illustrates the firing sequence, with thyristors in the same column fired simultaneously:

MAIN:	TH11	TH12	TH13	TH14	TH15	TH16
COMM:	TH25	TH26	TH21	TH22	TH23	TH24

Let us assume that TH11 and TH16 are conducting ($i_A = I_L$) and that capacitor voltage v^C_A is positive. At a selected instant, we desire to interrupt the flow of dc link current I_L through TH11, and route I_L through TH13. To achieve this, we fire TH21 and TH13 simultaneously. When auxiliary thyristor TH11 is fired, its associated main thyristor TH11 is turned off instantaneously and the auxiliary thyristor takes over the whole link current. v^C_A starts decreasing linearly at a rate dependent on I_L. TH11 is reverse biased until v^C_A becomes zero. TH11 must recover from its conducting state by this time because a forward voltage will be applied after v^C_A changes polarity. TH13 is reversed biased since the moment it was fired by the sum of capacitor and induced line voltages $v^C_A + v_{BA}$. The commutating capacitor has to build up a negative voltage equal to v_{BA} before TH13 can become forward biased and can start conduction. The time from the firing of TH21 until TH13 becomes forward biased is called linear charging period T1. With TH13 on, phase B is connected to the upper rail and a positive voltage spike of increasing amplitude appears at the motor terminals superimposed on v_{BA}. The voltage spike is due to a L-C resonant discharge of duration T2 involving two phase leakage inductances L and commutating capacitor C_A. By the end of this resonant period i_A has decreased to zero while i_B has built up to I_L. The .5 $i^2_A L$ energy stored in the motor leakage inductance is transferred to C_A in the form of a resonant (sinusoidal) voltage build-up $\hat{v}^C_A R$ superimposed on what was the capacitor voltage just before the resonant discharge started. At the end of period T2, by the time i_B becomes equal to I_L, TH21 is turned off by current starvation concluding the commutation. At this moment v^C_A has reached the value $-(v_{BA} + \hat{v}^C_A R)$ and this is the capacitor initial voltage for next commutation of TH11.

The "ideal" phase current, if we assume an infinitely short commutation time, has a so-called "quasi square wave" shape with 120° conduction and 60° dwell-time in a half period. The practical current waveshape with a standard induction motor and actual commutation time is shown in Figure 2 (i_A dashed line). i_{AF} approximate fundamental of i_A can also be seen drawn dashed on the i_A axis (i_{AF} amplitude is not scale).

In order to keep the voltage stresses at minimum, the commutating capacitance should be as large as possible. On the other hand, the sum of the varying T1 + constant T2 must not exceed a duration equivalent to 120° of one period; the main thyristor must be allowed to assume full conduction of line current for at least a very short time.

For given values of the capacitance, leakage inductance and motor terminal voltage, the dominant linear charging time is a sole function of the minimum value of link current which is equal to the magnetizing current (I1) of the motor.

Based on the above considerations, the relation among magnetizing current, maximum operating frequency and maximum value of commutating capacitance is given by (1) with a good approximation.

$$C \leq \left[.91 \frac{I1 \times F_n}{V_n F_{max}} \left(\sqrt{L + .202 \frac{V_n}{F_n I1}} - \sqrt{L} \right) \right]^2 \qquad (1)$$

where V_n is the peak motor nameplate line-to-line voltage [V]

I1 peak magnetizing current [A]

F_n is the motor nameplate frequency [Hz]

L leakage inductance of one phase [H] (stator + rotor)

C maximum value of commutating capacitance [F]

F_{max} upper limit of operating frequency [Hz]

The losses on equivalent stator resistances R_A, R_B, R_C have been neglected in the derivation of (1) which adds a small safety factor of about 7% to F_{max}.

Another restriction for successful commutation is that the resonant commutation angle, θ2, cannot exceed 60° of the cycle of the output current fundamental, at maximum operating frequency:

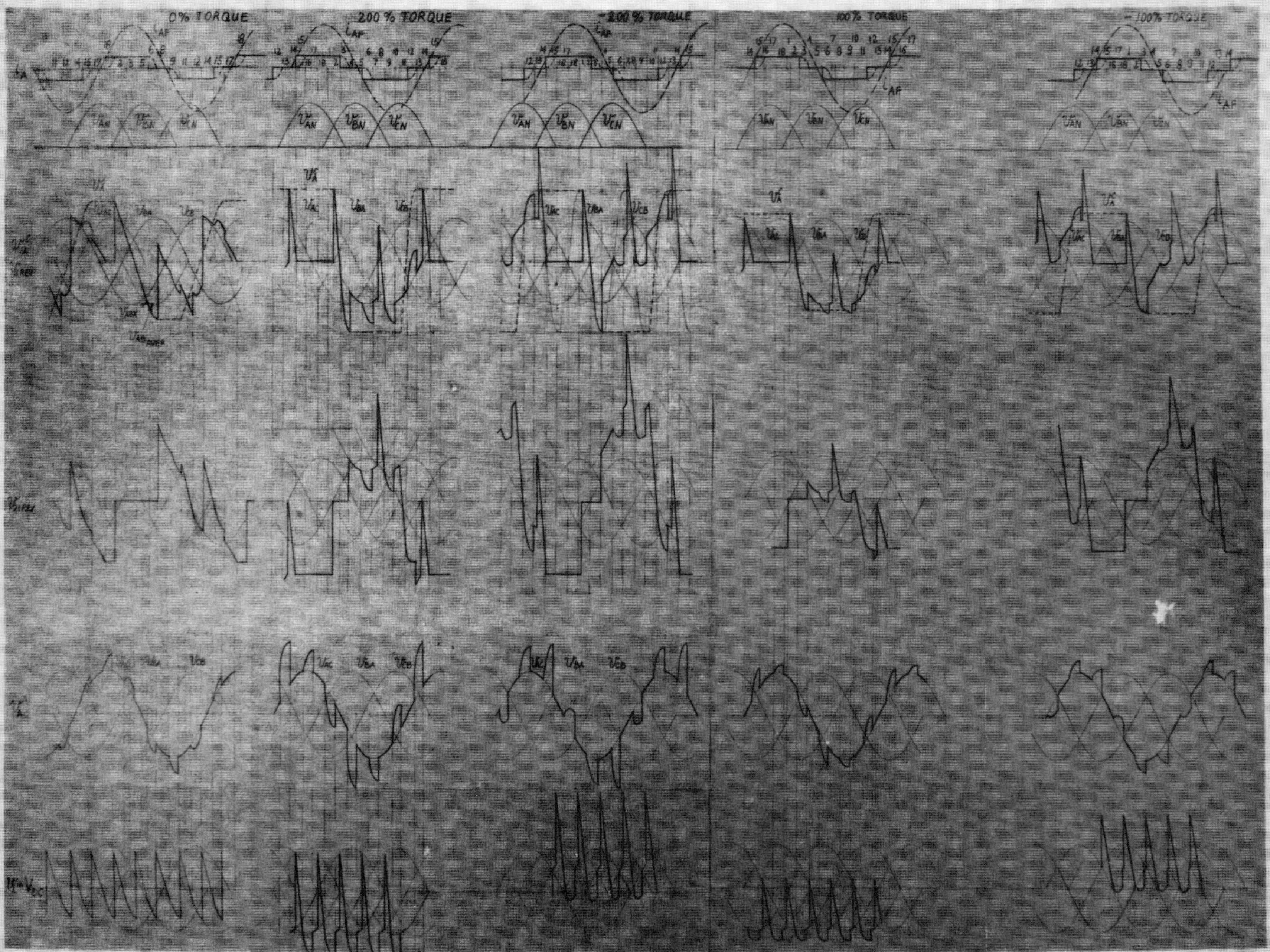

Figure 2. Voltage synthesis of TYPE 1 inverter.

$$T2 = \frac{\pi \sqrt{2LC}}{2} \leq \frac{1}{6F_{max}} \qquad (2)$$

$$\theta 2 = 360 \; T2 \; F_{max} \qquad (2a)$$

$\theta 2 = 26.5°$ for the standard 15 hp motor used in the analysis. C must be selected to satisfy the most restrictive of the two conditions.

The commutation capability of the auxiliary circuit must be retained down to zero frequency, where T1 and the reverse bias time is minimum for a given current. This is satisfied if T1, at zero frequency, exceeds the main thyristor recovery time t_{oo}. This conditions sets a lower limit to the value of the commutation capacitor C, given by

$$\sqrt{2LC} \; e^{-\frac{\pi}{4Q}} = t_{oo} \qquad (2b)$$

2.1 Waveform Synthesis During Steady State Operation (60 Hz)

In the following a synthesis is presented for obtaining the instantaneous voltages on the inverter components while the motor runs with steady state load and develops a given discrete -200, -100, 0, 100, or 200% of its rated torque. The purpose of the synthesis is to obtain voltages which are expressed in a scale graphical from (Figure 2) so that their peak values reveal the worst case stresses on the particular component at a given torque.

The graphical determination of the voltages across the various components of the inverter is implemented assuming a motor having the following parameters (Westinghouse Motor, Style No.:680B101G45):

Nameplate frequency: 60 Hz

Nameplate voltage: 230 V RMS line-line

Rated line current at 100% torque: 40.2 A_{rms}

Stator leakage inductance: .0008 H

Stator resistance: .188 ohms

Magnetizing inductance: .0203 H

Rotor leakage inductance: .00155 H

Rotor resistance: .104 ohms

It is assumed that the inverter drives the motor at 60 Hz and that the commutation capacitors are sized in such a manner that 60 Hz is the maximum frequency of operation at zero load.

The synthesis consists of the following steps:

a. Select on of the discrete torque values.

b. Determine the value of link current I_L which corresponds to the selected torque from Figure 3. The torque-current relationship is calculated from given motor data. The dashed line should be used which is corrected for the form factor of the "quasi square wave" shape of i_A

$$I_L = \frac{\pi}{2\sqrt{3}} \, i_{AF} \qquad (3)$$

where i_{AF} is the peak value of i_A fundamental.

c. Determine phase shift $\phi_{(V-I)}$ between v_{AN} and i_{AF} at the choosen torque from Figure 4. Use the curve for 60 Hz.

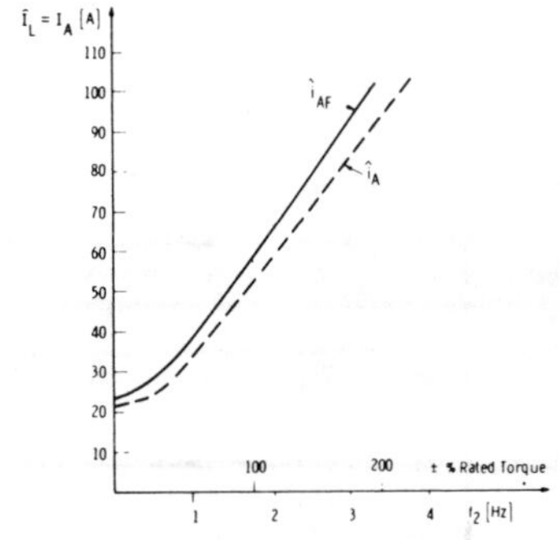

Fig. 3. Torque-current relationship at constant flux

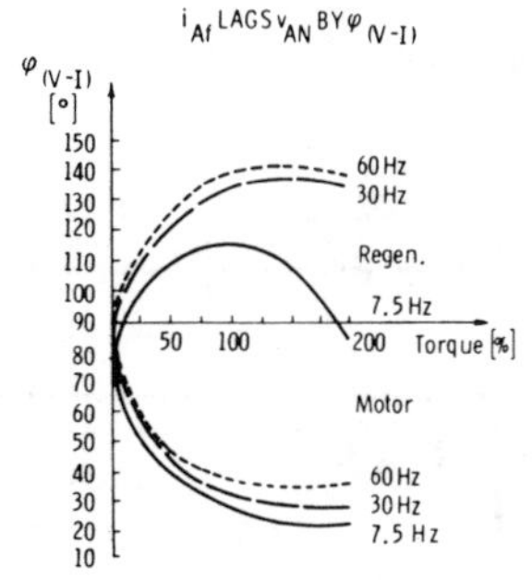

Fig. 4. Torque-$\varphi_{(V-I)}$ relationship at different speeds. 15 HP

d. Calculate $\theta 2$ from (2a), draw i_A as shown in Figure 2. The corners are sine wave quarters (0°-90°)

e. Draw the thyristor commutation sequence. The sequence for zero torque is shown in Figure 5. The sequence for a typical motoring (or generating) torque is shown in Figure 6.

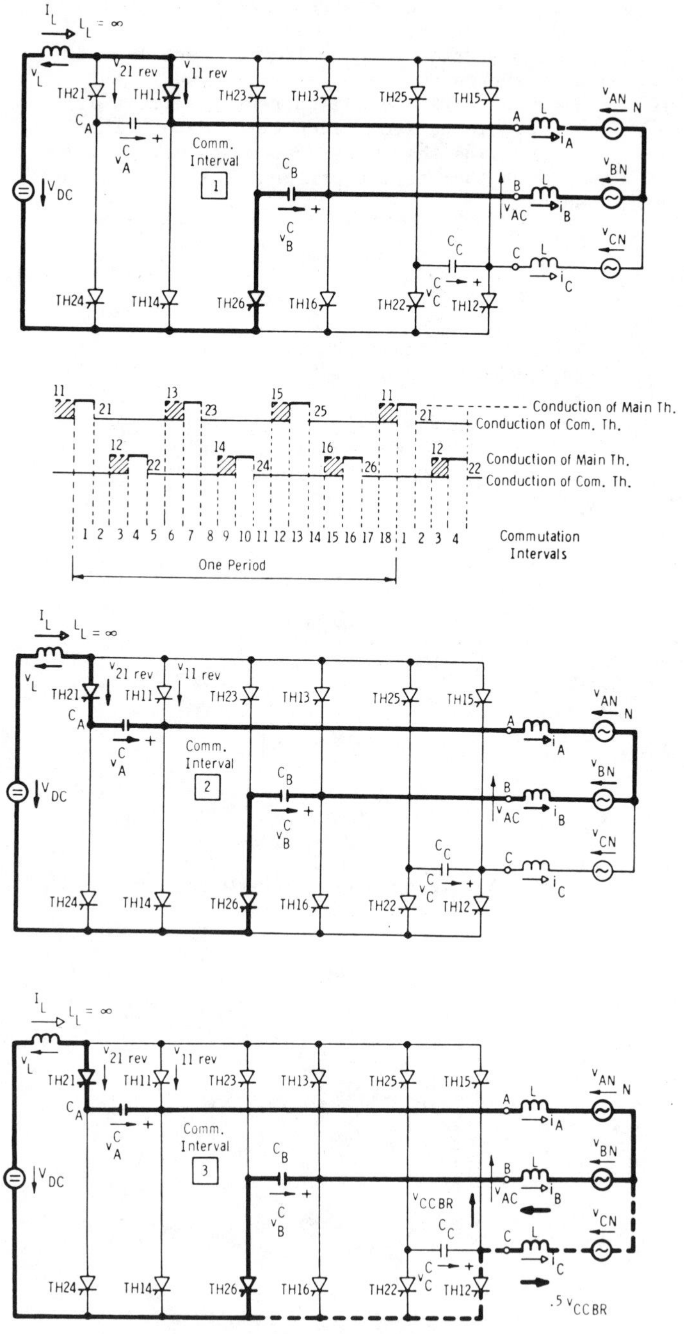

Fig. 5. Equivalent Circuits (zero torque)

Interpretation of commutation sequence diagram, typically Figure 5. One cycle of the inverter output waveform is divided in a number of commutation intervals by dashed vertical lines. The conduction of a commutation thyristor is represented by the thin continuous horizontal lines in the "conduction of com. th." rows. The corresponding commutation thyristor for each thin horizontal line is identified by its number at the beginning of the line. Let us recall that main and commutation thyristors are fired simultaneously according to the table in Chapter 2 and a commutation thyristor commences its conduction immediately after it is fired. Similarly, the conduction of main thyristors is represented by thick horizontal lines starting dashed and ending solid. A conducting main thyristor is identified by its number at the beginning of the dashed thick line. The dashed portion indicates that the corresponding main thyristor shares the link current with an outgoing commutation thyristor. The solid portion means that the corresponding main thyristor conducts the full link current. For example, considering interval 6, we see that thyristor TH22 is conducting, and that the conduction started at the beginning of interval 5. We also see that thyristor TH21 is conducting, and started its conduction at the beginning of interval 2. Finally we see that TH13 is conducting, and that the conduction started at the beginning of interval 6 (although TH13 was fired together with TH21). The dashed line indicates that TH13 shares the line current with TH21.

The duration of a particular interval varies in time and degrees of output cycle according to the operating frequency and torque conditions. At zero torque and maximum frequency, for instance, intervals 1, 4, 7... are infinitely short, since the main thyristors theoretically never carry the link current without sharing it with the outgoing commutation thyristor, that is:

$$T1 + T2 = \frac{1}{3\ F_{max}}$$

f. Designate each combination of thyristor conductions by a number within a period (18 commutation intervals) as shown.

g. Draw the equivalent circuit of each commutation interval. (As an example, the first three equivalent circuits are shown in Figure 5.)

h. Prepare Table 1 by applying Kirchoff's equations to each equivalent circuit ($\Sigma v = 0$).

NOTE: Use line to line voltages, e.g., v_{AC}. The resonant voltage build-up during commutation is divided equally between two leakage inductances (intervals 3, 6, 9,... etc.).

0% TORQUE TYPE 1 STRESSES

Commut. Interval	V_{AC}	V_{11}rev	V_{21}rev	v_A^C	$V_L + V_{DC}$
1	$\overline{V}_{AC}$	0	$-v_A^C$	NC	$\overline{V}_{BA} - v_B^C$
2	$\overline{V}_{AC}$	v_A^C	0	-+	$\overline{V}_{BA} + v_A^C - v_B^C$
3	$\overline{V}_{AC} + 1/2(v_{BR}^C)$	v_A^C	0	+	$-\overline{V}_{AC} + v_A^C - 1/2(v_{BR}^C)$
4	$\overline{V}_{AC}$	v_A^C	0	+	$-\overline{V}_{AC} + v_A^C$
5	$\overline{V}_{AC}$	v_A^C	0	+	$-\overline{V}_{CB} + v_A^C - v_C^C$
6	$\overline{V}_{AC} + 1/2(v_{AR}^C)$	v_A^C	0	+-	$\overline{V}_{CB} + 1/2(v_{AR}^C) - v_C^C$
7	$\overline{V}_{AC}$	$-\overline{V}_{BA}$	$-\overline{V}_{BA} - v_A^C$	NC	$\overline{V}_{CB} - v_C^C$
8	$\overline{V}_{AC}$	$-\overline{V}_{BA}$	$-\overline{V}_{BA} + v_B^C - v_A^C$	NC	$\overline{V}_{CB} + v_B^C - v_C^C$
9	$\overline{V}_{AC} - v_{CR}^C$	$-\overline{V}_{BA} + v_B^C - 1/2(v_{CR}^C)$	$-\overline{V}_{BA} + v_B^C - 1/2(v_{CR}^C) - v_A^C$	NC	$-\overline{V}_{BA} + v_B^C - 1/2(v_{CR}^C)$
10	$\overline{V}_{AC}$	$-\overline{V}_{BA} + v_B^C$	$-\overline{V}_{BA} + v_B^C - v_A^C$	NC	$-\overline{V}_{BA} + v_B^C$
11	$\overline{V}_{AC}$	$-\overline{V}_{BA} + v_B^C$	$-\overline{V}_{BA} + v_B^C - v_A^C$	-+	$-\overline{V}_{BA} + v_B^C - v_A^C$
12	$\overline{V}_{AC} + 1/2(v_{BR}^C)$	$\overline{V}_{AC} + 1/2(v_{BR}^C)$	$\overline{V}_{AC} + 1/2(v_{BR}^C) - v_A^C$	+	$\overline{V}_{AC} + 1/2(v_{BR}^C) - v_A^C$
13	$\overline{V}_{AC}$	$\overline{V}_{AC}$	$\overline{V}_{AC} - v_A^C$	+	$\overline{V}_{AC} - v_A^C$
14	$\overline{V}_{AC}$	$\overline{V}_{AC} + v_C^C$	$\overline{V}_{AC} + v_C^C - v_A^C$	+	$\overline{V}_{AC} + v_C^C - v_A^C$
15	$\overline{V}_{AC} + 1/2(v_{AR}^C)$	$\overline{V}_{AC} + 1/2(v_{AR}^C) + v_C^C$	$\overline{V}_{AC} + 1/2(v_{AR}^C) + v_C^C - v_A^C$	+-	$-\overline{V}_{CB} + v_C^C - 1/2(v_{AR}^C)$
16	$\overline{V}_{AC}$	$\overline{V}_{AC} + v_C^C$	$\overline{V}_{AC} + v_C^C - v_A^C$	NC	$-\overline{V}_{CB} + v_C^C$
17	$\overline{V}_{AC}$	$\overline{V}_{AC} + v_C^C$	$\overline{V}_{AC} + v_C^C - v_A^C$	NC	$-\overline{V}_{CB} + v_C^C - v_B^C$
18	$\overline{V}_{AC} - v_{CR}^C$	0	$-v_A^C$	NC	$\overline{V}_{BA} + 1/2(v_{CR}^C) - v_B^C$
1	$\overline{V}_{AC}$	0	$-v_A^C$	NC	$\overline{V}_{BA} - v_B^C$

Table 1

> ±100% TORQUE TYPE 1 STRESSES

Commut. Interval	V_{AC}	V_{11} rev	V_{21} rev	v_A^C	V_L
1	$\overline{V}_{AC}$	0	$-v_A^C$	NC	$\overline{V}_{CA}$
2	$\overline{V}_{AC}$	v_A^C	0	-+	$\overline{V}_{CA} + v_A^C$
3	$\overline{V}_{AC} + 1/2(v_{AR}^C)$	v_A^C	0	+-	$\overline{V}_{CB} + 1/2(v_{AR}^C)$
4	$\overline{V}_{AC}$	$\overline{V}_{AB}$	$\overline{V}_{AB} - v_A^C$	NC	$\overline{V}_{CB}$
5	$\overline{V}_{AC}$	$\overline{V}_{AB}$	$\overline{V}_{AB} - v_A^C$	NC	$\overline{V}_{CB} - v_C^C$
6	$\overline{V}_{AC} - v_{CR}^C$	$\overline{V}_{AB} - 1/2(v_{CR}^C)$	$\overline{V}_{AB} - 1/2(v_{CR}^C) - v_A^C$	NC	$\overline{V}_{AB}$
7	$\overline{V}_{AC}$	$\overline{V}_{AB}$	$\overline{V}_{AB} - v_A^C$	NC	$\overline{V}_{AB}$
8	$\overline{V}_{AC}$	$\overline{V}_{AB} + v_B^C$	$\overline{V}_{AB} + v_B^C - v_A^C$	NC	$\overline{V}_{AB} - v_B^C$
9	$\overline{V}_{AC} + 1/2(v_{BR}^C)$	$\overline{V}_{AC} + 1/2(v_{BR}^C)$	$\overline{V}_{AC} + 1/2(v_{BR}^C) - v_A^C$	NC	$\overline{V}_{AC} + 1/2(v_{BR}^C)$
10	$\overline{V}_{AC}$	$\overline{V}_{AC}$	$\overline{V}_{AC} - v_A^C$	NC	$\overline{V}_{AC}$
11	$\overline{V}_{AC}$	$\overline{V}_{AC}$	$\overline{V}_{AC} - v_A^C$	-+	$\overline{V}_{AC} - v_A^C$
12	$V_{AC} + 1/2(v_{AR}^C)$	$\overline{V}_{AC} + 1/2(v_{AR}^C)$	$\overline{V}_{AC} + 1/2(v_{AR}^C) - v_A^C$	+-	$\overline{V}_{BC} - 1/2(v_{AR}^C)$
13	$\overline{V}_{AC}$	$\overline{V}_{AC}$	$\overline{V}_{AC} - v_A^C$	NC	$\overline{V}_{BC}$
14	$\overline{V}_{AC}$	$\overline{V}_{AC}$	$\overline{V}_{AC} - v_A^C$	NC	$\overline{V}_{BC} + v_C^C$
15	$V_{AC} - v_{CR}^C$	0	$-v_A^C$	NC	$\overline{V}_{BA} + 1/2(v_{CR}^C)$
16	$\overline{V}_{AC}$	0	$-v_A^C$	NC	$\overline{V}_{BA}$
17	$\overline{V}_{AC}$	0	$-v_A^C$	NC	$\overline{V}_{BA} - v_B^C$
18	$\overline{V}_{AC} + 1/2(v_{BR}^C)$	0	$-v_A^C$	NC	$\overline{V}_{CA} - 1/2(v_{BR}^C)$
1	$\overline{V}_{AC}$	0	$-v_A^C$		$\overline{V}_{CA}$

Table 2

Interpretation of Table 1. The table is divided in five columns corresponding to five studied voltage stresses. These voltages are:

v_{AC} instantaneous line-line motor voltage

v_{11rev} reverse voltage across thyristor 11

v_{21rev} reverse voltage across thyristor 21

v_A^C voltage across commutating capacitor C_A

$v_L + V_{dc}$ instantaneous dc link voltage (ignoring converter ripple)

These voltage stresses are related to the motor induced line-line voltages ($\overline{V}_{AC}$, $\overline{V}_{CB}$, $\overline{V}_{BA}$) and to the commutating capacitor voltages by equations derived from Kirchoff's law. At a given commutation interval, the relationship between a given studied voltage and the other terms is expressed in the corresponding rows and columns. Example: At commutation interval 9, we have

$$V_{11} \text{ rev} = -\overline{V}_{BA} + V_B^C - 1/2(V_{CR}^C)$$

The terms V_{AR}^C, etc., indicate the resonant voltage built-up of the commutating capacitors. When there is no change in voltage across the commutating capacitor, this is indicated by NC in the V_A^C column. Arrows pointing down (up) indicate discharge (charge) of that capacitor.

i. Mark the intervals on i_A axis in Figure 2.

j. Estimate the position of i_{AF} (fundamental of i_A) relative to i_A and draw i_{AF}. (i_{AF} lags i_A by approximately .66 θ2 = 18°)

k. Draw v_{AN} with $\phi_{(V-I)}$ phase shift relative to i_{AF} (scale curve).

l. Draw line-to-line voltages on V_{11rev} axis (scale curve). (V_{AC} lags V_{AN} by 30°)

m. i_A is reduced to zero from its positive maximum in interval 6. The negative resonant charging of C_A starts at the beginning of the interval when $V_A^C = V_{AB}$. Estimate the average value of V_{AB} during the interval on the scale drawing. Mark the location of the average value V_{ANavg} on V_{AB}.

n. Calculate peak voltage V_{AR}^C with losses considered. The general expression for interval 6 is:

$$V_{AR}^C = \sqrt{2L\ I_L}\ \omega_o\ (\sin \omega_o t)\ e^{-\frac{\omega_o t}{2Q}} \tag{4}$$

The peak value of this sinusoidal wave is:

$$\hat{v}_{AR}^{C} = - I_L \sqrt{\frac{2L}{C}} \cdot e^{-\frac{\omega_o t_1}{2Q}} \qquad (5)$$

where $Q = \frac{X}{R} = 4$

$X = 2L\omega_o = 6$ ohm

$\omega_o = \frac{1}{\sqrt{2LC}} = 1256$ rad/sec

$\omega_o t_1 = \pi/2$

$R = R_1 + R_2' = 1.5$ ohm

The value of transient line-to-line resistance $R_1 + R_2$ at $f_1 = \omega_o/2\pi$ frequency is derived from motor data and from Reference [6] p. 159, where the skin effect dependence of the motor losses is analyzed.

For the present case of zero torque we obtain $\hat{v}_{AR}^{C} = -87$ V.

o. Add $\hat{v}_{AR}^{C} = -87$ V to $v_{AV_{aver}}$ at the marked location to obtain v_{AR}^{C}. ($v_{A-}^{C} = v_{AB_{aver}} + v_{AR}^{C}$, the negative peak voltage on the commutating capacitor.)

p. We can see from the commutation sequence that TH21 was fired at the beginning of interval 2. At that time C_A started its linear discharge from its positive peak voltage. Therefore a strating line between intersection of the beginning of interval 6 with v_{AB}(marked V_{ABX}), and the positive peak value of the capacitor voltage at the beginning of interval 2 gives v_A^C from interval 2 through interval 5. The sinusoidal resonant build up in interval 6 is nearly tangential to v_{A-}^C v_A^C remains at its negative peak until TH24 is fired in interval 11. Another linear discharge takes place followed by resonance in interval 15. By this time v_A^C reaches its positive peak which is preserved until TH21 is fired starting a new period. The other capacitor voltages v_B^C, v_C^C have the same magnitude as v_A^C but they are shifted 120° apart in the symmetrical 3-phase system.

Let us recall that all voltage stresses in Table 1 are derived from induced motor voltages and voltages of the commutating capacitors. Now we have all the components which appear in Table 1. We can obtain all the waveforms by linear superposition of instantaneous magnitudes while carefully observing their polarities.

q. When the troque is different from zero, e.g., 200% in Figure 2, the new thyristor commutation sequence must be found first. Draw i_A, i_{AF} and line to line voltages v_{AC}, v_{BA}, v_{CB}, as described in a to d. Note, θ2 (length of interval 3) is not load dependent.

r. Find the location of v_{ABaver} in interval 3 (Figure 3, 200% torque), calculate v_{AR}^C, v_A^C, as explained m, n, and o.

s. Calculate linear charge angle θ1 (interval 2) from (7)

$$T1 = C \frac{\left(\hat{v}_{AR}^{C} + v_{AB_{aver}} + v_{ABX}\right)_{\text{interval 6}}}{I_L} \qquad (6)$$

$$\theta 1 = 360\ T1\ F_{max} \qquad (7)$$

The conduction angle of TH21 is the sume of θ1 + θ2. With this data the commutation sequence is shown in Figure 6. It has been found that the commutation sequence does not change above 100% torque with this motor.

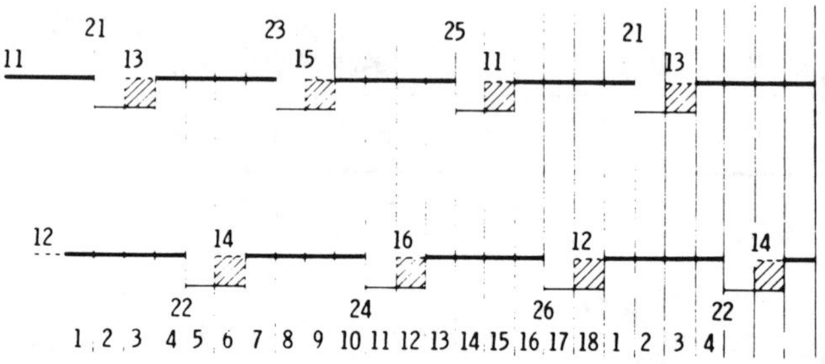

Fig. 6. Thyristor commutation sequence ±100% < torque < ±200%

t. Prepare Table 2 as explained in f to l and draw v_A^C on v_A^C axis (dashed line in Figure 2, 200% torque).

NOTE: The actual $\hat{v}_{AR}^{C}$ (recovery step) at the end of interval 3 is smaller than the calculated value because v_{AB} varies considerably during the resonance. The recovery step instead of $\hat{v}_{AR}^{C}$ should then be considered for the rest of the waveform synthesis.

2.2 Derivation of Worst Case Device Stresses as a Function of Motor Torque

By inspecting the peak amplitudes in Figure 2, we can determine the corresponding commutation intervals in which the peaks occur. The exact voltage equations then can be looked up in the corresponding table.

The calculated stress envelopes are given in Figure 7. The measured stresses are expected to be somewhat lower than calculated at regeneration due to the slight decrease of v_{LL} as regenerative torque is applied.

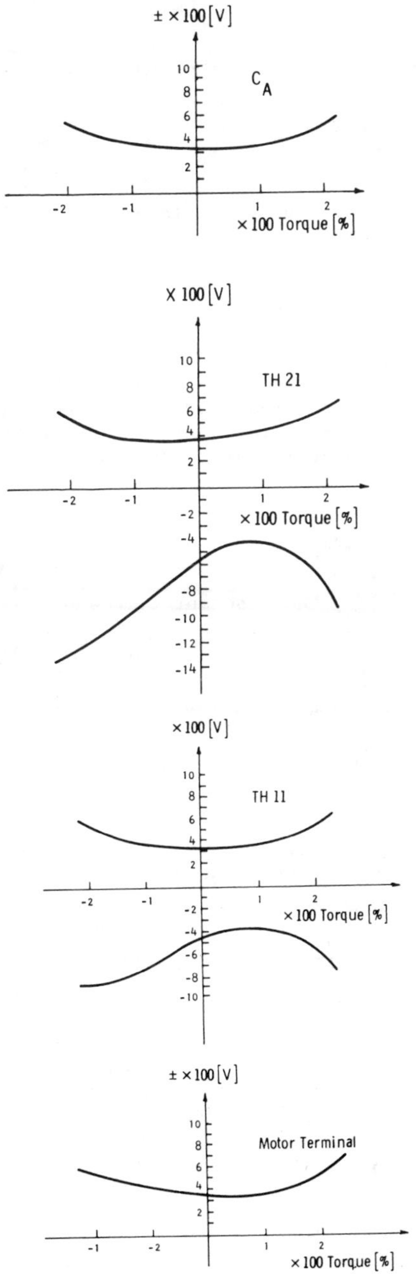

Fig. 7. Peak device stress envelopes (Type)

3. ACI INVERTER WITH COMMUTATING CAPACITORS AND DIODES (TYPE 2)

The operation of this inverter under various load conditions has been published in several papers. (References [2,3,4,5]). In these papers the commutation time is assumed instantaneous and the effects of line voltage variations during commutation are neglected. They also fail to give explicit data on device stresses.

In the following description of operation the same standard induction motor load is assumed as for the TYPE 1 inverter with optimized commutating capacitors for a selected speed range. Since motor voltage variations are significant during commutation they are considered in the derivation of peak stress envelopes of components.

The inverter circuit is shown in Figure 8. Each thyristor has a series isolating diode having the same number as the thyristor. The thyristors are numbered in the order of their firing sequence and they conduct through 120° of a full period. It is apparent from the circuit connection of Figure 8 that: The sum of voltages on the three equal capacitors in either group must be zero at any time, the equivalent capacitance of the delta connected capacitors is 1.5 C_{AB}, and the capacitors cannot be discharged by the motor due to isolating diodes D1-D6.

It could be shown similarly to the TYPE 1 case that the voltages on the commutating capacitors tend to converge to a single stable value depending on the motor nameplate voltage and instantaneous $\rho_{(V-I)}$ and I_L for a given motor.

It has been found that the inverter has two operating modes, both resulting in proper commutation. In mode 1 the commutating capacitor obtains its charge from the outgoing phase current only. In mode 2 a small charge is obtained through the "idling" phase which does not take part in that particular commutation.

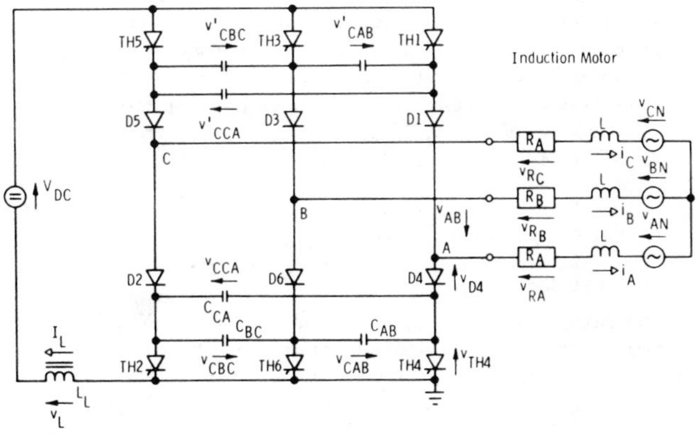

Fig. 8. Adjustable current link inverter with commutating capacitors and diodes (TYPE 2)

An exact mathematical analysis of mode 2 operation is not attempted here due to the complexity involved. It has been found that the peak voltage stress in mode 2 is less than in mode 1. The inverter operates practically in mode 1 with a standard motor developing +200% to -200% torque.

Let us go through a commutation cycle. When, e.g., TH4 and TH5 conduct, TH6 is the next to be fired in the sequence. Initially, $i_A = I_L$, the capacitor voltages stay constant because they are isolated from varying line voltages (e.g., v_{AB}). v_{CAB} must be negative for successful commutation. Assume $v_{CBC} = 0$ and $v_{CCA} = -v_{CAB}$. By firing TH6, v_{CAB} reverse biases TH4 and i_A is rerouted instantaneously through C_{AB} and TH6. At this point D6 is reverse biased by ($v_{CAB} + v_{AB}$). DC rail voltage ($v_L + v_{DC}$) is increased by a step of v_{CAB} magnitude. Since I_L is constant, C_{AB} is discharged at a constant rate resulting in a "linear charge" period similar to TYPE 1 commutation. The motor does not notice the current transfer if $L_L >> L$ in the series circuit. TH4 is reverse biased while $v_{CAB} < 0$. When $v_{CAB} = v_{AB}$, ($v_{AB} = v_{ABX}$) in Figure 10, D6 becomes forward biased. From that moment on, a resonant LC loop is closed, involving inductor L in phase B, inductor L in phase A and the capacitor bank, via D6 and D4. A resonant build up of i_B from $i_B = 0$ to $i_B = i_L$ occurs in a time corresponding to one quarter of the period of resonance, while the inductive energy stored in the leakage inductance of the outgoing phase is transferred to the commutating capacitor bank. (C_{BC} and C_{CA} are connected parallel to C_{AB}.) The resonant voltage build up is superimposed on the linear voltage build up prior to resonance C_{AB}. At this point, C_{AB} has the wrong polarity voltage for the next commutation.

Since the sum of capacitor voltages is always zero, while v_{CAB} is varied C_{BC} and C_{CA} must have gained half of v_{CAB} voltage respectively but with opposite polarities. There will be two commutations in the lower bank of the inverter before C_{AB} has to commutate again. Thus the initial commutation voltage with the right polarity (and the assumed initial magnitude), is re-established by the following two equal (but opposite polarity) voltage gains of C_{AB}. Let us mark the values of TYPE 2 components by a star. The maximum capacitance of a commutating capacitor is derived from (1) taking into account the delta connection of the capacitors:

$$C^* \leqq .666\left[.91\,\frac{I1 \times Fn}{V_n\,F_{max}}\left(\sqrt{L + .202\,\frac{V_n}{F_n\,I1}} - \sqrt{L}\right)\right]^2 \qquad (8)$$

The commutating capacitor resonant voltage build up time function is derived from (4)

$$v_{CABR} \leqq \sqrt{2LI_L}\,\omega_o\,(\sin \omega_o t)\cdot e^{-\frac{\omega_o t}{2Q}} \qquad (9)$$

where $\omega_o = \dfrac{1}{\sqrt{3LC^*}}$

and the build-up peak value from (5) and (9)

$$\hat{v}_{CABR} = I_L\sqrt{\frac{2L}{1.5\ C^*}}\ e^{-\frac{\pi}{4Q}} \qquad (10)$$

The commutating capacitor linear charge time is derived from (6), (6a):

$$T1^* = 1.5\ C^*\,\frac{\hat{v}_{CABR} + v_{AB_{aver}} + v_{ABX}}{I_L} \qquad (11)$$

where $v_{AB_{aver}}$ is the average motor line-to-line voltage during the resonant period and v_{ABX} is the value of v_{AB} at the beginning of interval 6. The corresponding linear charge angle is:

$$\theta 1^* = 360\ T1^*F_{max} \qquad (12)$$

From (2), (2a), the resonant time and the restriction on resonant charge angle $\theta 2^*$ becomes:

$$T2^* = \frac{\pi\sqrt{3LC^*}}{2} = \frac{L}{6\ F_{max}} \qquad (13)$$

$$\theta 2^* = 360\ T2^*F_{max} \qquad (14)$$

The star marked equations have the same dimensions as their TYPE 1 counterparts.

3.1 Waveform Synthesis During Steady State Operation (60 Hz)

Basically the same technique can be applied for obtaining the instantaneous steady state voltages as the one described in Chapter 2.1 in connection with the TYPE 1 circuit. The selected torque values are also the same for better comparison.

The commutation sequences for zero and equal or larger than ±100% torques are shown in Figure 9.

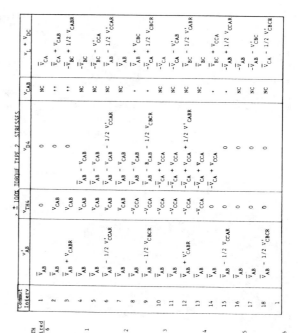

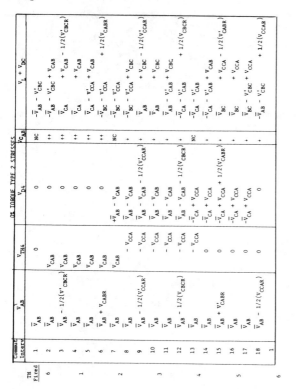

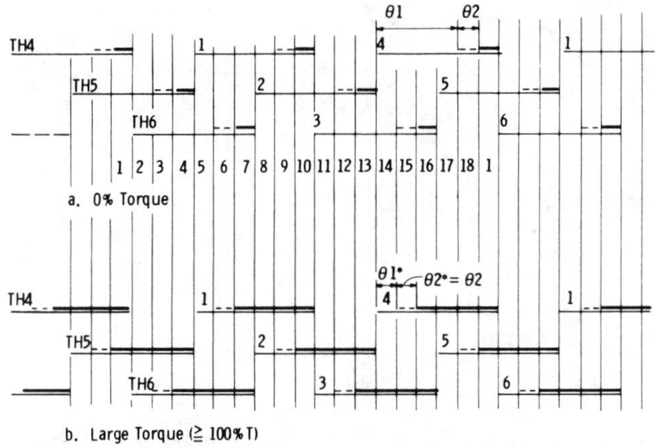

Fig. 9. Commutation sequence (TYPE 2)

The instant of a firing is marked with the corresponding thyristor number. A thin horizontal line below the number indicates the conduction of the thyristor. The dashed line means resonance, that is the corresponding diode to the thyristor commences conduction. The thick line indicates that the thyristor conducts its full associated phase current, e.g., $-i_A$ for TH4.

Data obtained from equivalent circuits of commutation intervals is tabulated for zero and larger than ±100% torques in Tables 3 and 4. The instantaneous voltage waveforms are shown in Figure 10. Peak device stress envelopes are given in Figure 11.

Table 3

0% TORQUE TYPE 2 STRESSES

TH Fired	Commut Interv	V'_{AB}	V_{TH4}	V_{D4}	$V_{C_{AB}}$	$V_L + V_{DC}$
6	1	$\bar{V}_{AB}$	0	0	NC	$-\bar{V}_{AB} - V'_{CBC}$
	2	$\bar{V}_{AB}$	V_{CAB}	0	++	$-\bar{V}_{AB} - V'_{CBC} + V_{CAB}$
	3	$\bar{V}_{AB} - 1/2(V'_{CBCR})$	V_{CAB}	0	++	$\bar{V}_{CA} + V_{CAB} - 1/2(V'_{CBCR})$
1	4	$\bar{V}_{AB}$	V_{CAB}	0	++	$\bar{V}_{CA} + V_{CAB}$
	5	$\bar{V}_{AB}$	V_{CAB}	0	++	$\bar{V}_{CA} - V'_{CCA} + V_{CAB}$
	6	$\bar{V}_{AB} + V_{CABR}$	V_{CAB}	0	++	$-\bar{V}_{BC} - V'_{CCA} + 1/2(V_{CABR})$
2	7	$\bar{V}_{AB}$	V_{CAB}	$+\bar{V}_{AB} - V_{CAB}$	NC	$-\bar{V}_{BC} - V'_{CCA}$
	8	$\bar{V}_{AB}$	$-V_{CCA}$	$\bar{V}_{AB} - V_{CAB}$	+	$-\bar{V}_{BC} - V'_{CCA} + V_{CBC}$
	9	$\bar{V}_{AB} - 1/2(V'_{CCAR})$	$-V_{CCA}$	$\bar{V}_{AB} - V_{CAB} - 1/2(V'_{CCAR})$	+	$\bar{V}_{AB} + V_{CBC} - 1/2(V'_{CCAR})$
3	10	$\bar{V}_{AB}$	$-V_{CCA}$	$\bar{V}_{AB} - V_{CAB}$	+	$\bar{V}_{AB} + V_{CBC}$
	11	$\bar{V}_{AB}$	$-V_{CCA}$	$\bar{V}_{AB} - V_{CAB}$	+	$\bar{V}_{AB} - V'_{CAB} + V_{CBC}$
	12	$\bar{V}_{AB} - 1/2(V_{CBCR})$	$-V_{CCA}$	$\bar{V}_{AB} - V_{CAB} - 1/2(V_{CBCR})$	+	$-\bar{V}_{CA} - V'_{CAB} + 1/2(V_{CBCR})$
4	13	$\bar{V}_{AB}$	$-V_{CCA}$	$-\bar{V}_{CA} + V_{CCA}$	NC	$-\bar{V}_{CA} - V'_{CAB}$
	14	$\bar{V}_{AB}$	0	$-\bar{V}_{CA} + V_{CCA}$	+	$-\bar{V}_{CA} - V'_{CAB} + V_{CAB}$
	15	$\bar{V}_{AB} + V'_{CABR}$	0	$-\bar{V}_{CA} + V_{CCA} + 1/2(V'_{CABR})$	+	$\bar{V}_{BC} + V_{CCA} - 1/2(V'_{CABR})$
5	16	$\bar{V}_{AB}$	0	$-\bar{V}_{CA} + V_{CCA}$	+	$\bar{V}_{BC} + V_{CCA}$
	17	$\bar{V}_{AB}$	0	$-\bar{V}_{CA} + V_{CCA}$	+	$\bar{V}_{BC} - V'_{CBC} + V_{CCA}$
	18	$\bar{V}_{AB} - 1/2(V_{CCAR})$	0	0	+	$-\bar{V}_{AB} - V'_{CBC} + 1/2(V_{CCAR})$
6	1					

Table 4

> ± 100% TORQUE TYPE 2 STRESSES

TH Fired	Commut Interv	v'_{AB}	v_{TH4}	v_{D4}	v_{CAB}	$v_L + v_{DC}$
6	1	$\bar{V}_{AB}$	0	0	NC	$\bar{V}_{CA}$
	2	$\bar{V}_{AB}$	V_{CAB}	0	++	$\bar{V}_{CA} + V_{CAB}$
	3	$\bar{V}_{AB} + V_{CABR}$	V_{CAB}	0	++	$-\bar{V}_{BC} + 1/2\ V_{CABR}$
1	4	$\bar{V}_{AB}$	V_{CAB}	$\bar{V}_{AB} - V_{CAB}$	NC	$-\bar{V}_{BC}$
	5	$\bar{V}_{AB}$	V_{CAB}	$\bar{V}_{AB} - V_{CAB}$	NC	$-\bar{V}_{BC} - V'_{CCA}$
	6	$\bar{V}_{AB} - 1/2\ V'_{CCAR}$	V_{CAB}	$\bar{V}_{AB} - V_{CAB} - 1/2\ V'_{CCAR}$	NC	$\bar{V}_{AB} - 1/2\ V'_{CCAR}$
2	7	$\bar{V}_{AB}$	V_{CAB}	$\bar{V}_{AB} - V_{CAB}$	NC	$\bar{V}_{AB}$
	8	$\bar{V}_{AB}$	$-V_{CCA}$	$\bar{V}_{AB} - V_{CAB}$	+	$\bar{V}_{AB} + V_{CBC}$
	9	$\bar{V}_{AB} - 1/2\ V_{CBCR}$	$-V_{CCA}$	$\bar{V}_{AB} - B_{CAB} - 1/2\ V_{CBCR}$	+	$-\bar{V}_{CA} + 1/2\ V_{CBCR}$
3	10	$\bar{V}_{AB}$	$-V_{CCA}$	$-\bar{V}_{CA} + V_{CCA}$	NC	$-\bar{V}_{CA}$
	11	$\bar{V}_{AB}$	$-V_{CCA}$	$-\bar{V}_{CA} + V_{CCA}$	NC	$-\bar{V}_{CA} - V'_{CAB}$
	12	$\bar{V}_{AB} + V'_{CABR}$	$-V_{CCA}$	$-\bar{V}_{CA} + V_{CCA} + 1/2\ V'_{CABR}$	NC	$\bar{V}_{BC} - 1/2\ V'_{CABR}$
4	13	$\bar{V}_{AB}$	$-V_{CCA}$	$-\bar{V}_{CA} + V_{CCA}$	NC	$\bar{V}_{BC}$
	14	$\bar{V}_{AB}$	0	$-\bar{V}_{CA} + V_{CCA}$	+	$\bar{V}_{BC} + V_{CCA}$
	15	$\bar{V}_{AB} - 1/2\ V_{CCAR}$	0	0	+	$-\bar{V}_{AB} + 1/2\ V_{CCAR}$
5	16	$\bar{V}_{AB}$	0	0	NC	$-\bar{V}_{AB}$
	17	$\bar{V}_{AB}$	0	0	NC	$-\bar{V}_{AB} - V'_{CBC}$
	18	$\bar{V}_{AB} - 1/2\ V'_{CBCR}$	0	0	NC	$\bar{V}_{CA} - 1/2\ V'_{CBCR}$
6	1					

So far it was assumed that the motor operates with constant nominal flux because a prescribed slip-current relationship is fulfilled according to Figure 3. The flux in a motor is nominal when it operates with nominal (nameplate) input voltage applied at zero torque. If, for any reason, the prescribed slip current relationship is not maintained the motor may become overexcited. Overexcitation results in longer than nominal motor terminal voltages which are seen by the ACI inverter.

Estimated "absolute worst case" peak stresses are also shown by dashed lines in Figure 11. The corresponding "maximum current available" is defined as the link current value at the highest torque a drive is designed to develop at nominal flux.

4. CONCLUSIONS OF COMPARISON OF THE TWO COMMUTATING CIRCUITS

A hybrid mathematical-graphical technique was developed and successfully applied for determination of instantaneous voltage waveforms on the components of two types of ACI inverters. The technique provides tables for quick determination of peak stresses on components and motor terminals in both inverters. Diagrams of worst case stresses were presented for a 6 pole 230 V, 15 hp standard induction motor operating at 60 Hz in the +200% to -200% nominal torque range. The results can be directly applied to motor drives of different nominal horsepower, frequency and line voltage from the one considered in the synthesis. A complete waveform synthesis is required, however, if the per unit values of the motor equivalent circuit are different from the "standard induction motor" given

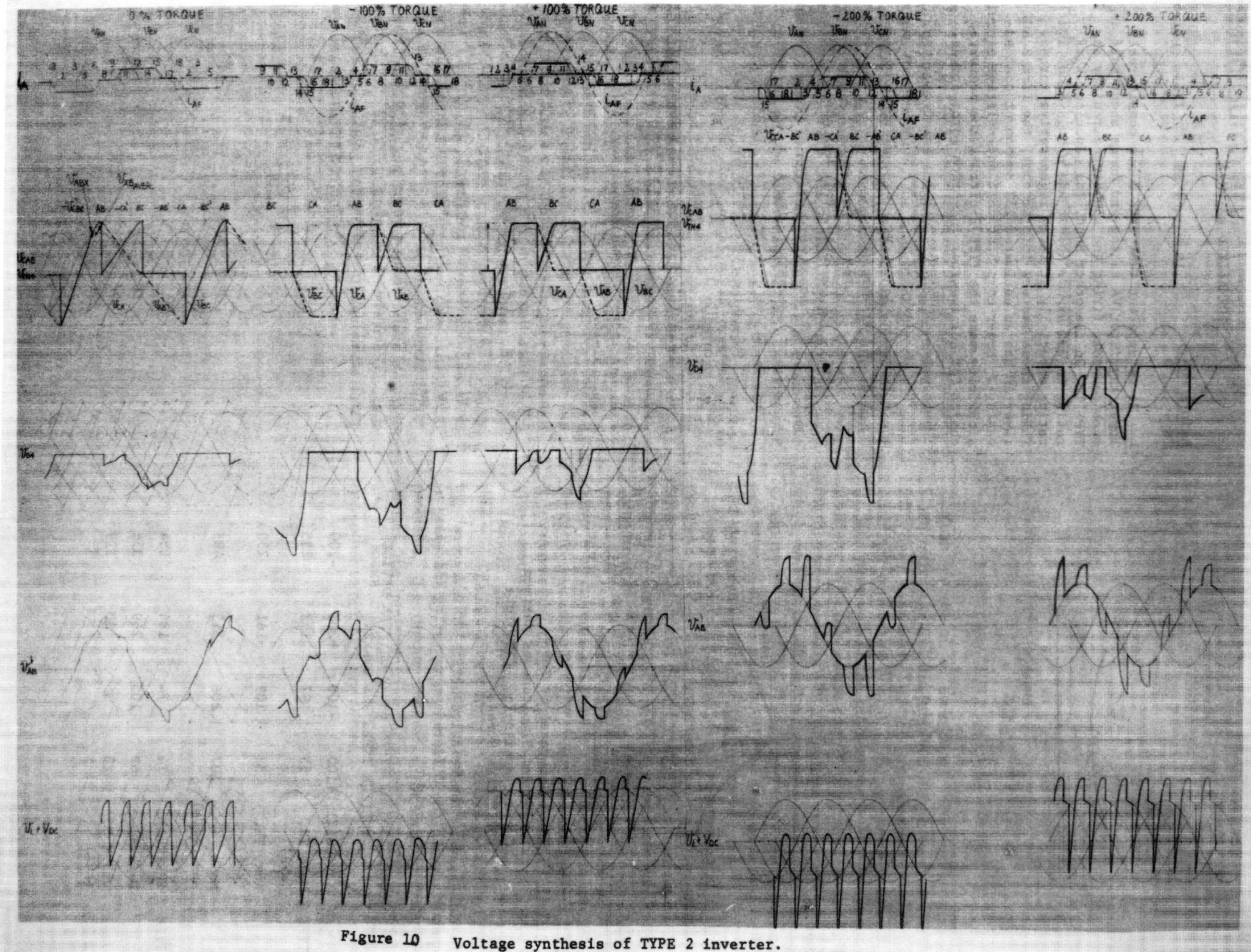

Figure 10 Voltage synthesis of TYPE 2 inverter.

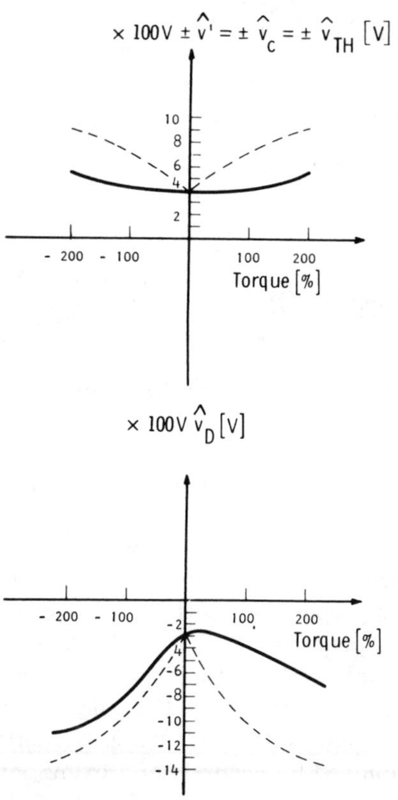

Fig. 11. Peak device stress envelopes (TYPE 2) at maximum rated speed

— Nominal Flux
--- Motor Saturated at Zero Torque and Simultaneous Maximum Current Available. (Absolute Worst Case Stress)

in this paper. The relative merits of inverters operating with the different commutating circuits are summarized below:

COMPARISON OF TYPES BASED ON PEAK DEVICE VOLTAGE TIMES RMS CURRENT

	Torque(%)	+100	-100	+200	-200
TYPE 1	Main TH. (kVA)	55	97	152	235
	Aux. TH. (kVA)	52	109	161	250
	Main. + Aux. (kVA)	107	206	313	485
TYPE 2	TH (kVA)	74	74	184	184
	D (kVA)	69	105	269	356
	.33 X D (kVA)	23	35	90	119

NORMALIZED RATINGS

$\frac{\text{Main \& AUX.}}{\text{TH + .33 D}}$	1.10	1.89	1.14	1.6

It is apparent that TYPE 2 has considerably lower thyristor VA rating and needs about half as much control, firing and snubbing circuitry as TYPE 1. On the other hand, its capacitor kVμF rating is 33% higher and it has a high diode kVA rating especially at large regenerative torque. Taking these factors into account, the TYPE 2 ACI inverter has an estimated 15% component cost advantage. Lower component cost and relative circuit simplicity make the TYPE 2 circuit the preferred choice for ACI inverter-induction motor drives.

REFERENCES

1. G. Backhaus, G. Moltgen: ETZ-A Bd. 90 (1969) H14, pp. 327-331.
2. N. Sato: EE in Japan, Vol. 84, No. 5 (1964) pp. 30-41.
3. S. Iida: EE in Japan, Vol. 90, No. 4 (1970) pp. 76-84.
4. H. Kasuno: EE in Japan, Vol. 90, No. 5 (1970) pp. 91-100.
5. K. P. Phillips: IEEE Conference Record of IGA, October 1971, pp. 385-392.
6. H. Largiader: Brown Boveri Rev. 4-70, pp. 152-167.

State-Variable Steady-State Analysis of a Controlled Current Induction Motor Drive

THOMAS A. LIPO, SENIOR MEMBER, IEEE, AND EDWARD P. CORNELL, MEMBER, IEEE

Abstract—**The exact equations defining steady-state operation of a controlled current induction motor drive system are derived by solving the system state equations in the stationary reference frame. These equations, which assume ideal current filtering, eliminate the difficulties involved in taking derivatives of discontinuous currents by defining a pair of pseudocurrent variables. Effects of saturation are included by using the slope ratio method. Electromagnetic torque and current pulsations are computed for various load conditions, and experimental confirmation of the calculated results is made. Similarities and differences to voltage controlled characteristics are presented. It is shown that normal open-loop operation occurs on the unstable side of the torque-slip characteristic necessitating the use of feedback control for stable operation.**

INTRODUCTION

Most INVERTERS in present use can be designated controllable-voltage adjustable-frequency sources, since the output terminal voltage is essentially independent of current. Recently, however, the useful features of inverters in which the current rather than voltage appears essentially as the independent variable have been recognized. Since a large portion of electrical equipment which utilizes ac power over a range of frequencies requires current of approximately constant amplitude, this type of inverter appears to have inherent advantages. Typical applications include ac motor drives which are controlled to develop constant motor torque over a fixed speed range. In addition, the simplicity of the inverter design, low cost, and regeneration capability make the controlled current source inverter (CCI) an attractive alternative to conventional controllable-voltage source inverters.

Although the principle of a controlled current source has been recognized for decades [1] and is, in fact, the principle behind operation of present-day HVDC links [2], the application to ac motor drives is of a much more recent nature [3]–[11]. Thus far, however, analyses of motor performance have been conducted by considering only the fundamental components of the ac line current [7], [8]. Although such an approach yields valuable information regarding quasi-stationary behavior, a more detailed solution is required if such important effects as motor heating and torque pulsations are to be accurately evaluated.

Paper TOD-75-45, approved by the Static Power Converter Committee of the IEEE Industry Applications Society for presentation at the IEEE 1974 Industry Applications Society Annual Meeting, Pittsburgh, Pa., October 7–10. Manuscript released for publication May 5, 1975.

The authors are with the General Electric Company, Schenectady, N. Y. 12301.

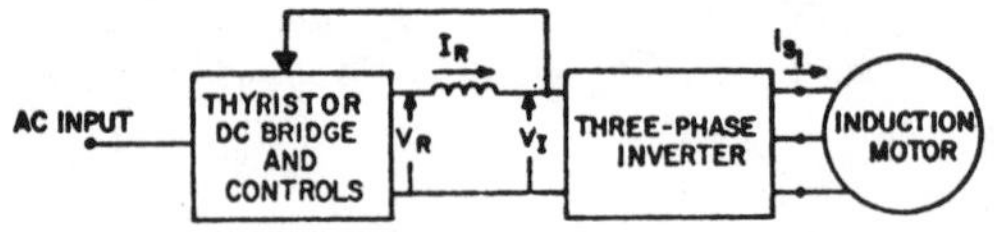

Fig. 1. Basic controlled current induction motor (CCI) drive system.

Probably the most popular approaches for analysis of ac motor drives which employ nonsinusoidal sources are harmonic superposition techniques using either symmetrical components [12] or multiple reference frames [13]. Although such methods are applicable to a CCI motor drive, the current in the lines of the motor are essentially "square wave" in nature. Numerous harmonics must be employed for accurate approximation of the input current waveshape.

This paper presents a simple alternative procedure for the steady-state analysis of an induction motor operating from a controlled current source. The method, which uses state variable techniques [14], [15], results in an exact closed form solution avoiding entirely the troublesome "Gibbs phenomena" or ringing in the solution which is characteristic of approaches using harmonic superposition. Since the solution is explicitly known, the effect of all parameters on motor performance can be rapidly and conveniently evaluated.

DESCRIPTION OF SYSTEM

A simplified diagram of the system considered in this paper is given in Fig. 1. Basically, the system consists of a controlled rectifier bridge, a dc link filter choke, a three-phase inverter, and an induction motor. The controlled rectifier bridge and choke together form a dc current source which supplies constant regulated dc current to the inverter. The inverter shown in Fig. 2 is the autosequential commutated inverter described by Ward [4] and appears to be the type in widest use. Although a number of alternative schemes are possible [3], [5], [8], they differ primarily in the means by which commutation is accomplished. Hence, the input–output terminal characteristics remain essentially the same. In Fig. 2, thyristors $T1$–$T6$ switch the load current at a rate established by the inverter control and establish the inverter output frequency. Capacitors $C1$–$C6$ provide the necessary com-

Reprinted from *IEEE Trans. Ind. Appl.*, vol. IA-11, pp. 704–712, Nov./Dec. 1975.

mutation energy while diodes $D1$–$D6$ isolate the capacitors from the load. A diagram illustrating the thyristor gating sequence and resulting line currents is shown in Fig. 3.

In this paper the following is assumed.

a) The thyristor bridge and dc choke may be considered as an ideal adjustable current source having infinite impedance. Hence, the dc link current I_R is assumed constant.

b) The inverter is considered as a zero impedance instantaneous switching device. That is, the effects of the commutating circuit will be neglected. Although assumptions a) and b) essentially eliminate the effect of source parameters on motor behavior, these assumptions have been found to be reasonably valid over low and medium frequency operating ranges.

c) The induction machine is considered an ideal machine in which the stator and rotor windings are distributed so as to always produce a single sinusoidal space distribution of MMF in the air gap.

d) The system is in the steady-state. In particular, the rotor speed is assumed to be constant. Although harmonic torques resulting from supply harmonics tend to produce speed oscillations, it is assumed that the rotor inertia is sufficiently large so as to minimize this effect.

e) All parameters of the machine are assumed to be constant.

f) The motor is assumed to be a wye-connected three-wire system. If the motor is delta-connected, it can be shown that if c) is valid, then this machine can be replaced by an equivalent wye-connection without altering its characteristics.

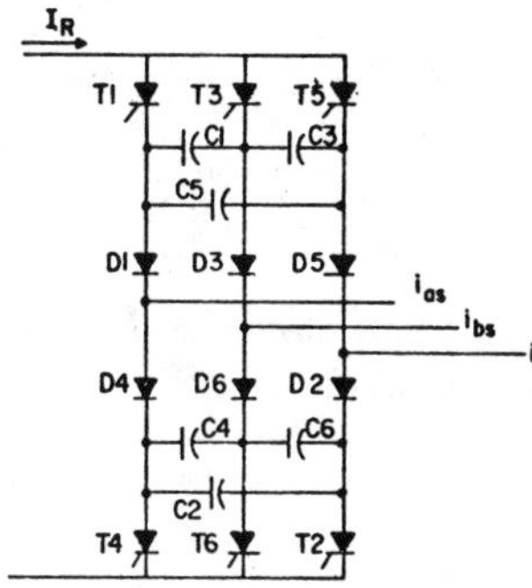

Fig. 2. Current-source inverter with autosequential commutation.

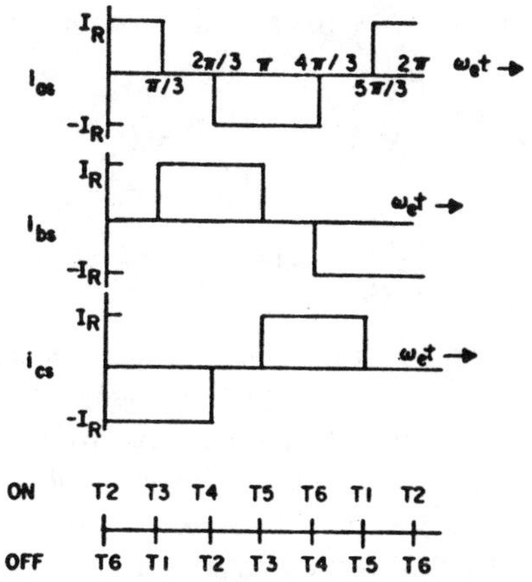

Fig. 3. Inverter gating sequence and resulting line currents.

SYSTEM EQUATIONS

The differential equations which describe transient behavior of an induction machine are conveniently expressed by transforming the stator and rotor phase variables to dq axes fixed either on the stator or rotor or rotating at synchronous speed. When the reference frame is fixed in the stator, the resulting equations are generally termed Stanley's Equations [16]. Using the notation of Krause and Thomas [17], these equations, expressed in per unit, are given in matrix form by

In these equations, the superscript s is employed to denote that the $d-q$ axes have been fixed in the stator. A p denotes the operator d/dt. Although six equations are generally required to completely define the machine response, the two zero-sequence equations have been omitted since the sums of the stator as well as rotor currents are zero. Also in (1), ω_b is the base electrical angular velocity used to obtain the per unit machine parameters, ω_r is the equivalent electrical angular velocity of the rotor, and r_s and r_r' are stator and referred rotor resistance. The quantities x_s, x_m, and x_r' are the stator self-, mutual, and rotor self-reactance referred to the stator, respectively.

$$\begin{bmatrix} v_{qs}^s \\ v_{ds}^s \\ 0 \\ 0 \end{bmatrix} = \begin{bmatrix} r_s + \frac{p}{\omega_b} x_s & 0 & \frac{p}{\omega_b} x_m & 0 \\ 0 & r_s + \frac{p}{\omega_b} x_s & 0 & \frac{p}{\omega_b} x_m \\ \frac{p}{\omega_b} x_m & -\frac{\omega_r}{\omega_b} x_m & r_r' + \frac{p}{\omega_b} x_r' & -\frac{\omega_r}{\omega_b} x_r' \\ \frac{\omega_r}{\omega_b} x_m & \frac{p}{\omega_b} x_m & \frac{\omega_r}{\omega_b} x_r' & r_r' + \frac{p}{\omega_b} x_r' \end{bmatrix} \times \begin{bmatrix} i_{qs}^s \\ i_{ds}^s \\ i_{qr}'^s \\ i_{dr}'^s \end{bmatrix} \qquad (1)$$

The voltages $v_{ds}{}^s$ and $v_{qs}{}^s$ are an equivalent set of voltages related to the phase voltages by the equations

$$v_{qs}{}^s = v_{as} \tag{2}$$

$$v_{ds}{}^s = \frac{1}{\sqrt{3}}(v_{cs} - v_{bs}). \tag{3}$$

The $d - q$ stator currents are similarly related to the stator phase currents by

$$i_{qs}{}^s = i_{as} \tag{4}$$

$$i_{ds}{}^s = \frac{1}{\sqrt{3}}(i_{cs} - i_{bs}). \tag{5}$$

The stator-referred $d - q$ rotor currents are expressed in terms of the rotor phase currents by

$$i_{qr}'^s = \frac{N_r}{N_s}\left[i_{ar}\cos\theta_r + \frac{1}{\sqrt{3}}(i_{cr} - i_{br})\sin\theta_r\right]. \tag{6}$$

$$i_{dr}'^s = \frac{N_r}{N_s}\left[-i_{ar}\sin\theta_r + \frac{1}{\sqrt{3}}(i_{cr} - i_{br})\cos\theta_r\right]. \tag{7}$$

In order to obtain (2)–(7), it is again necessary to assume that the sum of the stator currents and the sum of the rotor currents are zero. In (6) and (7), N_r/N_s is the effective rotor-to-stator turns ratio. θ_r denotes the relative displacement in electrical radians of the ar rotor axis with respect to the as axis, which have been assumed aligned at $t = 0$. That is, when speed is assumed constant, $\theta_r = \omega_r t$.

In addition to the motor current it is generally desirable to solve for the electromagnetic torque developed by the machine. When peak rated line-to-neutral voltage and peak rated line current are chosen as base quantities, the electromagnetic torque expressed in terms of the $ds - qs$ variables is

$$T_e = x_m(i_{qs}{}^s i_{dr}'^s - i_{ds}{}^s i_{qr}'^s). \tag{8}$$

Although (1)–(7) serve to completely define machine behavior regardless of the terminal conditions, it can be noted that voltages rather than currents appear as independent variables. When stator currents are known explicitly, it is apparent that only the third and fourth equation corresponding to the third and fourth row of (1) are independent in the solution. The first and second equations constitute two auxiliary relations which serve to define motor terminal voltage but which are not otherwise required in the solution. Since the stator currents can be considered as inputs to the rotor voltage equations, these terms can be transferred to the left side and the two rotor equations written as

$$\frac{\omega_r}{\omega_b}x_m i_{ds}{}^s - \frac{p}{\omega_b}x_m i_{qs}{}^s = \frac{p}{\omega_b}x_r' i_{qr}'^s + r_r' i_{qr}'^s - \frac{\omega_r}{\omega_b}x_r' i_{dr}'^s \tag{9}$$

$$-\frac{\omega_r}{\omega_b}x_m i_{qs}{}^s - \frac{p}{\omega_b}x_m i_{ds}{}^s = \frac{p}{\omega_b}x_r' i_{dr}'^s + r_r' i_{dr}'^s + \frac{\omega_r}{\omega_b}x_r' i_{qr}'^s. \tag{10}$$

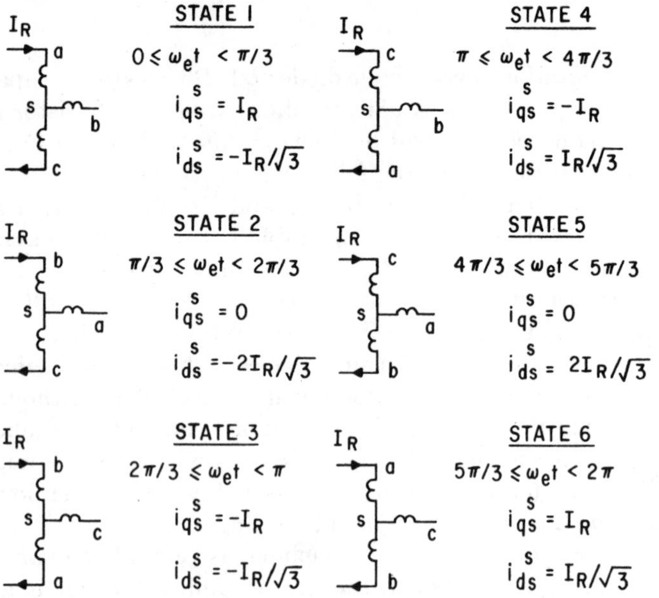

Fig. 4. Six connections for CCI motor operation and resulting $d-q$ stator currents.

In order to complete the problem definition the $d - q$ stator current inputs to the two system equations, (9) and (10), must be specified. In Fig. 4 the circuit connections corresponding to the six inverter states are summarized. The equations which define the $d - q$ stator currents for each of the six states follow directly from (4) and (5) and are given next to the appropriate sketch.

Equations (9) and (10) together with the equations which specify the input currents completely define system behavior when the dc link current is assumed constant and the inverter is symmetrically switched in accordance with Fig. 3. Moreover, these equations are equally valid for constant speed transients as well as steady-state. It can be observed from Fig. 4 that the input currents $i_{qs}{}^s$ and $i_{ds}{}^s$ are only piecewise continuous over a cycle. Although the first input term presents no problem, it is apparent that the second term contributes impulse functions at the switching instants $\omega_e t = 0, \pi/3, 2\pi/3$, etc. In order to avoid the difficulties encountered with impulse functions the derivative term can be removed by an appropriate change in variables.

It is useful to consider new pseudocurrent variables defined as

$$i_Q = i_{qr}'^s + \frac{x_m}{x_r'}i_{qs}{}^s \tag{11}$$

$$i_D = i_{dr}'^s + \frac{x_m}{x_r'}i_{ds}{}^s. \tag{12}$$

It should be noted that this change of variable is equivalent to specifying rotor flux linkages rather than currents as state variables. Equations (11) and (12) imply that

$$i_{qr}'^s = i_Q - \frac{x_m}{x_r'}i_{qs}{}^s \tag{13}$$

$$i_{dr}'^s = i_D - \frac{x_m}{x_r'}i_{ds}{}^s \tag{14}$$

$$\frac{p}{\omega_b} i_{qr}'^s = \frac{p}{\omega_b} i_Q - \frac{x_m}{x_r'} \frac{p}{\omega_b} i_{qs}^s \quad (15)$$

$$\frac{p}{\omega_b} i_{dr}'^s = \frac{p}{\omega_b} i_D - \frac{x_m}{x_r'} \frac{p}{\omega_b} i_{ds}^s. \quad (16)$$

Substituting (13)–(16) into (9) and (10) yields the following equations in terms of the modified rotor current variables:

$$\frac{r_r' x_m}{x_r'} i_{qs}^s = \frac{p}{\omega_b} x_r' i_Q + r_r' i_Q - \frac{\omega_r}{\omega_b} x_r' i_D \quad (17)$$

$$\frac{r_r' x_m}{x_r'} i_{ds}^s = \frac{p}{\omega_b} x_r' i_D + r_r' i_D + \frac{\omega_r}{\omega_b} x_r' i_Q. \quad (18)$$

It can be noted that the derivative terms have now been eliminated. Since the input does not now contain impulse functions it is evident that the modified current variables are clearly continuous functions even at the instants at which the stator currents are discontinuous. Hence, these equations are amenable to conventional state-variable techniques.

When (17) and (18) are solved for the derivative terms, they can be written in matrix form as

$$\frac{p}{\omega_b}\begin{bmatrix} i_Q \\ i_D \end{bmatrix} = \begin{bmatrix} \frac{-r_r'}{x_r'} & \frac{\omega_r}{\omega_b} \\ \frac{-\omega_r}{\omega_b} & \frac{-r_r'}{x_r'} \end{bmatrix} \begin{bmatrix} i_Q \\ i_D \end{bmatrix} + \frac{r_r' x_m}{(x_r')^2} \begin{bmatrix} i_{qs}^s \\ i_{ds}^s \end{bmatrix}. \quad (19)$$

Equation (19) can be written in conventional state-variable notation as

$$\frac{p}{\omega_b} \bar{\imath}_{QD} = \bar{A}\bar{\imath}_{QD} + b\bar{\imath}_{qds}^s \quad (20)$$

where

$$\bar{\imath}_{QD} = [i_Q, i_D]^t \quad (21)$$

$$\bar{\imath}_{qds}^s = [i_{qs}^s, i_{ds}^s]^t \quad (22)$$

and t denotes the transpose. Definitions of the quantities $\bar{A}$ and b are clearly implied by the context.

SYMMETRY RELATIONS

In general, because the $d - q$ axes input currents change six times per cycle, a closed form solution of (19) over a complete cycle is difficult. However, because of the symmetric nature of the inverter switching it can be shown that it is not generally necessary to obtain the solution over an entire cycle [15].

By reference to Fig. 3, it is clear that the three stator currents are half-wave symmetric. That is, for any two time instants relatively displaced by 180 electrical degrees

$$i_{as}(\omega_e t) = -i_{as}(\omega_e t + \pi) \quad (23)$$

$$i_{bs}(\omega_e t) = -i_{bs}(\omega_e t + \pi) \quad (24)$$

$$i_{cs}(\omega_e t) = -i_{cs}(\omega_e t + \pi). \quad (25)$$

Also, because of phase symmetry, the three currents are mutually displaced by 120 electrical degrees, or

$$i_{as}(\omega_e t + \pi/3) = i_{cs}(\omega_e t + \pi) \quad (26)$$

$$i_{bs}(\omega_e t + \pi/3) = i_{as}(\omega_e t + \pi) \quad (27)$$

$$i_{cs}(\omega_e t + \pi/3) = i_{bs}(\omega_e t + \pi). \quad (28)$$

Substracting (26)–(28) from (23)–(25), respectively, yields

$$i_{as}(\omega_e t + \pi/3) = -i_{bs}(\omega_e t) \quad (29)$$

$$i_{bs}(\omega_e t + \pi/3) = -i_{cs}(\omega_e t) \quad (30)$$

$$i_{cs}(\omega_e t + \pi/3) = -i_{as}(\omega_e t). \quad (31)$$

Equations (29)–(31) imply that if the current is specified at any time instant, then the currents 60 electrical degrees later are specified as well. Substituting (29)–(31) into (4) and (5) yields an equivalent relation in terms of the corresponding $d - q$ axes currents.

$$i_{qs}^s(\omega_e t + \pi/3) = \frac{1}{2} i_{qs}^s(\omega_e t) + \frac{\sqrt{3}}{2} i_{ds}^s(\omega_e t) \quad (32)$$

$$i_{ds}^s(\omega_e t + \pi/3) = \frac{-\sqrt{3}}{2} i_{qs}^s(\omega_e t) + \frac{1}{2} i_{ds}^s(\omega_e t). \quad (33)$$

Hence, a similar type of symmetry exists for the $d - q$ stator currents. Moreover, because i_{qs}^s and i_{ds}^s act as inputs to a set of linear equations it is apparent that in the steady-state, an equivalent set of relations apply for the $d - q$ pseudocurrents. That is, in matrix form

$$\bar{\imath}_{QD}(\omega_e t + \pi/3) = \bar{S}\bar{\imath}_{QD}(\omega_e t) \quad (34)$$

where

$$\bar{S} = \begin{bmatrix} \frac{1}{2} & \frac{\sqrt{3}}{2} \\ \frac{-\sqrt{3}}{2} & \frac{1}{2} \end{bmatrix}. \quad (35)$$

Equation (34) has considerable significance since it implies that if the solution is known over any 60° interval such as $0 \le \omega_e t < \pi/3$, then the solution is completely defined over the interval $\pi/3 \le \omega_e t < 2\pi/3$. Furthermore, the solution is also known over the remaining portion of an entire cycle by using this equation repetitively.

STATE-VARIABLE ANALYSIS

Since the input variables are constant over the interval $0 \le \omega_e t < \pi/3$, the formal solution of (20) is

$$\bar{\imath}_{QD}(\omega_e t) = \exp(\bar{A}\omega_b t)\bar{\imath}_{QD}(0) + \omega_b \int_0^t \exp[\bar{A}\omega_b(t-\tau)]b\bar{\imath}_1 \, d\tau \quad (36)$$

where from Fig. 4 during state 1,

$$\bar{\imath}_{qds}{}^{s} = \bar{\imath}_1 = \begin{bmatrix} I_R \\ -I_R/\sqrt{3} \end{bmatrix}. \tag{37}$$

In particular, from Fig. 3 at the time instant $t = \pi/3\omega_e = T$, since T, b, and $\bar{\imath}_1$ are constants,

$$\bar{\imath}_{QD}(\pi/3) = \exp(\bar{A}\omega_b T)\bar{\imath}_{QD}(0) + \exp(\bar{A}\omega_b T)\left[\int_0^T \exp(-\bar{A}\omega_b\tau)d(\omega_b\tau)\right]b\bar{\imath}_1. \tag{38}$$

Upon completing the integration, the pseudocurrents at the time instant $\omega_e t = \pi/3$ can be solved in terms of the initial currents at $t = 0$ and the input currents. Noting $T = \pi/3\omega_e$ and defining

$$f_R = \omega_e/\omega_b. \tag{39}$$

Equation (38) can be written

$$\bar{\imath}_{QD}(\pi/3) = \exp(\bar{A}\pi/3f_R)\bar{\imath}_{QD}(0) + [\exp(\bar{A}\pi/3f_R) - \bar{I}]\bar{A}^{-1}b\bar{\imath}_1 \tag{40}$$

where $\bar{I}$ is the identity matrix.

In general, the $d - q$ rotor currents are unknown at both time instants $\omega_e t = 0$ and $\pi/3$. However, because of symmetry the solution at these two time instants must be related by (34). That is

$$\bar{\imath}_{QD}(\pi/3) = \bar{S}\bar{\imath}_{QD}(0). \tag{41}$$

Solving (40) and (41), the initial condition $\bar{\imath}_{QD}(0)$ as a unique function of motor parameters and system inputs is

$$\bar{\imath}_{QD}(0) = [\bar{S} - \exp(\bar{A}\pi/3f_R)]^{-1} \cdot [\exp(\bar{A}\pi/3f_R) - \bar{I}]\bar{A}^{-1}b\bar{\imath}_1. \tag{42}$$

Having obtained the initial condition for the time interval the solution is known throughout the entire interval by (6). In particular

$$\bar{\imath}_{QD}(\omega_e t) = \exp(\bar{A}\omega_e t)\bar{\imath}_{QD}(0) + [\exp(\bar{A}\omega_e t) - \bar{I}]\bar{A}^{-1}b\bar{\imath}_1 \tag{43}$$

where $0 \leq \omega_e t < \pi/3$. The solution for the remaining five intervals is immediately known by virtue of symmetry, (34).

It should be noted that although the solution has been completely defined in terms of basic motor parameters, speed, and input current, the solution is only implicitly known by means of the defined quantities $\bar{A}$ and b. In some cases it would be desirable to obtain a solution wherein these variables appear explicitly. This can in fact be done by substituting explicitly for $\bar{A}$ and b in (42) and (43). Unfortunately, the resulting equations become quite cumbersome in spite of the fact that the system is second order and do not appear to lend any additional insight into the problem. Hence, the explicit form for these equations has not been included in this paper.

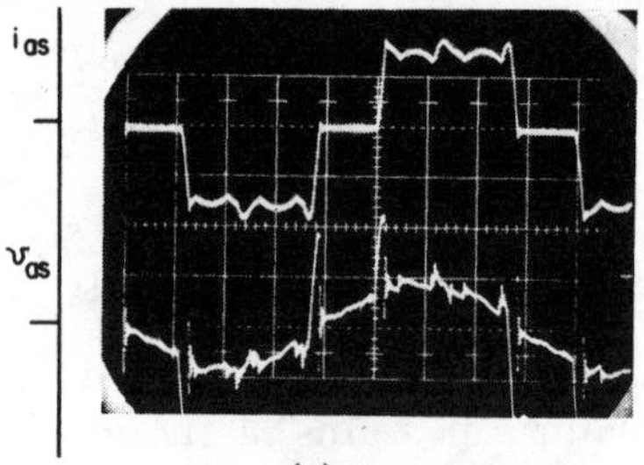

(a)

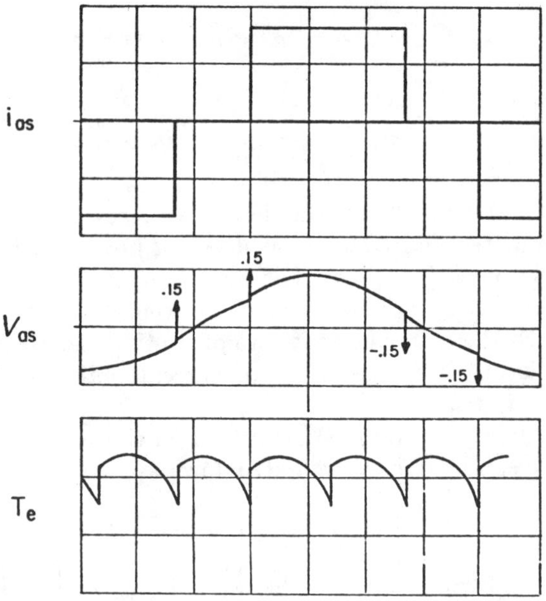

(b)

Fig. 5. (a) Experimental. (b) Calculated. Comparison of computed and measured results. I_R = 82 A; ω_e = 188.5 rad/s; slip = 0.04; T_L = 102 N·m SCALE: I_{as} – 50 A/div; V_{as} – 100 V/div; T_e – 50 N·m/div Time Axis – 4.17 ms (45°)/div.

In addition to the stator and rotor currents, the stator voltages are of considerable interest. The solution for the stator terminal voltages as well as for electromagnetic torque is given in Appendix I.

COMPARISON OF COMPUTED AND TESTED RESULTS

In order to verify the analysis, the equations which have been developed were evaluated using a digital computer, and the computed solution was compared to tested results. The induction machine used for purposes of comparison was a 230 V, 4 pole, 25 hp induction machine. A summary of the relevant parameters is given in Appendix II.

Fig. 5 shows a comparison of the computed results with measurements from the actual system. Although not measurable in the physical system, the instantaneous electromagnetic torque obtained from the computer solution is also plotted. For the case shown, the inverter frequency was set at one-half rated frequency (30 Hz) and the dc link current was adjusted until the fundamental component of motor line current was at rated value. The slip of the motor was fixed at 0.02. Although the computer program was written to compute per unit

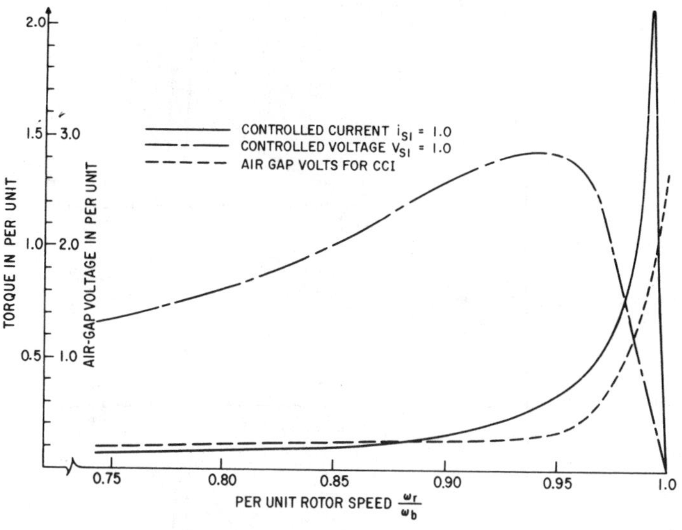

Fig. 6. Steady-state torque-speed characteristics for controlled current and controlled voltage operation.

quantities, all variables have been converted to normal units for purposes of comparison.

The location of the impulses computed from the digital computer solution is indicated with an arrow. The strength of the "area" of the impulse is adjacent to the arrow. It can be noted that the amplitudes of the voltages and locations of the pulses in the voltage waveshapes are in good agreement. In the actual system, of course, the voltage pulses are finite amplitude being limited by the commutating capability of the inverter. However, the "area" under the pulse remains essentially the same.

It can be noted that in addition to the voltage spikes, the instantaneous value of the terminal voltage changes slightly at each commutation. This effect can be attributed to two causes corresponding to the two resistive terms in (59) and (60). First, since the line current changes suddenly, the stator IR drop changes as well, resulting in a change in terminal voltage. Secondly, since the rotor flux linkages must be maintained constant when currents are switched in the stator circuits, the mutual flux linkage is not constant but changes slightly so as to maintain constant rotor flux linkage. This change is reflected in the air-gap voltage, and hence, the terminal voltage. The magnitude of the effect depends upon the relative values of rotor resistance and rotor leakage and magnetizing reactances.

Fig. 6 shows a typical torque versus speed curve at 60 Hz excitation wherein the dc link current I_R has been adjusted such that the rated fundamental component of current flows in the motor lines. The torque plotted in Fig. 6 is the average value of pulsating torque shown in Fig. 5(b). It can be noted that the torque remains very small until the motor approaches synchronous speed, then rises rapidly. Also shown on Fig. 6 for purposes of comparison is the torque for a conventional voltage source. It can be noted that the peak torque for the controlled current case is considerably higher than the voltage case suggesting that operation from a current source may be superior to operation from a voltage source. Further investigation, however, reveals that this is not the case. The high peak torque is, rather, a result of how saturation has been incorporated into the analysis.

The magnetizing reactance which has been used for the calculation of Fig. 6 was a saturated value (x_m = 2.68 pu) obtained from a conventional no-load test at nominal voltage. This value was then maintained constant in the computer program. Although the assumption of a constant (saturated) magnetizing reactance is reasonable when the motor is excited from a voltage source, operation with a fixed amplitude current results in operation over a wide range of flux conditions. Hence, the magnetizing reactance changes widely and plays a more dominant role in motor behavior.

In order to demonstrate this effect, the magnitude of the air-gap voltage has also been plotted in Fig. 6. At low speeds the flux level in the motor is very low so that the motor is essentially unsaturated. As the motor approaches synchronous speed, the flux begins to rise rapidly. When the speed exceeds the value corresponding to the intersection of the current source and voltage source characteristics, the air-gap flux begins to rise beyond 1.0 pu indicating a very saturated condition. It is apparent that operating points near synchronous speeds yield results which are highly in error when the saturation effect is not properly accounted for.

In order to more accurately account for saturation, the slope ratio method of deMello and Walsh was adopted [18]. This method assumes that the nonfundamental air-gap flux components resulting from saturation do not contribute to torque production and do not result in appreciable harmonic voltages at the machine terminals. Hence, space harmonics can still be neglected and saturation can be introduced into the analysis by adjusting the magnetizing reactance x_m so that $x_m i_m$ equals the air-gap flux ψ_m on the air-gap saturation curve. The per unit air-gap flux is obtained by subtracting the stator leakage impedance drop from the no-load terminal voltage. This voltage is used to define a factor $K(\psi_m)$ such that

$$K(\psi_m) = x_m(\text{sat})/x_m(\text{unsat}) \tag{44}$$

where x_m(unsat) is the magnetizing reactance corresponding to the air-gap line and where

$$x_m(\text{sat}) = \psi_m/i_m. \tag{45}$$

In (45), ψ_m equals the fundamental component of air-gap voltage and, i_m equals the magnetizing current. That is

$$i_m = [(i_{qs}{}^s + i_{qr}{}'^s)^2 + (i_{ds}{}^s + i_{dr}{}'^s)^2]^{1/2}. \tag{46}$$

In terms of variables defined over the interval $0 \leq \omega_e t < \pi/3$

$$i_m = [[I_R(x_r' - x_m)/x_r' + i_Q]^2 + [-I_R(x_r - x_m')/\sqrt{3}x_r' - i_D]^2]^{1/2}. \tag{47}$$

Fig. 7 gives the slope ratio curve for the tested motor.

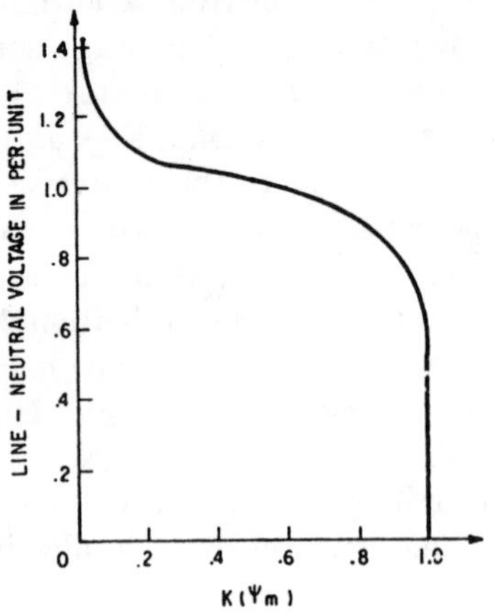

Fig. 7. Slope ratio curve.

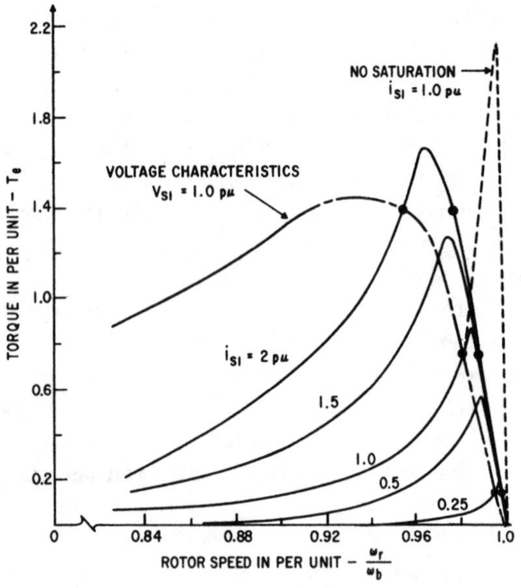

Fig. 8. Torque-speed curves for controlled current operation.

Although the solution with saturation is still represented in closed form, iteration must now be used to converge upon the proper value of $K(\psi_m)$ such that (45) is satisfied for the specified values of dc link current and motor speed. In Fig. 8 are shown a family of torque-speed curves wherein the effect of saturation has been included. The parameter i_{s1} is the fundamental component of motor line current rather than dc link current. In can be shown that i_{s1} is related to the dc link current by

$$i_{s1} = \frac{2\sqrt{3}}{\pi} I_R. \quad (48)$$

Also plotted for purposes of comparison is the characteristic for 1.0 pu current without saturation and for a conventional 1.0 pu voltage source. The large peak torque is clearly eliminated when saturation is accounted for.

MODES OF OPERATION

Examination of Fig. 9 suggests a number of possible modes of operation. a) The dc link current can simply be fixed at a value which ensures that breakdown torque

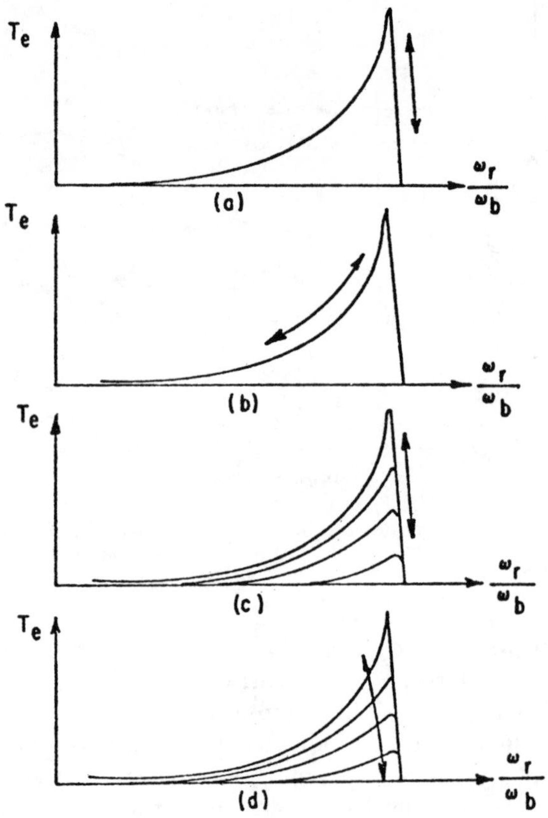

Fig. 9. Possible modes for controlled current operation.

is never exceeded. Operation is always on the stable negatively-sloped portion of the torque-speed characteristic. This mode would clearly result in a highly saturated condition and excessive heating in the motor, especially at light loads. b) The dc link current can be fixed at the same value as in a) and the motor operated on the positively-sloped portion of the curve. Although the motor is now unsaturated, motor heating is again high. Operation at light loads would be difficult. Since the positively-sloped portion of the curve is statically unstable, feedback would be necessary to ensure stable operation. c) The dc link current can be adjusted so that the motor always operates on the statically stable side of the curve at a speed slightly above the breakdown point. In this case stator heating would be less than in modes *a* and *b*, however, the motor would operate in a saturated condition. The dc link current and slip must be algebraically related so as to force this condition. d) The dc link current can be adjusted so that the motor operating points fall on the curve defined by the 1.0 pu voltage source. In this case performance would be closely analogous to operation from a voltage supply. Heating would again be minimal and the motor flux would be near its rated value. Link current and slip must again be related so as to force this condition. Since the motor operates on the statically unstable side of the motor torque-speed characteristic, a stabilization signal must be provided.

Since c) and d) appear to be the most desirable modes of operation, these two modes were compared at three typical operating conditions corresponding to light, rated,

TABLE I
PERFORMANCE CRITERIA (PER UNIT)

	Light Load i_{sl} = 0.25 pu		Rated Load i_{sl} = 1.0 pu		Heavy Load i_{sl} = 2.0 pu	
	Mode d	Mode c	Mode d	Mode c	Mode d	Mode c
Ave Torque	.13	.13	..76	..76	1.41	1.41
Pk-to-Pk Torque	.103	.132	.215	.566	.270	.888
Stator Loss	.0026	.0026	.042	.042	.167	.167
Rotor Loss	.0007	.0005	.017	.010	.070	.041

and heavy load. The six operating points selected are identified by the circles on Fig. 8. A summary of a number of performance criteria is given in Table I. It can be noted that the pulsating torque components are considerably higher for mode c) than for mode d). Since the stator current is the same in both cases the stator losses are, of course, the same. However, the rotor losses increase the case of mode d). The air-gap flux is considerably higher for mode c) suggesting also higher iron loss for this operating condition.

CONCLUSIONS

In this paper a detailed solution has been developed for the steady-state operation of an induction motor supplied from an ideal controlled current inverter source. Since the solution is obtained in closed form, iterative or superposition techniques, which approach the true solution only after numerous computational steps, are avoided entirely. The equations which have been derived are immediately applicable to transient solutions and can be extended, if necessary, to include the effects of finite source impedance. It was shown that saturation has an important effect on motor performance. The slope ratio method was incorporated to account for this effect. Investigation of feasible operating points indicates that if the motor is to remain unsaturated, feedback control is required in order to achieve stable operation.

APPENDIX I

The differential equations which define the voltages are given by the equations corresponding to the first two rows of (1)

$$v_{qs}^s = r_s i_{qs}^s + x_s \frac{p}{\omega_b} i_{qs}^s + x_m \frac{p}{\omega_b} i_{qr}'^s \tag{49}$$

$$v_{ds}^s = r_s i_{ds}^s + x_s \frac{p}{\omega_b} i_{ds}^s + x_m \frac{p}{\omega_b} i_{dr}'^s. \tag{50}$$

When the voltages are expressed in terms of the pseudocurrent variables, (49) and (50) become

$$v_{qs}^s = r_s i_{qs}^s + \frac{x_s x_r' - x_m^2}{x_r'} \frac{p}{\omega_b} i_{qs}^s + x_m \frac{p}{\omega_b} i_Q \tag{51}$$

$$v_{ds}^s = r_s i_{ds}^s + \frac{x_s x_r' - x_m^2}{x_r'} \frac{p}{\omega_b} i_{ds}^s + x_m \frac{p}{\omega_b} i_D. \tag{52}$$

Reference to Fig. 4 indicates that during state 1

$$i_{qs}^s = I_R \tag{53}$$

$$i_{ds}^s = \frac{I_R}{\sqrt{3}} - \frac{2I_R}{\sqrt{3}} u(t) \tag{54}$$

where $u(t)$ is the unit step function defined so that $u(t) = 0$ for $t < 0$, and $u(t) = 1$ for $t \geq 0$. Differentiating (53) and (54) it is apparent that over the specified interval

$$\frac{p}{\omega_b} i_{qs}^s = 0 \tag{55}$$

$$\frac{p}{\omega_b} i_{ds}^s = \frac{-2I_R}{\sqrt{3}\omega_b} \delta(t) \tag{56}$$

where $\delta(t)$ is the impulse function defined as the derivative of the unit step function.

Substituting these results into (46) and (47) yields

$$v_{qs}^s = r_s I_R + x_m \frac{p}{\omega_b} i_Q \tag{57}$$

$$v_{ds}^s = \frac{-r_s I_R}{\sqrt{3}} - \frac{x_s x_r' - x_m^2}{\omega_b x_r'} \frac{2I_R}{\sqrt{3}} \delta(t) + x_m \frac{p}{\omega_b} i_D. \tag{58}$$

The reactance factor $(x_s x_r' - x_m^2)/x_r'$ can be recognized as the stator transient reactance and is essentially equal to the sum of the stator plus rotor leakage reactance. The derivative terms $(p/\omega_b)i_Q$ and $(p/\omega_b)i_D$ are known from (19). Eliminating these terms from (57) and (58) yields the $d - q$ stator voltages uniquely in terms of the dc link current and $d - q$ pseudocurrents.

$$v_{qs}^s = \left[r_s + r_r' \frac{x_m^2}{(x_r')^2}\right] I_R - r_r' \frac{x_m}{x_r'} i_Q + \frac{\omega_r}{\omega_b} x_m i_D \tag{59}$$

$$v_{ds}^s = -\frac{1}{\sqrt{3}}\left[r_s + r_r' \frac{x_m^2}{(x_r')^2}\right] I_R - \frac{x_s x_r' - x_m^2}{\omega_b x_r'} \frac{2I_R}{\sqrt{3}} \delta(t) - r_r' \frac{x_m}{x_r'} i_D - \frac{\omega_r}{\omega_b} x_m i_Q. \tag{60}$$

The phase voltages are readily found from the inverse equations to (2) and (3)

$$v_{as} = \left[r_s + r_r' \frac{x_m^2}{(x_r')^2}\right] I_R - r_r' \frac{x_m}{x_r'} i_Q + \frac{\omega_r}{\omega_b} x_m i_D \tag{61}$$

$$v_{bs} = \frac{r_r' x_m}{x_r'}\left(\frac{i_Q}{2} + \frac{\sqrt{3}}{2} i_D\right) + \frac{\omega_r}{\omega_b} x_m \left(\frac{\sqrt{3}}{2} i_Q - \frac{1}{2} i_D\right) + \frac{x_s x_r' - x_m^2}{\omega_b x_r'} I_R \delta(t) \tag{62}$$

$$v_{cs} = -\left[r_s + r_r' \frac{x_m^2}{(x_r')^2}\right] I_R + \frac{r_r' x_m}{x_r'}\left(\frac{i_Q}{2} - \frac{\sqrt{3}}{2} i_D\right) - \frac{\omega_r}{\omega_b} x_m \cdot \left(\frac{\sqrt{3}}{2} i_Q + \frac{1}{2} i_D\right) - \frac{x_s x_r' - x_m^2}{\omega_b x_r'} I_R \delta(t). \tag{63}$$

Substituting (13) and (14) into (8), the electromagnetic torque expressed in terms of the stator currents and modified rotor currents is

$$T_e = x_m(i_{qs}{}^s i_D - i_{ds}{}^s i_Q). \quad (64)$$

Over interval 1 wherein $0 \leq \omega_e t < \pi/3$, the stator currents are given by (37), Hence, the electromagnetic torque over this interval is given by

$$T_e = x_m I_R \left[i_D + \frac{1}{\sqrt{3}} i_Q \right]. \quad (65)$$

APPENDIX II

Nameplate Motor Data	Base Quantities
18.65 kW	$V_{\text{Base}} = 187.8$ V
4-pole	$I_{\text{Base}} = 90.5$ A
3-phase Y-connected	$T_{\text{Base}} = 135.2$ N·m
$\omega_b = 377$ rad/s	
$V_{\text{rated}} = 230$ V rms line-to-line	
$I_{\text{rated}} = 64$ A rms	

Motor Parameters	SI Units	Per Unit
r_s	0.0788	0.0380
r_r'	0.0408	0.0197
x_s	5.75	2.77
x_r'	6.00	2.89
$x_{m(\text{sat})}$	5.54	2.68
$x_{m(\text{unsat})}$	9.33	4.50

REFERENCES

[1] A. H. Mittag, "Electric valve converting apparatus," U. S. Patent 1 946 292, Feb. 6, 1934.

[2] C. Adamson and N. G. Hingorani, *High Voltage Direct Current Power Transmission*. London: Garraway, 1960.

[3] M. Z. Khamudkhanov, *Frequency Control of Asynchronous Electric Drives with the Aid of Autonomous Inverters* (in Russian). Tashkent, Uzbek, SSR: Academy of Science, 1959.

[4] E. E. Ward, "Invertor suitable for operation over a range of frequency," *Proc Inst. Elec. Eng.*, vol. 111, pp. 1423–1434, Aug. 1964.

[5] F. Blaschke, H. Ripperger, and H. Steinkonig, "The control of induction motors with thyristor frequency converters for impressed stator current" (in German), *Siemens-Z.*, vol. 42, (9), pp. 773–777, 1968.

[6] M. Z. Khamudkhanov and A. A. Khashimov, *Theory and Methods of Calculation for Frequency Controlled Asynchronous Electric Drives for Asymmetrical Conditions* (in Russian). Tashkent, Uzbek, SSR: FAN, 1969.

[7] S. G. Zabrovski, G. B. Lazarev, A. V. Natalkin, and U. G. Tolstov, "The static characteristics of a frequency control system for an induction motor using current inversion" (in Russian), *Elektrichestvo*, pp. 38–41, Aug. 1971.

[8] P. G. Sperling, "The static converter fed induction motor operated with impressed square-wave current," *Siemens-Z.*, vol. 45, pp. 508–514, Aug. 1971.

[9] K. P. Phillips, "Current source inverter for ac motor drives," *IEEE Trans. Ind. Appl.*, vol. IA-8, pp. 679–683, Nov./Dec. 1972.

[10] M. B. Brennan, "A comparative analysis of two commutation circuits for adjustable current input inverters feeding induction motors," in *Conf. Rec. 1973 IEEE Power Electron. Specialists Conf.*, pp. 201–212.

[11] W. Farrer and J. D. Miskin, "Quasi-sine wave fully regenerative invertor," *Proc. Inst. Elec. Eng.*, vol. 120, pp. 969–976, Sept. 1973.

[12] W. V. Lyon, *Transient Analysis of Alternating Current Machinery*. New York: Technology M.I.T. and Wiley, 1954.

[13] P. C. Krause, "Method of multiple reference frames applied to the analysis of symmetrical induction machinery," *IEEE Trans. Power App. Syst.*, vol. PAS-87, pp. 218–227, Jan. 1968.

[14] D. Novotny and A. F. Fath, "The analysis of induction machines controlled by series connected semiconductor switches," *IEEE Trans. Power App. Syst.*, vol. PAS-87, pp. 597–605, Feb. 1968.

[15] T. A. Lipo, "The analysis of induction motors with symmetrically triggered thyristors," *IEEE Trans. Power App. Syst.*, vol. PAS-90, pp. 515–525, Mar./Apr. 1971.

[16] H. C. Stanley, "An analysis of the induction machine," *AIEE Trans.*, vol. 57, pp. 751–757, 1938.

[17] P. C. Krause and C. H. Thomas, "Simulation of symmetrical induction machinery," *IEEE Trans. Power App. Syst.*, vol. PAS 84, pp. 1038–1053, Nov. 1965.

[18] F. P. deMello and G. W. Walsh, "Reclosing transients in induction motors with terminal capacitors," *AIEE Trans. Power App. Syst.*, vol. 80, pp. 1206–1213, Feb. 1961.

CONTROL LOOP STUDY OF INDUCTION MOTOR DRIVES USING DQ MODEL

M.L. MacDonald
Powertonic Equipment Ltd.
Scarborough, Ontario, Canada

P.C. Sen
Department of Electrical Engg.
Queen's University
Kingston, Ontario, Canada

Abstract

This paper presents a systematic study of the various control loops in a current source inverter-induction motor drive and their effects on the dynamic response and stability of the system. A dq model is developed which incorporated the induction motor and the inverter power supply with current feedback. The model is first used to generate steady state curves to determine operating points. A linearised small signal model is developed to study stability and provide transfer functions for various control strategies. The stabilising effect of adding first slip speed control and then flux control is investigated. Two possible implementations of the flux control are compared, showing insignificant difference . The study reveals that with flux control, several pole-zero cancellations are possible, thereby simplifying the model and the design of other outer control loops such as speed or torque.

Introduction

The induction motor, in particular the squirrel cage type, has many inherent advantages for industrial applications. When compared to either dc or synchronous machines, the squirrel cage motor is lighter, more robust, less expensive, has higher torque-inertia ratio and is capable of much higher speeds. However, induction motors have generally been viewed as essentially constant speed machines. Developments in static power controllers have made reliable and flexible variable frequency supplies available. This enables use of the robust induction motors in high performance variable speed drives. For these drives, feedback loops are added for improved regulation and response and for protection purposes. Examples of these are speed sensing and current sensing. The induction motor is a difficult device to model because of the many nonlinearities. Besides, the addition of the inverter power supply adds other nonlinearities .

In order to assess or design outer control loops such as speed or torque controls, it is necessary to have a simple but reliable model for the inverter, the induction-motor, and any inner loops which might be considered as basic to the drive. This paper presents the systematic development of such a model and the simplification of that model such that it may be used for the design of further control loops. The example used for this study is a current source inverter-induction motor drive. This drive is modelled, its steady state characteristics determined and a linearized model used to assess stability of several feedbacks for performance improvement. The following control strategies have been investigated.

1. Independent current and frequency control.
2. Addition of slip speed control.
3. Addition of flux control.

The current source inverter (CSI) drive shown in Figure 1 has many advantages which compliment the induction motor. The CSI is a very rugged supply capable of recovery from short circuits or commutation faults. It offers inherent overcurrent protection when current feedback is used. It is a simple circuit which does not require fast turnoff thyristors. The CSI is capable of full regeneration with only 12 SCR's and 6 diodes. While it produces a square wave current supply, the motor voltage and hence flux is quasi sinusoidal. These numerous advantages have resulted in increasing use of the CSI for motor drives [1,2,3]

D-Q Model Development

The induction motor can be adequately modelled using a two axis representation developed from generalized machine theory. Such a representation is well documented [4,5,6,7]. Several assumptions are required in order to use this relatively elegant representation. These are: (i) the voltages, currents and impedances are symmetrical and balanced, (ii) the MMF distribution in the airgap is sinusoidal, (iii) no saturation occurs and (iv) there are no core losses. Generally, these assumptions do not present many restrictions while they simplify the model.

If a synchronously rotating reference frame is selected, then quantities at fundamental frequency are transformed into the two axis or dq model as dc values. Neglecting harmonics produced by the inverter power supply does not reduce the significance of the results in terms of stability and transient response [8,9]. Therefore only the fundamental frequency components are considered. This simplifies the model and enables faster computation.

The motor can then be described by a fourth order matrix equation of the form

$$\underline{v} = [R + G]\underline{i} + Lp\underline{i} \tag{1}$$

where p is the derivative operator and

$$R=\begin{bmatrix} r_s & 0 & 0 & 0 \\ 0 & r_s & 0 & 0 \\ 0 & 0 & r_r & 0 \\ 0 & 0 & 0 & r_r \end{bmatrix}, \quad G=\begin{bmatrix} 0 & \omega_s L_s & 0 & \omega_s M \\ -\omega_s L_s & 0 & -\omega_s M & 0 \\ 0 & \omega_{s\ell} M & 0 & \omega_{s\ell} L_r \\ -\omega_{s\ell} M & 0 & -\omega_{s\ell} L_r & 0 \end{bmatrix}$$

$$L=\begin{bmatrix} L_s & 0 & M & 0 \\ 0 & L_s & 0 & M \\ M & 0 & L_r & 0 \\ 0 & M & 0 & L_r \end{bmatrix}, \quad \underline{v}=\begin{bmatrix} v_{qs} \\ v_{ds} \\ v_{qr} \\ v_{dr} \end{bmatrix}$$

$$\underline{i}=\begin{bmatrix} i_{qs} \\ i_{ds} \\ i_{qr} \\ i_{dr} \end{bmatrix}.$$

The developed electrical torque for a p-pole motor is then

$$T_e = \frac{P}{2} M (i_{qs} i_{dr} - i_{ds} i_{qr}) \tag{2}$$

and the motor mechanical dynamics are described by

$$T_e = T_L + B\omega_r + Jp\omega_r \tag{3}$$

Now that the motor is modelled in the dq reference frame, it is necessary to transform the inverter, the dc link choke and the rectifier into the same reference frame. The harmonics produced by the current source inverter can be neglected and the inverter can be assumed to have no time delay. Then the CSI is an ideal supply capable of supplying sinusoidal currents and voltages. The fundamental component of the "a" phase current is

$$i_{as} = \frac{2\sqrt{3}}{\pi} I_{dc} \cos \omega_s t$$

where I_{dc} is the dc link current which is controlled by the controlled rectifier output voltage. If the

Reprinted from *Conf. Rec. IEEE/IAS 1978 Annual Meeting,* pp. 897-903, 1978.

q-axis is chosen to coincide with the axis of phase "a", then the dq currents are

$$i_{qs} = \frac{2\sqrt{2}}{\pi} I_{dc}$$

$$i_{ds} = 0$$

In order to maintain power invariance,

$$V_I I_{dc} = \frac{3}{2} v_{qs} i_{qs}$$

Then the q-axis voltage is

$$v_{qs} = \frac{\pi}{3\sqrt{2}} V_I$$

where V_I is the dc voltage at the inverter input. The dc link choke can now be transformed into the dq axis using the above relationships and the equation

$$V_R = V_I + (R_F + pL_F) I_{dc}$$

The quantities R_F and L_F are the choke resistance and inductance and V_R is the rectifier output voltage given by (for continuous current)

$$V_R = \frac{3\sqrt{3}}{\pi} V_S \cos\alpha - \frac{3}{\pi} X_{co} I_{dc}$$

where

V_S is the supply line voltage

X_{co} is the commutating reactance

and α is the firing angle

After transforming these to the dq axis, the first row of equation (1) can be rewritten as

$$v_c = (p(L'_F+L_S)+R'_F+X'_{co}+r_s)i_{qs}+pMi_{qr}+\omega_s M\, i_{dr}$$

where v_c=transformed rectifier output voltage

$$= \sqrt{3/2}\, V_S \operatorname{Cos}\alpha$$

$$L'_F = \pi^2/12\, L_F,$$

$$R'_F = \pi^2/12\, R_F,$$

and $X'_{co} = \pi/4\, X_{co}$

In operating a current source inverter, the objective is current control. Therefore feedback is normally used. Also current feedback provides over-current protection. If the firing angle control is such that v_c can be linearly controlled, and if proportional-integral error control is used, then

$$v_c = k_I\left(\frac{1+p\tau_1}{p}\right)(i_c - i_{qs})$$

where

K_I is the control gain

τ_I is the control time constant

and i_c is the current demand.

This expression can be accomodated in the matrix equation by introducing a new variable Q such that

$$pQ = K_I(i_c - i_{qs})$$

Then

$$v_c = Q(1+p\,\tau_I)$$

Because i_{ds} is zero and the equation for v_{ds} is redundant, the resulting matrix equation remains fourth order.

$$\begin{bmatrix} i_c \\ 0 \\ 0 \\ 0 \end{bmatrix} = \begin{bmatrix} p/K_I & 1 & 0 & 0 \\ -1-p\tau_I & r_s+R'_F+X'_{co}+p(L_s+L'_F) & PM & \omega_s M \\ 0 & PM & r_r+pL_r & \omega_{s\ell}L_r \\ 0 & -\omega_{s\ell}M & -\omega_{s\ell}L_r & r_r+pL_r \end{bmatrix} \begin{bmatrix} 0 \\ i_{qs} \\ i_{qr} \\ i_{dr} \end{bmatrix} \quad (4)$$

This expression describes the complete current source-induction motor drive in the two-axis or dq reference frame.

Steady State Characteristics

In steady state in the synchronous reference frame, all currents and voltages will be constant. Thus all the derivative or "p" terms are zero. For steady state, equation (1) and (4) can be written as

$$\underline{b} = A\underline{x}$$

The solution for $\underline{x}$ is

$$\underline{x} = A^{-1}\underline{b}$$

This general method can be applied to the steady state (p=0) versions of eqns. (1) and (4) . The resulting currents can be used to compute the developed torque from equation (2) and the flux linkages from

$$\underline{\lambda} = L\underline{i}$$

The matrix coefficients in (1) and (4) are dependant on motor speed and the applied frequency. A computer can be used to repeatedly compute the currents etc. for different speeds at a given frequency. The results for the voltage source and the current source supplied motor are shown in Figures 2 and 3. A comparison of the torque speed curves indicates that the CSI results in much less torque at low speeds but has a sharp peak closer to synchronous speed than the familiar voltage source. Another noteworthy difference is the sharp rise in flux linkage as the slip approaches zero. This results in saturation and high losses in a practical machine. Thus operation at full current at low slip is to be avoided.

Stability and Dynamic Response

In order to assess the stability and dynamic response, a number of operating points are considered. Figure 4 indicates those operating points used for this study. They represent different combinations of torque, frequency, slip speed and current. The equations describing the current source inverter-induction motor drive are linearized about each of these operating points. It can be seen that points such as B and E are statically unstable and will require feedback.

Linearized Model

In order to obtain a linearized model, equation (4) must first be put in the form $p\underline{x} = F(\underline{x},\underline{u})$. This can be done by rewriting the equation, separating the coefficient matrix, in the general form

$$\underline{u} = pL\underline{x} + RG\underline{x}$$

where L is a coefficient matrix of the "p" terms and RG the coefficient matrix of the "non-p" terms. This can then be rewritten as

$$p\underline{x} = L^{-1}(\underline{u} - RG\underline{x}) \quad (5)$$

For brevity this will not be expanded here. Equations (2)and (3) can be combined to form a fifth dynamic equation.

$$pw_r = \frac{1}{J}(2M\, i_{qs}\, i_{dr} - Bw_r - T_L) \quad (6)$$

In order to linearize these equations (5) and (6) all variables are considered to have a steady state value plus a small disturbance (eg. $\omega_r = \omega_{ro} + \delta\omega_r$). For simplicity, the δ symbol will be dropped. All steady state and second order terms are dropped, to give a linear state variable expression

$$p\underline{x} = A\underline{x} + B\underline{u} \qquad (7)$$

where
$$\underline{x} = \begin{bmatrix} i_{qs} \\ i_{qr} \\ i_{dr} \\ Q \\ \omega_r \end{bmatrix}, \quad \underline{u} = \begin{bmatrix} i_c \\ \omega_s \\ T_L \end{bmatrix}$$

the phase variable or controllable canonical form. Such a transformation is done using a recursive algorithm described by Tuel (11) and Rane (12). The algorithm provides a polynomial equation whose roots are the zeros of the transfer function. Thus, for a particular operating point and a particular choice of input and output, the complete transfer function can be found. This then defines stability and response. Linear feedback can be added and assessed in the same manner. The purpose of this is first to see the improvement provided by the particular feedback and secondly to simplify modelling of the total drive.

$$A = \begin{bmatrix} \frac{-L_r R_s}{L_\ell} & \frac{M r_r}{L_\ell} & \frac{-M L_r \omega_{ro}}{L_\ell} & \frac{L_r}{L_\ell} & \frac{-M L_r i_{dro}}{L_\ell} \\ \frac{M R_s}{L_\ell} & \frac{-(L_s+L'_F) r_r}{L_\ell} & \frac{L_r(L_s+L'_F)\omega_{ro}}{L_\ell} - \omega_{so} & \frac{-M}{L_\ell} & \frac{L_r(L_s+L'_F) i_{dro}}{L_\ell} \\ \frac{M(\omega_{so}-\omega_{ro})}{L_r} & (\omega_{so}-\omega_{ro}) & \frac{-r_r}{L_r} & 0 & \frac{-M i_{qso} - L_r i_{qro}}{L_r} \\ -k_I & 0 & 0 & 0 & 0 \\ \frac{2M i_{dro}}{J} & 0 & \frac{2M i_{qso}}{J} & 0 & -\frac{B}{J} \end{bmatrix}$$

and

$$B = \begin{bmatrix} \frac{k_I \tau_I L_r}{L_\ell} & 0 & 0 \\ \frac{-k_I \tau_I M}{L_\ell} & -i_{dro} & 0 \\ 0 & \frac{M i_{qso} + L_r i_{qro}}{L_r} & 0 \\ k_I & 0 & 0 \\ 0 & 0 & -\frac{1}{J} \end{bmatrix}$$

where $R_s = r_s + R'_F + X_{co} + K_I \tau_I$

$$L_\ell = L_r(L_s + L'_F) - M^2$$

This equation (7) describes the system for small disturbances about a steady state point which will determine the coefficients. Stability can be assessed knowing the poles or eigenvalues of the system. These can be found using various computer methods to first find the characteristic equation and then solve this for the roots or poles. In this study, the Levrrier-Faddeev (10) method was used to find the characteristic equation. This method avoids the need to invert the A matrix.

To define a complete transfer function, it is also necessary to determine the zeros and the steady state gain. These are unique to the choice of input and output while the poles are characteristics of the system. One method of obtaining the zeros involves transformation of the system equation to

Linear System Analysis

The computer methods, described above are used to find the poles and zeros at the operating points shown in Figure 4. Transfer functions are obtained for the following

1. independent current and frequency control
2. addition of slip speed control
3. addition of flux control

Results are presented in Table 1 for the transfer function ω_r/ω_s. Points B and E have poles with positive real components indicating instability. While

TABLE 1

Operation Point	Poles	Zeros	Gain
A	-194 ± j80 -3.7 ± j7 -5.5	-175 ± j117 -5.6	62
B	-193 ± j77 -7.8 ± j19 .78	-95 ± j187 50	7.2
C	-195 ± j80 -3.2 ± j78 -6.4	-175 ± j117 -6.4	70
D	-195 ± j80 -3.4 ± j3.3 -6.1	-162 ± j134 -5.8	23
E	-195 ± j80 -7.0 ± j19 .78	-175 ± j115 50	7.2
F	-195 ± j80 -4 ± j3.6 -4.7	-188 ± j94 -5.8	23

ω_r/ω_s, for Independent Current and Frequency Control

this is expected because of the positive slope in Figure 4, there is also a positive real zero. This means that attempting closed loop control of speed would not stabilize the system. However, when compared to point A which provides the same torque, point B is more desirable because of the saturation and high losses of operation at A.

Slip speed control is often used [13,14,15,16] to improve drive performance. Incorporating slip speed control as shown in Figure 5 requires a modification of the system equation. The input $\omega_{s\ell}$ replaces ω_s in the independent variable vector $\underline{u}$. Also the column of matrix A associated with ω_r is changed to correspond to $\omega_{s\ell} = \omega_s - \omega_r$. Table 2 presents the results for slip speed control at the same operating points as the previous case. The fact that all the poles are now in the left half plane, indicates the stabilizing effect of this type of control. Also, the pole at -0.1 caused by the mechanical time constant is now apparent and dominant.

TABLE 2

Operating Point	Poles	Zeros	Gain
A	-195 ± j80 -7.3 -5.7 -0.1	-90 ± j190 -5.6	62
B	-194 ± j78 -7.5 ± j18 -0.1	-95 ± j187 50	7.2
C	-195 ± j80 -7.3 -5.5 -0.1	-175 ± j117 -6.3	70
D	-195 ± j80 -7.6 -5.3 -0.1	-104 ± j183 -5.8	8.4
E	-195 ± j80 -6.5 ± j18.8 -0.1	-175 ± j116 50	7.2
F	-195 ± j80 -6.4 ± j1.6 -0.1	-188 ± j95 -5.8	23

Transfer Functions $\omega_r/\omega_{s\ell}$, for Independent Current and Slip Frequency Control

Flux Control

The use of motor flux control is suggested by a number of authors [3, 13, 17, 18] as a method of further improving performance of the drive. There are several possible methods: direct flux sensing, voltage sensing, and current-slip speed control. This last method is most suited for this application because both current and slip speed are already controlled.

The required relationship between current and slip speed can be most easily obtained by referring to the steady state characteristics. If the results for several current levels are calculated, the torque speed curves can be plotted as in Figure 6 . The points on each curve where rated flux linkage occurs are marked and joined by the broken line. Note that the resulting torque-speed curve (for constant flux operation) is linear. The current and slip speed for each point along this curve are also plotted in Figure 6. This is the required relationship between current and slip speed to maintain constant flux in the motor. Such operation can be obtained in two ways. The slip speed can be used to control the current reference as suggested by Phillips [1] and Maag [2] and shown in Figure 7a. The second method is to control the slip speed with the measured current level as proposed by Cornell and Lipo [3] and shown in Figure 7b. With either form of flux control, points A and C are no longer possible. The resulting transfer functions are given in Tables 3 and 4.

TABLE 3

Transfer Functions $\omega_r/\omega_{s\ell}$, for Flux Control Using Slip Speed to Control Current

Operating Point	Poles	Zeros	Gain
B	-194 ± j78 -7.5 ± j18.5 -.1	-10.3 ± j20.2 -129	375
D	-195 ± j80 -7.7 -5.3 -.1	-79 ± j61 -5.2	41
E	-195 ± j80 -6.5 ± 18.8 -.1	-137 -9.4 ± j20.3	375
F	-195 ± j80 -6.4 ± j1.6 -.1	-93 ± j34 -5.2	41

TABLE 4

Transfer Function ω_r/i_c, for Flux Control Using Current to Control Slip Speed

Operation Point	Poles	Zeros	Gain
B	-195 ± j88 -6.5 ± j18.4 -.1	-137 -9.6 ± j19.6	460
D	-195 ± j149 -6.4 -4.6 -.1	-137 -93 -6.3	150
E	-195 ± j82 -6.4 ± j18.7 -.1	-137 -9.6 ± j21.1	461
F	-195 ± j98 -6.1 ± j1.6 -.1	-137 -92 -6.5	150

A comparison of these two tables indicates little difference between the two schemes. By ignoring high frequency poles and zeros and making several pole-zero cancellations, it is possible to reduce the complexity of the transfer function. For example, points B and E can be reduced to a single pole associated with the mechanical time constant. Such a simple single pole model has been used previously [13, 18,19] but with little validation. The model is quite valid if both slip speed control and flux control are incorporated. Points D and F represent constant flux at low current levels. At these points, relatively low frequency poles exist which apparently can't be cancelled. This may indicate that constant flux operation may not be the best method for good dynamic response at low currents.

CONCLUSION

A current source inverter-induction motor drive has been modelled in order to develop the requirements for drives with good dynamic response and a simple model. It has been shown that the addition of slip speed control stabilizes the system, but does not simplify the model for the drive. A method for obtaining the required relationship between current and slip speed for constant flux from steady state curves has been demonstrated. Two implementations of this type of flux control were shown to have little difference . In general flux control has been shown to further improve dynamic performance and to enable simplification of the drive model. It has been demonstrated that the complete drive can be modelled as a single pole if both slip speed and flux control are added to the current feed back. This single pole is that associated with the motor mechanical time constant. Such simple model is particularly important when designing or assessing different outer control loops, such as speed or torque.

REFERENCES

1. Phillips, K.P.: Current source converter for ac motor drives, IEEE Transactions on Industry Applications, vol. IA-8, p. 679 (1972).
2. Maag, R.B.: Characteristics and application of current source slip regulated ac induction motor drives, IEEE Industry and General Applications Group Annual Meeting, p. 411 (1971).
3. Cornell, E.P., and Lipo, T.A.: Design of controlled current ac drive systems using transfer function techniques, IFAC Symposium on Control in Power Electronics and Electrical Drives, p. 133 (1974).
4. O'Kelly D., and Simmons, S.: Introduction to Generalized Electrical Machine Theory, McGraw-Hill (1968).
5. Adkins, B.: The Generalized Theory of Electrical Machines, Chapman & Hill Ltd. (1957).
6. Krause, P.C., and Thomas C.H.: Simulation of symmetrical induction machinery, IEEE Transactions on Power Apparatus & Systems, vol. PAS-93, p. 1410-1418 (1974).
7. De Sarker, A.K., and Berg, G.J.: Digital simulation of three-phase induction motors, IEEE Transactions on Power Apparatus & Systems, vol. PAS-89, p. 1031 (1970).
8. Krause, P.C., and Lipo, T.A. "Analysis and simplified representation of a rectifier-inverter induction motor drive, IEEE Transactions on Power Apparatus and Systems, vol. PAS-88, p. 588 (1969).
9. Sato, N., and Sawaki, N.: Steady-state and stability analysis of induction motor driven by current source inverter, IEEE Industry Applications Society Annual Meeting, p. 814 (1976).
10. Pennington, R.H. "Introductory Computer Methods and Numerical Analysis, MacMillan (1975).
11. Tuel, W.G. Jr.: On the transformation to (phase variable) canonical form, IEEE Transactions on Automatic Control, vol. AC-11, p. 607 (1966).
12. Rane, D.S.: A simplified transformation to (phase variable) canonical form, IEEE Transactions on Automatic Control, vol. AC-11, p. 608 (1966).
13. Sen, P.C., and McDonald, M,"Slip-Frequency Controlled Induction Motor Drives Using Digital Phase-Locked-Loop Control System". International Semiconductor Power Converter Conference Proceedings, 1977, pp. 413-419.
14. Amato, C.J.: Variable speed with controlled slip induction motor, IEEE Industrial Static Power Converter Conference, p. 181 (1965).
15. Slabiak, W., and Lawson, L.T.: Precise control of three phase squirrel cage induction motor using a practical cyclonconverter, IEEE Transactions on Industry and General Applications, vol. IGA-2, p. 274 (1966).
16. Mokrytzki, B.: The controlled slip static inverter drive, IEEE Transactions on Industrial and General Applications, vol. IGA-4, p. 312 (1968).
17. Abbondanti, A.: Method of flux control in induction motors driven by variable frequency, variable voltage supplies, Conference Record of International Semiconductor Power Converter Conference, p. 177 (1977).
18. Stefanovic, V.R.: Static and dynamic characteristics of induction motors operating under constant airgap flux control, IEEE Industrial Applications Society Annual Meeting, p. 436 (1976)
19. Moffat,R., Sen,P.C., Younker,R., and Bayoumi, M., "Digital Phase-Locked-Loop for Induction Motor Speed Control", IAS (IEEE) Annual Conference Proceedings, October, 1975, pp. 283-286.

APPENDIX

INDUCTION MOTOR PARAMETERS

The motor modelled in this study is a three phase, four pole, squirrel cage induction motor with the following parameters.

Magnetizing inductance	61 mH
Stator leakage inductance	1.9 mH
Rotor leakage inductance (referred to stator)	1.9 mH
Stator resistance	.6Ω
Rotor resistance (referred to stator)	.4Ω
Moment of inertia	.38 kg-m^2
Mechanical time constant	10 sec
Voltage	208V
Current	15A
Power	5 HP
Speed	1740 RPM
Frequency	60 Hz

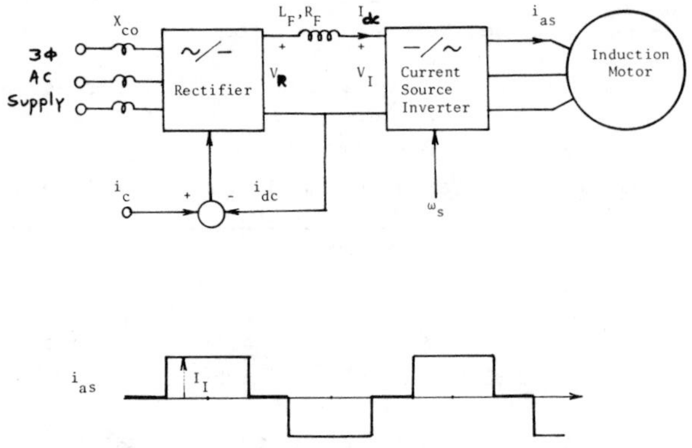

Figure 1 Current Source Inverter

a) block diagram
b) typical phase current waveform

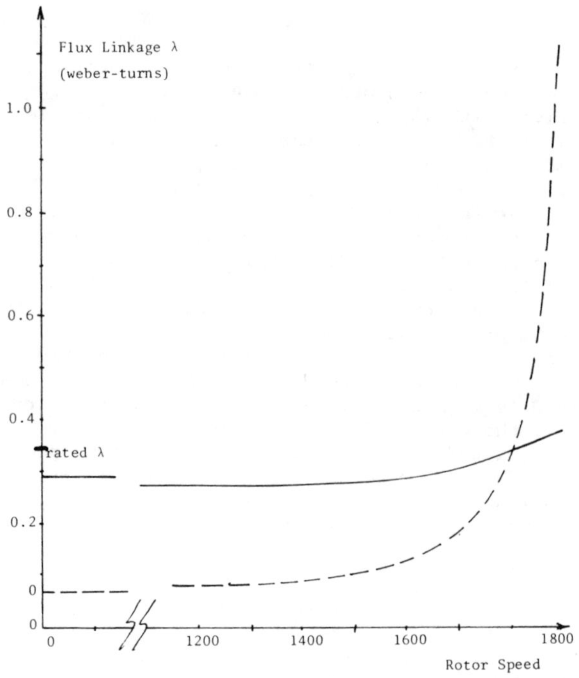

Figure 3 Flux Linkage Characteristics

a) - - - for Current Source Inverter
b) ——— for a Constant Voltage Source

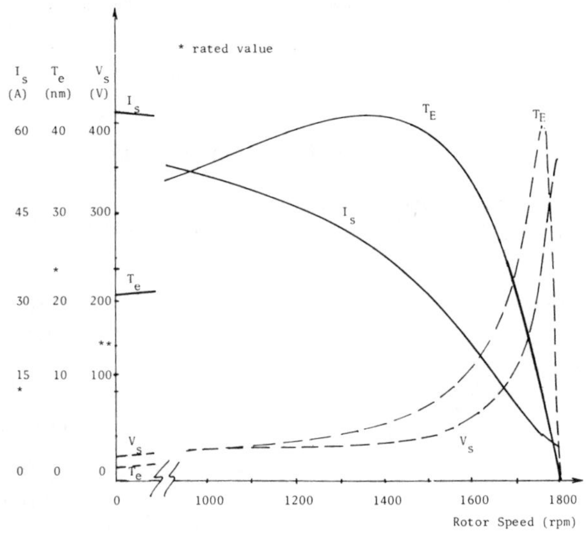

Figure 2 Developed Torque and Stator Voltage and Current Characteristics

- - - for Current Source Inverter

——— for Constant Voltage Supply

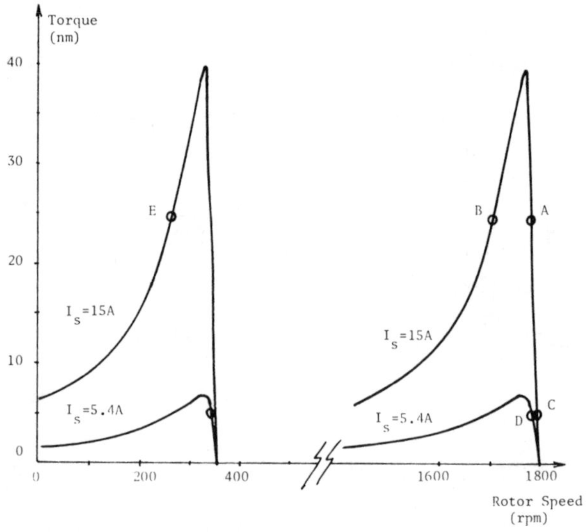

Figure 4 Operating Points Selected for Study

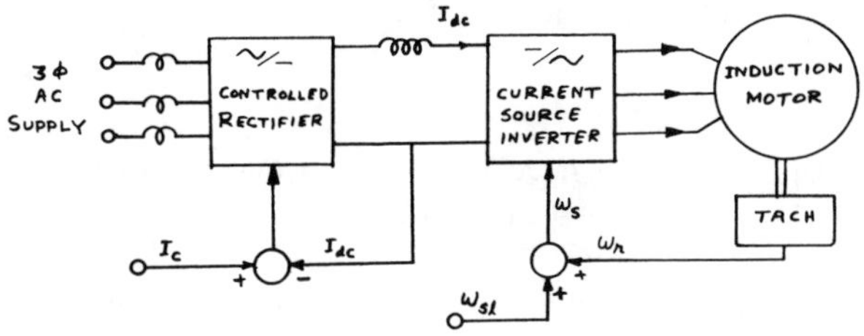

Figure 5 Independent Current and Slip Frequency Control

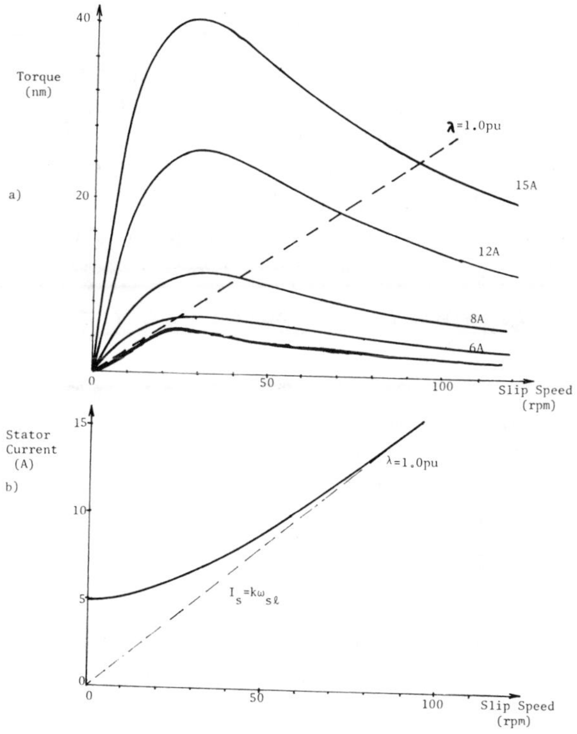

Figure 6 Characteristic Required for Constant Flux

a) Torque Speed Curves for Different Currents
b) Current Slip Curve Resulting from Constant Flux

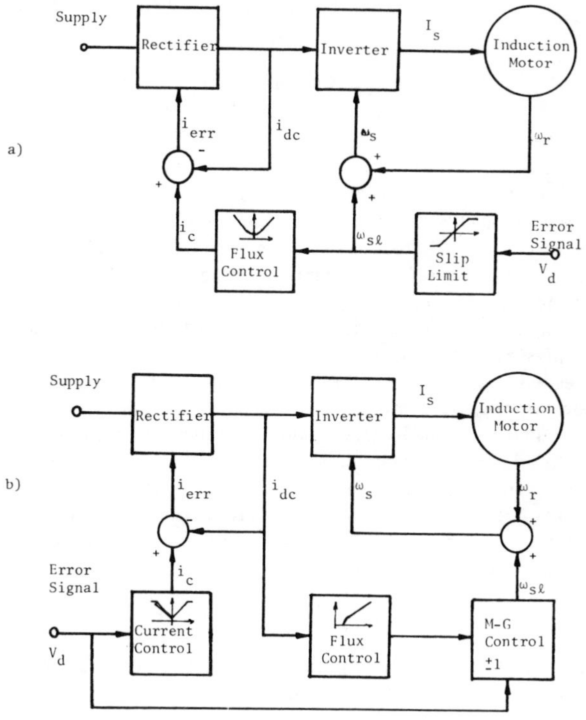

Figure 7 Flux Control Block Diagram

a) Slip speed controls current reference
b) Current controls slip speed

Modeling and Design of Controlled Current Induction Motor Drive Systems

EDWARD P. CORNELL, MEMBER, IEEE, AND THOMAS A. LIPO, SENIOR MEMBER, IEEE

Abstract—A dynamic model for current-controlled induction motor drives is developed, and a transfer function approach to the transient response investigation is formulated by means of *d-q* variables in the synchronously rotating reference frame. A sample control strategy is discussed, and transfer functions for various combinations of input and output variables are presented. It is shown that both dynamic and static instabilities exist for open-loop operation, but a well-damped closed-loop response is possible if slip frequency and current magnitude control are imposed. A comparison between the analytical transfer function transient performance predictions, the transient response predicted by the hybrid computer simulation, and actual laboratory tests is made using frequency response techniques. Bode plots are used to correlate the results between laboratory and analytical techniques. Hybrid computer and laboratory results are presented to show typical steady-state characteristics and waveforms.

INTRODUCTION

UNDER CERTAIN operating conditions a voltage/frequency-controlled induction motor drive operating either as a motor or generator can exhibit self-sustained oscillations about a steady-state operating point [1]. These oscillations are actual instantaneous rotor speed changes accompanied by variations in output torque, motor current, and input power. This operating point instability is directly related to the machine, load, and other system parameters and is not associated with the pulsating torques which normally accompany operation from inverter power sources. Related forms of instability are present in current/frequency-controlled induction motor drives. An investigation into the cause, method of analysis, and means for eliminating this instability problem is the subject of this paper.

It is shown that open-loop operation is unstable for most operating conditions, and control loops must be added to realize feasible operating points. The presence of open-loop instability makes design of the closed-loop control by purely laboratory techniques a difficult task. This paper presents an analytical design technique for finding the transfer function between a specific input command and a controlled output variable based on small-signal linearization. The frequency response corresponding to the appropriate transfer function is compared to the actual frequency response measured on a laboratory breadboard and also on a hybrid computer simulation of the system. The results demonstrate that transfer function techniques can be reliably used to synthesize the necessary slip frequency/current and speed control strategies.

Paper TOD-76-58, approved by the Static Power Converter Committee of the IEEE Industry Applications Society for presentation at the 1975 Tenth Annual Meeting of the IEEE Industry Applications Society, Atlanta, GA, September 28–October 2. Manuscript released for publication July 7, 1976.

The authors are with the General Electric Co., Schenectady, NY 12301.

INDUCTION MOTOR MODEL FOR CONTROLLED CURRENT OPERATION

A system block diagram for a controlled current induction motor drive fed from a three-phase source is shown as Fig. 1. In general, the system consists of an ac/dc-controlled rectifier bridge, a dc link smoothing reactor, a current-controlled inverter (CCI) and three-phase induction machine. The current source inverter is typically the auto-sequential type described by Ward [2] although other types are possible.

When supplied from a current-controlled inverter the motor phase currents are not sinusoidal but are rectangular in nature and flow for only 120 degrees of each half-cycle (neglecting commutation effects). Ideally, only two phases conduct at any instant of time resulting in six distinct modes of operation [3]. A diagram illustrating the resulting line currents is shown in Fig. 2. If I_R is the magnitude of the current in the dc link, these stepped currents exciting the three stator phases can be represented by the Fourier series expansions given by (1)-(3)

$$i_{as} = \frac{2\sqrt{3}}{\pi} I_R \left[\cos \omega_e t - \frac{1}{5}\cos 5\omega_e t + \frac{1}{7}\cos 7\omega_e t - \frac{1}{11}\cos 11\omega_e t + \cdots\right] \tag{1}$$

$$i_{bs} = \frac{2\sqrt{3}}{\pi} I_R \left[\cos(\omega_e t - 2\pi/3) - \frac{1}{5}\cos(5\omega_e t + 2\pi/3) + \frac{1}{7}\cos(7\omega_e t - 2\pi/3)\cdots\right] \tag{2}$$

$$i_{cs} = \frac{2\sqrt{3}}{\pi} I_R \left[\cos(\omega_e t + 2\pi/3) - \frac{1}{5}\cos(5\omega_e t - 2\pi/3) + \frac{1}{7}\cos(7\omega_e t + 2\pi/3)\cdots\right]. \tag{3}$$

When performing stability or transfer function analyses, it is convenient to view the system in a reference frame that rotates around the airgap in synchronism with the stator MMF at a speed corresponding to stator excitation frequency. Machine voltage, flux, and current variables become constant quantities during steady-state operation. These system equations represented in the synchronously rotating frame can thus be readily linearized around a particular steady-state operating point. Using the notation of Krause and Thomas [4], these

Reprinted from *IEEE Trans. Ind. Appl.*, vol. IA-13, pp. 321–330, July/Aug. 1977.

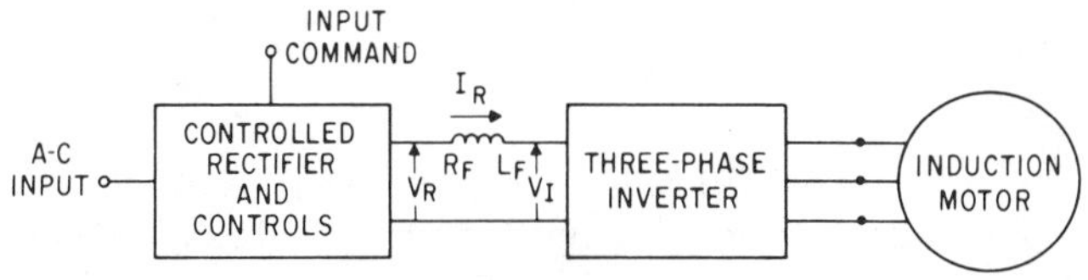

Fig. 1. Basic controlled current inverter induction motor drive system.

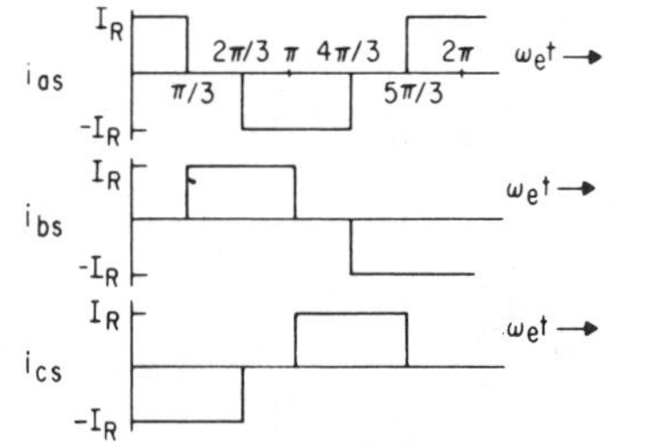

Fig. 2. Idealized motor line currents.

induction machine equations are given in matrix form

$$\begin{bmatrix} v_{qs}^e \\ v_{ds}^e \\ 0 \\ 0 \end{bmatrix} = \begin{bmatrix} r_s + \frac{p}{\omega_b} x_s & \frac{\omega_e}{\omega_b} x_s & \frac{p}{\omega_b} x_m & \frac{\omega_e}{\omega_b} x_m \\ -\frac{\omega_e}{\omega_b} x_s & r_s + \frac{p}{\omega_b} x_s & -\frac{\omega_e}{\omega_b} x_m & \frac{p}{\omega_b} x_m \\ \frac{p}{\omega_b} x_m & \frac{\omega_{sl}}{\omega_b} x_m & r_r' + \frac{p}{\omega_b} x_r' & \frac{\omega_{sl}}{\omega_b} x_r' \\ -\frac{\omega_{sl}}{\omega_b} x_m & \frac{p}{\omega_b} x_m & -\frac{\omega_{sl}}{\omega_b} x_r' & r_r' + \frac{p}{\omega_b} x_r' \end{bmatrix} \begin{bmatrix} i_{qs}^e \\ i_{ds}^e \\ i_{qr}'^e \\ i_{dr}'^e \end{bmatrix} \tag{4}$$

$$T_e = \frac{3}{2}\frac{P}{2}\frac{1}{\omega_b} x_m (i_{qs}^e i_{dr}'^e - i_{ds}^e i_{qr}'^e) = T_L + \frac{2J}{P}(p\omega_r).$$

In these equations, the superscript e is employed to denote that the d-q-axes are synchronously rotating. A p denotes the operator d/dt. All reactance values are referred to base frequency such that the operator p/ω_b always appears. Although six equations are generally required to completely define the machine response, the two zero-sequence equations have been omitted since the sum of stator as well as rotor currents are zero. Also, in (4), ω_b is the base electrical angular velocity used to obtain the per unit machine parameters, ω_r is the equivalent electrical angular velocity of the rotor, and $\omega_{sl} = \omega_e - \omega_r$ is the slip angular frequency. The parameters r_s and r_r' are stator, and referred rotor resistance referred to the stator. The quantities x_s, x_m and x_r' are the stator self, mutual, and rotor self-reactance referred to the stator, respectively.

Using (1)-(3) and applying the proper equations of transformation [4], the corresponding q- and d-axis currents in the synchronously rotating reference frame are

$$i_{qs}^e = \frac{2\sqrt{3}}{\pi} I_R \left(1 - \frac{2}{35}\cos 6\omega_e t - \frac{2}{143}\cos 12\omega_e t - \cdots\right) \tag{5}$$

$$i_{ds}^e = \frac{2\sqrt{3}}{\pi} I_R \left(-\frac{12}{35}\sin 6\omega_e t - \frac{24}{143}\sin 12\omega_e t - \cdots\right). \tag{6}$$

In (5) and (6) the q- and the as-axes are assumed aligned at time $t = 0$. For convenience, these currents can be expressed as

$$i_{qs}^e = I_R' g_{qs}^e \tag{7}$$

$$i_{ds}^e = I_R' g_{ds}^e, \tag{8}$$

where g_{qs}^e and g_{ds}^e are the switching or g functions defined as

$$g_{qs}^e = 1 - \frac{2}{35}\cos 6\omega_e t - \frac{2}{143}\cos 12\omega_e t - \cdots \tag{9}$$

$$g_{ds}^e = -\frac{12}{35}\sin 6\omega_e t - \frac{24}{143}\sin 12\omega_e t - \cdots \tag{10}$$

and

$$I_R' = \frac{2\sqrt{3}}{\pi} I_R. \tag{11}$$

It is important to note that (7) and (8) are valid even if I_R is not constant.

Assuming no power loss in the inverter, the power into and out of the inverter is identical, so that

$$V_I I_R = \frac{3}{2}(v_{qs}^e i_{qs}^e + v_{ds}^e i_{ds}^e). \tag{12}$$

The inverter voltage can be obtained by combining (7), (8), and (12) as

$$V_I = \frac{3\sqrt{3}}{\pi}(v_{qs}^e g_{qs}^e + v_{ds}^e g_{ds}^e) \tag{13}$$

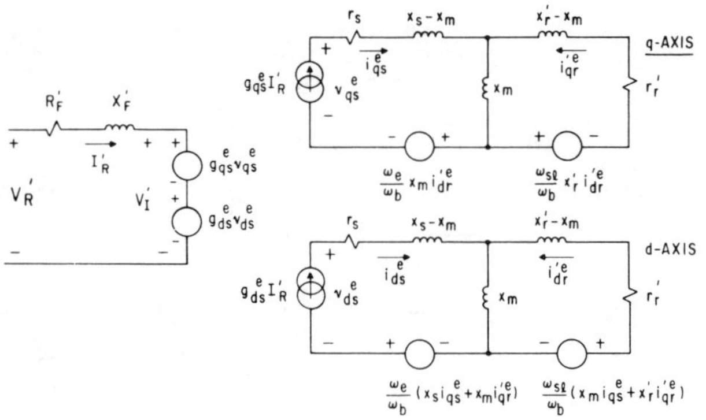

Fig. 3. *d-q* equivalent circuit of a CCI/induction motor drive in the synchronously rotating reference frame.

or simply

$$V_I' = v_{qs}^e g_{qs}^e + v_{ds}^e g_{ds}^e, \tag{14}$$

where

$$V_I' = \frac{\pi}{3\sqrt{3}} V_I. \tag{15}$$

Assuming continuous current in the smoothing reactor, the quantities I_R' and V_I' can be viewed as normalized dc link variables referred to the *d-q*-axes. The differential equation expressing the dc link variables can be expressed in terms of normalized quantities as

$$V_R' = V_I' + \left(R_F' + \frac{p}{\omega_b} X_F'\right) I_R', \tag{16}$$

where it has been convenient to define new normalized link parameters

$$R_F' = \frac{\pi^2}{18} R_F \tag{17}$$

$$X_F' = \frac{\pi^2}{18} X_F \tag{18}$$

and also

$$V_R' = \frac{\pi}{3\sqrt{3}} V_R. \tag{19}$$

Equations (4), (7), (8), (14), and (16) form the basic system equations in the synchronously rotating reference frame and can be used to form the system equivalent circuit shown in Fig. 3. In Fig. 3, and subsequently in Fig. 4, it is assumed that the time derivative operator is (p/ω_b).

Although the actual system operates with rectangular-wave excitation from a high equivalent impedance source, it is well-known that machine stability is determined primarily by the fundamental components of machine variables. If the effects of harmonics are ignored, the g functions become simply

$$g_{qs}^e \cong 1.0 \tag{20}$$

$$g_{ds}^e \cong 0, \tag{21}$$

whereby, from (7) and (8)

$$i_{qs}^e = I_R' \tag{22}$$

$$i_{ds}^e = 0. \tag{23}$$

Because of the normalization employed, the current in the stator q-axis, I_R', corresponds to the peak value of the fundamental component of motor phase current. The d-axis stator current is identically zero during both steady-state and transient conditions due to the positioning of the synchronously rotating reference frame axes. From (14), orientation of the *d-q*-axes also results in the identity that

$$v_{qs}^e = V_I', \tag{24}$$

with v_{ds}^e assuming the open-circuit value resulting from mutual coupling.

Neglecting harmonics, the detailed equivalent circuit of Fig. 3 reduces to the simplified equivalent circuit shown in Fig. 4.

Equations (7) and (8) can be combined with the equations of the induction machine in the synchronously rotating reference frame to yield the corresponding system equations, (25) and (26). Note that the ds equation in (25) has been omitted since i_{ds}^e is identically zero

$$\begin{bmatrix} V_R' \\ 0 \\ 0 \end{bmatrix} = \begin{bmatrix} r_s + R_F' + \frac{p}{\omega_b}(x_s + X_F') & \frac{p}{\omega_b} x_m & \frac{\omega_e}{\omega_b} x_m \\ \frac{p}{\omega_b} x_m & r_r' + \frac{p}{\omega_b} x_r' & \frac{\omega_{sl}}{\omega_b} x_r' \\ -\frac{\omega_{sl}}{\omega_b} x_m & -\frac{\omega_{sl}}{\omega_b} x_r' & r_r' + \frac{p}{\omega_b} x_r' \end{bmatrix} \times \begin{bmatrix} i_{qs}^e \\ i_{qr}'^e \\ i_{dr}'^e \end{bmatrix} \tag{25}$$

$$T_e = \frac{3P}{4\omega_b} x_m i_{qs}^e i_{dr}'^e = T_L + \frac{2J}{P} p\omega_r. \tag{26}$$

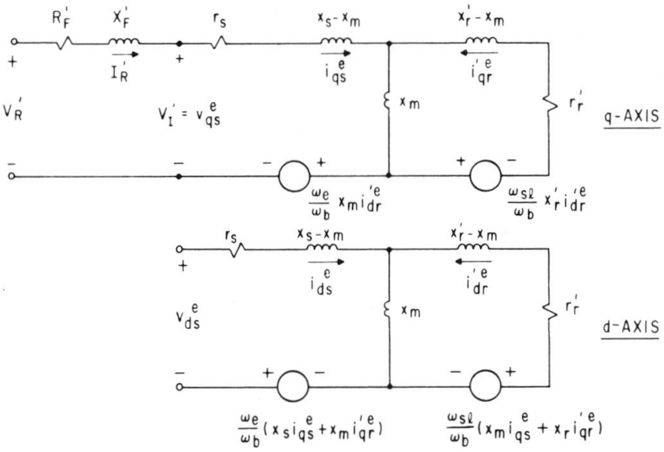

Fig. 4. Simplified *d-q* equivalent circuit.

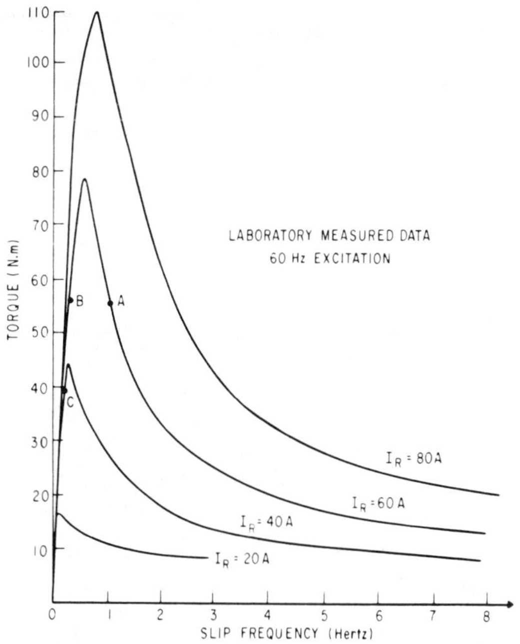

Fig. 5. Steady-state CCI motor characteristics.

TABLE I
OPEN-LOOP TRANSFER FUNCTION $\Delta I_R'/\Delta V_R'$ FOR f_e = 60 Hz

Point	T_e (N•m)	f_{sl} (Hz)	I_R (A)	DC Gain	Zeroes	Poles
A	57	1.2	60	−2.10	+8.25 −6.69 ± *j*15.7	+11.1 ± *j*14.7 −22.2 ± *j*24.1
B	57	.13	60	−1.48	−2.30 −1.41 ± *j*34.5	+41.3 −2.38 −30.5 ± *j*54.5
C	40	.34	40	−.42	−.76 −2.18 ± *j*17.8	+32.0 −.833 −26.6 ± *j*39.5

The steady-state slip-torque characteristics for a CCI/induction motor drive are shown in Fig. 5. The curves plotted in Fig. 5 were laboratory measured characteristics for an 18.5-kW machine. The parameters for the laboratory machine are given as Appendix I. These steady-state characteristics have been shown to correlate to within a few percent [3] of theoretical predictions using state-variable techniques.

Only a small portion of the slip-torque characteristics is shown. Examination of the characteristics indicates two regions of operation—one with a positive slope and the other with a negative slope. The positive sloped portions are well-known to be inherently unstable. This type of instability will be referred to as a static instability. It is also possible to have a dynamic instability on either portion of the characteristics. This is caused by negative electrical damping which is highly dependent on the selection of operating point and machine parameters. For CCI operation, the steady-state operating point can occur on either the upper or lower portion of the characteristic. For example, if operation is constrained such that the machine is never driven into a highly saturated condition, operation would be at point *A* for T_e = 57 N•m. Operation at point *B*, which yields the same output torque, corresponds to a highly saturated condition. For low torque and rated flux conditions, the steady-state operating point would be on the upper portion of a low current magnitude characteristic, for example, at point *C*. Point *C* is on the statically stable side of the I_R = 40 A characteristic. It is apparent that closed-loop control is necessary to insure stable operation over the entire load range of CCI drives.

CALCULATION OF CCI TRANSFER FUNCTIONS

One method available to study the transient performance of electric drives is small-signal linearization. Each variable is considered to be composed of a steady-state and small-time varying component. These components are substituted into the original nonlinear equations. All purely steady-state terms drop out and all second-order perturbations are ignored. The remaining equations are linear in the perturbation variables which allows the system designer to use linear control analytical techniques to synthesize the needed control. The linearized equations in matrix form are

$$\begin{bmatrix} X_F' + x_s & x_m & 0 & 0 \\ x_m & x_r' & 0 & 0 \\ 0 & 0 & x_r' & 0 \\ 0 & 0 & 0 & -\dfrac{2J\omega_b^2}{P} \end{bmatrix} \frac{p}{\omega_b} \begin{bmatrix} \Delta i_{qs}^e \\ \Delta i_{qr}'^e \\ \Delta i_{dr}'^e \\ \dfrac{\Delta\omega_r}{\omega_b} \end{bmatrix}$$

$$+ \begin{bmatrix} r_s + R_F' & 0 & \dfrac{\omega_e}{\omega_b} x_m & 0 \\ 0 & r_r' & \dfrac{\omega_{sl}}{\omega_b} x_r' & -x_r' i_{dro}'^e \\ \dfrac{\omega_{sl}}{\omega_b} x_m & -\dfrac{\omega_{sl}}{\omega_b} x_r' & r_r' & x_m i_{qso}^e + x_r' i_{qro}'^e \\ \dfrac{3P}{4\omega_b} x_m i_{dro}'^e & 0 & \dfrac{3P}{4\omega_b} x_m i_{qso}^e & 0 \end{bmatrix} \begin{bmatrix} \Delta i_{qs}^e \\ \Delta i_{qr}'^e \\ \Delta i_{dr}'^e \\ \dfrac{\Delta\omega_r}{\omega_b} \end{bmatrix}$$

$$= \begin{bmatrix} -x_m i_{dro}'^e \\ -x_r' i_{dro}'^e \\ x_m i_{qso}^e + x_r' i_{qro}'^e \\ 0 \end{bmatrix} \frac{\Delta\omega_e}{\omega_b} + \begin{bmatrix} 1 \\ 0 \\ 0 \\ 0 \end{bmatrix} \Delta V_R' + \begin{bmatrix} 0 \\ 0 \\ 0 \\ 1 \end{bmatrix} \Delta T_L. \tag{27}$$

If both sides of this matrix equation are premultiplied by the coefficient matrix of the derivative vector, the resulting equations can be arranged in state variable form

$$\frac{p}{\omega_b} \begin{bmatrix} \Delta i \\ \dfrac{\Delta\omega_r}{\omega_b} \end{bmatrix} = A \begin{bmatrix} \Delta i \\ \dfrac{\Delta\omega_r}{\omega_b} \end{bmatrix} + B\Delta \boldsymbol{u} \tag{28}$$

where

$$\Delta i = \begin{bmatrix} \Delta i_{qs}^e \\ \Delta i_{qr}'^e \\ \Delta i_{dr}'^e \end{bmatrix}; \qquad B = \begin{bmatrix} b_1^t \\ b_2^t \\ b_3^t \end{bmatrix};$$

$$\Delta \boldsymbol{u} = \begin{bmatrix} \dfrac{\Delta\omega_e}{\omega_b} \\ \Delta V_R' \\ \Delta T_L \end{bmatrix}. \tag{29}$$

A digital computer algorithm was used to find the transfer function [5] between an element of $\Delta \boldsymbol{u}$ and a specified output Δy of the form

$$\Delta y = \boldsymbol{c}^t \begin{bmatrix} \Delta i \\ \dfrac{\Delta\omega_r}{\omega_b} \end{bmatrix} + \boldsymbol{d}^t \Delta \boldsymbol{u}. \tag{30}$$

For a specific input two of the three elements of $\Delta \boldsymbol{u}$ are set equal to zero. The transfer function is returned in factored form with the associated dc gain. Stability and transient response is immediately apparent from the pole-zero locations.

If open-loop operation is attempted by applying an uncontrolled rectifier voltage and by commanding a constant inverter switching frequency (no current magnitude or slip control), the drive will accelerate into a saturated condition or slow down to zero speed. This can be predicted by calculating the transfer function $\Delta I_R'/\Delta V_R'$ and examining the pole-zero locations. The transfer function for open-loop operation corresponding to points A, B, and C on Fig. 5 are given as Table I [6].

It can be noted that the system is dynamically unstable at all four operating points. This type of instability results in one right-half plane pole in the linearized transfer function. In addition, for unsaturated conditions with sizable torque output, the system is also statically unstable which results in a second pole in the right-half plane at point A. These results are indicative of all operating points in both the motoring and regenerating region. If open-loop operation is attempted, the

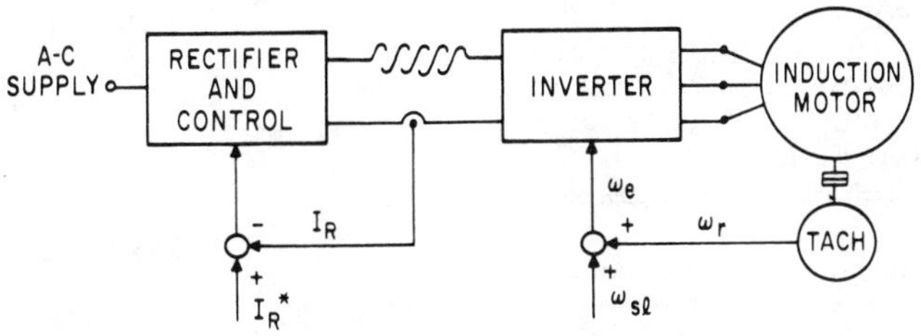

Fig. 6. Independent current and slip frequency control.

system will either slow down to zero-speed or speed up and operate in a very highly saturated condition. Closed-loop control is imperative for stable operation.

CLOSED-LOOP CONTROL

Examination of (27) suggests that two control variables are available for system stabilization. These system inputs are the rectifier voltage and the frequency command to the inverter. An elementary closed-loop control which results in stable operation for full motoring and regenerative operation is the independent current magnitude and slip frequency control shown in Fig. 6.

With slip frequency control, incremental changes in rotor speed are related to incremental changes in electrical frequency by the constraint

$$\Delta\omega_e = \Delta\omega_r + \Delta\omega_{sl}. \qquad (31)$$

Slip frequency control forces electrical frequency to change in response to rotor speed, which tends to maintain a constant angular displacement between rotor and stator MMF's during both steady-state and transient conditions. Although this type of control has a stabilizing effect, it is not capable of ensuring stable operation under all operating conditions. To obtain steady-state current control along with improved system transient response, the rectifier voltage must be constrained to respond to the error between a commanded value and the actual value of the dc link current. An integral plus proportional controller is used to give a satisfactory speed of response with zero steady-state error

$$\frac{\Delta V_R'}{(\Delta I_R'^* - \Delta I_R')} = \frac{K_c(1+\tau p)}{p}. \qquad (32)$$

To include the compensator in the analysis, a fifth state variable must be defined. ΔQ is defined to be the output of the integral controller, that is

$$\Delta Q = \frac{K_c}{p}(\Delta I_R'^* - \Delta I_R'). \qquad (33)$$

The rectifier voltage can then be expressed as

$$\Delta V_R' = \Delta Q(1 + \tau p). \qquad (34)$$

The system matrix equation including the slip frequency and current magnitude control is

$$\begin{bmatrix} r_s + R_F' + \frac{p}{\omega_b}(x_s + X_F') & \frac{p}{\omega_b}x_m & \frac{\omega_e}{\omega_b}x_m & -1-\tau p & x_m i_{dro}'^{e} \\ \frac{p}{\omega_b}x_m & r_r' + \frac{p}{\omega_b}x_r' & \frac{\omega_{sl}}{\omega_b}x_r' & 0 & 0 \\ -\frac{\omega_{sl}}{\omega_b}x_m & -\frac{\omega_{sl}}{\omega_b}x_r' & r_r' + \frac{p}{\omega_b}x_r' & 0 & 0 \\ K_c & 0 & 0 & p & 0 \\ \frac{3Px_m i_{dro}'^{e}}{4\omega_b} & 0 & \frac{3Px_m i_{qso}^{e}}{4\omega_b} & 0 & \frac{-2J\omega_b^2\left(\frac{p}{\omega_b}\right)}{P} \end{bmatrix} \begin{bmatrix} \Delta i_{qs}^{e} \\ \Delta i_{qr}'^{e} \\ \Delta i_{dr}'^{e} \\ \Delta Q \\ \frac{\Delta\omega_r}{\omega_b} \end{bmatrix}$$

$$= \begin{bmatrix} -x_m i_{dro}'^{e} \\ -x_r' i_{dro}'^{e} \\ (x_m i_{qso}^{e} + x_r' i_{qro}'^{e}) \\ 0 \\ 0 \end{bmatrix} \frac{\Delta\omega_{sl}}{\omega_b} + \begin{bmatrix} 0 \\ 0 \\ 0 \\ K_c \\ 0 \end{bmatrix} \Delta I_R'^* + \begin{bmatrix} 0 \\ 0 \\ 0 \\ 0 \\ 1 \end{bmatrix} \Delta T_L. \qquad (35)$$

TABLE II
TRANSFER FUNCTION $\Delta I_R'/\Delta I_R'^*$ WITH INDEPENDENT SLIP FREQUENCY AND CURRENT MAGNITUDE CONTROL FOR RATED ROTOR SPEED OPERATION WITH RATED AIRGAP FLUX

T_e (N•m)	f_{sl} (Hz)	I_R (A)	DC Gain	Zeroes	Poles
101	1.20	82	1.00	−8.03 −9.26 −2.55 ± *j*7.54	−3.71 −11.48 −4.01 ± *j*9.53 −229.
50	.45	45	1.00	−8.15 −9.17 −2.54 ± *j*2.85	−2.48 −11.04 −4.77 ± *j*3.94 −230.
0.1	.001	29	1.00	−2.54 ± *j*2.85 −8.10 −9.22	−2.45 −2.70 −7.97 −8.91 −231.
−50	−.45	45	1.00	−7.67 −.950 −2.61 ± *j*2.79	−6.52 −13.53 −1.54 ± *j*3.08 −230.
−101	−1.20	82	1.00	−8.11 −9.19 −2.55 ± *j*7.54	−6.01 −15.5 −1.00 ± *j*6.94 −229.

Since torque is proportional to the square of current, note that this type of control is essentially a torque control. In order to optimize performance, slip frequency must be adjusted with current to yield rated flux in the airgap. As can be seen by the pole-zero locations in Table II, good transient characteristics occur at all operating points from full motoring to full regenerating. This type of response is indicative of operation at all rotor speeds within the normal motor limitations.

CORRELATION WITH TEST RESULTS

In order to establish the validity of the transfer function approach two independent techniques were employed. The first technique utilized a hybrid computer system simulation in which the motor, the inverter, the rectifier, and the external feedback controls are all represented by simulating the appropriate algebraic and differential equations. The second technique involved frequency response measurements on actual laboratory hardware.

Fig. 7 shows a hybrid computer simulation recording of operation of an 18.5-kW CCI/induction motor drive used for laboratory evaluation. Included in this trace are motor current i_{as}, line-neutral motor volts v_{as}, inverter voltage V_I, rectifier voltage V_R, the actual current in the rotor of the machine i_{ar}, the q- and d-axis stator currents in the synchronously rotating reference frame i_{qs}^e, i_{ds}^e, and the developed electromechanical torque T_e. It should be noted that these waveforms corroborate the predictions made using state variable techniques [3].

Of special interest are the d- and q-axis stator current waveshapes in the synchronously rotating reference frame. These

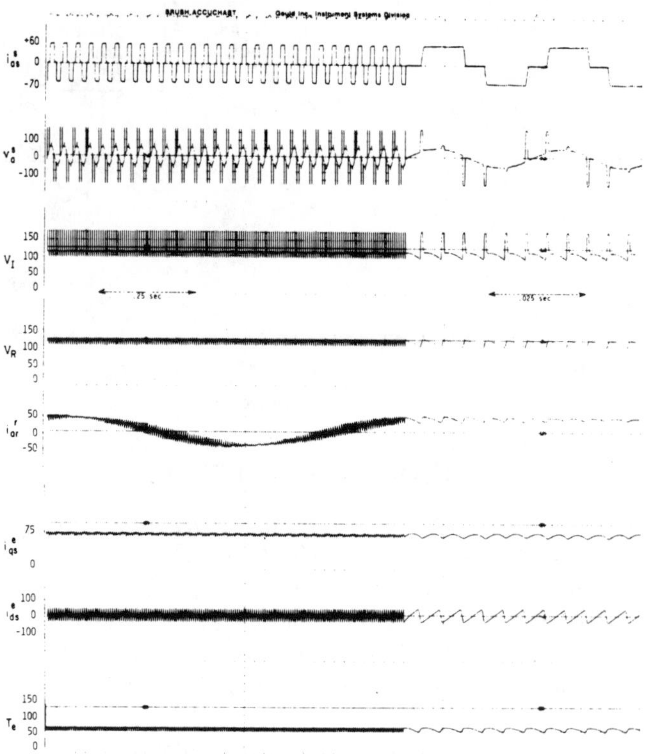

Fig. 7. Hybrid computer output of a CCI drive operating at half-speed and half-load.

currents are directly proportional to the switching functions presented as (9) and (10) with the proportionality constant I_R'. Neglecting all but the fundamental component in the transient response analyses results in $i_{ds}^e = 0$ and $i_{qs}^e = I_R'$. Using the simulation, it was established that system transients do not introduce any substantial changes in these assumed constraints.

In order to verify transfer function predictions of actual transient response, frequency response measurements were made on the computer simulation using a low frequency range servo-analyzer. Since the simulation is time scaled down to 1/20 of real time, and maximum output voltage is ±10 V, these tests must be run with extreme care to insure accuracy. It is well-known that speed variations add a pole and zero that nearly cancel for normal values of inertia. Therefore, to reduce the complexity of the frequency response output, speed variations were removed from the analysis by assuming infinite system inertia. This modification also has the effect of removing any static instability that may be present if control synthesis is being attempted using hybrid computer techniques.

Steady-state point A on Fig. 5 was selected as the sample operating point for correlation of various techniques. The transfer function of actual dc link current over commanded link current with constant slip frequency and constant speed operation was computed as

$$\frac{\Delta I_R}{\Delta I_R^*} = \frac{1.0(s/1.99 + 1)(s^2/63.41 + s/12.35 + 1)}{(s/1.52 + 1)(s/333 + 1)(s^2/80.0 + s/14.70 + 1)}. \tag{36}$$

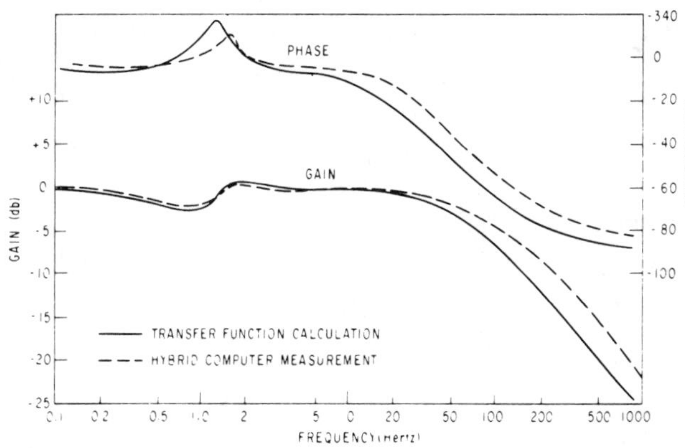

Fig. 8. Frequency response data for $\Delta I_R/\Delta I_R{}^*$ for $I_R = 60$ A, $f_{sl} =$ 1.2 Hz, $T_e = 51$ N·m, $n_r = 900$ rpm.

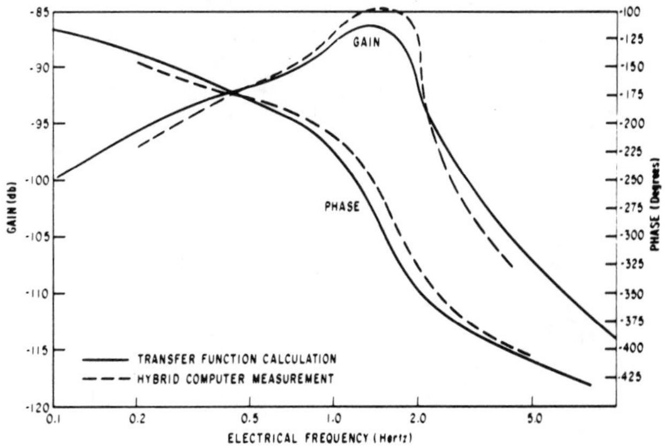

Fig. 9. Frequency response data for $\Delta I_R/\Delta\omega_{sl}$ for $I_R = 60$ A, $f_{sl} =$ 1.2 Hz, $T_e = 51$ N·m.

The frequency response corresponding to this transfer is plotted on Fig. 8 along with the frequency response measured directly from the hybrid computer simulation.

Fig. 9 corresponds to similar operating conditions as used for Fig. 8, but represents the transfer function of actual dc link current over slip frequency

$$\frac{\Delta I_R}{\Delta\omega_{sl}} = \frac{-2.9 \times 10^{-5}(s/4.95 \times 10^{-5} - 1)(s/18.67 - 1)}{(s/1.52 + 1)(s/333 + 1)(s^2/80.0 + s/14.70 + 1)}. \quad (37)$$

Note all transfer functions for the same operating condition have the same poles but different zeroes. In both cases correlation between hybrid computer measured and digital computer prediction of the frequency response was found to be very good.

As a second verification of the transfer function approach, actual laboratory frequency response measurements were conducted. The laboratory setup consists of an induction motor coupled to a dc load machine through a shaft torque transducer. The load machine is fed by an elementary Ward-Leonard system. In this case, speed cannot be held constant, so that the load dynamics must be included in the analysis. This was done in the analysis by constraining changes in load torque ΔT_L to be proportional to changes in rotor speed $\Delta\omega_r$.

The transfer function relating actual to commanded values of dc link current including speed variations is

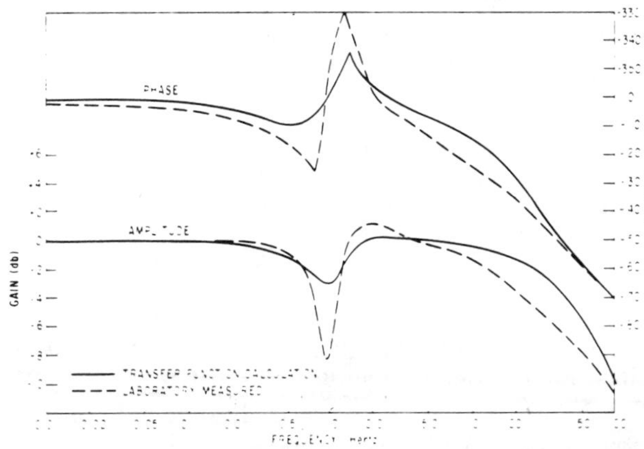

Fig. 10. Frequency response for $\Delta I_R/\Delta I_R{}^* - I_R = 60$ A, $f_{sl} = 1.2$ Hz, $n_r = 875$ rpm, $T_e = 54$ N·m.

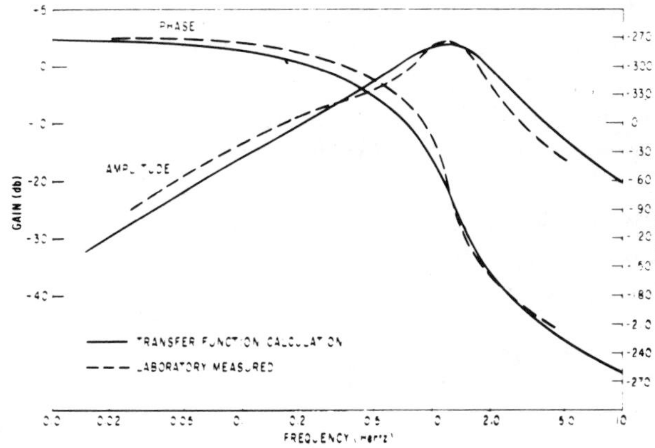

Fig. 11. Frequency response for $\Delta I_R/\omega_{sl} - I_R = 60$ A, $f_{sl} = 1.2$ Hz, $n_r = 875$ rpm, $T_e = 54$ N·m.

$$\frac{\Delta I_R}{\Delta I_R{}^*} = \frac{1.0(s/8.04 + 1)(s/9.25 + 1)(s^2/54.30 + s/10.61 + 1)}{(s/4.85 + 1)(s/10.95 + 1)(s/230 + 1)(s^2/73.65 + s/10.55 + 1)}. \quad (38)$$

This transfer function is correlated with laboratory measurements in Fig. 10.

The corresponding transfer function for changes in dc link current over changes in slip frequency is

$$\frac{\Delta I_R}{\Delta\omega_{sl}} = \frac{1.47 \times 10^{-4}(s/5.88 \times 10^{-4} - 1)(s/15.4 - 1)(s/9.87 + 1)}{(s/4.85 + 1)(s/10.95 + 1)(s/230 + 1)(s^2/73.65 + s/10.55 + 1)}. \quad (39)$$

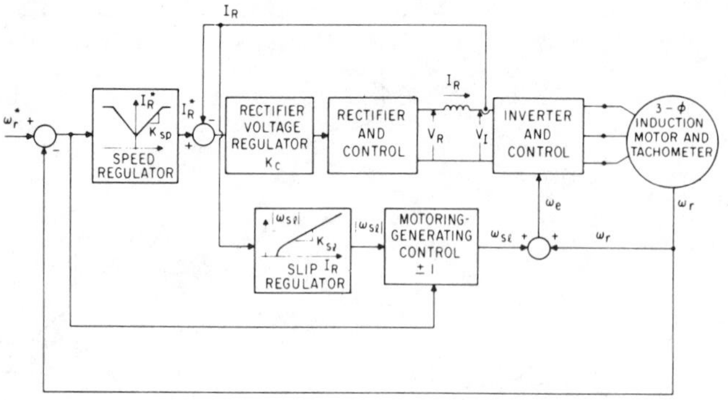

Fig. 12. Closed-loop CCI speed control.

Fig. 11 compares predicted response to actual laboratory measurements. These measurements indicate that the actual system response is somewhat less damped than the transfer function predictions. A more accurate representation of the load dynamics would be needed for better correlation of results.

OTHER CLOSED-LOOP CONTROLS

Transfer function techniques have also been used to synthesize more sophisticated controls for CCI operation. An interesting example has been reported in the literature [6], and a block diagram of the system is shown in Fig. 12.

In this system slip frequency is forced to respond to changes in current magnitude in order to maintain constant flux in the airgap during both steady-state and transient conditions. That is

$$\Delta\omega_{sl} = K_{sl}\Delta I_R, \tag{40}$$

where K_{sl} is chosen to maintain constant flux during a perturbation at a particular operating point. This constraint removes slip frequency as a system input. The two principle advantages of this constraint are that motor performance is optimized, and operation in the saturated rectifier voltage condition is possible (rectifier full on) because the slip frequency channel coupled to dc link current provided a stabilizing mechanism for the system. Speed control is obtained with a simple proportional controller

$$\Delta I_R{}^* = K_{sp}(\Delta\omega_r{}^* - \Delta\omega_r). \tag{41}$$

CONCLUSIONS

The feasibility of developing controlled current/induction motor drives using transfer function techniques has been established. A dynamic linearized model was used as the basis of the transient response study. Open-loop operation was shown to correspond to an unstable operating condition. It was shown that independent current magnitude and slip frequency control is capable of stabilizing the drive system for all operating points in the motoring and regenerating modes of operation.

Frequency response measurements made on a hybrid computer simulation and actual laboratory hardware establish the validity of this approach. An improved CCI drive was presented which provides additional system damping by constraining slip frequency to respond to changes in dc link current.

APPENDIX I

Nameplate motor data	Motor and filter parameters
18.6 kW	$r_s = 0.0788\ \Omega$
4-pole	$r_r' = 0.0408\ \Omega$
3-phase Y-connected	$x_s = 5.7518\ \Omega$
$\omega_b = 377$ rad/sec	$x_r' = 6.0028\ \Omega$
$J_{TOTAL} = 0.31$ Kg-m^2	$x_m = 5.54\ \Omega$
$V_{rated} = 230$ V rms	$X_F = 5.50\ \Omega$
$I_{rated} = 64$ A rms	$R_F = 0.091\ \Omega$

NOMENCLATURE

In general, subscripts have the following meaning

- o — steady state quantity as in $i_{dro}'^e$,
- d or q — equivalent 2-phase transformed variable as in $i_{ds}{}^e$,
- b — base quantity as in ω_b,
- e — electrical quantity as in ω_e,
- F — smoothing choke parameter as in R_F',
- I — inverter quantity as in V_I,
- m — mutual value as in x_m,
- l — leakage quantity,
- r — rotor quantity as in x_r,
- R — rectifier quantity as in V_R,
- s — stator quantity as in x_s,
- sl — slip quantity as in ω_{sl}.

Superscripts have the following meaning

- $*$ — command value as in $\omega_r{}^*$,
- $'$ — rotor quantity referred to the stator or as noted in text,
- e — quantities in synchronously rotating reference frame as $i_{dr}'^e$.

Variables have the following meaning

- Δ — perturbation variable,
- τ — voltage regulator zero location,
- ω — angular velocity,
- V — dc voltage,
- J — system inertia,
- K_{sp} — speed regulator gain,
- K_{sl} — slip channel gain,
- K_c — compensator gain,
- p — differential operator d/dt,
- P — machine poles,
- Q — output of integral controller.

REFERENCES

[1] T. A. Lipo and P. C. Krause, "Stability analysis of a rectifier-inverter induction motor drive", *IEEE Trans. on Power Apparatus and Systems,* Vol. PAS-87, No. 1, January 1968, pp. 227-234.

[2] E. E. Ward, "Invertor suitable for operation over a range of frequency", *Proceedings of IEE,* Vol. III, No. 8, August 1964, pp. 1423-1434.

[3] T. A. Lipo and E. P. Cornell, "State variable steady-state analysis of a controlled current induction motor drive", *IEEE Trans. Ind. Appl.*, vol. IA-11, No. 6, pp. 704–712, Nov./Dec. 1975.

[4] P. C. Krause and C. H. Thomas, "Simulation of symmetrical induction machinery", *IEEE Trans. on Power Apparatus and Systems,* Vol. PAS-84, November 1965, pp. 1038-1053.

[5] T. A. Lipo and A. B. Plunkett, "A novel approach to induction motor transfer functions", *IEEE Trans. on Power Apparatus and Systems,* Vol. PAS-93, No. 5, September/October 1974, pp. 1410-1418.

[6] E. P. Cornell and T. A. Lipo, "Design of controlled current ac drive systems using transfer function techniques", Conference Record of IFAC Symposium on Control in Power Electronics and Electrical Drives, Duesseldorf, October 7-9, 1974, Vol. I, pp. 133-147.

ANALYSIS AND CONTROL OF TORQUE PULSATIONS IN CURRENT FED INDUCTION MOTOR DRIVES

T.A. Lipo

General Electric Co.
Schenectady N.Y.

SUMMARY

Three types of feedback compensation schemes are described which can be used to reduce the torque pulsations which normally occur when induction machines are supplied from a current source inverter. The approach is verified both by a detailed computer simulation and also by test results from an actual system. Limitations of the method near zero slip frequency (no load) is discussed and a means suggested for elimination of the problem.

INTRODUCTION

Most inverters in present use can be designated as adjustable frequency voltage sources since the output terminal voltage is essentially independent of current. Recently, however, the useful features of current source inverters have been recognized in which the current rather than the voltage appears as the independent variable. The current source inverter drive is particularly appealing for use in four quadrant single motor drives. However, a serious drawback to the use of a current source inverter in such applications is the harmonic torque pulsations which exist at multiples of six times the output frequency of the inverter. In these applications shaft mechanical resonances often occur, generally in the range 30-100 Hz. These resonances are usually not of concern over most of the speed range. However, during starting or reversing operations the stator line frequency becomes sufficiently low that the sixth harmonic component of pulsating torque lies in the range of mechanical resonance. Operation continuously at such a resonance condition could result in abnormal wearing of gear teeth or shaft fatigue. In some cases, for example in machine tool drives, the presence of these torque pulsations are sufficiently large to affect the performance of the system even when not operating near a mechanical resonance.

When voltage source inverters are used in such applications, pulse-width-modulation is employed whereby the basic six step waveshape is modulated so as to eliminate the harmonic voltage components responsible for the sixth harmonic torque pulsation [1]. Unfortunately, application of pulse-width-modulation to current source inverters is more difficult [2]. Since the current source inverter employs 120 degree commutation only two of the three phases of the motor can be modulated at any instant. This results in a modulation scheme which is less desirable than that used in a voltage inverter which employs 180 degree commutation. Also the modulation algorithm is much more complex than the simple sine-triangle wave modulation typically used with voltage inverters. Losses in the machine are increased substantially and, although the lowest frequency torque pulsation can be eliminated higher harmonics are increased to the point where they too can create resonance problems.

This paper describes a different approach to the elimination of torque pulsations in current source induction motor drives by feedback control of the DC link current. Three different types of implementations increasing in complexity are described and the advantages and disadvantages of each are described. Practical application aspects of the control is discussed. The approach is verified both by a detailed computer simulation and on an actual system.

BASIC OPERATION

The basic three phase bridge configuration of a current source inverter drive is illustrated in Fig. 1, which omits the auxiliary circuitry that is required to force-commutate the thyristors. The inverter is fed with a controlled current i_d which, ideally, has negligible ripple and is sequentially switched from phase to phase of the motor load. Only two thyristors conduct at any given time, each carrying the impressed direct current for 120 degrees of the fundamental output period, except for commutation overlap. The resulting motor line currents have a waveform similar to the AC line current produced by a conventional six pulse voltage-fed rectifier as shown in Fig. 2.

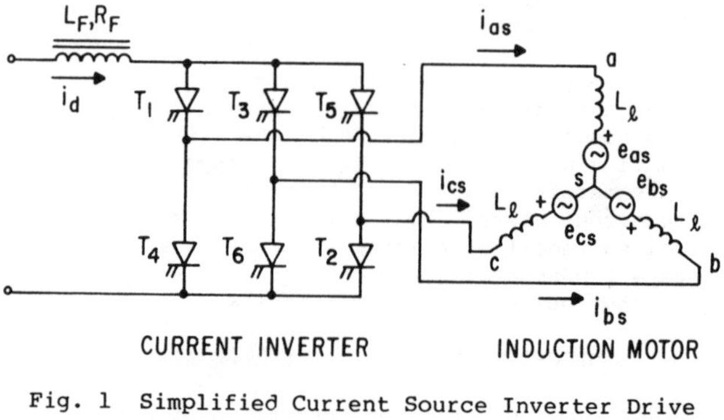

Fig. 1 Simplified Current Source Inverter Drive

Reprinted from *IEEE 1978 Power Elec. Spec. Conf.*, pp. 89-96, 1978.

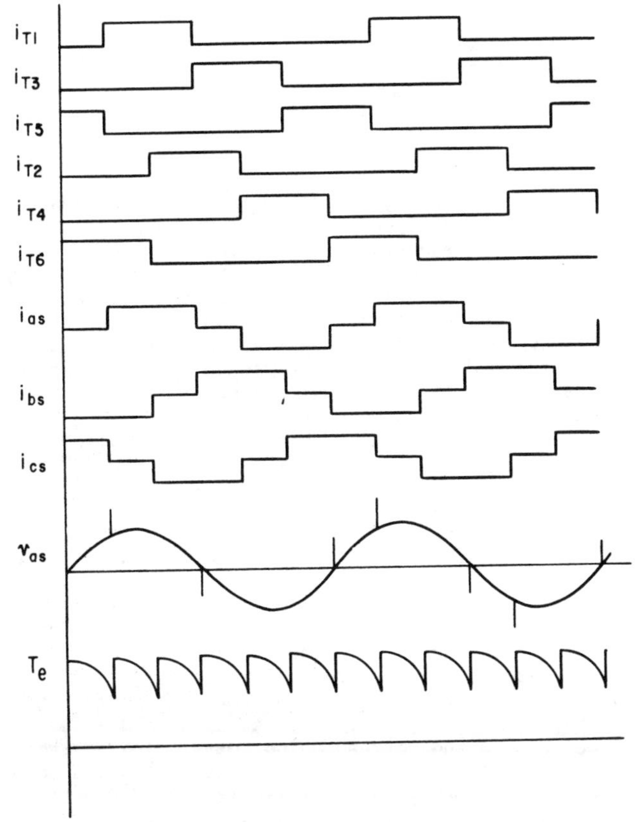

Fig. 2 Idealized Current, Voltage and Torque Waveforms.

During steady-state operation, an induction motor can be represented as an equivalent counter-emf in series with a small impedance. It can be noted that even though the currents are square wave in nature, the motor terminal voltage is essentially sinusoidal with voltage transients (spikes) superimposed at the instants of commutation. These transients appear across the motor leakage reactance and are generated by the commutating circuit. To produce instantaneous current transfer, as indicated by the ideal waveforms of Fig. 2, infinite impulse voltages would be necessary. In practice, however, the finite commutating voltage requires a non-zero time interval to force the current change through the leakage inductance.

The presence of a sinusoidal counter-emf implies that the air gap flux linkages are also nearly sinusoidal. Interaction of the sinusoidal air gap flux with the piece-wise constant stator currents result in a torque pulsation which resembles the output voltage of a conventional six pulse controlled rectifier bridge. This torque pulsation has a fundamental frequency equal to six times the inverter output frequency [3].

DESCRIPTION OF FEEDBACK APPROACH

A block diagram representation of a typical current source induction motor drive system is shown in Fig. 3 In general the system is equipped with a feedback control system which is employed to satisfy the system performance requirements. This control is represented simply as a box in Fig. 3. The control may have a variety of inputs depending upon the control scheme but typically employs stator current, air gap flux and rotor speed [4]. Many other feedback schemes are possible. However, regardless of the feedback variables only two system inputs are available for control, namely the frequency of the current source inverter and the phase delay angle of the rectifer bridge. In most cases a fast inner current loop is incorporated in the rectifier control system so that, effectively, the DC link current current can be considered as the second system input. The commanded values of the inverter angular frequency and rectifier output current determined by the control system are ω_e^* and α^* respectively. The inverter frequency is, in essence, the cause of the torque pulsation and use of this input to reduce torque pulsation would involve a type of pulse-width-modulation scheme mentioned previously. It is evident that the other control input might also provide a means for reducing torque pulsations if the DC link current can be "modulated" in the proper manner. The required modulation can be determined inherently if the undesirable quantity (torque pulsation) is detected and regulated to a minimum by means of feedback as illustrated in Fig. 3. Note that only the pulsating component of torque need be fed back to the summing junction since the average value is essentially fixed by the main control block.

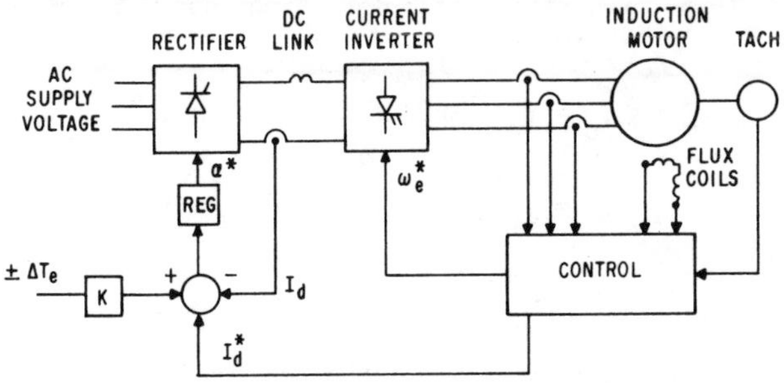

Fig. 3 Typical Control System Showing Added Input for Decogging Signal.

Method 1 - Using Motor Line Current and Rotor Speed

Because of the coupling which exists between phases of a three phase symmetrical induction machine it has been found useful to employ a transformation of variables resulting in an equivalent two phase machine [5]. In general, these equations are dependent upon the speed of the frame used to view the machine. However, when the reference frame is stationary or "fixed to the stator" the machine equations can be written [5]

$$v_{qs} = r_s i_{qs} + d\lambda_{qs}/dt \qquad (1)$$

$$v_{ds} = r_s i_{ds} + d\lambda_{ds}/dt \qquad (2)$$

$$0 = r_r i_{qr} + d\lambda_{qr}/dt - \omega_r \lambda_{dr} \quad (3)$$

$$0 = r_r i_{dr} + d\lambda_{dr}/dt + \omega_r \lambda_{qr} \quad (4)$$

$$T_e = \frac{3P}{4} (\lambda_{ds} i_{qs} - \lambda_{qs} i_{ds}) \quad (5)$$

The number of equations involved even with equivalent two phase variable is large and it is convenient to view the machine variables as components of vector quantities [6]. Let

$$\hat{v}_s = v_{qs}\hat{u}_q + v_{ds}\hat{u}_d \quad (6)$$

$$\hat{i}_s = i_{qs}\hat{u}_q + i_{ds}\hat{u}_d \quad (7)$$

$$\hat{\lambda}_s = \lambda_{qs}\hat{u}_q + \lambda_{ds}\hat{u}_d \quad (8)$$

$$\hat{\omega}_r = \omega_r \hat{u}_n \quad (9)$$

where $\hat{u}_d$ and $\hat{u}_q$ denote unit vectors in the two phase d- and q-axes respectively. Similar definitions apply for the rotor variables. Equations 1-5 can be written in vector form as

$$\hat{v}_s = r_s \hat{i}_s + d\hat{\lambda}_s/dt \quad (10)$$

$$\hat{0} = r_r \hat{i}_r + d\hat{\lambda}_r/dt - \hat{\omega}_r \times \hat{\lambda}_r \quad (11)$$

$$\hat{T}_e = \frac{3P}{4} \hat{\lambda}_s \times \hat{i}_s \quad (12)$$

By means of auxiliary equations relating the flux linkages and currents [5] it is possible to solve for $\hat{i}_r$ and $\hat{\lambda}_s$ in terms of $\hat{i}_s$ and $\hat{\lambda}_r$ as

$$\hat{\lambda}_s = (L_{\ell s} + k_r L_{\ell r})\hat{i}_s + k_r \hat{\lambda}_r \quad (13)$$

$$\hat{i}_r = \hat{\lambda}_r/(L_{\ell r} + L_m) - k_r \hat{i}_s \quad (14)$$

where

$$k_r = L_m/(L_m + L_{\ell r}) \quad (15)$$

When written only in terms of stator current and rotor flux linkage the equations which define transient behavior of the machine, Eqs. 10-12, can now be expressed as

$$\hat{v}_s = r_s \hat{i}_s + (L_{\ell s} + k_r L_{\ell r})\hat{i}_s + k_r d\hat{\lambda}_r/dt \quad (16)$$

$$k_r r_r \hat{i}_s = r_r \hat{\lambda}_r/(L_m + L_{\ell r}) + d\hat{\lambda}_r/dt - \hat{\omega}_r \times \hat{\lambda}_r \quad (17)$$

$$\hat{T}_e = \frac{3P}{4} k_r \hat{\lambda}_r \times \hat{i}_s \quad (18)$$

Note that Eq. 17 has been written so that the stator current appears on the left hand side. This form of the equation suggests that if the actual motor line currents which flow in the physical system can be measured then signals proportional to the two phase stator currents can be constructed and applied to an on-line model of the rotor circuits of the machine. The cross product of the measured stator current and calculated rotor flux linkage can be used to compute the electromagnetic torque. The pulsating value of torque can be obtained by eliminating the average component by means of a high pass filter.

Figure 4 shows a practical implementation of this control scheme. It can be noted that since actual stator current is used as inputs to the model, changes in stator resistance and stator leakage inductance due to temperature, frequency (skin effect) and tooth saturation do not affect its accuracy. However, measured values of magnetizing inductance, rotor leakage inductance and rotor resistance are used as constant coefficients in the model. Hence, changes in the corresponding actual machine parameters will affect the accuracy of the torque computation. Probably the most significant parameter change occurs in rotor resistance as the motor heats up during use. If necessary, this effect can be compensated by changing the modeled value of rotor resistance as a function of frame temperature. The order of improvement resulting from this simple compensation scheme is probably sufficient for most routine applications.

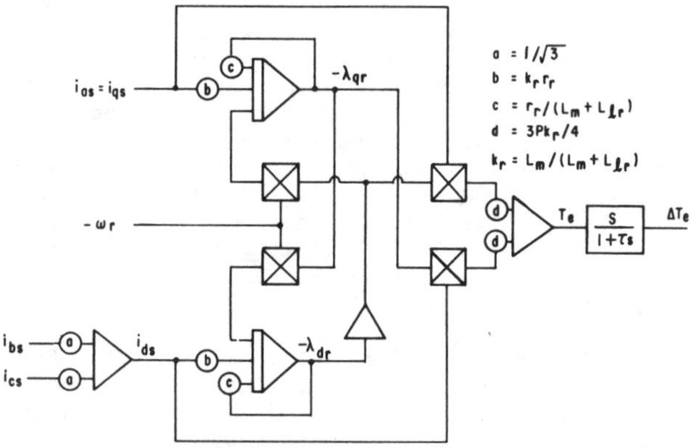

Fig. 4 Implementation of Method 1.

Method 2 - Using Terminal Voltage and DC Link Current

Although electromagnetic torque is most often calculated as the cross product of stator flux and stator current as in Eq. 12 many other equivalent forms for the torque equation exist. In particular, if stator flux linkage is expressed in terms of a leakage component plus an air gap component then

$$\hat{\lambda}_s = L_{\ell s}\hat{i}_s + \hat{\lambda}_m \qquad (19)$$

In terms of air gap flux linkage

$$\hat{T}_e = \frac{3P}{4}\hat{\lambda}_m \times \hat{i}_s \qquad (20)$$

since the cross product of any vector with itself is identically zero. It can be recalled from vector algebra that the cross product can be expressed equivalently as

$$T_e = \frac{3P}{4}|\hat{i}_s|[|\hat{\lambda}_m|\sin\theta] \qquad (21)$$

where the vertical bars imply the magnitude of the vector and θ is the instantaneous angle between vectors.

The quantity inside the brackets, namely $|\lambda_m|\sin\theta$, can be interpreted as the instantaneous component of the flux linkage vector normal to the current vector. This interpretation of the cross product suggests a means for computing the instantaneous pulsating torque from terminal measurements. Consider, for example, the time instant when the DC link current is directed into the c phase terminal and out of the b phase terminal with the a phase circuit open circuited or "floating" as shown in Fig. 5. From the definition of the d-q two phase variables it can be shown that the two components of the vector $\hat{\lambda}_m$ are

$$\lambda_{qm} = \lambda_{am} = \int v_{am}\,dt \qquad (22)$$

$$\lambda_{dm} = (\lambda_{cm} - \lambda_{bm})/\sqrt{3}$$
$$= (1/\sqrt{3})\int (v_{cm} - v_{bm})\,dt \qquad (23)$$

From Eq. 21 note that the torque producing component of flux linkage is, at this instant, the q-axis component since this component is normal to the current vector during the 60 degree portion of a cycle wherein this connection applies. The q-axis component is simply equal to the a phase component of flux linkage since the q-axis and a-phase axis are magnetically aligned by definition [5]. Since the a-phase current is zero over this interval it is clear that the voltage measured at the terminals of phase a is identically equal to the air gap voltage and the integral of this voltage is the flux linkages associated with phase a. Although one specific inverter connection was used as an example, these conclusions are general and can be applied to any of the six possible inverter connections. It is clear that the integration must always begin at each commutation so that the initial condition of the integral cannot be found without added information. Only the change in flux linkage over

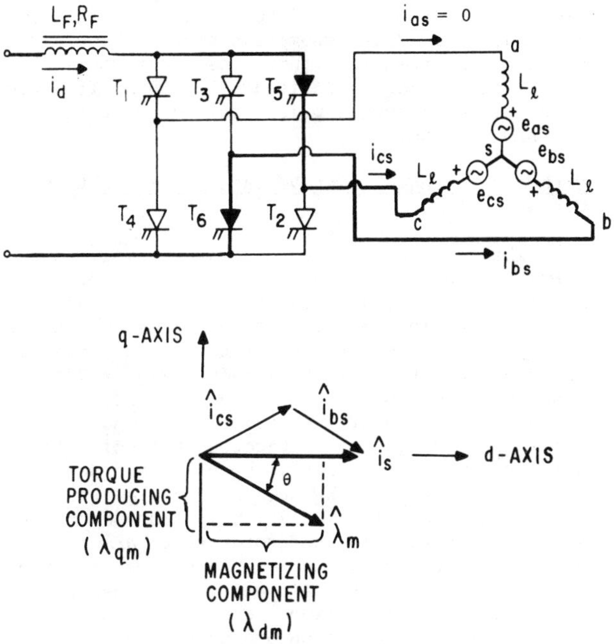

Fig. 5 Circuit Connection Diagram with Thyristors T5 and T6 Conducting.

the interval can be calculated. This component, however, is identically equal to the component which is responsible for the torque pulsation. Again, the unwanted average component of flux linkage can be removed with a high pass filter.

A mechanization of a torque pulsation measurement scheme employing this approach is shown in Fig. 6. The array of three switches are used to always connect the open circuited phase to the integrator. The other pair of switches is used to select the proper polarity of the open phase. The integrator is reset to zero at the start of each commutation. Since the average DC component of the integrator is not of concern this unwanted component is removed with a high pass filter. The product of the remaining, pulsating component of the integrator output times the DC link current is proportional to the pulsating torque.

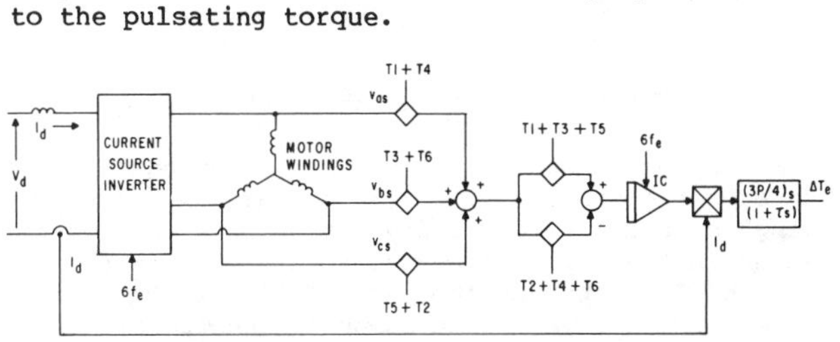

Fig. 6 Implementation of Method 2.

Because the torque producing component of flux linkage and current are measured exactly in Method 2, accuracy is not affected parameter changes due to saturation, temperature or other effects. However, the method assumes that commutation of the current source inverter is accomplished in negligible time compared to the basic 60 degree conduction interval. This assumption becomes less and less valid as the frequency increases. Nevertheless, a useful signal can be derived below 20 Hz line frequency (120 Hz torque pulsation) which is the region of primary concern.

Method 3 - Using Air Gap Voltage and Motor Line Current

The two previous methods which involve only terminal measurements are generally sufficiently accurate for most applications. However, when stringent restrictions are placed on torque pulsations it may be necessary to resort to internal measurements of the state of the machine. Figure 7 shows a scheme which employs search coils which are inserted in the top of the stator slots and are designed to measure the voltage corresponding to the air gap component of flux. The coils are designed to couple only with the useful, fundamental component of air gap flux. Other, unwanted components arise from saturation and rotor slot harmonics which do not produce useful torque. These components can be eliminated from the air gap voltage measurement by proper interconnection and weighting of the coil voltages [7]. Integration of the search coil voltages produces signals proportional to air gap flux linkages. The two phase stator currents can again be found algebraically from the three phase currents. Cross product multiplication of the two phase air gap flux linkage and stator current signals by means of Eq. 20 yields the motor torque. Again, if desired, the unneeded average value can be removed by a high pass filter.

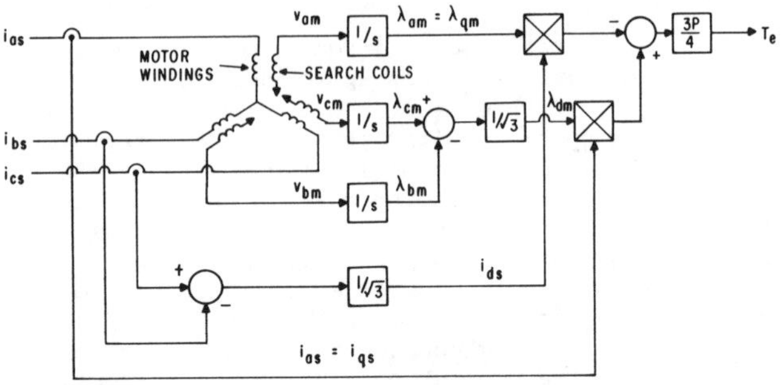

Fig. 7 Implementation of Method 3.

COMPUTED AND MEASURED RESULTS

Three methods for computing pulsating torque have been described which can be used as a "decogging" signal for modulating the DC link current as shown in Fig. 3. In order to verify the effectiveness of the approach the three feedback strategies were studied on a hybrid computer incorporating a detailed representation of motor and inverter [5,8]. For convenience the rectifier was modeled as an ideal power amplifier so that the desired output DC voltage was instantly obtained without sampling delay. The induction machine used for purposes of this study was a 230 V, 4 pole, 25 hp induction machine having the following parameters: r_s=0.0788Ω r_r=0.0408Ω, $x_{\ell s}$=0.2122Ω, $x_{\ell r}$=0.4362Ω, x_m=5.54Ω Base frequency used to compute the above reactances was 60 Hz. The DC link inductance was 29.2 mH.

Figure 8 shows a computer trace of a typical operating condition when the feedback decogging signal is out of service. The operating condition corresponds to a line frequency of 3.6 Hz, slip frequency of 1.2 Hz and link current of 60 A resulting in an average electromagnetic torque of 55 Nm or approximately one-half rated value. The large pulsating component superimposed on the average value of torque is apparent. In Fig. 9 the feedback decogging signal has been installed using control method 3. A large reduction in the pulsating torque is clearly evident. Similar results were also observed for the other two control methods since the identity of the feedback signal is essentially the same.

Figures 10 and 11 show scope traces taken from an actual experimental system. In Fig. 10 the decogging signal is out of service. It can be noted that the measured torque very closely resembles the torque pulsations predicted by the computer simulation. The basic sixth harmonic pulsation in the torque can be observed. As a practical matter much higher frequency components in the line current and torque can be observed which are not apparent in the simulation traces. These components arise from the 360 Hz ripple in DC link current resulting from harmonics contributed by the phase controlled rectifier. In Fig. 11 the feedback decogging has been installed using Method 3. It is evident that a substantial reduction in pulsating torque can be achieved in practice as well as in theory. Fourier Analysis of the torque traces in Figs. 10 and 11 indicate that a ten to one reduction in pulsating torque can be readily obtained.

PRACTICAL CONSIDERATIONS

Although the feedback scheme which has been described functions equally well in the regenerative mode as in motoring it is interesting that the polarity of the proper feedback signal changes sign. The needed sign change can be easily incorporated into the control but the proper feedback polarity becomes ambiguous at no load. Instability will result for either polarity at no load if the system gain is too high. A reduction in feedback gain can be avoided and torque pulsation kept at a low value if the air gap flux is reduced near the no load condition by use of the main system control. This stategy is often desirable for other purposes since the no load losses consisting mainly of iron losses will be reduced substantially at the same time. Figure 12 shows a series of computer traces illustrating a transition from motoring to regeneration by reducing the air gap flux. The pulsating torque clearly remains small during the entire transition.

Another practical matter involving implementation of such a control is speed of response. Certainly, when substantial changes in torque are required by the main system control it must not be interpreted as "pulsating torque" and eliminated by the decogging system. Since pulsating torque considerations are secondary when it is desired to move rapidly from one operating condition to another, one solution is to switch this signal out when the torque

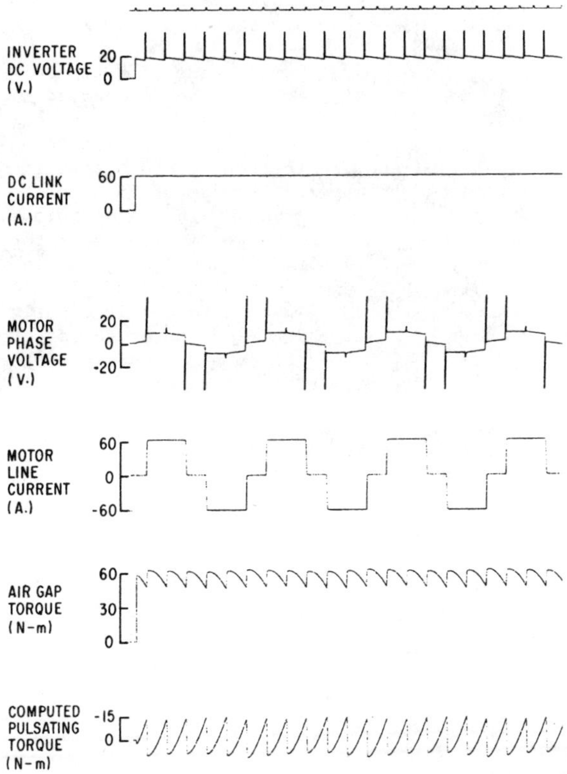

Fig. 8 Simulation Result with Decogging Signal Out of Service. Stator Frequency 3.6 Hz, Slip Frequency 1.2 Hz, DC Link Current 60 A.

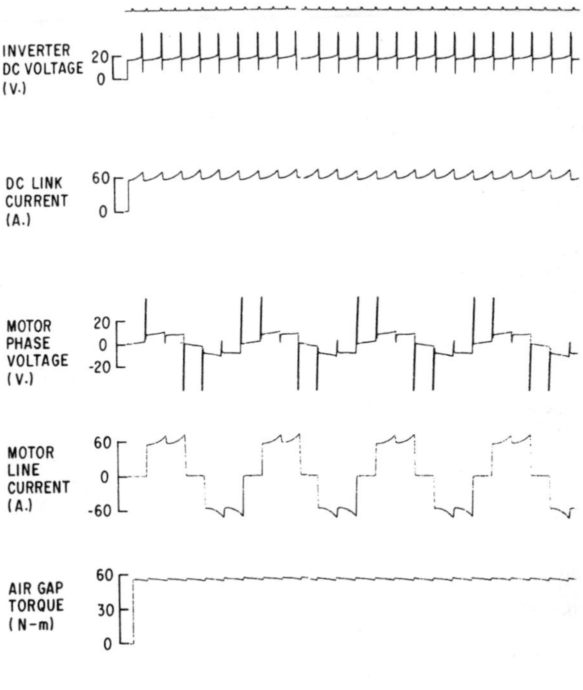

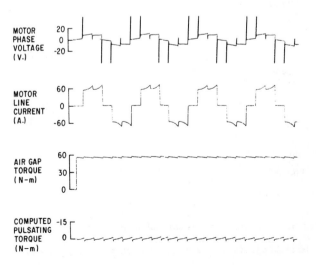

Fig. 9 Simulation Result with Decogging Signal In Service. Operating Conditions Same as Fig. 8.

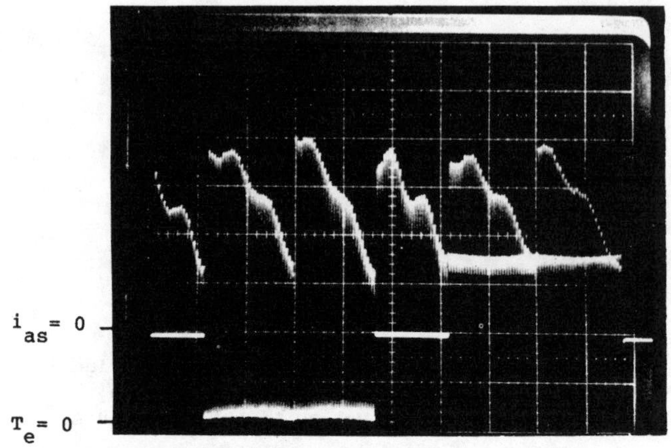

Fig. 10 Experimental Results from Actual Drive System with Decogging Signal Out of Service. Top Trace:Electromagnetic Torque, Scale 31 Nm/div., Bottom Trace:Motor Line Current, Scale 75 A/div., Time Scale 50 ms/div.

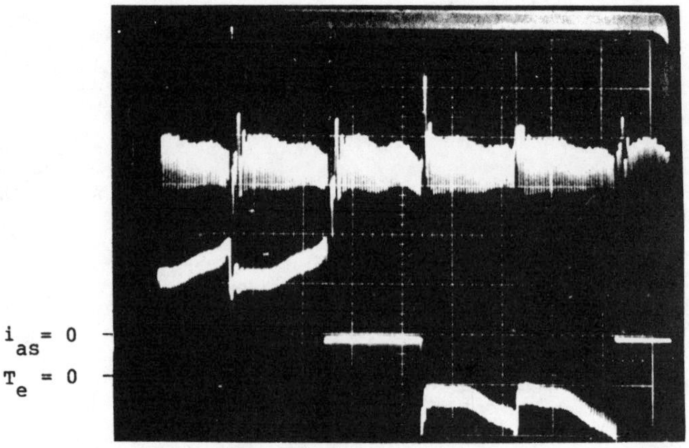

Fig. 11 Experimental Results with Decogging Signal in Operation. Scales same as Fig. 10.

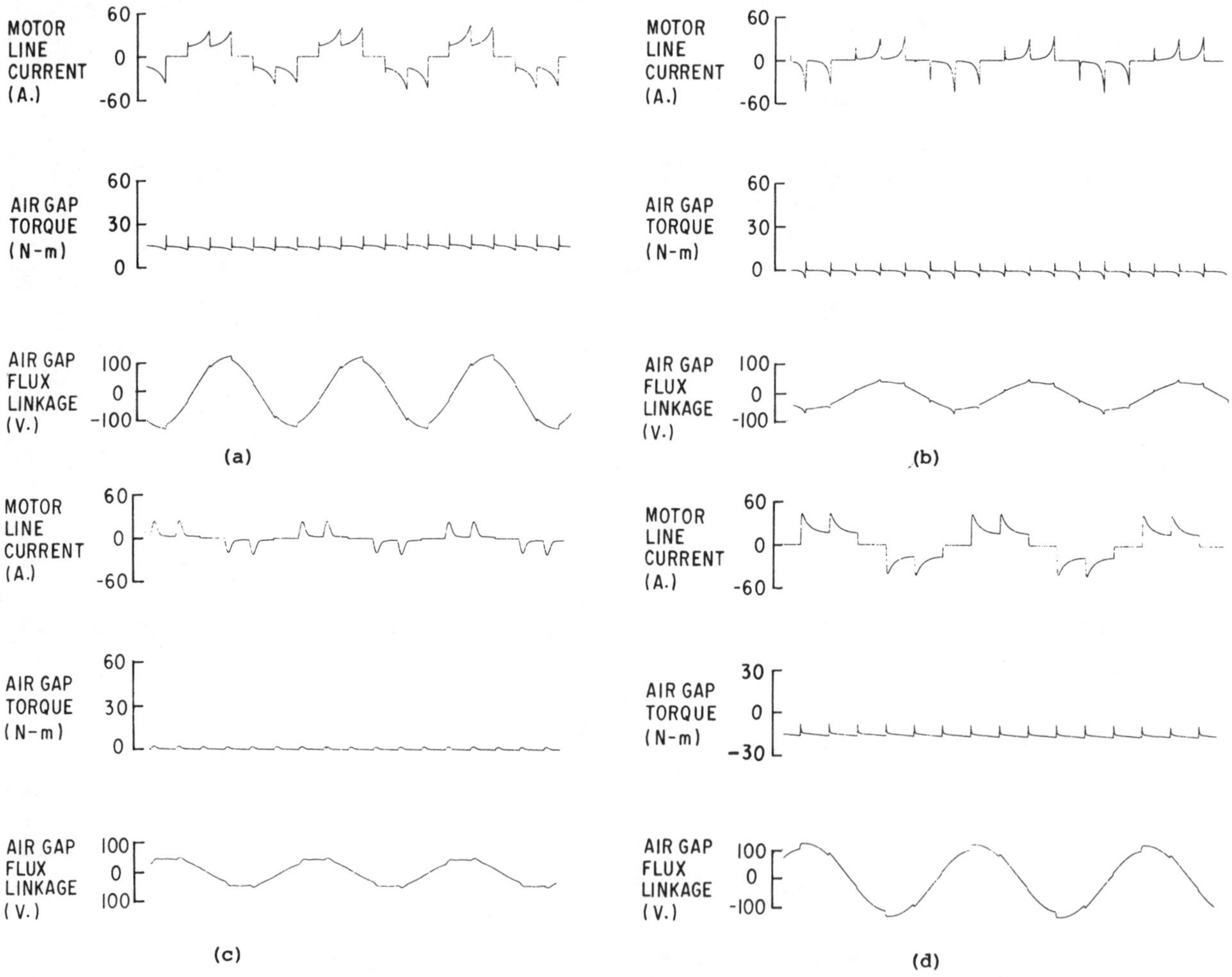

Fig. 12 Simulations Traces Showing Operation Through No-Load Using Weakened Air Gap Flux. a) Rated Flux, One-Eighth Rated Motor Load, b) Reduced Flux, No-Load, Positive Polarity Feedback, c) Reduced Flux, No Load, Negative Polarity Feedback, d) Rated Flux, One-Eighth Regenerative Load.

change exceeds some threshold. Alternatively, changes in operating condition can be sensed directly from the command signal i_d^*. Figure 13 describes a "riding gain" type of decogging control signal in which the feedback gain is inversely proportional to the change in command signal i_d^*.

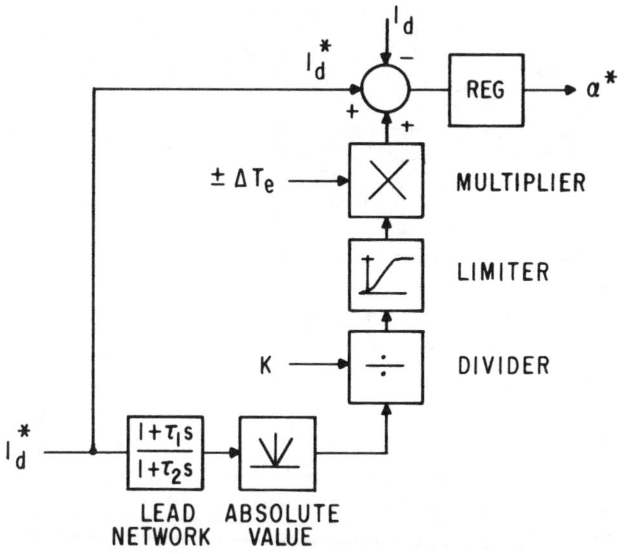

Fig. 13 Decogging Control System Employing Riding Gain.

CONCLUSION

This paper has presented three methods for feedback control of the pulsating torque which occurs when an induction motor is supplied from a current source inverter. Practical problems involved in the implementation of each have been discussed and it is demonstrated that a significant reduction in pulsating torque can be achieved in a practical system. Although this paper has concerned itself with induction motor drives it is clear the similar improvements can be attained by modulating the DC link of load commutated synchronous motor drives.

ACKNOWLEDGEMENT

The assistance of J.D. D'Atre and A.B. Plunkett during the experimental phase of this investigation are gratefully acknowledged.

REFERENCES

1. A. Schonung and H. Stemmler,"Static frequency changers with 'Subharmonic' control in conjunction with reversible variable speed AC drives", Brown Boveri Review, August/September 1964, pp. 555-577.

2. M. Blumenthal,"Pulse angle modulated operation of an ac machine fed by a current source inverter", Conference Record of the International Conference on Electrical Machines, 13-15 September, 1976, pp.I1-11 to I1-18.

3. T.A. Lipo and E.P. Cornell,"State-variable steady-state analysis of a controlled current induction motor drive", IEEE Trans. on Industry Applications, vol. IA-11, No. 6, November/December 1975, pp. 704-712.

4. A.B. Plunkett, J.D. D'Atre and T.A. Lipo,"Synchronous control of a static AC induction motor drive", Conference Record of the 1977 IEEE/IAS Annual Meeting, Oct. 26, 1977, pp. 609-615.

5. P.C. Krause and C.H. Thomas,"Simulation of symmetrical induction machinery", IEEE Trans. on Power Apparatus and Systems, vol. PAS-84, Nov. 1965, pp. 1038-1053.

6. K.P. Kovacs and I. Racz,"Transient behavior of AC machine",(book) Hungarian Academy of Science, Budapest, 1959 (In German).

7. T.A. Lipo,"Flux sensing and control of static AC drives by the use of flux coils", IEEE Trans. on Magnetics, vol. MAG-13, No. 5, September 1977, pp. 1403-1408.

8. T.A. Lipo,"Simulation of a current source inverter drive", Power Electronics Specialists Conference Record, June 14-16 1977, pp. 310-315.

Appendix

Analog Computer Symbols

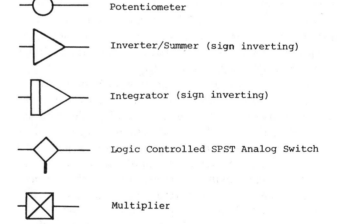

Potentiometer

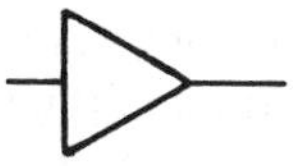
Inverter/Summer (sign inverting)

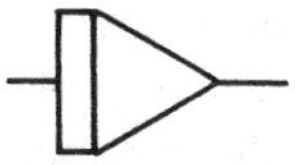
Integrator (sign inverting)

Logic Controlled SPST Analog Switch

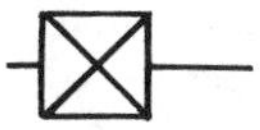
Multiplier

Synchronous Control of a Static AC Induction Motor Drive

ALLAN B. PLUNKETT, MEMBER, IEEE, JOHN D. D'ATRE, MEMBER, IEEE, AND THOMAS A. LIPO, SENIOR MEMBER, IEEE

Abstract—Conventional methods of controlling an induction motor utilize regulation of stator current and motor slip frequency in order to maintain system stability. This control strategy requires a shaft speed feedback and fast-response current regulation. An alternative method of controlling an induction motor is presented which achieves the necessary system stabilization by controlling only the motor frequency. The control inherently regulates the motor torque angle by properly adjusting the phase of the converter firing signals. By synchronizing the inverter firing pulses to the motor back electromotive force (EMF) possible adverse inverter operating modes are avoided. The concept of synchronous control eliminates the preprogrammed functional relations previously required and allows the control to adapt to any desired motor flux level. Any desired outer regulating loop can be incorporated to form a fast-response wide-range ac drive system.

INTRODUCTION

DURING recent years the development of static ac drives has opened up new fields of application for ac machines. These new applications have, in turn, necessitated the development of fast-response high-accuracy regulating systems. One regulating system which has prompted considerable attention is the current regulated induction motor drive. The basic configuration of such a drive is shown in Fig. 1. In this system ac or dc power is converted to variable amplitude dc power by means of a phase-controlled ac/dc rectifier or dc/dc chopper. The power is then converted to ac form by means of a dc/ac converter. A filter is inserted in the dc link so as to smooth the ripple currents inherent in rectifier or chopper operation. Alternatively, the ac to ac conversion could be accomplished in one step by use of a cycloconverter.

In order to realize practical operation of both motor and converter, stabilizing feedback must often be used to maintain normal motor flux and current levels [1]. Conventional control methods realize this stabilization by dynamic control of the motor current amplitude. Such strategy requires that the converter respond over the entire operating speed-torque range since system stability could be lost should the converter (rectifier, chopper, or cycloconverter) be phased fully on. The constraint severely limits the speed range over which such systems can be practically applied since operation in the "field weakening" or constant horsepower mode becomes difficult. In addition to current regulation, a function of current is used

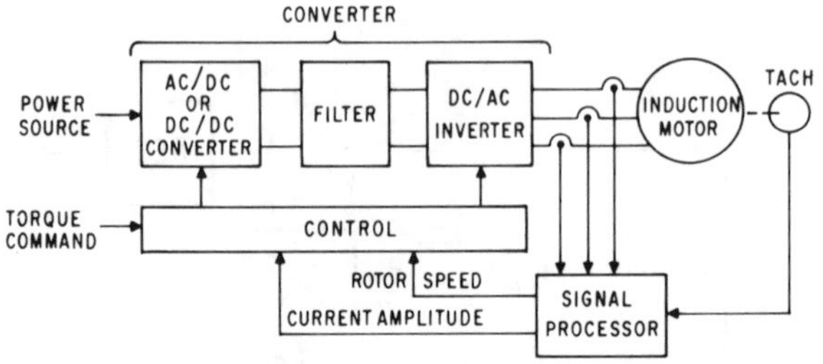

Fig. 1. Static ac drive with current and speed feedback.

to set the motor slip frequency which together with a motor speed feedback signal sets the inverter firing frequency [2], [3]. This functional relationship between link current and slip frequency is established to maintain, approximately, a constant flux in the motor as a function of load. The motor electrical slip frequency can also be calculated from terminal voltage and current [4]. However, accurate knowledge of motor parameters is required. Since motor parameters change with stator current, air gap flux, rotor frequency, and temperature, these regulation schemes are inherently difficult to operate over a wide range of speed and load. Stability can also be maintained by controlling power factor [5], but parameter changes again deteriorate performance.

In this paper an alternative method of control is presented. The scheme utilizes the inverter frequency as the necessary system stabilizing control, while the current amplitude merely adjusts the motor steady-state flux level. The stabilization is achieved by regulating the phase angle between motor current and motor flux. This approach causes the inverter firing pulses to synchronize to the motor counter electromotive force (CEMF) and is the dual of the voltage-fed induction motor wherein additional current is inherently provided to align the motor CEMF to the inverter [6].

The concept of synchronous control evolved from a desire to synchronize current flow in the inverter with respect to motor CEMF. It appears that an inverter with current regulation functions very much as a phase-controlled rectifier operating in the inverting mode. The firing angle of the inverter must be synchronized with respect to the motor voltage (or internal flux) in order to smoothly control power flow. Lack of synchronization results in an effect similar to the oscillation of a synchronous motor subjected to a sudden load change in which the rotor of the machine oscillates with respect to synchronous speed. Damping of rotor swings in a synchronous motor is obtained from short-circuited rotor windings which generate transient voltages, which in turn draw transient currents from the power supply to damp the oscillation. In the case of an induction motor supplied by a current

Paper ID 77-6, approved by the Industrial Drives Committee of the IEEE Industry Applications Society for presentation at the 1977 Industry Applications Society Annual Meeting, Los Angeles, CA, October 2–4.

A. B. Plunkett and T. A. Lipo are with Corporate Research and Development, General Electric Company, Schenectady, NY 12301.
J. D. D'Atre is with the Electric Utility System Engineering Department, General Electric Company, Schenectady, NY 12301.

Reprinted from *IEEE Trans. Ind. Appl.*, vol. IA-15, pp. 430–437, July/Aug. 1979.

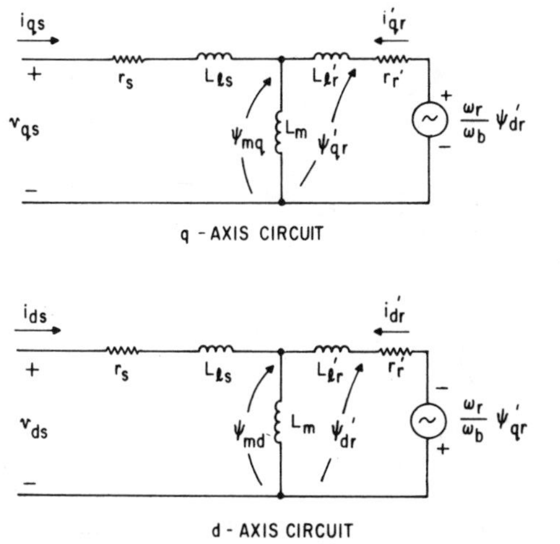

Fig. 2. Induction motor *d-q* euqivalent circuit in stationary reference frame.

source, a similar oscillating effect exists. However, the required damping current does not inherently flow from the power supply. By synchronizing the converter to the motor CEMF such hunting-type instabilities can be eliminated.

In addition to stability problems caused by hunting, difficulties are also introduced by converter commutation. Commutation introduces a time delay between the firing of a converter thyristor and the instant of actual current transfer in the motor lines. Ripple currents introduce additional random delays in the commutation time. Unless properly treated, these delays can cause difficulties in inverter firing and in subsequent loss of control. The concept of synchronous control allows the effects of variable inverter commutation delay to be attenuated by the gain within the regulation loop. Synchronous control also offers the possibility of tachometerless operation. The expense and mechancial problems associated with a tachometer can be eliminated.

GENERAL CONCEPTS

The conventional *d-q* axes equivalent circuit of a squirrel-cage induction motor expressed in a reference frame fixed in the stator [7] is given in Fig. 2. The quantities shown are the standard motor equivalent circuit parameters defined in the usual manner. The variables ψ_{mq}, ψ_{md}, ψ_{qr}', and ψ_{dr}' correspond to the *q*- and *d*-axis air gap and rotor flux linkages, respectively. These flux quantities carry units of voltage, are equal to the corresponding flux linkages λ times ω_b, the base angular frequency, and are defined by

$$\psi_{mq} = \omega_b L_m (i_{qs} + i_{qr}') \tag{1}$$

$$\psi_{md} = \omega_b L_m (i_{ds} + i_{dr}') \tag{2}$$

$$\psi_{qr}' = \omega_b [(L_m + L_{lr}') i_{qr}' + L_m i_{qs}] \tag{3}$$

$$\psi_{dr}' = \omega_b [(L_m + L_{lr}') i_{dr}' + L_m i_{ds}]. \tag{4}$$

In general, the superscript "*s*" is usually affixed to the *d-q* variables to signify the stator reference frame, but the practice is omitted here for convenience. The quantities $(\omega_r/\omega_b)\psi_{qr}'$ and $(\omega_r/\omega_b)\psi_{dr}'$ can be considered as equivalent to the generated CEMF of the motor and are cross-coupled between the *d* and *q* axes.

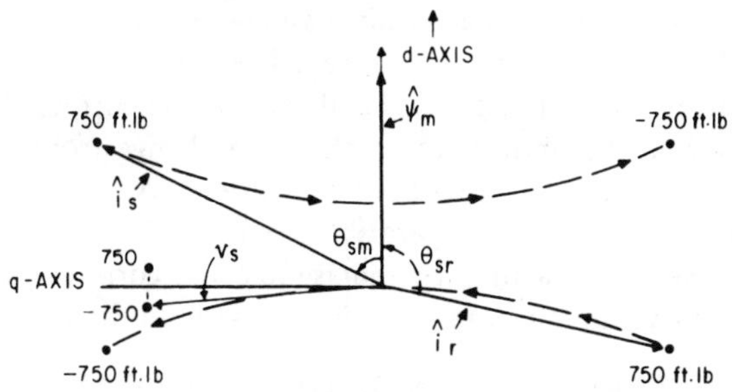

Fig. 3. Loci of voltage, current, and flux linkage vectors with load at rated frequency.

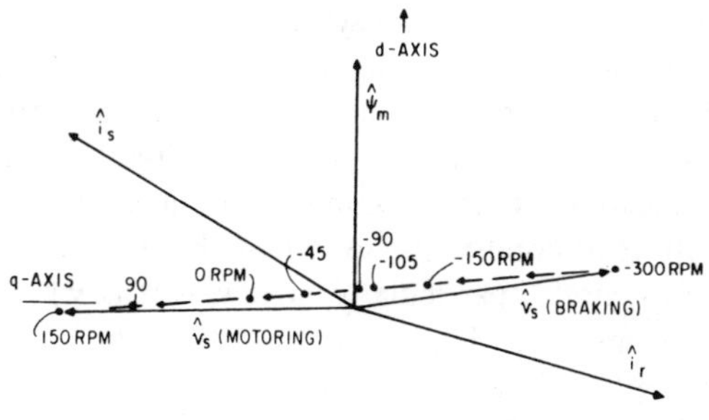

Fig. 4. Loci of voltage, current, and flux linkage vectors during speed reversal. Torque maintained constant.

Figs. 3 and 4 are vector diagrams portraying the *q*- and *d*-axis quantities as two components of a vector in order to show the interrelations which exist between variables. In particular, Fig. 3 shows the change in angular position of the motor current vectors as a function of load at a rated speed using the air gap flux linkage vector as a reference. The parameters of an experimental system given in the Appendix were used, wherein 750 ft•lb corresponds approximately to the rated load. At no load, the flux and current vector are in phase and 90° out of phase with the terminal voltage vector. When the motor load increases rotor current must be developed, which in turn requires a counteracting component of stator current.

It can be observed that the stator voltage vector $\hat{v}_s$ does not significantly change position with load, so that the frequency (or phase) of a voltage source need not vary with changes in load. The voltage source also offers an inherent stabilizing action by supplying damping current so that the motor is able to rapidly align to any new operating condition without assistance from the inverter. On the other hand, it can be noted that the stator current magnitude and angle change rapidly with load. The angular relation between current and motor flux now depends on the inverter firing and must be artificially provided by the control when a current source is employed. In addition, sufficient damping must be provided by the control for good transient behavior since the damping currents which normally flow are not inherently present.

A similar realignment effect occurs with voltage source inverters at low speeds, except it is now the terminal voltage vector that must be changed with respect to the flux vector. Fig. 4 shows a vector diagram illustrating a transition through zero speed while maintaining a torque constant. It should be noted that the current and flux linkage vectors remain relatively fixed while the terminal voltage vector varies widely in angle and magnitude with speed. The firing pulses of a voltage inverter source must now be aligned at the correct phase with respect to the motor flux such that a change from braking to motoring can occur without a transient. An angle control is now required when operating from a voltage source. Note, however, that with a current source the corresponding flux-current vector alignment is achieved without difficulty.

CALCULATION OF THE TORQUE ANGLE

In general, the calculation of the torque angle necessitates sensing motor electrical variables since this quantity cannot be directly measured as with a synchronous machine. One alternative is to measure the motor line current and terminal voltage and back-calculate the flux angular position from the per phase induction motor equivalent circuit. Unfortunately, motor parameters vary from motor to motor with temperature and with load conditions, so that such compensation may become inaccurate, particularly at low speeds where resistance drop becomes the largest portion of motor voltage. A preferred method is to sense the air gap flux directly [8], [9]. Since the air gap flux is the result of both stator and rotor currents the actual rotor circuit operating conditions are sensed, and the inaccuracies involved in terminal measurements can be eliminated.

It can be shown that when stator current and air gap flux linkages are written as vectors having d and q components, then the electromagnetic torque can be expressed as

$$\hat{T}_e = \left(\frac{3}{2}\right)\left(\frac{P}{2}\right)\frac{1}{\omega_b}\hat{i}_s x \hat{\psi}_m \tag{5}$$

where P is the number of motor poles and "x" denotes the cross product between the current and flux linkage vectors. Equation (5) can be written alternatively as

$$T_e = \left(\frac{3}{2}\right)\left(\frac{P}{2}\right)\frac{1}{\omega_b} |\hat{i}_s| |\hat{\psi}_m| \sin\theta_{sm} \tag{6}$$

where θ_{sm} is the angle between $\hat{i}_s$ and $\hat{\psi}_m$. Thus the torque angle can be computed from the equation

$$\sin\theta_{sm} = \frac{4\omega_b T_e}{3P|\hat{i}_s||\hat{\psi}_m|}. \tag{7}$$

In general, (7) is the most straightforward means for computing the torque angle. However, other forms for the torque equation exist which yield other possible flux-current relationships. It can be shown that (8)-(10) are alternative expressions for an instantaneous electromagnetic torque [10]:

$$\hat{T}_e = \left(\frac{3}{2}\right)\left(\frac{P}{2}\right)\frac{1}{\omega_b}\hat{\psi}_m x \hat{i}_r \tag{8}$$

$$\hat{T}_e = \left(\frac{3}{2}\right)\left(\frac{P}{2}\right)\frac{1}{\omega_b}\hat{\psi}_s x \hat{i}_r \tag{9}$$

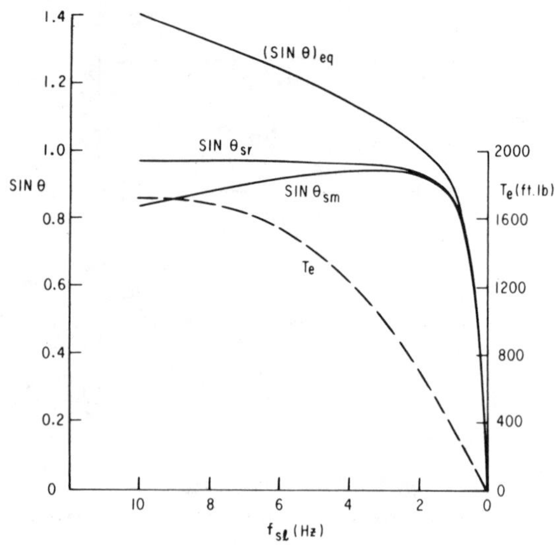

Fig. 5. Torque and torque angle as function of slip.

$$\hat{T}_e = \left(\frac{3}{2}\right)\left(\frac{P}{2}\right)\frac{1}{\omega_b}\hat{i}_s x \hat{\psi}_r \tag{10}$$

where $\hat{\psi}_s$ denotes the total stator flux linkage vector and $\hat{i}_r$ is the rotor current vector. Unfortunately the rotor current cannot be monitored directly, so that (8) and (9) are difficult to implement. However, from (1)-(4) the rotor and air gap flux linkages can be written as

$$\hat{\psi}_r = x_{lr}'\hat{i}_r + \hat{\psi}_m \tag{11}$$

$$\hat{\psi}_m = x_m(\hat{i}_s + \hat{i}_r) \tag{12}$$

where $x_{lr}' = \omega_b L_{lr}'$ and $x_m = \omega_b L_m$. Upon algebraically eliminating i_r, (11) can be written as

$$\hat{\psi}_r = \frac{x_r'}{x_m}\hat{\psi}_m - x_{lr}'\hat{i}_s \tag{13}$$

where $x_r' = x_{lr}' + x_m$. Hence the rotor flux linkage vector can be mechanized from the air gap flux linkages and stator current. The angle θ_{sr} between $\hat{i}_s$ and $\hat{\psi}_r$ corresponding to (10) can also be viewed as a torque angle. The angle θ_{sr} is defined by

$$\sin\theta_{sr} = \frac{4\omega_b T_e}{3P|\hat{i}_s||\hat{\psi}_r|} \tag{14}$$

and $\hat{\psi}_r$ is derived from (13).

The quantities $\sin\theta_{sm}$ and $\sin\theta_{sr}$ are plotted as functions of slip frequency in Fig. 5. Again the system parameters given in the Appendix were used. It can be noted that the quantity $\sin\theta_{sm}$ is a double-valued function of torque when the slip frequency increases from zero to breakdown. Since the angle control is located within a torque regulation loop, difficulties can be anticipated by this double-valued behavior. It can be noted that the angle between stator current and rotor flux is not double-valued but reaches a maximum of 90° when the slip frequency approaches breakdown. However, the angle

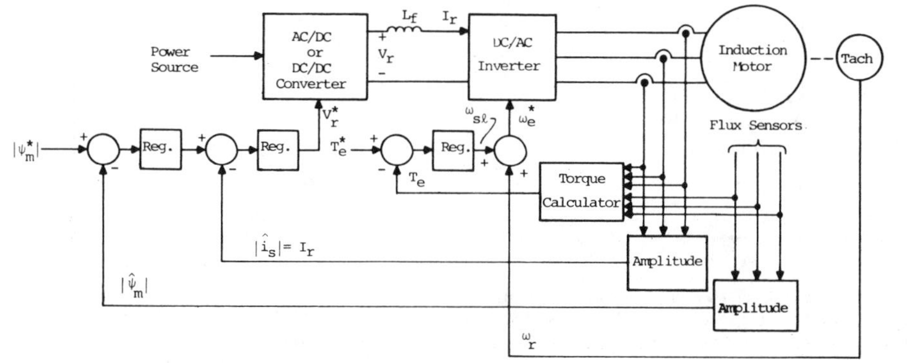

Fig. 6. Control scheme employing flux amplitude and torque regulation.

change is very small at high slips, indicating the low gain and consequent poor regulation.

One practical modification of this angle measurement technique is to basically provide a corrective signal to ensure that a monotonic relationship is maintained between torque and angle. Although (13) is a vector relationship it is interesting to consider the approximate relationship

$$|\hat{\psi}_r| \cong \frac{x_r'}{x_m} |\hat{\psi}_m| - x_{lr}' |\hat{i}_s|. \tag{15}$$

Equation (14) becomes

$$(\sin\theta)_{\text{equiv}} = \frac{4\omega_b T_e}{3P\left[\frac{x_r'}{x_m}|\hat{\psi}_m| - x_{lr}'|\hat{i}_s|\right]|\hat{i}_s|}. \tag{16}$$

The equivalent angle quantity $(\sin\theta)_{\text{equiv}}$ is also plotted in Fig. 5. Note that this modified variable takes on values greater than unity and thus does not strictly correspond to an actual angle. Nonetheless, the quantity increases monotonically over the entire operating region and thereby suggests a feasible variable for regulation.

TRANSIENT BEHAVIOR OF SYNCHRONOUS CONTROL

Although it has been demonstrated that a modified value of the torque angle has desirable steady-state properties, a detailed dynamic analysis is required in order to verify that the proposed angle regulation scheme is feasible. One analysis technique which has proven ideally suited to this task is the computation of the relevant induction motor transfer functions [11]. As part of this technique the *d-q* equations which express dynamic behavior of the inverter-motor system are defined, linearized for small changes about an operating point, and recast in state-variable form [3]. The relevant transfer function input is defined and the output is identified in terms of the state variables. The equations are then transformed to phase-variable canonical form wherein the desired transfer function can be obtained by inspection. Having obtained the desired transfer function, any of the usual control design techniques can be employed such as the Bode plot, the root-locus plot, the Nichols chart, etc.

It has already been mentioned that when the dc voltage source is active, the required system stabilization can be achieved by utilizing the dc link. One desirable control scheme which has been reported previously [9] is shown in Fig. 6. With this method the air gap flux magnitude and air gap torque are sensed. The dc voltage source is used to regulate the amplitude of the air gap flux and an inner current loop is provided to eliminate the time constant associated with the dc link inductor. In addition, slip frequency is adjusted to regulate the torque to the desired level.

In Fig. 7 the flux loop has been disabled and the dc link current loop has been closed through a simple gain G_i. That is,

$$V_r = G_i(I_r^* - I_r) \tag{17}$$

where the asterisk denotes the commanded value of the link current which is assumed constant. Fig. 7 shows the migration of the poles and zeros of the frequency loop transfer function between slip angular frequency and air gap torque $(\Delta T_e/\Delta\omega_{sl})$ as the gain G_i is increased in the dc link current regulator. The operating point considered is at rated frequency, rated terminal V/Hz, and a load of 750 ft·lb. The system parameters in the Appendix were again used. Note that although the system poles are in the right half-plane for zero gain, the poles move into the left half-plane with only a modest amount of feedback gain when the current regulator is in operation. Control of slip frequency is not required for stabilization. In Fig. 7 the effects of rotor speed changes have been neglected. Although rotor speed changes slightly modify the results of Fig. 7, the inertia will be assumed infinitely large throughout this study.

In Fig. 8 the behavior of the same transfer function is investigated. However, this time the current regulator is out of service, indicating that the dc voltage source is fixed. This condition corresponds to operation at high speeds in which the ac/dc rectifier or dc/dc chopper is phased full-on (fully conducting). Fig. 8 shows the location of the poles and zeros of $\Delta T_e/\Delta\omega_{sl}$ for a range of load points. Note that the system becomes unstable for loads above about 150 ft·lb. Moreover, the complex conjugate poles in the right half-plane are accompanied by a pair of zeros indicating that overall stability is not possible regardless of the feedback compensator used between the output ΔT_e and input $\Delta\omega_{sl}$. This behavior clearly points to the need for an improved control scheme during high-speed operation.

In Fig. 9 the sine of the angle between stator current and air gap flux linkage, $\sin\theta_{sm}$, has been implemented by means

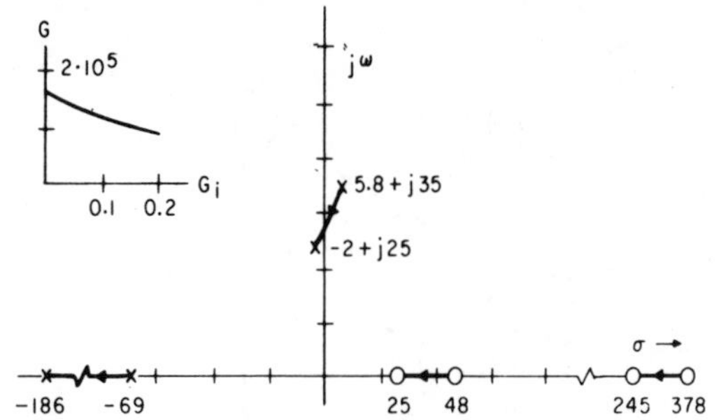

Fig. 7. Poles (x), zeros (0), and gain (G) of transfer function $\Delta T_e/\Delta\omega_{sl}$ for increasing link current regulator gain G_i. Initial point G_i = 0, final point G_i = 0.2. Operating condition 50 Hz, 210 V, 750 ft·lb.

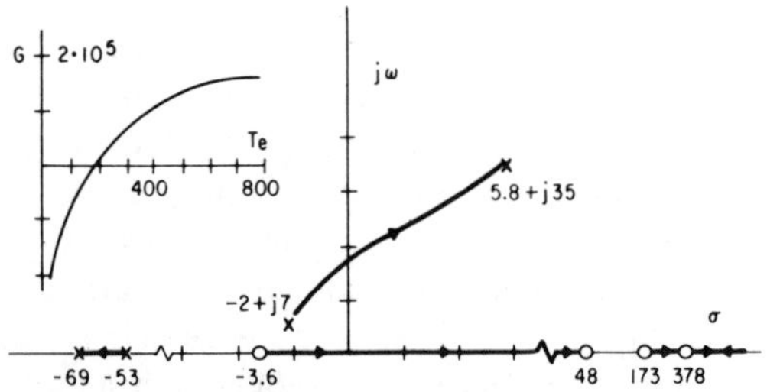

Fig. 8. Poles (x), zeros (0), and gain (G) of transfer function $\Delta T_e/\Delta\omega_{sl}$ with constant dc link voltage. Effect of increasing load torque with rated terminal voltage and frequency. Initial point T_e = 25, final point T_e = 750 ft·lb.

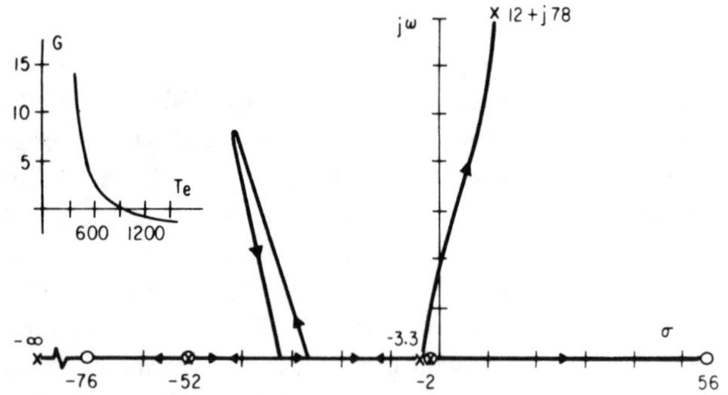

Fig. 9. Poles (x), zeros (0), and gain (G) of transfer function Δ sin $\theta_{sm}/\Delta\omega_{sl}$ with constant dc link voltage. Effect of increasing load torque with rated terminal voltage and frequency. Initial point T_e = 0 (no-load), final point T_e = 1675 ft·lb (breakdown).

of (7). The poles and zeros of $\Delta(\sin\theta_{sm})/\Delta\omega_{sl}$ are plotted over the same range of operating conditions as in Fig. 8. Since the transfer function pole locations are independent of input and output, the poles again enter the right half-plane at 150 ft·lb as load is increased. In this case the zeros originate at -2 and -52 at no load. The zeros move together, break off the real axis, and then return at a load of 500 ft·lb. Up to this point the zeros migrate only in the left half-plane indicating that stabilization remains feasible. However, as load continues to increase, one of the zeros moves to the right and finally enters the right half-plane beyond 1000 ft·lb. The point at which this zero changes sign corresponds to the inflection point of the plot of sin θ_{sm} versus f_{sl} in Fig. 5. In general, one of the right half-plane poles will be attracted to this zero as feedback gain is increased so that stabilizing feedback compensation appears impractical if not impossible.

Fig. 10 shows the pole-zero locations when the quantity sin θ_{sr} is measured rather than sin θ_{sm}. It can be noted that

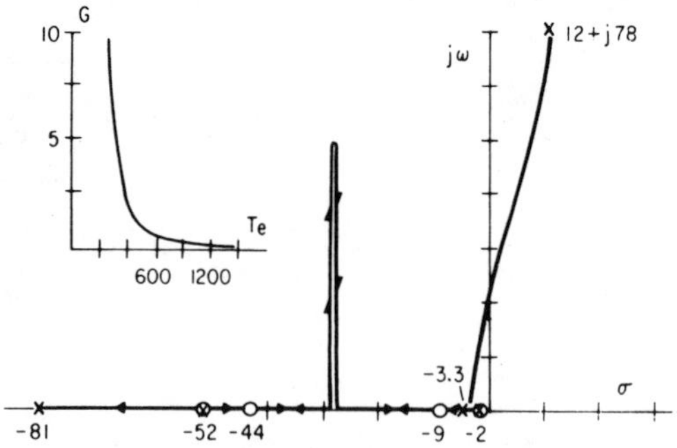

Fig. 10. Poles (x), zeros (0), and dc gain (G) of transfer function Δ sin $\theta_{sr}/\Delta\omega_{sl}$ with constant dc link voltage. Effect of increasing load torque with rated terminal voltage and frequency. Initial point T_e = 0, final point T_e = 1675 ft·lb.

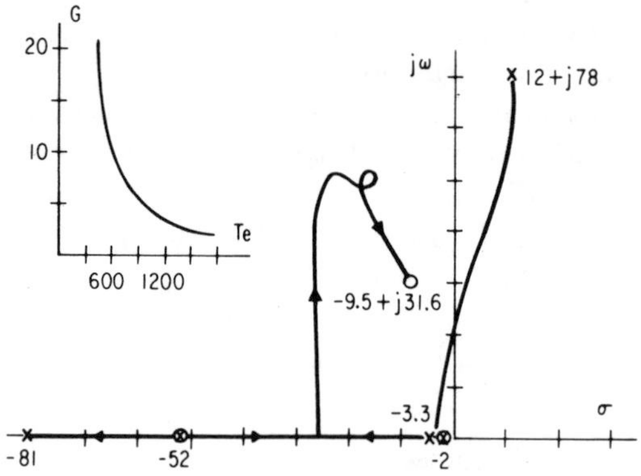

Fig. 11. Poles (x), zeros (0), and dc gain (G) of transfer function Δ (sin $\theta)_{\text{equiv}}/\Delta\omega_{sl}$ with constant dc link voltage. Effect of increasing load torque with rated terminal voltage and frequency. Initial point T_e = 0, final point T_e = 1675 ft·lb.

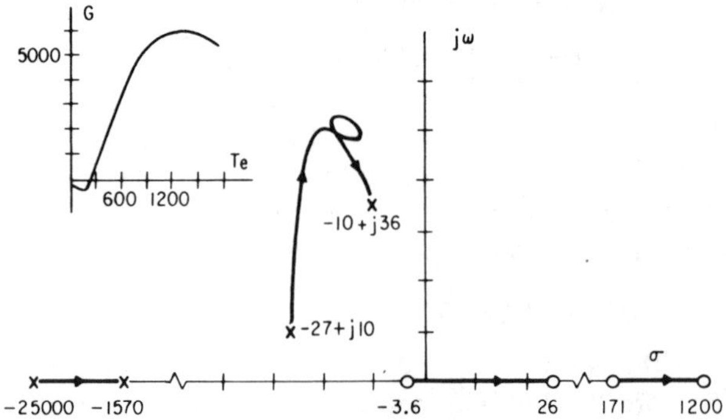

Fig. 12. Poles (x), zeros (0), and dc gain (G) of transfer function $\Delta T_e/\Delta$ (sin $\theta)^*$. Angle loop closed through gain G = 10. Initial point T_e = 25, final point T_e = 1675 ft·lb.

the zeros are better behaved and never enter the right half-plane. However, the system gain changes dramatically with load. Although it is possible to compensate with a feedback gain which increases inversely with load, such a technique is difficult to mechanize because practical considerations such as noise and sampling effects begin to affect performance.

In Fig. 11 the modified angle derived from (15) and (16) is considered as the feedback variable. Again, operation at 50 Hz is assumed. As noted previously, the open loop system poles are in the right half-plane for heavy loads. However, the system zeros remain in the left half-plane over the full range of operating conditions. Loop closure is now practical and system stability can be maintained for motoring loads. Fig. 12 indi-

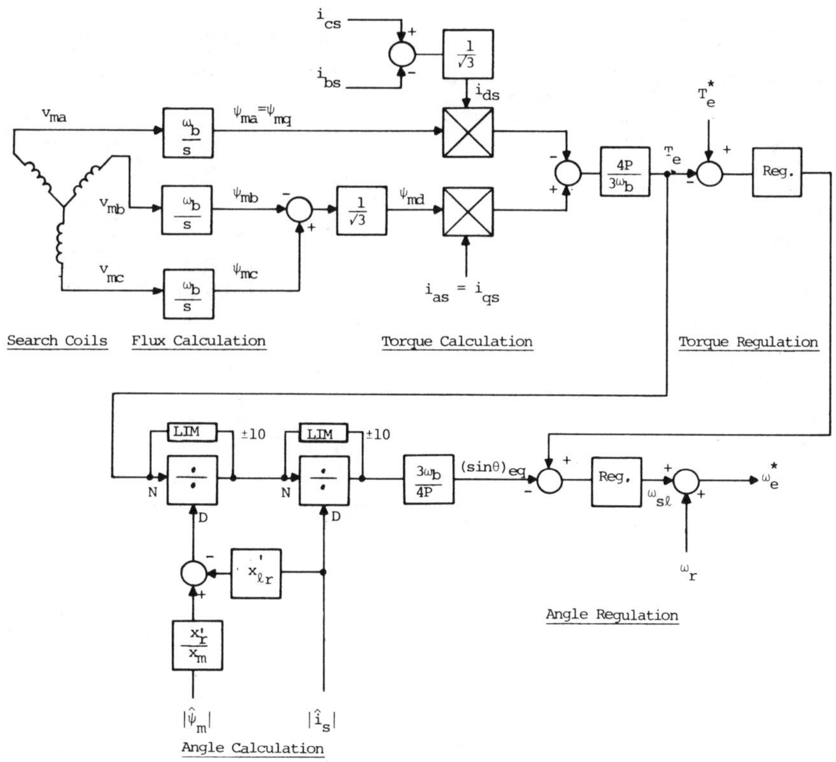

Fig. 13. Torque regulation scheme employing angle regulation.

cates the resulting pole-zero locations when the transfer function Δ (sin $\theta)_{\text{equiv}}/\Delta\omega_{sl}$ has been closed through a feedback gain $G_\theta = 10$. The poles and zeros in the subsequent outer loop $\Delta T_e/\Delta$ (sin $\theta)^*$ are shown. It is evident that system stability has been provided by the inner feedback loop, and closure of the torque regulating loop can now proceed without difficulty. Although rated line frequency has been assumed throughout this discussion, it can be shown that the system poles and zeros are not materially affected by operating frequency [3] so that these conclusions are valid over a wide speed range.

EXPERIMENTAL RESULTS

Fig. 13 depicts the overall method of torque angle calculation and regulation which has been successfully implemented. This scheme consists of calculating electromagnetic torque which is then normalized by magnitude of current and magnitude of flux linkage. The intermediate signals of torque, flux magnitude, and current magnitude are already required for use in other control loops which set the desired operating points for the motor [9] so that additional sensors are not required. In order to obtain a meaningful angle signal at start-up or under abnormal operating conditions, the values of $|\hat{i}_s|$ and $|\hat{\psi}_m|$ are fed to two dividers which have a minimum value set by a nonlinear limiter so that division by zero is avoided.

An alternative approach to the angle calculation which is somewhat restrictive in that saturation is not properly taken into account is shown in Fig. 14. In this case the flux and current signals are amplified, limited, and then connected to multipliers. The result is an angle signal which is not dependent on the magnitude of flux and current but only on their relative phase. For small amplitudes of current or flux, the calculation will properly yield a zero output. Division is not required. This method should lend itself well to digital implementation, although near zero speed continuous information will not be available from the digital phase detection circuit.

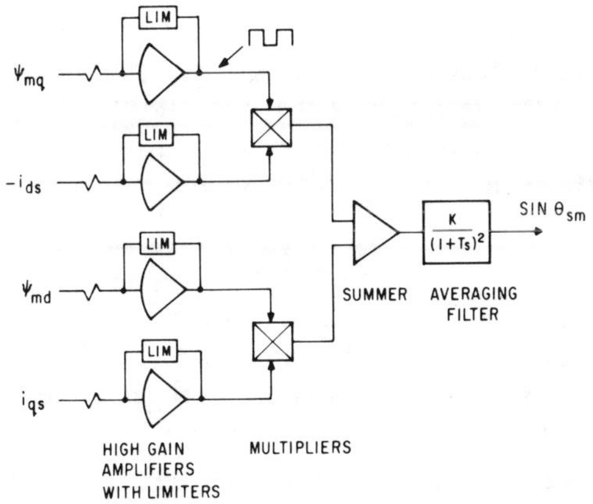

Fig. 14. Alternate angle computation.

The control system which has been described has been thoroughly instrumented and tested. Fig. 15 shows a representative system transient response for a step change in a commanded angle while operating in the motoring mode. The system parameters are again those summarized in the Appendix except that $L_f = 1.0$ mH. The dc source consists of a conventional six-pulse rectifier bridge supplied by fixed ac voltages. The chart traces display current magnitude, torque, (sin $\theta)_{\text{equiv}}$, and motor slip frequency. It can be noted that the angle follows the command very quickly (rise time ≅ 0.04 s), but it takes about 0.1 s for the torque and current to respond. The slow drift in torque (about 1.0 s time constant) is due to the change in the speed of the dc load motor as the torque level changes. The lower frequency ripple in the current is due to the beat between motor frequency and the rectifier

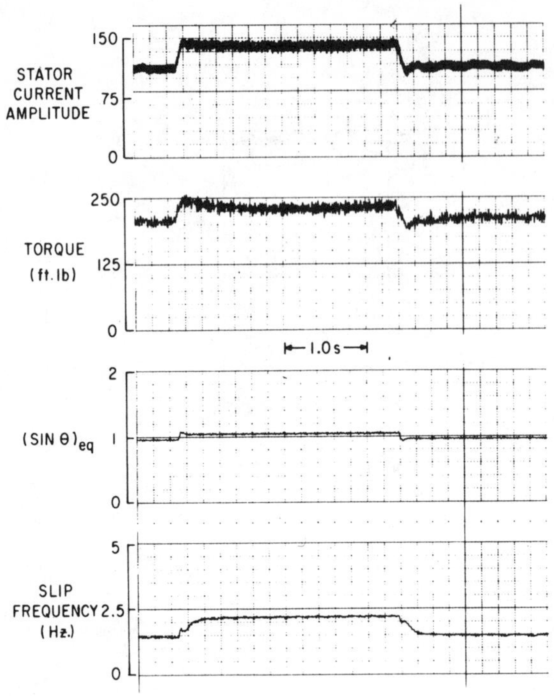

Fig. 15. Performance of prototype system for step changes in angle command with fixed dc source voltage.

ac source frequency. The high-frequency ripple is due to the pulsations in link current caused by inverter commutation. These results clearly show that accurate fast torque control can be achieved with synchronous control. Moreover, operation without current loop stabilization is entirely feasible.

CONCLUSION

The regulation of the angle between motor flux and current offers several direct benefits when an induction motor is supplied from a static converter. Most important, the necessary stabilizing force is exerted by inverter frequency alone, thus easing the requirements on the converter supply. Overall system stability and transient response have been shown to be enhanced by synchronous control. Moreover, the synchronization of inverter firing pulses to motor flux eliminates possible abnormal inverter commutation conditions. Problems involved with motor speed reversal with either current or voltage inverters is greatly reduced due to the vector alignment capability of the control. If necessary, mechanical sensing of shaft speed can be eliminated by controlling the torque angle since the system is self-synchronous. The basic control scheme should also be beneficial to synchronous motor drives since the control scheme which has been described does not require rotor position sensing nor armature reaction compensation.

APPENDIX

Motor Rating

Rated Voltage	210 V rms l-n.
Rated Frequency	50 Hz.
Breakdown Torque	1675 ft·lb.
Poles	4.

Motor Parameters at 50 Hz

$r_s = 0.0172\ \Omega.$

$x_{ls} = 0.0706\ \Omega.$

$r_r' = 0.0310\ \Omega.$

$x_{lr}' = 0.0903\ \Omega.$

$x_m = 2.8413\ \Omega.$

DC Link Parameters

$L_f = 1.824$ mH.

$r_f = 0.055\ \Omega.$

REFERENCES

[1] T. A. Lipo and E. P. Cornell, "State variable steady-state analysis of a controlled current induction motor drive," *IEEE Trans. Ind. Appl.*, vol. IA-11, pp. 704-712, Nov./Dec. 1975.

[2] R. B. Maag, "Characteristics and application of current source/slip regulated ac induction motor drives," in *IEEE/IAS Annu. Meeting Conf. Rec.*, pp. 411-417, Oct. 18-21, 1971.

[3] T. A. Lipo and E. P. Cornell, "Modeling and design of controlled current induction motor drive systems," in *IEEE/IAS Annu. Meeting Conf. Rec.*, pp. 612-620, Sept. 28-Oct. 2, 1975.

[4] A. Abbondanti and M. B. J. Brennen, "Control of induction motor drives by synthesis of a slip signal," *IEEE/IAS Annu. Meeting Conf. Rec.*, pp. 845-850, Oct. 7-10, 1974.

[5] G. R. Slemon, J. B. Forsythe, and S. B. Dewan, "Controlled power angle synchronous motor inverter drive system," *IEEE Trans. Ind. Appl.*, vol. IA-8, pp. 679-683, Nov./Dec. 1973.

[6] N. Remesh and S. D. T. Robertson, "Induction machine instability predictions–Based on equivalent circuits," *IEEE Trans. Power App. Syst.*, vol. PAS-92, pp. 801-807, Mar./Apr. 1973.

[7] P. C. Krause and C. H. Thomas, "Simulation of symmetrical induction machinery," *IEEE Trans. Power App. Syst.*, vol. PAS-84, pp. 1038-1053, Nov. 1965.

[8] F. Blaschke, "The method of field orientation for the control of induction machines," *Siemens Forsch. u. EntEntwick,.* pp. 184-193, Jan. 1972, (in German).

[9] A. B. Plunkett,"Direct flux and torque regulation in a PWM inverter-induction motor drive," in *IEEE/IAS Annu. Meeting Conf. Rec.*, pp. 591-597, Sept. 28-Oct. 2, 1975.

[10] K. P. Kovacs and I. Racz, *Transient Behavior of AC Machines* Budapest: Hungarian Academy of Science, 1959, (in German).

[11] T. A. Lipo and A. B. Plunkett,"A novel approach to induction motor transfer functions," *IEEE Trans. Power App. Syst.*, vol. PAS-93, pp. 1410-1418, Sept./Oct. 1974.

A High-Performance Controlled-Current Inverter Drive

LOREN H. WALKER, SENIOR MEMBER, IEEE, AND PAUL M. ESPELAGE

Abstract—**A production variable-frequency drive is described which uses an induction motor and a controlled-current inverter. The control loops of the drive have been arranged so that the performance of the drive is extended to zero speed and to the theoretical upper limit speed. Operating modes include torque smoothing by programmed dc link current, pulse-width modulated (PWM) current shaping, and flux-controlled constant horsepower operation.**

I. INTRODUCTION

CONTROLLED-current inverters (CCI) have acknowledged advantages over other variable-frequency drives for induction motors in that the CCI is very efficient, can regenerate readily, and is very rugged and simple in construction. The CCI has been subject to numerous limitations in response, however, which have prevented broad application in precision drives. This paper describes a controlled-current inverter drive which is a piece of production equipment intended to be free of all performance limitations. The goal in the development of this drive was to have performance matching the best dc drives.

The power circuit which is to be used in the CCI induction motor drive is shown in Fig. 1. This figure also shows the normal operating waveforms at input and output. The dynamic requirements imposed on this drive include:

- smooth performance without cogging or torque pulsations to zero speed;
- smooth reversals of direction of rotation under any torque condition;
- symmetrical four-quadrant operation;
- full torque from stall to rated speed, and constant horsepower at reduced flux to twice rated speed;
- ability to go from zero torque to peak torque in less than 0.2 s.

The production drives which have been built using this design are rated at 400 hp peak, 45 Hz base frequency, and 0–90 Hz frequency range.

The primary area of performance improvement in the development of this drive was the very low frequency region. The controlled-current inverter actually has some advantages over voltage-source inverters in this range in that:

- harmonic content of the current wave is independent of frequency;
- inasmuch as the CCI is a current source it does not respond to motor terminal voltage at all; thus, the controls are not "confused" by the fact that the *IR* drop of the motor is much larger than the induced back electromotive force (EMF) at very low frequencies.

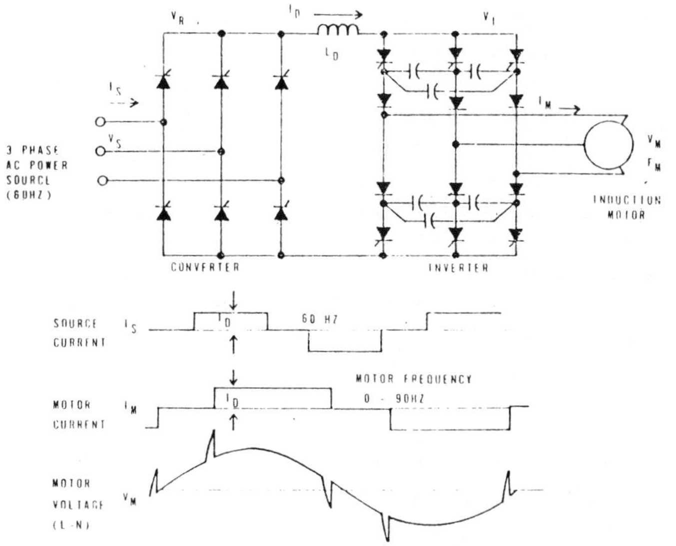

Fig. 1. CCI power circuit and waveforms.

The primary disadvantage of the CCI, when compared to the pulsewidth modulated (PWM) voltage source inverter, is control flexibility. The PWM can program a three-phase sinusoidal current at any phase and frequency. The CCI cannot do this, but in the drive described it will be found to approach it very closely.

The drive described incorporates two advanced concepts in low-speed operation: torque smoothing by programmed dc current and torque harmonic elimination by PWM current waveform. Neither of these features is built into the drive; rather, the control loops are chosen for optimum performance and the advanced "features" occur naturally.

The heart of the drive control is control of the angle between current and flux [1]. This control makes it possible to achieve fast transient response, stability at very high speeds and light loads, and flexible constant horsepower operation. The constant horsepower operating envelope is shown in Fig. 2. All of the performance improvements which are described are made without modification of the basic power circuit shown in Fig. 1.

II. BLOCK DIAGRAM AT LOW SPEED

The basic elements to be controlled in an induction motor drive are flux and torque. The controllable parameters are

Paper ID 79-19, approved by the Industrial Drives Committee of the IEEE Industry Applications Society for presentation at the 1979 Industry Applications Society Annual Meeting, Cleveland, OH, September 30–October 4. Manuscript released for publication November 7, 1979.

The authors are with the Drive Systems Department, General Electric Company, Salem, VA 24153.

Reprinted from *IEEE Trans. Ind. Appl.*, vol. IA-16, pp. 193–202, Mar./Apr. 1980.

current and frequency. In the approach used, frequency is used to control flux and current to control torque.

The block diagram at low speed is shown in Fig. 15. It will first be discussed in a general way, and then several specific features will be pointed out.

The speed regulator shown at the far left in Fig. 15 is conventional. The speed reference is compared to the tachometer feedback. The resultant error signal is passed through transfer function *G1* which is a proportional plus integral regulator in most applications. The output of block *G1* is the torque reference T^*, which controls the operation of the CCI. The speed regulator block is the only block in the diagram which contains time constants. The rest of the blocks labeled *G* are simple gain blocks. The torque reference is fed to three control paths, an upper path which controls current, a lower path which controls frequency, and a center path which controls flux.

The feedbacks which will be used in these paths are generated in the motor signals calculator shown in the lower right. This block accepts inputs of motor phase currents and motor flux. The motor flux signals are derived by integrating the output of voltage pickup coils located in the direct and quadrature axes of the stator. From these signals the motor signals calculator calculates instantaneous values of flux, torque, and angle, using the following expressions:

$$|\psi| = \sqrt{\psi_D{}^2 + \psi_Q{}^2} \tag{1}$$

$$T_e = K(I_Q \psi_D - I_D \psi_Q) \tag{2}$$

$$\sin \theta_e = \frac{K T_e}{|I|(|\psi| - K_1 |I|)}, \tag{3}$$

where

D and Q subscripts which denote the direct and quadrature components of each sensed parameter;
ψ the instantaneous airgap flux;
I the instantaneous stator current;
$|I|$ the stator current, three-phase rectified;
T_e the instantaneous electrical torque;
K a general constant of proportionality;
K_1 a constant related to motor parameters; and
θ_e the approximate angle between stator current and airgap flux.

This method of calculating motor parameters is as described by Plunkett, D'Atre, and Lipo in [1]. The value of K_1 is chosen by the method they describe.

Control of Current by Torque Regulation

To describe the upper control path on the block diagram, Fig. 15, we will neglect for the moment the inputs into this path from below. First, the absolute value of torque reference is compared to the absolute value of torque to generate a torque error signal at summing junction *J2*. This error signal is passed through the high-gain block *G5* to generate the voltage command to the ac to dc converter. Changes in the voltage at the dc link will cause changes in the dc link current,

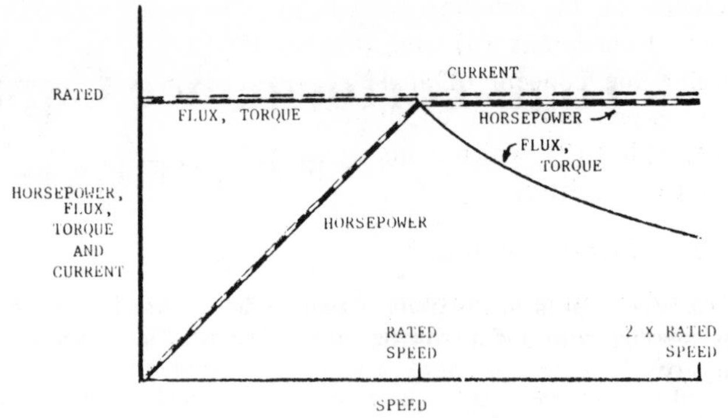

Fig. 2. Maximum drive performance envelope.

and hence, the motor current. Changes in motor current will cause changes in motor torque. Thus, the top control path is a closed-loop torque regulator acting through current control. Stability of this torque-regulating loop will be discussed later. For the present note only that this is a fast-response loop, with a crossover frequency of approximately 800 rad/s at full motor flux.

The feedback path from the dc link voltage through the ripple filter to summing junction *J4* is a positive feedback with a gain of unity. This feedback, by itself, causes the converter to put out a dc voltage which exactly matches the dc back EMF reflected by the inverter. This serves several purposes. When the reflected dc back EMF of the inverter changes suddenly, this unity gain feedback causes the converter output to track that change, so that the current remains approximately constant without a correction from *G5*. This also causes the signal level in *G5* to be more or less proportional to current only, independent of motor speed or voltage. This helps to maintain the dynamic performance of the drive uniform over very wide changes of operating point.

The current limit function is also applied to the rectifier voltage command by summing junction *J3*. If flux is transiently low when a high torque is suddenly commanded, the torque loop would raise current to very high levels. Thus, the current limit function is necessary to avoid excessive transient currents. The function of block *G7* will be discussed in conjunction with the flux control function.

Control of Frequency by Angle Regulation

The lower control path on the block diagram of Fig. 15 controls the frequency of the inverter output current. Its basic mechanism of control is to regulate θ_e, the instantaneous angle between stator current and airgap flux. This causes the current delivered by the current control function to be apportioned properly into the flux-producing and torque-producing axes of the induction motor. The current which is imposed in phase with the existing flux reinforces that flux; the current which is imposed in quadrature with flux produces torque. This concept can be expressed by the following expressions:

$$|\psi| = |I| X_m \cos \theta_e \tag{5}$$

$$T_e = K|I| \cdot |\psi| \sin \theta_e, \tag{6}$$

where terms are as previously defined, except that X_m is the reactance of the exciting branch of the motor equivalent circuit. Expressions (5) and (6) are very useful to understanding the concept of angle control, but are not precise. These expressions can be utilized to develop the control concepts, but (2), which is instantaneously valid, will be used to control the drive.

Concept of Optimum Angle

By substituting expression (5) into expression (6), a solution can be found for torque per ampere as a function of angle:

$$T_e = K|I|^2 X_m \cos\theta_e \sin\theta_e \tag{7}$$

$$\frac{T_e}{|I|^2} = K\frac{X_m}{2}\sin 2\theta_e. \tag{8}$$

Expression (8) shows that torque per ampere will be maximized if angle θ_e is maintained at a specific value of $\theta_e = \pm\pi/4$. The actual optimum angle is determined by computer analysis of the full equivalent circuit, including saturation. This optimum angle may be a function of operating torque level due to saturation effects. As a practical compromise, the angle may be optimized at the highest torque anticipated, and this same angle may be maintained at all lower torques.

Operation at constant angle is shown in Figs. 3 and 4. It is essentially the same as constant slip operation in its static characteristics. That is, both flux and current are proportional to the square root of torque. Constant angle is maintained for torques higher than 0.2 per unit (pu). At torque levels below 0.2 pu the angle is allowed to come to zero gradually to achieve a desired level of flux at zero torque. Angle control has significant dynamic advantages when compared to slip-controlled systems, as will be shown.

Angle Control Loop

Returning to Fig. 15 the lower control path is configured to carry out the constant angle concept. Starting at the left side of the lower control path, the torque reference *T** is modified by function *F1* to form an angle reference. The function *F1* is a symmetrically clamped gain stage to generate the function shown in Fig. 4.

Consider for the moment that the multiplier has a unity input at its upper input terminal so that it produces a gain of unity. Thus the angle reference is passed to summing junction *J6* where it is compared to the measure of actual instantaneous angle calculated by expression (3). The instantaneous angle error is amplified by gain *G4* and summed with the actual speed signal from the tachometer to form the frequency command to a voltage controlled oscillator (VCO).

The gain of the tachometer feedback into the VCO is set at unity, so that with no signal at the output of *G4* the frequency of the inverter will be the zero-slip frequency of the motor at any speed. This is similar in this respect to the feedback of dc link voltage into the converter input. It causes the output of *G4* to be proportional to angle θ_e but independent of speed. This again helps the drive to maintain uniform dynamic performance over a wide speed range.

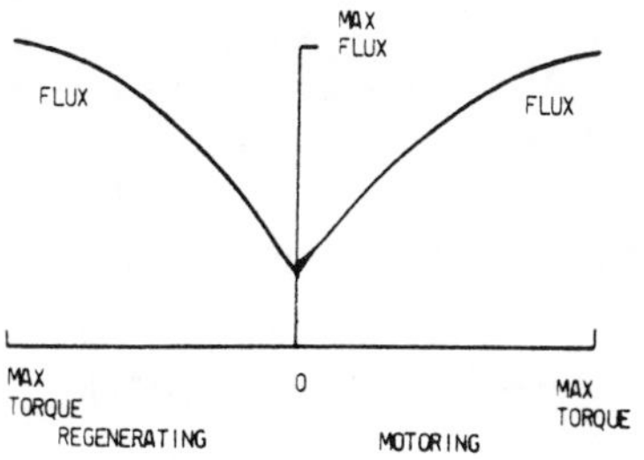

Fig. 3. Constant angle operation: flux versus torque.

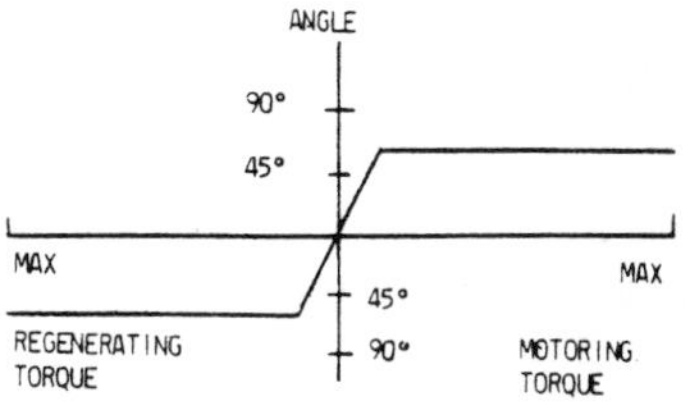

Fig. 4. Constant angle operation: angle versus torque.

Angle Loop Stability

The stability of the angle loop can be analyzed simply if it is assumed that the frequency and phase angle of motor flux do not change significantly during the transient responses of the angle loop. This is often true due to the mechanical inertia of the load and the electrical time constant of the rotor. If these assumptions are valid, the angle loop is defined by Fig. 5. It consists of *G4*, the VCO, the 6:1 countdown, the inverter, the current sensors, and the angle calculator. Note that the motor itself is not included in the loop. It provides only the flux as a phase reference to the loop. The loop, as defined, is a phase-locked loop and will be analyzed as such: its input is a phase command, it modifies a variable frequency to match a reference frequency (the frequency of the motor flux), and it is closed by a phase discriminator (the angle calculator).

The open loop gain is found by inspection of Fig. 5:

$$G(\omega) = G4\ \frac{(\text{volts})}{(\text{volt})} \cdot \frac{GF}{\omega}\ \frac{(\text{radian})}{(\text{second volt})} \cdot G_C\ \frac{(\text{volts})}{(\text{radian})}$$

$$G(\omega) = \frac{G4 \cdot GF \cdot G_C}{\omega}\ \frac{(1)}{(\text{seconds})} \tag{9}$$

The total phase shift associated with this gain is the sum of -90° due to the inherent integration and a transport delay associated with the discrete nature of the VCO and inverter. The VCO operates at a frequency six times the inverter fundamental frequency ω_s so the loop may be viewed as containing a maximum delay of 60° or $\pi/3$ rad at fundamental frequency, or an average delay to a random step in angle command of $\pi/6$ at fundamental frequency. The loop will be stable without further compensation if the gain drops below unity due to the inherent integration before the sum of the phase shifts due to the integration and the transport delay reaches 180°. This sum will reach 180° when the average transport delay reaches 90°. The frequency at which this occurs is found by

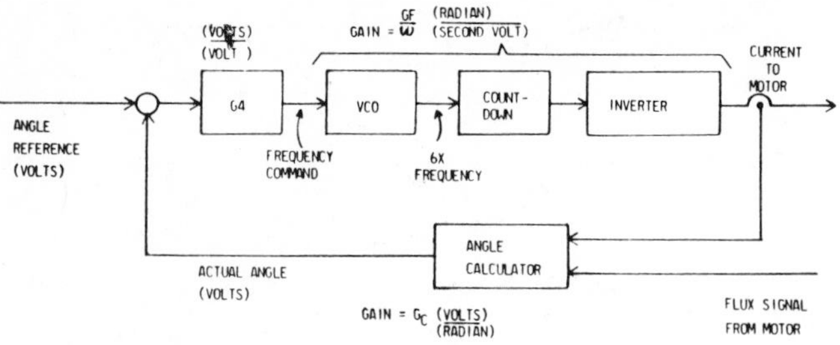

Fig. 5. Block diagram of angle control loop.

setting the average phase shift due to transport delay equal to 90° and solving for ω as a function of ω_s, the stator frequency:

phase shift due to transport delay = 90°

$$\frac{\pi}{6}\frac{\omega}{\omega_s}=\frac{\pi}{2} \tag{10}$$

$$\omega = 3\omega_s. \tag{11}$$

The maximum gain which can be stable in the angle loop is determined by setting gain to unity and ω equal to $3\omega_s$ in the expression (9) with the result:

$$(G4 \cdot GF \cdot G_C)\ \max = 3\omega_s. \tag{12}$$

Thus it is seen that with this simple stablization method and a fixed value of gain *G4*, the angle loop will be very stable at high motor frequencies and will go unstable at some low value of motor frequency. This characteristic, instability at low frequency, is viewed as an advantage of the angle loop, as will be discussed. In the actual drive, two values of gain are used in element *G4:* a very high gain above about 6 Hz motor frequency and a lower gain from 0-6 Hz. The lower value of gain causes the angle loop to become unstable below 0.5 Hz. The advantage of this instability will be discussed subsequently.

If the assumptions made to simplify this analysis are not valid, that is, if the phase of the flux does change during the transient response of the angle loop, the change in flux is in a direction to reduce the gain of the angle loop and so is generally a stablizing effect. The full dynamic model must be analyzed to determine the maximum stable gain in the angle loop, or inversely, the minimum stator frequency for stable operation at a given gain. As shown by expression (12), however, there is no value of gain which is stable at zero frequency, so the drive is designed to operate with the angle loop unstable below about 0.5 Hz.

Control of Flux

The center control path in Fig. 15 is for the control of flux. If the torque and angle loops were ideal, then the flux loop would be necessary only at zero torque. It is added to

1) determine the current level at zero torque;
2) correct the angle program to account for saturation and other imperfections;
3) provide an input to allow operation beyond rated speed in the constant horsepower region.

The function block *F3* converts the torque reference to a flux command ψ^*. The shape of this function is modeled after Fig. 3, the ideal parabola modified to a fixed level of flux at zero torque. This flux command ψ^* is compared at *J5* to the actual sensed flux (1) to form a flux error. This error is fed to the upper current control path through low gain *G7* in a sense to increase current when the flux is below the programmed value. This will convert the upper current control path to a flux regulator when torque and torque command are both near zero. This will be discussed in more detail later.

This flux error is also fed through offset function *F2* and low gain *G2* into the multiplier in the angle reference path. One purpose of function *F2* is to cause *G2* to generate an output of unity when flux error is zero. This causes the multiplier to operate as a gain of one to the angle reference. The sense of the signal injected by *G2* into the multiplier is to decrease the angle when flux is below the programmed value. This will act to divert more of the available current into the flux-producing axis.

When the torque command is zero the angle reference is zero, and *G2* cannot effect the output of the multiplier to introduce an angle modification. This allows the current loop to control flux through *G7* at zero torque without disturbing the angle.

Gain and Stability of the Torque and Flux Loops

The open-loop transfer function of the torque loop at constant flux and angle is given by (13):

$$G(s)=\frac{G5 \cdot G_V}{R_D} K_T \frac{e^{-t_d s}}{1+(L_D/R_D)s} \tag{13}$$

where

- *G5* the gain of block *G5* on Fig. 15;
- G_V the voltage gain of the ac/dc power converter;
- R_D the dc resistance of the dc link including the commutating resistance of the ac/dc converter;
- K_T the torque gain of the motor including the torque calculator in terms of *Te*/A dc;
- t_d the time delay introduced by the phase control in the three-phase full-wave converter.

The loop dynamics include a simple lag due to the dc inductor and a transport delay. This loop will be stable with no further modification if the gain is reduced to unity by the lag at a frequency below which the transport delay introduces an effective 90° phase shift. For such a loop, operating at 60 Hz input power, the maximum stable loop crossover is higher than 1100 rad/s. As was noted, the loop is operated at approximately 800 rad/s crossover at maximum motor flux.

Absent from expression (13) are the leakage inductances of the motor and the effective dc resistance of the motor and the inverter commutation. These items are removed from the loop by the positive feedback of dc link voltage to summing junction *J4*. They will appear in the expression as increases in R_D and L_D if the ripple filter break is not high with respect to the torque loop crossover frequency.

The gain K_T in expression (13) is not constant, but is a function of both flux and angle as expressed generally by (6). Thus if the flux increases with torque command as shown in Fig. 3 then the open-loop gain of the torque loop will also increase with torque command. Likewise, if an angle varies with torque command as shown in Fig. 4, then the gain of the torque loop due to angle will be constant over a wide range of torque, but this gain will drop to zero as the angle drops to zero near zero torque.

From the standpoint of stability of the torque regulating loop, all that is required to accommodate these gain changes is to assure that the torque loop gain at maximum flux and angle is within the range which can be stablized by available inductance. Since this torque loop is the dominant loop controlling the motor, however, the function of controlling current must be assumed by some other loop when the gain of the torque loop drops toward zero. This is accomplished, as was noted, by the path formed by *G7*, which allows the flux error to control the current. The gain of the flux loop by controlling current is given by (5). The inherent characteristic is the inverse of that of the torque loop: the flux loop has high gain when the angle θ_e is near zero and when the machine is unsaturated (X_m is large). Thus the flux loop gradually and smoothly takes control of the current as the torque and flux drop off and the angle moves toward zero.

Stability of the flux loops is provided by the rotor time constant. Both flux loops, controlling angle and controlling current, are operated at low gain well below the stability limits.

III. OPERATION OF THE DRIVE AT LOW SPEEDS

Low-speed operation of the drive is characterized by two modes of operation, with the transition between them occurring naturally. At frequencies from 0.5-6 Hz (approximately) the control operates in a "torque-smoothing" mode, producing a dc current waveform which is a series of exponentials or sections of a sine wave. Below approximately 0.5 Hz the drive operates in a pulsewidth modulated current mode to provide smooth control of torque and speed at stall conditions. Both of these modes of operation have been described by other authors [2]-[4], [7]. These authors have generally implemented these modes by adding control features. In this drive these modes are inherent in the control loops described above, without adding control features.

Torque Smoothing

Torque ripple is produced in a controlled-current drive by the interaction of the square-wave current with the sinusoidal flux. The expression for instantaneous torque (2) involves the cross product of square-wave or stepped-wave currents with sinusoidal fluxes. The resultant electrical torque has a waveform as shown in Fig. 6(c). The frequency of the ripple is six times the fundamental frequency of the motor current and flux. The waveshape of the ripple is a 60° section of a sine wave, with the section selected being determined directly by the angle θ_e. The amplitude of the ripple is determined by the product of the flux level and exciting current of the motor.

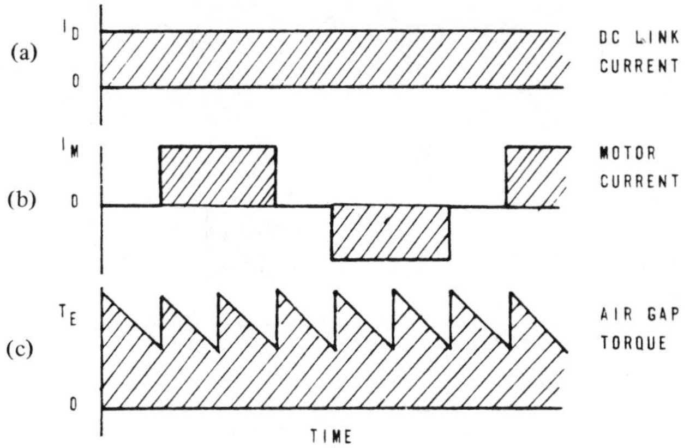

Fig. 6. Inherent torque ripple.

Several authors [2], [3], and [7] have shown that this torque can be reduced (theoretically) to zero by introducing a modulation of the dc current which is the inverse of the torque ripple. This effect is shown in Fig. 7. In Fig. 7(a) the current is given a ripple which is the inverse of the inherent torque ripple shown in Fig. 6(c). The resulting motor current waveform is shown in Fig. 7(b). The electrical torque which results is sketched in Fig. 7(c). The torque ripple has been removed, and a series of spikes remain. The spikes occur at each commutation of the inverter and are present because the dc current cannot be changed instantaneously. The downward steps in current shown in Fig. 7(a) must have a finite slope due to the presence of the dc inductor. During this fall time the torque will be out of control instantaneously until the current can assume the proper value for the new value of flux it faces after each commutation.

Torque smoothing in this drive is accomplished by the action of the torque loop. The actual torque T_e, having an inherent waveform as shown in Fig. 6(c), is compared at summing junction *J2* (Fig. 15) to the torque reference T^* which is a smooth dc level. The resultant torque error contains not only a dc component due to errors in average value of torque, but a large ripple signal. The high-gain fast-torque loop acts to reduce this torque error toward zero. This forces the dc current into the sawtooth waveform shown in Fig. 7(a) in a closed-loop manner. The result is shown in Fig. 8. In this dynamic recording, the motor current is shown in the top trace. The torque command and airgap torque are shown (inverted) in the second and third traces. Note that the actual torque contains a small amount of the original sinusoidal waveform due to less than infinite gain in the torque loop and exhibits the spikes due to limited rate of change of current.

Other writers [3] have discussed difficulty in keeping the proper amplitude and phase of ripple injected into the dc current program. This difficulty does not arise here since the torque smoothing is entirely by closed-loop action.

It is evident that torque smoothing of this type would be effective only when the torque ripple frequency is much lower than the crossover frequency of the torque regulating loop. With loop crossover of 800 rad/s (127 Hz) excellent torque smoothing can be achieved up to a ripple frequency of 36 Hz (6 Hz fundamental frequency), and useful smoothing

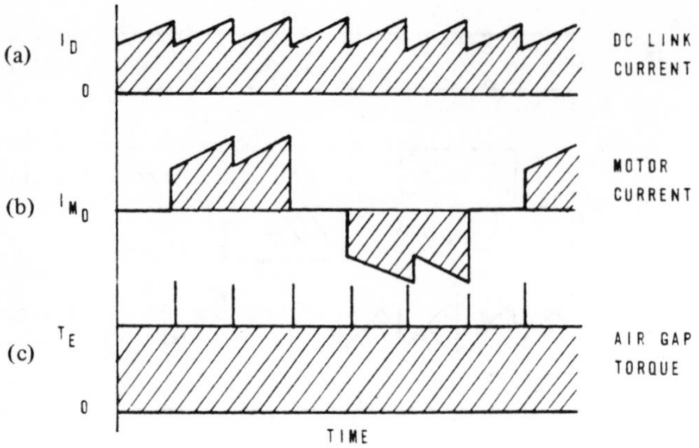

Fig. 7. Current and torque waveforms with torque smoothing.

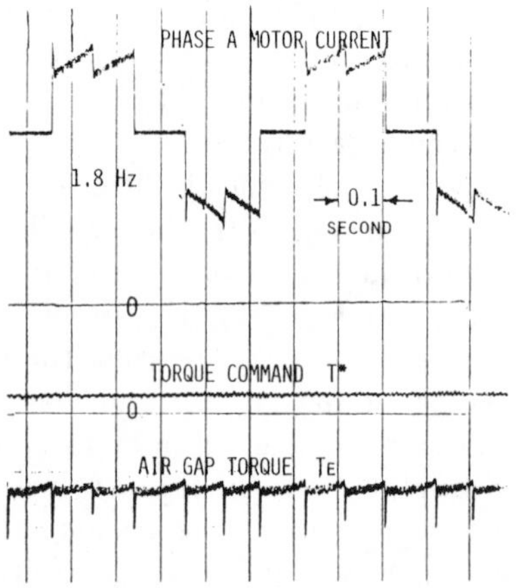

Fig. 8. Dynamic traces of torque smoothing.

can be achieved up to 60 Hz ripple frequency (10 Hz fundamental frequency). Since most major mechanical resonances are well below 36 Hz, the torque-smoothing performance should be adequate.

The action of the torque-smoothing function could be improper near zero torque. When the average torque is zero the ripple is first positive then negative. The absolute value circuit which precedes summing junction *J2* (Fig. 15) will rectify this ripple on T_e, and the torque loop will attempt to drive the dc current to zero to null the perceived excess dc value of torque. In practice this is no problem, due to the inherent effect of constant angle operation on the torque loop. As torque approaches zero the flux drops. Since torque ripple is proportional to the product of exciting current and flux, it is reduced approximately as flux squared. The low flux also reduces the loop gain in the torque loop. Thus, as torque approaches zero, the torque-smoothing gain becomes very small, but the torque ripple also becomes very small. Fig. 9 shows the torque ripple under both loaded and zero torque conditions.

Dynamic Response of the Drive at Low Speed

The 800 rad torque loop allows the drive to deliver fast torque response even at very low operating frequencies. Fig. 9 shows the drive at 3 Hz in speed control, initially at zero

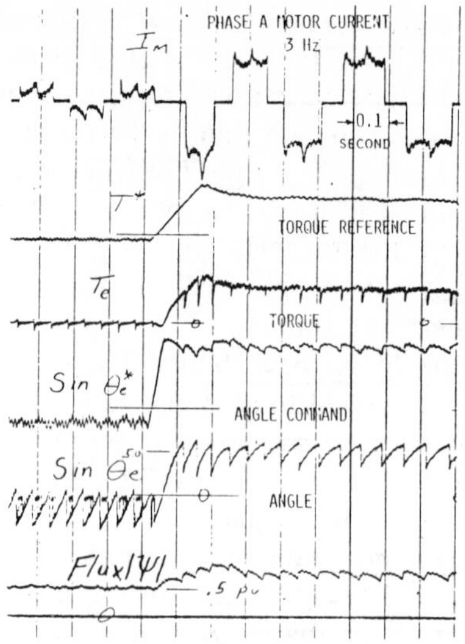

Fig. 9. Response to torque reference change at low speed.

torque and suddenly called upon to deliver a large regenerative torque (to resist a suddenly applied overhauling load). The torque reference is shown in the second trace. Its shape is determined by the mechanical inertia and by the dynamic response of the speed regulator loop. The actual airgap torque is shown on the next trace. Note that the torque follows the reference precisely, even though the command changes a large amount between firings of the inverter. Note also that the motor reaches full torque in about 0.15 s and could have responded more rapidly had not the command rate been limited.

Also shown in Fig. 9 are the angle command sin θ_e* and the calculated angle sin θ_e. Note the characteristic six-pulse ripple on the sin θ_e signal. The sudden change in angle which is required in order to step from zero torque to rated torque is accomplished by transiently lowering the frequency. This is evident both in the sin θ_e signal and in the phase *A* motor current signal.

The angle command signal sin θ_e* would be a clean signal except for the small influence of the flux loop. The small six-pulse ripple which is evident on the angle command is due to the flux error acting through the multiplier on the angle loop. This flux loop also provides some transient forcing on the angle command until the flux assumes its steady-state value.

Pulsewidth Modulation of Current

Another means of torque ripple control which has been described in the literature [3], [4] is the pulsewidth modulation of the motor current by multiple commutations of the inverter in each half cycle. This approach, sketched in Fig. 10, forms a motor current waveform which is notched at specific points to eliminate certain harmonics using the method first analyzed by Turnbull [5] in 1964. The effect is to convert the square waveform of the motor current into an equivalent sine wave by pulsewidth averaging.

Each notch added to the current waveform allows the elimination of one harmonic. Thus the double notched waveform at each quarter cycle in Fig. 10(b) allows the elimination

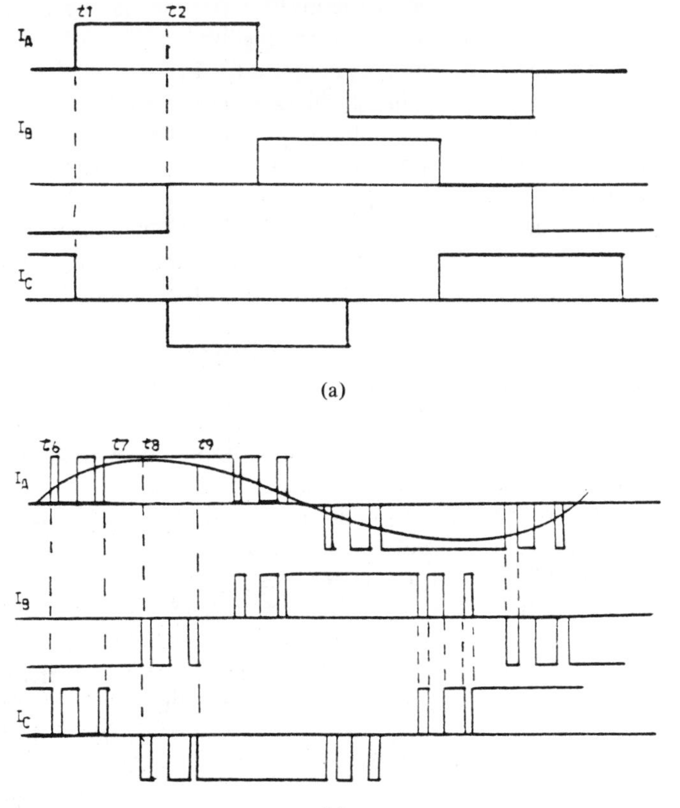

Fig. 10. (a) Normal motor current waveforms. (b) Motor current waveforms with pulsewidth modulation for torque ripple reduction.

of the fifth and seventh harmonics of current and thus eliminates the sixth harmonic of torque. There are a family of current waveforms which can be generated, each having one more notch per quarter cycle, and each eliminating one more current harmonic. An ideal controller might generate a single notched waveform below 4 Hz, a double notched waveform below 2 Hz, a triple notched waveform below 1 Hz, etc., to maintain a constant minimum frequency of torque pulsation.

The mechanism of producing these waveforms in the power circuit is by multiple switching of the inverter. The dc current is maintained constant in generating the idealized waves of Fig. 10. The inverter switching is the same as the normal mode. Refer to Fig. 10(a) and note that each termination of positive current in one phase is accompanied by an intiation of positive current in another phase. Consider time *t1* when positive current is commutated from phase *C* to phase *A*, or time *t2* when negative current is commutated from phase *B* to phase *C*. All commutations are of this type. Now refer to Fig. 10(b). At time *t6* the positive current is commutated from phase *C* to phase *A*. On the next four switchings the positive current is passed back and forth between phases *A* and *C* until it finally remains in phase *A* at time *t7*. At time *t8* the negative current is passed from phase *B* to phase *C*. Four more times it is passed back and forth until it remains in phase *C* at time *t9*. Thus all three phase currents are tapered by notching until they approach sine waves which produce no torque ripple.

The ability to operate in this mode is inherent in the power stage. In the example shown in Fig. 10(b) each inverter leg commutates five times as often as in the normal case of Fig. 10(a). The narrowest pulse (shortest time between commutations) is approximately 1/20 of that in Fig. 10(a). Thus if the motor were operating at a fundamental frequency of 3 Hz in Fig. 10(b), the power stage is switching at 15 Hz rate or at 60 Hz pulsewidth. Since a typical ac motor drive is operated at a maximum fundamental frequency of at least 60 Hz, it could operate in this PWM mode at fundamental frequency of 3 Hz with no unusual stress on the power stage. In this drive the PWM mode is used only at fundamental frequencies below 0.5 Hz.

Controlling the PWM Operation

The means which have been described by previous authors for implementing PWM patterns similar to that in Fig. 10(b) usually include a logic waveform generator operating from a high-frequency clock phase locked to the motor current fundamental frequency. The waveform generator may be enabled below a certain frequency, and a second more complex waveform generation may be enabled below a second frequency which is lower yet. This pattern must be integrated into the drive control responsive to the frequency control loop.

The means which is used in this drive control to generate the ideal PWM pattern is the angle loop which was described in connection with Fig. 5. The VCO and ring counter used in the angle loop are designed to be capable of bidirectional operation so that the motor can change smoothly from forward rotation to reverse rotation. Thus the countdown is a bidirectional ring counter, and the VCO outputs either "count forward" or "count reverse" pulses to the countdown. A simplified block diagram of the VCO is shown in Fig. 11. The operation of the VCO is of an integrating nature; it integrates the frequency command until it reaches a defined positive or negative value. It then produces an output pulse and resets the integrator. The sense of the integral determines the "count forward" or "count reverse" sense of the output pulse.

As was stated earlier, the overall angle loop as shown in Fig. 5 will go unstable due to the transport lag below a certain fundamental frequency. The mode of instability into which it goes is the PWM pattern which is desired. The resulting motor current in this mode of operation is shown in the dynamic traces of Fig. 12. This figure shows both the current pattern and the instantaneous value of sin θ_e at a fundamental frequency of 0.3 Hz.

The reason the pattern generated is very nearly the ideal sine wave PWM pattern is the ideal nature of the unstable angle loop. The loop acts as a switching regulator, advancing or reversing the sense of rotation of the discrete current vector to minimize the integrated error from the desired angle. Examination of the trace of sin θ_e in Fig. 12 shows that the attempt of the angle loop to minimize the difference of the angle signal from its dc value results it the injection of a current which closely approximates the desired pattern.

The gain of block *G4* in Fig. 15 determines both the fundamental frequency at which the PWM operation starts

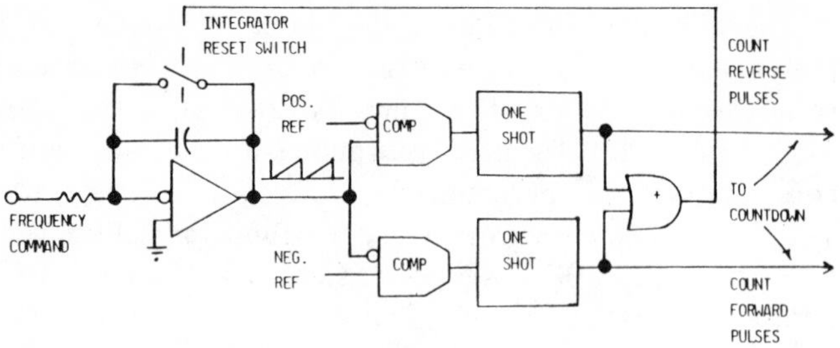

Fig. 11. Block diagram of bidirectional VCO.

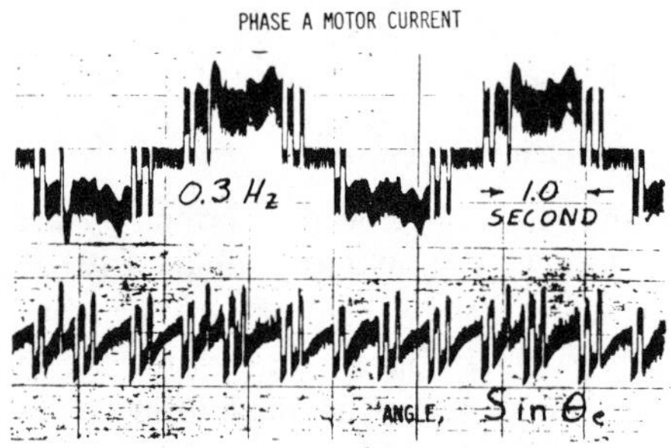

Fig. 12. Inherent PWM operation.

and the number of notches in the pattern. Expression (12) showed that the minimum stable frequency as a linear (not switching) angle regulator is proportional to loop gain in the angle loop. Once the loop is unstable (operating in the switching mode) the integral of each notch is determined by the loop gain. Thus when the gain *G4* is set for PWM operation to begin at 0.5 Hz, the complexity of the pattern at each frequency from 0.5 Hz to 0 Hz is determined. The PWM pattern in operation tends to maintain an average chopping frequency of approximately nine commutations per second at all frequencies below 0.5 Hz. One advantage of this mode of operation is that it keeps the inverter commutating at very low frequencies rather than stopped. This keeps the commutation capacitors from losing their charge.

IV. CONTROL IN THE HIGH-SPEED REGION

The control block diagram at very high speeds needs two features not shown on the low-speed block diagram:

1) a close control of motor current, particularly at zero torque;
2) a method of modifying the angle program and flux program to accomplish a constant horsepower mode of operation.

Control of Current

The close control of motor current is necessary to achieve the maximum possible performance at high speed and light load. As discussed by Lineau [6], the commutation time of the inverter current stretches abnormally at high speed. This causes the motor current to be much smaller than the dc current due to current which charges the commutation capacitors but does not pass through the motor. It also causes the phase angle of motor current to become almost unrelated to the time of firing the inverter thyristors. This is because the commutation time of the inverter current may be as large as 120 electrical degrees and is highly variable with slight variations of motor voltage or current level.

The approach to this problem in this drive is to have a high-gain high-speed angle loop to control the angle of actual motor current. As long as the motor current magnitude is kept stable, the angle loop can sense the angle of that current and control it by varying the thyristor firing time. This control can be maintained even though commutation times are long and variable so long as a minimum motor current is maintained.

The angle loop is already in place on the low-speed block diagram of Fig. 15. Its gain *G4* is increased by 4:1 once the motor leaves the torque-smoothing speed range (0–6 Hz) to give the most precise control of current angle when very high speeds are reached.

The precise control of motor current is accomplished by substituting the magnitude of motor current ($|I|$ from expression (4)) for T_e as the feedback quantity in the torque control loop at the top center of Fig. 15. This causes the motor to have a current which is proportional to torque command plus a constant minimum current bias, rather than a torque proportional to torque command. At no load at low speed, the current control mechanism degenerated into a low-gain flux controller. At high speed it remains as a high-gain current controller at all loads to maintain a firm minimum on motor current.

The current proportional to torque command does not produce a linear system. Since constant angle control is still used, torque is not proportional to current but is approximately proportional to current squared. This nonlinearity can be easily eliminated by interposing a nonlinear function in the current feedback path. This is not always done because the nonlinear characteristic is desirable in some applications.

The combination of the high-gain control of both magnitude and angle of motor current allows the drive to operate stably with commutation delays of up to 120 electrical degrees between thyristor firing and extinction of current in the previous motor phase. This approaches the maximum theoretical limit calculated by Lineau [6].

Constant Horsepower Operation

Constant horsepower operation is obtained by two features: a modification of the peak value of the flux program *F3* and a bias on the angle command which is a function of both speed and torque. The level of the maximum flux command out of block *F3* is made proportional to the inverse of speed past rated speed. Rated speed is defined as the speed at which motor voltage at maximum torque required by the load matches the maximum voltage the inverter and converter can produce.

The bias on angle command is a function of both speed and torque command. It is injected at the input of block *F2* on Fig. 15. It increases the angle when the combination of torque commanded and speed would cause a motor voltage higher than the inverter and converter can supply. Since the constant angle control produces low flux at low torque, the drive can operate at moderate torque to twice rated speed without any bias needed on the angle.

Dynamic Response at High Speed

Fig. 13 shows the waveforms of motor current and torque as the drive accelerates through the transition from torque smoothing to current control at approximately 6 Hz. The current waveform changes from the double sawtooth waveform to the controlled amplitude flat top. The torque waveform changes from smooth with spikes to the inherent sawtooth. These relatively large pulsations in electrical torque are at a 36 Hz rate, too high in frequency to be transmitted by the mechanical coupling to the load.

Fig. 14 shows dynamic traces during a step application and removal of mechanical torque at 25 Hz in the speed-regulated mode. It should be compared to Fig. 9 which shows a similar transient at 3 Hz. The torque is in a motoring sense in Fig. 14, but other than that difference, the transient response of the drive is very similar at high and low speeds.

V. CONCLUSION

Without modifying the basic power configuration of the controlled-current inverter, the control strategy described gives the CCI induction motor drive the smooth, precise control of a dc drive.

ACKNOWLEDGMENT

Basic concepts utilized in this drive originated from several people. Particular acknowledgment is made to David L. Lippitt and William G. Wright (deceased) of General Electric Company, Schenectady, NY.

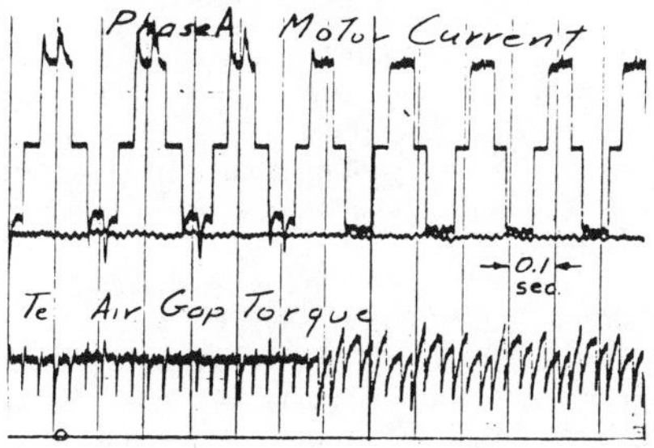

Fig. 13. Current and torque waveforms when accelerating through switch from torque control to current control.

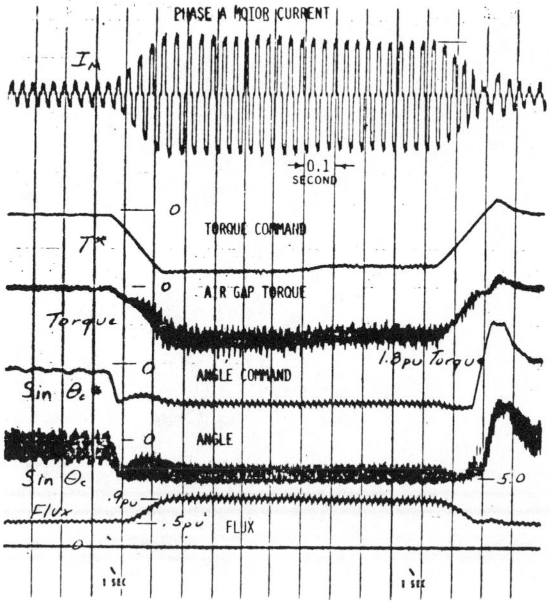

Fig. 14. Dynamic traces at 25 Hz.

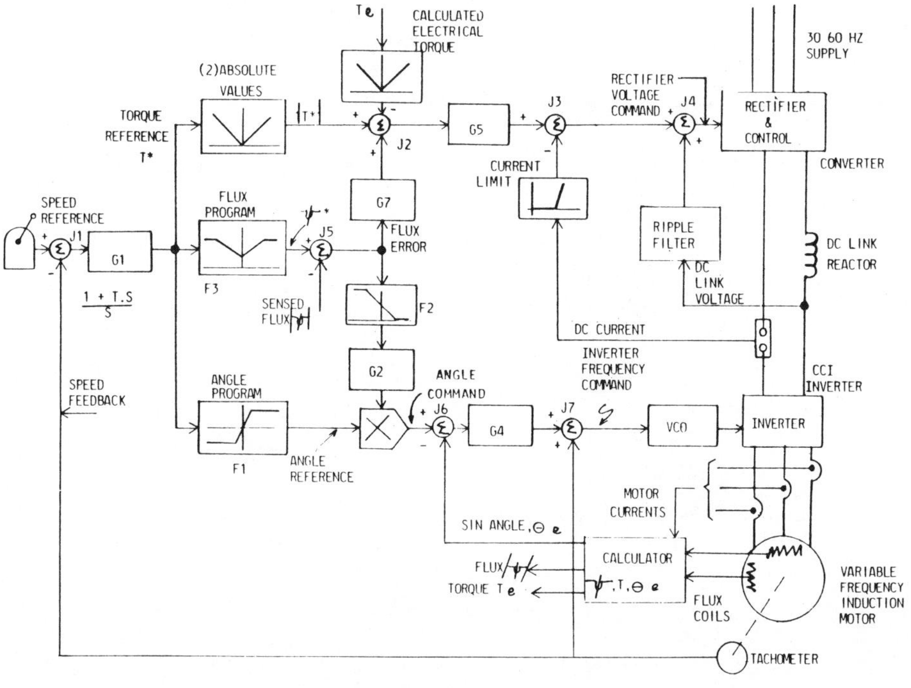

Fig. 15. Control block diagram at low speed.

REFERENCES

[1] A. B. Plunkett, J. D. D'Atre, and T. A. Lipo, "Synchronous control of a static AC induction motor drive," in *1977 IEEE Ind. Appl. Soc. Annual Meeting Conf. Rec.*, pp. 609–615.

[2] T. H. Chin, "A new controlled current type inverter with improved performance," in *Int. Semiconductor Power Converter Conf. Rec.*, pp. 185–192, 1977.

[3] J. Zubek, "Evaluation of techniques for reducing shaft cogging in current fed AC drives," in *1978 Ind. Appl. Soc. Annual Meeting Conf. Rec.*, pp. 517–524.

[4] W. Lineau, A. Muller-Hellmann, and H. C. Skudelny, "Power converters for feeding asynchronous traction motors of single phase AC vehicles," in *Int. Semiconductor Power Converter Conf. Rec.*, pp. 295–304, 1977.

[5] F. G. Turnbull, "Selected harmonic reduction in static DC/AC inverters," *IEEE Trans. Commun. Electron.*, vol. 83, no. 73, pp. 374–378, July 1964.

[6] W. Lienau, "Commutation modes of a current source inverter," in *2nd IFAC Symp. Preprints on Control in Power Electronics and Electrical Drives*, pp. 219–229, 1977.

[7] T. A. Lipo, "Analysis and control of torque pulsations in current fed induction motor drives" in *IEEE Power Elec. Spec. Conf. Rec.*, p. 89–96, 1978.

A NEW MULTIPLE CURRENT-SOURCE INVERTER

A. Nabae, T. Shimamura and R. Kurosawa
Tokyo Shibaura Electric Co., Ltd.
Yokohama, Japan

SUMMARY

The authors have developed a new multiple current-source inverter for driving an induction motor at variable speed. Its output current shows multiple step waveform and gives better torque ripple than that of the previously used system. This new system also solves the parallel running problem of current-source inverters. This paper describes the principle of the new multiple current-source inverter, including its circuit and torque ripple analysis.

INTRODUCTION

Many current-source inverter-induction motor drive systems are going to be used more widely in the general industrial application field, because of the feasibility of their power regeneration to AC source and their speed control economical merits. Notwithstanding these merits, two major problems should be solved for wider application. One concerns transient response and stability problem and the other is torque ripple problem in low frequency range, caused by the current square waveform.

The authors have developed a new slip frequency controlling system, in which the effect of slip frequency transient term is considered. It gives a solution to the former problem.[1]

This paper treats the latter problem. The current-source inverter (CSI) supplies "square wave" output current, including many higher harmonics. The induction motor driven by CSI, therefore, has high torque ripple of 6 f_0 frequency, where f_0 is the fundamental frequency of the output current. In wide range variable speed drive, this high torque ripple may cause irregular revolution and sometimes resonance oscillating, especially in the low speed range. Therefore, it was highly desirable to develop a CSI which renders better low torque ripple characteristics and high ripple frequency.

A new multiple CSI developed for this purpose is explained in the following.

PRINCIPLES OF NEW MULTIPLE CURRENT-SOURCE INVERTERS

Presently, the most encouraging current-source inverter is called "Series Diode Type" CSI,[(2)-(4)] and it is yet under improvement by many investigators.

The multiple CSIs, described herein, are composed of two or more "unit CSI"s. All types of CSI can be applied to the component "unit CSI" of this multiple CSI.

The induction motor driven by sinusoidal wave current basically includes no appreciable output torque ripple component. CSI is often connected in multiple stage, giving a multiple step current waveform very similar to a sinusoidal waveform.

Figure 1 shows a main circuit of the conventional multiple CSI. In the figure, two variable DC sources are shown by E_{ds1} and E_{ds2}. Unit inverters are shown by INV 1 and INV 2, respectively. Figure 2 shows the voltage and current waveforms whose commutation overlap angle is neglected. INV 1 and INV 2 are phase-shifted to each other by θ.

DC input voltage E_{d1} and E_{d2} are expressed as follows.

$$E_{d1} = \frac{3\sqrt{2}}{\pi} V_M \cos(\phi - \frac{\theta}{2}) \quad (1)$$

$$E_{d2} = \frac{3\sqrt{2}}{\pi} V_M \cos(\phi + \frac{\theta}{2}) \quad (2)$$

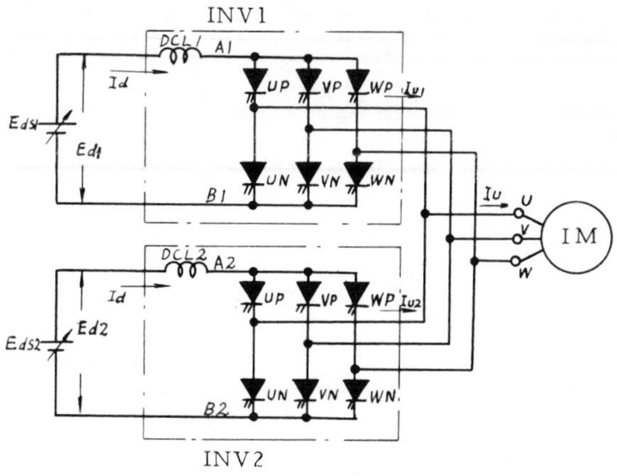

Fig. 1 Conventional Multiple CSI (Duplex)

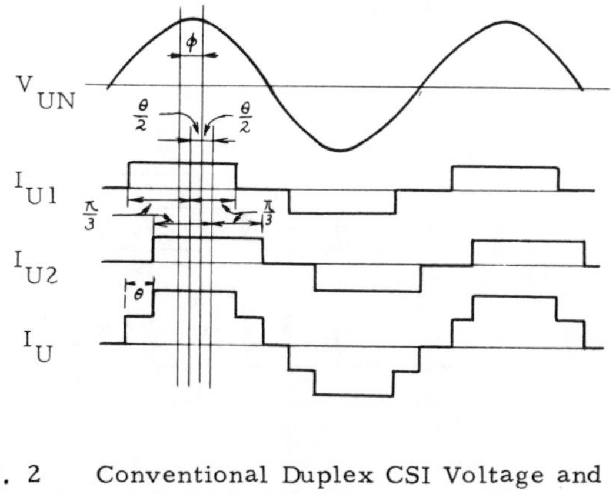

Fig. 2 Conventional Duplex CSI Voltage and current Waveforms

Reprinted from *1977 IEEE/IAS Intl. Semi. Power Conv. Conf.*, pp. 200-204, 1977.

where V_M is r.m.s. value of motor terminal voltage. When the induction motor operates in lag power factor and ϕ is not equal to zero, $E_{d1} \neq E_{d2}$. Therefore, E_{ds1} and E_{ds2} are unable to be supplied from one common converter and controlled by one common controller; in other words, E_{ds1} and E_{ds2} must be equipped separately, controlled individually and their converter ratings must necessarily differ. The new multiple inverter improves these demerits by equalizing input DC voltages of each unit CSI. Figure 3 shows the timing chart of this new duplex CSI, in which current duration of each switching element and output current waveform are corresponded and compared to each other. Current duration of each switching element repeats $(2/3\pi + \theta)$ and $(2/3\pi - \theta)$ in both positive (UP, VP, WP) and negative (UN, VN, WN) sides.

In this way, input DC voltages of INV 1 and INV 2 become equal and the values are as follow:

$$E_{d1} = E_{d2} = \frac{3\sqrt{2}}{\pi} V_M \cos\phi \cos\frac{\theta}{2} \tag{3}$$

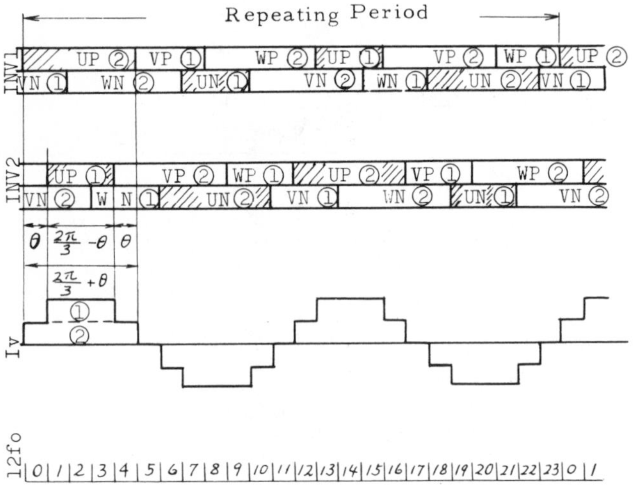

Fig. 3 New Duplex CSI Timing Chart

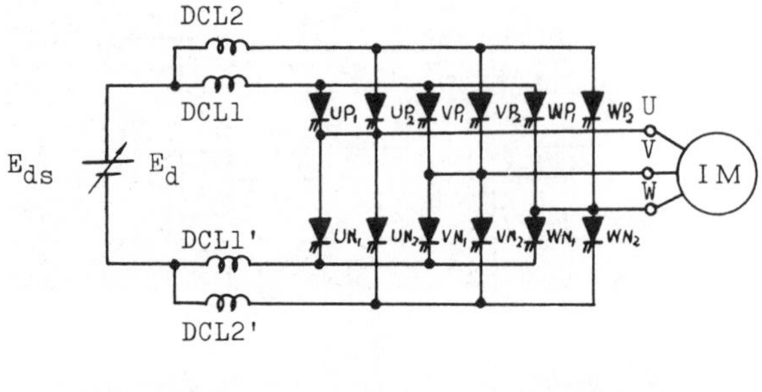

Fig. 4 New Duplex CSI Diagram

In this duplex CSI, positive and negative bus potentials of converter output to motor neutral point are all equal in both INV 1 and INV 2. Hence, the DC source can supply in common from one converter and is controlled by one controller as shown in Fig. 4. Figure 5 shows the logic circuit of this duplex CSI controller, in which the sharing circuits of gate signals Pup through Pwn are used to drive thyristors UP through WN. In Fig. 5, input of the MOD 24 ring counter is $12 \cdot f_0$ clock-pulse, where f_0 designates the inverter output frequency. Outputs of the ring counter are logical sum of main thyristor conduction pulses. Figure 6 shows the timing chart of the quadruple CSI. In this quadruple CSI, the DC source can be composed of two converters, one for INV 1 and INV 2 and the other for INV 3 and INV 4, and are controlled by one controller.

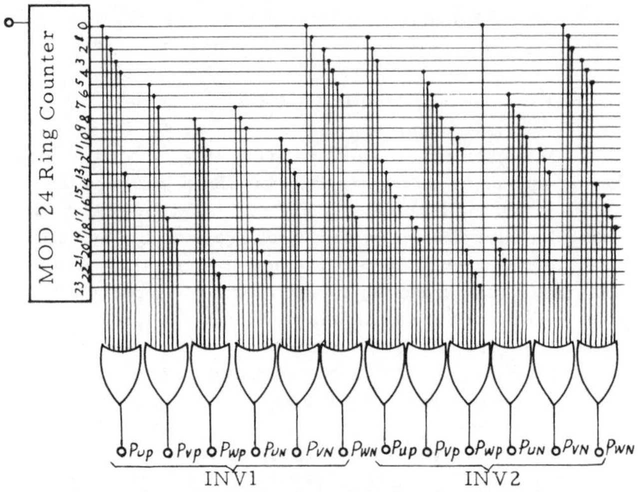

Fig. 5 New Duplex CSI Logic Circuit

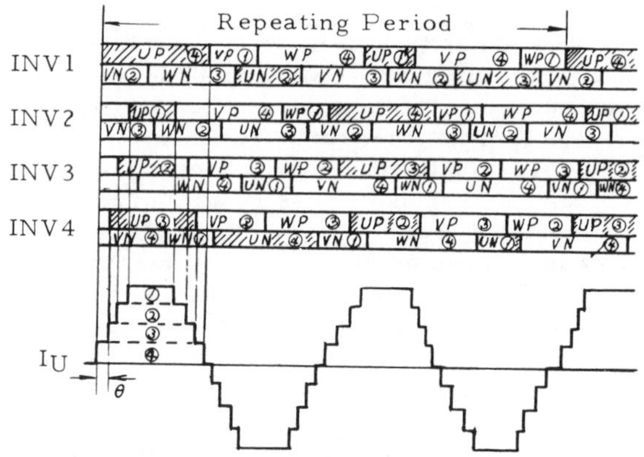

Fig. 6 New Quadruple CSI Timing Chart

ANALYSIS OF CURRENT BALANCING IN PARALLELED UNIT CSI

Equivalent circuit of the paralleled unit CSI is shown in Fig. 7. Symbols in Fig. 7 represent the following;

$$L_1 = L_1' + L_1'' , \quad R_1 = R_1' + R_1''$$

L_1' : DC reactor inductance
L_1'' : Induction motor leakage inductance (sum of primary and secondary leakage inductance per phase)

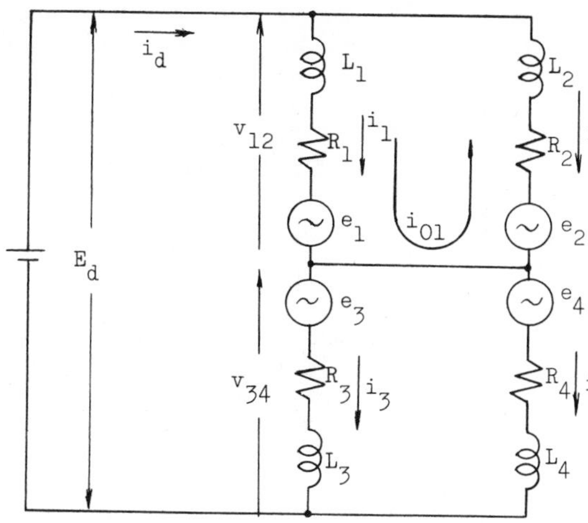

Fig. 7 Equivalent Paralleled Unit CSI Circuit.

R_1' : DC reactor resistance
R_1'' : Induction motor resistance (sum of primary and secondary resistance per phase)
V_M : Induction motor phase to phase voltage effective value
$e_1 \sim e_4$: Induced voltage per phase, selected by main thyristor
ϕ : Induction motor power factor
θ : Output current step width (refer to Fig. 2 and Fig. 3)

The circuit equation is expressed as,

$$v_{12} = R_1 i_1 + L_1 \frac{di_1}{dt} + e_1 = R_2 i_2 + L_2 \frac{di_2}{dt} + e_2 \qquad (4)$$

Expressing average values of e, i for a half cycle as $\bar{e}$, $\bar{i}$,

$$\bar{v}_{12} = R_1 \bar{i}_1 + L_1 \overline{\frac{di_1}{dt}} + \bar{e}_1 = R_2 \bar{i}_2 + L_2 \overline{\frac{di_2}{dt}} + \bar{e}_2 \qquad (5)$$

where

$$L_1 \overline{\frac{di_1}{dt}} = L_2 \overline{\frac{di_2}{dt}} = 0$$

$$\bar{e}_1 = \bar{e}_2 = \bar{e}_3 = \bar{e}_4 = \frac{\sqrt{2}\, V_M}{\sqrt{3}} \cdot \frac{3}{4\pi} \left\{ \int_{\phi-(\frac{\pi}{3}+\frac{\theta}{2})}^{\phi+(\frac{\pi}{3}+\frac{\theta}{2})} \cos\omega t\, d\omega t + \int_{\phi-(\frac{\pi}{3}-\frac{\theta}{2})}^{\phi+(\frac{\pi}{3}-\frac{\theta}{2})} \cos\omega t\, d\omega t \right\}$$

$$= \frac{3\sqrt{2}}{2\pi} V_M \cos\phi \cdot \cos\frac{\theta}{2}$$

Then,

$$\bar{i}_1 R_1 = \bar{i}_2 R_2$$

in the same way, $\bar{i}_3 R_3 = \bar{i}_4 R_4$

$$\text{Therefore, } E_d = R_d \bar{i}_d + \frac{3\sqrt{2}}{\pi} V_M \cos\phi \cos\frac{\theta}{2} \qquad (6)$$

$$\text{where} \quad R_d = \frac{R_1 R_2}{R_1 + R_2} + \frac{R_3 R_4}{R_3 + R_4}$$

Thus, it is clarified that the average current sharing ratio between two unit CSIs is determined by the resistance ratio, and that DC voltage is expressed by the same expression as that of the single inverter, as shown in Eq. (6).

Next, cross current i_{01} shown in Fig. 7, will be discussed. Assuming that $L_1 = L_2 = L_3 = L_4 = L$ and Ri drop is negligibly small, compared with L di/dt, referring to Fig. 8, we obtain

$$2L \frac{di_{01}}{dt} = e_2 - e_1 = V_M \sin(\omega t - \frac{\pi}{3})$$

$$\Delta i_{01} = \frac{V_M}{2L} \int_{\frac{(\phi+\frac{\pi}{3}-\frac{\theta}{2})}{\omega}}^{\frac{(\phi+\frac{\pi}{3}+\frac{\theta}{2})}{\omega}} \sin(\omega t - \frac{\pi}{3}) dt = \frac{V_M \sin\phi \sin\frac{\theta}{2}}{\omega L} \qquad (7)$$

When $\theta = \frac{\pi}{6}$,

$$(\Delta i_{01})_{\theta=\frac{\pi}{6}} = \frac{V_M \sin\phi}{2\omega L} \qquad (7)'$$

where Δi_{01} means the peak to peak value of i_{01}, as shown in Fig. 8(b). Figure 8 (c) shows output current I_u, calculated by the above mentioned equations.

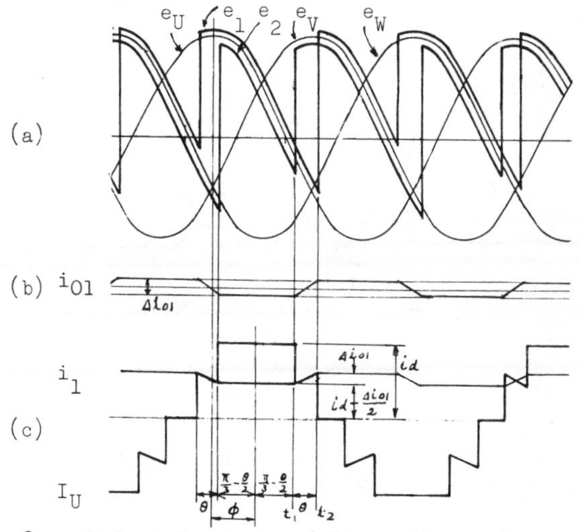

Fig. 8 Output Current and Cross Current Analytical Waveforms
a) Motor induced voltage,
b) Cross current, c) Output current

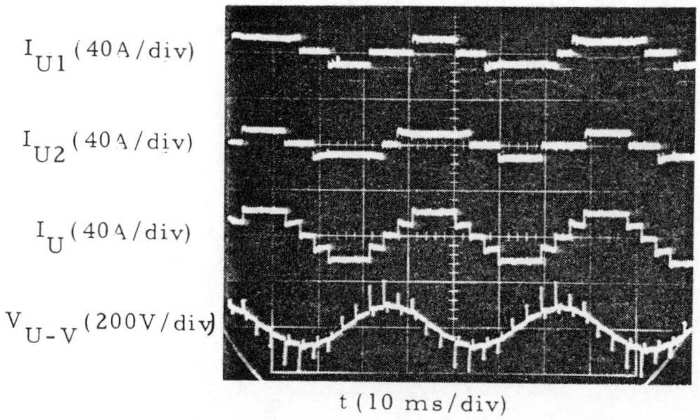

Fig. 9 Experimental Results — Output Current (I_u) Waveform.

Figure 9 shows output current waveform measured by the experimental circuit of the duplex CSI, shown in Figs. 3 and 4. The analytical result shown in Fig. 8(c) agrees well with experimental results, shown in Fig. 9. As is clear from Eq. (7), the value of Δi_{01} is kept low, when an adequate DC reactor value is adopted.

HARMONIC ANALYSIS AND TORQUE RIPPLES IN CSI-INDUCTION MOTOR DRIVE SYSTEMS

Figures 10 and 11 show harmonic components of duplex and quadruple CSI, respectively. In these figures, horizontal axes are current step width θ. It is preferable to choose θ as $\pi/6$ for duplex and θ as $\pi/12$ for quadruple CSI, for the sake of logic circuit simplicity. Figure 12 shows calculated results of instantaneous output torque in single, duplex ($\theta = \pi/6$) and quadruple ($\theta = \pi/12$) CSIs. It is clear that torque ripple values decrease approximately in inverse proportion to numbers of multiples, and torque ripple frequencies increase in proportion to numbers of multiples.

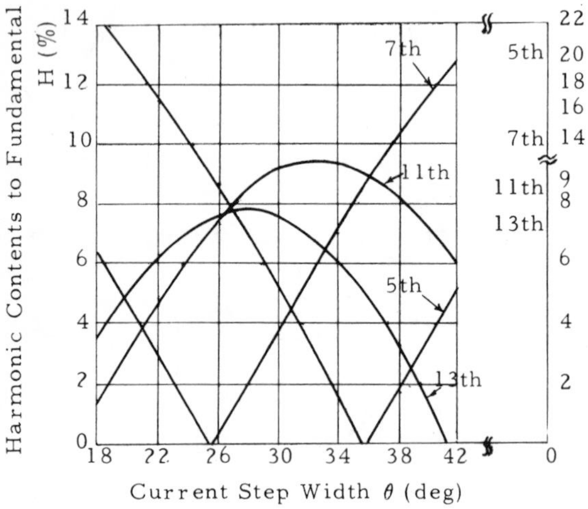

Fig. 10 Duplex CSI Harmonic Contents.

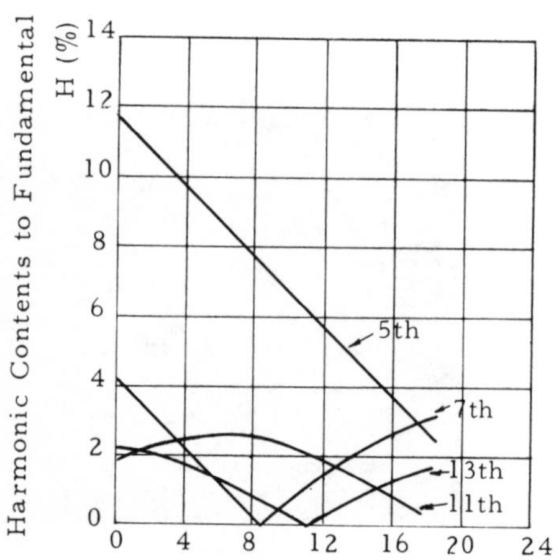

Fig. 11 Quadruple CSI Harmonic Contents.

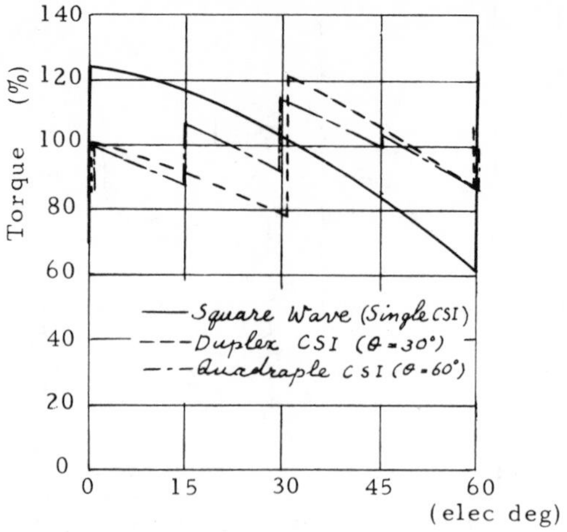

Fig. 12 Instantaneous Output Torque.

CONCLUSION

The new multiple CSI current contains less harmonics and gives better torque ripple characteristics in the CSI-IM drive system, compared with those of a single CSI.

The new multiple CSI can be supplied from a single DC source; one converter and one controller for duplex CSI and two converters and one controller for quadruple CSI.

The authors also have discussed cross current problem in paralleled unit CSIs and torque ripple problem.

The new multiple CSI was found to furnish a superior thyristor drive induction motor system.

REFERENCES

1. A. Nabae, R. Kurosawa, T. Shimamura: Toshiba Review, Vol. 31, No. 7, pp. 599-604 (1976).
2. N. Sato: EE in Japan, Vol. 84, No. 5, pp. 30-41 (1964).
3. M. B. Brennen: PESC 73 Record pp. 201-212 (1973).
4. W. Farrer, J. D. Miskin: PIEE, Vol. 120, No. 9, pp. 969-976 (1973).

Section 5
Slip Power Controlled Drives

Induction Motor Speed Control with Static Inverter in the Rotor

A. LAVI, SENIOR MEMBER, IEEE, AND R. J. POLGE, MEMBER, IEEE

Abstract—**A speed regulating scheme using a wound rotor induction motor and a static synchronous inverter is investigated. The purpose of the inverter is to receive the slip power from the rotor and to deliver it to the ac line.**

The steady state and the transient performances of the system are studied and compared with experimental results. The dependence of system performances upon the design of the inverter and the commutating filter inductance is presented and design criteria are developed.

The resulting motor inverter system has a linear torque current relation independent of the speed. In this respect, the system has the characteristic of a separately excited dc motor.

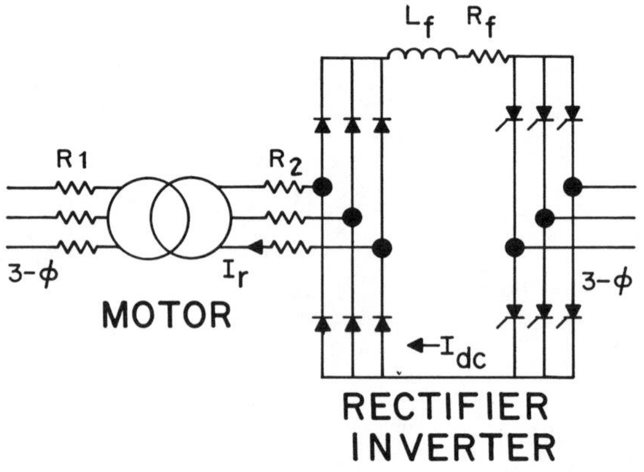

Fig. 1. Inverter speed-control system.

INTRODUCTION

THE SPEED CONTROL of ac motors has been of interest to engineers for some time. For a wound rotor induction motor, continuous speed control can be obtained by variation of the frequency of the ac source, by variation of an impedance in series with stator or rotor, or by the insertion of a counter EMF in the rotor circuit. The Scherbius system is a good example of counter EMF control. Its efficiency is good because most of the slip power is fed back to the ac source. With the recent progress in silicon diodes and silicon controlled rectifiers (SCRs), it becomes advantageous to replace the auxiliary machines of the Scherbius system by a bridge rectifier and a static inverter (see Fig. 1). A study of a similar drive has been given by Meyer with particular reference to large power applications [1]. However, the problem is still open to a comprehensive analysis. Elaboration on the influence of the factors of design upon the performance of a speed regulating system is desirable.

In this work, the study of the speed control of a wound rotor induction motor in conjunction with a static synchronous inverter is presented. Steady-state static relationships between torque, slip, and inverter voltage are derived. The transient performance of the system is also investigated.

Paper 31 TP 65–724, recommended and approved by the Rotating Machinery Committee of the IEEE Power Group for presentation at the IEEE Summer Power Meeting, Detroit, Mich., June 27–July 1, 1965. Manuscript submitted May 18, 1964; made available for printing April 27, 1965.

This is part of a dissertation submitted by R. J. Polge in partial fulfillment of the requirement for the Ph.D. degree, Carnegie Institute of Technology, Pittsburgh, Pa.

A. Lavi is with the Carnegie Institute of Technology, Pittsburgh, Pa.

R. J. Polge is with the University of Alabama, Huntsville, Ala.

Reprinted from *IEEE Trans. Power App. Syst.*, vol. PAS-85, pp. 76–84, Jan. 1966.

Basic Steady-State Relations of the System

Under the assumptions of negligible leakage reactances and ideal filtering ($L_f = \infty$), the rotor currents are alternating square pulses of $2\pi/3$ radians duration as shown in Fig. 2(a). The rms value of the rotor current I_r is $\pi/3$ times the rms value of the fundamental component I_{22}. From power considerations, a rotor equivalent circuit can be derived from which the fundamental component I_{22} and the total i^2R losses can be determined.

The inverter produces an average counter EMF equal to $1.35\ V \cos \alpha$. Here V is the line-to-line voltage and α is the firing angle (defined in Fig. 2). The inverter absorbs energy from the rotor when $\pi/2 < \alpha < \pi$ and can deliver power when $\alpha < \pi/2$.

The balance of power in each rotor phase gives

$$E_{22}I_{22} = r_{22}I_r^2 + \frac{1}{3}R_fI_{dc}^2 - \frac{1}{3}(1.35\ V\cos\alpha - W) \times I_{dc} + P_{\text{mech}} \qquad (1)$$

where W is the voltage drop in the semiconductors.

The power dissipation in R_f is $1/3R_fI_{dc}^2$. This is equivalent to the power dissipation caused by the flow of an rms current I_r in a resistance $(R_f/2)$ in each rotor phase, because $I_{dc}^2 = (3/2)I_r^2$; $(1/3)R_fI_{dc}^2 = R_f(3/2) \times 3I^2 = (1/2)R_fI_r^2$. Assuming that the mechanical torque is caused only by the fundamental component I_{22}, the mechanical power of the rotor is P_{mech}

$$P_{\text{mech}} = \left[\left(r_{22} + \frac{R_f}{2}\right)I_{22}^2 - (1.35\ V\cos\alpha - W)\frac{I_{dc}}{3}\right]\frac{1-\mathcal{S}}{\mathcal{S}} \qquad (2)$$

where $\mathcal{S}$ is the slip.

Hence

$$E_{22}I_{22} = \left(\frac{\pi^2}{9} - 1\right)\left(r_{22} + \frac{R_f}{2}\right)I_{22}^2 + \left[\left(r_{22} + \frac{R_f}{2}\right)I_{22}^2 - (1.35\ V\cos\alpha - W)\frac{\pi}{3\sqrt{6}}I_{22}\right]\frac{1}{\mathcal{S}}. \qquad (3)$$

This suggests the per phase equivalent circuit of Fig. 3(a), where the effect of the stator resistance r_1 is included. X_M is the magnetizing reactance. $R_l = (r_{22} + R_f/2)(\pi^2/9 - 1)$ represents the losses caused by the harmonic content of I. $R_b = [r_{22} + R_f/2 - (1.35\ V\cos\alpha - W)\pi/3\sqrt{6}\ I_{22}]$ takes into account the effect of the inverter.

Figure 3(a) can be modified into Fig. 3(b) where the inverter back EMF is represented by an ac voltage source opposing the flow of current. $R_t = (a^2/\mathcal{S})(r_{22} + R_f/2)$ and $E' = (a/\mathcal{S})(-E\cos\alpha + \pi W/3\sqrt{6})$, (a) being the transformation ratio of the motor.

Note: While Fig. 3(b) gives the magnitude of I_1, I_2, and all powers (assuming no leakage inductance), it does not give the proper current waveforms.

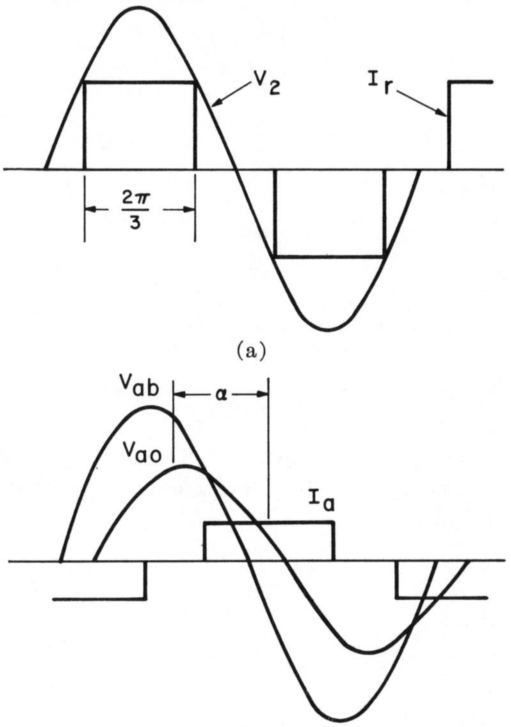

Fig. 2. (a) Rotor current. (b) Inverter current.

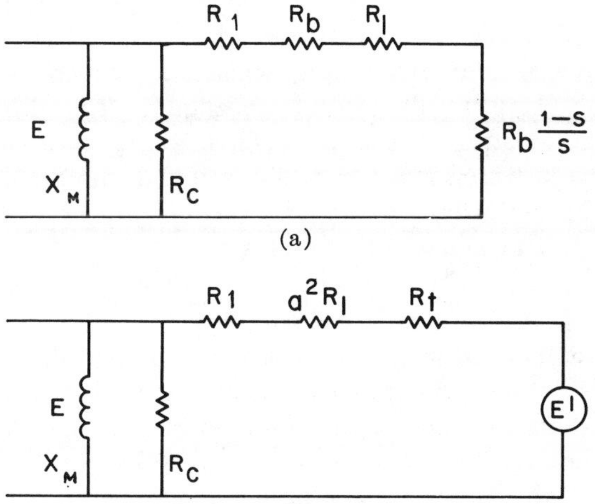

Fig. 3. (a) Equivalent circuit. (b) Equivalent circuit with voltage source.

Neglecting the semiconductor drop W,

$$I_2 \equiv \frac{I_{22}}{a} = \frac{E\left(1 + \dfrac{a\cos\alpha}{\mathcal{S}}\right)}{r_1 + \dfrac{a^2}{\mathcal{S}}r_{22}} = \frac{E(\mathcal{S} + a\cos\alpha)}{\mathcal{S}r_1 + a^2r_{22}} \qquad (4)$$

($\cos\alpha$ is normally negative).

The developed torque T is the fundamental power in the rotor divided by the synchronous speed N_s

$$T = \frac{-(E/\mathcal{S})\cos\alpha + r_{22}I_{22}/\mathcal{S}}{N_s}I_{22}. \qquad (5)$$

Combining (4) and (5), the torque becomes

$$T = \frac{E}{aN_s} \frac{1 - a \cos \alpha \dfrac{r_1}{a^2 r_{22}}}{1 + \mathrm{S} \dfrac{r_1}{a^2 r_{22}}} I_{22} \tag{6}$$

or

$$T = \frac{EI_{22}}{aN_s}\left(1 - \frac{a \cos \alpha\, r_1}{a^2 r_{22}}\right)\left[1 - \frac{\mathrm{S} r_1}{a^2 r_{22}} + \frac{(\mathrm{S} r_1)^2}{(a^2 r_{22})^2} \cdots\right]. \tag{7}$$

Hence

$$T \approx \frac{EI_{22}}{aN_s}\left[1 - \frac{(a \cos \alpha + \mathrm{S}) r_1}{a^2 r_{22}}\right]. \tag{8}$$

In (8), the term $r_1/a^2 r_{22}$ is at most unity. From (4), because $E/(\mathrm{S} r_1 + a^2 r_{22})$ is equal to at least the short-circuit current, $(\mathrm{S} + a \cos \alpha)$ must be a quantity much smaller than one. Hence, the torque becomes approximately proportional to I_{22},

$$T \approx \frac{E}{aN_s} I_{22}. \tag{9}$$

With this simplified model, some interesting results become apparent. The torque developed is proportional to the fundamental component of rotor current I_r. The current depends upon the difference between the rotor voltage at the particular slip and the inverter voltage. Hence, at fixed firing angle of the inverter, the torque–speed characteristic is almost linear. It follows that, at fixed torque, the inverter voltage must vary linearly with the slip. These results require negligible $R_r I_r$ drop. The characteristics of the motor-inverter system resembles very much that of a separately excited dc motor.

The motor-inverter system can operate at $\mathrm{S} < 0$ and $\mathrm{S} > 1$. The balance of powers is

$$\underset{\text{stator}}{TN_s} = \underset{\text{elec. rotor}}{N_s T \mathrm{S}} + \underset{\text{mechanical}}{N_s T(1 - \mathrm{S})}. \tag{10}$$

It results that, if the motor is driven by a prime mover at speeds above synchronous ($\mathrm{S} < 0$), power is absorbed from the shaft and delivered to the ac source through the stator or the inverter. ($T < 0$, $N_s T < 0$, $N_s T(1 - \mathrm{S}) < 0$.) If $\mathrm{S} > 1$, the rotor acts as a brake; a large power is fed back to the ac source through the inverter. This power comes from the shaft and the stator. ($T > 0$, $N_s T(1 - \mathrm{S}) < 0$.)

Since only the fundamental component of the rotor current is assumed to contribute to the torque, the same value of the fundamental component I_{22} is necessary to obtain a given torque, whether the current is sinusoidal or pulsating. Hence, the rotor losses at fundamental frequency are equal to $r_{22} I_{22}^2$ independently of the rotor current waveform. The Fourier analysis of alternating square pulses of $2\pi/3$ radians duration shows that the rms value I_m of the mth harmonic component is I_{22}/m and m equals $6l + 1$ (l integer). These harmonic components produce additional $i^2 R$ losses. Neglecting the variation of rotor resistance with frequency, the rotor losses for alternating square pulses are

$$3 r_{22} I_{22}^2 \left[1 + \sum_{l=1}^{\alpha} \frac{1}{(6l \pm 1)^2}\right] = 3 r_{22} I_{22}^2 \times \frac{\pi^2}{9}, \tag{11}$$

i.e., $\pi^2/9 = 1.09$ times that of a purely sinusoidal current. It results that, under the assumption of negligible leakage reactances and ideal filtering, the motor must be derated by at least 10 percent or so.

Electrodynamic Effects of the Harmonics

The presence of the rectifier bridge in the rotor circuit causes harmonic currents in the rotor and in the stator which produce harmonic torques in addition to the harmonic torques already present in conventional operation.

The influence of the revolving fields produced by the rotor harmonics upon the stator currents is investigated, assuming that the rotor currents are square pulses. Consider, for example, the fifth harmonic rotor current. The rotor rotates at a speed $N = N_s(1 - \mathrm{S})$ and the fifth harmonic rotor current is at the frequency $5\mathrm{S}f$. These 3-phase harmonic rotor currents set up a magnetic field in the air gap which revolves with respect to the rotor at the speed $-5\mathrm{S}N_s$. Since the rotor is at speed $N_s(1 - \mathrm{S})$, the resulting speed of this fifth harmonic wave with respect to the stator is $N_s(1 - 6\mathrm{S})$ and the induced stator current is at frequency $(1 - 6\mathrm{S})f$.

The frequency of the secondary circuit (that of the stator) is now divided by $(1 - 6\mathrm{S})/5\mathrm{S}$ resulting in the equivalent circuit of Fig. 4(a). If X_1 and X_M are, respectively, the stator leakage reactance and the magnetizing reactance at fundamental frequency f, then at the fifth harmonic frequency $X_{(5\mathrm{S})} \approx 5\mathrm{S}X_1$; $X_{M(5\mathrm{S})} = 5\mathrm{S}X_M$.

The magnitude of the current induced in the stator by the fifth harmonic rotor current is

$$I_{(5\mathrm{S})} = \frac{I_{22}}{5} \frac{X_M}{\sqrt{(r_1/1 - 6\mathrm{S})^2 + (X_1 + X_M)^2}}. \tag{12}$$

This current is maximum at stall, null at $\mathrm{S} = 1/6$, and increases from there again for smaller slips. The reasoning is, of course, valid for the other rotor harmonics. One concludes that the harmonic currents induced in the stator are smaller than the corresponding harmonic currents in the rotor and their frequency is not, in general, commensurate with the line frequency. Thus, the stator current is nearly sinusoidal, although not periodic in general.

The harmonics forced into the rotor by the rectification process produce parasitic asynchronous torques. Again, take the case of the fifth harmonic of rotor current. Using the equivalent circuit of Fig. 4(b), the total power transferred to the stator is

$$P_{(5\mathrm{S})} = \frac{r_1 I_{(5\mathrm{S})}^2}{\mathrm{S}_{(5\mathrm{S})}}. \tag{13}$$

The harmonic slip is

$$\mathrm{S}_{(5\mathrm{S})} = \frac{1 - 6\mathrm{S}}{5\mathrm{S}} \tag{14}$$

and the synchronous harmonic speed is $5\mathrm{S}N_s$, so that the harmonic torque is

$$T_{(5\mathrm{S})} = \frac{r_{22} I_{(5\mathrm{S})}^2}{\mathrm{S}_{(5\mathrm{S})} N_{(5\mathrm{S})}} = \frac{r_{22} I_{(5\mathrm{S})}^2}{(1 - 6\mathrm{S}) N_s}. \tag{15}$$

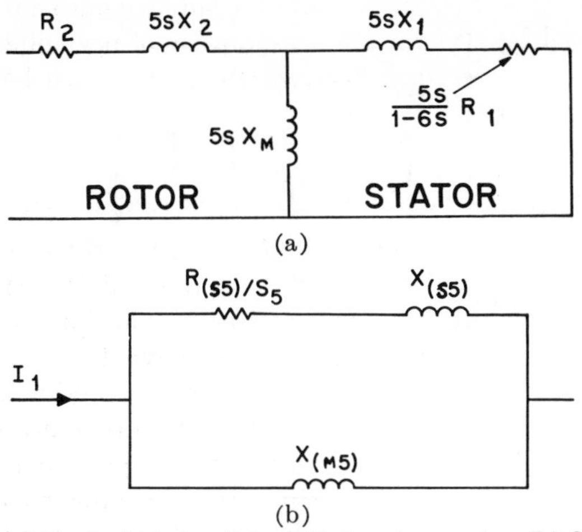

Fig. 4. (a) Equivalent circuit for fifth time harmonic. (b) Constant current circuit of fifth space harmonic.

The value of $I_{(5S)}$ obtained in (12) is substituted in (15)

$$T_{(5S)} = \frac{r_1}{(1-6S)N_s}\left(\frac{I_{22}}{5}\right)^2 \frac{X_M{}^2}{\left(\dfrac{r_1}{1-6S}\right)^2 + (X_1+X_M)^2}. \tag{16}$$

The value of I_{22} depends mainly on the equivalent circuit at fundamental frequency,

$$I_{22} \approx \frac{E}{\sqrt{(r_1+R_{b/S})^2+(X_1+X_2)^2}} \tag{17}$$

where R_b is the equivalent resistance of the inverter for the given steady state. The fundamental torque is approximately

$$T \approx \frac{E^2}{N_s}\,\frac{r_1+R_{b/S}}{(r_1+R_{b/S})^2+(X_1+X_2)^2}. \tag{18}$$

The ratio of the fifth harmonic to fundamental torque is

$$\frac{T_{(5S)}}{T} = \frac{1}{25}\,\frac{r_1X_M{}^2}{(1-6S)\{[r_1/(1-6S)]^2+(X_1+X_M)^2\}}\cdot\frac{S}{r_1S+R_b}. \tag{19}$$

The maximum of this ratio occurs near $S = 1/6$. The ratio is positive at $S > 1/6$ and negative at $S < 1/6$. Realizing that $r_1 \leq R_b$ and that the other term is always less than unity, the ratio is never larger than 0.04. The torques produced by the other harmonics can be obtained similarly and are even smaller.

The interaction of rotor current harmonics and stator space harmonics produce synchronous torques at specific speeds, but they are very small in a good motor design where the magnitude of space harmonics is minimized [2].

Thus, all harmonic torques can be neglected and it is sufficient to consider the torque to be caused only by the fundamental component of rotor current.

Effects of the Leakage Reactances

Because of the presence of leakage reactances in the phases of the induction motor, the commutation of current among the phases does not occur instantaneously.

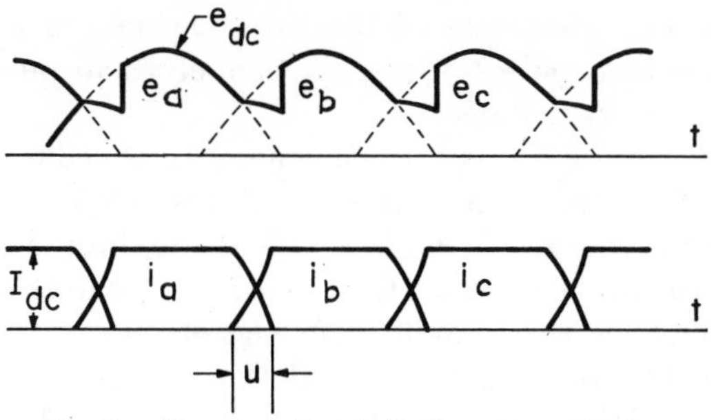

Fig. 5. Commutation in 3-phase ac generator.

There is a period of current overlap whereby two phases carry current simultaneously. The commutation in the induction motor is similar to that in a transformer with a rectified load, except that the internal resistance to consider is now $R_e = r_1 + r_{22}/S$. This resistance is not negligible at very small slips. Assuming perfect filtering, the sum of the phase currents is constant. With the notations of Fig. 5,

$$i_a + i_b = I_{dc} \tag{20}$$

and, from Kirchhoff's law,

$$e_a - e_b = X_e\left(\frac{di_a}{d\theta} - \frac{di_b}{d\theta}\right) + R_e(i_a - i_b). \tag{21}$$

The duration of commutation is u (angle of overlap) and the equations of the current pulse can be obtained from (20) and (21). By Fourier analysis, the fundamental component I_{22}, the phase shift of the fundamental φ, and the rms value I_r are calculated. When R_e is neglected, the rise and decay of the current pulses are arcs of sine wave and closed form equations can be obtained; otherwise numerical techniques must be employed.

When $S > 0.1$ (R_e negligible)

$$u = \cos^{-1}\left(1 - \frac{\sqrt{2}X_e}{\sqrt{3}\,E}I_{dc}\right) \tag{22}$$

$$\varphi = \tan^{-1}\frac{(2u - \sin 2u)}{(1-\cos 2u)}. \tag{23}$$

When $S < 0.1$ (R_e large), u is defined by

$$\sin\gamma\, e^{-(R_e/X_e)u} - \sin(\gamma - u) = \frac{I_{dc}}{2i_{sh}}[1 + e^{-(R_e/X_e)u}] \tag{24}$$

where

$$\tan\gamma = \frac{X_e}{R_e},\qquad i_{sh} = \frac{\sqrt{3}\,E}{\sqrt{2}\,\sqrt{R_e^2+X_e^2}}$$

then

$$\varphi \approx \frac{u}{2}.$$

Typical order of magnitude, at rated current: $u = 30°$, $\varphi = 20°$.

Concluding, the presence of leakage reactances produces the following effects:

1) The direct voltage of the bridge rectifier is reduced by a factor $\cos^2(u/2)$.

2) The waveform of the rotor current is a series of alternating pulses round in shape, each pulse lasting $(2\pi/3 + u)$ radians.

3) The fundamental component I_{22} is reduced slightly and displaced by an angle φ with respect to the phase to neutral voltage (u and φ depend on torque and slip). Hence, for a given I_{dc}, the torque is less than the value obtained from the simplified analysis of (9).

4) The rms current I_r and the harmonic content are reduced; the rms value of the mth harmonic becomes less than I_{22}/m.

Performances of the System

The linear equivalent circuit of Fig. 3 cannot be modified to take the leakage reactances into account. This is because the equations of the rotor currents are more complex; the currents I_r, I_{22}, I_{dc} and the angle of overlap u are interdependent, and all are unknown to start with. Since the linear circuit analysis is not applicable, a numerical procedure based on the flow of power is employed. The powers are normalized, the total fundamental rotor power at rated torque being equal to 100.

For a given slip $\mathcal{S}$ and per unit torque k, the performance of the system can be calculated. If k_m is the per unit torque corresponding to the mechanical losses, the delivered torque is $k' = k + k_m$. The fundamental rotor power is $100k'$. This divides into mechanical output $P_{\text{mech}}' = 100k'(1 - \mathcal{S})$ and electrical output at fundamental frequency $100k'\mathcal{S} = 3r_{22}I_{22}^2 + P_4$ where P_4 is the electric power out of the rotor.

In order to find the losses everywhere, the currents I_r, I_{22}, and I_{dc} must be calculated by an iterative procedure. After an initial guess for u and I_r, the quantities I_{22} and I_{dc} and the powers everywhere can be evaluated, resulting in more precise values for u and I_r. I_r = apparent power/$\sqrt{3}\,V$. These values are used in a new series of computations until the value obtained for I_r differs from the value used to compute it by a prescribed percent error.

The power fed back to the line through the inverter is $P' = P_4 -$ (losses in diodes, SCR, and filter). The direct voltage at the inverter is $1.35\,V \cos\alpha = P'/I_{dc}$; this relation defines $\cos\alpha$.

The flow of powers is shown in Fig. 6. The total power P_t into the system is equal to the difference between the power absorbed by the stator P_i and the power fed back by the inverter. The total reactive power Q_t into the system is the sum of the reactive powers absorbed by the stator and the inverter; the total harmonic power H_t is the sum of the harmonic powers absorbed by the bridge rectifier and by the inverter. The total apparent power A_t is equal to

$$A_t = \sqrt{(P_t^2 + Q_t^2 + H_t^2)}. \tag{25}$$

The efficiency η of the system is the ratio of the useful power on the shaft $100k(1 - \mathcal{S})$ to the total power consumed, namely,

$$\eta = \frac{100k(1 - \mathcal{S})}{P_t}. \tag{26}$$

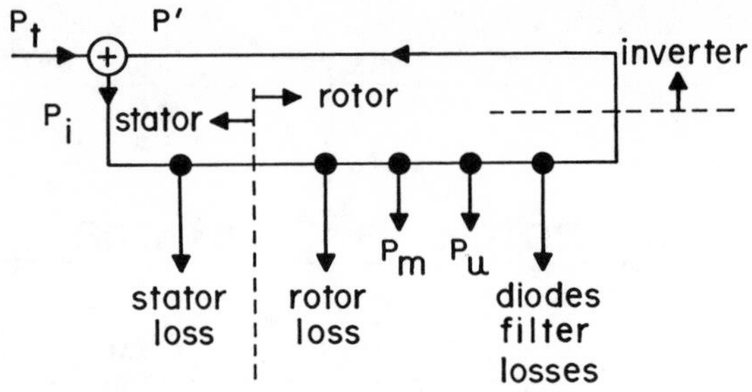

Fig. 6. Power flow diagram.

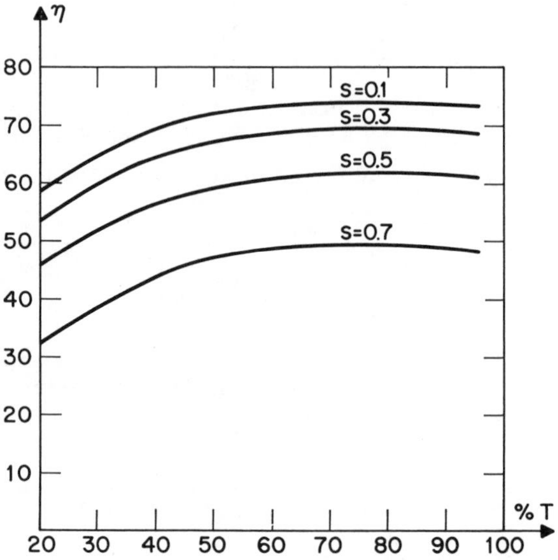

Fig. 7. Calculated efficiency for experimental machine.

The apparent factor F_p of the system is, by definition, the ratio of the total active power to the total apparent power, namely,

$$F_p = \frac{P_t}{A_t}. \tag{27}$$

The calculations have been checked against experimental results for a specific machine and the agreement is very good. These results show also that the simplified analysis neglecting the leakage reactances describes rather well the basic operation of the motor-inverter system.

The numerical computations show the efficiency (Fig. 7) of the specific motor inverter system used in the experimentation is quite good; in fact, near rated speed and rated torque, the overall efficiency is 76 percent while the efficiency of the machine with the short-circuited rotor is 80 percent.

This is, by far, better than what can be expected in a Scherbius system of similar power ratings; in the Scherbius system the power fed back has to flow through three machines, resulting in a poor efficiency.

The power factor (Fig. 8) of the rotor circuit is near unity, but the power factor of the motor inverter system is rather poor. This is so because the inverter absorbs a reactive power from the ac line, which is a function of the load torque and the slip. At rated speed and rated torque,

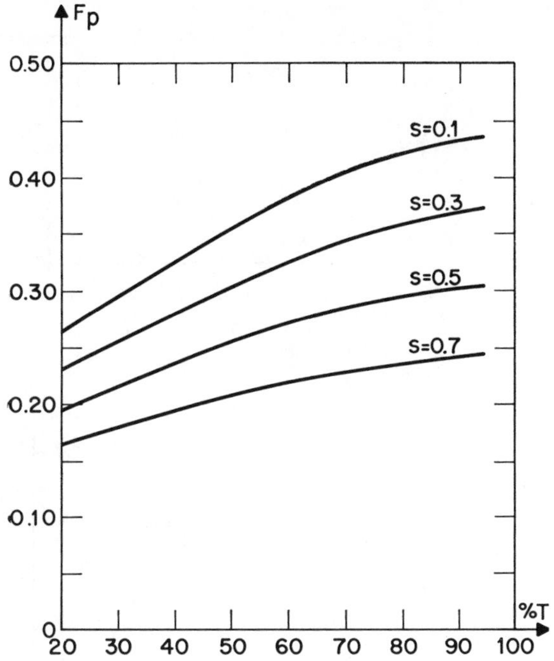

Fig. 8. Calculated apparent power factor for experimental machine.

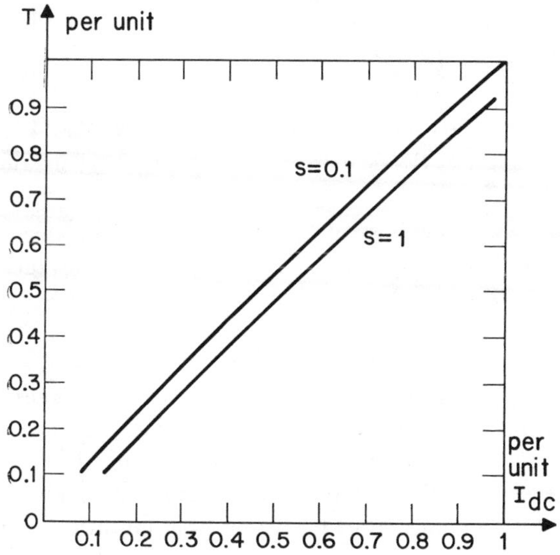

Fig. 9. Torque vs. I_{dc} (experimental).

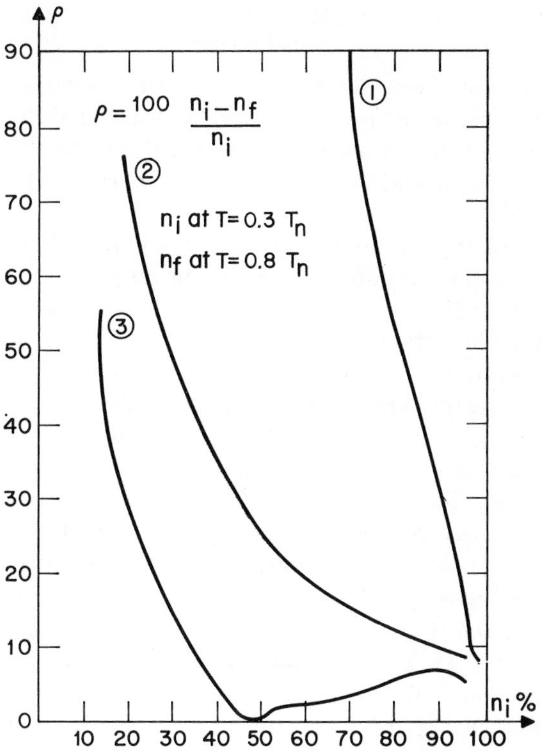

Fig. 10. Steady-state speed regulation: 1) Variable resistance. 2 Inverter. 3) Inverter with *RI* compensation.

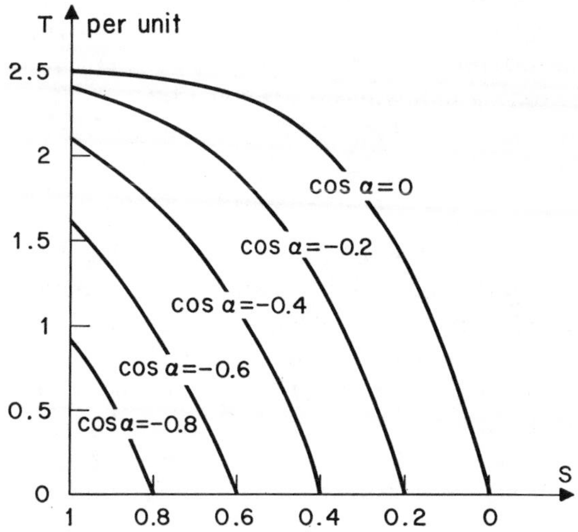

Fig. 11. Torque vs. slip for fixed firing angle.

the power factor reaches 0.5. However, the power factor can be improved if the range of speed control is limited and the proper transformation ratio is selected. Meyer reports a power factor equal to 0.7 at rated torque for a range of speed control limited between synchronous and 0.75 synchronous.

The proportionality between torque and direct current at stall (starting torque) has been confirmed experimentally. This is a very important feature in that high-starting torques can be obtained with reasonable currents. This is a consequence of the "high-power factor" of the rotor circuit. The proportionality between torque and direct current depends very little upon the slip as shown in Fig. 9.

The speed regulation, that is, the percent variation of speed with change in torque, at fixed inverter EMF is much better than for the case of resistance control. However, because of the large *RI* drops in the rotor and filter, the regulation is inferior to that of a dc machine. *RI* compensation, similar to that used for dc machines, can be added to improve the speed regulation. For a change of load torque from 30 to 80 percent of rated, the regulation, above 35 percent of rated speed, is better than 8 percent on the small machine used in the experiment. In a larger machine, the per unit drop is reduced and better results are to be expected. Figure 10 shows the speed regulation of the machine with resistance control and with inverter control (with and without *RI* compensation). Finally, Fig. 11 gives the torque–speed characteristic at various values of α.

DESIGN

In the design of this motor drive system, there are three important parameters at the disposal of the designer. They are the filtering inductance, the motor transformation ratio a, and the dependence of inverter EMF upon the signal.

The filtering inductance is required in the system to provide commutation of inverter current and to limit the losses in the rotor windings. With a finite inductance the rotor currents are not flat topped, but have ripples produced by the harmonic voltages of the bridge rectifier and the inverter. These ripples produce additional losses in the rotor. The first harmonic voltage of the inverter V_6', $\pi/2$ (at frequency $6f$) is maximum for $\alpha = \pi/2$ and equal to 0.33 V. The first harmonic voltage of the bridge rectifier V_6 (at frequency $6f$) is equal to 0.08 aV for a typical machine at rated current. These two harmonic voltages have the same frequency at stall and may add up arithmetically resulting in a harmonic voltage $V_{6t} = 0.41\ V$ (for $a = 1$). The maximum harmonic current I_{6t} resulting from insufficient filtering is obtained using the equivalent circuit of Fig. 12. This circuit consists of a voltage source V_{6t} in series with twice the leakage reactance of the machine and the filtering inductance. The resistances being relatively small are, thus, neglected. The increase in the rotor losses caused by the ripples I_{6t} on the direct current I_{dc} is equal to $(1 + I_r^2/I_{dc}^2)$. If the allowed increase of losses resulting from insufficient filtering is μ^2 percent and $a = 1$, the filtering inductance L_f is given by

$$L_f = \frac{0.44\ V}{\pi f \mu I_{dc(\text{rated})}} - \frac{X_e}{\pi f} \tag{28}$$

Thus, a large filtering inductance is desirable. On the other hand, the filtering inductance L_f determines the transient performance of the machine and, therefore, a compromise must be made between motor derating and fast response. In general, however, it is more important to limit the losses using a large inductance, because the speed of response can be improved through negative feedback. This design criterion for the filtering inductance is suggested.

In order to obtain a good power factor for the motor inverter system, the rotor voltage should be made as large as possible; excluding the use of a transformer on the inverter side, the motor transformation ratio must be made equal to $a = S_{\max}$. $S_{\max}$ is the highest controllable slip. If speed control between slip one and zero is desired, a must be one; if the motor is to provide a braking torque at S larger than one, an even larger transformation ratio is needed. If the maximum slip is limited to a value less than one and the transformation ratio is selected so as to maximize the power factor, the motor must be started by means of an auxiliary resistor, because the inverter cannot provide sufficient back EMF at high slips.

The variation of inverter EMF with slip, at constant torque, is linear, as seen from Fig. 13. This indicates that, if cos α of the inverter is proportional to an external signal, linear control will result. In other words, in order to obtain a linear control system, the inverter EMF voltage must be proportional to the net control signal.

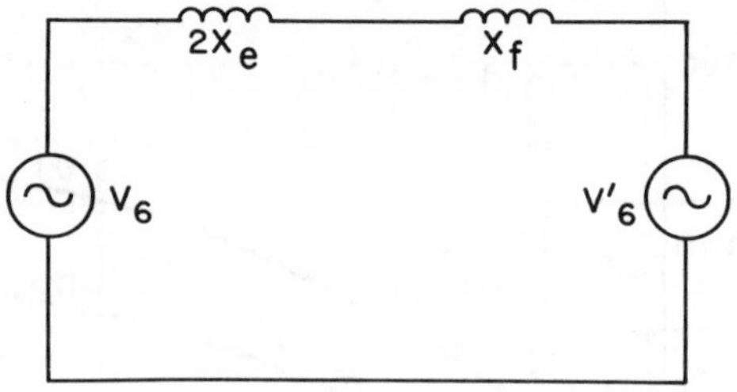

Fig. 12. Simplified circuit for voltage harmonic.

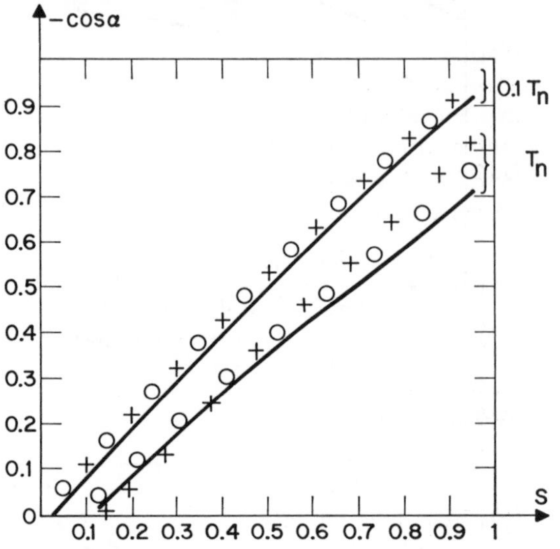

Fig. 13. Evaluation of cos a for given load and slip.

TRANSIENT ANALYSIS

The complete transient analysis of a conventional induction motor is quite difficult. The presence of SCR switching certainly does not simplify matters. Thus, an analysis for slow changes of speed about an operating point is attempted. Three major assumptions are introduced: 1) the time constant of the dc part of the circuit is much larger than the time constant of the ac; 2) the speed of the motor does not vary appreciably during one-sixth of a cycle of the ac line, because of the motor and load inertia; 3) the motor torque during transient is proportional to the current in the dc circuit. The constant of proportionality found in the steady state is assumed valid for a small perturbation about an initial operating point.

Under these assumptions, a simplified circuit is derived (Fig. 14) which consists of two unidirectional voltage sources (bridge rectifier and inverter) in series with the rotor, filter, and diode resistances and with the filtering inductances. The two voltage sources are periodic at different frequencies; thus, the direct current is not constant, even at constant slip. i_L is defined by the differential equation

$$v(t) - v'(t) = (2r_{22} + R_f + 2R_d)i_L(t) + L_f \frac{di_L}{dt}. \tag{29}$$

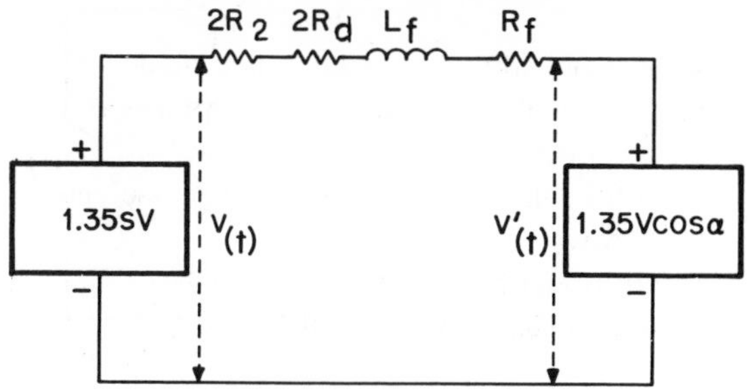

Fig. 14. Simplified rotor inverter circuit.

Because of the motor and load inertia, the speed does not vary considerably within a small fraction of a cycle. Thus, integration of (29) from one firing instant α_{m-1} to another α_m and averaging over one-sixth of a cycle gives the difference equation

$$\bar{V}(m) - \bar{V}'(m) = R\bar{I}_L(m) + 6fL_f[i_L(m) - i_L(m-1)]$$

$$R \equiv 2r_{22} + R_f + 2Rd \qquad (30)$$

where the notation over the variable means one-sixth of a cycle average and the subscript (m) designates the mth inverter firing interval; $i_L(m)$ is the instantaneous value of i_L at the end of the mth interval.

Since one-sixth of a cycle is a short time in comparison to the possible duration of a transient, the difference equation can be replaced by a first-order differential equation. Letting

$$\bar{V}(m) = 1.35\ V\bar{\mathcal{S}},\ \bar{V}'(m) = -1.35\ V\ \overline{\cos\alpha},$$

$$[i_L(m) - i_L(m-1)]6f \approx \frac{d\bar{I}_L}{dt}, \qquad (31)$$

then

$$1.35\ V(\bar{\mathcal{S}} + \overline{\cos\alpha}) = R\bar{I}_L(t) + L_f\frac{d\bar{I}(t)}{dt}$$

where $\bar{I}_L(t)$ is the time variation of the filter current averaged over a reasonably long interval.

From Newton's second law, assuming the torque is proportional to the direct current,

$$T = K_T\bar{I}_L$$

$$J\frac{dN(t)}{dt} + T_L = T \qquad (32)$$

where J = moment of inertia and T_L = load torque.

The two differential equations are combined into a second-order differential equation which describes the system for small signal variations. Since $N = N_s(1 - \mathcal{S})$ and

$$\frac{d\overline{N(t)}}{dt} = -N_s\frac{d\overline{\mathcal{S}(t)}}{dt},$$

$$\frac{d^2\overline{\mathcal{S}(t)}}{dt^2} + \frac{R}{L_f}\frac{d\overline{\mathcal{S}(t)}}{dt} + \frac{1.35\ VK_T}{N_sL_fJ}\mathcal{S}(t) = \frac{RT_L}{L_fJN_s} - \frac{1.35\ VK_T\overline{\cos\alpha}}{N_sJL_f}. \qquad (33)$$

For large up transients, only piecewise linearization is possible; the constants of the differential equation must be modified to take account of the reduction of the rotor induced EMF and of the reduction of the torque–current constant. Equation (33) can be Laplace transformed and, thus, a transfer function relating the motor speed to the command signal (cos α) is obtainable which can be used in the synthesis of a complete speed regulating system incorporating the motor-inverter.

Appendix I

Nomenclature

A_t	=	total apparent voltamperes
a	=	stator to rotor transformation ratio, also phase subscript
b	=	phase subscript
c	=	phase subscript
e_a, e_b, e_c	=	instantaneous EMF in rotor phases
e	=	subscript for equivalent
E_{22}	=	rms value of rotor EMF at fundamental frequency per phase
E'	=	equivalent inverter back EMF referred to stator per phase
E	=	per phase voltage
f	=	frequency, also subscript for filter
F_p	=	apparent power factor
H_t	=	total harmonic power
I, i	=	current
I_1	=	rms stator current sinusoidal
I_2	=	rms rotor current sinusoidal referred to stator
I_{22}	=	rms rotor current sinusoidal referred to rotor
I_r	=	total rms rotor current
I_{dc}	=	filter current, dc average
I_L	=	transient current in inductance
i_a, i_b, i_c	=	instantaneous phase currents in rotor
J	=	moment of inertia
k	=	per unit torque, k' per unit delivered torque, k_m per unit friction torque, K_T torque constant
L	=	subscript for load
L_f	=	filter inductance
l	=	integer
m	=	counting integer, one-sixth of cycle identifier
M	=	subscript magnetization
N	=	speed, N_s synchronous speed
P	=	power, P' inverter power, P_4 electric rotor power
P_t	=	total power
P_{mech}	=	mechanical power, P_m loss, P_u useful
Q_t	=	total reactive power
r	=	subscript for rotor
r_1	=	stator resistance per phase
r_2	=	rotor resistance referred to stator per phase
r_{22}	=	actual rotor resistance per phase
R_f	=	filter resistance
R_e	=	equivalent resistance
R	=	total resistance

R_b = equivalent resistance of inverter
R_d = dynamic resistance of rectifier junction
s = slip, subscript for synchronous
t = subscript for total
T = torque
T_L = load torque
u = angle of overlap
V = line to line voltage, v instantaneous voltage of bridge
V' = inverter voltage, v' instantaneous voltage of inverter
W = semiconductor voltage drop
X = reactance
X_1 = stator leakage reactance per phase
X_2 = rotor leakage reactance per phase referred to stator
X_{22} = rotor leakage reactance per phase referred to rotor
X_e = equivalent reactance
X_M = magnetization reactance

References

[1] V. M. Meyer, "Über die untersynchrone stromrichterkaskade," *Elektrotech. Z.*, Ausgabe A, September 1961.
[2] Robert J. Polge, "Speed control of a wound rotor induction motor with static inverter in the rotor circuit," Ph.D. dissertation, Carnegie Institute of Technology, Pittsburgh, Pa., 1963.

Slip Power Recovery in an Induction Motor by the Use of a Thyristor Inverter

WILLIAM SHEPHERD, MEMBER, IEEE, AND JACK STANWAY

Abstract—The low-speed efficiency of an induction motor is improved by rectifying slip-frequency power, inverting this to line frequency, and injecting it back into the supply directly (line feedback) or through auxiliary stator windings (stator feedback). Torque–speed curves then have the nature of a variable-speed drive. The low power factor and nonsinusoidal supply current of the line feedback connection are improved by use of the stator feedback method but the improvement of efficiency is then much less. Line feedback with a two-phase induction motor eliminates the need for a variable control voltage source.

I. Introduction

A THREE-PHASE induction motor fed from a constant voltage, constant frequency source is inherently inefficient at low speeds. For operation at less than one-half synchronous speed, more than a half of the power crossing the air gap is dissipated in the rotor windings and the external rotor resistors.

Many methods have been suggested whereby slip-frequency power is extracted from the induction motor (rotor) brushes, rectified, and used actively rather than dissipated in resistors. The rectified rotor current has been used in two basic ways: 1) by connection to the armature of a dc motor mechanically coupled to the induction motor shaft [1]–[4], 2) by inversion to line frequency, alternating current which is injected back into the supply [5]–[8]. In addition, various individual inventions have been described that involve rectification of the rotor currents [9]–[12]. Erlicki has described a rectifier-inverter scheme [13] for feeding the rotor with slip-frequency electromotive forces (EMF) using the principle of the stator-fed shunt commutator motor.

In the present paper, the efficiency-speed characteristic for conventional induction motor operation with a closed rotor circuit is established. Conventional operation is then contrasted with that obtainable by the use of an induction motor combined with a line commutated (constant frequency) inverter for utilizing slip power. A novel connection is introduced by which the power factor of the usual induction motor–inverter combination may be improved.

Paper TOD 112-67, SPC 67-23, approved by the Static Power Converter Committee for publication in this TRANSACTIONS. Manuscript received September 4, 1968.

W. Shepherd is with the University of Bradford, Bradford, Yorkshire, England.
J. Stanway is with the Royal Military College of Science, Shrivenham, Swindon, Wiltshire, England.

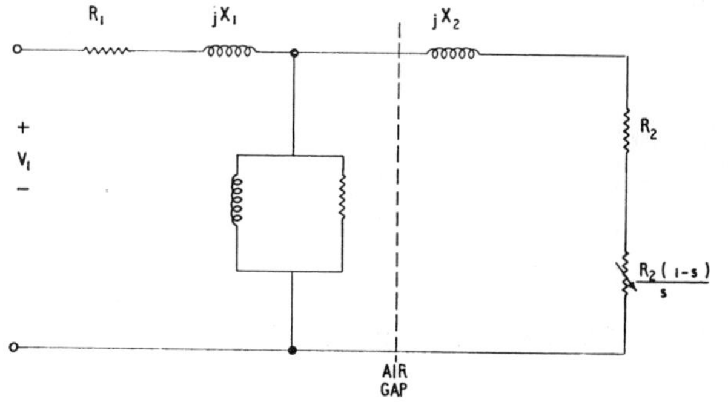

Fig. 1. Per-phase equivalent circuit of three-phase induction motor with sinusoidal supply.

II. Efficiency Speed for Normal Induction Motor Operation

Of the total power per phase crossing the air gap in an induction motor, that portion dissipated as rotor copper loss may be considered to be dissipated by the resistor R_2 in Fig. 1. The remaining rotor power (i.e., that dissipated in resistor $R_2(1-s)/s$ in Fig. 1) represents the mechanical power developed by the rotor which is spent in overcoming its own friction loss and in driving the load. Let the motor operate at an angular velocity w mechanical rad/s. Then if the synchronous angular velocity is w_s, the per-unit slip s is defined by

$$w = w_s(1 - s). \tag{1}$$

If the developed torque is T N·m/phase, the (internal) mechanical power developed per phase, in watts, is

$$\begin{aligned} P_{\text{developed}} &= Tw \\ &= Tw_s(1 - s). \end{aligned} \tag{2}$$

But the power delivered to the air gap from the stator windings is given [14] in

$$\text{power into rotor} = Tw_s \quad \text{W/phase}. \tag{3}$$

The rotor power ratio, given in (4), is shown in Fig. 2.

$$\frac{\text{mechanical power developed}}{\text{power into rotor}} = 1 - s. \tag{4}$$

That portion of rotor input power represented by the upper triangle in Fig. 2 is dissipated as copper loss in the rotor windings and external rotor resistors.

Reprinted from *IEEE Trans. Ind. Gen. Appl.*, vol. IGA-5, pp. 74–82, Jan./Feb. 1969.

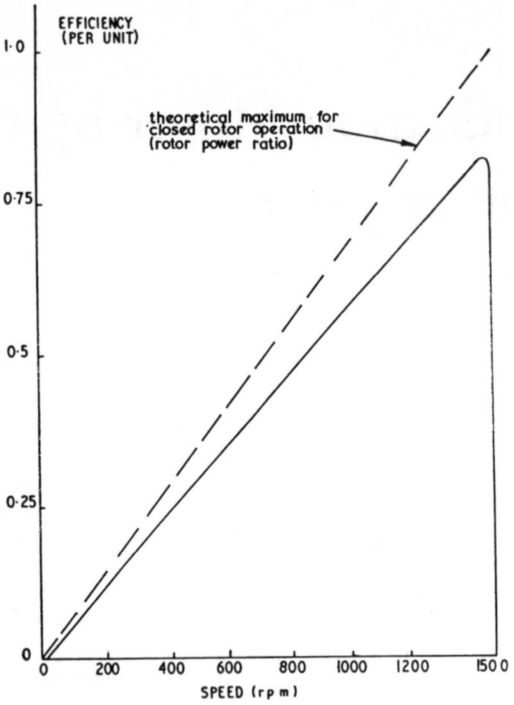

Fig. 2. Efficiency–speed for three-phase induction motor with sinusoidal supply and rotor winding closed.

Now due to stator copper losses, core losses, and friction loss the motor efficiency is always less then the linear relationship defined by (4).

$$\frac{\text{mechanical power delivered to the load}}{\text{power into motor}} < 1 - s. \quad (5)$$

A typical form of motor efficiency-speed characteristic, indicated in Fig. 2, displays the inherent inefficiency of low-speed sinusoidal induction motor operation, except where some method of slip-power recovery is employed. At 30 percent synchronous speed, for example, the maximum efficiency obtainable by conventional operation is less than 30 percent.

III. Performance with Inverted Slip Power Fed Back Directly into Supply

A. Three-Phase Motor Arrangement

A schematic arrangement of an induction motor–inverter combination, given in Fig. 3, is hereafter called the "line feedback" arrangement. The average open-circuit value of the direct voltage V_D is known [15] to be

$$|V_D| = \frac{3}{\pi} \sqrt{2} \sqrt{3} \, |V_p| = 2.34 \, |V_p| \quad (6)$$

where V_p is the per phase EMF of the star-connected rotor. The rectified voltage sV_D is smoothed by inductor L and fed to the input of a three-phase full-wave line-commutated inverter whose output is connected directly to the supply. Operation of this type of inverter is extensively described in the literature [16].

When the equal firing angles of the six inverter thyristors are greater than 90 degrees (measured from the crossover points of two successive supply phase voltages) and when the applied, rectified voltage is high enough to overcome the opposing EMF of the inverter, inversion occurs and power is recycled from the rotor back into the supply. The magnitude of the inverted current depends on the difference between the rectified voltage and the opposing EMF presented by the inverter. For the case of infinitely smoothed direct voltage and neglecting the ac source impedance, the opposing EMF presented by the inverter depends on the firing angle and the supply voltage as defined [16] by

$$V_{\text{opposing}} = 1.35 \, V \cos \alpha \quad (7)$$

where α is the thyristor firing angle and V is the root mean square (rms) line-to-line supply voltage. With any fixed firing angle α, inversion requires that the direct input voltage be greater than the corresponding value V_{opposing} from (7).

Measured torque-speed characteristics are given in Fig. 4 for a 3-hp 50-Hz induction motor, controlled by the line feedback system, Fig. 3. Tests were carried out at one-half rated voltage and the upper limits of the torque characteristics were constrained by thermal winding ratings. It is seen that direct feedback of current into the supply results in the system becoming a fairly good variable-speed drive with thyristor firing angle as the speed control parameter. Rather surprisingly, inversion continued slightly below 90-degree firing angle. Continuous control was achieved through zero speed and into the "overhauling" or braking quadrant II, Fig. 4. Inversion ceased at −300 r/min with 170-degree firing angle.

Oscillograms of the current waveforms are given in Figs. 5 and 6. Under all conditions the inverted current retains a conduction angle of 120 degrees and the same waveshape, while the motor winding current is substantially sinusoidal so that a distorted current waveform is drawn from the supply. At low speeds, with higher firing angles, the rotor current waveform becomes roughly rectangular as indicated in Fig. 6. The system currents remained balanced under all operating conditions. The variation of efficiency for operation at a range of fixed firing angles is given in Fig. 7. Low-speed efficiencies well in excess of the theoretical maxima for closed-rotor operation were obtained. In order to realize maximum operating efficiency over the whole speed range, the firing angle should preferably be adjusted at speeds coinciding with the peaks in Fig. 7. A more practicable criterion, however, is to smoothly adjust the triggering angle so as to provide a specified torque at each speed.

The variation of input power factor to the motor was obtained by the use of two dynamometer wattmeters, standard moving-iron ammeters, and a voltmeter. Power factor curves corresponding to the torque-speed curves of Fig. 4 are given in Fig. 8. It is seen that the order of the power factor is low compared with that obtainable with

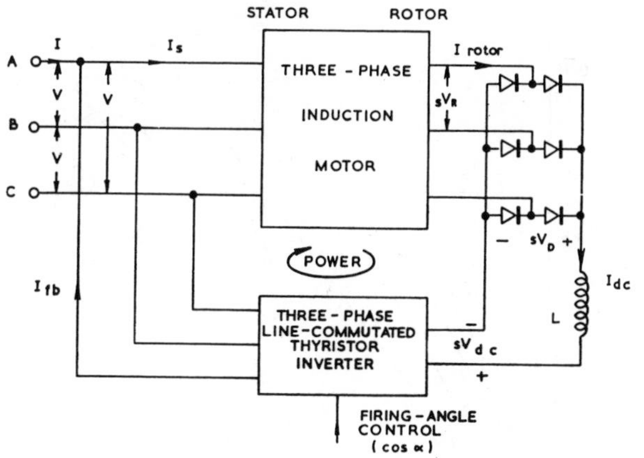

Fig. 3. Schematic arrangement of three-phase line feedback scheme.

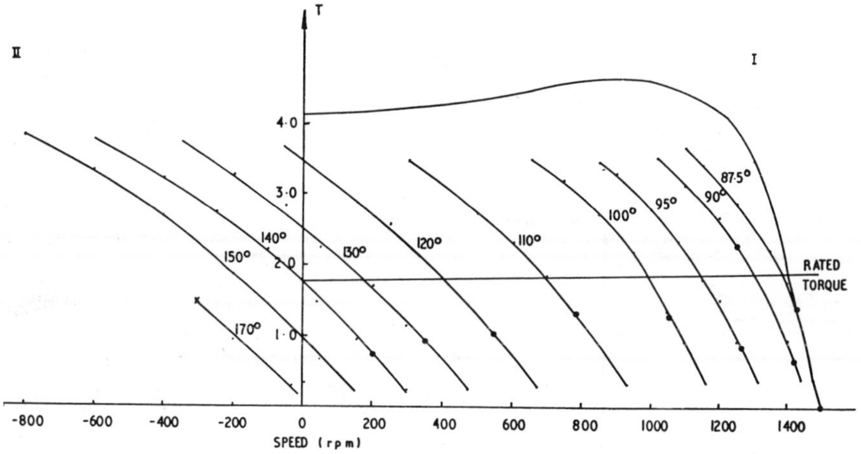

Fig. 4. Measured torque–speed characteristics of three-phase line feedback system with thyristor firing angle as control parameter. 3 hp, 200 volts, 50 Hz.

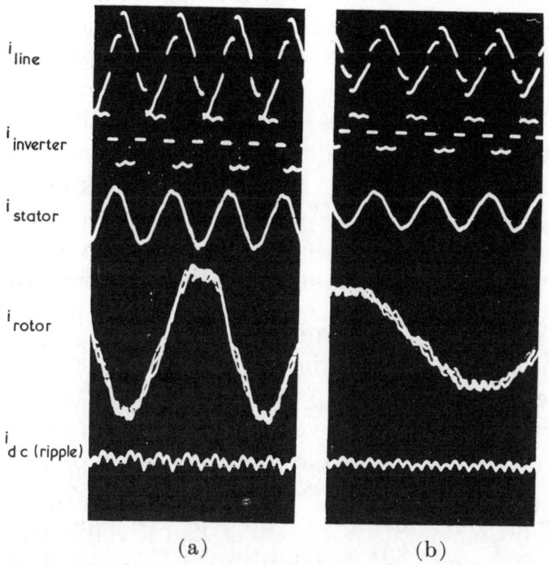

(a) (b)

Fig. 5. Current waveforms of three-phase line feedback system. Firing angle 90°. (a) 900 r/min. (b) 1200 r/min.

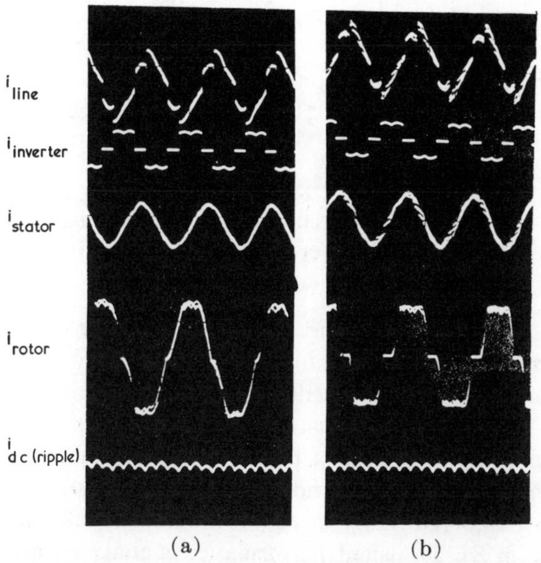

(a) (b)

Fig. 6. Current waveforms of three-phase line feedback system. Speed 400 r/min (s = 0.734). (a) Firing angle 110°. (b) Firing angle 130°.

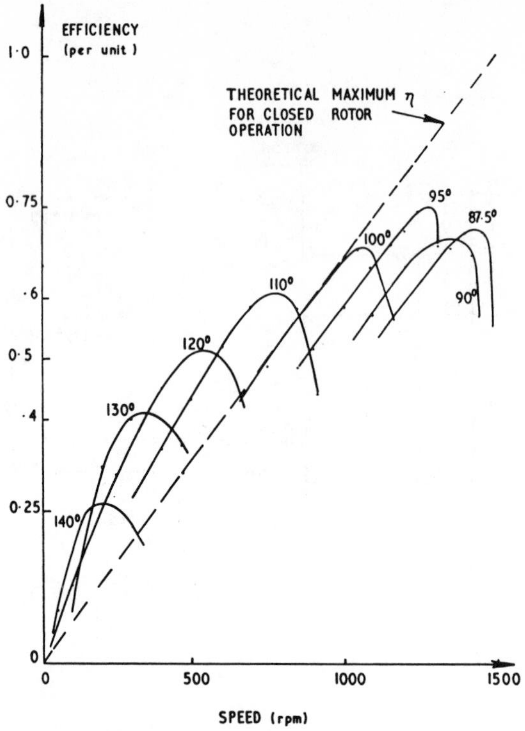

Fig. 7. Measured efficiency–speed characteristics of three-phase line feedback system with thyristor firing angle as control parameter. 3 hp, 200 volts, 50 Hz.

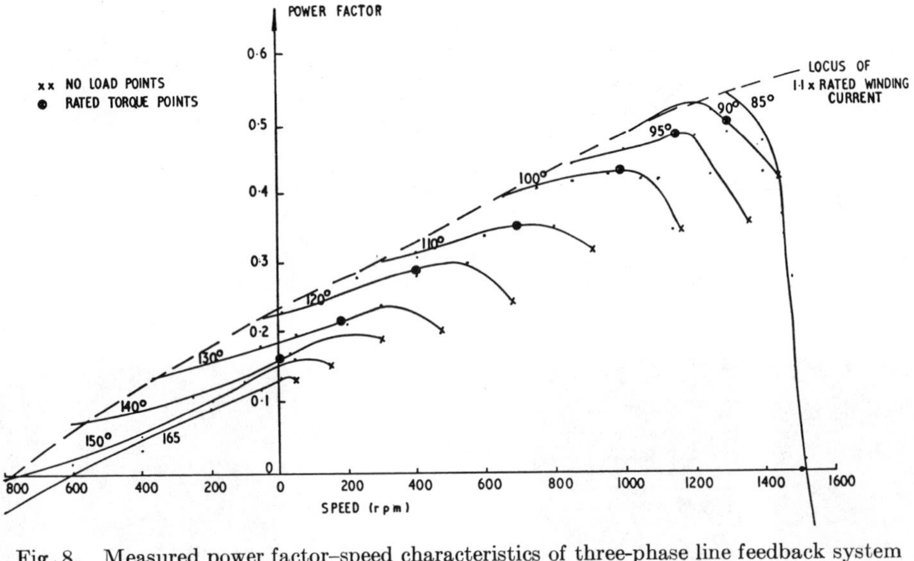

Fig. 8. Measured power factor–speed characteristics of three-phase line feedback system with thyristor firing angle as control parameter. 3 hp, 200 volts, 50 Hz.

sinusoidal control. In Fig. 9 the power factor obtained by an excursion over the forward speed range with rated torque using the line feedback method is contrasted with that obtained by adjustment of the rotor resistances. The poor power factor of the line feedback method offsets, to some extent, the superior torque speed curves of Fig. 3.

An extensive analysis of the line feedback method is given in [8]. Transient performance is analyzed and a discussion of harmonic losses due to nonsinusoidal currents is included.

B. Two-Phase Motor Arrangement

The line feedback principle can be applied to a two-phase induction motor by use of a single-phase thyristor inverter. Two constant voltage electrical supplies in time quadrature are required as indicated in Fig. 10, and these may be obtained from Scott-connected transformers. A 200-volt, 3-hp motor was used at the reduced voltage $V_{A'A} = V_{B'B} = 80V$ to limit the current. The set of measured torque-speed characteristics, Fig. 11, shows that adjustment of the thyristor firing angles produces a range of

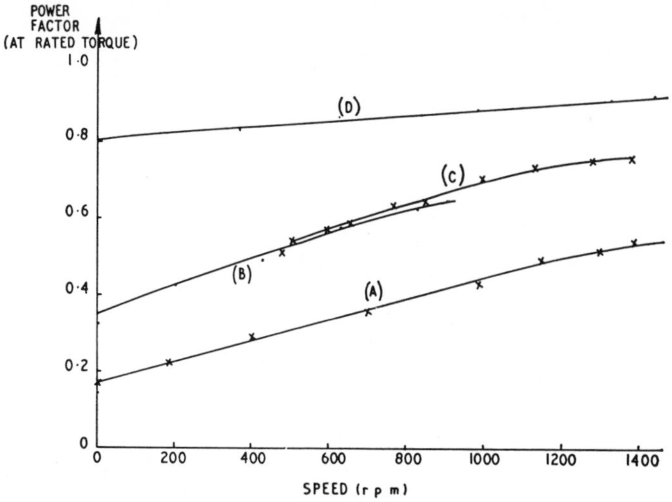

Fig. 9. Power factor (at rated torque)–speed characteristics for 3-hp 50-Hz three-phase induction motor, $V = 200$ volts. (A)—line feedback; (B)—stator feedback, without transformer; (C)—stator feedback, with 1 to 2 transformer between inverter and stator; (D)—rotor resistance control.

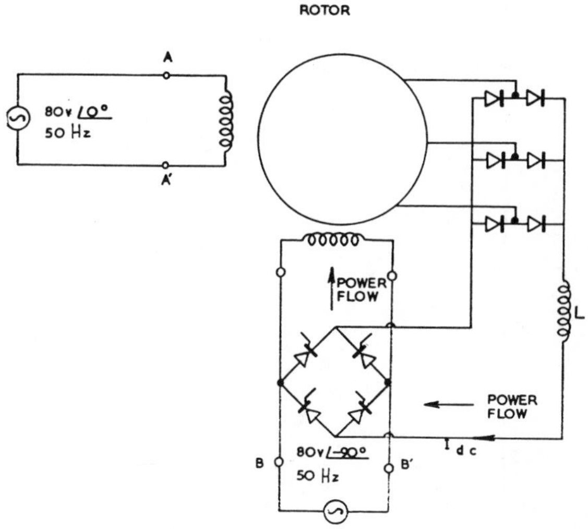

Fig. 10. Schematic arrangement of two-phase line feedback scheme.

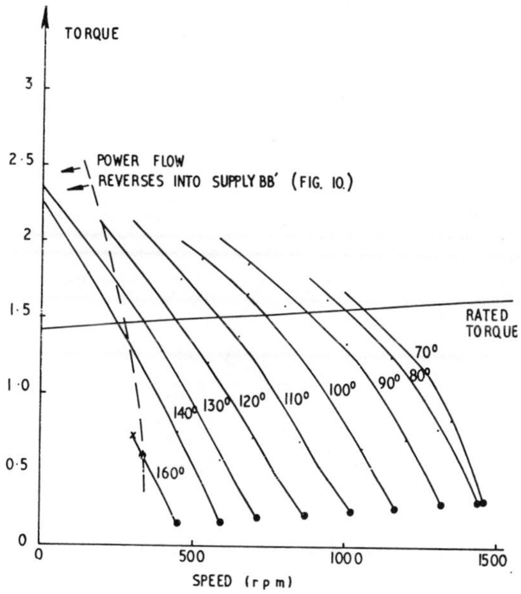

Fig. 11. Measured torque–speed characteristics of two-phase line feedback system with thyristor firing angle as control parameter. 3 hp, $V_{AA'} = V_{BB'} = 80$V, 50 Hz.

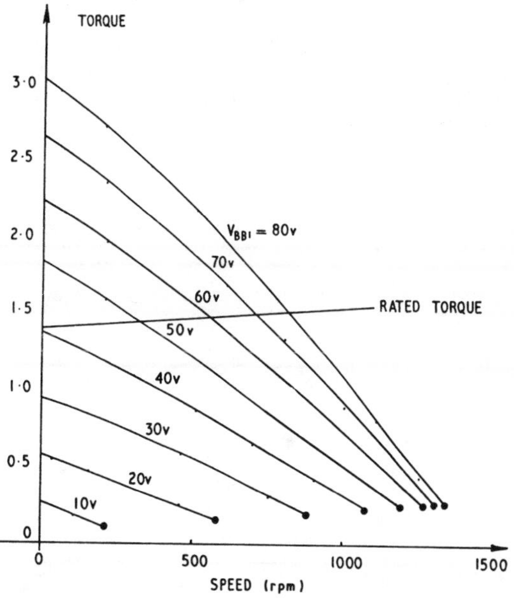

Fig. 12. Measured torque–speed characteristics of two-phase induction motor with sinusoidal control. 3 hp, $V_{AA'} = 80$V, $V_{BB'}$ variable, 50 Hz.

curves similar to that, Fig. 12, obtainable by varying the magnitude of $V_{B'B}$ with, for example, sinusoidal control.

It was not possible to get low-speed control for this particular machine because of the low transformation ratio from stator to rotor voltages. In the low-speed region, the wattmeter in line B indicated a reversal of power flow. The various wattmeters showed that power was entering terminals $A'A$, circulating through the rotor and rectifier–inverter combination, maintaining unidirectional flow in winding BB', but entering supply BB' (rather than leaving it as at higher speeds).

The line feedback arrangement for a two-phase motor is seen, Fig. 13, to once again lead to substantial increase of efficiency over that obtainable with the rotor closed. Unlike the three-phase case the motor currents are unbalanced, except at 0-degree firing angle. Winding AA' operates at a power factor greater than 0.8 lag over the whole range whereas the power factor looking into terminals BB', Fig. 10, does not exceed 0.25 lag at the highest speed attained.

In summary, the two-phase, line feedback scheme, Fig. 10, gives superior torque-speed characteristics at higher efficiency than by closed rotor control. The need for a variable voltage source is eliminated since thyristor firing angle becomes the control parameter. Nonsinusoidal current is drawn from the supply into the feedback winding and the motor currents are usually unbalanced, producing asymmetrical heating. The action of the inverter triggering is to retard the phase of the current so that the power factor of the input to the feedback winding is very poor.

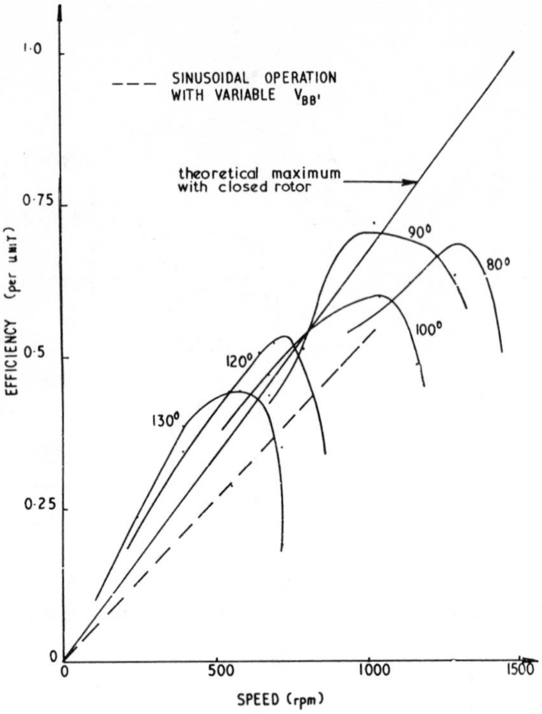

Fig. 13. Measured efficiency–speed characteristics of two-phase line feedback system with thyristor firing angle as control parameter. 3 hp, $V_{AA'} = V_{BB'} = 80$ V, 50 Hz.

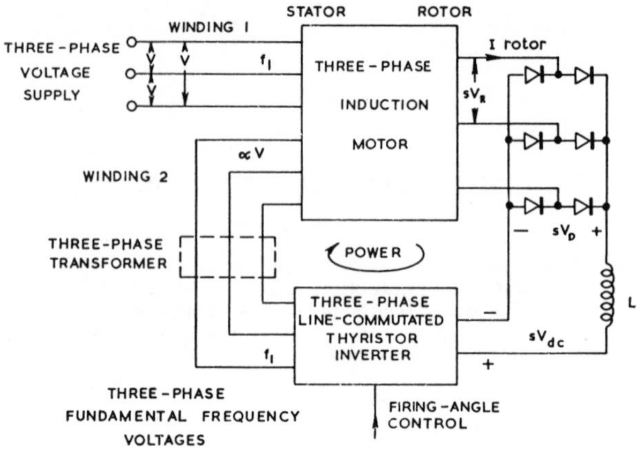

Fig. 14. Schematic arrangement of three-phase stator feedback scheme.

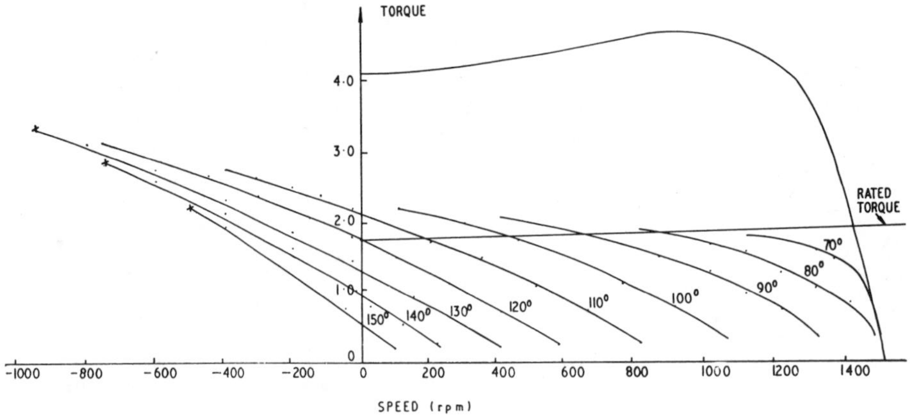

Fig. 15. Measured torque–speed characteristics of three-phase stator feedback system with thyristor firing angle as control parameter, no coupling transformer. 3 hp, 200 volts, 50 Hz.

IV. Performance with Inverted Slip Power Fed Back to Auxiliary Stator Windings

An alternative scheme to the line feedback method is to feed the rectified and inverted power back into auxiliary windings on the stator, as shown in Fig. 14. This arrangement is hereafter known as the "stator feedback" scheme [17], [18]. Magnetic coupling between the two electrically independent three-phase stator windings maintains a fixed open-circuit voltage, proportional to the supply voltage, between the terminals of the auxiliary stator winding. Power is transferred to the supply by transformer action in the stator and some regulation of the auxiliary terminal voltage occurs on load.

The measured torque-speed performance of the 3-hp motor used in Section III-A is given in Fig. 15 for the case of unity turns ratio between the two stator windings. Comparison of performance between Fig. 4 for the line feedback method and Fig. 15 for the stator feedback method, for the same motor operating at the same (reduced) applied voltage, shows that line feedback is superior, giving higher torque with less speed regulation. The torque output of the stator feedback method can be increased and the speed regulation decreased by increasing the induction motor transformation ratio of rotor voltages to stator voltages. Alternatively, the effective motor transformation ratio may be modified by connecting a transformer between the inverter and the auxiliary stator winding, as shown in Fig. 14, or between the rotor slip rings and the

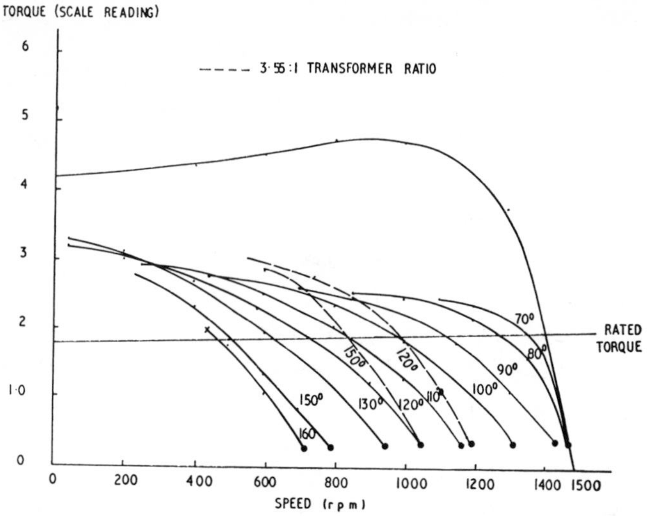

Fig. 16. Measured torque–speed characteristics of three-phase stator feedback system with thyristor firing angle as control parameter. Coupling transformer connected. 3 hp, 200 volts, 50 Hz, 2 to 1 transformation ratio.

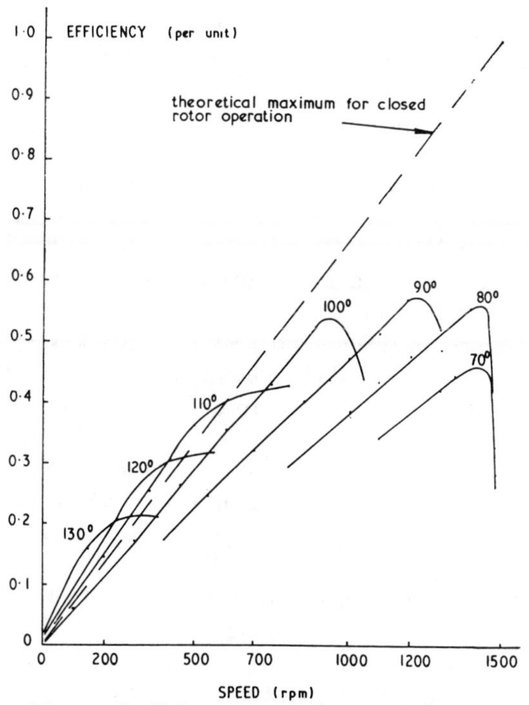

Fig. 17. Measured efficiency–speed characteristics of three-phase stator feedback system with thyristor firing angle as control parameter. No coupling transformer. 3 hp, 200 volts, 50 Hz.

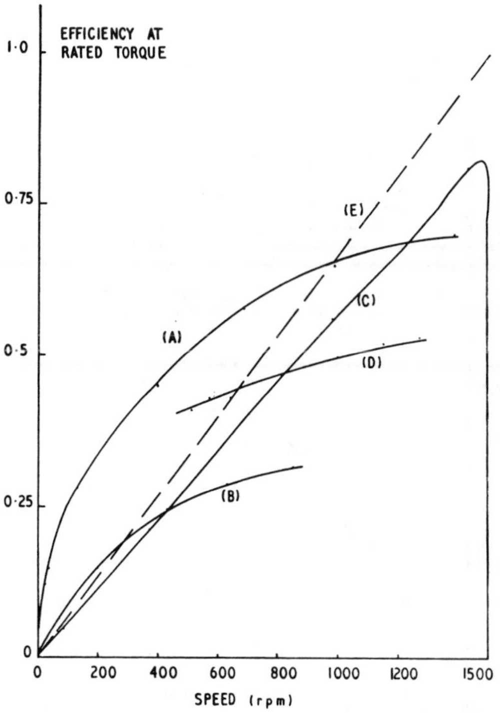

Fig. 18. Measured efficiency (at rated torque)–speed characteristics for three-phase 3-hp induction motor, $V = 200$ volts; (A)—line feedback; (B)—stator feedback; (C)—rotor resistance control; (D)—stator feedback with 1 to 2 transformer between inverter and stator; (E)—theoretical maximum with closed rotor.

full-wave diode bridge rectifier. The transformer should be connected, in Fig. 14, with its high-voltage side coupled to the stator so that the voltage "seen" by the inverter is reduced below the supply voltage. Measured torque-speed curves for stator feedback with different stepup transformer ratios are given in Fig. 16. As the transformer turns ratio is increased the range of speed control attainable is decreased. When the transformer connection that produced the curves of Fig. 16 is reversed to give voltage stepdown between the inverter and the stator, a set of torque-speed curves is obtained that is restricted to the low- and reverse-speed regions.

Efficiency versus performance corresponding to the torque-speed curves of Fig. 15 is given in Fig. 17. The increase of efficiency over closed-rotor performance is less than for corresponding control with the line feedback arrangement. A comparison of the efficiency at rated torque for different control methods over the forward speed range is given in Fig. 18. Stator feedback with a 1 to 1 stator transformation ratio is seen to give very little advantage

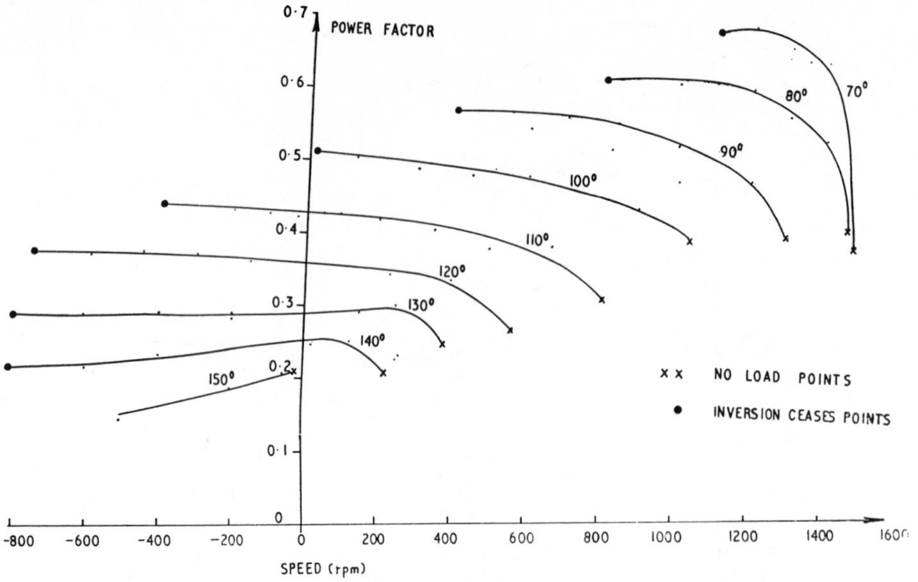

Fig. 19. Measured power factor–speed characteristics for three-phase stator feedback system with thyristor firing angle as control parameter. 3 hp, 200 volts, 50 Hz.

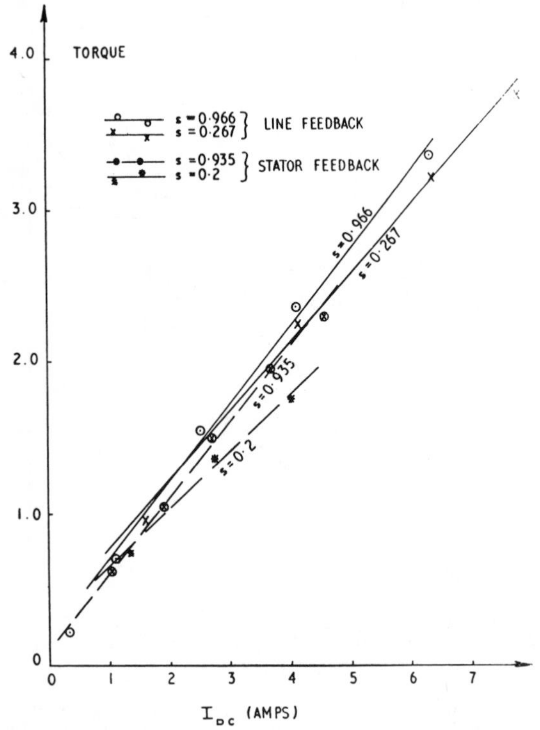

Fig. 20. Measured torque–rectified current for three-phase slip power recovery schemes. 3 hp, 200 volts, 50 Hz.

over conventional rotor resistance control. Considerable improvement of efficiency results from the use of stator feedback with a transformer, but the line feedback arrangement is clearly the best alternative.

Power factor versus speed for the stator feedback system is given in Fig. 19 and is seen to be markedly superior to the corresponding curves (Fig. 8) for line feedback control. Comparison of the power factor versus speed performance with rated torque, given in Fig. 9 for the different control methods, shows that stator feedback gives a power factor about halfway between line feedback and rotor resistance control.

Current waveforms for the stator feedback method are similar to those of Figs. 5 and 6 for line feedback, except that the distortion of the supply current is a little less severe due to the inductive filtering of the stator windings.

In the steady state, the torque developed at fixed speed is proportional to the direct current into the inverter and therefore roughly proportional to the rotor current for both the line feedback and stator feedback systems as shown by the measured characteristics of Fig. 20. In this respect and in the form of the torque-speed curves, the systems have the nature of a separately excited dc motor, and contrast with customary induction motor performance in which torque is proportional to the square of rotor current.

V. Conclusions

The inherent inefficiency of low-speed induction motor operation with the rotor closed can be overcome by rectifying the slip-frequency power, inverting this to line frequency, and injecting it back into the supply directly (line feedback) or through auxiliary stator windings (stator feedback). Motor torque-speed curves then have the nature of a variable-speed drive or a separately excited dc motor, with much better speed regulation than is attainable by stator voltage control or rotor resistance control. Line feedback control results in much improved efficiency at low speeds but causes operation at low-power factor and draws nonsinusoidal current from the supply. The power factor and supply current distortion are less degraded by the use of stator feedback, but the improvement of low speed efficiency is markedly less.

Line feedback with a two-phase induction motor eliminates the need for a variable control voltage source since thyristor firing-angle is the controlling variable.

References

[1] "Improvements in or relating to control arrangements for slip ring induction motors," General Electric Co., Ltd., British Patent Specification 868 763, May 1961.

[2] W. von Brutsche, "Drehstrom-Gleichstrom-Kaskaden—eine wirtschaftliche Lösung der Drehzahlverstellung von Asynchronmotoren," *Siemens Z.*, pp. 710–714, October 1962.

[3] R. J. Bland, N. N. Hancock, and R. W. Whitehead, "Considerations concerning a modified Kramer system," *Proc. IEE* (London), vol. 110, pp. 2228–2232, December 1963.

[4] I. M. Macauley, W. Wood, J. S. Michael, and H. Waxeley, "High efficiency, variable-speed alternating current motor equipments," *AEI Engrg.*, vol. 4, pp. 237–242, September/October 1964.

[5] "Improvements in and relating to variable-speed alternating current motor equipments," British Thompson-Houston Co., Ltd., British Patent Specification 842 624, July 1960.

[6] M. Teissie-Solier and C. Curie, "A control and automatic speed regulation system for an asynchronous electric induction motor," British Patent Specification 920 259, March 1963.

[7] C. Curie, "The thyrasyntrol—an electronic device for a variable speed induction motor," *L'Electricien*, pp. 295–302, December 1964.

[8] A. Lavi and R. J. Polge, "Induction motor speed control with static inverter in the rotor," *IEEE Trans. Power Apparatus and Systems*, vol. PAS-85, pp. 76–84, January 1966.

[9] W. H. Lee, "Induction motor speed control," British Patent Specification 841 342, July 1960.

[10] H. von Gallistl, "Ein Gittersteuersatz für vielseitige Anwendung," *BBC Nach.*, pp. 675–681, November/December 1961.

[11] G. Hausen, "Compensated dynamic braking of three-phase wound-rotor induction motors," *Trans. Engrg. Inst. Canada*, Paper 31, 1962.

[12] E. Golde, "Asynchronmotor mit elektronischer Schlupfregelung," *AEG Mitt.*, no. 11–12, pp. 666–671, 1964.

[13] M. S. Erlicki, "Inverter rotor drive of an induction motor," *IEEE Trans. Power Apparatus and Systems*, vol. PAS-84, pp. 1011–1016, November 1965.

[14] A. E. Fitzgerald and C. Kingsley, *Electric Machinery*, 1st ed. New York: McGraw-Hill, 1952, ch. 9.

[15] J. Schaefer, *Rectifier Circuits*. New York: Wiley, 1964.

[16] B. D. Bedford and R. G. Hoft, *Principles of Inverter Circuits*. New York: Wiley, 1965, sec. 3.4.

[17] W. Shepherd and J. Stanway, "Improvements in or relating to electric motors," U.K. Patent 39865/64, 1964.

[18] N. O. Kenyon, D. C. Reay, and M. J. C. Waggett, "Inverter control of an induction motor," 18th Tech. Staff Course (EI Group), Royal Military College of Science, Shrivenham, England, Design Exercise Rept., April 1966.

COMMUTATORLESS KRAEMER CONTROL SYSTEM FOR LARGE-CAPACITY INDUCTION MOTORS FOR DRIVING WATER SERVICE PUMPS

TAKAAKI WAKABAYASHI — Tokyo Metropolitan Water Works Bureau, Tokyo, Japan

TAKAMASA HORI — Hitachi Research Laboratory, Hitachi Ltd. Ibaraki-Ken, Japan

KOUSAKU SHIMIZU — Hitachi Works, Hitachi Ltd. Ibaraki-Ken, Japan

TAKAYUKI YOSHIOKA — Ohmika Works, Hitachi Ltd. Ibaraki-Ken, Japan

ABSTRACT

As a means of speed control of large-capacity induction motors, a commutatorless Kraemer control system that supplies secondary slip power of the induction motor to an inverter type commutatorless motor directly coupled mechanically to the induction motor was developed and applied commercially for driving large-capacity pumps.

Since this speed control system is of the Kraemer structure, very little reactive current and higher harmonic current flow into the power source; moreover, even when power source voltage has dropped for a short period of time (0 to 0.6 seconds) in a supply voltage depression, operation can be continued without trouble. Thus, this speed control system offers high reliability.

This paper describes the speed control method by the commutatorless Kraemer control system, as well as the motor characteristics and the operating characteristics at supply voltage depression.

1. INTRODUCTION

Kraemer control and Scherbius control in which a static frequency converter is used on the secondary side of a wound-rotor induction motor offer high efficiency and good controllability, as well as a wide speed control range, so that they have been employed extensively in speed control of large-capacity pumps.[1)-3)] Table 1 shows the development of speed control systems for induction motors using such control systems and also the technical targets and social background that induced the development.

In recent years, water service pump facilities have been built in larger capacities and with increasing automatic control features, to meet the growing demand for water which has necessitated higher efficiency and labor-saving. Also, to ensure smooth regulation of flow rate and pressure, high-performance speed control systems are being introduced. Such a speed control system must satisfy the following requirements:

(a) High reliability and safety
Specifically, back-up operation must be possible in the event of failure of the control system; operation must not be interrupted by a supply voltage depression due to lightning surge or other causes. In this way, stoppage of water supply and generation of turbid water must be prevented, so that steady water supply and better service can be assured.

(b) Good maintenability
Ease of maintenance must be secured so that labor can be saved.

(c) Little disturbance of power source
Higher harmonic current and reactive current must be prevented from flowing into the power source as far as possible.

Recently, mechanical commutators, which are a weak point in the maintenance of DC motors, have been replaced by thyristors and other contactless switching devices in commutatorless motors, and such commutatorless motors have come to be used in general industrial fields, offering higher reliability and better maintenability than conventional DC motors.

Taking advantage of the features of the commutatorless motor, we developed a commutatorless Kraemer control system (hereinafter referred to as "CL-Kraemer control system") in which the DC motor in the conventional Kraemer control system is replaced by a commutatorless motor. The new control system satisfies all the requirements listed above. Two units of 1,900kW capacity and three of 1,450kW are now in operation with excellent results at the Misono Purification Plant of the Tokyo Metropolitan Waterworks Bureau.

The operation and characteristics of the CL-Kraemer control system developed and commercialized by the authors will be described below.

2. SPEED CONTROL BY CL-KRAEMER CONTROL SYSTEM

2.1 Circuit configuration

Fig. 1 shows a skeleton diagram of the main circuit and control circuit of the CL-Kraemer control system. A thyristor frequency converter and commutatorless motor (CLM) are electrically connected to the secondary circuit of the induction motor (IM). Moreover the induction motor and commutatorless motor are directly coupled mechanically, to utilize the secondary slip power of the induction motor as the driving shaft power. The thyristor frequency converter consists of a diode-rectifier circuit (REC) which rectifies the secondary voltage of the induction motor, a current smoothing reactor (DCL), and a thyristor inverter (INV). The inverter frequency is controlled to synchronize with the signal frequency from the distributor (DIST) mounted on the driving shaft. The liquid resistor (ST.R) in Fig. 1 is a starting rheostat to accelerate the motor to the speed control range. Resistor EX.R is for recovery from a speed drop at supply voltage depression and for suppression of surge voltage at recovery from supply voltage depression. SH is a secondary short circuiter for back-up operation with the induction motor alone in the event of failure of the inverter or control system of the commutatorless motor.

Reprinted from *IEEE/IAS 1976 Annual Meeting*, pp. 822–828, 1976.

Since the higher harmonic current and reactive current generated by the inverter flow into the commutatorless motor, such currents flowing into the power source in this control system are less than in a thyristor Scherbius control system. This obviates the need for a filter for absorbing the higher harmonic current and saves the capacitors for reactive current compensation.

In the following description, the components will be referred to by the abbreviations used in Fig. 1.

2.2 Speed control method

(1) Speed control range

At low speed, INV commutation is impossible because the electromotive force of the CLM is small. Since the IM secondary voltage is proportional to slip, the voltage at low speed is greater than the CLM electromotive force (no-load voltage), and voltage equilibrium between the REC and INV cannot be achieved. In the CL-Kraemer control system, therefore, a starting rheostat is used to accelerate the motor up to a speed where DC voltage can be equilibrated, then the starting rheostat is cut off and the CLM is connected to the IM secondary circuit, to start Kraemer operation.

Fig. 2 shows the speed control range of the CL-Kraemer control system. The ideal speed control range is from the speed at the intersection of the curve for IM secondary voltage E_{2s} and that for electromotive force E_{t0} at no-load voltage of CLM to the speed in the neighborhood of IM synchronous speed ($s \simeq 0$). When one considers the voltage drop due to armature reaction with a load on the CLM and INV commutation, the minimum controllable speed would be a speed slightly higher ($s = s_m$) than where E_{t0} and E_{2s} become equal, and the maximum controllable speed would be a speed slightly lower than that an IM secondary short.

(2) Speed control method

Since the REC is connected to the IM secondary circuit in a three-phase bridge connection, REC DC voltage E_{di} will be given as below when the IM and REC voltage drop is denoted by $R_{im}I_d$:

$$E_{di} = (3\sqrt{2}/\pi)\, s E_2 - R_{im}I_d \qquad (1)$$

Since the INV is also three-phase bridge connected, INV DC voltage E_{ds} will be given as follows, in which the CLM internal voltage is denoted by E_{ts}, INV control angle advance by γ , and CLM voltage drop by $R_{sm}I_d$:

$$E_{ds} = (3\sqrt{2}/\pi)\, E_{ts} \cos\gamma + R_{sm}I_d \qquad (2)$$

Considering the voltage drop R_dI_d in a DC circuit, from Eqs. (1) and (2) we get :

$$s E_2 = E_{ts} \cos\gamma + R_mI_d \qquad (3)$$

where

$$R_m = (\pi/3\sqrt{2})\, (R_{im} + R_{sm} + R_d) \qquad (4)$$

Since E_{ts} is proportional to the product of rotating speed N and field flux Φ , we get :

$$E_{ts} = K\Phi N \qquad (5)$$

And slip s will be given as follows where the IM synchronous speed is denoted by N_0 :

$$s = (N_0 - N) / N_0 \qquad (6)$$

From Eqs. (3) through (6), we get :

$$N = (E_2 - R_mI_d) N_0 / (E_2 + K\Phi N_0 \cos\gamma) \qquad (7)$$

In Eq. (7), E_2, K and N_0 are constant, so that it will be seen that rotating speed N can be controlled by regulating Φ or γ . Since Φ is a function of the field current of the CLM, speed control by field current would involve the following inconveniences :

(a) At low field current, armature reaction will increase.

(b) For operation at $s \simeq 0$, field current must be brought close to zero, so that INV commutation voltage will be lacking, with the result that commutation failure will occur.

(c) As field time constant is large, quick-response control is difficult.

Therefore, the authors used a speed control method in which principally the control angle is regulated. Nevertheless, field current control is necessary for limiting the CLM terminal voltage and reducing the withstand voltage of thyristors that make up the INV, rather than for speed control.

(3) Configuration of control circuit

The gate pulse generator (GPG) in Fig. 1 operates by receiving synchronizing signals from the DIST mounted on the CLM driving shaft. Even though the synchronizing signal frequency may vary within the speed control range, control angle γ is continuously controlled by control voltage.

The speed control instruction and speed feedback signal enter the automatic speed regulator (ASR) and the ASR output serves as the instruction for the automatic current regulator (ACR). As the current regulation instruction and current feedback signal enter the ACR, the ACR generates the control voltage for the GPG. When acceleration is required, the regulator operates so as to decrease the INV DC voltage E_{ds} (increase γ) and increase current.

3. COMMERCIALIZATION OF CL-KRAEMER CONTROL SYSTEM

3.1 Specifications for CL-Kraemer control system

Table 2 shows the specifications for the CL-Kraemer control system commercialized by the authors. This system is used to drive the pumps installed at the Misono Purification Plant of the Tokyo Metropolitan Waterworks Bureau. The plant purifies the raw water introduced from the River Arakawa and the purified water is pumped
to the Nerima Pumping Plant about 4km away and
to the Itabashi Pumping Plant about 11km away.

3.2 Characteristics and construction of motor

Fig. 3 shows the construction of the motor in the 1,900kW CL-Kraemer control system. To simplify the motor construction, a common yoke, common shaft structure was adopted. For the 1,900kW and 1,450kW motors, the same outside diameter and slot shape and dimensions were used for the sake of standardization of fabrication. Also, consideration was given to the IM secondary voltage so that the same kind of CLM could be used.

In order that the sliprings for both the IM and CLM may be inspected and serviced in one operation, all sliprings are provided at the top. The distributor is also provided at the top for ease of inspection, and consideration was given to end play.

Fig. 4 is a photo of a CL-Kraemer motor as installed at the Misono Purification Plant, Tokyo Metropolitan Waterworks Bureau. Five CL-Kraemer sets are arranged.

3.3 Resonance with natural frequency of mechanical system

Since CLM is used in the CL-Kraemer control system, there is a torque ripple (with a waveform similar to INV DC voltage E_{ds}) at intervals of 60 degrees electrical angle, as shown in Fig. 5. This torque ripple is maximum at 100% speed and acts as a torsional torque on the driving shaft. Hence torsional vibration comes into question.

The transmission ripple torque that drives the pump through the driving shaft is not the electrical ripple torque generated by the CLM; its magnitude varies with the moment of inertia of the motor and pump, and with the spring stiffness coefficient and damping coefficient of the shaft. The most serious problem here occurs when the ripple frequency of the electrical torque approaches the torsional natural frequency of the driving shaft, to cause resonance, which can lead to local damage of the machine.

In this system, however, as shown in Table 3, the torsional resonance frequency is removed from these frequencies, so that there is no possibility of resonance.

3.4 Maximum current at acceleration and field current control

The INV has an allowable maximum current at which no commutation failure occurs, and this maximum current is a function of field current, speed, and control angle.

Fig. 6 shows the allowable maximum voltage and current calculated under the following conditions in consideration of thyristor withstand voltage and INV commutation respectively: that the CLM rated maximum terminal voltage will be held below 750V and the no-load voltage (voltage applied to INV at zero main circuit current) below 1,100V.

At steady-state operation, main circuit current is equal to pump load current; but at acceleration, the ACR instruction value is limited, to hold the acceleration current within the allowable maximum current, because the main circuit current can exceed the allowable maximum current.

4. OPERATING CHARACTERISTICS OF CL-KRAEMER CONTROL SYSTEM

4.1 Operating characteristics

Fig. 7 shows the input-output characteristics and total efficiency of the CL-Kraemer control system. The calculated and measured values are in good agreement, indicating that the desired characteristics have been achieved. The total efficiency is just about as high as that of a thyristor Scherbius control system.

Fig. 8 shows the speed-load characteristics at varied loads under condition of a constant speed instruction. A constant speed is maintained by means of the automatic speed regulator and ideal shunt characteristics are obtained. Smooth operation is guaranteed at various speed instructions and loads.

4.2 Voltage and current waveforms

Fig. 9 shows voltage and current waveform of the CLM at a steady-state operation. The current has a rectangular waveform with a 120 degree period peculiar to inverters. The voltage waveform has a dip synchronized with inverter commutation.

5. OPERATING CHARACTERISTICS AT SUPPLY VOLTAGE DEPRESSION

An outstanding feature of this CL-Kraemer control system is that it operates normally even at supply voltage depression.

5.1 Speed drop at supply voltage depression and its effect

When, during pump operation, pump speed falls suddenly as driving force is lost due to a supply voltage depression, a resultant pressure drop in the pipeline will cause a water hammer, which can lead to damage of the pipeline. Therefore, water hammer must be studied in advance and, if necessary, a surge tank must be provided or other measures taken. For the pipeline system to which this control system was applied, it was required that the speed drop at supply voltage depression should not exceed 20% at 100% speed operation and that at minimum speed operation the speed should not drop to under 60% at supply voltage depression.

In the CL-Kraemer control system, the IM and CLM are mechanically coupled, so that there is a great inertia, and speed drop at supply voltage depression is minimal. Usually, a low inertia force of the driving motor is an advantage; but in this control system, a high inertia force is a benefit.

According to past field data, supply voltage depression normally lasts from 0.1 to 0.5 seconds. In view of this, this control system is so designed that operation can be continued uninterrupted even during supply voltage depression of up to 0.6 seconds.

In the 1,900kW machine, for instance, a supply voltage depression lasting 0.6 seconds will reduce speed by 13% at 100% operation, and by 6% at 70% operation. These speed drops easily meet the conditions for preventing water hammer.

5.2 Operation of main circuit and control circuit at supply voltage depression

Measures for supply voltage depression can be taken more readily in the CL-Kraemer control system than in a thyristor-Scherbius control system for the following reasons: In the latter system, the inverter is connected to the AC power source so that inverter commutation voltage will fall along with a decrease in power supply voltage and this makes a commutation failure of inverter liable; in the CL-Kraemer control system, on the other hand, even if the power supply voltage of the CLM field circuit falls, the large field time constant (of second order) enables field current to continue flowing, with the result that electromotive force of the CLM is generated and inverter commutation is carried out in a stable manner.

In order to ensure continued operation during supply voltage depression, the circuit must first detect the power source failure, then (a) apply to the ACR a control signal for decreasing the main circuit current, and (b) connect resistor EX.R to the IM secondary circuit and at the same time open the circuit breaker between the IM and REC. In this way the IM secondary transient voltage at recovery from supply voltage depression is suppressed and the voltage is prevented from being applied to the REC. After the secondary transient voltage has decreased to a certain extent and the IM has been accelerated by the EX.R, it is confirmed that the speed is within the speed control range, then the REC is connected to the IM and at the same time the EX.R is cut off and normal operation is resumed. By the above procedure, stable operation can be continued regardless of an instantaneous power source failure.

If the control power source voltage falls during a supply voltage depression, the control circuit would of course be unable to perform stably. Therefore, the control power source is supplied by a battery-fed inverter power source.

The series of operation at supply voltage depression described above is illustrated in Fig. 10. It will be seen that normal operation is maintained with no overcurrent flowing in the main circuit, even at the start of and recovery from a supply voltage depression.

6. CONCLUSIONS

The operation and characteristics of a CL-Kraemer control system have been described in the foregoing. Major features of this control system are that stable operation can be continued even at supply voltage depression and that very little higher harmonic current and reactive current are generated. A bright future is promised for this system as a speed control system for large-capacity induction motors for driving water service pumps.

The authors wish to express their deep gratitude to Mr. H. Shimazaki, Chief of the Electrical Machinery Designing Section of the Tokyo Metropolitan Waterworks Bureau and other persons for the valuable guidance and cooperation they provided in developing and commercializing this control system.

7. REFERENCES

1) T. Hori, Y. Hiro, "The Characteristics of an Induction Motor Controlled by a Scherbius System (Application to Pump Drive)", IEEE IAS 7th Annual Meeting Rec., pp.775-782, Oct. 1972.
2) J. Noda, Y. Hiro, T. Hori, "Brushless Scherbius Control of Induction Motors", IEEE IAS 9th Annual Meeting Rec., pp.111-118, Oct. 1974.
3) K. Honda, Y. Aso, Y. Hiro, "6,200kW Raw Water Transfer Pumping Equipment of the Asaka Purification Plant, Tokyo Metropolitan Water Works Bureau", Hitachi Review 15 (2) pp.60-70, Feb. 1966.

Table 1. Development of speed control system for induction motors for large-capacity water service pump drive

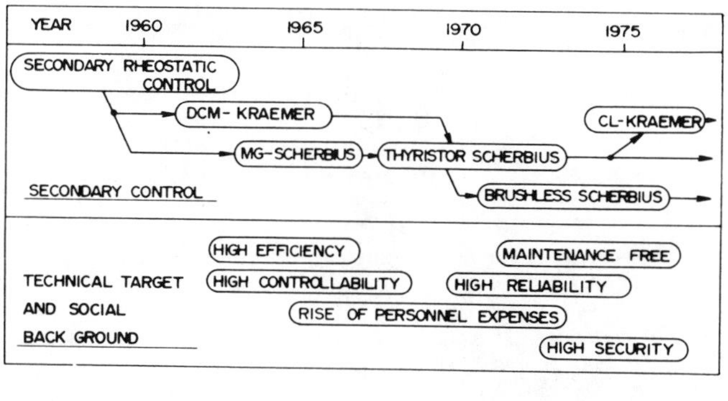

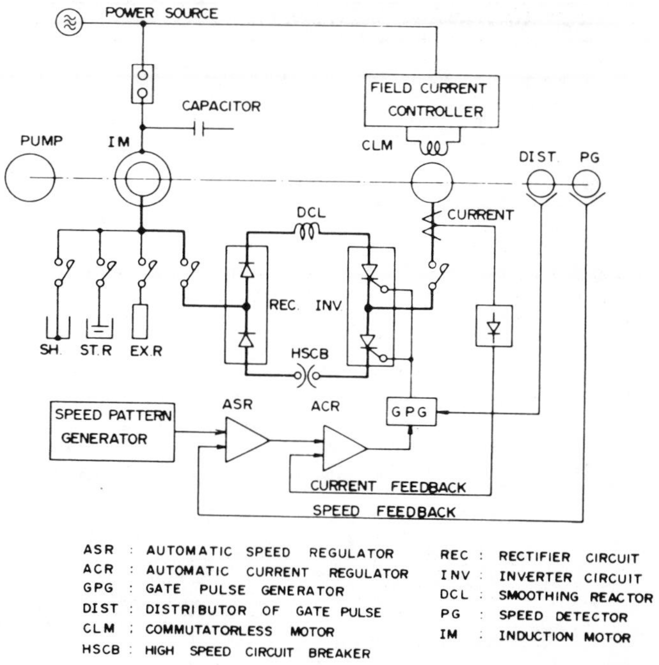

Fig. 1 Basic construction of CL-Kraemer control system.

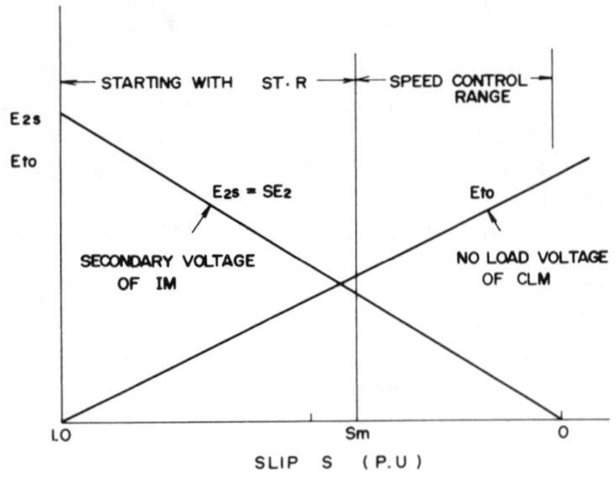

Fig. 2 Speed control range of CL-Kraemer system.

Table 2. Specifications for CL-Kraemer control system

ITEM \ FEED PUMP TO~		NERIMA PUMPING PLANT	ITABASHI PUMPING PLANT
PUMP	HEAD	60 m	50 m
	FLOW	146 m^3/min	132 m^3/min
	TYPE	VERTICAL SINGLE SUCTION CENTRIFUGAL PUMP	
NO. OF PUMPS		2	3
OUTPUT kW		1900 kW (AT 404 rpm)	1450 kW (AT 400 rpm)
VOLTAGE OF IM		3000V, 50HZ, 14POLES	
SPEC. OF CLM		RATED MAX. TERMINAL VOLTAGE 750 V RATED MAX. ARMATURE CURRENT 1115 A 14 POLES	
SPEED CONTROL RANGE		404 ~ 283 rpm (100 ~ 70 %)	400 ~ 260 rpm (100 ~ 65 %)

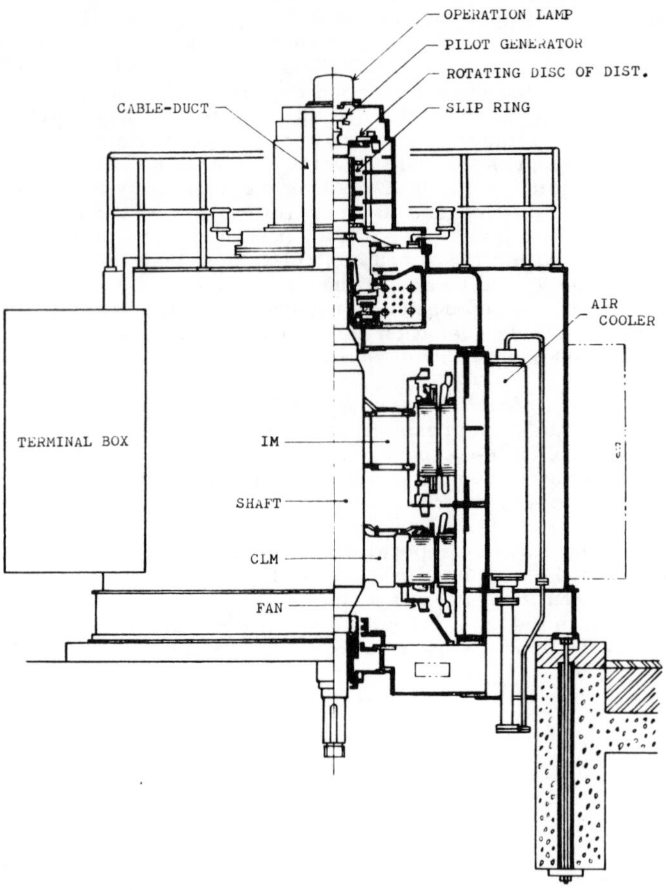

Fig. 3 Construction of 1,900kW CL-Kraemer motor.

Fig. 4 Photo of CL-Kraemer motor for Misono Pumping Plant, Tokyo Metropolitan Waterworks Bureau.

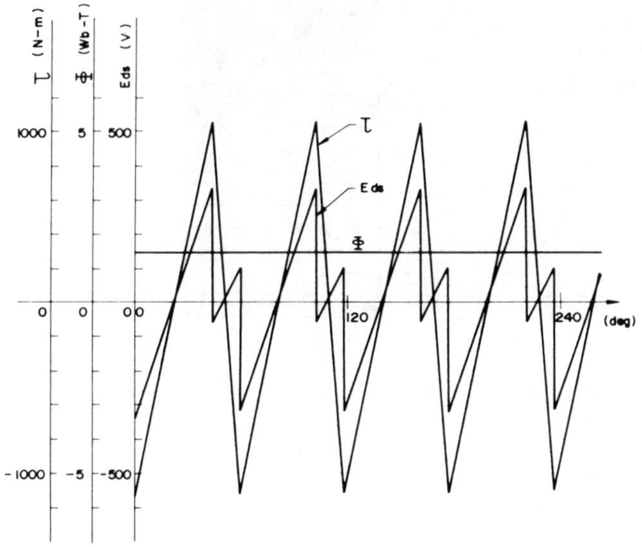

Fig. 5 Torque (τ), DC voltage (E_{ds}), field flux (Φ) of CLM at 100% load and speed (1,900 kW Machine).

Table 3. Torsional natural frequency of 1,900kW CL-Kraemer control system

TORSIONAL RESONANCE FREQUENCY	FIRST	SECOND	THIRD
	15.5 HZ	84.2 HZ	166 HZ
VOLTAGE FREQUENCY (f_M) OF CLM IN CONTROL RANGE	47.1 ~ 33 HZ (100 ~ 70% SPEED)		
TORQUE RIPPLE FREQUENCY ($6f_M$) OF CLM	283 ~ 198 HZ		
ROTATING SPEED FREQUENCY	6.73 ~ 4.72 HZ (404 ~ 283 rpm)		

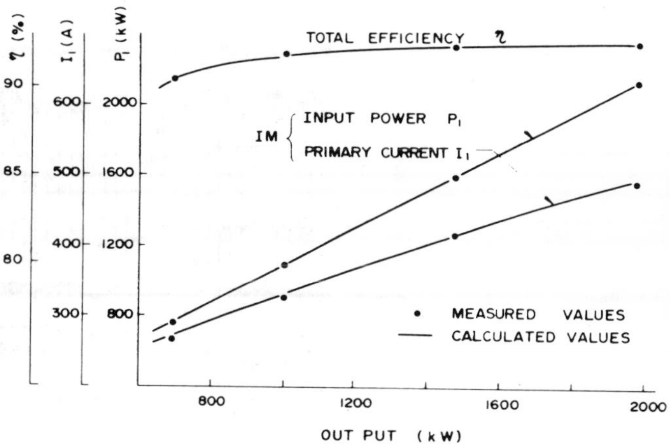

Fig. 7 Output-input, current characteristics and total efficiency (1,900kW Machine).

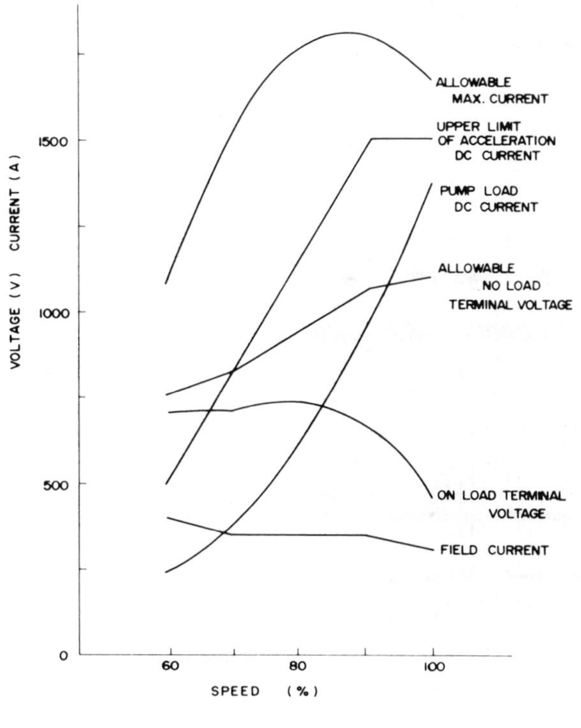

Fig. 6 Allowable maximum current, voltage and field current pattern of CLM (1,900kW Machine).

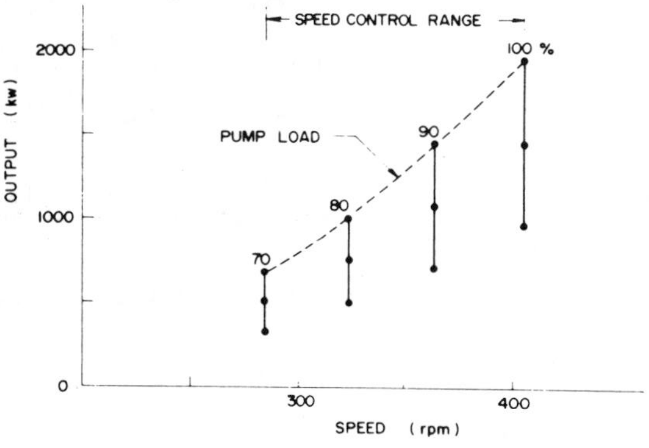

Fig. 8 Speed-load characteristics at constant speed instructions (1,900kW Machine).

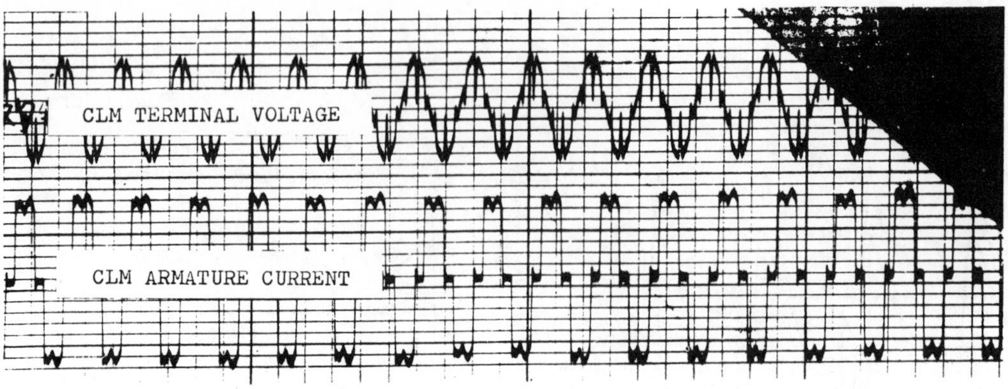

Fig. 9 Voltage and current waveform of CLM.

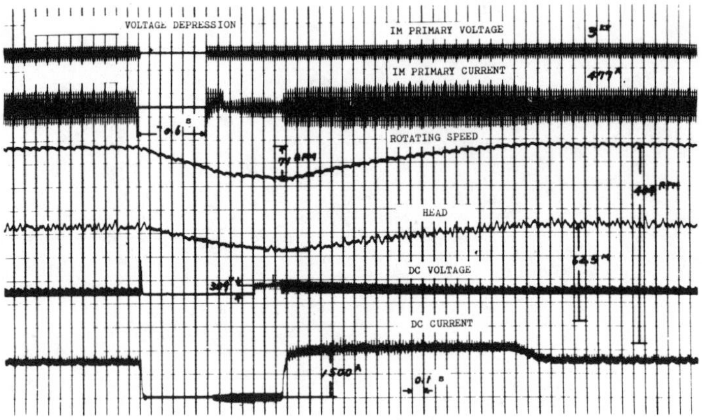

(a) at 100% speed

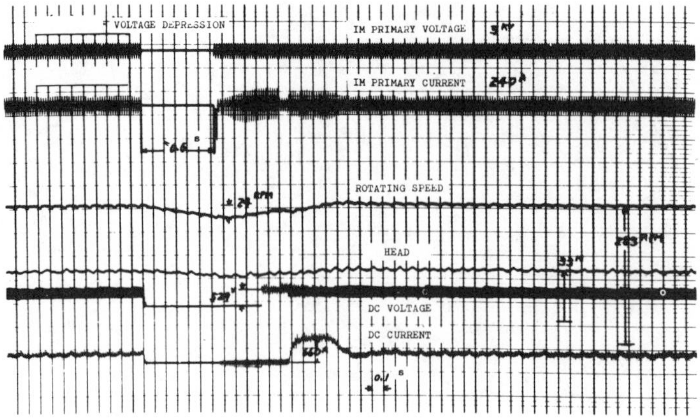

(b) at 70% speed

Fig. 10 Operation at supply voltage depression.

Rotor Chopper Control for Induction Motor Drive: TRC Strategy

PARESH C. SEN, SENIOR MEMBER, IEEE, AND K. H. J. MA

Abstract—**A solid-state induction motor speed regulating scheme using a rotor chopper-controlled external resistor is investigated. This control scheme provides continuous and contactless variation of rotor resistance and thereby eliminates the undesirable features of the conventional rotor resistance control method. A thorough analysis of the steady-state performance of the system is presented. Both dc and ac circuit models are derived to describe the performance characteristics when the chopper operates under the time-ratio control (TRC) strategy. Effects of machine parameters on performance characteristics are studied. Theoretical results from the model are verified by comparison with experimental results. The torque-speed characteristic of this speed control system is essentially linear for a particular time ratio.**

INTRODUCTION

THE SIMPLEST speed control scheme for wound-rotor induction motors is achieved by changing the rotor resistance. It has been established that this rotor resistance control method can provide high starting torque with low starting current and variation of speed over a wide range below the synchronous speed of the motor. Moreover, the power factor is generally improved. Thus, it is extensively used where a high starting current may cause serious line disturbances and when the simplicity of operation is desired. It is also used when the load is of an intermittent nature, requiring high starting torque and relatively rapid acceleration and retardation, such as foundry or steel mill hoists and cranes. Conventionally, the rotor resistance is altered manually and in discrete steps. This mechanical operation is undesirable because the time response is slow and speed variation is not smooth. With the recent progress in power semiconductor technology, these undesirable features of the conventional rheostatic control scheme can be eliminated by using a three-phase rectifier bridge and a chopper-controlled external resistance as shown in Fig. 1.

A chopper is a power switch electronically monitored by a control module. When the chopper is in the ON mode all the time, the fictitious resistance R^* connected to X and Y is zero. When the chopper is in the OFF mode all the time, R^* will be equal to the external resistance R_{ex}. If the chopper is periodically regulated so that, in each chopper period, it is ON for some time but is OFF for the rest, it is possible to obtain variation of R^* between zero and R_{ex}. Thus, the chopper electronically alters the external resistance R_{ex} in a continuous and contactless manner. Also, the rectified current builds up during the ON time interval but decays during the OFF time interval.

In the time-ratio control (TRC) strategy, the period of chopper T_{ch} (consisting of time duration of ON and OFF together) is kept constant but the duty cycle or time ratio α, defined by

$$\alpha = t_{on}/T_{ch}$$

where t_{on} is the time interval of the ON mode, is varied. A typical waveform of the rectified current for the TRC is shown in Fig. 2.

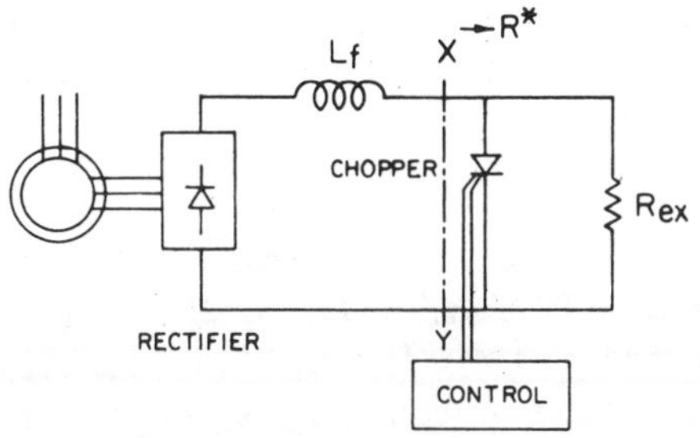

Fig. 1. Schematic for rotor chopper control for induction motor drive.

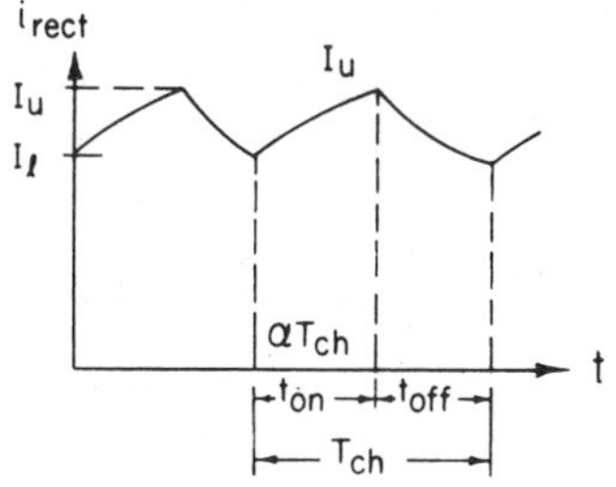

Fig. 2. Waveform of rectified current for TRC.

ANALYSIS AND DERIVATION OF CIRCUIT MODELS

In this development, a per-unit notation is adopted. The base values (see Appendix) are as follows:

voltage = V_{base} = rated stator phase voltage, V,
current = I_{base} = rated stator phase current, A,
speed = ν_{base} = synchronous angular speed, r/s,
impedance = z_{base} = V_{base}/I_{base}, Ω,
power = 3 $V_{base}I_{base}$, W,
torque = T_{base} = P_{base}/ν_{base}, N·m.

This per-unit system has certain advantages. The numerical values of the per-phase power is the same as that

Paper TOD-73-144, approved by the Industrial Control Committee of the IEEE Industry Applications Society for publication in this TRANSACTIONS. Manuscript released for publication August 15, 1974.

P.C. Sen is with the Department of Electrical Engineering, Queen's University, Kingston, Ont., Canada K7L 3N6.
K.H.J. Ma is with Federal Pioneer Electric Company, Winnipeg, Man., Canada.

Reprinted from *IEEE Trans. Ind. Appl.*, vol. IA-11, pp. 43–49, Jan./Feb. 1975.

of the total power in per unit. Besides, the unit torque corresponds to unit air-gap power. This is in contrast to the usual arbitrary choice of the rated full-load shaft torque as a unit base.

Exact analysis is tedious, involving the phasor calculations for motor fundamental and harmonic quantities and step-by-step analysis of nonlinearities in the rectifier-chopper circuitry. However, it has been found possible to develop circuit models from which a good prediction of the performance characteristics can be made.

DEVELOPMENT OF DC CIRCUIT MODEL

Fig. 3(a) shows the per-phase circuit model with the stator impedance referred to the rotor side. The dc model will be obtained for the three-phase system. Due to both rectification and chopping processes, the rotor current is no longer sinusoidal. If the chopper frequency is high and/or smoothing inductor is large, then the rectified current is essentially a dc with negligible superimposed ripple. Thus the rotor current is approximately composed of alternating square pulses of $2\pi/3$ duration [1]. The average rectified current I_{dc} is related to rotor rms current I_2 by

$$I_2^2 \simeq 2I_{dc}^2/3. \tag{1}$$

The power loss in the stator and rotor resistances for all three phases is $3I_2^2(sR_1' + R_2)$ or $2I_{dc}^2(sR_1' + R_2)$ if expressed in terms of I_{dc}. Hence, $3(sR_1' + R_2)$, transferred across the rectifier bridge, appears as $2(sR_1' + R_2)$ in the dc side.

Due to the leakage reactances sX_1' and sX_2, the commutation of current between diodes in the rectifier bridge is no longer instantaneous. There is a period of current overlap whereby two phases carry current simultaneously. This causes a voltage reduction V_R from the terminals of the rectifier bridge which is given by [1]

$$V_R = 3s(X_1' + X_2)I_{dc}/\pi.$$

If the magnetizing current is not specifically required and the diode drops are neglected, the system can be represented by the dc equivalent circuit as shown in Fig. 3(b), where

$$V_{dc} = s(3(6)^{1/2}E_s'/\pi).$$

A set of equations for the ON and OFF modes of the chopper can be derived. In the period that starts at mT_{ch} (m being a dummy variable) the chopper is ON for $mT_{ch} \le t \le (m + \alpha)T_{ch}$ and OFF for $(m + \alpha)T_{ch} \le t \le (m + 1)T_{ch}$. Let the current flowing during the ON and OFF mode be $i_{on}(t)$ and $i_{off}(t)$, respectively. The equations for $i_{on}(t)$ and $i_{off}(t)$ are: for $mT_{ch} \le t \le (m + \alpha)T_{ch}$

$$\frac{di_{on}(t)}{dt} + \frac{1}{\tau_{on}} i_{on}(t) = \frac{V_{dc}}{L_f} \tag{2}$$

and for $(m + \alpha)T_{ch} \le t \le (m + 1)T_{ch}$

$$\frac{di_{off}(t)}{dt} + \frac{1}{\tau_{off}} i_{off}(t) = \frac{V_{dc}}{L_f} \tag{3}$$

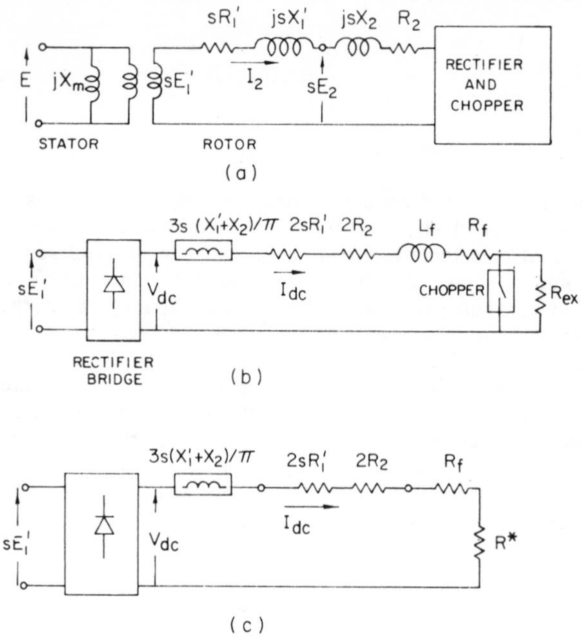

Fig. 3. Development of dc circuit model.

where

$$\tau_{on} = L_f/[R_m(s) + R_f] \tag{4}$$

$$\tau_{off} = L_f/[R_m(s) + R_f + R_{ex}] \tag{5}$$

and

$$R_m(s) = [2R_1' + 3(X_1' + X_2)/\pi]s + 2R_2. \tag{6}$$

The solutions for (2) and (3) are

$$i_{on}(t) = (I_l - I_{sh}) \exp[-(t - mT_{ch})/\tau_{on}] + I_{sh} \tag{7}$$

and

$$i_{off}(t) = (I_u - I_{op}) \exp\{-[t - (m + \alpha)T_{ch}]\tau_{off}\} + I_{op} \tag{8}$$

where

$$I_{sh} = V_{dc}/[R_m(s) + R_f] \tag{9}$$

$$I_{op} = V_{dc}/[R_m(s) + R_f + R_{ex}]. \tag{10}$$

I_l and I_u are the initial current of the ON and OFF modes, respectively.

In the steady state, the following equations must be satisfied:

$$i_{on}(mT_{ch}) = i_{off}[(m + 1)T_{ch}]$$

and

$$i_{on}[(m + \alpha)T_{ch}] = i_{off}[(m + \alpha)T_{ch}].$$

Since α and T_{ch} are known in the TRC, from the preceding two conditions, I_l and I_u are found to be

$$I_l = \frac{I_{sh}(1 - \lambda)\mu + I_{op}(1 - \mu)}{1 - \mu\lambda} \tag{11}$$

and

$$I_u = \frac{I_{sh}(1 - \lambda) + I_{op}(1 - \mu)\lambda}{1 - \mu\lambda} \tag{12}$$

where

$$\lambda = \exp[-\alpha T_{ch}/\tau_{on}]$$

and

$$\mu = \exp[-(1-\alpha)T_{ch}/\tau_{off}].$$

The average current is given by

$$I_{dc} = \frac{1}{T_{ch}} \int_{mT_{ch}}^{(m+\alpha)T_{ch}} i_{on}(t)\,dt + \int_{(m+\alpha)T_{ch}}^{(m+1)T_{ch}} i_{off}(t)\,dt$$

$$= \frac{1}{T_{ch}} \{(I_l - I_{sh})\tau_{on}(1-\lambda) + (I_u - I_{op})\tau_{off}(1-\mu)\}$$

$$+ \alpha I_{sh} + (1-\alpha)I_{op}. \tag{13}$$

The rms current is given by

$$I_{rms} = \left[\frac{1}{T_{ch}}\left(\int_{mT_{ch}}^{(m+\alpha)T_{ch}} i_{on}^2(t)\,dt + \int_{(m+\alpha)T_{ch}}^{(m+1)T_{ch}} i_{off}^2(t)\,dt\right)\right]^{1/2}$$

$$= [\alpha I_{sh}^2 + (1-\alpha)I_{op}^2$$
$$+ (1/T_{ch})\{0.5(I_l - I_{sh})^2\tau_{on}(1-\lambda^2)$$
$$+ 0.5(I_u - I_{op})^2\tau_{off}(1-\mu^2)$$
$$+ 2[I_{sh}(I_l - I_{sh})\tau_{on}(1-\lambda)$$
$$+ I_{op}(I_u - I_{op})\tau_{off}(1-\mu]\}]^{1/2}. \tag{14}$$

The rotor copper loss is given by

$$P_{cu} = [V_{dc} - 3s(X_1' + X_2)I_{dc}/\pi]I_{dc} - 2R_1' s I^2{}_{rms}.$$

The developed torque at slip s becomes

$$T = P_{cu}/s = (1/s)\{[V_{dc} - 3s(X_1' + X_2)I_{dc}/\pi]I_{dc} - 2R_1' s I_{rms}^2\}. \tag{15}$$

If the chopper frequency is high then I_{rms} is essentially equal to I_{dc}. Thus, (15) may be approximated by

$$T \simeq (1/s)\{[V_{dc} - 3s(X_1' + X_2)I_{dc}/\pi]I_{dc} - 2R_1' s I_{dc}^2\}. \tag{16}$$

Also, if the chopper frequency is high and/or the smoothing inductor is sufficiently large, then

$$x = \alpha T_{ch}/\tau_{on} \ll 1$$

$$y = (1-\alpha)T_{ch}/\tau_{off} \ll 1$$

and λ and μ can be approximated by the first two terms of the Taylor's expansion of the exponent, i.e.,

$$\lambda = \exp[-x] \simeq 1 - x \tag{17}$$

$$\mu = \exp[-y] \simeq 1 - y. \tag{18}$$

Also the cross-product term xy can be neglected. The simplified expressions for I_l and I_u are

$$I_l \simeq \frac{I_{sh}x(1-y) + I_{op}y}{1-(1-x)(1-y)} \simeq \frac{I_{sh}x + I_{op}y}{x+y}$$

$$I_u \simeq \frac{I_{sh}x + I_{op}y(1-x)}{1-(1-x)(1-y)} \simeq \frac{I_{sh}x + I_{op}y}{x+y}.$$

Since the approximate expression for I_l and I_u are equal, the average current

$$I_{dc} \simeq \frac{I_{sh}x + I_{op}y}{x+y} = \frac{\text{num}}{\text{den}}$$

where

$$\text{num} = \frac{V_{dc}}{R_m(s) + R_f}\frac{\alpha T_{ch}}{L_f}[R_m(s) + R_f]$$
$$+ \frac{V_{dc}}{R_m(s) + R_f + R_{ex}}\frac{(1-\alpha)T_{ch}}{L_f}$$
$$\cdot[R_m(s) + R_f + R_{ex}]$$

and

$$\text{den} = \frac{\alpha T_{ch}}{L_f}[R_m(s) + R_f]$$
$$+ \frac{(1-\alpha)T_{ch}}{L_f}[R_m(s) + R_f + R_{ex}].$$

Therefore,

$$I_{dc} = \frac{V_{dc}}{R_m(s) + R_f + R_{ex}(1-\alpha)}.$$

Hence, the equivalent resistance appearing across the terminals of the rectifier bridge is

$$R_{eq} = [R_m(s) + R_f + R_{ex}(1-\alpha)].$$

The fictitious resistance for the chopper-controlled external resistance is

$$R^* = R_{ex}(1-\alpha). \tag{19}$$

By adjusting the time ratio α in the range $0 \leq \alpha \leq 1$, R^* will vary in the range $0 \leq R^* \leq R_{ex}$. Consequently, an induction motor speed control scheme in which the rotor resistance can be continuously and contactlessly adjusted by electronic means is possible. The expression for R^* (19) simplifies the dc model to a significant extent as shown in Fig. 3(c).

DEVELOPMENT OF AC CIRCUIT MODEL

The power loss in the dc side of the rectifier bridge, under per-phase consideration, becomes $I_{dc}^2(R^* + R_f)/3$. From (1), this is equivalent to the power dissipation caused by the flow of I_2 in a resistance $0.5(R_f + R^*)$ in each rotor phase. The balance of power in each rotor phase gives

$$E_2 I_{21}\cos\theta_1 = R_2 I_2^2 + 0.5(R_f + R^*)I_2^2 + P_{mech}$$

where I_{21} is the fundamental of the rotor current in rms, θ_1 is the angle between E_2 and I_{21}, and P_{mech} is the mechanical power. If the mechanical torque is caused by the rotor fundamental current then P_{mech} is given by

$$P_{mech} = [R_2 + 0.5(R_f + R^*)]I_{21}^2(1-s)/s.$$

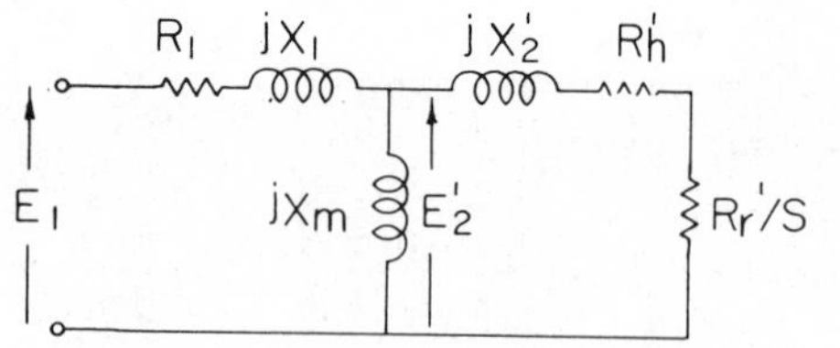

Fig. 4. Per-phase ac circuit model.

Hence,

$$E_2 I_{21} \cos \theta_1 = (\pi^2/9 - 1)[R_2 + 0.5(R_f + R^*)]I_{21}^2 + [R_2 + 0.5(R_f + R^*)]I_{21}^2/s$$

since [2]

$$I_{21} = 3I_2/\pi.$$

The per-phase ac circuit model is shown in Fig. 4, where $R_r' = [R_2 + 0.5(R_f + R^*)]n^2$ and $R_h = R_r'(\pi^2/9 - 1)$. The latter, R_h' represents the harmonic loss due to the rectification (R_r' and R_h' are referred to the stator).

PERFORMANCE CHARACTERISTICS

To investigate the effects of the chopper frequency, the torque–slip curves for three different frequencies, 200Hz, 50 Hz, and 20Hz, with the time ratio fixed at 0.6 are evaluated by 1) using the expression given by (15), 2) using the expression given by (16), and 3) using the model shown in Fig. 3(c).

The results are shown in Fig. 5. If the chopper frequency is high (200 Hz), all three methods give essentially the same results. When the chopper operates at 50 Hz, the values of torque based on (15) lie on those at 200 Hz but are lower than those given by (16). The error in using (16) instead of (15) becomes more pronounced when the frequency is 20 Hz. Hence the model of Fig. 3(c) can be confidently used when the chopper frequency is high and the smoothing inductor is large.

The torque–slip characteristics predicted from the dc circuit model [Fig. 3(c)] and from the per-phase ac circuit model [Fig. 4] are shown in Figs. 6 and 7, respectively. Different values of α would yield a family of torque–speed curves. These curves are essentially linear. Since the two sets of results are in close agreement, either model can adequately describe the torque–speed characteristic of TRC control strategy. The dc circuit model is more convenient to use if only the torque–speed relation is desired because it involves simple calculation.

For the purpose of comparison, a typical set of torque–slip curves from the experiment and different theoretical circuit models is shown in Fig. 8. The dc model for an ideal machine (with all the machine parameters ignored) gives results that are unrealistic because the torque developed at a particular speed (curve a) is too high. Besides, maximum torque exceeds the pull-out torque of the machine. When the internal power losses have been considered by including the machine resistances R_1' and R_2 in the model, the torque developed (curve b) becomes realistic. When the harmonic resistance R_h is neglected in the ac model, the torque (curve c) is higher than that when R_h is included. The discrepancy between the theoretical calculations (curves d and e) and the experimental results is less than 10 percent. Thus, both dc and ac circuit models with all the machine parameters considered give satisfactory predictions of the torque–slip characteristic of TRC strategy.

When other performance characteristics such as the input current, power factor, and efficiency are also required, then ac circuit model should be used because these can be evaluated along with the calculation of torque. A

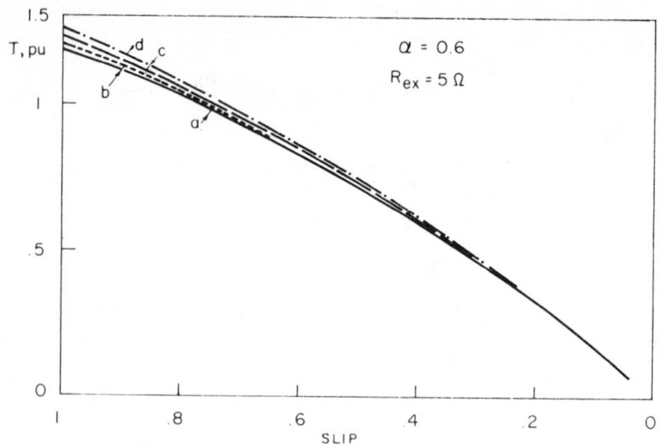

Fig. 5. Effects of chopper frequency on torque–speed characteristic: a—for 200Hz = based on (15), (16), and Fig. 3(c); b—for 20Hz = based on (15); c—for 50Hz = based on (15); d—for 20Hz = on (16).

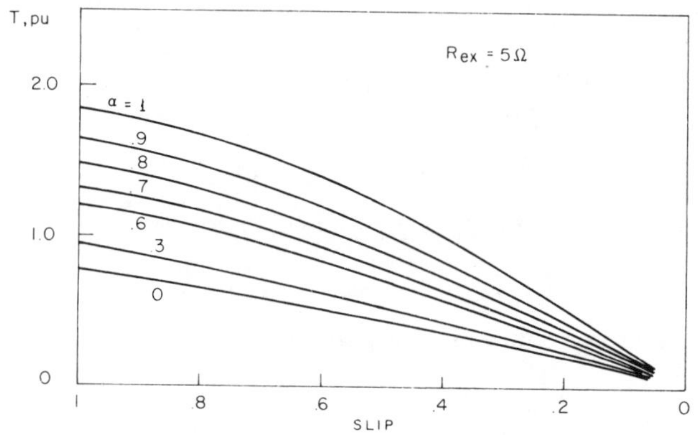

Fig. 6. Torque–slip curves based on dc circuit model.

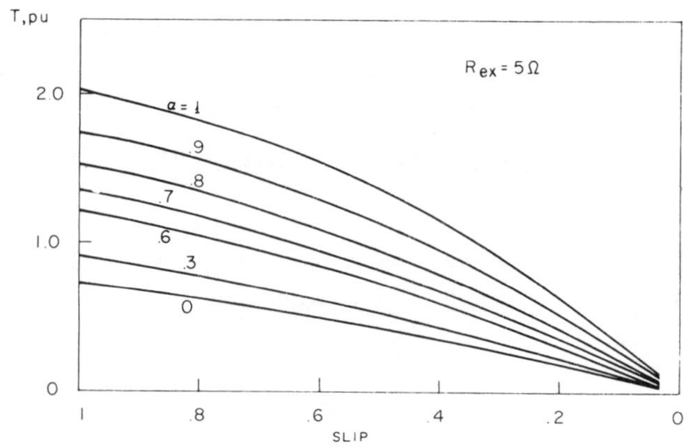

Fig. 7. Torque–slip curves based on ac circuit model.

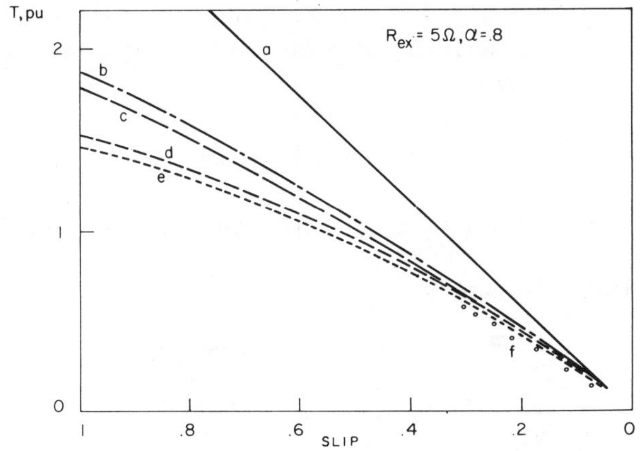

Fig. 8. Comparison of results from model study and experiment: *a*—ideal machine; *b*—X_1 and X_2 neglected in Fig. 3(c); *c*—R_h' neglected in Fig. 4; *d*—ac circuit model (Fig. 4); *e*—dc circuit model (Fig. 3(c)); *f*—experimental.

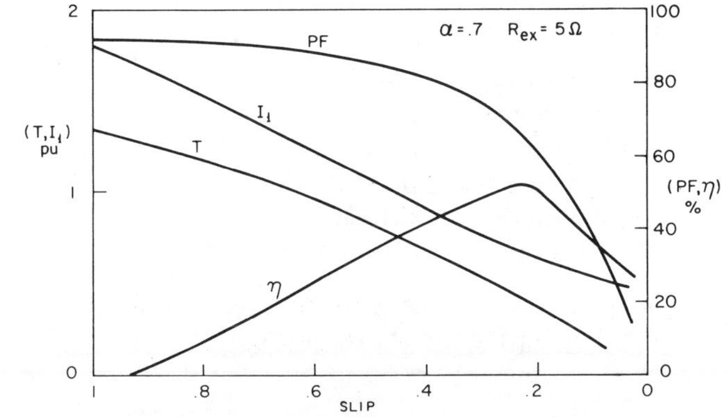

Fig. 9. Performance characteristics for TRC (α = 0.7, R_h considered).

typical set of these performance characteristics is shown in Fig. 9 for a particular duty cycle of the chopper ($\alpha = 0.7$). Power factor is improved at lower speeds. Starting current is also reduced. The maximum efficiency is low because the machine has high rotor circuit resistance and harmonic losses are involved.

EXPERIMENTAL SETUP AND OSCILLOGRAMS

The power and control circuits are simple. The chopper is the conventional Jones circuit [3]. The control unit is essentially an astable multivibrator. Its output frequency and duty cycle are adjustable.

The experimental torque–slip relation is shown in Fig. 8 by discrete circles. The oscillograms for the rectified current, voltage across the external resistor, and rotor and stator currents are illustrated in Fig. 10. The effect of commutation overlap is evident from the waveform of the rotor current (Fig. 10(a)). Also the rotor current is not sinusoidal but is rich in harmonics. However, the stator current is essentially sinusoidal (shown in Fig. 10(b)), which is a desirable feature because harmonics in the supply may produce adverse effects on the supply system.

Fig. 10(c) illustrates the waveforms for rectified current and the voltage across the external resistor.

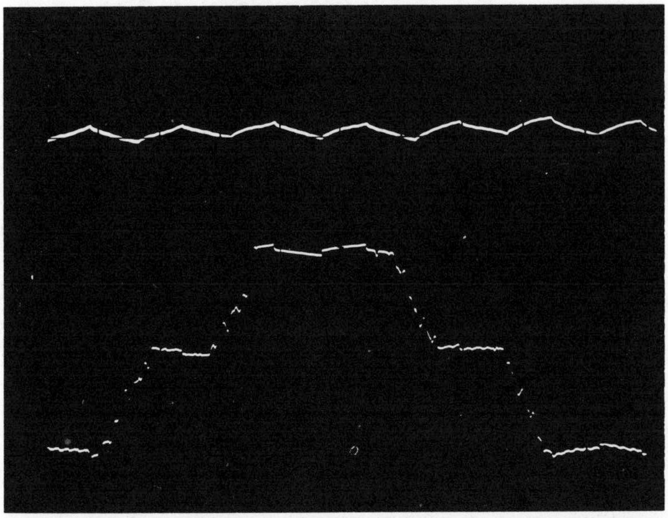

(a)

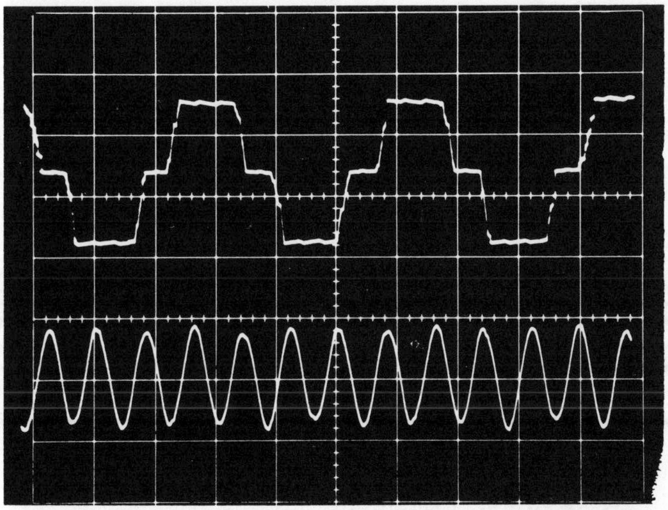

(b)

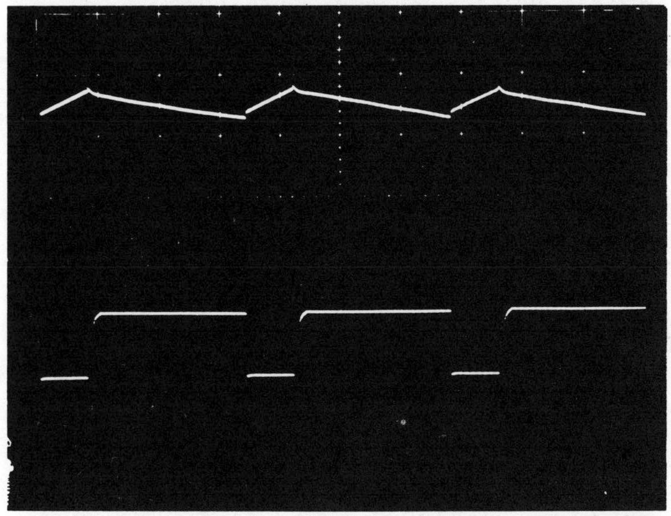

(c)

Fig. 10. Oscillograms for TRC. (a) Rectified current and rotor current. (b) Rotor current and stator current. (c) Rectified current and voltage across R_{ex}.

CONCLUSION

The solid-state speed control scheme for induction motor drive studied in this paper is simple but elegant. The scheme provides continuous and contactless adjustment of the rotor resistance by electronic means. The ac and dc circuit models derived in this study are important tools for predicting performance characteristics of the system. The torque–slip characteristic of the TRC is essentially linear. The feasibility of the system and verification of theoretical results are demonstrated by experimental results. It is anticipated that such simple and elegant control scheme will find applications in many industrial drive system.

NOMENCLATURE

α	Ratio of t_{on} to T_{ch}.
E_1	Per-phase supply voltage in rms.
E_1'	Per-phase supply voltage referred to rotor.
I_2	Rms rotor current.
I_{21}	Rms value of rotor fundamental current.
I_{dc}	Average rectified current.
I_l	Lower current limit.
I_u	Upper current limit.
I_{op}	Steady-state rectified current when the chopper is OFF all the time.
I_{sh}	Steady-state rectified current when the chopper is ON all the time.
i_{on}	Rectified current in ON mode.
i_{off}	Rectified current in OFF mode.
L_f	Smoothing inductance.
n	Stator to rotor turn ratio.
P_{cu}	Rotor copper loss.
P_{mech}	Mechanical power.
R_1	Stator resistance.
R_2	Rotor resistance.
R_1'	Rotor-referred stator resistance.
R_2'	Stator-referred rotor resistance.
R_f	Resistance of smoothing inductor.
R_r'	Effective rotor resistance referred to stator.
R_{ex}	External resistance.
R^*	Fictitious resistance of the chopper-controlled resistance.
$R_m(s)$	$= [2R_1' + 3(X_1' + X_2)/\pi]s + R_2$.
R_h'	Harmonic resistance referred to stator.
s	Slip.
sE_2	Induced rotor voltage at slip s.
T	Developed torque.
T_{ch}	Chopper period.
t_{on}	ON time interval.
t_{off}	OFF time interval.
τ_{on}	$= L_f/[R_m(s) + R_f]$.
τ_{off}	$= L_f/[R_m(s) + R_f + R_{ex}]$.
V_{dc}	$= s\,3(6)^{1/2}E_1'/\pi$.
X_1	Stator reactance at supply frequency.
X_2	Rotor reactance at supply frequency.
X_m	Stator magnetizing reactance.
X_1'	Rotor-referred stator reactance at supply frequency.
X_2'	Stator-referred rotor reactance at supply frequency.

APPENDIX

MOTOR PARAMETERS AND BASE VALUES

Motor Parameters

Quality	Magnitude
R_1	0.75 Ω
R_2	0.36 Ω
X_1	0.73 Ω
X_2	0.195 Ω
X_m	20.0 Ω
n	1.937

Windage and frictional loss (total) = 310 W.

Base Values

Base voltage	120 V
Base current	9 A
Base speed	188.5 r/s
Base impedance	13.33 Ω
Base power	3240 W
Base torque	17.189 N·m

The performance characteristics of the machine under normal operation are shown in Fig. 11.

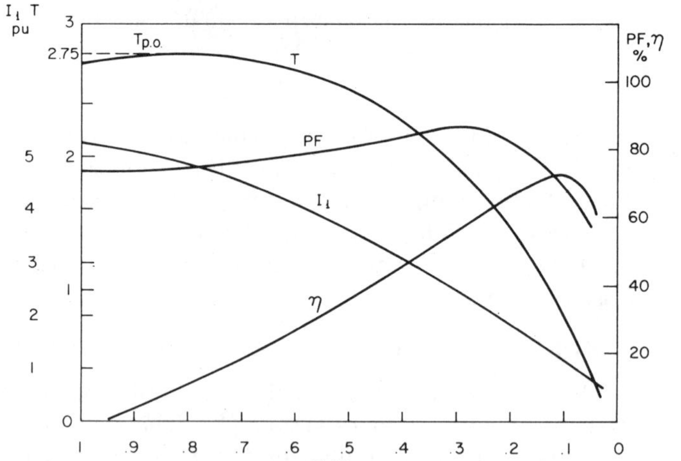

Fig. 11. Performance characteristics of machine under test.

Miscellaneous Data

L_f	210 mh
R_f	1 Ω
R_{ex}	5 Ω

Unless specified otherwise, the chopper frequency is 200Hz.

REFERENCES

[1] B. Bedford and R. Hoft, *Principles of Inverter Circuits.* New York: Wiley, 1964.

[2] J. Schaefer, *Rectifier Circuits: Theory and Design.* New York: Wiley, 1965.

[3] General Electric *SCR Manual*, 4th ed. General Electric Co., 1967.
[4] A. Lavi and R. Polge, "Induction motor speed control with static inverter in the rotor," *IEEE Trans. Power App. Syst.*, vol. PAS-85, pp. 76–84, Jan. 1966.
[5] A. Kusko, "State of arts: Solid state ac and dc motor drives in industry," presented at the IEEE Int. Semiconductor Power Converter Conf., Baltimore, Md., May, 1972.
[6] N. Hayashi, "Speed control of wound-rotor inductor motors by through-pass inverters," *Proc. Inst. Elec. Eng.*, Japan, vol. 90, no. 6, 1970.

Adjustable Speed AC Drive Systems for Pump and Compressor Applications

HERBERT W. WEISS, SENIOR MEMBER, IEEE

Abstract—**The paper covers the application and performance of a cycloconverter type frequency converter and a doubly fed wound rotor ac motor as applicable to adjustable speed drive systems for pump and compressor loads. The paper covers the subject by discussion of the following: 1) basic theory and operation of a cycloconverter and a doubly fed wound rotor motor; 2) motor and converter ratings to meet the performance requirements of typical pump and compressor applications; 3) control schemes for starting, running, and process control; 4) efficiency and power factor; and 5) summation, including advantages of adjustable speed ac drive systems and static frequency conversion equipments.**

Introduction

THE TRANSPORTATION of liquid or gas through a pipeline system utilizes either contrifugal pumps or compressors to produce the required flow. The various and sometimes continually changing operating conditions are a function of the throughput requirements and the physical properties of the product and the pipeline system. When designing the pipeline system, the number of stations, quantity of units per station, and the prime mover for the pump or compressor units must be selected on the basis of performance requirements and installation and operating cost.

Since the process requires changes in fluid flow rate and control of flow rate, the pumping stations must include some provision to adjust and control flow. The basic methods used to accomplish these functions are: 1) constant speed prime mover and throttle valve; 2) quantity of small constant speed units per station with selectivity of units to control flow; and 3) adjustable speed prime mover. Depending on the pipeline system requirements, any one or combination of the three basic methods may provide the best operating system.

The constant speed drive represents the lowest installed cost, but may not provide the flexibility or operating efficiency which can be attained from an adjustable speed drive. An adjustable speed drive can be accomplished by prime movers such as: 1) internal combustion engine; 2) gas turbine; 3) constant speed electric drive with slip coupling; 4) dc motor with adjustable voltage or adjustable field control; and 5) ac motor with adjustable voltage or adjustable frequency control.

The engine and gas turbine have been applied to gas line systems because of the economic availability of fuel. However, these prime movers do not have as high an operating efficiency as can be attained by some of the electric drives which are available. Included as one of the adjustable speed electric drives is a doubly fed wound rotor ac motor.

Paper TOD-72-166, approved by the Petroleum and Chemical Industry Committee of the IEEE Industry Applications Society for presentation at the Petroleum and Chemical Industry Technical Conference, Denver, Colo., September 18–20, 1972. A version of this paper was published in the *Oil and Gas Journal*, December 11, 1972. Manuscript released for publication December 29, 1972.

The author is with the Drive Systems Product Department, General Electric Company, Salem, Va.

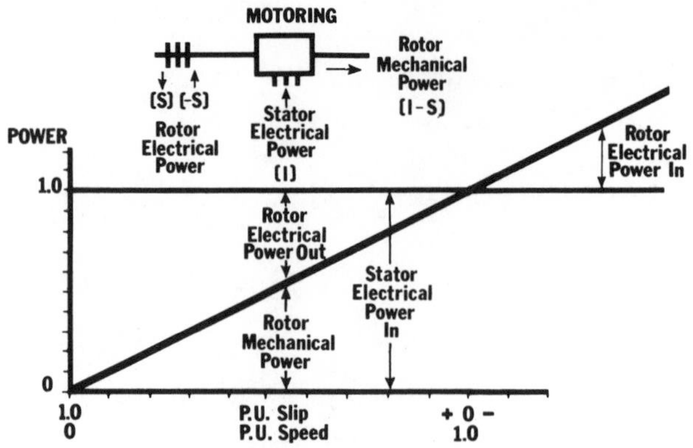

Fig. 1. Power flow in wound rotor motor.

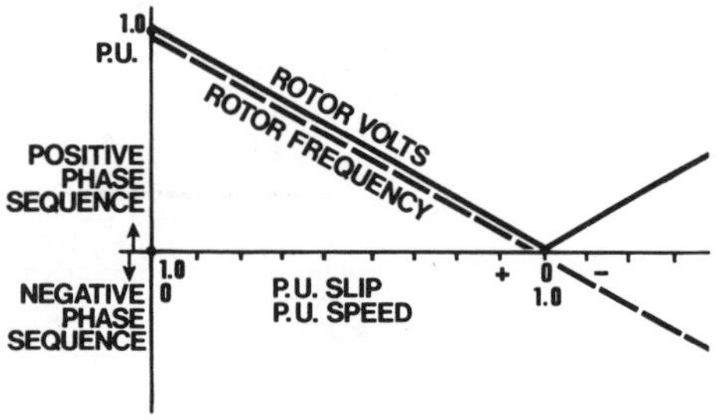

Fig. 2. Rotor voltage.

The application of wound rotor motors on adjustable speed drives is not new, nor is the application of doubly fed wound rotor motors. Schemes using secondary resistance are quite common but are not practical on large drives where efficiency is important. Schemes using rotating machines in the rotor circuit to recapture rotor power, such as the Clymer, modified Kraemer, and Sherbius systems have been used quite successfully on large drive systems.

With the rapid advance and acceptance of solid-state technology over the last decade, it is now possible to apply static power conversion equipment in the rotor circuit of a wound rotor machine to produce an adjustable speed drive with high efficiency.

Basic Theory and Operation

The wound rotor ac motor provides a means to adjust the drive speed by changing the rotor excitation current. The shaft output power is the difference between the stator input

Reprinted from *IEEE Trans. Ind. Appl.*, vol. IA-10, pp. 162–167, Jan./Feb. 1974.

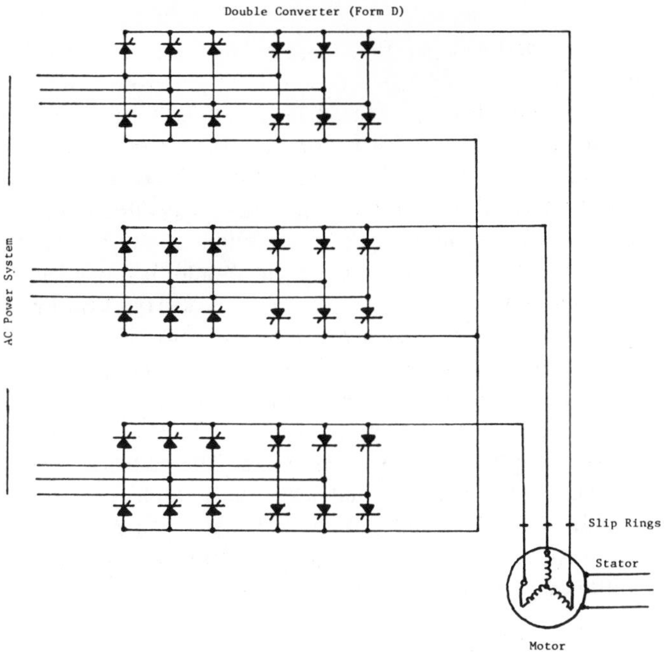

Fig. 3. Basic cycloconverter circuit.

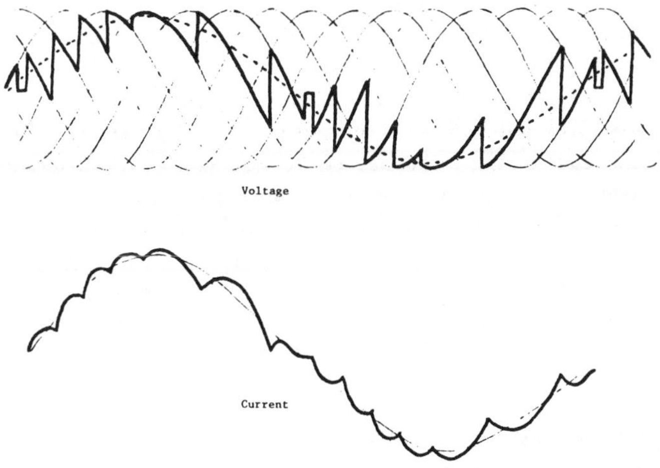

Fig. 4. Cycloconverter waveforms.

power and the rotor output power. Fig. 1 shows the power flow for a wound rotor motor. In a singly fed machine, the rotor power is always in one direction, out of the machine. In a doubly fed machine, a voltage source is connected to the rotor circuit which is capable of putting power into or taking power out of the rotor. Thus, the shaft output power may be increased above 1 per unit by feeding power into the machine through the stator *and* the rotor. The speed of the machine is a function of the difference between the stator and rotor frequencies, $N = (1 - S)$ per unit, where S is the per unit slip frequency. If the rotor is excited in opposite phase sequence, S becomes negative, and the machine will operate above its synchronous speed.

Fig. 2 shows the magnitude and frequency of the internal rotor voltage as a function of motor speed. An external voltage source connected to the rotor circuit must be capable of operating at the rotor internal voltage over the speed range required by the drive. For this type drive, a frequency converter is required to convert and adjust conventional power system voltage and frequency to match that developed by the machine rotor.

When considering this type of adjustable speed drive for a pipeline application, several factors indicate it to be a favorable selection.

1) The range of adjustable speed required is rather small; usually from 70 to 100 percent of maximum speed is sufficient. Therefore, the frequency converter rating does not have to match the motor rating.

2) Since the motor is doubly fed, its base rating (at synchronous speed) is less than the maximum horsepower required for the drive.

3) A cycloconverter type frequency converter has the inherent capability of transferring power in either direction. Thus, a converter rated approximately 15 percent of the motor horsepower can provide ±15-percent power transfer from the rotor to cover a 30-percent speed range at constant torque.

4) A cycloconverter produces a good waveform at the lower frequencies.

5) A doubly fed machine has characteristics similar to that of a synchronous machine. Thus some degree of power factor correction can be achieved by overexciting the rotor.

6) Static converter equipments have very low losses, thus increasing operating efficiency.

Cycloconverter

The cycloconverter is a power electronic equipment designed to convert constant voltage, constant frequency ac power to adjustable voltage, adjustable frequency ac power without utilizing a dc link. A basic cycloconverter circuit which utilizes the three-phase bridge thyristor converter is shown in Fig. 3. Note that the power conversion circuits are connected to the ac power system through isolation transformers, thus providing electrical isolation between the power system and the motor and between output phases of the cycloconverter. The transformers also provide the correct voltage to the converter circuit, as required, to meet the highest voltage available at the motor slip rings throughout the adjustable speed range.

The basic cycloconverter operation utilizes the technology of the dc thyristor converter, by phase-controlling the output of a double (reversing) converter to produce an alternating voltage. This operation is shown in Fig. 4 for a unity power factor load. When two or three of these double converters are connected to the motor windings, the outputs are phase-shifted 90° for a two-phase cycloconverter system and 120° for a three-phase cycloconverter system. Since the cycloconverter output is derived directly from the ac power system, the maximum output frequency is limited to a fraction of the power system frequency (typically 1/3) to maintain an acceptable waveform with low harmonic content. Output voltage control is also accomplished by phase-controlling the converters to produce the desired voltage level of the output ac voltage wave. By proper manipulation of the control and thyristor gating circuits, the output phase sequence can be readily changed from positive to negative.

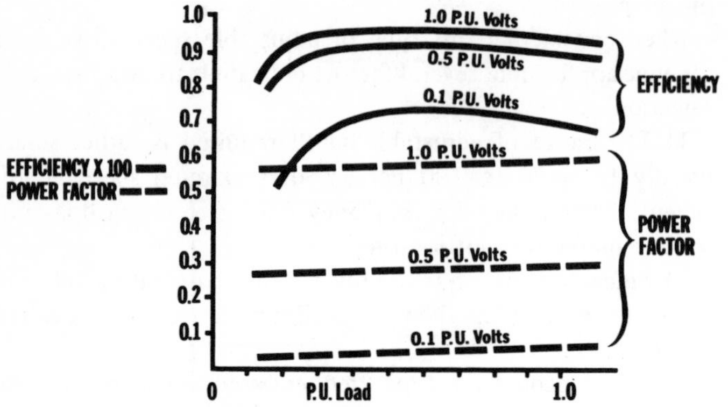

Fig. 5. Cycloconverter efficiency and power factor.

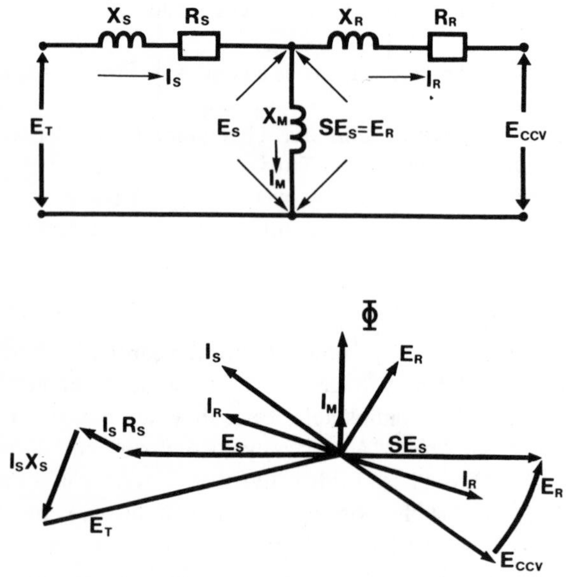

Fig. 6. Equivalent circuit and vector diagram.

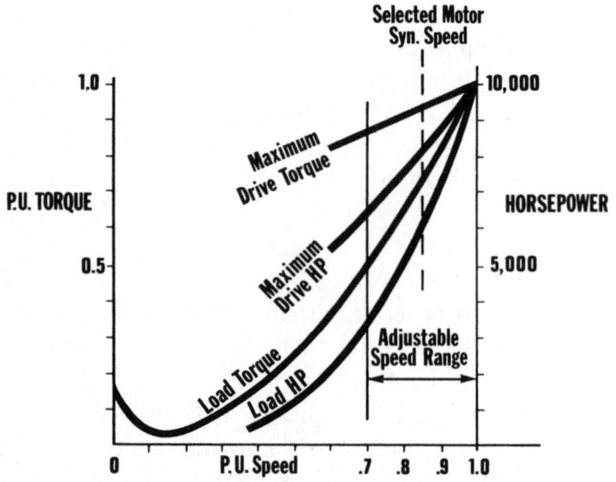

Fig. 7. Typical speed–torque–horsepower curve for pump and compressor applications.

The converter operation continually transfers the load current from one thyristor to another, and from positive to negative current flow; thus the shape and polarity of the output ac voltage is produced. The current carrying capacity of the thyristor devices is based on this continual switching, which requires that certain output frequencies be avoided to prevent excessive heating on a particular grouping of thyristors. The most important frequencies which should be avoided for continuous operation are 0, 30, 45, and 60 Hz. Continuous operation at these frequencies may be possible at reduced load, but should be recognized when rating and applying the equipments to the drive system requirements.

The efficiency of the cycloconverter is high due to the use of static components such as thyristors. The no-load losses are less than $\frac{1}{2}$ percent of rated kVA, and full-load losses are approximately 4 percent of rated kVA. The approximate cycloconverter efficiency and power factor as a function of output is shown in Fig. 5 for an 0.8-pF load.

Wound Rotor Motor

The wound rotor machine may be treated exactly like a squirrel-cage induction motor when singly fed, but exhibits somewhat different characteristics when doubly fed. Fig. 6 shows the equivalent circuit diagram and vector diagram for the wound rotor machine. Note that the rotor is subjected to two voltages. One is the voltage generated by the rotating stator field and is proportional to per unit slip. The second is the output of the frequency converter. The vector difference between these two voltages produces the current in the rotor circuit. The rotor current is reflected into the stator and when vectorially added to the magnetizing current (*Im*) becomes the current drawn from the power system. It is possible to "overexcite" the rotor, thus supplying the magnetizing current and improving the power factor of the system.

Pump or Compressor Application

This type of application is a variable torque load, where the load torque varies approximately as the square of motor speed and horsepower varies approximately as the cube of motor speed. Fig. 7 shows a typical speed-load torque and horsepower curve with a proposed speed range.

Assuming maximum horsepower requirement of the drive to be 10 000 hp at 1.0 per unit speed and adjustable speed range of 0.7 to 1.0 per unit speed, the motor synchronous speed may be selected near 0.85 per unit of the drive requirement. The torque capability of the motor must equal the maximum torque requirement of the drive which is at 1.0 per unit speed. This would mean a motor rating of $1.0 \times 0.85 \times 10\,000 = 8500$ hp at motor synchronous speed of 0.85 per unit. Since the drive requirement at 0.85 per unit speed is $(0.85)^2 \times 0.85 \times 10\,000 = 6150$ hp, it would be desirable to reduce the motor horsepower even more by taking advantage of the more efficient cooling of self-ventilated and totally enclosed fan-cooled (TEFC) machines at increased motor speed. Assuming the motor design allows a 6-percent increase in torque for an 18-percent increase in speed, the motor rating at 0.85 per unit speed would be approximately 0.94 × 0.85 × 10 000 = 8000 hp, with 106 percent continuous torque capability at 1.0 per unit drive speed.

The frequency converter must be capable of increasing the drive horsepower to 10 000 hp by supplying power into the machine through the rotor. Therefore, the frequency converter would be rated to supply 10 000 − 8500 = 1500 hp to the rotor. This maximum torque and horsepower capability of the selected drive is shown in Fig. 7, compensated for the increase or decrease in rating due to cooling efficiency and speed. Note that the drive horsepower and torque capability exceeds the load requirement below the maximum design point. This capability is available to handle any process requirements requiring high torque at reduced flow if the pump or compressor is sized to include this capability.

The frequency converter must be capable of delivering 1500 hp into the rotor, but needs to be further defined. Assuming the motor secondary open circuit secondary voltage to be 2000 V at standstill ($S = 1.0$), the rotor EMF at 1.0 per unit speed is $SE = S \times 2000 = 0.1765 \times 2000 = 355$ V. Rotor frequency is equal to slip frequency, $S = 0.1765$. The frequency converter must be rated for operation at 0.1765 × 60 Hz = 10.65 Hz maximum, adjustable from 0 to 10.65 Hz, positive or negative phase sequence. The rated rotor current is related to motor horsepower and rotor voltage and is approximately

$$I = \frac{\text{hp} \times 746}{\sqrt{3} \times V \times D}$$

where D is the factor to compensate for voltage drop = 0.95 and

$$I = \frac{(8000 \times 1.06) \times 746}{1.7314 \times 2000 \times 0.95} = 1920 \text{ A}.$$

Frequency converter output kVA = $\sqrt{3} \times 1920 \times 355 = 1180$ kVA. Therefore, a motor rated 8000 hp at 0.85 per unit drive speed and a frequency converter rated 1180 kVA, 355 V, 1920 A, 10.65-Hz would drive a pump or compressor rated 10 000 hp with a 30-percent speed range, in the doubly fed rotor motor system.

Control Schemes

Drive Start

There are advantages and disadvantages associated with starting a doubly fed machine with a static frequency converter. The major advantage is the ability to soft start the drive, thus eliminating high inrush currents to the motor and the resultant voltage drop on the power system.

The disadvantages arise from the fact that the frequency converter rating is based on limited speed range and cannot match the rotor voltage at the high slip conditions during a start. To compensate for this, some additional control equipment is required to get the machine up to the speed at which the frequency converter can take over.

There are several methods to start the drive, the selection of which should be based on the particular drive requirements and economics. Assuming that the drive may be started at reduced load, the torque required at 70-percent drive speed would be approximately 20 percent of maximum rated, and 25 percent of the motor rated torque. Three starting methods are shown in Fig. 8.

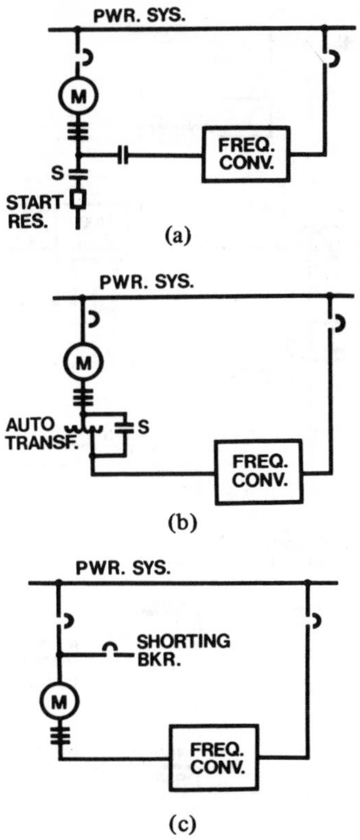

Fig. 8. Starting methods.

Fig. 8(a) uses resistors in the rotor circuit to accelerate the drive to 70-percent speed, synchronize the frequency converter to the rotor voltage, then apply the frequency converter and disconnect the starting resistor.

Fig. 8(b) uses an auto transformer to reduce the rotor voltage by the ratio of 2000:355 to match rotor voltage to the frequency converter voltage at standstill. With the autotransformer in the circuit, the frequency converter must be capable of supplying 20 percent of rated drive torque which requires 0.2 × 1920 × 2000/355 = 2160 A.

This current magnitude represents only a small overload to the frequency converter, which on a short-time basis would cause no problems. The frequency converter output wave form would not be good at high slip, but since the drive torque requirements at the high slip, low speed are quite low, the performance should be acceptable. After the drive is accelerated to 70-percent speed, the auto transformer is removed from the circuit for normal drive operation.

Fig. 8(c) uses an "inside-out" starting sequence. A circuit breaker is used to short circuit the stator winding. The frequency converter is used to energize the rotor and accelerate the drive as an induction motor. As an induction motor, the machine torque varies as the square of the applied voltage, on a constant volts/hertz basis. At 70-percent speed, the machine requires approximately 70 × (60 + 10.5) = 49.5 Hz, and 49.5/10.65 × 355 = 1650 V to produce rated torque.

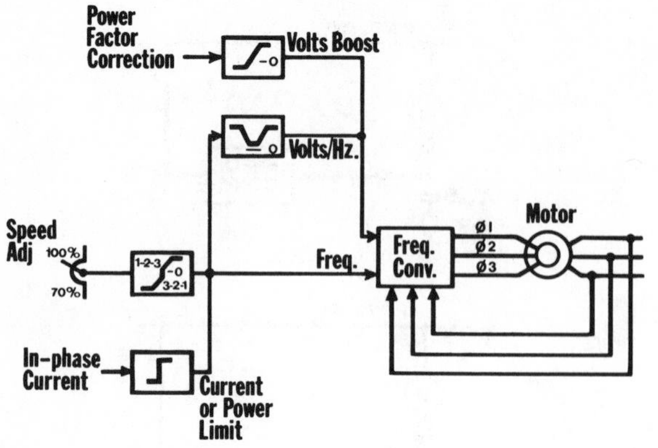

Fig. 9. Typical control scheme.

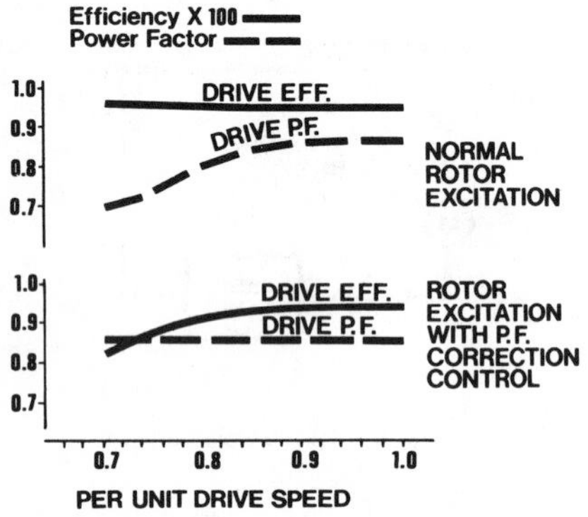

Fig. 10. Estimated efficiency and power factor.

Since the frequency converter capability is limited to 355 V, the available motor torque would be $(355)^2/(1650)^2 = 4.65$ percent.

The machine pull-out torque might be $2 \times 4.65 = 9.3$ percent, but this would still not be adequate to accelerate the drive. For this particular example, this starting method could not be used, but a drive with a wider speed range could result in favorable conditions for this method. The complete starting cycle requires acceleration as an induction motor on both sides, closing the stator line breaker, and synchronizing and closing in the frequency converter on the rotor. This sequence could be completed in a very short time.

Drive Control

The speed of the drive is the difference between the stator and rotor frequency $N = 60\,f(1 - S)/P$. Since f is constant and P is fixed by machine design, the speed may be adjusted by changing S, which is the frequency applied to the rotor. Therefore, speed control of the drive may be quite accurately accomplished by adjusting the frequency of the converter. More accurate speed control can be accomplished by the use of a speed feedback signal and a closed loop regulator.

Fig. 9 shows a typical control scheme. Motor speed is set by frequency, and the rotor voltage is directly related to frequency. Therefore, the control must also adjust the frequency converter output voltage to maintain the proper rotor excitation for the drive load requirements. Since the internal rotor voltage changes on a linear volts/hertz basis, the control must also establish a similar relationship for the frequency converter. Additional control is required to maintain rotor excitation near synchronous speed where frequency is almost zero, and a direct volts/hertz relationship would not provide enough voltage to overcome the rotor circuit *IR* drop. Depending on the drive system requirements, additional control functions may be necessary or desirable to enhance performance, such as: 1) current or load limit; 2) stabilizing networks; and 3) power factor adjustment.

Since continuous drive operation at motor synchronous speed is sometimes not desirable, the control can be set up to recognize this condition and establish a deadband in the speed adjustment to prevent continuous operation below approximately $\frac{1}{4}$-Hz rotor frequency. This deadband is quite small, and should not present any great operational disadvantage in most process flow adjustments.

Drive System Performance

The drive system efficiency and power factor can be quite good. Fig. 10 shows estimated performance values for the load horsepower requirements as shown in Fig. 7. The top curve shows the expected performance for "normal" rotor excitation, where rotor currents are maintained at the value required for the motor load. Drive efficiency remains good throughout the adjustable speed range, but power factor falls off below motor synchronous speed.

The bottom curve shows the expected performance when the drive control scheme is designed to maintain constant power factor by overexciting the rotor below the maximum speed point. Since the load requirement is variable torque, the motor and the cycloconverter will take the overexcitation currents without exceeding their rating. Note that drive efficiency falls off at the lower speeds due to the losses created by the overexcitation currents.

To achieve power factor correction at the maximum power factor point would require an oversized motor and cycloconverter to handle the overexcitation currents and would decrease drive efficiency.

Summation

The doubly fed ac motor drive system utilizing a cycloconverter type frequency converter is one of many methods available to provide the adjustable speed required by a process. For the particular application of driving a pump or compressor as would be typical for the transportation of gas or liquid through a pipeline, the doubly fed machine represents an attractive choice. Some of the features of the doubly fed drive which should be considered are as follows:

1) stepless speed control;
2) high operating efficiency;
3) low installed cost due to reduced motor horsepower and plus-minus synchronous speed control accomplished with a small frequency converter;
4) proven reliability of cycloconverter power and control circuits and devices in over 2 million kW of dc motor drive applications;

5) some degree of power factor control;
6) low installation costs of power electronic equipments;
7) production by cycloconverter of a good waveshape with low harmonic content at low frequencies;
8) proven high reliability and low maintenance of ac motor.

The need for an adjustable speed drive must be determined by the pipeline system designer based on overall system performance and economics. The doubly fed wound rotor motor utilizing a cycloconverter provides the system engineer with an excellent tool to do the job.

HIGH RESPONSE CONTROL OF STATOR WATTS AND VARS FOR LARGE WOUND ROTOR INDUCTION MOTOR ADJUSTABLE SPEED DRIVES

C.B. Mayer, Sr. Member, IEEE
Analytical Engineering, General Electric Co.
Schenectady, New York

ABSTRACT

High response control of stator watts and vars has been obtained on a cycloconverter controlled, doubly-fed, double-range, 15,000 Hp wound rotor induction machine with an orthogonal control scheme which linearizes the machine equations and combines both feedforward and feedback error signals. Leading and lagging power factor and positive or negative stator power flow can be smoothly and rapidly controlled over a speed range in excess of $\pm 35\%$ of the induction motor synchronous speed.

This doubly fed drive has been appropriately named the "Scherbiustat Drive" because the wound rotor induction motor secondary power conversion equipment is a cycloconverter which is the static equivalent of the Scherbius machine. Likė the Scherbius drive, it is capable of controlling both motor power and vars at shaft speeds below and above synchronism, with smooth transition through synchronism. The Sherbiustat drive, like the Scherbius drive, does not employ a DC link in the motor secondary power conversion equipment. It should not be confused with a Kraemerstat drive which employs a DC link in the conversion. The Kraemerstat drive uses rectifiers and inverters and is so named because of its similarity to the Kraemer drive which uses a synchronous converter and a DC motor in the secondary power conversion process.

INTRODUCTION

The subject of this paper is the control of a 22 Pole, 60 Hz, doubly-fed 15,000 HP wound rotor induction machine. The machine is part of an energy storage and supply system for powering an experimental fusion reactor. The overall system is depicted in Figure 1.

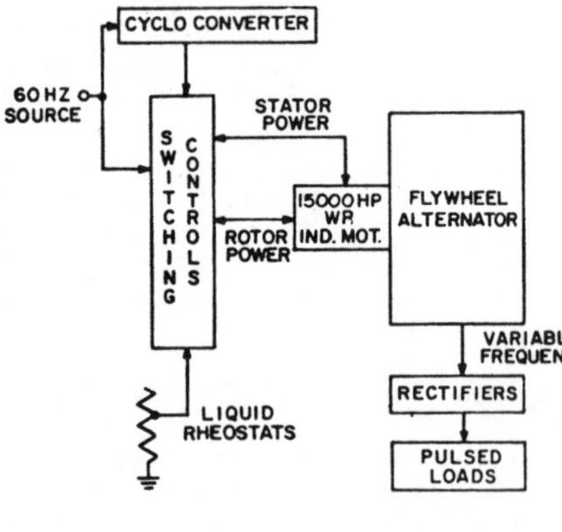

Figure 1 Overall System Diagram

The function of the induction machine is to accelerate the huge "flywheel alternator" to a top speed of 375 RPM. Load is applied for a period of about 10 seconds by rectifying the output of the flywheel alternator utilizing the stored rotational energy. During this period, the system is decelerated to a minimum speed of 257 RPM. Normal operation is to then reaccelerate the system from this speed to its top speed in preparation for the next load pulse. The minimum/maximum speed corresponds to 70.6% to 114.6% of synchronous speed for the induction machine. It takes the 15,000 HP machine 4.5 minutes to reaccelerate the system. After completion of experiments or upon emergency stop signal, the wound rotor machine must decelerate the system to zero speed in a minimum time. This is accomplished by pumping power back into the utility supply system and/or dissipating power in the liquid rheostats within the maximum ratings of the components of the drive system involved.

DEFINITION OF OPERATING MODES

To perform these functions along with the initial startup, there are three basic operating control modes of the drive system as depicted in Figure 2.

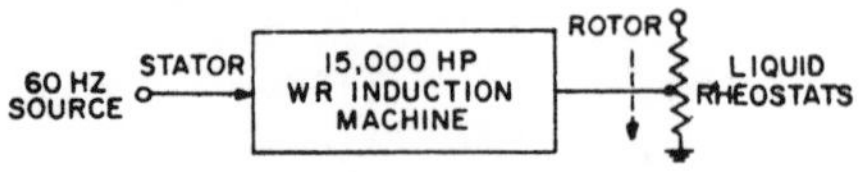

Figure 2a Mode 1, Startup

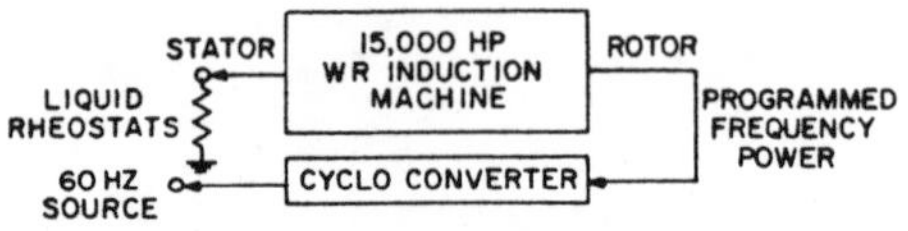

Figure 2b Mode 2, Normal Operation

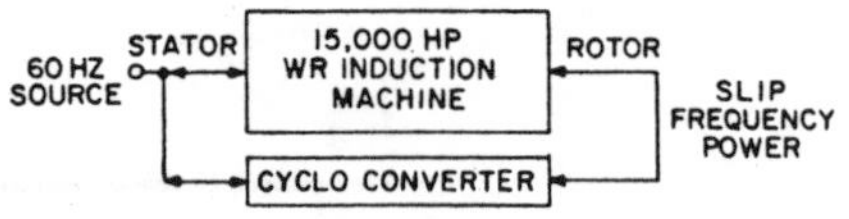

Figure 2c Mode 3, Stop

Mode #1 is the initial acceleration mode from zero speed to 257 RPM. In this mode, the liquid rheostat is connected to the rotor slip rings and a convention control is employed to accelerate the drive at a maximum rate within the rated capacity of the machine and/or liquid rheostat.

Mode #2 is the normal operating condition. In this mode, the liquid rheostat is replaced by a cyclo convertor which converts 60 Hz line frequency power directly to slip frequency power. This form of drive is the modern static version of an "ancient" drive developed before 1920 by A. Scherbius which is described in some early literature around that time.[1,2,3] The Scherbius drive along with the Kraemer/Kramer drive, which contains a DC link between the conversion, are described in many AC machine texts throughout the intervening history.[4,5,6,7,8,9] The modern version described here has been appropriately termed the Scherbiustat.

The requirement of the controller is to maintain the proper magnitude and phase of the applied slip ring voltage to achieve constant stator watts and vars and preventing upsets to the utility supply during the long

Reprinted from *Conf. Rec. IEEE/IAS 1979 Annual Meeting,* pp. 817-823, 1979.

acceleration or the rapid deceleration resulting from the tremendous overhauling torque produced by the fusion experiment. This control mode is the primary subject of the paper.

A speed regulator must also be provided to maintain speed if the load is not immediately applied upon reaching top speed.

When stopping, deceleration under motor control is accomplished by commanding and regulating negative stator watts until the drive reaches the minimum speed point for the normal mode (regenerative braking).

Mode #3 is utilized in decelerating the drive from the minimum speed of Mode #2 (257 RPM) to rest. In this mode, the cyclo convertor remains connected to the rotor and converts generated rotor power of programed frequency vs. speed to 60 Hz power and pumps it back into the line at the maximum rate permitted by the cyclo convertor current rating (current regulated). Simultaneously maximum power is pumped into the liquid rheostat which is set to maximum resistance and is connected to the stator in place of the normal 60 Hz line voltage.

Drive system control in each of the above modes was developed and evaluated along with transition controls between modes by means of a hybrid computer[1] simulation of the drive system. In this simulation, the customary two axis representation was used for the induction machine,[10,11] the 60 Hz supply system, and the secondary supply system. The cyclo convertor voltages were modeled as ideal single frequency sine waves produced in response to the control signals and transfer function of the power convertor.

The main subject of this paper is the control approach for the normal operating mode of the drive (Mode #2). The development of the unique orthogonal regulator which evolved after unsuccessful results with the initial approaches is discussed herein.

DEVELOPMENT OF CONTROL APPROACH

Definition of Terminology - Motor/Cyclo Convertor Model

A balanced three phase system has been assumed hence in discussing the control approach, it is only necessary to consider the relationships which exist in a single phase. The single phase equivalent circuit is shown in Figure 3.

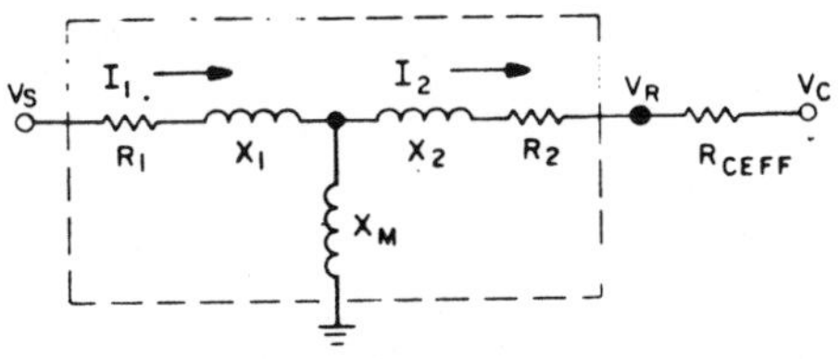

Figure 3 Single Phase Equivalent Circuit

The reference frame[12] has been fixed in the rotor and hence all voltages and currents are at slip frequency. The resistance R_{ceff} represents the effective internal impedance of the cyclo converter. This comes about by virtue of the negative current feedback within the power convertor in addition to the high gain voltage loop with a crossover frequency of around 1000 radians/second. As such, it does not represent a true system loss. Therefore, the voltage V_c represents the input voltage to the cyclo convertor's power convertor while V_r represents the true cyclo convertor output voltage or rotor slip ring voltage. V_s represents the stator voltage as seen in the rotor reference frame. Figure 4 is a typical vector diagram of the voltage and current relationships at a particular operating point.

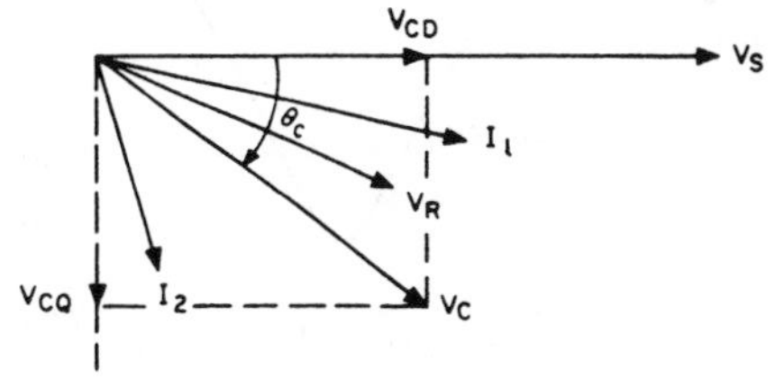

Figure 4 Typical Vector Diagram

Initial Control Approach

It is instructive before describing the orthogonal regulator approach which was finally developed to consider the initial approach which was attempted and to understand the reasons for the difficulties which arise from that approach.

The initial approach was founded on the philosophy that control of the rotor voltage magnitude (Control Channel 1) would basically vary the machine excitation and hence the vars while variation of the phase of the rotor voltage via frequency (Control Channel 2) would control the power angle and hence stator watts to the machine. While this philosophy is adequate and reasonably correct for large values of slip, it proved to be inadequate for control over the entire operating speed range of $\pm$ 20% slip.

After considerable difficulty, a suitable control was developed using this basic approach with some degree of cross coupling between channels. This control performed well during the acceleration from minimum to maximum speed which is quite slow and takes place over a period of several minutes. However, under the conditions of "rapid" deceleration by the overhauling fusion experiment load the control could not maintain constant stator watts and vars when passing thru zero slip.

Inherent Problem of the Magnitude/Phase Approach

The reason for the difficulty with the magnitude/phase approach outlined above can be seen for Figure 5. This shows a typical plot of the cyclo convertor control voltage (magnitude and phase) which must be fed to the power convertor as a function slip. The conditions are for rated stator watts and .4 per unit vars lagging.

It will be noted that while the "V" shaped voltage magnitude function appears quite reasonable for a control regulator to achieve, the rapid phase transition required near zero slip in order to maintain the constant stator watts and vars is a near insurmountable task when the speed changes rapidly. Increasing regulator gain sufficiently to permit tracking of the phase channel was impossible to achieve with a stable control.

[1] Combination of digital and analog computers with A/D and D/A converters to provide interface between analog and digital variables.

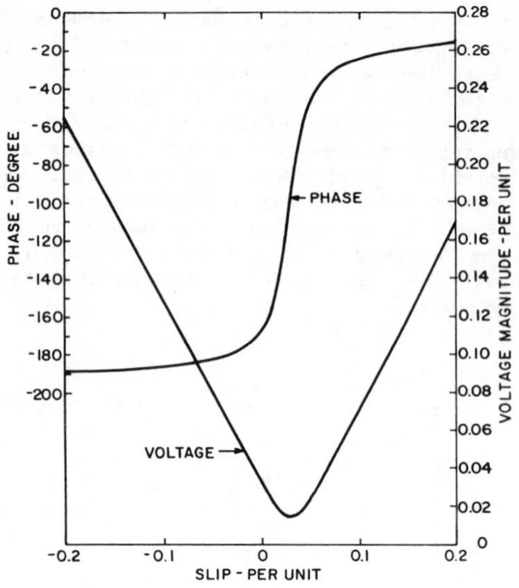

Figure 5 Magnitude and Phase of Cyclo Convertor Control Voltage vs. Slip for R_{ceff} = .0246 PU ohms, Watts = .86 PU, Vars = .4 PU, Stator Watts = 1.0 PU and VARS = 0.0 PU

Basis for the Orthogonal Regulator - Linearization of the Control Problem

The concept of the orthogonal regulator was to utilize two independent voltage amplitude channels of either polarity which were 90° displaced in time phase. Both of the amplitude channels were to be phase synchronized with respect to the rotor slip frequency. The first thought was to use the rotor current as the reference. However, after examining the in phase and quadrature voltage components required to maintain constant watts and vars vs. slip it, was found that the control was still non-linear. However, when the stator voltage was chosen as the reference and the inphase and quadrature components were computed, a linear set of equations was obtained in terms of stator watts, vars and slip. Specifically the equations are of the form.

(1) $V_{cq} = C_1 + C_2 W + C_3 V + S (C_4 W + C_5 V)$

(2) $V_{cd} = C_2 V - C_3 W + S (C_4 V - C_5 W + C_6)$

Where V_q is the per unit voltage component 90° displaced from the stator voltage
V_d is the per unit voltage component in phase with the stator voltage
W is the per unit stator watts
V is the per unit stator vars (lagging = +)
S is the per unit slip

The coefficients C_1 - C_6 are a function of machine and power convertor parameters and are therefore constant. They are defined by the equations.

(3) a) $C_1 = -(R_2 + R_{ceff})/X_m = -R_{2TOT}/X_m$

b) $C_2 = -C_1 R_1$

c) $C_3 = -C_1 (X_1 + X_m)$

d) $C_4 = -(X_1 + X_2 + X_1 X_2/X_m)$

e) $C_5 = R_1 (1 + X_2/X_m)$

f) $C_6 = (1 + X_2/X_m)$

The locus of the voltage vector in the D-Q plane can easily be plotted using the equations 1 and 2. Figures 6 and 7 show parametric plots obtained for the 15000 HP machine with an effective resistance of the cyclo convertor of .0246 pu ohms.

Relating these plots back to Figure 5, the reason for the rapid phase change in the vector sum of V_d and V_q is apparent due to the skirting of the origin near zero slip. The effect of increasing the "resistance" of the cyclo convertor, R_{ceff}, by means of increased current feedback can be seen in Figure 8. This pushes the locus of the control voltage away from the origin for a given condition of stator watts and vars. If one were to use sufficient current feedback in the power convertor, then it should be feasible to avoid rapid phase changes near zero slip and thereby develop a control along the lines originally attempted. However, the orthogonal approach has a far greater performance potential.

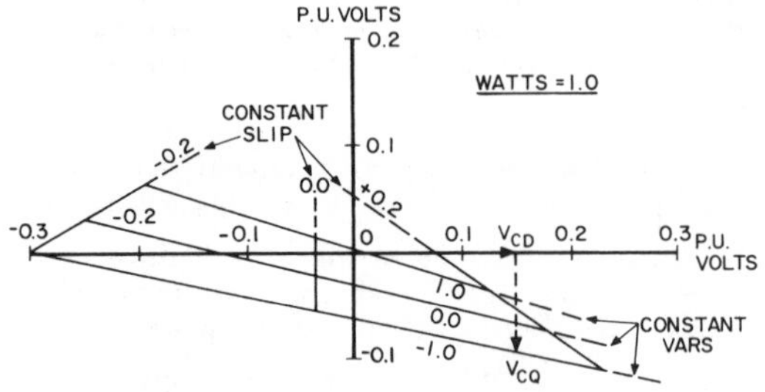

Figure 6 Parametric Locus Plot of Cyclo Converter Reference Voltage for Constant Stator Watts = 1.0 PU with R_{ceff} = .0246 PU ohms

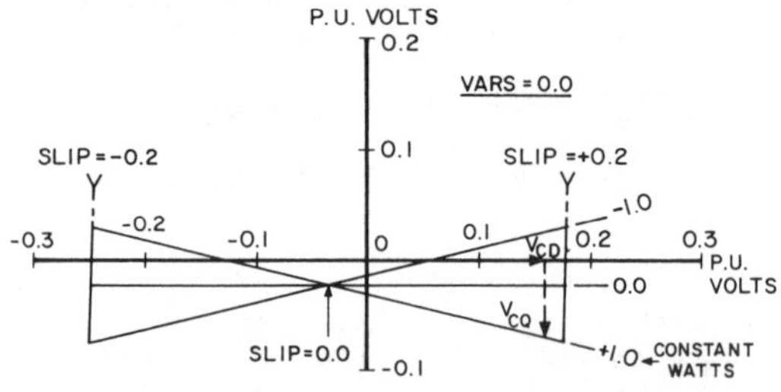

Figure 7 Parametric Locus Plot of Cyclo Converter Reference Voltage for Constant Stator Vars = 0.0 PU, with R_{ceff} = .0246 PU ohms

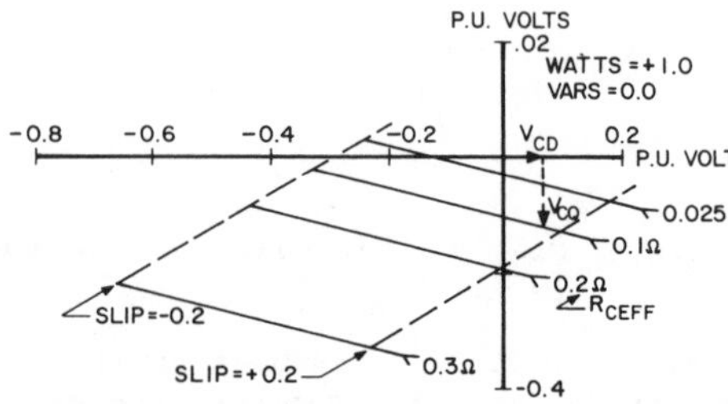

Figure 8 Parametric Locus Plot of Cyclo Converter Reference Voltage vs Effective Cyclo Converter Resistance, R_{ceff}, for Constant

Orthogonal Regulator Concept

Neglecting for the moment the problem of maintaining phase synchronization of the D, Q component with the stator voltage, let us develop the basic regulator concept. Since equations 1 and 2 define the magnitude of the components in terms of the desired watts and vars, the only additional variable needed is to measure the slip which can be derived from the tachometer signal. Thus, in principle the machine could be operated open loop with the feed forward controller depected in Figure 9. To allow for minor variation in machine, cyclo convertor and system control parameters, a regulator is needed. The approach chosen is shown in Figure 10.

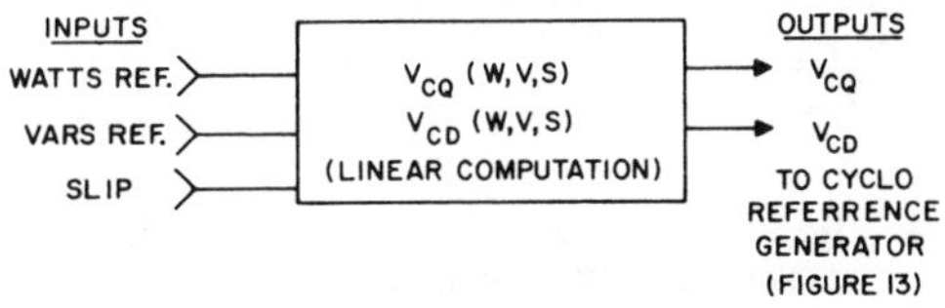

Figure 9 Open Loop Feedforward Orthogonal Control of Motor Watts and Vars Using Equations 1 and 2

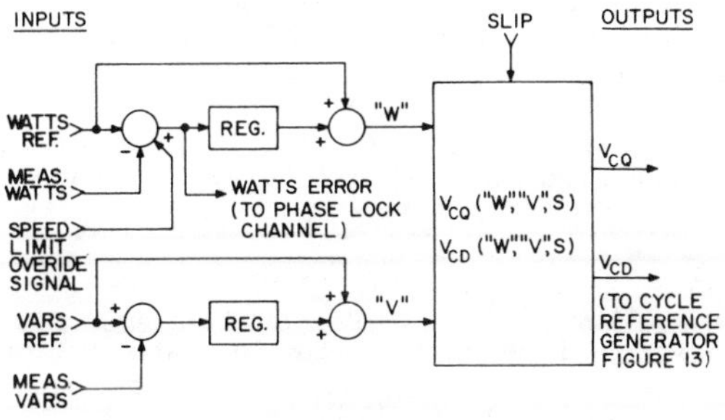

Figure 10 Orthogonal Controller with Watts and Vars Regulator Channels

Here a watts and vars regulator channel have been inserted ahead of the orthogonal controller and the channel output signals derived from the watts/vars error signals are added to the watts and vars reference signals to form the dynamic inputs to the controller based upon equations 1 and 2. With the addition of the regulators it is feasible to eliminate the second order terms in the orthogonal controller. Specifically the coefficients C_2, C_5 and C_7 can be neglected without significant degradation in performance. A regulator gain of 16/1 and a crossover frequency in excess of 100 radians was readily achieved in both channels. (See Orthogonal Regulator Design below)

Phase Synchronization

As indicated above, the orthorgonal control approach requires phase synchronization between the rotor voltage channels and the stator voltage as seen in the rotor reference frame. One method of producing a phase synchronization is by mechanical means. This approach would utilize some form of a shaft position sensor to produce a phase locked oscillator signal running at slip frequency. This could be obtained from the difference frequency between two separate locked oscillators, one synchronized to the line frequency, the other synchronized to the rotor speed.

The actual method employed did not require a mechanical shaft position sensor, but rather developed the frequency and phase synchronization electronically by means of the analog tachometer speed signal and the watts error signal from the orthogonal regulator described above. The analog slip frequency signal is derived by subtracting the speed signal from a fixed voltage source equal to the speed signal voltage at synchronous motor speed. The basis for using the watts error signal can be seen in Figure 11. Shown here is the per unit watts error obtained per degree of phase error in the applied rotor voltages. This was obtained by resolving equations 1 and 2 for Watts and Vars, and performing the following steps:

1. Choose a value for W, V, S and compute V_{cq}, V_{cd} from equations 1 and 2.
2. Choose the phase error in degrees, then rotate the V_{cq} and V_{cd} vectors by the phase error and compute V_{cq}', V_{cd}' (components in phase and in quadrature to stator reference voltage).
3. Using the resolved equations, determine the stator watts and vars with V_{cq}', V_{cd}' applied to the rotor.
4. Calculate the watts error by subtracting from the original value chosen.

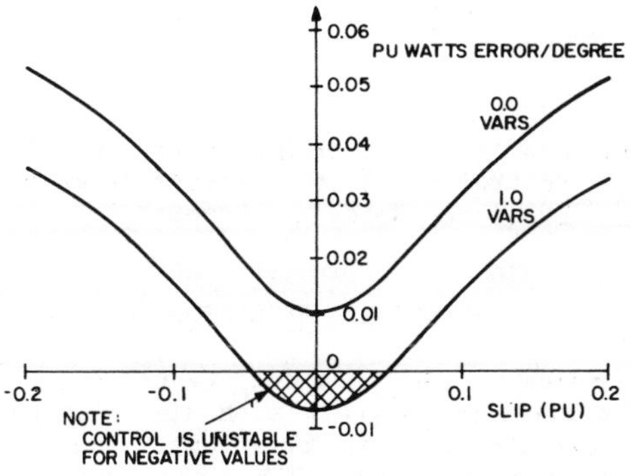

Figure 11 Plot of Per Unit Watts Error vs Slip. Parameter = Stator Vars

This analysis showed that unless conditions call for greater than around .6 PU watts lagging power factor at zero slip (see unstable region Fig. 11) the watts error signal would always be of the same polarity although the sensitivity per degree of phase error would vary with slip and stator Vars. The variations in sensitivity proved to be inconsequential and the region defined as unstable is not within the normal desired operating conditions of a machine. Hence the phase locked regulator channel shown in Figure 12 performed the function of maintaining phase synchronization quite adequately. The phase lock synchronization signals and slip frequency command signals are summed to produce the total frequency command signal. These, along with the orthogonal voltage command signal, are fed to the 3 phase cyclo converter sine wave reference generator (Figure 13) which produces the quantized reference sine waves which control the power converter firing circuits. These in turn fire the thyristor cells and produce the slip frequency voltages applied to the rotor slip rings. The initial phase synchronization was achieved by a separate phase synchronizing circuit while operating under liquid rheostat control during startup.

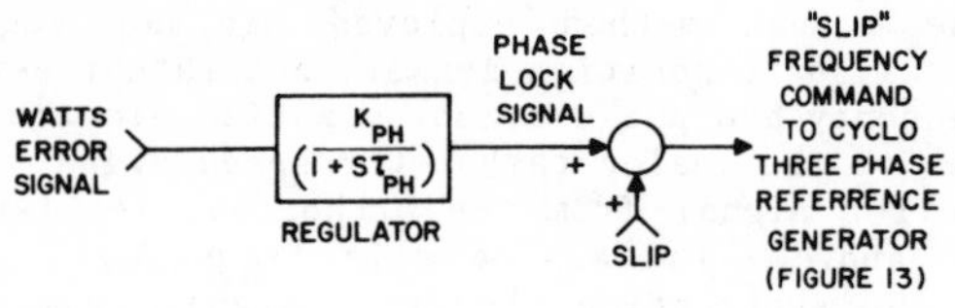

Figure 12 Phase Lock Regulator Channel

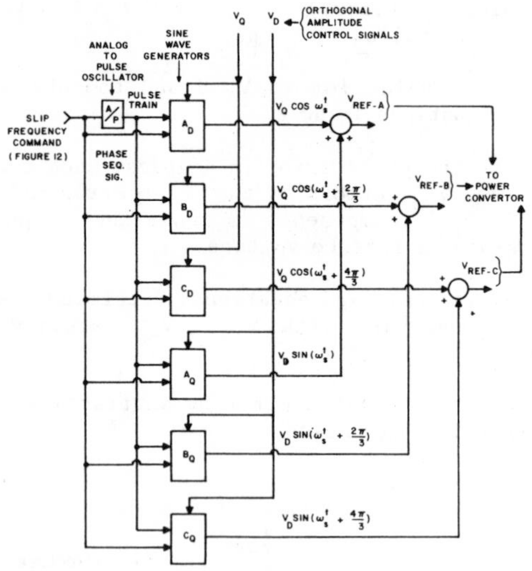

Figure 13 Functional Diagram of Orthogonal Amplitude Control of Cyclo Convertor Three Phase Slip Frequency Reference Generator

ORTHOGONAL REGULATOR DESIGN

The poles of the open loop transfer function for the watts and vars signals with the orthogonal feed forward linearization, via equations 1 and 2, are well damped and occur around 120 rad/sec. This is illustrated in the Bode amplitude response of Figure 14 which shows a slightly peaked response (1.6 db).

Regulator design consisted of a single down break around 6 rads/sec followed by a double up break near 120 radians/sec to compensate for the phase shift resulting from the complex poles of the motor. Zero frequency gain of 24 db produces a crossover in the 150 rad/sec (see Figure 14) range resulting in a very high performance system. Further improvement might be obtained by more careful design of the compensation circuitry if required. Performance achieved with the present design was more than adequate for the present application. The rapid response capability of the system should make it feasible to not only prevent large disturbances to the power system but to serve as an active damper for improving power system stability.

PERFORMANCE SIMULATION

Simulation of all three operating modes was carried out on an EAI Pacer 100 hybrid computer system to verify system response. Second order coefficients C_2, C_5 and C_7 (equations 1 and 2) have been set to zero. Figure 15 shows simulated response of the drive as it approaches top speed and then goes into speed limit operation due to the speed limit override signal (Figure 10). This is followed by the initiation of the fusion experiment overhauling load which reduces the speed to its minimum value in 7 or 8 seconds and then the long reacceleration to top speed. Normal operation would initiate the fusion experiment just prior to reaching the top speed and thus eliminate the disturbance to the power system caused by speed limiting function. Under these conditions, stator watts and vars will remain constant throughout the operation cycle.

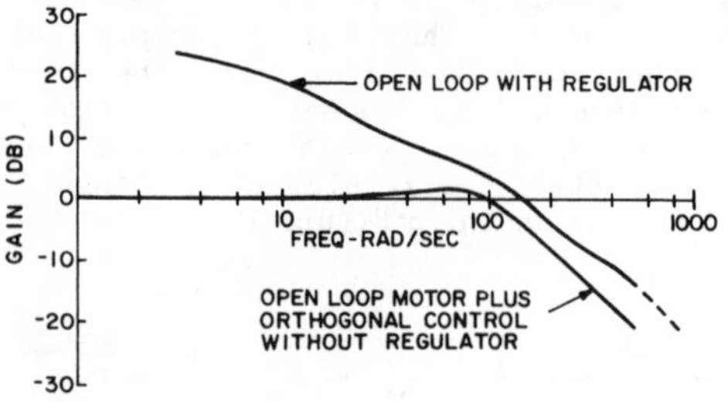

Figure 14 Open Loop Bode Plot of Watts Channel for Doubly Fed Motor with Orthogonal Control with and without Regulator

Figure 16 shows the simulation of the transition from liquid rheostat control (Mode #1) to the doubly fed cyclo converter control. During this transition, the cyclo converter is first synchronized to the rotor voltage prior to connection to the slip rings. Following connection of the cyclo converter the liquid rheostat is ramped to its maximum resistance value and then disconnected. The stator vars are approximately .84 PU while operating under liquid rheostat control. This is regulated to .4 PU following the transition to Mode #2.

Figure 17 is a simulation of the transition from Mode #2 to Mode #3 during the stopping sequence. Here the sequence of events is to initiate the stop by calling for minus rated stator watts and maintaining that condition until the transition speed is reached, at which time zero stator watts and vars are commanded. After decay of the stator current to zero, the stator breaker is opened and the cyclo convertor suicided. The liquid rheostat is then connected to the stator and the cyclo convertor is switched to current regulator operation. Applying 1 per unit current reference to the cyclo converter completes the transition. The cycloconverter frequency is programmed as a function of motor speed so that the cycloconverter regenerates power back into the 60 Hz system.

SUMMARY

An approach has been developed which permit extremely rapid control of stator watts and vars of a wound rotor induction machine of large capacity. The orthogonal regulator concept by which this has been accomplished has been described and simulation results of the drive system given. Application to a 15,000 HP machine for a flywheel generator system to power a fusion reactor experiment are in progress and testing of the drive should be conducted in the near future. The application of the drive system to act as a power system stabilizer would appear fruitful and should be explored.

ACKNOWLEDGEMENT

The author wishes to acknowledge the significant contributions of Mr. M. Horton, General Electric Company to the conception and encouragement in the development of the Orthogonal Control approach described herein.

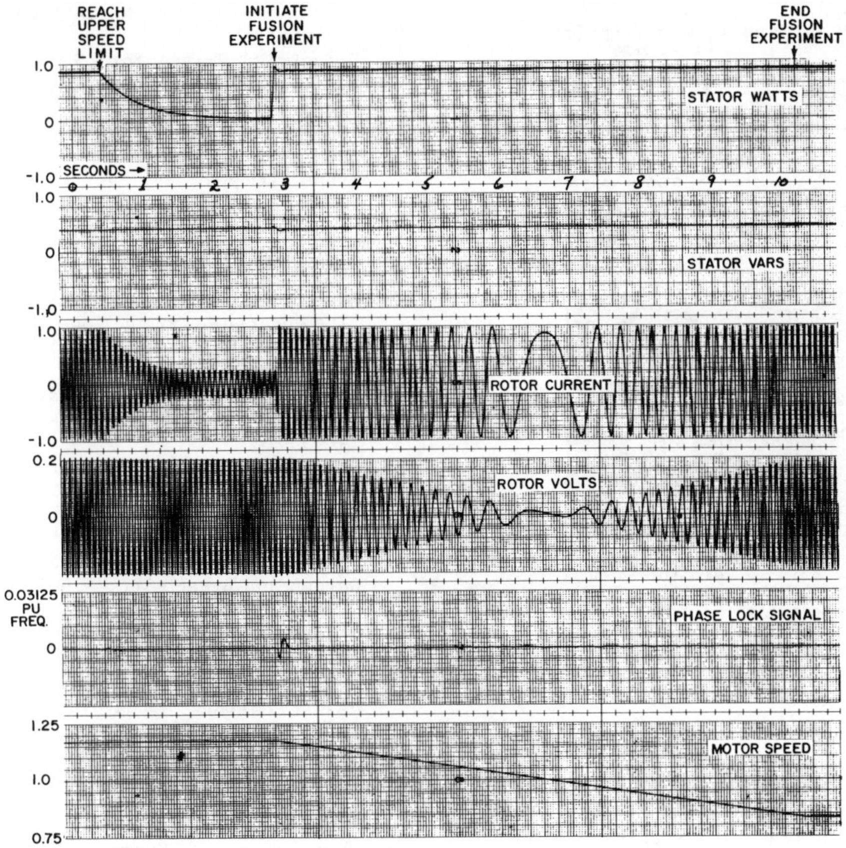

Figure 15 Simulated System Response of Speed Limiting Followed by Drive System Deceleration Due to Fusion Experiment

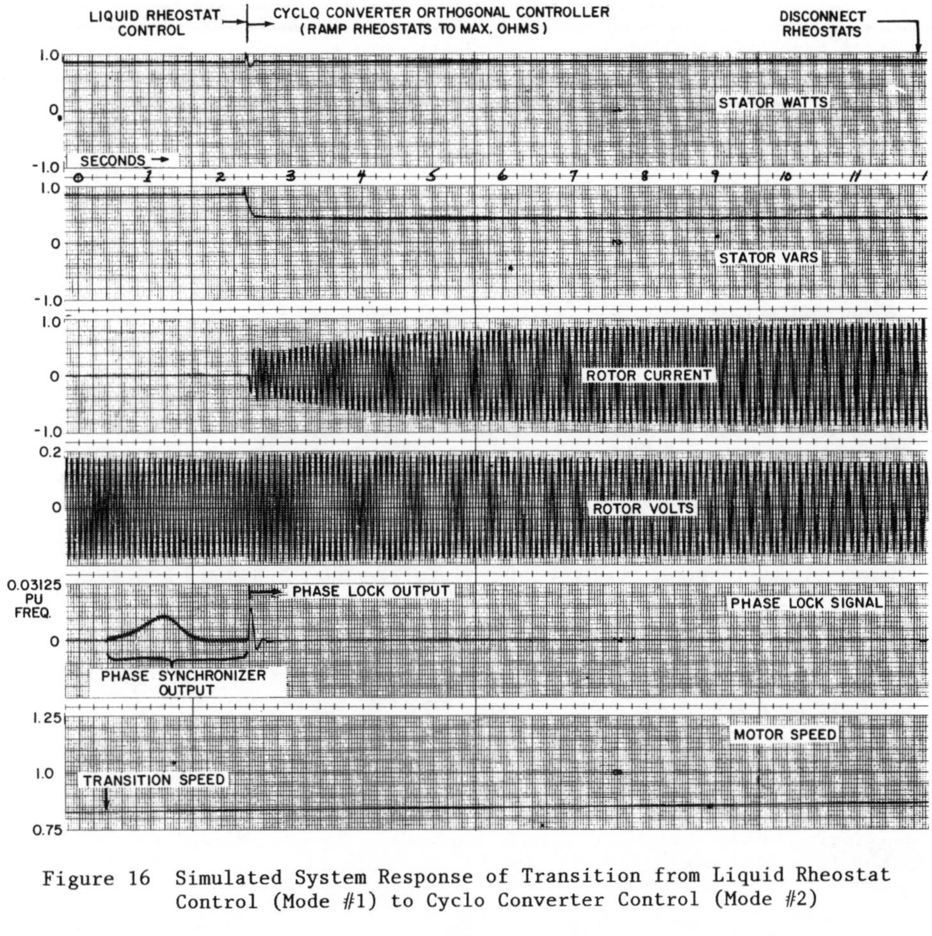

Figure 16 Simulated System Response of Transition from Liquid Rheostat Control (Mode #1) to Cyclo Converter Control (Mode #2)

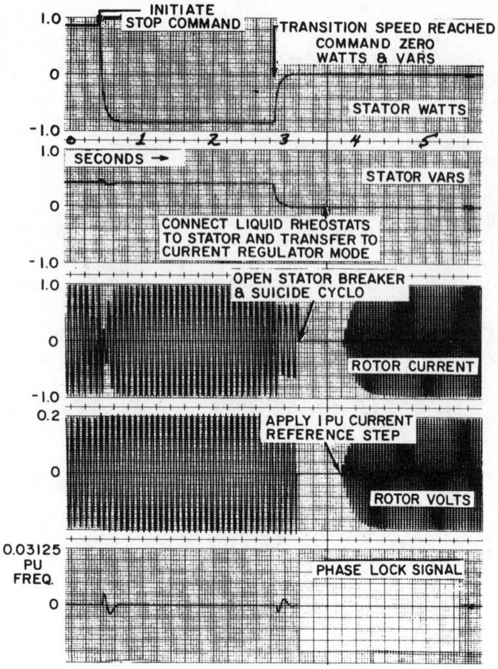

Figure 17 Simulated System Response of Transition from Normal Operation (Mode #2) to Current Regulator (Mode #3) for Stopping Drive System

REFERENCES

1. "Electric Drive for Steel Main Rolls," K.A. Pauly, G.E. Review, Vol 22, No. 5, May 1919.
2. "Theory of Speed and Power Factor Control of Large Induction Motors by Neutralized Polyphase Alternating-Current Commutator Machines," J.I. Hull, Trans AIEE, Vol 39, Part II, 1920, pp 1135-1177.
3. "Some Methods of Obtaining Adjustable Speed with Electrically Driven Rolling Mills, K.A. Pauly, G.E. Review, Vol 24, 1921, pg 422.
4. "The Induction Motor and Other Alternating Current Motors," B.A. Behrend, McGraw Hill, 1921.
5. "The Control of the Speed and Power Factor of Induction Motors," M. Walker, Ernest Benn Ltd., 1924, pp 118-120.
6. "AC Machines," A.F. Puchstein, T.C. Lloyd, A.G. Conrad, John Wiley & Sons, 1936, 1942, 1954, pp 342-350.
7. "Principles of AC Machinery," R.R. Lawrence, McGraw Hill, 1940, pp 531-535.
8. "AC Machinery," B.F. Bailey, J.S. Gault, McGraw Hill, 1951, pp 260-264.
9. "The Nature of Induction Machines," P. Alger, Gordon and Breach Inc., 1965, pp 318-320.
10. "Two Reaction Theory of Synchronous Machines, Generalized Method of Analysis -Part I," R.H. Park, Trans. AIEE, Vol 48, July 1929, pp 716-730.
11. "An Analysis of the Induction Machine," H.C. Stanley, Trans AIEE, PAS, Vol 57, 1938, pp 751-757.
12. "Method of Multiple Reference Frames Applied to the Analysis of Symmetrical Induction Machinery," P.C. Krause, IEEE Trans PAS, Vol PAS-87, January 1968, pp 218-227.
13. "Simulation of Symmetrical Induction Machinery, P.C. Krause, C.H. Thomas," IEEE Trans PAS, Vol PAS-84, November 1965, pp 1038-1053.

Part II
Synchronous Motor Drives

Section 6
Current Fed Inverter Drives

Synchronous Motor Drive with Current-Source Inverter

GORDON R. SLEMON, SENIOR MEMBER, IEEE, SHASHI B. DEWAN, MEMBER, IEEE, AND JAMES W. A. WILSON, MEMBER, IEEE

Abstract—**Steady-state properties of a variable-speed drive using a synchronous motor fed by a controlled current-source inverter are derived from an equivalent circuit model, including saturation, and are confirmed by experiment. The system is shown to be stable under all operating conditions. Control strategy for open-loop operation is discussed.**

INTRODUCTION

FOR CERTAIN applications, ac motor drives using inverter-fed synchronous machines provide some features that make them preferable to induction motor drives [1], [2]. For example, where precise simultaneous speed control of a number of motors is required, a system using synchronous motors offers a practical approach. As a further example, a synchronous machine responds more quickly than an induction machine to change in torque. Also, a variable-frequency synchronous motor drive offers the possibility of simple precise position control.

Most inverters used in ac drives are controllable voltage sources of variable frequency. Recently, the advantages of simplicity, greater controllability, regenerative capability, and ease of protection provided by the current-source inverter have become widely recognized [3], [4].

This paper examines some of the properties of a drive using a synchronous motor fed by a controlled current-source inverter and operating in the open-loop mode. The torque–angle characteristics of the drive are derived from an equivalent circuit model that includes the effects of magnetic saturation. The accuracy of the model is confirmed by experiment. The stability of the drive is then examined using perturbation techniques. Finally, the control strategy for this drive is discussed, and comparisons are made between synchronous motor drives using voltage and current sources.

DESCRIPTION OF SYSTEM

The current-fed synchronous motor drive is shown in block diagram form in Fig. 1. A six-pulse phase-controlled rectifier, operating from a three-phase supply, is controlled in such a way as to produce a unidirectional current of controllable magnitude in a smoothing inductor. This current is then switched in a thyristor inverter to provide variable-frequency three-phase current to a standard synchronous motor.

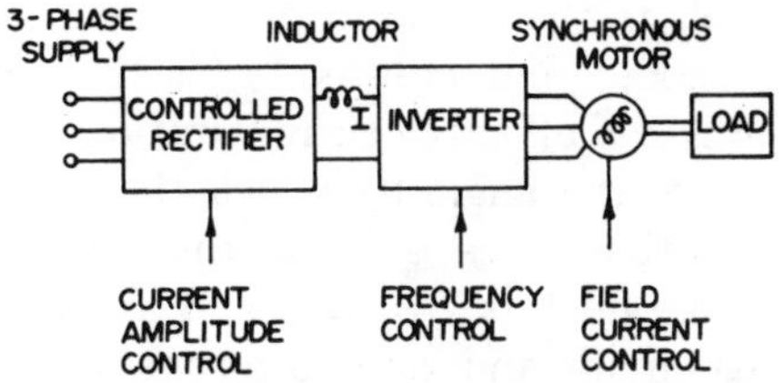

Fig. 1. Drive configuration for open-loop current-fed synchronous motor control.

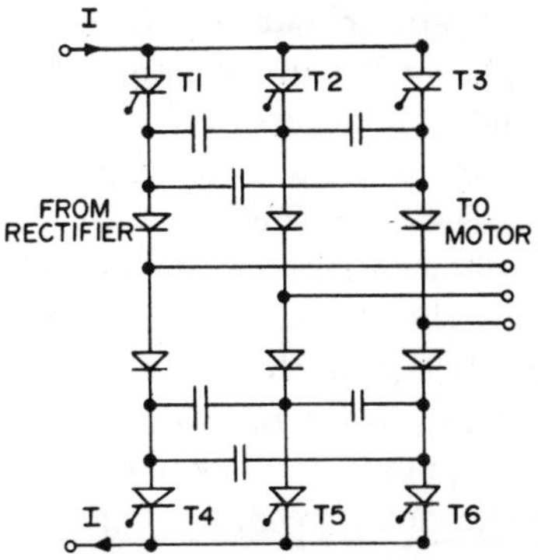

Fig. 2. Inverter circuit.

The inverter circuit [3] is shown in Fig. 2. It contains six thyristors, T1 to T6, each of which conducts for one-third of each cycle. Six capacitors provide commutation energy, and six diodes are included to prevent interaction of the commutation capacitors with the motor load and to prevent interaction between the upper and lower commutation circuits.

The current waveform of the inverter is of the six-step type. The voltage waveform is dependent on the properties and loading of the synchronous motor.

Precise speed control can be achieved through control of the inverter frequency. Reversal of rotation is achieved by reversal of the phase sequence of thyristor switching as the frequency goes through zero value. Reversal of power flow is achieved by reversing the polarity of the voltage in the direct-current link between the rectifier and the inverter. Thus, full four-quadrant operation of the drive is provided. In its simplest form the drive can be operated with the magnitude of the stator current set at rated value and with the field current held constant. For a multiple-motor drive, the stator windings of all motors are connected in series.

Paper TOD-73-89, approved by the Static Power Converter Committee Industry Applications Society for presentation at the 1973 Eighth Annual Meeting of the IEEE Industry Applications Society, Milwaukee, Wis., October 8–11. Manuscript released August 15, 1973.

G. R. Slemon and S. B. Dewan are with the Department of Electrical Engineering, University of Toronto, Toronto, Ont. Canada M5S1A4.

J. W. A. Wilson is with the Research and Development Center for Electronics, Reliance Electric Co., Ann Arbor, Mich.

Reprinted from *IEEE Trans. Ind. Appl.*, vol. IA-10, pp. 412–416, May/June 1974.

STEADY-STATE PERFORMANCE

The equivalent circuit of Fig. 3 has been used successfully in predicting the performance of voltage-fed synchronous machines [2]. In this section let us examine the extent to which this circuit is applicable for current-fed machines. In the process, some of the special characteristics of current-fed machines will be demonstrated.

In Fig. 3, R_s is the stator resistance per phase, L_{Ls} is the stator leakage inductance, L_{ms} is the magnetizing inductance, n is the effective current ratio, i_f is the field current, and β is the angle by which the field axis lags the magnetomotive force axis of the stator winding. Values of the parameters L_{Ls}, L_{ms}, and n can be derived from the open-circuit and zero-power-factor-lagging load test data [5] plotted in the graph of Fig. 3. The equivalent circuit is based on assumptions of negligible saliency and negligible effects of stator current harmonics.

With constant stator current I_s and constant field current i_f, the phasor loci are shown in Fig. 4. The stable zero torque condition occurs when the stator and field magnetomotive forces are aligned at $\beta = 0$. As the motor is loaded β increases, causing the effective field current phasor I_f to follow a circular locus. The magnetizing current I_{ms}, being the phasor sum of I_s and I_f, also follows a circular locus. If the machine were magnetically linear, the air-gap voltage phasor E_{ms} would also follow a circular locus. The torque would be equal to the air-gap power per unit of synchronous speed, i.e., for a p-pole machine,

$$T = \frac{3p}{2\omega} (I_s E_{ms}^*) \text{ real part}$$

$$= \frac{3p}{2} \mid I_s \mid \mid I_f \mid L_{ms0} \sin \beta \qquad (1)$$

where L_{ms0} is the unsaturated value of magnetizing inductance.

With magnetic saturation the air-gap voltage ceases to be linearly dependent on the magnetizing current. For any value of I_{ms}, the appropriate value E_{ms} may be found from the open-circuit curve of Fig. 3, choosing the value corresponding to $i_f = I_{ms}/n$. Fig. 4 shows the locus of E_{ms} including saturation. The loci of Fig. 4 apply for any value of speed and inverter frequency, except that the magnitude of the air-gap voltage is linearly proportional to frequency ω. The effect of saturation is independent of frequency.

A torque–angle relationship for a current-fed machine is shown in Fig. 5. If saturation is ignored, this has the sine form of (1). With saturation, the system is less stiff and reaches its maximum value of torque at a field angle β greater than Π/2.

Test Results

Load tests were performed on a 5-hp laboratory machine with a cylindrical rotor. The machine parameters (per unit) were $R_s = 0.039$ and $L_{Ls} = 1.3$ (at rated air-gap

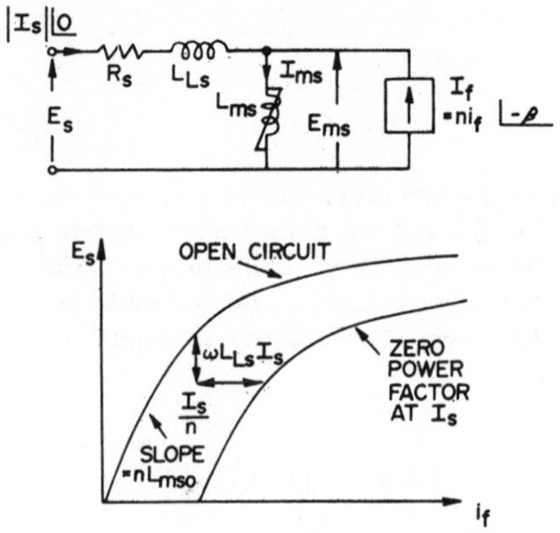

Fig. 3. Equivalent circuit of synchronous machine together with open-circuit and zero-power factor test curves.

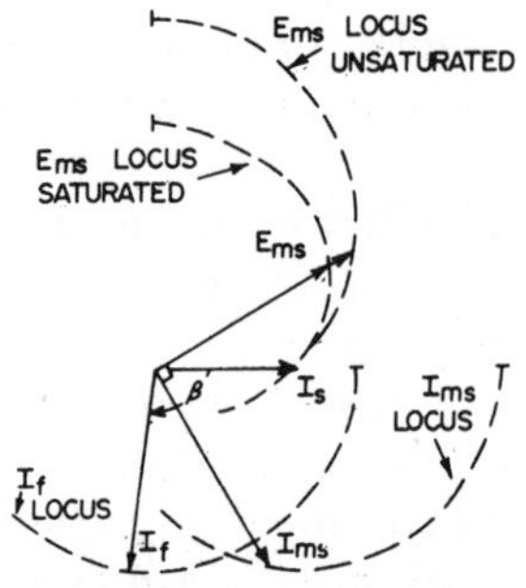

Fig. 4. Loci of effective field current I_f, magnetizing current I_{ms}, and air-gap voltage E_{ms}.

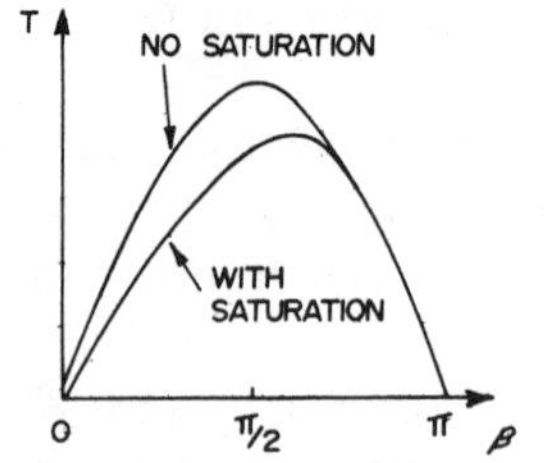

Fig. 5. Torque T as a function of field angle β showing the effect of magnetic saturation.

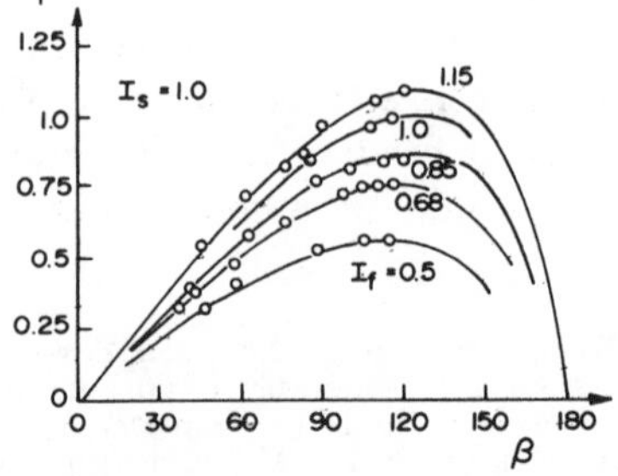

Fig. 6. Torque characteristic of laboratory machine. ——— Calculated. ○○○ Experimental. All values in per unit.

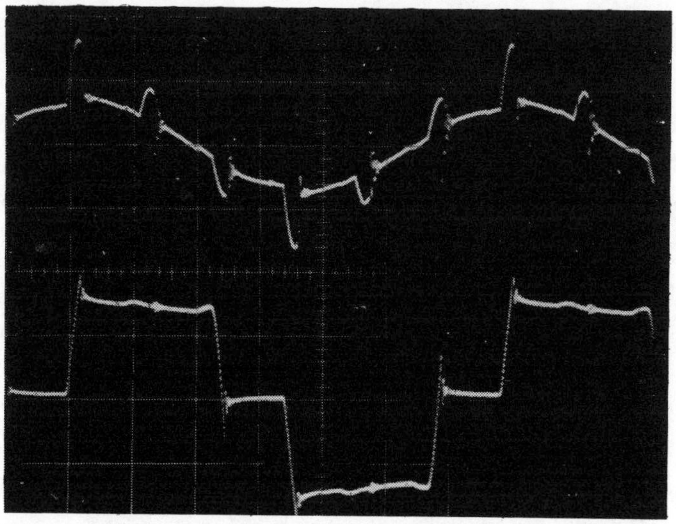

Fig. 7. Typical waveforms of stator voltage (top) and stator current (bottom).

flux). The stator current was maintained constant at rated value, and torque–angle measurements were made in the stable range for several values of field current. The results, shown in Fig. 6, demonstrate good agreement between measured and computed values.

It is noted that the peak torque in Fig. 6 occurs at a field angle β of about 120° with rated current in both stator and field windings.

The input current to the stator is of the simple stepped form containing no even harmonics, and in the limiting case of an ideal waveform, only 20 percent and 14 percent, respectively, of fifth and seventh harmonics. The stator voltage is approximately sinusoidal but contains transient pulses resulting from the high rates of change of stator current at the switching intervals. Typical current and voltage waveforms are shown in Fig. 7.

STABILITY

Voltage-fed synchronous machines may experience instability or continuous oscillation under some operating conditions, particularly with low values of supply voltage and frequency [6]. The stability of the current-fed synchronous motor has been examined using a conventional *d-q* model for the motor and the method of perturbations [7]. The analysis, given in the Appendix, shows that the machine should exhibit no spontaneous sustained oscillations when operated within its steady-state torque limits. This is shown to be true, even if the machine has saliency. The field winding provides damping of forced oscillations under loaded conditions. The analysis in the Appendix shows that a quadrature axis damper winding is, however, required to provide damping at no load.

CONTROL STRATEGY

In the simplest control system, the stator current is set at a constant amplitude, usually of rated value. The field current is also held at a constant value. The speed is controlled by varying the inverter frequency. As the load torque changes, the field angle β adjusts itself accordingly.

The choice of an appropriate value for the field current involves a compromise between field heating limitations and maximum torque capability. In a machine that is magnetically linear, the maximum torque would be proportional to the field current (1). The experimental torque curves of Fig. 6 show that the peak torque is less than proportional to field current.

At no load, the motor operates with the stator and rotor fields aligned, i.e., at $\beta = 0$. Fig. 4 shows that this is normally a highly saturated condition. As load is added, the field angle β increases and the air-gap voltage follows the locus shown in Fig. 4. The machine is normally just emerging from saturation as the peak gap power and peak torque condition is reached. With constant stator and rotor currents, the resistive losses are constant. The iron losses are at a minimum at the maximum torque condition and increase as load torque is decreased. The increase of the iron losses at high speed is compensated by the increased cooling capability of the machine at high speed. It may, however, be necessary to set the stator and field currents somewhat below normal rated values to restrict overall losses to acceptable limits for cooling.

Fig. 6 shows that, with stator and field currents set at rated values, the peak torque is equal to the normal rated value of the motor when fed by a rated voltage supply. It is not practicable to operate at this peak value of torque since any small perturbation in torque would cause the motor to lose synchronism. The maximum design value of load torque must be limited to a value T_L lower than $T_{\max}$ as shown in Fig. 8.

One factor determining the required margin between $T_{\max}$ and T_L is the maximum torque impulse that the drive is expected to withstand at rated load without loss of synchronism. Suppose a shock torque of magnitude $\hat{T}$ occurs, in addition to the normal T_L, for a short time interval Δt when the motor is revolving at speed ν. The kinetic energy taken from the rotor inertia by the impulse is

$$\Delta W_{\text{kin}} = \hat{T}\nu\Delta t \qquad (2)$$

assuming the change in the speed to be relatively small. The rotor angle will subsequently swing to the right of the initial operating point of Fig. 8. Stability will be retained only if the shaded area of Fig. 8 is greater than ΔW_{kin} [5] in (2). It is noted from (2) that the shock torque constraint is most severe at maximum speed.

Another factor influencing the required torque margin is the rate at which the drive speed can be changed. Let J be the polar moment of inertia of the motor and its load. For a load torque T_L, the acceleration is given by

$$\frac{d\nu}{dt} = \frac{1}{J}(T - T_L) \qquad (3)$$

where the developed torque T has a maximum value $T_{\max}$ as shown in Fig. 8.

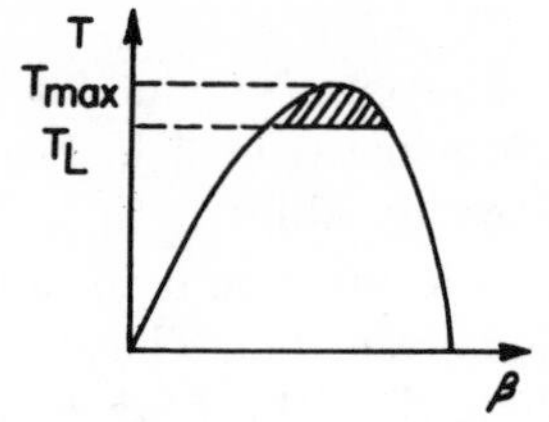

Fig. 8. Normal operating torque T_L of motor. Shaded area shows maximum shock energy which can be absorbed stably.

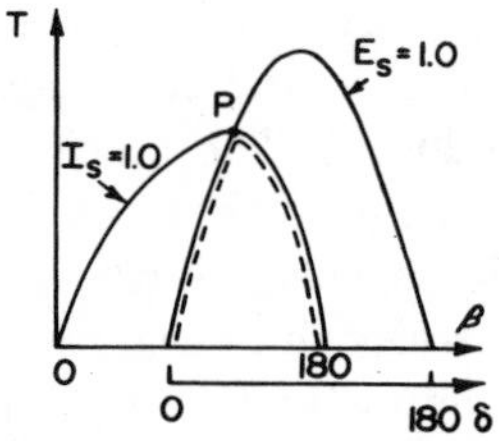

Fig. 9. Torque characteristics for current-fed motor with $I_s = 1.0$ and voltage-fed motor with $E_s = 1.0$. Dotted curve applied for voltage-fed motor with current limit of 1.0.

When a load torque limit T_L has been chosen, the rate of increase of inverter frequency must be limited to maintain synchronism. For a p-pole machine, the stator frequency f has a maximum increase rate of

$$\frac{df}{dt} = \frac{P}{4\pi J}(T_{\max} - T_L). \tag{4}$$

Under normally loaded conditions, deceleration may occur stably at a considerably faster rate than acceleration because $T_{\max}$ in (4) is negative while T_L is still positive. Under deceleration conditions the rectifier operates in the inverter mode, returning power to the supply.

If pull-out occurs under any excessive load condition, the system is self-protecting since the source inverter limits the stator current. In that event, however, it is necessary to provide means of adjusting the inverter frequency to restore synchronism.

Comparison with Voltage-Fed Motor

A comparison of the torque–angle characteristics of a synchronous motor with constant stator current and with constant stator voltage is shown in Fig. 9. The latter is plotted on the load angle δ between the rotor axis and the stator flux axis. At point P the terminal conditions are identical.

It is evident from Fig. 9 that the voltage-fed machine has a larger torque limit and more stability margin than the current-fed machine. This larger peak torque is, however, achieved by allowing the stator current to increase considerably, typically to about twice rated value. If the inverter on the voltage-fed drive had been fitted with a current limiter set for rated stator current ($I_s = 1.0$) the torque–angle locus would have been the dotted curve shown in Fig. 9. The voltage-fed machine would then be subject to an even greater restriction on the acceptable value of load torque T_L than the current-fed machine in the presence of shock torque. The current limit of a voltage-fed motor would normally be set at a value somewhat greater than rated value since only a short-term overload would be anticipated. Thus, there is probably little difference in the torque ratings with the two types of drive.

In a voltage-source inverter, the current limit is normally imposed by reduction of voltage when the current tends to exceed the limit. While this approach works well with induction machines it may lead to instability if used with synchronous machines in which the stator current may increase as a consequence of a stator voltage reduction. This suggests that a direct means of current control such as that of Fig. 1 may be necessary for a voltage-fed system and that using a current-source system throughout the complete load and speed range is relatively simpler.

A feedback control loop may be applied to the system of Fig. 1 to control the current amplitude to a value that will limit the motor voltage to a maximum of approximately 1.1 per unit. During light load periods the stator current will be considerably less than rated value. The maximum stator current can then be increased, possibly to 1.5 per unit without overheating during short-term peak load periods. The maximum torque is thereby increased.

The system of Fig. 1 can be fitted for constant field angle operation by using a sensor on the motor shaft to provide firing pulses for the inverter [2]. Fig. 6 shows that the optimum field angle setting for such a drive is in the region of 120°.

CONCLUSION

A current-fed synchronous motor operates in a saturated mode for most of its torque range, emerging from saturation as maximum torque is approached. A steady-state analysis based on a nonlinear equivalent circuit is shown to provide a good prediction of the relation between torque and field angle. The field angle is shown to be greater than 90° at the condition of maximum torque.

Operating open-loop with constant stator current and constant field voltage, the drive is shown to be free of the spontaneous oscillations which characterize voltage-fed synchronous motors at low voltage and frequency.

APPENDIX

Stability Analysis

In the following analysis, saturation and damper windings are ignored but provision is made for inclusion of saliency. Following Parks' transformation, a constant stator current of value I_s at a field angle β can be represented by direct- and quadrature-axis currents as

$$i_d = I_s \cos\beta \tag{5}$$

and

$$i_q = I_s \sin\beta. \tag{6}$$

The field circuit is represented by the equation

$$e_f = (R_f + pL_f)i_f + pL_{md}i_d \tag{7}$$

where $p = d/t$ and L_{md} is equal to the inductance L_{ms0} in Fig. 3 if the field variables e_f and i_f are appropriately referred to the stator using the current ratio n. If the load torque is T_L (including mechanical losses), the torque expression is

$$T_L = L_{md} i_f i_q + (L_{md} - L_{mq}) i_d i_q - J\, d\nu/dT \qquad (8)$$

where L_{mq} is the quadrature-axis magnetizing inducance.

Designating steady-state values by the subscript 0 and small perturbations by Δ, the perturbation equations of the machine are

$$\Delta i_d = -I_s \sin \beta_0 \Delta\beta \qquad (9)$$

$$\Delta i_q = I_s \cos \beta_0 \Delta\beta \qquad (10)$$

$$\Delta e_f = (R_f + pL_f)\Delta i_f + pL_{md}\Delta i_d \qquad (11)$$

$$\Delta T_L = L_{md}(i_{f0}\Delta i_q + i_{q0}\Delta i_f) + (L_{md} - L_{mq}) \cdot (i_{d0}\Delta i_q + i_{q0}\Delta i_d) + Jp^2\Delta\beta. \qquad (12)$$

With a constant field voltage, Δe_f is equal to zero. Substitution of (9)–(11) into (12) gives, after some manipulation,

$$\frac{\Delta\beta}{\Delta T_L} = (R_f + pL_f)/D \qquad (13)$$

where

$$D = p^3(JL_f) + p^2(JR_f) + pI_s[L_{md}^2 I_s \sin^2 \beta_0 + L_f L_{md} i_{f0} \cos \beta_0 + L_f(L_{md} - L_{mq}) I_s \cos 2\beta_0] + R_f I_s[L_{md} i_{f0} \cos \beta_0 + (L_{md} - L_{mq}) I_s \cos 2\beta_0].$$

The bracketted part of the constant term in D is equal to $dT_L/d\beta$ from (8). It is therefore positive as long as the motor is operating with an angle β_0 in the positive slope portion of the torque/angle curve (Fig. 5). This condition is necessary for retention of synchronism. If this term is positive, the coefficient of p in D will also be positive. Applying the Routh criterion, the system will be stable if all coefficients of D are positive and if

$$L_{md}^2 I_s \sin^2 \beta_0 > 0 \qquad (14)$$

a condition met everywhere except for $\beta_0 = 0$.

At the no-load condition with $\beta_0 = 0$, (13) simplifies to

$$\frac{\Delta\beta}{\Delta T_L} = \frac{1}{p^2 J + [L_{md} i_{f0} I_s + (L_{md} - L_{mq}) I_s^2]}. \qquad (15)$$

This second-order relation is undamped. Quadrature-axis dampers are required if the motor is to have adequate damping of oscillations due to load disturbances near no load.

REFERENCES

[1] A. Bellini, A. de Carli, and M. Murgo, "Speed control of synchronous machines," *IEEE Trans. Ind. Gen. Appl.*, vol. IGA-7, pp. 332–338, May/June 1971.

[2] G. R. Slemon, J. B. Forsythe, and S. B. Dewan, "Controlled-power-angle synchronous motor inverter drive system," *IEEE Trans. Ind. Appl.*, pp. 216–219, Mar./Apr. 1973.

[3] K. P. Phillips, "Current source converter for ac motor drives," *IEEE Trans. Ind. Appl.*, vol. IA-8, pp. 679–683, Nov./Dec. 1972.

[4] A. Habock and D. Kollensperger, "State of development of converter-fed synchronous motors with self-control," *Siemens Rev.*, vol. 38, no. 9, pp. 390–392, 1971.

[5] G. R. Slemon, *Magnetoelectric Devices.* New York: Wiley, 1966.

[6] T. A. Lipo and P. C. Krause, "Stability analysis for variable frequency operation of synchronous machines," *IEEE Trans. Power App. Syst.*, vol. PAS-87, pp. 227–234, Jan. 1968.

[7] B. Adkins, *The General Theory of Electrical Machines.* New York: Wiley, 1957.

Simplified Model and Closed-Loop Control of a Commutatorless DC Motor

ALAIN JAKUBOWICZ, M. NOUGARET, AND ROBERT PERRET

Abstract—**A simplified dynamic model of a current-fed, self-commutated synchronous motor, operating with a constant angle between the motor counter electromotive force (CEMF) and the input current. The model, which is justified experimentally, includes only two time constants, associated with the rotor inertia and the dc link inductance. This permits the design of the current and speed controlling loops along the same principles used for classical armature-controlled dc motor drives. In this way, the experience which is widely available in industry can be applied, at least with a good approximation, to design of the commutatorless dc motor drives. Various control configurations are discussed. Control loops are designed and closed around the basic motor-inverter block. An experimental result is obained for each control configuration, thus validating the proposed model.**

I. INTRODUCTION

VARIABLE-SPEED drives use in most cases motors powered with dc current and fed by static converters using rectifier bridges or choppers. The use of these drives has a number of mechanical and electrical limitations caused by the necessity of fitting a mechanical commutator [1].

The construction of the mechanical commutator has to be extremely delicate and finally the actual power is limited in the region of 10 MW for a speed of 1000 r/min and of 500 kW for 5000 r/min. The use of these machines is absolutely impossible in explosive, corrosive, or dust-filled atmosphere. The limitations of use of dc machines and the hope that in the near future they can be replaced by less fragile equipment at a lower cost has led to the development of equipment which use a static converter in conjunction with an ac machine.

As soon as the power passes 500 kW the forced commutation converters are no longer able to compete (the size of the condensers assisting commutation coupled with the number of components becoming necessary). The solution achieved in this range of power is a current-fed self-commutated synchronous machine. This equipment is the most simple in use, particularly because of its ability to work with natural commutation [2]-[4]. With the development of motors fitted with permanent magnets, it can well be imagined that in the near future this type of associated equipment will also be used within the range of current power at around 50-500 kW.

Paper ID 79-35, approved by the Industrial Drives Committee of the IEEE Industry Applications Society for presentation at the 1979 Industry Applications Society Annual Meeting, Cleveland, OH, September 30–October 4. Manuscript released for publication November 2, 1979.

The authors are with the Laboratoire d'Electrotechnique, E.R.A./C.N.R.S 534, ENS IEG, Institut National Polytechnique, Grenoble, France.

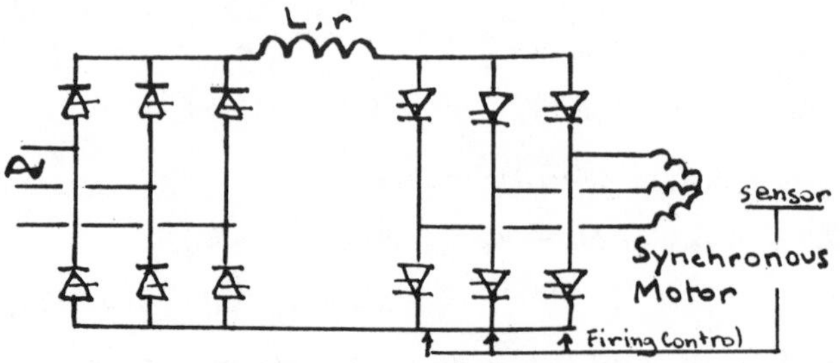

Fig. 1. Basic principle.

II. DESCRIPTION OF THE EQUIPMENT

The equipment used is described in Fig. 1. The motor power is 13 kW. The current source is connected with the assistance of a rectifier bridge controlled in series with an inductor [5].

The static frequency converter is a six-thyristor Graetz bridge, piloted by a position sensor mounted on the motor shaft. The converter creates a square wave current in the stator windings. The current waves as registered experimentally in the machine is shown Fig. 2. The spread of conductors is such that the tension wave is reduced to its fundamental term as a first approximation [6], the phase angle between the tension v_a and the current i is called ϕ, and ψ_a is the phase angle between the electromotive force (EMF) E and the pulse.

III. REPRESENTATION OF A SELF-CONTROLLED SYNCHRONOUS MACHINE

A model has to be constructed which can take into account a constant steady state and the various transients in the functioning zones required in order to meet the wished-for requirements (dc current loop, speed, etc.). A considerable number of studies have been undertaken to take into account the transient evolutions and the processes of control. However, these models are too complex and, therefore, difficult to use for the purpose of demonstrating the self-controlled synchronous machine in regard to all its envisaged requirements [7], [8]. Thus an extremely simplified model has been built, which is experimentally valid in all the functioning areas of importance, allowing a certain number of hypothesis [5].

A. The Motor Model for Control

To establish the model, the main theme has been as follows: the system contains a large number of time constants which

Reprinted from *IEEE Trans. Ind. Appl.*, vol. IA-16, pp. 165–172, Mar./Apr. 1980.

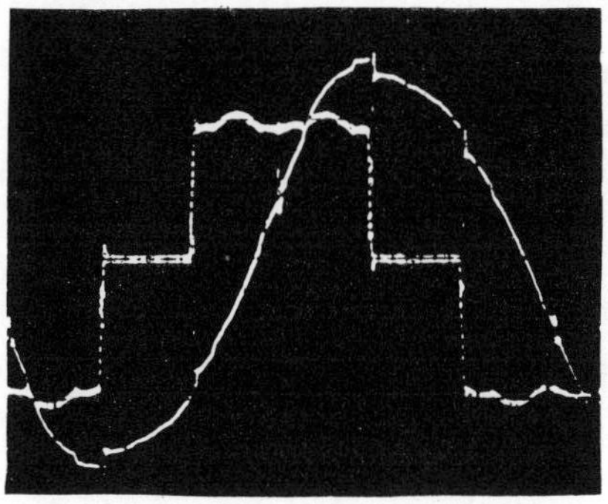

Fig. 2. Experimental register of current and tension waves.

can be classified in two groups, those taking into account rapid phenomena and those dealing with slow phenomena. In the field of operation being considered, it is possible to consider these two sections separately and separate rapid phenomena from slow phenomena [9].

Average values have been considered, solely leaving out all internal time constants of the machine caused by group inertia and smoothing inductors [10]. In order to establish the model, four steps will be envisaged.

1) The equations were established registering the evolution of the system during time interval dt.

Mechanical equation:

$$J\frac{d\Omega}{dt} + f_1\Omega = \Gamma(t) - \Gamma_r$$

where J is the group inertia, Ω the speed, f_1 the friction coefficient, Γ the electromechanical torque, and Γ_r the resistant torque. In this case a dc generator supplying a resistor bank is used as a load for the motor; the rule for resistant torque is $\Gamma_r = k\Omega$:

$$J\frac{d\Omega}{dt} + f\Omega = \Gamma(t),$$

where $f = f_1 + k$.

Electrical equations:

$$\Gamma(t) = k\phi_m I_0 \sin(\omega t + \psi_a)$$

$$rI_0(t) + L\frac{dI_0(t)}{dt} = u_0(t) - u_0'(t)$$

where I_0 is the current in the dc current loop, r and L are the resistance and the inductance of the smoothing inductor, and u_0 and u_0' are the tensions before and after the inductance (Fig. 1).

2) The average value of the differential equations is taken describing the system during a time interval Δt sufficiently large to cope with rapid electrical phenomena. In the meantime, Δt can be considered as a differential element in connection with slow electromechanical phenomena.

In such timing conditions, an average value can be used for $\Gamma(t)$, $u_0(t)$, and $u_0'(t)$; there is no time for the speed to develop, and thus it remains constant. Practically, it must be considered that at the moment t the system is at operating point Γ_0, Ω_0 (in steady state $f\Omega_0 = \Gamma_0$). At moment $t + \Delta t$, the speed is passed to $\Omega_0 + \Delta\Omega$ and the torque to $\Gamma_0 + \Delta\Gamma$. Thus

$$J\frac{\Delta\Omega}{\Delta t} + f(\Omega_0 + \Delta\Omega) = \Gamma_0 + \Delta\Gamma$$

may be

$$\left(\frac{J}{f} + \Delta t\right)\frac{\Delta\Omega}{\Omega_0} = \frac{\Delta\Gamma}{\Gamma_0}\Delta t.$$

For the scheme of separation of rapid and slow phenomena to be valid, two conditions must be assumed:

$$\Delta t \gg dt$$

$$J/f \gg \Delta t.$$

In the case studied, $J/f = 5$ s, and if one takes $\Delta t = \pi/\omega$ corresponding to the speed 150 r/min, $\Delta t = 100$ ms, which validates our working hypothesis. Based on an average Δt, the following equations are obtained:

$$J\frac{d\Omega}{dt.} + f\Omega = \langle\Gamma\rangle$$

$$\langle\Gamma\rangle = kI_0\phi_m \cos\psi, \qquad \text{with } \psi = \psi_a + \pi/6$$

$$rI_0 + \frac{LdI_0}{dt} = \langle u_0(t)\rangle - \langle u_0'(t)\rangle$$

$$\langle u_0(t)\rangle = \frac{3\sqrt{3}}{\pi}V\sqrt{2}\cos\alpha$$

$$\langle u_0'(t)\rangle = \frac{3}{\pi}\sqrt{3}E\sqrt{2}\cos\psi$$

where α is the delay angle of the rectifier bridge and V the supply voltage. The relation between E and Ω is

$$E = k\Omega_m.$$

3) An operational structure can be adopted for the average equations based on Δt:

$$Jp\Omega + f\Omega = \Gamma$$

$$rI_0 + LpI_0 = \frac{3\sqrt{3}}{\pi}V\sqrt{2}\cos\alpha - \frac{3\sqrt{3}}{\pi}E\sqrt{2}\cos\psi.$$

4) A working sketch showing the self-controlled synchronous machine modeled for its control can be established as a translation of these equations (Fig. 3). It is now necessary to validate this model by comparing its theoretical answer with the practical experiments.

B. Validity of the Model

In order to study the model, it was decided to work to a size of output I_0. This could be translated to schematic diagram as shown in Fig. 4.

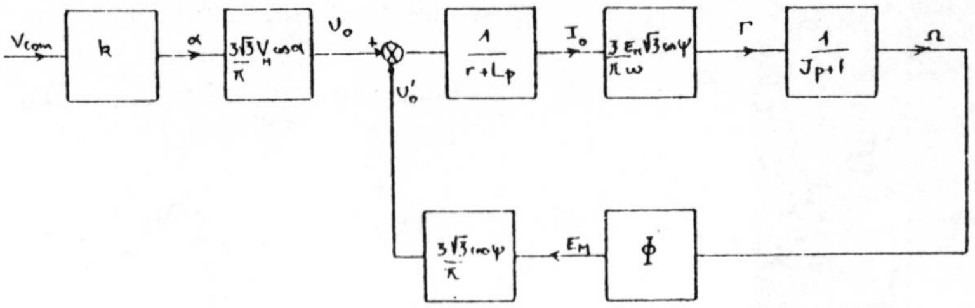

Fig. 3. Functional plan of the self-controlled synchronous machine.

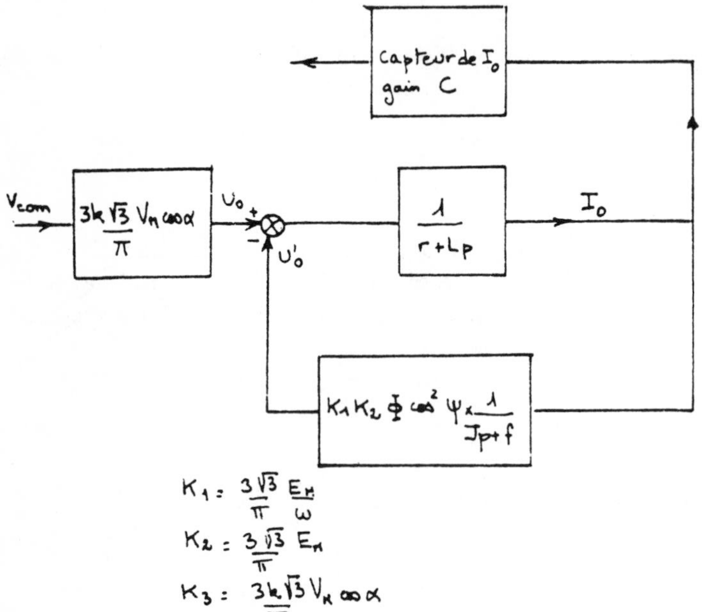

Fig. 4. Schematic diagram with I_0 output.

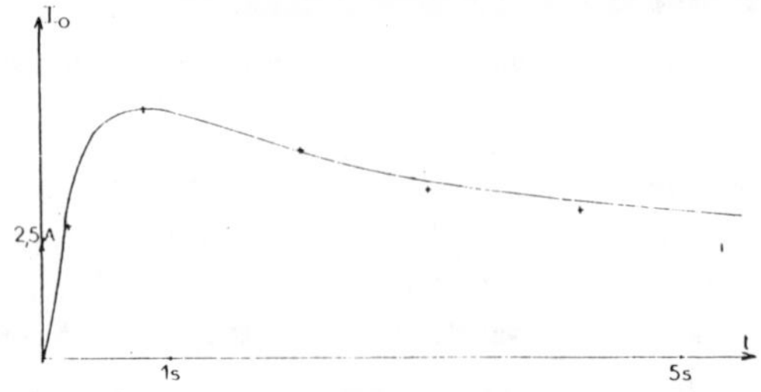

Fig. 5. The simulated theoretical response.

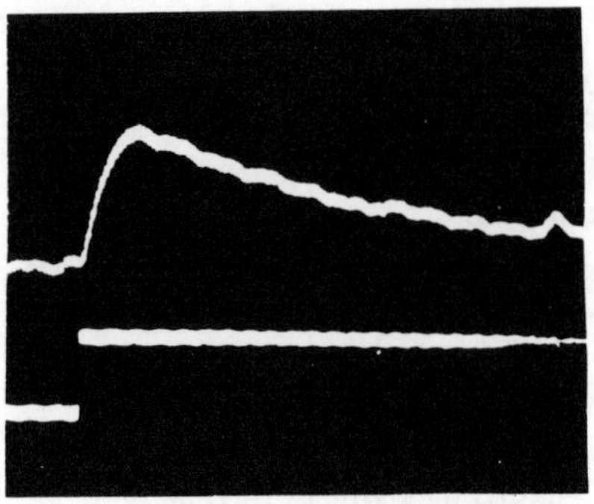

I_o = 2.5A/C

V_c = 0.1V/C

1 s/C

Fig. 6. Experimental response.

The transfer I/v_c, where I shows the output of the current sensor and V_c shows the control tension of the rectifier bridge, is

$$F(p) = \frac{I}{V_c} = K_s \frac{1+\tau_1 p}{1+\dfrac{2z}{\omega n}p+\dfrac{p^2}{\omega_n{}^2}}$$

where

$$K_s = \frac{K_3 df}{rf + A\cos^2\psi}$$

$$\tau_1 = J/f \qquad \omega_n = \sqrt{\frac{K_3 cf}{JLK_s}}$$

$$z = \frac{1}{2}\sqrt{\frac{K_s}{K_3 cf}} + \frac{rj+Lf}{\sqrt{JL}}.$$

The experimental conditions are as follows

$$J/f = 5\ \text{s}; \qquad \frac{1}{\tau_1} = 0.2\ \text{rad/s}$$

$$L = 27.6\ \text{mH}; \qquad r = 120\ \text{m}\Omega$$

$$\psi_a = 145'; \qquad K_s = 2; \qquad K_3 = 4.13; \qquad C = 0.13.$$

Thus

$$F(p) = 2\frac{1+5p}{(2p+1)(0.24p+1)}.$$

1) Simulated Theoretical Response: In order to have a theoretical response to a current step, $F(p)$ has been simulated. The result obtained is shown in Fig. 5.

2) Experimental Response: A self-controlled synchronous machine has been used in exactly the same experiment; the current growth and the sensor output is shown in the oscillogram of Fig. 6.

3) Harmonic Response: A sinusoidal tension $v_e(t)$ of constant amplitude has been superimposed on the control tension V_c, and thus the frequence is varied. In Fig. 7, the asymptotes in the plan of Bode of the theoretical transfer function are shown with their registered values in experiment.

C. Conclusion

A very clear agreement can be seen between the simulated response and the experimental response just as shown in the harmonic experiment between the theoretical response and the experimental response. Thus the approximations that were used are justified by the experiments made and it can be

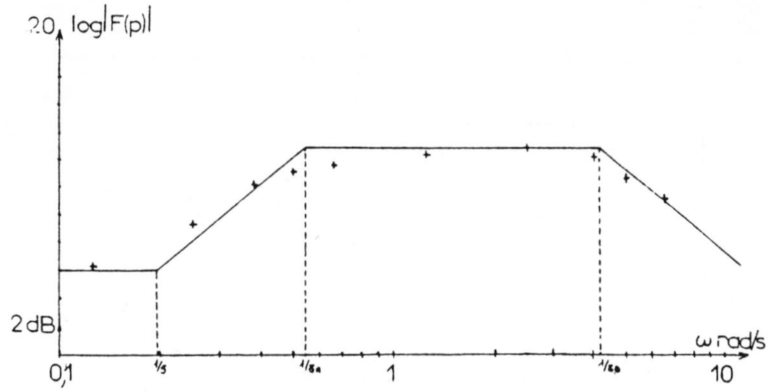

Fig. 7. Theoretical and experimental harmonic respones of the system.

considered that the proposed simplified model demonstrates correctly the theory of the system in the kinds of operation involved.

IV. CONTROL OF CURRENT IN THE DC LOOP

A. Necessity for Regulating Current in the DC Loop

It was found that the source of supply to the commutator acted exactly as a source of dc current. Thus it was found that the system had to be desensitized with regard to disturbances of load and in order to make certain that the current was maintained constantly and precisely as required in the dc loop.

B. Aims of this Regulation

We have seen the response form of the current at a reference-tension step. The response time may be taken as 2 s and the open loop system has a transient which is too slow. The system, therefore, must be corrected to obtain a proper dynamic benefit, in fact greater accuracy. Using the control model, the regulation structure will be easy to establish.

C. Regulation Structure

Taking into account time constants of current filtering, the transfer function in the open loop is as follows:

$$F(p) = 2\,\frac{5p+1}{(2p+1)(0.24p+1)(0.027p+1)}\,.$$

With the aim of having a nil error, an integrator was placed in the feedback path which made a gain and thus compensated the time constant $(1+2p)$ and which slows down the system.

The structure of the corrector was such that the gain worked out as

$$C(p) = \frac{1+2p}{\tau p}\,.$$

Assuming $F_c(p)$ the transfer function of the corrected system in closed loop is shown in Fig. 8, with

$$F_c(p) = \frac{F(p)C(p)}{1 + C(p)\cdot F(p)}$$

$$F_c(p) = \frac{N_c(p)}{D_c(p)}$$

with

$$-N_c(p) = 1 + 5p$$

$$-D_c(p) = \left|1 + \left(5 + \frac{\tau}{2}\right)p\right| \left|\frac{p^2}{\omega_{nc}^2} + \frac{2z_c}{\omega_{nc}}p + 1\right|.$$

The natural pulsation becomes

$$\omega_{nc} = 12.42\sqrt{1 + \frac{10}{\tau}}\ \text{rad/s}.$$

The damping factor is equal to

$$z_c = \frac{1.68}{\sqrt{1 + \frac{10}{\tau}}}\,.$$

The gain of the corrected closed loop becomes, therefore,

$$F_c(p) = \frac{1}{\frac{p^2}{\omega_{nc}^2} + \frac{2z_c}{\omega_{nc}}p + 1}\,.$$

The corrected system in the closed current circuit worked almost like a second-order process. The best response was obtained when the damping factor z_c was equal to 0.4 and for $\tau = 0.6$. The natural pulsation becomes then $\omega_{nc} = 52.2$ rad/s. The theoretical findings provide the values which enabled the adoption of the type of current loop shown in Fig. 9 and which was used for the experiment.

D. Regulation Experiment

1) Step Response: The current response to a step of reference was registered as in Fig. 10. For the predetermined value of τ, an overshoot of 28 percent was observed. The response time at five percent is 110 ms. The theoretical

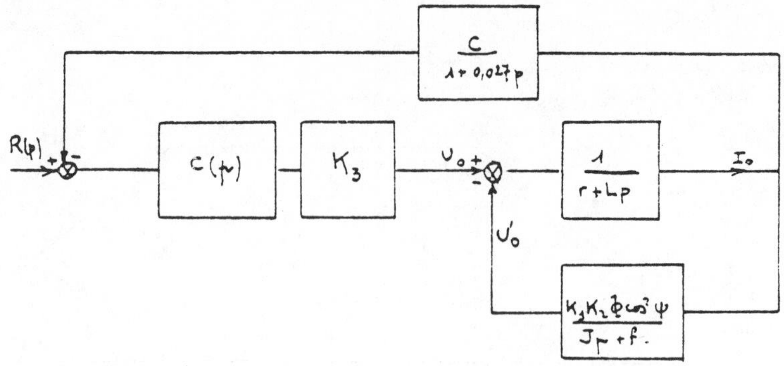

Fig. 8. The corrected system in closed-current loop.

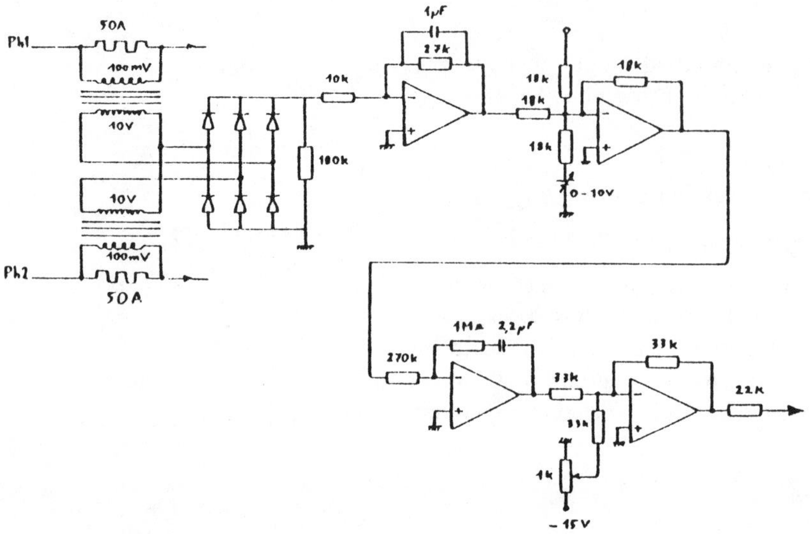

Fig. 9. Practical realization of the control of current.

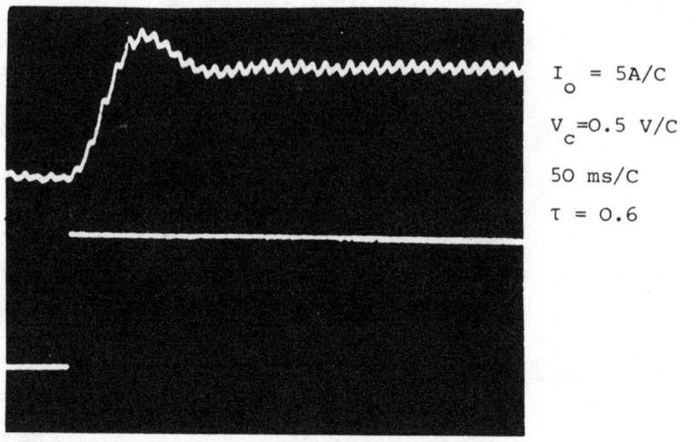

Fig. 10. Step response.

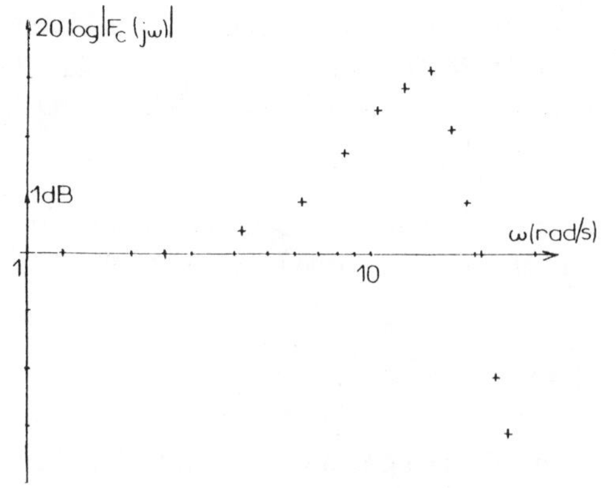

Fig. 11. Harmonic response in closed-current circuit.

calculation given for t_r was

$$t_r = \frac{2\pi}{0.9\omega_{nc}} = 133 \text{ ms.}$$

2) Harmonic Experiment: Making a superposition of sinusoidal tension control of constant amplitude and variable frequency in the Bode plan, the response in Fig. 11 was obtained.

E. Conclusion

The modelization proposed for control is very satisfying. It has allowed regulation of current within the dc loop such that the practical response corresponds very well with the theoretical response, and the simplicity of the model has allowed a direct approach to the problem.

V. REGULATION OF THE SYSTEM SPEED

A. Aims of Speed Regulation

The system has a very slow reaction when the output speed is considered. The response time varies from 5-20 s whether the system is set up in open loop or closed circuit. However, the speed is not controlled (it varies with the resistant torque applied). It has been proposed, therefore, to

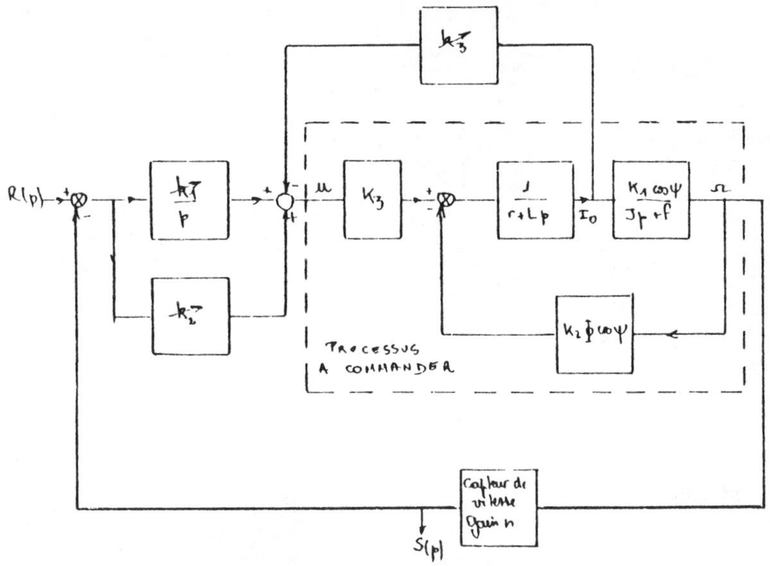

Fig. 12. Functional scheme for the speed control system.

make a speed regulation of which the properties will be as follows:

- response time at least five times faster than that in the open speed loop,
- well muffled response ($d = 20$ percent),
- linearity and fidelity in regard to the reference, and
- desensitization agreement with disturbances.

The study will be based upon the control model which has been established, and calculations will be made around a functioning point $\psi_a = 145°$.

B. Control Structure

The control process has been reduced to two times (J/f and L/r). In order to compensate for the influence of disturbances, to make certain of linearity between the input (reference) and the output (speed), and to have a nil static error, an integrator will be introduced into the system. Thus the total system will be in a third order.

It is being found that, in the structure of the mounting, the current in the dc open loop is a state variable. Thus the solution for control of state variable feedback becomes necessary. The system being in the third order, it is going to be necessary to fit three adjustable gains bearing on the three variable feedback states in order to control the system dynamics.

The variable feedbacks are

- I_0 to be the current in dc loop,
- the error, and
- the integral of the error.

The functional scheme is shown in Fig. 12.

C. Theoretical Calculations of Control

1) Transfer Function: The transfer function is $H(p) = S(p)/R(p)$ where $S(p)$ is the speed output after the sensor and $R(p)$ the reference-tension.

$$\frac{S(p)}{u(p)} = \frac{nK_1K_3 \cos \psi}{(Jp+f)(r+L_p)+A \cos^2 \psi}$$

$$u(p) = \left(\frac{k_1}{p} + k_2\right)[R(p) - S(p)] - k_3 I_0(p).$$

Therefore, $H(p)$ can be put in the form $N(p)/D(p)$ with

$$N(p) = \frac{k_2}{k_1} p + 1$$

$$D(p) = \frac{0.1}{k_1} p^3 + \left(\frac{0.147 + 4.95k_3}{k_1}\right) p^2$$

$$+ \left(\frac{0.2 + 3.05k_1 + k_2}{k_1}\right) p + 1;$$

k_1, k_2, and k_3 figure in linear fashion as the coefficients of $D(p)$.

$D(p)$ will be identified to $D_s(p)$ (the denominator of the specific response; that is to say, the response which it is desirable to obtain).

2) Specific Response: The response is required of the type

$$D_s(p) = (1 + \tau_s p)$$

$$\cdot \left(1 + \frac{2z_s}{\omega_{ns}} p + \frac{p^2}{\omega_{ns}^2}\right).$$

A response time at five-percent $t_r = 1$ s and a damping factor z_s at 0.4 are fixed. In this case

$$\omega_{ns} = \frac{2\pi}{0.9t_r} = 6.98 \text{ rad/s}.$$

The choice of τ_s is made because of the fact that it does not alter the response time at five percent ($\tau_s = 0.1$ s is to be

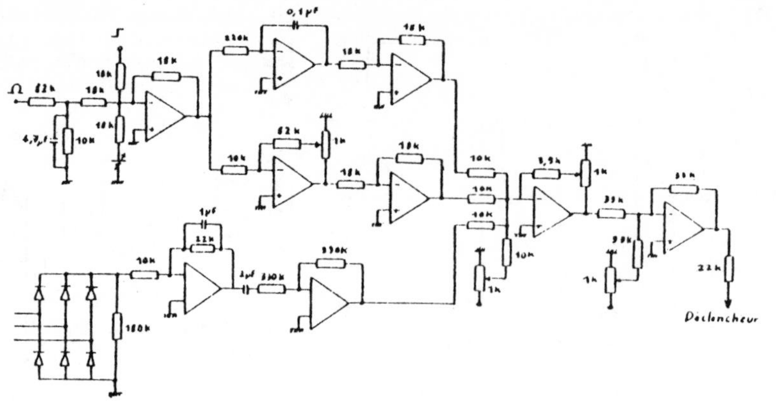

Fig. 13. Speed regulation scheme.

taken). Thus $D_s(p)$ is easily determinable:

$$D_s(p) = (1 + 0.1p)\left(1 + \frac{0.8}{6.98}\,p + \frac{p^2}{(6.98)^2}\right).$$

It only remains to identify the coefficients of p, p^2, and p^3 which become

$$k_1 = 48.72; \qquad k_2 = 9.4;$$

$$k_3 = 2.19; \qquad \frac{k_2}{k_1} = 0.19.$$

$N(p) = (1 + 0.19p)$ is of the same order as $(1 + \tau_s p)$. The response will be very close to a second order. A practical situation is shown in Fig. 13.

D. Experiments and Control Performances

1) Step Response: The theoretical gains of the corrector and the control system have been put into position. The response is given in Fig. 14. The response time at five percent measured is t_r = 1.3 s. Fig. 15 shows the error signal. The static error can be stated to be null. The response of the error signal corresponds correctly to the theoretical dynamic.

2) Linearity: The variation of the speed has been measured for a variant reference of 0–10 V. The response is correctly linear at 150 r/min to 1500 r/min as shown in the Fig. 16.

VI. CONCLUSION

A simplified dynamic model of a current-fed self-commutated synchronous motor has been easily established. This model has been justified experimentally. It has permitted the design of the current and speed controlling loops along the same principles used for classical armature-controlled dc motor drives. For each control configuration an experimental result has been obtained, validating with a good approximation the proposed model. The simplicity of this model makes the calculation of the controlling loop very easy, and so the experience which is widely available in industry

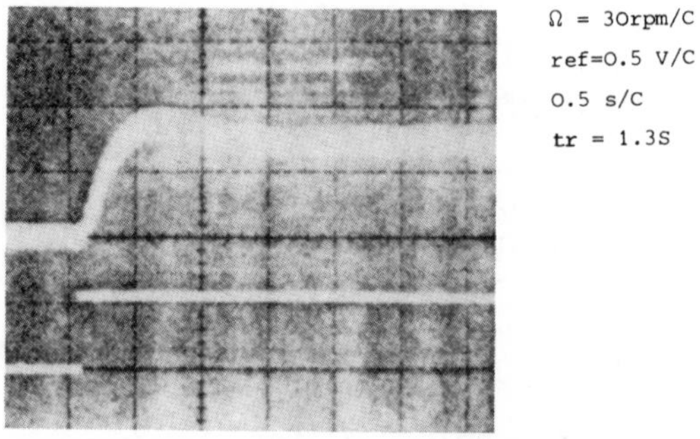

Fig. 14. Speed response to a step of the reference.

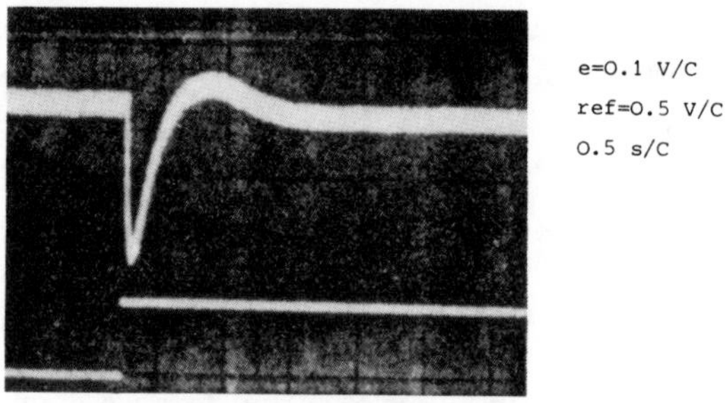

Fig. 15. Error signal response.

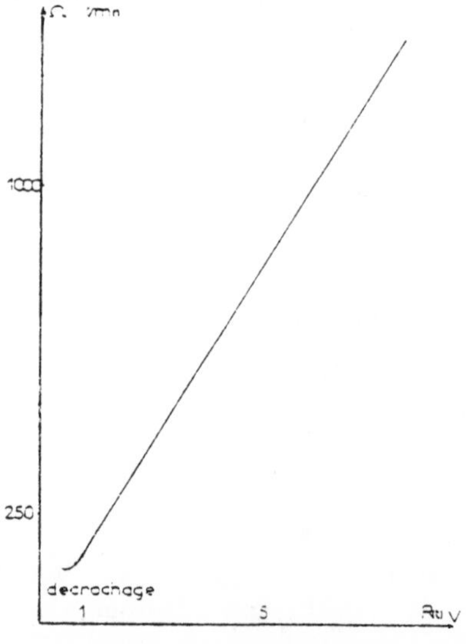

Fig. 16. Control linearity.

can be applied to design of the commutatorless dc motor drives.

REFERENCES

[1] L. Pierrat, "Les entraînements à vitesse variable par variateurs statiques de fréquence," *J. d'Etudes,* SEE—9 & 10, Mar. 1978.

[2] G. Rojat, "Machine synchrone autopilotée alimentée par un convertisseur statique à commutation assistee," Thèse de Spécialité, Toulouse, July 12, 1974.

[3] B. DeFornel, "Machines à courant alternatif alimentées a fréquence variable par convertisseurs statiques," Thèse de Doctorat ès Sciences, Toulouse, Apr. 3, 1976.

[4] M. Lajoie-Mazenc and B. Trannoy, "Quelques aspects de l'étude du remplacement du collecteur par un commutateur statique," *R.G.E.T.*, 81, 1972.

[5] A. Jakubowicz, "Etude d'une machine synchrone autopiloteé: Réalisation, modélisation et asserissements," Thèse de Spécialité, Grenoble, 1978.

[6] A. Abdel-Razek, "Contribution à l'étude des régimes transitories déséquilibres des machines synchrones dans deux cas: Courts-circuits brusques et alimentation par convertisseur statique," Thèse de Doctorat ès Sciences, Grenoble, Dec. 7, 1976.

[7] T. Lipo and E. Cornell, "State variable steady state analysis of a controlled current induction motor drive," *IEEE Trans Ind. Appl.*, Nov. Dec. 1975.

[8] V. Stefanovic, "Variable frequency induction motor drive dynamics," McGill University, Montreal, PQ, Canada, 1975.

[9] M. Nougaret: "Sur une méthode de synthèse, par retur d'état, de la commande des procédés physiques," Thèse de Doctorat ès Sciences, Grenoble, 1972.

[10] E. Toutain, "Etude d'unm hacheur à accumulation capacitive à transistors. Réalisation, modélisation et asservissements," Thèse de Spécialité, Grenoble, 1978.

Dynamic Performance of Self-Controlled Synchronous Motors Fed by Current-Source Inverters

FUMIO HARASHIMA, MEMBER, IEEE, HARUO NAITOH, AND TOSHIMASA HANEYOSHI

***Abstract*—Dynamic characteristics of self-controlled synchronous motors are analyzed on the basis of the state-space method and Runge-Kutta-Gill method. As the results of this analysis, the effects of dc chokes, damper windings, and saliency of the motor on these characteristics are quantitatively clarified. Transfer function models of this type of motors are also proposed.**

INTRODUCTION

CURRENT-SOURCE inverter-fed synchronous motors, hereafter called commutatorless motors, have come into wide use in industry as a kind of variable speed motors, which can be controlled with good performance like dc motors. Furthermore, this system has a rigid structure and is easy to maintain like ac motors. Fig. 1 shows the fundamental construction of a commutatorless motor-drive system. It is clear from Fig. 1 that this system is an ac motor drive system with variable voltage and variable frequency-power source.

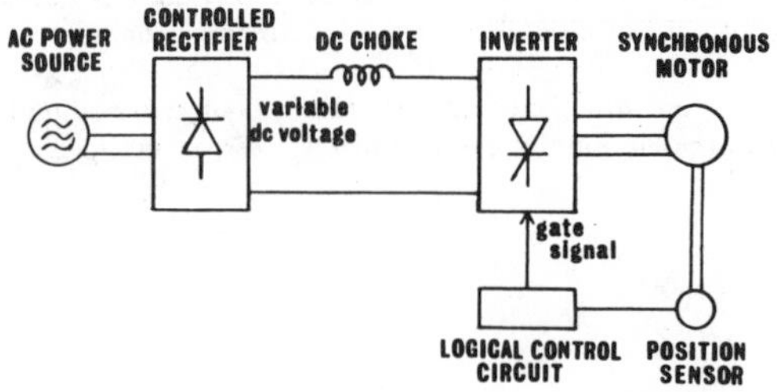

Fig. 1. Fundamental construction of commutatorless motor drive systems.

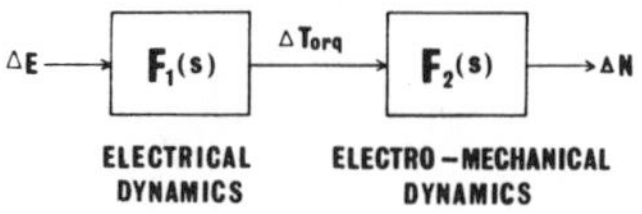

Fig. 2. Block diagram for dynamic performance of commutatorless motors.

When this system is used as a variable-speed drive, it is an important problem to consider dynamic characteristics as well as steady-state ones. Two kinds of dynamic characteristics, that is, electrical dynamics and electro-mechanical dynamics, should be taken into consideration. Since the response of electrical dynamics is faster than electro-mechanical dynamics, these two characteristics can be divided into two parts as is shown in Fig. 2. $F_2(s)$ in Fig. 2, which represents electro-mechanical dynamics, is easily obtained from torque-speed curves of the motor and torque characteristics of loads. However, $F_1(s)$ in Fig. 2, which represents electrical dynamics, cannot be easily obtained, since it depends on complicated electrical phenomena.

However, $F_1(s)$ becomes a very important factor when commutatorless motor systems are required to be controlled with high performance like dc drives. Thus the electrical dynamics is mainly considered in this paper.

Up to now, many authors have analyzed this system assuming that the input dc current I_0 is constant [1]. In the practical cases, however, the dc chokes have finite values and produce great effects on performance of commutatorless motor systems. In particular, for analyzing the dynamic characteristics, it is indispensable to consider the fact that the dc choke has a finite value. Recently, Harashima *et al.* analyzed a commutatorless motor system with finite size of dc choke under a steady-state condition using the state-space method [2]. They clarified quantitatively the ripples of dc input current, instantaneous waveforms of output torque, effects of damper windings, and so forth.

In this paper, first, the method to obtain rigorous solutions of this system by digital simulation is explained. Second, some remarks on the results of the digital simulation are shown. Third, the approximate models of this system are proposed and compared with the precise models by the digital simulation. Finally, some experimental results demonstrate the adequacy of this method.

RIGOROUS SOLUTION BY DIGITAL SIMULATION

Circuit Configuration

In Fig. 1, the output voltage of the controlled rectifier is controlled by the gate signal of the thyristors. In this case, although the output voltage includes ripples, they do not have serious influences on the characteristics of the motors since ripple frequencies are generally high and smoothed by dc choke L_0 for the most part. Therefore, the controlled rectifier can be considered as a dc power source with variable output voltage. The inverter is a natural commutation six-pulse bridge-inverter that is commutated by the induced voltage of the synchronous motor. Then, Fig. 1 can be replaced by Fig. 3.

Coordinate System Used in the Analysis

In the present analysis, the coordinate system based on the two-axis theory is used. Fig. 4 shows the vector representa-

Paper SPCC 77-27, approved by the Static Power Converter Committee of the IEEE Industry Applications Society for publication in this TRANSACTIONS. Manuscript released for publication July 18, 1978.

The authors are with the Institute of Industrial Science, University of Tokyo, Tokyo 106, Japan.

Reprinted from *IEEE Trans. Ind. Appl.*, vol. IA-15, pp. 36–46, Jan./Feb. 1979.

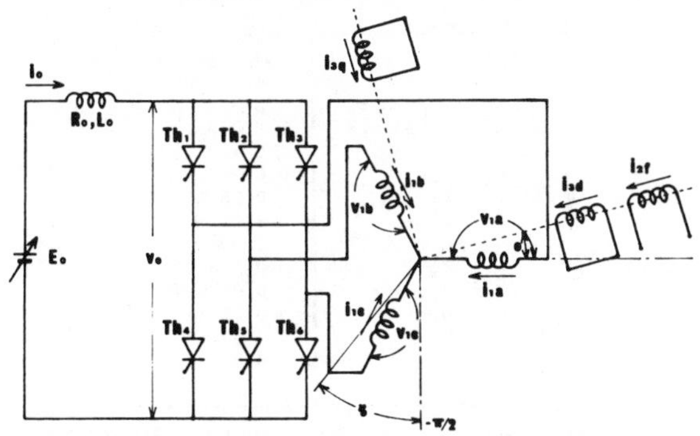

Fig. 3. Circuit diagram of salient-pole commutatorless motor.

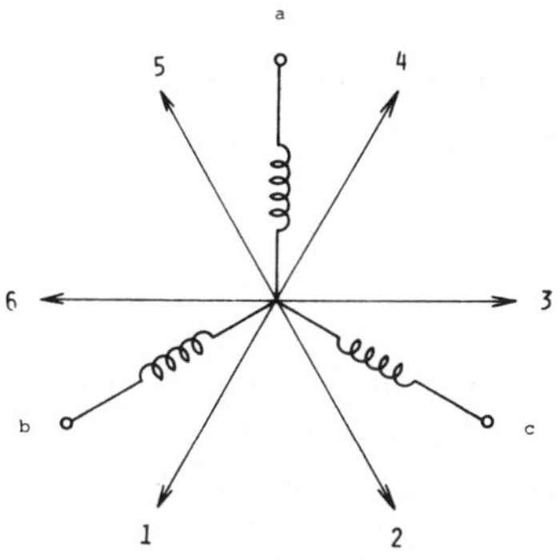

Fig. 4. Vector representation of armature currents.

tions of armature current. When thyristors Th_1 and Th_5 in Fig. 3 are conducting, the current flows from winding *a* through winding *b*. In this case, the direction of armature current vector is that of vector 1. When thyristor Th_6 turns on, the commutation starts. During the commutation period the current begins to flow through winding *c*, and the current, which is flowing through winding *b*, is reduced to zero. Meanwhile, the armature-current vector direction rotates from vector 1 to 2. When the current of winding *b* becomes zero, the commutation is finished, and then, the current vector direction remains in that of vector 2 until the next commutation starts. If the coordinates shown in Fig. 5 are used, the whole system can be analyzed by considering the period from commutation starting time to next commutation starting time. This coordinate system is stationary to the armature windings except for the instant of commutation start. This coordinate system is called "α–β coordinates," and hereafter the commutatorless motor system is analyzed using α–β coordinates.

Circuit Equation of a Synchronous Motor Based on the α–β Coordinate System

Generally, a three-phase machine can be transformed into a two-phase machine based on two-axis theory. In this paper, using the α–β coordinate system fixed to the armature windings, three armature windings are transformed into two windings, and the rotor windings, that is, the field winding and the damper windings, are transformed on the armature side.

The α–β coordinate system is fixed to the armature windings except for the moments of commutation starting. Therefore, the circuit equations based on this coordinate system are

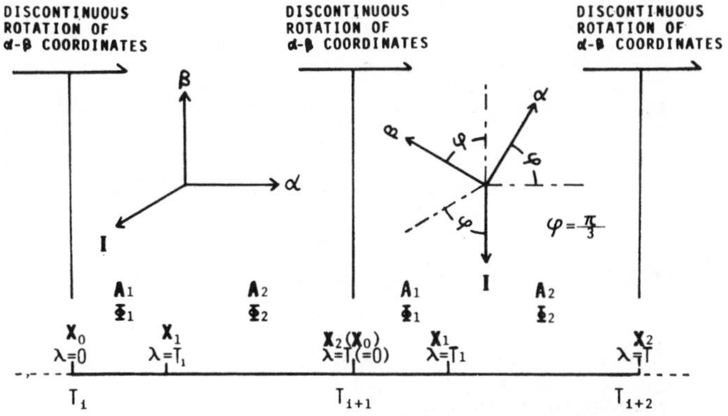

I : VECTOR OF ARMATURE WINDINGS
A_1, A_2 : COEFFICIENT MATRICIES
Φ_1, Φ_2 : STATE TRANSITION MATRIDIES
X_0, X_1, X_2 ; STATE VARIABLE VECTORS

Fig. 5. Discontinuous rotation of α–β coordinates and mode analysis.

the same forms as the two-axis equations fixed to the armature windings, and they are [3] :

$$V = Z' \cdot I. \tag{1}$$

Classification of Operation Mode

There are two fundamental operation modes in the commutatorless motor system. One mode corresponds to the period during which the current flows through three armature windings. The other mode corresponds to the period during which the current flows through two armature windings. The former mode is called "commutation mode," and the latter mode is called "single mode."

When the motor is running in steady-state, the sum of the duration of a commutation mode T_1 and the duration of a single mode T_2 is constant. This sum is denoted by T.

Circuit Equations for Each Mode

The field current I_f is assumed to be constant. From Fig. 3, the following relations are given:

$$E_0 - (R_0 + pL_0)I_0 = V_0 \tag{2}$$

$$V_{3\alpha} = V_{3\beta} = 0 \tag{3}$$

$$i_{2\alpha} = I_f \cos\theta, \qquad i_{2\beta} = I_f \sin\theta. \tag{4}$$

Equation (4) is changed into (5):

$$pi_{2\alpha} = -\dot{\theta} i_{2\beta}, \qquad pi_{2\beta} = \dot{\theta} i_{2\alpha}, \qquad (p = d/dt) \tag{5}$$

where θ is the angle between α-axis and field-current vector. Here, the relation between α-axis current $i_{1\alpha}$ and direct current I_0 is given as

$$i_{1\alpha} = \sqrt{\frac{3}{2}} I_0. \tag{6}$$

Then, the relation for each mode is considered in the case where the current flows from winding *a* through winding *c* after commutation.

For the commutation mode, the following relations are given:

$$i_{1\alpha} = -i_{1c} = I_0, \qquad i_{1b} = 0$$

$$V_0 = V_{1a} - V_{1c} = V_{1a} - V_{1b}. \tag{7}$$

Thus in α-β coordinates, (2) and (7) are changed into

$$V_{1\alpha} = -\frac{2}{3}(R_0 + pL_0)i_{1\alpha} + \frac{2}{3}E_0 \tag{8}$$

$$V_{1\beta} = 0. \tag{9}$$

For the single mode,

$$i_{1a} = I_0, \qquad i_{1a} + i_{1b} + i_{1c} = 0$$

$$V_0 = V_{1a} - V_{1c}, \tag{10}$$

and then,

$$\sqrt{3}V_{1\alpha} + V_{1\beta} = -2(R_0 + pL_0)i_{1\beta} + \sqrt{2}E_0 \tag{11}$$

$$i_{1\alpha} = \sqrt{\frac{3}{2}}I_0, \quad i_{1\beta} = \frac{1}{\sqrt{2}}I_0. \tag{12}$$

Finally, from (1)-(12) and (29), the total state equations are obtained as

$$pX = A_1 \cdot X: \quad \text{for the commutation mode}$$

$$pX = A_2 \cdot X: \quad \text{for the single mode} \tag{13}$$

where

$$X = \text{col}\,(i_{1\alpha}, i_{1\beta}, i_{2\alpha}, i_{2\beta}, i_{3\alpha}, i_{3\beta}, E_0): \quad \text{state vector} \tag{14}$$

$$A_1, A_2: \quad 7 \times 7 \text{ matrix}.$$

The elements of matrices A_1 and A_2 are constant for non-salient pole machines and time-variable for salient pole machines.

Connection of Each Mode and Solution for Steady-State Running

Mode analysis of the commutatorless motors system is shown in Fig. 5. If the motor is a nonsalient pole machine, the state transition matrices Φ_1 and Φ_2 for these two modes are given as

$$\Phi_1(\lambda) = \exp(A_1 \cdot \lambda): \quad \text{for the commutation mode}$$

$$\Phi_2(\lambda) = \exp(A_2 \cdot \lambda): \quad \text{for the single mode}$$

$$A_1, A_2: \quad \text{See Appendix}. \tag{15}$$

Then, state vectors X_0, X_1, and X_2, shown in Fig. 5, are subject to the following relations [4]:

$$X_1 = \Phi_1(T_1) \cdot X_0$$

$$X_2 = \Phi_2(T_2) \cdot X_1$$

$$X_0 = BX_2. \tag{16}$$

TABLE I
CONSTANTS OF MOTOR

Power	3.7 kw
Pole number	4
Phase number	3
Rated voltage	100.0 V
R_0	0.0 ~ 0.43 Ω
L_0	0.0 ~ 0.123 H
R_1	0.22 Ω
L'_1	8.22 mH
L_{1s}	2.30 mH
R_3	0.366 Ω
R'_{3s}	0.0877 Ω
L'_3	8.22 mH
L'_{3s}	2.30 mH
M'_{12}	0.35 H
M'_{13}	7.17 mH

where matrix B expresses the roation of the α-β coordinate system. The details of the matrix B is shown by (30) in the Appendix.

From (16) the following relation is derived under steady-state condition [5] [6]:

$$X_0 = B \cdot \Phi_1(T_1) \cdot \Phi_2(T_2)X_0. \tag{17}$$

Equation (17) gives seven relations for eight unknown variables, that is, seven initial components of X_0 and commutation period $T_1(T_1 = T - T_2)$. So, another equation is required, which represents the relation between $i_{1\alpha}$ and $i_{1\beta}$ at time $\lambda = T_1$. In the α-β coordinates, this relation is given as

$$i_{1\beta}(T_1) = i_{1\alpha}(T_1)/\sqrt{3}. \tag{18}$$

The solutions of (17) and (18) determine the state vector X_0 and the time T_1. Then, the following relations give the solutions of state variables at any instant:

$$X_1(\lambda) = \Phi_1(\lambda)X_0, \qquad 0 < \lambda \leqslant T_1,$$

$$X_2(\lambda) = \Phi_2(\lambda)X_1(T_1), \qquad T_1 < \lambda \leqslant T. \tag{19}$$

Solution for Electrical Dynamics of the Commutatorless Motor System

The solution of electrical dynamics such as step response can be obtained by solving (13) by the Runge-Kutta-Gill method using the solutions of steady-state running of the commutatorless motors as initial values. Fig. 6 is a flowchart for both steady-state and dynamic solutions.

NUMERICAL EXAMPLES

The Dynamic Characteristics

The numerical examples of dynamic characteristics of the commutatorless motor system are shown in Fig. 7 and Fig. 8. Fig. 7 shows the step response of I_0 when E_0 is changed to $E_0 + \Delta E$. Fig. 8 shows the step response of output torques under the same condition in Fig. 7. Table I shows the constants of the motor of this system.

Some Remarks on Rigorous Solution

Fig. 9 shows some examples of time constant for step response of the system. In Fig. 9, $\overline{I}_{st}$, $\overline{I}_0$, $\overline{I}_\infty$ denote the initial,

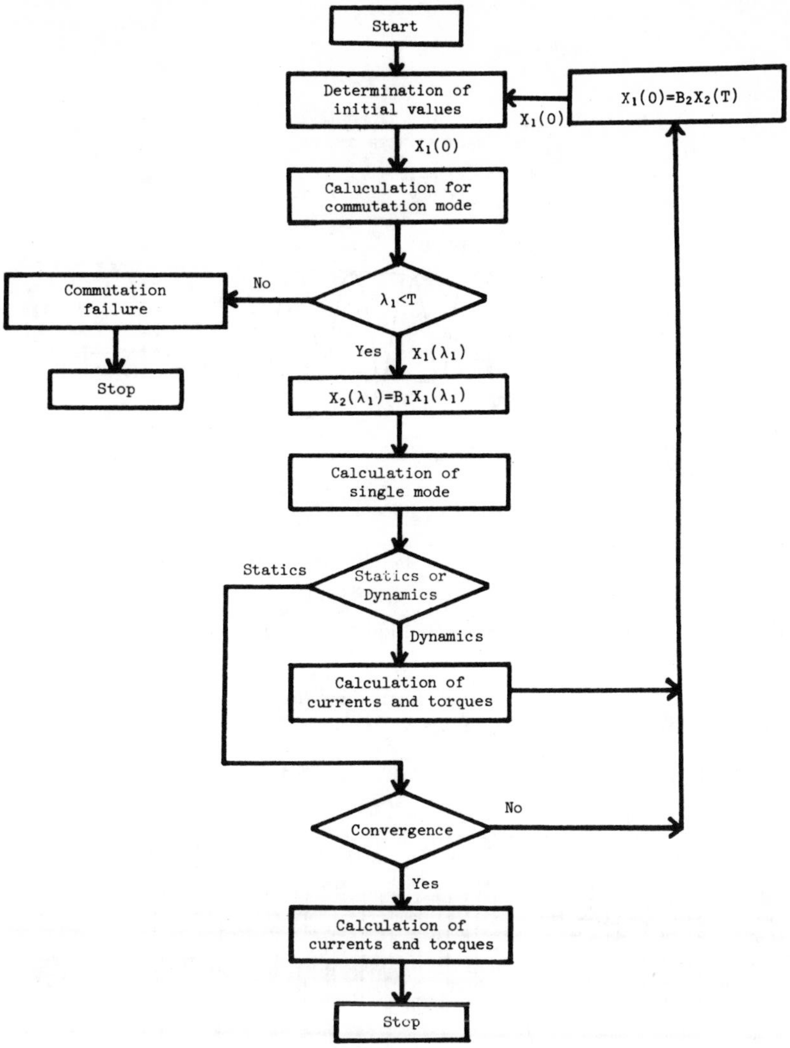

Fig. 6. Flowchart for steady-state and dynamic solution.

transient, and final average dc imput current, respectively, and each time constant is given as the reciprocal number of the gradient of each line. From Fig. 9, dc choke L_0 and damper windings act to make the time constant larger.

Fig. 10 shows the comparison of response speeds of step responses in the case where the couppling coefficient k is varied. From Fig. 10, it is more evidently clarified that damper windings make the response of the motor slower. The reason why damper windings act in such a manner is that damper windings reduce the commutating inductance L_c and consequently the effective resistance R_e (see (27)) is also reduced.

Fig. 11 shows the comparison of response speeds of step responses in the case where the time constant of damper windings T_{dw} is varied. In Fig. 11, T_{dwo} is the time constant of damper windings of the experiment machine, and T_{dwo} = 22.5 ms. It is observed that the response speed becomes faster with the increase of T_{dw}.

DERIVATION OF TRANSFER FUNCTION MODELS

Here, the approximate transfer function models are proposed. As mentioned in the Introduction, only electrical dynamics are considered. The speed of the motor is therefore assumed to be constant. As the input signal, small variations of dc voltage E_0 are used, and as the output signal, dc imput currents or output torques are used.

The Fundamental Model

First, a model for a nonsalient pole machine without damper windings is derived. For this type of machine, the most important problem is how to deal with overlapping phenomena, that is, commutation mode.

There exist single modes and commutation modes. The circuit equations for each mode are different from each other. In order to derive the transfer function of the whole system, these two equations have to be combined.

A line-commutated phase-controlled converter can be replaced by a variable dc source and a resistance on the average [7]. The equivalent circuit is given in Fig. 12(a). In this figure, the resistance $3\omega L_c/\pi$ expresses the effects of overlapping phenomena equivalently, and the amplitude of the variable dc source is given by $\tilde{E}\cos\gamma_0$, where $\tilde{E}$ is the induced voltage of the motor when γ_0 is equal to zero and γ_0 is the leading angle of the gate control. By replacing the overlapping phenomena by the equivalent resistances in this way, it is possible to regard that only single mode exists throughout the operation of this system.

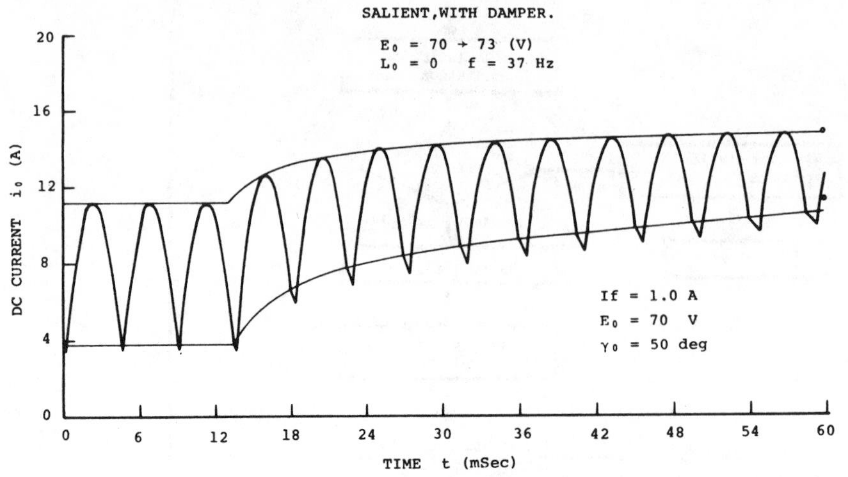

Fig. 7. Example of input dc current.

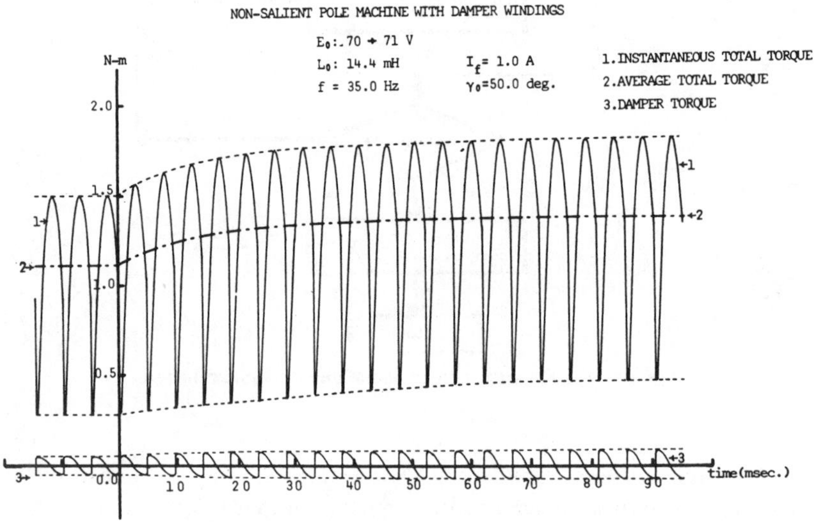

Fig. 8. Step response of output torques.

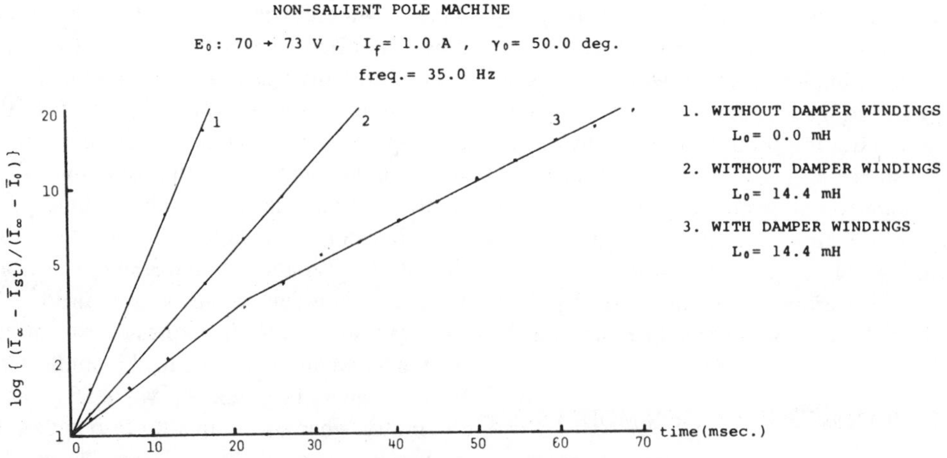

Fig. 9. Derivation of time constants.

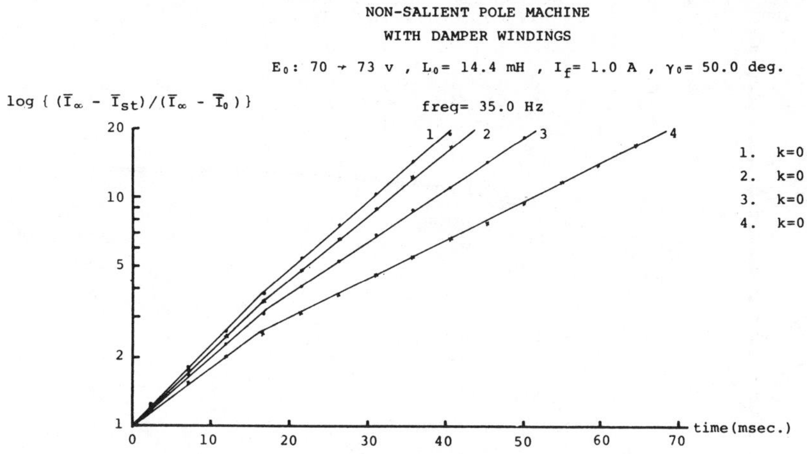

Fig. 10. Effects of coupling coefficients on response speed.

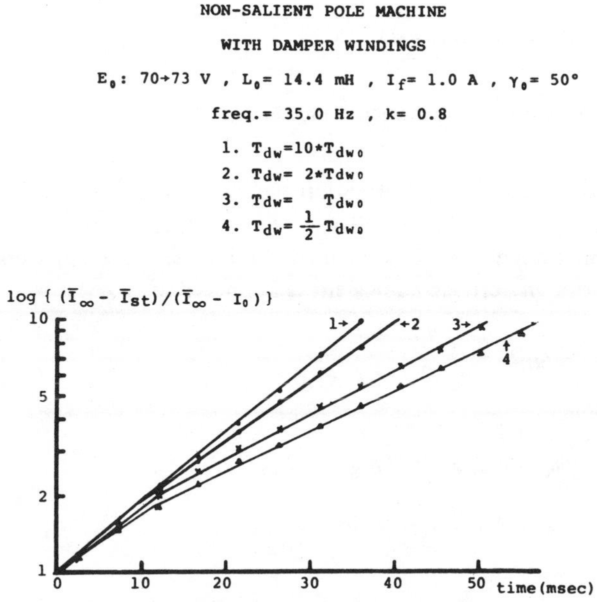

Fig. 11. Effects of time constant of damper windings on response speed.

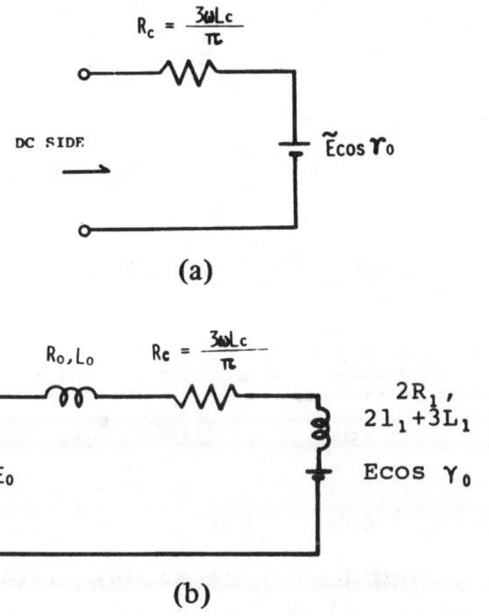

Fig. 12. Equivalent circuits. (a) Inverter-synchronous motor systems observed from dc side. (b) Fundamental model.

Therefore, the equivalent circuit of the whole system becomes as shown in Fig. 12(b). From this figure, the transfer function G_{1E}, which denotes the relation between the voltage increment ΔE_0 and the average circuit current increment $\Delta \bar{I}_0$ is given as

$$G_{1E}(s) = \frac{\Delta \bar{I}_0}{\Delta E_0} = \frac{1}{(Ls + R)} \tag{20}$$

where

$$L = L_0 + 2l_1 + 3L_1$$

$$R = R_0 + 2R_1 + \frac{3\omega}{\pi} L_c$$

$L_c = l_1 + \frac{3}{2} L_1$ L_c: commutating inductance per phase.

However, the torque equation of synchronous motors is given as

$$\text{torq} = M_{12}'(i_{2\alpha} i_{1\beta} - i_{1\beta} i_{1\alpha}). \tag{21}$$

From (3) and (9), (21) becomes (22):

$$\text{torq} = 2M_{12}'I_f \cdot i_0 \cos\left(\theta + \frac{\pi}{3}\right) \tag{22}$$

where

$$\theta = \omega t - \frac{\pi}{2} - \gamma_0 \; (0 < t < T).$$

Integrating (22) from $t = 0$ through $t = T$, using the relation that the correlation between i_0 and $\cos(\theta + (\pi/3)$ is relatively small, the following relation is obtained:

$$\overline{\text{torq}} = \frac{3\sqrt{2}}{\pi} M_{12}'I_f \cos \gamma_0 \cdot \bar{I}_0 . \tag{23}$$

Equations (20) and (23) give the transfer function G_{1T}, which denotes the relation between ΔE_0 and the average

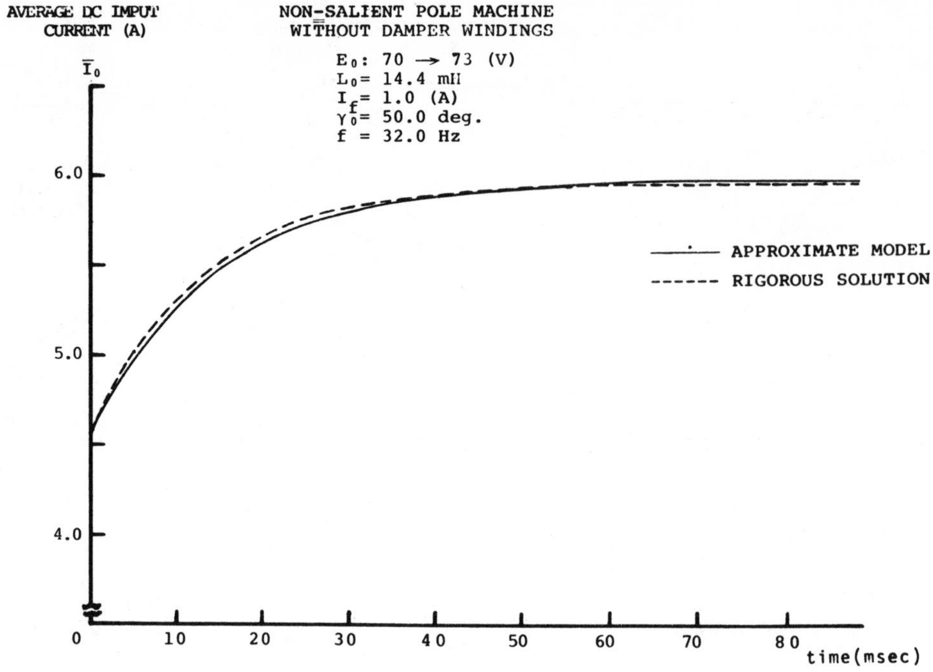

Fig. 13. Step response of average dc input current of fundamental model.

torque increment $\Delta\,\overline{\text{torq}}$:

$$G_{1T}(s) = \frac{\Delta\,\overline{\text{torq}}}{\Delta E_0} = \frac{3\sqrt{2}M_{12}'I_f\cos\gamma_0}{\pi(Ls+R)}. \tag{24}$$

The Practical Model

Here, a nonsalient pole machine with damper windings is considered. Generally, salient pole machines with damper windings are the most practical. The saliency, however, is not considered for the sake of simplicity, since the effects of saliency are less important than that of damper windings. In the Appendix, it is shown briefly how to deal with the saliency.

As with the fundamental model, the effects of overlapping phenomena can also be represented by the equivalent resistance. In this model, it is important to consider the effects of the coupling between armature windings and damper windings. By the effects of this coupling, the commutating inductance L_c is reduced to σL_c, where σ is the leakage factor and is given by

$$\sigma = \left(1 - \frac{3M_{13}^2}{2L_1 \cdot L_3}\right). \tag{25}$$

The transfer function of this system can be derived in the following manner. The basic differential equation of this system is given as

$$pI = A_3^{-1}B_2I + A_3^{-1}D_2 \tag{26}$$

where

$$I = \text{col}\,(\bar{i}_{1\beta}, \bar{i}_{3\alpha}, \bar{i}_{3\beta})$$

A_3, B_2, D_2: (see Appendix).

The Laplace transformation of (29) gives the transfer function $G_{2E}(=\Delta\bar{I}_0/\Delta E_0)$ of this system. Here, the effects of overlapping phenomena and damper windings are taken into consideration in deriving equivalent resistance Re:

$$G_{2E}(s) = \frac{L_3s + R_3}{a_2s^2 + a_1s + a_0} \tag{27}$$

where

$$a_2 = (L_0 + 2L_1')L_3 - 2M_{13}'^2$$

$$a_1 = R_{\text{eff}}L_3 + R_3(L_0 + 2L_1')$$

$$a_0 = R_e \cdot R_3$$

$$R_e = R_0 + 2R_1 + \frac{3\omega\sigma}{\pi}L_c$$

Then, the transfer function $G_{2T}(=\Delta\overline{\text{torq}}/\Delta E)$ is considered. Damper currents also generate output torque, that is, damper torque. The torque by damper currents is, however, much smaller than that of armature currents (see Fig. 11). Therefore, in neglecting the damper torque, the torque equation in this model also becomes (23). Then, the following relation is derived:

$$G_{2T}(s) = \frac{\Delta\,\overline{\text{torq}}}{\Delta E_0} = \frac{3\sqrt{2}}{\pi}M_{12}'I_f\cos\gamma_0 G_{2E}. \tag{28}$$

Comparison of Approximate Models and Rigorous Solution

Fig. 13 shows the comparison of the fundamental model and the corresponding rigorous solution. Fig. 14 shows the Bode diagrams of the fundamental model and the corresponding rigorous solution. Fig. 15 shows the comparison of the practical model and the corresponding rigorous solution. Fig. 16 shows the Bode diagrams of the practical model and the

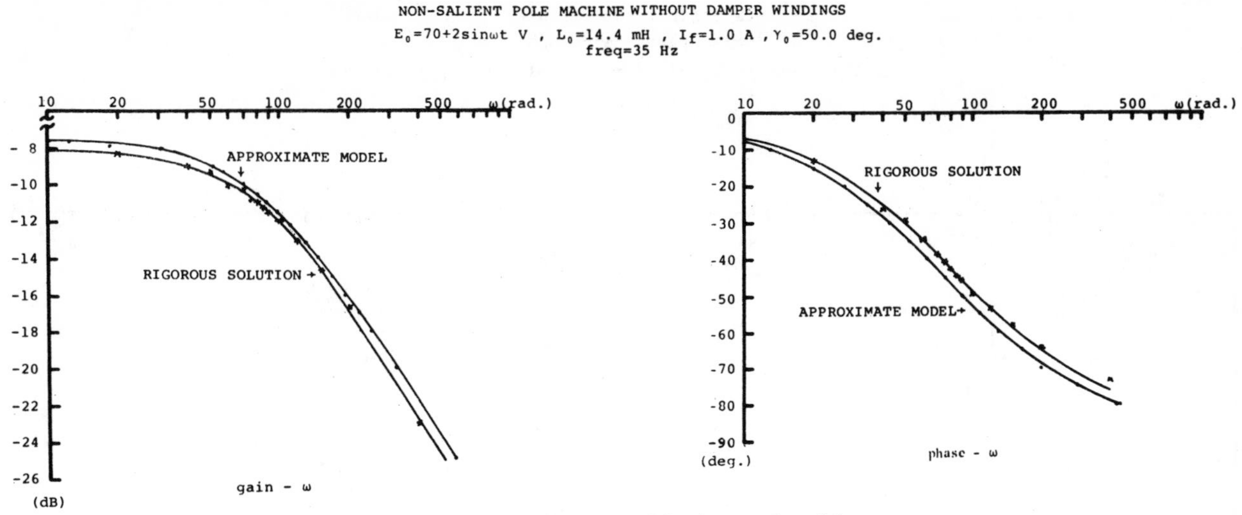

Fig. 14. Bode diagram of fundamental model.

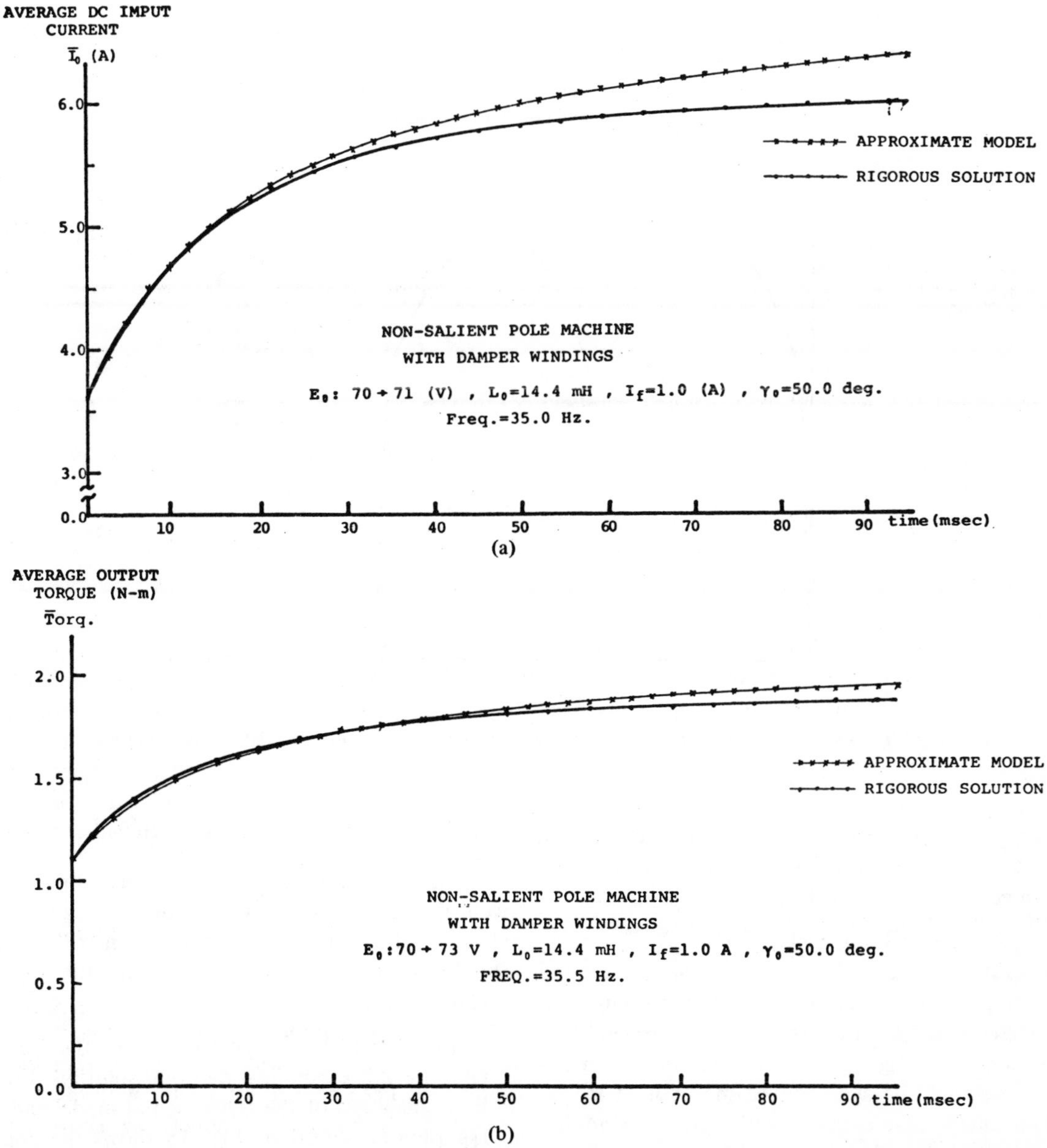

Fig. 15. (a) Step response of average dc input current of practical model. (b) Step response of average output torque of practical model.

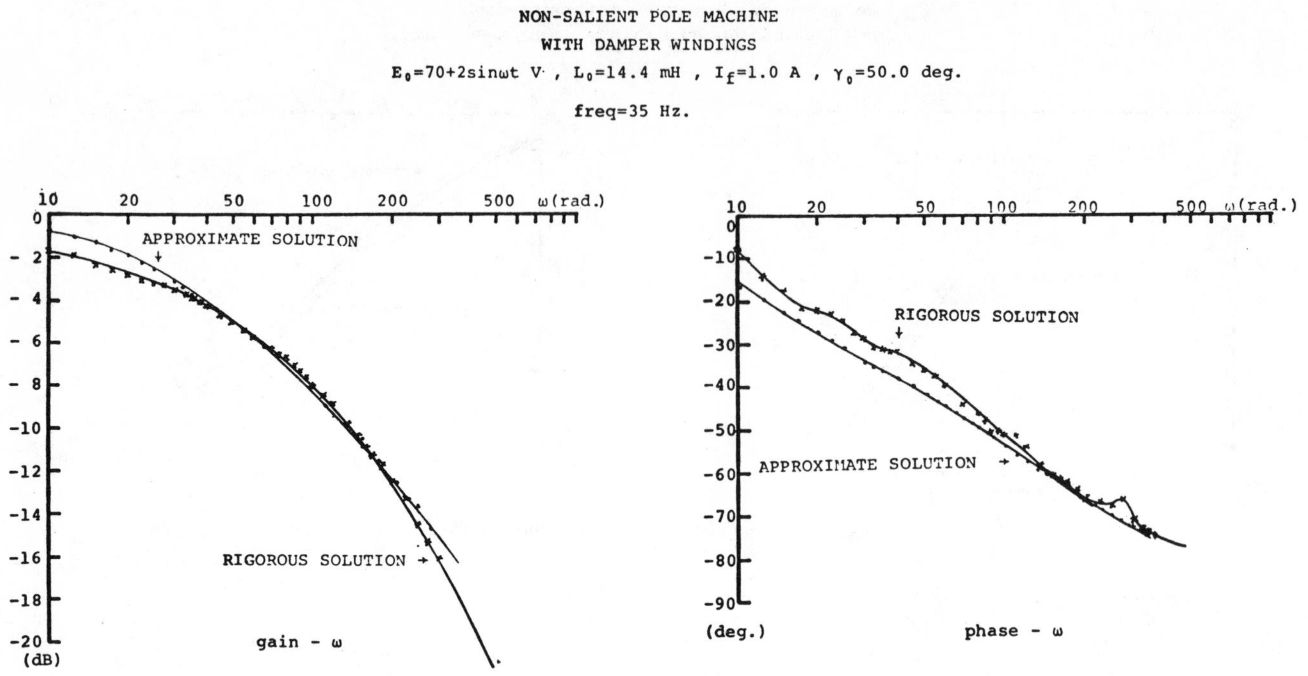

Fig. 16. Bode diagram of practical model.

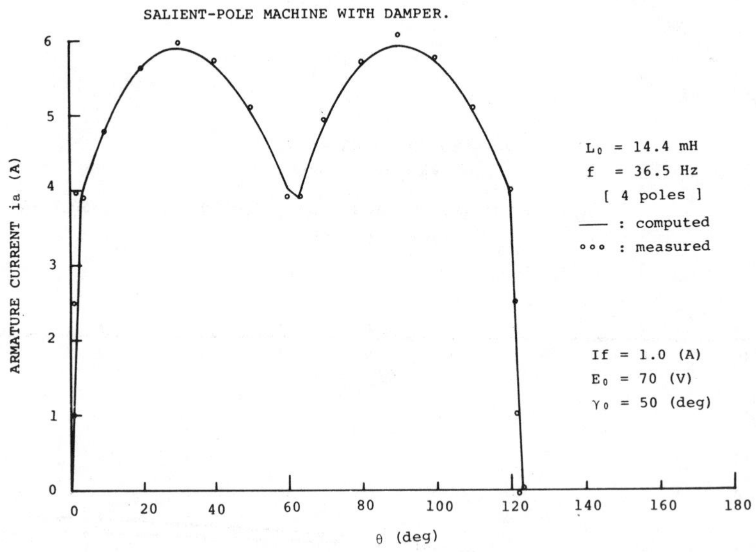

Fig. 17. Comparison of experimental result and rigorous solution under steady-state condition.

corresponding rigorous solution. From the above Figs. 13-16, it is confirmed that the approximate models that we have proposed are in good correlation with the rigorous solutions.

EXPERIMENTAL RESULTS

Fig. 17 shows the comparison of an experimental result and the corresponding rigorous solution by digital simulation under steady-state condition. Fig. 18 shows experimental results and the corresponding rigorous solution for a step response of this system. The synchronous motor used in our experiment is a salient pole machine with damper windings. From this figure, it is clear that our analysis is powerful even in dynamic analysis.

CONCLUSION

As the results of this study, the dynamic characteristics as well as static ones of self-controlled synchronous motors fed by a current source inverter have been rigorously analyzed on the basis of state-space method and the Runge-Kutta-Gill method.

The approximate models of this system under dynamic conditions are also proposed. The transfer functions of these systems have been derived by dealing with the dynamic phenomena on the average. These models have shown good correlation with rigorous solution by digital simulation and experimental results.

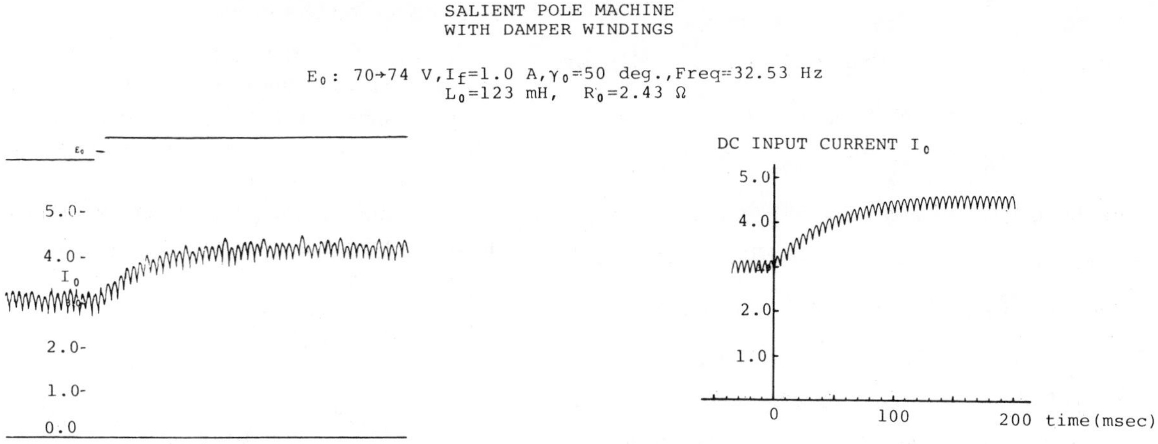

(a) (b)

Fig. 18. Comparison of experimental result and rigorous solution for step response. (a) Experimental result. (b) Rigorous solution.

APPENDIX

Impedance Matrices

Impedance matrix Z is transformed into matrix Z' on the basis of two-axis theory:

$$Z' = \begin{bmatrix} R_1 - 2L_{1s}'\dot{\theta}\cos 2\theta & 2L_1'\dot{\theta}\cos 2\theta & -2M_{13s}'\dot{\theta}\cos 2\theta & 2M_{13s}'\dot{\theta}\cos 2\theta \\ 2L_{1s}'\dot{\theta}\cos 2\theta & R_1 + 2L_{1s}'\dot{\theta}\cos 2\theta & 2M_{13s}'\dot{\theta}\cos 2\theta & 2M_{13s}'\dot{\theta}\cos 2\theta \\ -M_{13s}'\dot{\theta}\cos 2\theta & \dot{\theta}(M_{13}' + M_{13s}'\cos 2\theta) & R_3 + R_{3s}\cos 2\theta - L_{3s}\dot{\theta}\cos 2\theta & R_{3s}\cos 2\theta + \dot{\theta}(L_s + L_{3s}\cos 2\theta) \\ \dot{\theta}(-M_{13}' + M_{13s}\cos 2\theta) & M_{13s}'\dot{\theta}\cos 2\theta & R_{3s}\cos 2\theta + \dot{\theta}(-L_3 + L_{3s}\cos 2\theta) & R_3 - R_{3s}\cos 2\theta + L_{3s}\dot{\theta}\cos 2\theta \end{bmatrix}$$

$$+ \begin{bmatrix} L_1' + L_{1s}'\cos 2\theta & L_{1s}'\cos 2\theta & M_{13}' + M_{13s}'\cos 2\theta & M_{13s}'\cos 2\theta \\ L_{1s}'\cos 2\theta & L_1' - L_{1s}'\cos 2\theta & M_{13s}'\cos 2\theta & M_{13}' - M_{13s}'\cos 2\theta \\ M_{13}' + M_{13s}'\cos 2\theta & M_{13s}'\cos 2\theta & L_3 + L_{3s}\cos 2\theta & L_{3s}\cos 2\theta \\ M_{13s}'\cos 2\theta & M_{13}' - M_{13s}'\cos 2\theta & L_{3s}\cos 2\theta & L_3 - L_{3s}\cos 2\theta \end{bmatrix} \cdot P \tag{29}$$

where

$$L_1' = \frac{3}{2}L_1 + l_1, \quad L_{1s}' = \frac{3}{2}L_s, \quad M_{12}' = \sqrt{3/2}M_{12},$$

$$M_{13}' = \sqrt{3/2}M_{13}, \quad M_{13s}' = \sqrt{3/2}M_{13s},$$

$$M_{13} = \frac{M_{13d} + M_{13q}}{2}, \quad M_{13s} = \frac{M_{13d} - M_{13q}}{2},$$

$$L_3 = \frac{L_{3d} + L_{3q}}{2}, \quad R_3 = \frac{R_{3d} + R_{3q}}{2}, \quad R_{3s} = \frac{R_{3d} + R_{3q}}{2},$$

$$L_{3s} = \frac{L_{3d} - L_{3q}}{2} \quad (L_d = L_1' + L_{1s}', \quad L_q = L_1' - L_{1s}').$$

The suffixes 1, 2, and 3 denote that the circuit element that has one of these suffixes is relative to armature, field, and damper windings, respectively. Also

R_i resistance of i winding per phase,
L_i inductance of i winding per phase,
M_{ij} mutual inductance between i and j windings ($i, j = 1$, 2, or 3),
l_1 leakage inductance of armature winding per phase,
L_{1s} inductance of armature winding per phase due to saliency,

$$\theta = \dot{\theta}t - \frac{\pi}{2} - \gamma_0.$$

The matrix Z' is for salient pole machines. For nonsalient pole machines where the rotor coils have equal impendances, R_{3s}, L'_{1s}, L_{3s}, and M'_{13s} of matrix Z' become zero.

Matrices A_1, A_2

$$A_1 = \begin{bmatrix} a_{11} & a_{12} & & a_{14} & a_{15} & a_{16} & a_{17} \\ a_{21} & a_{22} & a_{23} & & a_{25} & a_{26} & \\ & & & -\dot{\theta} & & & \\ & & \dot{\theta} & & & & \\ a_{51} & a_{52} & & a_{54} & a_{55} & a_{56} & a_{57} \\ a_{61} & a_{62} & a_{63} & & a_{65} & a_{66} & \end{bmatrix}$$

$$a_{11} = -(2R_0 + 3R_1')/\Delta_3, \quad a_{12} = 3\dot{\theta}M_{13}{}^{2\prime}/L_3/\Delta_3,$$
$$a_{14} = 3\dot{\theta}M_{12}'/\Delta_3, \quad a_{15} = 3R_3'M_{13}'/L_3'\Delta_3, \quad a_{16} = 3\dot{\theta}M_{13}'/\Delta_3,$$
$$a_{17} = \sqrt{6}/\Delta_3, \quad a_{21} = -\dot{\theta}M_{13}{}^{2\prime}/L_3'\Delta_2, \quad a_{22} = -R_1'/\Delta_2,$$
$$a_{23} = -\dot{\theta}M_{12}', \quad a_{25} = -\dot{\theta}M_{13}'/\Delta_2, \quad a_{26} = R_3'M_{13}'/L_3'\Delta_2,$$
$$a_{51} = M_{13}'(2R_0 + 3R_1')/L_3\Delta_3, \quad a_{52} = -\dot{\theta}M_{13}'(2L_0 + 3L_1')/L_3'\Delta_3,$$
$$a_{54} = -R_3'(2L_0 + 3L_1')/L_3'\Delta_3, \quad a_{55} = -R_3'(2L_0 + 3L_1')/L_3'\Delta_3,$$
$$a_{56} = -\dot{\theta}(2L_0 + 3L_1')/\Delta_3, \quad a_{57} = -\sqrt{6}M_{13}'/L_3'\Delta_3,$$
$$a_{61} = \dot{\theta}L_1'M_{13}'/L_3'\Delta_2, \quad a_{62} = R_1'M_{13}'/L_3'\Delta_2,$$
$$a_{63} = \dot{\theta}M_{12}'M_{13}'/L_3'\Delta_2, \quad a_{65} = \dot{\theta}L_1'/\Delta_2, \quad a_{66} = -R_3'L_1'/L_3'\Delta_2,$$
$$\Delta_2 = L_1' - M_{13}'^2/L_3', \quad \Delta_3 = 2L_0 + 3L_1' - 3M_{13}{}^{2\prime}/L_3'.$$

$$A_2 = \begin{bmatrix} a_{11}' & a_{13}' & a_{14}' & a_{15}' & a_{16}' & a_{17}' \\ a_{21}' & a_{23}' & a_{24}' & a_{25}' & a_{26}' & a_{27}' \\ & & -\dot{\theta} & & & \\ & \dot{\theta} & & & & \\ a_{51}' & a_{53}' & a_{54}' & a_{55}' & a_{56}' & a_{57}' \\ a_{61}' & a_{63}' & a_{64}' & a_{65}' & a_{66}' & a_{67}' \end{bmatrix}$$

$$a_{11}' = -(R_0 + 2R_1')/\Delta_1, \quad a_{13}' = -3\dot{\theta}M_{12}'/2\Delta_1,$$
$$a_{14}' = 3\dot{\theta}M_{12}'/2\Delta_1, \quad a_{15}' = M_{13}(3R_3'/L_3' - 3\dot{\theta}/2\Delta_1,$$
$$a_{16}' = M_{13}'(3\dot{\theta} + 3R_3'/L_3')/2\Delta_1, \quad a_{17}' = 3/2\Delta_1,$$
$$a_2' = a_{1_i}/3 (i = 1 \sim 7),$$
$$a_{51}' = M_{13}'\{-\dot{\theta}/3 + (R_0 + 2R_1')/\Delta_1\}/2\Delta_1,$$
$$a_{53}' = 3\dot{\theta}M_{12}'M_{13}'/2L_3'\Delta_1, \quad a_{54}' = -3\dot{\theta}M_{12}'M_{13}'/2L_3'\Delta_1,$$
$$a_{55}' = -\{M_{13}{}^{2\prime}(3R_3'/L\theta' - 3\dot{\theta})/2\Delta_1 + R_3'\}/L_3',$$
$$a_{56}' = -\{M_{13}{}^{2\prime}(3\dot{\theta} + 3R_3'/L_3')/2\Delta_1 + \dot{\theta}L_3'\}/L_3',$$
$$a_{57}' = -3M_{13}'/2L_3'\Delta_1,$$
$$a_{61}' = M_{13}'\{\dot{\theta} + (R_0 + 2R')'3\Delta_1\}/L_3', \quad a_{63}' = \dot{\theta}M_{12}'M_{13}'/2L_3'\Delta_1,$$
$$a_{64}' = -3\dot{\theta}M_{12}'M_{13}'/2L_3'\Delta_1,$$
$$a_{65}' = \{\dot{\theta}L_0' + M_{13}'^2(-3R_3'/L_3' + \dot{\theta})/2\Delta_1/L_3',$$
$$a_{66}' = \{-R_3' - M_{13}{}^2(3\dot{\theta} + R_3'/L_3')/2\Delta_1\}/L_3',$$
$$a_{67}' = -M_{13}'/3L_3'\Delta_1$$
$$\Delta_1 = L_0 + 2L_1' - 2M_{13}{}^{2\prime}/L_3'.$$

Matrix B

$$B = \begin{bmatrix} B_0 & & & \\ & B_0 & & \\ & & B_0 & \\ & & & 1 \end{bmatrix} \quad B_0 = \begin{bmatrix} \cos\psi & \sin\psi \\ -\sin\psi & \cos\psi \end{bmatrix} \tag{30}$$

where $\psi = \pi/3$.

About Saliency of Motors

In order to obtain rigorous solutions for salient pole machines, first, (17) and (18) are solved by regarding the salient pole machines as the corresponding nonsalient pole ones. Then, using the above solutions as initial values, (13) is solved by Runge-Kutta-Gill method.

The approximate models of the salient pole machines integrate the terms of matrix Z' which includes $\dot{\theta}$ from $t = 0$ through $t = T(\omega T = \pi/3)$.

Matrices A_3, B_2 and D_2

$$A_3 = \begin{bmatrix} L_0 + 2L_1' & \sqrt{\frac{3}{2}}M_{13}' & \frac{1}{\sqrt{2}}M_{13} \\ \sqrt{\frac{3}{2}}M_{13}' & L_3 & 0 \\ \frac{1}{\sqrt{2}}M_{13}' & 0 & L_3 \end{bmatrix} \tag{31}$$

$$B_2 = \begin{bmatrix} R_0 + 2R_1 + R_e & 0 & 0 \\ 0 & R_3 & 0 \\ 0 & 0 & R_3 \end{bmatrix} \tag{32}$$

$$D_2 = \begin{bmatrix} E_0 - \tilde{E}\cos\gamma_0 \\ 0 \\ 0 \end{bmatrix}. \tag{33}$$

Then, $1/T\int_0^T (2I_f M_{12}'\dot{\theta}\sin(\theta - \pi/6)\,dt = \sqrt{2}\tilde{E}\cos\gamma_0$, where $\tilde{E} = (1/\sqrt{2})\cdot(3/\pi)M_{12}'\dot{\theta}I_f$.

REFERENCES

[1] I. Schmit, "Analysis of converter-fed synchronous motors," in *IFAC Symp. Preprints,* vol. 1, pp. 571-585, Oct. 1974.

[2] F. Harashima *et al.*, "Analysis of thyristor commutatorless motors with finite DC reactors, " *Elec. Eng. Japan,* vol. 94-B, no. 11, pp. 551-558, Nov. 1974.

[3] W. J. Gibbs, *Electric Machine Analysis Using Matrices.* London, England: Pitman, 1962.

[4] J. T. Tou, *Modern Control Theory.* New York: McGraw-Hill, 1964.

[5] W. Charlton, "Matrix approach to steady state analysis of inverter-fed induction motor, " *Electron. Lett.,* vol. 6, no. 14, pp. 451-452, July 9, 1970.

[6] T. A. Lipo *et al.*, "State variable steady state analysis of a controlled current induction motor drive," *IEEE Trans. Industry App.,* vol. IA-11, pp. 704-712, Nov. / Dec. 1975.

[7] W. McMurray, *The Theory and Design of Cycloconverters* Cambridge, MA: 1972.

STABILITY BEHAVIOR OF A SYNCHRONOUS-RELUCTANCE MACHINE SUPPLIED FROM A CURRENT SOURCE INVERTER

C.M. Ong
Purdue University
West Lafayette, IN

T.A. Lipo
General Electric Co.
Schenectady, NY

Summary

The development of an equivalent circuit model for a current source inverter/reluctance machine drive is described. The dynamic behavior of the basic drive with dc voltage or current source excitation is examined using linearized equations based on this model. Two forms of instability are identified which tend to limit high speed operation when dc voltage source excitation is employed. The results demonstrate the need for closed-loop control for practical implementation.

Introduction

In a related two-part paper, a review of the literature on the current source inverter (CSI) and its application potential in adjustable speed drives was presented (1). These two papers set forth the steady-state behavior of a synchronous-reluctance machine operated from a CSI. In the application of ac drives, steady-state characteristics are not sufficient since the transient behavior of the drive is equally important to ensure reliable performance.

The stability behavior of reluctance machines fed from adjustable frequency voltage sources is well-known (2-4). In general, instability of any ac machine supplied from a voltage source results either from inherent instability of the machine or from its interaction with the source for a given operating condition (2,5). Machine parameters, source parameters, load characteristics and operating conditions all affect system stability. For a voltage source inverter reluctance machine drive, the influence of these parameters on stability has been investigated using a variety of methods. In the case of conventional voltage source inverters it has been found that source parameters have an adverse effect on system stability (5). In particular, it has been established that the dc side filter capacitor must be carefully selected.

In this paper, the open-loop stability of a current source inverter reluctance machine drive is investigated. In particular, two types of dc source characteristics are investigated, namely a constant dc current and a constant dc voltage source. These two source characteristics correspond to idealized representations of a thyristor ac/dc bridge operating in the controlled-current mode and in the zero phase delay (full rectify) mode respectively. A simple equivalent circuit model of the rectifier, filter, CSI and reluctance machine is developed. The dynamic response of the system is investigated based on a linearized model derived from this equivalent circuit. Although idealized source characteristics is perhaps an oversimplification it is the intent of this paper to portray the open-loop behavior of the CSI-reluctance machine system in as simplified a manner as possible. A detailed closed-loop design approach which utilizes this open-loop model is the subject of a companion paper (6).

Description of the System

The basic system configuration of a controlled-current source inverter reluctance motor drive is shown in Fig. 1. The system consists of a controlled-rectifier dc source supplied from an ac voltage source, a dc link with a series choke, a three-phase current inverter (CSI) and a three-phase synchronous-reluctance machine. The basic circuit arrangement of the CSI using auto-sequential commutation described by Ward (7) is shown in Fig. 2. With a constant input current, the idealized output current waveform from the CSI is quasi-square (1).

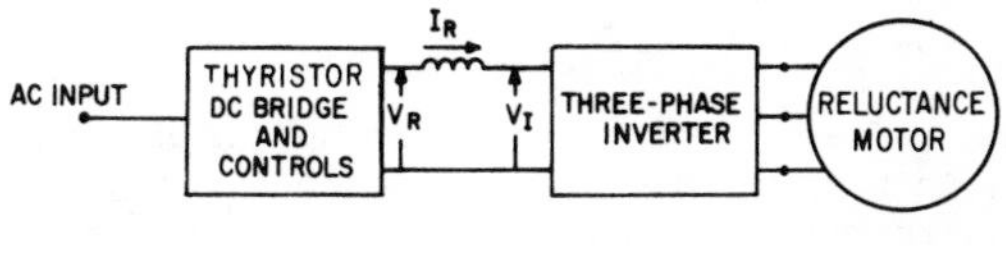

Fig. 1 Current source inverter/reluctance motor drive.

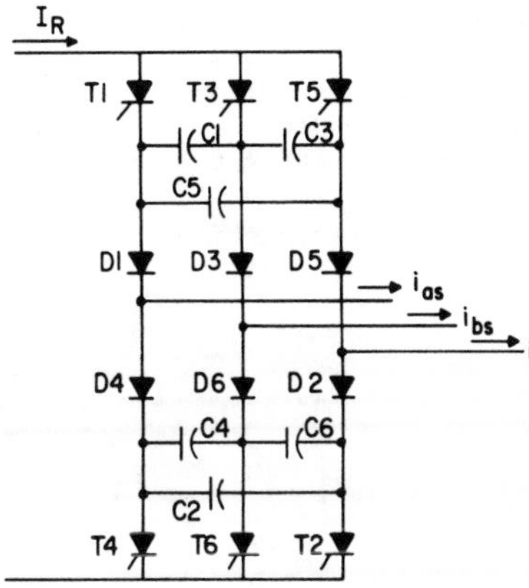

Fig. 2 Current source inverter with auto-sequential commutation.

In this paper it is assumed that:

- The CSI is a zero impedance switching device and the duration of the commutation interval can be assumed to be negligibly small.
- The reluctance machine can be represented as an idealized machine having a set of uniformly distributed stator windings. The rotor of the machine can be adequately represented by one equivalent short-circuited rotor winding in both direct and quadrature axes. The stator windings of the machine are assumed wye-connected. If necessary, delta-connected stator windings can be represented as an equivalent wye connection having identical terminal characteristics.
- The effects of saturation in the machine can be accounted for by appropriate values of saturated mutual reactance (8). For this study, the machine parameters are assumed constant for a given operating point.
- The harmonics of the dc link current can be ignored and the resulting dc component of the dc link current has sufficiently smooth fluc-

Reprinted from *Conf. Rec. IEEE/IAS 1975 Annual Meeting*, pp. 484–493, 1975.

tuations such that it can be approximated as constant throughout any inverval between two switching instants of the inverter. The validity of these simplifying approximations are clearly influenced by the inductance of the choke and the harmonics present in the dc rectifier voltage for a given operating condition.

Machine Representation

It is common in machine analysis to express the variables of an ac machine in dq0 components referred to one or more of the following reference frames,

- the stationary reference frame
- the synchronously rotating reference frame
- the reference frame fixed on the rotor.

The time reference of these reference frames is generally taken with respect to the instant the d-axis of the synchronously rotating reference frame is aligned with the axis of the phase 'a' winding of the machine. For machines with asymmetrical rotor structures, simlification is gained by referring the variables of the machine to the reference frame fixed on the rotor. Following the usual convention, the d-axis of that reference frame is aligned with the path of minimum reluctance on the rotor. However, since current rather than voltage is considered as the independent variable, it is convenient in this case to orient the d-axes of the above-mentioned reference frames with the axis of phase 'a' as shown in Fig. 3[1].

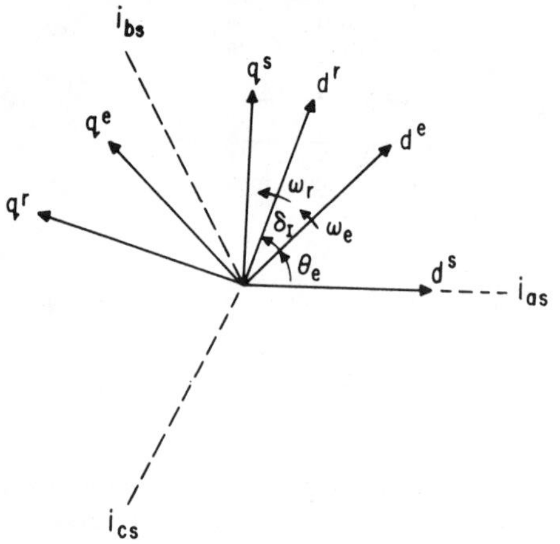

Fig 3 Axes of reference.

When a wye-connected three-phase stator winding is supplied from a balanced three-phase, three-wire ac source, the instantaneous phase voltages and currents respectively sum to zero.

$$v_{as} + v_{bs} + v_{cs} = 0 \tag{1}$$

$$i_{as} + i_{bs} + i_{cs} = 0 \tag{2}$$

As a result, the zero sequence components of these variables are not present. The transformation of the dq components of voltages and currents from the stationary reference frame to the synchronously rotating reference frame is given by the matrix $\underline{T}(\theta_e)$ where $\theta_e=\omega_e t$[1]

$$\underline{T}(\theta_e) = \begin{bmatrix} \cos\theta_e & \sin\theta_e \\ -\sin\theta_e & \cos\theta_e \end{bmatrix} \tag{3}$$

Also, from Fig. 3, the transformation of dq variables in the synchronously rotating reference frame to the reference frame fixed on the rotor $\underline{T}(\delta_I)$ is given by the same transformation matrix but with δ_I replacing θ_e in the corresponding trigonometric terms of Eq. 3. It should be noted that the angle δ_I as defined by Fig. 3, although related to torque, is different than the conventional torque angle for a voltage source. The voltage and torque expressions which describe the three-phase synchronous-reluctance machine having two mutually orthogonal rotor windings are those set forth in Refs. 1 and 2 and will not be repeated here.

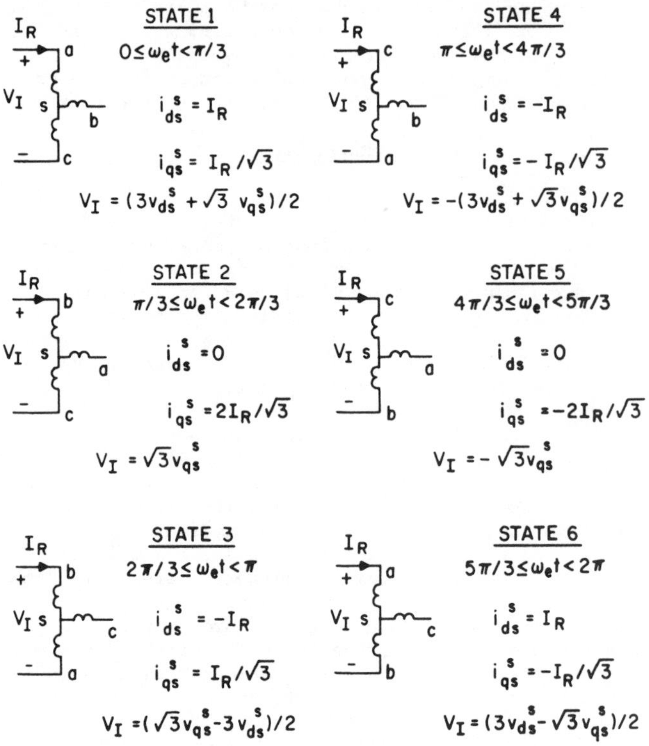

Fig. 4 The six inverter states and resulting d-q axes voltages and currents.

CSI Terminal Relations

Assuming that the CSI is an ideal lossless switching device with negligible commutation time, the sequence of current paths through the stator windings of the machine over a complete switching cycle is shown in Fig. 4. For each conduction state shown the ac phase voltages and currents can be expressed in terms of the inverter terminal voltage V_I and the rectifier current I_R. When expanded in a Fourier series these currents can be expressed

$$i_{ds}^s = \frac{2\sqrt{3}}{\pi} I_R \left[\cos\omega_e t - \frac{\cos 5\omega_e t}{5} + \frac{\cos 7\omega_e t}{7} - \frac{\cos 11\omega_e t}{11} + \dots\right] \tag{4}$$

$$i_{qs}^s = \frac{2\sqrt{3}}{\pi} I_R \left[\sin\omega_e t + \frac{\sin 5\omega_e t}{5} + \frac{\sin 7\omega_e t}{7} + \frac{\sin 11\omega_e t}{11} + \dots\right] \tag{5}$$

Note that Eqs. 4 and 5 are general and valid even if I_R is not a constant.

For brevity, it is convenient to define the switching functions

$$g_{ds}^{s} = \left[\cos\omega_e t - \frac{\cos 5\omega_e t}{5} + \frac{\cos 7\omega_e t}{7} - \frac{\cos 11\omega_e t}{11} + \ldots\right] \quad (6)$$

$$g_{qs}^{s} = \left[\sin\omega_e t + \frac{\sin 5\omega_e t}{5} + \frac{\sin 7\omega_e t}{7} + \frac{\sin 11\omega_e t}{11} + \ldots\right] \quad (7)$$

and also

$$I_R' = \frac{2\sqrt{3}}{\pi} I_R \quad (8)$$

The stator currents i_{ds}^s and i_{qs}^s can then be written in the form

$$i_{ds}^{s} = I_R' g_{ds}^{s} \quad (9)$$

$$i_{qs}^{s} = I_R' g_{qs}^{s} \quad (10)$$

The inverter terminal voltage can also be related to the switching functions. An inspection of the expressions for inverter terminal voltage V_I summarized in Fig. 4 will reveal that

$$V_I' = \left(v_{ds}^{s} g_{ds}^{s} + v_{qs}^{s} g_{qs}^{s}\right)$$

where

$$V_I' = \frac{\pi}{3\sqrt{3}} V_I \quad (11)$$

Since the power flow into the inverter is expressed as the product of the rectifier current and the inverter terminal voltage

$$V_I I_R = \frac{3\sqrt{3}}{\pi} I_R \left(v_{ds}^{s} g_{ds}^{s} + v_{qs}^{s} g_{qs}^{s}\right) \quad (12)$$

Substituting for $I_R g_{ds}^s$ and $I_R g_{qs}^s$ in the above expression, results in

$$V_I I_R = \frac{3}{2}\left(v_{ds}^{s} i_{ds}^{s} + v_{qs}^{s} i_{qs}^{s}\right) \quad (13)$$

Note that the left-hand side of Eq. 13 is the dc power into the CSI and the right-hand side is the power flow into the machine, thus, the result is consistent with the assumption that the CSI is a lossless switching device. Referring to Fig. 2 it can be shown that the normalized rectifier voltage V_R' may be expressed in terms of the inverter voltage V_I' and the current I_R' as

$$V_R' = V_I' + \left(R_F' + \frac{p}{\omega_b} X_F'\right) I_R' \quad (14)$$

Where ω_b is the base frequency, p denotes the operator $\frac{d}{dt}$ and

$$V_R' = \frac{\pi}{3\sqrt{3}} V_R,\ R_F' = \frac{\pi^2}{18} R_F,\ X_F' = \frac{\pi^2}{18} \omega_b L_F .$$

An Equivalent Circuit Model

Frequently in the analysis of uniform airgap machines, i.e., induction machines, it is found convenient to refer the variables of the machine to the synchronously rotating reference frame. In general, the synchronously rotating frame is defined as that reference frame which rotates at the angular velocity of the fundamental component of the excitation vector.

Applying the transformation $\underline{T}(\theta_e)$, Eq. 3, to the stator currents i_{ds}^s and i_{qs}^s given by Eqs. 9 and 10, the corresponding dq currents in the synchronously rotating reference frame i_{ds}^e and i_{qs}^e are

$$i_{ds}^{e} = I_R' g_{ds}^{e} \quad (15)$$

$$i_{qs}^{e} = I_R' g_{qs}^{e} \quad (16)$$

where

$$\begin{bmatrix} g_{ds}^{e} \\ g_{qs}^{e} \end{bmatrix} = \underline{T}(\theta_e) \begin{bmatrix} g_{ds}^{s} \\ g_{qs}^{s} \end{bmatrix} \quad (17)$$

Similarly, it can be shown that

$$V_I' = v_{ds}^{e} g_{ds}^{e} + v_{qs}^{e} g_{qs}^{e} \quad (18)$$

When expressed explicitly as a function of time, the functions g_{ds}^e and g_{qs}^e are given by

$$g_{ds}^{e} = 1 - \frac{2}{35}\cos 6\omega_e t - \frac{2}{143}\cos 12\omega_e t - \ldots \quad (19)$$

$$g_{qs}^{e} = \frac{12}{35}\sin 6\omega_e t + \frac{24}{143}\sin 12\omega_e t + \ldots \quad (20)$$

It is clear that the constant term of g_{ds}^e in Eq. 19 corresponds to the fundamental component of the resultant voltage or current waveform in the stationary reference frame. The 'g' functions defined by Eqs. 6, 7, 19 and 20 are plotted in Fig. 5. When I_R is constant it is clear that these functions also effectively define the waveshape of the d-q stator currents.

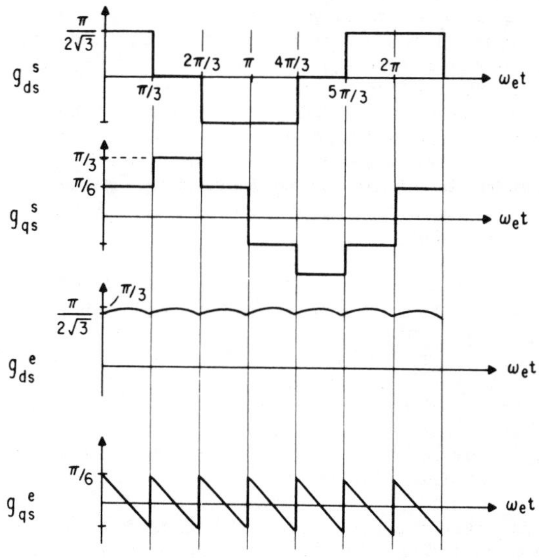

Fig. 5 The d- and q- axes switching functions.

Because the airgap of a reluctance machine is not uniform, the airgap flux density is not proportional to the MMF at every point in the gap. Hence, the self and mutual inductances of the stator windings are functions of angular position of the rotor. Voltage and current variables of the machine must be referred to a reference frame fixed on the rotor in order that the equations relating these variables become independent of rotor position. A transformation of variables from the synchronously rotating reference frame to a reference frame fixed on the rotor can be accomplished with the aid of the transformation matrix $\underline{T}(\delta_I)$. An equivalent circuit model of both current source inverter and reluctance machine is shown in Fig. 6. Consistent with previous papers,[10] it is assumed that the time derivative operator in Fig. 6 is p/ω_b. Note that since the synchronously rotating axes have been located relative to the stator current, the torque angle δ_I is consistent with Fig. 3.

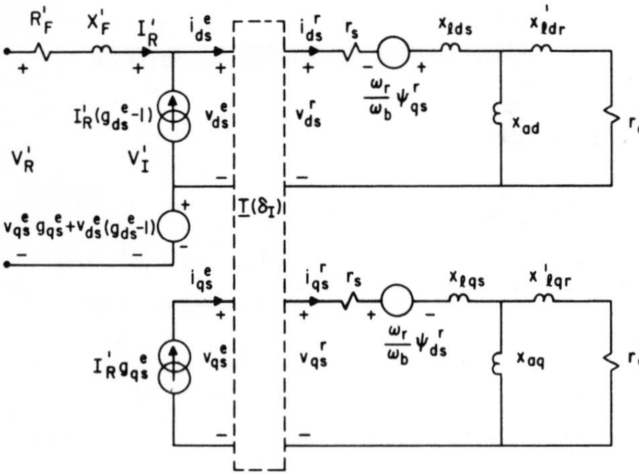

Fig. 6 A d-q axes equivalent circuit of a current source inverter/reluctance motor drive.

Small Signal Stability Behavior

The reluctance machine is, in general, described by a set of nonlinear differential equations for which no analytical solution is yet available. For the purpost of a stability investigation, a linearized system model is generally employed. Clearly, an analysis based on linearization of the set of nonlinear equations of the machine is valid only when small perturbations about an operating point is considered. In spite of this limitation, however, such linear analyses have been found to be invaluable in predicting small-disturbance transient response and in establishing guidelines for design modifications.

It was observed from the steady-state solution that harmonic components are present in the current and torque waveforms, whereas the waveforms of the phase voltages are nearly sinusoidal[1]. Reports of previous stability investigations of ac machines supplied from voltage source inverters indicate that the harmonic components do not significantly affect stability behavior[2-5]. It has been shown that the resultant contribution from the harmonics to energy conversion is small, even though the torque pulsations may be considerable. Although harmonics do contribute some additional damping, no satisfactory quantitive theory is yet available so that provision for refinement must ultimately be based on an experimental study. Hence, for simplicity the effects of harmonics will be neglected in this analysis and also in the subsequent analytical design[6].

Neglecting harmonics the 'g' functions defined by Eqs. 19 and 20 are, approximately

$$g_{ds}^{e} \cong 1 \tag{21}$$

$$g_{qs}^{e} \cong 0 \tag{22}$$

Thus, Eqs. 15 and 16 and 18 reduce to the form

$$i_{ds}^{e} \cong I_R' \tag{23}$$

$$i_{qs}^{e} \cong 0 \tag{24}$$

Replacing g_{ds}^e and g_{qs}^e in Fig. 5 by their approximate values leads to the simplified equivalent circuit shown in Fig. 7. This equivalent circuit representation clearly demonstrates the mechanism of power transfer through the dc link to the machine which, for unidirectional rectifier current, is dependent on the inverter terminal voltage, $V_I' = v_{ds}^e$, where v_{ds}^e assumes positive values for motor action and is negative for generator operation. System stability information may now be obtained by perturbing the differential equation defining Fig. 7. The inverter terminal voltage v_{ds}^e and also the currents i_{ds}^r and i_{qs}^r are dependent on the load angle δ_I. The perturbation equations, given in Appendix I, may be established from the system equations in a straightforward manner. These linear, time-invariant equations can be arranged into the state variable matrix form as given by Eq. 33. The small-signal stability behavior of the drive is investigated by evaluating the eigenvalues of this linear system of equations. To facilitate a comparison between the stability behavior exhibited by a reluctance machine supplied by a voltage inverter source and from a CSI, the machine parameters employed are similar to those used in previous studies.[1,2] These machine parameters are summarized in Appendix II.

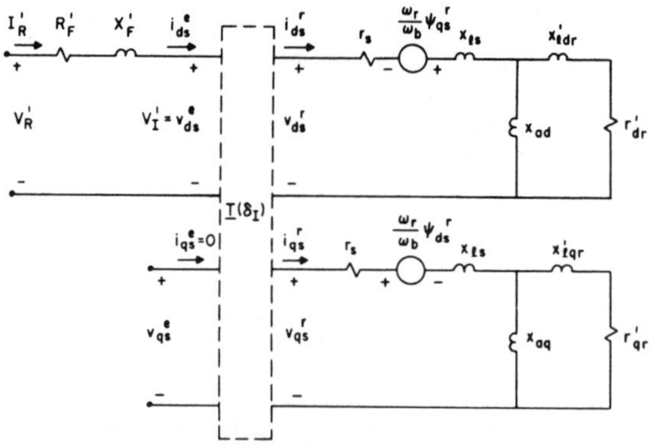

Fig. 7 Simplified d-q axes equivalent circuit.

Constant Current Source Operation

When source harmonics are neglected it can be noted from Fig. 7 that $i_{qs}^e = 0$. Hence, its perturbation Δi_{qs}^e is always zero so that the order of the system becomes one less than for an equivalent voltage source inverter. Also, for a constant current rectifier source, $\Delta I_R' = \Delta i_{ds}^e = 0$. The order of the system is therefore four when the motor is supplied from a fixed current source. System stability can be determined by calculating the system eigenvalues using Eq. 33. A typical set of system eigenvalues consists of two real and one complex pair of roots. In general, when dc link current is held constant, the machine has been found to be stable for all motoring and generating conditions which do not exceed the pull-out torque. The result agrees with previous investigators who indicated that a conventional synchronous motor is stable for fixed sinusoidal current excitation.[9] In addition, it has been observed from the eigenvalue analysis that identical sets of eigenvalues are always obtained for equal motoring and generation torque indicating that transient response in the motoring and generating regions are identical.

Because many important motor and system parameters can often be specified by the designer, the influence of system parameters and operating conditions on stability is often an invaluable aid to design. Although space limitations prevent treating the subject exhaustimely, comparison of ovserved trends with results from a study of reluctance machines fed from voltage inverter sources are of interest.[2] The most significant parameters affecting transient behavior are:

- A reduced level of current excitation results in a higher overshoot but better damping.
- An increase in rotor inertia tends to lower the transient response frequency and, for the operating conditions investigated an improvement in the damping occurs. This result is in contrast to previous work involving voltage excitation.[3]
- Changes in the stator winding resistance and leakage reactance do not affect the transient response behavior, as can be anticipated with a constant current source excitation. Even in the case of an approximate current source, the effects due to changes in value of these stator parameters will normally be masked by the large source impedance.
- A higher resistance to leakage reactance ratio (r/x_ℓ) in the d- or q-axis rotor windings tends to improve the damping when the machine is on load. However, for no-load conditions, a higher (r/x_ℓ) ratio in the d-axis rotor winding does not affect damping. Nevertheless, a higher (r/x_ℓ) ratio in the q-axis rotor winding provides additional damping even when the machine is operating on no-load.
- A decrease in rotor saliency (x_{ad}/x_{aq}) results in a lower damped natural frequency together with improved damping.
- Finally, an increase in loading on the machine always tends to improve damping.

In order to verify these conclusions, an analog computer study was performed wherein all of the relevant system non-linearities were preserved and the CSI represented in complete detail. Computer traces of key system variables are shown in Fig. 8 for a step change in applied torque. Figures 8a and 8b indicate response for step changes in load torque from no-load to 0.2 pu motoring and 0.2 pu generating operation respectively. Note that, as predicted, the transient responses are identical except for the changes in sign. Since transient responses are identical in both regions, only the transient response in the motoring region are shown in subsequent traces.

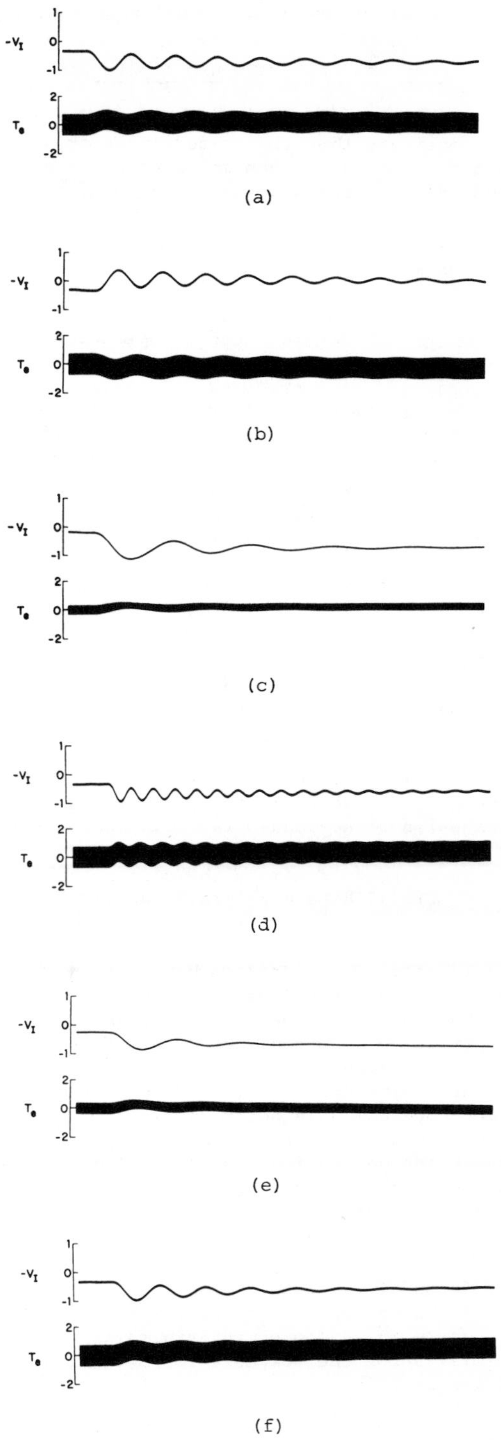

Fig. 8 Step response for constant current source excitation, I_R=0.8 (except c), f_R= 1.0, T_L=±0.2. Motor parameters are in Appendix II except as noted; (c) I_R=0.5, (d) H = 0.1 s., (e) x_{ad}=1.5, (f) r_{qr}'=0.045.

Figure 8c illustrates the conclusion that for reduced current excitation, a higher overshoot occurs followed by a better damped response. Smaller inertia results in a higher transient response frequency and noticeable speed fluctuations as portrayed in Fig. 8d. Hence, a larger inertia enhances relative stability, although this may not be true for all operating conditions. Figure 8e shows the traces for the case of reduced rotor saliency, which has the equivalent effect of a higher load angle operation. It is observed that the response is relatively less oscillatory and because of the smaller saliency ratio, the torque pulsation is also correspondingly smaller. Figure 8f illustrates the effect of higher rotor resistances. In particular, note that a higher q-axis rotor winding resistance results in greater damping. Since these analog computer traces have been obtained with a detailed representation of the reluctance machine with non-linearities and harmonics taken into account they clearly justify the assumptions made in the small signal eigenvalue analysis.

Constant Voltage Source Operation

In many practical applications, limitations in KVA rating necessitate that the ac-dc thyristor bridge be permitted to "saturate" when motor speed exceeds a specified value of operation above a certain inverter output frequency. This mode of operation corresponds to constant horsepower operation and is analogous to the field weakening mode in conventional dc motor drives. Since operation from a voltage source is typical for many wide speed applications, it is of interest to examine the consequences of operation under such a condition. In this analysis the voltage variation of the ac supply is not considered. It will be assumed herein that for the drive under consideration, the assumption of an infinite ac busbar is justified so that $\Delta V_R = 0$. Under such conditions, the rectifier is assumed to operate without phase angle delay. If necessary, the effects of finite source impedance can be lumped with the filter choke parameters.

Stability of the CSI/reluctance motor drive operating from a voltage source is again defined by Eq. 33 where, in this case, $\Delta i_{ds}^e \neq 0$. In contrast to operation from a current source, the reluctance motor exhibits regions of unstable behavior when the CSI is supplied from a voltage source. Figures 9a and 9b show the boundaries between stable and unstable regions of operation as a function of the per unit frequency f_R. The operating condition assumes a fixed value of rectifier voltage which, for each frequency, provides 0.8 pu current excitation at no-load. In Fig. 9(a), an arbitrarily large value of choke inductance, 50 pu, has been selected so as to approximate the current source condition. In Fig. 9(b) a more practical value, X_F=1.2 pu has been chosen. It is interesting to note that the two widely different values of choke inductance has no discernable effect on stability behavior in the motoring region. Subsequent investigation of steady-state behavior has indicated that these boundary curves can also be predicted from considerations of steady-state power transfer (Appendix III). Hence, the result is an inherent characteristic of the system and indicates that stable operation is not possible without voltage control of the ac/dc thyristor bridge. The instability boundary in the generating region is clearly dependent on filter parameters and corresponds to a dynamic instability analogous to those which occur for conventional voltage source operation.

The forms of instability for the operating conditions labled 'b' and 'c' in Fig. 9(b) are illustrated by traces from an analog computer simulation study. Figure 10(a) shows the response following a step change of load torque from an initial (stable) operating point 'a' (T_{eo} = 0, f_R= 1.0) to an operating point labeled 'b' (T_{eo} = 0.1, f_R= 1.0). The decay of the rectifier current and torque indicates that a stable steady-state operating condition does not exist at point 'b'. Similarly, Fig. 10(b) is for a step change of load torque from operating point 'a' to a generating condition at point 'c'(T_{eo} = -0.1, f_R= 1.0). Operating point 'c' is outside the region of stable operation where net system damping becomes negative and the response diverges. Note that the speed of the machine oscillates about synchronous speed instead of the gradual decay observed for motor operation, point 'b'. The operating conditions in Fig. 9 are for fixed values of rectifier voltage curresponding to rated excitation current. In this case the rectifier output voltage and region of stability is small. Nevertheless, these results are the same regardless of the voltage level.

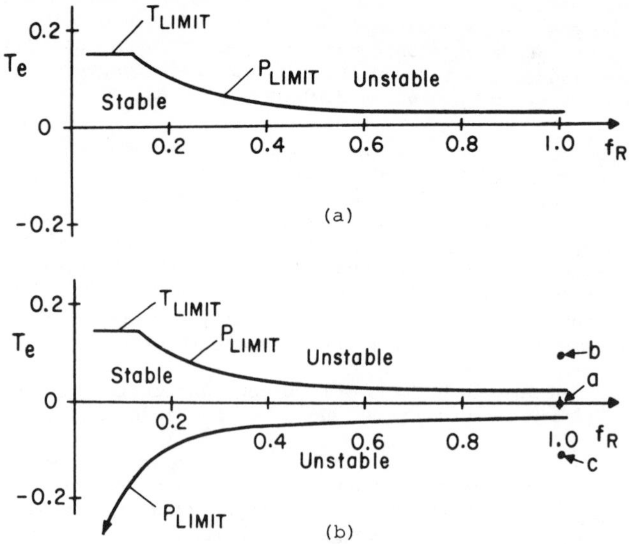

Fig. 9 Stability boundary of CSI/reluctance motor drive with constant dc rectifier voltage, (a) X_F =50, (b) X_F=1.2.

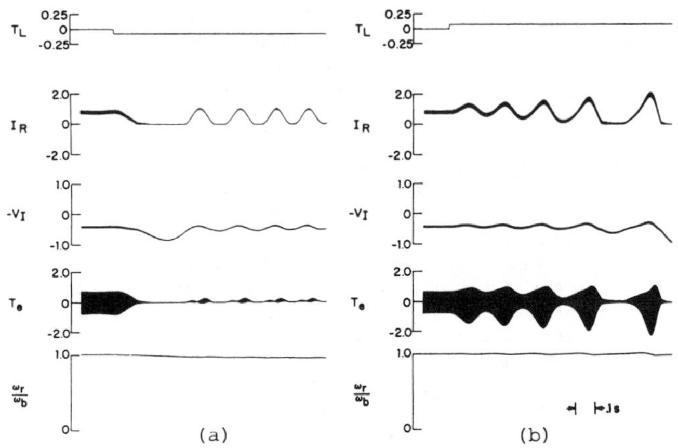

Fig. 10 Load torque switching for constant dc voltage operation, f_R = 1.0. Step change in load torque, (a) 0 to 0.1 pu, (b) 0 to -0.1 pu.

Although difficult to achieve practically, an examination of other operating conditions are also of interest. Figure 11 shows a larger region of stable operation when the rectifier voltage is adjusted so as to produce a fixed value of current excitation, I_R= 0.8 pu, independent of steady-state load and operating frequency. Hence, this mode of operation can be considered as the open loop equivalent of a voltage source adjusted to appear identical, on a steady-state basis, to a current source. It is apparent that dynamic behavior is entirely different from a conventional voltage source even though the steady-state operating point is identical. Again, a real pole moves into the right-haof plane in the motoring region. The mechanism leading to this instability is again the limitation in power transfer from rectifier to machine although this limit now occurs at somewhat higher values of torque (power). Again, a pair of complex poles moves into the right-half plane for generator operation, predicting growing oscillations. Presence of the unstable generating region is again due to insufficient damping.

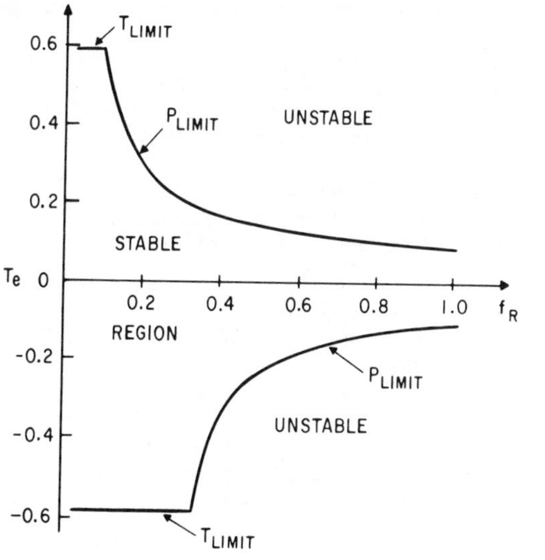

Fig. 11 Stability boundary with fixed dc rectifier voltage and steady-state current $I_R = 0.8$.

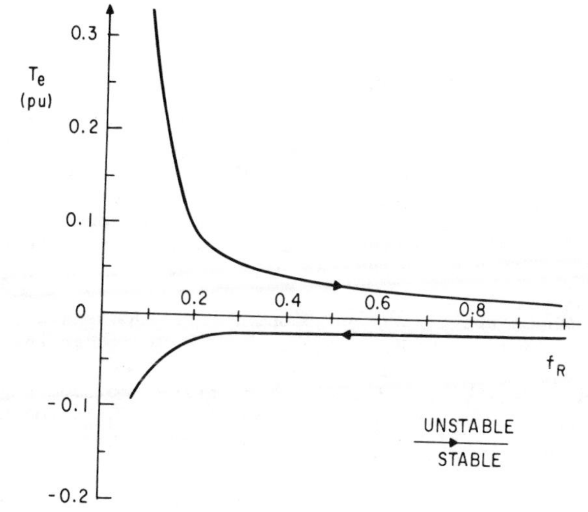

Fig. 12 Stability boundary with fixed dc rectifier voltage. Air gap flux maintained at one per unit.

Another typical operating constraint is to maintain air gap flux constant independent of load and operating frequency. Figure 12 shows the regions of stable and unstable operation when rectifier voltage is adjusted and then held constant so as to maintain rated stator flux at the specified load condition. Similar observations about the mechanisms of instability can be made.

In contrast to conventional voltage inverter operation[2] it has been established that changes in motor parameters have only relatively minor effects on the two instability modes. A study of parameter variations similar to Ref. 2 was not deemed necessary. These results point to the need for closed loop control to provide the necessary system damping. The closed loop design of such a system is the subject of a companion paper.

Conclusion

In this paper, a derivation of an equivalent circuit model of a current source inverter (CSI)/reluctance motor drive has been presented. The introduction of switching functions (g functions) have been shown to facilitate the identification of fundamental and harmonics components generated by the switching operations of the CSI, thus providing a ready means of isolating their respective contributions. When the effects of harmonic components are ignored, the equivalent circuit model leads directly to a simplified equivalent circuit model from which small displacement equations may be obtained by linearization.

Many drive system require feedback control to ensure stability. However, before design of the necessary feedback loops a stability analysis of the basic open-loop system is useful since it can often be used to establish guidelines for improving system response. Although idealized constant current and constant voltage dc source characteristics have been investigated a number of interesting results have been obtained. In particular it was found that the dc source characteristic has a dominant effect on stability and transient behavior. With a stiff current source, the CSI and reluctance machine system is unconditionally stable, whereas with a weak current source (voltage source) the same system has a much poorer stability profile.

From a stability point of view a CSI/reluctance motor drive clearly operates best when supplied from a constant current source. However, when the drive is considered as a whole, a source characteristic of this type is not necessarily desirable despite the attractiveness of an unconditionally stable system. Based on steady-state characteristics,[1] fixed dc link current operation leads to excessive voltage stresses on the inverter thyristors at light load and has the operational disadvantage of overexcitation and consequent higher torque pulsations and losses. On the other hand, open loop operation of the basic system with constant rectifier voltage has limited practical utility because of the restricted region of stable operation. It is apparent that the characteristics of a properly regulated dc current source would be a reasonable compromise. The application of the results of this paper to the closed-loop design of an adjustable speed CSI/reluctance motor drive are described in a companion paper.[6]

List of Symbols

All quantities are in per unit unless noted

Symbol	
V_R	rectifier dc terminal voltage
I_R	rectifier dc output current
V_I	inverter terminal voltage
v_{as}	phase a voltage of machine
i_{as}	phase a current of machine
v_{ds}, v_{qs}	stator voltages in d- and q-axis circuits
i_{ds}, i_{qs}	stator currents in d- and q-axis circuits
i_{dr}, i_{qr}	rotor currents in d- and q-axis circuits
ψ_{ds}, ψ_{qs}	stator flux linkages of d- and q-axis circuits
ψ_{dr}, ψ_{qr}	rotor flux linkages of d- and q-axis circuits
T_e	electromagnetic torque

T_L	load torque
ω_b	base electrical angular frequency in rad/s
ω_e	electrical angular frequency of the inverter (fundamental component) in rad/s
ω_r	rotor electrical angular frequency in rad/s
f_R	frequency ratio ω_e/ω_b
δ_I or δ	load angle for current source in rad
H	inertia constant of rotor in sec
x_{ds}, x_{qs}	synchronous reactances in the d- and q-axis circuits at base frequency
x_{ad}, x_{aq}	stator-rotor mutual reactances in the d- and q-axis at base frequency circuits
$x_{\ell s}$	stator leakage reactance at base frequency
$x_{\ell dr}$, $x_{\ell qr}$	rotor circuit leakage reactances in d- and q-axis circuits at base frequency
r_s	stator resistance
r_{dr}, r_{qr}	rotor resistances in the d- and q-axis circuits
L_F, R_F	filter inductance and resistance
p	differential operator d/dt
s	Laplace variable
t	time

Superscripts

s	to indicate a variable referred to the stationary reference frame
e	to indicate a variable referred to the synchronously rotating reference frame
r	to indicate a variable referred to the reference frame fixed to the rotor
'	to indicate a rotor parameter referred to the stator or as otherwise defined in the text

Subscript

o	to indicate the steady-state portion of the variable

Prefix

Δ( )	to denote a small change of the variable

References

(1) Ong, C.M., Lipo, T.A., "Steady-State Analysis of a Current Source Inverter/Reluctance Motor Drive, Part I - Analysis and Part II - Experimental and Analytical Results", to be presented at the 1975 IEEE-IAS Annual Meeting, Sept. 28-Oct. 2, 1975.

(2) Lipo, T.A. and Krause, P.C., "Stability Analysis of a Reluctance Synchronous Machine", IEEE Trans. PAS, Vol. 86, 1967, pp. 825-834.

(3) Lawrenson, P.J. and Bowes, S.R., "Stability of Reluctance Machines", Proc. IEE, Vol. 117, No. 2, Feb. 1971, pp. 356-369.

(4) Cruickshank, A.J.O., Anderson, A.F., and Menzies, R.W., "Theory and Performance of Reluctance Motors with Axially Laminated Anisotropic Rotors", Proc. IEE, Vol. 118, No. 7, July 1971.

(5) Lipo, T.A. and Krause, P.C., "Stability Analysis of a Rectifier-Inverter Induction Motor Drive", IEEE Trans. on Power Apparatus and Systems, Vol. PAS-88, No. 1, January 1969, pp. 55-66.

(6) Ong, C.M., and Lipo, T.A., "An Approach to Closed Loop Design of a Curreht Source Inverter/Reluctance Motor Drive System", to be presented at the 1975 IEEE-IAS Annual Meeting, Sept. 28-Oct. 2, 1975.

(7) Ward, E.E., "Inverter Suitable for Operation over a Range of Frequency", Proc. IEE, Vol. III, No. 8, Aug. 1964, pp. 1423-1434.

(8) Williamson, A.C., "Calculation of Saturation Effects in Segmented-Rotor Reluctance Machines", Proc. IEE, Vol. 121, No. 10, Oct. 1974, pp. 1127-1133.

(9) Slemon, G.R., Dewan, S.R. and Wilson, J.W.A., "Synchronous Motor Drive with Current-Source Inverter", IEEE-IAS, Conf. Record, 1973, pp. 875-879.

(10) Krause, P.C. and Lipo, T.A., "Analysis and Simplified Representations of Rectifier-Inverter Reluctance-Synchronous Motor Drives", IEEE Trans. PAS, Vol. 88, June 1969, pp. 962-940.

Appendix I

The linear equations describing the small excursion from a nominal load condition may be obtained by considering small perturbation of the system equations. The subscript 'I' is omitted from the torque angle δ_I. The subscript 'o' is used to denote the steady-state or "operating point" value of the variable.

The perturbation equations for an inductance machine have been derived in Ref. 2. Note that in Eqs. 26-33, the rotor currents are expressed in a rotor reference frame whereas the stator voltage and current has been referred to the synchronously rotating frame. Assuming $\Delta i_{qs}^e = 0$ the perturbation equations are

$$\begin{aligned}\Delta v_{ds}^e = {} & (x_{ds}\cos^2\delta_o + x_{qs}\sin^2\delta_o)\frac{p}{\omega_b}\Delta i_{ds}^e \\ & - (x_{ds} - x_{qs})\, i_{dso}^e \frac{\sin 2\delta_o}{2}\frac{p}{\omega_b}\Delta\delta \\ & + x_{ad}\cos\delta_o \frac{p}{\omega_b}\Delta i'^{r}_{dr} - x_{aq}\sin\delta_o \frac{p}{\omega_b}\Delta i'^{r}_{qr} + r_s \Delta i_{ds}^e \\ & - (x_{ds} - x_{qs})\, i_{dso}^e \frac{\sin 2\delta_o}{2}\frac{\Delta\omega_r}{\omega_b} \\ & - \frac{\omega_{ro}}{\omega_b}(x_{ds} - x_{qs})\frac{\sin 2\delta_o}{2}\Delta i_{ds}^e \\ & - \frac{\omega_{ro}}{\omega_b}(x_{ds} - x_{qs})\, i_{dso}^e \cos 2\delta_o\, \Delta\delta \\ & - \frac{\omega_{ro}}{\omega_b} x_{aq}\cos\delta_o\, \Delta i'^{r}_{qr} - \frac{\omega_{ro}}{\omega_b} x_{ad}\sin\delta_o\, \Delta i'^{r}_{dr}\end{aligned} \tag{25}$$

$$\begin{aligned}\Delta v_{qs}^e = {} & (x_{ds} - x_{qs})\frac{\sin 2\delta_o}{2}\frac{p}{\omega_b}\Delta i_{ds}^e \\ & + x_{ad}\sin\delta_o \frac{p}{\omega_b}\Delta i'^{r}_{dr} + x_{aq}\cos\delta_o \frac{p}{\omega_b}\Delta i'^{r}_{qr}\end{aligned}$$

$$- (x_{ds}\sin^2\delta_o + x_{qs}\cos^2\delta_o)\, i^e_{dso} \frac{p}{\omega_b}\Delta\delta$$

$$+ (x_{qs}\sin^2\delta_o + x_{ds}\cos^2\delta_o)\, i^e_{dso} \frac{\Delta\omega_r}{\omega_b}$$

$$+ \frac{\omega_{ro}}{\omega_b}(x_{qs}\sin^2\delta_o + x_{ds}\cos^2\delta_o)\Delta i^e_{ds}$$

$$+ \frac{\omega_{ro}}{\omega_b}(x_{qs} - x_{ds})\, i^e_{dso}\sin 2\delta_o\, \Delta\delta$$

$$+ \frac{\omega_{ro}}{\omega_b} x_{ad}\cos\delta_o \Delta i'^r_{dr} - \frac{\omega_{ro}}{\omega_b} x_{aq}\sin\delta_o \Delta i'^r_{qr} \qquad (26)$$

$$0 = -x_{ad} i^e_{dso}\sin\delta_o \frac{p}{\omega_b}\Delta\delta + x_{ad}\cos\delta_o \frac{p}{\omega_b}\Delta i^e_{ds}$$

$$+ x'_{dr}\frac{p}{\omega_b}\Delta i'^r_{dr} + r'_{dr}\Delta i'^r_{dr} \qquad (27)$$

$$0 = -x_{aq} i^e_{dso}\cos\delta_o \frac{p}{\omega_b}\Delta\delta - x_{aq}\sin\delta_o \frac{p}{\omega_b}\Delta i^e_{ds}$$

$$+ x'_{qr}\frac{p}{\omega_b}\Delta i'^r_{qr} + r'_{qr}\Delta i'^r_{qr} \qquad (28)$$

$$\Delta T_e = -(x_{ad} - x_{aq})\, i^e_{dso}\sin 2\delta_o\, \Delta i^e_{ds}$$

$$-(x_{ad} - x_{aq})(i^e_{dso})^2\cos 2\delta_o\, \Delta\delta$$

$$- x_{ad} i^e_{dso}\sin\delta_o\, \Delta i'^r_{dr} - x_{aq} i^e_{dso}\cos\delta_o\, \Delta i'^r_{qr} \qquad (29)$$

$$\Delta V'_R = \Delta v^e_{ds} + (R'_F + X'_F \frac{p}{\omega_b})\Delta i^e_{ds} \qquad (30)$$

$$2Hp\frac{\Delta\omega_r}{\omega_b} = \Delta T_e - \Delta T_L \qquad (31)$$

$$p\Delta\delta = \Delta\omega_r - \Delta\omega_e \qquad (32)$$

These equations can be written in matrix form as

$$\begin{bmatrix} (X'_F + x_{ds}\cos^2\delta_o + x_{qs}\sin^2\delta_o) & x_{ad}\cos\delta_o & -x_{aq}\sin\delta_o & 0 & 0 \\ x_{ad}\cos\delta_o & x'_{dr} & 0 & 0 & 0 \\ -x_{aq}\sin\delta_o & 0 & x'_{qr} & 0 & 0 \\ 0 & 0 & 0 & 1 & 0 \\ 0 & 0 & 0 & 0 & -2H\omega_b \end{bmatrix} \frac{p}{\omega_b} \begin{bmatrix} \Delta i^e_{ds} \\ \Delta i'^r_{dr} \\ \Delta i'^r_{qr} \\ \Delta\delta \\ \frac{\Delta\omega_r}{\omega_b} \end{bmatrix} =$$

$$\begin{bmatrix} [r_s + R'_F - \frac{\omega_{ro}}{\omega_b}(x_{ds} - x_{qs})\frac{\sin 2\delta_o}{2}] & -\frac{\omega_{ro}}{\omega_b}x_{ad}\sin\delta_o & -\frac{\omega_{ro}}{\omega_b}x_{aq}\cos\delta_o & -i^e_{dso}(x_{ds} - x_{qs})\cos 2\delta_o & -i^e_{dso}(x_{ds} - x_{qs})\sin 2\delta_o \\ 0 & r'_{dr} & 0 & 0 & -i^e_{dso}\, x_{ad}\sin\delta_o \\ 0 & 0 & r'_{qr} & 0 & -i^e_{dso}\, x_{aq}\cos\delta_o \\ 0 & 0 & 0 & 0 & -1 \\ -i^e_{dso}(x_{ds} - x_{qs})\sin 2\delta_o & -i^e_{dso}x_{ad}\sin\delta_o & -i^e_{dso}x_{aq}\cos\delta_o & -(i^e_{dso})^2(x_{ds} - x_{qs})\cos 2\delta_o & 0 \end{bmatrix} \cdot \begin{bmatrix} \Delta i^e_{ds} \\ \Delta i'^r_{dr} \\ \Delta i'^r_{qr} \\ \Delta\delta \\ \frac{\Delta\omega_r}{\omega_b} \end{bmatrix} -$$

$$\begin{bmatrix} -1 & 0 & (x_{ds} - x_{qs})\, i^e_{dso}\frac{\sin 2\delta_o}{2} \\ 0 & 0 & i^e_{dso}\, x_{ad}\sin\delta_o \\ 0 & 0 & i^e_{dso}\, x_{aq}\cos\delta_o \\ 0 & 0 & 1 \\ 0 & -1 & 0 \end{bmatrix} \cdot \begin{bmatrix} \Delta V'_R \\ \Delta T_L \\ \frac{\Delta\omega_e}{\omega_b} \end{bmatrix} \qquad (33)$$

Appendix II

Motor Parameters	Per Unit Value
r_s	0.045
r'_{qr}	0.015
r'_{dr}	0.030
$x_{\ell s}$	0.100
$x'_{\ell dr}$	0.100
$x'_{\ell qr}$	0.100
x_{aq} (unsat)	0.500
x_{ad} (unsat)	2.000
H (inertia constant)	0.400 (S)
R_F	0.100
X_F	1.2

Both quantities are peak-to-peak values of the machine phase voltage and line current(10). Base frequency, $\omega_b/2\pi$, is 60 Hz.

Appendix III

Conditions for Maximum Steady-State Torque

When rectifier voltage is held constant, dropping the 'o' subscript for simplicity

$$V'_R = I'_R R'_F + I'_R \left[r_s - f_R(x_{ds} - x_{qs}) \frac{\sin 2\delta_I}{2} \right] \quad (34)$$

The average electromagnetic torque developed by the machine is given by

$$T_e = -I'^2_R (x_{ds} - x_{qs}) \frac{\sin 2\delta_I}{2} \quad (35)$$

Combining these two equations

$$V'_R I'_R = I'^2_R (R'_F + r_s) + f_R T_e \quad (36)$$

Since $I_R > 0$, Eq. 36 can be solved as

$$I'_R = \frac{V'_R + \sqrt{V'^2_R - 4(R'_F + r_s) f_R T_e}}{2(R'_F + r_s)} \quad (37)$$

The dc link current I_R must also be real. Hence, the maximum torque for a given operating frequency is

$$T_{LIM} = \frac{V'^2_R}{4 f_R (R'_F + r_s)} \quad (38)$$

at which point

$$I'_R = \frac{V'_R}{2(R'_F + r_s)} \quad (39)$$

Equation 38 defines the upper limit to the region of permissible steady-state operation for a fixed rectifier voltage and corresponds to the negatively sloped portion of the stability regions in Figs. 9, 11 and 12. If Eq. 36 is differentiated with respect to I'_R it can be shown that Eq. 38 corresponds to a maximum power limitation. At sufficiently low frequencies the upper limit becomes defined by a torque limit rather than a power limit. The transition point occurs when the maximum torques in the two regions are equated. Setting $\delta_I = 45°$ in Eq. 35 and equating to 38 we have

$$I'^2_R \frac{(x_{ds} - x_{qs})}{2} = \frac{V'^2_R}{4 f_R (R'_F + f_s)} \quad (40)$$

However, the transition point I'_R is also defined by Eq. 39 so that Eq. 40 reduces to

$$\frac{x_{ds} - x_{qs}}{8(R'_F + r_s)^2} \quad \frac{1}{4 f_R (R'_F + r_s)} \quad (41)$$

Hence, the transition point between torque limited and power limited operation is defined by

$$f_R = \frac{2(R'_F + r_s)}{x_{ds} - x_{qs}} \quad (42)$$

Control and Simulation of a Current-Fed Linear Inductor Machine

BIMAL K. BOSE, SENIOR MEMBER, IEEE, AND THOMAS A. LIPO, SENIOR MEMBER, IEEE

***Abstract*—A self-controlled, current-fed inverter excited linear inductor machine is described, in which the field current and inverter switching angles are programmed such that the machine operates at constant air-gap flux and unity displacement factor under all operating conditions. The machine can operate in wide speed range and constant predetermined torque is developed during both motoring and regeneration. The control incorporates overlap angle compensation for the inverter and permits modulation of inverter firing angle with resolution of a fraction of a degree. The dynamic *d-q* axis model of the machine with the power converters and the proposed control strategy has been simulated on a hybrid computer, and static and dynamic performance have been studied in detail. The control circuit has been designed and tested with a 112-kW prototype inductor machine. The experimental results give good agreement with the theory and with simulation.**

I. INTRODUCTION

THE CURRENT-SOURCE inverter fed synchronous machine drives with a self-control feature (also known as commutatorless machine) have attracted increasing interest in recent years [1]. In this class of drives, the power conversion circuit is simple for bidirectional power control, and load commutation can be implemented if the machine is permitted to operate at a certain minimum leading displacement factor. The self-controlled synchronous machine behaves as a "modified" dc machine which can be operated in wide speed and torque ranges with very little stability problems and with no fear of pulling out of step. An inductor-type machine has the added advantage of higher speed capability because the rotor dc field is excited from the stator side. Therefore, the machine is attractive for applications such as flywheel energy storage, aircraft generators, linear motor propulsion, and other high-speed applications.

In this paper a speed control system is described for a homopolar, sector wound inductor machine [2]. This sector-type machine is, in fact, a laboratory model of linear synchronous motor propulsion system which is presently under test in a high speed rail propulsion program sponsored by the U.S. Department of Transportation. In the present control strategy the field current and inverter firing angles are regulated such that the machine always operates at constant air-gap flux and the terminal displacement angle is zero under all operating conditions. Maintaining constant rated air-gap flux allows optimum utilization of the magnetics, and unity displacement factor permits efficient and economical design of the inverter and machine as a system [3].

Paper ID 78-1, approved by the Industrial Drives Committee of the IEEE Industry Applications Society for presentation at the 1978 Industry Applications Society Annual Meeting, Toronto, ON, Canada, October 1-5. Manuscript released for publication November 8, 1978.

B. K. Bose is with the Corporate Research and Development Center, General Electric Company, Schenectady, NY 12301.

T. A. Lipo was with General Electric Company, Schenectady, NY. He is now with Purdue University, Lafayette, IN.

II. DESCRIPTION OF CONTROL SYSTEM

Fig. 1 shows the power circuit of the drive system and Fig. 2 gives the idealized voltage and current waves at the machine terminal for unity displacement factor. Three-phase 60-Hz ac power at the input is converted to dc by a phase-controlled bridge rectifier, and is then inverted by an auto-sequential type force-commutated inverter to generate three-phase variable-frequency variable-current power for the inductor machine. Ignoring the forced commutation circuit of the inverter, both sides of the power circuit are symmetrical about the dc link. Since the machine displacement factor is constrained to be unity during both motoring and regeneration, the inverter firing angle with respect to machine phase voltage is 0° during motoring mode but switches to 180° during regeneration.

The general control block diagram of the system is shown in Fig. 3. It can be noted that the machine has two main feedback control loops, a speed control in the outer loop and a torque control in the inner loop. The command speed ω^* in the system is compared with the feedback speed ω_r, and the resulting error signal generates the armature current command I_a^* of the machine. Under the condition of constant air-gap flux and unity terminal displacement factor, it can be shown that the developed torque of the machine is proportional to armature current amplititude I_a. Since the rectifier input current amplitude I_a' has essentially the same amplitude as I_a it has been used as feedback signal because it is derived from currents having a constant 60-Hz frequency. The absolute value circuit maintains a positive polarity for the commanded value of armature current I_a^*, which is symmetrical with minimum and peak values clamped during both motoring and regeneration.

The I_a^* signal is used to generate the field current command I_f^* and the armature current phase angle command ϕ^* through the respective function generator as shown in the figure. The polarity of ϕ^* signal depends on the polarity of speed error which in turn governs the motoring or regeneration mode. The derived phase angle is then subtracted from the bias angle of 180° to generate the switching delay angle α_d. The self-control circuit of the machine consists of a set of position sensors, digital angle controller, and the thyristor firing circuit which will be described later.

Overlap Angle Compensation

The actual phase current wave of the machine deviates from the ideal six-step wave due to commutation effect as shown in

Reprinted from *IEEE Trans. Ind. Appl.*, vol. IA-15, pp. 591–600, Nov./Dec. 1979.

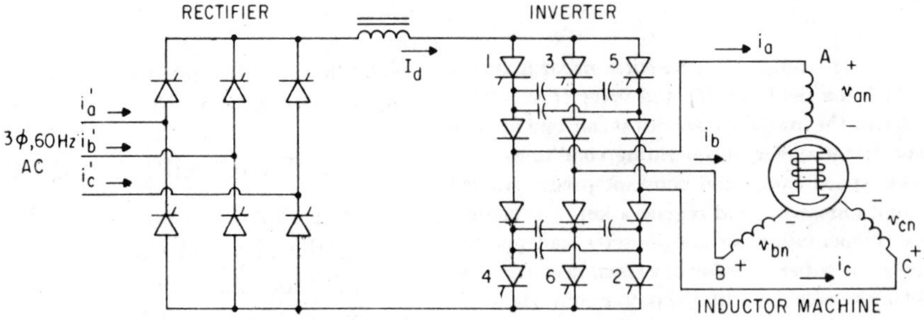

Fig. 1. Power circuit of the drive system.

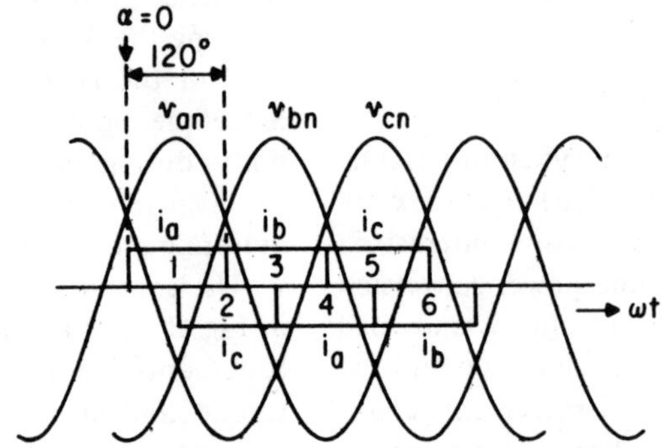

Fig. 2. Idealized voltage and current waves at unity.

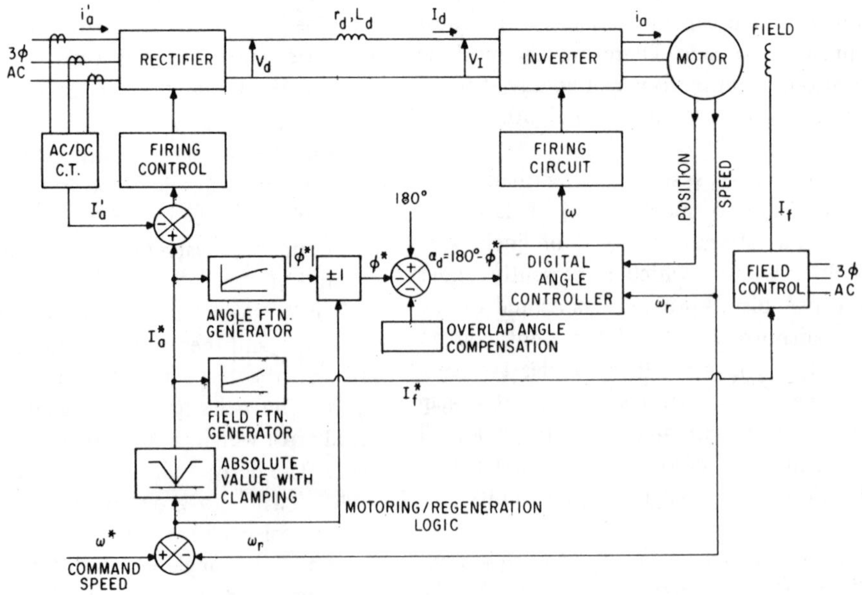

Fig. 3. General control block diagram.

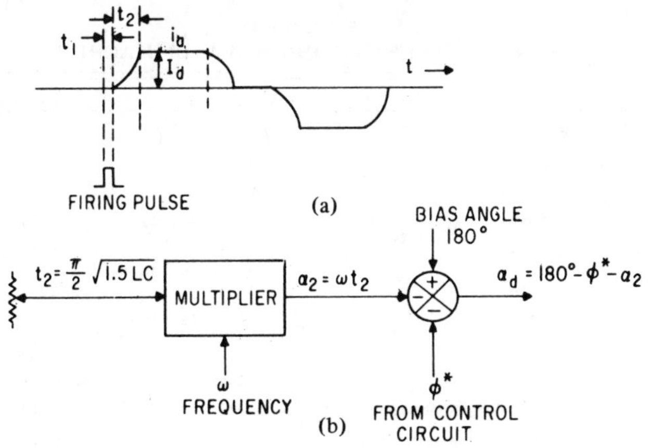

Fig. 4. Overlap angle compensation.

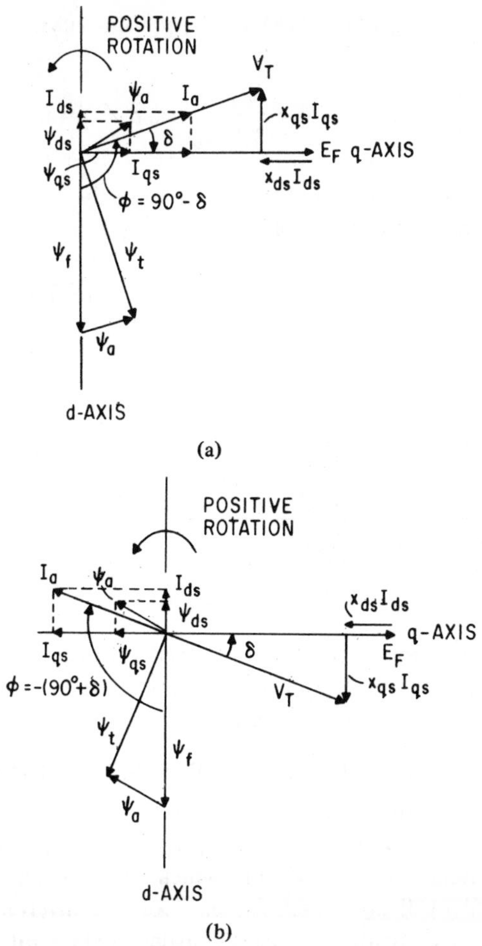

Fig. 5. Phasor diagram. (a) Motoring mode. (b) Regeneration mode.

Fig. 4(a). From the instant of firing the SCR, the phase current rises to maximum value I_d after a delay $t_1 + t_2$, where t_1 = constant current charging time of commutation capacitors and t_2 = delay corresponding to overlap angle α_2. In the practical operating range, t_1 can be neglected and $t_2 = \pi/2\sqrt{1.5LC}$, where L and C are commutating inductance and capacitance, respectively. Since L and C are constant parameters, the phase angle delay due to commutation can be compensated by the circuit as shown in Fig. 4(b). The signal t_2 is multiplied by angular frequency ω to construct α_2 which is then subtracted from α_d to give the equivalent phase lead of inverter switching angles.

III. DERIVATION OF ϕ^* AND I_f^* SIGNALS

Fig. 5 shows the phasor diagrams of the machine for both motoring and regeneration at unity displacement angle, and Fig. 5 shows the corresponding waveforms. The stator resistance has been neglected for simplicity.

Relating to the phasor diagram for motoring, the following expressions can be written in standard symbols:

$$X_{ds} = \omega_b L_{ds} \tag{1}$$

$$X_{qs} = \omega_b L_{qs} \tag{2}$$

$$X_{ad} = \omega_b L_{ad} \tag{3}$$

$$\psi_{ds} = X_{ds} I_{ds} = X_{ds} I_a \sin\delta \tag{4}$$

$$\psi_{qs} = X_{qs} I_{qs} = X_{qs} I_a \cos\delta \tag{5}$$

$$\psi_f = X_{ad} I_f' \tag{6}$$

$$I_f = \frac{I_f'}{N} \tag{7}$$

$$E_a = \frac{\omega}{\omega_b} E_A = \frac{\omega}{\omega_b} \psi_f \tag{8}$$

$$V_t = \frac{\omega}{\omega_b} V_T = \frac{\omega}{\omega_b} \psi_t \tag{9}$$

$$\phi = 90° + \delta \tag{10}$$

$$\bar{\psi}_t = \bar{\psi}_f + \bar{\psi}_a \tag{11}$$

$$\tan\delta = \frac{X_{qs} I_{qs}}{E_f - X_{ds} I_{ds}}. \tag{12}$$

In (1)-(12), ω_b is a constant base or design frequency used to define the machine reactances, and ω is the operating frequency. The voltage V_t is the machine terminal volts at operating angular frequency ω, whereas V_T is the terminal voltage at base frequency. Similarly, E_a and E_A are the Thevenin internal EMF's at ω and ω_b, respectively.

From (11),

$$\psi_t^2 = (\psi_f - \psi_{ds})^2 + \psi_{qs}^2. \tag{13}$$

Combining (13) with (1)-(6) and (9),

$$V_T^2 = (X_{ad} I_f' - X_{ds} I_a \sin\delta)^2 + X_{qs}^2 I_a^2 \cos^2\delta. \tag{14}$$

Combining (3), (6), (8), and (12)

$$\tan\delta = \frac{X_{qs} I_{qs}}{X_{ad} I_f' - X_{ds} I_{ds}}. \tag{15}$$

Substituting (4) and (5) in (15) and simplifying

$$\left(\frac{X_{ds}}{X_{qs}} - 1\right) \sin^2 \delta - \frac{X_{ad}}{X_{qs}} \frac{I_f'}{I_a} \sin \delta + 1 = 0. \tag{16}$$

Equations (14) and (16), respectively, can be written in normalized form as follows:

$$\left(\frac{I_f'}{I_a} - \frac{X_{ds}}{X_{ad}} \sin \delta\right)^2 + \frac{X_{qs}^2}{X_{ad}^2} (1 - \sin^2 \delta) = \frac{V_T^2}{X_{ad}^2 I_a^2} \tag{17}$$

and

$$\frac{I_f'}{I_a} = \frac{1 + \left(\frac{X_{ds}}{X_{qs}} - 1\right) \sin^2 \delta}{\frac{X_{ad}}{X_{qs}} \sin \delta}. \tag{18}$$

Equations (7), (10), (17), and (18) can be solved numerically with the given machine parameters in Table I to express ϕ^* and I_f^* as function of I_a^*.

IV. ROTOR POSITION SENSING PRINCIPLE

An optical interruption method of position sensing is used in the present system as illustrated in Fig. 6. There are six stationary optical sensors which generate parallel and independent firing pulses for each of the inverter SCR's. Each sensor consisting of a light emitting diode, phototransistor and pulse amplifying circuit is positioned at 60 electrical degrees interval on one side of the rotor drum. A round interrupting stud is mounted on the rotor at the center line of each north pole. As the machine rotates in the direction shown, the rotor position encoded pulses are generated in the sequence 1-2-3-4-5-6 which are then fed at the input of the digital angle controller circuit.

V. DIGITAL ANGLE CONTROLLER

The function of the angle controller is to delay the set of position encoded pulses proportional to the analog signal α_d which then generate the corresponding firing pulses of the inverter SCR's. This digital control method has the advantage that it eliminates offset and drift errors and prevents any asymmetry in SCR firing.

Since the ideal machine phase current flows in 120° pulses, the SCR for the corresponding phase is switched on 60° earlier to satisfy the fundamental frequency wave relations as shown in Fig. 7. The rotor position encoded pulse for a particular phase polarity is generated with a 240° lead angle with respect to positive peak value of ψ_f wave so that the SCR switching delay angle is given by $\alpha_d = 180° - \phi$ where ϕ is positive for motoring and negative for regeneration. In the present design, $\phi_{max} = 128°$ (corresponding to $\delta_{max} = 38°$) which keeps a margin of 52° for overlap angle compensation.

A schematic of the angle controller is shown in Fig. 8 and its operation is explained in Fig. 9. There are six identical parallel channels each of which receives position encoded

TABLE I
PARAMETERS OF THE SIMULATED MACHINE

	Simulated	Constructed
Power	112 kW	77 kW
Armature Current	281 Amps	370 Amps
Phase Voltage	140 Volts	87 Volts
Wheel Speed	1526 RPM (111 m/s)	1526 RPM
Frequency	394 Hz (ω_b = 2476 rad/s)	394 Hz
Rotor Poles	30	30
Stator Poles	3.5 (Effective)	5.0
δ_{max}	38°	38°
Stator Resistance r_s	0.0258 Ω	0.013 Ω
Stator Leakage Reactance $X_{\ell s}$	0.208 Ω	0.14 Ω
d-Axis Stator Magnetizing Reactance X_{ad}	0.25 Ω	0.17 Ω
q-Axis Stator Magnetizing Reactance X_{aq}	0.222 Ω	0.14 Ω
d-Axis Stator Self-Reactance $X_{ds} = X_{\ell s} + X_{ad}$	0.458 Ω	0.31 Ω
q-Axis Stator Self-Reactance $X_{qs} = X_{\ell s} + X_{aq}$	0.43 Ω	0.28 Ω
Field Leakage Reactance $X'_{\ell f}$ (Referred to Stator)	0.2371 Ω	0.47 Ω
Field Resistance r_f' (Referred to Stator)	0.229 Ω	0.2125 Ω
Field/Stator Effective Turns Ratio N	4.06	9.0

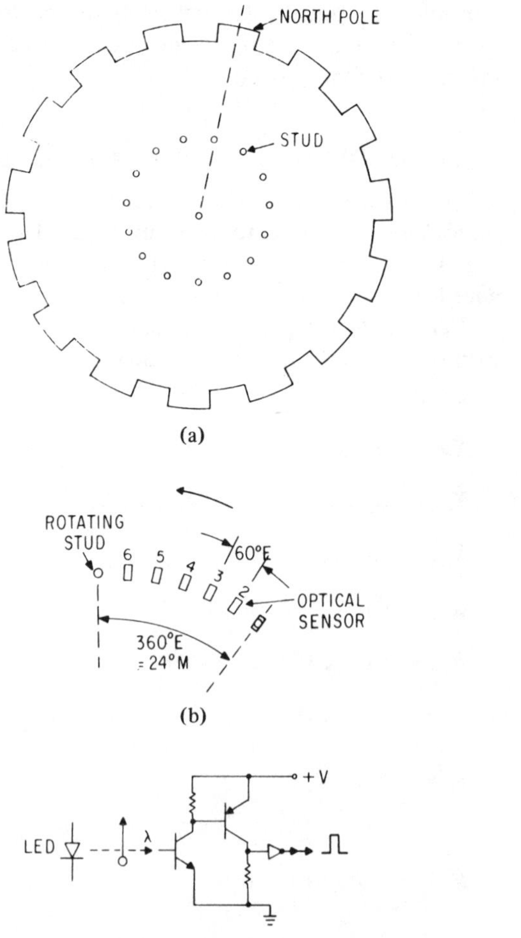

Fig. 6. (a) Rotor end view showing light interrupting studs. (b) Optical sensor configuration. (c) Sensor circuit.

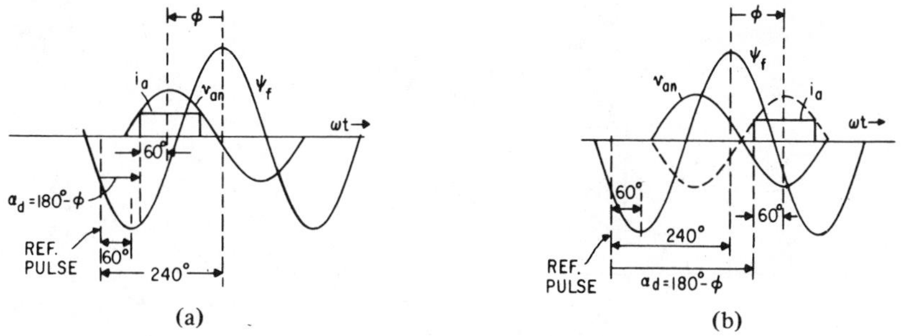

Fig. 7. Idealized waveforms. (a) Motoring mode. (b) Regeneration mode.

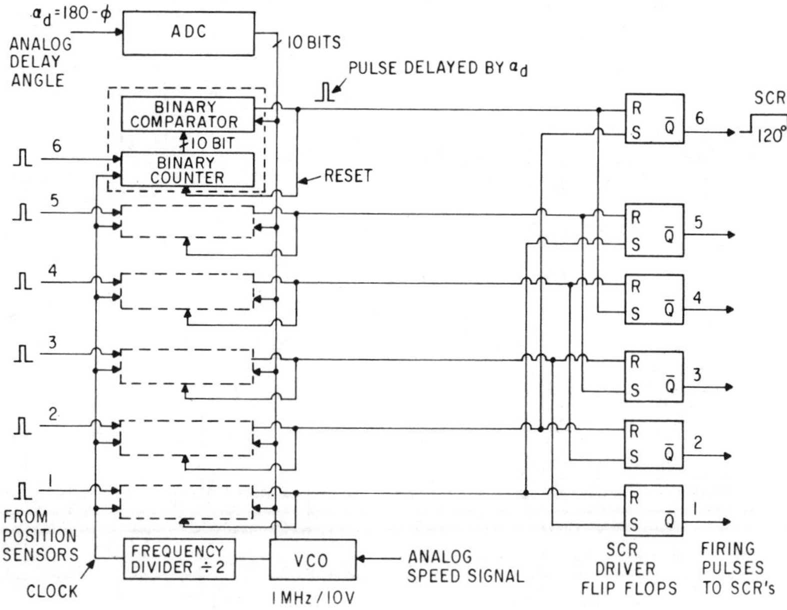

Fig. 8. Schematic of angle controller.

pulses at the input and generates the firing pulse for the corresponding SCR at delay angle α_d. The circuit uses the digital ramp comparison principle and generates the time delay $t_d = \alpha_d/\omega$, where ω is the inverter angular frequency. The ADC box converts the analog signal into the equivalent digital word. A clock proportional to the inverter frequency is generated from the speed signal and fed to the binary counter. The position encoded pulse 'enables' the counter to ramp up with a slope proportional to ω as shown in Fig. 9. The binary comparator compares the counter and ADC outputs and generates a pulse at the crossover point which corresponds to the delay angle α_d. If α_d is fixed, but speed is reduced, the time delay extends proportionately to maintain same α_d. The comparator output also resets the counter which is enabled again after time period $T = 2\pi/\omega$. The output pulses in parallel are combined in driver flip-flops to generate 120° wide firing pulses.

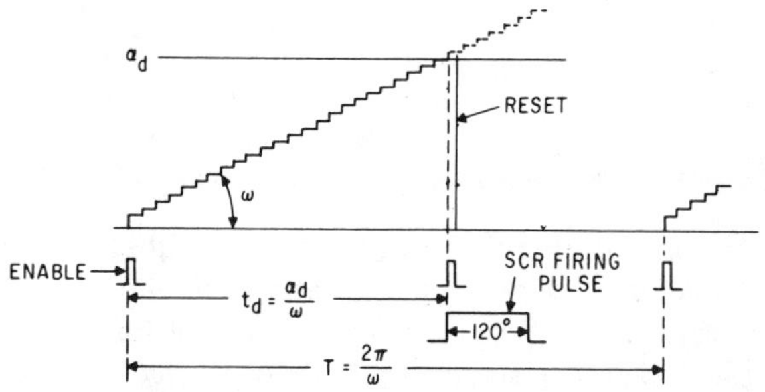

Fig. 9. Operation of angle control circuit.

VI. HYBRID COMPUTER SIMULATION

In order to gain insight into the effectiveness of the control scheme the entire system including current source inverter, filter, inductor machine, and associated control system was simulated on the hybrid computer. Such an analysis phase is extremely useful in the design of any converter system since it makes available a "breadboard" for the adjustment of those regulator gains, time constants, and maximum and minimum limits [4] which can best be accomplished by experimentation. Since this phase can be completed well in advance of construction, modifications to the control can be easily incorporated without delays in the assembly of the hardware. The system can also be exercised over all modes of operation and a high degree of confidence established before the design is put to hardware.

Fig. 10 shows the basic simulation that was used to model the four-pole inductor-type synchronous machine. The actual machine is a sector-type motor rated at 150 hp (112 kW). Details of the machine parameters are given in Table I. Since

$$a = 1/(1 + x_{aq}/x_{\ell s})$$
$$b = \omega_b r_s / x_{\ell s}$$
$$c = \omega_b$$
$$d = 1/(1 + x_{\ell s}/x_{\ell f} + x_{ad}/x_{\ell f})$$
$$e = 1/(1 + x_{\ell f}/x_{\ell s} + x_{ad}/x_{\ell s})$$
$$f = \omega_b r_f / x_{\ell f}$$
$$g = \omega_b r_f / x_{ad}$$
$$h = 1/x_{\ell s}$$
$$j = 1/x_{\ell f}$$
$$k = 3P/4\omega_b$$
$$\ell = P/2J$$

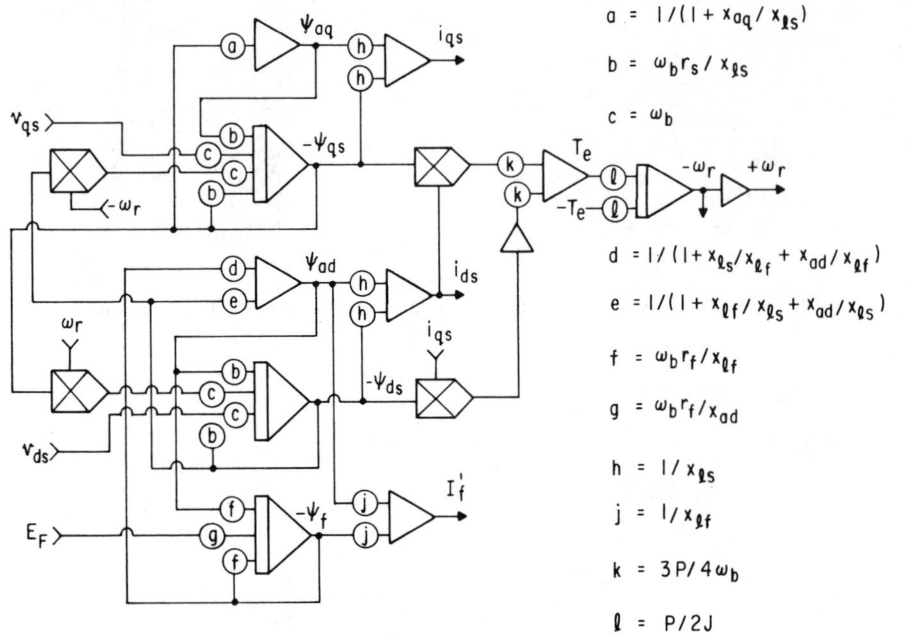

Fig. 10. Simulation diagram of the machine.

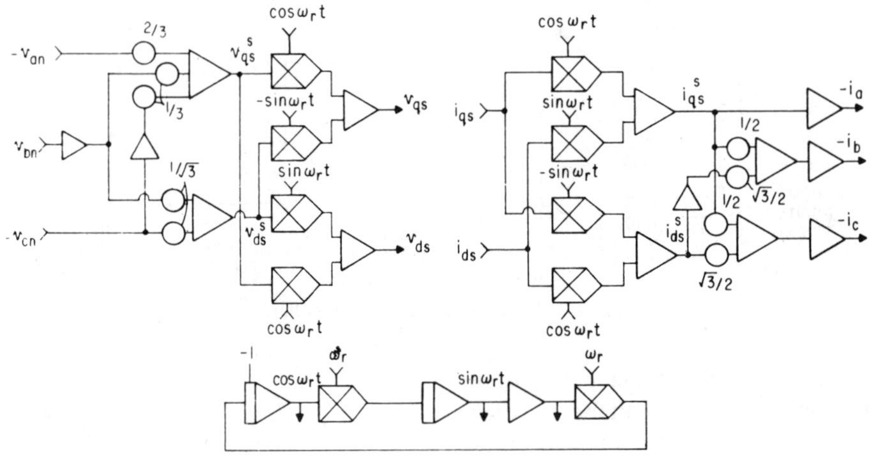

Fig. 11. Simulation of transformation equations.

the machine is essentially a linear motor with a finite length the equations describing its operating behavior are not well documented. However, a detailed analysis of this machine has indicated that the static entry and exit effects can be considered as negligible [5]. Hence the machine operates essentially as an ideal cylindrical four-pole inductor machine which can be modeled adequately by Park's equations [6], [7]. Also, the rotor saliencies are constructed of finely laminated steel without damper windings. Hence the eddy currents in the rotor resulting from rotation (dynamic end effects) can be ignored. Correspondingly, simulation of the machine need not contain extra rotor circuits to account for amortisseur bars or eddy currents.

In Fig. 10 the quantities ψ_{qs}, ψ_{ds}, ψ_f are the total instantaneous flux linkages of the q-axis stator, d-axis stator, and field winding, respectively, in the rotor reference frame; ψ_{aq} and ψ_{ad} are the air-gap flux linkages. Similarly, v_{qs}, v_{ds}, e_f, and i_{qs}, i_{ds}, and i_f are the instantaneous q-axis stator, d-axis stator, and field winding voltages and currents, respectively. The

torques T_e and T_l are the developed electromagnetic torque and load torque, and ω_r is the rotor speed in equivalent electrical rad/s. The parameters of the machine which appear as potentiometer settings are identified in Table I.

In Fig. 11 the equivalent two-phase d-q voltages and currents in both the stationary and rotating reference frames are developed from the three-phase variables. The superscript 's' has been used in Fig. 11 in order to distinguish the two-phase d-q variables expressed in the stationary reference frame from those referred to the rotor. Also, it can be noted that the zero sequence component has been omitted since the machine is wye-connected. The currents, i_a, i_b, and i_c are the machine line currents while the voltages v_{an}, v_{bn}, and v_{cn} are the motor phase voltages.

Fig. 12 shows the implementation of the current-fed inverter bridge. In the figure the logic-controlled switches on the right-hand side serve to direct the dc link current to one of the three thyristors in the upper half of the bridge and route a return path through one of the three lower-half thyris-

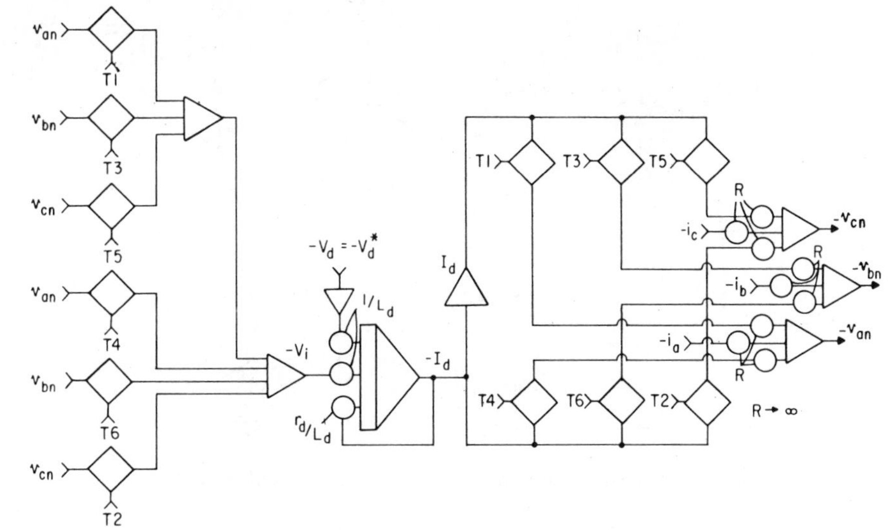

Fig. 12. Simulation of current-fed inverter.

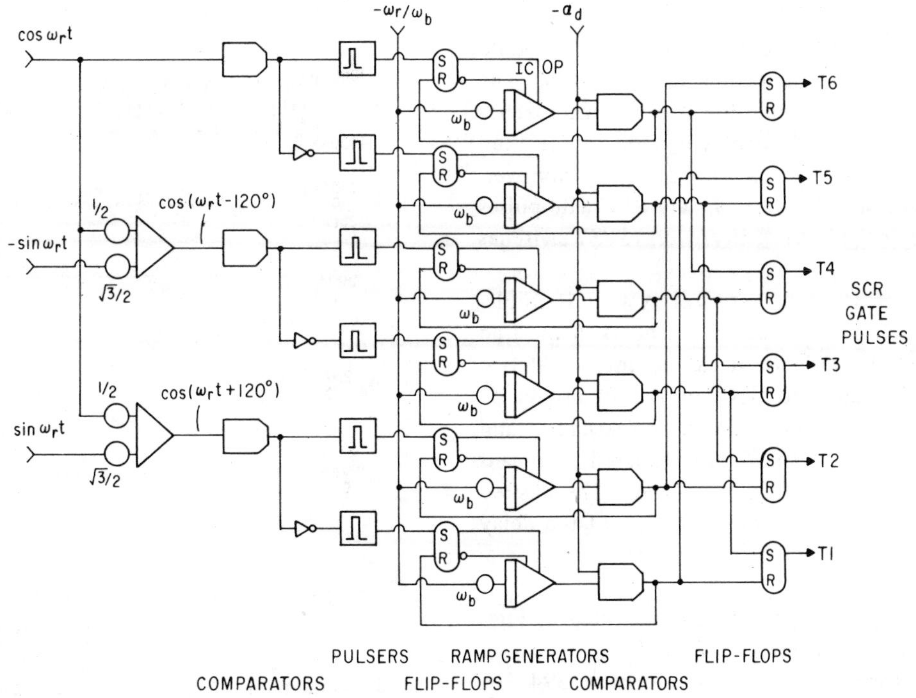

Fig. 13. Simulation of angle control scheme.

tors. The same firing pulses are fed to the left-hand array of switches which function to select the proper voltages corresponding to the two conducting thyristors, summing the result with the proper polarity to make up the dc inverter voltage. The difference between the inverter voltage and the applied dc link voltage V_d is used to compute the dc link current I_d.

It can be observed that the inverter is modeled as an "ideal" current source inverter having negligible commutation time. While it is recognized that commutation effects are more severe in current than in voltage source inverters the time delay due to commutation is predictable and can be readily compensated. If desired, a sampling delay which is a function of link current and air-gap flux (CEMF) can be introduced. Alternatively, more detailed models have been developed which can be used [8]. For purposes of this study, however, such additional accuracy was not deemed necessary. Also, the details of the input ac/dc rectifier are not of interest for this study so that it is assumed that the actual dc link voltage V_d is identically equal to the commanded (desired) value V_d^* derived from the control system.

In Fig. 13 the 120° logic pulses used to drive the inverter are developed. Proceeding from left to right, it can be noted that a three-phase set of sinusoids are developed from the

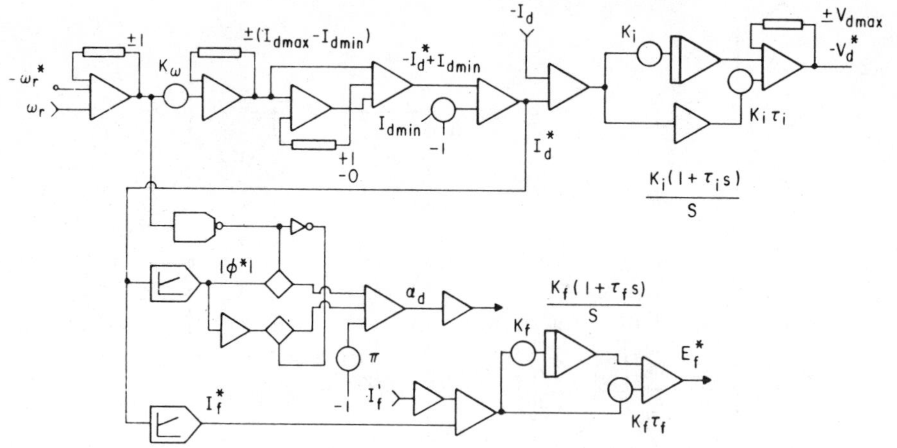

Fig. 14. Simulation of unity displacement factor control system.

variable frequency oscillator which is also used for the voltage and current transformations needed for the simulation of Park's equations. The zero crossings of the sine waves are detected and six clock pulses are produced per 360 electrical degrees rotation and hence model the position encoded pulses of the optical sensors. These clock pulses are used to initiate timed ramps having a slope proportional to rotor speed. The ramp is terminated when the amplitude of the ramp reaches α_d. The outputs of the six right-hand comparators are then interconnected through flip-flops so as to ensure nonoverlapping gate pulses. The signals $T_1, \cdots, T_6$ are 120° gate pulses which are used to trigger the inverter thyristors (SPST switches in the simulation).

The unity displacement factor control system is implemented in Fig. 14. In all cases the asterisk denotes the desired (commanded) value of the variable. It can be noted that two function generators are used to develop the functions ϕ^* versus $I_a{}^*$ and $I_f{}^*$ versus $I_a{}^*$ corresponding to (10), (17), and (18). The desired values of I_d and I_f are regulated by means of the dc link voltage V_d and field voltage E_f. The desired torque angle δ is obtained as shown by the use of the delay angle α_d which is used as the bias signal for the ramp generators in Fig. 13.

Typical computer simulation results are illustrated in Figs. 15 and 16 using the parameters given in Table I. Fig. 15 shows the steady-state waveforms of system variables at 394 Hz, corresponding to a maximum surface (peripheral) speed of 200 mi/h for the prototype linear motor. The machine is operating near rated load (0.85 pu). The computer traces clearly show the "notches" in the machine voltages typical of operation from a current source inverter. In Fig. 16 the machine is subjected to a step increase from an unloaded condition to the same load condition as Fig. 15. It can be noted that the machine responds smoothly to the transient with no tendency towards oscillation or hunting. Similar type of behavior was observed over the entire range of operation from 50 to 394 Hz.

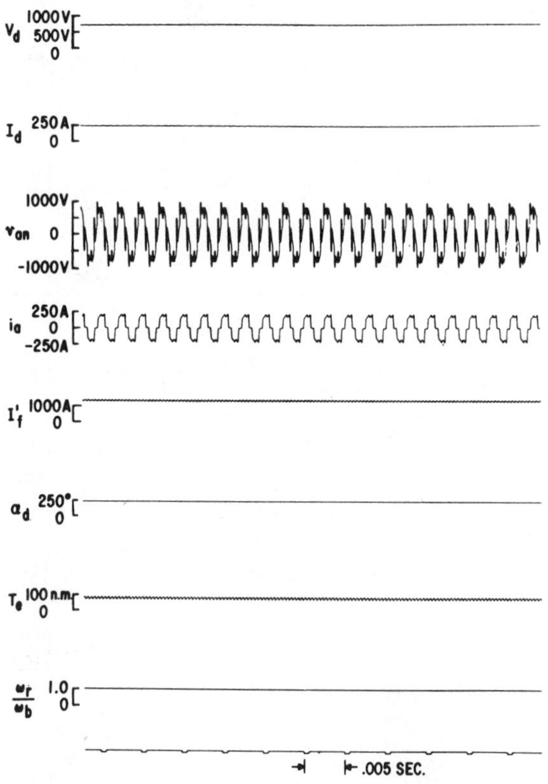

Fig. 15. Simulation of steady-state waveform at rated torque (79.1 N·m) and rated speed (1526 r/min).

VII. DESIGN AND EXPERIMENTATION

The complete control system including the position sensing circuit was designed and tested in the laboratory with the prototype 112-kW inductor machine. A photograph of the machine is shown in Fig. 17 which shows sector wound stator with the field coil at the bottom and the rotor light interrupting studs on the right. The module consisting of optical sensors and associated electronics is located on the shaft supporting bracket. A dc separately excited load machine is coupled to the same shaft for test purpose. The rectifier and inverter of the inductor machine and the load machine with its converters are a part of the general purpose laboratory facility.

In the design of control circuits, the angle and field function generators were approximated by straight lines for simplicity. The digital angle controller was designed to provide 0.33° resolution of SCR firing angle and no appreciable un-

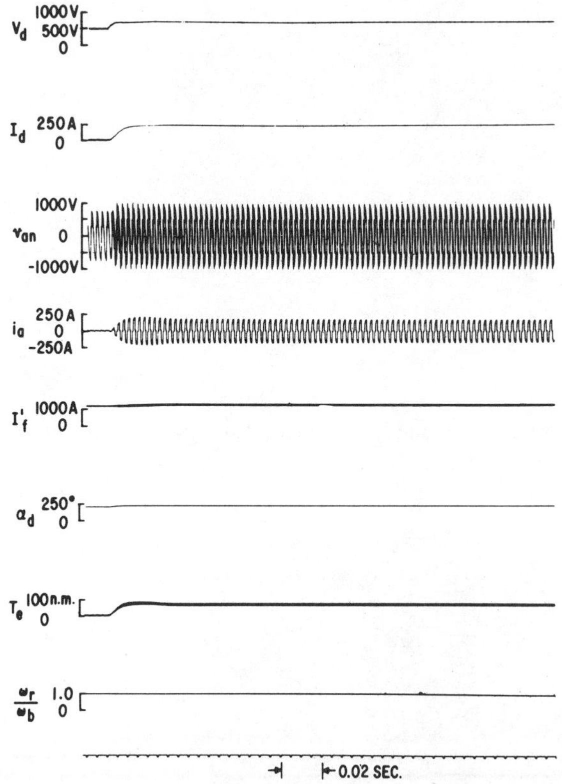

Fig. 16. Simulation response for step increase in load torque from no load to 79.1 N·m 394 Hz.

Fig. 17. Experimental 112-kW inductor machine in the laboratory showing position sensing assembly.

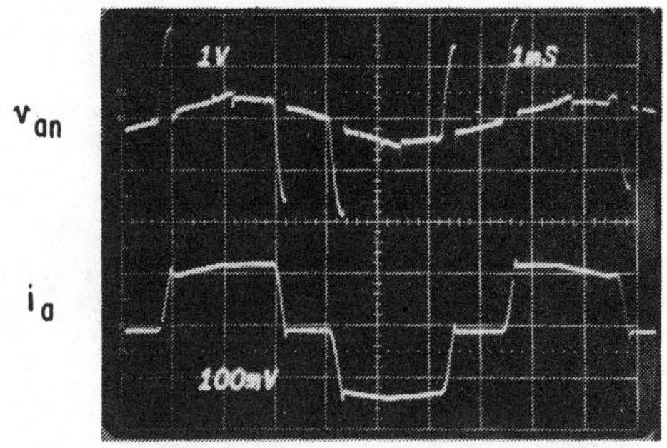

Fig. 18. Scope trace of phase voltage and current at 150 Hz with 200-A dc link current. Vertical scale: 100 V/cm 200 A/cm, respectively. Horizontal scale: 1 ms/cm.

balance was noticed. The overlap angle compensation was fine tuned until desired operation was obtained.

The inductor machine was tested for static and dynamic performance in wide speed and torque ranges and the experimental results agree well with the theory and simulation results. The displacement factor was found to vary between 0.95 and unity because of cumulative effect of several imperfect conditions. Fig. 18 shows the typical phase voltage and current waves at the machine terminal in motoring condition. The current commutation effect and the resulting voltage transient is evident from the figure. Fig. 19 shows the transient response of the machine at constant command speed when step load torque is applied and subsequently removed. This trace can be compared qualitatively to Fig. 16. Note again the well-damped response similar to Fig. 16.

Fig. 20 shows the transient response for step changes in speed command under a constant load torque condition. Note that the armature current responds almost immediately to the sudden error in speed command. Response to a step decrease in speed command is similar, however, in this case a speed decrease is accomplished by regenerative braking. Note that when the rotor speed approaches the desired operating speed the machine automatically switches over to motoring operation. The transition point between braking and motoring is clearly apparent.

VIII. CONCLUSION

A control strategy for a self-controlled sector wound linear inductor machine has been described which permits operation at constant air-gap flux and unity displacement factor under all operation conditions. The machine is capable of motoring and regeneration in wide speed and torque ranges and well balanced firing pulses with 0.33° resolution is obtained through a digital angle controller. The control incorporates an overlap angle compensation scheme which helps to restore unity displacement factor.

The *d-q* axis dynamic model of the machine with the control and converter circuits has been simulated on a hybrid computer and performances have been studied in detail. The control circuit with the position sensors have been designed and tested in the laboratory with a newly designed 112-kW machine, and the experimental results show good agreement with the theory and simulation results.

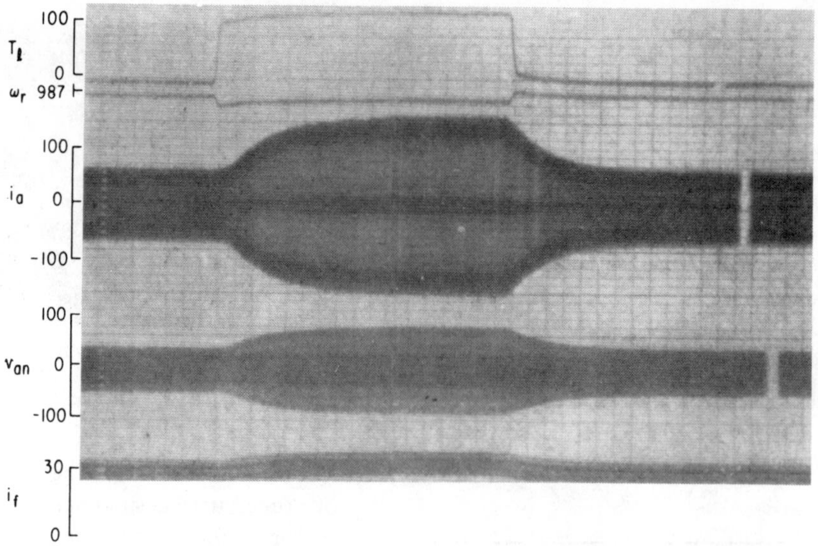

Fig. 19. Transient response with step changes in load torque near 0.4 per unit speed. Speed before application of load: 609 r/min; after: 596 r/min.

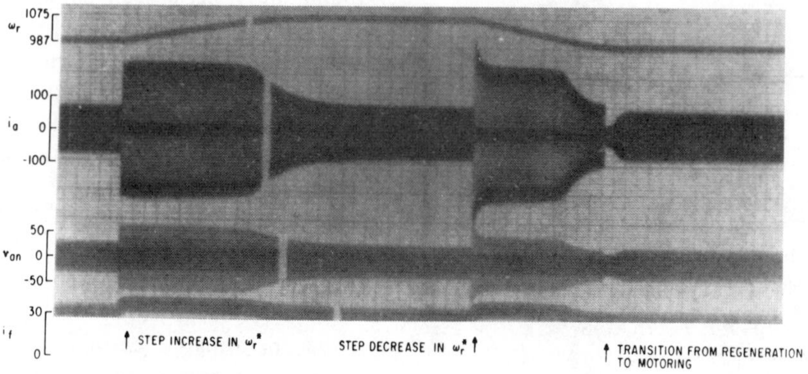

Fig. 20. Transient response with step changes of speed command near 0.4 per unit speed.

ACKNOWLEDGMENT

The authors gratefully acknowledge the help of W. Mischler during the experimental part of the project. Thanks are also due to M. Guarino of DOT-FRA who has guided the exploratory development of the inductor motor.

REFERENCES

[1] A. Häbock and D. Köllensperger, "State of development of converter-fed synchronous motors with self control," *Siemens Rev.*, XXXVIII, pp. 390–392, 1971.

[2] E. Levi, "Linear synchronous motors for high-speed ground transport," *IEEE Trans. Magn.*, vol. MAG-9, pp. 242–248, Sept. 1973.

[3] H. Stemmler, "Drive systems and electronic control equipment of the gearless tube mill," *Brown Boveri Rev.*, vol. 57, pp. 120–128, Mar. 1970.

[4] C. B. Mayer and T. A. Lipo, "Use of simulation in the design of an inverter drive," in *Conf. Rec., 1972 IEEE Ann. Meeting*, pp. 745–752.

[5] B.-T. Ooi, "Homopolar linear synchronous motor dynamic equivalents," *IEEE Trans. Magn.*, vol. MAG-13, pp. 1424–26, Sept. 1977.

[6] R. H. Park, "Two-reaction theory of synchronous machinery, I—Generalized method of analysis, *Trans. AIEE*, vol. 48, pp. 716–730, Jul. 1929.

[7] P. C. Krause and K. Carlsen, "Analysis of a homopolar inductor-alternator," in *Conf. Rec. IEEE Ind. Gen. Appl. Ann. Meeting*, pp. 117–125, 1968.

[8] T.A. Lipo, "Simulation of a current source inverter drive," *Conf. Rec. 1977 PESC Conf.*, pp. 310–315.

COMMUTATORLESS DC DRIVE FOR STEEL ROLLING MILL

Yuko Shinryo
Masahiro Kataoka

Mitsubishi Electric Corporation
Nagasaki, Japan

Isamu Hosono
Masahiko Akamatsu

Mitsubishi Electric Corporation
Amagasaki, Japan

Keijiro Syoji

Mitsubishi Electric Corporation
Kobe, Japan

ABSTRACT

This paper describes the fundamental principles of operation and control of a new commutatorless d-c drive for steel rolling mill. In particular it is shown that inertia of motor becomes to be small to cancel the armature reaction with the compensating field windings, and torque ripple of motor is decreased by using multiphase motor with d-c link.

The regulating characteristics of this system is similar to that of Thyristor Leonard system. This paper includes the corroborating experimental results obtained from the field installation of a 250KW drive for steel rolling mill, which is already in operation since October of 1976.

INTRODUCTION

Since its introduction some ten years ago, the thyristor converter has become firmly established as the static electrical power conversion equipments of many types, ranging in power ratings from a few hundred watts to tens megawatts.

Thus, at one end of the scale, the thyristor converter finds use in low power equipments, such as speed controllers for fractional horsepower d-c motors and the like. At the other end of the scale, the thyristor converter has now become the unquestioned successor to the high power grid controlled mercury arc converter, as well as to the rotating AC to DC converter - particular the classical Ward Leonard system - in applications such as high power variable speed drives for steel rolling mill, having ratings of many megawatts.

The d-c machine has, in the past, been generally preferred over other types, in most industrial applications requiring a variable speed drive. This is because the speed of the d-c machine can be controlled relatively easily, by control of either its armature voltage, or its field voltage, or both, depending upon the desired performance characteristics of the drive. It does, however, have the practical disadvantage of employing a commutator and brushes, which is relatively costly, and also requires periodic maintenance. Moreover, the use of a commutator makes operation difficult with voltages or currents beyond a certain practical limit ; furthermore, the commutator makes a scale limit as a single armature.

The a-c machine, on the otherhand, does not require a commutator. As a result, it is generally considerably less expensive than the d-c machine, as well as being much more robust mechanically and, therefore, much less in need of maintenance. However, the a-c motor is not generally regarded as being a variable speed machine, because its speed is a function of its applied frequency, which is normally fixed.

Of course, various means for controlling the speed of a-c machines, connected to a fixed frequency supply, have been devised. These methods, although satisfactory within their own limitations, either are inefficient, or are not generally applicable to a drive which is required to provide a high performance over a wide range of load and speed, with perhaps, the added requirement for reversing.

An exception to this rule is the a-c commutator motor ; but here, of course, because of the use of the commutator, the machine is similar in construction to the d-c motor, and does not therefore have the mechanical, or economical, advantages of the commutatorless a-c machine.

In point of fact, if a supply of variable frequency, variable voltage, power is connected to a commutatorless a-c machine, then it is possible, by means of appropriate control of the frequency and amplitude of the voltage, to provide an efficient variable speed drive, with a performance characteristics which is, for all practical purposes, equivalent to that of high performance 4-quardrant armature voltage controlled d-c drive. [1]

The ideal variable speed motor should be easy to control and maintenance.

Under the above circumstances we have worked on the development of commutatorless a-c or d-c motors fed by rectangular waveform converter and supplied them for the speed control drive in a variety of industrial applications, such as tunnel blower drives, exhaust fan drives, pump drives, extruder drives, printing press drives, piling cranes, paper winders, and so on since 1970. [2] [3]

Although it has numerous excellent features, there are still some difficult problems to be solved. For example the inertia of its motors is seem to be larger than that of conventional d-c motors, and torque ripple is generated by the harmonic components of armature current.

However we have overcome these difficulties with development of a new type commutatorless d-c drive system and built a 250KW drive for steel rolling mill drive. And this is already to operation since October of 1976.

Reprinted from *IEEE/IAS 1977 Annual Meeting*, pp. 263–271, 1977.

This paper is intended to describe the fundamental principles of operation and control of this drive system and experimental results obtained from the field installation of a 250KW drive for steel rolling mill as follows.

SYSTEM OUTLINE

A new commutatorless d-c drive system is a multiphase motor with d-c link composed of the static power converter and synchronous motor. Fig. 1 shows the block diagram of commutatorless d-c drive system. This motor is the same structure as that of the conventional rotary field synchronous motor with respect to the principle. Its stator has two sets of three phase armature windings which have a 30 electrical degree phase shift between them. Its rotor is cylindrical and has field windings like conventional synchronous machines do, but the compensating field windings are also integrated in addition to the main field windings. The main field current is decreased according to the field weakening pattern like that of a d-c motor as shown in Fig. 2, and the compensating field current is controlled in order to cancel the armature reaction.

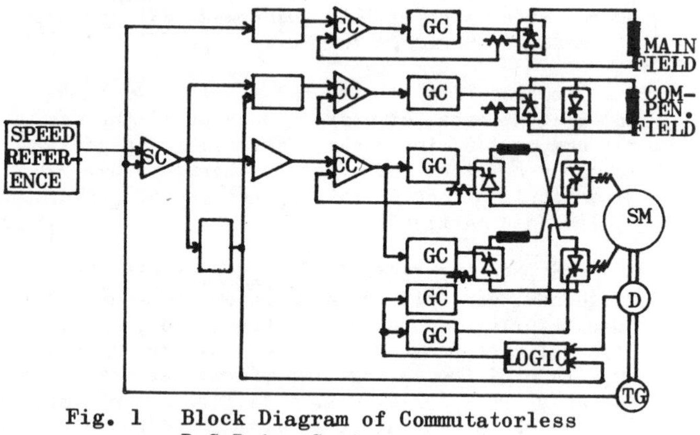

Fig. 1 Block Diagram of Commutatorless D-C Drive System

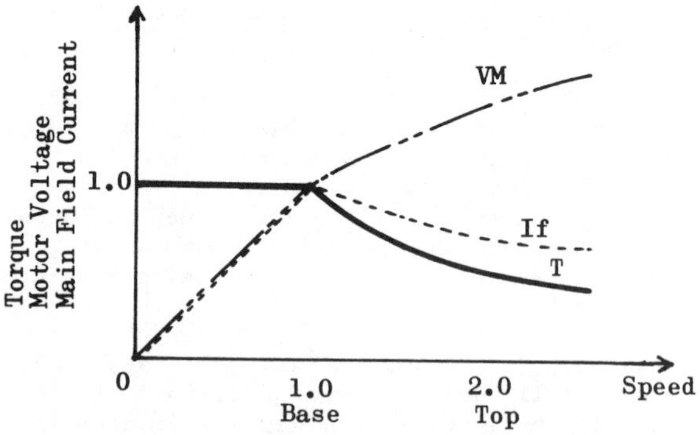

Fig.2 Field Weakening Pattern

This static power converter is composed of the series connection of the line side converter (rectifier) and the machine side converter (inverter) consisting of a smoothing reactor and 2×12 thyristors. This type of the static power converter is attractive because of its simplicity, and because its commutation can be securely made by induced emf of the motor without using any complicated auxiliary commutating circuit.

The task of the static power converter is to take power drawn from the supply system of fixed voltage and frequency in order to change speed of a synchronous motor.

This drive system is provided with a current control loop inside the speed control loop and the speed of motor is controlled by the phase control of the line side converter, which is almost the same technique as the Thyristor Leonard system.

As before, the desired performance characteristics of steel rolling mill is constant torque from zero to base speed and constant power from base to top speed. Therefore, the main field current is kept constant at the range of constant torque and decreased at the range of constant power as shown in Fig. 2.

THEORY AND CHARACTERISTICS

A basic circuit and winding composition of the multiphase motor with d-c link is illustrated in Fig. 3 and 4.

The distributor detects the position of the rotor and the logic circuit selects the most effective thyristor (UD11, VD11, WD11, ...) of the inverter to generate torque. Fig. 5 shows the principle and output signals of the distributor.

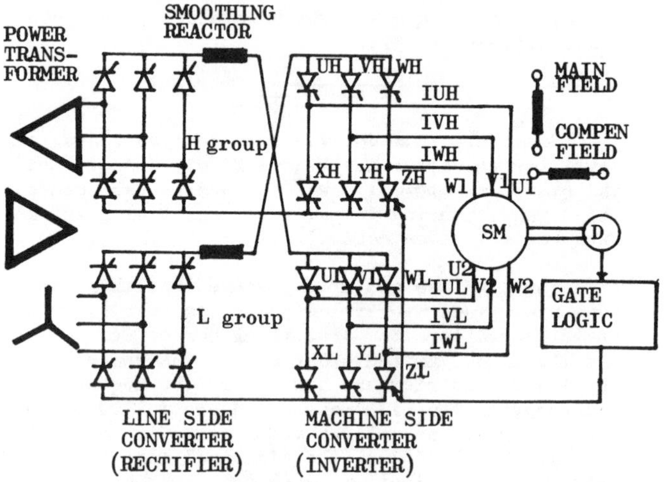

Fig. 3 Basic Circuit of Multiphase Motor with D-C Link

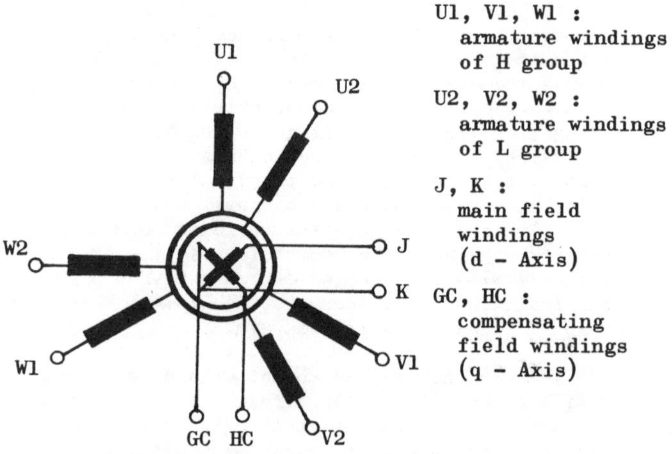

Fig. 4 Composition of Windings

There are 2×6 modes of the thyristor combination as shown in Fig. 6. In each mode the selected thyristor gives the inverter output current of rectangular waveform to the armature windings of the multiphase motor with d-c link in order, U1-U2-V1-V2-W1-W2.

As a result, the flow chart of the armature current for the multiphase motor with d-c link is shown in table 1.

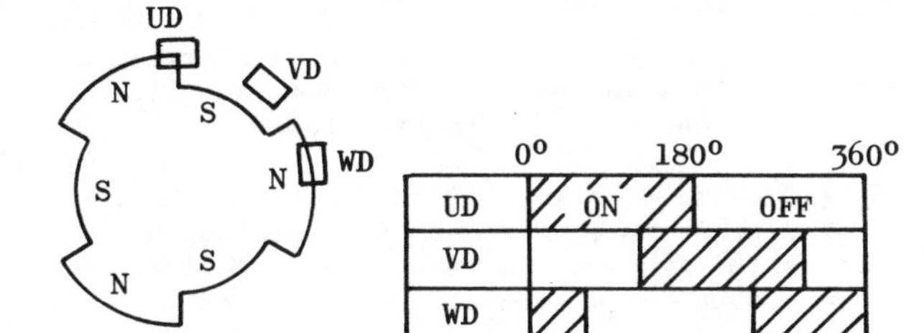

(For 6 pole machine)

Fig. 5 Distributor Output Signals

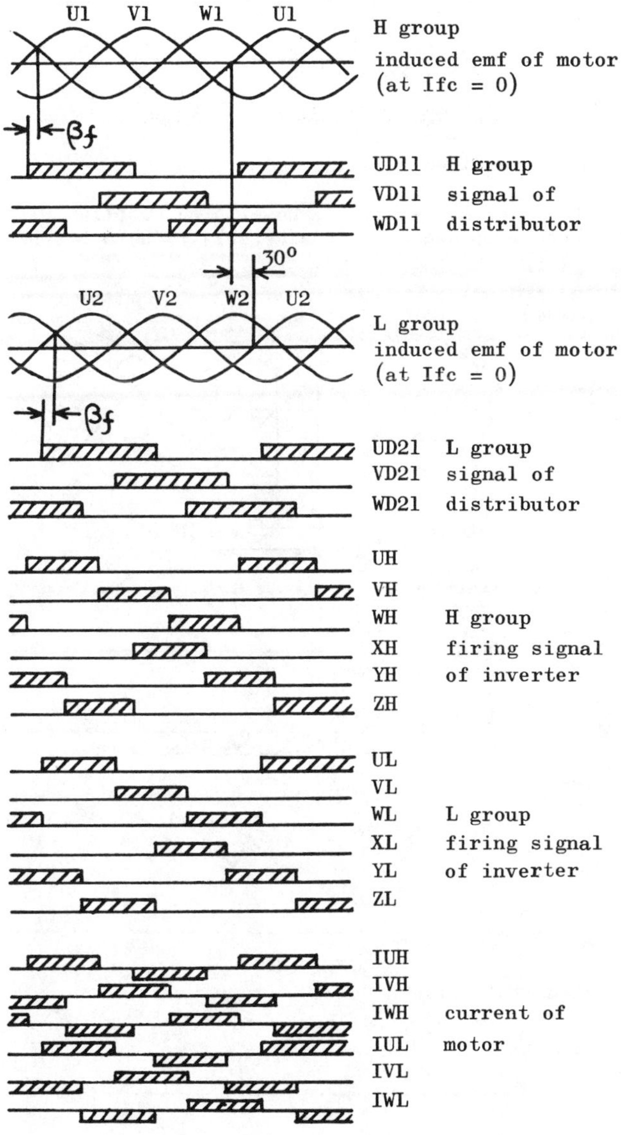

Fig. 6 Modes of Thyristor Combination

→ Power transformer (Y connection)
→ Positive thyristor of L group rectifier
→ Smoothing reactor of L group
→ Negative thyristor of H group inverter
→ Armature winding (U1, V1, W1)
→ Positive thyristor of H group inverter
→ Negative thyristor of H group rectifier
→ Power transformer (Δ connection)
→ Positive thyristor of H group rectifier
→ Smoothing reactor of H group
→ Negative thyristor of L group inverter
→ Armature winding (U2, V2, W2)
→ Positive thyristor of L group inverter
→ Negative thyristor of L group rectifier

Table 1. Flow Chart of Armature Current for Multiphase Motor with D-C Link

The armature current flows in series. The rectifier and inverter are the 12-pulse thyristor converter circuit with two 6-pulse commutating groups each other. A stable operation of this motor is kept with good commutation of inverter achieved by means of induced emf commutation.

As mentioned above, there are two complete sets of three phase armature windings, which have a 30 electrical degree phase shift between them, on the stator. The one is connected to L group inverter, the other is connected to H group inverter. On the other hand, there are two complete sets of field windings, which have a 90 electrical degree phase shift between them, on the rotor. The one is a main field winding which produces main flux, and the other is a compensating winding which cancels the armature reaction.

To grasp its steady state characteristics, consider the vector diagram of the multiphase motor with d-c link as shown in Fig. 7.

In Fig. 7, it is possible to consider the voltages currents, and fluxes as vectors in a plane having d and q co-ordinate axes mutually perpendicular and oriented in exactly the same way as the d and q axes of the machine itself as given by Fig.4. [4]

The fundamental component of the armature current Ia is lagging to $\frac{u}{2}$ with respect to the start point of the commutation, where u is the commutation overlap angle. Therefore the control firing angle βf must be set almost equal to $\frac{u}{2}$ at the full load in order that Ia and q axis become to be in phase. The compensating field current If points in the opposite direction of Ia in order to cancel the component of the armature reaction Xaq·Ia.

On the other hand, the safety margin angle γ to commutate is obtained in order that the induced emf Ea is lagging to $\Delta\phi c$ with respect to q axis by the over compensating field current Ifc'. The effective margin angle γ' to commutate should not be estimated by the vector diagram, but it must be estimated by the instantaneous waveform as shown in Fig. 9.

At each mode of operation, the control firing angle βf is set as shown in Fig. 8. At the regenerating operation that is leading to 180° with respect to the phase of Ia at the motoring operation. Thus the compensating field current is quickly reversed by suitable control of the dual converter connected to the compensating field windings.

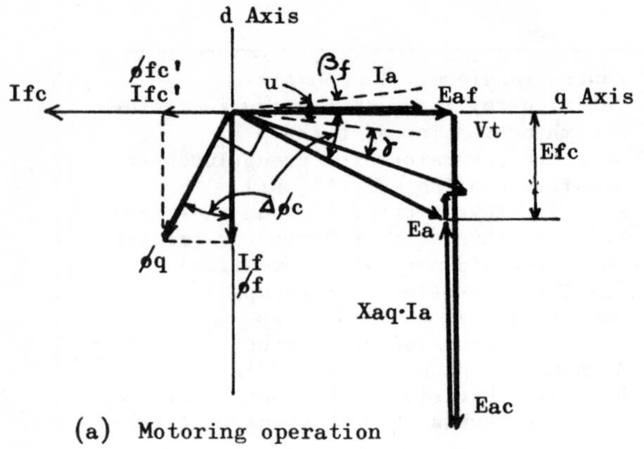

(a) Motoring operation

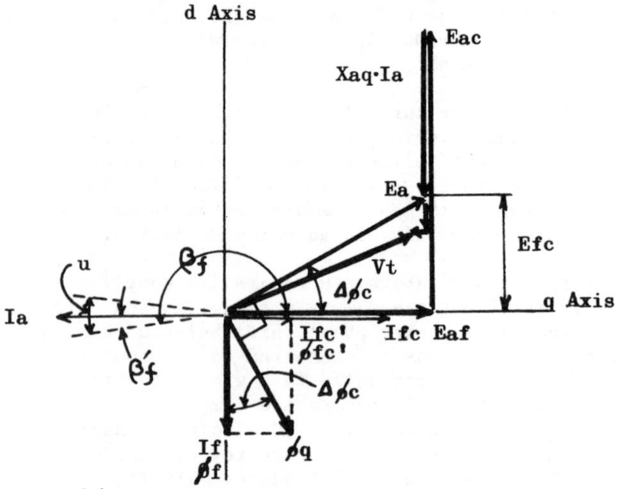

(b) Regenerating operation

Fig. 7 Vector Diagrams of Multiphase Motor with D-C Link

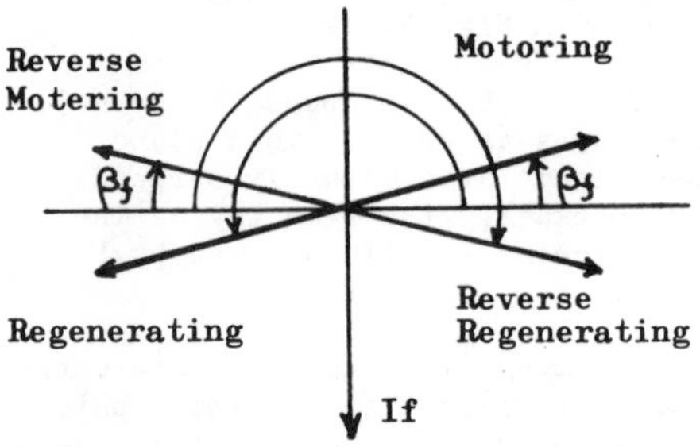

Fig. 8 Control Firing Angle βf

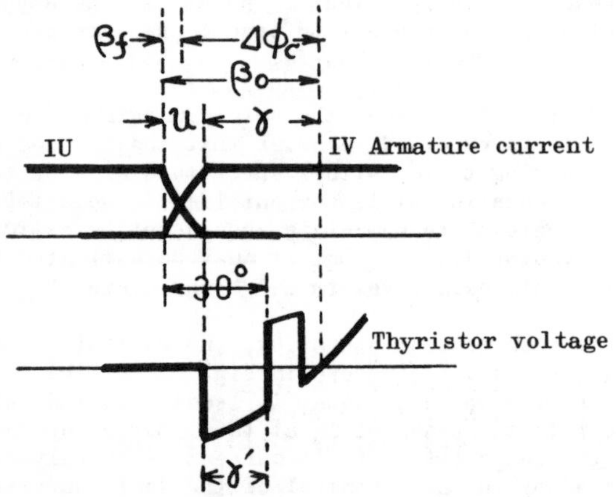

Fig. 9 Voltage Waveform During Commutation

The instantaneous torque is strictly calculated by the fundamental equations as described later. But it can be discussed approximately by the following assumption.

Assumption :

(1) The induced emf is sinusoidal waveform.

(2) The armature current is trapezoidal waveform.

The instantaneous torque

$$T(\theta) = \frac{P}{2\ fM} \sum_{j=1}^{m} e_j(\theta) \cdot i_j(\theta) \quad .... (1)$$

By substituting H group of induced emf and armature current (Fig. 10-a and b) into the equation (1), torque is obtained as shown in Fig. 10-c.

By expanding equation (1) in Fourier series, it is found that superimposed on the mean torque produced by the fundamental wave of the armature and the field are harmonic torques of the order

$$\nu = km \qquad (k = 1,2,3, ...)$$

Fig. 11 shows the torque pulsation ΔT of 12th and 24th order in β_o and u parameters with 12-pulse connections.

For $\beta_o = 35^\circ$, $u = 10^\circ$ at full load, for example, the amplitude of the harmonic torque of 12th and 24th order is

$$\frac{\Delta T12}{Tmean} = \frac{T12\ max. - T12\ min.}{2 \times T\ mean} = 0.085$$

$$\frac{\Delta T24}{T\ mean} = \frac{T24\ max. - T24\ min.}{2 \times T\ mean} = 0.025$$

as can be seen from Fig. 11.

When special conditions have to be complied with, the torque ripple can be considerably reduced by the use of 12-pulse connections.

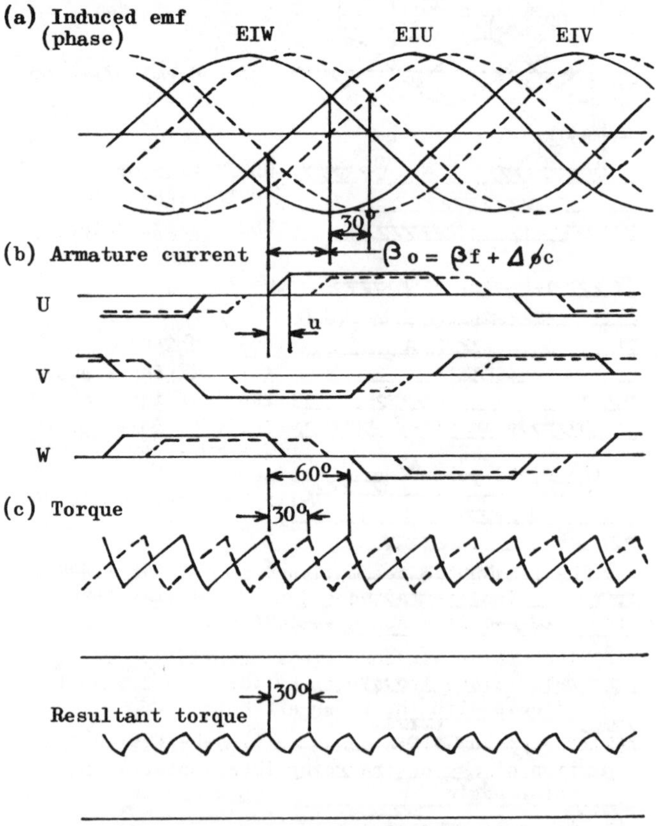

Fig.10 Torque Ripple of Multiphase Motor with D-C Link

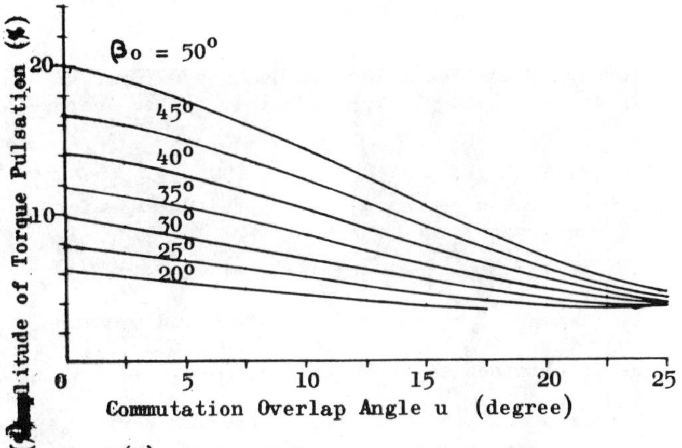

(a) Torque Pulsation of 12th order

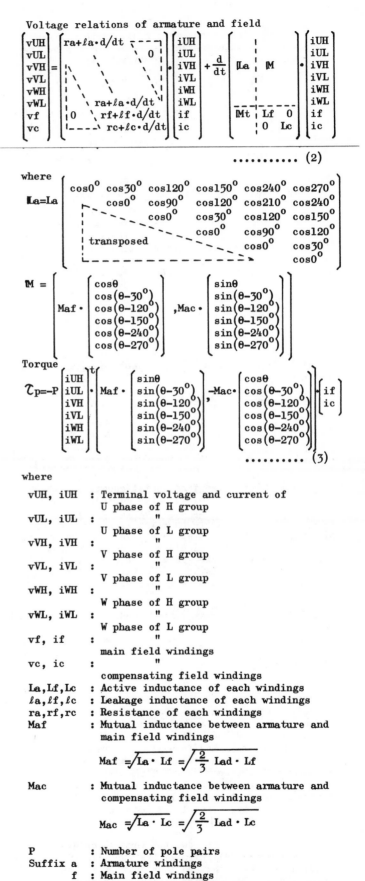

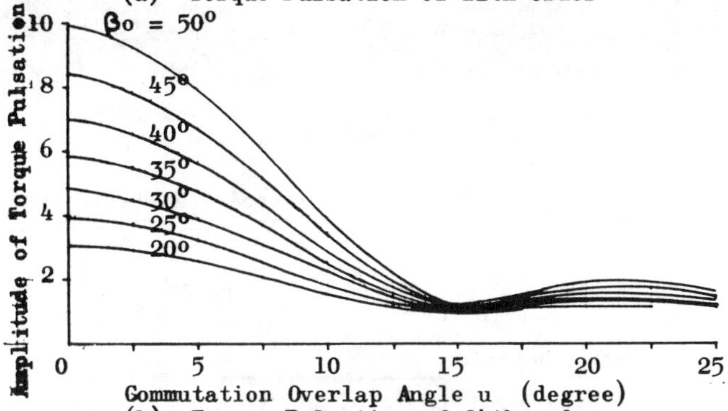

(b) Torque Pulsation of 24th order

Fig.11 Amplitude of Torque Pulsation

To obtain the instantaneous waveform, consider the analytical mode of the multiphase motor with d-c link as follows. [5]

Assumption :

(1) Motor
- (a) Rotor construction : cylindrical
- (b) Winding configuration :
  - Armature windings : two sets of three phase star connection
  - Field windings : main and compensating field windings
  - Damper windings : nothing

(2) Static power converter
- (a) Main power source : voltage power source with 12-pulse commutated ripple
- (b) Main inverter : two sets of three phase bridge inverter
- (c) Field power source : voltage power source with 6-pulse commutated ripple

The reasonable assumptions are made in order to derive the fundamental equation of the multi-phase motor with d-c link shown as follows.

(1) The stator windings are sinusoidally distributed along the airgap as far as all mutual effects with the rotor are concerned.

(2) The stator slots cause no appreciable variation of any of the rotor inductances with rotor angle.

(3) Saturation is neglected

The electrical performance of the multiphase motor with d-c link may now be described by the following fundamental equations as shown in Table 2.

Voltage relations of armature and field

$$\begin{bmatrix} vUH\\ vUL\\ vVH\\ vVL\\ vWH\\ vWL\\ vf\\ vc \end{bmatrix} = \begin{bmatrix} ra+\ell a\cdot d/dt & & & & & & & 0\\ & \ddots & & & & & & \\ & & & & & & & \\ & & & & & & & \\ & & & & & & & \\ & & & & & ra+\ell a\cdot d/dt & & \\ & & & & & & rf+\ell f\cdot d/dt & \\ 0 & & & & & & & rc+\ell c\cdot d/dt \end{bmatrix} \cdot \begin{bmatrix} iUH\\ iUL\\ iVH\\ iVL\\ iWH\\ iWL\\ if\\ ic \end{bmatrix} + \frac{d}{dt}\begin{bmatrix} \mathbb{L}a & \mathbb{M} & \\ \mathbb{M}t & Lf & 0\\ & 0 & Lc \end{bmatrix} \cdot \begin{bmatrix} iUH\\ iUL\\ iVH\\ iVL\\ iWH\\ iWL\\ if\\ ic \end{bmatrix}$$

.... (2)

where

$$\mathbb{L}a=La\begin{bmatrix} \cos 0^{\circ} & \cos 30^{\circ} & \cos 120^{\circ} & \cos 150^{\circ} & \cos 240^{\circ} & \cos 270^{\circ}\\ & \cos 0^{\circ} & \cos 90^{\circ} & \cos 120^{\circ} & \cos 210^{\circ} & \cos 240^{\circ}\\ & & \cos 0^{\circ} & \cos 30^{\circ} & \cos 120^{\circ} & \cos 150^{\circ}\\ & \text{transposed} & & \cos 0^{\circ} & \cos 90^{\circ} & \cos 120^{\circ}\\ & & & & \cos 0^{\circ} & \cos 30^{\circ}\\ & & & & & \cos 0^{\circ} \end{bmatrix}$$

$$\mathbb{M} = \left[Maf\cdot\begin{bmatrix} \cos\theta\\ \cos(\theta-30^{\circ})\\ \cos(\theta-120^{\circ})\\ \cos(\theta-150^{\circ})\\ \cos(\theta-240^{\circ})\\ \cos(\theta-270^{\circ}) \end{bmatrix}, Mac\cdot\begin{bmatrix} \sin\theta\\ \sin(\theta-30^{\circ})\\ \sin(\theta-120^{\circ})\\ \sin(\theta-150^{\circ})\\ \sin(\theta-240^{\circ})\\ \sin(\theta-270^{\circ}) \end{bmatrix} \right]$$

Torque

$$\tau p=-P\begin{bmatrix} iUH\\ iUL\\ iVH\\ iVL\\ iWH\\ iWL \end{bmatrix}^{t} \cdot \left[Maf\cdot\begin{bmatrix} \sin\theta\\ \sin(\theta-30^{\circ})\\ \sin(\theta-120^{\circ})\\ \sin(\theta-150^{\circ})\\ \sin(\theta-240^{\circ})\\ \sin(\theta-270^{\circ}) \end{bmatrix}, -Mac\cdot\begin{bmatrix} \cos\theta\\ \cos(\theta-30^{\circ})\\ \cos(\theta-120^{\circ})\\ \cos(\theta-150^{\circ})\\ \cos(\theta-240^{\circ})\\ \cos(\theta-270^{\circ}) \end{bmatrix} \right] \cdot \begin{bmatrix} if\\ ic \end{bmatrix}$$

.... (3)

where

vUH, iUH	:	Terminal voltage and current of U phase of H group
vUL, iUL	:	" U phase of L group
vVH, iVH	:	" V phase of H group
vVL, iVL	:	" V phase of L group
vWH, iWH	:	" W phase of H group
vWL, iWL	:	" W phase of L group
vf, if	:	" main field windings
vc, ic	:	" compensating field windings
La, Lf, Lc	:	Active inductance of each windings
$\ell a, \ell f, \ell c$	:	Leakage inductance of each windings
ra, rf, rc	:	Resistance of each windings
Maf	:	Mutual inductance between armature and main field windings
Mac	:	Mutual inductance between armature and compensating field windings
P	:	Number of pole pairs
Suffix a	:	Armature windings
f	:	Main field windings
c	:	Compensating field windings

$$Maf = \sqrt{La\cdot Lf} = \sqrt{\frac{2}{3}\,Lad\cdot Lf}$$

$$Mac = \sqrt{La\cdot Lc} = \sqrt{\frac{2}{3}\,Lad\cdot Lc}$$

Table 2. Fundamental Equations of Voltage and Torque

On the other hand, main field current If and compensating field current Ic is determined by the following procedure. (in per-unit form)

$$Ef = \frac{1}{\sqrt{1+(Eco/Ef)^2}} \quad \cdots\cdots\cdots\cdots (3)$$

$$Eco = \left(\frac{Eco}{Ef}\right)\times\frac{1}{\sqrt{1+(Eco/Ef)^2}} \quad \cdots\cdots (4)$$

therefore

$$If = \frac{\sqrt{2}}{\omega Maf}\cdot Ef \quad \cdots\cdots\cdots\cdots (5)$$

$$Ico = \frac{\sqrt{2}}{\omega Mac}\cdot Eco \quad \cdots\cdots\cdots\cdots (6)$$

Thus the component of the compensating field current which is proportional to the armature current is represented by using compensating factor Δ Kc.

$$Ica = \frac{\sqrt{2}\,Laq}{Mac}\cdot ia\cdot\Delta Kc \quad \cdots\cdots\cdots\cdots (7)$$

where

$$Laq = Lad = \frac{3}{2}\cdot La$$

$$La = Lf = Lc = Maf = Mac$$

By substituting the above relations in equation (7), Ica becomes.

$$Ica = \frac{3}{\sqrt{2}}\cdot ia\cdot\Delta Kc \quad \cdots\cdots\cdots\cdots (8)$$

After all, the compensating field current is of the form.

$$Ic = Ico + Ica \quad \cdots\cdots\cdots\cdots (9)$$

From the derived fundamental equations, the instantaneous values of voltage, current, torque and commutation overlap angle of the multiphase motor with d-c link are numerically calculated by the digital computer. We have developed the proper program "SCAP-M" to calculate the instantaneous characteristics of the system including electrical machines and thyristors. Its details will be reported at a later date if possible.

CONSTRUCTION

MOTOR

A commutatorless d-c motor is composed of the motor itself, distributor, tachogenerator and forced ventilating fan.

There are many possible types of commutatorless d-c motor distinguished by the type of rotor construction and the method of motor excitation. The former is homopole, clawpole, salient pole, or cylindrical, and the latter is the method with rotating-rectifier and induction generator exciter, or with an excitation transformer having a rotating secondary connected to a rotating-rectifier. To avoid the necessity for any maintenance the commutatorless d-c motor without brushes is usually employed in a wide variety of industrial applications.

Nevertheless the new type commutatorless d-c motor for steel rolling mill has two sets of field windings on the rotor and they must be instantaneously controlled by the static field converter. Therefore the excitation with two sets of sliprings is adopted for a 250KW drive. Fig. 12 shows external view and cross section of a 250KW motor.

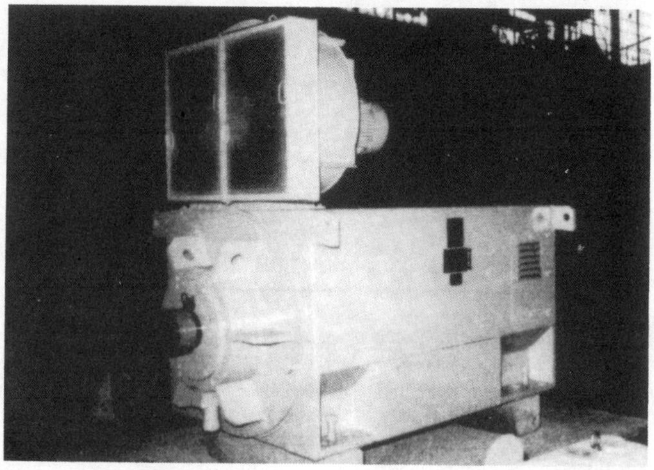

(a) External View

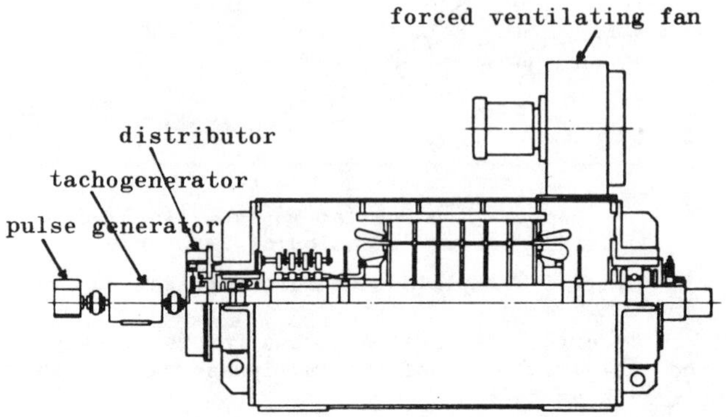

(b) Cross Section

Fig.12 External View and Cross Section of a 250KW Drive for Steel Rolling Mill

DISTRIBUTOR

The distributor is composed of a toothed disc on a free motor shaft end and two sets of three position detecting sensors. The function of the distributor is to detect the rotor position. The detecting sensors are of the molded oscillation type to prevent problems even under the bad ambient condition. Fig.13 illustrates construction of distributor which is ablè to adjust the position of detecting sensors.

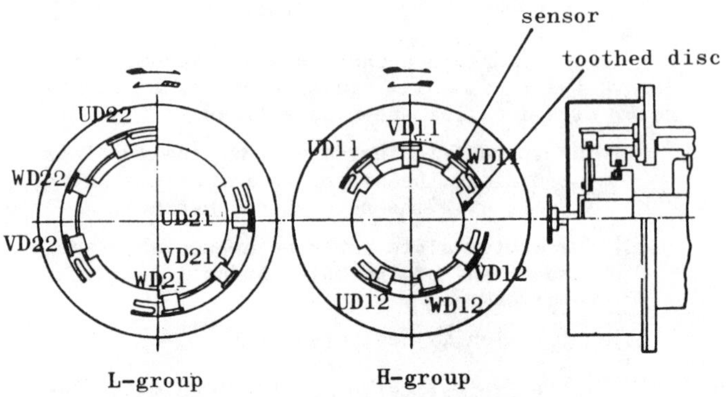

Fig. 13 Construction of Distributor

CONTROL DEVICE

The control devices are composed of the following principal blocks.

(1) Multiphase static power converter
(2) Main field control converter
(3) Compensating field control converter
(4) Motor control panel
(5) Smoothing reactor
(6) Power transformer

This multiphase static power converter has the complete function for steel rolling mill drive, but has not unique construction because of its small capacity. If the large capacity of static power converter is built at the next stage, paticular attentions must be paid in its construction, for example Freon boiling cooled thyristor and so on. Fig. 14 and 15 show the external view of the control devices and the inside view of the static power converter. [6][7]

Fig.14 External View of Static Power Converter

(a) Logic Units

(b) Thyristor Units

Fig.15 Inside View of Static Power Converter

EXPERIMENTAL RESULTS

Tests were made on a multiphase motor with d-c link at works and site. The specifications of the motor are given in Table 3.

Application	steel rolling test mill
Number of phase	$2\times3\phi$
Rated output	250KW
Rated voltage	370/560V
Rated current	240/180A
Rated frequency	29.75/65HZ
Number of poles	6P
Rated speed	595/1300rpm
Rating	1 hour
Overload rating	220% 10sec
Class of insulation	F
Enclosure	drip proof
Rotor construction	cylindrical rotor with sliprings
Number of thyristor elements	24
Type of thyristor	FT-500A-50
Composition of thyristor	1S1P 6Arms 4Groups

Table 3 Specifications of Multiphase Motor with D-C Link

The conventional and paticular tests are made for this motor. The conventional tests are defined by JEC 114. This rule is established for a conventional synchronous machine, so it is not sufficient for a commutatorless d-c motor. But any other rule has never been established for a commutatorless d-c motor. Therefore the parts of it, which are suitable for a commutatorless d-c motor, and the particular tests are applied.

(1) Tests of motor itself
 (a) No-load saturation test
 (b) Short circuit test
 (c) Mechanical overspeed test

(2) Combined tests of motor and static power converter
 (a) No-load characteristics test
 (b) Load characteristics test
 (c) Temperature rise test
 (d) Start-stop test
 (e) Step response test
 (f) Impact drop test
 (g) Jogging test
 (h) Protection test

Among them the principal test results are described.

Fig.16 shows load characteristics of the multiphase motor with d-c link at steady state operation, Fig.17 shows the waveform of induced emf with respect to distributor signals. This waveform includes some harmonics because of main field and compensating field current pulsation. Fig.18, 19 and 20 show the dynamic characteristics which are obtained from the field tests at the customer's site, that is, the oscillograms of start-stop test, step response test and impact drop test. These oscillograms show that the regulating characteristics of this system is similar to Thyristor Leonard system (static Ward Leonard system).

The features of this system are summarized in Table 4.

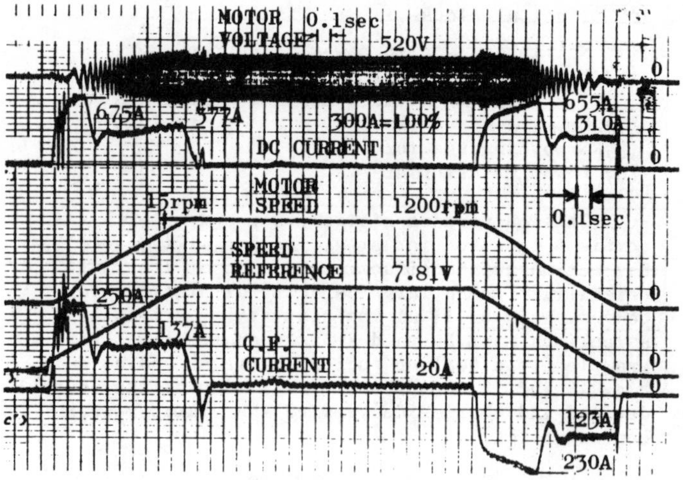

Fig.18 Start-Stop Test of Multiphase Motor with D-C Link

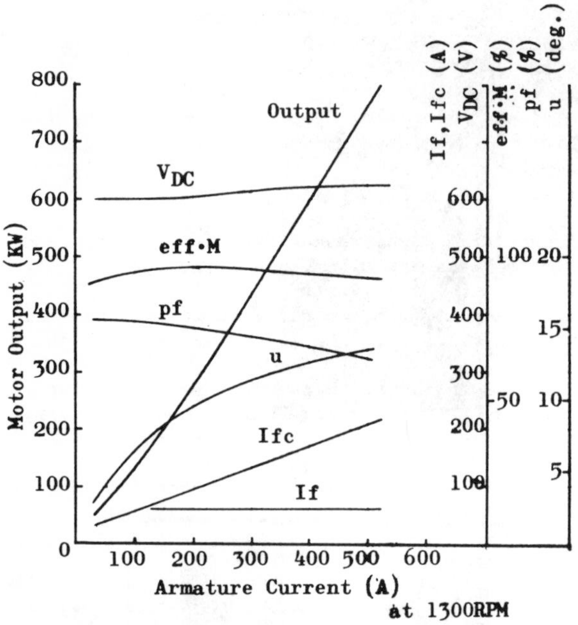

Fig.16 Load Characteristics of Multiphase Motor with D-C Link

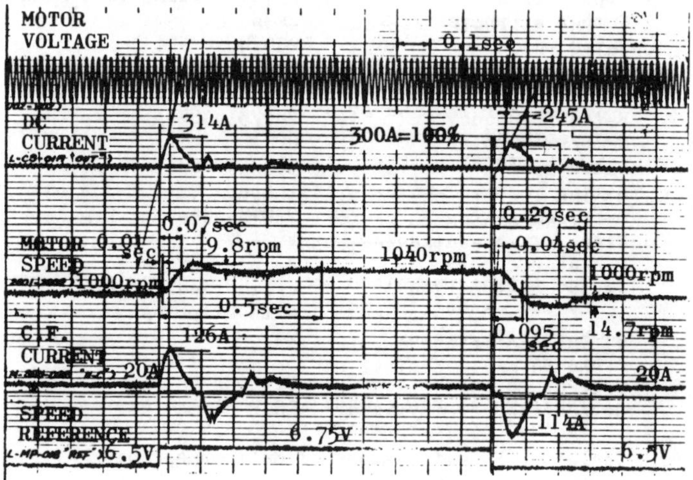

Fig.19 Step Response Test of Multiphase with D-C Link

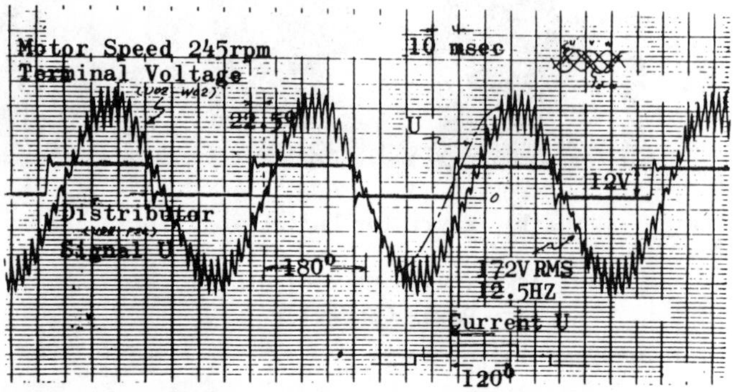

Fig.17 Waveform of Terminal Voltage and Distributor Signals

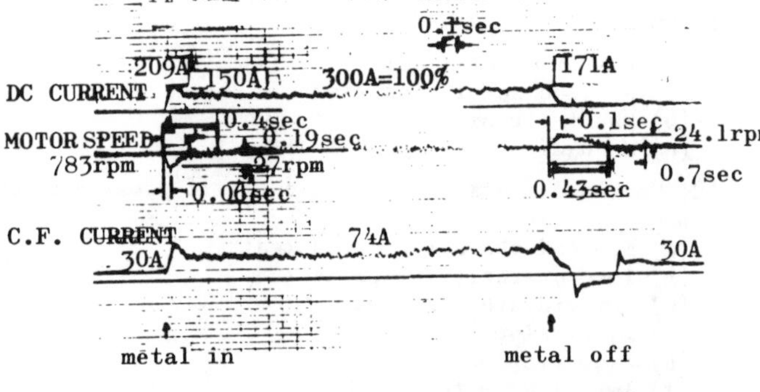

Fig.20 Impact Drop Test of Multiphase Motor with D-C Link

Type of motor	Multi-phase with d-c link
Shape of motor current	Rectangular
Commutation method	Pulse operation at low speed Induced emf commutation at high speed
Power factor of motor	0.8 ~ 0.9
Inertia of motor	Almost same with d-c motor
Torque pulsation	Smaller pulsation with more number of motor phases
Power factor of line	Almost same with Thyristor Leonard system
Harmonics of line side	Almost same with Thyristor Leonard system

Table 4. Features of Commutatorless D-C Drive

CONCLUSION

The paper presents a new type commutatorless d-c drive system for steel rolling mill, which is a combination of new technique of static power converter and rotating machine.

The following conclusions may be drawn from the results :

(1) The performance characteristics of this system is calculated and simulated by the vector diagrams and fundamental equations of the multiphase motor with d-c link.

(2) The inertia of the multiphase motor with d-c link becomes to be small to cancel the armature reaction by the compensating field current.

(3) The torque pulsation can be considerably reduced by the use of the multiphase motor.

(4) This motor can be operated in the constant power by suitable control of the main field current.

(5) This system has almost the same characteristics as that of Thyristor Leonard system.

From the facts described above, we may conclude that in near future a commutatorless d-c drive might be considered to be a substitute for conventional d-c drive.

ACKNOWLEDGEMENTS

Our thanks are due to Mr. S. Araki, Nippon steel Corporation, for encouragement and advice during the course of this accomplishment.

We also express our sincere appreciation for the kindness rendered by all other persons concerned.

Our thanks are due to Mrs. Nakamura, who has borne the brunt of the typing of the manuscript, and has uncomplainingly and successfully carried out the often difficult task of deciphering the original.

REFERENCES

(1) B. R. Pelly, "Thyristor Phase-Controlled Converters and Cycloconverters : Operation, Control, and Performance", John Wiley, New York, 1971.

(2) R. Yamashita and others, Mitsubishi A.C. Commutatorless Motors (Type CS Thyristor Motors)", Mitsubishi Denki ENGINEER, pp.16-22, September 1972.

(3) M. Kataoka and others, "Driving Motor and Control Equipment for Highway Tunnel Blower", Mitsubishi Denki Giho, vol. 49, No.12, pp.767-772, December 1975.

(4) C. Concordia, "Synchronous Machines : Theory and Performance", John Wiley, New York, 1951.

(5) D.C. white and H. H. Woodson, "Electromechanical Energy Conversion", John Wiley, New York, 1958.

(6) S. Otani and others, "High-power Semiconductor Rectifier Equipment using Boiling and Condensing Heat Transfer", IAS 1975 Annual Meeting.

(7) M. Yano and others, "Static Converter Starting of Large Synchronous Motors", IAS 1976 Annual Meeting.

Analysis of a Novel Forced-Commutation Starting Scheme for a Load-Commutated Synchronous Motor Drive

ROBERT L. STEIGERWALD AND THOMAS A. LIPO, SENIOR MEMBER, IEEE

Abstract—A load-commutated inverter synchronous motor drive system employing a simple auxiliary commutation circuit for machine startup is analyzed, and results of a hybrid computer simulation are presented. The commutation circuit employs a single commutation capacitor connected to the neutral of the machine and two auxiliary thyristors, which are used only during machine starting. A practical operating scheme is developed for the forced commutated inverter, which insures commutation over all load currents by actively allowing the commutation capacitor to charge to a voltage proportional to load current. Results of key computer runs are given including inverter waveforms, transient waveforms during transition from forced to load commutation, as well as the effect of forced commutation and load commutation on pulsating torque. The forced-commutation circuit is used only for synchronous machine startup. However, due to its simplicity it also is an attractive alternative to be considered for other types of current-fed inverter ac drives.

INTRODUCTION

ONE OF THE factors which has prevented the widespread use of ac drive systems has been the relative expense of the forced-commutated inverter compared to their ac–dc or dc–dc converter counterparts. When employing a synchronous motor, the field winding can be overexcited such that a leading power factor is presented to the inverter. In this case natural or load commutation of the inverter thyristors can be obtained. Load commutation considerably simplifies the inverter and is a highly desirable mode of operation. Unfortunately, such a drive cannot operate at low speeds where the machine counter electromotive force (EMF) is insufficient to commutate the inverter thyristors. Even in applications where the machine must operate over a limited speed range, some method must be employed to start the machine. One technique which has been used to commutate the inverter at low speed is to interrupt the dc link current by proper control of the phase-controlled rectifier feeding the inverter [1]. Since the phase-controlled rectifier must drive the dc link current to zero six times per cycle, this method can only be used at relatively low motor speeds. When operating from a dc source, such as in a traction application, a force-commutated chopper circuit must be used to utilize this commutation scheme.

In this paper an alternate method of starting a synchronous machine is presented employing a simple forced-commutation circuit. The circuit utilizes a single commutation capacitor and only two auxiliary thyristors for the entire inverter. Since the forced commutation part of the inverter need operate only over low speeds where the inverter is unable to load commutate, the rating of the circuit components need only be a fraction of the full machine rating. Fig. 1 shows a circuit diagram of the commutation scheme. It can be noted that the commutation capacitor is connected to the neutral of the machine. Since the fundamental component of the commutation capacitor current is three times the inverter fundamental frequency, an inverter employing this scheme may be termed a third harmonic auxiliary commutated inverter. Such a commutation scheme has been proposed in the past for use in HVdc transmission systems to obtain forced commutation of the inverter (rather than the usual line commutation) and thus reduce the amount of reactive power consumed from the ac system. The arrangement is attributed to Kaganor and Saba in [2], where further references dating from 1940 are given. Recently the technique has been adapted for use in applying forced commutation to a thyristor cycloconverter [3].

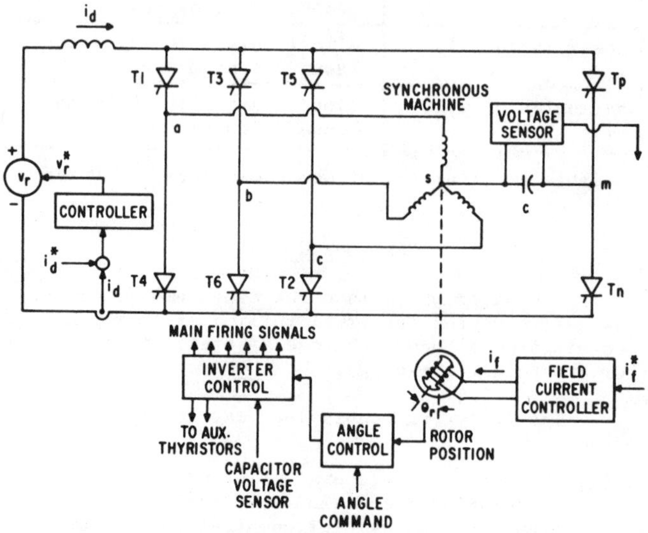

Fig. 1. Load-commutated synchronous motor drive with third harmonic auxiliary commutation.

This paper analyzes the third harmonic commutated inverter specifically for use in starting synchronous machines. However, the approach also offers the possibility of a simple economical forced-commutated current-fed inverter for induction-motor drives. The behavior of the commutation circuit, including effects on motor performance, are analyzed with the aid of a hybrid computer. Operating problems of the drive system in the load-commutated mode including transients occurring

Paper ID 77-11, approved by the Industrial Drives Committee of the IEEE Industry Applications Society for presentation at the 1977 Industry Applications Society Annual Meeting, Los Angeles, October 2–4. Manuscript released for publication October 9, 1978.

The authors are with the Corporate Research and Development Center, General Electric Company, Schenectady, NY 12301.

Reprinted from *IEEE Trans. Ind. Appl.*, vol. IA-15, pp. 14–24, Jan./Feb. 1979.

during the transition from forced to load commutation are addressed. System control strategies including proper thyristor gating control are developed.

SYSTEM DESCRIPTION

Fig. 1 shows a simplified version of the drive system to be studied. This figure also served to define the nomenclature which will be used throughout this paper. In this system a controllable dc link current i_d is fed to a three-phase inverter which in turn drives a wound-field synchronous machine. The link current is controlled by adjusting the voltage source v_r in response to an error signal generated by the difference of actual link current i_d and the commanded link current i_d*. In practice the voltage source can be any type of ac/dc rectifier or dc/dc chopper, but is not considered in detail here. The inverter main thyristors *T*1-*T*6 are gated in the usual fashion sequentially as numbered. The thyristors are fired directly by the output of the angle control block, which consists of a set of six pulses for each electrical revolution of the machine. The pulses occur at a known location relative to the shaft position by employing a shaft position sensor which measures the shaft angle and then phase shifting this signal by a controllable amount as determined by the angle command. The thyristors are thus fired at preselected rotor locations, and therefore the angle between the field magneto motive force (MMF) and the stator MMF is controlled.

The machine field current i_f is also assumed controllable in the same fashion as the dc link current in response to a command signal i_f*. Since the stator current, rotor angle, and field current are controllable, the power factor can be controlled if desired, and hence load commutation can be accomplished. However, at standstill or low speed, the machine has negligible counter EMF, and the inverter cannot load-commutate. Therefore, an auxiliary method of commutating the thyristors is needed until the machine gains sufficient speed to load-commutate. The commutation circuit shown in Fig. 1 is used to accomplish the necessary inverter commutation. The commutation capacitor voltage sensor is used to actively control the peak capacitor voltage, and its operation will be explained below. Two different types of thyristor gating can be implemented and warrant separate consideration.

PRINCIPLES OF COMMUTATION–SIMPLE SEQUENTIAL GATING

In order to provide an adequate starting point for the discussion to follow, the commutation principle will be discussed briefly. In analyses of this type the machine can be represented by a counter EMF of peak value E_M in series with a commutating inductance L_k. Consider the commutation from main thyristor *T*2-*T*4. The equivalent circuit for this commutation is shown in Fig. 2 with the initial capacitor voltage as indicated. The waveforms for this interval are sketched in Fig. 3. Examination of Fig. 3 reveals that commutation takes place in three stages.

Stage 1

Commutation is initiated by firing auxiliary thyristor *Tn*, which effectively places capacitor *C* in parallel with e_3 and L_k

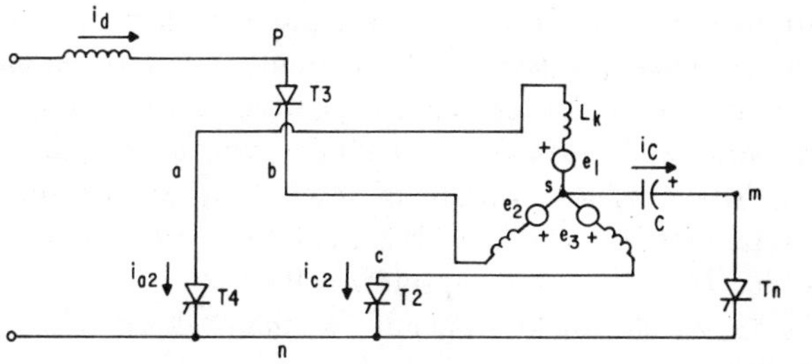

Fig. 2. Equivalent circuit for commutation from *T*2 to *T*4.

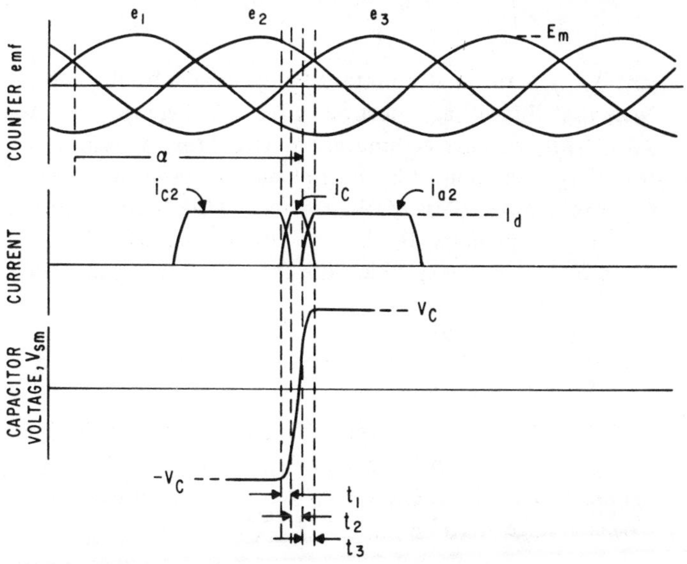

Fig. 3. Waveforms for commutation from *T*2 to *T*4.

(motor phase *c*, line to neutral). For proper operation, the initial capacitor voltage v_{ms} is higher than e_3, and consequently the current in phase *C* immediately starts to decrease, while the current in the commutating capacitor begins to build up. The sum of these currents is the constant direct current I_d. There is gradual commutation of current from phase *c* to the commutating capacitor. During this current transfer, the capacitor voltage v_{sm} becomes more positive. Stage 1 lasts until the entire load current has been commutated to *Tn* and the capacitor, at which point *T*2 goes off. This first stage lasts for a time t_1 as indicated on Fig. 3.

Stage 2

Since the capacitor is now carrying all of the load current, the capacitor voltage v_{sm} rises linearly until it equals the counter EMF generated in phase *a* (i.e., until $v_{sm} = -e_1$). The second stage lasts for a time t_2 as seen in Fig. 3. Note that during stage 2, the current in phases *a* and *c* is zero. Thus the actual motor currents will be less than the usual 120° duration.

Stage 3

When the capacitor voltage v_{sm} overcomes the counter EMF of phase *a* ($-e_1$) current begins to transfer from capacitor *C* to thyristor *T*4. It is assumed that *T*4 is fired as soon as forward voltage is applied to it or, alternatively, that *Tn* and *T*4 are fired simultaneously and the gate pulse to *T*4 remains throughout stages 1 and 2 and is present when the voltage on $T4$ becomes positive. Since current i_{a2} is increasing, the capacitor current is decreasing, and the rate of rise of capacitor voltage decreases until commutation is complete ($i_{a2} = I_d$). The capacitor voltage is now of the proper polarity to commutate $T3$ off when Tp is fired. From symmetry it is concluded that the initial capacitor voltage was $-V_c$ as shown in Fig. 3. As shown in Appendix B, the peak capacitor voltage V_c is given by

$$V_c = I_d \sqrt{\frac{L_k}{C}} - E_M \sin\ (\alpha + \pi/6) \tag{1}$$

where E_M is the peak counter EMF generated by the machine. Note that this voltage must be sufficiently high to commutate $T3$ off for the next commutation. This situation will only be true if e_3, the counter EMF of phase c, is lower than $-e_1$, the negative counter EMF of phase a, during the commutation. As proven in Appendix B, this condition essentially limits the range of angles α over which the commutation circuit is effective to

$$90° < \alpha < 270°.$$

The particular case of zero counter EMF illustrates the potential problem with the commutation circuit. Following Appendix B with the counter EMF zero, the capacitor charges to a peak voltage given by

$$V_c = I_d \sqrt{L_k/C}. \tag{2}$$

During the next commutation the current in the offgoing thyristor is given by

$$i_{c2} = I_d - \frac{V_c}{\sqrt{L_k/C}} \sin\ (t/\sqrt{L_k C}), \tag{3}$$

while the capacitor voltage is given by

$$v_{ms} = V_c \cos\ (t/\sqrt{L_k C}). \tag{4}$$

The time when current reaches zero in the offgoing thyristor is found by setting (3) equal to zero. This results in

$$t/\sqrt{L_k C} = \sin^{-1}\ (I_d \sqrt{L_k/C}/V_c). \tag{5}$$

Substituting V_c from (2) gives

$$t/\sqrt{L_k C} = \sin^{-1} 1 = \pi/2. \tag{6}$$

Incorporating this expression in (4) yields

$$v_{ms} = 0, \tag{7}$$

that is, the capacitor has discharged completely to zero in a quarter cycle oscillation, while driving the current in the offgoing thyristor to zero.

Since the capacitor now carries the dc link current, it immediately reverses, reapplying forward voltage to the thyristor which results in zero turn-off time. It is clear that for practical circuits, which will contain some losses, the thyristor will not turn off. From the above simplified discussion it is clear that at zero speed, the capacitor will not charge high enough at the end of a commutation to insure the next commutation. To overcome this problem, the technique of delayed gating has been employed as described below.

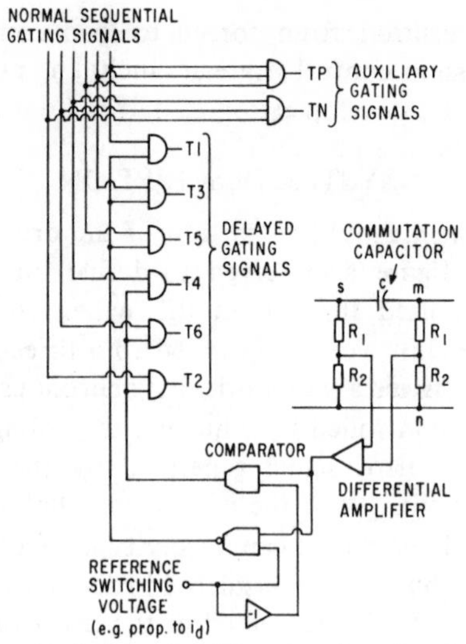

Fig. 4. Technique for delayed gating of main thyristors.

PRINCIPLES OF COMMUTATION–DELAYED GATING

It is useful to now refer to Fig. 4, which illustrates the delayed gating principle. In general, the sequential gating signals for the main thyristors are delayed until the capacitor charges to a predetermined voltage level as sensed by a suitable capacitor voltage sensor (differential amplifier and comparators). The capacitor is thus assured of charging to a predetermined voltage level and commutation is guaranteed.

In order to illustrate the delayed commutation principle, consider a commutation of current from thyristor $T2$–$T4$ (phase c to phase a), and assume the capacitor voltage v_{ms} is positive. To turn $T1$ off, auxiliary thyristor Tn is fired, and C is placed across phase c in such a direction as to drive i_{c2} to zero. When i_{c2} is driven to zero, $T2$ turns off, and the entire link current is in C, which therefore reverses its voltage linearly. When the voltage on capacitor C is of sufficient value to forward bias $T4$, the firing of $T4$ is delayed by the inhibit signals being applied to the AND gates. $T4$ is allowed to be fired only after C has charged to a sufficiently high voltage to insure the next commutation. This voltage level is determined by comparing the signal voltage indicative of capacitor voltage v_{sm} with a desired threshold level (reference switching voltage). The output of the comparator inhibits the gating of the main thyristors until the capacitor has charged to the reference level. The voltage dividers, consisting of $R1$ and $R2$, allow the capacitor voltage to be sensed using signal level circuitry (i.e., the differential amplifier).

It can be noted that the reference switching voltage may be constant value or be controlled to follow a signal indicative of some other circuit parameter. For example, the threshold level can be proportional to link current, thus assuring a higher commutating voltage as the link current increases. In this manner the circuit commutating ability can be made to track that required by the load.

If the main thyristor gating is delayed until the capacitor voltage charges to a reference value V_{ref}, then the final capacitor voltage will be given by

$$V_c = \sqrt{[V_{ref} + E_M \sin (\alpha + \pi/6)]^2 + I_d^2 L_k/C} - E_M \sin (\alpha + \pi/6). \quad (8)$$

Precharging of Commutation Capacitor

When it is desired to load commutate, firing of the auxiliary thyristors is simply inhibited. To initially charge the capacitor for startup, an auxiliary thyristor and an opposite main thyristor (e.g., *Tn* and *T*1) may be fired causing the capacitor to ring up through the link inductor L_d and phase *a* toward twice the link voltage. Then, as operation begins the capacitor will pump up as determined by the reference switching voltage. Alternatively, the capacitor may be initially charged through a resistor from an auxiliary power supply.

SYSTEM SIMULATION

In order to properly evaluate system behavior, a hybrid computer simulation of the third harmonic commutated inverter was implemented, including a detailed model of both inverter and machine. The synchronous machine was simulated in the usual manner as an equivalent two-phase machine with the reference frame for voltage and current vectors fixed in the rotor [4]. The thyristors were simulated as perfect switches with series *di*/*dt* reactors. The simulation of the commutating circuit is of particular interest and is included in Appendix A. The remainder of the inverter was simulated in similar fashion. The dc link as shown in Fig. 1 was also simulated as well as the rotor position sensor and necessary voltage and current transformation (three-phase to two-phase transformation and viceversa).

The synchronous motor chosen for this study corresponds to a 20 kVA, eight-pole homopolar-inductor type of machine. The maximum rotor speed is 7500 r/min corresponding to a line frequency of 500 Hz. Rated terminal line voltage is 90 V at this frequency. These parameters correspond to those of an onboard inductor-motor flywheel energy-storage device for an electric vehicle application [5]. The parameters of the machine and the rest of the system parameters are as follows

Synchronous machine (homopolar-inductor motor) parameters:

$r_s = 0.0139\ \Omega$,	base 1 = *n* voltage = 71.8 V (peak),
$x_{ls} = 0.07228\Omega$,	base frequency = 500 Hz,
$x_{md} = 0.215\ \Omega$,	base kVA = 20,
$x_{mq} = 0.1197\ \Omega$,	poles = 8,
$x_{lfr}' = 0.7458\ \Omega$,	inertia = J = 0.0109 kg·m^2,
$r_{fr}' = 0.00309\Omega$.	

Inverter and dc link filter parameters:

$C = 160\ \mu F$,
$l = 10\ \mu H$,
$L = 0.358$ mH, ⎫
$RL = 0.0085\ \Omega$, ⎭ Figs. 6–8.

SIMULATION RESULTS

Third Harmonic Commutation

Fig. 5 shows the inverter operating at a fundamental frequency of 250 Hz. This condition corresponds to high-frequency operation of the inverter, which would probably not be encountered in the event that load commutation is used for high-speed operation, but serves to illustrate the commutation waveforms. Note that the motor voltage and current waveforms are similar, but not identical to those obtained for a conventional current-source inverter [6]. In particular, the currents in the main thyristors *T*1, *T*2, ···, T6 are slightly less than 120°. Notches in the phase voltage now occur in pairs corresponding to the turn-on and turn-off of the auxiliary thyristor currents.

Since the commutation circuit voltage must exceed the counter EMF of the machine, the commutation time depends greatly on the motor winding EMF which appears in series with the commutation capacitor at the commutation instant. Note that the "rate of rise" of the thyristor current *Tp* is significantly greater than the subsequent "rate of decrease." In the limiting case, the thyristor currents i_{Tp} and i_{Tn} are each on for half the time, and the main thyristor currents become 60° rather than 120° blocks of current. Also, the capacitor voltage becomes triangular, and the reverse recovery voltage on the thyristor is very small indicating that thyristors *Tp* and *Tn* are on the verge of shorting the dc link. This situation represents the highest operating frequency attainable for the given load and circuit parameters. For the present system, the limiting frequency was 400 Hz at a dc link current of 150 A.

The waveforms shown in Fig. 5 were taken for an ideal dc link current source (infinite link inductor). The waveforms for a finite dc link inductor (0.358 mH) are nearly identical to those of Fig. 5 except that small ripples appear in the dc link current during the commutation periods as shown in Fig. 6.

Effect of Third Harmonic Commutation on Electromagnetic Torque

In general, the pulses of current that flow through the thyristors *Tp* and *Tn* contribute a zero sequence component that is a nontorque producing component of current. As a result it can be shown that the torque is essentially reduced by one-half during the commutation interval. Fig. 7 illustrates this effect for the 50 Hz case. Fig. 8 shows the torque waveform for four higher line frequencies. The "notches" that occur in the torque waveform are clearly evident. Although the duration of the notches is effectively the same for all operating frequencies, as the frequency increases, the notches become a greater and greater portion of a cycle. At 100 Hz the notching has resulted in the average torque being depleted by approximately one-eighth, at 200 Hz by one-fourth, and at 400 Hz by one-half.

Load Commutation

Fig. 9 displays the same operating condition as Fig. 5, except in this case the commutation circuit was disabled and the six-pulse thyristor bridge was forced to operate under load commutation (back EMF commutation). Note that the com-

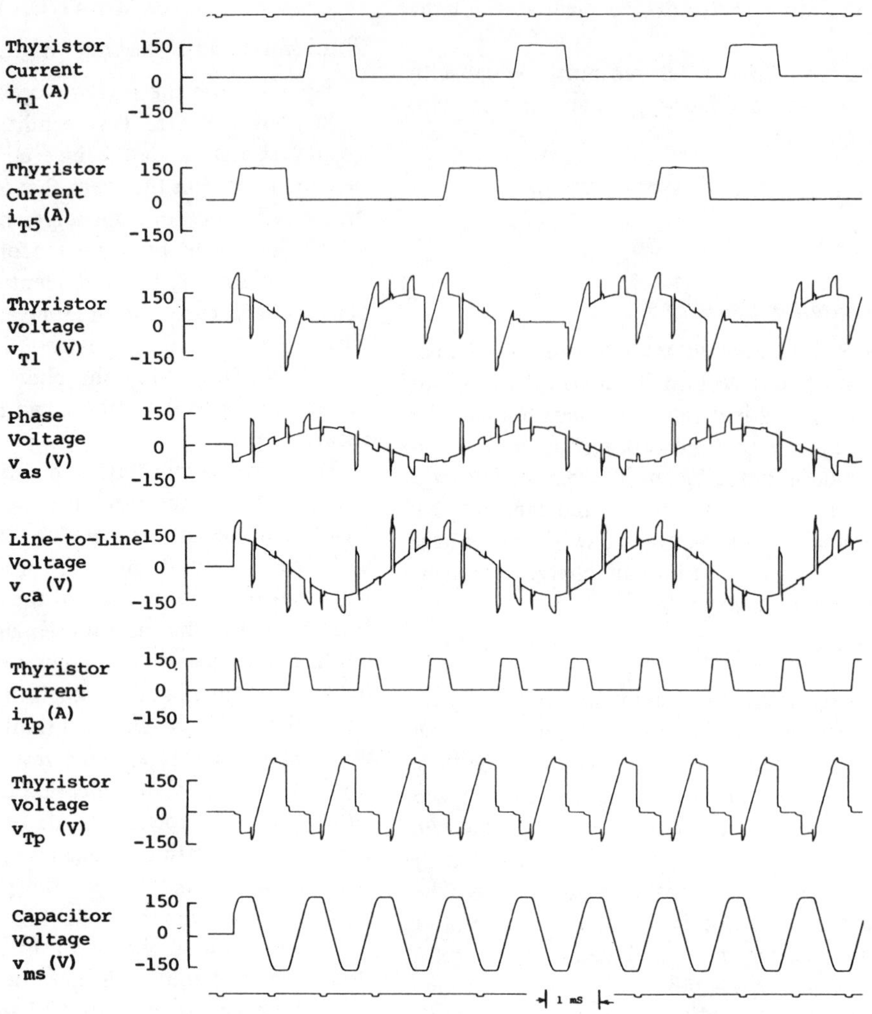

Fig. 5. Steady-state operation under third harmonic auxiliary commutation at 250 Hz with constant dc link current.

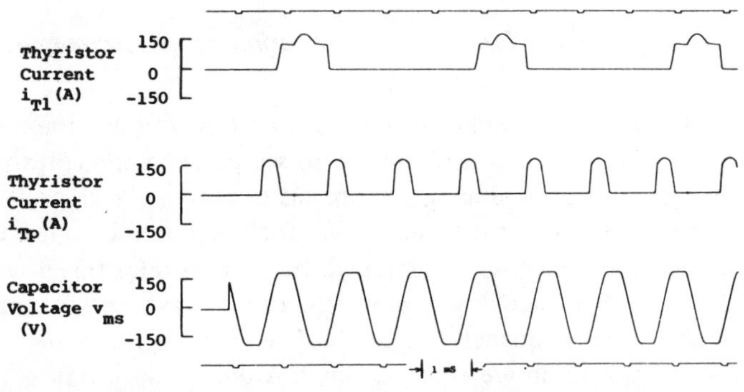

Fig. 6. Effect of finite link inductor on operation under third harmonic commutation at 250 Hz.

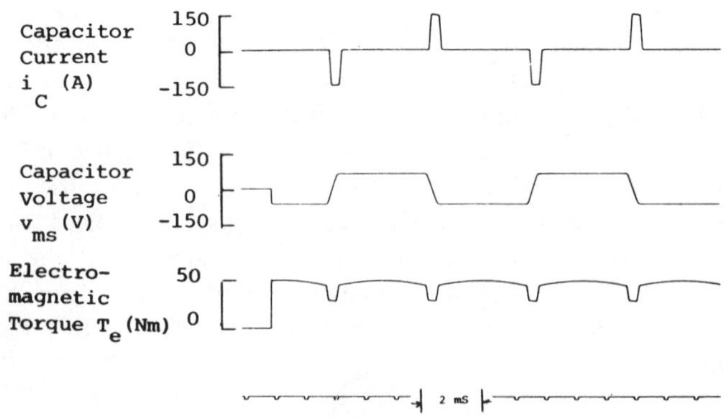

Fig. 7. Effect of commutation on electromagnetic torque at 50 Hz.

mutation notches are much larger (longer) than for the corresponding forced commutation case. The large spikes which appear on the motor windings are no longer present.

Effect of Delayed Gating

In Fig. 10 the proper main thyristor has been gated at the same instant as the auxiliary thyristor (no commutation delay). The dc link current is slowly reduced to zero to observe the effect of load on commutation. Since the commutation capacitor voltage is a function of the link current (8), it decreases as i_d is decreased. Note that $\alpha \cong 160°$ for this case, and thus delayed gating is not needed since $90° < \alpha < 270°$ as explained earlier. However, the method is unable to commutate upon starting or for other values of α without delayed gating.

In Fig. 11 the firing of the main thyristors has been delayed. In particular, the oncoming main thyristor has been delayed until the capacitor voltage exceeds a preset threshold

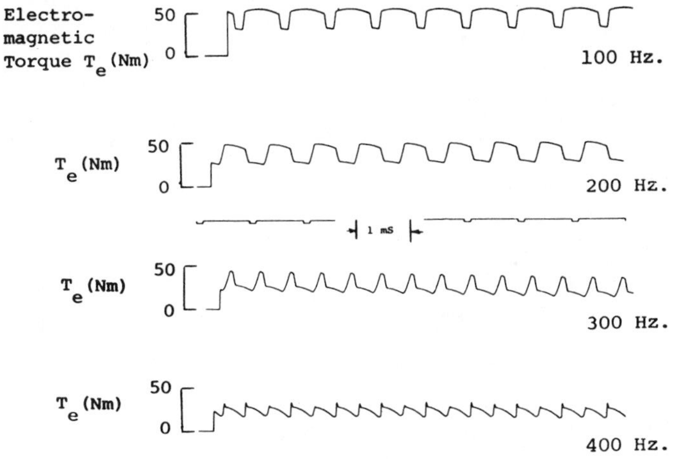

Fig. 8. Effect of commutation on electromagnetic torque for progressively higher frequencies.

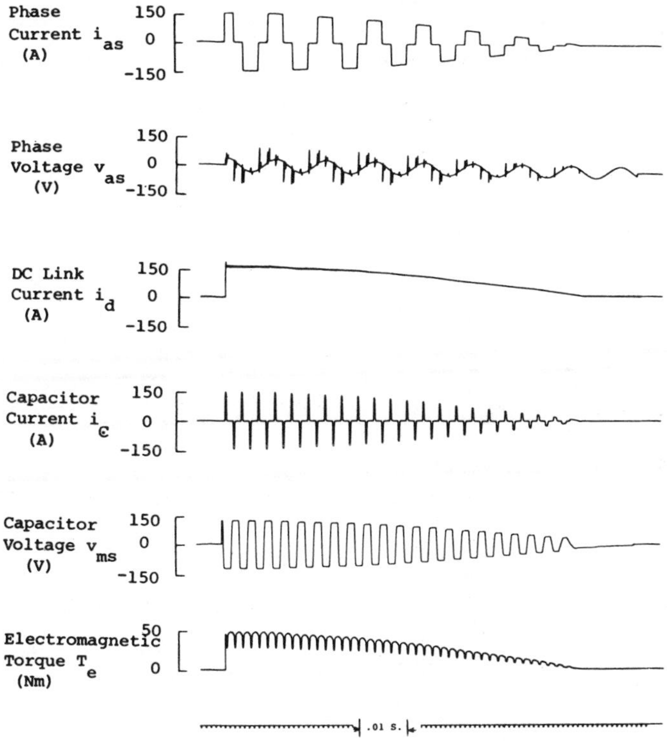

Fig. 10. Performance of commutation circuit as dc link current is decreased. No delay in commutation time.

of 150 V. Note that commutation continues successfully until the link current is reduced to a sufficiently small value that the capacitor voltage does not exceed the prescribed threshold before the opposite auxiliary thyristor is fired. At this point a commutation failure occurs as the auxiliary thyristors shunt the dc link. It is apparent that the commutation time increases for decreasing i_d resulting in widening "notches" in the electromagnetic torque.

Fig. 12 shows the effect of making the threshold voltage for commutation delay a function of dc link current. Note that the capacitor voltage again decreases as link current is decreased. The result is similar to Fig. 10. However, in this case, the commutation time is under control and can be adjusted at will, and the inverter can be operated at all values of α as well

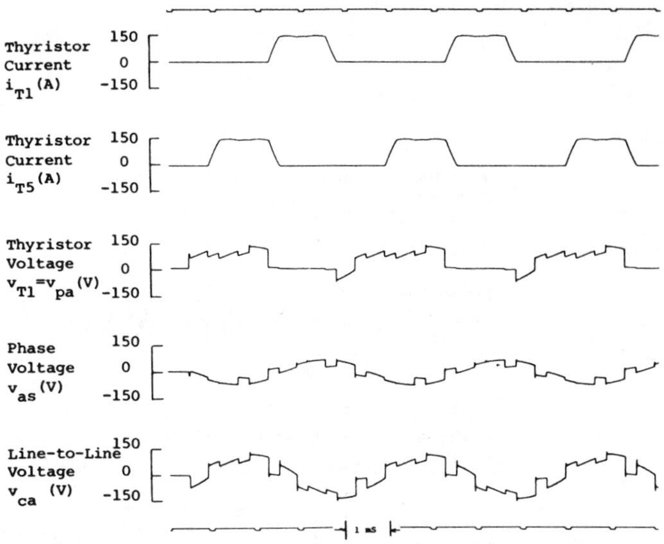

Fig. 9. Steady-state operation under load commutation at 250 Hz, $\alpha = 0°$.

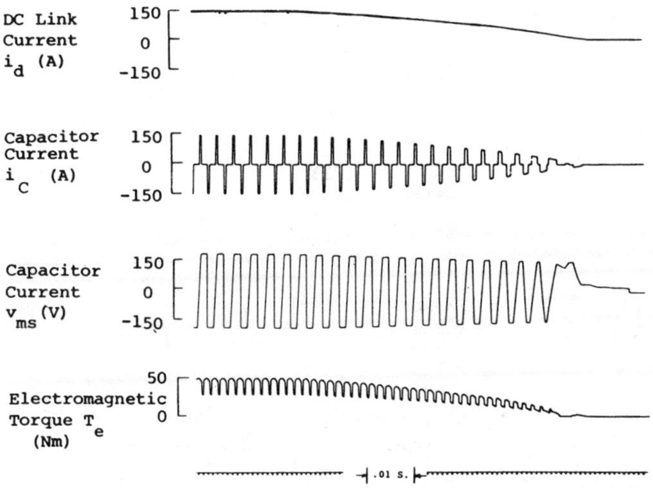

Fig. 11. Performance of commutation circuit as dc link current is decreased. Delay in commutation time maintaining constant commutation voltage.

as at motor standstill. If desired, a lower limit can be provided in the commutation time for very light load conditions.

Transient Behavior During Transition from Forced to Load Commutation

Fig. 13 displays a typical transition from forced to load commutation. In particular, the MMF angle is set at 60° and the motor is adjusted for leading power factor. For this run, the dc link current as well as the speed were held constant. It can be observed that the torque increases markedly after the transition to load commutation.

Effect of Link Current and Line Frequency on Load Commutation

Two of the system variables which have a dominant effect on load commutation are the motor line frequency and dc link current. In Fig. 14 the dc link current has been increased gradually so as to observe the effect on commutation. It can

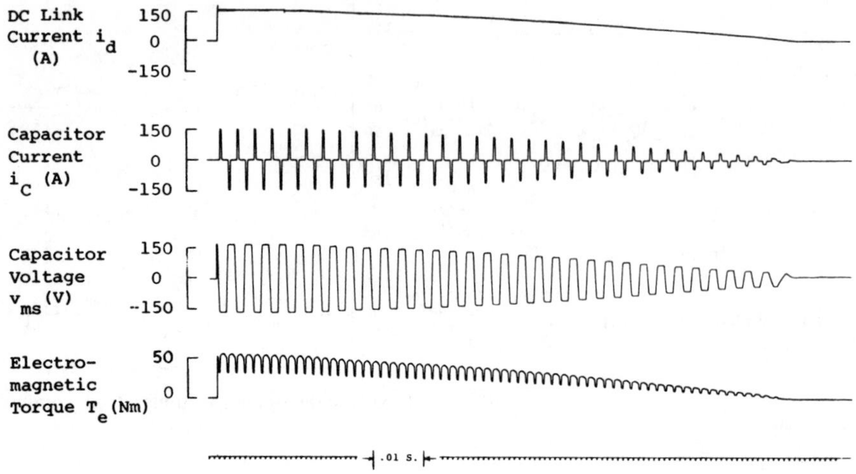

Fig. 12. Performance of commutation circuit as dc link current is decreased. Threshold level proportional to dc link current.

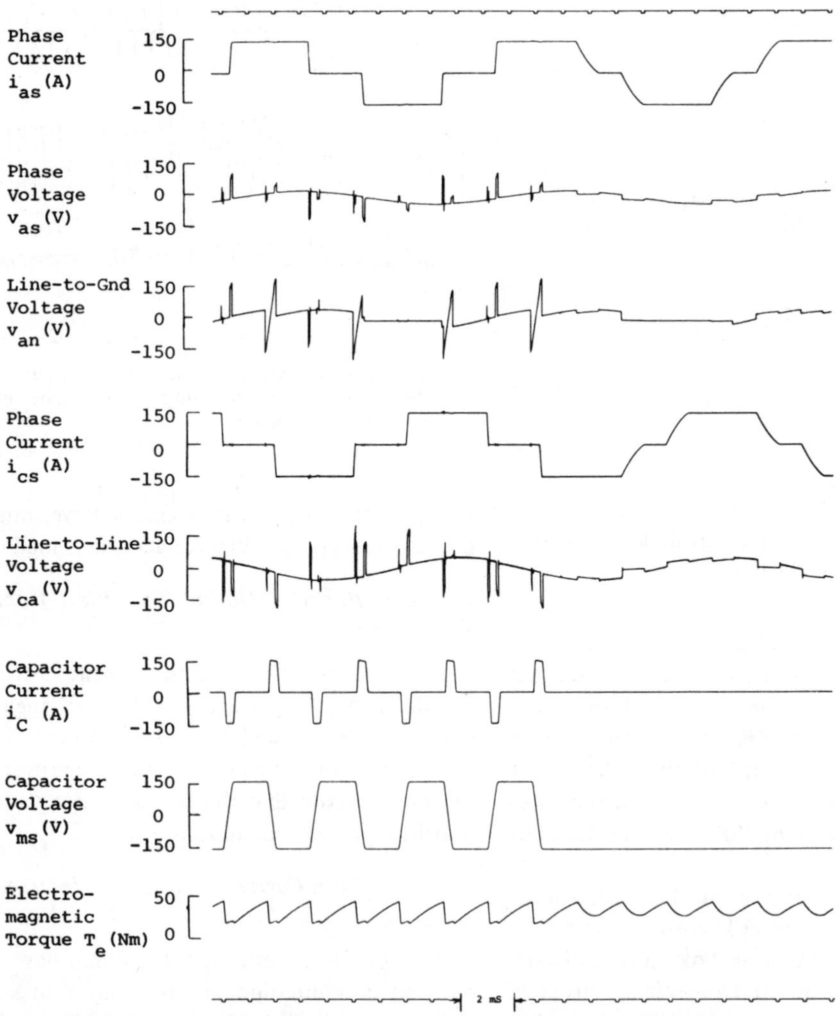

Fig. 13. Transient behavior of system during transition from forced to load commutation, $\alpha = 60°$, 100 Hz.

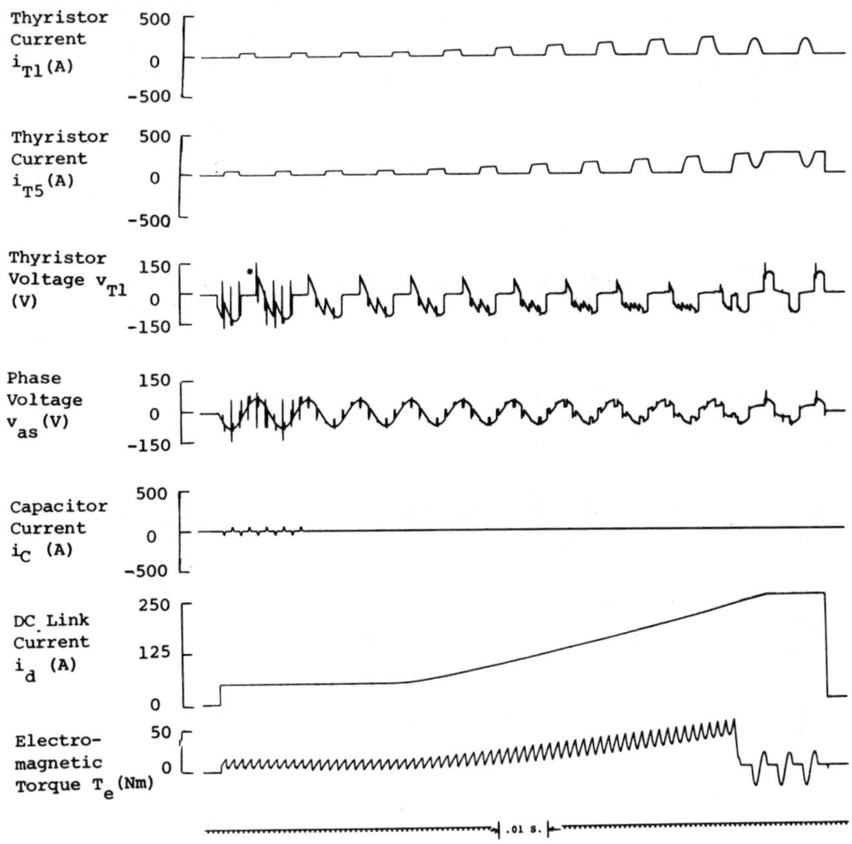

Fig. 14. Effect of increasing dc link current on load commutation at 50 Hz.

be noted that the turn-off time decreases in proportion to the link current so that ultimately thyristor *T5* is unable to commutate to *T1* and shoot-through occurs.

In Fig. 15 the dc link current has been held constant, but the motor speed reduced slowly. In general, successful transition from force to load commutation could not be achieved below 0.1 pu speed (50 Hz). However, once having achieved load commutation much lower speeds were possible. In Fig. 15 a transition to load commutation is made at 50 Hz. and the line frequency then slowly decreased. It can be noted that commutation failure does not occur until the machine reaches a frequency of 0.03 per unit (15 Hz).

CONCLUSIONS

A load-commutated inverter drive employing a simple forced commutation circuit for machine startup has been analyzed and shown to be an excellent candidate for a practical ac drive system. Since the commutation circuit consists of a single capacitor and only two auxiliary thyristors, the basic simplicity of the load-commutated inverter power circuit is maintained. By delaying the gating of the incoming main thyristor, the forced commutation can be maintained over all inverter delay angles. Using this technique, the commutating capacitor peak voltage can be actively controlled to optimize circuit operation. Transition from forced to load commutation and vice versa is accomplished without any difficulty. While the analysis presented here has centered specifically around a synchronous machine drive, the commutation scheme can also be used for other current-fed ac drives (i.e., synchronous-reluctance, permanent magnet, or induction motor drives). Third harmonic commutation shows promise as a simple reliable commutation technique for low-cost ac motor drives.

APPENDIX A

SIMULATION OF COMMUTATION CIRCUIT

The simulation diagram for the commutation circuit will be developed with reference to Fig. 16. The phase currents and dc link voltage are available from other parts of the simulation and are generated in conventional fashion. It can be noted that a large resistance is placed from neutral to ground to develop the neutral voltage, while the *di/dt* inductors are used to generate the auxiliary thyristor currents. Referring to Fig. 16, the neutral to ground voltage is

$$v_{sn} = R_{sn}(i_{as} + i_{bs} + i_{cs} + i_{Tp} - i_{Tn}).$$

Assuming that the respective thyristor is conducting,

$$i_{Tp} = \frac{1}{l} \int (v_{pn} - v_{ms} - v_{sn})\, dt \geqslant 0 \tag{9}$$

and

$$i_{Tn} = \frac{1}{l} \int (v_{ms} + v_{sn})\, dt \geqslant 0. \tag{10}$$

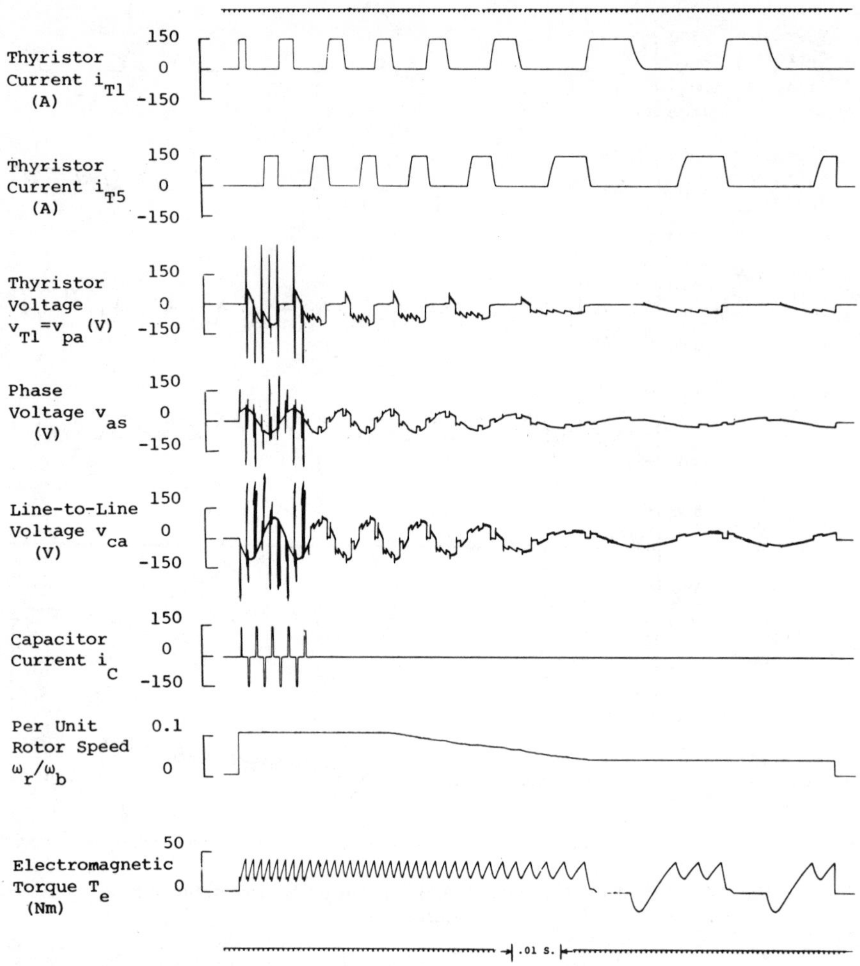

Fig. 15. Effect of decreasing rotor speed on load commutation with fixed dc link current.

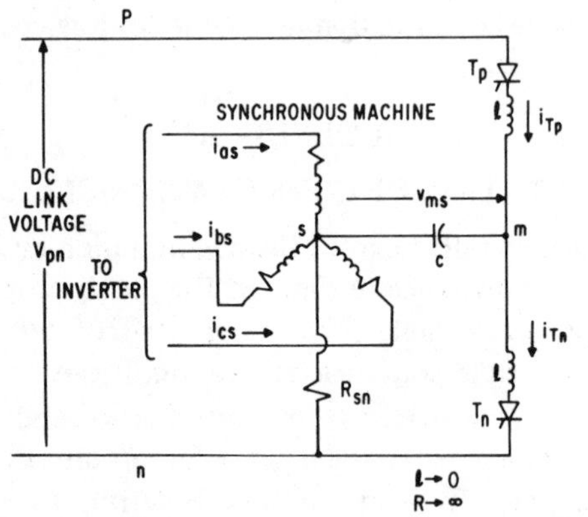

Fig. 16. Equivalent circuit for commutation circuit simulation.

The voltage across the commutation capacitor is computed as

$$v_{ms} = \frac{1}{C}\int (i_{Tp} - i_{Tn})\,dt. \tag{11}$$

These equations are simulated in Fig. 17. The integrators for i_{Tp} and i_{Tn} are gated in the simulation from the identical arrangement logic circuits illustrated in Fig. 3. Additional gating logic is not needed since the auxiliary thyristor currents naturally go to zero and remain at zero as the commutating capacitor charges. The limiters around the integrators ensure unidirectional auxiliary thyristor current.

APPENDIX B
ANALYSIS OF COMMUTATION

In this analysis it is assumed that the counter EMF changes a negligible amount during the commutation interval. Thus the counter EMF can be assumed constant for a given α.

Stage 1

Thyristor $T2$ is conducting, and $T4$ does not conduct during this interval. Referring to Fig. 2

$$L_k\frac{di_{c2}}{dt} - \frac{1}{C}\int i_c\,dt - e_3 = 0. \tag{12}$$

Solution of this equation under the constraint

$$i_c + i_{c2} = I_d$$

and the initial conditions

$$i_{c2}(0) = I_d;\ v_{ms}(0) = V_c$$

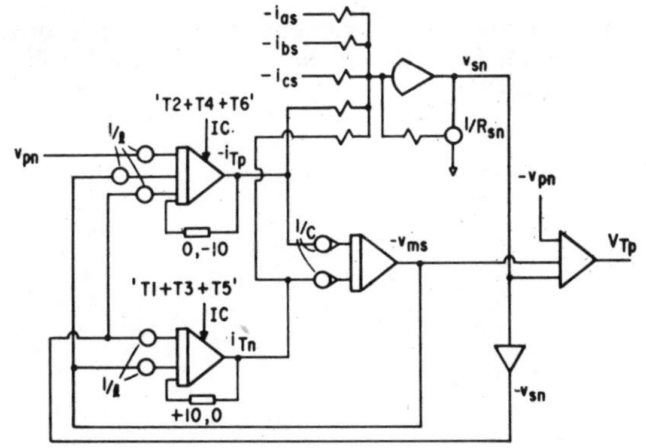

Fig. 17. Simulation diagram for third harmonic auxiliary commutation circuit.

results in

$$i_{c2} = I_d - \frac{V_c - e_3}{\sqrt{L_k/C}} \sin\ (t/\sqrt{L_kC}) \tag{13}$$

where $e_3 = -E_M \sin(\alpha - \pi/6)$. The duration of Stage 1 is found by solving 13 for $i_{c2} = 0$:

$$t_1 = \sqrt{L_kC}\ \sin^{-1}\left[\frac{I_d\sqrt{L_k/C}}{V_c + E_M\ \sin\ (\alpha - \pi/6)}\right]. \tag{14}$$

The capacitor voltage at the end of Stage 1 is found from

$$v_{sm}(t_1) = -V_c + \frac{1}{C}\int_0^{t_1}\left(\frac{V_c - e_3}{\sqrt{L_k/C}}\right)\cdot \sin\frac{t}{\sqrt{L_kC}}\,dt \tag{15}$$

where t_1 is given by (14).

Stage 2

Stage 2 starts when the entire link current is in the commutation capacitor C and lasts until the capacitor voltage equals the counter EMF e_1 (no delayed gating) or until *T*4 is fired (delayed gating), whichever is longer. During Stage 2 the capacitor voltage changes linearly

$$\frac{dv_{sm}}{dt} = \frac{I_d}{C}. \tag{16}$$

Note that the reapplied *dv/dt* to T2 is

$$\frac{dv}{dt} = \frac{I_d}{C} + \frac{de_3}{dt} = \frac{I_d}{C} - 2\pi f E_M\ \cos\ (\alpha - \pi/6) \tag{17}$$

where *f* is the frequency of the counter EMF.

Stage 3

During Stage 3, *T*2 is off, and *T*4 is conducting. Referring to Fig. 2,

$$L_k\frac{di_{a2}}{dt} - e_1 - \frac{1}{C}\int_0^t i_C\,dt = 0. \tag{18}$$

Under the constraint $i_{a2} + i_C = I_d$ and the initial conditions $i_{a2}(0) = 0$, $v_{ms}(0) = e_1$ (no delayed gating), the solution is

$$i_C = I_d\ \cos\ (t/\sqrt{L_kC}) \tag{19}$$

$$i_{a2} = I_d[1 - \cos\ (t/\sqrt{L_kC})]. \tag{20}$$

Stage 3 lasts until $i_C = 0$. The duration of Stage 3 is, from (19),

$$t_3 = \frac{\pi}{2}\sqrt{L_kC}. \tag{21}$$

The capacitor voltage is

$$v_{sm} = -E_M\ \sin\ (\alpha + \pi/6) + \frac{1}{C}\int_0^t I_d\ \cos\ (t/\sqrt{L_kC})\,dt$$

$$= -E_M\ \sin\ (\alpha + \pi/6) + \frac{I_d}{C}\sqrt{L_kC}\ \sin\ (t/\sqrt{L_kC}). \tag{22}$$

Substituting (21) into (22) gives the peak capacitor voltage V_c as

$$V_c = -E_M\ \sin\ (\alpha + \pi/6) + I_d\sqrt{\frac{L_k}{C}}\ \text{(no delayed gating).} \tag{23}$$

For delayed gating a similar analysis applies, but with the initial condition

$$v_{sm}(0) = V_{ref} > -e_1. \tag{24}$$

In (24) V_{ref} is the capacitor threshold voltage at which *T*4 is fired. The peak capacitor voltage becomes

$$V_c = \sqrt{[V_{ref} + E_M \sin(\alpha + \pi/6)]^2 + I_d^2(L_k/C)}$$

$$-E_M\ \sin\ (\alpha + \pi/6)\quad \text{(delayed gating).} \tag{25}$$

The ranges of α for which delayed gating is not required can be found by observing that for commutation to be successful the minimum value of i_{c2} as given in (13) must be less than or equal to zero. This is equivalent to the constraint

$$I_d - \frac{V_c - e_3}{\sqrt{L_k/C}} < 0. \tag{26}$$

Substituting $-E_M \sin(\alpha - \pi/6)$ for e_3 and (22) for V_c in this inequality, results in

$$\sin\ (\alpha - \pi/6) > \sin\ (\delta + \pi/6), \tag{27}$$

which has the solution

$$90° < \alpha < 270°. \tag{28}$$

REFERENCES

[1] T. Peterson and K. Frank, "Starting of large synchronous motor using static frequency converter", *IEEE Trans. Power App. Syst.,* vol. PAS-91, pp. 172–179, Jan./Feb. 1972.

[2] V. P. Bakharerskii and A. M. Utevskii, "A circuit for two-stage artificial commutation of an inverter," *Direct Current,* pp. 153–159, June 1957.

[3] E. J. Stacey, "An unrestricted frequency changer employing force commutated thyristors," 1975 Power Electronics Specialists Conf. *Conf. Record,* pp. 165–173.

[4] C. H. Thomas, discussion of "Analogue computer representations of synchronous generators in voltage-regulation studies," by M. Riaz, *AIEE Trans. (Power App. Syst.),* vol. 75, pp. 1178–1184, Dec. 1956.

[5] A. B. Plunkett and F. G. Turnbull, "Load commutated inverter/synchronous motor drive without a shaft position sensor," 1977 IEEE/IAS Annual Meeting, *Conf. Record.*

[6] T. A. Lipo, "Simulation of a current source inverter drive," *Conf. Record of the Power Electronics Specialists Conf.*, Palo Alto, CA, June 14–16, 1977.

Static Variable Frequency Starting and Drive System for Large Synchronous Motors

Beat Mueller
Member, IEEE
Brown Boveri Corp.
U.S.A.

Thomas Spinanger
Member, IEEE
Brown Boveri Corp.
U.S.A.

Dieter Wallstein

Brown Boveri & Co.
Switzerland

Abstract - Large synchronous motors and generators used in connection with compressors, gas turbines, pumped storage plants, and synchronous compensators, in most cases, need the help of an electrical drive for starting. The purpose and the technical reasons for a starting system, as well as different types are summarized.

A new technique, the static variable frequency starting system (VFSS), is increasingly employed in power and industrial plants. The technical features of this system are illustrated in detail by discussion of one of the largest such systems in the world (with a rating of 2 x 30 MW) which is currently under construction for the starting of large compressor drives.

I. INTRODUCTION

Starting a synchronous motor has always been a problem and a variety of systems have been developed over the years. The choice of the methods employed depends very much on the particular requirements and conditions. With gas turbines, for instance, the rotating exciter or separate D.C. motor has been used. For pumped storage plants, on the other hand, it was normal to mount an induction type pony motor on the main shaft of the machine group. Both systems are quite specific for a certain application. Recently, however, a general system has been developed, the static starting device, which is adaptable to any specific requirement. In addition, the static starting device is not only capable of starting the synchronous motor up to a predetermined speed, but also of driving the motor at variable speeds.

The general purpose of a starting system is to accelerate a rotating machine up to a speed at which it can begin to run on its own or at which it can be synchronized to the main supply network. A gas turbine for example, must be run up to about 60% of its rated speed before it is capable of producing enough power by itself to continue accelerating the set. Synchronous motors for compressors, pumps, and synchronous condensers, on the other hand, have to be accelerated to 100% rated speed in order to be synchronized with the network.

To draw comparisons between conventional starting methods and static starting devices, we summarize very briefly the most important starting systems, the material involved and the behavior and rating capacity of each kind. (See Fig. 1)

a. Starting with Starter Motor

Each motor is fitted with a flange mounted starter motor which accelerates the rotor of the synchronous motor to the required speed. The starter motor is either an induction motor with starting resistors or a D.C. motor. The starting motor has to be mounted on the free shaft end of the synchronous motor and thus making the group longer. Braking is possible with additional equipment. This method is mainly used, when the opposing torque and therefore the rating of the starting system is not more than about 5% of the motor rating.

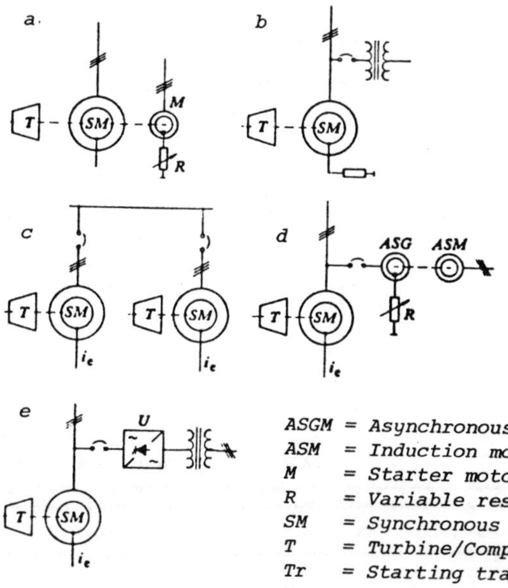

ASGM = Asynchronous generator
ASM = Induction motor
M = Starter motor
R = Variable resistor
SM = Synchronous machine
T = Turbine/Compressor
Tr = Starting transformer
U = Frequency changer
i_e = Excitation current

Fig. 1 - Basic circuits for starting systems

a: Starting with starter motor
b: Starting with transformer or direct connection
c: Frequency starting with synchronous auxiliary machine
d: Frequency starting with asynchronous generator (Unger connection)
e: Frequency starting with static frequency changer

b. Asynchronous Starting

The synchronous motor is either connected directly or through a transformer to the power supply. This method results in short start-up times but will draw a high reactive current (6 to 7 times the rated current) and therefore drops the voltage in the supply network. The additional transformer reduces the voltage drop but increases the start-up time. Braking is again possible with additional equipment. (See reference [1]).

c. Frequency Starting with Existing Machine Group

The synchronous motor can be started with an existing turbine - generator set. The two machines are electrically coupled together at standstill or at low speed and then run up at variable frequency. Since this met-

Reprinted from *Conf. Rec. IEEE/IAS 1979 Annual Meeting,* pp. 429-438, 1979.

hod is used primarily for pumped storage, the last generator cannot be used as a pumping motor, unless it is started from a neighboring power station. (See reference [2]).

d. Unger Connection

The synchronous motor is started without a large power consumption from the supply system by using an auxiliary rotating group, consisting of an induction motor and an asynchronous generator. The accelerating power is governed by the starting resistor in the rotor circuit of the asynchronous generator. Problems occur, however, on starting from zero frequency to a certain speed, because the system requires a minimum frequency.

e. Static Variable Frequency Starting System (VFSS)

The synchronous machine is started from standstill by applying a phase synchronized, variable frequency generated by solid-state power conversion equipment. The starting system can be located remotely from the motor with one start-up system applied for several motors. The system can also be used for driving, braking, or reversing large synchronous machines. A full description of this system is given in this paper.

SUMMARY (See Table 1)

Systems a, c, and d need rotating auxiliary devices, which need space, maintenance, and limit the freedom of installation. System B draws a high reactive current and results in a voltage drop on the power supply. None of the four systems have breaking or variable speed capability.

System E, on the other hand, needs a synchronous motor with static - or special rotating - excitation and static power electronic devices and controls, which however, are no problem today. In addition, harmonics in the line and in the motor must be considered.

	Starter Motor (a)	Asynchronous Starting (b)	Frequency Starting (c)	Unger Conn. (d)	Static Start Device (e)
Can be installed remotely from motor	No	Yes	Yes	Yes	Yes
Only one start up system for several motors	No	Yes	Yes	Yes	Yes
Controlled braking capability	No	No	No	No	Yes
Variable speed capability	No	No	No	No	Yes
Minimal voltage drop in the supply network during start up	Yes	No	Yes	Yes	Yes
Power train with only solid state conversion	No	No	No	No	Yes
Can use rotating excitation	Yes	Yes	Yes	Yes	*
Minimal harmonic disturbance to line and motor	Yes	Yes	Yes	Yes	**
Relative cost	medium	low	low	med.	***

Table 1 - Comparison of different Starting System

* *Additional equipment needed.*
** *Dependent on pulse - number of converters*
*** *Dependent on proportion of number of starting systems to number of motors*

II. STATIC VARIABLE FREQUENCY STARTING SYSTEM/VFSS

A. Features and Applications

The static variable frequency starting system draws power from the supply system at constant voltage and frequency, and converts it into power of variable voltage and frequency, in order to start and accelerate a synchronous motor. (See Fig. 2).

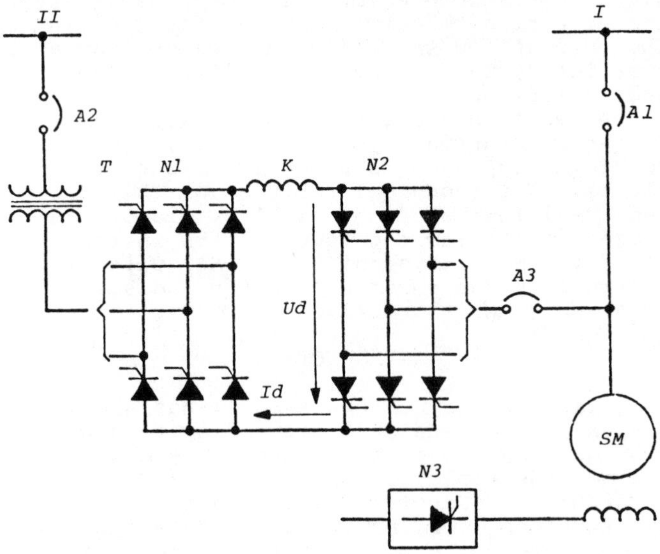

Fig. 2 - Static frequency changer

Basic circuit for starting system

I = Main supply system
II = Auxiliary supply system
A1,A2,A3 = Circuit breakers
T = Transformer
N1 = Rectifier
N2 = Inverter
N3 = Converter for excitation
K = Reactor

The conversion is performed by two thyristor converters in three-phase bridge connection and a D.C. buffer circuit. The D.C. link (with reactor) decouples the different three-phase bridge frequencies and smooths the rectified current. During start-up or while driving, the converter on the A.C. side is connected through a transformer to the main supply system, while the other converter is connected directly, or through a transformer, to the synchronous motor.

The power required for starting the motor is determined by the following factors:

- .. Magnitude and curve-shape of opposing torque (load torque) of driven machine group (Mg in Fig. 3)
- .. Moment of inertia of machine group
- .. Run up time and speed

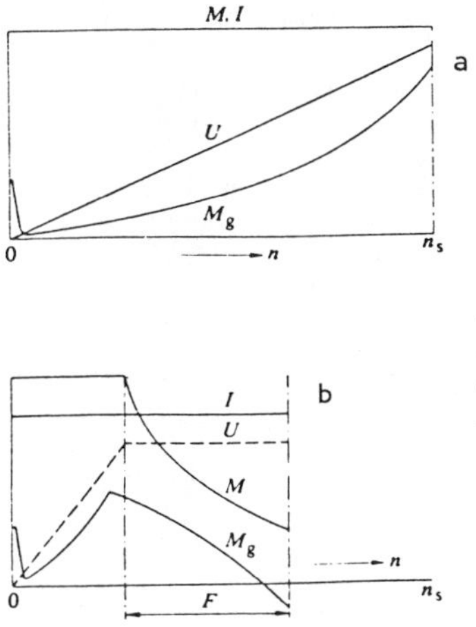

F = Field weakening
I = Motor current
M = Motor torque
M_g = Opposing torque
n_s = Rated speed
U = Motor voltage

Fig. 3 - Schematic of torque and motor quantities as a function of speed n

a: Starting hydro motors or compressor drives.
b: Starting gas turbine sets

There are two main different operations of starting a synchronous motor. One is at constant machine flux and the other is in field weakening operation. The opposing torque curve of a gas turbine, for instance, allows the operation of the starting device in the field weakening mode (Fig. 3b). This means the synchronous machine can be run up at constant voltage over a wide range of speed and the frequency changer is better utilized than with constant flux operation for the same run up time. When starting a hydro machine or a large compressor set, on the other hand, the opposing torque does not allow a field weakening operation. (Fig. 3a). Voltage is increased proportional to the speed of the motor.

The current is in both cases constant and irrespective of speed. This means also a constant current loading on the frequency changer.

B. Operating Principle (See Fig. 2)

In the driving mode, the bridge *N1* on the input side is functioning as a rectifier. Through the action of its current-regulation system it forces a D.C. current Id in the D.C. buffer circuit. Bridge *N2* on the motor side operates as a line commutated inverter. When braking, the function of converters *N1* and *N2* are interchanged.

The changeover from one mode to the other, (that is from motoring to generating modes) is effected by moving the firing angle, alpha, of converter *N2* into the rectifier mode. As a result, the buffer circuit voltage, *Ud*, reverses and the current regulator of Bridge *N1* resets *N1* automatically to the inverter mode.

The current of Bridge *N2* is commutated by the motor voltage. The motor current, therefore, lags behind the voltage and the synchronous machine is overexcited. Motor voltage and motor frequency are proportional to each other for a certain motor flux. At zero frequency, the amplitude of the motor voltage is therefore zero and commutation of the current in Bridge *N2* is not possible.

For this reason, at low frequencies the motor is driven by pulse operation. Commutation is then not controlled by the motor voltage but by the supply side bridge *N1*. (For details of operating principles in both the motor controlled operation and the pulse operation, see Appendix I).

C. Synchronization

Several procedures of synchronizing a synchronous motor to the supply system are possible, depending on operating and load conditions.

.. The motor is accelerated until a preset oversynchronous speed (motor frequency higher than the system frequency) is reached. The VFSS is then disconnected and the motor speed falls according to the mechanical torque. Paralleling takes place when the conditions detected by synchronizing and voltage matching devices are met. This procedure is used when the opposing torque is small, allowing the use of field weakening to obtain a stator voltage which is a fraction of the rated voltage.

.. The motor is accelerated by the VFSS until the rated speed is almost reached. Then the automatic synchronizer takes over for the final speed adjustment. The set slip limit enables the paralleling detector to release the breaker closing order in advance, taking into account the breaker closing time. The VFSS is still driving the motor and thus paralleled with the supply system. The VFSS is only blocked after actual closing of the synchronizing breaker. This procedure is employed for a large opposing torque and requires a specialized synchronizing control.

D. Use of VFSS as Drive System

As already pointed out before, a VFSS is not only capable of starting a synchronous motor, but also of driving it at variable speed. For this continuous operation, special attention should be paid to the power factor and the harmonic content of the output voltage, which results in a certain heating of the motor windings.

For a start up system with its relatively short start up time, it is usually sufficient to oversize the excitation system. For a drive system, however, the design of the motor has also to be given special consideration. This is above all, the case, when the rating of the drive system and of the motor are about of the same order. The whole system and especially the construction of the motor have therefore to be coordinated very thoroughly.

E. Reliability

.. Brown Boveri developed and designed as early as 1936 a variable frequency drive for synchronous motors. (See reference [3]), but only the development of static power devices and integrated circuits in the late 1960's led to a breakthrough of this technique.

- .. Experience gained with more than 40 start-up and drive systems in operation (rated 0.7 MW - 30 MW), as well as several HVDC systems, substantiate the reliability of the system.

- .. Due to this long experience and a sophisticated design, the Mean Time to Failure (MTTF) is high and the Mean Time to Repair (MTTR) is low.
  This results in a high degree of availability of the VFSS. (See reference [6]).

III. USAF - ASTF PROJECT

The USAF is building an Aeropropulsion Systems Test Facility (ASTF), at their Arnold Engineering Development Center in Tennessee. When completed, this will be one of the largest engine test facilities in the world.

The ASTF engine test facility is designed to simulate actual subsonic and supersonic flight environments for aircraft propulsion testing. The ASTF will consist of an air supply unit, air heaters, a refrigeration unit, two test cells (28' in diameter and 85' long) and an exhaust unit. The compressors for the air supply and exhaust side are driven by large synchronous motors in the range of 20 MW to 40 MW. Each compressor and motor is rigidly coupled, with the motor starting under substantial load. The required starting system rating, to meet the duty cycle is 40,000 hp (30 MW).

Before specifying synchronous motors and VFSS, the USAF made a detailed economic study of eleven different drive and starting systems. (See reference [4]). The evaluation was made for a 25 year life and took into account such factors as:

- .. Detailed capital costs including the costs of the buildings for the various alternatives.
- .. Detailed operating costs.
- .. Detailed expected maintenance costs over the life.

The study considered the following combinations:

- .. Synchronous motor drives with wound rotor starting motors (2 different combinations).
- .. Synchronous motor drives with VFSS (3 different combinations).
- .. Gas turbine drives.
- .. Gas turbine/electric drives (2 different combinations).
- .. Gas/steam turbine drives.
- .. Steam turbine drives.
- .. Diesel engine drives.

While the exact system chosen did not have the lowest capital costs, it did, however, have the lowest life cycle cost. In addition, in all comparible cases, the synchronous motor/VFSS combination had the lowest capital cost for this application.

To obtain redundancy in motor starting, two similar assemblies, one for the air supply and one for the exhaust side, will be installed. Each assembly of itself, however, has to be capable of starting all 18 motors in one hour.

A. General Design, Single-Line Diagram

The USAF-ASTF specification has the following requirements:

- .. 2 similar systems, one for the air supply, the other for the exhaust side.
- .. 161 kV supply system, input transformer.
- .. 12-pulse rectifier.
- .. 24-pulse inverter.
- .. Optional output transformer.
- .. Start 18 motors, (one after the other) in one hour, repeat after one hour, with one assembly.
- .. Synchronize motors with 13.8 kV running supply system.
- .. Overcome a -10% rated speed at beginning of start.

The above mentioned requirements result in the following general design of the ASTF-VFSS and are shown in Figure 4 and discussed below.

- .. The input transformer ***T1*** has two secondary windings, one wye, one delta, each connected to a 6-pulse rectifier.
- .. The 12-pulse rectifier is formed of (2) 6-pulse rectifiers, ***N1*** and ***N2*** connected in series. By using a 12-pulse rectifier, the content of harmonic currents to the supply system is reduced considerably. (See reference [7], [9]).
- .. In the D.C. link there is a reactor ***K*** which absorbs the difference of the instantaneous voltage magnitudes between rectifier and inverter D.C. voltage and thus decouples them.
- .. A protection circuit, the free-wheeling circuit ***F***, absorbs the energy stored in the D.C. reactor under converter fault conditions and thus avoids overvoltage on the converters.
- .. The inverter consists of (4) 6-pulse bridges connected in series, ***M1, M2, M3, M4***, forming a 24-pulse system. The 24-pulse system reduces the harmonic currents to the motor and avoids overheating the motor during start-up.
- .. The output transformer is required to match and combine the (4) 3-phase inverter outputs to the 13.8 kV synchronous motor input. The transformer is split up in two transformers in one enclosure, ***T2***

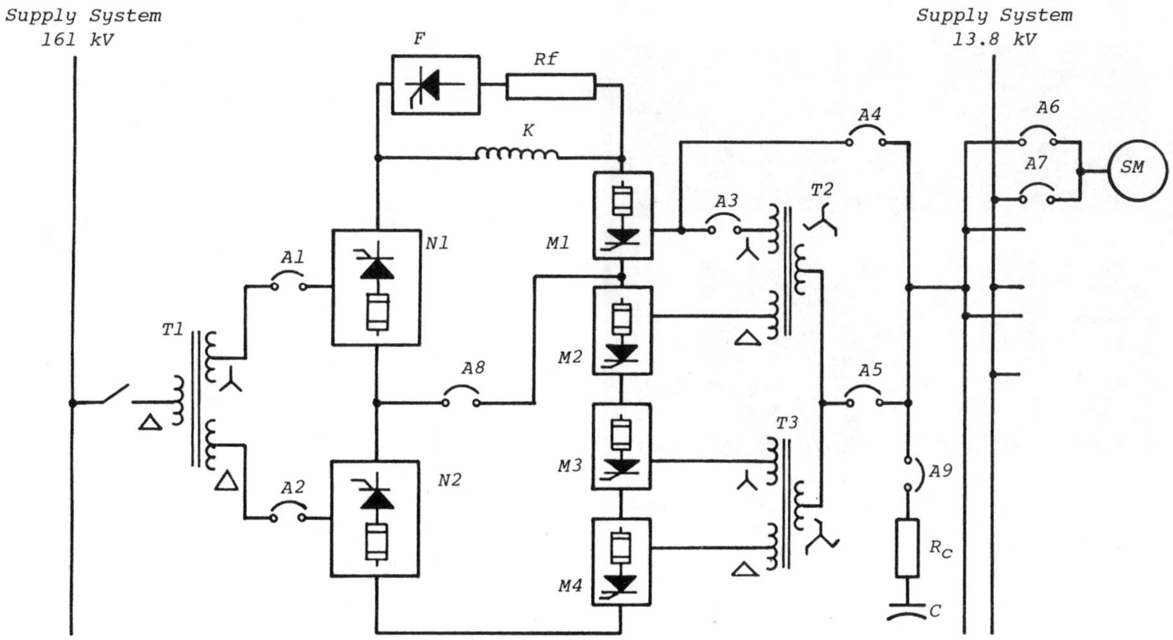

Fig. 4 - Tullahoma VFSS/Single Line Diagram

A1-A9	= *Circuit Breakers*	*F*	= *Freewheeling circuit*
T1,T2,T3	= *Transformer*	*Rf*	= *Nonlinear resistor*
N1,N2	= *Rectifier*	*C*	= *Capacitor*
M1,M2	= *Inverter*	*Rc*	= *Damping resistor*
M3,M4		*SM*	= *Synchronous motor*

& *T3*, each forming a 12-pulse system with two inverters. Each transformer has a delta and a wye-zigzag secondary winding. One zigzag has a phase shift of +7.5°, and the other of -7.5°. The two transformers connected in parallel, together with the 4 inverters, form therefore a 24-pulse system.

.. The synchronous motor cannot produce the required reactive power for commutation during low frequency operation (0 to about 6 Hz). In this frequency range, the commutation of the inverters is insured by pulse-operation (see Appendix I). The auxiliary voltage for inverter commutation is derived from a rotor position encoder mounted on the motor shaft. During the low frequency operation, the output transformer is by-passed and only 6-pulse operation is possible. (Breakers *A3* and *A5* open, breaker *A4* closed). At the same time there is also 6-pulse operation on the rectifier. (Breaker *A8* closed). When changing over from pulse operation to motor-commutation operation, breakers *A4* and *A8* are opened and breakers *A3* and *A5* are closed.

.. The capacitor bank *C* (with damping resistor *Rc*) serves as a commutating aid for the inverter. In some operation points the permissible commutation time may otherwise be exceeded due to high transformer and motor reactances.

B. Ratings and Construction

The specified ratings are a 161 kV supply system, and the required 13.8 kV, 30 MW output value.

To meet these requirements and to optimize the converter circuits, a D.C. voltage of 11.5 kV was chosen, resulting in 5180V/2200A/19.7MW per 6-pulse rectifier on the supply side and 2730V/2200A/10.4MW per 6-pulse inverter on the motor side. All converters are water cooled with a closed deionized water circuit. (See Fig. 5 & 6). Both transformers, (input and output) are oil filled, self-cooled, and rated 40 MVA. The circuit breakers are of the air-magnetic type and are rated to withstand short circuit currents occurring in worst case conditions. All connections between the different installation parts are enclosed bus duct, rated 15kV/2000A, and all equipment is metal enclosed, to meet the EMC requirements.

Fig. 5 - Tullahoma/VFSS-6-pulse Rectifier 5180 V/2200A/19.7 MW

Fig. 6 - Tullahoma/VFSS-One leg of the 6-pulse Rectifier, with impulse transformers, snubber circuit and indicating lights, water cooled thyristor stack.

C. Operating Principle/Synchronizing

Only one motor is started at the same time. Once the motor wanted is preselected, the sequence logic will start and take the motor to rated speed and synchronize it with the running system. After synchronization, the VFSS is disconnected and ready to start the next motor.

Fig. 7 & 8 show two typical start-up schematics. One motor is started following a speed ramp, that means not with full torque, but with constant acceleration (Fig. 7). The other motor is started under current limitation and full torque. (Fig. 8).

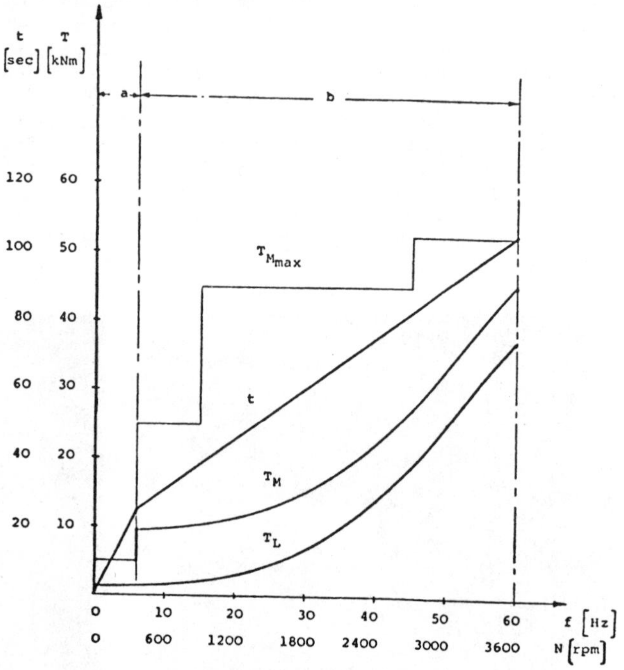

Fig. 7 - Schematic of Start Up following a speed ramp 52,000hp motor

t = *Time*
T_L = *Load torque*
T_M = *Motor torque*
a : *Pulse controlled operation*
b : *Motor controlled operation*

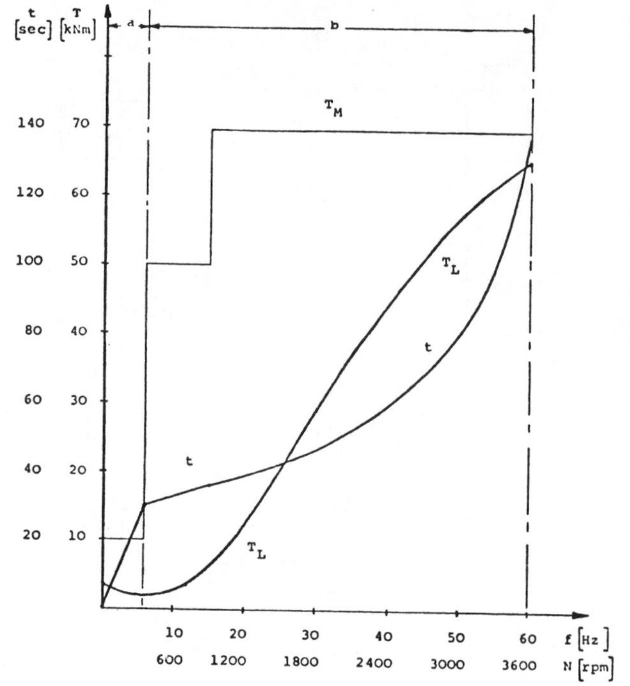

Fig. 8 - Schematic of Start Up under current limitation (Maximum VFSS torque) 44,000 hp motor

Taking into account that the motor has to be synchronized under load, the following synchronizing method has been chosen:

The motor is run up by the VFSS until rated speed is nearly reached. Then the automatic synchronizing device takes over and matches the voltage of the VFSS output to the running line voltage. As soon as the two values correspond, the slip signal, which is derived within the synchronizer, is checked, and if below a certain value, the frequency matching initiated. The frequency of the VFSS is increased by following a preset ramp, which insures that the slip remains beneath a certain value.

During all this time, the VFSS is still supplying the motor with a leading power factor and therefore paralleling has to take place at a determined angle, avoiding any paralleling with a power factor not suitable to the inverters. As soon as paralleling with the running bus is activated and the running breaker closed, the VFSS is immediately shut down by shifting the firing angle of the thyristor gate pulses.

Considering the special synchronizing conditions, manual synchronizing is prohibited. To insure a proper operation of the automatic synchronizer, two similar devices are connected in parallel. The actual synchronizing command is only given when both devices are within 5 electrical degrees.

D. Control and Protection

There are two control systems and one overall protection system; all three are autonomous but interacting with each other:

- An overall control logic, which works as an autonomous system within the VFSS, controls the programmed start sequence and its timing, the switching in and out of the circuit breakers, the converters and the excitation system, as well as the coordination between the two starting systems, the air supply and the exhaust systems. The control system is a standard device (Decontic K) which was developed for the automation of power plants.

- A regulation control, which controls strictly the converters by means of special CT's, PT's and control loops.

- Both control systems are influenced by the protection system. The protection functions include overcurrent, stator earth fault, overload, differential and others.

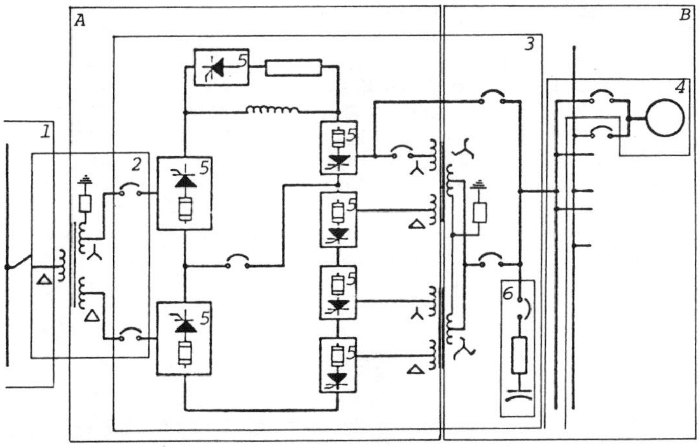

Fig. 9 - Tullahoma/VFSS - Protection Zones

1 = Bus Differential
2 = Transformer Differential
3 = Converter Differential
4 = Overcurrent/Motor
5 = Fuse and Overvoltage/Converter
6 = Time Overcurrent/Capacitor Bank
A = Earth Fault/VFSS
B = Earth fault/Motor

Fig. 9 shows the main protection zones.

The differential zones (1, 2, 3) are overlapping and thus reach from the supply system to the output of the VFSS. Each converter is protected by its own overvoltage protection and each converter phase by a fuse.

While the VFSS is in operation, the normal motor protection system is blocked, because the motor protection relays are designed for rated frequency and voltage and, at low frequencies, may give erroneous measurements and false tripping signals.

Only static type relays are used and special attention is given to the current wave form and the variable frequency. The protection system is divided into two separate subsystems, each subsystem capable of insuring the main protection functions for the VFSS by itself, and therefore, allowing the shut down of one part for testing or revision.

E. Mathematical Model

The whole starting system is programmed on a digital computer. Models have been developed for the VFSS power system, the VFSS control system and the synchronous motor with its excitation system. The different parameters can easily be changed and therefore different types of motors, regulators, and sequences optimized, and different start operations tested. The protection system can also be tested under different fault conditions.

F. Reliability

The availability and reliability of the system are held at a high level by the following precautions.

.. All electronic and power electronic parts pass a 100% incoming functional test.

.. All converters have a (n_s-1) design, that means the converter is still operable after failure of one series thyristor per leg.

.. Due to a sophisticated protection scheme with several levels of protection, it is almost impossible that a failure, inside or outside the VFSS, can result in a damage to the equipment.

SUMMARY

When comparing various start-up systems, the lower losses, higher flexibility, smaller space requirements and breaking and drive capabilities of the VFSS have to be taken into account. The VFSS is not limited in power output and reduces the mechanical part of the drive system to a minimum.

.. For large compressor drives or under load starting in general, the VFSS often is the only feasible way of starting. In such an application starting motors would need a rating in the order of the actual motor.

.. The VFSS allows the system designer to optimize the synchronous compensator to the supply network without degrading performance during start up.

.. For pumped storage plants, the VFSS combines starting, braking, and reversing capabilities in one.

.. For gas turbines, the VFSS reduces the length of the machine group considerably.

.. With the appropriate design considerations, the VFSS makes it possible to use the synchronous motor at variable speed.

On the other hand, the VFSS implies several requirements which must be coordinated with the whole system:

.. Leading power factor and harmonics, which have to be considered when designing the motor and the excitation system with reference to heating.

.. Special protection coordination with respect to D.C. components on ground fault current, variable frequency and voltage, special current and potential transformers.

APPENDIX I

A. 3-Phase, 6 Pulse Converter

Depending on the firing delay angle α, the following rectified D.C. voltages are obtained.

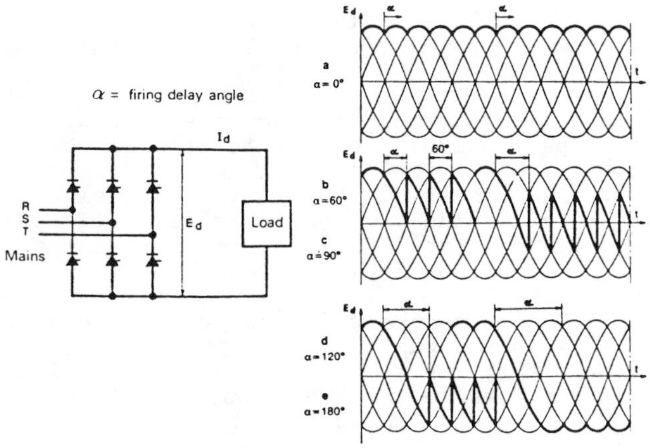

Fig. 10 - 3-Phase, 6 Pulse Converter

a,b,(c) : Rectifier Mode
(c),d,e : Inverter Mode

If α is between 0° and 90°, the converter is working as a rectifier; if α is between 90° and 180°, the converter is working as an inverter.

(For more details, see references [8], [9].

B. Operating Principle of VFSS (See reference [5]).

The firing angle of the motor side converter is chosen to obtain the best possible utilization of the frequency changer and of the motor. This is the case when current and voltage are in phase, that means with power factor cos φ = 1. The power factor on the motor side converter is however limited to approximately 0.85 owing to the commutation angle and the minimum extinction angle of the motor side converter. For proper operation, the motor voltage is used as the reference voltage for the thyristor firing sequence.

B.1 Motor Controlled Operation (See Fig. 11)

At higher frequencies, the motor itself determines the output frequency of the converter by means of its own stator voltage. This voltage also commutates the inverter. The synchronous motor is thus self-controlled and behaves exactly the same way as a separately excited D.C. motor; it can neither hunt nor slip from the impressed stator current.

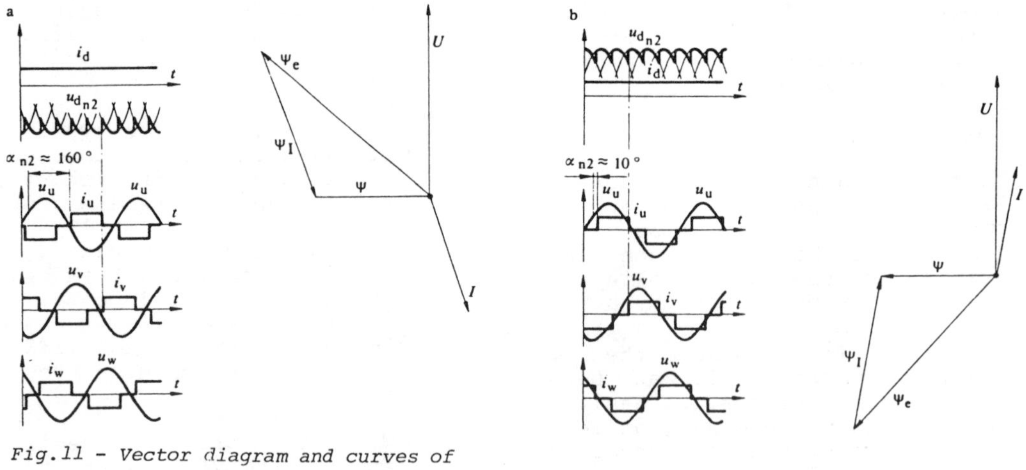

Fig.11 - Vector diagram and curves of currents and voltages when

a: Driving
b: Braking

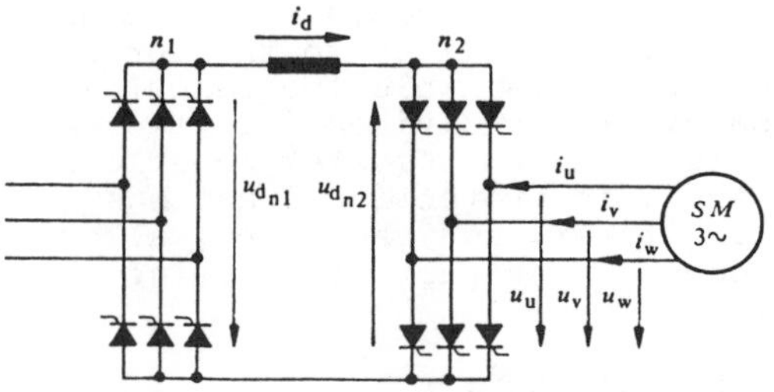

i_d	=	*Direct current in buffer circuit*
i_u, i_v, i_w	=	*Machine phase currents*
n_1, n_2	=	*Converters*
Udn1,Udn2	=	*Direct voltages in buffer circuit*
Uu,Uv,Uw	=	*Machine phase voltages*
Ψ	=	*Resultant flux in stator*
Ψe	=	*Flux in stator created by rotor current*
ΨI	=	*Flux in stator created by stator current (armature reaction)*

B.2 Pulse Operation

At very low frequencies, when the amplitude of the motor voltage is too small for proper commutation, a pulse mode of operation is employed. (See Fig. 12). At very low frequencies, a three phase auxiliary signal voltage is derived either from an external voltage or a rotor position encoder. As soon as the motor voltage can be used as a reference, the auxiliary signal voltage is derived from the motor voltage. This auxiliary voltage controls the firing pulses for the motor side converter and through a logic, the pulses of the supply side converter. Commutation in this mode is not performed by the motor, however, but by the supply side converter. Each time the current in the motor side converter is required to commutate e.g., from motor winding "u" to motor winding "v", the D.C. current is reduced to zero by the supply side converter. After a short time, which is long enough to allow the thyristors feeding "u" to regain their blocking capability, the supply side converter is again pulsed along with the thyristor feeding "v".

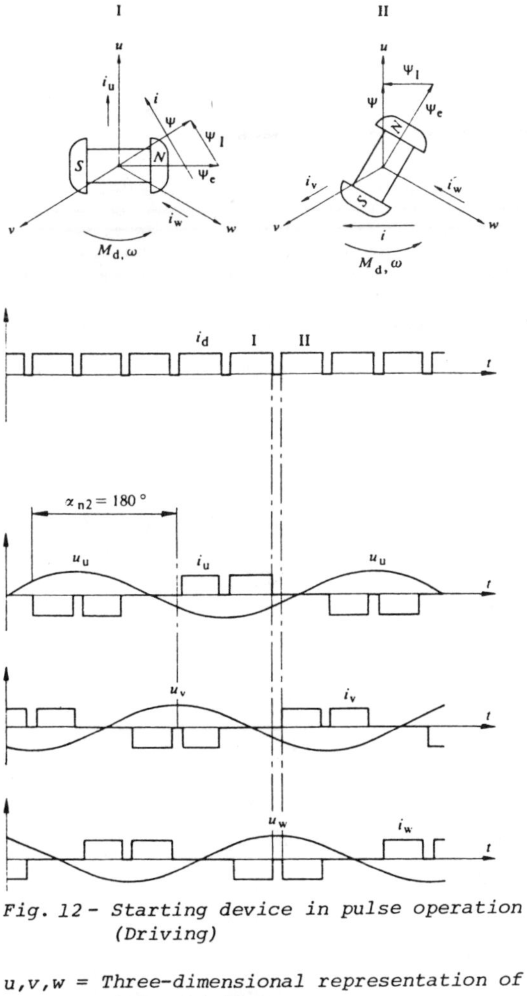

Fig. 12 - Starting device in pulse operation (Driving)

u,v,w = Three-dimensional representation of stator windings
Md = Torque of synchronous machine
w = Angular velocity

Other symbols as Fig. 7

I = Windings u and w carrying current
II = Windings v and w carrying current

C. Mathematical Formula of Average Torque (See reference [5]).

The synchronous motor develops a torque which can be described by the following formula (losses and influence of the harmonics of flux and current are disregarded):

$$Md = K \times I_d \times \frac{U}{w} \times \cos\varphi$$

Md = Average torque

K = Constant

I_d = Average current in D.C. buffer circuit

U = Voltage at terminals of synchronous motor

w = Angular velocity of synchronous motor

φ = Phase difference between current and voltage in synchronous motor.

The amount of torque is determined by the value of the direct current (I_d). The torque direction is determined by the sign selected for the power factor. The motor flux (U/w) and the power factor (cos φ) are given values which determine the optimum size and design of the VFSS. Current (I_d) is regulated by a control loop acting on the supply side converter. The limit of (I_d) is specified by the VFSS load capacity. The motor voltage is regulated in such a way, that flux is held at a constant value, that means, motor voltage varies proportional to the frequency.

D. Oscillograph of Actual Starting Sequence

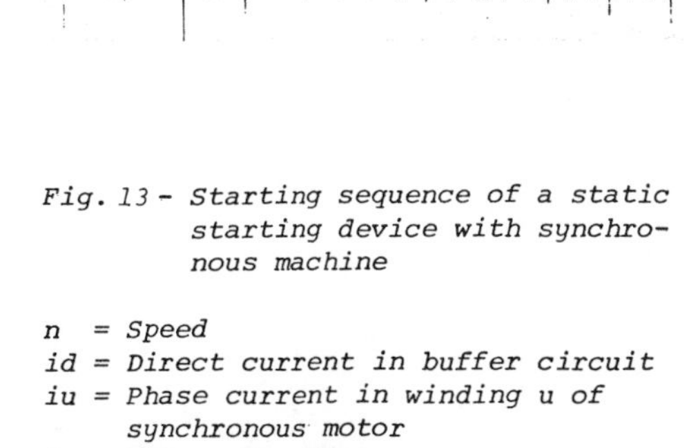

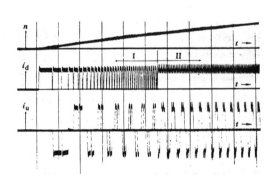

Fig. 13 - Starting sequence of a static starting device with synchronous machine

n = Speed
id = Direct current in buffer circuit
iu = Phase current in winding u of synchronous motor
I = Pulse operation
II = Motor-controlled operation

REFERENCES

[1] M. Canay, "Asynchronous starting of a 230 MVA synchronous machine in "Vianden 10" pumped storage station", Brown Boveri Rev. 61 1974 (7) pp. 313-318

[2] M. Canay, "Partial frequency starting in pumped storage plants", Brown Boveri Rev. 61 1974 (7) pp. 319-326

[3] E. Keller, "Starting synchronous machines by means of mutators with special reference to turbo-generators used as phase-advancers", Brown Boveri Rev. 23, 1936 (6) pp. 173-175

[4] DMJM/Norman Engineering Co., "ASTF, Prime Mover Study", prepared for AEDC, Tullahoma, Tn., Vol. III-3.2/June 1973

[5] F. Peneder, R. Lubasch, A. Vonmard, "Static equipment for starting pumped-storage plant, synchronous condensers and gas turbine sets", Brown Boveri Rev. 61 1974 (9/10) pp. 440-447

[6] K. Gamlesaeter and F. Peneder, "Large Synchronous Compensators with Static Excitation and Starting System", Brown Boveri Rev. 65 1978 (5) pp. 303-311

[7] A. Kusko, "Power Electronics Applications in Power Systems", IEEE paper EH0135-4/78/0000-0027, presented at the 1979 IEEE Winter Power Meeting, New York, N.Y.

[8] V.R. Stefanovic, "Thyristor Rectifiers and Synchronous Inverters", IEEE paper EH0135-4/78/0000-0005, presented at the 1979 IEEE Winter Power Meeting, New York, N.Y.

[9] "Silizium Stronrichter Handbuch", BBC Baden, Switzerland, 1971

[10] F. Peneder and V. Suchanek, "Standardized Starting Equipment", Brown Boveri Rev. 65, 1978 (3) pp. 607-613

Utilization and Rating of Machine Commutated Inverter-Synchronous Motor Drives

JOHN ROSA

Abstract—For a dc-link type variable frequency inverter-synchronous motor drive, the relationship between the operation of the machine commutated thyristor inverter and the characteristics of the synchronous motor are analyzed. The factors of maximum motor utilization, the required rating of the machine-side converter (inverter), and the rating and input power factor of the line-side converter (rectifier) are established. The effect of four different firing angle control principles applied to the machine converter on rating and performance is analyzed.

I. INTRODUCTION

AMONG solid-state variable-speed ac-motor drives machine commutated inverter-synchronous motor drives constitute a special group. In these drive systems the thyristor inverters, which feed the motors, operate with natural commutation, made possible by the electromotive force (EMF) of the synchronous machine. Consequently, these inverters do not require the complex and costly force commutating circuits which form an integral part of solid-state variable-speed induction motor drives. Since, however, operation of the inverter is tied to the behavior of the synchronous motor, the design and utilization of such systems are dependent on certain motor characteristics.

This paper deals with the relationship between inverter operation and machine characteristics and analyzes the factors of system utilization and rating with the objective of maximizing the economic benefits offered by this type of drive system.

The result is a set of equations which can be used to calculate the maximum power obtainable from a given motor (characterized by its nameplate data) and the necessary operating criteria to extract the same. The required rating of the machine-side converter (inverter) and the rating and input power factor of the line-side converter (rectifier) to handle maximum power can be calculated. The effect on utilization and rating of four alternative firing angle control schemes is also taken into account. A numerical example illustrates the use of the outlined method.

II. CONVERTER-MACHINE INTERFACE

The basic circuit diagram of the system is shown in Fig. 1. The line converter, together with inductor L, acts as a dc current source. Its output I_{dc} is impressed at the dc input of the

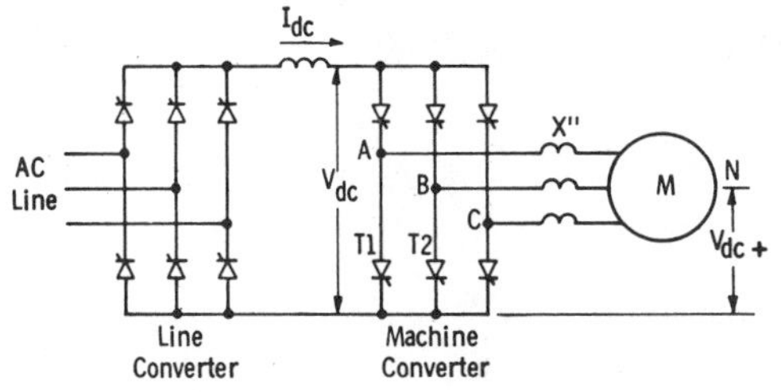

Fig. 1.

machine converter operating as an inverter (assuming motoring operation). Natural commutation of the machine converter is made possible by the fact that its ac terminals are connected to the synchronous machine, which is "seen" by the converter as a three-phase ac source of terminal voltage V (line-to-neutral) and source reactance X'' (per phase). The commutation process will be discussed later. First, it is necessary to describe the interaction between the converter and the synchronous machine. At this point, it suffices to know that dc link current I_{dc} is cyclically distributed into the three ac lines in the form of rectangular ac currents of I_{dc} amplitude. The fundamental components I of these three ac line currents are 120° displaced; they lead their respective V line-to-neutral motor terminal voltages by displacement angle $\beta' + 180°$. The line currents which include harmonic components of $h = 6k \pm 1$ order ($k = 1, 2, 3, \cdots$) result in voltage drops across the machine reactances; consequently, the loaded terminal voltage of the machine differs from its no-load terminal voltage E. The problem of determining the loaded terminal voltage is somewhat complicated by the fact that different currents "see" different reactances: the voltage drop caused by the fundamental line current I is IX_s where X_s is the machine's synchronous reactance, the voltage drop caused by a harmonic line current I_h is $I_h(\omega_h/\omega)\,X''$ where X'' is the machine's subtransient reactance at the fundamental ω frequency; and ω_h is the given harmonic frequency. The corresponding equivalent circuit diagram is shown in Fig. 2(a). The bandpass filter across X'' permits I to bypass the subtransient reactance and the high-pass filter across X_s permits the harmonic currents to bypass the synchronous reactance. This diagram is too cumbersome for our purpose because it requires reconstruction of the terminal voltage from fundamental and harmonic components.

In a more convenient approach, the harmonic components are handled together with the fundamental. In order to permit this, the diagram is rearranged as shown in Fig. 2(b). Since the fundamental component I now also flows through X'', the re-

Paper SPCC 78-1, approved by the Static Power Converter Committee of the IEEE Industry Applications Society for presentation at the 1978 Industry Applications Society, Toronto, ON, Canada, October 1-5. Manuscript released for publication October 24, 1978.

The author is with the Power Electronics Laboratory, Electronics Research and Development Division, Westinghouse Electric Corporation, 1310 Beulah Road, Pittsburgh, PA 15235.

Reprinted from *IEEE Trans. Ind. Appl.*, vol. IA-15, pp. 155–164, Mar./Apr. 1979.

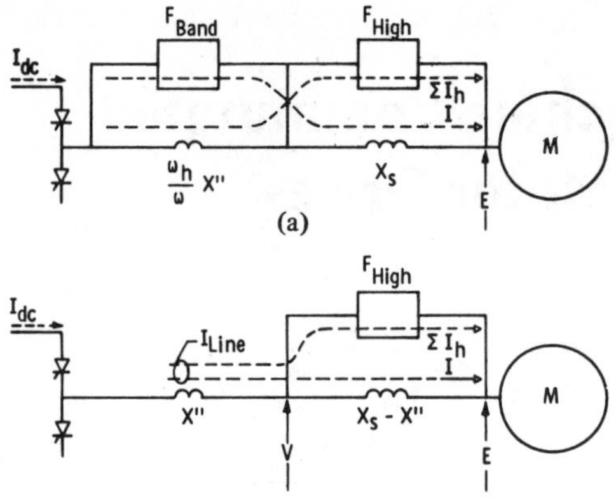

(b)

Fig. 2.

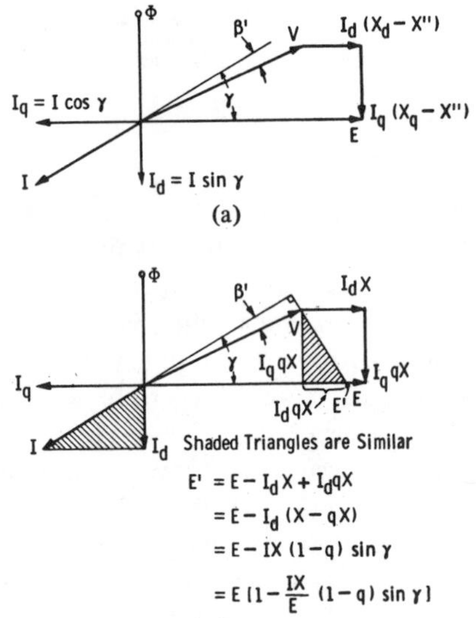

(b)

Fig. 3.

actance carrying I alone must be reduced to $X_s - X''$. Thus the total effect of I is unchanged, and the harmonic currents are combined with the fundamental in their path through the subtransient reactance.

With this equivalent circuit diagram, the fundamental current I results in a sinusoidal voltage drop $I(X_s - X'')$ across the "adjusted" synchronous reactance resulting in sinusoidal voltage V "behind the subtransient reactance," where the interface between machine and converter can be assumed to lie.

As described in greater detail in [1] and illustrated by the vector diagram of Fig. 3(a), loaded terminal voltage V (line-to-neutral) can be derived from no-load terminal voltage E (line-to-neutral), which in turn is induced by pole flux ϕ. The assumption is made that the resistive voltage drop IR across the winding resistance R of the machine is negligible in comparison with the reactive voltage drop across the machine's synchronous reactance.

With this assumption, terminal voltage V is sinusoidal and can be obtained as the vector difference between E (lagging 90° behind pole flux ϕ) and reactive voltage drop $I_d(X_d - X'')$ and $I_q(X_q - X'')$, each leading the respective current component I_d and I_q by 90°. I_d and I_q are the direct and quadrature axis components of I with respect to ϕ (direct axis). X_d and X_q, the direct and quadrature axis synchronous reactances, are machine constants. In the analysis that follows, it will be further assumed that the direct and quadrature axis values of X'' are identical. For a given pole excitation and motor speed, ϕ and E are also constant. For a given line current I, leading terminal voltage V at a given supplementary displacement angle β', V will assume a certain amplitude and lead E by angle $\gamma - \beta'$, where angle γ is the supplementary displacement angle of I with respect to E.

The relationship between V, E, I_d, I_q, X_d, and X_q is described by

$$V^2 = [E - I_d(X_d - X'')]^2 + [I_q(X_q - X'')]^2. \tag{1}$$

Using notations

$$X_d - X'' = X \qquad X_q - X'' = qX \tag{2}$$

and

$$I_d = I \sin\gamma \qquad I_q = \cos\gamma, \tag{3}$$

the ratio of V to E is given by

$$\frac{V}{E} = \left[1 - 2\frac{IX}{E}\sin\gamma + \left(\frac{IX}{E}\right)^2 \cdot (\sin^2\gamma + q^2\cos^2\gamma)\right]^{1/2}. \tag{4}$$

The relationship between angles γ and β' is also defined by the vector diagram:

$$\cos(\gamma - \beta') = \frac{E - IX\sin\gamma}{V}$$

$$\sin(\gamma - \beta') = \frac{qIX\cos\gamma}{V}. \tag{5}$$

Solving these equations for β' and by substituting (4), we get

$$\cos\beta' = \frac{E}{V}\cdot\left[1 - \frac{IX}{E}\cdot(1-q)\sin\gamma\right]\cdot\cos\gamma = \frac{E'}{V}\cos\gamma$$

$$\cos\beta' = \frac{\left[1 - \frac{IX}{E}(1-q)\sin\gamma\right]\cos\gamma}{\left[1 - 2\frac{IX}{E}\sin\gamma + \left(\frac{IX}{E}\right)^2(\sin^2\gamma + q^2\cos^2\gamma)\right]^{1/2}}. \tag{6}$$

Equation (6) introduces a new voltage vector E':

$$E' = E\left[1 - \frac{IX}{E}(1-q)\sin\gamma\right]. \tag{7}$$

The geometrical and electrical meaning of E' is illustrated in Fig. 3(b). The vector diagram represents a salient pole machine, in which $X_d > X_q$; thus $q < 1$. E' is derived from terminal voltage V with the assumption that both the direct and quadrature axis synchronous reactances have the value X_q. The vector E' thus constructed satisfies two requirements.

i) It satisfies the power equation $P_{ac} = 3\ VI\ \cos\beta' = 3\ E'I\cos\gamma$.

ii) The vector difference between E' and V is perpendicular to I, which it should be since this difference is a reactive voltage drop. It is therefore E' that qualifies as "the voltage behind the synchronous reactance". The introduction of E' helps to extend the analysis to salient pole machines as well.

The ac and dc side relationships of the converter and the commutation phenomena are discussed in conjunction with Fig. 4, which shows a commutation event from thyristor $T1$ in line A to thyristor $T2$ in line B (see also the circuit diagram in Fig. 1). The instant of initiating the commutation is defined by "firing angle" α, or the supplementary "commutation lead angle" β. Prior to the instant defined by these angles, thyristor $T1$ is conducting $I_{A\ \text{line}}$ current in line A. The amplitude of this current equals the amplitude of dc link current I_{dc}. When thyristor $T2$ is fired, conduction in line B is initiated, and commutation from $T1$ to $T2$ starts. Due to the presence of "source" reactance X'', however, conduction in $T1$ and in line A does not stop immediately. During the interval defined by "commutation overlap angle" u, the current in $T2$ and in line B gradually rises to the level of I_{dc}, while the current in $T1$ and in line A correspondingly decreases from the level of I_{dc} to zero, forced by the voltage difference $V_{AB} = V_{A-N} - V_{B-N}$. At any moment $I_{A\ \text{line}} + I_{B\ \text{line}} = I_{dc}$. At the instant defined by "turn-off angle" δ, $T1$ ceases to conduct and becomes reverse-biased ($V_{T1} = V_{AB}$ is negative) for the duration defined by angle δ. Thus although the commutation was initiated at lead angle β, the time of recovery available to the thyristor prior to supporting forward voltage is defined by angle δ; the remaining time is taken up by the commutation overlap defined by angle u.

The relationship between angles β, u, δ, and quantities V, V_{dc}, I_{dc}, X'' are found in [2].

$$V_{dc} = \frac{3}{\pi}(\sqrt{6}V\cos\beta + I_{dc}X'') \tag{8}$$

$$\cos(\beta - u) = \cos\delta = \cos\beta + \sqrt{\frac{2}{3}}\cdot\frac{I_{dc}X''}{V}. \tag{9}$$

Thus the converter operation is defined by β, whereas the machine terminal voltage depends on β' or γ. In order to define the interaction between the converter and the synchronous machine, it is necessary to find the relationship between commutation lead angle β and the corresponding supplementary phase displacement angle γ. This relationship can be derived from the observation that the dc power $P_{dc} = V_{dc}\cdot I_{dc}$ fed to the inverter must equal the real ac power $P_{ac} = 3E'I\cdot\cos\gamma$ fed to the motor. (The losses of the converter are negligible and are ignored). The relationship between V and V_{dc} is defined by (8). I is the fundamental component (rms value) of the rectangular wave ac line current of amplitude I_{dc}:

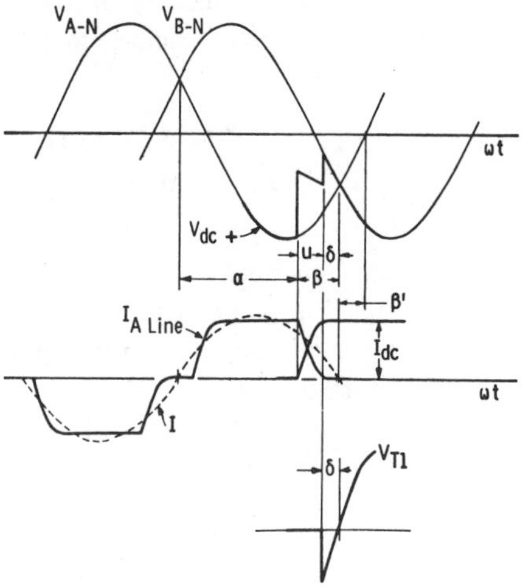

Fig. 4.

$$I = \frac{\sqrt{6}}{\pi}I_{dc}. \tag{10}$$

Thus with the use of (8) and (10),

$$P_{dc} = 3VI\cos\beta + \frac{\pi}{2}I^2X'' = 3E'I\cos\gamma = P_{ac}. \tag{11}$$

Solving for $\cos\beta$,

$$\cos\beta = \frac{\left[1 - \frac{IX}{E}(1-q)\sin\gamma\right]\cos\gamma - \frac{\pi}{6}\cdot\frac{IX''}{E}}{\frac{V}{E}}. \tag{12}$$

With notation,

$$X'' = pX, \tag{13}$$

and substituting (4) for V/E, we get

$$\cos\beta = \frac{\left[1 - \frac{IX}{E}(1-q)\sin\gamma\right]\cos\gamma - \frac{\pi}{6}p\frac{IX}{E}}{\left[1 - 2\frac{IX}{E}\sin\gamma + \left(\frac{IX}{E}\right)^2(\sin^2\gamma + q^2\cos^2\gamma)\right]^{1/2}}, \tag{14}$$

which states the relationship between the commutation lead angle β and the resultant phase displacement angle γ for a given normalized line current IX/E. Substituting (14) into (9), the relationship between γ and turn-off angle δ is also obtained:

$$\cos\delta = \frac{\left[1 - \frac{IX}{E}(1-q)\sin\gamma\right]\cos\gamma + \frac{\pi}{6}p\frac{IX}{E}}{\left[1 - 2\frac{IX}{E}\sin\gamma + \left(\frac{IX}{E}\right)^2(\sin^2\gamma + q^2\cos^2\gamma)\right]^{1/2}}. \tag{15}$$

It is interesting to note the symmetry between (14), (6), and (15), which define the converter machine interface.

III. CALCULATION OF POWER OBTAINABLE FROM THE MACHINE

Assuming an ideal lossless machine, the mechanical power obtainable from the machine is the same as the electrical power fed to the machine, expressed in (11). It is convenient to calculate this power in normalized form as per units of the machine's short circuit "power":

$$P_{\mathrm{sh}} = \frac{3E^2}{X}. \tag{16}$$

The normalized machine power then becomes

$$\frac{P}{P_{\mathrm{sh}}} = \frac{3E'I \cos\gamma}{3\dfrac{E^2}{X}} = \frac{IX}{E} \cdot \frac{E'}{E} \cos\gamma. \tag{17}$$

To maximize the power obtainable with a given current, angle γ should be chosen as small as possible. γ, however, must not be less than the value defined by (15) in which the value of turn-off angle δ is allowed the minimum safe value for the given design. Thus for any normalized current IX/E, angle γ can be calculated by iteration from (15), which will result in maximum power at that current while also satisfying the commutation safety requirements. This maximum power can be calculated after substituting the value of γ into (17). Note that E'/E is also a function of γ, unless $q = 1$ (see (7)).

Fig. 5(a) shows a plot of γ versus IX/E for a machine having a cylindrical rotor ($q = 1$) and a subtransient reactance ratio $X''/X = p = 0.2$. The turn-off angle is kept at $\delta = 15°$. The corresponding normalized power P/P_{sh} is plotted against IX/E normalized current in Fig. 5(b). It can be seen that with increasing current, the power first increases and then declines, after having reached a maximum value of $(P_M/P_{\mathrm{sh}}) = 0.295$ at $(IX/E)_M = 0.5$. (The subscript M is used hereon to label quantities associated with the maximum power).

In Fig. 6, P_M/P_{sh} is plotted against subtransient reactance ratio p, with turn-off angle δ as a parameter and with $q = 1$. This graph clearly illustrates the effect of the subtransient reactance and of the required recovery angle on the maximum power obtainable from the machine.

Calculations made for salient pole machines show that there are only slight differences in obtainable maximum power and the current value at which this power is available is the same.

Equation (17) and the resultant graphs, such as the one shown in Fig. 5(b), are generally valid, regardless of the state of excitation of the machine. In a specific case, it is necessary to establish the decrease in rating of a given machine when fed from an inverter. For this purpose, we shall calculate the machine's maximum power handling ability in inverter operation and under nominal excitation and express it in per units of nominal, rated power, terminal voltage, and reactance (which are given for unity power factor operation on the nameplate):

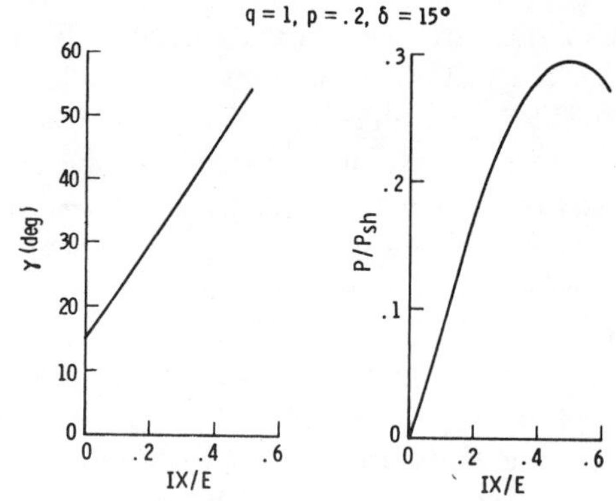

(a) (b)

Fig. 5.

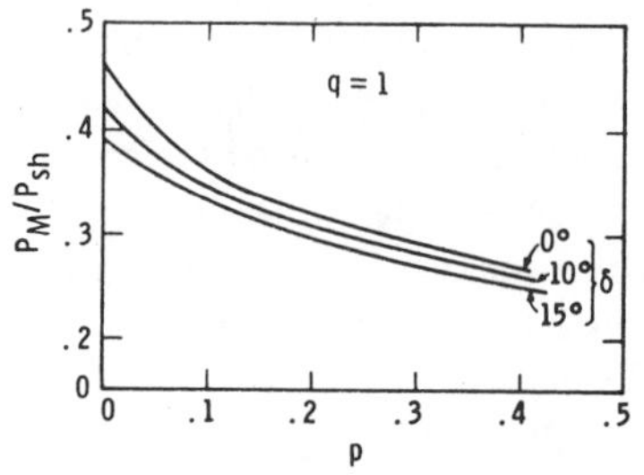

Fig. 6.

$$\frac{P_{Mn}}{P_n} = \frac{P_{Mn}}{P_{\mathrm{shn}}} \cdot \frac{P_{\mathrm{shn}}}{P_{\mathrm{shn}}'} \cdot \frac{P_{\mathrm{shn}}'}{P_n}. \tag{18}$$

The quantity P_{shn}' is the machine's "short circuit power" obtained with the original X_d reactance. Thus

$$\frac{P_{\mathrm{shn}}}{P_{\mathrm{shn}}'} = \frac{3\dfrac{E_n^2}{X}}{3\dfrac{E_n^2}{X_d}} = \frac{X_d}{X_d - X''} \tag{19}$$

and

$$\frac{P_{\mathrm{shn}}'}{P_n} = \frac{3\dfrac{E_n^2}{X_d}}{3V_nI_n} = \frac{\left(\dfrac{E_n}{V_n}\right)^2}{\dfrac{I_nX_d}{V_n}}. \tag{20}$$

(Note that subscript n indicates the given variable's value under nameplate conditons.)

To calculate E_n/V_n, (5) is used; sin γ and cos γ are calculated from Fig. 3; keeping in mind that for unity power factor $\beta' = 0$ and that the reactance is X_d rather than X,

$$\frac{E_n}{V_n} = \frac{1 + s\left(\dfrac{I_nX_d}{V_n}\right)^2}{\sqrt{1 + \left(s\dfrac{I_nX_d}{V_n}\right)^2}} \tag{21}$$

and

$$\frac{P_{shn}'}{P_n} = \frac{\left[1 + s\left(\frac{I_n X_d}{V_n}\right)^2\right]^2}{\left[1 + \left(s\frac{I_n X_d}{V_n}\right)^2\right]\frac{I_n X_d}{V_n}}. \tag{22}$$

Substituting different $I_n X_d/V_n$ and $sI_n X_d/V_n$ normalized direct axis and quadrature axis synchronous reactance values into (22), one finds that the P_{shn}'/P_n ratio is virtually constant (= 2) over a wide range of reactances. Thus for the condition of nominal excitation and speed, the approximate (within about 5 percent) maximum theoretically available normalized power obtained from (17) can be converted to per unit values of the nameplate rating by multiplication with 2 and with the $X_d/(X_d - X'')$ ratio.

In the more general case, the P/P_{sh} normalized power (maximum or not) at any excitation and speed can be converted to per unit power as indicated by

$$\frac{P}{P_n} = \frac{P}{P_{sh}} \cdot \frac{P_{sh}}{P_{shn}'} \cdot \frac{P_{shn}'}{P_n} = \frac{P}{P_{sh}} \cdot \frac{P_{shn}'}{P_n} \cdot \left(\frac{E}{E_n}\right)^2 \cdot \frac{X_d}{X_d - X''} \tag{23}$$

where the factors are the normalized power, the nominal short circuit power-to-nominal power ratio (see 21), the square of the actual-to-nominal no-load voltage ratio, and the ratio of direct axis to "adjusted" synchronous reactance. Similarly, the IX/E normalized current can be converted to per unit current as defined by

$$\frac{I}{I_n} = \frac{IX}{E} \cdot \frac{E}{E_n} \cdot \frac{E_n}{V_n} \cdot \frac{V_n}{I_n X_d} \cdot \frac{X_d}{X_d - X''}. \tag{24}$$

The factors are the normalized current, the actual-to-nominal no-load voltage ratio, the nominal excitation voltage-to-terminal voltage ratio (see (21)), the reciprocal of per unit direct axis synchronous reactance, and the ratio of direct axis to "adjusted" synchronous reactance. The V/E normalized voltage (see (4)) can be translated to per unit voltage:

$$\frac{V}{V_n} = \frac{V}{E} \cdot \frac{E}{E_n} \cdot \frac{E_n}{V_n} \tag{25}$$

with factors described above.

IV. FIRING ANGLE CONTROL OF THE MACHINE CONVERTER

The function of the firing angle control is to maintain turn-off angle δ at a safe value so that $\delta \geqslant 180\omega t_{off}/\pi$, where t_{off} is the turn-off time deemed safe for the given thyristor and ω is the machine's angular frequency at the given operating speed. Four plausible approaches of firing angle control will be investigated.

A. δ = constant

If δ is maintained at the minimum safe value compatible with the given thyristor, the machine will operate at the highest possible power factor and yield the highest possible power with the given current. Thus a control scheme which always maintains δ at the permissible minimum level results in the most efficient use of the machine.

Implementing a firing angle control which maintains δ = constant at a given speed, however, is cumbersome because δ is a quantity that "happens" after the firing (at some angle β) and the commutation overlap interval, the angle u of which is a dependent variable itself as shown by (9). Thus a δ = constant type control would have to operate by prediction, based on sensing the actual δ obtained previously.

B. β = constant

If commutation lead angle β is kept constant at a value such that the resultant δ turn-off angle is at the safe minimum level when the machine is delivering maximum power, then the turn-off angle at lighter loads will be larger and thus fall in the safe range. The required value of $\beta = \beta_M$ can be readily calculated from (14), after substitution of $(IX/E)_M$ and γ_M.

Sensing of β generally requires the availability of a signal representing voltage V. Since this voltage is "behind the subtransient reactance," it will have to be either "reconstructed" from the converter ac terminal voltage or obtained by filtering same.

C. ϵ = constant

For the convenience of analysis, the instant of commutation has been arbitrarily defined by angle β measured from the 330° point of the V sinusoid. To implement the firing angle control, the instant of commutation could equally be well defined by an angle ϵ, related to the rotor position. The relationship between β "electrical firing angle" and ϵ "mechanical firing angle" is simply

$$\epsilon = \beta - \beta' + \gamma. \tag{26}$$

Again, the required value of ϵ is determined on the basis that the resultant δ turn-off angle is at the safe minimum level when the machine is delivering maximum power. At lighter loads, the turn-off angle will be larger and thus fall in the safe range. The required value of $\epsilon = \epsilon_M$ can be readily calculated from (26), after substituting γ_M into (26) and γ_M and $(IX/E)_M$ into (6) and (14) to obtain β_M' and β_M.

Sensing of ϵ requires a rotor position transducer which delivers a pulse to the firing angle control at the six ϵ angles associated with the three phases. Since ϵ is measured in electrical degrees, there must be six pick-up points associated with each pole pair.

D. V_{dc}/E = constant

As shown by (8), the V_{dc} voltage developed at the machine converter's dc terminal is a measure of electrical firing angle β. A firing angle control which is based on the principle of maintaining V_{dc}/E at a constant level is described by the following relationship. From (8),

$$\frac{V}{E} \cdot \cos \beta = \frac{\pi}{3\sqrt{6}} \cdot \frac{V_{dc}}{E} - \frac{\pi}{6} p \frac{IX}{E}. \tag{27}$$

By substituting the product of (4) and (14) for $(V/E) \cos \beta$, we get the relationship between angle γ and normalized dc voltage V_{dc}/E:

$$\frac{V_{dc}}{E} = \frac{3\sqrt{6}}{\pi} \cdot \left[1 - \frac{IX}{E}(1-q) \sin \gamma \right] \cos \gamma. \tag{28}$$

It is V_{dc}/E that is kept constant according to this firing angle control technique. In order to ensure minimum permissible turn-off time at maximum power, this constant must be at $(V_{dc}/E)_M$ level, also resulting in γ_M angle. It can be seen that in a cylindrical rotor machine ($q = 1$), the value of γ remains constant, regardless of the current IX/E. In a salient pole machine, γ will slightly but monotonously increase from value γ_M as the current decreases from $(IX/E)_M$ to 0.

To implement this control would generally require a smooth signal proportional to V_{dc}. To provide this presents no particular technical problem, except that again (like in the case of δ = constant) control of the firing angle is carried out by "prediction" based on sensing (with some time delay due to filtering) the previously obtained V_{dc}.

Figs. 7-9 illustrate the effect of the firing angle control schemes on the different quantities associated with the converter-motor operation of a cylindrical pole machine having $q = 1, p = 0.2$, with the minimum turn-off angle at maximum power set to $\delta_M = 15°$. A brief study of Figs. 7-9 leads to the following observations.

Fig. 7(a) shows that the power versus current relationship is linear in the V_{dc} = constant scheme and becomes increasingly nonlinear in this order: ϵ = constant, β = constant, δ = constant. More importantly, for a given $IX/E < (IX/E)_M$ normalized current, the δ = constant scheme yields the highest power. The other schemes utilize the machine less effectively in this order: β = constant, V_{dc} = constant, ϵ = constant.

Fig. 7(b) shows that, as one might expect, the δ = constant scheme maintains the turn-off angle at a fixed value as the current decreases from $(IX/E)_M$ to 0. The other schemes result in an increasingly generous safety margin of the turn-off angle in this order: β = constant, V_{dc} = constant, ϵ = constant.

Fig. 8 shows the effect of the different control techniques on current displacement angles β' and γ over the $(IX/E)_M$ to zero current range. The smallest angles (and thus, the highest power factor) at reduced loads are produced by δ = constant. The other schemes become increasingly poorer in this order: β = constant, V_{dc} = constant, ϵ = constant.

Fig. 9 shows that, according to its defined principle, the V_{dc} = constant scheme maintains V_{dc} constant over the $(IX/E)_M$ to zero current range. The ϵ = constant scheme results in a decreasing dc voltage, the other schemes result in increasingly higher dc voltage at reduced loads, in this order: β = constant and δ = constant.

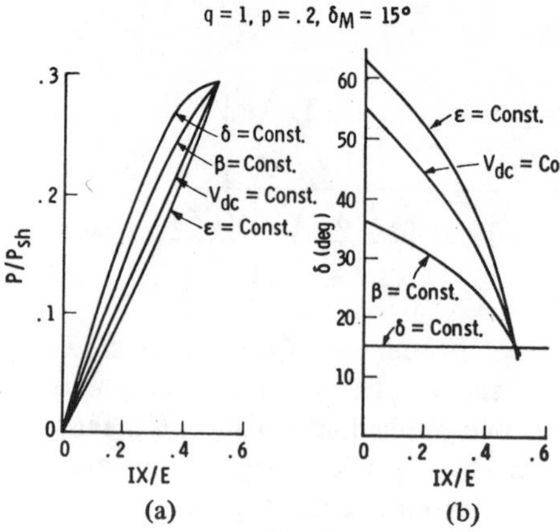

Fig. 7.

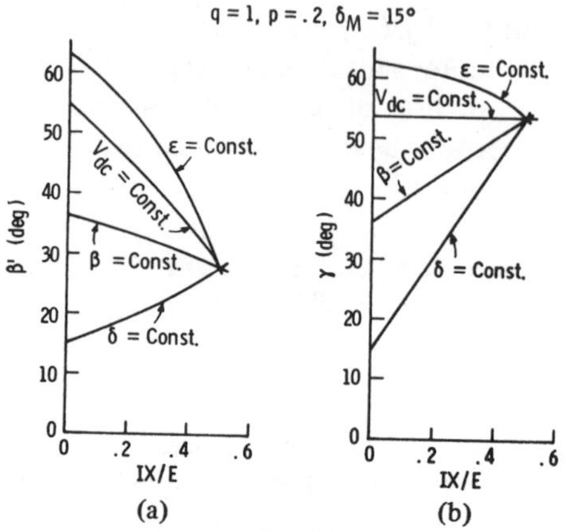

Fig. 8.

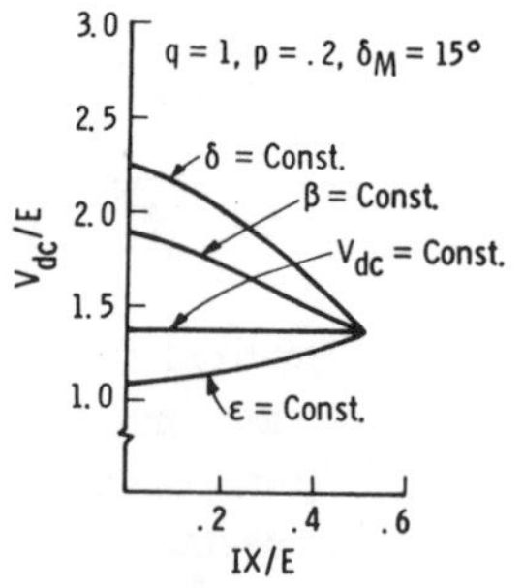

Fig. 9.

V. THE RATING OF THE MACHINE CONVERTER

The normalized power rating of the machine converter is defined as the ratio of the maximum fundamental ac volt-ampere versus the motor short circuit volt-ampere at highest excitation.

$$\frac{P_{mc}}{P_{sh}} = \frac{3I_M E_0}{3\frac{E_M{}^2}{X}} = \left(\frac{IX}{E}\right)_M \cdot \frac{E_0}{E_M}. \tag{29}$$

I_M is the highest ac current the converter delivers (i.e., the current at maximum motor power) and E_0 is the highest ac voltage the converter is exposed to (i.e., the terminal voltage at no-load). If the machine excitation remains fixed during operation, the E_0/E_M term is unity. This obviously results in a serious penalty in converter rating which can be avoided if the excitation is controlled to keep V constant over the entire $(IX/E)_M$ to 0 current range. Since the V/E ratio is $(V/E)_M$ at the maximum power and is unity at 0 current, keeping V constant via excitation control means gradually reducing E in the same $(V/E)_M$ proportion; while the current decreases from $(IX/E)_M$ to 0, E is reduced from E_M to $E_0 = E_M(V/E)_M$. If such control is applied, the value of the E_0/E term in (29) becomes $(V/E)_M$.

It should be noted that E itself becoming a variable if excitation control is applied does not affect the validity of the dimensionless expressions derived previously. It simply means that in the normalized terms IX/E, V/E, etc., E is not a fixed quantity, and in order to convert these terms to amperes and volts, the appropriate value of E must be used. Luckily, at the two extremes of the IX/E range, even this is unnecessary. At 0 current, $IX/E = 0$, $V/E = 1$, and at maximum power $IX/E = (IX/E)_M$, $V/E = (V/E)_M$, regardless of the value of E. At the maximum power point E has the value selected to obtain the desired power. If no excitation control is applied, this value is maintained over the entire load range; with excitation control, E is gradually decreased with decreasing load, as discussed above.

VI. THE RATING OF THE LINE CONVERTER

The normalized power rating of the line converter can be expressed as

$$\frac{P_{1c}}{P_{sh}} = \frac{I_{dcM} V_{dc\max}}{3\dfrac{E_M{}^2}{X}} = \frac{\pi}{3\sqrt{6}} \cdot \left(\frac{IX}{E}\right)_M \cdot \left(\frac{V_{dc}}{E} \cdot \frac{E}{E_M}\right)_{\max}. \tag{30}$$

I_{dcM} is the highest dc link current (experienced at maximum power). $V_{dc\max}$ is the highest dc input voltage to the machine converter. Ignoring the voltage drop across the dc line reactor, the same voltage appears on the dc side of the line converter. $V_{dc\max}$ may or may not occur at maximum load, depending on the selected firing angle control and excitation control schemes. With fixed excitation, $E = E_M =$ constant, and the maximum value of the term in parentheses in (30) is reached where V_{dc}/E is maximum. As was shown in Fig. 9, with the δ = constant or β = constant firing angle control schemes, the maximum of V_{dc}/E occurs at $IX/E = 0$ and its value is $(3\sqrt{6}/\pi)\cos\delta$ and $(3\sqrt{6}/\pi)\cos\beta$, respectively. With the ϵ = constant scheme, the maximum of V_{dc}/E most likely occurs at $(IX/E)_M$ and can be calculated with (28). With very low values of p (see (13)), however, the maximum may occur at $IX/E = 0$ in which case its value is $(3\sqrt{6}/\pi)\cos\epsilon_M$. With the V_{dc} = constant firing angle scheme $V_{dc}/E = (V_{dc}/E)_M$ over the entire load range and can be calculated with (28).

With excitation control (which maintains V constant at V_M level), $V_{dc\max}$ is reached when $V_{dc}/E_M = (V_{dc}/E)\cdot(E/E_M)$ product reaches maximum. Checking the two extremes of the load range will give an indication on the location and value of the maximum. At $(IX/E)_M$ maximum load $V_{dc}/E_M = (V_{dc}/E)_M$ which can be calculated from (28). At $IX/E = 0$, $E/E_M = (V/E)_M$ obtained from (4) and V_{dc}/E is obtained from (28) with substitution of $IX/E = 0$ and $\gamma = \delta_M, \beta_M$ or ϵ_M (depending on the type of firing angle control used in the machine converter). For the V_{dc} = constant firing angle scheme, the value V_{dc}/E_M is obviously constant over the entire load range.

Thus the line converter rating is affected by the selected firing angle control scheme applied to the machine converter and, eventually, by the absence or presence of machine excitation control.

VII. THE INPUT DISPLACEMENT FACTOR OF THE LINE CONVERTER

In order to operate the machine at maximum speed over the entire load range, the line converter must produce a dc voltage V_{dc} ranging from $V_{dc\max}$ to some $V_{dc\min}$. This is accomplished by firing angle control of the line converter which in turn manifests itself in a variable displacement angle of the ac line current. In the ideal case, when $V_{dc\max}$ dc output voltage of the line converter is obtained at 0° firing angle, the displacement factor at a given V_{dc} output voltage is

$$\cos\phi = \frac{V_{dc}}{V_{dc\max}} = \frac{\dfrac{V_{dc}}{E_M}}{\dfrac{V_{dc\max}}{E_M}} = \frac{\dfrac{V_{dc}}{E} \cdot \dfrac{E}{E_M}}{\dfrac{V_{dc\max}}{E_M}}. \tag{31}$$

As discussed previously, without excitation control, $E/E_M = 1$. With excitation control, $E/E_M = (V/E)_M/(V/E)$. The factors affecting $V_{dc\max}/E_M$ were discussed. It can be seen from (31) that the input power factor is also influenced by the type of firing angle control applied to the machine converter and by the absence or presence of machine excitation control.

VIII. NUMERICAL EXAMPLE

A hypothetical synchronous machine has the following nameplate data: 920 kW, 1.0 PF, 4260 V (l-l), 125 A, 1200 r/min, 60 Hz, $X_d = 15.84\ \Omega$, $X_q = 10.56\ \Omega$, $X'' = 2.64\ \Omega$.

In our nomenclature the above data translate to

P_n	$= 920\,000$ W,
V_n	$= 4260\sqrt{3} = 2460$ V,
I_n	$= 125$ A,
x	$= 15.84 - 2.64 = 13.2\ \Omega$,
q	$= (10.56 - 2.64)/13.2 = 0.6$,
p	$= 2.64/13.2 = 0.2$,
s	$= 10.56/15.84 = 0.667$,
$I_n X_d/V_n$	$= 0.80$ per unit.

No-load terminal voltage at nominal excitation (from (21)):

$$E_n = 1.26 \times 2460 = 3100\ V.$$

Assume the inverter utilizes conventional "slow" thyristors of a maximum turn-off time of 200 μs. At 60 Hz, this corresponds to 4.32 electrical degrees. To provide ample safety margin, a minimum turn-off angle of $\delta = 10°$ is selected.

Using the above quantities and IX/E as independent variable, the iterative solution of (15) gives corresponding values of γ. Substituting IX/E and γ pairs into (7) and (17), values of P/P_{sh} are obtained which reach a maximum of $P_M/P_{sh} = 0.316$ at $(IX/E)_M = 0.51$ and $\gamma_M = 43.79°$. From (4), the terminal voltage at maximum power is $(V/E)_M = 0.684$.

From (6), the effective displacement factor of the machine at maximum power is $\cos \beta' = 0.906$. With the 1.1 distortion factor of the motor line current this corresponds to a power factor of 0.824.

Above information translates into dimensionless and electrical units. Maximum power at nominal excitation ((18), (19), (22)):

$$P_{Mn}/P = 0.316 \times 1.98 \times (15.84/13.2) = 0.75 \text{ per unit}$$

$$P_{Mn} = 0.75 \times 920 = 690 \text{ kW}.$$

Current for maximum power (24):

$$I_{Mn}/I_n = 0.51 \times 1 \times 1.26 \times (1/0.8) \times (15.84/13.2)$$

$$= 0.96 \text{ per unit}$$

$$I_{Mn} = 0.96 \times 125 = 120 \text{ A}.$$

Since the line current has a distortion factor of 1.1, total rms line current at maximum power (nominal excitation):

$$1.1 \times 120 = 132 \text{ A } (1.06 \text{ per unit}).$$

Terminal voltage at maximum power ((20), (25)):

$$V_{Mn}/V_n = 0.684 \times 1 \times 1.26 = 0.86 \text{ per unit}$$

$$V_{Mn} = 0.86 \times 2460 = 2120 \text{ V}.$$

We can now proceed to calculate the converter ratings.

Machine Converter Rating:

a) Without excitation control (29),

$$E_0/E_M = 1 \qquad P_{mc}/P_{sh} = 0.51.$$

b) With excitation control (29),

$$E_0/E_M = (V/E)_M = 0.684; P_{mc}/P_{sh} = 0.51 \times 0.684 = 0.349.$$

With $P_{mc} = P_{mcn}$, $P_{sh} = P_{shn}$, and $P_{shn}'/P_n = 1.98$, the figures translate into dimensionless and electrical units.

a) Without excitation control (18),

$$P_{mcn}/P_n = 0.51 \times 1.98 \times (15.84/13.2) = 1.21 \text{ per unit}$$

$$P_{mcn} = 1.21 \times 920 = 1115 \text{ kVA}.$$

b) With excitation control (18),

$$P_{mcn}/P_n = 0.349 \times 1.98 \times (15.84/13.2) = 0.83 \text{ per unit}$$

$$P_{mcn} = 0.83 \times 920 = 763 \text{ kVA}.$$

Firing Angle Control for the Machine Converter:

With the $\delta = 10°$ minimum turn-off angle, the four firing angle control schemes described would maintain the following controlled parameters at a constant level:

$$\left.\begin{array}{l} \delta = \text{constant } \delta = \delta_M = 10° \\ \beta = \text{constant } \beta = \beta_M = 34.07° \text{ (from (12))} \\ \epsilon = \text{constant } \epsilon = \epsilon_M = 52.88° \text{ (from (26))} \\ V_{dc} = \text{constant } V_{dc}/E_M = 1.45° \text{ (from (28))} \end{array}\right\} \begin{array}{l} \text{using } (IX/E)_M \\ = 0.51 \text{ and } \gamma_M \\ = 43.79°. \end{array}$$

Line Converter Rating:

a) Without excitation control (30), (28):

with δ = constant $P_{lc}/P_{sh} = 0.51 \times \cos 10° = 0.50$

with β = constant $P_{lc}/P_{sh} = 0.51 \times \cos 34.07° = 0.42$

with ϵ = constant $P_{lc}/P_{sh} = (\pi/3\sqrt{6}) \times 0.51$

$\times 1.45 = 0.32$

with V_{dc} = constant $P_{lc}/P_{sh} = (\pi/3\sqrt{6}) = 0.51$

$\times 1.45 = 0.32.$

b) With excitation control: To establith the maximum (V_{dc}/E_M), the value of $(V_{dc}/E)(E/E_M)$ is calculated at no-load and maximum load for each of the four firing angle control schemes. With $(V/E)_M = 0.684$ (from (4)) and $(V_{dc}/E)_M = 1.45$ (from (28)) being identical for all four schemes, at $IX/E = 0.51$ maximum load $V_{dc}/E_M = 1.45$ for all four schemes. At $IX/E = 0$,

for δ = constant = 10°,

$$V_{dc}/E_M = 0.684 \times (3\sqrt{6}/\pi) \times \cos 10° = 1.58$$

for β = constant = 34.07°,

$$V_{dc}/E_M = 0.684 \times (3\sqrt{6}/\pi) \times \cos 34.07° = 1.33$$

for ϵ = constant = 52.88°,

$$V_{dc}/E_M = 0.684 \times (3\sqrt{6}/\pi) \times \cos 52.88° = 0.97$$

for V_{dc}/E_M = constant, $\qquad V_{dc}/E_M = 1.45.$

Using the appropriate maximum for each scheme, we get

with δ = constant $P_{lc}/P_{sh} = (\pi/3\sqrt{6}) \times 0.51 \times 1.58 = 0.35$

with β = constant $P_{lc}/P_{sh} = (\pi/3\sqrt{6}) \times 0.51 \times 1.45 = 0.32$

with ϵ = constant $P_{lc}/P_{sh} = (\pi/3\sqrt{6}) = 0.51 \times 1.45 = 0.32$

with V_{dc} = constant $P_{lc}/P_{sh} = (\pi/3\sqrt{6})$

$\times 0.51 \times 1.45 = 0.32.$

With $P_{lc} = P_{lcn}$; $P_{sh} = P_{shn}$; $P_{shn}'/P_n = 1.98$, the above figures translate into dimensionless and electrical units.

a) Without excitation control:

$$\text{with } \delta = \text{constant } P_{\text{lcn}}/P_n = 0.5 \times 1.98 \times (15.84/13.2)$$

$$= 1.19 \text{ per unit}; \qquad P_{\text{lcn}} = 1095 \text{ kVA}$$

$$\text{with } \beta = \text{constant } P_{\text{lcn}}/P_n = 0.42 \times 1.98 \times (15.84/13.2)$$

$$= 1.0 \text{ per unit}; \qquad P_{\text{lcn}} = 920 \text{ kVA}$$

$$\text{with } \epsilon = \text{constant } P_{\text{lcn}}/P_n = 0.32 \times 1.98 \times (15.84/13.2)$$

$$= 0.76 \text{ per unit}; \qquad P_{\text{lcn}} = 699 \text{ kVA}$$

$$\text{with } V_{\text{dc}} = \text{constant } P_{\text{lcn}}/P_n = 0.32 \times 1.98 \times (15.84/13.2)$$

$$= 0.76 \text{ per unit}; \qquad P_{\text{lcn}} = 699 \text{ kVA}.$$

b) With excitation control:

$$\text{with } \delta = \text{constant } P_{\text{lcn}}/P_n = 0.35 \times 1.98 \times (15.84/13.2)$$

$$= 0.83 \text{ per unit}; \qquad P_{\text{lcn}} = 764 \text{ kVA}$$

$$\text{with } \beta = \text{constant } P_{\text{lcn}}/P_n = 0.32 \times 1.98 \times (15.84/13.2)$$

$$= 0.76 \text{ per unit}; \qquad P_{\text{lcn}} = 699 \text{ kVA}$$

$$\text{with } \epsilon = \text{constant } P_{\text{lcn}}/P_n = 0.32 \times 1.98 \times (15.84/13.2)$$

$$= 0.76 \text{ per unit}; \qquad P_{\text{lcn}} = 699 \text{ kVA}$$

$$\text{with } V_{\text{dc}} = \text{constant } P_{\text{lcn}}/P_n = 0.32 \times 1.98 \times (15.84/13.2)$$

$$= 0.76 \text{ per unit}; \qquad P_{\text{lcn}} = 699 \text{ kVA}.$$

Input Displacement Factor:

Figures will be calculated for $(IX/E)_M$ to obtain values at full load and for $IX/E = 0$ to show the trend with decreasing loads. Values of $(V_{\text{dc}}/E)\cdot(E/E_M)$ are obtained from the calculations made for the line converter. Similarly, figures for

$$\frac{V_{\text{dc}}}{E} \cdot \frac{E}{E_M} = \frac{V_{\text{dc}}}{E} \cdot \left(\frac{V}{E}\right)_M \bigg/ \frac{V}{E} = 0.684 \times (V_{\text{dc}}/E)/(V/E)$$

are obtained from there and tabulated for substitution into (32), with δ, β, ϵ, and V_{dc} equal constant.

a) Without excitation control:

V_{dcmax}/E_M	2.30	1.94	1.45	1.45
V_{dc}/E at $(IX/E)_M$	1.45	1.45	1.45	1.45
0	2.30	1.94	1.41	1.45
$\cos\phi$ at $(IX/E)_M$	0.63	0.75	1.00	1.00
0	1.00	1.00	0.97	1.00

b) With excitation control:

V_{dcmax}/E_M	1.58	1.45	1.45	1.45
$(V_{\text{dc}}/E)\cdot(E/E_M)$ at $(IX/E)_M$	1.45	1.45	1.45	1.45
0	1.58	1.33	0.97	1.45
$\cos\phi$ at $(IX/E)_M$	0.92	1.00	1.00	1.00
0	1.00	0.92	0.67	1.00

TABLE I

Machine Power (kW)	920 nameplate/690 inverter op.			
Terminal Voltage ($V_{\ell\text{-}\ell}$)	4260 nameplate/3672 inverter op.			
Line Current (A)	125 nameplate/132 inverter op.			
Machine Power Factor	1.0 nameplate/.824 inverter op.			
Machine Converter Rating				
w/o excitation control (kVA)	1115			
w excitation control (kVA)	763			
Firing Angle Control Variable	δ	β	ϵ	V_{dc}
Line Converter Rating				
w/o excitation control (kVA)	1095	920	699	699
w excitation control (kVA)	764	699	699	699
AC Line Displacement Factor				
w/o excitation control FL/NL	.57/.91	.68/.91	.91/.88	.91/.91
w excitation control FL/NL	.84/.91	.91/.84	.91/.61	.91/.91

The input power factor is obtained by dividing the displacement factor by the distortion factor of the input line current. Assuming a six-pulse line converter, this distortion factor is 1.1. Table I summarizes the results. The results in Table I clearly indicate that a significant derating (to 75 percent) of the motor is unavoidable if the machine excitation at maximum power is kept at nominal nameplate level. The available rating can be increased by increasing the motor excitation and thereby E, subject to the following constraints:

i) thermal rating of the field winding,
ii) saturation characteristics of the magnetic structure,
iii) thermal rating of the stator winding.

Using the normalized expressions derived above, the designer can select a suitable motor and establish the set of operating parameters which result in an optimum design for the given drive requirements.

IX. CONCLUSIONS

Machine commutated inverter-synchronous motor drives display operating characteristics which are mathematically well definable. Most important among these characteristics is the maximum power which can be extracted from the motor. Low subtransient-to-synchronous reactance ratio and short thyristor turn-off time (in this order) are conducive to higher per unit power. The value of the synchronous reactance has minimal effect. (Of course, the same electromagnetic structure having lower synchronous reactance would have higher nominal power).

The specific scheme selected for firing angle control of the machine converter has no effect on the obtainable maximum power (as long as the control is properly adjusted for the maximum power point). It does affect, however, the power factor—and thus the efficiency—of the motor at reduced loads and, more importantly, the rating and the input power factor of the line converter. Excitation control, which maintains the machine terminal voltage (behind X'') constant is most effective to reduce both converter ratings and improve the input power factor. If excitation control is not possible (e.g., permanent magnet machine) or not practical, the ϵ = constant and the V_{dc} = constant schemes may prove to be superior. While they have no effect on the machine converter rating, they result in low line converter rating and high input power factors. Such performance can be matched by the other two schemes only if excitation control is applied.

NOMENCLATURE

- E — The rms value of line-to-neutral no-load terminal voltage (excitation voltage).
- E' — The rms value of line-to-neutral voltage "behind the synchronous reactance" (see 7)).
- E_0 — The rms value of the highest line-to-neutral ac voltage seen by the machine converter.
- I — The rms value of the fundamental component of the machine line current.
- I_d — Direct axis component of I.
- I_q — Quadrature axis component of I.
- I_{dc} — Average value of the dc link current.
- M — Index, indicating a variable at maximum power available from the motor at the given speed and excitation.
- n — Index, indicating a variable at nominal (nameplate) value.
- P, P_{ac} — Real ac power fed to the machine.
- P_{dc} — dc power fed to the machine converter.
- P_{lc} — Rating of line converter.
- P_{mc} — Rating of machine converter.
- P_{sh} — Short circuit (through X) volt-ampere of machine at given excitation.
- P_{sh}' — Short circuit (through X_d) volt-ampere of machine at given excitation.
- q — Ratio of adjusted quadrature-to-direct axis synchronous reactance.
- s — Ratio of quadrature-to-direct axis synchronous reactance.
- u — Commutation overlap angle in electrical degrees.
- V — The rms line-to-neutral machine terminal voltage "behind the subtransient reactance."
- V_{dc} — Average dc input voltage to the machine converter.
- X — "Adjusted" direct axis synchronous reactance ($X_d - X''$).
- X_d — Direct axis synchronous reactance.
- X_q — Quadrature axis synchronous reactance.
- X'' — Subtransient reactance (no distinction made between direct and quadrature axis).
- α — Machine converter firing angle in electrical degrees.
- β — Machine converter "electrical" commutation lead angle, with respect to V, in electrical degrees.
- β' — Supplementary displacement angle of I, with respect to V, in electrical degrees.
- γ — Supplementary displacement angle of I with, respect to E, E', in electrical degrees.
- δ — Thyristor turn-off angle in electrical degrees.
- ϵ — Machine converter "mechanical" commutation lead angle with respect to quadrature axis in electrical degrees.
- ϕ — Line converter input displacement angle in electrical degrees.
- ω — Machine angular frequency at given speed.

REFERENCES

[1] Fitzgerald, Kingsley, and Kusko, *Electric Machinery.* New York: McGraw Hill, 1961.

[2] B. R. Pelly, *Thyristor Phase-Controlled Converters and Cycloconverters.* New York: Wiley-Interscience, 1971.

Section 7
Cycloconverter Controlled Drives

AC COMMUTATORLESS AND BRUSHLESS MOTOR

TAKUZO MAENO and **MASAKAZU KOBATA**
Toyo Electric Mfg., Co. Ltd.
Yokohama, Japan

ABSTRACT

We have developed a type of ac commutatorless motor which is a combination of a motor and thyristor controller. The rotor of the motor has neither a commutator, slip rings nor brushes. The motor can therefore be designed rugged for a high speed control. The motor may be most suitable for use in such an environment as with acid or alkaline gas and at locations where maintenance-free motors are most urgently required.

We have introduced the theoretical treatment of the commutation angle together with the character of such a motor, its construction and its performances.

INTRODUCTION

There has arisen in recent years a remarkable many number of instances[2,3,4] of commutatorless or brushless motor such as an induction motor fed from a variable frequency source of an inverter or a cycloconverter system, or a synchronous motor combined with a similar controller. We have developed a synchronous type motor deprived of slip rings and provided with a distributor, combined with a cycloconverter supplied from a three-phase power source of commercial frequency.

Such a motor together with the controller enables a completely satisfying start without any auxiliary apparatus and a high speed running over a synchronous speed which is determined by the source frequency and number of poles. We have manufactured and delivered about 200 sets of such motors of the capacity in the range of 11kW ~ 500kW.

We have treated in this paper the composition of such a motor, its construction, performance and its theoretical basis for the commutation phenomena as well as the equation for calculating the commutation angle. Our result concerning the commutation angle is largely different from the precedent author's.[1] The above equation is reduced from the fundamental equations of a salient pole synchronous machine together with the boundary conditions in the commutation period of such a motor. We will give the theoretical treatment in Appendix(II).

Fig.1. Appearance of 15kW BL Motor.

COMPOSITION

Figs.1 and 2 show the external appearance of our ac commutatorless and brushless motor and Fig.3 shows its composition. We call this motor a BL motor in simplified term hereunder. In Fig.3 a three-phase ac power source of commercial frequency is supplied through smoothing reactors L_1 and L_2 and a cycloconverter composed of 18 thyristors to a three-phase armature winding of the BL motor.

The BL motor is a claw pole type synchronous motor as shown in Fig.4 with field windings mounted on brackets which are excited from a dc power source and its armature coils are also on the stator side. The rotor is of a rigid construction with solid steel poles of claw type with neither coils nor slip

Fig.2. Appearance of Control Equipment.

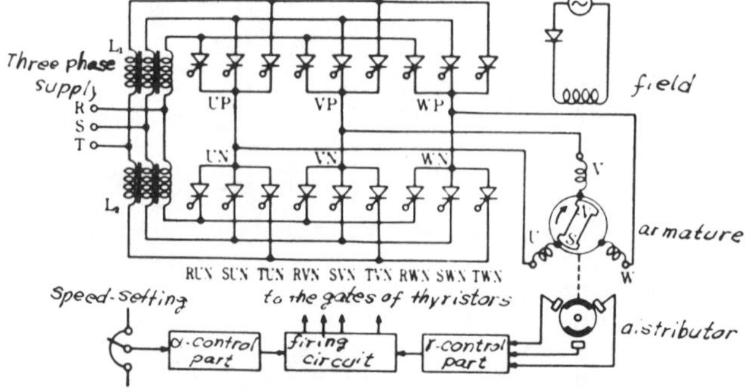

Fig.3. Composition of the BL Motor.

Paper 71 C 1-IGA, recommended and approved by the Rotating Machinery Committee of the IEEE Power Engineering Society for presentation at the 1971 Sixth Annual Meeting of the IEEE Industry and General Applications Group, Cleveland, Ohio, October 18-23, 1971. Manuscript submitted August 11, 1970; made available for printing March 17, 1972.

Reprinted from *IEEE Trans. Power App. Syst.*, vol. PAS-91, pp. 1476–1484, July/Aug. 1972.

rings.

The main flux proceeds as shown by a dotted line in Fig.5.

The rotor shaft is constructed of non-magnetic material in order that a shunting flux path is not produced. The rotor shaft is provided with a distributor on its end and three sensors respectively corresponding to three-phase armature coils mounted on brackets pick up the rotor position.

Fig.4. Rotor of the BL Motor.

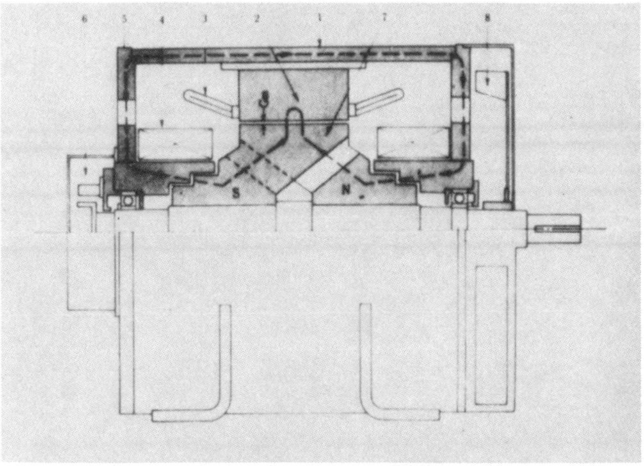

① Frame ③ Armature coils ⑤ Bracket ⑦ Field core
② Armature core ④ Exciting coils ⑥ Distributor ⑧ Fan

Fig.5. Construction of the BL Motor.

These signals can be preset to determine the lead angle γ_0 of the armature phase current against the corresponding armature induced voltage while the switching phase angle α in relation to the three-phase supply voltage is adjusted by the speed setting rheostat.

The firing signals are composed of logical products of the above two sets of signals one of which γ_0 is composed of six signals having rotating frequency and the other α composed of six signals having supply frequency. The characteristic of superior performance of the BL motor as a synchronous motor over a wide range of speed is based upon the mechanism of firing and commutating thyristors by the option of the distributors to direct which phase of the armature winding in synchronism with rotation should be conducted negatively or positively. Therefore the output current

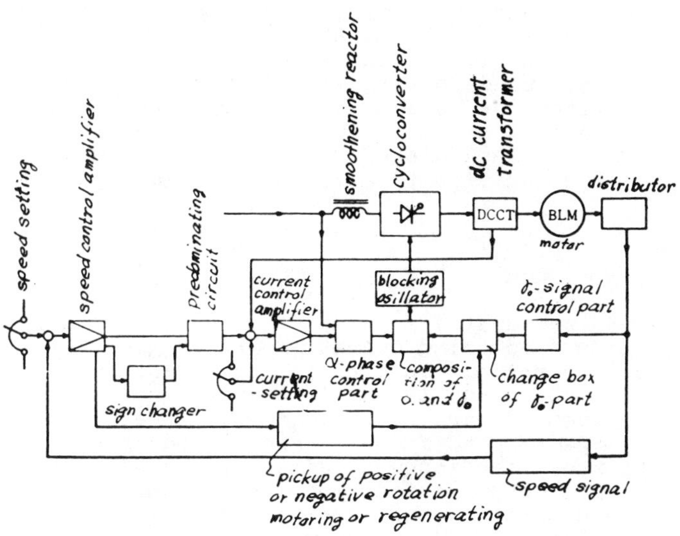

Fig.6. Control Block Diagram.

of the cycloconverter i.e. the motor current has a fixed phase difference in relation to the motor induced voltage with a frequency bound by the distributor and consequently there arises no chance of being pulled out of synchronism. The machine regulated as such is so called a self-controlled commutatorless motor of a synchronous type.

The sensors of the distributor utilize phototransistors combined with photodiodes as optical sources, or magnetodiodes which are excited by a distributor made of a permanent magnet having sectors of the same number of poles as the motor. As for commutation the so called load commutation utilizing the motor induced voltage is exerted in a speed range higher than one third of the synchronous speed (determined by the source frequency and number of poles of the motor) and in a lower speed range the so called supply voltage commutation enters which backs up the insufficient commutating voltage of the former.

This mechanism of commutation necessitates no commutating condensers and offers the most simplified and efficient type for practical use. And as shown in Fig.6 the speed signal from the distributor is compared with the preset speed signal and the difference between these two is led to control the α -signal combined with the γ_0 -signal from the distributor to actuate a closed-loop for control, which is assured in a stable state by the attachment of the minor loop of current control, for the four quadrant operation of torque and speed.

THE PRINCIPLE OF THE PERFORMANCE

In Fig.7, corresponding to the case of motoring, are shown the performance modes of the BL motor, the waveforms of currents and induced voltages. The total of 18 thyristors are decomposed into 6 groups, each of 3 ones denoted UP,UN,VP,VN,WP and WN respectively. A train of firing signals of motor frequency, corresponding to the above six groups, each having 120 degrees electrical angle width as shown on the top in Fig.7 are denoted as γ_0 signal. The motor current is thus specified to have a conduction period of 120 degrees and a recess period of 60 degrees electrical angles per half cycle and approaches the rectangular wave in the case of the reactor having sufficient inductance.

The speed-setting signal α operates so as to determine which of 3 thyristors of that group that has been just selected by the distributor should be fired.

The equation to give the rotating speed is

$$n = \frac{60\left(\frac{3\sqrt{6}}{\pi} E_S \cos\alpha - R_d I_S\right)\frac{2}{3\sqrt{3}}}{\frac{2p}{2a} k_p k_d Z \Phi \cos(\gamma_0 - \frac{u}{2}) \cos\frac{u}{2}} \qquad (1)$$

See Appendix(I).

In a motoring operation the angle α will be controlled in a range from zero up to 90 degrees, and in a regenerating one, say, between 90 and 160 degrees, speed being changed approximately proportional to cos α. As above mentioned in spite of the similarity in its construction to a synchronous motor, the BL motor has the advantage of being driven in a wide speed range independent from the supply frequency with a constant torque characterstic much the same as a dc shunt motor.

COMMUTATION[1,2]

In case of the BL motor supplied from a three-phase alternating current system, the following two kinds of commutation occur. One is that which utilizes the induced electromotive force of the armature coil, the so called " load commutation " and the other is that which utilizes the electropotential difference between two phases of power supply.

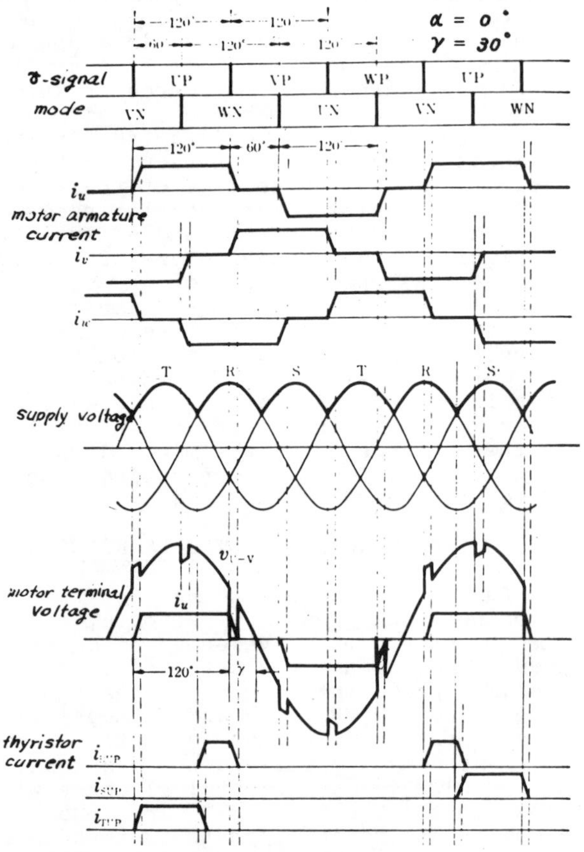

Fig.7. Waveforms of Respective Parts.

This latter is so called " supply voltage commutation ". At starting and low-speed running the former type of commutation i.e. load commutation is difficult because of a scanty electromotive force, and fortunately in these ranges the supply voltage commutation enters in substitution. And at low speed the maximum delay time for the latter commutation corresponds to 120 degrees electrical angle of supply frequency or 6.6 ms in case of 50Hz which is short enough compared to the total conduction period to cause no trouble such as the unbalance between motor phase currents.

Load commutation will be explained as follows: In Fig.8 commutation arises from U phase to V phase, that is, when thyristor RUP and TWN are conducting current thyristor RVP becomes energized and on account of electromotive force ($e_u - e_v$) in a closed circuit A→B→U→V→C→A, a rush current flows in a reverse direction to i_u.

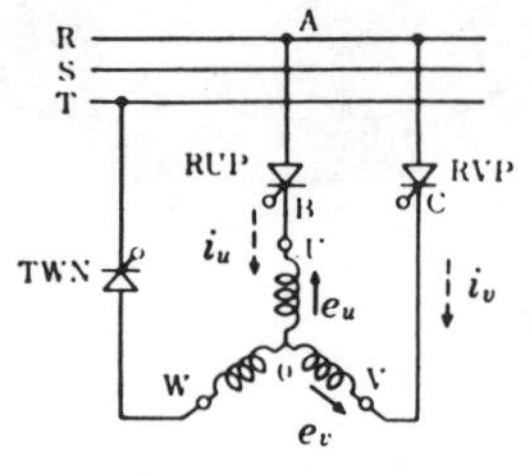

Fig.8. Mechanism of Commutation.

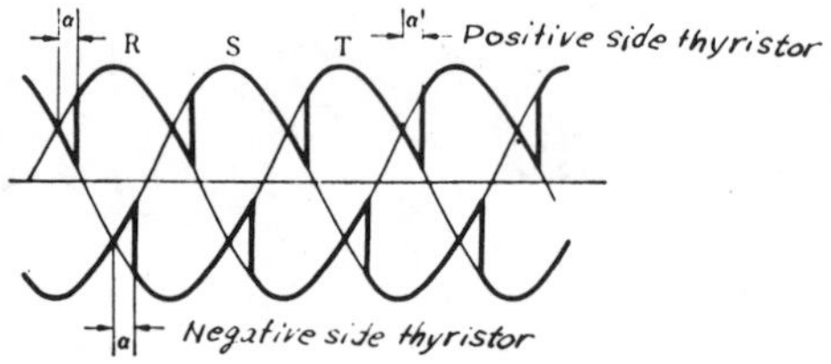

(a)

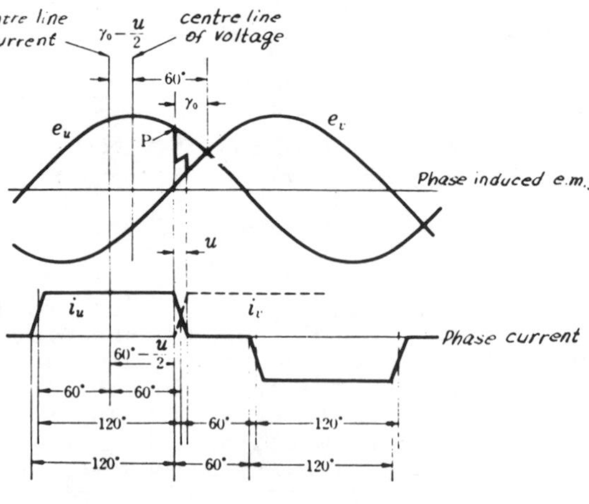

(b)

Fig.9. (a) Phase Control of Power Source Voltage.
(b) Phase Control of Phase Voltage and Current of Motor.

Then i_u commences to decrease with i_v increasing. i_u finally decreases to zero value effecting commutation from RUP to RVP to be finished. In Fig.9(b) the commutation from

U phase to V phase begins at point P, and the leading angle of the current i_u to the motor phase voltage e_u is shown to be $(\gamma_0-\frac{u}{2})$; where u is the commutation period in electrical phase angle. The value γ_0 is selected to be from 30 to 60 electrical degrees.

The vector diagram of the BL motor is as shown in Fig.10 where the internal phase angle is denoted as δ and the phase difference of motor current in relation to motor terminal voltage is $(\gamma_0-\frac{u}{2}-\delta)$ or $(\gamma-\frac{u}{2})$. As above mentioned, the commutation can be accomplished successfully when γ is larger than u and this condition also specifies the maximum torque the motor can bear.

The commutation period converted to a phase angle u is as a measure calculated by the following Eq.(2) in case of γ_0 being 60 degrees.

$$\frac{X_{af}I_{fav}}{I_s/\sqrt{3}} = \frac{X_d\sin(\frac{\pi}{3}-\frac{u}{2})\sin\frac{3}{2}u+X_d'\cos(\frac{\pi}{3}-\frac{u}{2})\cos\frac{3}{2}u+X_q[\frac{3}{2}-2\sin u\sin(\frac{\pi}{3}-u)]}{\cos(\frac{\pi}{3}-u)-\frac{1}{2}} \qquad (2)$$

See Appendix(II).

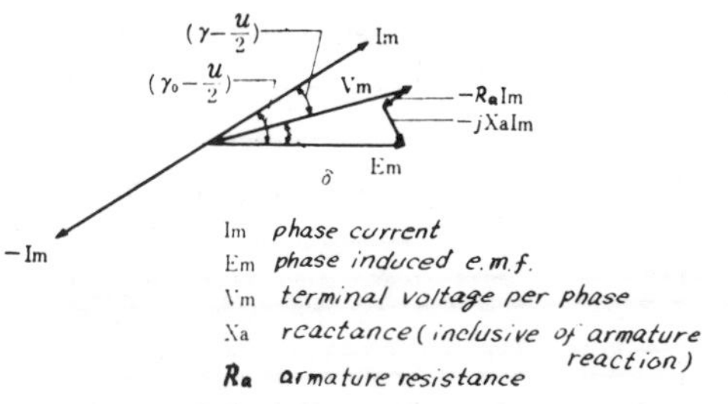

Fig.10. Vector Diagram of Motor.

SMOOTHING REACTOR

The smoothing reactors have a role to keep the fluctuation of the motor current as small as possible, hence making the initial motor current in commutation at a lower value to shorten the commutation period. Another role is to suppress a rush current in the case of a sudden increase of supply voltages, incorporating with the minor loop for current control. The magnetic coupling is assured between the three-phase windings of reactors as shown in Fig.11, in order to almost annul their impedances during commutation between the supply lines.

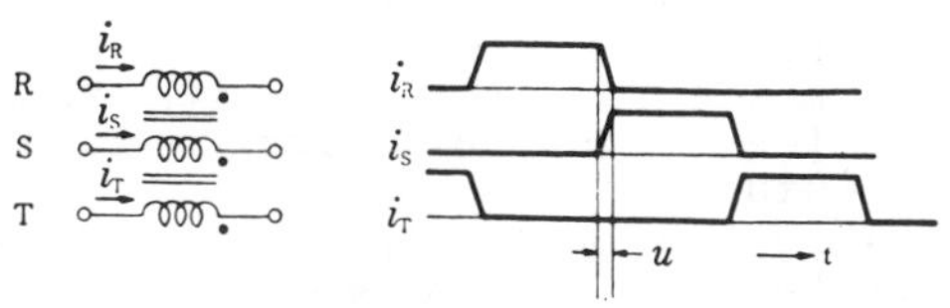

Fig.11. Coupling of Reactor.

TYPES OF THE BL MOTOR

The representative types of the BL motors manufactured by us are 4-pole 15kW, 22kW, 37kW; 6-pole 55kW, 132kW, and 8-pole 200kW, 250kW, 315kW, 375kW with the speed from zero up to 2,800rpm. We have manufactured about 200 sets or more of the BL motors in the above range.

We have also produced the BL motors, 8-pole 500 kW 1,600rpm. The number of poles of the BL motor is selected to give the most efficient design for armature coil space or rotor pole structure preferably limitted to a lower value with the consideration for electronics parts.

CHARACTERISTICS[3,5,6]

Hereunder characteristics of 4-pole 15kW 2500 rpm BL motor are given in Figs.12~18.

In Fig.12 torque-speed characteristics with γ_0 and field current held constant and α as a parameter is shown. By controlling the α signal the speed is changed in a wide range. With γ_0 increasing the speed increases and at the same time the speed regulation becomes greater.

In Fig.13 the efficency, power-factor, current, leading angle, and commutation angle corresponding to the commutation period etc. are shown in ordinates in relation to the torque in abscissa. In Fig.14 or 15, $\gamma=\gamma_0-\delta$,

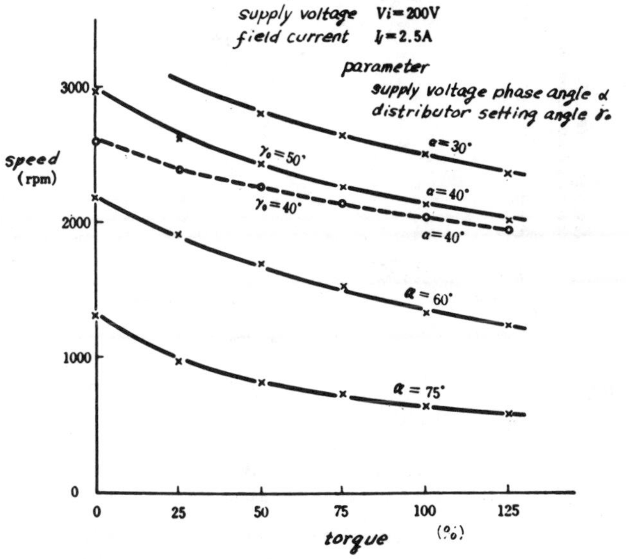

Fig.12. Characteristic of Rotating Force versus Speed.

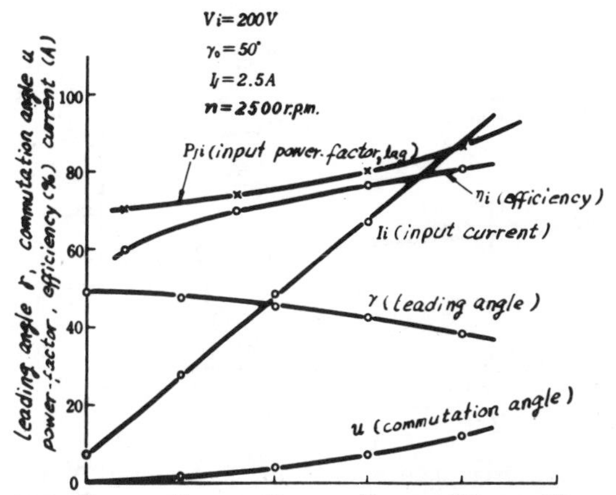

Fig.13. Speed Constant Characteristic.

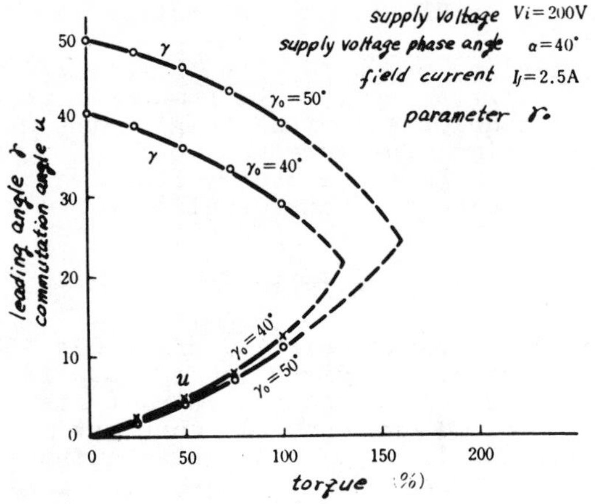

Fig.14. Characteristic of Rotating Force versus γ, u.

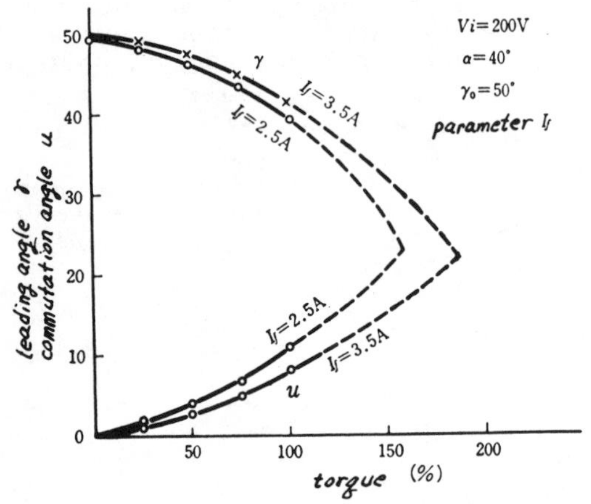

Fig.15. Characteristic of Rotating Force versus γ, u.

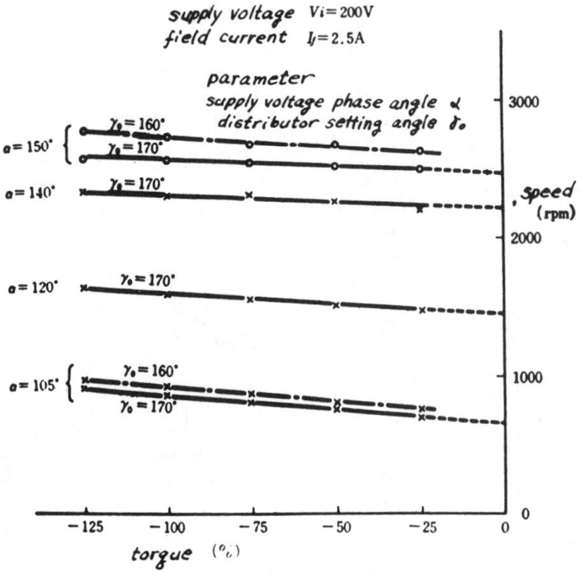

Fig.16. Characteristic of Rotating Force versus Speed (Regenerative Brake Side).

decreases with δ or the load current increasing. Moreover angle u corresponding to the commutation period increases to effect the phase difference of the load current unto motor terminal voltage to decrease. The limitation of the overload corresponds to the cross points of torque- γ & torque-u curves in Fig.14 or 15.

The efficiency η_i expresses the overall value inclusive of thyristors. In Fig.15 torque-(γ , u) characteristic is shown with a parameter of field current.

In Fig.16 torque-speed characteristic with field current and γ_o held constant is shown in the regenerative quadrants with no closed loop for speed control.

In Fig.17 for the four quadrant operation torque-speed characteristics with a closed speed control loop are shown, in which the speed regulation is given to be about 0.5%.

The minor loop for current control shown in Fig.6 is essential for the quick response in such a case of a sudden change of the supply voltage.

In Fig.18 is shown an oscillogram for motor drive side.

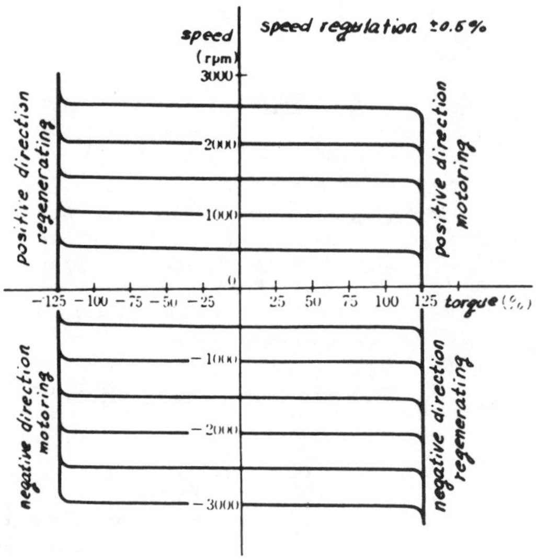

Fig.17. Speed Characteristic in case of fitting Speed Stabilizer.

SUMMARY

The characteristic performance and feature of the BL motor are as follows.

(1) Maintenance free except bearing, because of non-brush, non-commutator, and non-slip ring.

(2) Atmosphere even with special gas as SO_2 etc. does not prohibit to use the BL motor.

(3) The space for the BL motor itself is not greater than that for dc motor or ac commutator motor.

(4) High speed and high capacity machines can be readily produced because of no restriction of the overspeed caused by a commutator.

(5) Good efficiency on account of speed control by thyristors.

(6) Performance for the four quadrant running is competitive to dc machine controlled by solid state totally adjustable reversing control and superior in the simplicity of the control.

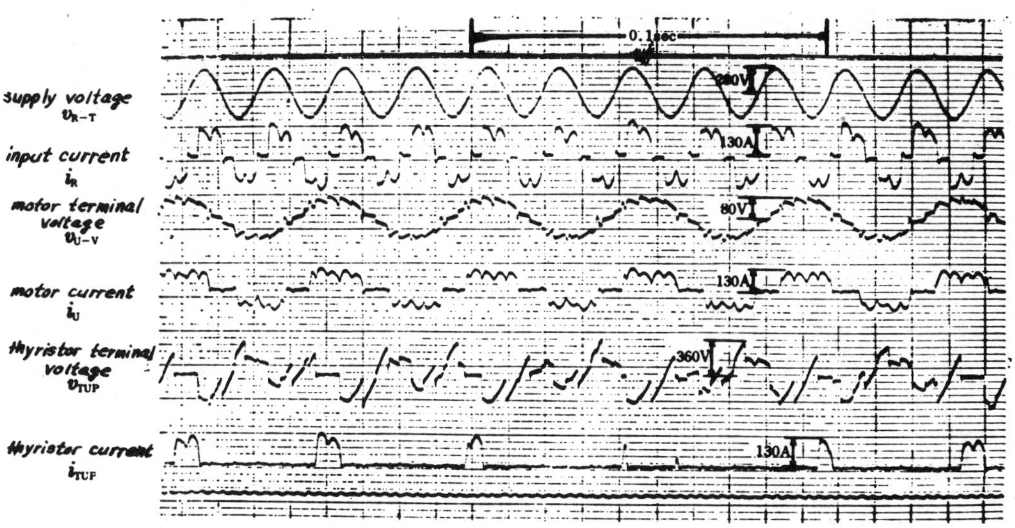

Fig.18. Oscillogram of Motor Drive Side.

NOMENCLATURES

2a: number of parallel circuits per phase of the armature winding of motor

α : firing electric angle respective to supply voltage (radians)

γ_o: the preset angle of the distributor which corresponds to the leading angle of the armature current in relation to the armature induced voltage at no load (radians)

γ: the leading angle of the armature current in relation to the armature terminal voltage at no load (radians)

E_m: the armature induced voltage per phase (volts)

E_s: the source voltage reduced to one phase of motor (volts)

δ : the internal phase angle between the induced voltage and the terminal voltage (radians)

I_m: the motor armature phase current in r.m.s. value (amps)

I_s: the peak value of the motor armature phase current (amps)

i_d: the direct-axis armature current (instantaneous value)

i_q: the quadrature-axis armature current (instantaneous value)

i_f: the field current (instantaneous value)

i_u, i_v, i_w: per phase motor current for u, v & w phase respectively (instantaneous value)

k_p: the short pitch coefficient of the armature winding

k_d: the distribution coefficient of the motor armature winding

$L_d (X_d)$: the direct-axis synchronous inductance (reactance)

$L_d' (X_d')$: the transient direct-axis synchronous inductance (reactance)

$L_q (X_q)$: the quadrature-axis synchronous inductance (reactance)

$L_{af} (X_{af})$: the mutual inductance (reactance) between field coil and armature direct-axis

$L_{ff} (X_{ff})$: self inductance(reactance) of field coil

2p: number of poles of the motor

n : rotating speed (rpm)

Φ : flux per pole of motor at no load (Wb)

R_a: resistance of the armature coil per phase (ohms)

R_d: resistance inclusive of the two phases of motor armature coils and smoothing reactors (ohms)

R_f : the field coil resistance (ohms)

u : phase angle corresponding to commutation period (radians)

V_m: terminal voltage of the motor armature winding (volts)

v_d; the direct-axis armature terminal voltage (instantaneous value)

v_q; the quadrature-axis armature terminal voltage (instantaneous value)

v_f; the field coil terminal voltage (instantaneous value)

v_u, v_v, v_w; per phase motor terminal voltage for u, v and w phase respectively (instantaneous value)

X_a: armature reactances inclusive of the armature reaction and the leakage reactance per phase (ohms)

ω : angular velocity in electrical phase angle (radians/s)

Z : total no. of conductors of the armature winding

Note: All inductances are in H and all reactances in ohms.

APPENDIX (I)[2]

In Fig.19 the waveform of the motor terminal voltage is shown and the absolute average value of the instantaneous terminal voltage is calculated as shown hereunder , when the leading angle of the distributor is set as causing the motor current to lead the motor terminal voltage by an angle γ . During the commutation period the terminal voltage is kept at the average value of the two terminal voltages, i.e. $(v_u - v_w)$ and $(v_v - v_w)$ for instance as you see in Fig.19.

The following three equations express the instantaneous values of the phase voltages of star-connected armature windings.

$$v_u = \sqrt{2}\, V_m \sin \omega t = \sqrt{2}\, V_m \sin \theta$$

$$v_v = \sqrt{2}\, V_m \sin(\omega t - \tfrac{2}{3}\pi) = \sqrt{2}\, V_m \sin(\theta - \tfrac{2}{3}\pi)$$
$$v_w = \sqrt{2}\, V_m \sin(\omega t + \tfrac{2}{3}\pi) = \sqrt{2}\, V_m \sin(\theta + \tfrac{2}{3}\pi) \qquad (3)$$

The absolute average value V_{av} of the motor terminal voltage is expressed by the sum of the three integrals as follows.

$$V_{av} = \frac{3}{\pi}\int_{\frac{\pi}{2}}^{\frac{5}{6}\pi-\gamma} (v_u - v_w)\,d\theta + \frac{3}{\pi}\int_{\frac{5}{6}\pi-\gamma}^{\frac{5}{6}\pi-\gamma+u} \frac{1}{2}\left[(v_u-v_w)+(v_v-v_w)\right]d\theta + \frac{3}{\pi}\int_{\frac{5}{6}\pi-\gamma+u}^{\frac{5}{6}\pi} (v_v - v_w)\,d\theta$$

$$= \frac{3\sqrt{6}}{\pi} V_m \cos(\gamma - \tfrac{u}{2}) \cos\tfrac{u}{2} \qquad (4)$$

Whereas the speed electromotive force E_m per phase will be expressed by the following Eq.(5).

$$E_m = \frac{\sqrt{2}\pi}{2} f K_p K_d \frac{Z}{2a} \Phi$$

$$= \frac{\pi}{2\sqrt{2}} \cdot \frac{2p}{2a} \cdot \frac{n}{60} \cdot K_p K_d \Phi Z \qquad (5)$$

As an approximation the following connection in Eq.(6) holds from the vector diagram in Fig.10, disregarding the resistance drop.

$$E_m \cos(\gamma_o - \tfrac{u}{2}) = V_m \cos(\gamma - \tfrac{u}{2}) \qquad (6)$$

From Eqs.(4), (5) and (6) the following equation will be introduced.

$$E_m \cos(\gamma_o - \tfrac{u}{2}) = \frac{\pi}{3\sqrt{6}} \frac{V_{av}}{\cos\frac{u}{2}}$$

or

$$\frac{n}{60} = \frac{V_{av}\frac{2}{3\sqrt{3}}}{\frac{2p}{2a} K_p K_d Z \Phi \cos(\gamma_o - \frac{u}{2}) \cos\frac{u}{2}} \qquad (7)$$

If the firing angle with respect to the supply voltage is α, the average value of the motor terminal voltage is expressed as follows in approximation.

$$V_{av} = \frac{3\sqrt{6}}{\pi} E_s \cos\alpha - R_d I_s \qquad (7a)$$

Eq.(7) combined with Eq.(7a) will be reduced to Eq.(1).

APPENDIX (II)

The fundamental equations for the synchronous machine can be expressed by the following equations(8) by the two reaction theory.

$$\frac{d}{dt}(L_d i_d + L_{af} i_f) - \omega (L_q i_q) + R_a i_d = v_d$$

$$\frac{d}{dt}(L_q i_q) + \omega (L_d i_d + L_{af} i_f) + R_a i_q = v_q \qquad (8)$$

$$\frac{d}{dt}(L_{ff} i_f + \tfrac{3}{2} L_{af} i_d) + R_f i_f = v_f$$

We treat the commutation period in which the current i_w decays to zero from the initial value I_s and at the same time the current i_u increases from zero up to that value I_s --- we assume the constant resultant current i.e. I_s flowing throughout the commutation period.

When the commutation begins, the position of the rotor pole head N lags behind the u-coil axis by the electrical phase angle of $(\gamma_o + \frac{5}{6}\pi)$ as shown in Fig.20. The phase voltage v_u lags behind the phase current i_u by the electrical phase angle $(\gamma - \frac{u}{2})$ as shown in Fig.21, hence the instantaneous value of v_u can be expressed by the following equation (9), and v_v and v_w can also be expressed by Eqs.(10) and (11) respectively.

$$v_u = \sqrt{2}\, V_m \cos(\omega t - \tfrac{\pi}{3} - \gamma) \qquad (9)$$

$$v_v = \sqrt{2}\, V_m \cos(\omega t - \tfrac{\pi}{3} - \gamma - \tfrac{2}{3}\pi) \qquad (10)$$

$$v_w = \sqrt{2}\, V_m \cos(\omega t - \tfrac{\pi}{3} - \gamma + \tfrac{2}{3}\pi) \qquad (11)$$

With the aid of d-q transformation for the armature voltage v_d and v_q may be expressed as follows.

$$v_d = \frac{2}{3}\left[v_u \cos\theta + v_v \cos(\theta - \tfrac{2}{3}\pi) + v_w \cos(\theta + \tfrac{2}{3}\pi)\right] \qquad (12)$$

$$v_q = -\frac{2}{3}\left[v_u \sin\theta + v_v \sin(\theta - \tfrac{2}{3}\pi) + v_w \sin(\theta + \tfrac{2}{3}\pi)\right] \qquad (13)$$

Considering, during commutation, the terminal voltages at U and W terminals should equal to each other and be half of the sum of v_u and v_w which are the instantaneous values respectively.

Then, we have by using Eqs.(9), (10), (11), (12) and (13),

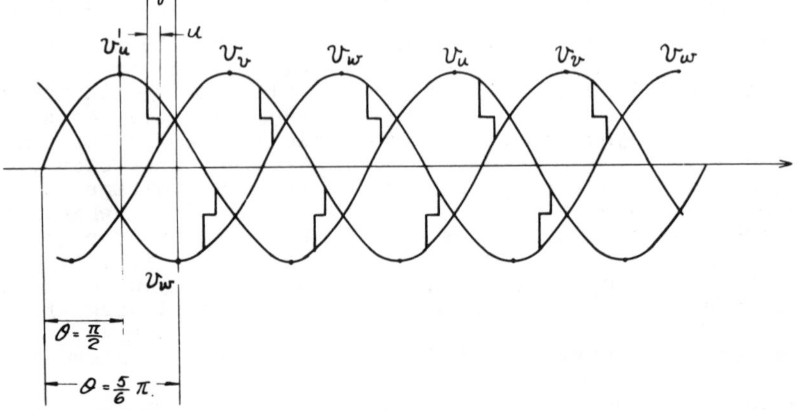

Fig.19. Waveform of Motor Terminal Voltage.

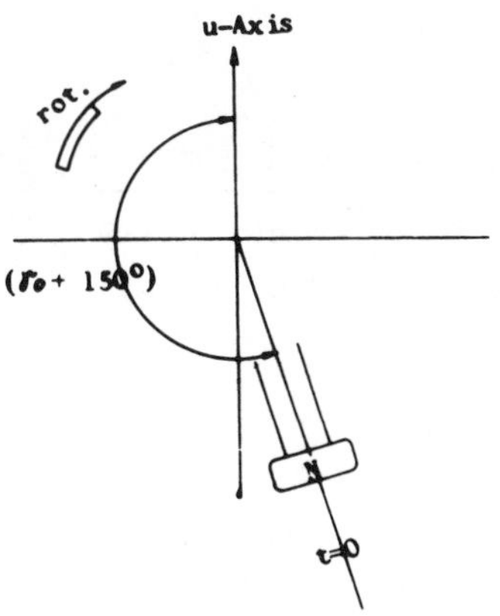

Fig.20. Rotor Position when Commutation begins.

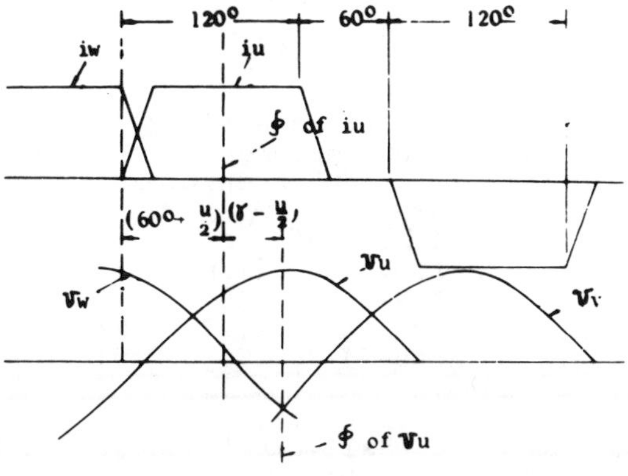

Fig.21. Relation of Phase Angle of the Voltage and Current.

$$v_d = \frac{2}{3}\left[-\frac{1}{2}(v_u+v_w)\left\{\cos\theta+\cos(\theta+\frac{2}{3}\pi)\right\}+v_v\cos(\theta-\frac{2}{3}\pi)\right]$$

$$=\frac{V_m}{\sqrt{2}}\left[\cos(2\omega t+\theta_0-\gamma+\frac{\pi}{3})+\cos(\theta_0+\gamma+\frac{\pi}{3})\right] \qquad (14)$$

$$v_q = -\frac{2}{3}\left[-\frac{1}{2}(v_u+v_w)\left\{\sin\theta+\sin(\theta+\frac{2}{3}\pi)\right\}+v_v\sin(\theta-\frac{2}{3}\pi)\right]$$

$$=-\frac{V_m}{\sqrt{2}}\left[\sin(2\omega t+\theta_0-\gamma+\frac{\pi}{3})+\sin(\theta_0+\gamma+\frac{\pi}{3})\right] \qquad (15)$$

where

$$\omega t+\theta_0=\theta \qquad (16)$$

$$\theta_0=-(\frac{5}{6}\pi+\gamma_0) \qquad (17)$$

Therefore by using the relation in Eq.(16) and (17) the initial value V_{do} and V_{qo} can be reduced to the next equations.

$$V_{do}=v_d\Big|_{t=0}=-\frac{V_m}{\sqrt{2}}\left[\sin(\gamma_0-\gamma)+\sin(\gamma_0+\gamma)\right] \qquad (18)$$

$$V_{qo}=v_q\Big|_{t=0}=\frac{V_m}{\sqrt{2}}\left[\cos(\gamma_0-\gamma)+\cos(\gamma_0+\gamma)\right] \qquad (19)$$

i_d and i_q can as well be given by the expression in Eqs.(20) and (21) respectively with the aid of d-q transformation similar to the above equations (12), (13) for v_d and v_q, referring to the conditions in Eqs.(22) and (23).

I_s is the initial value of i_w and $-i_v$.

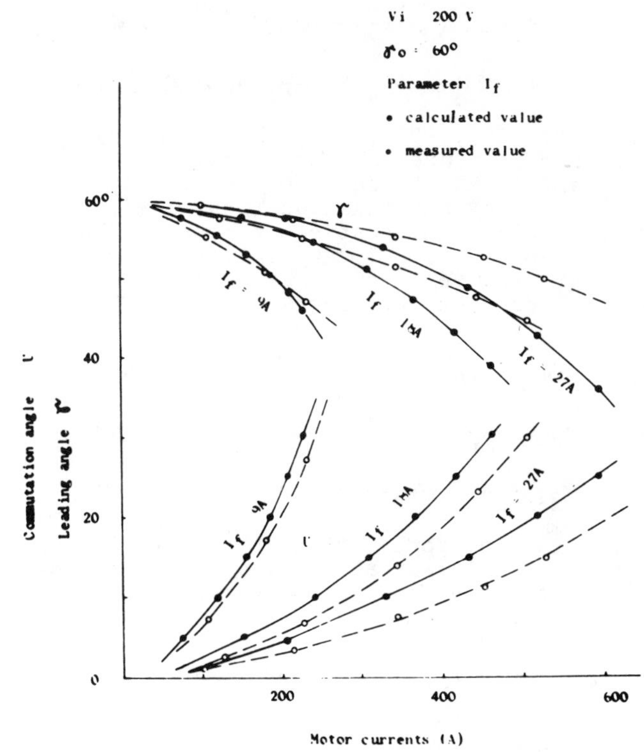

Fig.22. Comparison between calculated and measured Values for 125kVA BL Motor.

$$i_d=\frac{2}{\sqrt{3}}\left[i_u\sin(\frac{\pi}{3}+\theta)-I_s\sin\theta\right] \qquad (20)$$

$$i_q=\frac{2}{\sqrt{3}}\left[i_u\cos(\frac{\pi}{3}+\theta)-I_s\cos\theta\right] \qquad (21)$$

$$i_w=I_s-i_u \qquad (22)$$

$$i_v=-I_s \qquad (23)$$

From Eqs.(20) and (21) and referring to the condition $\theta=\theta_0=-(\frac{5}{6}\pi+\gamma_0)$ the initial value of i_d and i_q can be reduced to

$$I_{do}=\frac{2}{\sqrt{3}}I_s\sin(\frac{\pi}{6}-\gamma_0) \qquad (24)$$

$$I_{qo}=\frac{2}{\sqrt{3}}I_s\cos(\frac{\pi}{6}-\gamma_0) \qquad (25)$$

From Eq.(8), considering the initial conditions given by Eqs.(18), (19), (24), (25) and $i_f=I_{fo}$ at t=0 and disregarding the $R_a i_d$, $R_a i_q$, and $R_f i_f$ terms in Eq.(8), we can by Laplace transformation method easily attain the following solutions.

$$X_d' I_{du}=\left[X_d'-X_d(1-\cos u)\right]I_{do}+X_q I_{qo}\sin u$$

$$-X_{af}I_{fo}(1-\cos u)-\frac{X_{af}}{X_{ff}}\cdot V_f\cdot u$$

$$-2\sqrt{2}V_m\cos(\gamma-\frac{u}{2})\sin\frac{u}{2}\sin(\gamma_0-u) \qquad (26)$$

$$X_q I_{qu}=-X_d I_{do}\sin u+X_q I_{qo}\cos u-X_{af}I_{fo}\sin u$$

$$+2\sqrt{2}V_m\cos(\gamma-\frac{u}{2})\sin\frac{u}{2}\cos(\gamma_0-u) \qquad (27)$$

$$\frac{2}{3}\frac{X_{ff}X_d'}{X_{af}}I_{fu}=X_d I_{do}(1-\cos u)-X_q I_{qo}\sin u$$

$$+ X_{af} I_{fo} \left(\frac{2}{3} \frac{X_{ff} X_d}{X_{af}^2} - \cos u \right) + \frac{2}{3} \frac{X_d}{X_{af}} \cdot V_f \cdot u$$

$$+ 2\sqrt{2} V_m \cos(\gamma - \frac{u}{2}) \sin\frac{u}{2} \sin(\gamma_o - u) \qquad (28)$$

In Eqs.(26), (27), (28), I_{du}, I_{gu} and I_{fu} respectively represent the final values of i_d, i_g and i_f at the end of the commutation period, where the relation $i_w=0$, $i_u=I_s$ and $i_v=-I_s$ holds and enables the reduction of Eqs.(30) and (31). In approximation we can also have the relation in Eq.(29) concerning the field current the average value of which is I_{fav}.

$$\frac{1}{2}(I_{fo} + I_{fu}) = I_{fav} \qquad (29)$$

$$I_{du} = -\frac{2}{\sqrt{3}} I_s \cos\left(\frac{\pi}{3} - \gamma_o + u\right) \qquad (30)$$

$$I_{gu} = \frac{2}{\sqrt{3}} I_s \sin\left(\frac{\pi}{3} - \gamma_o + u\right) \qquad (31)$$

We usually apply this type of the BL motor where γ_o is set at 60 electrical degrees and the boundary conditions can be expressed by simpler equations i.e.

$$I_{do} = -\frac{I_s}{\sqrt{3}}, \quad I_{go} = I_s, \quad I_{du} = -\frac{2}{\sqrt{3}} I_s \cos u$$

and

$$I_{gu} = \frac{2}{\sqrt{3}} I_s \sin u$$

In this case we can acquire the following relations in Eqs.(2), (32) and (33).

$$\frac{X_{af} I_{fav}}{I_s/\sqrt{3}}$$

$$= \frac{X_d \sin(\frac{\pi}{3} - \frac{u}{2}) \sin\frac{3}{2}u + X_d' \cos(\frac{\pi}{3} - \frac{u}{2}) \cos\frac{3}{2}u + X_g\left[\frac{3}{2} - 2\sin u \sin(\frac{\pi}{3} - u)\right]}{\cos(\frac{\pi}{3} - u) - \frac{1}{2}} \qquad (2)$$

$$\frac{X_{af} I_{fav}}{\sqrt{2} V_m \cos(\gamma - \frac{u}{2})}$$

$$= \frac{X_d \sin(\frac{\pi}{3} - \frac{u}{2}) \sin\frac{3}{2}u + X_d' \cos(\frac{\pi}{3} - \frac{u}{2}) \cos\frac{3}{2}u + X_g\left[\frac{3}{2} - 2\sin u \sin(\frac{\pi}{3} - u)\right]}{X_d' \cos\frac{u}{2}(2\cos u - 1) + X_g \sin\frac{u}{2}(\sqrt{3} + 2\sin u)} \qquad (32)$$

$$I_{fo} = I_{fav} - \frac{3}{2} \frac{X_{af}}{X_{ff}} \left(\cos u - \frac{1}{2}\right) I_s \qquad (33)$$

By using these equations we have calculated values for 3 phase 125 kVA synchronous motor operating as the BL motor and in Fig.22 is shown the comparison between tested and calculated values. The current I_m is set to be 0.82 I_m from the experiments and in Fig.22 I_m-u curves give a better conformity rather than I_m-γ curves. We must introduce another new method if a better result is requested.

REFERENCES

(1) M. Stöhr, " Die Typenleistung Kollektorloser Stromrichtermotoren bei verbesserten Motorshaltungen" Arch.Elektrotech 32, No.11, 12, 1938.

(2) N. Sato, " A Study for Commutatorless Motor " J.I.E.E.J. (Japan) Vol.84-8, No.911, pp.1249-1257, 1964.

(3) T. Maeno and T.Kitazawa, " BL Motor (A.C. Brushless Motor) ", Toyo Review, No.11 pp.2-8, 1969.

(4) T. Maeno and M. Kobata, " Brushless Motor for Electric Rail-car ", Toyo Review, No.11, pp.9-14, 1969.

(5) S. Ogino,"Application of Brushless Motor" DENKI KEISAN, Vol.37, No.4, pp.724-728, 1969.

(6) M. Kobata, "Thyristor-motor into Application Era ", SHINDENKI, Vol.24, No.3, pp.28-32, 1970.

SYNCHRONOUS MOTOR RAILCAR PROPULSION

FRANK J. BOURBEAU
Delco Electronics Division
General Motors Corporation
Santa Barbara, California

ABSTRACT

Development of ac motor drives for rail transit car propulsion has centered on the induction motor with PWM inverter control. Interest in the induction motor as a replacement for the series dc traction motor stems from the simplicity of the squirrel cage rotor of the induction motor. In this article, the shortcomings of PWM inverter-induction motor transit car drive are examined. It is shown that the synchronous, or brushless dc, motor drive can provide performance exceeding both the PWM inverter-induction motor and the conventional dc motor in the tranist car application.

INTRODUCTION

The rapid transit railcar has experienced revived interest in recent years. The dominant features of this mode of passenger transport are the utilization of wayside electrical power for traction, and guidance by means of the flanged steel wheel on steel rail. Wayside power pickup is a means of utilizing abundant energy from coal reserves for passenger transport. The steel wheel and rail interface provides the lowest possible rolling resistance. Formerly, the flanged wheel's lateral guidance and levitation capability were taken for granted. These factors now become significant in view of the current interest in wheelless linear motor propulsion using electrical or pneumatic power for levitation.

The advantages of the rapid transit car have in the past been offset by passenger crowding, high noise levels, extremes of temperature and humidity and poor ride quality — factors which have contributed to a gradual decline in ridership. The Urban Mass Transportation Administration's (UMTA) Rapid Rail Vehicles and Systems Program, with Boeing Vertol as systems manager, is aimed at alleviating these negative aspects of rail rapid transit.

The UMTA rapid rail program consists of State of the Art Car (SOAC) and Advanced Concept Train (ACT) phases. Chopper controlled separately excited dc motors power the pair of SOAC cars which are undergoing revenue service trials. In the ACT-I program, the energy savings potential of an onboard flywheel to store braking energy will be evaluated. In this paper, a synchronous motor propulsion system is described which promises reduced operating cost and improved performance over existing dc or induction motor transit car propulsion. A new type of power converter, termed the capacitor-coupled-cycloinverter, was developed to match the special characteristics of the synchronous motor. A review of existing transit car propulsion systems places the synchronous motor system in proper perspective.

DC MOTOR TRACTION

The self-propelled rail transit car has long depended on the series dc motor for motive power. This motor form has excellent overload capability and immunity to line voltage variations, and provides good torque sharing with motors connected to wheels of different diameters. Over the years, vehicle speeds have increased and dynamic braking capability has assumed importance to the extent that motors are sized by the required braking performance[1] The peak electrical power output of contemporary rapid transit traction motors operating as braking generators is up to three times the motoring power. The ac motor, to be considered as a viable candidate to succeed the dc motor, must equal or exceed this formidable braking performance.

Resistor Control

Tractive effort control with conventional series dc motor propulsion is accomplished with a motor driven switch which supplies the proper combinations of armature current limiting and field shunting resistors, and series-parallel motor connections. The power loss in the current limiting resistors is a serious liability with this control method. Also, the magnitude and rate of rise of starting current imposes a large burden on the dc supply fault protection sensor since, under some conditions, the surge current when starting a train of cars can resemble a fault current profile[2] . The change in tractive effort, or jerk, occurring when resistor or motor connections are changed has been reduced to a tolerable level at the expense of increased controller complexity.

In addition to the problems of efficiency, system fault protection, and jerk, the slow response of the motor driven controller does not permit adequate wheel slide protection. Also, because of efficiency and resistor power rating limitations, continuous operation is practical only on those switch settings which place the traction motors in series or in parallel across the dc supply. The close speed regulation necessary for advanced automatic train control is thus difficult to achieve. Regenerative braking, with its great potential for energy saving, is also not practical with the resistor controlled series dc traction motor.

Chopper Control

The difficulties associated with resistor control of tractive effort are relieved with a dc-dc converter, or chopper. This approach, which is functionally equivalent to a dc-dc transformer, is utilized in the Bay Area Rapid Transit (BART) system and in the previously mentioned SOAC developmental cars. With chopper control, dc input current and power, and motor voltage, increase with speed to base speed. Above base speed, the chopper is full on, connecting the motors directly to the dc supply. Constant power operation to maximum speed is then accomplished through the use of contactor-controlled field shunts with series wound motors (as in BART), or by programmed field current in separately excited motors (as in SOAC).

Chopper control eliminates the initial power loss due to starting resistance, minimizes jerk, and eliminates the problem of discriminating between train starting current and fault current. Chopper control can also provide regenerative braking, regulated speed operation, and suppression of wheel slide.

DC Motor Limitations

The chopper-controlled dc traction motor, though a significant improvement, still retains liabilities associated with the dc motor which adversely affect the cost and performance of transit car operation. For example, even though commutator bar and brush wear rates are extremely low, commutator condition must be frequently monitored to detect the onset of deterioration which rapidly leads to catastrophic failure. Motor size and weight, while reduced in recent years through the use of high temperature insulation system and commutator refinements, is a principal factor limiting transit car speed.

INDUCTION MOTOR TRACTION

Much development effort has been expended on controlled-slip induction motor drives for off-highway vehicles [3] , battery

Reprinted from *IEEE/IAS 1974-Part I*, pp. 533–540, 1974.

powered autos [4], and rail vehicles [5,6]. The ruggedness and simplicity of the induction motor are invariably cited to justify its selection. Another factor influencing motor choice may be the ease with which a standard induction motor may be adapted for use in the initial phase of a drive development program. A standard synchronous motor, on the other hand, needs considerable modification to make it suitable for use in a development program. This goes far to explain the dearth of effort on synchronous motor drives even though, as described below, the synchronous motor offers inherent performance advantages in some applications.

Two examples of induction motor rail drives have reached the prototype stage. A European diesel electric locomotive [5] employs a single pulse width modulated (PWM) inverter energized from an alternator-rectifier dc link and driving four traction motors in parallel. Similarly, a U.S. transit car drive [6] operates from the nominal 600 Vdc third rail and uses a dual PWM inverter to power four parallel connected traction motors.

The PWM inverter, shown in Figure 1, combines voltage modulation and dc-to-3 phase inversion in one set of thyristor-diode switches to eliminate the separate dc chopper modulation stage that is required for the simple, six-step, variable input voltage

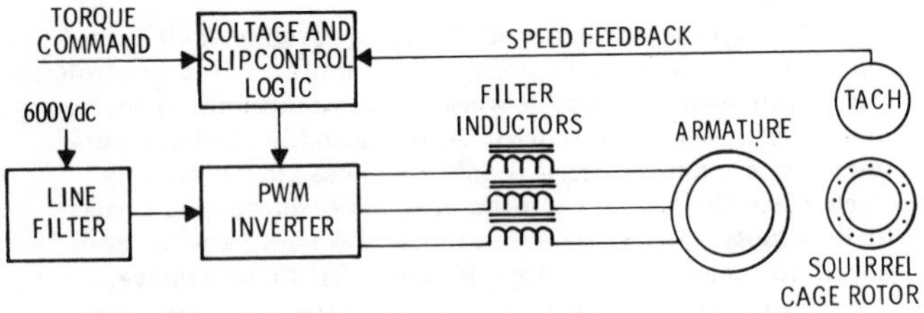

Figure 1. PWM Inverter - Induction Motor

inverter. As described below, the apparent simplicity of the PWM approach is obtained at the expense of weak field operation which leads to significant motor oversizing and extreme sensitivity of torque to wheel size variations.

Acceleration Mode

During acceleration, the full performance motor torque profile consists of constant torque from zero to base speed and torque decreasing inversely with speed from base to maximum speed. The motor torque level in the first phase is derived from the design adhesion limit, while the motor power level in the second phase is constrained by the capacity limit of the power distribution system. In the constant torque region, flux is held near the saturation level by raising the motor voltage. In the constant power region, voltage modulation is not available for the reasons discussed below. Motor voltage is then proportional to the dc supply voltage and flux decreases inversely with speed.

The base motor frequencies of the locomotive and transit car drives [5,6] cited above are 33 Hz and 60 Hz, respectively. This frequency limitation is due principally to the relatively long turn-off time required for production grade thyristors with voltage and current ratings adequate for this high power inverter service.

Located in the Appendix is an analysis of the PWM modulation cycle which shows how a 100-microsecond turn-off time specification can dictate a maximum modulation frequency of 63 Hz, or a 1900 r/min base speed with a four-pole motor (two-pole motors are not generally considered in this application because of their poorer material utilization).

The induction motor is characterized by breakdown, or pullout, torque which must exceed the load torque at all speeds. The relationship of breakdown torque to $(\text{flux})^2$ is readily demonstrated. Also, except at low speed and within the saturation limit, flux varies directly with voltage and inversely with synchronous speed. The actual speed is close to synchronous speed, so that the variation of breakdown torque with voltage and speed can be stated as

$$T_b \;\alpha\; (V/N)^2$$

For constant voltage operation, breakdown torque becomes a function of dc supply voltage as

$$T_b \;\alpha\; (E_{DC}/N)^2$$

With constant power, the per-unit load torque decreases inversely with speed as

$$T_\ell = 1/N$$

Because breakdown torque decreases more rapidly with speed than load torque, it will match the load torque at a maximum per-unit pullout speed of M. Thus, the breakdown torque normalized to base speed load torque can be written as

$$T_b = M/N^2$$

This equation shows that the motor must be sized for a breakdown torque of M times the base speed load torque in order to avoid pullout at high speed.

The induction motor operated with controlled slip is able to provide load torque up to 80% of breakdown torque [7]. Thus, for the case of a pullout speed of M = 4, the constant voltage, variable speed-controlled slip motor has an excess of breakdown torque of 0.8M = 3.2 over that required for a fixed speed-controlled slip motor. Since motor weight is proportional to breakdown torque, the weight of the variable speed motor exceeds that of the constant speed motor by a factor approaching 3.2.

Torque Sharing

The two PWM inverter-induction motor rail drives described above [5,6] operate four traction motors in parallel. This eliminates inverter interaction on the dc side, simplifies the motor control system, and gives inherent wheel slide protection. However, these advantages are offset by a severe torque sensitivity to differences in wheel diameters, as shown below.

Induction motor torque normalized to breakdown torque at any speed can be expressed in terms of normalized slip [8] as

$$T = \frac{2S}{S^2 + 1}$$

where slip frequency normalized to slip frequency at breakdown is

$$S = s/s_b$$

The actual motor torque is

$$T_m = TT_b$$

Combining the expressions given above for breakdown torque and normalized torque gives the actual motor torque as

$$T_m = \frac{M}{N^2}\left(\frac{2S}{S^2+1}\right)$$

The torque sensitivity to wheel diameter tolerance can be seen from the following example. Assume two wheelsets with nominal diameters, a third set with wheels oversized by a fraction $\Delta d/d$, and a fourth set with wheels undersized by the same fraction. The slip control system will average the outputs of the four motor tachometers, forming a speed signal which closely approximates the true rotational frequency of the two motors driving the nominal diameter wheelsets. In response to a torque command, the control system adds a slip frequency to the tachometer derived average rotational frequency to form the inverter output frequency. The rotational frequencies of the motors connected to the large and small wheelsets will be fractionally low and high, respectively, by $\Delta d/d$. The normalized slip of these motors in terms of a programmed slip, S, and an error term, DN, is then

$$S\,(\text{lg whl}) = S + DN$$

$$S\,(\text{sm whl}) = S - DN$$

where

$$D = \frac{\Delta d/d}{s_b/f_B}$$

represents the ratio of fractional wheel diameter to fractional breakdown slip at base frequency. The torques of the motors with non-standard wheel diameters are then

$$T_m(\text{lg whl}) = \frac{2M}{N^2}\left[\frac{S + DN}{(S + DN)^2 + 1}\right]$$

$$T_m(\text{sm whl}) = \frac{2M}{N^2}\left[\frac{S - DN}{(S - DN)^2 + 1}\right]$$

The load torque is the sum of the torques of the four motors, so that

$$T_\ell = 2T_m\,(\text{nom whl}) + T\,(\text{lg whl}) + T\,(\text{sm whl})$$

$$T_\ell = \frac{2M}{N^2}\left[\frac{2S}{S^2 + 1} + \frac{S + DN}{(S + DN)^2 + 1} + \frac{S - DN}{(S - DN)^2 + 1}\right]$$

For constant power operation, load torque varies with speed as

$$T_\ell = 4/N$$

Figure 2 is a plot of individual motor torques and load torque for the case of wheel tolerance and breakdown slip ratios of

$$\Delta d/d = 0.75\text{ cm}/75\text{ cm} = 0.01$$

$$s_b/f_B = 3\text{ Hz}/60\text{ Hz} = 0.05$$

$$D = 0.01/0.05 = 0.2$$

We note that even the relatively small wheel diameter tolerance of ± 0.75 cm (5 cm is a typical spread between new and worn wheel diameters) results in a torque reversal (braking torque) on the motor driving the undersize wheelset. This is offset by a motoring torque on the motor driving the large wheelset of 2.4 times normal at base speed. The wheel-rail interface of course could not support this torque and wheel slide would result.

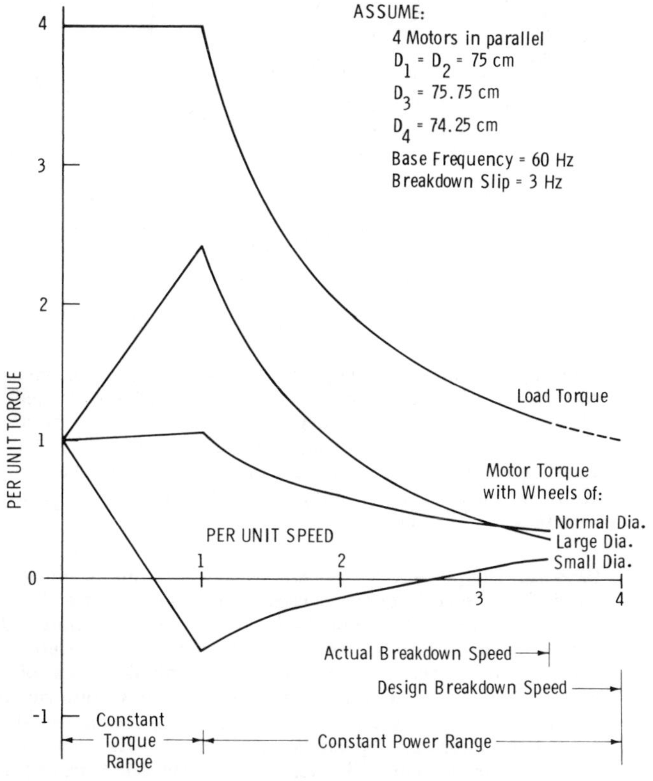

Figure 2. Induction Motor Torque Unbalance

In addition to wheel slide, other formidable problems appear because of unequal wheel size. Motoring efficiency will be greatly reduced if a motor is braking. Power factor will also suffer because of higher slip operation on the slower turning motors. The sensitivity to wheel size can be reduced by using higher slip motors, but this approach may reduce efficiency to a level well below that obtained with dc motor propulsion. The somewhat surprising sensitivity of torque-to-wheel size at low speed is a corollary to the motor oversizing discussed earlier. Thus, this excess torque capability must be accommodated by very low slip operation in the vicinity of base speed. Writing the motor torque equation again as

$$T_m = \frac{M}{N^2}\left(\frac{2S}{S^2 + 1}\right)$$

and noting that $N = 1$ and $T_m = 1$ at base speed, the required base speed slip is

$$S = M\left[1 - (1 - 1/M^2)^{1/2}\right] \cong 1/2\,M$$

For a typical pullout speed of $M = 4$, $S \cong 1/8$ of the slip frequency to produce breakdown torque. In the example cited above, the motor designed for 3.0 Hz breakdown slip must operate at 3/8 Hz slip at base speed. The effect of a small error in determining the true synchronous speed, whether from wheel size variations, tachometer offset, or other sources, is magnified in the base speed region to produce a large torque difference in the parallel connected motors.

Braking Mode

The braking performance of the PWM inverter-induction motor transit car drive mirrors the motoring performance [6]. By contrast, braking torque exceeds motoring torque at high speed by as much as 3/1 in modern dc motor drives. High torque capability at high speed requires a high flux level. This flux is not available because of the state-of-the-art PWM inverter is unable to modulate motor voltage at high speed.

The inability of the PWM inverter-induction motor drive to handle the kinetic energy of the transit car in braking has two implications. Most obvious is the added load placed on the friction brakes and the attendant increase in maintenance costs. Also, in transit systems which are receptive to regenerative braking, deficient torque capability at high speed means that a large portion of the kinetic energy is not recoverable when high deceleration levels are required.

SYNCHRONOUS MOTOR TRACTION

The synchronous ac motor functions as commutator-less dc motor when the phase of the stator (armature) current is controlled by a rotor position sensor. This is a direct analog of the commutator motor, with the rotor postion sensor, converter, and stationary armature performing the same function as the mechanical commutator and rotating armature. The concept of the self-controlled, or self-synchronous, motor appears to have originated in the early 1930's with Alexanderson and Mittag [9]. The advent of reliable high power thyristors now makes the self-synchronous motor a viable competitor to the commutator machine and to the inverter-fed induction motor in many applications.

Self-synchronous drives have received more emphasis in Europe and Japan than in the U.S. [10, 11, 12]. These foreign applications include a 1,500 kW sodium pump drive in a nuclear station and numerous gearless cement mill drives with ratings to 6,000 kW. In the U.S., the only known thyristor-controlled self-synchronous traction drive is a military truck conversion[13].

This vehicle uses the standard IC engine to drive a high frequency alternator and six cycloconverter controlled wheel motors. The motors are homopolar inductor machines.

Machine Structures

In the past, the maintenance and reliability penalties of slip rings restricted self-synchronous traction drives to the use of the inductor machine. This brushless double airgap structure lacks the efficient materials utilization obtained with wound rotors. The resulting weight penalty is only partially offset by the high speed capability of the solid steel rotor. The advent of the reliabile silicon diode rotating rectifier, coupled with a rotary transformer [14], now makes the brushless wound rotor machine a practical reality. No longer is there a need to compromise performance to obtain a brushless machine.

The brushless wound rotor machine offers an obvious weight saving potential over the dc machine. Assuming equal current and flux densities and active material dimensions, elimination of the commutator significantly shortens the length of the machine (the rotary transformer and rectifier occupying much less space than the commutator) and also allows higher speed operation with a corresponding reduction in weight. Finally, commutator-less operation permits liquid cooling which raises the permissible current density. The 400 Hz, 12,000 r/min aircraft alternators in current use embody these three weight saving factors to achieve an impressive power density of up to 4.5 kW/kg. By contrast, high performance dc traction machines produce a power density approaching 0.65 kW/kg in dynamic braking.

It is apparent that a significant potential for weight reduction exists with the synchronous ac traction machine. This weight reduction will permit higher vehicle speed because of better truck stability and will reduce track wear because of lower dynamic normal force loads. Perhaps,most significantly, the synchronous motor approach enhances the feasibility of the monomotor bogie. Here, one reasonably sized ac motor replaces two large conventional dc motors. The result is a reduction in the number of traction motors per car by half and improved truck stability and adhesion [15].

Power Converter Approaches

All of the self-synchronous drives referenced above operate from low frequency utility power, or from a high frequency alternator [9, 10, 11, 12, 13]. Thus, there is no established approach to follow in adapting the self-synchronous ac motor to operate from dc third rail power.

The nature of the brushless synchronous machine dictates two requirements that a converter must meet:

1. Provide a current-source characteristic to minimize flux undulations.
2. Provide a source of high frequency excitation for the rotary transformer field supply.

Three candidate converter approaches are evaluated below. The PWM inverter, described above in connection with induction motor rail drives, and the dc chopper-variable current input (VCI) inverter both have serious limitations. A new approach, the capacitor-coupled-cycloinverter, is shown to be well matched to the requirements of the synchronous motor.

PWM Inverter. A block diagram of a PWM inverter-synchronous motor drive is shown in Figure 3. This system resembles the induction motor drive with the addition of a separate inverter for excitation of the primary of the rotary transformer.

Because of the thyristor switching speed limitations described above, the inverter must provide constant voltage above base speed. Constant power operation is then obtained by programming the field supply inverter to weaken the field inversely with speed, as in shunt dc motor operation.

There are a number of disadvantages in the PWM inverter-synchronous motor approach. Perhaps the most important is the need for relatively large inductors in series with the motor leads. Enough reactance must be supplied at the motor frequency to prevent flux undulations which would result in excessive damper winding loss and field winding induced voltage.

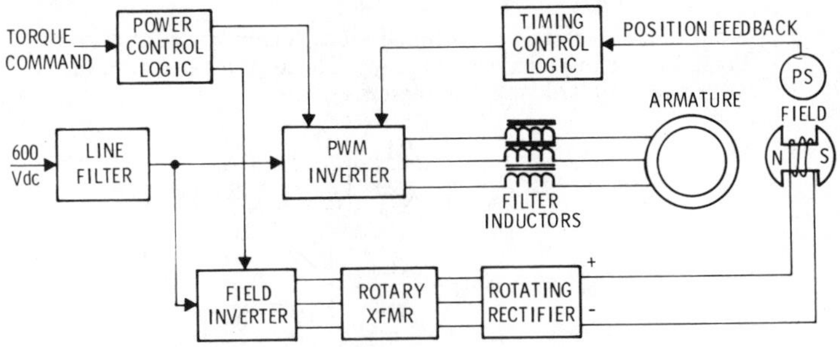

Figure 3. PWM Inverter - Synchronous Motor

Stated another way, the inverter should have a quasi-current source characteristic to permit the machine to operate with an approximately sinusoidal voltage waveform. In addition to the obvious weight penalty of the three filter inductors, the effective reduction in motor power factor must also be considered.

Other areas of concern with the PWM inverter-synchronous motor are the reduced efficiency implied by high armature current (relative to that with variable voltage operation) at high speed, and the large torque angle required with weak flux operation. A potential stability problem (speed tending to increase with torque) may also exist with the equivalent shunt dc motor operation [16].

VCI Inverter. The block diagram of the variable current input inverter is shown in Figure 4. The required current source characteristic is provided by the reactance of the coupling inductor, with no inductors required in series with the motor. As with the PWM inverter, a separate high frequency inverter must be provided for field excitation.

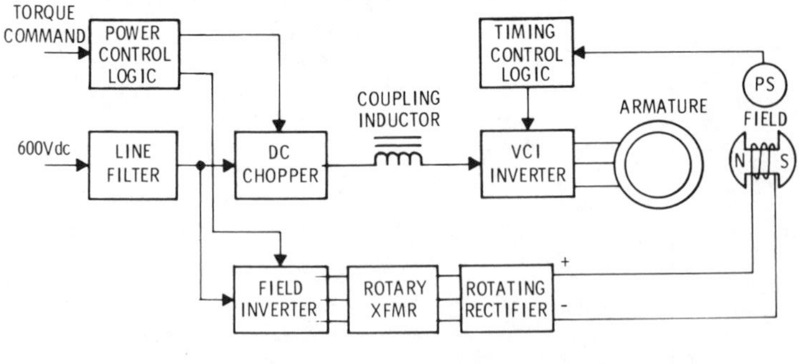

Figure 4. VCI Inverter - Synchronous Motor

This approach allows efficient unity power factor motor operation. In principle, it also offers the possibility of modulation above base speed to avoid the liabilities associated with the weak field operation over a wide speed range. In practice, this "full time" modulation may require use of a complex polyphase chopper to avoid the continuous high RMS filter capacitor current that would result.

Capacitor-Coupled-Cycloinverter. Generically, the cycloinverter consists of a high frequency inverter and cycloconverter [17] Control of the amplitude of the low frequency ac output voltage has been through control of the dc input to the inverter, PWM control of the inverter voltage, or by phase control of the cycloconverter. In the capacitor-coupled-cycloinverter, shown in Figure 5, motor current is controlled by frequency modulation of the interstage capacitive reactance. This reactance effectively acts as a controlled lossless impedance in series with the motor terminals, serving both to control motor current and to provide the converter with a current source characteristic.

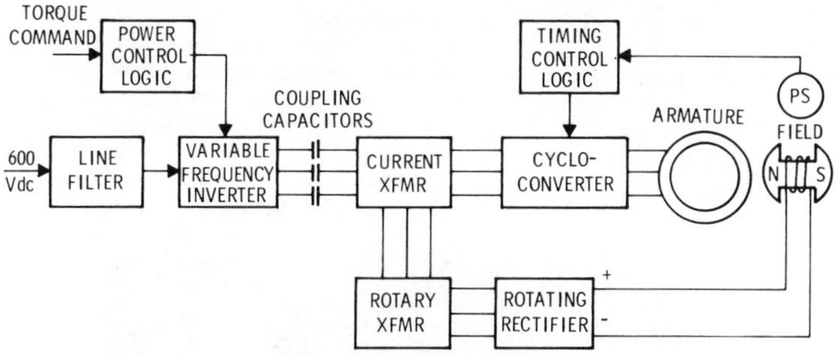

Figure 5. Capacitor-Coupled-Cycloinverter

The machine terminal voltage is then essentially sinewave for efficient constant flux operation. Field excitation is by means of the high frequency interstage current transformer which drives the motor field winding via the rotary transformer and rotating rectifier, with dc current proportional to armature current. The self-synchronous ac machine thus takes on the characteristics of a series dc motor.

Propulsion System Description

The capacitor-coupled-cycloinverter was developed for synchronous motor transit car propulsion primarily because of the inherent current source output characteristic and the simple means of obtaining motor field excitation. The inherent reliability of thyristor commutation in the input and output stages was also a factor in selecting this unconventional converter for development. Adaptability to a bi-modal commuter car operation was still another consideration. Here, the railcar is powered by a turbo-alternator-cycloconverter in rural areas and by noise and pollution free dc third rail-inverter-cycloconverter in urban areas.

Control Considerations. The transit car traction motor must have equal performance in both directions. With dc motors, brushes must be set at the neutral angle. Similarly, in a self-synchronous motor, the rotor postion sensor must be fixed so that its output signals have the same variation with angle of rotation in both directions. This bidirectional performance requirement dictates the need for a change in the timing of the applied current with load to achieve the desired power factor. This change is termed commutation angle advance. Referring to the phasor diagram of Figure 6a where the simplifying assumptions of smooth rotor construction and negligible resistance are made, the quantities are:

A = Armature MMF	V = Terminal voltage
F = Field MMF	δ = Torque angle
R = Resultant of A and F	θ = Power factor angle
Φ = Resultant flux	α = Commutation advance angle

From [8] the torque is

$$T = \frac{\pi}{2}\left(\frac{\text{poles}}{2}\right)^2 \Phi F \sin\delta$$

With the further assumption of a linear magnetic circuit, the flux is proportional to the resultant MMF so that torque in terms of a constant K_1 is

$$T = K_1 RF \sin\delta$$

From Figure 6a it is apparent that with no commutation angle advance ($\alpha = 0$), the power factor and torque angles are identical.

Commutation angle advance in the drive under discussion is obtained by blending a signal derived from the motor terminal voltage with a signal developed from the variable reluctance type rotor position sensor. Except at very low speed, the motor current is maintained approximately in-phase with the terminal voltage. Figure 6b shows the unity power factor case where commutation angle advance is equal to the torque angle. In this case, the resultant MMF is

$$R = \sqrt{F^2 - A^2}$$

Because of the current transformer field excitation, field MMF is proportional to armature MMF and the torque angle is constant. The resultant is

$$R = A\sqrt{K_2^2 - 1}$$

The torque thus becomes

$$T = K_1 K_2 A^2 \sqrt{K_2^2 - 1} \sin\delta = K_3 A^2$$

Power is constant above base speed so that torque varies with speed as

$$T = P/N$$

The armature current must thus be programmed to vary with speed to obtain

$$A = \sqrt{T/K_3} = \sqrt{P/K_3 N} = K_4/\sqrt{N}$$

Also, for constant power, the product of A and V is constant so that terminal voltage must vary above base speed as

$$V = K_5 P/A = K_6\sqrt{N}$$

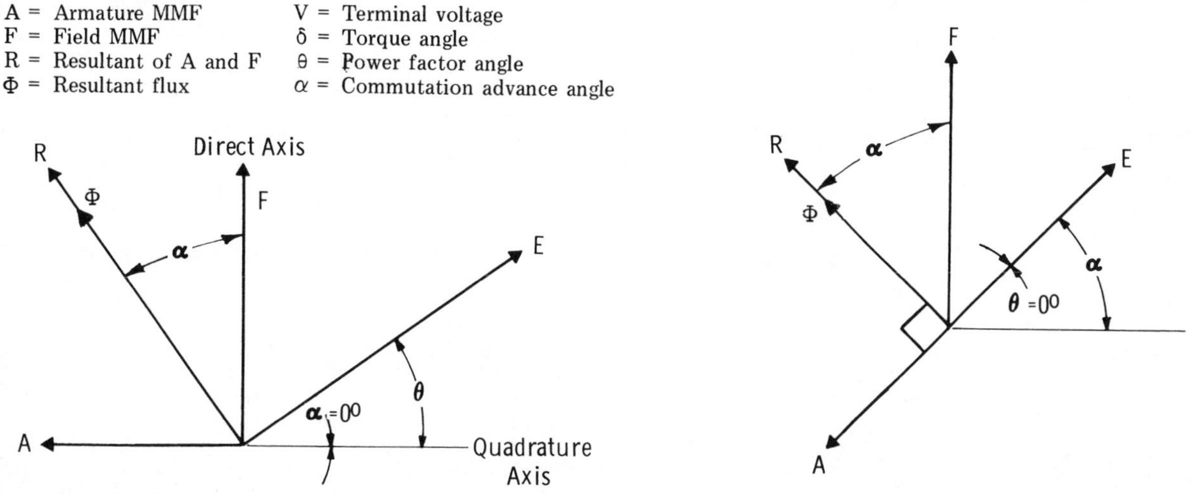

a) Zero Commutation Angle Advance

b) Commutation Angle Advanced for Unity Power Factor

Figure 6. Phasor Diagrams

High Frequency Inverter. The inverter stage of the propulsion system consists of three modified Mapham inverters [18] which are gated to produce a three phase output voltage over a frequency range of approximately 300 Hz to 1,200 Hz. This sine wave series inverter provides lower thyristor di/dt and longer turn-off time than parallel inverters, and simpler and more reliable commutation than force commutated inverters. The peak thyristor blocking voltage approaches twice that of the dc supply voltage. The circuit diagram of Figure 7 shows how the three multisection capacitors combine the functions of commutating, coupling and dc blocking.

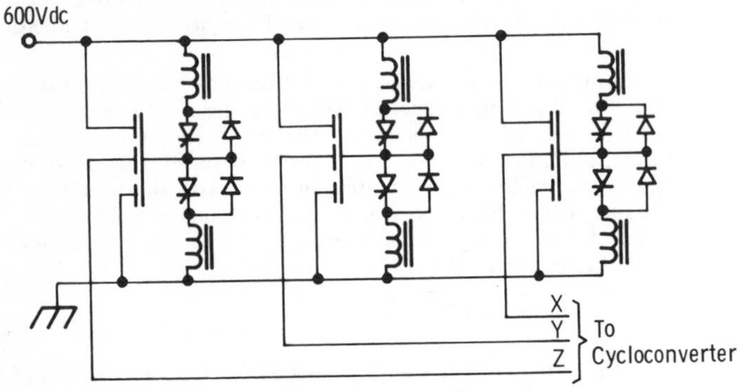

Figure 7. Input Inverter Stage

The feasibility of the capacitor-coupled-cycloconverter hinged on the development of efficient resonating inductors and the commercial availability of high current capacitors. Adequate inductors, with Q factors in excess of 150, were developed. Forced convection cooled capacitors of the type used in the induction heating industry were adapted. Typical capacitor specifications are: 102 μF, 1200 Hz, 1250 V rms, 960 A rms, 1200 kvar, 0.003 dissipation factor.

Field Excitation. The current transformer-rotary transformer field supply maintains the approximately constant ratio of field to armature MMF. At very low inverter frequencies, the magnetizing current of the cascaded transformers become significant. This effect is put to use to "fade out" the field current at very low power settings to permit starting with very low jerk. At higher inverter frequencies, leakage inductance is the dominant, but manageable, transformer characteristic. The current transformer leakage inductance serves a useful purpose in controlling the di/dt of the cycloconverter thyristors.

Cycloconverter. A circuit diagram of the 18 thyristor cycloconverter is shown in Figure 8. The manner in which the synchronous motor characteristics affect cycloconverter design and operation are described below.

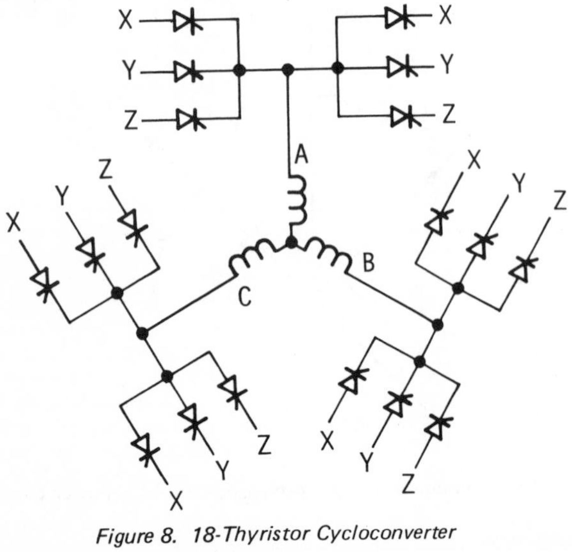

Figure 8. 18-Thyristor Cycloconverter

The well-defined counter EMF of the self-controlled synchronous motor forces phase current to fall rapidly and predictably upon removal of the phase current logic command. Hence the motor conduction angle is only moderately greater than the duration of the logic signal command, or trigger angle. This allows the use of 120° trigger angle which is readily developed from the rotor position sensor outputs. Figure 9 shows oscillograms of motor current and voltage response with 120° trigger angle. The conduction angle is observed to be approximately 130°. The square wave current waveform and sine wave voltage waveform verify the current source characteristic of the capacitor-coupled cycloinverter.

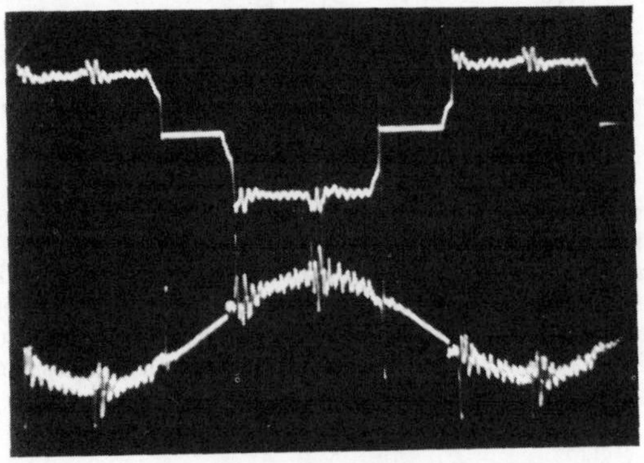

Figure 9. Phase Voltage and Current - 200A/cm, 200 V/cm, 5 ms/cm

The nominally 120° wide current waveform is very advantageous. Harmonic current is minimized and the phases can be connected to form a neutral to eliminate zero sequence current components [19]. Interconnecting the phases permits use of an 18-thyristor cycloconverter instead of the 36-thyristor cycloconverter used in induction motor drives [3,20].

Braking. Field excitation of the traction machine, operating as an alternator, controls the braking effort. This field excitation is obtained using the current transformer described above by continuously gating the cycloconverter to place a line-to-line short circuit at the input to the cycloconverter. Current from the variable frequency inverter can flow through the coupling capacitors and primary windings of the current transformer. The current transformer secondary winding then energizes the primary of the rotary transformer.

For dynamic braking, the machine voltage can be applied to a three phase resistor bank with an electromechanical or static switch, or can be rectified and applied to a dc load resistor. Regenerative braking is obtained by applying the rectified machine output to the dc supply line, raising the supply voltage to a maximum level which determines the limit of regenerated current receptivity. With a nominal 600 Vdc supply voltage, regeneration is possible down to the machine base speed. With a low line voltage, indicating heavy current drain by nearby accelerating cars, regeneration is available to a lower speed. For the case of deceleration from a maximum speed of 129 km/h (80 mi/h) to a typical base speed of 36 km/h, and assuming no losses in the system, a fraction

$$1 - (36/129)^2 = 0.92$$

of the total kinetic energy can be recouped to a receptive source.

Cooling System. The synchronous motor drive system is forced convection cooled using a non-flammable phosphate-ester lubricating coolant. The motor cooling design is derived from aircraft alternator practice. A separate lower temperature cooling circuit is provided for the converter. Thyristors, diodes, and heat sinks are placed in containers filled with coolant at essentially atmospheric pressure. Coolant passes first through the heat sinks and then into the container where additional heat transfer takes place from the exterior of the heat sink. This arrangement gives a thermal resistance from the puck thyristor case to inlet coolant of typically 0.06^oC/watt. Circulating liquid cooling, as opposed to air cooling, brings the entire thermal mass of the power converter into play to limit component transient temperature fluctuations during the typically intermittent power operation of the transit car. The inevitable dirt buildup on the exterior of the oil-to-air heat exchanger can be conveniently removed with conventional transit car cleaning equipment.

Reliability Considerations. In a statically controlled motor drive, the reliability of the thyristor commutation process dominates overall system reliability. With the dc chopper or force-commutated PWM inverter, a pre-charged commutation capacitor must be discharged by firing an auxiliary thyristor to turn off the conducting main thyristor. Fewer components are involved in the commutation process of the inverter stage of the cycloinverter. In this circuit, a thyristor, once gated conductive, turns off naturally by resonant current reversal. Commutation is also highly reliable in the line commutated cycloconverter stage. At low speed, reversals of the input line voltage turn off motor phase current when the 120° wide gate drive is removed from a group of thyristors. At high speed, the motor EMF appearing across the cycloconverter output line forces thyristor current to zero at the end of the normal conduction period.

The provision of separate converters for each monomotor bogie is the most significant reliability factor in the propulsion system. This permits operation at 50% performance level with one power converter disabled. The probability of a stranded car with the dual propulsion system is extremely remote compared to the case of a car equipped with the usual single dc chopper or PWM inverter.

Test Results. The capacitor-coupled-cycloinverter was tested with a standard 170 kW synchronous motor converted to self-synchronous operation by the addition of a rotor position sensor. The internal fan was removed and separate ventilation was provided for cooling at low speed.

Curves showing the control of locked rotor phase current and torque with inverter frequency are plotted in Figure 10. The dc supply current which accounts for power converter and motor losses is also shown. For inverter frequencies above 600 Hz, phase current (a dc quantity at zero speed) varies linearly with frequency while the curve of torque vs. frequency increases more rapidly with frequency, as expected with series field excitation.

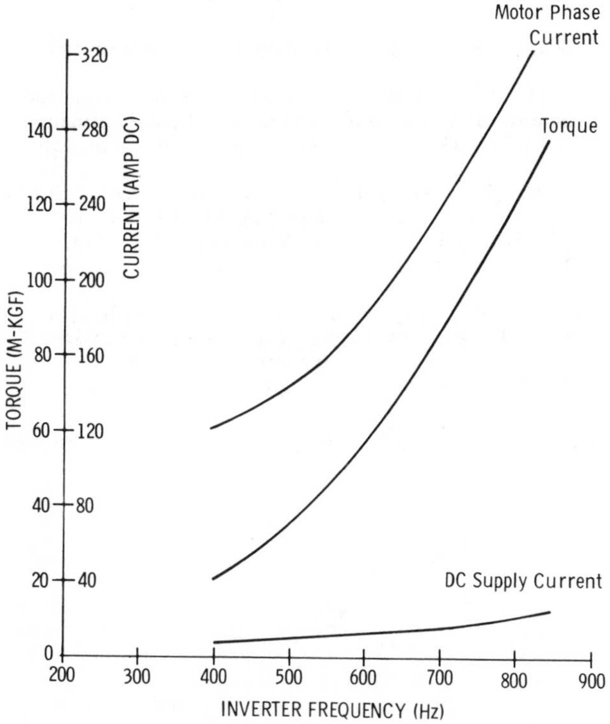

Figure 10. Locked Rotor Test Results

The dc-dc transformer action of the converter is revealed by low level of dc supply current over the frequency range. Here the high dc input voltage (approximately 600 Vdc in this test) and low input current, transform to high motor current and low motor voltage (the motor IR drop with R = 0.04 ohm). At the maximum torque level (limited in this test by the torque transducer), the dc current transformation ratio is 340/25 = 13.6.

Torque and efficiency vs. speed for constant power operation is shown in Figure 11. The efficiency shown is mechanical output power (speed x torque) over converter input dc power. Losses in the line filter inductor were not included. At the 50 kW level, efficiency varies from 84% to 90% over the speed range of 750 to 1,800 r/min. At 115 kW, dynamometer limitations increase the minimum speed to 1460 r/min. Efficiency ranges from 90% to 91% at the higher power level.

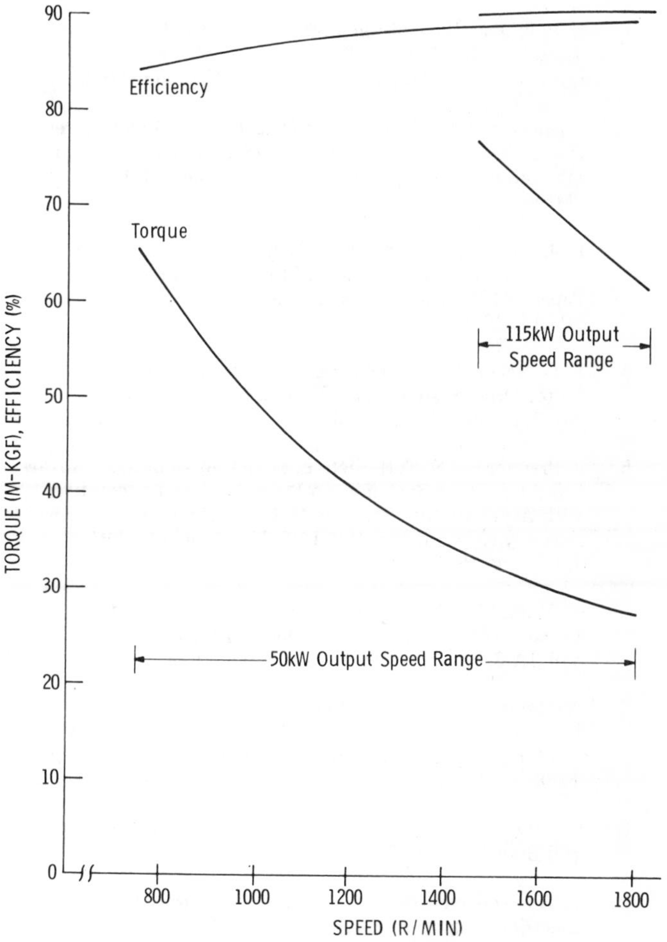

Figure 11. Torque and Efficiency Test Results

SUMMARY

The ac motor rail transit car propulsion system described above represents a departure from convention in both the motor and the converter. The motor is a wound rotor synchronous machine, made brushless by means of a rotary transformer-rectifier and self-synchronous by the addition of a rotor position sensor. Rotary transformer excitation is derived from a current transformer to produce dc field current proportional to armature current, as in a series dc motor. The synchronous motor was selected over the squirrel cage induction motor because of this series dc motor characteristic and because of its ability to provide adhesion limited braking torque over the entire vehicle speed range.

The special requirements of the synchronous motor led to the development of a new type of dc-to-three phase ac converter termed the capacitor-coupled-cycloinverter. In this converter, the reactance of interstage coupling capacitors is frequency modulated over a typical 300 Hz to 1,200 Hz range to control the synchronous motor armature current. Motor voltage, approximately sine wave, is established by flux and speed. A current transformer in series with the coupling capacitors supplies the necessary high frequency excitation for the rotary transformer-rectifier in the motor. This form of cycloinverter was chosen over conventional converter approaches because of its natural current source characteristic, simple means of series field excitation, low ripple current demand on the input filter, and simple control circuitry. Ready adaptability to bimodal (turbine-3rd rail) propulsion was also a selection factor.

Tests have shown a typical system efficiency of 90% over a wide speed range in spite of the two-stage power conversion. This is due to the use of low loss inductors and capacitors in the inverter stage and the relatively high efficiency of the synchronous motor.

REFERENCES

1. E. P. Priebe. "Propulsion Motor Requirements for Mass Transportation." IEEE Transactions, Industry Applications, Vol. IA-8, No. 3, pp. 310 - 315, May/June 1972.

2. J. Stewart, J. Waldron. "Fault Detection on Direct Current Rapid Transit Systems." IEEE - IAS Conference Record. 8th Annual Meeting, Oct. 8-11, 1973, Paper LT-THU-PMI-895.

3. L. J. Lawson, R. Borland. C. Puchy, " Optimal Control System Performance for an AC Electric Vehicle Drive." Paper 66135. Automotive Engineering Congress, Jan. 10 - 14, 1966.

4. P. D. Agarwal. "The GM High-Performance Induction Motor Drive System." IEEE Transactions, Vol. PAS-88, No. 2, pp. 86-93, Feb. 1969.

5. J. Brenneisen, E. Futterlieb, E. Muller, M. Schultz. " A New Converter Drive System for a Diesel Electric Locomotive with Asynchronous Traction Motors." IEEE Transactions, Industry Applications, Vol. IA-9, No. 4, pp. 482-291, Jul. / Aug. 1973.

6. N. Vutz. "PWM Inverter Induction Motor Transit Car Drives." IEEE Transactions Industry Applications, Vol. IA-8, No. 1, pp. 89-91, Jan. / Feb. 1973.

7. A. Humphrey. " Constant Horsepower Operation of Induction Motors." Paper SPC-TUE-2-743, IEEE - IGA Conference Record, 2nd Annual Meeting, Sept. 29/ Oct. 3, 1968.

8. A. Fitzgerald, C. Kingsley, Electric Machinery., McGraw-Hill Book Co., 2nd Ed., 1961.

9. E. Alexanderson, A. Mittag. " The Thyratron Motor." Electrical Engineering (AIEE), pp. 1517 - 1523, Nov. 1934.

10. A. Habock and D. Kollensperger. " State of Development of Converter-Fed Synchronous Motors with Self-Control." Siemens Review, No. 9, pp. 390 - 392, 1971.

11. H. Stremmler, " Drive System and Electronic Control Equipment of the Gearless Tube Mill." Brown Boveri Review, Vol. 57, No. 3, pp. 120 - 128, March 1970.

12. J. Inagaki, M. Kuniyoshi, S. Tadakuma. "Commutators get the Brushoff." IEEE Spectrum, Vol. 10, No. 6, pp. 52 - 58, June 1973.

13. G. Collins, W. Slabiak. "Brushless Synchronous Propulsion Motor." Paper 680455 presented at SAE Mid-Year Meeting, Detroit, May 1968.

14. G. Kracke. "Rotating-Rectifier Excitation for Synchronous Motors, Synchronous Condensers, and Converter-Fed Synchronous Motors." Siemens Review, No. 10, pp. 530 - 534, 1970.

15. F. Nouvion. "Diesel and Electric Railroad Operation in France." Paper No. C-72-936-8-IA, presented at the Joint ASME/IEEE Railroad Conference Jacksonville, Florida, March 14-15, 1972.

16. E. Cornell, D. Novotny. "Theoretical Analysis of the Stability and Transit Response of Self-Controlled Synchronout Machines." Paper 73-144-2, 1973, IEEE Power Engineering Society Winter Meeting, New York, N.Y., Jan. 28 - Feb. 2, 1973.

17. SCR Manual, 5th Ed., General Electric Co., p. 396, 1972.

18. N. Mapham. "An SCR Inverter with Good Regulation and Sine Wave Output." IEEE Transactions Industry Applications, Vol. IGA-3, No. 2, pp. 176 - 187, March/Apr. 1967.

19. L. Jacovides. "Analysis of Induction Motor Drives with a Nonsinusoidal Supply Voltage using Fourier Analysis." IEEE Transactions, Industry Applications, Vol. IA-9, No. 6, pp. 741 - 747, Nov./Dec. 1972.

20. L. Jacovides. "Analysis of a Cycloconverter - Induction Motor Drive System Allowing for Stator Discontinuities." IEEE Transactions, Industry Applications, Vol. IA-9, No. 2, pp. 206 - 215, March/April 1973.

Electrical Aspects of the 8750 hp Gearless Ball-Mill Drive at St. Lawrence Cement Company

JOHN A. ALLAN, MEMBER, IEEE, W. A. WYETH, GORDON W. HERZOG, SENIOR MEMBER, IEEE, AND JOHN A. I. YOUNG, MEMBER, IEEE

Abstract—**The first large gearless ball-mill in North America has been successfully installed at the Mississauga, Ontario, plant of the St. Lawrence Cement Company. The motor rating is 8750 hp (6500 kW) at 14.5 r/min (4.84 Hz). The electrical aspects considered of interest are discussed in the paper.**

INTRODUCTION

THE CONCEPTS of gearless ball-mill drives were presented at the 1968 Cement Industry Conference (IEEE), while some of the installations in Europe have been reported on at the 1970 and 1972 Conferences. This paper discusses the electrical aspects of the first North American installation of a mill of this type. The mill is powered by a synchronous motor/cycloconverter drive in the self-controlled mode of operation. The mill was supplied by Aerofall Mills. The motor, power conversion, and control were designed and manufactured by the Canadian General Electric Company. The drive was installed at the Mississauga, Ontario, plant of St. Lawerence Cement, with the technical services supplied by Holtec (Holderbank Technical Services Ltd.). Cement was first produced on August 14, 1973.

GENERAL

As mill and motor design drawings were being developed by Aerofall Mills and Canadian General Electric, respectively, foundation and electrical designs were being simultaneously prepared by Holtec. Close cooperation between the three participants was established early in the project and consequently rapid progress was made.

The foundation for the mill discharge end bearing and motor stator was designed as one base slab of concrete with piers extending from it to support the bearing and stator. The stator support piers including the slab thickness were designed to meet the stiffness specification set by the motor requirements. Preliminary sizing of the major components of the gearless drive system and of the principal power switching equipment enabled structural design and construction of the building to be completed on schedule. The layout of these components and equipment is shown in Fig. 1.

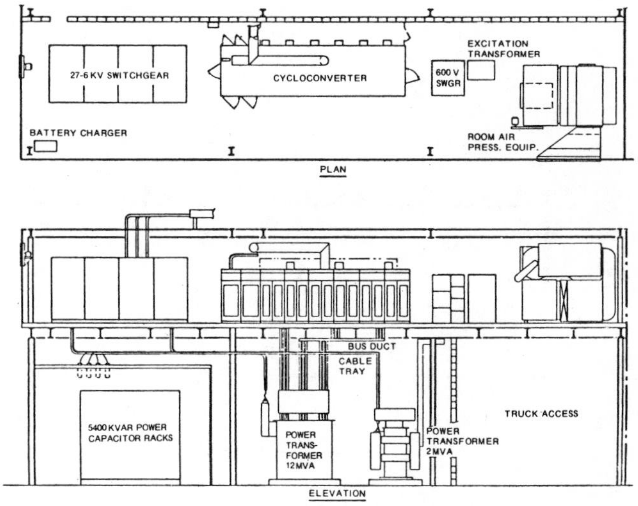

Fig. 1.

Air Conditioning of Electrical Room

A 24 000 ft^3/min (11.33 m^3/s) air conditioning unit was installed to provide clean moisture-free air of uniform temperature. This unit was equipped with cooling coils and 130 kW of heating elements and was capable of maintaining the required room temperature whether the equipment was operating or not, regardless of season. In addition, the unit filtered the air as well as maintained a positive pressure in the room. A small portion of the room air was used for makeup for the motor closed air system.

Dust Resistant Enclosures

All electrical equipment terminals outside of the electrical rooms were enclosed in cable boxes. A totally enclosed bus duct was used for all major power interconnections. The transformer secondaries were connected to the cycloconverter via a miniphase bus duct while the output power was fed to the stator by means of an isolated phase bus duct.

The capacitor bank was housed separately from other equipment. It was divided so that a two-branch filter could be added if required.

Paper TOD-74-107, approved by the Cement Industry Committee of the IEEE Industry Applications Society for presentation at the 1974 IEEE Cement Industry Technical Conference, Mexico City, Mexico, May 13–16. Manuscript released for publication January 27, 1975.

J. A. Allan is with Holderbank Technical Services, Ltd., Clarkson, Ont., Canada.

W. A. Wyeth, G. W. Herzog, and J. A. I. Young are with Canadian General Electric Company, Ltd., Peterborough, Ont., Canada K9J 7B5.

Reprinted from *IEEE Trans. Ind. Appl.*, vol. IA-11, pp. 681–687, Nov./Dec. 1975.

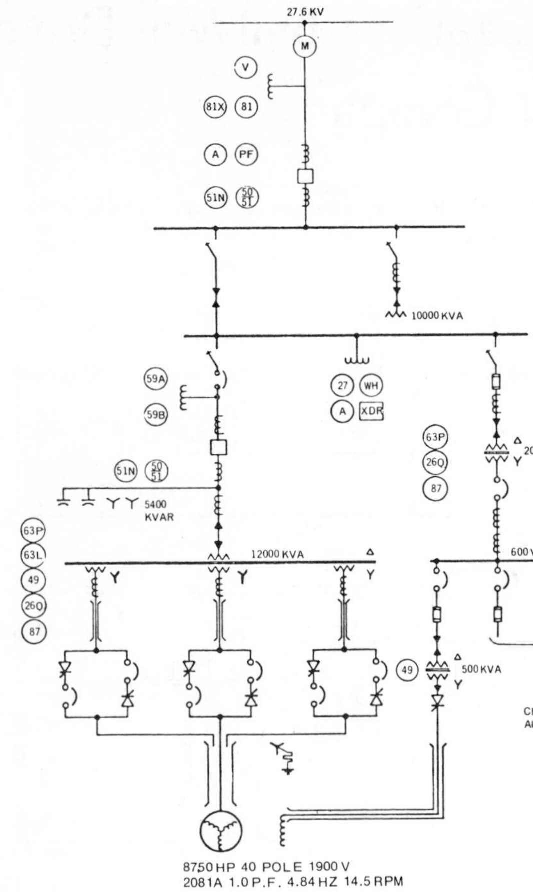

Fig. 2. Gearless drive power system.

Power System Protection

The power system was designed to provide the protection required with the minimum amount of equipment. Protection included such features as capacitor unit failure detection, transformer differential, overcurrent, undervoltage, overtemperature, underfrequency and other normal requirements. The one-line diagram of the power system for the drive and auxiliaries is shown in Fig. 2. Motor and converter protection is discussed later.

Control

The control system for the mill was designed to supervise the start-up of the complete mill circuit, including the automatic staged start-up of the mill itself. This supervisory feature permitted the plant central operator to know, prior to initiating the mill circuit sequencing start-up, that all components were in a "Go" state and thus avoiding an abortive exercise due to an activated safety or protective device. Approximately 200 alarm points were monitored by the computer which annunciated and logged any incoming alarm.

Drive Layout

The side elevation of the drive shown in Figs. 3, 4(a) and 4(b) are photographs of the motor and mill, while Fig. 5 is a picture of the cycloconverter static power conversion and control room.

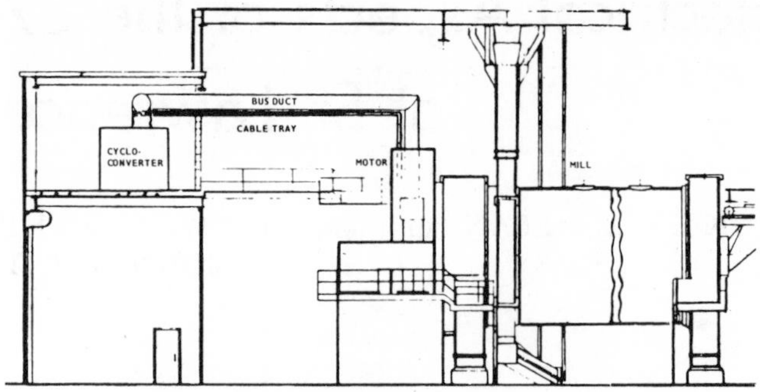

Fig. 3. Side elevation with mill foreshortened.

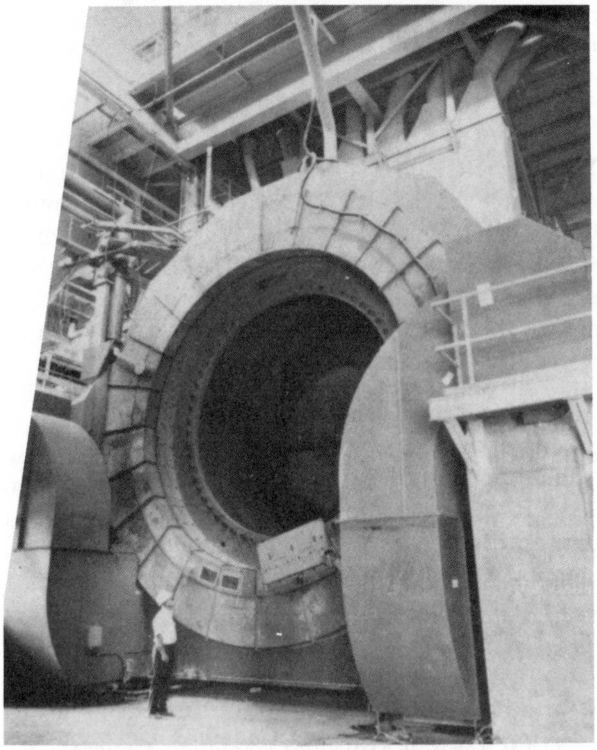

(a)

(b)

Fig. 4. (a) Motor. (b) Motor and mill.

Fig. 5. Control room.

Fig. 6. Rotor prepared for shipment.

MOTOR

Rating and Enclosure

The motor was rated 8750 hp, 1.0 PF, 6850 kVA, 14.5 r/min, 1900 V, 4.84 Hz, 40 Poles, Class B, 1.0 Service Factor. A totally enclosed structure with water to air heat exchangers and with air circulating fans was supplied. The surface air coolers were mounted on each side of the lower portion of the frame on the opposite mill side. Ducts carried the air from both ends of the frame to enclosing covers for double end type of ventilation.

This arrangement concentrated the motor services in one location, minimizing the mount of air ducts, while preserving accessibility to the motor.

The collector was in a separate compartment of the enclosing covers, separately ventilated to keep carbon dust out of the main motor enclosure. Dust seals were provided at the interface between the rotor spider and the enclosing covers. The entire enclosure was pressurized to minimize the ingress of dust.

Mounting

Figs. 4(a) and 4(b) of the installed motor show the overhung mounting arrangement, which is a variation of the overhung arrangement in [2].

The motor was mounted axially from the mill discharge end bearing with the rotor overhung on a tube extension of the mill. This torque tube, designed and supplied by Aerofall Mills, supports the rotor and transmits torque, and attenuates any out of round deflections, or changes in the mill attachment.

This arrangement removed the attendant problems of heat conduction from the mill to the rotor, and differential expansion between the mill and the rotor spider, when the motor was mounted in between the mill bearings. Furthermore, this arrangement provided accessibility to both the mill and the motor separately, while maintaining an acceptable distance between the centers of the mill bearing and the motor.

The rotor spider was mounted on a centering spigot on the torque tube flange and bolted with sufficient pre-load on the bolts to drive the mill by friction between the mating flange faces.

Construction

The mounting arrangement and the proximity of the manufacturing plant to the user's site, enabled a one-piece rotor to be supplied. Some of the poles were removed for shipment to meet the shipping clearances as shown in Fig. 6.

This construction enabled the use of a shrink fit between the rotor rim and the rotor spider central web.

An amortisseur winding was not necessary since the self-controlled mode offered the bonus of a speed regulator design that was stable without the need for amortisseur damping. This also minimized losses from harmonics in the stator current.

The stator frame was required to be split for shipment. This frame, suitable for horizontal mounting, was designed to withstand the weight, thermal expansion, and the unbalanced magnetic forces from a radially displaced rotor. No reduction in the magnetic forces was used in the frame design, even though a multiplicity of stator winding circuits was chosen. The motor can experience rotor displacements at zero speed and at the spotting speed when the counter effect from different stator circuit currents is nil or very low.

The effect of frame distortion, rotor spider, and torque tube deflections, on the peripheral air gap unbalanced magnetic force was solved in an iterative manner to obtain the air gap change for an initial rotor displacement. A major portion of any horizontal directed unbalanced

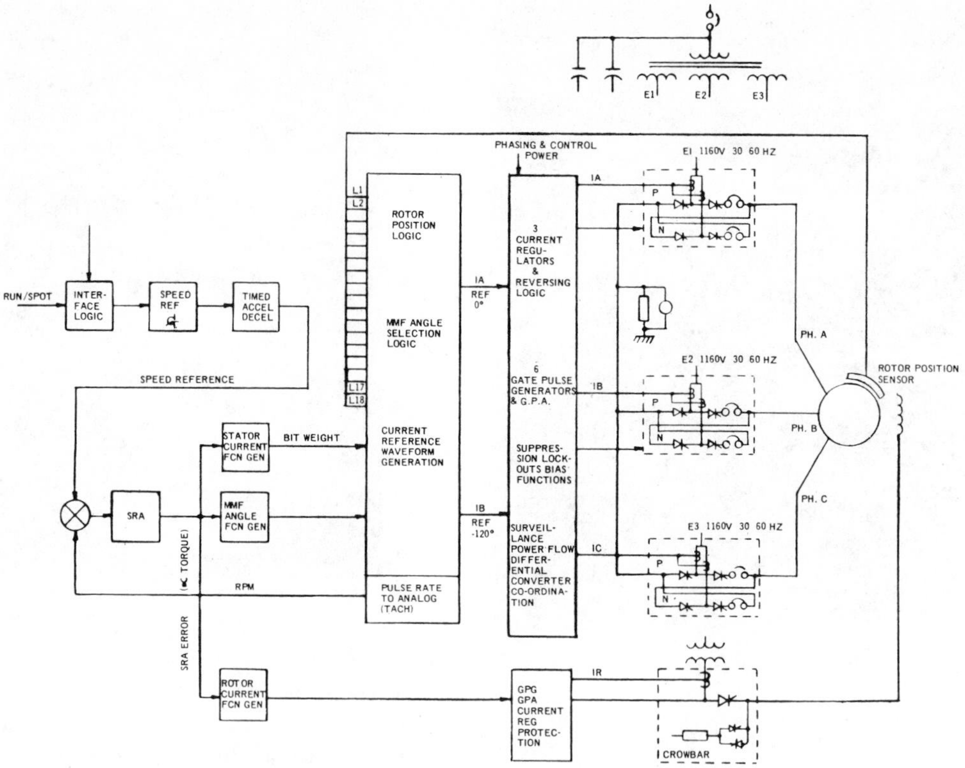

Fig. 7. Block diagram.

magnetic force must be carried by one stator foot pier, because the frame is less stiff, (when subjected to a horizontal air gap force), than its foundation piers. A specification of acceptable pier horizontal stiffness was given to the purchaser (which has been referred to earlier in the paper).

The vertical deflection of the rotor from both static and dynamic mill loads, together with the differential temperature expansions between the stator and the mill bearings, were used to establish the air gap setting for the field installation.

Class B insulation was provided on both the stator and the rotor. The stator coils were insulated with MicaMatt tape, and impregnated by a vacuum-pressure process with epoxy resin. All of the coils except those required to close the spans were installed in the factory. The stator coil end head bracing consisted of lashing to a support ring with peripheral blocks between arms by a material which conformed to the coil arm shape.

Bearing currents were eliminated by the overhung mounting arrangement so that it was not necessary to insulate the bearing pedestals. Rotor spider grounding brushes were provided to conduct any capacitive current that flowed from the thyristor rotor supply.

STATIC POWER CONVERSION AND CONTROL

Fig. 7 is the block diagram that identifies the major components of the drive system. These are the motor, the stator and rotor power converters, rotor position sensor, control, protection, and reactive compensation.

Stator Power Converter

Other than the motor, the largest component was the static power converter frequency changer. Since the maximum machine frequency of 4.84 Hz was less than 1/12 of the supply frequency, the cycloconverter was selected as the best choice for this function. Although cycloconverters have been known and well described for many years, there are some features worth identifying.

The transformer has a single core with 3 isolated secondaries which allows the cycloconverter to be Y connected with advantages of economy and reduced harmonic generation. The motor is also Y connected, and there is no connection between the two neutrals, so that the cycloconverter can be operated in the trapezoidal (flat-topped) voltage mode without the penalty of triplen currents flowing in the motor. This improves the power factor at rated speed.

Since the neutrals are floating, it is necessary to multiplex simultaneously the thyristor gating pulses to all appropriate cells in the active banks in each cycloconverter output phase. This is accomplished by the use of the Silpac® type bias-cosine gate pulse generation control system which operates mainly in the logic domain and is interfaced with the bank selection for each motor phase. This selection is controlled by unique reversing logic circuitry that receives information from solid-state thyristor conduction state sensors. This allows positive control

® Registered trademark.

of the currents even though there is no neutral tie between the motor and the cycloconverter.

Rotor Position Sensor

The motor is in the self-controlled mode, with the stator frequency directly controlled by rotor speed. Therefore the motor cannot fall out of step and thus has operating characteristics similar to a dc machine. This mode of control is accomplished by an unique rotor position sensor which provides signals that are decoded for each electrical cycle. Thus the rotor position relative to the stator is known at all times. At standstill it is known within ±5° electrical and when running ±0.2° for this particular design. Two quasi-sine wave references are generated in digital to sine wave converters that are controlled by logic functions. These waveforms form the references for the stator current regulators. The symmetry and spacing of these waveforms is very accurately maintained, and the sine wave rms distortion is negligible resulting in a high quality 3-phase reference waveform generator.

Control

The outer control loop is a speed regulator with the speed regulating amplifier (SRA) error voltage calibrated in proportion to the required machine torque needed to maintain the set speed. There is a rotor-stator MMF angle selection circuit function that allows the 3-phase reference set to be shifted over the necessary control range in response to a dc signal. Thus the stator rms current magnitude is proportional to the bit weight reference to the digital to sine converter and its frequency is controlled by the rotor speed. Also, the stator flux has a definite spatial relationship to the rotor flux and further the stator flux can be varied by the MMF control circuitry mentioned. The tachometer signal is derived from the rotor position sensor information.

There is a logic interface between the mill control and drive control via a timed acceleration and deceleration function generator. This results in smooth starting, stopping, and permits stepless vernier speed changes.

DRIVE PROTECTION

In addition to the power system protection already described, the following protective devices were among those provided for the devices.

Motor Protection

1) Loss of differential air pressure across the fans.
2) Fans exhaust air temperature.
3) Temperature detectors embedded in stator windings.
4) Condensation control via space heaters and a water temperature regulating valve as a function of air temperature.
5) Rotor radial displacement sensors that will de-energize the motor and prevent a motor start should abnormal conditions occur.
7) Stator ground detector.
8) Rotor ground detector.

Conversion and Control Protection

1) Loss of cooling air.
2) Differential protection around cycloconverter.
3) Cell loss annunciation.
4) Overcurrent; circulating currents; conduction through.
5) Loss of air.
6) Incomplete starting sequence.
7) Overspeed.
8) Current imbalance.

Diagnostics and Fault Finders

The status of all the key functions can be determined by light indicator cards in the control pages. Further, the significant control points were wired to diagnostic cards so that it was possible to connect oscilloscope channels to about 200 circuit locations simply by selecting thumbwheel switch locations.

A further diagnostic aid was an electronic rotor position simulator that allowed exercising all the drive signal electronics (with the thyristor voltages removed) as though the motor was actually moving.

These were extremely powerful maintenance aids.

DRIVE PERFORMANCE

The following section contains some of the performance data gathered under field conditions. The performance met or exceeded the design criteria and has been most encouraging.

Acceleration to "Run" Mode

Fig. 8 is a recording of an acceleration from "standstill" to "run" that was set for approximately 16 s to 12.64 r/min. The current built up until the mill started to move. The mill broke away and accelerated in a very smooth manner. The rotor current (I_R) was zero in the standby mode, and it built up to the value requested by the regulator.

Steady-State Parameters

Fig. 9 shows the 3 motor stator currents (I_A, I_B, I_C) and the rotor current (I_R) at 13.1 r/min. The stator currents were about 0.90 pu and 1 machine cycle is displayed. Fig. 10 shows the same waveforms with the phase A line to neutral voltage. This was at 13.5 r/min.

MMF Angle Control

Fig. 11 is a composite picture of a portion of phase A reference for three different lag settings relative to logic signal SCP10. The north-pole center line will be exactly in the center of the phase A belt when SCP10 first goes from a "0" to a "1." SCP10 refers to the "stator position logic signal decoded relative to the N-pole for the region 0° to 10°."

Reactive Compensation

As can be seen in Fig. 10 at 93 percent rated speed, the drive is in the trapezoidal mode. The system power factor

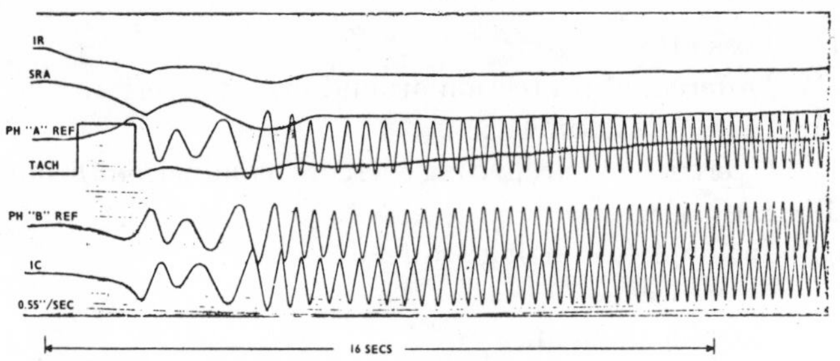

Fig. 8. Run acceleration.

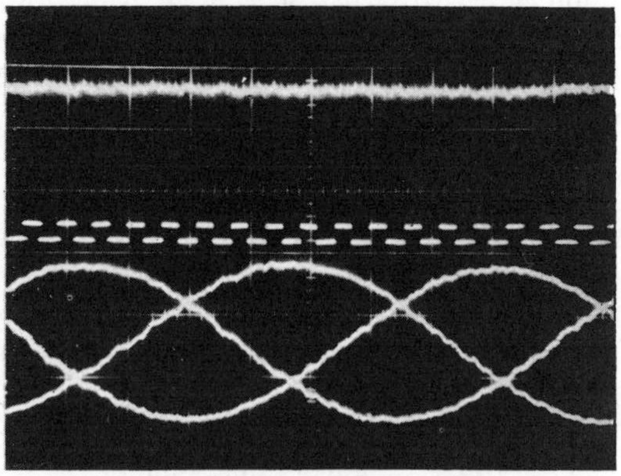

Fig. 9. Current waveforms.

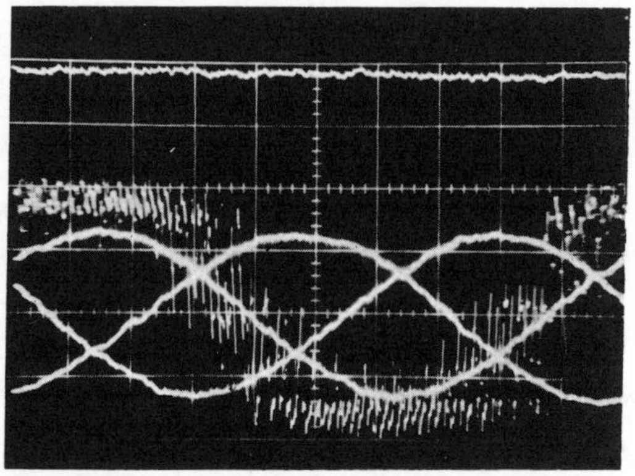

Fig. 10. Voltage and current waveforms.

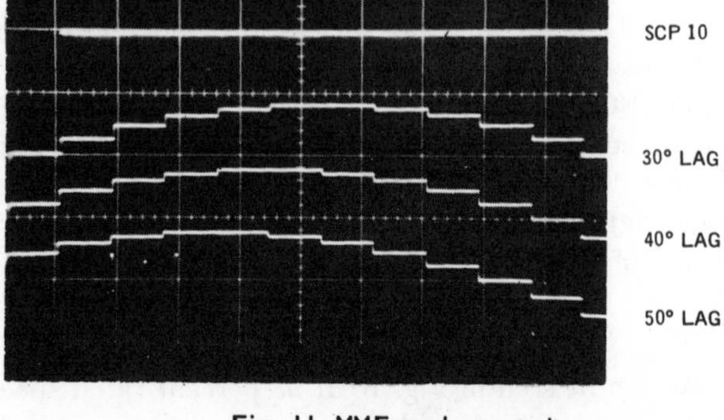

Fig. 11. MMF angle control

is approximately 96 percent leading since the lagging cycloconverter vars are still less than the leading vars of the capacitor bank.

Drive Simulation

To check the overall concepts and the control design strategy, an analytical study was made of the drive system. The study consisted of a digital analysis followed by an analogue computer simulation of the gearless motor and the cycloconverter regulators. As a second stage to the study the motor simulation was connected to the actual cycloconverter control including a simulated rotor position sensor for full closed-loop operation. This second stage as well as allowing a better check on regulator responses permitted a more comprehensive test and "burn in" time on the control hardware.

The field experience has established excellent correlation between the computer predictions and actual drive performance. The control has the ability to inject a vernier MMF angle reference that will either add to, or subtract from, the programmed value from the regulator. It is thus possible to vernier the stator amps and volts plus rotor amps to the desired values with the drive running. This was demonstrated during the computer study.

CONCLUSIONS

Apart from the elimination of gears with their attendant problems, there are other advantages of a gearless drive that were theorized and now are quite evident. The drive characteristics are similar to a dc motor. Starting current peak is considerably lower than conventional geared drives, actually reaching a little over 100 percent peak prior to cascade. This was particularly important in view of the low capacity feeder available.

A wide range of speed control is inherently available, which has proven very useful in balancing the process for maximum efficiency particularly during start-up. Very low speed for spotting without switching or coupling other equipment is another inherent feature. Normal stops are powered down by a timed deceleration so that there is very little rollback. Reverse rotation operation is feasible, if this function is required.

The overall length of the gearless mill is appreciably less than for conventional geared drives. This is an important consideration in planning a new mill.

The technical feasibility and production advantages predicted in [1] have been verified. We are confident that its viability for substantially larger ratings and broader applications have been established.

REFERENCES

[1] E. A. E. Rich, "Concepts of gearless ball-mill drives," *IEEE Trans. Ind. Gen. Appl.*, vol. IGA-5, pp. 13–17, Jan./Feb. 1969.
[2] "Gearless rotary mill," U.S. Patent 3 272 444, Sept. 13, 1966.
[3] R. A. Hamilton and G. R. Lezan, "Thyristor adjustable frequency power supplies for hot strip mill run-out tables," *IEEE Trans. Ind. Gen. Appl.*, vol. IGA-3, pp. 168–175, Mar./Apr. 1967.
[4] W. S. Chow, J. Duckworth, G. Hausen, and J. A. I. Young, "The role of cycloconverters in solid state power conditioning for ac drives," *1972 IEEE Intercon Digest*, 72CH0581-9 IEEE.

[5] H. U. Wurgler. "The world's first gearless mill drive," *IEEE Trans. Ind. Gen. Appl.*, vol. IGA-6, pp. 524–527, Sept./Oct. 1970.
[6] R. Zins, "Gearless drive for a finish-grind mill with a capacity of 160 t/h," *IEEE Trans. Ind. Appl.*, vol. IA-9, pp. 21–24, Jan./Feb. 1973.
[7] A. S. Cornford, "The design of grinding mills for mechanical reliability," Society of Mining Engineers of AIME, Preprint 72-B-335.

Drive System and Electronic Control Equipment of the Gearless Tube Mill

H. Stemmler

621.313.323:621.926.54
621.926.54–83
621.926.54–523.8

The article considers the electrical aspects of the gearless tube mill drive. It is shown that the self-controlled converter-fed synchronous motor employed has basically the same control properties as a d.c. commutator motor. It cannot fall out of step, its torque is not limited by a pull-out torque and there is no tendency to hunt. Unlike a d.c. commutator motor, however, the synchronous motor is not subject to limitations imposed by the mechanical commutator.

Introduction

Over the past few years there has been intensive activity in the development of converter-fed three-phase drives. The areas where such drives are particularly applicable are now becoming defined, and include

– industries in which a large number of similar variable-speed drives are required and where the emphasis is on minimum maintenance and mechanical ruggedness, examples being roller tables and the synthetic-fibres industry [1, 2, 3]
– applications requiring high speeds and drive outputs where the requirements cannot be met with commutator motors [4].

An interesting example of a high-output converter-fed three-phase motor is the cement-mill drive described in this issue [5]. The drive has an output of 8800 hp and a speed variable between zero and 15 rev/min.
The main components of the system can be seen in Fig. 1. The synchronous motor II is arranged round the cement mill in the form of a ring, the mill being of the tube type with a drum 16·5 m long and about 5 m in diameter. Power, of variable frequency, is obtained from the controlled converter I by way of electronic control system III.
The severe operating conditions impose heavy demands not only on the purely mechanical side of the installation, though these we need not consider here, but also on the electrical equipment. The reasons can be summarized as follows.
The mill contains steel balls which are carried up the drum wall as the mill turns, and then run back, crushing the clinker. In accordance with the requirements of the mill designer, the drive is constructed for a starting torque equal to 1·6 times the rated torque, a value which provides an ample reserve compared with the starting torque actually needed.
Under normal operating conditions it must be possible to vary the mill speed within a small range about the rated speed. The system power factor must then not drop below 0·86.
To change the ball charge and replace the steel cladding it must be possible to turn the mill very slowly for positioning purposes.

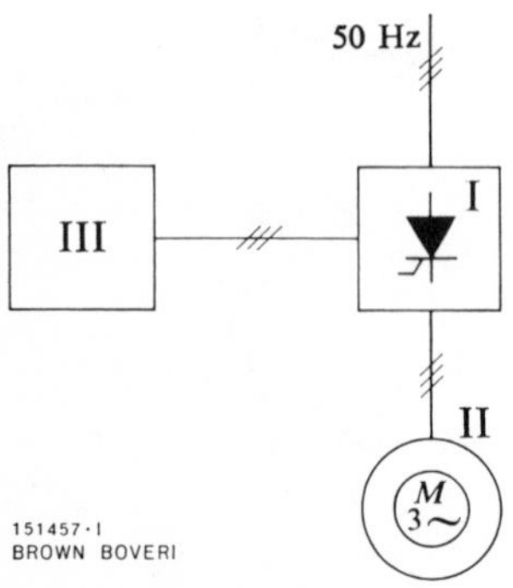

Fig. 1 – Drive arrangement for the gearless cement mill

I = Converter
II = Synchronous motor
III = Electronic control system

Key to Symbols

d = Real axis in the coordinate systems
$\underline{E}$ = Vector of stator e.m.f.
f = Stator frequency of synchronous motor
$\underline{I}$ = Vector of stator current
$\underline{I}_e$ = Vector of excitation current
L = Total inductance of rotating stator field
M_e = Coupling inductance of rotating field
P = Power supplied to stator
p = Number of pole pairs
q = Imaginary axis in the coordinate systems
R = Resistance of stator
$\underline{U}$ = Vector of stator voltage
U_N = Amplitude of system voltage u_N
U_T = Amplitude of timing signal u_T
U_l = Amplitude of rotor voltage u_l of three-phase slipring motor
T = Electrical torque
α_0 = Angular position of rotor relative to stator phase R at time $t = 0$
φ_{IE} = Angle between $\underline{I}$ and $\underline{E}$
φ_{IU} = Angle between $\underline{I}$ and $\underline{U}$
φ_I = Angle between $\underline{\Psi}_e$ and $\underline{\Psi}_I$
$\underline{\Psi}$ = Vector of resultant stator flux linkage
$\underline{\Psi}_e$ = Effective flux linkage created in stator by excitation current $\underline{I}_e$

Reprinted with permission from *Brown Boveri Review,* pp. 120-128, Mar. 1970.

$\underline{\Psi}_I$ = Effective flux linkage created in stator by stator current $\underline{I}$
ω = Angular frequency in stator

Reference Quantities

E_n = Amplitude of rated stator e.m.f.
I_n = Amplitude of rated stator current
T_n = Rated electrical torque
U_n = Amplitude of rated stator voltage
Ψ_n = Amplitude of rated stator flux linkage
ω_n = Rated angular frequency in stator
$\varphi_{IE} = 0$ = Phase displacement between stator current and voltage under rated operating conditions

Referred Quantities

$\underline{U}' = \underline{U}/E_n$
$\underline{E}' = \underline{E}/E_n$
$\underline{I}' = \underline{I}/I_n$
$\underline{I}'_e = \underline{I}_e/I_n$
$\underline{\Psi}'_I = \underline{\Psi}_I/\Psi_n$ where $(\underline{\Psi}_I = \underline{I}L)$
$\underline{\Psi}'_e = \underline{\Psi}_e/\Psi_n$ where $(\underline{\Psi}_e = \underline{I}_e M_e)$
$\underline{\Psi}' = \underline{\Psi}/\Psi_n$
$T' = T/T_n$
$\omega' = \omega/\omega_n$
$R' = R \,.\, I_n/E_n$
$X' = L \,.\, I_n/\Psi_n$
$X'_e = M_e \,.\, I_n/\Psi_n$

Relationships between Reference Quantities

$$E_n = \omega_n \,.\, \Psi_n$$
$$U_n = E_n + I_n \,.\, R$$
$$T_n = \frac{3}{2} p \,.\, \Psi_n \,.\, I_n \cos\varphi_{IE} \quad (\cos\varphi_{IE} = 1)$$

Construction and Operation

The principle of the converter and motor is explained below, followed by a description of the electronic control system linking them.

The Converter

Both the stator and the rotor of the synchronous motor are fed through converters. The d.c. excitation current for the rotor is supplied by a controlled rectifier, while the stator converter (Fig. 2a) is a line-commutated, controlled static frequency changer. Each of the three phases contains two six-pulse bridges connected in antiparallel, the output voltage of which is controlled with variable sinusoidal frequency and amplitude. One bridge supplies current i_+ for the positive half-wave of the stator current, the other i_- for the negative half-wave.
The half of the converter not carrying current is blocked off by the control system to prevent an internal short circuit in the converter. As in the case of all three-phase drives with variable-frequency supply, the amplitude and

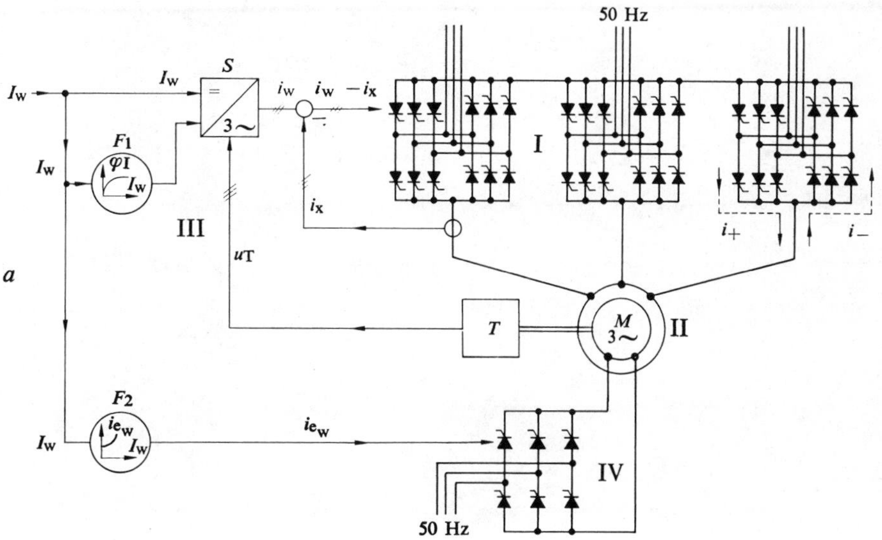

Fig. 2a – Basic arrangement of converter and control system for a self-controlled synchronous motor

I = Converter for stator
II = Synchronous motor
III = Electronic control system
IV = Converter for rotor
F_1, F_2 = Function generators
S = Stator current desired value unit
u_+, u_-, i_+, i_- = Voltages and currents of positive and negative half-waves

Subscripts:
T = Frequency generator
w = Desired value
x = Actual value

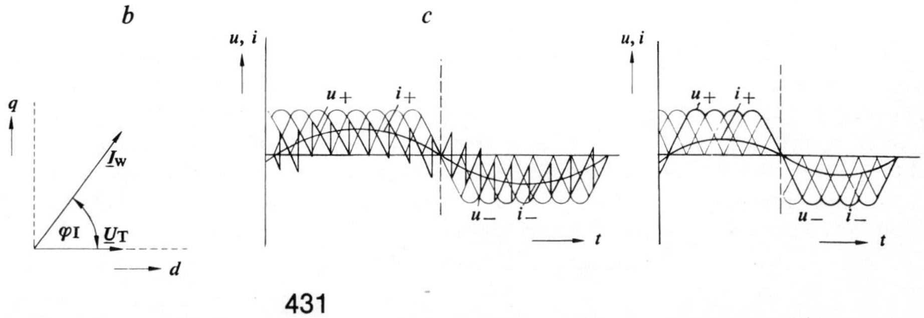

Fig. 2b – Desired value unit S acts on timing signal u_T as a phase/amplitude adjuster and produces the stator current desired value i_w

Fig. 2c – Output voltage and current in relation to time, for two different frequencies
u_+, u_- = Voltage of positive and negative half-wave

Fig. 3 – Stator of synchronous motor for the gearless cement-mill drive

frequency of the voltage have to be made approximately linear to one another.

The way in which the frequency changer operates can be seen from the behaviour of the output voltage. This is illustrated in Fig. 2c for two different frequencies and amplitudes. The diagram shows how the converter cuts separate sections out of the voltages from the 50 Hz system and then combines them into a new voltage of which the mean value is sinusoidal and the frequency is low compared with the power frequency. If one wishes to produce an output voltage of low amplitude, as in the case of the left-hand oscillogram, the sections are chopped from the flanks of the sinusoidal power voltage. If the output voltage is to be increased together with the frequency, the sections are shifted progressively towards the peaks of the power voltage. The highest output voltage amplitude is obtained when the output voltage is composed of the crests of the system voltage (Fig. 2c, right). One tries to maintain this condition under normal operation, as the converter then draws the least reactive power from the network.

The curves in Fig. 2c show the output voltage and output current for the case where the power factor of the machine $\cos \varphi_{IU}$ is unity. However, the converter permits any phase position. The drive is thus suitable for four-quadrant operation, i.e. driving and braking in both directions.

It should also be noted that the converters, like the machine, are connected in star to keep converter voltages of the same phase away from the machine winding. The third harmonic of the converter voltage, which is particularly strong in the trapezoidal mode, is thus unable to cause any current harmonics.

The Synchronous Motor

The stator of the gearless cement mill drive is illustrated in Fig. 3. The picture gives an indication both of the construction and of the size.

Equations

The characteristics and operation of the synchronous machine differ appreciably from the usual case where the machine is connected direct to the 50 Hz supply. Although the machine actually has a salient-pole rotor, the following discussion is limited to a motor with a smooth rotor. The equations are simpler in consequence, but nevertheless provide important information.

It is convenient to consider the stator converter not as a voltage source, but as a current source which impresses a current, produced in three-phase sinusoidal form by the current control system, on the stator of the motor. The exciter rectifier also has a current control system and impresses the d.c. excitation current on the rotor.

The resultant stator flux linkage $\underline{\Psi}$ (including leakage) is composed of the magnetic flux linkages $\underline{\Psi}_I$ and $\underline{\Psi}_e$ created in the stator by the impressed stator current $\underline{I}$ and excitation current $\underline{I}_e$:

$$\underline{\Psi} = \underline{\Psi}_I + \underline{\Psi}_e \qquad \underline{\Psi}_I = \underline{I}L;\ \underline{\Psi}_e = \underline{I}_e M_e \tag{1}$$

The relationship between the resultant stator flux linkage $\underline{\Psi}$ and the terminal voltage $\underline{U}$ applied to the stator is in fact of no immediate interest in the present context, because we are considering the converter as a current source. It is given here, however, for the sake of completeness:

$$\underline{U} = \underline{I}R + \underline{E} \tag{2}$$

$$\underline{E} = j\omega\underline{\Psi} \tag{3}$$

The applied voltage is thus divided into the resistive stator voltage drop $\underline{I}R$ and the stator e.m.f. $\underline{E}$, which is proportional to ω and $\underline{\Psi}$.

Next we must derive an expression for the electrically developed torque T. For the power supplied P less the resistive stator losses $3\,(I/\sqrt{2})^2R$, we have

$$P - 3\left(\frac{I}{\sqrt{2}}\right)^2 R = 3\,\frac{E}{\sqrt{2}}\,\frac{I}{\sqrt{2}}\cos\varphi_{IE} \tag{4}$$

If the iron losses are disregarded, power P and torque T are related in the following manner:

$$T\frac{\omega}{p} = P - 3\left(\frac{I}{\sqrt{2}}\right)^2 R \tag{5}$$

With Eq. (3) and (4) it follows that

$$T = \frac{3}{2}\,p\,\Psi\,I\cos\varphi_{IE} \tag{6}$$

It is convenient to convert Eq. (1), (2), (3) and (6) to dimensionless form by referring them to the rated values. If we also introduce a rotating system of coordinates, the real axis d of which coincides with the pole axis of the rotor, while the imaginary axis leads by 90°, Eq. (1), (2), (3) and (6) yield:

$$\underline{\Psi}' = \underline{\Psi}'_I + \underline{\Psi}'_e;\ \underline{\Psi}'_I = \underline{I}'X';\ \underline{\Psi}'_e = \underline{I}'_e X'_e \tag{7}$$

$$\underline{U}' = \underline{I}'R' + \underline{E}' \tag{8}$$

$$\underline{E}' = j\omega'\,\underline{\Psi}' \tag{9}$$

$$T' = \Psi'\,I'\cos\varphi_{EI} \tag{10}$$

Vector Diagram

These equations are shown in the form of a vector diagram in Fig. 4. On the left is the motor with the impressed specific current loadings of the rotor and stator, and also the associated magnetic flux linkages $\underline{\Psi}'_e$ and $\underline{\Psi}'_I$, which are shown as vectors rotating at angular velocity ω. The rotating system of coordinates (d and q axes) is also shown. Fig. 4b illustrates how the two magnetic fields in the stator are combined to yield the resultant stator flux linkage $\underline{\Psi}'$ (corresponding to Eq. (7)). Disregarding the resistive voltage drop, the vector of the stator voltage $\underline{U}'$ leads the vector of the stator flux linkage $\underline{\Psi}'$ by 90° (corresponding to Eq. (8) and (9)), and the stator current

Fig. 4a – Specific current loading of stator and rotor

Fig. 4b – Vector diagram of the synchronous motor

$\underline{\Psi}' = \underline{\Psi}'_I + \underline{\Psi}'_e$
$\underline{U}' \approx j\omega\,\underline{\Psi}'$
$T' = \Psi'\,I'\cos\varphi_{IE}$ (corresponds to area of triangle $\underline{\Psi}'$, $\underline{\Psi}'_I$ and $\underline{\Psi}'_e$)

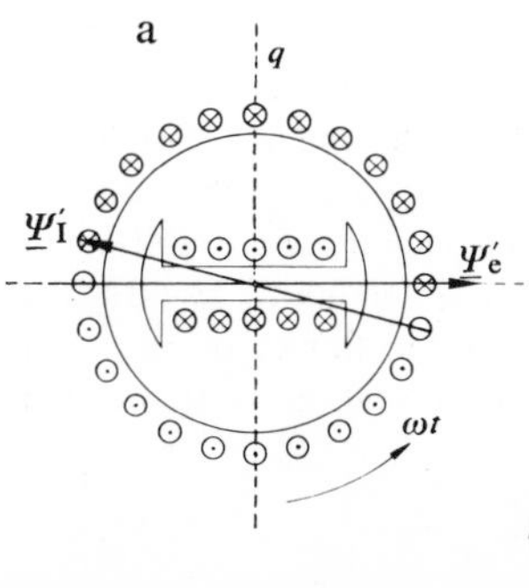

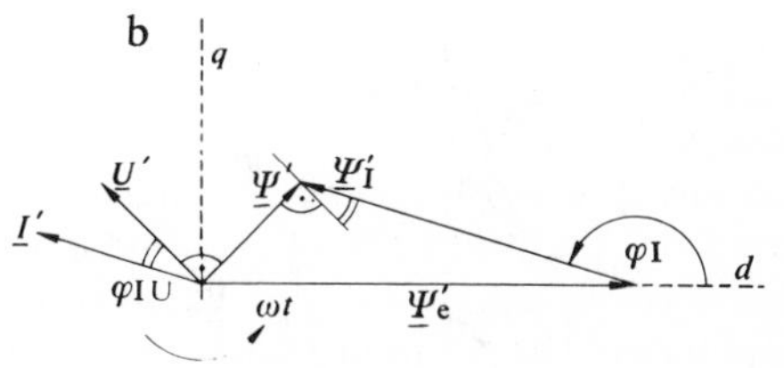

vector $\underline{I}'$ has the same direction as the magnetic flux linkage $\underline{\Psi}'_\mathrm{I}$ created by it. Torque T can also be shown in this vector diagram. According to Eq. (10) it is proportional to the area of the triangle formed by the vectors of the magnetic flux linkages $\underline{\Psi}'_\mathrm{I}$, $\underline{\Psi}'_\mathrm{e}$ and $\underline{\Psi}'$.

Principle of the Control System

Controlled Variables

When planning a speed regulation system, the first question to consider is what controlled variables are available. In the present case, where the converters function as current sources, the choice (Fig. 4) consists of

- excitation current I_e
- stator current amplitude I, and
- angle φ_I between the vector of the stator current m.m.f. and the rotor, i.e. the d-axis, termed the current angle in the following.

It is perhaps at first surprising that the stator angle φ_I, and not the stator frequency, is used as the controlled variable for the speed. A brief look at the d.c. commutator motor will explain why, however, and also show the close relationship between this kind of machine and a self-controlled synchronous motor. The controlled variables for adjusting torque and speed are basically the same for the d.c. commutator motor as for a self-controlled synchronous machine, namely,

- the field winding excitation current, which corresponds to the excitation current of the rotor in a synchronous machine
- the armature current, corresponding to the stator current amplitude in a synchronous machine
- the angle between the vector of the armature current m.m.f. and the axis of the field winding, which in the case of the synchronous machine corresponds to current angle φ_I.

Whereas with a synchronous machine the current angle φ_I can be set by the electronic control system described below, in the case of a d.c. commutator motor the corresponding angle between the axis of the field winding and the vector of the armature current m.m.f. is determined by the design of the machine and the brush setting.

Of the three controlled variables of the synchronous motor, the amplitude I of the stator current can be considered the final controlled variable, which is then used to vary the torque, and hence the speed. The two other controlled variables I_e and φ_I are adjusted in relationship to the set amplitude I of the stator current in such a way that

- magnetic stator flux linkage Ψ' always stays as its rated value $\Psi' = 1$ (except during operation under field-weakening conditions), because the motor is designed for this flux linkage
- current I and voltage U of the stator are always in phase (i.e. $\varphi_\mathrm{IE} = \varphi_\mathrm{IU} = 0$), because neither the motor nor the converter then need to carry reactive current and can be made smaller accordingly.

The vector diagram for this case is shown in Fig. 5. As the stator current amplitude I' is varied, the vector of the stator flux linkage $\underline{\Psi}'$ turns on a circle of radius 1 (first condition above, i.e. $\Psi' = 1$), while the magnetic flux linkage $\underline{\Psi}'_\mathrm{I}$ of the stator current $\underline{I}'$ always remains perpendicular to the resultant stator flux linkage $\underline{\Psi}'$ (second condition above, i.e. $\varphi_\mathrm{IE} = \varphi_\mathrm{IU} = 0$). Mathematically, the vector diagram of Fig. 5 immediately yields the equations

$$\varphi_\mathrm{I} = \arctan \frac{1}{I'X'} \quad (11)$$

$$I'_\mathrm{e}X'_\mathrm{e} = \sqrt{1 + (I'X')^2} \quad (12)$$

$$\text{for} \begin{cases} \Psi' = 1 & (13) \\ \varphi_\mathrm{IU} = \varphi_\mathrm{IE} = 0 & (14) \end{cases}$$

Thus if, in accordance with Eq. (11) and (12), the angle φ_I between the vector of the stator current m.m.f. and the d-axis of the rotor and also the excitation current I'_e are adjusted as a function of the final controlled variable (stator current amplitude I'), the stator flux linkage always stays at its rated value $\Psi' = 1$, and the converter and motor carry pure active current ($\varphi_\mathrm{IE} = \varphi_\mathrm{IU} = 0$). Then, from Eq. (10), (13) and (14) we have for the electrical torque:

$$T' = I' \quad (15) \quad \text{for} \begin{cases} \Psi' = 1 & (13) \\ \varphi_\mathrm{IU} = \varphi_\mathrm{IE} = 0 & (14) \end{cases}$$

According to Eq. (15), torque T' is proportional to stator current I' and is not limited by a pull-out torque.

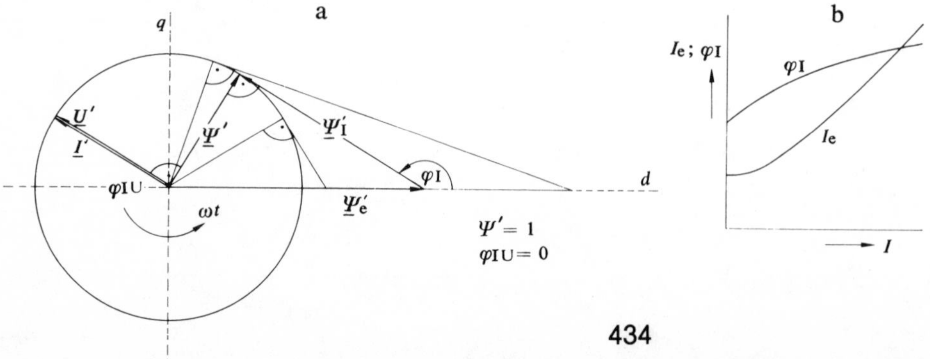

Fig. 5a – Vector diagram of synchronous motor for constant stator flux linkage $\Psi' = 1$ and purely active current $\cos \varphi_\mathrm{IU} = 1$

Fig. 5b – Values to be adjusted for excitation current I_e and angle φ_I between the vector of the stator current m.m.f. and rotor in relation to set stator current amplitude I (Eq. 11 and 12)

Fig. 6 – Electronic control cubicles for the gearless cement mill drive

Design of the Control System

The basic arrangement of the control system is shown in Fig. 2a. Since the angle between the vector of the stator current m.m.f. and the rotor position is used as a controlled variable, a frequency generator T is required to detect the rotor position (d-axis). The frequency generator provides a three-phase sinusoidal signal u_T of constant amplitude whereby the vector of phase R is in phase with the d-axis of the rotor. The way the unit functions is described in the next section.

This three-phase signal u_T is passed to the stator current desired value unit S where it is varied in amplitude by input value I_w (presented as a d.c. voltage) and altered in phase relative to the d-axis by a second d.c. input voltage "φ_I". The desired value unit S thus acts on the frequency generator signal u_T as a phase/amplitude adjuster (Fig. 2b) and converts it into the desired stator current i_w, which is also of three-phase sinusoidal shape. This is considered in greater detail in the next section.

The desired value of the stator current i_w is in turn compared with the measured actual value i_x, and the difference $i_w - i_x$ regulates the converter I (Fig. 2a) so that it impresses on the motor II a stator current i_x which follows the desired value i_w as closely as possible.

Thus by "turning" the phase/amplitude adjuster it is possible to vary the angular position of the stator's specific current loading relative to the rotor position, and employ the "stretching" or "squeezing" action to increase or reduce its intensity.

All this is subordinated to the speed-regulation loop which is not shown in Fig. 2. The difference between its desired and actual values provides the desired value I_w of the stator current amplitude. This value acts direct on the "amplitude" input of the desired value unit S and, via function generators F_1 and F_2, which are represented by Eq. (11) and (12) and ensure that conditions (13) and (14) are maintained, on the "angle" input (φ_I) of desired value unit S. It also acts on the excitation current via exciter transformer IV.

The hinged-frame cubicles containing the control equipment are illustrated in Fig. 6.

Operation of the Frequency Generator

The frequency generator T (Fig. 2a) consists of a small machine and a computing unit (Fig. 7). Its purpose is to generate a three-phase sinusoidal signal of constant amplitude which, interpreted as a rotating vector, turns synchronously and in phase with the d-axis of the rotor.

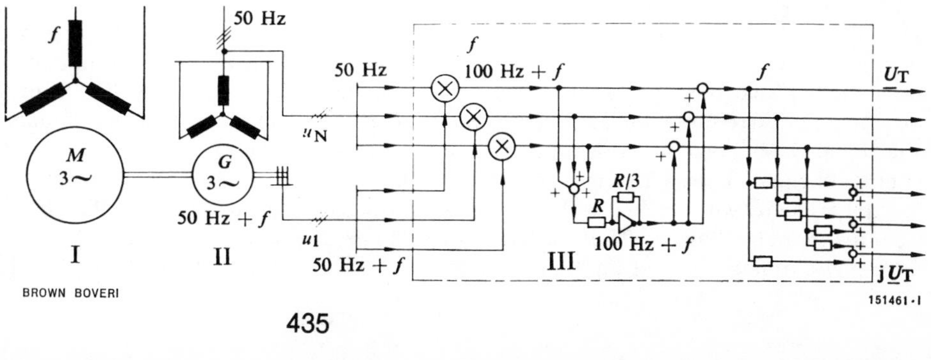

Fig. 7 – The shaft of synchronous motor I has coupled to it frequency generator II which modulates the 50 Hz power frequency with motor frequency f

Computer unit III acts as a demodulator and provides the timing signal $\underline{U}_T$ (frequency f) and phase position of rotor.

A three-phase tacho-generator would not be suitable as it produces no voltage output at standstill. However, the purpose is served by a small three-phase slipring rotor II (Fig. 7) coupled to the shaft, the stator receiving the voltage of the 50 Hz system. This voltage, with frequency 50 Hz $+ f$ modulated by the speed of the synchronous motor I (Fig. 7), is picked off again at the rotor.
The rotor voltage is passed to the computing unit III (Fig. 7) together with the three-phase 50 Hz system voltage. This unit functions as a demodulator and provides the required signal of frequency f and with the phase position of the rotor.
The 50 Hz system voltage thus functions as a carrier frequency which is modulated by speed, or stator frequency f, of the synchronous motor. In principle it could be replaced by a three-phase voltage of some other frequency.
Generation of the timing signal can be described mathematically in the following manner. According to the angular position $2\pi f t + \alpha_0$ of the rotor (d-axis) relative to phase R of the synchronous motor's stator, the required timing signal is

$$u_T = U_T \cos(2\pi f t + \alpha_0 + \vartheta) \qquad (16)$$

$\vartheta = 0°$el for phase R, —120°el for phase S and +120°el for phase T

The system voltage

$$u_N = U_N \cos(2\pi \; 50 \text{ Hz}\; t + \vartheta) \qquad (17)$$

is frequency-modulated by the three-phase slipring rotor (Fig. 7) in such a way that the rotor voltage is described by the equation

$$u_1 = U_1 \cos[2\pi (50 \text{ Hz} + f)\, t + \alpha_0 + \vartheta] \qquad (18)$$

$$U_1 = U_N \frac{50 \text{ Hz} + f}{50 \text{ Hz}}$$

In the demodulator (Fig. 7) the three-phase voltage u_1, with reversed phase sequence ($-\vartheta$ instead of $+\vartheta$), is multiplied by the three-phase system voltage u_N to produce the voltage u_2:

$$u_2 = u_1 \, u_N = \text{const.} \qquad (19)$$

$$= U_T \begin{Bmatrix} + \cos[2\pi f t + \alpha_0 - \overset{\widehat{=} + \vartheta}{\widetilde{2\vartheta}}] + \\ + \cos[2\pi (100 \text{ Hz} + f)\, t + \alpha_0] \end{Bmatrix}$$

$$U_T = \frac{1}{2} U_N^2 \cdot \frac{50 \text{ Hz} + f}{50 \text{ Hz}} = \text{const.}$$

This voltage consists not only of the required three-phase (ϑ) timing signal according to Eq. (16), but also an unwanted component (no ϑ) equal in phase and with the summation frequency 100 Hz $+ f$. This can easily be eliminated. If the three phases of voltage u_2 are added together, the three-phase voltages cancel each other out and the sum obtained comprises the three voltage components of equal phase. These are passed through a phase-inverter amplifier with a gain of $-1/3$ and again subtracted from the output voltage u_2. Thus finally only the required three-phase timing signal of Eq. (16) remains.

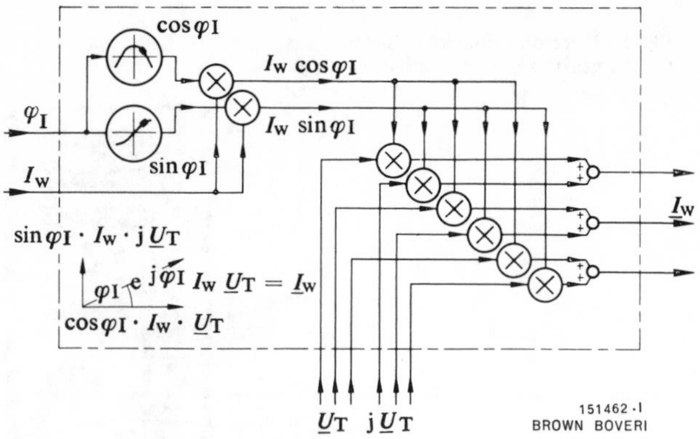

Fig. 8 – Desired value unit for stator current

This acts as a phase/amplitude adjuster on timing signal $\underline{U}_T$, turns it through angle φ_I and extends it by the factor I_w.

Its constant amplitude const. $^1/_2\, U_N^2$ (50 Hz $+ f$)/50 Hz is present even when the drive is stopped, i.e. when $f = 0$. It varies only very slightly in relation to the speed of the drive because its frequency f is always low compared with 50 Hz. Even these small deviations are regulated by means of an amplitude-control loop (not shown in Fig. 7). The three-phase output voltage displaced by 90° j$\underline{U}_T$, which is required by the desired value unit S, is obtained by adding each pair of phases of the three-phase output voltage $\underline{U}_T$.
If one of the two three-phase input signals were to be replaced by a single-phase signal, two superposed three-phase voltages with the sum and difference frequencies of both input signals would appear at the output of the demodulator. It would be difficult, if not impossible, to separate the two signals clearly.

Operation of the Stator Current Desired Value Unit

The function of the desired value unit S (Fig. 2) is to modify the phase position and amplitude of the three-phase sinusoidal output voltages of the frequency generator, which indicate the position of the rotor, and thence form the three-phase sinusoidal desired value for the stator current. This value controls the intensity of the stator's specific current loading and its angular position relative to the rotor. This angle has to be controlled at every operating frequency and at standstill. The unit cannot therefore employ any RC or LC phase-shifting elements because these are dependent on frequency.
The unit operates as a phase/amplitude adjuster, as mentioned earlier, and can be considered as a representation of the (extended) Euler equation

$$I_w \underline{U}_T \cos\varphi_I + I_w \underline{U}_T \, j \sin\varphi_I = I_w \underline{U}_T \, e^{\,j\varphi_I} \qquad (20)$$

Vector $\underline{U}_T$ is "turned" by varying φ_I, while the "stretching" or "squeezing" effect is obtained by modifying I_w.
Construction and operation of the unit can be explained with the aid of Fig. 8. From the value φ_I, represented in the form of a d.c. voltage signal, the upper of the two function generators produces the d.c. signal $\cos\varphi_I$ and the lower function generator the d.c. signal $\sin\varphi_I$. Both signals are multiplied by d.c. signal I_w, thus yielding d.c. signals of $I_w \cos\varphi_I$ and $I_w \sin\varphi_I$. The first of these is then multiplied by each phase of the frequency generator

Fig. 9 – Currents of phases R, S and T and one frequency changer voltage on running up at constant maximum torque

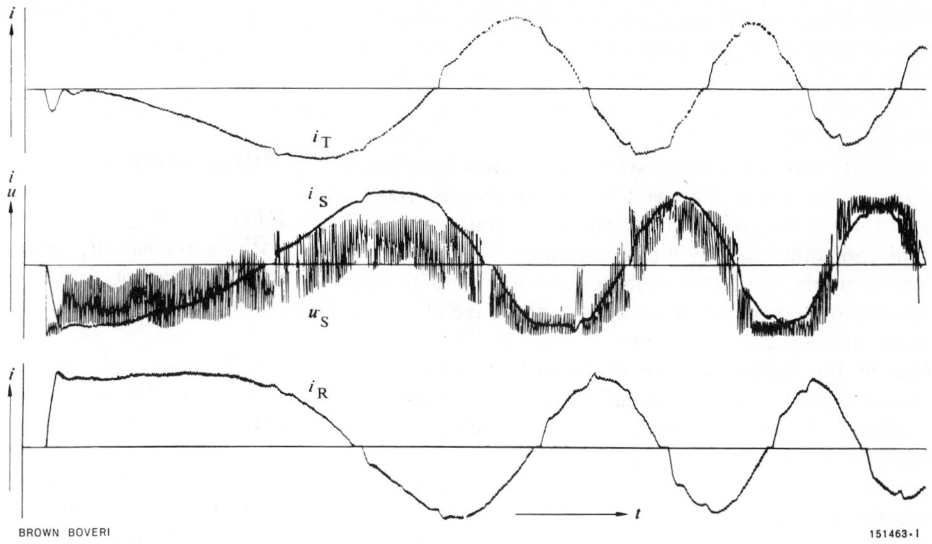

sine voltage $\underline{U}_T$, and the second with each phase of the frequency generator voltage (displaced by 90°) $j\underline{U}_T$. This then yields the two three-phase a.c. voltages $\underline{U}_T I_w \cos \varphi_I$ and $j\underline{U}_T I_w \sin \varphi_I$, which are added at the output to give the desired value of the three-phase sinusoidal stator current $\underline{I}_w$.

$$\cos \varphi_I I_w \underline{U}_T + \sin \varphi_I I_w j\underline{U}_T = e^{j\varphi_I} I_w \underline{U}_T = \underline{I}_w \qquad (21)$$

The amplitude of the desired stator current $\underline{I}_w$ can thus be varied by means of d.c. signal I_w, and the phase position with d.c. signal φ_I.

The Chief Electrical Features of the Self-Controlled Synchronous Motor

A synchronous motor controlled in this way has operating characteristics quite different from those of a machine connected direct to a 50 Hz supply. The latter has a pull-out torque, it can fall out of step, and shows a tendency to hunt if subjected to shock loads.

These unwelcome features can be avoided with the self-controlled synchronous machine supplied through a converter as described here. With a self-controlled motor it is possible to obtain any required intensity of the stator current loading and any required angular position between this and the rotor, regardless of speed and shock loads. It is therefore impossible for the machine to pull out of synchronism. There is also none of the usual tendency of a synchronous machine to hunt in the event of shock loads. Hunting is caused by regular fluctuations in amplitude and phase position of the stator current loading relative to the rotor. Torque is dependent solely on the stator current amplitude, which can be varied at will within limits set by the design of the machine and the frequency changer (Eq. 15). Thus any desired torque can be applied, both at standstill and on starting. In so doing, the machine p.f. can always be kept at unity. Consequently, both the converter and the motor need be designed only for the purely active power. At low speeds the system p.f. is also low, but in the rated speed range of 13 to 15 rev/min it is possible to achieve a system p.f. $\geqslant$ 0·86.

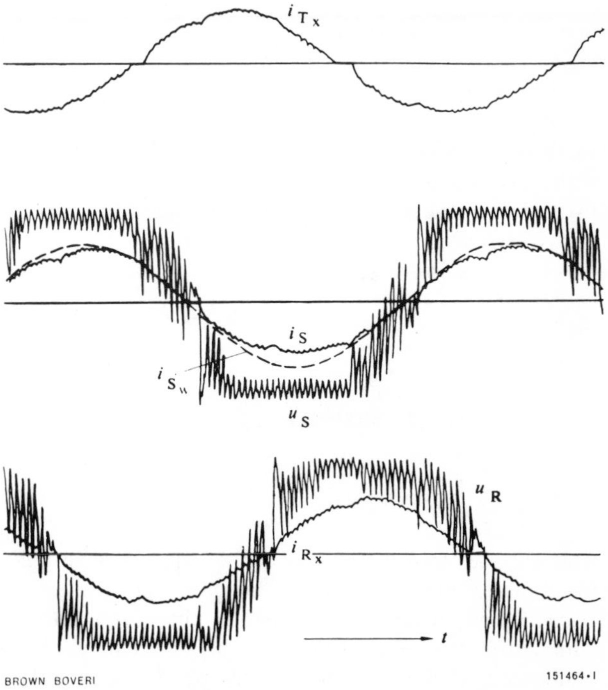

Fig. 10 – Currents i and voltage u of converter and motor in phases R, S and T under steady-state conditions

Subscripts:
R, S, T = Phases R, S and T
w = Desired value
x = Actual value

With the type of control system described here, a synchronous machine has control properties as good as those of a d.c. commutator machine, but without the limitations imposed by the commutator.
The currents of the three phases and one voltage of the frequency changer when running up the drive at constant maximum torque are shown in Fig. 9. The converter loading increases progressively, according to Eq. (7) and (8), as the frequency rises, until it reaches the maximum (trapezoidal curve) at the last two half-cycles. It can be seen, both here and in Fig. 10, which shows the same quantities for the steady state, that current and voltage are always in phase.
The low current harmonics content is to a large extent due to the fact that with the control principle employed the damper winding, which would sharply reduce the reactance on which harmonics depend, can be omitted without any risk of hunting.

Bibliography

[1] *A. Schönung, H. Stemmler:* Static frequency changers with "subharmonic" control in conjunction with reversible variable-speed a.c. drives. Brown Boveri Rev. 1964 *51* (8/9) 555–577.

[2] *J. Franke, A. Schönung:* Steuerung statischer Umformer zum Speisen der Antriebe von Chemiefaser-Spinnmaschinen. Elektrotech. Z. – B 1968 *20* (21) 616–621.

[3] *R. A. Hamilton, G. R. Lezan:* Thyristor adjustable frequency power supplies for hot strip mill run-out tables. IEEE Trans. Ind. and gen. Appl. 1967 *3* (2) 168–175.

[4] *H. Stemmler:* Speisung einer langsamlaufenden Synchronmaschine mit einem direkten Umrichter. VDE-Fachtagung 1969, Energieelektronik, Hannover, 177–189.

[5] *E. Bläuenstein:* The first gearless drive for a tube mill. Brown Boveri Rev. 1970 *57* (3) 96–105.

Selected Bibliography

Section 2

1. P. C. Krause, "A constant frequency induction motor speed control," *Proc. Nat. Elec. Conf.*, vol. 20, pp. 361–365, 1964.
2. D. A. Paice, "Speed control of large induction motors by thyristor converters," *IEEE Trans. IGA,* vol. IGA-5, no. 5, pp. 545–551, Sept./Oct. 1969.
3. W. Shepherd and J. Stanway, "The polyphase induction motor controlled by firing angle adjustment of silicon controlled rectifiers," *IEEE Int. Conf. Rec.,* vol. 12, no. 4, pp. 135–154, 1964.
4. W. Shepherd, "On the analysis of the three-phase induction motor with voltage control by thyristor switching," *IEEE Trans. IGA,* vol. IGA-4, no. 3, pp. 304–311, May/June 1968.
5. W. Shepherd and J. Stanway, "An experimental closed loop variable speed drive incorporating a thyristor driven induction motor," *IEEE Trans. IGA,* vol. IGA-3, no. 6, pp. 559–565, Nov./Dec. 1967.
6. S. L. Benn, "Practical phase control of linear induction motors," *IEE Int. Conf. on Electrical Variable Speed Drives,* pp. 30–33, Sept. 1979.
7. D. W. Novotny and A. F. Fath, "The analysis of induction machines controlled by series connected semiconductor switches," *IEEE Trans. PAS,* vol. PAS-87, no. 2, pp. 597–605, Feb. 1968.
8. R. E. Bedford and V. D. Nene, "Voltage control of the three-phase induction motor by thyristor switching: a time domain analysis using the $\alpha-\beta-0$ transformation," *IEEE Trans. IGA,* vol. IGA-6, no. 6, pp. 553–562, Nov./Dec. 1970.

Section 3

1. U. Putz, "Thyristor industrial drives," *IEEE/IAS Int. Sem. Power Conv. Conf. Rec.,* pp. 340–354, 1977.
2. R. L. Risherg, "A wide speed range inverter fed induction motor drive," *IEEE/IAS Annual Meeting Conf. Rec.,* pp. 629–633, 1969.
3. R. P. Veres, "New inverter supplies for high power drives," *IEEE Trans. IGA,* vol. IGA-6, no. 2, pp. 121–127, Mar./Apr. 1970.
4. E. M. Sabbagh and W. Shewan, "Characteristics of an adjustable speed polyphase induction machine," *IEEE Trans. PAS,* vol. PAS-87, no. 3, pp. 613–624, Mar. 1968.
5. I. M. MacDonald, "A static inverter wide range adjustable speed drive," *IEEE Int. Conv. Rec.,* vol. 12, no. 4, pp. 34–41, 1964.
6. R. W. Stokes, "High voltage transistor inverters for ac traction drives," *IEEE/IAS Int. Sem. Power Conv. Conf. Rec.,* pp. 270–294, 1977.
7. P. Lataire, "Induction motor drives with power transistor inverters," *IEEE/IAS Annual Meeting Conf. Rec.,* pp. 763–769, 1978.
8. P. C. Krause and C. H. Thomas, "Simulation of symmetrical induction machinery," *IEEE Trans. PAS,* vol. PAS-84, no. 11, pp. 1038–1053, Nov. 1965.
9. T. A. Lipo and F. G. Turnbull, "Analysis and comparison of two types of square wave inverter drives," *IEEE Trans. IA,* vol. IA-11, no. 2, pp. 137–147, Mar./Apr. 1975.
10. K. D. Lach, "A study of the effect of variable frequency operation on induction motor stability," *Proc. Nat. Elec. Conf.,* pp. 376–380, 1967.
11. K. Y. G. Li, "Analysis and operation of an inverter fed variable speed induction motor," *Proc. IEE,* vol. 116, no. 4, pp. 1571–1580, Sept. 1969.
12. G. W. McLean, G. F. Nix, and S. R. Alwash, "Performance and design of induction motors with square wave excitation," *Proc, IEE,* vol. 116, no. 8, 1405–1411, Aug. 1969.
13. D. M. Mitchell and C. J. Triska, "An investigation of an SCR inverter drive for an induction motor," *IEEE/IAS Annual Meeting Conf. Rec.,* pp. 81–90, 1967.
14. P. C. Krause and L. T. Woloszyk, "Comparison of computed and test results of a static ac drive system," *IEEE Trans. IGA,* vol. IGA-4, no. 6, pp. 583–588, Nov./Dec. 1968.
15. H. Largiader, "Design aspects of induction motors for traction applications with supply through static frequency changers," *Brown Boveri Review*, vol. 57, pp. 152–167, April 1970.
16. A. Nabae, "Performance of slip frequency controlled induction machines," *IEEE/IAS Annual Meeting Conf. Rec.,* pp. 852–856, 1975.
17. M. J. Youn and R. G. Hoft, "Variable frequency induction motor bode diagrams," *IEEE/IAS Int. Sem. Power Conv. Conf. Rec.,* pp. 124–136, 1977.
18. A. R. Miles and D. W. Novotny, "Transfer functions of the slip controlled induction machine," *IEEE Trans. IA,* vol. IA-15, no. 1, pp. 54–62, Jan./Feb. 1979.
19. M. Ramamoorty, "Steady state analysis of inverter driven induction motors using harmonic equivalent circuits," *IEEE/IAS Annual Meeting Conf. Rec.,* pp. 437–440, 1973.
20. N. Vutz, "PWM inverter induction motor transit car drives," *IEEE Trans. IA,* vol. IA-8, no. 1, pp. 89–91, Jan./Feb. 1972.
21. F. G. Turnbull, "Selected harmonic reduction in static dc-ac inverters," *IEEE Trans. Comm. Elec.,* vol. 83, no. 7, pp. 374–378, July 1964.
22. A. Schonung and H. Stemmler, "Static frequency changers with subharmonic control in conjunction with reversible variable speed drives," *Brown Boveri Review,* vol. 51, pp. 555–577, Aug./Sept. 1964.
23. G. C. Jain, "The effect of wave shape on the performance of a 3-phase induction motor," *IEEE Trans. PAS,* vol. PAS-83, no. 6, pp. 561–566, June 1964.
24. E. A. Klingshirn and H. E. Jordan, "Polyphase induction motor performance and losses on nonsinusoidal voltage sources," *IEEE Trans. PAS,* vol. PAS-87, no. 3, pp. 624–631, Mar. 1968.
25. B. J. Chalmers and B. R. Sarker, "Induction motor losses due to nonsinusoidal supply waveforms," *Proc. IEE,* vol. 115, no. 12, pp. 1777–1782, Dec. 1968.
26. T. A. Lipo, P. C. Krause, and H. E. Jordan, "Harmonic torque and

speed pulsations in a rectifier-inverter induction motor drive," *IEEE Trans. PAS,* vol. PAS-88, no. 5, pp. 579–587, May 1969.
27. F. N. Klein and W. Frederich, "PWM inverter induced harmonics effects in ac motors," *IEEE/IAS Annual Meeting Conf. Rec.,* pp. 1007–1015, 1974.
28. J. W. A. Wilson and J. A. Yeamans, "Intrinsic harmonics of idealized PWM systems," *IEEE/IAS Annual Meeting Conf. Rec.,* pp. 967–973, 1976.
29. J. W. A. Wilson, "Adaptation of pulse-width modulation theory for use in ac motor drive inverters," *IEEE/IAS Int. Sem. Power Conv. Conf. Rec.,* pp. 193–199, March 1977.
30. G. S. Buja and G. B. Indri, "Optimal pulse-width modulation for feeding ac motors," *IEEE Trans. IA,* vol. IA-13, no. 1, pp. 38–44, Jan./Feb. 1977.
31. G. B. Kliman and A. B. Plunkett, "Development of a modulation strategy for a PWM inverter drive," *IEEE Trans. IA,* vol. IA-15, no. 1, pp. 72–79, Jan./Feb. 1979.
32. S. Sone and Y. Hori, "Harmonic elimination of microprocessor controlled PWM inverter for electric traction," *IEEE/IECI Conf. Rec.,* pp. 278–283, 1979.
33. J. M. D. Murphy, R. G. Hoft, and L. S. Howard, "Controlled slip operation of an induction motor with optimum waveforms," *IEE Int. Conf. Rec. on Electrical Variable Speed Drives,* pp. 157–160, Sept. 1979.
34. B. Mokrytzki, "Pulse width modulated inverters for ac motor drives," *IEEE Trans. IGA,* vol. IGA-3, no. 6, pp. 493–503, Nov./Dec. 1967.
35. A. J. Humphrey, "Constant horsepower operation of induction motors," *IEEE Trans. IGA,* vol. IGA-5, no. 5, pp. 552–557, Sept./Oct. 1969.
36. J. B. Forsythe and S. B. Dewan, "Output current regulation with PWM inverter induction motor drives," *IEEE Trans. IA,* vol. IA-11, no. 5, pp. 517–525, Sept./Oct. 1975.
37. A. Kusko, "Application of microprocessors to ac and dc electric motor drive systems," *IEEE/IAS Annual Meeting Conf. Rec.,* pp. 1079–1081, 1977.
38. R. Moffat, P. C. Sen, R. Younker, and M. M. Bayoumi, "Digital phase-locked loop for induction motor speed control," *IEEE Trans. IA,* vol. IA-15, no. 2, pp. 176–182, Mar./Apr. 1979.
39. P. C. Sen and M. MacDonald, "Slip frequency controlled induction motor drives using digital phase locked loop control systems," *IEEE/IAS Int. Sem. Power Conv. Conf. Rec.,* pp. 412–419, 1977.
40. P. C. Sen and M. L. MacDonald, "Stability analysis of induction motor drives using phase locked loop control system," *IEEE/IAS Annual Meeting Conf. Rec.,* pp. 681–689, 1978.
41. S. B. Dewan and S. A. Mirbod, "Slip speed control in an induction motor drive with a phase locked loop," *IEEE/IAS Annual Meeting Conf. Rec.,* pp. 952–955, 1979.
42. A. Abbondanti and M. B. Brennen, "Variable speed induction motor drives use electronic slip calculator based on motor voltages and currents," *IEEE Trans. IA,* vol. IA-11, no. 5, pp. 483–488, Sept./Oct. 1975.
43. I. D. Landau, "Wide range speed control of three-phase squirrel-cage induction motors using static frequency converters," *IEEE Trans. IGA,* vol. IGA-5, no. 1, pp. 53–60, Jan./Feb. 1969.
44. T. A. Lipo, "Flux sensing and control of static ac drives by the use of flux coils," *IEEE Trans. Magnetics,* vol. MAG-13, no. 5, pp. 1403–1408, Sept. 1977.
45. A. B. Plunkett, "Direct flux and torque regulation in a PWM inverter induction motor drive," *IEEE Trans. IA,* vol. IA-13, no. 2, pp. 139–146, Mar./Apr. 1977.
46. A. B. Plunkett and T. A. Lipo, "New methods of induction motor torque regulation," *IEEE Trans. IA,* vol. IA-12, no. 1, pp. 47–55, Jan./Feb. 1976.
47. V. R. Stefanovic, "Static and dynamic characteristics of induction motors operating under constant airgap flux control," *IEEE/IAS Annual Meeting Conf. Rec.,* pp. 436–444, 1976.
48. V. R. Stefanovic and T. H. Barton, "The speed torque transfer function of electric drives," *IEEE Trans. IA,* vol. IA-13, no. 5, pp. 428–436, Sept./Oct. 1977.
49. K. H. Bayer, H. Waldmann, and M. Weibelzahl, "Field oriented closed loop control of synchronous machine with the new TRANSVEKTOR control system," *Siemens Review,* vol. 34, pp. 220–223, 1972.
50. T. L. Grant and T. H. Barton, "Control strategies for PWM drives," *IEEE/IAS Annual Meeting Conf. Rec.,* pp. 780–784, 1979.
51. A. Nabae, K. Otsuka, H. Uchino, and R. Kurosawa, "An approach to flux control of induction motors operated with variable frequency power supply," *IEEE/IAS Annual Meeting Conf. Rec.,* pp. 890–896, 1978.
52. B. DeFornel, J. M. Farines, and J. C. Hapiot, "Numerical estimation of the speed of an asynchronous machine supplied by a static converter," *IEEE/IAS Annual Meeting Conf. Rec.,* pp. 956–962, 1979.
53. L. J. Garces, "Parameter adaption for the speed controlled static ac drive with squirrel cage induction motor," *IEEE/IAS Annual Meeting Conf. Rec.,* pp. 843–850, 1979.
54. M. Abbas and D. W. Novotny, "Stator referred equivalent circuits for inverter driven electric machines," *IEEE/IAS Annual Meeting Conf. Rec.,* pp. 828–835, 1978.
55. A. R. Miles, D. W. Novotny, and J. Betro, "The effect of voltz/hertz control of induction machine dynamic response," *IEEE/IAS Annual Meeting Conf. Rec.,* pp. 802–809, 1979.
56. F. G. G. DeBuck, "Leakage reactance and design considerations for variable-frequency inverter-fed induction motors," *IEEE/IAS Annual Meeting Conf. Rec.,* pp. 757–769, 1979.
57. T. Konishi, K. Kamiyama, Y. Miyahara, T. Ohmae, and N. Sugimoto, "An application technology of microprocessor to adjustable speed motor drives," *IEEE/IAS Annual Meeting Conf. Rec.,* pp. 669–675, 1978.
58. E. A. Skogholm, "Efficiency and power factor for a square wave inverter drive," *IEEE/IAS Annual Meeting Conf. Rec.,* pp. 463–469, 1978.
59. A. Ludbrook, "Basic economics of thyristor adjustable speed drive systems," *IEEE/IGA Annual Meeting Conf. Rec.,* pp. 591–596, 1969.
60. S. B. Dewan and D. L. Duff, "Optimum design of an input commutated inverter for ac motor control," *IEEE Trans. IGA,* vol. IGA-5, no. 6, pp. 699–705, Nov./Dec. 1969.
61. H. Kielgas and R. Nill, "Converter propulsion systems with three-phase induction motors for electrical traction vehicles," *IEEE/IAS Int. Sem. Power Conv. Conf. Rec.,* pp. 304–319, 1977.
62. P. K. Sattler, "Suitable choice of an inverter fed induction motor system for several applications," *IEEE/IAS Annual Meeting Conf. Rec.,* pp. 798–805, 1978.
63. T. M. Jahns, "Improved reliability in solid state ac drives by means of multiple independent phase-drive units, *IEEE/IAS Annual Meeting Conf. Rec.,* pp. 770–780, 1978.
64. K. Mauch and M. R. Ito, "A multimicroprocessor ac drive controller," *IEEE/IAS Annual Meeting Conf. Rec.,* pp. 634–640, 1980.

Section 4

1. R. B. Magg, "Characteristics and application of current source slip regulated ac induction motor drives," *IEEE/IGA Annual Meeting Conf. Rec., pp.* 411–413, 1971.
2. M. P. Kazmierkowski, M. Nowak, and R. Barlik, "Induction motor drive systems controlled by frequency converter with current source inverter," *IEEE/IAS Annual Meeting Conf. Rec.,* pp. 1067–1072, 1978.
3. T. H. Chin, "A new controlled current type inverter with improved performance," *IEEE/IAS Int. Sem. Power Conv. Conf. Rec.,* pp. 185–192, 1977.
4. A. Nabae, "Pulse amplitude modulated current source inverters for ac drives," *IEEE/IAS Annual Meeting Conf. Rec.,* pp. 539–545, 1978.
5. T. A. Lipo, "Simulation of a current source inverter drive," *IEEE Power Elec. Spec. Conf. Rec.,* pp. 310–315, 1977.
6. W. McMurray, "The performance of a single phase current fed inverter with counter EMF-inductive load," *IEEE Trans. IA,* vol. IA-14, no. 4, pp. 319–329, July/Aug. 1978.

7. F. Harashima and H. Hayashi, "Dynamic performance of current source inverter fed induction motors," *IEEE/IAS Annual Meeting Conf. Rec.*, pp. 904–909, 1978.
8. S. Nobuhiko and N. Sato, "Steady state and stability analysis of induction motor driven by current source inverter," *IEEE Trans. IA*, vol. IA-13, no. 3, pp. 244–253, May/June 1977.
9. A. Joshi and S. B. Dewan, "Comparison of ac traction drives using voltage and current source inverters," *IEEE/IAS Annual Meeting Conf. Rec.*, pp. 781–788, 1978.
10. B. Kampschulte, "Comparison of an asynchronous machine drive with current-source and voltage-source inverter," *IEEE/IAS Annual Meeting Conf. Rec.*, pp. 945–951, 1979.
11. M. L. MacDonald and P. C. Sen, "Control loop study of induction motor drives using DQ model," *IEEE/IAS Annual Meeting Conf. Rec.*, pp. 897–903, 1978.
12. E. P. Cornell and T. A. Lipo, "Design of controlled current ac drive systems using transfer function techniques," *IFAC Symposium Conf. Rec.*, vol. 1, pp. 133–147, 1974.
13. R. H. Nelson and T. A. Rodomski, "Design methods for current source inverter/induction motor drive systems," *IEEE/IECI Conf. Rec.*, pp. 141–145, 1975.
14. R. Ueda, T. Mochizuki, and S. Takata, "Steady state stability analysis and synthesis of current source inverter driven induction motor system," *IEEE/IAS Annual Meeting Conf. Rec.*, pp. 864–872, 1979.
15. W. Lineau, "Torque oscillations in traction drives with current fed asynchronous machine," *IEE Int. Conf. Rec. on Electrical Variable Speed Drives*, pp. 102–107, 1979
16. R. Palaniappan, J. Vithayathil, and S. K. Dutta, "Principle of a dual current source converter for ac motor drives," *IEEE Trans. IA*, vol. IA-15, no. 4, pp. 445–452, July/Aug. 1979.
17. J. D. Van Wyk and H. R. Holtz, "A simple and reliable four-quadrant variable frequency ac drive for industrial application up to 50 Kw," *IEE Int. Conf. Rec. on Electrical Variable Speed Drives*, pp. 34–37, 1979.
18. S. B. Dewan and G. R. Slemon, "Multi-axle ac traction drives using thyristor converters," *IEE Int. Conf. Rec. on Electrical Variable Speed Drives*, pp. 108–111, 1979.
19. P. W. Hammond, "Power factor correction of current source inverter drives with pump loads," *IEEE/IAS Annual Meeting Conf. Rec.*, pp. 520–529, 1980.
20. E. R. Laithwaite and S. B. Kuznetsov, "Development of an induction machine commutated thyristor invertor for traction drives," *IEEE/IAS Annual Meeting Conf. Rec.*, pp. 580–585, 1980.
21. E. R. Laithwaite and S. B. Kuznetsov, "Natural commutation of current source thyristor inverter by cage rotor induction machines," *IEEE/IAS Annual Meeting Conf. Rec.*, pp. 897–902, 1980.
22. F. Harashima, T. Haneyoshi, and H. Inaba, "Operating performances of inverter fed induction motors considering magnetic saturation," *IEEE/IAS Annual Meeting Conf. Rec.*, pp. 586–591, 1980.
23. E. F. Sousa, H. Tokieda, T. Itoh, R. Ueda, and S. Takata, "Steady state analysis of current fed inverter induction motor drive including harmonic components," *IEEE/IAS Annual Meeting Conf. Rec.*, pp. 592–597, 1980.
24. S. B. Dewan and A. Joshi, "Modified steady state analysis of current source inverter squirrel cage motor drive," *IEEE/IAS Annual Meeting Conf. Rec.*, pp. 598–604, 1980.
25. R. Krishnan and V. R. Stefanovic, "Control principles in current source induction motor drives," *IEEE/IAS Annual Meeting Conf. Rec.*, pp. 605–617, 1980.
26. W. S. Mok and P. C. Sen, "Induction motor drives with microcomputer control system," *IEEE/IAS Annual Meeting Conf. Rec.*, pp. 653–662, 1980.
27. A. Hirata, S. Saito, M. Miyazaki, and T. Nakagawa, "New applications of current type inverters," *IEEE/IAS Annual Meeting Conf. Rec.*, pp. 891–896, 1980

Section 5

1. W. Shepherd and J. Stanway, "Slip power recovery in an induction motor by the use of a thyristor inverter," *IEEE Trans. IGA*, vol. IGA-5, no. 1, pp. 74–83, Jan./Feb. 1969
2. M. S. Erlicki, "Inverter rotor drive of an induction motor," *IEEE Trans. PAS*, vol. PAS-84, no. 9, pp. 1011–1016, Nov. 1965.
3. Y. Wallach, M. S. Erlicki, and J. Ben-Uri, "Transients in a Kramer cascade," *IEEE Trans. IGA*, vol. IGA-2, no. 2, pp. 158–162, Mar./Apr. 1966.
4. H. Kazuno, "A wide range speed control of an induction motor with static Scherbius and Kramer systems," *J. Elec. Eng. Japan*, vol. 89, no. 2, pp. 10–19, 1969.
5. T. Hori and Y. Hiro, "The characteristics of an induction motor controlled by a Scherbius System," *IEEE/IAS Annual Meeting Conf. Rec.*, pp. 775–782, 1972.
6. J. Noda, Y. Hiro, and T. Hori, "Brushless Scherbius control of induction motors," *IEEE/IAS Annual Meeting Conf. Rec.*, pp. 111–118, 1974.
7. A. Kusko and C. B. Somuah, "Speed control of a single-frame cascade induction motor with slip power pump back," *IEEE Trans. IA*, vol. IA-4, no. 2, pp. 97–105, Mar./Apr. 1978.
8. P. C. Sen and K. H. Ma, "Constant torque operation of induction motors using chopper in rotor circuit," *IEEE Trans. IA*, vol. IA-14, no. 5, pp. 408–414, Sept./Oct. 1978.
9. M. Ramamoorty and N. S. Wani, "Dynamic model for a chopper controlled slip ring induction motor," *IEEE Trans. IECI*, vol. 25, no. 3, pp. 260–266, Aug. 1978.
10. W. F. Long and N. L. Schmitz, "Cycloconverter control of the doubly fed induction motor," *IEEE Trans. IGA*, vol. IGA-7, no. 1, pp. 95–100, Jan./Feb. 1971.
11. A. K. Chattopadhyay, "An adjustable speed induction motor drive with a cycloconverter type thyristor-commutator in the rotor," *IEEE Trans. IA*, vol. IA-14, no. 2, pp. 116–122, Mar./Apr. 1978.
12. A. K. Chattopadhyay, "Digital computer simulation of an adjustable speed induction motor drive with a cycloconverter type thyristor commutator in the rotor, *IEEE Trans. IECI*, vol. IECI 23, no. 1, pp. 86–92, Feb. 1976.

Section 6

1. E. Ohno, T. Kishimoto, and M. Akamatsu, "The thyristor commutatorless motor," *IEEE Trans. Magnetics*, vol. MAG-3, no. 9, pp. 236–240, Sept. 1967.
2. G. E. M. S. Fahmy, "On commutatorless motor (CLM) drives," *IEEE/IAS Annual Meeting Conf. Rec.*, pp. 419–426, 1974.
3. A. H. Hoffman, "Brushless synchronous motors for large industrial drives," *IEEE Trans. IGA*, vol. IGA-5, no. 2, pp. 158–162, Mar./Apr. 1969.
4. C. S. J. Lamb, "Commutatorless alternating voltage fed variable speed motor," *Proc. IEE*, vol. 110, no. 12, pp. 2221–2227, 1963.
5. P. Grumbrecht and F. Hentschel, "Controlled drives with converter-fed synchronous machines," *IEEE/IAS Annual Conf. Rec.*, pp. 806–811, 1978.
6. A. Habock and D. Köllensperger, "State of development of converter fed synchronous motors with self control," *Siemens Review*, vol. 38, no. 9, pp. 390–392, 1971.
7. G. R. Slemon, J. B. Forsythe, and S. B. Dewan, "Controlled power angle synchronous motor inverter drive system," *IEEE Trans. IA*, vol. IA-9, no. 2, pp. 216–219, Mar./Apr. 1973.
8. F. Harashima, H. Taoka, and H. Naitoh, "A microprocessor based PLL speed control system of converter-fed synchronous motor," *IEEE/IECI Conf. Rec.*, pp. 272–277, 1979.
9. E. L. Lustenader, R. H. Guess, E. Richter, and F. G. Turnbull, "Development of a hybrid flywheel/battery drive system for electric vehicle applications," *IEEE Trans. Veh. Tech.*, vol. VT-26, no. 2, pp. 135–143, May 1977.
10. A. B. Plunkett and F. G. Turnbull, "Load commutated inverter/synchronous motor drive without a shaft position sensor," *IEEE Trans. IA*, vol. IA-15, no. 1, pp. 63–71, Jan./Feb. 1979.
11. A. B. Plunkett and F. G. Turnbull, "System design method for a load commutated inverter synchronous motor drive," *IEEE/IAS Annual Meeting Conf. Rec.*, pp. 812–819, 1978.
12. T. Okuyama, T. Hori, N. Morino, T. Miyata, and I. Shimizu,

"Effects of machine constants on steady state and transient characteristics of commutatorless motors," *IEEE/IAS Annual Meeting Conf. Rec.*, pp. 272–279, 1977.

13. F. C. Brockhurst, "Performance equations for dc commutatorless motors using salient pole synchronous type machines," *IEEE/IAS Annual Meeting Conf. Rec.*, pp. 861–868, 1978.
14. C. M. Ong and T. A. Lipo, "Steady-state analysis of a current source inverter-reluctance motor drive. Part I: Analysis. Part II: Experimental and analytical results," *IEEE/IAS Annual Meeting Conf. Rec.*, pp. 841–851, 1975.
15. I. Schmit, "Analysis of converter fed synchronous motors," *IFAC Symposium,* vol. 1, pp. 571–585, 1974.
16. T. A. Lipo and P. C. Krause, "Stability analysis for variable frequency operation of synchronous machines," *Trans. PAS,* vol. PAS-87, no. 1, pp. 227–234, Jan. 1968.
17. T. A. Lipo and P. C. Krause, "Stability analysis of a reluctance synchronous machine," *IEEE Trans. PAS,* vol. PAS-86, no. 8, pp. 825–834, Aug. 1967.
18. C. M. Ong and T. A. Lipo, "An approach to closed loop design of a current source inverter/reluctance motor drive," *IEEE/IAS Annual Meeting Conf. Rec.*, pp. 494–500, 1975.
19. T. A. Lipo, "Performance calculations of a reluctance motor drive," *IEEE/IAS Annual Meeting Conf. Rec.*, pp. 558–567, 1976.
20. N. Sato and V. V. Semenov, "Adjustable speed drive with a brushless dc motor," *IEEE Trans. IGA*, vol. IGA-7, no. 4, pp. 539–543, July/Aug. 1971.
21. F. Harashima, H. Naitoh, and T. Haneyoshi, "Dynamic performance of self-controlled synchronous motors fed by current source inverters," *IEEE Trans. IA,* vol. IA-15, no. 1, pp. 36–47, Jan./Feb. 1979.
22. H. Naitoh, T. Haneyoshi, and F. Harashima, "Analysis of thyristor circuits with time-dependent parameter loads," *IEEE Trans. IECI,* vol. IECI-25, no. 3, pp. 285–291, Aug. 1978.
23. T. Peterson and K. Frank, "Starting of large synchronous motor using static frequency converter," *IEEE Trans. PAS,* vol. PAS-91, no. 1, pp. 172–179, Jan./Feb. 1972.
24. H. G. Meyer, "Static frequency changers for starting synchronous machines," *Brown Boveri Review,* vol. 51, pp. 526–530, Aug./Sept. 1964.
25. I. Hosono, K. Katsuki, and M. Yano, "Static converter starting of large synchronous motors," *IEEE/IAS Annual Meeting Conf. Rec.*, pp. 536–551, 1976.
26. K. Imai, "New applications of commutatorless motor systems for starting large synchronous motors," *IEEE/IAS Int. Sem. Power Conv. Conf. Rec.*, pp. 237–246, 1977.
27. Y. Hirane and A. Kumamoto, "A double speed commutatorless motor," *IEEE/IAS Annual Meeting Conf. Rec.*, pp. 922–927, 1979.
28. G. S. Buja, "Optimum control of current inverter synchronous motor drives," *IEEE/IAS Annual Meeting Conf. Rec.*, pp. 812–816, 1979.
29. H. Le-Huy, R. Perret, and D. Roye, "Microprocessor control of a current fed synchronous motor drive," *IEEE/IAS Annual Meeting Conf. Rec.*, pp. 873–880, 1979.
30. H. Le-Huy, "A self controlled synchronous motor drive using terminal voltage sensing," *IEEE/IAS Annual Meeting Conf. Rec.*, pp. 562–569, 1980.
31. A. V. Gumeste and G. R. Slemon, "Steady state analysis of a permanent magnet synchronous motor drive with voltage fed inverter," *IEEE/IAS Annual Meeting Conf. Rec.*, pp. 618–625, 1980.
32. T. Kataoka and S. Nishikata, "Transient performance analysis of self controlled synchronous motors," *IEEE/IAS Annual Meeting Conf. Rec.*, pp. 626–633, 1980.

Section 7

1. T. Tsuchiya, "Basic characteristics of cycloconverter type commutatorless motors," *IEEE Trans. IGA,* vol. IGA-6, no. 4, pp. 349–356, July/Aug. 1970.
2. E. P. Cornell and D. W. Novotny, "Commutation by armature induced voltage in self controlled synchronous machines," *IEEE/IAS Annual Meeting Conf. Rec.*, pp. 143–157, 1973.
3. E. P. Cornell and D. W. Novotny, "Theoretical analysis of the stability and transient response of self-controlled synchronous machines," *IEEE/PES Winter Meeting,* pp. 15–23, 1973.
4. P. T. Finlayson and D. C. Washburn, "Cycloconverter controlled synchronous machines for load compensation on ac power systems," *IEEE Trans. IA,* vol. IA-10, no. 6, pp. 806–813, Nov./Dec. 1974.
5. F. J. Bourbeau, "Wide speed range cycloconverter ac motor drives operating from utility frequency power," *IEEE/IAS Int. Sem. Power Conv. Conf. Rec.*, pp. 319–326, 1977.
6. P. M. Espelage and B. K. Bose, "High frequency link power conversion," *IEEE Trans. IA,* vol. IA-13, no. 5, pp. 387–394, Sept./Oct. 1977.
7. H. N. Wurgler, "The world's first gearless mill drive," *IEEE Trans. IGA,* vol. IGA-6, no. 5, pp. 524–527, Sept./Oct. 1970.
8. R. Zims, "Gearless drive for a finish-ground mill with a capacity of 160 t/h," *IEEE Trans. IA,* vol. IA-9, no. 1, pp. 21–24, Jan./Feb. 1973.
9. E. A. E. Rich, "Concepts of gearless ball mill drives," *IEEE Trans. IGA,* vol. IGA-5, no. 1, pp. 13–17, Jan./Feb. 1969.
10. T. Okuyama, T. Shibata, T. Konishi, T. Sukegawa, N. Morino, and K. Iwata, "Cycloconverter-fed synchronous motor drive for steel rolling mill," *IEEE/IAS Annual Meeting Conf. Rec.*, pp. 820–827, 1978.
11. L. J. Jacovides, M. F. Matuoka, and D. W. Shimer, "A cycloconverter synchronous motor drive for traction applications," *IEEE/IAS Annual Meeting Conf. Rec.*, pp. 549–561, 1980.

Author Index

W

Y

Z

Subject Index

Editor's Biography

Dr. Bimal K. Bose received the B.E. degree from Calcutta University, Calcutta, India in 1956, the M.S. degree from the University of Wisconsin, Madison, WI in 1960, and the Ph.D degree from Calcutta University in 1966.

From 1956 to 1958, he was an electrical engineer at the Tata Power Company, Bombay, India. Then, in 1960, he joined as a faculty member the Bengal Engineering College (Calcutta University) where he organized the industrial electronics program and did research in the areas of magnetic amplifiers, instrumentation, and power electronics applied to drives. He was awarded the Premchand Roychand Scholarship in 1968 and the Mouat Gold Medal in 1970 by Calcutta University for research contributions. In 1971, he joined Rensselaer Polytechnic Institute, Troy, NY as an Associate Professor of Electrical Engineering and served there for five years. In R.P.I., Dr. Bose was responsible for organizing undergraduate and graduate power electronics programs which included an advanced course on adjustable speed ac drives. He was a consultant for several industries which included the General Electric Company, Bendix Corporation, Lutron Electronics, and PCI Ozone Corporation. In 1976 he joined the General Electric Research and Development Center in Schenectady, NY and did various projects on converter fed machine drive systems, electric and hybrid vehicle microcomputer control, etc. Currently, his research interests are power conversion systems and microcomputer control of ac drives. Dr. Bose has published 38 papers and holds a number of U.S. patents. He is continuing as Adjunct-Associate Professor at Rensselaer Polytechnic Institute. He is chairman of Transactions Review of Static Power Converter Committee, IEEE, a member of Microcomputer Control Committee of Industrial Electronics and Control Instrumentation Group, IEEE, and a member of Scientific Committee of International Conference on Digital Control of Electrical Machines. Dr. Bose is a senior member of IEEE.